BIOLOGY OF THE LOBSTER

Homarus americanus

This book is dedicated to Dr. John M. Anderson (Cornell University) and to the memory of Dr. Priscilla F. Pollister (Brooklyn College), who introduced me to the invertebrates.

BIOLOGY OF THE LOBSTER

Homarus americanus

Edited by

Jan Robert Factor

Division of Natural Sciences
Purchase College
State University of New York
Purchase, New York

Dear Fred — Many thanks for your help and advice along the way and with this volume. With my warmest regards!

Academic Press

San Diego New York Boston London Sydney Tokyo Toronto

This book is printed on acid-free paper. ♾

Academic Press, Inc.
A Division of Harcourt Brace & Company
525 B Street, Suite 1900, San Diego, California 92101-4495

United Kingdom Edition published by
Academic Press Limited
24-28 Oval Road, London NW1 7DX

Library of Congress Cataloging-in-Publication Data

Biology of the lobster Homarus americanus / edited by Jan Robert Factor.
p. cm.
Includes bibliographical references and index.
ISBN 0-12-247570-4
1. American lobster. I. Factor, Jan Robert.
QL444.M33B566 1995
595.3'841--dc20 94-25271
CIP

PRINTED IN THE UNITED STATES OF AMERICA
95 96 97 98 99 00 EB 9 8 7 6 5 4 3 2 1

Contents

CHAPTER

4

Postlarval, Juvenile, Adolescent, and Adult Ecology

Peter Lawton and Kari L. Lavalli

CHAPTER 5

Fishery Regulations and Methods

Robert J. Miller

CHAPTER 6

Populations, Fisheries, and Management

Michael J. Fogarty

CHAPTER 7

Interface of Ecology, Behavior, and Fisheries

J. Stanley Cobb

CHAPTER 8

Aquaculture

D. E. Aiken and S. L. Waddy

CHAPTER

9

Reproduction and Embryonic Development

P. Talbot and Simone Helluy

CHAPTER
10
Control of Growth and Reproduction

S. L. Waddy, D. E. Aiken, and D. P. V. de Kleijn

CHAPTER
11
Neurobiology and Neuroendocrinology

Barbara Beltz

CHAPTER
12
Muscles and Their Innervation

C. K. Govind

CHAPTER

13

Behavior and Sensory Biology

Jelle Atema and Rainer Voigt

CHAPTER

14

The Feeding Appendages

Kari L. Lavalli and Jan Robert Factor

CHAPTER

15

The Digestive System

Jan Robert Factor

CHAPTER

16

Digestive Physiology and Nutrition

Douglas E. Conklin

CHAPTER

17

Circulation, the Blood, and Disease

Gary G. Martin and Jo Ellen Hose

CHAPTER

18

The Physiology of Gas Exchange, Circulation, Ion Regulation, and Nitrogenous Excretion: An Integrative Approach

Brian R. McMahon

Contributors

Numbers in parentheses indicate the pages on which the authors' contributions begin.

D. E. Aiken (153, 217) Biological Station, Department of Fisheries and Oceans, St. Andrews, New Brunswick, Canada E0G 2X0

Jelle Atema (313) Boston University Marine Program, Marine Biological Laboratory, Woods Hole, Massachusetts 02543

Barbara Beltz (267) Department of Biological Sciences, Wellesley College, Wellesley, Massachusetts 02181

J. Stanley Cobb (139) Department of Zoology, University of Rhode Island, Kingston, Rhode Island 02881

Douglas E. Conklin (441) Department of Animal Science, University of California at Davis, Davis, California 95616

D. P. V. de Kleijn (217) Department of Animal Physiology, University of Nijmegen, Toernooiveld, 6525 ED Nijmegen, The Netherlands

G. P. Ennis (23) Science Branch, Department of Fisheries and Oceans, St. John's, Newfoundland, Canada A1C 5X1

Jan Robert Factor (1, 349, 395) Division of Natural Sciences, Purchase College, State University of New York, Purchase, New York 10577

Michael J. Fogarty (111) National Oceanic and Atmospheric Administration, National Marine Fisheries Service, Northeast Fisheries Science Center, Woods Hole, Massachusetts 02543

C. K. Govind (291) Life Sciences Division, Scarborough College, University of Toronto, Scarborough, Ontario, Canada M1C 1A4

Simone Helluy (177) Department of Biological Sciences, Wellesley College, Wellesley, Massachusetts 02181

Jo Ellen Hose (465) Department of Biology, Occidental College, Los Angeles, California 90041

Kari L. Lavalli (47, 349) Marine Science Center, Northeastern University, Nahant, Massachusetts 01908

Peter Lawton (47) Biological Station, Department of Fisheries and Oceans, St. Andrews, New Brunswick, Canada E0G 2X0

Gary G. Martin (465) Department of Biology, Occidental College, Los Angeles, California 90041

Brian R. McMahon (497) Department of Biological Sciences, University of Calgary, Calgary, Alberta, Canada T2N 1N4

Robert J. Miller (89) Department of Fisheries and Oceans, Halifax, Nova Scotia, Canada B3J 2S7

P. Talbot (177) Department of Biology, University of California at Riverside, Riverside, California 92521

Rainer Voigt (313) Boston University Marine Program, Marine Biological Laboratory, Woods Hole, Massachusetts 02543

S. L. Waddy (153, 217) Biological Station, Department of Fisheries and Oceans, St. Andrews, New Brunswick, Canada E0G 2X0

Austin B. Williams (13) National Marine Fisheries Service Systematics Laboratory, National Museum of Natural History, Smithsonian Institution, Washington, D.C. 20560

Preface

The American lobster, *Homarus americanus*, is one of the more ecologically and economically important invertebrates inhabiting the coastal waters of North America. The publication of this volume marks the centennial of the first collection of scientific knowledge dedicated to this species. Francis Hobart Herrick's monographs (The American Lobster: A Study of Its Habits and Development, *Bull. U.S. Fish Comm.* 15, 1–252, 1895, and Natural History of the American Lobster, *Bull. U.S. Bur. Fish.* 29, 149–408, 1909) gathered a wealth of information at a time of great interest in the biology, management, and culture of lobsters. Recognizing the accumulation of information by the late 1970s, J. Stanley Cobb and Bruce F. Phillips produced a two-volume set on the various species of both clawed and spiny lobsters (*The Biology and Management of Lobsters*, Academic Press, New York, 1980). The recent volume by B. F. Phillips, J. S. Cobb, and J. Kittaka (*Spiny Lobster Management*, Fishing News Books, Oxford, 1994) considers fisheries biology, management, and aquaculture of commercially important spiny lobsters from around the world.

Familiar to scientists and the public alike, few invertebrates have attracted as much attention as *Homarus americanus*. Both as a research organism and as a valuable commercial food species, the lobster has been the subject of a vast array of studies and research reports. The need to gather information about this species into a single, accessible form was obvious to Herrick in 1909, when he wrote "Our knowledge of the lobster has increased to such an extent during the past fifteen years that in all probability there is no marine invertebrate in the world which is now better known." One hundred years later, after ever so much more research and accumulated information, the need for an updated volume is obvious.

It is no longer possible for one author to write authoritatively on all aspects of the biology of *Homarus americanus*. The modern version of Herrick's monographic approach requires the expertise of specialists in the varied fields of lobster biology. This volume strives to be more than a summary of recent advances in research. It is intended to bring together a wide assortment of interrelated topics in a compendium of knowledge that provides both overview and historical perspective, allowing it to serve as a reference text for all those interested in the lobster. I hope it will be useful to a wide audience, including lobster biologists seeking a synthesis of their own areas of specialty or detailed information outside their specialties, researchers who work with spiny and European lobsters, carcinologists, and invertebrate zoologists. I think it will also be a significant resource for the wider world of marine biologists, ecologists, environmentalists, and policymakers.

This project began with a conversation with Stan Cobb, to whom I am grateful for his immediate enthusiasm and heartening encouragement. I am also thankful to Jelle Atema, C. K. Govind, Kari Lavalli, Mary Rice, and Susan Waddy for their early and continuing advice. My special thanks go to Andrea Anthony for her encouragement and support throughout this project.

I also express my sincere appreciation to those who freely gave their time and insight by providing me with valuable comments and suggestions concerning portions of this book: John M. Anderson, Jelle Atema, Peter Auster, Warren Burggren, J. Stanley Cobb, Douglas Conklin, Diane Cowan, Charles Derby, Gerald Ennis, Michael Fogarty, C. K. Govind, Gareth Harding, L. B. Holthuis, Christiane Hudon, Glenn Jefferson, Kevin Johnson, Phyllis Johnson, Mark Kirk, Jay Krouse, Kari Lavalli, Peter Lawton, Linda Mantel, Gary Martin, Robert Miller, Keith Nelson, Charles Page, Douglas Pezzack, Greg Roach, Allen Selverston, Prue Talbot, and Dale Tshudy. If only it had been possible to take their valuable advice in every instance.

I thank Chuck Crumly, Cheryl Uppling, and the staff at Academic Press for their expert advice and good work in bringing this volume to successul completion.

Above all others, this volume is made possible by the hard work and graceful cooperation of the contributing authors.

Jan Robert Factor

CHAPTER

1

Introduction, Anatomy, and Life History

JAN ROBERT FACTOR
Division of Natural Sciences
Purchase College
State University of New York
Purchase, New York 10577

I. Introduction

It has been 100 years since the publication of the first monograph on the American lobster, *Homarus americanus* H. Milne Edwards, 1837. In 1895,[1] Francis Hobart Herrick presented the sum of knowledge that had accumulated during the 19th century (and earlier) in his monograph "The American Lobster: A Study of Its Habits and Development." Again in 1909,[2] Herrick summarized the state of contemporary understanding of this important species in "Natural History of the American Lobster." The intense interest in the American lobster during this period was stimulated by the economic importance of the fishery, concern over the declining state of the fishery, and the potential of aquaculture, as well as by the general advance of biology, the interest in American species, and the application of contemporary techniques. These same factors continue to stimulate our interest in this important species.

In the intervening years, burgeoning fields such as ecology, population biology, physiology, and neurobiology have added to the growth of information about lobsters in general. Significant bibliographies of homarid lobsters have been produced by R. D. Lewis (1970) and W. S. W. Nowak (1972). By the late 1970s, J. Stanley Cobb and Bruce F. Phillips (1980a,b) recognized the need to review our understanding of the various clawed and spiny lobsters from around the world.

Few invertebrates have attracted as much attention for such a long period of time as *Homarus americanus*. The more recent explosion of information, when added to the generally superb earlier work, presents a bewildering array of research papers, facts, and opinions. This volume is intended to impose a degree of order and overview on this vast literature and to provide a modern perspective on the American lobster.

Individual chapters summarize our understanding of particular fields of research from a modern viewpoint, yet with historical perspective. Bringing

[1] Published as an article in *Bull. U.S. Fish Comm.*, Vol. 15, pp. 1–252 + plates A–J and 1–54, 1895. Issued February 5, 1896, by the Government Printing Office, Washington, D.C. Document No. 300 of the U.S. Commission of Fish and Fisheries. Variously cited as 1895 and 1896. Also issued as a separately bound monograph, apparently prior to the issuance of the journal series.

[2] Published as an article in *Bull. U.S. Bur. Fish.*, Vol. 29, pp. 149–408 + plates XXVIII–XLVII, 1909. Issued July 13, 1911, by the Government Printing Office, Washington, D.C. Document No. 747 of the U.S. Bureau of Fisheries. Variously cited as 1909 and 1911. Also issued as a separately bound monograph; however, there appears to be no evidence that this work was published before 1911. A reprint of this monograph was published in 1977 by Arno Press (A New York Times Co.), New York, as part of their "History of Ecology" series.

BIOLOGY OF THE LOBSTER *Homarus americanus*
Jan Robert Factor, Editor

together such a variety of topics in one place also serves to highlight the diversity of our interests in *Homarus americanus*. Areas of overlap among related chapters, although generally considered from somewhat different perspectives, point out the natural interrelatedness among many topics. The liberal use of cross-referencing will guide the reader to related material found elsewhere in the volume.

It is inherently problematic to break up the naturally integrated biology of an animal into discrete subjects. From an ecological perspective, for example, it is difficult to draw the traditional line between the premetamorphic larval stages (stages I–III) and the postmetamorphic stages (beginning with the postlarva, stage IV). Rather, Ennis (Chapter 3) considers the ecology of the planktonic stages (larvae and the early, presettlement postlarva), while Lawton and Lavalli (Chapter 4) deal with the ecology of the benthic phases (beginning with the settling postlarva and continuing through the adult). Lawton and Lavalli (Chapter 4) have brought together physiology, ecology, and behavior into a coherent rationale for the terms applied to the life history phases of *Homarus americanus*.

The interrelationships among population biology, genetics, and fishery management are elucidated by Miller (Chapter 5) and Fogarty (Chapter 6), and the integration of fisheries with the ecology and behavior of *Homarus americanus* is discussed by Cobb (Chapter 7). The relationship of these fields, as well as reproductive and nutritional biology, to the longstanding interest in lobster aquaculture is made clear by Aiken and Waddy (Chapter 8).

The neurobiology of *Homarus americanus* is consid-

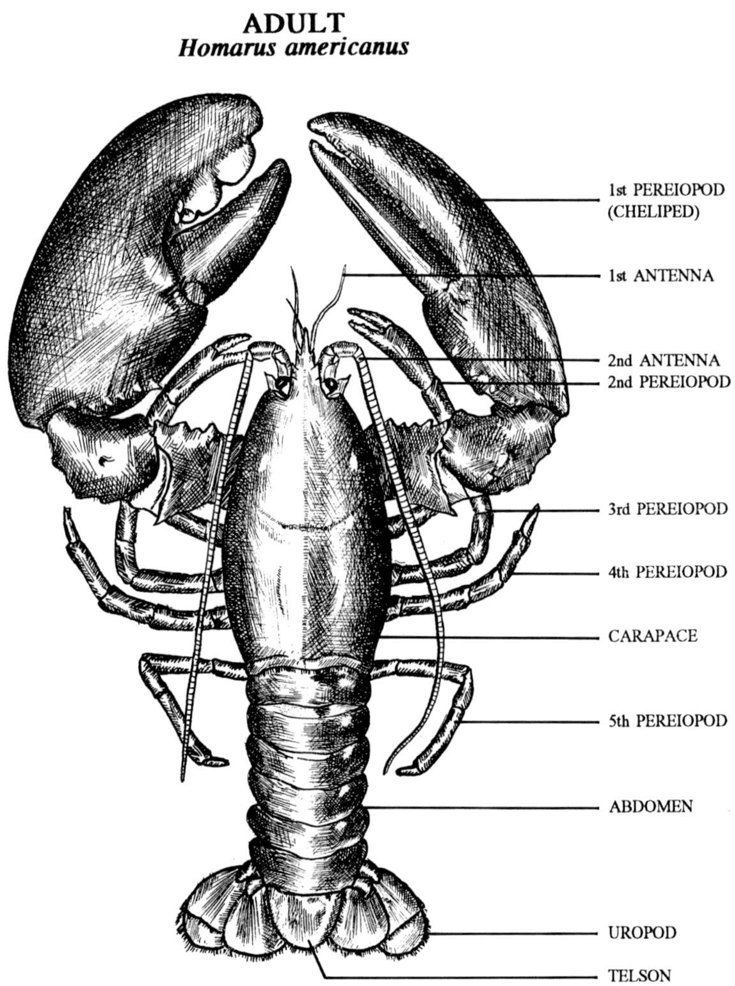

FIGURE 1 Adult lobster, *Homarus americanus*. Dorsal view. Length of specimen, 18.5 inches total length (TL); weight, 11.75 pounds; age, "probably about 16 years." (Drawn from life by Philip B. Hadley, 1906; labels added.)

ered in three related chapters on neurobiology and neuroendocrinology (Beltz, Chapter 11), muscles and their innervation (Govind, Chapter 12), and behavior and sensory biology (Atema and Voigt, Chapter 13).

The complex processes of feeding and digestion illustrate the intimate connections among ecology (Ennis, Chapter 3, and Lawton and Lavalli, Chapter 4), sensory biology (Atema and Voigt, Chapter 13), functional morphology of appendages (Lavalli and Factor, Chapter 14), structure and organization of the digestive system (Factor, Chapter 15), and physiology of digestion and nutrition (Conklin, Chapter 16).

Additional links are also obvious, such as those between mating behavior (Atema and Voigt, Chapter 13) and the physiology of reproduction (Talbot and Helluy, Chapter 9), between the behavioral and physiological mechanisms of reproduction and the exogenous and endogenous control mechanisms of reproductive processes (Waddy, Aiken, and de Kleijn, Chapter 10), and between the nature of the circulatory system and blood cells (Martin and Hose, Chapter 17) and the integrated physiology of *Homarus americanus* (McMahon, Chapter 18).

This monograph is focused as tightly as possible on the American lobster, *Homarus americanus*. It is not intended to be a wide-ranging review of crustacean or decapod biology. Limited comparisons with closely related species, however, point out particularly interesting points of similarity or difference, or serve to fill in gaps in our understanding of *H. americanus*. It is not assumed that the closely related, yet distinct (Williams, Chapter 2), European lobster (*H. gammarus*) is identical to the American lobster (*H. americanus*) in any aspect of its biology, unless there is evidence to support specific instances.

When considered together, the chapters in this volume provide a level of overview and synthesis that is rarely possible in research reports or reviews of limited aspects of the literature.

II. Anatomy

The segmented body of *Homarus americanus* is organized into three regions: head, thorax, and abdomen. Each segment bears a pair of appendages which are modified for particular functions within and among regions. The head (cephalon) bears sensory and feeding appendages, antennae and mouthparts; the thorax bears feeding and locomotory appendages, mouthparts and pereiopods ("walking legs"); and the abdomen ("tail") bears locomotory and reproductive appendages, pleopods ("swimmerets") and the tail fan. The head and thorax are fused into a cephalothorax, which is covered dorsally and laterally by a carapace that partially obscures the segmentation of the body. The body of a mature adult lobster is illustrated in Fig. 1; although not all of the appendages are visible in this dorsal view, they are listed in Fig. 2, along with their regional and functional organization.

The anteriormost head appendages, the first and second antennae, serve sensory functions. The mouthparts comprise the last three appendages of the head region (mandible and first and second maxillae) and the first three thoracic appendages (first through third maxillipeds). The first pereiopods are exaggerated into the great claws (chelipeds, great chelae), and the second through fifth pereiopods serve as walking legs. In addition to the mouthparts, the pereiopods are also used in feeding, especially the first three pairs, which are chelate (Lavalli and Factor, Chapter 14). Among the abdominal appendages, the first pleopod is sexually dimorphic; it forms the hardened sexual intromittent organ of the male (gonopod) and is reduced in the female. The second through fifth pleopods serve as swimmerets and carry the fertilized eggs of the ovigerous female. The sixth abdominal segment bears paired, biramous, flattened

Appendages of *Homarus americanus*

ANTERIOR

HEAD		1st ANTENNAE - antennules
		2nd ANTENNAE - antennae
	MOUTHPARTS	MANDIBLES
		1st MAXILLAE - maxillules
		2nd MAXILLAE - maxillae
THORAX		1st MAXILLIPEDS
		2nd MAXILLIPEDS
		3rd MAXILLIPEDS
	WALKING LEGS	1st PEREIOPODS - chelipeds (great claws)
		2nd PEREIOPODS - chelate
		3rd PEREIOPODS - chelate
		4th PEREIOPODS
		5th PEREIOPODS
ABDOMEN	SWIMMERETS	1st PLEOPODS - ♂ sexual organ
		2nd PLEOPODS
		3rd PLEOPODS
		4th PLEOPODS
		5th PLEOPODS
		6th PLEOPODS - uropods
		TELSON

POSTERIOR

FIGURE 2 The appendages of *Homarus americanus*.

uropods, which appear to represent highly modified pleopods. The telson, the posterior terminus of the body, may or may not represent a true body segment or somite (discussed by Schram, 1986). The uropods, together with the telson, form the "tail fan" used in escape behavior when the abdomen is flexed rapidly (Atema and Voigt, Chapter 13). All of the appendages are fundamentally biramous (with both endopodite and exopodite), but the exopodites of the pereiopods are lost at metamorphosis.

McLaughlin (1980) and Felgenhauer (1992) consider the general internal anatomy of decapod crustaceans. The best overview of the internal anatomy of *Homarus americanus*, specifically, remains Herrick's (1909) account. His cutaway view shows the position and gross shape of the major organs and is reproduced as Fig. 3. Additional details of internal anatomy are presented in the chapters of this volume that cover the various organ systems.

III. Life History

The life history of *Homarus americanus* can be divided into a series of developmental phases based on morphological, physiological, behavioral, and ecological considerations (see Lawton and Lavalli, Chapter 4). Embryonic, larval, postlarval, shelter-restricted juvenile, emergent juvenile, vagile juvenile, adolescent, and adult phases complete the life cycle.

Newly fertilized eggs are enclosed in egg envelopes and attached to the pleopods, where embryonic development occurs under the care and protection of the female (see Talbot and Helluy, Chapter 9). Early cleavage, gastrulation, and organogenesis result in a naupliar stage. This "egg-nauplius," so-called because the naupliar stage is passed within the egg, is characterized by a median eye and three pairs of appendages (first and second antennae and mandibles). Further development, including addition of the postmandibular appendages, is accompanied by embryonic molts and results in the postnauplius (or metanauplius). The stage that is finally released from the egg envelopes at eclosion is called the prelarva (or prezoea).

Soon after hatching, the prelarva undergoes a molt that results in the first larval stage (stage I). The three larval stages, sometimes called mysid larvae, are considered to be the equivalent of the zoeal stages of other decapods because they use thoracic appendages for locomotion (Gurney, 1942; Williamson, 1982).

Several external morphological features, which are easily observed through a dissecting microscope, are used to distinguish the early, free-living stages of *Homarus americanus* (Figs. 4–7). The recognition characters are generally details of the appendages and body form (Hadley, 1905, 1906; Herrick, 1909). Although the measurements of the early stages may vary with the study and conditions of growth, some

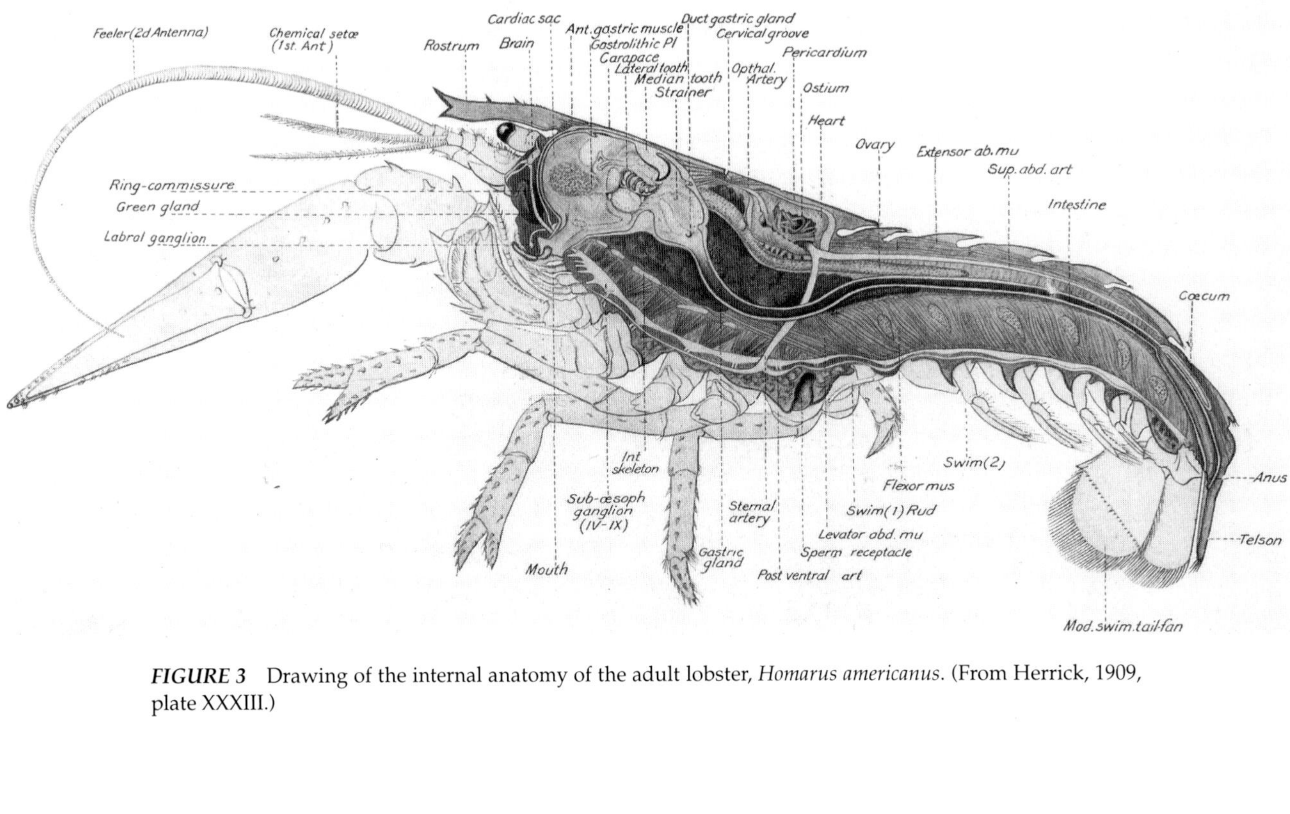

FIGURE 3 Drawing of the internal anatomy of the adult lobster, *Homarus americanus*. (From Herrick, 1909, plate XXXIII.)

are included as guidelines. Carapace lengths are from Gruffyd *et al.* (1975), as cited by Charmantier *et al.* (1991), based on 25 lobsters in each stage; total lengths are from Herrick (1895, Table 25), based on 2–79 lobsters in each stage; weights are from Charmantier and Aiken (1987), as cited by Charmantier *et al.* (1991), based on 40–50 lobsters in each stage.

Stage I larvae (Fig. 4) are distinguished by the presence of exopodites on the pereiopods, which are used for swimming, and by the absence of abdominal pleopods and uropods. The telson is simple and fork shaped. Carapace length averages 1.74 mm and total length averages 7.84 mm (range 7.50–8.03).

Stage II larvae (Fig. 5) appear similar to the first stage. The major difference, however, is that the second through fifth abdominal segments now bear rudimentary pleopods (pleopod buds) without setae. Carapace length averages 2.64 mm, total length averages 9.20 mm (range 8.3–10.2), and weight averages 6.5 mg.

In stage III (Fig. 6), thoracic exopodites are still

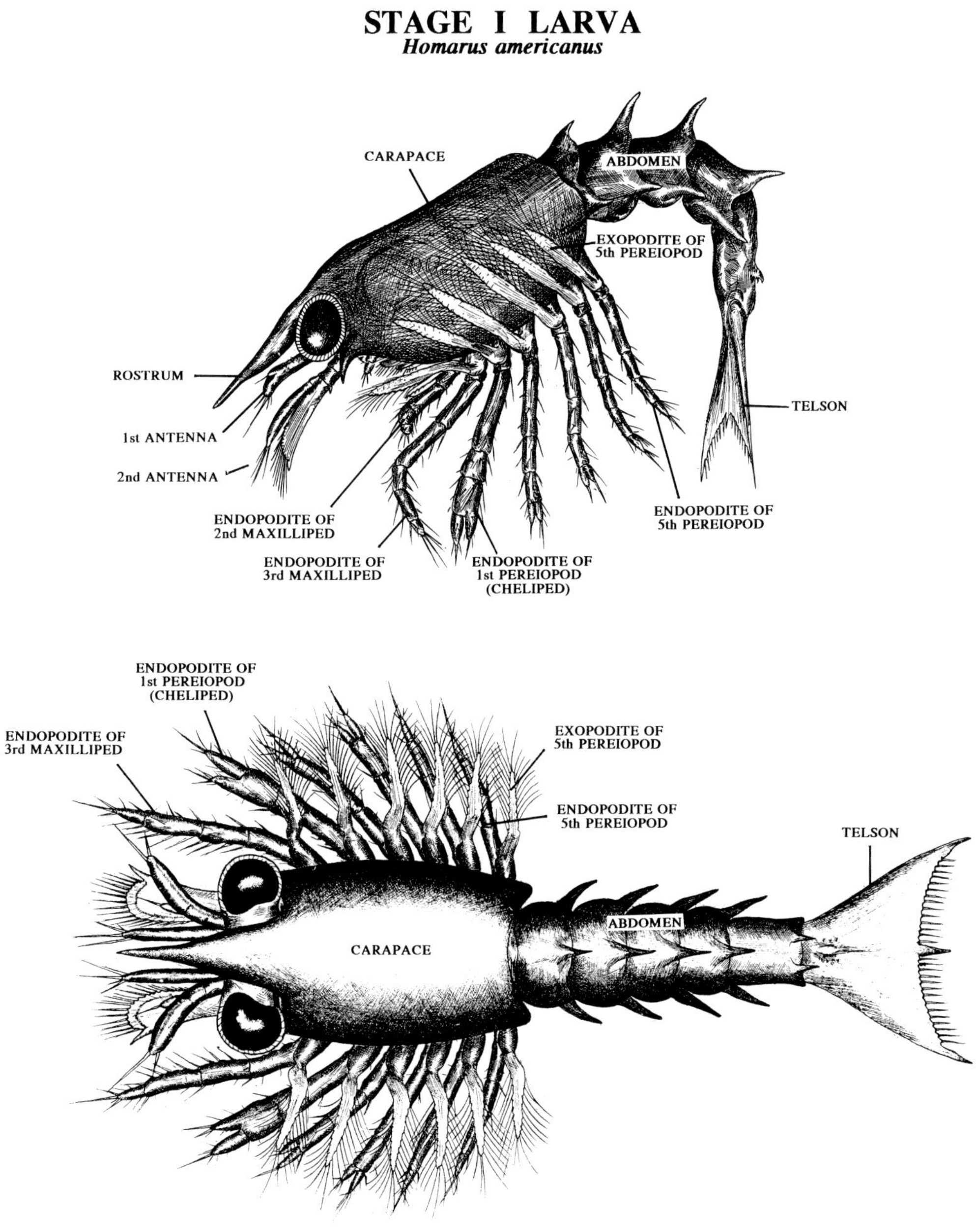

FIGURE 4 Stage I larva, *Homarus americanus*. Lateral and dorsal views. Length of specimen, 8 mm TL; age, 3 days. (Drawn from life by P. B. Hadley, 1906; labels added.)

present, but the pleopods on abdominal segments two through five now bear a delicate fringe of setae. In addition, the telson and newly developed uropods (appendages of the sixth abdominal segment) together constitute a broad tail fan. Carapace length averages 3.20 mm, total length averages 11.1 mm (range 10–12), and weight averages 14.5 mg.

The molt of the stage III larva into the postlarva (stage IV) is generally considered to represent metamorphosis in *Homarus americanus* (Charmantier *et al.*, 1991). The carapace length of stage IV averages 3.75 mm, total length averages 12.6 mm (range 11–14), and weight averages 24.5 mg. The postlarva begins to look like a miniature adult (Fig. 7), but represents a transitional stage in several ways (see Lawton and Lavalli, Chapter 4). The postlarva is the equivalent of the megalopal stage (or glaucothoe) of other decapods. Williamson (1982, p. 50) prefers the term megalopa for "the first stage with functional pleopods in any group"; postlarva, however, is

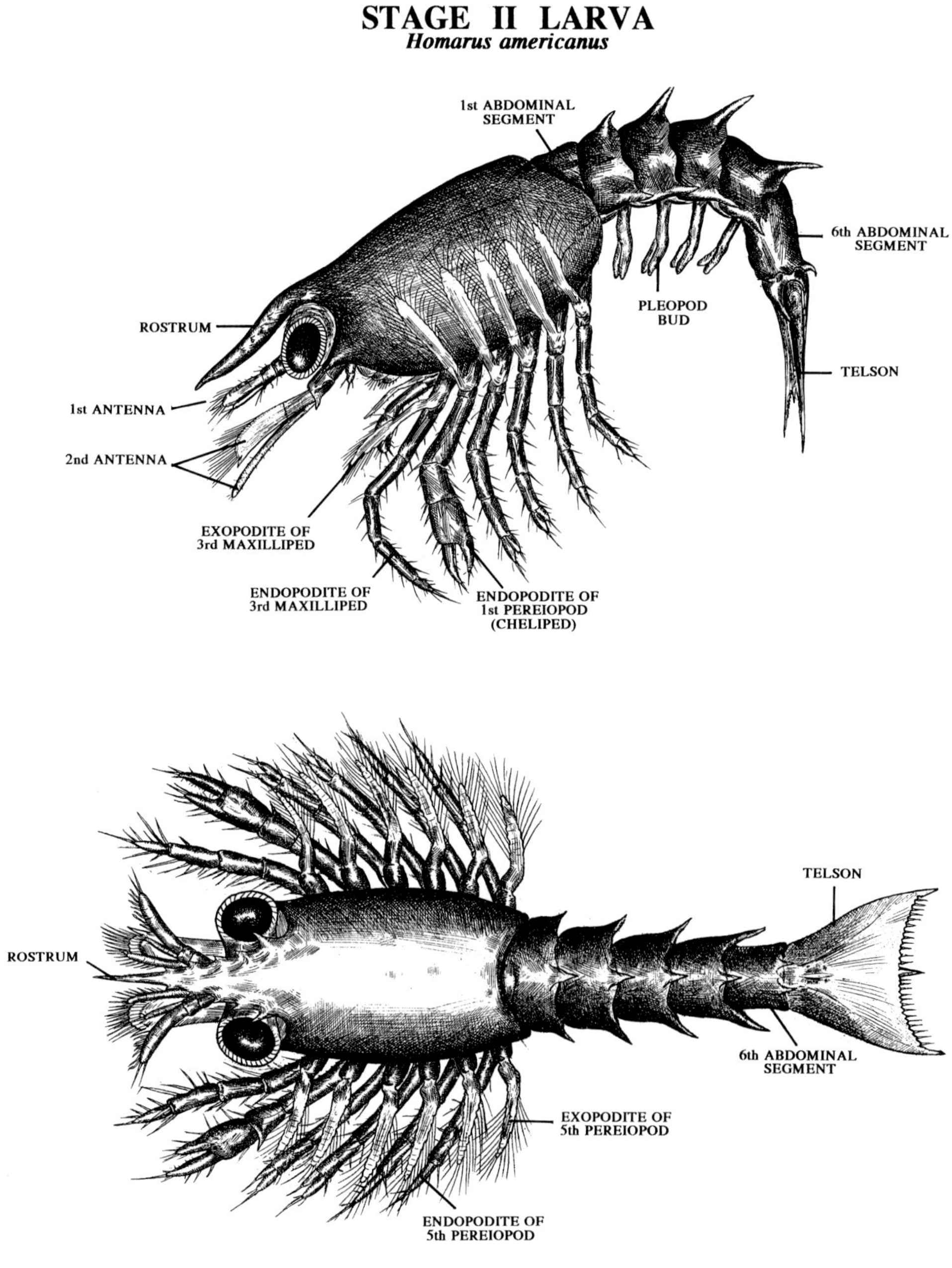

FIGURE 5 Stage II larva, *Homarus americanus*. Lateral and dorsal views. Length of specimen, 9.5 mm TL; age, 6 days. (Drawn from life by P. B. Hadley, 1906; labels added.)

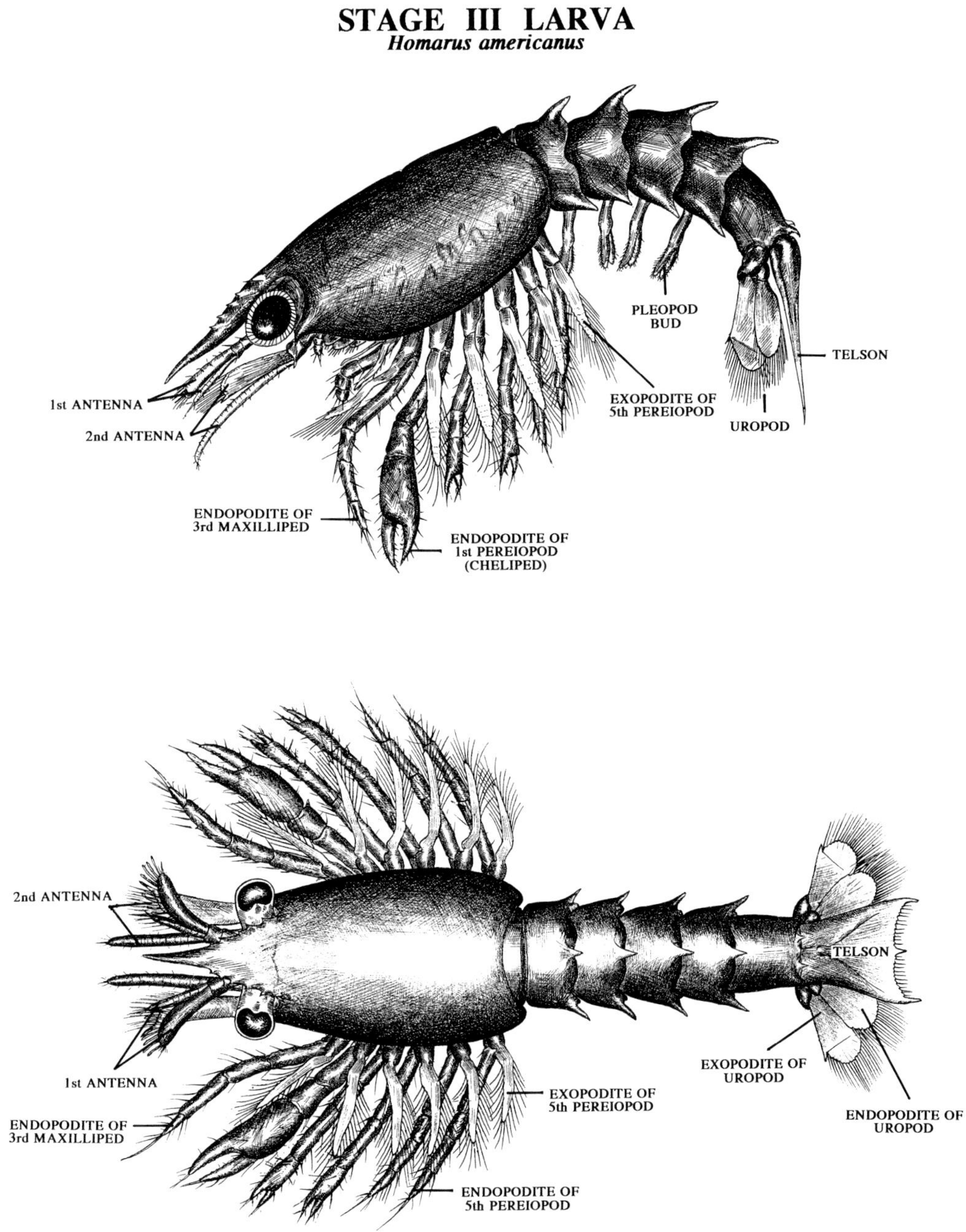

FIGURE 6 Stage III larva, *Homarus americanus*. Lateral and dorsal views. Length of specimen, 11.5 mm TL; age, 9 days. (Drawn from life by P. B. Hadley, 1906; labels added.)

already in wide use for lobsters. While the postlarva of *H. americanus* clearly represents a metamorphic transformation from the third stage, it is not dramatically different in general form from the early juvenile stages that follow (beginning with stage V); this distinguishes it from the megalopa of a brachyuran, for example, which has a markedly different body form from the subsequent "first crab" stage, which looks like a miniature adult crab.

Although morphological changes occur throughout the development of the lobster, a variety of notable external and internal anatomical changes mark the metamorphosis to stage IV (compare Figs. 6 and 7 and Figs. 8a and 8b). Externally, the second antennae develop long, whip-like endopodites. The thoracic appendages lose the swimming exopodites

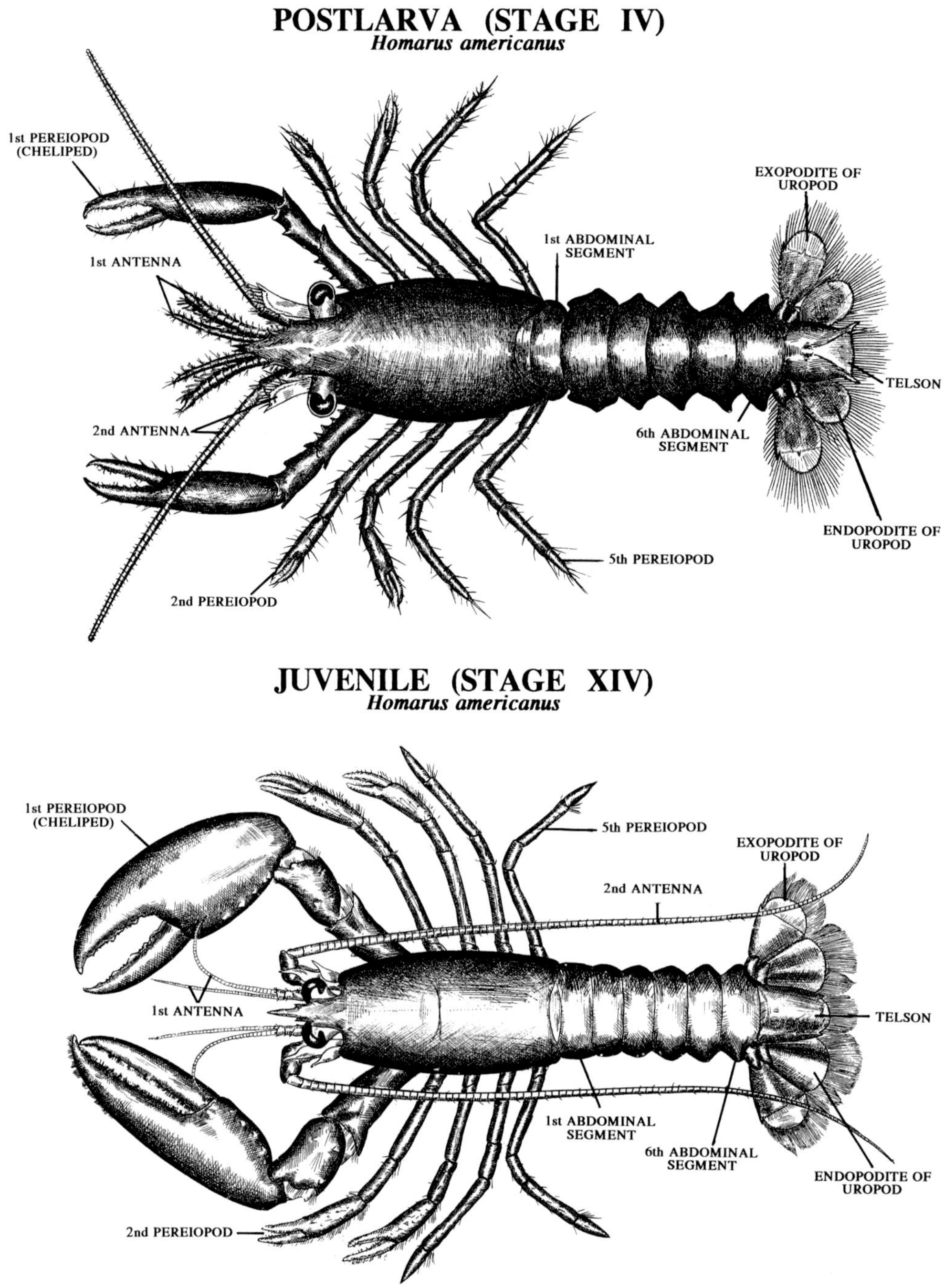

FIGURE 7 Stage IV postlarva and juvenile, *Homarus americanus*. Dorsal views. Length of stage IV specimen, 14.8 mm TL; age, 14 days. Length of juvenile male specimen, 65 mm TL; age, 14 months. (Drawn from life by P. B. Hadley, 1906; labels added.)

characteristic of the three larval stages; enlarged and strengthened pleopods now assume the function of swimming, which shifts from the thoracic to the abdominal appendages at metamorphosis. The uropods become equal in length to the telson.

The larval stages are strictly planktonic, while stage IV postlarval lobsters begin to take up the benthic existence typical of all subsequent stages. This change in habitat is brought about by behavioral responses to light, pressure, and gravity. The behaviors of larval lobsters keep them in the plankton and changes that begin in stage IV cause lobsters to seek

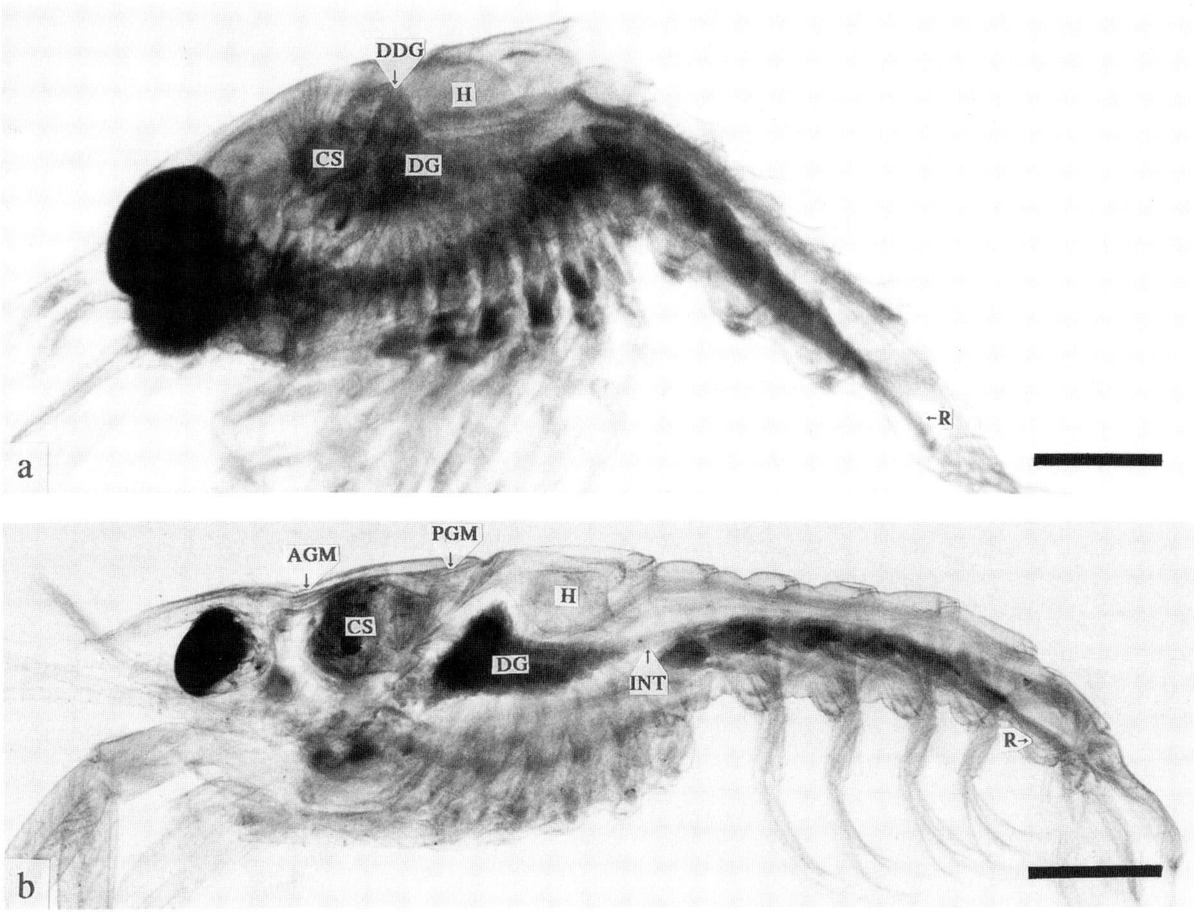

FIGURE 8 Stage I larva (a) and stage IV postlarva (b), *Homarus americanus*. Aspects of internal anatomy are visible (several landmarks are labeled). Photomacrographs of whole-mounted specimens, transmitted illumination. AGM, Anterior gastric muscle; CS, cardiac stomach; DDG, dorsal lobe of digestive gland; DG, digestive gland; H, heart; INT, intestine; PGM, posterior gastric muscle; R, rectum. Scale bars: (a) 0.5 mm; (b) 1.0 mm.

the bottom (see Ennis, Chapter 3, and Lawton and Lavalli, Chapter 4). The nephrosacs of the antennal glands and the statocysts apparently become functional in stage IV (Neil *et al.*, 1976; Waite, 1899). The function of swimming shifts from the thoracic exopodites in the larva to the abdominal pleopods in the postlarva; the well-developed pleopods confer considerable swimming ability on the postlarva, which can also walk on the bottom. Ecologically and behaviorally, the postlarva can be considered a transitional stage, with considerable swimming and bottom-testing behavior to find suitable habitat. By late stage IV (or perhaps stage V, as settlement can be delayed if necessary), the lobster assumes the benthic habitat. Locomotion again becomes a thoracic function as the lobster uses the endopodites of the pereiopods to walk on the bottom.

Dramatic changes, both external and internal, in the feeding apparatus (see Lavalli and Factor, Chapter 14, and Factor, Chapter 15) are coordinated with the behavioral and locomotory changes. Interesting developmental changes occur in the mouthparts (particularly the mandibles and third maxillipeds) at metamorphosis and the development of a gastric mill with heavily cuticularized teeth enables the lobster to grind food into minute particles appropriate for final digestion. Physiological changes in nutrient storage also accompany metamorphosis (Biesiot, 1982, 1986).

Together, the developmental changes that occur at metamorphosis (reviewed by Charmantier *et al.*, 1991) prepare the lobster for the transition to the benthic environment.

A series of juvenile stages begins with stage V, and

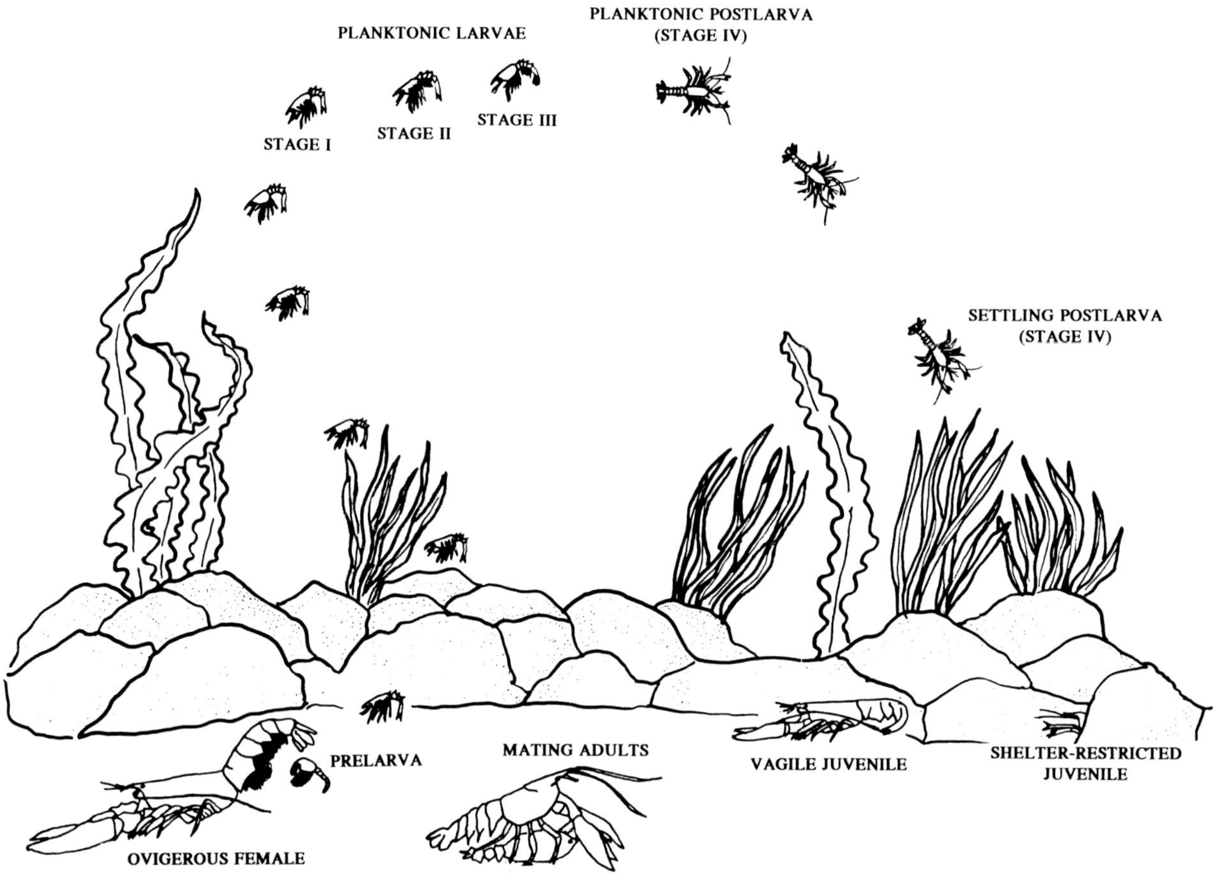

FIGURE 9 Diagrammatic summary of the life history of the American lobster, *Homarus americanus*. (Redrawn with permission by K. L. Lavalli, based on a drawing by R. E. Duggan as published in Harding, 1992.)

these can be grouped into several life history phases (see Lawton and Lavalli, Chapter 4). The movements of the recently settled shelter-restricted juvenile (carapace length ~4–14 mm) appear to be limited within its subterranean habitat. It is a suspension feeder on planktonic organisms, as well as a browser on food found within the shelter and at its entrance. The next phase, the emergent juvenile (carapace length ~15–25 mm), is mostly confined to the shelter, but undertakes limited forays in the vicinity of the shelter. The vagile juvenile (carapace length ≳25 mm) is a shelter user, as are all subsequent phases, but makes more extensive movements out of the shelter for food.

Adolescent-phase lobsters (carapace length ~50 mm) are marked by physiological, but not functional, sexual maturity. They are mostly nocturnal and may participate in seasonal movements with reproductive animals. The adult phase (carapace length ≳50 mm) begins with the onset of functional sexual maturity. In males, functional maturity begins with the capability of mating with and inseminating a female, given a reasonable opportunity. In females, functional maturity is clearly indicated when external eggs are present; individuals not carrying eggs, however, may also be mature. Mating apparently occurs in the shelters of male lobsters (Atema and Voigt, Chapter 13).

The life history of *Homarus americanus* is summarized diagrammatically in Fig. 9. Most life history phases are imperfectly understood, and some have been recognized only recently. The long-standing mystery of the first year of benthic life is finally yielding as understanding of early benthic habits grows and several phases of juvenile life are differentiated.

The mysteries of the biology of any complex organism do not yield easily to a brief consideration, or even to a serious overview chapter. Ultimately, the reader must assemble a personal synthesis of this species based on a realistic understanding of the complexities of the subject. The chapters that follow will assist the reader in this goal by presenting important details and knowledgeable syntheses of the various topics that together constitute the biology of the lobster, *Homarus americanus*.

Acknowledgments

I thank Drs. John M. Anderson, Michael Fogarty, and Austin Williams for their bibliographic detective work on my behalf and John M. Anderson and Kari Lavalli for offering helpful comments about the manuscript.

References

Biesiot, P. (1982). Development of the hepatopancreas in early stages of the American lobster *Homarus americanus*. *Am. Zool.* 22, 871.

Biesiot, P. M. (1986). "Changes in Midgut Gland Morphology and Digestive Enzyme Activities Associated with Development in Early Stages of the American Lobster *Homarus americanus*." Ph.D. thesis. Massachusetts Institute of Technology–Woods Hole Oceanogr. Inst. Joint Program, Woods Hole, Massachusetts.

Charmantier, G., and Aiken, D. E. (1987). Intermediate larval and postlarval stages of *Homarus americanus* H. Milne Edwards, 1837 (Crustacea, Decapoda). *J. Crust. Biol.* 7, 525–535.

Charmantier, G., Charmantier-Daures, M., and Aiken, D. E. (1991). Metamorphosis in the lobster *Homarus* (Decapoda): A review. *J. Crust. Biol.* 11, 481–495.

Cobb, J. S., and Phillips, B. F., eds. (1980a). "The Biology and Management of Lobsters," Vol. I, "Physiology and Behavior." Academic Press, New York.

Cobb, J. S., and Phillips, B. F., eds. (1980b). "The Biology and Management of Lobsters," Vol. II, "Ecology and Management." Academic Press, New York.

Felgenhauer, B. E. (1992). Internal anatomy of the Decapoda: An overview. *In* "Microscopic Anatomy of Invertebrates" (F. W. Harrison, ed.), Vol. 10, "Decapod Crustacea" (F. W. Harrison and A. G. Humes, eds.), pp. 45–75. Wiley-Liss, New York.

Gruffyd, L. D., Riser, R. A., and Machin, D. (1975). A comparison of growth and temperature tolerance in larvae of the lobsters *Homarus gammarus* and *Homarus americanus*. *Crustaceana* 28, 23–32.

Gurney, R. (1942). "Larvae of Decapod Crustacea." Ray Soc., London.

Hadley, P. B. (1905). Changes in the form and color in successive stages of the American lobster (*Homarus americanus*). Preliminary report. *Annu. Rep.—R. I. Comm. Inland Fish.* 35, 44–80.

Hadley, P. B. (1906). Regarding the rate of growth of the American lobster (*Homarus americanus*). *Annu. Rep.—R. I. Comm. Inland Fish.* 36, 153–235 + plates XXVI–XXXVII.

Harding, G. C. (1992). American lobster (*Homarus americanus* Milne Edwards): A discussion paper on their environmental requirements and the known anthropogenic effects on their populations. *Can. Tech. Rep. Fish. Aquat. Sci.* No. 1887.

Herrick, F. H. (1895). The American lobster: A study of its habits and development. *Bull. U.S. Fish Comm.* 15, 1–252 + plates A–J and 1–54.

Herrick, F. H. (1909). Natural history of the American lobster. *Bull. U.S. Bur. Fish.* 29, 149–408 + plates XXVIII–XLVII.

Lewis, R. D. (1970). A bibliography of the lobsters, genus *Homarus*. *U.S., Fish Wildl. Serv., Spec. Sci. Rep.—Fish.* No. 591.

McLaughlin, P. A. (1980). "Comparative Morphology of Recent Crustacea." W. H. Freeman and Co., San Francisco.

Milne Edwards, H. (1837). "Histoire Naturelle des Crustacés, Comprenant l'Anatomie, la Physiologie, et la Classification de ces Animaux." Vol. II. Librairie Encyclopédique de Roret, Paris.

Neil, D. M., Macmillan, D. L., Robertson, R. M., and Laverack, M. L. (1976). The structure and function of thoracic exopodites in the larvae of the lobster *Homarus gammarus* (L.). *Philos. Trans. R. Soc. London* 274B, 53–68.

Nowak, W. S. W. (1972). The lobsters (Homaridae) and the lobster fisheries: An interdisciplinary bibliography. "Marine Science Research Laboratory Technical Reports" No. 6, Memorial Univ. of Newfoundland, St. John's.

Schram, F. R. (1986). "Crustacea." Oxford Univ. Press, New York.

Waite, F. C. (1899). The structure and development of the antennal glands in *Homarus americanus* Milne Edwards. *Bull. Mus. Comp. Zool.* 35, 151–210.

Williamson, D. I. (1982). Larval morphology and diversity. *In* "The Biology of Crustacea" (D. E. Bliss, ed.), Vol. 2, "Embryology, Morphology, and Genetics" (L. G. Abele, ed.), pp. 43–110. Academic Press, New York.

CHAPTER

2

Taxonomy and Evolution

AUSTIN B. WILLIAMS
National Marine Fisheries Service Systematics Laboratory
National Museum of Natural History
Smithsonian Institution
Washington, D.C. 20560

I. Introduction

This chapter introduces the American lobster (*Homarus americanus*) in its systematic setting and summarizes what is known concerning evolution of the species. It seems appropriate to undergird this discussion by listing part of the systematic hierarchy that includes lobsters and lobster-like decapod crustaceans. Vernacular names help to distinguish selected categories in this outline, and ranges of fossil records for the major groups set forth their antiquity. Synonymies of generic and specific names for the American lobster precede diagnostic features for the species, and its geographic distribution is given along with that of its immediate relatives (Glaessner, 1969; Holthuis, 1991). A phylogenetic discussion concludes the chapter.

II. Taxonomy and Systematic Hierarchy

The genus *Homarus* Weber, 1795, contains two currently recognized living species: *H. americanus* H. Milne Edwards, 1837, the American lobster, and *H. gammarus* (Linnaeus, 1758), the European lobster. *Homarinus capensis* (Herbst, 1792), the poorly known Cape lobster (see Kornfield *et al.*, 1995), has been removed from *Homarus*.

The systematic position of lobsters and lobster-like forms in the crustacean order Decapoda follows, with selected vernacular names, ranges of fossil records, and geographic distributions (Glaessner, 1969; Holthuis, 1991).

Order Decapoda Latreille, 1803. Permotriassic–Recent.
- Suborder Pleocyemata Burkenroad, 1963. Permotriassic–Recent.
  - Infraorder Astacidea Latreille, 1803. Permotriassic–Recent.
    - Superfamily Nephropoidea Dana, 1852 (clawed marine lobsters). Permotriassic-Recent.
      - Family Thaumastochelidae Bate, 1888. Recent.
      - Family Nephropidae Dana, 1852. Mid-Jurassic–Recent.
        - Subfamily Neophoberinae Glaessner, 1969. Mid-Jurassic–Recent.
        - Subfamily Thymopinae Holthuis, 1974. ?Palaeocene–Recent.
        - Subfamily Nephropinae Dana, 1852. Lower Cretaceous-Recent.
          - *Homarinus capensis* (Herbst, 1792). Recent. South Africa, from Table Bay to Transkei (Holthuis, 1991; Kornfield *et al.*, 1995).
          - *Homarus americanus* H. Milne Edwards, 1837. Recent. Atlantic coast of North America between the Strait of Belle Isle, Newfoundland, Canada, and North Carolina, rarely southward to off

BIOLOGY OF THE LOBSTER *Homarus americanus*
Jan Robert Factor, Editor

Miami, Florida (Holthuis, 1974, 1991; Cofer-Shabica and Nielsen, 1988).
Homarus gammarus (Linnaeus, 1758). Recent. Eastern North Atlantic from northwestern Norway (Lofoten Islands) to the Azores and coast of Morocco; Mediterranean Sea, except in extreme eastern part east of Crete; and along northwest coast of Black Sea. Not present in Baltic Sea (Holthuis, 1991).
Superfamily Astacoidea Latreille, 1803 (freshwater crayfishes of Northern Hemisphere). Upper Jurassic or Lower Cretaceous-Recent.
Superfamily Parastacoidea Huxley, 1879 (freshwater crayfishes of Southern Hemisphere). Pleistocene-Recent.
Infraorder Palinuridea Latreille, 1803 (spiny lobsters and allies). ?Lower Triassic, Mid-Triassic–Recent.
Superfamily Eryonoidea De Haan, 1841. Upper Triassic-Recent.
Superfamily Glypheoidea von Zittel, 1885. ?Lower Triassic, Mid-Triassic–Recent.
Superfamily Palinuroidea Latreille, 1803. Lower Jurassic–Recent.
Family Palinuridae Latreille, 1803. Lower Jurassic-Recent.
Family Synaxidae Bate, 1881. Recent.
Family Scyllaridae Latreille, 1825. Lower Cretaceous–Recent.
Subfamily Ibacinae Holthuis, 1985. ?Upper Cretaceous, ?Oligocene–Recent.
Subfamily Arctidinae Holthuis, 1985. Recent.
Subfamily Scyllarinae Holthuis, 1985. ?Lower Cretaceous, ?Lower Eocene–Recent.
Subfamily Theninae Holthuis, 1985. Recent.
Infraorder Thalassinidea Latreille, 1831 (lobster, ghost, and mud shrimps).
Superfamily Thalassinoidea Latreille, 1831.
Family Thalassinidae Latreille, 1831. Pleistocene-Recent.
Family Axianassidae Schmitt, 1924. Recent.
Family Axiidae Huxley, 1879. Lower Jurassic-Recent.
Family Laomediidae Borradaile, 1903. Lower Miocene–Recent.
Family Callianideidae Kossmann, 1880. Recent.
Family Callianassidae Dana, 1852. Lower Cretaceous–Recent.
Family Ctenochelidae Manning and Felder, 1991. Recent.
Family Upogebiidae Borradaile, 1903. Upper Jurassic–Recent.

From this listing, it can be seen that the only group of direct interest in this discussion is the family Nephropidae, and narrowly the subfamily Nephropinae, but these and other categories in the hierarchy are subjects of current revision as new information becomes available.

Homarus Weber, 1795

Homarus Weber, 1795:94. Name 494, placed on *Official List of Generic Names in Zoology* (Hemming and Noakes, 1958), Opinion 104 (originally published in 1928). Type species: *Astacus marinus* Fabricius, 1775 (=*Cancer gammarus* Linnaeus, 1758).

Homarus Guérin-Méneville, 1825, 10:768. Type species, by original designation and monotypy: *Cancer gammarus* Linnaeus, 1758.

Homarus H. Milne Edwards, 1837, 2:333. Type species: *Homarus vulgaris* H. Milne Edwards, 1837 (=*Cancer gammarus* Linnaeus, 1758).

Holthuis (1974:815, 1991:57) explained this synonymy in great detail, pointing out that the name *Homarus* was independently chosen for this genus by three different authors, each of the homonyms having different nominal species selected as their types.

Homarus americanus H. Milne Edwards, 1837

Homarus americanus H. Milne Edwards, 1837, 2:334.
Astacus marinus Say, 1817:165 (not Fabricius, 1775).
Astacus americanus Stebbing, 1893:203, Figs. 17–19 (larvae from S. I. Smith).
Homarus mainensis Berrill, 1951:238, 1956:224 (suggested substitute name).

The type locality of *Astacus marinus* Say and *Homarus americanus* H. Milne Edwards is "Longbranch, part of the coast of New Jersey" (Say, 1817:166). (Holthuis, 1991).

The preceding synonymy of scientific names for the American lobster includes only original descriptions, not the numerous biological references to the species.

The descriptive characters for *Homarus americanus* that follow are from Williams (1984), paraphrasing Holthuis (1974:815, including Fig. 24).

Essentially smooth, chelate lobster. Rostrum bearing dorsal teeth on lateral margin and usually a ventral tooth. Carapace with postorbital and antennal spines present; distinct median dorsal groove extending from rostrum to posterior margin, subdorsal carinae low. A set of other connecting grooves as follows: postcervical groove distinct in upper part, lower part faint and almost entirely replaced by intercervical groove connecting with cervical groove; urogastric, cervical, and antennal grooves distinct; gastroorbital and hepatic grooves obscure. Thoracic sternum narrow; sternites between first to third legs with narrow single or double median or submedian ridges; that between fourth legs with two posteriorly divergent ridges in male, bearing sperm receptacle in female.

First chelipeds prominent, naked, rather smooth, and asymmetrical; major crusher with gaping fingers bearing molariform teeth in addition to smaller teeth; minor cutter without gape, cutting edges straight and armed with numerous nearly uniform small denticles, larger teeth if present neither molariform nor on cutting edge.

Abdomen smooth, no carina separating tergites from pleura. Telson narrowing posteriorly to convex terminal margin, short transverse proximal ridge and pair of posterolateral spines present. Exopods of uropods with transverse suture. Sternites of second to fifth segments with median spine in male, spines usually absent or greatly reduced in female. Male with first pleopods modified into rigid copulatory stylets.

Color dark bluish green to brownish olive mottled with very dark greenish black spots, often almost black; pleura with reddish tips, orange to whitish below (from Williams, 1987).

In contrast, the European lobster is often bluish to bluish black dorsally, with white tracings or mottlings on the carapace and the abdomen dorsally. Chelipeds have white tubercles; chelae may be suffused with orange tints and walking legs are lighter blue. The whole body may be much lighter in color, and underparts may appear yellowish or white. Recently preserved Cape lobsters were light yellowish orange when photographed.

Morphological differences among *Homarus americanus*, *H. gammarus*, and *Homarinus capensis* are well summarized by Holthuis (1991) in an illustrated key as: *H. gammarus*, nearly identical to *H. americanus*, but with no ventral tooth on the rostrum; *H. capensis*, differing from both of these in its small size, attaining a total length of 10 cm, and having palms of large first chelipeds covered with setae, especially near the lower margin.

III. Evolution

Within the lineage of astacideans, whose fossil record extends from the Permotriassic to the Recent, the family Nephropidae has a record extending from the Mid-Jurassic to the Recent. The family to which *Homarus americanus* belongs thus originated early in the evolutionary development of the lobster hierarchy. The evidence is at least twofold: (1) fossil presence in the rock column establishes geologic ages and (2) ornamentation on the carapace of fossil lobster remains, the most consistently preserved large part of the lobster body, provides a set of grooves and eminences that form time-linked patterns. Ornamentations on other major exoskeletal parts of fossil lobsters, such as abdominal segments and chelipeds, tend to be incomplete and variable, thus precluding their usefulness in comparative treatments of evolutionary change even though they contribute to descriptive accounts of species. Lobsters did not fossilize abundantly, but their slender record is complete enough to establish a trail through time.

Paleontological descriptive literature is replete with analyses of the ornamental patterns on the nephropid lobster carapace. The transversely oblique grooves represent, according to most authorities, external evidence for the fundamental, but variously modified, segmentation of this region of the body. Other grooves, ridges, and tubercles represent additional homologous structures. Summaries of these patterns have been written by Mertin (1941), Glaessner (1960, 1969), Secretan (1960a,b, 1964), Förster (1966, 1967), and Holthuis (1974). On the basis of these patterns, Glaessner (1969) accepted the idea of Mertin (1941) that fossil and recent nephropids are divisible into two subfamilies, the Nephropinae and the Homarinae, but Holthuis (1991) could not substantiate this division on the basis of the morphology of living lobsters and placed the Homarinae in synonymy. To make these arguments intelligible, the main outlines of carapace ornamentation need explanation in order to bring forth evidence for both taxonomic decisions and postulation of evolutionary sequence.

Glaessner (1969:R403, R410, and R626) rendered a great service in illustrating primary segmentation of the lobster carapace that summarized metameric conventions developed by several authors, deriving much of the information from Förster (1966, 1967), and he linked these figures to labeled outlines of selected fossil and living genera. Glaessner pointed out that in several Triassic and many Jurassic decapods "the carapace shows clearly three trans-

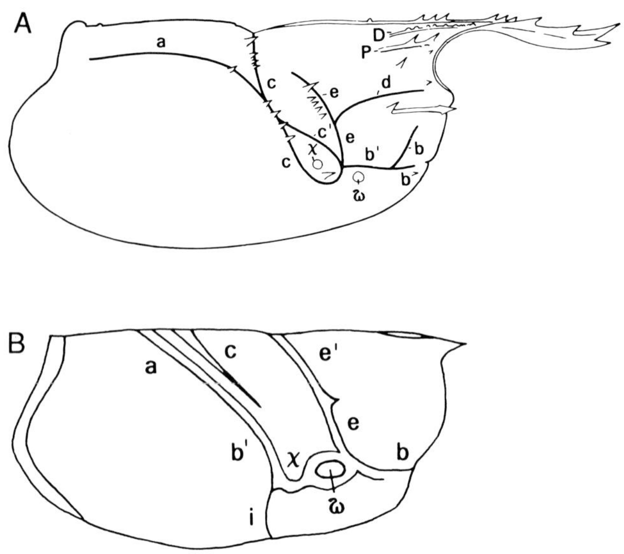

FIGURE 1 Schematic lateral views of the idealized nephropid carapace showing selected grooves, ridges, and tubercles. a, Branchiocardiac groove; b, antennal groove; b′, hepatic groove; c, postcervical groove; c′, intercervical groove; d, gastroorbital groove; e-e′, cervical groove; i, inferior groove; D, subdorsal carina; P, supraorbital carina; ω, external point marking articulation of the mandible on its endosternum; χ, point marking the place of attachment of the musculus dorsoventralis posterior. Both the cervical and branchiocardiac grooves are continued to the middorsal line in *Eryma* and other fossil lobsters. [Adapted and simplified from (A) Holthuis (1974) and (B) Glaessner (1969).]

verse grooves termed **cervical, postcervical,** and **branchiocardiac**" (Fig. 1). These grooves were regarded by both Glaessner (1960) and Secretan (1960a,b, 1964) as remnants of somite boundaries, although Albrecht (1981) regarded the grooves as secondary structures resulting from muscle attachments and therefore useful for systematic purposes but of little phylogenetic significance. Nevertheless, the major grooves are persistent features. The grooves in living lobsters are not all boundaries of muscle attachments, nor do they serve for attachment of membranes, but in primitive lobsters the posteriormost groove, the branchiocardiac, marks the anterior boundary of the branchial chamber. Successively more anterior are the postcervical and cervical grooves. Along with other grooves, this grouping is recognizable in the Penaeidea (penaeid shrimps) and is modified in the Caridea (caridean shrimps) by disappearance of the (intermediate) postcervical groove. In lobsters, the postcervical groove later takes the place and appearance of the cervical groove in dorsal aspect, connected laterally with the cervical groove by the intercervical groove, and (confusingly) is often called the cervical groove.

Aside from these prominent grooves, Glaessner (1960, 1969) and Holthuis (1974) noted other landmarks recognizable on the carapace of lobsters that are of systematic or evolutionary significance. Pertinent in this discussion are the subdorsal carina, behind the rostrum and below the middorsal line; the supraorbital carina, behind the orbital margin and below the subdorsal carina; the antennal carina, behind the antennal spine; the gastroorbital groove, a groove arcing backward from near the orbital margin to connect with the cervical groove; and the hepatic groove, continuing backward from the posterior end of the antennal groove, uniting through a looplike curve with lower ends of the cervical and postcervical grooves, and continuing ventrally as the inferior groove. Within and/or lying near loops formed by the hepatocervical–postcervical grooves are small protuberances that may or may not be well developed. Area ω covers the external articulation of the mandible on its endosternum "and is a valuable pointer to the homologies of carapace grooves" (Glaessner, 1969:R406). Behind ω, area χ marks the place of attachment of the musculus dorsoventralis posterior, near the lower end of the postcervical groove.

The fossil record of the genus *Homarus* and that of similar forms give clues to their probable relationships. The assemblage existed in various ages from the Early Cretaceous Period to the Recent, and the record for *Homarus* was stated by Glaessner (1969) to embrace this entire interval. *Palaeohomarus* Mertin, 1941, treated as a subgenus of *Homarus* by Glaessner (1969), has a fossil record extending from the Mid-Cretaceous (Albian Epoch) to the end of that period. The fossil record for *Hoploparia* McCoy, 1849, considered by some workers to be congeneric with *Homarus* (discussed by Secretan, 1964), extends from the Early Cretaceous to the end of the Eocene. Paleontologists who worked with traditional systematic methods regarded these genera as forming a line of descent (subfamily Nephropinae) separate from, but paralleling, a line leading to modern freshwater crayfishes (superfamily Astacoidea) (Mertin, 1941; Secretan, 1964; Glaessner, 1969; Hasiotis, 1990); the stem from which these lines descended has origins shared by the subfamily Eryminae (including the genus *Eryma* von Meyer, 1840), whose fossil record reaches from the Upper Triassic to the Upper Cretaceous or Lower Tertiary (Förster, 1966, 1967).

Mertin (1941), who recognized the Homarinae as a distinct subfamily, viewed *Hoploparia* as the form ancestral to *Palaeohomarus*, in the Upper Cretaceous, and to *Homarus*, in the Upper Tertiary (Fig. 2).

Secretan (1964) also regarded *Hoploparia*, numer-

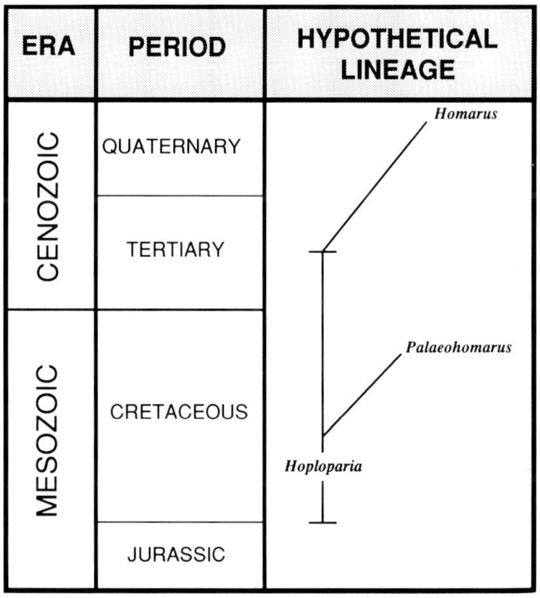

FIGURE 2 Abridged hypothetical phylogeny of *Homarus*, with associated genera, proposed by Mertin (1941).

ous and widely distributed in the Cretaceous and Lower Tertiary, as an ancestral form, but perhaps more central in nephropid lobster phylogeny than had Merton. She thought that, within this large genus, species with attributes trending toward nephropians were numerous and widely distributed in Europe, the Americas, and the Southern Hemisphere; those with homarine tendencies might have been intermediate (Fig. 3) between a group that preserved traces of fairly complete carapace segmentation (i.e., those that resemble *Nephrops*), and still others having greater carapace fusion that might be more nearly related to the freshwater crayfishes, the Astacidae. The latter were viewed as specialized in the Southern Hemisphere in the Lower Cretaceous and later, when they were distributed in Europe around the opening of the Eocene as well as in America. The genus *Homarus* appeared, in Secretan's view, to succeed the cohort with homarine tendencies.

Glaessner (1969) had the advantage of having more material available for developing his evolutionary hypotheses, and he visualized a more diversely branched radiation based on ornamentation of the carapace (Fig. 4). He altered the radiating pattern of carapace form shown in his Fig. 227 to that proposed by Förster (1966, 1967) based on Permian *Protoclytiopsis* Birstein, 1958, from which sprang two lines of descent. One of these passed through Triassic *Clytiella* Glaessner, 1931, into Jurassic *Eryma* to end in Cretaceous *Enoploclytia*. The other line was postulated to pass from Permian *Protoclytiopsis* into Jurassic

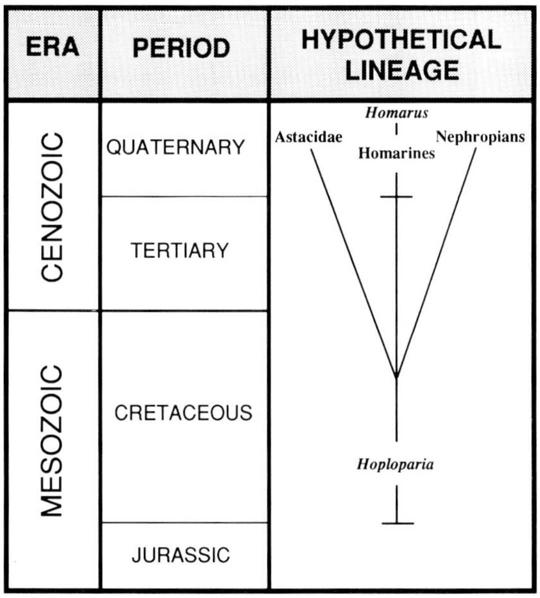

FIGURE 3 Abridged hypothetical phylogeny of *Homarus*, with associated groups, proposed by Secretan (1964).

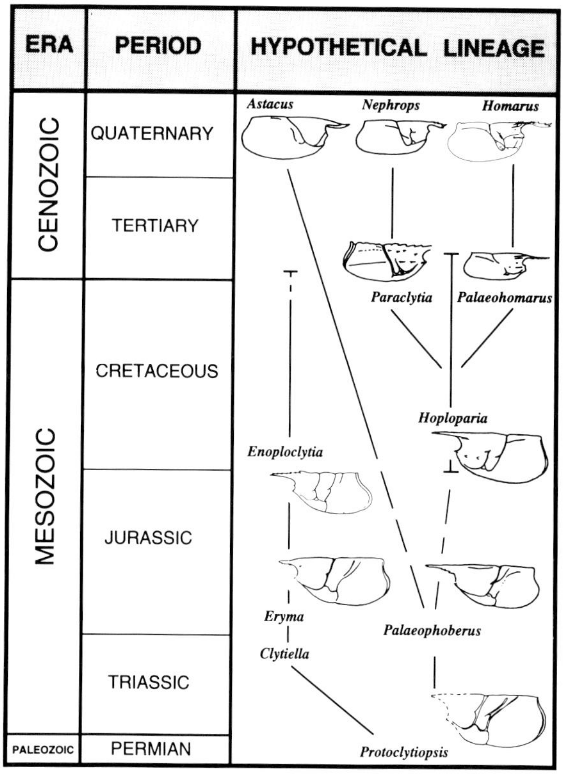

FIGURE 4 Abridged hypothetical phylogeny of *Homarus*, with associated genera, proposed by Glaessner (1969).

Palaeophoberus, which in turn led to modern *Astacus* on one hand, and into Cretaceous *Hoploparia* on the other. *Hoploparia*, in turn, split into two lines, Cretaceous *Paraclytia* leading to Recent *Nephrops* on one hand, and Cretaceous *Palaeohomarus* leading to Recent *Homarus* on the other.

All of the fossil material is fragmentary, but the basic outlines of carapace ornamentation suggested that a cladistic approach to analysis of this series might yield more objective insights into the line of descent for *Homarus* than had the traditional systematic methods, although it is important to remember that any such approach is hypothetical. Any set of fossils is only a sample of reality, and therefore if one attempted to simplify the analysis by restricting attention to Glaessner's (1969) suggested lines leading from lower Mesozoic fossil lobsters to modern *Homarus*, what evolutionary tree would appear? A suite of characters that can be found on *Homarus*, *Palaeohomarus*, and *Hoploparia*, the line that seems to lead from the Mesozoic ancestors, was used for analysis (Tables 1 and 2). *Eryma* was chosen as an outgroup, based on Glaessner's (1969) figures. This set of characters provided a single tree (Fig. 5) with a consistency index of 100%. The entire clade is defined by seven synapomorphies. Five autapomorphies define *Hoploparia*. Three and four autapomorphies, respectively, define *Palaeohomarus* and *Homarus*. The latter two genera are distinguished from *Hoploparia* by a single synapomorphy. The evidence thus supports the conclusions of Mertin (1941), Secretan (1964), and Glaessner (1969) that *Hoploparia*, though closely related to *Homarus*, is distinct from it. Tshudy (1993) has refined this approach in a more extensive analysis.

The slight morphological differences between Recent *Homarus americanus* (American lobster) and *H. gammarus* (European lobster) have been confirmed by biochemical genetics. Rather low levels of genetic variability are present at 44 loci encoding electrophoretically detectable proteins among eight populations of *H. americanus*, the average proportion of heterozygous loci per individual being 3.8% (paraphrasing Tracey *et al.*, 1975). Genetic variation is concentrated at only eight loci, with just five loci having proportions of heterozygotes greater than 20%.

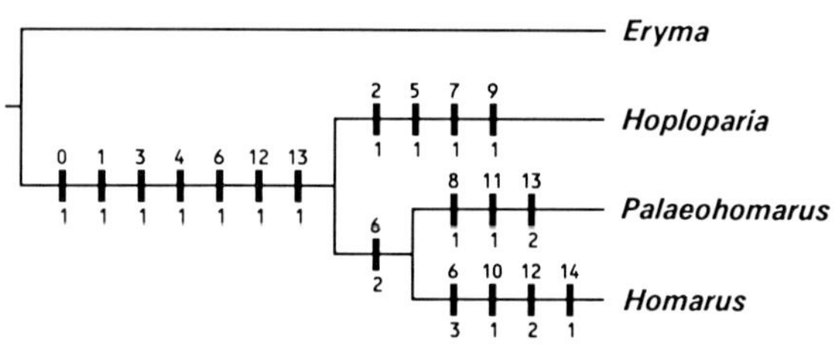

FIGURE 5 An estimated phylogeny among four genera of nephropid lobsters produced with the aid of Hennig86, Version 1.5 (Farris, 1986), and Clados, Version 1.2 (Nixon, 1992). Upper numbers of hash marks on internodes represent characters; lower numbers represent the state of each. Tree length, 19; consistency index, 100; average consistency index, 100.

TABLE 1 **Characters on Carapace of Fossil Lobsters and Living *Homarus* Used for Analysis**

1. Transverse grooves
   - Nearly parallel
   - Not parallel
2. Longitudinal spine rows
   - Without
   - With
3. Cervical groove
   - Deep, moderately inclined, not strongly sinuous
   - Shallow but variable, inclined, sinuous
4. Cervical, postcervical, and branchiocardiac grooves
   - Reaching dorsal midline
   - Only postcervical reaching midline
5. Postcervical groove
   - Not connected directly with hepatic groove
   - Connected directly with hepatic groove
6. Hepatic groove
   - Confluent under χ and ω
   - Confluent under χ
   - Confluent under ω
   - None of these
7. Branchiocardiac groove
   - Not connected with postcervical groove ventrally
   - Connected with postcervical groove ventrally
8. Subdorsal and suborbital carinae
   - Not present
   - Present
9. Cervical groove
   - Not terminated by a λ-shaped sulcus
   - Terminated by a λ-shaped sulcus
10. Suborbital region
    - Produced anterolaterally
    - Oblique
11. Suborbital margin
    - Smooth
    - Spined
12. Tubercles
    - χ and ω distinct
    - Only ω distinct
    - χ and ω indistinct
13. Rostrum
    - Short
    - Moderate length
    - Long and attenuate
14. Rostral lateral carinae
    - Smooth or lightly dentate
    - Strongly toothed

TABLE 2 **Data Matrix for Analysis of Lobsters**

	Transformation series														
Taxon	0	1	2	3	4	5	6	7	8	9	10	11	12	13	14
Outgroup	0	0	0	0	0	0	0	0	0	0	0	0	0	0	0
Palaeohomarus	1	1	0	1	1	0	2	0	1	0	0	1	1	2	0
Hoploparia	1	1	1	1	1	1	1	1	0	1	0	0	1	1	0
Homarus	1	1	0	1	1	0	3	0	0	0	1	0	2	1	1

Genetic identity is above 0.99 in all but three comparisons. Differentiation occurs only at the malic enzyme locus, but the degree of this differentiation supports a previous suggestion from migration and morphological studies that *H. americanus* is divided into a number of more or less geographically isolated inshore and offshore populations, although these populations are genetically similar.

Extension of this approach to *Homarus gammarus* from the coast of Norway and the Irish Sea, similarly examined with respect to 17 functionally different proteins, has shown that 41 loci encoding these proteins are homologous with loci studied in *H. americanus* (paraphrasing Hedgecock *et al.*, 1977). Again, although the different populations have different allelic frequencies at several polymorphic loci, gene frequencies appear quite similar. The average amounts of genetic variability within European and American lobster populations appear to be equivalent. More than one allele is detected at 20% of the loci, the average detected per locus is 1.2, and the average proportion of loci heterozygous per individual is 4.0%.

At 30 loci, *Homarus gammarus* is monomorphic for the common *H. americanus* allele, but polymorphic loci show various degrees of interspecific divergence. Average genetic identity and average genetic distance, compared to values for conspecific population comparisons, indicate that a small but significant amount of divergence separates the European and American lobsters.

It is speculated that *Homarus americanus* and *H. gammarus* were isolated during the Pleistocene Epoch. The apparent weakness of reproductive isolating barriers suggests that these species evolved allopatrically.

The Cape lobster, *Homarinus capensis*, is enigmatic because it had been collected only a few times until recently. Populations living cryptically in shallow waters of South Africa recently have been rediscovered and studied from both morphological and biochemical points of view (Kornfield *et al.*, 1995). The distinction of this species from species of *Homarus* in the Northern Hemisphere is clear-cut.

IV. Summary

The American lobster, *Homarus americanus*, and the European lobster, *H. gammarus*, are included among the Nephropidae along with the recorded time span for all lobster families. The synonymy of *H. americanus* is presented and its distinguishing characters are given. The lineage of the genus *Homarus* is reviewed, as determined by traditional methods based on succeeding changes in the pattern of ornamentation on the carapace, the most consistently fossilized large part of the lobster body. A cladistic analysis shows essential agreement with the earlier conclusions: from an Eryma-like ancestor in the Mid-Jurassic, *Homarus* seems to have emerged along with successive but similar forms, *Hoploparia* and *Palaeohomarus*. *Homarus americanus* and its very close relative, *H. gammarus*, have been shown to exhibit a small but significant amount of genetic difference, and it is speculated that these species were isolated during the Pleistocene Epoch.

Acknowledgments

J. D. Thomas and R. Hershler assisted with the cladistic analysis and R. E. Gibbons provided computer graphics. The manuscript was critically reviewed by G. A. Bishop, B. B. Collette, L. B. Holthuis, R. B. Manning, and D. Tshudy.

References

Albrecht, H. (1981). Zur Deutung der Carapaxfurchen der Astacidea (Crustacea, Decapoda). *Zool. Scr.* 10, 265–271.

Bate, C. S. (1881). On *Synaxes*, a new genus of Crustacea. *Ann. Mag. Nat. Hist., Ser.* 5 7, 220–228 + plate 14.

Bate, C. S. (1888). Report on the Crustacea Macrura Collected by *H.M.S. Challenger* during the Years 1873–76. *Rep. Sci. Res. Voyage H.M.S. Challenger, Zool.* 24.

Berrill, N. J. (1951). "The Living Tide." Dodd, Mead, New York.

Berrill, N. J. (1956). "The Living Tide." Fawcett World Library, New York.

Birstein, J. A. (1958). Ein Vertreter der ältesten Ordo der Crustacea Decapoda, *Protoclytiopsis antiqua* gen. nov. sp. nov. aus dem Perm West-Siberiens. *Dokl. Akad. Nauk SSSR* 122, 477–480. (In Russian.)

Borradaile, L. A. (1903). On the classification of the Thalassinidea. *Ann. Mag. Nat. Hist., Ser. 7* 7, 534–551.

Burkenroad, M. D. (1963). The evolution of the Eucarida, (Crustacea, Eumalacostraca), in relation to the fossil record. *Tulane Stud. Geol.* 2, 2–17.

Cofer-Shabica, S. V., and Nielsen, R. B., Jr. (1988). First record of an American lobster, *Homarus americanus*, from off southern Florida. *Bull. Mar. Sci.* 43, 315–317.

Dana, J. D. (1852). Conspectus crustaceorum, &c. Conspectus of the Crustacea of the Exploring Expedition under Capt. Wilkes, U.S.N., Vol. 6: Including the Paguridea continued, the Megalopidea, and the Macroura. *Proc. Acad. Nat. Sci. Philadelphia* 6, 6–28.

De Haan, W. (1841). Crustacea. *In* "Fauna Japonica Sive Descriptio Animalium, Quae in Itinere per Japoniam, Jussu et Auspiciis Superiorum, Qui Summum, in India Batava Imperium Tenent, Suscepto, Annis 1823–1830 Collegit, Notis, Observationibus et Adumbrationibus Illustravit" (P. F. von Siebold, ed.), Fasc. 5, pp. 109–164 + plates 33–37, 39–42, and 47.

Fabricius, J. C. (1775). "Systema Entomologiae, Sistens Insectorum Classes, Ordines, Genera, Species, Adiectis Synonymis, Locis, Descriptionibus, Observationibus." Libraria Kortii, Flensburg and Leipzig, Germany.

Farris, J. S. (1986). "Hennig86, Version 1.5," computer program. State University of New York at Stony Brook.

Förster, R. (1966). Über die Erymiden, eine alte konservative Familie mesozoischer Dekapoden. *Palaeontographica* 125(A), 61–175 + plates 13–20.

Förster, R. (1967). Die reptanten Dekapoden der Trias. *Neues Jahrb. Geol. Palaeontol., Abh.* 128, 136–194 + plates 9–12.

Glaessner, M. F. (1931). Eine Crustaceenfauna aus den Lunzer Schichten Niederösterreichs. *Jahrb. Geol. Bundesanst. (Austria)* 81, 467–486 + plates 15–17.

Glaessner, M. F. (1960). The fossil decapod Crustacea of New Zealand and the evolution of the order Decapoda. *N. Z. Geol. Surv. Paleont. Bull.* 31, 1–63 + plates 1–7.

Glaessner, M. F. (1969). Decapoda. *In* "Treatise on Invertebrate Paleontology" (R. C. Moore, ed.), Part R, Arthropoda 4, Vol. 2, pp. R399–R628. University of Kansas, Lawrence, Kansas, and Geological Society of America, Boulder, Colorado.

Guérin-Méneville, F. E. (1825). Homard, *Homarus*. Lat. (*Fam. natur.*). *In* "Histoire Naturelle. Entomologie, ou Histoire Naturelle des Crustacés des Arachnides et des Insectes" (M. Latreille, ed.), Encyclopédie Méthodique, etc., Zoologie, Vol. 10, p. 768. Mme. Veuve Agasse, Imprimeur-Librairie, Paris.

Hasiotis, S.T. (1990). "Paleontology, Ichnology, and Paleoecology of the Upper Triassic Chinle Formation of the Canyonlands, Southeastern Utah," M. A. thesis. State University of New York at Buffalo.

Hedgecock, D., Nelson, K., Simons, J., and Shleser, R. (1977). Genic similarity of American and European species of the lobster *Homarus*. *Biol. Bull., (Woods Hole, Mass.)* 152, 41–50.

Hemming, F., ed., and Noakes, D., asst. (1958). "Official List of Generic Names in Zoology. First Instalment: Names 1–1274." International Trust for Zoological Nomenclature, London.

Herbst, J. W. F. (1792). "Versuch einer Naturgeschichte der Krabben und Krebse, nebst einer systematischen Beschreibung ihrer verschiedenen Arten," Vol. 2, Heft 2, pp. 49–78 + plates 30–33. Bey Gottlieb August Large, Berlin and Stralsund, Germany.

Holthuis, L. B. (1974). The lobsters of the superfamily Nephropidea of the Atlantic Ocean (Crustacea: Decapoda). *Bull. Mar. Sci.* 24, 723–884.

Holthuis, L. B. (1985). A revision of the family Scyllaridae (Crustacea: Decapoda: Macrura). I. Subfamily Ibacinae. *Zool. Verh.* 218, 1–130.

Holthuis, L. B. (1991). FAO Species Catalogue. Marine Lobsters of the World. An Annotated and Illustrated Catalogue of Species of Interest to Fisheries Known to Date. *FAO Fish. Synopsis 125*, 13.

Huxley, T. H. (1879). "The Crayfish. An Introduction to the Study of Zoology," Int. Sci. Ser., Vol. 28. Appleton, New York. (Many subsequent editions of this book were issued.)

Kornfield, I., Williams, A. B., and Steneck, R. S. (1995). Assignment of *Homarus capensis*, the Cape lobster of South Africa, to *Homarinus* new genus (Decapoda: Nephropidae). *Fish. Bull.*, 93, 97–102.

Kossmann, R. (1880). "Zoologische Ergebnisse einer Reise in die Küstengebiete des Roten Meeres. Zweite Hälfte, erst Lieferung: III, Malacostraca (2. Theil: Anomura)," pp. 67–140 + plates 4–15. von Wilhelm Engelmann, Leipzig, Germany.

Latreille, P. A. (1803). "Histoire Naturelle, Générale et Particulièr, des Crustacés et des Insects," Vol. 5. Dufart, Paris.

Latreille, P. A. (1825). *In* "Genre de Crustacés. Encyclopédie Méthodique. Histoire Naturelle. Entomologie, ou Histoire Naturelle des Crustacés, des Arachnides et des Insects," Vol. 10. Panckoucke and Laporte, Paris.

Latreille, P. A. (1831). "Cours d'Entomologie; ou, de l'histoire Naturelle des Crustacés des Arachnides, des Myriapodes et des Insects, à l'usage des Élèves de l'Ecole du Muséum d'Histoire Naturelle," Ann. I. Publisher unknown, Paris.

Linnaeus, C. (1758). "Systema Naturae per Regna tria Naturae, secundum Classes, Ordines, Genera, Species, cum Characteribus, Differentiis, Synonymis, Locis," 10th ed., Vol. 1. Laurentii Salvii, Holmiae (Stockholm).

Manning, R. B., and Felder, D. L. (1991). Revision of the American Callianassidae (Crustacea: Decapoda: Thalassinidea). *Proc. Biol. Soc. Washington* 104, 764–792.

McCoy, F. (1849). On the classification of some British fossil Crustacea, with notices of new forms in the University Collection at Cambridge. *Ann. Nat. Hist. Ser. 2* 4, 161–179.

Mertin, H. (1941). Decapode Krebse aus dem Subhercynen und Braunschweiger Emscher und Untersenon, sowie Bermerkungen uber einige verwandte Formen in der Oberkreide. *Nova Acta Leopold.* n.f. **10**(68), 149–264. (Translated from German by the Smithsonian Institution, Washington, D.C., 1981, TT 81-58208.)

Milne Edwards, H. (1837). "Histoire Naturelle des Crustacés, Comprenant l'Anatomie, la Physiologie et la Classification de ces Animaux," Vol. 2. Librairie Encyclopédique de Roret, Paris.

Nixon, K. C. (1992). "Clados, Version 1.2," computer program. L. H. Bailey Hortorium, Cornell University, Ithaca, New York.

Say, T. (1817). An account of the Crustacea of the United States. *J. Acad. Nat. Sci. Philadelphia* 1, 57–169.

Schmitt, W. L. (1924). The macruran, anomuran, and stomatopod Crustacea. *Bijdr. Dierkd.* 23, 61–81 + plate 8.

Secretan, S. (1960a). Essai d'interprétation des sillons des Crustacé Décapodes Astacoures. *C. R. Hebd. Seances Acad. Sci.* 250, 168–169.

Secretan, S. (1960b). Observations relatives au processus d'evolution des sillons chez les Crustacés Décapodes Macroures. *C. R. Hebd. Seances Acad. Sci. (Paris)* 251, 1551–1553.

Secretan, S. (1964). Les Crustacés décapodes du Jurassique supérieur et du Crétacé de Madagascar. *Mem. Mus. Natl. Hist. Nat., Ser. C* 14, 1–224.

Stebbing, T. R. R. (1893). "A History of Crustacea. Recent Malacostraca," Int. Sci. Ser., Vol. 74., Paul, London.

Tracey, M. L., Nelson, K., Hedgecock, D., Shleser, R. A., and Ressick, M. L. (1975). Biochemical genetics of lobsters: Genetic variation and the structure of American lobster (*Homarus americanus*) populations. *J. Fish. Res. Board Can.* 32, 2091–2101.

Tshudy, D. M. (1993). Taxonomy and Evolution of the Clawed Lobster Families Chilenophoberidae and Nephropidae. Ph.D. dissertation, Kent State University. *Diss. Abstr. Int.* 54(8), 4055-B.

von Meyer, H. (1840). "Neu Gattungen fossiler Krebs aus Gebilden von bunten Sandstein bis in die Kreide," pp. 1–28 + 4 plates. Publisher unknown, Stuttgart, Germany.

von Zittel, K. A., . (1885). "Handbuch der Palaeontologie," Vol. 1, No. 2, pp. 679–721. Oldenbourg, Munich, Germany.

Weber, F. (1795). "Nomenclator Entomologicus Secundum Entomologiam Systematicum Ill. Fabricii, Adjectis Speciebus Recens Detectis et Varietatibus...," Apud C. E. Bohn, Kiel, Germany.

Williams, A. B. (1984). "Shrimps, Lobsters, and Crabs of the Atlantic Coast of the United States, Maine to Florida." Smithsonian Institution Press, Washington, D.C.

Williams, A. B. (1987). Spiny Lobsters—Identification, World Distribution, and U. S. Trade. *Mar. Fish. Rev.* 48 (2), 1–36.

CHAPTER

3

Larval and Postlarval Ecology

G. P. ENNIS
Science Branch
Department of Fisheries and Oceans
St. John's, Newfoundland, Canada A1C 5X1

I. Introduction

The 6–8 weeks of planktonic existence is arguably the most complex and least understood phase of the complex life cycle of *Homarus americanus*. This planktonic phase includes three larval stages plus a postlarval stage, during which the critical transition from pelagic to benthic lifestyle occurs. Understanding the factors that influence the distribution, growth, and survival of the planktonic stages is fundamental to understanding recruitment mechanisms and establishing a practical relationship between parental stock and recruitment to the population (see Cobb, Chapter 7, and Fogarty, Chapter 6).

The last summary of the state of our knowledge of larval and postlarval *Homarus americanus* was included in a broad review of the larval ecology of clawed, spiny, slipper, and coral lobsters (Phillips and Sastry, 1980). At that time, the authors concluded that a great many questions remain unanswered, particularly concerning larval behavior and the effects of environmental variables on their planktonic existence. Further, a detailed understanding of many facets of the distribution and abundance of clawed lobster larvae had been hampered by relatively small and highly variable numbers in plankton sampling. Resolution of these issues would require much-improved techniques and a level of sampling entailing enormous logistical problems.

Since 1980, there has been a substantial growth in the literature on larval and postlarval lobsters and an international workshop that focused on recruitment (various authors, 1986). While the older literature provides the foundation for existing knowledge, recent contributions present significant new insights. The objective of this chapter is to provide a comprehensive summary of the present state of our knowledge of the planktonic life of *Homarus americanus*.

II. Hatching and Larval Release

A. Distribution of Ovigerous Females

Hatching and larval release occur following a 9- to 12-month period of embryonic development, during which the eggs are protected and maintained by the ovigerous female. (See Talbot and Helluy, Chapter 9, on reproduction and embryonic development.) Recaptures of ovigerous females caught, tagged, and released in shallow water (<20 m) near Grand Manan, Bay of Fundy, reveal migrations greater than 20 km into deeper water (>200 m), exposing the developing eggs to the maximum temperature available during the winter months (Campbell, 1986). Females return to shallow water the following summer to hatch their eggs when the surface temperature is high (Campbell, 1986), thereby conferring a survival advantage to the pelagic larvae by decreasing their development time to the benthic stage (Caddy, 1979; Hudon and Fradette, 1988). Ovigerous females are sometimes densely aggregated in certain shallow-

BIOLOGY OF THE LOBSTER *Homarus americanus*
Jan Robert Factor, Editor

water areas of Grand Manan at the time of hatching (Campbell, 1990). Some offshore lobsters undergo extensive seasonal migrations from the slope of the continental shelf to shallower shelf waters to take advantage of higher temperatures for molting and reproduction (Cooper and Uzmann, 1980). In these deepwater lobsters, there is no evidence of a migration of ovigerous females to coastal waters, comparable to that near Grand Manan, and the presence of stage I larvae in surface waters indicates that hatching occurs over a broad expanse in offshore waters (Rogers *et al.*, 1968; Lund and Stewart, 1970; Harding *et al.*, 1987a; Harding and Trites, 1988; Katz *et al.*, 1994). However, recaptures of tagged, ovigerous females that were displaced from Veatch Canyon to Narragansett Bay, Rhode Island, indicate that they remained within the bay until their eggs were hatched and then returned to the general offshore area from which they were originally captured (Saila and Flowers, 1968). Coastal lobsters in other areas undergo small-scale, shallow–deep movements (Bergeron, 1967; Munro and Therriault, 1983; Ennis, 1984). In the case of Newfoundland lobsters, these movements are an avoidance response to severe, storm-induced wave action in the nearshore area during autumn and offer no advantage in terms of exposure to higher temperature in slightly deeper water (Ennis, 1984). Larval lobsters are present in surface waters throughout Atlantic Canada in the summer. This indicates that in cold-water areas, such as the northeast coast of Newfoundland, embryonic development still occurs within 12 months, allowing hatching to take place in the summer (see Lawton and Lavalli, Chapter 4).

B. Seasonality of Hatching

The time of onset of hatching in the summer and the duration of the hatching period, as indicated by the presence of stage I larvae in the plankton (stage I duration is 2 days at 22°C), vary from year to year and over the geographical range of *Homarus americanus*. In coastal waters of southern New England, stage I larvae have been taken from late May to mid-August (Lund and Stewart, 1970; Bibb and Hersey, 1979; Fogarty *et al.*, 1983), with peak hatching from late June through early July (Lund and Stewart, 1970; Bibb *et al.*, 1983; Fogarty *et al.*, 1983). In other studies in the same general area, stage I larvae were usually present beyond early to mid-July (Collings *et al.*, 1983; Lux *et al.*, 1983). Stage I larvae were found in Cape Cod Bay from mid-June to mid-August (Lawton *et al.*, 1983). In coastal waters of central Maine, very few stage I larvae were found during June; they were present throughout July, with the greatest numbers during the second half, and up to mid-August (Sherman and Lewis, 1967). Farther north in the Gulf of Maine, hatching occurred from July to September, but peaked during the second half of August at Grand Manan (Campbell, 1986). During 15 years of sampling in Northumberland Strait, stage I larvae were first encountered from mid-June to early July, with peak numbers during July and well into August in some years; they were last encountered during the second half of September (Scarratt, 1964, 1973). On the south coast of Newfoundland, the first occurrence of stage I larvae in annual sampling conducted since 1972 has usually been during the first half of July, but can be as late as the end of July or early August; their last occurrence has varied from the end of July to mid-September (G. P. Ennis, unpublished data). On the northeast coast of Newfoundland, the only lobster larvae found in extensive sampling from mid-May to late September appeared during early to mid-July (Ennis, 1983).

Overall, hatching in *Homarus americanus* takes place during a 4-month period from late May through much of September. The hatching season tends to begin earlier and continue somewhat longer in the southern part of the lobster's range. In an area, both the start and the duration of the season can vary between years by several weeks.

Over a 10-year period at the Massachusetts State Lobster Hatchery and Research Station, the lowest temperature at which hatching occurred was 12.2°C, but hatching usually began at 15.0°C and was most intensive at approximately 20°C (Hughes and Matthiessen, 1962). In the field, bottom temperatures at first occurrence of stage I larvae in plankton samples ranged from 11.0° to 13.6°C south of Cape Cod, from 9.0° to 12.7°C in Cape Cod Bay (the southern Gulf of Maine), from 7.9° to 13.9°C farther north in the Gulf of Maine (Fogarty and Lawton, 1983; Campbell, 1986), from 4.2° to 10.6°C in Northumberland Strait (Scarratt, 1964), and from 10.0° to 13.8°C along the coast of Newfoundland (G. P. Ennis, unpublished data).

The timing of first larval appearance in the plankton tends to coincide with a narrow range of surface temperatures averaging around 12.5°C [see Harding *et al.* (1983), who also provide dates of first appearance and corresponding surface and bottom temperatures covering most of the lobster's range]. Peak abundance of stage I larvae around the Îles de la Madeleine is synchronized with a rapidly increasing surface temperature during July, which induces rapid growth and shortens planktonic life (Hudon and Fradette, 1988; see Section IV,A).

C. Mechanisms of Hatching and Larval Release

Attaining the planktonic larval phase from fully developed eggs involves two steps: the actual escape from egg membranes and the subsequent release of newly hatched larvae by the ovigerous female (Templeman, 1937a). Hatching (eclosion) occurs when the outer membrane of the egg ruptures under the internal pressure generated as water is absorbed by the developing embryo (Davis, 1964). A prelarval stage (prezoea), enveloped in a cuticle and incapable of swimming, emerges from a delicate inner egg membrane that is torn open by the weight of the embryo and swimmeret movements by the female (see Talbot and Helluy, Chapter 9). The first-stage larva emerges from the prelarval molt either in conjunction with rupture of the egg membranes or up to 24 hours later (Aiken, 1980), and remains attached to the cuticle. Release of free-swimming stage I larvae is accomplished by extremely vigorous pleopod beating by the female, usually lasting less than 1 minute (Templeman, 1937a; Ennis, 1975a), and occurs most frequently at night, shortly after darkness. Hatching and prelarval molting go on throughout the 24-hour period between successive larval releases. From one or two up to 2000 larvae may be released at any one time, and the time required to hatch and release a full clutch of eggs can vary from 15 to 31 days (Ennis, 1975a). Immediately upon release, larvae swim upward and swarm within a few centimeters of the surface (Herrick, 1895; Templeman, 1937a).

III. Larval Development and Metamorphosis

Beginning with the newly hatched, prelarval stage, lobster larvae undergo a series of four molts in their development to the postlarval stage, during which settlement onto the bottom occurs.

The main external morphological and anatomical characteristics of stage I larvae, which are about 8 mm long, include a segmented body with functional appendages on all but the abdominal segments, large eyes, a conspicuous rostral spine, long dorsal spines on the abdominal segments, and a triangular telson. Swimming capability is provided primarily by six pairs of exopodites on the third maxillipeds and five thoracic limbs.

Stage II larvae are only slightly larger, approximately 9 mm long, and closely resemble those of the first stage. The main distinguishing feature is the presence of four pairs of nonsetose "swimmerets" (pleopods) on the second through fifth abdominal segments. Other structural changes include the appearance through the transparent cuticle of rudimentary uropods at the base of the tail fan and noticeably larger chelae on the first pereiopods, which are destined to become the great chelae.

The general appearance of stage III larvae is similar to that of the second stage, but they are readily distinguished by their larger size of approximately 11 mm, the presence of the completed tail fan with uropods, larger swimmerets fringed with short setae, and larger claws.

Stages I–III are zoeal stages. During the molt to the next stage, larval lobsters undergo a metamorphosis in which the anatomical characteristics of the larval stages are replaced by those of a juvenile and the animal is recognizable as a lobster. Stage IV is the postlarval stage. Exopodites of the biramous thoracic appendages of the first three stages are reduced to rudiments, leaving uniramous appendages representing the endopodites only. The swimmerets are much larger and more setose and provide the postlarvae with a greatly improved swimming ability. The dorsal abdominal spines are lost, a true statocyst is developed, the chelipeds are much enlarged and extend forward rather than hang downward, and the antennules and antennae become much larger. The postlarval lobster resembles a miniature adult, although the proportions differ, and it is now ready to make the transition from a pelagic to a benthic lifestyle. (See Factor, Chapter 1, and Lawton and Lavalli, Chapter 4, for discussions of life history.)

Intermediate stages resulting from incomplete metamorphosis from stage III to postlarva occur in several forms (Templeman, 1936a; Charmantier and Aiken, 1987). These are usually associated with exposure to toxic substances or other unfavorable conditions and are quite common after eyestalk ablation, but also occur at low frequencies under apparently favorable rearing conditions.

Various other morphological and anatomical as well as ecological, ethological, and physiological characteristics are significantly altered during larval development and metamorphosis (reviewed by Charmantier *et al.*, 1991). Development of the mouthparts and the digestive system (Factor, 1978, 1981, Chapter 15; Hinton and Corey, 1979; Lavalli and Factor, Chapter 14) and of the neuromuscular system (Lang *et al.*, 1977; King and Govind, 1980; Stephens and Govind, 1981; Costello *et al.*, 1981; reviewed by Govind, 1982, Chapter 12) has been described in detail.

The X-organ and the sinus glands are present in the eyestalks of larvae (Pyle, 1943). The latter have been found to contain different types of neurosecretory granules in *Homarus gammarus* larvae, suggesting that they are at least partly functional in larval stages (Rotllant *et al.*, 1991). Further details of various aspects of larval development and metamorphosis are provided in the following sections.

IV. Growth and Survival

Size increase in lobsters is a discontinuous process achieved by molting. It is a function of the incremental increase in size at molting as well as molt frequency. Physiologically, however, growth is a continuous process as a lobster proceeds through an ongoing series of molt cycles of increasing duration over its life span. Under favorable conditions, lobsters undergo three molts in very rapid succession during their planktonic life.

Our direct knowledge of the molt cycle of *Homarus americanus* and its molting physiology and endocrinology is based primarily on studies of adults and much of our understanding is based on studies of other species. The limited information available for larval and postlarval stages indicates similarities with adult lobsters (reviewed by Charmantier *et al.*, 1991; Waddy *et al.*, Chapter 10, provide a comprehensive treatment of molting).

A. Temperature

There is an inverse relationship between temperature and the time required from hatching to attain the postlarval stage (Templeman, 1936b). Larvae take as few as 11 days to reach the postlarval stage at 22°C and as many as 54 days at 10°C. At these temperatures, time from hatching to the end of the postlarval stage ranges from 22 to 103 days, respectively. Stage duration increases with development and, with decreasing temperature, successive stages become progressively longer. As the temperature decreases from 22° to 10°C, the stage duration increases from 2 to 14 days for stage I, from 4 to 15 days for stage II, from 5 to 25 days for stage III, and from 11 to 49 days for the postlarva (Templeman, 1936b; similar results were obtained by MacKenzie, 1988). The postlarval stage is attained at temperatures below 10°C, but molting to stage V does not occur.

The ratio of the duration of a particular stage to total larval development time is constant among temperature treatments (i.e., equiproportional development). Those that develop faster or slower through a given stage show the same deviation, proportionally, in the next stage (MacKenzie, 1988).

Sizes (as expressed by carapace length) attained at the various stages by larvae reared individually tend to be quite variable, but larvae that are large at the time of hatching usually remain larger than average during development (see Section IV,C). Although the smallest size at any stage is generally achieved in larvae reared at the highest temperature (22°C), temperature has relatively little effect on the incremental increase in carapace length at the molt from each larval stage to the next. However, stage V (first juvenile) lobsters reared at 15° and 18°C have significantly greater dry weights and carapace lengths than those reared at 10°, 12°, and 22°C, indicating that both the weight and carapace length increments at the molt from the postlarval stage are affected by temperature (MacKenzie, 1988).

Growth curves derived from mean carapace length at each stage and mean stage duration illustrate the effect of temperature on the growth of laboratory-reared larvae and postlarvae (Fig. 1) (MacKenzie, 1985). Mean carapace lengths for larvae obtained from the wild (Templeman, 1948; Wilder, 1953; Hudon and Fradette, 1988) are within the ranges of carapace lengths in Fig. 1 for stages I and II; however, wild-caught stage III and postlarval lobsters are generally larger, by up to 0.5 and 1.0 mm, respectively. Carapace length molt increments (derived from mean carapace lengths for successive stages) are also greater for wild-caught larvae than for laboratory-reared larvae, at 34–36% for stages I–II, 32–33% for stages II–III, and 30–34% for stage III to postlarva. A growth curve (Fig. 1) produced from estimated molt increments and stage durations for larvae caught near Îles de la Madeleine from late July through early August, when growth is faster than earlier in July and later in August (Hudon and Fradette, 1988), indicates that quite rapid growth can be achieved in the wild. There is, however, considerable spatial and temporal variability in both stage duration and molt increment (Hudon and Fradette, 1988).

Estimated growth rates (in milligrams of protein per day) during the postlarval stage for postlarvae caught near Rhode Island are substantially higher than the highest rate obtained for postlarvae reared in the laboratory under the best possible conditions. Observed variability in postlarval growth rates during the summer in this area appear to be unrelated to temperature or availability of prey (Juinio and Cobb, 1994; see Section IV,E).

Temperature has little effect on stage-specific survival of lobsters in larval stages I and II (>60%), but survival rates in stage III and the postlarval stage are reduced to <26% at 10°C, compared with >75% among those reared at higher temperatures (MacKenzie, 1988). Total cumulative survival to stage V (first juvenile) is 4, 56, 64, 68, and 47% at 10°, 12°,

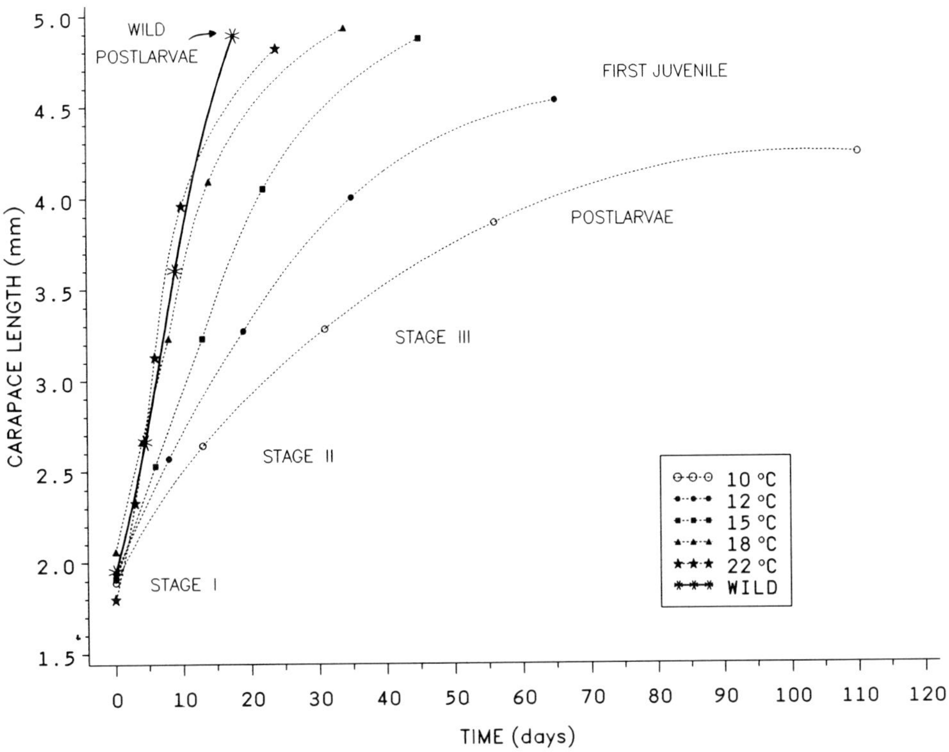

FIGURE 1 Growth of larval and postlarval lobsters in relation to temperature. The five data points for each curve represent the mean carapace length at the accumulated mean stage duration for larvae (stages I–III), the postlarva (stage IV), and the first juvenile (stage V). The curves were drawn by eye based on data from MacKenzie (1985). The curve for larvae in the wild was estimated from data of Hudon and Fradette (1988).

15°, 18°, and 22°C, respectively. Survival in nature depends very much on the time of hatching in relation to the summertime portion of the annual surface temperature cycle (see Section II,B). Even though hatching in Northumberland Strait occurs over a 3-month period in most years, survival to the postlarval stage is greatest for larvae hatched earlier in the season. Overall survival is greatest at an intermediate mean surface temperature (11.1°C) during the period from the onset of vernal warming to the end of stage I production (Caddy, 1979). Early hatching favors attainment of the postlarval stage, which is more dependent on high temperature for completion than the earlier stages (Caddy, 1979), by the time peak surface temperature is reached (Hudon and Fradette, 1988). This induces rapid growth overall, which enhances survival by reducing the duration of exposure to predators in the plankton (Hudon and Fradette, 1988) and ensures successful completion of the postlarval stage before autumn cooling is sufficiently advanced to cause significant delay (Caddy, 1979).

Temperature is the most important factor affecting growth and survival of larval and postlarval lobsters. Reduced temperature prolongs the duration of all stages quite considerably and at low temperature, survival of stage III and the postlarval stage is much reduced. Incremental growth at the molt from one stage to the next is little affected by temperature except at the end of the postlarval stage, in which greater growth is achieved at intermediate temperatures. In the wild, the highest survival rate is associated with hatching earlier in the season, when surface temperature is increasing rapidly. This results in the most rapid growth overall and the shortest duration of planktonic life.

B. Temperature and Salinity

The effect of temperature on the duration of larval development and survival is modified by salinity; tolerance to salinity extremes varies with temperature.

At rearing temperatures of 15° to 17.5°C, the time required to reach the postlarval stage is unaffected by salinity in the 21- to 32-ppt range (Templeman, 1936b). Survival, however, is reduced from ~83% at 31–32 ppt to ~63% at 21–22 ppt. At salinities lower than 20 ppt, survival to the postlarval stage is greatly reduced (<10% at 19–20 ppt), and at 17 ppt few survive even to stage II. Size attained by postlarvae is not affected by salinity in the 21- to 32-ppt range.

Larvae have been cultured at 20 combinations of temperature (10°, 15°, 20°, and 25°C) and salinity (15, 20, 25, 30, and 35 ppt) (Sastry and Vargo, 1977). At 15°C, development is completed to the postlarval stage at salinities from 20 to 35 ppt; at 20°C, the postlarval stage is attained at salinities from 15 to 30 ppt. At 15°C, survival is highest at 35 ppt; at 20°C, however, it is highest at 20–30 ppt. It appears that high temperature reduces tolerance to high salinity, but increases tolerance to low salinity over the 15- to 35-ppt range.

Larval and postlarval lobsters have considerable capacity to osmoregulate in both reduced and increased salinities (Charmantier *et al.*, 1988). Larval stages are nearly isosmotic in molt stage A (postmolt) and hyperosmotic in stage C and stage D (premolt) over a wide range of salinity. (See Waddy *et al.*, Chapter 10, for a discussion of molt stages and the molting cycle.) Postlarvae are hyperosmotic, but their ability to hyperregulate at low salinities increases considerably from postmolt (stage A) to premolt (stage D). Tolerance to low salinity decreases during larval development. The lethal salinity for 50% of the animals after 24 hours of exposure (24-hour LS_{50}) increases from 14 ppt for stage I larvae in molt stage C to 17 ppt at metamorphosis (i.e., premolt stage III larvae and postmolt postlarvae). Tolerance increases in postlarvae in molt stage C (24-hour LS_{50} = 11.6 ppt), but decreases slightly in molt stage D.

C. Light Intensity and Photoperiod

Survival to the postlarval stage is higher for larvae reared in almost complete darkness (with a very brief photoperiod at 1- to 3-day intervals) than for larvae reared in a natural light–dark (LD) cycle (94 versus 78%) (Templeman, 1936b), and similarly (55 versus 45%) for those reared in a very short photoperiod (LD 01:23) compared to a very long photoperiod (LD 23:01) (Aiken *et al.*, 1981). However, for larvae reared in continuous darkness, survival to the postlarval stage is substantially lower (29 versus 54%) than for those reared at LD 12:12 and low light intensity (Eagles *et al.*, 1986). More rapid development to the postlarval stage (80 versus 55% after 14 days at 20°C) occurs at LD 01:23 than at LD 23:01 and in continuous darkness than in LD 12:12 (98 versus 84% after 14 days at 20°C) (Eagles *et al.*, 1986). At LD cycles from 01:23 to 23:01, however, the greatest carapace length molt increments during larval development and the largest mean size of postlarvae occur at the longest photoperiod (Aiken *et al.*, 1981); this is inconsistent, however, with larger postlarvae occurring in near-darkness compared to a natural light cycle (Templeman, 1936b), and in continuous darkness compared to LD 12:12 (Eagles *et al.*, 1986). At three different photoperiods, there is a significant negative correlation between the carapace length of postlarvae and the number of days required to reach the postlarval stage—the largest postlarvae are consistently the first to reach this stage (see Section IV,A).

The effect of photoperiod is variable in larvae reared each month from March through September (Aiken *et al.*, 1982). Survival is greatest at LD 01:23 in all months except May and June, when the best survival occurs at LD 12:12. In the spring and early summer, the postlarval stage is achieved faster at a short photoperiod (LD 01:23), whereas in September, development is faster at a long photoperiod (LD 23:01). The mean carapace length of postlarvae is largest at LD 23:01 in February–March; in April–June, the mean carapace length is comparable at LD 23:01 and LD 01:23, but is shorter at LD 12:12; but in July–August, it is largest at the short photoperiod (LD 01:23).

Light intensity also influences development (Eagles *et al.*, 1986). Larvae reared at the LD 12:12 photoperiod and low light intensity exhibit greater survival to the postlarval stage (44 versus 38%) and larger postlarval size than those reared at higher light intensity (30 times higher, but still less than full sunlight). Development time is unaffected by light intensity.

The general effect of photoperiod on larval growth is variable. Inconsistent, sometimes contradictory, results can be obtained, especially with experiments conducted during different seasons. Nevertheless, the general pattern in experiments during the normal summertime larval period is one of greater survival and faster development to the postlarval stage, as well as larger postlarvae, when they are reared in darkness or a very short photoperiod. Also, greater survival and larger postlarvae are obtained at low rather than high light intensity, while development time is unaffected.

D. Density

Stocking density of laboratory-reared larvae can be expressed as volume of water per individual; density of postlarvae and juveniles can be expressed as bottom surface area per individual. In this way, density can be determined for lobsters raised communally or individually.

There is no significant difference in survival between larvae reared to early juvenile stages at two densities that vary by a factor of 2: 30 ml per larva and 13 cm^2 per postlarva; and 60 ml per larva and 26

cm^2 per postlarva (Sastry and Zeitlin-Hale, 1977). Lobsters reared communally and individually at the same densities exhibit comparable survival rates through the first larval stage (range, 67–78%). For individually reared animals, survival through each subsequent stage increases to 100% through the postlarval stage. Among communally reared larvae, however, survival drops sharply to only 30% through stage II, then increases to about 45% through stage III and 62% through the postlarval stage.

There is no connection between mortality and the time of molting among animals reared individually. Among those reared communally, however, mortality occurs primarily during the period of molting. Within each stage, mortality is directly related to the degree of asynchrony in molting.

Molting and newly molted lobsters are particularly vulnerable to cannibalism, which probably accounts for most of the increased mortality among communally reared animals. However, survival of communally held postlarvae is higher than for the second and third stages, even though the postlarval molt period is longer. This may be associated with the behavioral changes that result in decreased aggression among postlarval individuals in a group, hence reduced mortality through the establishment of dominance hierarchies.

When postlarvae are held in pairs, the first one to molt does so after an average of 12 days, which is the same as for postlarvae held individually. The second in the pair, however, molts after an average of 16 days. A dominance relationship in the interaction between individuals is a hypothetical cause of the delay of molt (Cobb, 1970).

E. Nutrition and Bioenergetics

Reducing the food supply results in reduced survival and increased development time. Halving the usual level of feeding of mostly live copepods reduces survival to the postlarval stage from 60 to 20% and increases the time required to reach the postlarval stage from 25–30 days to 50–55 days (Templeman, 1936b). With a further halving of the food supply, few larvae reach stage II and none reach stage III. A more modest reduction in the food supply (frozen brine shrimp) has very little effect on survival or development time to the postlarval stage or on the size of postlarvae (Eagles *et al.*, 1986). However, poor-quality food (broken and fragmented brine shrimp versus intact adults) reduces survival (20 versus 36%), increases development time (17.6 versus 14.8 days), and reduces size (5.2 versus 6.2 mg dry weight). Live adult brine shrimp are taken more readily by larval lobsters and the growth of these lobsters is significantly faster compared to larvae fed frozen brine shrimp (Carlberg and Van Olst, 1976; Van Olst *et al.*, 1980). Postlarval lobsters have a high resistance to starvation. Most can survive at least 12 days of food deprivation, during which protein catabolism is the main source of energy. They are quite tolerant as well to reduced food supply and can maintain the same stage duration on low rations as with *ad libitum* feeding (Juinio *et al.*, 1992).

Nutritional and bioenergetic aspects of development can be elucidated by examining the various changes in biochemical and physiological processes associated with developmental transitions from the egg through the juvenile stages. Larvae from eggs that develop at a high temperature have relatively larger energy reserves at the time of hatching, possibly providing an important adaptive advantage to newly hatched larvae by delaying the necessity of feeding and prolonging the period of starvation resistance (Sasaki *et al.*, 1986). This indicates that the annual thermal regime to which lobsters are exposed influences larval survival. If sufficient yolk reserves remain at the time of hatching, unfed larvae survive and molt to stage II, although the duration of stage II is prolonged (Anger *et al.*, 1985). Without sufficient reserves, however, early feeding is essential for survival through stage I. Eggs produced by large females tend to be larger and to have a higher energy content (Attard and Hudon, 1987). This indicates a survival advantage for the larvae they produce.

The wet weight of larvae increases 70–80% at each molt, but remains relatively stable between molts. Ash level increases at each molt, particularly in the postlarval stage, in which it represents a greater proportion of the total weight. This is associated with the development of a more heavily calcified exoskeleton following metamorphosis. Protein is the largest biochemical constituent throughout the larval and postlarval stages. In terms of caloric equivalents, protein increases from ~4 calories per individual in early stage I larvae to ~29 calories per individual in late postlarvae, lipid increases from just over 1 to ~16 calories, and carbohydrate remains at a low but stable level throughout (<2 calories). As a percentage of all biochemical components, however, protein decreases during the postlarval stage as the lipid level increases (Sasaki *et al.*, 1986). Under good conditions in the laboratory, postlarvae add protein at a rate of 0.26 mg/day, but higher rates (up to 0.6 mg/day) are achieved in the wild (Juinio and Cobb, 1994; see Section IV,A).

The weight-specific respiration rate increases at each stage. It is highest in early postlarvae, but drops sharply at the end of the postlarval stage. The ammonia excretion rate is similar for stages I and II, increases at stage III and the postlarval stage, but, as with the respiration rate, drops at the end of the postlarval stage. Increases in ammonia excretion, evident in the early postmolt periods and through much of the postlarval stage, may be related to the production of chitin for the exoskeleton. The oxygen–nitrogen ratio declines slightly over the larval period, but decreases quite substantially toward the end of the postlarval stage, indicating a reduction in metabolic activity prior to settlement (Sasaki *et al.*, 1986).

The premetamorphic larval stages show no evidence of lipid storage and utilize all biochemical constituents in their metabolism. The postmetamorphic postlarval stage, however, shows increased dependence on protein and accumulates lipid stores. This probably confers considerable advantage in adapting to a benthic habitat by enabling newly settled lobsters to rely temporarily on stored reserves as they make the transition from planktonic to benthic existence (Sasaki *et al.*, 1986).

F. *Substrate*

It seems unlikely that substrate type per se would affect growth and survival of the planktonic larval lobsters; however, this has not been investigated. Experimentation on the effects of substrate during development has focused primarily on aspects of the behavior of postlarvae associated with substrate selection and settlement.

Bottom-seeking behaviors begin 2–6 days after molting into the postlarval stage (Cobb *et al.*, 1989a; see Section V,E,3). For settlement, a strong preference for substrate with preformed crevices and macroalgal cover is evident (Botero and Atema, 1982; Johns and Mann, 1987). When early postlarvae (less than 1 day after molting) are introduced into aquaria, each with a single substrate, they all settle within 34 hours when released over a substrate of sand overlaid by rock with algal cover, within 38 hours over sand with scattered rocks and pebbles, and within 62 hours over mud. Over a flat sand substrate, however, less than 40% settle within 62 hours and even 2 weeks later most of the postlarvae are still swimming in the water column; their size and the absence of exuviae indicate that they had probably not molted to stage V (Botero and Atema, 1982).

The duration of the postlarval stage is 11 days at 22°C. (The foregoing observations were made at 18–23°C.) In addition to delaying settlement when unsuitable substrate is available, it appears that postlarvae may also delay molting to stage V (the first juvenile stage).

Advanced postlarvae (>10 days postmetamorphosis), which settle readily when suitable substrate is available, are extremely vulnerable to predation by cunners when released over sand substrate devoid of cover 30 minutes before the fish are introduced (Johns and Mann, 1987). When cover is available on the same sand substrate, their vulnerability is reduced but nevertheless remains high.

Delaying settlement to avoid unsuitable substrate risks prolonged exposure to predation in the water column; however, the reduced vulnerability to predation after settling on suitable substrate represents a survival advantage over the long term.

G. *Prey, Food Preference, and Natural Diet*

Most of what is known about the prey and food preferences of larval and postlarval lobsters results from analysis of stomach contents. Lobster postlarvae held in hatching bags suspended in the wild had a preference for copepods and diatoms that were available naturally, rather than the clam meat that was provided as food (Williams, 1907). Other larva-rearing experiments have been carried out using live plankton obtained from the wild as food (Templeman, 1936b). In culture facilities, all stages take live brine shrimp more readily than the frozen product (Carlberg and Van Olst, 1976). Aiken and Waddy (Chapter 8) provide a thorough review of aquaculture.

The natural diet of larval and postlarval lobsters includes the wide variety of phytoplankton and zooplankton available to them. While it seems unwise to generalize from studies of stomach contents of lobsters from a particular region, when taken together they produce the best picture available of natural diets.

In studies of lobsters from the southern Gulf of St. Lawrence, a variety of copepod species are reported to be a large part of the diet of larval and postlarval stages; cladocerans are found regularly in the stomachs of larvae and, to a lesser extent, postlarvae; *Cancer* spp. zoeae and megalopae occur regularly in the stomachs of stage III and postlarval lobsters, respectively; and gastropod larvae are quite important in the diet of postlarvae (Harding *et al.*, 1983). Stage III larvae and particularly postlarvae prey preferentially on larger zooplankton, but nevertheless consume a broad spectrum of prey sizes. The smaller copepod species dominate the plankton community in relatively warm inshore waters and may contribute to larval survival by providing an abundant supply of suitably sized prey for the first two stages

(Harding *et al.*, 1983).

Stomach contents of larval stages I–III from another area of the southern Gulf of St. Lawrence are reported to consist mostly of unidentifiable, amorphous, organic and inorganic material (Varma, 1977). The identifiable material includes eggs, copepod fragments, and several phytoplankters, including diatoms, dinoflagellates, and filamentous algae. [This contrasts with the observations of Harding *et al.* (1983), who made no mention of phytoplankton.]

In coastal waters of Rhode Island, postlarvae feed primarily on larvae of several decapod crustaceans and a variety of copepods (Juinio and Cobb, 1992). Insect parts and fish eggs also occur regularly in stomach contents, along with filamentous algae, diatoms, chaetognaths, larvaceans, cladocerans, and mysids as minor components. Among the decapod larvae, megalopae of *Cancer* spp. and unidentified megalopae are particularly prevalent. These postlarvae also prey selectively on the larger zooplankton available to them (Juinio and Cobb, 1992).

The high growth rates estimated for larval stages near Îles de la Madeleine (see Section IV,A) indicate good nutritional conditions. Undernourished animals compose less than 6% of the postlarvae taken near Rhode Island. They are generally much better nourished than those reared under *ad libitum* feeding conditions in the laboratory. The better nutritional condition and the higher growth rate (see Section IV,A) of wild postlarvae are probably due to a higher quality of prey in the field and indicate that starvation or poor nutrition is unlikely to be a major source of natural mortality (Juinio and Cobb, 1994).

Lobster larvae and postlarvae are omnivorous, opportunistic feeders. There is a strong preference for live prey and all stages consume a broad spectrum of prey types and sizes. Smaller zooplankton appear to be important to those in the first two stages, but later stages prefer larger species. Evidence indicates that food limitation, resulting in poor nutrition or prolonged stage duration, is not a significant factor in their ecology.

H. Predators

Very little is known about predators of lobster larvae and postlarvae. Sources of information include analyses of the stomach contents of predators and behavioral observations.

In the Gulf of Maine, "young lobsters" have been found in 15% of the herring gull (*Larus argentatus*) and 2% of the common tern (*Sterna hirundo*) stomachs examined (Mendall, 1934, as cited by Mills, 1957). No larvae or postlarvae were found in 36 herring gull stomachs and only one postlarva was found in one of the 15 common tern stomachs examined from the southern Gulf of St. Lawrence, when lobster larvae were abundant (Mills, 1957).

Unspecified decapod larvae are reported from more than 20% of the feeding juvenile herring examined from coastal waters of Maine (Sherman and Perkins, 1971), and herring held in large tanks have been observed to capture lobster larvae (Battle *et al.*, 1936). Decapod larvae have also been found among the stomach contents of mackerel in Newfoundland waters (Moores *et al.*, 1975).

Experimental observations suggest that postlarvae may be particularly vulnerable to predation during substrate selection and settlement (Johns and Mann, 1987). The presence of a fish predator (the cunner, *Tautogolabrus adspersus*) in the same tank, even though separated by a mesh barrier, induces a conspicuous change in the bottom locomotion of late-stage postlarvae, which results in considerable delay in reaching shelters. Without the protection of a barrier, mortality attributable to predation by cunners ranges from 68 to 96%, depending on the type of shelter available to the postlarvae (see Section IV,F). Postlarvae are also sensitive to cunner metabolites and display avoidance responses when exposed to a cunner odor plume (Boudreau *et al.*, 1993a). Observations of stage I larvae released into a naturalistic observation pool indicate that cunners and other fish contribute significantly to larval mortality (Herrick, 1895). Observations made by divers on laboratory-reared postlarvae released into the wild indicate that they are quite vulnerable to predation by fish, especially cunners, when they approach the bottom (Ennis, 1975b; Cobb *et al.*, 1983).

Together, such studies indicate that a variety of pelagic and demersal fish species (and some birds) are significant predators of the larvae and postlarvae of *Homarus americanus*.

I. Water Quality

Lobster larvae are tolerant of relatively wide ranges of temperature, salinity, and dissolved oxygen, but are tolerant of narrow ranges of other water-quality variables, such as pH, ammonia, and especially pollutants. Precise values for most variables for optimal culture conditions have been determined (Van Olst *et al.*, 1980). In nature, suboptimal conditions probably contribute to reduced survival indirectly by prolonging development.

Larvae have very limited tolerance to temperatures lower than 5° or greater than 30°C (Huntsman, 1924) and the optimal temperature is 20–22°C

(Hughes *et al.*, 1972; Van Olst *et al.*, 1980). At a salinity of 11.6 ppt, very few stage I larvae survive longer than 24 hours and none molt; at 42.5 ppt, they may survive for several days, but the few that attempt to molt die in the process. The optimal salinity is 30–31 ppt (Templeman, 1936b; Van Olst *et al.*, 1980).

Dissolved oxygen concentrations <1 mg/liter and above saturation and pH levels <5.0 and >9.0 are lethal. Optimal conditions are 6.4 mg/liter of dissolved oxygen and pH 8.0 (Van Olst *et al.*, 1980). Postlarvae are quite sensitive to un-ionized ammonia (NH_3-N), which has an incipient LC_{50} value of 1.4 mg/liter and is particularly toxic at high temperatures and pH. Preventing its accumulation is a major concern in a culture facility, where concentrations <0.14 mg/liter are optimal (Delistraty *et al.*, 1977). Under natural conditions, the dissolved oxygen, pH, and ammonia levels are unlikely to be suboptimal to an extent that would affect larval survival.

Larval survival is quite likely affected by pollution of the marine environment by human activity, especially in coastal areas.

A number of heavy metals are quite toxic. Mercury, copper, and cadmium, with 48-hour LC_{50} values of 10, 75, and 430 μg/liter, respectively, are the most lethal for larvae (Connor, 1972; Dorband *et al.*, 1976).

Crude and other petroleum oils, each with its own mixture of hydrocarbons and metals, are also toxic. Venezuelan crude oil, for example, has a 96-hour LC_{50} of 0.86 mg/liter for stage I larvae and 4.9 mg/liter for stage III larvae and postlarvae (Wells and Sprague, 1976). Polycyclic aromatic hydrocarbons enter the marine environment from a variety of sources, including creosote, which is quite toxic and has a 96-hour LC_{50} of 0.02 mg/liter for stage I larvae at 20°C (McLeese and Metcalfe, 1979).

Another group of toxic pollutants includes various organochlorines (pesticides and industrial compounds), chloramines, and chlorine. Chloramine is much more toxic than free chlorine and is also more toxic at high temperatures. At 20°C, chloramine has a 48-hour LC_{50} value of 4.08 mg/liter for stage I larvae, compared to 0.56 mg/liter at 30°C (Capuzzo *et al.*, 1976). A variety of common pesticides are extremely toxic to crustaceans generally, and undoubtedly to larval lobsters. Very low concentrations of a number of organochlorines are lethal to sand shrimp. Several pyrethroids are lethal for adult lobsters and presumably also for larvae at low concentrations (McLeese *et al.*, 1980). Fenitrothion (an organophosphate) has a 96-hour LC_{50} value of ~1.0 μg/liter for larval lobsters (McLeese, 1974).

Drilling fluids ("muds") have varying toxicities. The most toxic of five tested has a 96-hour LC_{50} of 74 mg/liter for stage I larvae (Derby and Capuzzo, 1984). The chemical components of these fluids, which vary, and not the physical properties appear to be primarily responsible for their toxic effects.

Quite conceivably in many coastal areas, poor water quality related to pollution could be a direct cause of lobster larval mortality. In addition, sublethal effects of exposure to various pollutants appear likely to contribute to larval mortality indirectly.

Development is retarded when larvae are exposed to crude oil at a concentration of 0.14 mg/liter and the incidence of "intermediate" larval stages increases (see Section III); food consumption decreases at 0.19 mg/liter (Wells and Sprague, 1976). Sublethal concentrations of crude oil alter the lipid metabolism and energetics of larvae, resulting in less available energy and causing delayed molting and reduced growth (Capuzzo *et al.*, 1984). Chloramine causes respiratory stress at concentrations of 0.05 mg/liter and, after 48 hours of exposure, reduced respiration rates (Capuzzo *et al.*, 1976). Sublethal concentrations of drilling fluids reduce respiration rates of larvae, alter energetics, and impair growth and development (Derby and Capuzzo, 1984).

J. Disease

Under high-density culture conditions, especially if physiological stress associated with poor water quality or inadequate nutrition is a factor, lobster larvae are susceptible to a number of disease-causing organisms (see Aiken and Waddy, Chapter 8, on aquaculture).

Infestation of larvae by a variety of noninvasive, filamentous microorganisms (epibionts) is a common occurrence. The most prevalent is the bacterium *Leucothrix mucor*, but other bacteria, algae, and stalked protozoa are found as well (Nilson *et al.*, 1975). The filamentous growth on external surfaces hinders feeding and may cause death by entanglement during molting. Gill surfaces are sometimes infested sufficiently to cause a degree of anoxia. The onset of prolonged preening behavior in postlarvae enables them to remove epibionts from external surfaces, thereby considerably reducing their susceptibility (Nilson *et al.*, 1975).

Two phycomycetous fungi also infest larvae. Both are invasive organisms that cause death by destroying internal tissues. *Lagenidium* sp. occurs in association with filamentous epibionts that appear to aid infection by providing lodging on the exoskeleton for encysted spores (Nilson *et al.*, 1975; Fisher *et al.*, 1978). *Haliphthoros milfordensis* occurs primarily on postlarvae, but larvae are presumed to be susceptible; it is usually found alone, without association with epibionts (Fisher *et al.*, 1975; see Martin and Hose, Chapter 17).

Larvae and postlarvae are also affected by shell disease caused by chitinolytic bacteria that erode the exoskeleton. Death results most frequently for larvae, their exoskeletons presumably providing little protection (Fisher *et al.*, 1976, 1978; see Waddy *et al.*, Chapter 10, on the exoskeleton).

Pathogens have not been reported for larvae or postlarvae taken from the wild and the extent to which disease contributes to their mortality is unknown.

V. Behavior

A. Modes of Locomotion

Herrick (1909) noted that one of the most striking habits of *Homarus americanus* larvae is their incessant activity and apparently aimless swimming immediately after hatching. He also observed, however, that they can direct their movements with a certain degree of precision when necessary. In the three larval stages, swimming propulsion is provided primarily by the beating of six pairs of exopodites of the biramous third maxillipeds and the five pereiopods, which are thoracic appendages (see Section III). The exopodites are flattened and fringed with long, featherlike setae and beat in rapid, vibratory strokes (Hadley, 1908a; Herrick, 1909; see Lavalli and Factor, Chapter 14, on larval appendages). The position of the exopodites can be shifted anteriorly, resulting in movement backward and upward, or posteriorly, resulting in movement forward and upward (Hadley, 1908a). A change in orientation accomplishes forward or backward movement in any direction. Larvae occasionally use short bursts of very quick tail flexes for rapid backward movement.

Nonsetose swimmerets, which are present only as minute buds beneath the cuticle in stage I, appear on the second through fifth abdominal segments of stage II larvae. These are fringed with short, rudimentary setae in stage III and their motion assists in preventing the larvae from sinking when apparently at rest (Herrick, 1909). At metamorphosis, the exopodites of the thoracic appendages are reduced to functionless stumps and the swimmerets develop into highly setose, fully functional, swimming appendages.

Accompanying the change in mode of locomotion at metamorphosis is a remarkable development of swimming ability (see Section III). Postlarvae swim with much greater agility, precision, and speed than do the earlier stages (Herrick, 1909). They tend to swim steadily and strongly forward (Cobb *et al.*, 1983). Statocysts, or balancing organs located at the base of the first antennae, are fully developed and functional and the reeling, uncertain gait of the larval stages is no longer present. Postlarvae glide along in a forward direction, propelled by action of the swimmerets with the cephalothorax and abdomen straight and the claws extended straight in front of the head and held either close together or apart (Herrick, 1909; Cobb *et al.*, 1983). The claws-together mode is used for fast, directional swimming, during which postlarvae tend to disregard everything else. If an object is encountered, they stop, back away slightly, and swim around it, continuing in the same direction. Measured in the field, swimming speeds of wild, free-ranging postlarvae averaged 18 cm/sec in the claws-together mode, with a maximum speed of 24 cm/sec (Cobb *et al.*, 1989b). The claws-apart mode is used when investigating objects or feeding or when "relaxing" in the water column, moving their pleopods slowly and showing no directionality of movement (Botero and Atema, 1982; Cobb *et al.*, 1983; Rooney and Cobb, 1991). As in earlier stages, rapid backward movement is accomplished by bursts of abdominal flexes.

B. Feeding Behavior

Herrick (1909) noted that the preying and fighting instincts and voracity are particularly well-developed habits of newly hatched larvae. Under crowded conditions they are very readily cannibalistic whenever the food supply is insufficient or unsuitable. Although at close range they possess sharp vision for small floating particles and often pursue and capture copepods and other members of the plankton, they seem to be dependent on their incessant and aimless swimming to bring them into contact with food that is suspended or also swimming in the water column (Herrick, 1909). Larvae appear to be adapted to feeding on suspended food material and are clearly responsive to chance contact with potential food items (Hinton and Corey, 1979). Although they lack precise discrimination with respect to food selection and often ingest inorganic particles, they do have a clear preference for live food (see Section IV,G). Work with *Homarus gammarus* larvae suggests that food capture is unselective, but ingestion depends on chemical cues; additionally, conditioning must play an important role in ensuring that larvae select nutritious and readily digestible items (Kurmaly *et al.*, 1990).

Postlarvae are particularly adept and voracious predators. Williams (1907) saw a swarm of postlarvae seize, drag under, and devour a full-grown cricket that had fallen into their tub. Yet, even after fasting, they would not attack each other unless one was newly molted. A group of postlarvae resting at the bottom of a container responded to clam juice by immediately flocking to the surface, where they continued active swimming for some moments. This indicates a chemosensory involvement in food detection. However, live megalopae are generally not captured until postlarvae bump into them, suggesting that tactile stimulation is also necessary for prey capture (Juinio and Cobb, 1992). (See Atema and Voigt, Chapter 13, on sensory biology.)

In the wild, when swimming near the surface in the claws-apart mode, postlarvae pause to investigate patches of foam and floating particulate matter they encounter by picking with their claws in an apparent feeding motion (Cobb *et al.*, 1983). They are opportunistic feeders and exercise a degree of selection for larger prey items. They have been observed in the wild carrying crab megalopae and winged insects; however, their diet includes a wide variety of prey groups and sizes, indicating a nonselective feeding mode as well (Juinio and Cobb, 1992). Although postlarvae forage successfully throughout the day, feeding activity is greatest during the night, when visual perception is assumed to be reduced or absent; this seems to indicate that prey detection is not primarily visually mediated (Juinio and Cobb, 1992), as has been suggested by laboratory (Herrick, 1909) and field observations (Cobb *et al.*, 1983). The increased feeding activity of postlarvae at night is associated with a higher incidence of copepods as newly ingested prey and may be attributed to the increased abundance of this prey group near the surface during this time (Juinio and Cobb, 1992).

C. *Responses to Physical Environmental Stimuli*

1. Orientation and Geotaxis

The normal orientation of larvae is with head down, cephalothorax–abdomen articulation uppermost, with the cephalothorax inclined at an angle of about 30° from the horizontal; the axis of the abdomen is about 60° from the longitudinal axis of the cephalothorax, orienting it approximately vertically in the water (Hadley, 1908a). While larvae are exposed to light, this general equilibrium position is regularly interrupted by apparently haphazard swimming involving all sorts of erratic movements and swimming positions (Hadley, 1908a; Herrick, 1909). In very subdued light or when they are blinded, swimming becomes delicate and regular, with larvae hovering for many seconds in a single position and moving backward or forward with equal ease. They revert instantly to haphazard swimming if light is reintroduced (Hadley, 1908a).

Barely visible as shallow depressions of the integument in second- and third-stage larvae, fully functional statocysts (gravity receptors) appear in postlarvae. The general instability of the larval stages is replaced with well-balanced, stable equilibrium with the longitudinal axis generally maintained in a horizontal orientation. The statocyst is lined with sensory hairs and contains numerous fine sand grains, without which postlarval movements become uncertain (Herrick, 1909; see Atema and Voigt, Chapter 13, on sensory perception).

The response of larval and postlarval lobsters to gravity does not include vertical movement in either direction (i.e., geotaxis). However, gravity does provide a basis for orientation and maintaining equilibrium that is particularly well developed in postlarvae, but appears also to play a role in larval stages (Hadley, 1908b), albeit to a considerably lesser extent.

2. Phototaxis, Photopathy, and Photokinesis

Larval and postlarval lobsters display a broad spectrum of responses to light. Movement toward or away from a light source (positive or negative phototaxis, respectively) is particularly well developed, but selection of a region of optimal light intensity along a gradient (photopathy) and changing the level of locomotory activity with changing light intensity (photokinesis) are also included in their behavior.

In unidirectional light, larvae turn and head away from the source and tend to maintain the longitudinal plane of the body parallel to the direction of the light rays and to approach or recede from the source in that position. Postlarvae turn and head toward the source, although the tendency to maintain the longitudinal plane of the body parallel to the light rays is not as strong as in larvae (Hadley, 1908a,b).

In larvae, the mechanics of orientation to unidirectional light involves unequal stimulation of the two eyes. The swimming appendages on the same side as the eye that is more intensely stimulated by the light vibrate more rapidly. This turns or rotates the body until the longitudinal plane is parallel and both eyes are stimulated equally, at which point they vibrate at the same rate on both sides. The intensity of light striking both eyes equally determines whether the swimming appendages tilt forward or backward, moving the larva toward or away from the light source (Hadley, 1908a).

Postlarvae blinded in one eye swim forward in an

arc curving away from the side of the uninjured eye (Hadley, 1908b). Presumably, this is due to greater activity of the swimmerets on the uninjured side and suggests that the mechanics of their orientation to unidirectional light is different from that in larvae, since they orient with the head toward the source. The difference likely relates to the much greater stability of postlarvae and their use of gravity in orientation.

In the absence of directionality to the lighting, larvae and postlarvae select, from the range provided, a region of optimal or preferred light intensity (Hadley, 1908a).

The nature and intensity of the response to light are quite varied, both within and among stages; however, a general pattern can be seen in Hadley's (1908a) very extensive series of experimental observations.

During the first few hours after hatching, larvae are strongly positively phototactic and also positively photopathic (i.e., they select areas of brightest illumination). They react even to very slight differences in the intensity of illumination. As stage I larvae get older, the response becomes less definite; a greater light intensity is required to elicit a comparable response, and there is more individual variation in the nature of the response to a given stimulus. During the second or third day, many become negatively phototactic, especially to very bright light (e.g., sunshine). As they approach the molt, the response becomes somewhat indefinite; although most are positively phototactic at low light intensity, many remain negatively phototactic at high intensity. The sign of the taxis can be changed from positive to negative by a sudden increase in light intensity, but larvae usually adjust quickly and revert to positive. A gradual increase to the same intensity, however, does not elicit the negative response. Throughout stage I, the photopathic response remains positive (Hadley, 1908a).

Newly molted stage II larvae are quite sensitive to slight differences in light intensity and their phototactic response to light of nearly all intensities is negative. As the molt approaches, there is a shift to a positive phototactic response, especially at low intensity, although many continue to respond negatively at high intensity. The photopathic response also remains positive throughout stage II (Hadley, 1908a).

Newly molted stage III larvae display a negative phototactic response that becomes increasingly pronounced as intensity is increased. In daylight, practically all respond negatively. Later in stage III, many become positively phototactic at low light intensity, but revert to a negative response as intensity is increased. In larvae about to molt, positive phototaxis is more pronounced, especially at low intensity, and many remain positive even in high-intensity daylight. The photopathic response, although less definite than in the first two stages, is mostly positive, especially just after the molt (Hadley, 1908a).

In early postlarvae, the phototactic response is predominantly negative, although in high-intensity light (e.g., bright sunlight) many display a positive response. By midstage, the negative response is quite definite at all intensities and remains so throughout. The photopathic response is positive in early postlarvae, but the tendency to select more brightly illuminated areas diminishes and the response is generally (but not completely) negative toward the end of the stage (Hadley, 1908a).

In all stages, the phototactic response overrides the photopathic response. Larvae and postlarvae respond phototactically to unidirectional light introduced horizontally and move to an area of light intensity that otherwise would be avoided (Hadley, 1908a).

While Hadley (1908a) did not address photokinesis per se, it is clear from his observations of very quiescent swimming by larvae in very dim light, comparable to the swimming of blinded larvae, that exposure to light stimulates a greater intensity of locomotory activity. It is not clear, however, to what extent increasing the intensity of light stimulates greater activity. Blinded postlarvae are also quiescent swimmers (Hadley, 1908b) and normal individuals swim more during the day than at night (Rooney and Cobb, 1991), indicating a photokinetic response.

Hadley's (1908a) observations of phototactic responses were made using light introduced horizontally. In unidirectional light from above, however, phototactic responses in terms of vertical movements (see Section V,D) are poorly defined and generally very weak (Ennis, 1975b). After 10 minutes in darkness, stage I larvae respond to overhead light with very slow upward swimming toward the light source during the first 5 minutes, followed by slow downward movement during an additional 30 minutes of exposure. The same pattern is evident over a range of light intensities and the rate of downward movement increases when larvae are returned to darkness following the 35-minute exposure to light. When light intensity is changed during the period of exposure, there is downward movement as intensity decreases and usually upward movement as it increases. Stage II larvae are only slightly less responsive, but stage III larvae are almost nonresponsive. Although postlarvae are very mobile in the observation column, their movements do not appear to be in response to light.

3. Barokinesis

Larvae and postlarvae respond to instantaneous increases in hydrostatic pressure as small as 1.0 psi (equivalent to a depth increase of ~68 cm) and as great as 20 psi by swimming upward (Ennis, 1975b). The response is instantaneous and involves short bursts of rapid tail flexes, as well as increased beating rate of the swimming appendages. Responsiveness of stage I larvae, in terms of percentage swimming near the surface in a column after a 5-minute exposure, increases linearly to near 100% at 10 psi. They continue to respond during a series of small, stepped increases in pressure and also retain their sensitivity and responsiveness through a series of pressure pulses as great as 20 psi. Responsiveness, especially to smaller increases, is greater in overhead lighting than in darkness. Following a period of exposure to increased pressure, larvae respond to even a partial decrease by moving downward. There is some downward movement during a period of sustained pressure increase, indicating a degree of accommodation to the pressure, which is much more prevalent at lower pressure and among older larvae. Stage II larvae are less responsive to smaller pressure increases than are stage I larvae, but are quite responsive to larger increases. Although sensitivity to pressure changes is retained throughout, stage III larvae and postlarvae are less responsive than are stage I and II larvae (see Section V,D,2).

4. Rheotaxis

In flowing water, larval swimming is always uncertain and desultory (Hadley, 1906). Stage III larvae, however, often display a tendency to retain equilibrium in the water, but are nevertheless borne along unresistingly in the current. Immediately upon molting into the postlarval stage, however, they head directly into the current and swim against it. While they often make headway, they are displaced backward if the current is too strong (see Section V,E). This rheotactic response in postlarvae can be elicited by an optical stimulus, a cardboard cylinder with slits that rotates around a motionless glass dish containing the postlarvae (Hadley, 1906). If light is introduced at an angle to the direction of flow, the phototactic response overrides rheotaxis in postlarvae. This is especially so at night, when the rheotactic response is most definite. Even in daylight, the rheotactic response is most pronounced when postlarvae are shaded from direct light.

Although it is weak, a rheotactic response is also present in the larval stages (Ennis, 1986a). In a flow tank, some larvae of all three stages are able to remain swimming and avoid impingement on a downstream screen at a current velocity of 2 cm/sec over a 30-minute period. At a velocity of 5 cm/sec, however, their capabilities are very limited and very few remain swimming after 5 minutes. Larvae that continue swimming in flowing water tend to orient in the upstream direction.

With their vastly superior swimming ability, postlarvae display a capacity to remain swimming at current velocities up to 9 cm/sec and their swimming performance improves during the postlarval stage (Ennis, 1986a). Laboratory-reared postlarvae have been shown to be capable of making over-the-bottom headway up to 5 cm/sec in water flowing at a velocity of 14 cm/sec, which translates into a through-the-water swimming speed of 19 cm/sec. Time spent swimming during a 30-minute period decreases appreciably with an increase in water velocity from 8 to 14 cm/sec, especially at night (Rooney and Cobb, 1991).

5. Thermokinesis

Direct evidence of the effect of temperature on larval and postlarval behavior is limited. Swimming speeds increase at higher temperature in various decapod larvae (Sulkin, 1984). Some respond to a temperature decrease by ascending and to a temperature increase by descending, and sensitivity to temperature change can decrease with larval development to the point of unresponsiveness in later stages (Forward, 1990).

When wild-caught postlarvae are placed in a flow tank with water velocity of 14 cm/sec for 30 minutes, they swim very little (<1% of the time) at 15°C and 12% of the time at 21°C (Rooney and Cobb, 1991). Larval and postlarval lobsters generally remain in the warmer water above the thermocline in a vertical temperature gradient (Boudreau *et al.*, 1991; see Section V,D). While much of the nature of their behavioral responses to changes in temperature remain to be elucidated, thermokinesis is certainly a factor in their responses to environmental stimuli.

D. Vertical Distribution and Movements

1. Depth Distribution and Daily Vertical Movements

Despite the inherent difficulties associated with plankton sampling at discrete depths, early work in coastal areas demonstrated that larval and postlarval lobsters are highly concentrated at or very near the surface and are taken only occasionally, in comparatively very small numbers, in nets towed at depths greater than 1–2 m. The high concentrations observed to depths of 9 m in the Cape Cod Canal (Collings *et al.*, 1983) and to 12 m near the eastern end of the canal in Cape Cod Bay (Matthiessen and Scherer, 1983) are unusual occurrences attributable to the turbulent mixing of water flowing through the canal (a list of field studies relevant to the vertical distribution of *Homarus americanus* larvae and postlarvae is provided in Harding *et al.*, 1987a). A pattern of daily vertical movements controlled by light can be inferred from variation in catches at the surface. When larvae are present in the water column, some are taken at the surface throughout the day; however, they tend to avoid bright sunshine by moving downward. They seem to be strongly attracted to the surface in dull daylight. During a 24-hour period, larvae are abundant at the surface in the early daylight hours, but move down in the stronger light of midday. Many are attracted back to the surface late in the day as light intensity declines to an optimal level, but then disperse downward as night approaches and light intensity drops below the minimum level required to keep them near the surface. They are attracted back to the surface as light intensity increases near dawn (Templeman, 1937b, 1939; Templeman and Tibbo, 1945).

Use of multicompartment nets to discretely sample narrow depth zones has confirmed this general pattern and demonstrated that these semidiurnal vertical movements are largely confined to the upper 2–3 meters of the water column and involve mostly stage I larvae. Daytime catches of stage I larvae in the upper 60 cm are consistently much greater than at night, but for later stages the difference is small and inconsistent between sampling periods (Scarratt, 1973). All stages, particularly stage I, respond to changing light conditions during daytime with small-scale vertical movements close to the surface. Under low light intensity conditions, such as at dawn, dusk, and under overcast skies, stage I larvae are much more highly concentrated in the upper 80 cm than under sunny skies. Later stages, especially postlarvae, tend to be even more highly concentrated near the surface than stage I larvae under all daytime light conditions, but they, too, are less concentrated in the upper 80 cm under high light intensity conditions (Harding *et al.*, 1982; Hudon *et al.*, 1986).

Earlier sampling in offshore areas had indicated very few larvae below the surface layer (Stasko, 1977; Stasko and Gordon, 1983). However, sampling at 5-m intervals from the surface to 30 m, and occasionally at 40 and 50 m, with a large net that could be opened and closed at depth, revealed stage I larvae to be most abundant at the 15- to 30-m depths with very few at the surface during the day. At night, most are at the surface and from 5 to 10 m, but very few are deeper (Harding *et al.*, 1987a). Integrated throughout the water column, on average <1% of stage I larvae occur above 2.5-m depth during daylight, compared to ~28% after dark. Stage II and III larvae are taken over a broad depth range as well; most are in the upper 20 m, but some stage II larvae are taken as deep as 30 m. There is no indication of shifts in vertical distribution between day and night. Postlarvae are present in 75% of the surface samples representing 90% of the total catch, but some are taken as deep as 30 m. The daytime and nighttime catches at the surface are similar.

In coastal waters, the vertical distribution and movements of lobster larvae and postlarvae are confined to the upper 2–3 m of the water column, whereas in offshore waters they appear to be unrestricted by depth within the upper 30 m. In both areas, only stage I larvae undergo daily vertical movements. The semidiurnal pattern in coastal waters tends to concentrate stage I larvae near the surface during early morning and late afternoon/early evening. This contrasts with the diurnal pattern, which tends to concentrate them near the surface in offshore waters during the night. This indicates that something other than changing light conditions is the main mediator of these vertical movements in the offshore area (Harding *et al.*, 1987a). In neither area do later stages undergo vertical movements in response to the day–night cycle and common to both areas is the tendency for postlarvae, to a much greater extent than larvae, to be highly concentrated near the surface.

The variable observations on the vertical distribution and movements of the planktonic stages of the lobster in the field are a general reflection of their variable behavioral responsiveness to environmental stimuli, as shown under experimental conditions, individually as well as within and between stages (see Section V,C). While their behavioral repertoire provides for good depth-regulatory capabilities, there appear to be no well-defined mechanisms which serve to restrict or control their vertical distribution or movements within narrow limits.

2. Depth Regulation

The precise mechanisms of depth regulation and how they differ between coastal larvae, which tend to remain within a very narrow depth range near the surface, and offshore larvae, which occupy a comparatively broad range of depth, are not clear. The small-scale vertical movements of stage I larvae in the inshore area can be explained fairly readily by the changing light intensity hypothesis, but this is not so with the 20-m migrations observed offshore (Harding *et al.*, 1987a).

Stage I larvae are much more responsive than

other stages, in terms of depth-regulatory swimming behavior, to changes in light and hydrostatic pressure; they are, therefore, more likely to regulate depth in response to changing conditions. Yet, in coastal areas, the later stages appear to be more restricted in terms of vertical movements. Possibly their lower sensitivity to light and pressure changes allows them to respond more strongly to some other factor. Although a geotactic response has not been demonstrated in postlarvae, they clearly rely on gravity for orientation and equilibrium (see Section V,C,1). Negative geotaxis may be involved in their concentration near the surface. Even in stage I larvae, responsiveness to changes in light and pressure varies with age and at any given time the stage I population in the wild will be composed of a mixture of ages from newly hatched to ready to molt. Considerable individual variability could account for the absence of an all-or-nothing type of response to changing environmental conditions. It seems likely that responses to light and pressure serve more as general orientation cues rather than as mechanisms for regulating depth within precise limits, especially in later larval and postlarval stages.

Larvae that originate in offshore waters may have been conditioned to high hydrostatic pressure and low light intensity during their embryonic development. Upon hatching, they face a very different environment and a much longer journey to the surface than larvae originating in coastal areas. Offshore larvae may be responsive to light at a much lower intensity, hence their attraction to the surface at night, and may be generally unresponsive to changes in pressure.

Exceptions to the general pattern of variation in the abundance of larvae at the surface, such as the occasional high concentration at the surface during bright sunshine, have been observed. Attempts to explain such exceptions have examined the effects of turbidity, which alters light penetration and possibly reduces the responsiveness of larvae to variation in light intensity above the surface (Templeman and Tibbo, 1945). However, summer light attenuation inshore in the Gulf of St. Lawrence (Harding *et al.*, 1987b) and offshore over Browns Bank (Harding *et al.*, 1987a) are similar, with 1% of surface radiation reaching 18–20 m. Factors other than light intensity and light penetration, such as prey–predator interactions, may be responsible for the differences in the scale of vertical distribution and movements between inshore and offshore larvae. While the possibility has not been investigated for lobster larvae, the vertical migrations of other zooplankters are influenced by the changing depth distribution of their predators and prey (Janssen and Bradt, 1980; Ohman *et al.*, 1983; Harding *et al.*, 1986).

Vertical temperature and salinity profiles and the depth of the thermocline and halocline affect the vertical distribution of larvae. In an experimental chamber, newly hatched larvae swim upward in high-salinity seawater (31–32 ppt), but normally do not pass into an overlying freshwater layer. Those that pass through the interface either react vigorously and swim down or become inactive, sink into the seawater, and recover (Scarratt and Raine, 1967). Further, larvae avoid swimming upward into seawater diluted to 21.4 ppt, although they readily swim upward into 26.7-ppt seawater and are noticeably more active at the higher salinity (Scarratt and Raine, 1967). Salinity gradients and low surface salinity may be factors in vertical distribution in nearshore areas where runoff is extensive or in localized areas where heavy rain is mixed downward by onshore wind. In offshore areas with relatively high salinity extending from the surface to deep water, salinity is not likely to be a factor.

The vertical distribution of larvae and postlarvae has been observed in a 190-cm vertical column in relation to a thermal gradient with and without a strong thermocline (Boudreau *et al.*, 1991). In a 20°C (top) to 15°C (bottom) temperature gradient, all stages distribute throughout the depth of the column, although the later larval and postlarval stages tend to occupy the lower portion. However, in the presence of a thermocline (~17°C above and 9°C below) at about 140-cm depth, with the exception of small numbers of stage III larvae and postlarvae, almost all distribute above the thermocline. In neither case is the vertical distribution of larvae and postlarvae affected by the day–night cycle.

In offshore waters, the structure and slope of the vertical temperature profile sometimes change dramatically with the tidal cycle, but temperature typically declines gradually from ~16°C at the surface to ~10° at 30 m. While some are present in the cold water at 30–40 m, most larvae and postlarvae are distributed in the warmer upper layers (Harding *et al.*, 1987a). A thermocline, regardless of its depth, will generally restrict lobster larvae and postlarvae to the warmer-water layer above. A temperature gradient, while it is much less restrictive, also tends to limit their vertical distribution to shallow, warm layers.

Depth regulation in larvae of many decapod species involves feedback mechanisms in which the sign of a taxis (orientation) and the direction of change in activity with a change in the intensity of stimulation (i.e., high or low kinesis) depend on the intensity of another stimulus (reviewed by Sulkin,

1984). Taxes and kineses often change during ontogeny, as does the sign of a given feedback mechanism. Positive feedback reinforces a taxis or a kinesis and results in movement to the surface or to the bottom, whereas negative feedback is counteractive and results in depth regulation somewhere in midwater. Common feedback mechanisms include reversals in the sign of geotaxis or phototaxis as a result of sudden changes in temperature or salinity (Sulkin, 1984; Forward, 1989, 1990). Such feedback mechanisms have not been clearly established for the larvae of *Homarus americanus*; however, observations of their reactions to reduced salinity (Scarratt and Raine, 1967) and of their avoidance of low temperature (Boudreau *et al.*, 1991) indicate the likely involvement of feedback mechanisms in depth regulation.

E. Horizontal Distribution, Movements, Dispersal, and Settlement

1. Horizontal Distribution and Dispersal

A considerable reduction in catches of larvae at the surface has been noted shortly after the wind changes direction from onshore to offshore (Templeman, 1937b). This, along with their strong affinity for the surface layer, has led to the generally accepted conclusion that larvae are passive drifters and their dispersal from the parental stock location is determined by surface currents. There are strong cases for surface drift as a key mechanism in the dispersal of lobster larvae from offshore to inshore areas of southern New England (Katz *et al.*, 1994) and the Gulf of Maine (Harding and Trites, 1988) and around the Îles de la Madeleine in the Gulf of St. Lawrence (Hudon and Fradette, 1993). Further, it is hypothesized that thermoclines are perceived by settling postlarvae as a negative physical cue that tends to keep them near the surface, thereby facilitating the location of shallow-water coastal areas, where good settling substrate is found (Boudreau *et al.*, 1992). This is supported by a negative correlation between the frequency of strong winds during the settlement season, which presumably disrupt the thermocline, and lobster landings 8 years later in the Îles de la Madeleine (Boudreau *et al.*, 1991). On the other hand, strong thermal stratification is associated with less favorable conditions for overall survival of the lobster's planktonic stages (Caddy, 1979).

It is also well established that larvae of a variety of other decapod species move vertically and utilize currents moving in different directions at different depths in their dispersal (Phillips, 1981; Sulkin, 1984). Vertical movements have also been used to explain variability in lobster larval abundance in nearshore areas in relation to wind direction. Detection of turbulence ("noise") induced by wind or surf close to shore may trigger vertical movements that enable larvae to utilize subsurface countercurrents to avoid long-distance displacement by surface currents and maintain their position near parental grounds (Squires, 1965, 1970; Squires *et al.*, 1971).

Field studies in two areas of the southern Gulf of St. Lawrence defined centers of abundance for each of the four planktonic stages. In one study, the center for postlarvae was 20–40 km to the west of that for earlier stages, despite a prevailing southwesterly wind (Caddy, 1979). This suggests that stage III larvae and postlarvae at least make use of behavioral and hydrodynamic mechanisms to maintain position and move upwind. They may move vertically, presumably at night, to a depth where a countercurrent results in upstream transport in relation to the surface. There is little evidence for drift of larvae into or out of the survey area and the larvae present appear to be predominantly the result of local spawning.

In another area, the three larval stages were found to be strongly aggregated most frequently near the mouth of a small bay (Hudon *et al.*, 1986; Hudon and Fradette, 1993). On the other hand, postlarvae were found more frequently in nearshore stations and inside the bay. Postlarvae were found predominantly near the surface during both day and night, suggesting that their distribution results from active movement. Similarly, over the southern New England continental shelf, postlarvae have been more abundant closer to shore than larvae (Rogers *et al.*, 1968; Katz *et al.*, 1994).

Larvae are captured more frequently from the surface layer if patches of floating seaweed are present; the small-scale patchiness of both larvae and seaweed may be caused by Langmuir circulation (Harding *et al.*, 1982). Larvae and postlarvae are also concentrated in downwellings characteristic of shallow sea fronts, where water masses converge (Cobb *et al.*, 1983). (These are evident at the surface by lines of foam 200–250 m long and 1–3 m wide, which often orient parallel to the direction of tidal current flow.) They are found in the slightly colder (as much as 2°C) and clearer water on one side of the front, where bits of flotsam and pieces of seaweed collect. Postlarvae actively feed in these fronts, which are ephemeral and appear or disappear with localized changes in currents or wind (Cobb *et al.*, 1983).

Although surface drift is very much involved in their horizontal distribution, lobster larvae are not simply passive current drifters. Evidence indicates that they move vertically to utilize currents moving in different directions as a means of remaining near parental grounds and avoiding long-distance displacement. Postlarvae, on the other hand, appear to determine their horizontal distribution by means of active, directed movement.

2. Directional Swimming and Orientation

Although the rheotactic response is present (see Section V,C,4), the three larval stages are relatively weak swimmers in terms of maintaining position or making headway in flowing water. In contrast, postlarvae are remarkable swimmers and display a considerable capacity for rapid, directed swimming (see Section V,A).

In laboratory studies, wild-caught postlarvae were faster swimmers than those reared in the laboratory, possibly because of their larger size and better nutritional state (Rooney and Cobb, 1991). A high proportion (66%) of the postlarvae observed in the wild in one study were swimming in a northerly to easterly direction, clearly indicating a capability for oriented swimming (Cobb *et al.*, 1989b). If sustained, directional swimming in the claws-together mode, with speeds up to 24 cm/sec (see Section V,A), could allow movement on the order of tens of kilometers. Late postlarvae have accumulated metabolic reserves sufficient to survive up to 5 days without feeding (Sasaki *et al.*, 1986). Assuming sustained swimming is physiologically possible, in 5 days postlarvae traveling at 15 cm/sec could move 65 km (Cobb *et al.*, 1989b).

Mechanisms by which postlarvae orient for directional swimming are not known, but several possibilities have been suggested. Many arthropods use celestial cues for orientation and navigation. Postlarvae show a preference for daytime swimming, suggesting the possibility of a visual orientation cue, such as the sun, to guide their directional swimming (Rooney and Cobb, 1991). Visual reference to the land mass or the sky–land interface (Ennis, 1986a) or some difference in the quality of light over land and ocean (Cobb *et al.*, 1989b) may also provide orientation cues. Orientation to polarized light (polarotaxis) is common among animals. Sensitivity to polarized light has been demonstrated in some crustaceans: some decapod larvae are sensitive to polarized light (Via and Forward, 1975; Bardolph and Stavn, 1978); orientation to polarized light is involved in the vertical migratory behavior of the copepod *Cyclops vernalis* (Umminger, 1968); and the littoral mysid *Mysidium gracile* is capable of orientation to polarized light (Bainbridge and Waterman, 1957). Polarotaxis may conceivably be a factor in the oriented swimming of lobster postlarvae. Chemical cues originating in the nearshore habitat might provide directional information for the postlarvae of *Homarus americanus* (Hudon *et al.*, 1986; Boudreau *et al.*, 1993a) and sound produced by waves breaking on shore may provide the orientation that enables puerulus of *Panulirus longipes* to return to the coastal areas of Western Australia after a long period at sea (Phillips and McWilliam, 1986).

3. Substrate Selection and Settlement

Although postlarvae display a strong affinity for swimming near the surface until they settle, settling behavior and behavior appropriate to the bottom-dwelling lifestyle develop very early (Berrill and Stewart, 1973; Botero and Atema, 1982). The duration of the postlarval stage is 11 days at 22°C (see Section IV,A) and settlement may occur from early to late in the stage (Cobb *et al.*, 1989a; Incze and Wahle, 1991). Diving and bottom-testing behavior begins 2–6 days after molting to the postlarval stage (Cobb *et al.*, 1989a). Descents to the bottom involve directed swimming in the claws-together mode with the body oriented vertically, as well as passive sinking in which swimming stops, the claws are apart, and the mid-body is arched downward. Ascents from the bottom usually involve directed, claws-together, vertical swimming. These are clearly recognizable behavioral events that become increasingly frequent after 2 days of postlarval life. Dives to the bottom followed by ascents within 30 seconds ("touchdowns") and ascents from the bottom followed within 30 seconds by returns to the bottom ("liftoffs") also increase significantly after 2 days. This appears to be true bottom-testing behavior (Cobb *et al.*, 1989a). For settlement, substrate with preformed crevices and macroalgal cover is strongly preferred; when such is provided for postlarvae less than 1 day after molting, all settle and burrow within 34 hours (Botero and Atema, 1982). Settlement is delayed quite considerably when unsuitable substrate is provided and molting to the first juvenile stage may also be delayed (see Section IV,F). As postlarvae age and when they are released in water previously conditioned by cunners, they tend to settle more quickly and choose shelters less selectively (Boudreau *et al.*, 1993b). In the presence of a thermocline or thermal gradient, settlement will also be delayed and even settled postlarvae will respond to a reduction in ambient temperature by leaving the bottom (Boudreau *et al.*, 1992).

There is considerable variability in the distribution of molt stages among wild-caught postlarvae from one year to the next and one area to another (Cobb *et al.*, 1989a; Incze and Wahle, 1991). Observations indicate that settlement may occur anytime from quite early to quite late in the postlarval period. By the time postlarvae reach molt stage D_0, they have accumulated enough metabolic reserves to last up to 5 days without further feeding (Sasaki *et al.*, 1986). Molt stage D_0 through D_3 (late premolt) postlarvae are commonly taken in the wild. The foregoing indicates that postlarvae have considerable flexibility in responding to different and changing environmental conditions associated with choosing the time and place of settling. (See Section IV,F, and Lawton and Lavalli, Chapter 4.)

When considered together, studies clearly show that postlarvae are very well-adapted behaviorally for locating bottom that is suitable for settling in terms of enhancing the survival of the early benthic stages. Successful settlement is a preoccupation of postlarvae rather than a matter of chance and is an important element of recruitment.

VI. Directions for Further Research

Throughout the extensive geographic range of *Homarus americanus*, which likely encompasses several stocks (Ennis, 1986b), there is a very broad range of both physical and biological environmental conditions to which larvae and postlarvae are exposed. The challenges to successful settlement and recruitment to localized populations probably vary greatly between areas and probably also between years in any given area. Over the lobster's range, the origin of local recruitment may range from egg production that is completely within the local population to egg production that is completely outside and may include all combinations of the two possibilities. Unknowns regarding the foregoing represent serious limitations to effective management and conservation of the lobster resource (Fogarty, Chapter 6).

A detailed understanding of all the processes involved in determining where the larvae that hatch in a given area eventually settle is essential to understanding the extent to which egg production in an area contributes to recruitment in the same area or elsewhere. This is basic to establishing a relationship between egg production and recruitment to the population and understanding recruitment fluctuations and whether these are caused by variation in egg production or in environmental conditions affecting survival of larvae and postlarvae or successful postlarval settlement.

Over the past 15 years, there have been significant advances in our knowledge of larval and postlarval ecology due to highly focused field studies and laboratory experimentation. Further advances will certainly result from continued research applying each approach. However, progress toward resolving the following questions would contribute most significantly to knowledge and understanding of the recruitment processes and mechanisms.

1. Why are catches of postlarvae virtually restricted to the surface layer when they spend considerable time descending and ascending during bottom testing prior to settlement? Do thermoclines or thermal gradients reduce or eliminate postlarval bottom-testing behavior? Is representative, quantitative sampling possible, or does changing vulnerability to nets from one stage to the next preclude comparing their relative abundance and mortality? Gear avoidance behavior may be much more effective at depth than at the surface. The extent to which reduced vulnerability to gear towed at depth may account for the general absence of larvae and postlarvae from subsurface tows must be evaluated.
2. In nearshore areas, how do larvae and postlarvae avoid being cast ashore by onshore, wind-generated wave action? Is their disappearance from the surface when the wind changes to offshore due to downwind surface dispersal, or do they swim down from the surface layer? To what extent is dispersal from parental grounds by hydrodynamic processes controlled or influenced by larval/postlarval activity and behavior as opposed to passive drift in the near-surface layer? Comprehensive studies are needed of the horizontal and vertical distribution of larvae and postlarvae and their movements in response to changing environmental conditions and especially in relation to small- and large-scale oceanographic conditions, both physical and biological.
3. Can larval and postlarval abundance and survival through their planktonic existence be measured and the causes of year-to-year fluctuations be determined?
4. What cues do postlarvae use for directional swimming? Is their capacity for long-distance, directional swimming designed to relocate parental grounds or locate the nearest suitable settling bottom?
5. Why is it so difficult to sample postlarvae quantitatively after settlement to the benthic environment? A method of sampling is needed that would provide a basis for a quantitative measure of postlarval settlement that could be used to develop time series of annual estimates for different areas.

VII. Summary

Upon hatching in early summer from eggs carried externally by ovigerous females, larvae swim immediately to the surface to begin the 6- to 8-week planktonic phase of their life cycle. Development includes three larval stages (stages I–III) and a postlarval stage (stage IV), which results from a metamorphosis during the molt to stage IV.

Growth and survival during the planktonic phase are influenced by a variety of abiotic and biotic factors. Other factors being adequate, temperature exerts the greatest effect on stage duration, with the time required to attain and complete the postlarval stage ranging from 22 to 103 days at 22° and 10°C, respectively. Overall survival is quite low at 10°C and highest at temperatures around 18°C. Temperature has relatively little effect on sizes attained by larvae and postlarvae, but the size increment during the molt at the end of the postlarval stage is much reduced at low temperature. In the wild, stage III and postlarval lobsters tend to be larger than laboratory-reared animals and overall growth rates comparable to those under optimal rearing conditions can be achieved. Hatching early in the season, when surface temperature is increasing rapidly, favors greater survival through the planktonic phase. Stage duration is not affected by salinity from 21 to 32 ppt; however, survival is much reduced at the low end of the range. At high temperature, there is increased tolerance to low salinity, but reduced tolerance to high salinity. Inadequate food supply reduces survival and increases development time. Larvae and postlarvae are tolerant of narrow ranges of some water-quality variables, such as pH, ammonia, and especially pollutants. Suboptimal conditions contribute to reduced survival directly, or indirectly by prolonging exposure to predation by various pelagic and benthic fish. Duration of the postlarval stage may be prolonged in the absence of substrate suitable for settlement.

In the larval stages, swimming is accomplished with vibratory strokes of the exopodites on the third maxillipeds and the thoracic appendages. Although very active and capable, larvae tend to be haphazard and aimless swimmers. At metamorphosis, the swimmerets become fully functional, as do the statocysts, and postlarvae are transformed into very proficient and purposeful swimmers. Postlarvae possess well-balanced, stable equilibrium and they swim with much greater agility, precision, and speed than do larvae.

Larvae and particularly postlarvae are voracious predators. Their natural diet includes phytoplankton and a broad spectrum of zooplankton types and sizes. In coastal waters especially, feeding conditions are good and it appears that food limitation is not a significant factor in their ecology.

Behavioral responses of larvae and postlarvae to physical stimuli, particularly light, gravity, hydrostatic pressure, and currents, are highly variable within and between stages and often depend on stimulus intensity and interaction among stimuli. Their behavioral repertoire is extensive and provides a basis for responding to a broad range of continuously changing conditions, rather than for predictable, all-or-nothing responses.

In coastal waters, the vertical distribution and movements of larvae and postlarvae are confined to the upper 2–3 m of the water column, but offshore they appear to be unrestricted by depth within the upper 30 m. A semidiurnal pattern of vertical movements in coastal waters tends to concentrate stage I larvae nearer the surface during the day, whereas offshore a diurnal pattern concentrates them near the surface at night. In neither area do later stages undergo vertical movements in response to the day–night cycle and common to both areas is a tendency for postlarvae, to a much greater extent than larvae, to be highly concentrated near the surface. While larvae and postlarvae are behaviorally well equipped for depth regulation, highly variable observations in the field and in the laboratory indicate no well-defined mechanisms that control their vertical distribution or movements within narrow limits.

Although currents strongly influence their horizontal distribution, larvae are not passive surface drifters. They are capable of moving vertically and possibly utilize currents moving in different directions as a means of avoiding long-distance displacement. Postlarvae appear to determine their horizontal distribution by means of active, directed movement. They have a very strong affinity for swimming near the surface; however, bottom-testing and behavior appropriate to the bottom-dwelling lifestyle develop early in the stage. With their remarkable swimming ability and a capacity for directional orientation, postlarvae are capable of long-distance travel to locate bottom that is suitable for settling in terms of enhancing the survival of the early benthic stages. Successful settlement is achieved by a strong preference for substrate that provides good shelter, combined with an ability to settle very early in the postlarval stage or to delay settlement until appropriate substrate is found. Recruitment to the population is by design and is not a matter of chance.

Although there is a general understanding of recruitment processes, detailed knowledge of all factors determining where larvae that originate in a given area eventually settle is lacking. The absence of a clear relationship between egg production and recruitment to the population, together with a lack of understanding of the causes of recruitment fluctuation, limits conservation of *Homarus americanus*.

References

Aiken, D. E. (1980). Molting and growth. *In* "The Biology and Management of Lobsters" (J. S. Cobb and B. F. Phillips, eds.), Vol. 1, pp. 91–163. Academic Press, New York.

Aiken, D. E., Martin, D. J., Meisner, J. D., and Sochasky, J. B. (1981). Influence of photoperiod on survival and growth of larval American lobsters (*Homarus americanus*). *J. World Maric. Soc.* 12, 225–230.

Aiken, D. E., Rowe, W. J., Martin-Robichaud, D. J., and Waddy, S. L. (1982). Seasonal differences in the effect of photoperiod on survival and development of larval American lobsters (*Homarus americanus*). *J. World Maric. Soc.* 13, 287–293.

Anger, K., Storch, V., Anger, V., and Capuzzo, J. M. (1985). Effects of starvation on moult cycle and hepatopancreas of stage I lobster (*Homarus americanus*) larvae. *Helgol. Wiss. Meeresunters.* 39, 107–116.

Attard, J., and Hudon, C. (1987). Embryonic development and energetic investment in egg production in relation to size of female lobster (*Homarus americanus*). *Can. J. Fish. Aquat. Sci.* 44, 1157–1164.

Bainbridge, R., and Waterman, T. H. (1957). Polarized light and the orientation of two marine crustacea. *J. Exp. Biol.* 34, 342–364.

Bardolph, M., and Stavn, R. H. (1978). Polarized light sensitivity in the stage I zoea of the mud crab, *Panopeus herbstii*. *Mar. Biol.* 46, 327–333.

Battle, H. I., Huntsman, A. G., Jeffers, A. M., Jeffers, G. W., Johnson, W. H., and McNairn, N. A. (1936). Fatness, digestion and food of Passamaquoddy young herring. *J. Biol. Board Can.* 2, 401–429.

Bergeron, J. (1967). Contribution à la biologie du homard, *Homarus americanus* M.-Edw., des Îles-de-la Madeleine. *Nat. Can. (Que.)* 94, 169–207.

Berrill, M., and Stewart, R. (1973). Tunnel-digging in mud by newly-settled American lobsters, *Homarus americanus*. *J. Fish. Res. Board Can.* 30, 285–287.

Bibb, B. G., and Hersey, R. L. (1979). "Distribution and Abundance of Lobster Larvae in Block Island Sound, 1978," report prepared by Raytheon Company for the New England Power Company, Westborough, Massachusetts.

Bibb, B. G., Hersey, R. L., and Marcello, R. A., Jr. (1983). Distribution and abundance of lobster larvae (*Homarus americanus*) in Block Island Sound. *NOAA Tech. Rep., NMFS SSRF* 775, 15–22.

Botero, L., and Atema, J. (1982). Behavior and substrate selection during larval settling in the lobster *Homarus americanus*. *J. Crust. Biol.* 2, 59–69.

Boudreau, B., Simard, Y., and Bourget, E. (1991). Behavioural responses of the planktonic stages of the American lobster *Homarus americanus* to thermal gradients, and ecological implications. *Mar. Ecol. Prog. Ser.* 76, 13–23.

Boudreau, B., Simard, Y., and Bourget, E. (1992). Influence of a thermocline on vertical distribution and settlement of post-larvae of the American lobster *Homarus americanus* Milne-Edwards. *J. Exp. Mar. Biol. Ecol.* 162, 35–49.

Boudreau, B., Bourget, E., and Simard, Y. (1993a). Behavioural responses of competent lobster postlarvae to odor plumes. *Mar. Biol.* 117, 63–69.

Boudreau, B., Bourget, E, and Simard, Y. (1993b). Effect of age, injury, and predator odors on settlement and shelter selection by lobster *Homarus americanus* postlarvae. *Mar. Ecol. Prog. Ser.* 93, 119–129.

Caddy, J. F. (1979). The influence of variations in the seasonal temperature regime on survival of larval stages of the American lobster (*Homarus americanus*) in the southern Gulf of St. Lawrence. *Rapp. P.-V. Reun., Cons. Int. Explor. Mer* 175, 204–216.

Campbell, A. (1986). Migratory movements of ovigerous lobsters, *Homarus americanus*, tagged off Grand Manan, eastern Canada. *Can. J. Fish. Aquat. Sci.* 43, 2197–2205.

Campbell, A. (1990). Aggregations of berried lobsters (*Homarus americanus*) in shallow waters off Grand Manan, eastern Canada. *Can. J. Fish. Aquat. Sci.* 47, 520–523.

Capuzzo, J. M., Lawrence, S. A., and Davidson, J. A. (1976). Combined toxicity of free chlorine, chloramine and temperature to stage I larvae of the American lobster *Homarus americanus*. *Water Res.* 10, 1093–1099.

Capuzzo, J. M., Lancaster, B. A., and Sasaki, G. C. (1984). The effects of petroleum hydrocarbons on lipid metabolism and energetics of larval development and metamorphosis in the American lobster (*Homarus americanus* Milne Edwards). *Mar. Environ. Res.* 14, 201–228.

Carlberg, J. M., and Van Olst, J. C. (1976). Brine shrimp (*Artemia salina*) consumption by the larval stages of the American lobster (*Homarus americanus*) in relation to food density and water temperature. *Proc. Annu. Meet.—World Maric. Soc.* 7, 379–389.

Charmantier, G., and Aiken, D. E. (1987). Intermediate larval and postlarval stages of *Homarus americanus* H. Milne Edwards, 1837 (Crustacea: Decapoda). *J. Crust. Biol.* 7, 525–535.

Charmantier, G., Charmantier-Daures, M., Bouaricha, N., Thuet, P., Aiken, D. E., and Trilles, J.-P. (1988). Ontogeny of osmoregulation and salinity tolerance in two decapod crustaceans: *Homarus americanus* and *Penaeus japonicus*. *Biol. Bull. (Woods Hole, Mass.)* 175, 102–110.

Charmantier, G., Charmantier-Daures, M., and Aiken, D. E. (1991). Metamorphosis in the lobster *Homarus* (Decapoda): A review. *J. Crust. Biol.* 11, 481–495.

Cobb, J. S. (1970). Effects of solitude on time between fourth and fifth larval molts in the American lobster (*Homarus americanus*). *J. Fish. Res. Board Can.* 27, 1653–1655.

Cobb, J. S., Gulbransen, T., Phillips, B. F., Wang, D., and Syslo, M. (1983). Behavior and distribution of larval and early juvenile *Homarus americanus*. *Can. J. Fish. Aquat. Sci.* 40, 2184–2188.

Cobb, J. S., Wang, D., and Campbell, D. B. (1989a). Timing of settlement by postlarval lobsters (*Homarus americanus*): Field and laboratory evidence. *J. Crust. Biol.* 9, 60–66.

Cobb, J. S., Wang, D., Campbell, D. B., and Rooney, P. (1989b). Speed and direction of swimming by postlarvae of the American lobster. *Trans. Am. Fish. Soc.* 118, 82–86.

Collings, W. S., Cooper-Sheehan,C., Hughes,S. C., and Buckley,J. L. (1983). The spatio-temporal distribution of American lobster, *Homarus americanus*, larvae in the Cape Cod Canal and approaches. *NOAA Tech. Rep., NMFS SSRF* 775, 35–40.

Connor, P. M. (1972). Acute toxicity of heavy metals to some marine larvae. *Mar. Pollut. Bull.* 3, 190–192.

Cooper, R. A., and Uzmann, J. R. (1980). Ecology of juvenile and adult *Homarus*. *In* "The Biology and Management of Lobsters" (J. S. Cobb and B. F. Phillips, eds.), Vol. 2, pp. 97–142. Academic Press, New York.

Costello, W. J., Hill, R., and Lang, F. (1981). Innervation patterns of fast and slow motor neurones during development of a lobster neuromuscular system. *J. Exp. Biol.* 91, 271–284.

Davis, C. C. (1964). A study of the hatching process in aquatic invertebrates. XIII. Events of eclosion in the American lobster, *Homarus americanus* Milne-Edwards (Astacura, Homaridae). *Am. Midl. Nat.* 72, 203–210.

Delistraty, D. A., Carlberg, J. M., Van Olst, J. C., and Ford, R. F. (1977). Ammonia toxicity in cultured larvae of the American lobster (*Homarus americanus*). *Proc. Annu. Meet.—World Maric.*

Soc. 8, 647–672.

Derby, J. G. S., and Capuzzo, J. M. (1984). Lethal and sublethal toxicity of drilling fluids to larvae of the American lobster, *Homarus americanus*. *Can. J. Fish. Aquat. Sci.* 41, 1334–1340.

Dorband, W. R., Van Olst, J. C., Carlberg, J. M., and Ford, R. F. (1976). Effects of chemicals in thermal effluent on *Homarus americanus* maintained in aquaculture systems. *Proc. Annu. Meet.—World Maric. Soc.* 7, 391–414.

Eagles, M. D., Aiken, D. E., and Waddy, S. L. (1986). Influence of light and food on larval American lobsters, *Homarus americanus*. *Can. J. Fish. Aquat. Sci.* 43, 2303–2310.

Ennis, G. P. (1975a). Observations on hatching and larval release in the lobster *Homarus americanus*. *J. Fish. Res. Board Can.* 32, 2210–2213.

Ennis, G. P. (1975b). Behavioral responses to changes in hydrostatic pressure and light during larval development of the lobster *Homarus americanus*. *J. Fish. Res. Board Can.* 32, 271–281.

Ennis, G. P. (1983). The Effect of Wind Direction on the Abundance and Distribution of Decapod Crustacean Larvae in a Newfoundland Nearshore Area. *Can. Tech. Rep. Fish. Aquat. Sci.* 1138.

Ennis, G. P. (1984). Small-scale seasonal movements of the American lobster *Homarus americanus*. *Trans. Am. Fish. Soc.* 113, 336–338.

Ennis, G. P. (1986a). Swimming ability of larval American lobsters, *Homarus americanus*, in flowing water. *Can. J. Fish. Aquat. Sci.* 43, 2177–2183.

Ennis, G. P. (1986b). Stock definition, recruitment variability, and larval recruitment processes in the American lobster, *Homarus americanus*: A review. *Can. J. Fish. Aquat. Sci.* 43, 2072–2084.

Factor, J. R. (1978). Morphology of the mouthparts of larval lobsters, *Homarus americanus* (Decapoda: Nephropidae), with special emphasis on their setae. *Biol. Bull.* (Woods Hole, Mass.) 154, 383–408.

Factor, J. R. (1981). Development and metamorphosis of the digestive system of larval lobsters, *Homarus americanus* (Decapoda: Nephropidae). *J. Morphol.* 169, 225–242.

Fisher, W. S., Nilson, E. H., and Shleser, R. A. (1975). Effect of the fungus *Haliphthoros milfordensis* on the juvenile stages of the American lobster *Homarus americanus*. *J. Invertebr. Pathol.* 26, 41–45.

Fisher, W. S., Rosemark, T. R., and Nilson, E. H. (1976). The susceptibility of cultured American lobsters to a chitinolytic bacterium. *Proc. Annu. Meet.—World Maric. Soc.* 7, 511–523.

Fisher, W. S., Nilson, E. H., Steenbergen, J. F., and Lightner, D. V. (1978). Microbial diseases of cultured lobsters: A review. *Aquaculture* 14, 115–140.

Fogarty, M. J., and Lawton, R. (1983). An overview of larval American lobster, *Homarus americanus*, sampling programs in New England during 1974–79. *NOAA Tech. Rep., NMFS SSRF* 775, 9–14.

Fogarty, M. J., Hyman, M. A., Johnson, G. F., and Griscom, C. A. (1983). Distribution, relative abundance, and seasonal production of American lobster, *Homarus americanus*, larvae in Block Island Sound in 1978. *NOAA Tech. Rep., NMFS SSRF* 775, 23–28.

Forward, R. B., Jr. (1989). Behavioral responses of crustacean larvae to rates of salinity change. *Biol. Bull. (Woods Hole, Mass.)* 176, 229–238.

Forward, R. B., Jr. (1990). Behavioral responses of crustacean larvae to rates of temperature change. *Biol. Bull. (Woods Hole, Mass.)* 178, 195–204.

Govind, C. K. (1982). Development of nerve, muscle and synapse. *In* "The Biology of Crustacea" (D. E. Bliss, ed.), Vol. 3 of "Neurobiology: Structure and Function" (H. L. Atwood and D. C. Sandeman, eds.), pp. 37–68. Academic Press, New York.

Hadley, P. B. (1906). The relation of optical stimuli to rheotaxis in the American lobster, *Homarus americanus*. *Am. J. Physiol.* 17, 326–343.

Hadley, P. B. (1908a). The behavior of the larval and adolescent stages of the American lobster (*Homarus americanus*). *J. Comp. Neurol. Psychol.* 18, 199–301.

Hadley, P. B. (1908b). The reaction of blinded lobsters to light. *Am. J. Physiol.* 21, 180–199.

Harding, G. C., and Trites, K. W. (1988). Dispersal of *Homarus americanus* larvae in the Gulf of Maine from Browns Bank. *Can. J. Fish. Aquat. Sci.* 45, 416–425.

Harding, G. C., Vass, W. P., and Drinkwater, K. F. (1982). Aspects of larval American lobster (*Homarus americanus*) ecology in St. Georges Bay, Nova Scotia. *Can. J. Fish. Aquat. Sci.* 39, 1117–1129.

Harding, G. C., Drinkwater, K. F., and Vass, W. P. (1983). Factors influencing the size of American lobster (*Homarus americanus*) stocks along the Atlantic coast of Nova Scotia, Gulf of St. Lawrence, and Gulf of Maine: A new synthesis. *Can. J. Fish. Aquat. Sci.* 40, 168–184.

Harding, G. C., Vass, W. P., Hargrave, B. T., and Pearre, S., Jr. (1986). Diel vertical movements and feeding activity of zooplankton in St. Georges Bay, N.S., using net tows and a newly developed passive trap. *Can. J. Fish. Aquat. Sci.* 43, 952--967.

Harding, G. C., Pringle, J. D., Vass, W. P., Pearre, S., Jr., and Smith, S. J. (1987a). Vertical distribution and daily movements of larval lobsters *Homarus americanus* over Browns Bank, Nova Scotia. *Mar. Ecol. Prog. Ser.* 41, 29–41.

Harding, G. C., Hargrave, B. T., Vass, W. P., Sheldon, R. W., and Pearre, S., Jr. (1987b). Vertical flux of particulate matter by sedimentation and zooplankton movements in St. Georges Bay, the southern Gulf of St. Lawrence. *Biol. Oceanogr.* 4, 323–357.

Herrick, F. H. (1895). The American Lobster: A Study of Its Habits and Development. *Bull. U.S. Fish Comm.* 15, 1–252.

Herrick, F. H. (1909). Natural history of the American lobster. *Bull. U.S. Bur. Fish.* 29, 149–408.

Hinton, D. J., and Corey, S. (1979). The mouthparts and digestive tract in the larval stages of *Homarus americanus*. *Can. J. Zool.* 57, 1413–1423.

Hudon, C., and Fradette, P. (1988). Planktonic growth of larval lobster (*Homarus americanus*) off Îles de la Madeleine (Québec), Gulf of St. Lawrence. *Can. J. Fish Aquat. Sci.* 45, 868–878.

Hudon, C., and Fradette, P. (1993). Wind-induced advection of larval decapods into Baie de Plaisance (Îles de la Madeleine, Québec). *Can. J. Fish. Aquat. Sci.* 50, 1422–1434.

Hudon, C., Fradette, P., and Legendre, P. (1986). La répartition horizontale et verticale des larves de homard (*Homarus americanus*) autour des Îles de la Madeleine, Golfe du Saint-Laurent. *Can. J. Fish. Aquat. Sci.* 43, 2164–2176.

Hughes, J. T., and Matthiessen, G. C. (1962). Observations on the biology of the American lobster, *Homarus americanus*. *Limnol. Oceanogr.* 7, 414–421.

Hughes, J. T., Sullivan, J. J., and Shleser, R. (1972). Enhancement of lobster growth. *Science* 177, 1110–1111.

Huntsman, A. G. (1924). Limiting factors for marine animals. 2. Resistance of larval lobsters to extreme temperature. *Contrib. Can. Biol. Fish.* 2, 91–93.

Incze, L. S., and Wahle, R. A. (1991). Recruitment from pelagic to early benthic phase in lobsters *Homarus americanus*. *Mar. Ecol. Prog. Ser.* 79, 77–87.

Janssen, J., and Bradt, S. (1980). Feeding ecology and vertical migration of adult alewives (*Alosa pseudoharengus*) in Lake Michigan. *Can. J. Fish. Aquat. Sci.* 37, 177–184.

Johns, P. M., and Mann, K. H. (1987). An experimental investigation of juvenile lobster habitat preference and mortality among habitats of varying structural complexity. *J. Exp. Mar. Biol. Ecol.*

109, 275-285.

Juinio, M. A. R., and Cobb, J. S. (1992). Natural diet and feeding habits of the postlarval lobster *Homarus americanus*. *Mar. Ecol. Prog. Ser.* 85, 83–91.

Juinio, M. A. R., and Cobb, J. S. (1994). Estimation of recent growth of field-caught postlarval American lobsters, *Homarus americanus*, from RNA:DNA ratios. *Can. J. Fish. Aquat. Sci.* 51, 286–294.

Juinio, M. A. R., Cobb, J. S., Bengston, D., and Johnson, M. (1992). Changes in nucleic acids over the molt cycle in relation to food availability and temperature in *Homarus americanus* postlarvae. *Mar. Biol.* 114, 1–10.

Katz, C. H., Cobb, J. S., and Spaulding, M. (1994). Larval behavior, hydrodynamic transport, and potential offshore-to-inshore recruitment in the American lobster *Homarus americanus*. *Mar. Ecol. Prog. Ser.* 103, 265–273.

King, J. A., and Govind, C. K. (1980). Development of excitatory innervation in the lobster claw closer muscle. *J. Comp. Neurol.* 194, 57–70.

Kurmaly, K., Jones, D. A., and Yule, A. B. (1990). Acceptability and digestion of diets fed to larval stages of *Homarus gammarus* and the role of dietary conditioning behavior. *Mar. Biol.* 106, 181-190.

Lang, F., Govind, C. K., Costello, W. J., and Greene, S. I. (1977). Developmental neuroethology: Changes in escape and defensive behavior during growth of the lobster. *Science* 197, 682–685.

Lawton, R., Kouloheras, E., Brady, P., Sides, W., and Borgatti, M. (1983). Distribution and abundance of larval American lobsters, *Homarus americanus* Milne-Edwards, in the western inshore region of Cape Cod Bay, Massachusetts. *NOAA Tech. Rep., NMFS SSRF* 775, 47–52.

Lund, W. A., Jr., and Stewart, L. L. (1970). Abundance and distribution of larval lobsters, *Homarus americanus*, off the coast of southern New England. *Proc. Natl. Shellfish. Assoc.* 60, 40–49.

Lux, F. E., Kelly, G. F., and Wheeler, C. L. (1983). Distribution and abundance of larval lobsters (*Homarus americanus*) in Buzzards Bay, Massachusetts, during 1976–79. *NOAA Tech. Rep., NMFS SSRF* 775, 29–33.

MacKenzie, B. R. (1985). "Temperature Considerations of Larval Lobster (*Homarus americanus* Milne Edwards) Ecology in Waters off Southwestern Nova Scotia," M.S. thesis. Dalhousie University, Halifax, Nova Scotia, Canada.

MacKenzie, B. R. (1988). Assessment of temperature effects on interrelationships between stage durations, mortality, and growth in laboratory-reared *Homarus americanus* Milne Edwards larvae. *J. Exp. Mar. Biol. Ecol.* 116, 87–98.

Matthiessen, G. C., and Scherer, M. D. (1983). Observations on the seasonal occurrence, abundance, and distribution of larval lobsters (*Homarus americanus*) in Cape Cod Bay. *NOAA Tech. Rep., NMFS SSRF* 775, 41–46.

McLeese, D. W. (1974). Olfactory responses and fenitrothion toxicity in American lobsters (*Homarus americanus*). *J. Fish. Res. Board Can.* 31, 1127–1131.

McLeese, D. W., and Metcalfe, C. D. (1979). Toxicity of creosote to larval and adult lobsters and *Crangon* and its accumulation in lobster hepatopancreas. *Bull. Environ. Contam. Toxicol.* 22, 796–799.

McLeese, D. W., Metcalfe, C. D., and Zitko, V. (1980). Lethality of permethrin, cypermethrin and fenvalerate to salmon, lobster and shrimp. *Bull. Environ. Contam. Toxicol.* 25, 950–955.

Mendall, H. L. (1934). The relationship of certain sea birds to the fishing industry of the state of Maine. *Bienn. Rep.—Comm. Sea Shore Fish., Maine* 8, 37–64.

Mills, D. H. (1957). Herring gulls and common terns as possible predators of lobster larvae. *J. Fish. Res. Board Can.* 14, 729–730.

Moores, J. A., Winters, G. H., and Parsons, L. S. (1975). Migrations and biological characteristics of Atlantic mackerel (*Scomber scombrus*) occurring in Newfoundland waters. *J. Fish. Res. Board Can.* 32, 1347–1357.

Munro, J., and Therriault, J.-C. (1983). Migrations saisonnières du homard (*Homarus americanus*) entre la côte et les lagunes des Îles de la Madeleine. *Can. J. Fish. Aquat. Sci.* 40, 905–918.

Nilson, E. H., Fisher, W. S., and Shleser, R. A. (1975). Filamentous infestations observed on eggs and larvae of cultured crustaceans. *Proc. Annu. Meet.—World Maric. Soc.* 6, 367–378.

Ohman, M. D., Frost, B. W., and Cohen, E. B. (1983). Reverse diel vertical migration: An escape from invertebrate predators. *Science* 220, 1404–1407.

Phillips, B. F. (1981). The circulation of the southeastern Indian Ocean and the planktonic life of the western rock lobster. *Oceanogr. Mar. Biol.* 19, 11–39.

Phillips, B. F., and McWilliam, P. S. (1986). The pelagic phase of spiny lobster development. *Can. J. Fish. Aquat. Sci.* 43, 2153–2163.

Phillips, B. F., and Sastry, A. N. (1980). Larval ecology. *In* "The Biology and Management of Lobsters" (J. S. Cobb and B. F. Phillips, eds.), Vol. 2, pp. 11–57. Academic Press, New York.

Pyle, R. W. (1943). The histogenesis and cyclic phenomena of the sinus gland and X-organ in Crustacea. *Biol. Bull. (Woods Hole, Mass.)* 85, 87–102.

Rogers, B. A., Cobb, J. S., and Marshall, N. (1968). Size comparisons of inshore and offshore larvae of the lobster, Homarus americanus, off southern New England. *Proc. Natl. Shellfish. Assoc.* 58, 78–81.

Rooney, P., and Cobb, J. S. (1991). Effects of time of day, water temperature, and water velocity on swimming by postlarvae of the American lobster, *Homarus americanus*. *Can. J. Fish. Aquat. Sci.* 48, 1944–1950.

Rotllant, G., Charmantier-Daures, M., Trilles, J.-P., and Charmantier, G. (1991). Ultrastructure de la glande du sinus chez les larves de *Homarus gammarus* (L.): Premiers résultats. *C. R. Hebd. Seances Acad. Sci.* 313, 201–206.

Saila, S. B., and Flowers, J. M. (1968). Movements and behavior of berried female lobsters displaced from offshore areas to Narragansett Bay, Rhode Island. *J. Cons., Cons. Int. Explor. Mer* 31, 342–351.

Sasaki, G. C., Capuzzo, J. M., and Biesiot, P. (1986). Nutritional and bioenergetic considerations in the development of the American lobster *Homarus americanus*. *Can. J. Fish. Aquat. Sci.* 43, 2311–2319.

Sastry, A. N., and Vargo, S. L. (1977). Variations in the physiological responses of crustacean larvae to temperature. *In* "Physiological Responses of Marine Biota to Pollutants" (F. J. Vernberg, A. Calabrese, F. P. Thurberg, and W. B. Vernberg, eds.), pp. 401–423. Academic Press, New York.

Sastry, A. N., and Zeitlin-Hale, L. (1977). Survival of communally reared larval and juvenile lobsters, *Homarus americanus*. *Mar. Biol.* 39, 297–303.

Scarratt, D. J., (1964). Abundance and distribution of lobster larvae (*Homarus americanus*) in Northumberland Strait. *J. Fish. Res. Board Can.* 21, 661–680.

Scarratt, D. J. (1973). Abundance, survival, and vertical and diurnal distribution of lobster larvae in Northumberland Strait, 1962–1963, and their relationships with commercial stocks. *J. Fish. Res. Board Can.* 30, 1819-1824.

Scarratt, D. J., and Raine, G. E. (1967). Avoidance of low salinity by newly hatched lobster larvae. *J. Fish. Res. Board Can.* 24, 1403–1406.

Sherman, K., and Lewis, R. D. (1967). Seasonal occurrence of larval lobsters in coastal waters of central Maine. *Proc. Natl. Shellfish.*

Assoc. 57, 27–30.

Sherman, S., and Perkins, H. C. (1971). Seasonal variations in the food of juvenile herring in coastal waters of Maine. *Trans. Am. Fish. Soc.* 100, 121–124.

Squires, H. J. (1965). Decapod Crustaceans of Newfoundland, Labrador and the Canadian Eastern Arctic. Manuscript Rep. *Fish. Res. Board Can.* 810.

Squires, H. J. (1970). Lobster (*Homarus americanus*) fishery and ecology in Port au Port Bay, Newfoundland, 1960–65. *Proc. Natl. Shellfish. Assoc.* 60, 22–39.

Squires, H. J., Tucker, G. E., and Ennis, G. P. (1971). Lobster (*Homarus americanus*) in Bay of Islands, Newfoundland, 1963–65. Manuscript Rep. *Fish. Res. Board Can.* 1151.

Stasko, A. B. (1977). Lobster Larvae on the Scotian Shelf. *Can. Atl. Fish. Sci. Advis. Comm. Res. Doc.* 77/31.

Stasko, A. B., and Gordon, D. J. (1983). Distribution and Relative Abundance of Lobster Larvae off Southwestern Nova Scotia 1977–1978. *Can. Tech. Rep. Fish. Aquat. Sci.* 1175.

Stephens, P. J., and Govind, C. K. (1981). Peripheral innervation fields of single lobster motoneurons defined by synapse elimination during development. *Brain Res.* 212, 476–480.

Sulkin, S. D. (1984). Behavioral basis of depth regulation in the larvae of brachyuran crabs. *Mar. Ecol. Prog. Ser.* 15, 181–205.

Templeman, W. (1936a). Fourth stage larvae of *Homarus americanus* intermediate in form between normal third and fourth stages. *J. Biol. Board Can.* 2, 349–354.

Templeman, W. (1936b). The influence of temperature, salinity, light and food conditions on the survival and growth of the larvae of the lobster (*Homarus americanus*). *J. Biol. Board Can.* 2, 485–497.

Templeman, W. (1937a). Egg-laying and hatching postures and habits of the American lobster (*Homarus americanus*). *J. Biol. Board Can.* 3, 339–342.

Templeman, W. (1937b). Habits and distribution of larval lobsters (*Homarus americanus*). *J. Biol. Board Can.* 3, 343–347.

Templeman, W. (1939). Investigations into the life history of the lobster (*Homarus americanus*) on the west coast of Newfoundland, 1938. *Newfoundland, Dep. Nat. Resour., Res. Bull. (Fish.)* 7, 1–52.

Templeman, W. (1948). Growth per moult in the American lobster. *Newfoundland, Dep. Nat. Resour., Res. Bull. (Fish.)* 18, 26–48.

Templeman, W., and Tibbo, S. N. (1945). Lobster investigations in Newfoundland 1938 to 1941. *Newfoundland, Dep. Nat. Resour., Res. Bull. (Fish.)* 16, 1–98.

Umminger, B. L. (1968). Polarotaxis in copepods. I. An endogenous rhythm in polarotaxis in *Cyclops vernalis* and its relation to vertical migration. *Biol. Bull. (Woods Hole, Mass.)* 135, 239–251.

Van Olst, J. C., Carlberg, J. M., and Hughes, J. T. (1980). Aquaculture. *In* "The Biology and Management of Lobsters" (J. S. Cobb and B. F. Phillips, eds.), Vol. 2, pp. 333–384. Academic Press, New York.

Various authors (1986). International Workshop on Lobster Recruitment. *Can. J. Fish. Aquat. Sci.* 43, 2062–2402.

Varma, C. P. (1977). Natural food of the lobster larvae (*Homarus americanus* Milne-Edwards) in the Northumberland Strait, New Brunswick, Canada. *Biol. Membr.* 4, 206–213.

Via, S. E., and Forward, R. B., Jr. (1975). The ontogeny and spectral sensitivity of polarotaxis in larvae of the crab *Rhithropanopeus harrisi* (Gould). *Biol. Bull. (Woods Hole, Mass.)* 149, 251–266.

Wells, P. G., and Sprague, J. B. (1976). Effects of crude oil on American lobster (*Homarus americanus*) larvae in the laboratory. *J. Fish. Res. Board Can.* 33, 1604–1614.

Wilder, D. G. (1953). The growth rate of the American lobster (*Homarus americanus*). *J. Fish. Res. Board Can.* 10, 371–410.

Williams, L. W. (1907). The stomach of the lobster and the food of larval lobsters. *Annu. Rep.—R.I. Comm. Inland Fish.*, 37, 153–180.

CHAPTER

4

Postlarval, Juvenile, Adolescent, and Adult Ecology

PETER LAWTON
Biological Station
Department of Fisheries and Oceans
St. Andrews, New Brunswick, Canada E0G 2X0

KARI L. LAVALLI
Marine Science Center
Northeastern University
Nahant, Massachusetts 01908

I. Introduction

The complex life cycle of the American lobster, *Homarus americanus*, is dominated by a benthic period that may extend to more than 30 years, even under contemporary patterns of commercial exploitation. During this period, the lobster's physical attributes and its movement potential increase over several orders of magnitude: <5 mm to >200 mm in carapace length (CL); <100 mg to >9 kg in live weight; <1 m to >1 km in daily activity range; <1 km to >100 km in annual displacement. Relaxed predation pressure and the increased metabolic demands accompanying greater body size result in dramatic ontogenetic shifts in realized niche as the lobster progresses from a shelter-restricted habit to a more migratory existence. Throughout the benthic period, the lobster seeks out shelter-providing habitat, although there is a relaxation from obligate need following initial recruitment to the benthos to facultative use of shelter in later life. Habitat architecture may thus be influential in molding local lobster population demography and, on a regional scale, habitat limitation in the early benthic life history period may constitute an important factor regulating population size.

This chapter reviews the benthic life history of *Homarus americanus*, including aspects of postlarval supply and transition to the benthos that are influential in determining initial benthic distribution. A new life history classification is presented that divides the benthic portion of the life cycle into five distinct phases: three juvenile phases (shelter-restricted, emergent, and vagile), an adolescent phase, and an adult phase. Patterns in the distribution and abundance, shelter usage and feeding, movement potential, social interaction, and community role of these successive benthic life history phases are surveyed. Given the status of *H. americanus* as a major fisheries resource, this review also delves into the interaction between fisheries science and ecology. Emphasis is on the last decade of research, which has disclosed the *mode de vie* of lobsters in their first few months and years of benthic existence. New insight has also been obtained into adult ecology through the use of enhanced animal tracking and *in situ* observational techniques, which have partially overcome the historical reliance on commercial fisheries sampling to assay the distribution and abundance of this final protracted life history phase.

II. Life History Phases

The life history of *Homarus americanus* has long been divided into several developmental phases, each consisting of one or a series of stages. A history of the various schemes is depicted in Fig. 1. Several important trends are apparent: (1) the recognition of pronounced morphological, physiological, and behavioral changes accompanying the metamorphic molt into the fourth stage; (2) acknowledgment of behavioral differences occurring within the early years of benthic existence, as the lobster grows from ~5- to 40-mm CL; and (3) consideration of the impact of reproductive maturation on lobster movement and social interaction.

Existing schemes all seem inadequate in one way or another. For example, Herrick (1895) used *larvae* for the first four stages; three larval stages are now generally recognized, followed by a metamorphosis that results in a postlarva (stage IV; see Section IV,A).

Hudon (1987) adopted *postlarvae* for lobsters that are cryptic and flee when exposed (<25-mm CL). This term does not adequately distinguish among the pelagic postmetamorphic stage, the settling postlarva, and the new benthic recruit. Furthermore, use of *juvenile* for lobsters of 25- to 73-mm CL does not account for the observation that physiological maturity can be reached at 40-mm CL and ignores the typical use of *juvenile* for immature animals.

Barshaw and Bryant-Rich (1988) distinguished between cryptic *early* juveniles, with undifferentiated claws, and more vagile *late* juveniles, with differentiated claws. Their descriptions are not precise, however, as the claws begin to differentiate externally at stage VI (Govind and Lang, 1978), but the muscle fibers do not complete differentiation until stage VIII for the cutter (Govind and Pearce, 1989) and stage XVI (approximately 1 year of age) for the crusher (Govind and Lang, 1978; Costello and Lang, 1979).

Wahle and Steneck (1991) introduced *early benthic phase* to describe a range of sizes (from 5- to between 20- and 40-mm CL) of cryptic juveniles typically found in shelter-providing habitats, notably cobble. They applied *adolescent phase* to the more conspicuous, inshore, prereproductive lobsters and *reproductive phase* to sexually mature lobsters. *Early benthic phase* stresses the demographic similarity of lobsters between 5- and 40-mm CL, yet deemphasizes the behavioral transitions within this size range (Hudon, 1987; Barshaw and Bryant-Rich, 1988). Cobb and Wahle (1994) modified *adolescent phase* to include lobsters that are vagile and capable of habi-

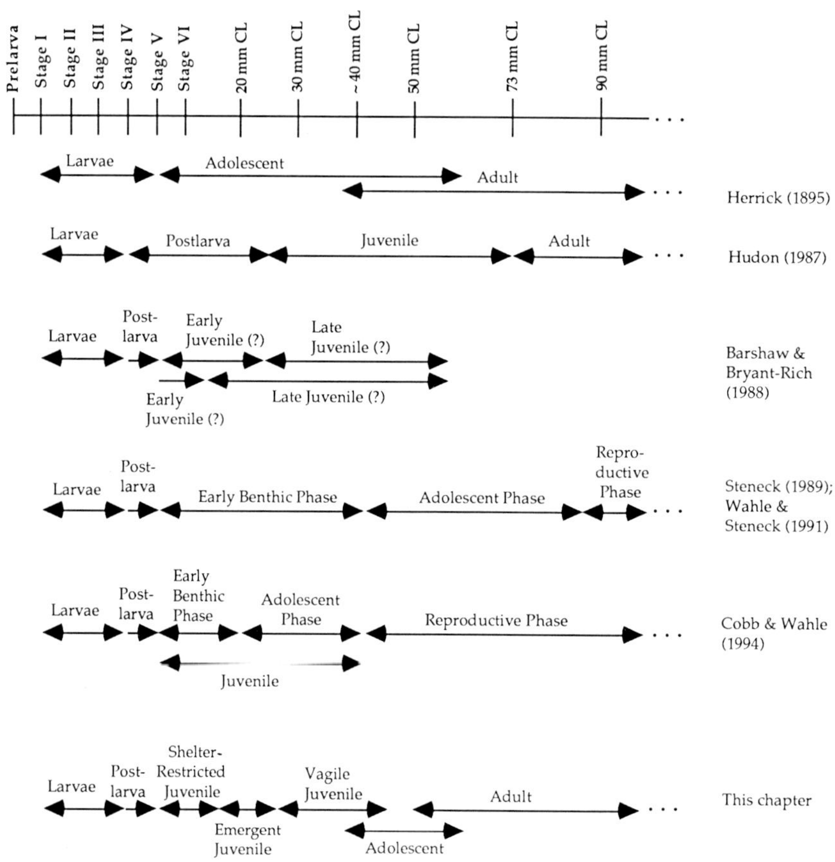

FIGURE 1 Evolution of the terminology used for various life history phases of the American lobster (*Homarus americanus*).

tat shifts (~20- mm CL), but this ignores the typical use of *adolescent* for physiologically maturing individuals.

The life history phases proposed in Table 1 integrate the ecological ontogeny of lobsters more explicitly with developmental, biological, and individual behavioral attributes. Lobsters that have recently settled to the benthos constitute the *shelter-restricted juvenile phase*. These retain developmental and behavioral legacies from the postlarval phase, such as the capacity for suspension feeding and rapid, highly effective tail-flipping. With further growth and morphological differentiation of the claws and the assumption of a fully benthic foraging mode, yet only limited movement outside of the shelter, the lobster enters the *emergent juvenile phase*. The subsequent *vagile juvenile phase* is marked by a progressive change from movements restricted to the immediate vicinity of shelter to a wider-ranging, surface-active foraging mode (*sensu* Lawton, 1987), and the potential for habitat shifts (Hudon, 1987).

With the distinction between emergent and vagile juvenile phases, the *adolescent phase* is more clearly linked to reproductive development and is marked by physiological (but not functional) sexual maturity. While changes in activity pattern and movement between the successive juvenile phases may be linked to relaxation of predation pressure, it is within the adolescent phase that reproductive maturation and associated behavioral attributes (i.e., social interactions) begin to exert their dominant influence on seasonal movements and local demography of lobster populations. The adolescent phase also marks the onset of clear, size-related, sexual differentiation. Males may become physiologically mature at a smaller size than females. The final and most protracted life history phase, the *adult phase*, begins with the onset of functional maturity (see Section VI,B).

These terms are used throughout this chapter, as defined above and in Table 1, which provides explicit criteria. They clearly distinguish among lobsters that differ in morphological attributes, behavior, the external abiotic and biotic factors of the environment to which they are exposed, and, thus, realized niche. Body size at transition between all phases, particularly in the later adolescent and adult phases, will vary geographically, with significant demographic implications (see Section VI,B). Thus, lobster life history phases should be geographically indexed to estimates of size at maturity. For certain purposes, several phases may be combined, as Wahle and Steneck (1991) have done in their use of the term *early benthic phase*.

III. General Distribution Patterns of Lobsters

A. Geographic Range and Distribution

The American lobster is found along the continental shelf and upper slope of the northwestern Atlantic, from the Straits of Belle Isle, Newfoundland and Labrador, to Cape Hatteras, North Carolina. Occasional records are reported as far south as Rich Inlet, Wilmington, North Carolina (Squires, 1990). While lobsters are encountered intertidally, the principal depth distribution is from the sublittoral fringe to 50 m based on *in situ* density estimates and landings from the commercial fisheries (Pringle and Burke, 1993). However, lobsters are fished to depths of 700 m on the edge of the continental shelf (Cooper and Uzmann, 1971). Within this general range, the major inshore fisheries are found between Rhode Island and Newfoundland. Offshore, lobsters are found in highest concentrations on Browns Bank, Georges and Crowell basins in the Gulf of Maine, and near submarine canyons from Corsair Canyon, eastern Georges Bank, to Norfolk Canyon off the coast of Virginia. The terms *inshore* and *offshore* are used broadly in this chapter, as definitions vary between fisheries management agencies, as does popular usage in fishing communities along the geographical range. Miller (Chapter 5) and Fogarty (Chapter 6) provide comprehensive accounts of *Homarus americanus* fisheries.

B. Habitats Used

Homarus americanus is found in a wide variety of habitats (Table 2) (Cooper and Uzmann, 1980). Inshore populations of vagile juvenile, adolescent, and adult lobsters are found on mud, cobble, bedrock, peat reefs, and eelgrass beds and (in certain locations) within sandy depressions (Thomas, 1968; Cooper, 1970; Cobb, 1971; Cooper *et al.*, 1975; Hudon, 1987; Able *et al.*, 1988; Heck *et al.*, 1989; Wahle and Steneck, 1991; Lawton and Robichaud, 1992). Offshore populations are found on similar substrates, as well as on clay, which makes up much of the outer continental shelf (Cooper and Uzmann, 1980). Shelter-restricted and emergent juvenile phases seem to be principally distributed among cobble, rocks on sand, and peat reefs (Hudon, 1987; Able *et al.*, 1988; Wahle and Steneck, 1991).

C. Environmental Factors Determining Distribution

Within its geographical range, the lobster inhabits regions where temperatures can be as low as 5°C or as

TABLE 1 **Life History Phases of *Homarus americanus* with Associated Behaviors and Ecological Processes**[a]

Phase	Size (mm CL)[b]	Activity pattern	Foraging mode	No. of shelters occupied	Realms (maximum abundance)	Ecological processes
Larval (stages I–III)	~2 – 4	Vertical migrator; poor swimmer; passive (?) drifter	Raptorial feeder	N/A	Pelagic (offshore to inshore, depending on area)	Dispersal; development
Postlarval (stage IV)	~4 – 5	Strong surface swimmer; benthic settler	Raptorial/ suspension feeder	Selects preexisting shelter or excavates new one	Pelagic (inshore) to benthic	Settlement; predation while settling or shortly thereafter
Shelter-restricted juvenile	~4 – 14	Recent settler; remains under cover; subterranean movements within spatially complex habitats	Suspension feeder; browser within shelter; ambusher at shelter entrance	Usually one, sometimes several within a contiguous shelter space	Benthic (shallow)	Recruitment; predation
Emergent juvenile	~15 – 25	Mostly shelter confined; limited movements outside of shelter, but remains in close vicinity	Browser; ambusher	One to several	Benthic (shallow)	Predation
Vagile juvenile	~25 – size of physiological maturity (~40)	Shelter user, but more extensive movements out of shelter for food	Ambusher; pursuer; searcher	One to several	Benthic (shallow)	Competition
Adolescent	Physiological maturity, but not functional maturity (~50)	Active, mostly nocturnal; may participate in seasonal movements with reproductive animals	Pursuer; searcher	One to many, depending on seasonal movement	Benthic (shallow to deep)	Competition; indirect fishing mortality due to retention in traps
Adult	Functional maturity (>50)	Active, mostly nocturnal; seasonal, reproductively mediated movements	Pursuer; searcher	One to many, depending on seasonal movement	Benthic (shallow to deep)	Mate selection; reproductive success; direct fishing mortality

[a]Expanded from the scheme presented by Steneck (1989) by inclusion of characteristic activity pattern, foraging mode (*sensu* Hughes, 1980), and shelter usage (at any particular time) for each defined life history phase. Some statements are based on current hypotheses, rather than on direct observation or quantification. See Section II for further details.

[b]Sizes at transition between all phases may vary geographically. For adolescent and adult lobsters, the sizes specified are the minimum carapace lengths (CLs) for entry into these life history phases. Size at maturity varies geographically, such that functional maturity may not be reached until ~100-mm CL in some areas.

high as 20°C (Aiken and Waddy, 1986). The thermal tolerance of lobsters is broad, from −1° to 30.5°C, and they can survive abrupt temperature increases of 16°C or decreases of 20°C (Harding, 1992). Where annual warm-water conditions last longer, such as in the southern Gulf of St. Lawrence (Templeman, 1936a) and Long Island Sound (Briggs and Mushacke, 1979), growth is accelerated and sexual maturity occurs at smaller sizes.

Proper synchrony of molting and reproductive

TABLE 2 **Inshore and Offshore Habitats Frequented by *Homarus americanus*[a]**

Substrate description	Place found	Associated flora	Associated fauna	Size of lobsters present
Sand base with rock: shelters formed by excavating sand under a rock(s) to form U-shaped, shallow tunnels	Inshore; offshore	Sea lettuce; nori; Irish moss; dulse; sea wrack; rockweeds; kelps; sea colander; encrusting corallines	Sponges; polychaetes; crabs; shrimps; mollusks; brachiopods; ascidians; echinoderms; inshore fishes (flounder, rocklings, gunnels, shannies, rock and conger eels, sculpins, tautogs, cunners, cusks, small skates, cod, goosefish); offshore fishes (dogfish, cod, wolffish, tilefish)	All
Cobble: shelters formed in the interstitial spaces between heterogeneous rocks, pebbles, and boulders making up the bed	Inshore	Algal species generally uncommon except for encrusting corallines	Crabs; shrimps; polychaetes; echinoderms; mussel beds; small fishes such as rocklings, gunnels, shannies, grubbies, cunners, rock eels, juvenile flounder	Shelter-restricted, emergent, and vagile juvenile phases; some adolescents
Bedrock base with rocks and boulders: shelters formed by rock overhangs or crevices, but burrowing is generally not possible	Inshore only	Algal species same as in sand base with rock	Crabs and fish species same as in sand base with rock, although fewer species present	Vagile juveniles, adolescents, adults
Mud base with burrows: shelters formed from excavations in soft substrate	Inshore, in estuaries and harbors; not commonly found offshore	Generally absent	Anemones; shrimps; crabs; certain fishes, such as flounder, hake, cusk, hagfish, rockling, snake and conger eels, ocean pout, sculpins, tautog, cunner, redfish, goosefish, wolffish, squirrelfish, dogfish	Adults
Peat reefs: shelters formed from excavations that cut deep into peat; often obscured by dense algal growth; reef forms from blocks of salt marsh peat that break and fall into adjacent marsh creeks and channels	Inshore along salt marshes	*Spartina alterniflora*; green fleece; sea lettuce; green weeds	Crabs; fishes, particularly cunners, American eel, rock gunnel, hake, shannies, pollock, grubbies	All life history phases may be present
Eelgrass meadows: shelters formed from excavations into rhizomes of the eelgrass	Inshore	*Zostera*; sea lettuce; red weeds	Crabs; shrimps; hermit crabs; fishes such as mummichog sticklebacks, flounder, grubbies, pipefish, sand lance, tautog, tomcod, cod, scup	Shelter-restricted, emergent, and vagile juveniles, adolescents
Clay or sand base with burrows and depressions: shelters consist of bowl-like depressions	Offshore	No algal species	Anemones; polychaetes; crabs; shrimps; some species of mollusks; echinoderms; same fishes as on mud substrates	Adolescents, adults
Mud–clay base with anemones: depressions formed around base of anemone tubes; thin layer of silt present	Offshore	No algal species	Crabs; polychaetes; shrimps; echinoderms; same fishes as on mud substrates	Adolescents, adults
Submarine canyon clay wall with burrows: shelters formed from excavations into compacted clay walls; depressions may be shallow, half-moon-like, or deep, cave-like holes; some described as a "pueblo village" which is an extensively bioeroded clay exposure supporting a rich megafaunal community	Offshore canyons	No algal species	Same species as on sand- and clay-based substrates	Adolescents, adults

[a] Data are based on an initial classification by Cooper and Uzmann (1980) and subsequent observations (Hudon, 1987; Able *et al.*, 1988; Barshaw and Lavalli, 1988; Heck *et al.*, 1989; Wahle and Steneck, 1991; Wahle, 1992a; K. L. Lavalli and P. Lawton, unpublished observations).

cycles may require particular annual thermal regimes (Aiken and Waddy, 1986; see Waddy *et al.*, Chapter 10, on control of growth and reproduction). Toward the southern limit of distribution, winter water temperatures must decline to approximately 8–10°C (Aiken and Waddy, 1986). At the northern limit of distribution, summer water temperatures may not rise high enough for successful spawning, as occurred with a group of male and female lobsters (81- to 114-mm CL) transplanted from the northeast coast of Newfoundland to St. Michael's Bay, southern Labrador, about 200 km beyond the known northern limit (Boothroyd and Ennis, 1992).

Lobsters have historically been considered to be stenohaline marine organisms, limited to coastal and offshore habitats where salinities are typically >25 ppt (Dall, 1970). However, even with their limited ability to osmoregulate, adolescent and adult lobsters exploit estuarine areas, perhaps engaging in seasonal movements to avoid low salinity or high summer temperatures (Thomas, 1968). Settled postlarvae and shelter-restricted phase juveniles do not appear to live deep within the upper reaches of estuaries (Wahle, 1993). The lower lethal limit of adult lobsters exposed to various dilutions of seawater is generally between 8 and 14 ppt, depending on temperature, oxygen, and acclimation conditions (McLeese, 1956). Larvae and molting individuals are more sensitive to reduced salinity than are intermolt benthic lobsters (Charmantier *et al.*, 1988). Salinities of ≤10 ppt cause significant physiological changes that are extremely stressful, even if the animals manage to survive short-term exposure (McLeese, 1956; Jury *et al.*, 1994a). Heavy mortalities have been reported after extreme spring runoffs in estuarine systems (Thomas and White, 1969). Despite the stress, some lobsters may live in estuaries year-round. Increased temperatures facilitating earlier and more frequent molts, greater food availability, and reduced competition may promote movement into estuaries; upper estuarine areas may represent an underexploited, albeit suboptimal, habitat for subdominant lobsters (Jury *et al.*, 1994a).

Lobsters can survive in waters with low levels of dissolved oxygen (hypoxia), with the possible exception of locations experiencing severe organic enrichment. The lower lethal oxygen level for juveniles and adults ranges from 0.2 mg O_2/liter at 5°C to 1.2 mg O_2/liter at 25°C in 30-ppt salinity (Harding, 1992). Other significant environmental factors affecting lobster distribution include light, winds, and currents, which are considered in detail in the sections that follow. Because severe storms have stranded large numbers of lobsters along certain coasts, the onset of winter storms has been implicated in the seasonal movement of lobsters to deeper water (Cooper *et al.*, 1975; Ennis, 1984a; P. Lawton, unpublished observations).

IV. Postlarval Lobsters

A. Historical Perspective

Lobster developmental stage IV was long considered a planktonic larval form, rather than a postlarval (or benthic-like) form. Part of this confusion stemmed from Herrick's (1895, p. 174) observation that "the young lobster may remain at the surface of the ocean, even after the sixth molt," and his demarcation of the larval period as the duration of pelagic life, which he considered "at a close by the end of the fifth stage." Later, Herrick (1909, p. 348) restated this position, but noted "... a record by Meek of the capture by surface net of a young specimen of the European lobster, which measured 20.5 millimeters.... Such a lobster should be in the sixth or seventh stage."

These ideas were endorsed by Hadley (1908, p. 254), who noted that stage IV lobsters might construct burrows in sand and gravel bottoms, but they did not remain in them, continuing to swim actively near the water surface. Conversely, stage V lobsters not only immediately began burrowing in such substrates, but remained within their shelters, rarely swimming. Hadley (1908) concluded, as did Herrick (1895), that stage V lobsters were the settling stage, and that stage IV lobsters should be regarded as larvae.

Templeman (1940) and Templeman and Tibbo (1945) noted few stage V or VI larvae in their field plankton tows. Recent field studies by Scarratt (1973) and Cobb *et al.* (1989) revealed that stage IV lobsters in later molt stages (from D_1 onward) were absent from the plankton, leading to the conclusion that settlement takes place sometime before the end of stage IV. Cobb (1988), Factor (1989), and Charmantier (1989) briefly discussed appropriate terms for stage IV, concluding that *postlarva* should be reserved for this stage in recognition of a true metamorphosis between stages III and IV (historical review and detailed justification provided by Charmantier *et al.*, 1991; see Factor, Chapter 1, for an overview of life history).

B. Settlement

1. Relationship between Larval Supply and Benthic Recruits

Both laboratory observations (Cobb *et al.*, 1989) and field plankton surveys (Scarratt, 1973; Cobb *et al.*,

1989) suggest that postlarvae may settle before or around the middle of stage IV, yet the time of settlement is not precise and may be influenced by environmental conditions. Ennis (Chapter 3) discusses movements of larvae and planktonic postlarvae, settlement, and substrate selection.

Despite intensive sampling, much information is lacking on how postlarval planktonic abundance relates to recruitment into the benthos. In a study along the coast of Maine, around Damariscove Island, planktonic postlarval abundance in August was positively correlated to benthic recruits (7- to 8-mm CL) in September; however, no consistent relationship was found between the patterns of planktonic postlarval abundance and the density of the benthic recruits (Incze and Wahle, 1991). The island seems to cast a "recruitment shadow," intercepting postlarvae carried in wind-driven currents (Wahle and Incze, 1993). Settlement is apparently a gradual process and, as such, transient pulses in planktonic postlarval abundance do not seem to affect final recruit densities. More important factors may include the presence of suitable settlement habitat, the presence of predators, temperature influences on larval development time, density-dependent interactions between new recruits and the previous year's cohort, and hydrographic conditions (Incze and Wahle, 1991).

Abundance and spatial association of consecutive stages of lobster larvae in the Baie de Plaisance, Îles de la Madeleine, Quebec, suggests that advective processes may strongly influence the distribution and supply of earlier larval stages (Hudon and Fradette, 1993). Where such events are predictable, yearly occurrences, postlarval supply tends to be stable; where these events occur randomly from year to year, postlarval supply and settlement will be more variable. Because of their strong swimming ability, postlarvae can spread out within a region (Katz *et al.*, 1994); however, advective processes determine where they are initially located (Hudon and Fradette, 1993; Hudon, 1994).

2. Factors Affecting Settlement and Postsettlement Distribution

Historically, recently settled postlarvae have been nearly impossible to locate in the field; thus, our understanding of the settlement process is based on laboratory studies with cultured postlarvae. Ennis (Chapter 3) discusses bottom-testing behavior and the ability of planktonic postlarvae to delay settlement to find appropriate benthic habitats. Environmental cues that may be influential in determining initial postsettlement distribution include reactions to predator odors and thermal gradients (Boudreau *et al.*, 1992; 1993a), phototactic responses (Botero and Atema, 1982), and chemical cues from benthic conspecifics and/or substrates (Hudon *et al.*, 1986; Boudreau *et al.*, 1993b).

One of the most important settlement cues may be the presence or absence of a thermal gradient through which the postlarvae must swim to encounter the benthos. In the laboratory, postlarvae are reluctant to pass through thermal gradients of a 4–5°C difference (Boudreau *et al.*, 1992). If expressed in the field, such behavior could lead the postlarvae to warmer, shallower, inshore areas, where suitable habitat such as cobble and peat is available (Hudon, 1987; Wahle and Steneck, 1991). Thermal cues may explain why settled postlarvae were not found in areas of coastal Maine colder than 15°C (Wahle and Steneck, 1991; R. S. Steneck, unpublished data), as lower temperatures in addition to pronounced thermal gradients are thought to inhibit settlement (Huntsman, 1923). However, recent *in situ* sampling of new benthic recruits in the Fundy Isles region of the Bay of Fundy (Lawton and Robichaud, 1992) suggest that Huntsman's (1923) original speculations on factors limiting larval distribution in the Bay of Fundy, as well as current efforts at synthesizing regional settlement patterns, are somewhat premature.

The change in phototaxis at stage IV (Hadley, 1908) has been proposed as a mechanism that may lead postlarvae to seek dark areas (i.e., the bottom) (Botero and Atema, 1982), but, in controlled laboratory settings, postlarvae have not been seen to change their vertical position during light and dark conditions (Boudreau *et al.*, 1992). Further investigation is needed, as nocturnal settlement could have important ramifications for predator avoidance.

Chemical cues may provide settling postlarvae with information about predation risk, substrate type, and survivability of conspecifics. Postlarvae are attracted to water conditioned by adult lobsters and macroalgae (Boudreau *et al.*, 1993b)—odors that might indicate a structurally complex habitat capable of supporting a lobster population. In contrast, they avoid water conditioned by cunners, which can be significant predators (Lavalli and Barshaw, 1986; Barshaw and Lavalli, 1988; Wahle and Steneck, 1992). Surprisingly, postlarvae did not prefer water from sand or rock substrates over sterile, filtered seawater; however, in choice tests, postlarvae preferred rock-conditioned water to sand-conditioned water (Boudreau *et al.*, 1993b). Despite these encouraging results, ablation studies should be conducted to determine how important chemical cues are to the overall benthic habitat selection process.

Settled postlarvae and shelter-restricted phase

juveniles are not randomly distributed in nature, being most dense on cobble substrates (Wahle and Steneck, 1991), rocks on sand (Hudon, 1987), and peat reefs (Able *et al.*, 1988). These new benthic recruits have also been collected from eelgrass beds, cobble/boulder substrates (Heck *et al.*, 1989; Wahle and Steneck, 1991), and mud flats (MacKay, 1929; Cooper and Uzmann, 1980), although densities are typically lower; they are not found on exposed bedrock or sand substrates. Such a distribution pattern could arise from active habitat selection, from random settlement followed by differential predator-induced mortality, or from a combination of both processes. Field support for active habitat selection has been obtained by following the behavior of postlarvae released in shallow water (<2 m) in Narragansett Bay, Rhode Island, and Buzzards Bay, Massachusetts (Cobb *et al.*, 1983). Further corroboration comes from planktonic (Hudon *et al.*, 1986) and benthic sampling efforts in the Gulf of St. Lawrence (Hudon, 1987). Postlarval production is estimated at 5800 individuals per square kilometer (Hudon *et al.*, 1986) and, based on substrate availability, this would lead to an expected density of 0.48 individuals per square meter if postlarvae preferentially chose shallow rocky bottoms (Hudon, 1987). With predation rates of 30% factored in, the expected densities would drop to 0.317 individuals per square meter. Hudon (1987) found densities of 0.18 stage V lobsters per square meter, indicating that substrate selection seems to occur in nature. However, because predation is extremely important (see Section IV,C), deciphering the settlement process from sampled postsettlement distributions is problematic.

3. Burrowing Ability

Postlarvae, juveniles, and adolescents of three species of clawed (nephropid) lobsters (*Homarus americanus*, *H. gammarus*, and *Nephrops norvegicus*) create remarkably similar burrows with one, two, or more openings [see the work of Cobb (1971), Berrill and Stewart (1973), and Botero and Atema (1982), on *H. americanus*]. The burrows begin as U-shaped tunnels and may be expanded later into additional openings. The "entrance" tunnel is usually a craterlike, or wide-mouthed, depression with a mound of sediment at one end. The "rear exit" and additional openings simply open onto the flat surface of the sediment. Upon selection of a site, a lobster typically faces the area where the hole (later to become the entrance) will be made and digs into the substrate with the first two to three pairs of pereiopods, with pleopod fanning to blow the sediment away (Fig. 2), resulting in a pile of sediment outside the hole (Fig. 3). Some of this sediment may later be bulldozed away from the mouth of the burrow, but a mound will remain.

Ventilation is necessary to prevent low oxygen and high carbon dioxide stress during burrow occupancy. Thus, newly settled postlarvae, which seem to be shelter bound, must either expend energy to ventilate the burrow actively, via their pleopod or respiratory currents, or excavate the burrow initially in such a manner as to exploit passive flow through it (Atkinson and Taylor, 1988). The construction of lobster burrows and their openings probably promotes passive ventilation (Lavalli, 1992; K. L. Lavalli, unpublished observations). Fluid will move through the burrow from the end where the flow is slower (smaller, blunter, or lower opening) to the end where the flow is faster (larger, sharper, or higher opening) due to Bernoulli's principle (Fig. 3) (Vogel and Bretz, 1972; Vogel, 1977). For lobster burrows in rock substrates, currents will cause water to flow between the interstitial spaces, thereby providing the necessary ventilation. Shelter-restricted phase juveniles are found in the greatest densities in 2–10 m of water (Hudon, 1987; Able *et al.*, 1988; Wahle and Steneck, 1991; Incze and Wahle, 1991), depths which typically experience high-velocity water flow (Howard and Nunny, 1983).

While postlarvae are more tolerant to ammonia than are larvae, they are less tolerant than adults. Lethal concentrations (LC_{50}s; the level of toxicant lethal to 50% of test individuals) have been reported

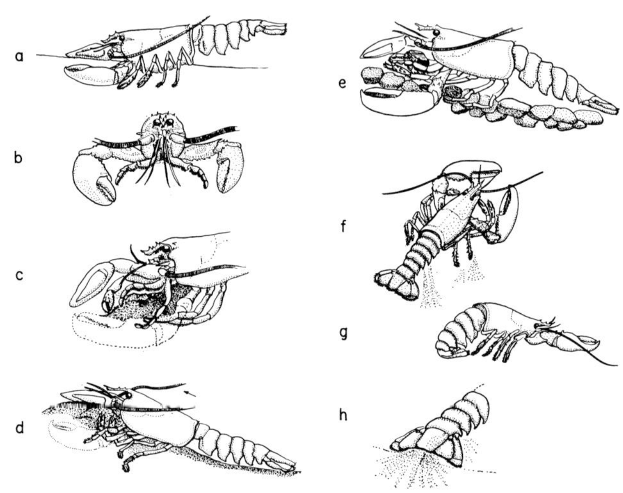

FIGURE 2 Burrowing behavior of the American lobster: (a–d) bulldozing; (e) rock moving; (f and g) "backward dig"; and (h) pleopod fanning. (From "The shelter related behavior of the lobster, *Homarus americanus*" by J. S. Cobb, *Ecology*, 1971, **52**, 108–115. Copyright © 1971 by Ecological Society of America. Reprinted by permission.)

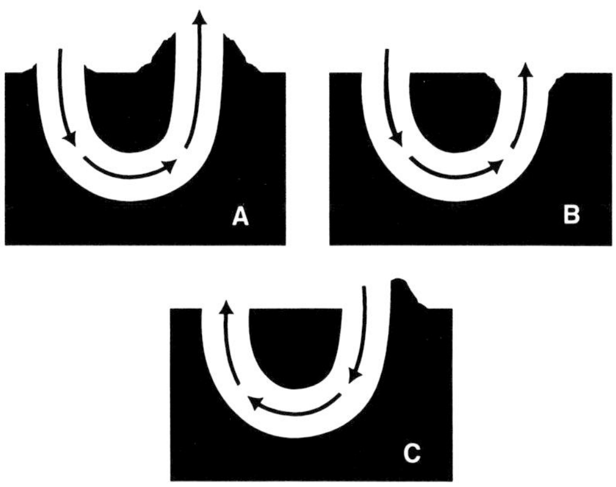

FIGURE 3 Burrow designs that would promote passive flow of water through the chamber: (A) water flows from the lower opening (where pressure is higher, but velocity is slower) to the higher opening (where pressure is lower, but velocity is higher); and (B) water flows from the more sheltered, blunter opening to the less sheltered, wider opening. (*sensu* Vogel, 1974; both designs are used by lobsters.) (C) Water flows from the opening with a mound of sediment on one side only to the more sheltered, blunter opening. (From Lavalli, 1992, with permission.)

for postlarvae held at 20°C as low as 4.8 mg of NH_3/liter over 24 hours and 2.36 mg/liter for 96 hours. Incipient LC_{50} (50% mortality under continued exposure for periods after which acute lethal action has ceased) for postlarvae is 1.4 mg/liter (Sastry and Laczak, 1975; Delistraty *et al.*, 1977; Young-Lai *et al.*, 1991). Tolerance increases to 12.14 mg of NH_3/liter in stage VI lobsters and to 121 mg/liter in juveniles exposed for 24 hours. Hypoxic conditions may increase the toxic effects of ammonia, but increased scaphognathite beating, as seen in the response of *Homarus gammarus* to hypoxia (Thomas, 1954), may alleviate this problem. No measurements of oxygen conditions within *H. americanus* burrows have been made, although conditions within *Nephrops norvegicus* burrows do not become as severely hypoxic as in those of other burrowing decapods (Atkinson and Taylor, 1988). Newly settled postlarvae and shelter-restricted phase juveniles spend approximately 20% of their time pleopod fanning within the shelter (Barshaw and Bryant-Rich, 1988). While perhaps most important for food acquisition (see Section IV,D), pleopod fanning, along with proper hydrodynamic burrow design, respiratory currents, and other movements of the lobster, may adequately exchange burrow water and eliminate wastes.

4. Substrate Availability

The availability of habitat most suitable for settling postlarvae is relatively unknown along the range of the American lobster. Only four geographically extensive areas and one estuarine gradient have been surveyed. Eleven percent of the substrates surveyed along 60 km of Maine coastline was cobble, 24% was sediment, and 65% was ledge (Wahle and Steneck, 1991). Censuses of other shelter-providing habitats, such as eelgrass and *Spartina* peat reefs, have not yet been conducted. Nonetheless, because of the low proportion of cobble habitat, which is clearly more protective (see Section IV,C), Wahle and Steneck (1991) suggested that availability of shelter-providing habitat may limit postlarval recruitment to the benthos. Of the substrates surveyed along 240.47 km^2 of the Baie de Plaisance, Gulf of St. Lawrence, 91.2% were sand, 7.6% were deep (>5 m) rocky bottom, and 1.2% were shallow (<5 m) rocky bottom (Hudon, 1987). In contrast, 50% of the shoreline of Nova Scotia is bordered by rocky subtidal, of which 28% is bedrock and the remaining 72% is boulder and cobble bottom [see Miller *et al.* (1992), summarizing earlier surveys by Moore and Miller (1983) and Moore *et al.* (1986)].

The 448-km shoreline of the mainland and large islands that form the Quoddy region of the Bay of Fundy consists of bedrock (35.2%), coarse sedimentary (55.1%) and mud (9.2%) shores, and isolated salt marshes (0.5%) (Thomas *et al.*, 1983). A synoptic survey of inshore lobster distribution in this region evaluated siting conflicts between salmon aquaculture and the local fishery (briefly described in Lawton and Robichaud, 1992). The predominant substrates sampled from near-intertidal to 15-m depth were bedrock and cobble/boulder bottom in which new benthic recruits and later life history phases were encountered. From 15- to 25-m depth, sedimentary substrates comprised >80% of the sampled habitats and supported only vagile juvenile, adolescent, and adult lobsters.

Based on direct benthic censuses along a 22-km length of the Narragansett Bay estuary, new benthic recruits were absent from featureless sedimentary habitats that form the majority of the bottom in this shallow bay (generally <10 m deep). Rocky habitat and cobble/boulder habitat supported both new benthic recruits and older lobsters; mechanisms for an apparent restriction of recruitment at upper-bay sites were suggested to be reduced larval supply and physiological stress (Wahle, 1993).

5. Optimal Settlement Models

Few attempts have been made to dynamically model the trade-offs facing a postlarva trying to settle to the benthos. By choosing to hold off settlement until it finds a high-quality substrate, the lobster decreases its chance of postsettlement predation directly by preventing predator access, and indirectly by providing a shelter that is less likely to collapse and reexpose the lobster. However, delaying settlement can also increase predation risk because continued diving to test substrates, as well as the actual investigatory behavior, repeatedly exposes the postlarva to both pelagic and benthic predators. Furthermore, the longer a postlarva waits to settle, the greater the chance that it will molt into stage V without a shelter, and thus become highly vulnerable to predators.

Barshaw (1988) proposed a dynamic optimization model that assumed that (1) stage IV lobsters can delay molting to stage V for a discrete amount of time only; (2) if a stage IV lobster molts to stage V prior to settlement, it has a small probability of surviving; (3) the probability of surviving to reproduction differs for different substrates; (4) the probability of surviving to reproduction is constant for a particular substrate; and (5) a lobster does not resume substrate testing once it has settled. Assumption 4 is clearly not the case, as the probability of surviving to reproduction probably increases, the older a lobster becomes, since it outgrows many of its predators (Wahle and Steneck, 1992). Assumption 5 is also an oversimplification, as stage V lobsters have been found in the water column, albeit infrequently (Herrick, 1895; Templeman, 1940; Templeman and Tibbo, 1945; Barshaw, 1988). These assumptions, however, make the model tractable and result in the baseline output reproduced in Fig. 4. By varying the mortality rates for each substrate, or by varying the relative availability of substrates, the slope of the curve is modified and the postlarva should be either more or less selective in its settlement choices.

In fact, postlarvae do seem to base decisions on some of the above factors. Older postlarvae are progressively less selective (Cobb *et al.*, 1989; Boudreau *et al.*, 1993a), as are injured postlarvae, which may already have a decreased chance of survival (Boudreau *et al.*, 1993a). Environmental cues may act sequentially to influence postlarval decision-making (Boudreau *et al.*, 1993b): the presence or absence of a thermocline distributes settling lobsters into shallow or unstratified waters; chemical cues near the bottom provide substrate information; and the presence of light drives the postlarvae into darker microhabitats. Information is still lacking on the costs of repeatedly sampling substrates, the probabilities of survival to reproduction for each substrate type, and the availability of substrates along the entire range of the American lobster. Microtag–recapture studies on *Homarus gammarus* juveniles indicate that after a size of 10- to 15-mm CL, survival to reproductive age is fairly high (Bannister *et al.*, 1991; Van der Meeren and Næss, 1991; Burton, 1992; C. A. Burton, unpublished

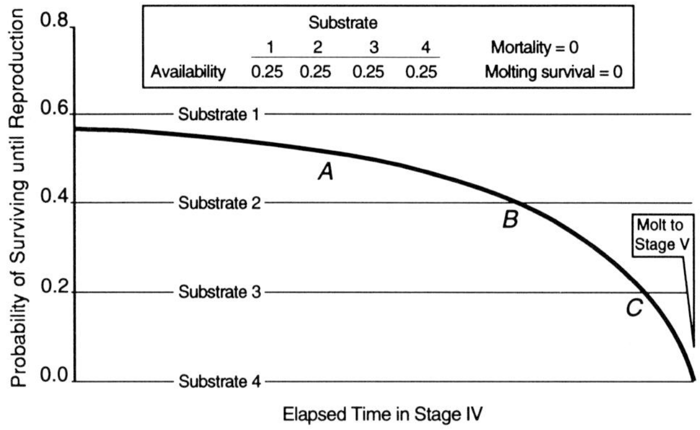

FIGURE 4 Hypothetical probability of surviving until reproduction in four substrates of different quality, and while testing bottom, during stage IV. Availability, Probability of encountering each substrate on the bottom; mortality, percentage of lobsters that are preyed on each time they test the substrate; molting survival: percentage of lobsters that survive after molting to stage V without settling. In region *A*, a lobster should settle into substrate 1 if it encounters it; otherwise, it should continue testing substrates. At point *B*, a lobster should settle into either substrate 1 or 2, if encountered, or should continue testing if it only encounters substrate 3 or 4. At point *C*, the lobster should accept any of the previous substrates, only rejecting substrate 4. (Adapted from Barshaw, 1988, with permission.)

data); large-scale, long-term studies have yet to be conducted on *H. americanus*.

Another model of benthic recruitment (Incze and Wahle, 1991) assumes that (1) only competent postlarvae will initiate diving behavior; (2) all competent postlarvae will dive with the same frequency and will settle if they encounter suitable habitat; (3) diving postlarvae will be subject to the same rate of predation; and (4) diving frequency and successful recruitment to the benthos will not vary on a diel basis. Using the final average density of recruits, as well as planktonic postlarval abundance, Incze and Wahle (1991) determined that as the frequency of diving excursions increases, the proportion of successfully recruited postlarvae decreases; they suggested that differences in diving frequency might explain site-specific differences in the final average density of new benthic recruits, rather than short-term differences in planktonic postlarval abundance.

C. Predation Pressures

Most of our information on the predation risks facing postlarval and juvenile lobsters following benthic settlement comes from laboratory experiments using cultured stock. Initial studies investigating substrate preferences assumed that if settling lobsters could find an appropriate substrate and build a shelter, they would be free from predation (Atema *et al.*, 1982;

Botero and Atema, 1982; Pottle and Elner, 1982). However, it is now apparent that settled postlarvae are highly vulnerable to attack by benthic fishes and crabs and the rate of attack is dependent on the substrate type. Sand is the least protective substrate, since lobsters are unable to build shelters and remain exposed within shallow depressions (Lavalli and Barshaw, 1986). Cohesive mud substrates also provide little protection against fish and crab predators that are able to locate shelters by sight and destroy them (field observations by Roach, 1983; laboratory trials by Barshaw and Lavalli, 1988). Eelgrass beds, while structurally complex, do not provide any more protection than mud substrates (Barshaw and Lavalli, 1988), but peat reefs (see Section III,B) provide better protection against fishes and crabs, presumably because lobsters blend in with the root structures of the associated plants (Barshaw *et al.*, 1994). If burrowed into cobble substrates, postlarvae remain vulnerable to burrowing crabs (Lavalli and Barshaw, 1986; Johns and Mann, 1987; Barshaw and Lavalli, 1988), but may be less susceptible to fish predators such as sculpins or cunners (Roach, 1983; Lavalli and Barshaw, 1986). Cobble substrates also provide immediate access to interstitial spaces, which reduces exposure time during burrowing activities. However, when predators were added to substrates of sand, peat, or cobble prior to the introduction of lobsters, there was no significant decrease in survival compared to experiments in which predators were added after lobster settlement (Barshaw *et al.*, 1994). Thus, the more important factor determining the protective value of a substrate may be whether it is easily manipulated by predators.

In field experiments with tethered lobsters (5- to 7-mm CL), cobble environments seem to be most protective, while mud and bare substrates are least protective (Wahle and Steneck, 1992). However, tethering interferes with the ability to gain and maintain shelter in mudlike substrates (Barshaw and Able, 1990), and it is unclear whether less complex substrates are as unprotective as sand (Wahle and Steneck, 1992). Nonetheless, sand, mud, eelgrass, and peat reef substrates provide little protection against predators, with survival rates of 0–27% (Barshaw *et al.*, 1994). Survival rates are not much different for untethered postlarvae released in more complex substrates; survival was 42% after 1 day and 20% after 9 days in a caged environment of oyster shells on sand (Kittaka *et al.*, 1983). After 4 weeks, survival of untethered postlarvae was significantly lower when predators were present in caged substrates of rock and vegetation or mud (17.6 versus 34.5% in nonpredator cages) (Roach, 1983).

D. Natural Diets

1. Feeding Studies on Natural Foods

Although there are no field studies of stomach contents of recently settled lobsters, some laboratory studies have highlighted potential natural sources of food. Stage IV lobsters are capable of surviving on plankton derived solely from an unfiltered water supply (Emmel, 1908). Postlarvae held in canisters submerged in a bay with only fouling organisms for food survived and grew as well as those fed brine shrimp in the laboratory (D'Agostino, 1980). Barnacle larvae, copepods, mysids, crab zoeae, and unidentified plankton (152 μm to 1 mm) have provided sufficient nutrition for the growth and development of postlarvae and shelter-restricted phase juveniles (Daniel *et al.*, 1985; Barshaw, 1989; Lavalli, 1991). Whether or not postlarvae routinely function as detritivores, consuming the fecal deposits of bivalves, remains to be demonstrated (Bryden, 1973).

Prior to settlement, postlarvae are known to consume a wide range of planktonic organisms (Table 3). There is some speculation that postlarvae accumulate enough lipid reserves by molt stage D_0 to cease feeding until after they molt into stage V (Sasaki *et al.*, 1986); however, it is not clear whether the reserves are merely sufficient for normal metabolic activity or they provide enough energy for bottom-testing behavior (see Ennis, Chapter 3). In fact, the postlarvae may possibly continue to feed via the techniques described in Section IV,D,3, although at a lower rate, as implied by the reduction in incremental growth at the molt from stage IV to stage V (Hudon, 1987).

2. Physiological Constraints on Feeding

While the postlarvae are morphologically similar to adult lobsters, their claws are symmetrical and relatively small (Lang *et al.*, 1977). It is not until the lobster reaches 30–40 mm in total length (TL) that the mechanical advantage of the crusher claw reaches twice that of the cutter claw and is capable of crushing foods such as small mollusks and gastropods (Costello and Lang, 1979; see Lavalli and Factor, Chapter 14). Due to their small claws, their vulnerability to predators when exposed, and the ineffectiveness of the claws in capturing organisms, it seems likely that newly settled postlarvae feed on suspended particles and items within their shelters.

3. Food Capture Techniques

Postlarval lobsters are capable of removing particles of 1 mm to at least 70 μm from the water column

TABLE 3 **Food of Postlarval, Juvenile, and Adult *Homarus americanus* Based on Stomach Content Analysis**

Age class	Food items found	References
Postlarvae (stage IV), pelagic	Copepods, diatoms, bacteria, crustacean remains, decapod larvae (particularly crab zoea and megalops), amphipods, algae, insects/insect pieces, fish eggs, gastropod larvae, echinoderms, worms, molluscan larvae, sand	Herrick (1895, 1909), Williams (1907), Templeman and Tibbo (1945), Harding *et al.* (1983), Gunn (1987), Juinio and Cobb (1992)
Juveniles (>12 mm CL)	Echinoderms (sea urchins[a], brittle stars[a], seastars), mollusks (mussels[a], periwinkles), decapods (rock crabs, lobsters), polychaetes (polynoids[a], nereids[a]), algae[a], copepods[a], hydrozoans[a], bryozoans, fish, tunicates, eelgrass, detritus, pebbles and sand	Scarratt (1980), Carter and Steele (1982b)
Adults	Fish, isopods, copepods, decapods (*Cancer irroratus*[a], *Carcinus maenus*[a]), *Hyas* sp., *Pagurus* sp., spider crabs, amphipods, lobster shells[a], barnacles, nematodes, polychaetes[a], insects, mollusks (limpets[a] and bivalves[a], particularly mussels, clams, scallops), gastropods[a] (periwinkles[a], whelks, dogwinkles), echinoderms (sea urchins[a], sea stars[a]), sea anemones, foraminiferans, jellyfish, sponges, bryozoans, ascidians[a], hydroids, ectoprocts, algae, eelgrass[a], detritus, small stones, sand	Herrick (1909), Squires and Ennis (1968), Weiss (1970), Ennis (1973), Reddin (1973), Scarratt (1980)
	Never eaten: coelenterates, sponges, algae	Scarratt (1980)

[a]Most frequently found in the stomachs.

(Lavalli and Barshaw, 1989); these organisms are captured differently depending on their size (Lavalli, 1992). Small organisms are carried in currents created by pleopods, exopodites, and mouthpart movements, particularly the third and second maxillipeds; the prey is then captured behind a mesh of setae along the medial edge of the segments of the third maxillipeds, transferred to the second maxillipeds, and passed back to the mouth. This type of food capture is known as suspension feeding (*sensu* Hughes, 1980). Larger organisms are captured by raptorial feeding techniques (Lavalli, 1992). Lavalli and Factor (Chapter 14) survey the functional morphology of the feeding appendages and discuss particle retention mechanisms.

Suspension and raptorial feeding on planktonic organisms does not preclude other means of feeding. Crnkovic (1968) suggested that one of the reasons for the continual construction of new openings in *Nephrops norvegicus* burrows is to search for food within the sediment. Both *Homarus gammarus* and *H. americanus* lunge out of their burrows to grab food near the entrance (Berrill, 1974; Barshaw and Bryant-Rich, 1988). Planktonic and benthic organisms in and around their shelters may thus support continued growth of vulnerable postlarvae and shelter-restricted phase juveniles (Barshaw, 1988; Lavalli, 1991); however, quantitative field assessments of prey availability have yet to be undertaken.

E. Behavior

1. Antipredator Behaviors

Recently settled postlarvae are extremely vulnerable to predators if exposed (Lavalli and Barshaw, 1986; Johns and Mann, 1987; Barshaw and Lavalli, 1988; Wahle and Steneck, 1992), as are shelter-restricted phase juveniles (Cooper and Uzmann, 1980; Roach, 1983; Kittaka *et al.*, 1983; Hudon, 1987; Barshaw and Bryant-Rich, 1988). If these vulnerable new recruits can subsist on shelter-based food, there is little reason for them to leave the protection of their shelters.

Postlarval lobsters in protective laboratory environments such as cobble remain sheltered 100% of the time (Lavalli *et al.*, 1995), yet they are capable of responding appropriately to predators if necessary (Johns and Mann, 1987; Wahle, 1992b). If unsheltered, or under sterile laboratory conditions, postlarval and shelter-restricted phase juveniles respond to the presence or the odor of fish predators by significantly increasing their use of unnatural shelters of PVC pipe (Wahle, 1992b). If exposed when predators are present, postlarvae increase the frequency of "retreating" or "freezing" behavior. Retreats are slow, locomotory movements that do not terminate until shelter is found. Freezing seems to be an adaptive response to visual predators, and occurs mostly in response to moving objects (Johns and Mann, 1987). Herrick (1895) described this phenomenon as a "death-feigning habit," but he could not rationalize (p. 186) how "in the environment of these animals, where so many of their enemies are scavengers or omnivorous, it could be of much service to its possessor when finally established on the bottom." Predators, however, appear to be ephemeral components in particular habitats (Wahle and Steneck, 1992), so freezing until the threat is gone may be an effective way to avoid predation.

2. Social Interactions

While larger vagile juvenile, adolescent, and adult lobsters are known for their aggressive natures, little is known about the agonistic behavior of newly settled postlarvae (reviewed by Atema and Voigt, Chapter 13). Paired postlarval lobsters influence the timing of one another's molt in such a way that the dominant animal of the pair molts first, while the subordinate molts an average of nearly 5 days later (Cobb, 1970). In addition, singly held postlarvae can have subtle effects on each other, most notably in reducing near-neighbors' growth rates. These inhibitory effects are thought to be chemically mediated; however, when lobsters are held on opposite sides of partitions, which allow both visual and chemical communication but eliminate physical contact, molt delays are eliminated (Cobb *et al.*, 1982). Molt delays are also eliminated in communally reared lobsters if the claws are removed (Cobb *et al.*, 1982). Postlarvae in the field, as well as in the laboratory, are known to cohabit in shelters (Cobb, 1971; Boudreau *et al.*, 1992, 1993a; D. F. Cowan, unpublished data) or to live in close association with each other, but in separate shelters under the same rock (K. L. Lavalli and P. Lawton, unpublished observations). Whether they affect the growth rates of their nearest neighbors in the wild is unknown.

V. Juvenile and Adolescent Lobsters

A. *Nursery Area Dynamics*

1. Habitat Types and Patterns of Abundance

Shelter-restricted, emergent, and vagile juveniles occupy a subset of the habitats described in Table 2, specifically those selected by settling postlarvae (see Section IV,B). As benthic development proceeds and lobsters begin to forage more widely, they disperse from these primary settlement sites and encounter a wider array of habitat types. Until dispersal, however, these primary settlement sites may constitute discrete "nursery areas." Given the pace of coastal development along much of the lobster's inshore range, there is some urgency for the identification of such nursery areas (Elner and Hamet, 1984; Harding, 1992; de Lafontaine *et al.*, 1992; Lawton and Robichaud, 1992).

Integrating knowledge of the structural or architectural aspects of benthic habitats with the density and size structure of associated lobster populations is currently a focus of *Homarus americanus* research. [See Howard (1980) on *H. gammarus*.] Shelter-restricted and emergent juveniles seem most abundant on cobble substrates [up to 12.4 individuals per square meter (Wahle and Steneck, 1991)] at depths of 2–7 m. They are also found in *Spartina alterniflora* peat reefs [up to 5.7 individuals per square meter (Able *et al.*, 1988)] or under rocks embedded in sand [0.18 individuals per square meter (Hudon, 1987)], but at lower densities. Larger stones with algal cover provide shelter for both emergent and vagile phase juveniles at densities of 1.2–3.8 individuals per square meter. These larger juvenile phases are found infrequently in substrates of eelgrass and sand (fewer than 0.04 and 0.02 individuals per square meter, respectively) (Hudon, 1987). Early benthic phase lobsters (*sensu* Wahle and Steneck, 1991) (Fig. 1) are rarely found on featureless soft or bedrock substrates (fewer than 0.2 individuals per square meter), but if bedrock is colonized by kelp–mussel beds, then densities of these juveniles are nearly the same as in cobble (1–2.9 individuals per square meter). Vegetation, however, does not enhance densities on sediment substrates (Wahle and Steneck, 1991). At greater depths in the Bay of Fundy, juvenile lobsters are most common in cobble/boulder habitats [up to 16 individuals per square meter (P. Lawton, unpublished data)]. The average field densities on the habitats thus far sampled are lower than those at which juveniles can apparently coexist (16 individuals per square meter), at least for a period of 6 months, and laboratory indications suggest that even higher densities are possible (27 individuals per square meter) (Van Olst *et al.*, 1975).

Shelter-restricted, emergent, and vagile phase juveniles have also been hand-captured from intertidal stations in parts of Maine, New Hampshire, Massachusetts, Rhode Island, and Connecticut (Krouse and Nutting, 1990a,b; Kirby, 1994; D. F. Cowan, unpublished data; J. S. Krouse, Maine Department of Marine Resources, unpublished data). Densities of these intertidal lobsters vary on a month-to-month basis, yet can be as high as 4.33 individuals per square meter (D. F. Cowan, unpublished data). These lobsters range from 3- to 42-mm CL, are often found cohabiting with both conspecifics and green crabs, and can be in salinities as low as 0 ppt during winter snow runoff (D. F. Cowan, unpublished data).

Intertidal lobsters (14.5- to 87-mm CL) have been found consistently in large numbers over a 16-year period at Pratt's Island Beach, Maine (J. S. Krouse, unpublished data). At all of these sites, males outnumber females, usually in a ratio of 2:1 or greater. Similarly, 86 emergent and 176 vagile phase juveniles were found under *Fucus*-covered rocks, as well as in mud, sand, and eelgrass at low tide in a bay of Prince Edward Island (MacKay, 1920).

The population densities above are based on direct *in situ* sampling by divers or field workers; however, population size estimates derived from mark–recapture techniques may be comparable to those obtained by direct sampling (Bernstein and Campbell, 1983). Population densities for juvenile lobsters ≥20-mm CL ranged from 6.37 ± 0.78 (SE) to 4.68 ± 0.62 individuals per 50 m^2 over a 3-month sampling period at a shallow subtidal area with a substrate of boulders overlying sand and gravel off McNutt Island, Nova Scotia. This direct census generated estimates of the total population size for the 4.5-ha site of 4932 ± 257 lobsters. The corresponding population estimate from mark–recapture data was 4745 ± 2759 lobsters. Bernstein and Campbell (1983) noted the close correspondence between the two estimates, yet the lower error term for direct census, which they concluded to be the more cost-effective approach.

Direct *in situ* sampling by divers appears to be the best method available for surveying shallow inshore nursery areas, but may, over time, reduce the quality of an area as a settlement site. Population densities and size ranges may fluctuate as the habitat deteriorates (Campbell, 1991). Directed trapping methodologies for juvenile lobsters are being reevaluated in an attempt to develop techniques for the assessment of abundance of lobster recruits 1–2 years prior to their entry to the commercial fishery (R. J. Miller, Department of Fisheries and Oceans, Halifax, personal communication) and may prove useful for sampling deeper or turbid habitats. Clearly, further work is required to calibrate results among these diverse approaches to the inventory of juvenile lobster abundance.

2. Carrying Capacity Concepts

The possibility that lobster populations may be subject to density-dependent controls during the early benthic life history period emerged from a reanalysis of a long-term time series of larval abundance and lobster stock size in Northumberland Strait (Fogarty and Idoine, 1986; original time series by Scarratt). From a different perspective, Caddy (1986) suggested the prospect of a recruitment "bottleneck" in the early benthic juvenile phases related to the obligate nature of crevice use. In conceptualizing his approach, which was based on the theory of fractal surfaces, Caddy argued for a return to a more limited, three-dimensional definition of *niche* as "that unit of bounded space which when occupied by an individual organism, improves that organism's chance of survival." Subsequently, Wahle and Steneck (1991) interpreted field data on the restriction of early benthic phase lobsters to shelter-providing habitats (cobble substrate) in terms of potential demographic bottlenecks. A vigorous debate on the prospects for testing this hypothesis on a regional scale is still evolving (Addison and Fogarty, 1992; Wahle, 1993).

The bottleneck hypothesis, while strengthening the resilience of the fishable stock, may also limit lobster recruitment; the recent large-scale increase in landings suggests an environmentally driven production system (as in many shrimp species), rather than density-dependent mechanisms that impact the juvenile phases (Elner and Campbell, 1991). Increased water temperatures may have improved larval and juvenile survival through increased growth and food availability. In addition, low fishing mortality in some areas in the late 1970s may have allowed more females to produce eggs, swelling recruitment into the fishery 5–8 years later. Differential growth then may have spread these cohorts over several years, sustaining the landings (Elner and Campbell, 1991). Fogarty (Chapter 6) discusses the lobster production dynamics underlying this recent landings pulse.

An earlier attempt to determine the carrying capacity of inshore lobster grounds approached the question from a community trophodynamics perspective (Miller *et al.*, 1971). Estimates of production of lobster prey species in a subtidal kelp community in St. Margaret's Bay, Nova Scotia, exceeded that needed to support the local lobster population by a factor of 10, suggesting that food production was not a limiting factor. Part of this production may not be available to lobsters because of prey size, population density, or loss to competitors. Instead, food quality, predators, parasites, space, and recruitment may be more important in setting lobster abundance levels (Miller *et al.*, 1971).

Some of these concerns are being addressed through experimental manipulation of habitat structural elements and lobster population demography (Wahle, 1991; Bologna and Steneck, 1993; R. A. Wahle, University of Rhode Island, personal communication). Linked with these field studies are attempts to evaluate the prospects for enhancement of local carrying capacity through release of hatchery-reared juveniles into existing habitats or placement of artificial habitats (Scarratt, 1968; Sheehy, 1976; Conan, 1986; Caddy and Stamatopoulos, 1990; Bannister,

1993). It is clear that both physical niches (*sensu* Caddy, 1986) and trophic dimensions of the classic Hutchinsonian niche will need to be considered if lobster recruitment mechanisms and habitat carrying capacities are to be understood.

B. Shelter-Related Behavior

1. Physical Attributes of Shelters

Shelter-restricted and emergent phase juveniles have the same substrate preferences as do postlarval lobsters. Shelter-restricted phase juveniles (9- to 15-mm CL; stages VII–X) choose a substrate and begin shelter excavation within 15 minutes of introduction into experimental tanks (Pottle and Elner, 1982). Significantly more lobsters choose to settle in gravel (~20-mm size) after exploring the substrate by probing crevices with their chelipeds (Pottle and Elner, 1982; Wahle, 1992a). Lobsters frequently construct more than one shelter and are seen in different locations at different times (Pottle and Elner, 1982). In the field, emergent and vagile phase juveniles (stages XI–XV; 50- to 90-mm TL) settle within 30 seconds under blocks on sand; within 24 hours, their burrows are well shaped (Kittaka *et al.*, 1983).

In nature, *Homarus americanus* prefers shelters where the height is less than the width (Cobb, 1971), a preference also expressed in laboratory choice experiments with shelter-restricted and emergent phase juveniles (>11-mm CL) (Boudreau *et al.*, 1990). [Similar height–width relationships are seen in tunnels constructed in mud by *H. gammarus* juveniles (Howard and Bennett, 1979).] With increasing size, lobsters prefer larger shelters, but expression of this preference is dependent on the dimensions of available shelters (Cobb, 1971; R. S. Steneck, unpublished data). Generally, the relationship between shelter size and lobster size is very close for shelter-restricted and emergent phase juveniles, but begins to relax in the vagile phase as the lobsters grow above 30-mm CL and become more wide-ranging in their movements (Fig. 5). [*Homarus gammarus* also selects shelters proportional to its size (Dybern, 1973).]

Lobsters prefer opaque shelters to transparent ones; this may be related to their preference for lower-profile shelters, which would presumably be darker (Cobb, 1971).

2. Hydrodynamic Limitations

The association of lobsters with shelter-providing habitats purportedly provides protection against predators, especially during vulnerable molt periods. However, it may also be related to the need to avoid tidal currents and wave action (Howard and Nunny, 1983). *Homarus gammarus* adolescents (50-mm CL) act normally in current speeds of 5 cm/sec, but as currents increase to 10–15 cm/sec, walking is impaired and control over antennae is lost. As the current is increased to 21–43 cm/sec, some lobsters slip downstream, while others remain standing still. These reactions are the same on rock and gravel substrates. Emergent phase juveniles (16-mm CL) also behave normally in currents of 5 cm/sec, but as soon as they encounter a place where they can do so, they immediately enter or begin to construct a shelter.

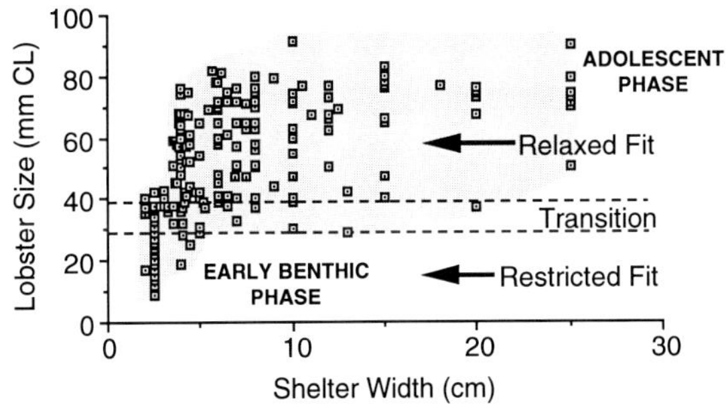

FIGURE 5 American lobster size–shelter size relationship. Most lobsters (88.4 ± 6%) were found in shelters composed of rock substratum. Lobster body size corresponds to shelter size; the fit is tight for lobsters up to 30-mm CL, but more relaxed for larger body sizes. Data are from five sampling regions distributed along 400 km of the Maine coast. Terminology of life history phases is per Wahle and Steneck (1991) (see also Fig. 1). (From R. S. Steneck, unpublished data, used with permission.)

Oscillatory flow from wave action may also create problems and reduce the benefit of posturing to minimize drag, suggesting that critical velocities (those eliminating all movement) might be even lower than predicted from the above studies (Miller, 1990). Sheltering, or occupying substrates that have irregular topography, may protect lobsters against exhaustion or damage (Howard and Nunny, 1983).

3. Shelter Competition

The level of intraspecific shelter competition is unclear, but field studies suggest that it may be less frequent than inferred from laboratory studies (O'Neill and Cobb, 1979; Karnofsky *et al.*, 1989a,b). Only 33% of aggressive encounters between adolescent lobsters in a shallow cove off Buzzards Bay involved shelter defense (Karnofsky *et al.*, 1989b). In all cases, the evicted lobsters usually left the immediate area after several days (Karnofsky *et al.*, 1989b).

Population densities of lobsters increase when artificial shelters are placed at 0.5-m intervals in a field setting (Fig. 6A); however, the proportion of empty shelters increases due to numerous evictions (Fig. 6B) (Steneck, 1989; R. S. Steneck, unpublished data). Moreover, the size of lobsters occupying the shelters decreases (Fig. 6C). As spacing between shelters is increased, population density decreases, but fewer shelters are empty and larger lobsters tend to predominate. Where spacing is held constant and shelter size is varied, lobsters occupy shelters proportional to their body size. When adolescent and adult lobsters (30- to 90-mm CL) are placed in a shelter "arena" to observe competitive interactions, larger individuals typically evict smaller ones from their shelters, but only if they are above 60-mm CL. Lobsters below 60 mm CL are never observed competing for space (Steneck, 1989; R. S. Steneck, unpublished data).

Despite these observations of intraspecific competition, nonmating lobsters are known to share shelters, both in the field and in the laboratory (Stewart, 1972; Cooper *et al.*, 1975; Sheehy, 1976; Cooper and Uzmann, 1977; O'Neill and Cobb, 1979; Pottle and Elner, 1982; D. F. Cowan, unpublished data; K. L. Lavalli and P. Lawton, unpublished observations). Generally, cohabitation occurs between animals of different sizes, when one or both lobsters has missing claws, when shelter is rare, or when water temperature is low.

Nondecapod interactions have also been observed. Eels are often seen occupying lobster shelters in eelgrass beds (Karnofsky *et al.*, 1989a) and are known for their aggressive encounters with adolescent and adult lobsters (Stein *et al.*, 1975). Thus, they might be responsible for decreases in lobster densities within eelgrass beds over a seasonal period (Karnofsky *et al.*, 1989a). Fish, particularly cunners, will occupy lobster shelters or will investigate fresh molt shells within a shelter, but interactions between larger juvenile lobsters and cunners do not appear overly aggressive (Stein *et al.*, 1975). While cunners can displace emergent and vagile phase lobsters from their shelters, they do not appear to interact beyond the eviction (Lavalli *et al.*, 1995). Interactions between interspecifics and shelter-restricted, emergent, or vagile phase juveniles have not yet been adequately investigated.

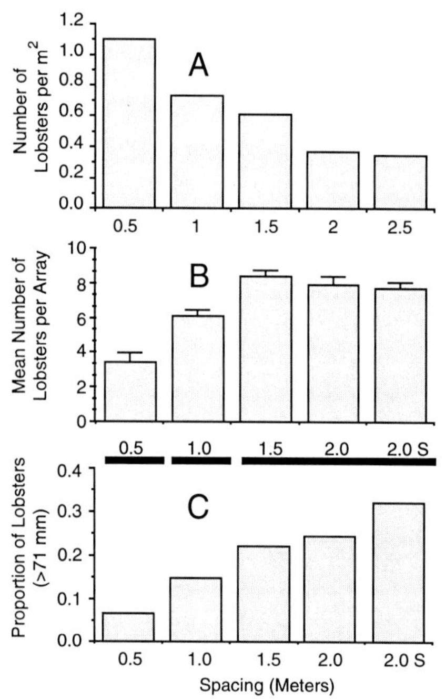

FIGURE 6 Demographic consequences of shelter competition between American lobsters: (A) lobster population density as a function of intershelter spacing; (B) mean number of lobsters per shelter spacing array, illustrating the effects of competition as shelters are placed more closely together; (C) proportion of larger lobsters (>71-mm CL) per shelter spacing array, showing fewer larger lobsters when shelters are closely spaced. Shelter arrays were double rows of 5 shelters, except 2.0 S (single row of 10 shelters). Data not significantly different at $P = 0.05$ (Student–Newman–Keuls test) are represented with a contiguous horizontal bar. (From R. S. Steneck, unpublished data, used with permission.)

C. Foraging Behavior

1. Chemosensory Behavior

Lobsters are sensitive to waterborne food compounds, suggesting that prey selection may, in part, be chemically mediated (Derby and Atema, 1981; Atema and Voigt, Chapter 13). Juvenile, adolescent, and adult lobsters express preferences for particular prey species (Evans and Mann, 1977; Elner, 1980; Carter and Steele, 1982a; Elner and Campbell, 1987; Hudon and Lamarche, 1989). After feeding on one food source, lobsters become more sensitive to metabolites of that food source, while requiring higher concentrations of metabolites of other foods before eliciting behavioral responses (Derby and Atema, 1981; Daniel and Bayer, 1987a). Naive, shelter-restricted phase juveniles, however, seem to require a learning period to express food search responses to extracts and metabolites of prey they might encounter in the wild (Daniel and Bayer, 1987b), suggesting a lack of preprogrammed food preferences.

2. Physiological Considerations

Beginning at stage VI, a 47% increase in mechanical advantage occurs for the crusher claw due to differential development (Costello and Lang, 1979). This increase is rapid through stage VIII, in which the mechanical advantage of the crusher becomes twice that of the cutter claw, then continues at a reduced rate until the adult condition is reached (Costello and Lang, 1979; Govind, Chapter 12). Because of this developmental pattern, it is doubtful that the claws of shelter-restricted and emergent phase juveniles are used to crush prey. They may be used for prey seizure, but their dexterity and coordination do not seem well developed for efficient transfer of food to the chelate walking legs and mouthparts until some time after stage VI (Lavalli, 1992; Section IV,D). Unfortunately, previous gut content analyses have not discriminated between different sizes of benthic juveniles, so it is unknown whether claw development is reflected in the food preferences of small juveniles.

3. Natural Diet

The natural diet of emergent and vagile phase juveniles (≥15-mm CL) is better known than that of lobsters in the shelter-restricted phase. Plankton provides an adequate diet for the growth and survival of shelter-restricted and emergent phase juveniles (Emmel, 1908; D'Agostino, 1980; Daniel *et al.*, 1985; Barshaw, 1989; Lavalli, 1991). While longer intermolt periods and lower growth increments were noted for laboratory-held, plankton-fed, shelter-restricted phase juveniles (Barshaw, 1989; Lavalli, 1991), these effects may not have occurred if the lobsters had access to benthic organisms (amphipods, polychaetes, etc.), as they would have in the sediment of their shelters. Benthic invertebrates, such as amphipods, have proven to be excellent diets for juvenile lobsters, supporting high survival, rapid growth, and wild-type coloration (Good *et al.*, 1982; Daniel and Bayer, 1987a). Jatzke (1970) noted that a 6-month-old *Homarus gammarus* grew as well as lobsters maintained in the laboratory on artificial diets, and survived for 17 months in a "trichterkreisel" with only plankton and coelenterates, small crustaceans, and polychaetes as food.

Both stage V and VI *Homarus americanus* are capable of the suspension and raptorial modes of feeding described for postlarvae (see Section IV,D,3). Juvenile lobsters may retain this ability for some time, as suggested by the length of the setae on their mouthparts (Lavalli and Factor, 1992); new evidence for *H. gammarus* suggests that even adults may be capable of suspension feeding (Loo *et al.*, 1993). Laboratory evidence suggests that emergent phase juveniles feed periodically throughout the day, perhaps as a function of their gut clearance rate (Bordner and Conklin, 1981). Such periodic food needs could be met by a shelter-based food supply of benthic zooplankton and invertebrates, and would also allow the emergent phase juveniles to reduce out-of-shelter exposure until they attain a size less vulnerable to predation (Lavalli and Barshaw, 1989; Wahle and Steneck, 1992; Wahle, 1992b).

Stomach contents of juvenile, adolescent, and adult *Homarus americanus* (38- to 76-mm CL, Weiss, 1970; 12- to 73-mm CL, Carter and Steele, 1982b) suggest that while adolescent lobsters consume the same type of prey as adult lobsters, the relative proportion of the prey items taken is dependent on the size of the lobster (Table 3). Smaller lobsters consume more hydroids, gastropods, nonreptantian crustaceans, polychaetes, and brittle stars than do larger lobsters (Weiss, 1970). As juvenile lobsters use only their chelae to crush mollusks about the hinge (Elner and Jamieson, 1979), size selectivity of the prey items is evident with large mussels, scallops, oysters, and sea urchins escaping predation either by exceeding the chela spread of the lobsters or by being too thick to crush (Carter and Steele, 1982b; Elner and Lavoie, 1983). Use of perimeter tactics to open bivalves might be advantageous for adolescent lobsters, as it is for *Carcinus maenas* and *Cancer irroratus*, but it appears that lobsters lack the dexterity necessary for such tactics (Moody and Steneck, 1993; but see Section VI,F,2). Crusher chela force increases linearly with lobster body size (Elner and Campbell, 1981), whereas molluscan shell strength is exponentially related to shell thickness (Boulding, 1984), mass (Curry, 1988), and length (Griffiths and Seiderer, 1980). Thus, prey strength increases more rapidly than lobster chelae strength, as demonstrated by the size of lobster needed to crush sea scallops (*Placopecten magellanicus*). Lobsters of 20.8-, 25-, and 29-mm CL can crush scallops of 20- to 25-, 25- to 30-, and 30- to 35-mm shell height, respectively; however, lobsters must be larger than 42.2-mm CL to crush scallops greater than 35 mm in height (Elner and Jamieson, 1979).

Plant material appears to be a consistent component of the natural diet of lobsters, as evidenced by available stomach content analyses (reviewed by Elner and Campbell, 1987), even in habitats such as sea urchin barrens, where macroalgal abundance is greatly reduced. Thus, plants may not be incidentally ingested, but actively selected, forming a functional nutritional component of the diet. (Conklin, Chapter 16, discusses nutrition.) Seaweeds feature in the diets of many other decapod crustaceans, but until recently their potential role has been somewhat neglected. Conversely, the significance of other components of the natural diet of lobsters (and other decapods) may have been inflated. The bias in stomach content analyses toward those prey that leave indigestible skeletal structures is well appreciated. However, certain prey may be only partially ingested. Juvenile lobsters may successfully prey on juvenile intermolt *Cancer* crabs up to approximately 40% of their own body size, but as the predator–prey size ratio increases, they consume a decreasing fraction of the biomass of the prey, such that overall meal intake remains relatively constant (P. Lawton, unpublished observations). Thus, previous studies on natural diet and prey selectivity (Evans and Mann, 1977) tend to overestimate the dietary contribution of decapod prey (Lawton, 1985).

4. Selective Feeding

The diet of juvenile lobsters varies seasonally such that they consume more limpets, crustacean matter, filamentous algae, bryozoans, pebbles, and sand in early summer (June to mid-August), but more sea urchins, periwinkles, chitons, bivalves, crabs, lobster exuviae, and polynoids in late August and mid-September; by fall (October and November), only mussels, periwinkles, and crabs are eaten in high frequencies (Weiss, 1970; Ennis, 1973; Scarratt, 1980; Carter and Steele, 1982b). This pattern might reflect the seasonal availability of prey (Weiss, 1970; Scarratt, 1980; Carter and Steele, 1982b; Karnofsky *et al.*, 1989a) or might be related to lobster size (Scarratt, 1980).

Lobsters begin feeding shortly after molting and eat mostly items high in calcium for remineralization of the exoskeleton (Weiss, 1970; Scarratt, 1980). Thus, seasonal variations in diet may be due more to the need for calcium-rich diets during the molting season (Ennis, 1973; Carter and Steele, 1982b), such that natural dietary intake follows a specific regime governed by the molt cycle, rather than availability at any one time (Leavitt *et al.*, 1979). Nutrient analyses of stomach contents have shown that hard-shelled lobsters (molt stage C_1–D_3) have significantly higher crude protein and gross energy levels than soft-shelled animals (molt stage A_1–A_2), but significantly lower serum levels of total minerals and calcium.

Despite habitat differences, diet is fairly consistent for shelter-restricted, emergent, and vagile phase juveniles (7.8- to 49.9-mm CL) and is dominated by mussels, lobsters, Atlantic rock crabs, gastropods, and ectoprocts (Hudon and Lamarche, 1989). Such consistency implies that the lobster is a selective feeder, maintaining its particular feeding habits despite variations in prey availability in the various habitats it occupies (Elner and Campbell, 1987). Therefore, the contention that lobsters are scavengers (Herrick, 1895, 1909), unspecialized feeders (Scarratt, 1980), or opportunistic omnivores (Squires, 1970; Miller *et al.*, 1971) no longer appears tenable (Elner and Campbell, 1987; Hudon and Lamarche, 1989).

5. Foraging Activities

Models proposed by Barshaw (1988) and Wahle (1992b) predict that if shelter-restricted phase juveniles use both raptorial and suspension feeding modes exclusively within their shelters, they can avoid the predation risk that would accompany foraging outside their shelters. However, as ingestion rates and nutritional requirements increase with body size, the lobster must choose between the potential risk of predation while foraging and a reduced growth rate. As long as a shelter-based food supply exceeds the energetic needs of the lobster, it can remain in the shelter; once these needs exceed supply, the lobster must begin foraging outside of its shelter. Increased growth also reduces predation risk. As size increases and risk decreases, the lobster may expand its foraging area. Thus, emergent phase lobsters can be considered central place foragers (Lawton, 1987; Wahle, 1992b). There is some evidence that energetic needs are no longer met by diets composed exclusively of zooplankton for juveniles between stages VIII and IX (approximately 11-mm CL) (Lavalli, 1991). If the supply of shelter-based benthic organisms is adequate, lobsters may be able to prolong their sheltered existence and grow even larger before emerging to forage. However, more information on growth rates on natural diets is needed before the size at which lobsters should begin foraging can be predicted.

Emergent and vagile phase lobsters engage in foraging bouts outside their shelters (38- to 76-mm CL, Weiss, 1970; 20- to 46-mm CL, Lawton, 1987), commencing shortly after dusk and continuing for about 5 hours. Conversely, while adolescent and adult lobsters may be active throughout a 24-hour period, they become most active 2–3 hours after sunset; activities then abruptly decline 2–3 hours before sunrise (Ennis, 1984b; similar results are reported by Conan *et al.*, 1984, and Karnofsky *et al.*, 1989a). In the laboratory, activity levels gradually decrease from 24:00 on (Lawton, 1987). Food is carried back to the shelter for consumption—in some cases, repeated trips to the food patch will be made until as many as 10 mussels are collected. Infrequent and briefer feeding bouts occur in the open. When starved for increasing periods of time (24, 48, and 72 hours), lobsters alter their activity patterns in ways that increase predation risk, i.e., increased foraging during daytime (Lawton, 1987).

Risk-prone foraging behavior is also present in vagile juveniles and adolescent lobsters (40- to 53-mm CL) (McKenzie, 1989). Exposure to fish (tautog) causes a switch from consuming mussels primarily at the mussel patch to acquiring mussels and retreating to the shelter for consumption. Predation risk also causes lobsters deprived of food for 12 hours to reduce the number of trips to the food patch, but such is not the case for lobsters deprived of food for 60 hours. Nonetheless, with food deprivation and predator risk, overall consumption of mussels is reduced. There is no significant shift, however, to daytime feeding for lobsters subjected to the longer starvation periods. McKenzie (1989) suggested that lobsters may respond to movements of predators via sound or pressure-field detection. However, lobsters are also highly chemosensitive and shelter-restricted and emergent phase juveniles (9- to 15- and 18- to 26-mm CL) respond to both sculpins and sculpin-conditioned water (Wahle, 1992b). Vagile phase juveniles (30- to 38-mm CL) do not increase shelter usage greatly, but respond to the fish predator with aggressive displays both during its presence and during a 6-hour postpredator period.

D. Seasonal Movements and Annual Displacement

As the benthic period of life history progresses, lobsters undergo temporally organized, spatially oriented movements of successively increasing scale. Herrnkind (1980) defined three types of movement patterns in spiny lobsters: *migration*, movement of individuals or populations over considerable distances with a return to the original area; *nomadism*, wandering of individuals over large areas without clear-cut start and end points; and *homing*, periodic, often daily, excursions from a shelter to some nearby area with subsequent return to that shelter or others nearby. [The terms *transient behavior* and *resident behavior*, used with respect to short- and long-term shelter site fidelity, respectively, by Stewart (1972) and Ennis (1984b), are functionally equivalent to *nomadism* and *homing*.]

1. Seasonal Movements and Overwintering

Emergent and vagile juveniles and, to some degree, adolescent *Homarus americanus* are likely to be nontransient (or resident) in specific shelter sites during the winter due to inactivity at low water temperatures (Ennis, 1984b; Karnofsky *et al.*, 1989a). While several other field studies have included winter population censuses (e.g., Cooper *et al.*, 1975; Elner and Hamet, 1984; D F. Cowan, unpublished data), there is a paucity of information for vagile juveniles that would indicate a clear seasonal movement to deeper shelter sites to overwinter.

Lobsters in Bonavista Bay, Newfoundland, undergo seasonal movements to deeper water in autumn, returning to shallow water in the spring and early summer (Ennis, 1984a,b). Movement to deeper water in the fall is probably a response to increased turbulence due to storm conditions. As temperatures increase in the spring, increased activity and nomadic movement result in reoccupation of the shallow-water shelters during the spring. In the summer, lobsters are restricted to a narrower depth band (in shallower water) than in winter, perhaps due to the presence of a fairly strong thermocline at 5–10 m. While there is a clear seasonal depth change, some lobsters remain in shallow water throughout the winter.

In demographic studies of lobster populations (5- to 90-mm CL) in the Boothbay region of Maine (at depths of 6–12 and 18–24 m), there is little evidence of offshore movement in midsummer and midwinter, but lobsters occupying shallow burrows (~10 m or less) move to greater depths during stormy periods with strong vertical turbulence in the water (Cooper *et al.*, 1975). These movements generally involve horizontal distances of 100 m or less and increases in depth of only 6–10 m. Only 2–4% of shelters examined during the winter were shared by two, occasionally three, lobsters (40- to 90-mm CL), which usually differed considerably in size (Cooper and Uzmann, 1977).

In Bideford River, Prince Edward Island, a different seasonal movement pattern is found (Thomas, 1968). Adolescent and adult lobsters (average, 64-mm CL) move from deeper (>4-m depth), cooler stations within the estuary (that they occupy over the summer) to shallower water along the sides of the estuary, where they excavate burrows into the muddy bottom. Accumulation of fine silt on the lobsters occupying the burrows indicates negligible activity during the winter (Thomas, 1968).

2. Daily Movements and Annual Displacement

There are few published field studies on the daily foraging ranges of emergent and vagile juvenile lobsters and only anecdotal references to the scale of annual displacement. While homing, nomadism, and migration are exhibited by adolescent and adult lobsters (see Section VI,C), movements of the vagile juvenile phase are probably restricted to homing, with a trend toward nomadism, as shelter locations are exchanged perhaps in response to local food resource depletion or to physically outgrowing the shelters. Nomadism, or transient use of shelter sites, may lead to an annual displacement (or dispersion) from primary settlement sites. Horizontal movements of emergent and vagile phase juveniles may be on the order of several meters or less, while adolescent lobsters (40- to 45-mm CL) may range up to 300 m

(Cooper and Uzmann, 1977). This represents a considerable ontogenetic shift in behavior and may signal the period at which dispersion occurs. Present understanding of the field activity patterns of emergent and vagile juvenile lobsters is compromised by the need to destructively sample substrates to determine shelter occupancy and individual lobster identity.

In sampling of shallow-water, rocky-bottomed habitats in the Îles de la Madeleine from May through September, a large mode of juveniles (15- to 30-mm CL), present at the start of sampling, did not progress to larger size classes, as would be expected from growth of a cohort (Hudon, 1987); this is ascribed to the simultaneous effects of differential growth, mortality, and emigration occurring over this size range. Lobsters >30-mm CL could often be seen walking outside their burrows, even in daytime. Fluctuations in their abundance suggest that they may have engaged in migrations closer to shore or toward deeper water at different times during the summer, although the horizontal scale of these movements is not specified (Hudon, 1987).

Long-term monitoring studies at specific coastal nursery sites in Maine have estimated annual displacements of vagile juveniles to be no more than several hundred meters, including a shift to deeper water in the winter (R. S. Steneck, unpublished data). Field studies involving the release of hatchery-reared color morphs (Wahle, 1991) and microtagged juveniles (reviewed by Krouse and Nutting, 1990b; D. F. Cowan, unpublished data) into specific microhabitats, as opposed to enclosures (Roach, 1983), should eventually allow a better definition of growth rates, survival, and dispersal rates of juvenile lobsters from initial settlement sites.

E. *Predation Pressures*

1. Physiological Considerations

Shelter-restricted and emergent phase juveniles rely on a tail flip with a low behavioral threshold to escape harm. Because the abdomen forms a substantial portion of the animal's total mass, the ratio of mass to force developed during flexion is high, resulting in an effective propulsion when the lobster is small (Lang *et al.*, 1977). As the lobster grows, this ratio decreases and the tail flip becomes less efficient. However, the ratio of abdomen to animal mass does not change until the TL of the lobster becomes 40–60 mm (~20 mm CL), when the rest of the animal grows at a higher rate than the abdomen. Concomitantly, the claws make up a small proportion of the animal's mass. With subsequent growth, claw length increases more rapidly than CL until the lobster reaches 50 mm TL, and claw weight increases more rapidly than total weight (Lang *et al.*, 1977).

Thus, emergent phase juveniles (31- to 52-mm TL) respond to stimuli with a tail flip 98% of the time, compared to uninjured adult lobsters (170- to 250-mm TL), which tail flip only 18% of the time. With growth, the escape reflex is partially replaced by defensive claw displays (Lang *et al.*, 1977). These allometric changes have important implications, particularly for different-sized juveniles under predation risk.

2. Influence of Habitat Structure

Sediment-based substrates, such as mud or eelgrass, do not significantly protect shelter-restricted phase juveniles from either fish or crab predators (Roach, 1983; Barshaw and Lavalli, 1988). However, in the field, lobster survival increases with body size: shelter-restricted phase juveniles (5- to 7-mm CL; stages IV and V) suffer 100% mortality when exposed to predators for 24 hours; emergent phase juveniles (15-to 20-mm CL) suffer 87% mortality; and vagile phase juveniles (30- to 40-mm CL) suffer only 13% mortality (Wahle and Steneck, 1992). In shorter periods (6.5 hours), lobsters of 5- to 7-mm CL suffer 91% mortality, while those of 8- to 25-mm CL suffer 59% mortality; half of the lobsters preyed on are lost within the first 15 minutes of exposure to predators. The time needed to gain shelter under rocks is considerably less than 15 minutes (Berrill, 1974; Kittaka *et al.*, 1983), whereas more time is needed to construct shelters in soft sediments (Atema *et al.*, 1982; Roach, 1983) and may contribute to the differential survival (but see Barshaw *et al.*, 1994, for an alternative view). As is true in laboratory studies, survival in the field is greatest on cobble substrates (88%) and lowest on bare and mud substrates (42%) (Wahle and Steneck, 1992).

With untethered lobsters and longer periods in the field, the number of animals remaining in an area seems to be lower. The percentage of emergent phase lobsters (stages XI–XV; 50- to 90-mm TL) remaining on a substrate of blocks on sand after 11 days ranged from 8 to 25.5% (Kittaka *et al.*, 1983). If the blocks were placed near habitats where fishes were more abundant, only 2.7% of the stocked lobsters remained. It is not known whether these numbers represent survival rates, effects of dispersion, or a combination of both. Only crab predation was observed on these emergent phase juveniles (Kittaka *et al.*, 1983).

3. Predators and Antipredator Behavior

Inshore predators of juvenile lobsters include crabs (*Carcinus maenas* and *Cancer* spp.), blenny-like fishes, sculpins, flounders, cunners, lobsters, hermit crabs, and shrimp (*Crangon septemspinosa*) (Hudon, 1987; Wahle and Steneck, 1992). The most frequent attacks on shelter-restricted and emergent phase juveniles (5- to 7-mm and 8- to 25-mm CL) are by cunners and sculpins. *Carcinus maenas* attacks are also frequent, but these crabs concentrate on the shelter-restricted phase juveniles. Attacks by the remaining predator species are infrequent and also focus on the shelter-restricted phase juveniles (Wahle and Steneck, 1992). All predators, except sculpins, attack the shelter-restricted phase juveniles more quickly than they attack emergent phase juveniles (within the first 15 minutes, as opposed to 45 minutes). In the laboratory, mud crabs [*Dyspanopeus (=Neopanope) sayi*] also prey on shelter-restricted phase juveniles (stage V), but at lower rates than fish (Lavalli and Barshaw, 1986; Barshaw and Lavalli, 1988). Based on historical reports and *in situ* video observations (Herrick, 1909; Wahle and Steneck, 1992), cunners seem to be a significant predator of shelter-restricted phase juveniles; however, lobsters are only found infrequently in their stomachs and are more commonly found in long- and short-horned sculpin stomachs (Ojeda, 1987). This may reflect differences in mouth size in the two fishes, such that sculpins can consume a greater size range of juveniles.

Historical reviews have reported that cod and pollock are significant predators of small lobsters (Herrick, 1895, 1909; Bigelow and Schroeder, 1953; Scott and Scott, 1988). Despite these anecdotal reports, there is little direct evidence that cod and other groundfish greatly impact juvenile, and thus adult, lobster populations. Nonetheless, some investigators believe that historical patterns of overfishing of these groundfish stocks in coastal areas may be responsible for current regional disparities in fish predation intensity between coastal and offshore sites (see Wahle and Steneck, 1992, for predators of lobsters; Witman and Sebens, 1992, for predators of *Cancer* crabs; and Addison and Fogarty, 1992, for counterarguments).

Because of the relative risks involved, it seems that shelter-restricted phase juveniles behave differently from emergent and vagile phase conspecifics in terms of shelter usage patterns and exposure time. Prior to introduction of a caged sculpin, shelter usage is low (3–5%) and there is no difference in nocturnal shelter usage of three groups of juvenile lobsters (9- to 15-, 18- to 26-, and 30- to 38-mm CL) (Wahle, 1990, 1992b). However, both during and following introduction of the predator, all sizes of juveniles increase their shelter usage, but the increase shows a strong inverse relationship to the size of the lobster. Shelter-restricted phase juveniles remain in their shelters nearly 100% of the time, while the vagile phase juveniles tend to respond with increased mobility and visual displays, rather than sheltering. These surprising results indicate that there are no differences in the sheltering behaviors of shelter-restricted, emergent, and vagile phase juveniles prior to predator introduction—contrary to repeated observations that shelter-restricted phase juveniles are never seen out of their shelters (Roach, 1983; Kittaka *et al.*, 1983; Barshaw and Bryant-Rich, 1988). However, the conditions of this laboratory experiment were sterile and consisted of experimental plastic trays with PVC tubes for shelters. Additionally, the lobsters were fed at some distance from their shelters, which encouraged them to leave the PVC tubes. Expanded versions of these experiments (Lavalli *et al.*, 1995) using naturalistic cobble substrates and unfiltered sea water (presumably providing a food supply) show that shelter-restricted phase juveniles (<10-mm CL) never leave their shelters during any of the three treatments. Emergent phase juveniles (11- to 25-mm CL) use their shelters approximately 50–80% of the time, whereas vagile phase juveniles (28- to 40-mm CL) use their shelters about 30–50% of the time and do not increase shelter usage in response to the predators; they do, however, visually display more frequently. Thus, it may be that lobsters respond to predators based on the structural and trophic quality of the habitat in which they reside and perhaps even the type of predator encountered; however, more research is needed in this area.

F. Social Behavior

1. Territoriality

In laboratory experiments, adolescent lobsters (46- to 56-mm CL) appear to defend territories that consist of the immediate area surrounding their shelters (Jacobson, 1977). These shelters/territories are consistently occupied for 1 week or more and are actively defended, even against animals of higher dominance status (Jacobson, 1977; Jacobson and Atema, 1977). Because of differences in the types of dominance hierarchies established and the ability of subordinates to hold territories when lobster density is increased, Jacobson (1977) suggested that adolescent lobsters may be territorial only when they exist in low densities. At higher densities, despotic dominants arise and often prevent subordinates from obtaining shelters, and thus territories. Conversely, video observations of lobsters occupying artificial shelters in the Gulf of Maine suggest that adolescent lobsters display little site fidelity and rarely remain within the same shelter, or general area, from day to day (R. S. Steneck, unpublished data). If these lobsters are more nomadic in nature, this would suggest that they do not, in fact, maintain and defend territories for extended periods. Alternatively, the lobsters observed in the artificial shelters may have shelters/territories elsewhere, and may be displaying a type of exploratory behavior when the artificial shelters were added into their normal foraging range. Then their lack of fidelity to these artificial shelters would be due not to a nomadic nature, but to the cessation of exploratory behavior.

2. Aggression and Dominance

Postlarvae have subtle effects on each other's molting patterns and growth rates, but do not seem to be overly agonistic toward each other (Section IV,E,2). In contrast, juvenile lobsters begin to manifest adultlike patterns of aggression after stage VII (per J. Mitchell, cited in Atema and Cobb, 1980). In laboratory settings, when a pair of emergent phase juveniles, equally matched in terms of size, are introduced into an arena, agonistic encounters can escalate into fights in which appendages may be damaged or lost (Huber and Kravitz, 1995). In the short term (24–48 hours), the subordinate animal remains subordinate and will not engage the formerly fought dominant or any other newly presented animal, including another subordinate (Huber and Kravitz, 1995). This has important ramifications for a subordinate, as dominant lobsters control food resources when active, consume food at greater rates, and control shelter if it is limiting (Cobb and Tamm, 1975; Lawton, 1987). Subordinates are forced to spend more time foraging during the day, which may increase their risk of predation (Lawton, 1987).

In the long term, the subordinate delays ecdysis so that the dominant animal molts first (Cobb and Tamm, 1975). The dominant lobster then assumes subordinate postures, while the subordinate animal assumes the dominant status. However, if the formerly dominant animal survives its molt (i.e., is not cannibalized), it regains its dominant status. When the formerly subordinate lobster molts, it is typically killed during subsequent agonistic encounters (Cobb and Tamm, 1975). These changes in aggression levels and dominance status are related to the molt cycle; animals in molt stage C are dominant over those in stages A, B, and D_3 (immediate premolt) (Tamm and Cobb, 1978). Encounters are typically of an approach–avoid kind, but if fights ensue, lobsters in molt stage A (immediate postmolt) typically lose appendages or are killed. Shell hardness seems to play an important role in such encounters, as lobsters in molt stages D_1 and D_2 are dominant or equal in status to those in stage C. Encounters between two molt stage C animals are highly ritualized and hotly contested (Tamm and Cobb, 1978; Huber and Kravitz, 1995). This overall pattern is also found in groups of three emergent phase juveniles (13- to 24-mm CL) (Zeitlin-Hale and Sastry, 1978) and five shelter-restricted phase juveniles (6.5- to 9.5-mm CL) (Sastry and Ehinger, 1980), but aggression and activity levels decrease over time.

Establishment of dominance hierarchies under laboratory conditions suggests that individuals within populations of closely spaced juvenile lobsters can have profound effects on growth rate, access to food and shelter, and injury rate. However, whether such hierarchies are established in wild populations remains to be seen (see Section V,B,3). Nonetheless, growth rates in natural populations appear to be affected by density of lobsters present on a particular substrate, with lower densities promoting greater overall growth increases of the population after 4–6 weeks (Roach, 1983). Growth rates between individual lobsters also appear to vary greatly, with the differences between the largest and smallest lobsters becoming greater over time (Roach, 1983). These ioral interaction, whereby certain individuals inhibit the growth rates of other lobsters by controlling more resources or interfering with their activities.

VI. Adult Lobsters

A. Historical Perspective

In the absence of heavy commercial exploitation, the adult phase of *Homarus americanus* life history may exceed 30 years. Historically, American lobsters have reached live weights in excess of 19 kg (Wolff, 1978); even under contemporary exploitation, lobsters up to 217-mm CL (~9 kg) may be encountered in shallow water during the summer months (Lawton and Robichaud, 1992). Such lobsters may be in excess of 20 years old, if male, or over 30 years old in the case of females (Campbell, 1983).

Much of our information on adult lobsters comes from fishery-related studies, using traps or trawls as sampling tools, and deals mainly with size distribution, seasonal patterns of molting, movement as revealed by mark–recapture studies, etc. Reliance on trap and trawl surveys, prior to the development of *in situ* approaches (e.g., scuba diving or submersibles), significantly colored our perception of adult lobster ecology. (Cobb, in Chapter 7, discusses the interrelationships among ecology, behavior, and fisheries.)

B. Size at Maturity

Size at sexual maturity has implications for understanding lobster population demography and for evaluating impacts of particular regulations in the commercial fishery (Ennis, 1984c; Miller, Chapter 5; Fogarty, Chapter 6). In males, *physiological* maturity is indicated by the presence of mature spermatozoa, and in females, by developed ovaries (reviewed by Aiken and Waddy, 1980). Based on these criteria, physiological maturity occurs at smaller sizes in males than in females (Krouse, 1973; Briggs and Mushacke, 1980; Aiken and Waddy, 1980). The adolescent phase for lobsters has been defined as the period between the onset of physiological and *functional* maturity (Table 1; Section II).

Functional maturity in males occurs when, given a reasonable opportunity, the male is capable of mating with and inseminating a female (Aiken and Waddy, 1980). While there are distinct morphometric correlates of maturity in males (Aiken and Waddy, 1980; Elner and Campbell, 1981), this definition has behavioral dimensions that are difficult to apply to the field situation. Nonetheless, in areas where large adult lobsters form a significant proportion of the landed catch (e.g., the Bay of Fundy), concern has been expressed that differential removal of large males may reduce the probability of mating for large females (Campbell, 1992).

In females, functional maturity is clearly indicated when external eggs are present. However, because individuals not carrying eggs may also be mature, the proportion of ovigerous (berried) lobsters in the population, usually inferred from trap samples, is an equivocal index of maturity (Aiken and Waddy, 1980; Pezzack and Duggan, 1989). Pleopod cement glands are not found on male or immature female lobsters; they develop in the pubertal female when undeveloped (white) ovaries undergo vitellogenesis, providing a convenient determination of female maturity (Aiken and Waddy, 1982) which is easily field-validated (Ennis, 1984d).

Regardless of the criteria used, size at sexual maturity varies markedly over the geographical range of *Homarus americanus* (Aiken and Waddy, 1980) and has significant demographic consequences (see Fogarty, Chapter 6). An intriguing hypothesis states that agonistic behavior of lobsters increases as they mature (Steneck, 1989). However, rather than engage constantly in agonistic interactions with smaller conspecifics, lobsters maturing in high-density coastal recruitment sites disperse to lower-density areas, leading to a "preharvest" decline in abundance, or "diffusion" into deeper-water habitats. Where size at sexual maturity corresponds closely to, or exceeds, the minimum legal size in the fishery, such behavioral influences will be difficult to distinguish from harvesting pressures. However, where size at maturity falls below harvestable size [e.g., in Long Island Sound (Briggs and Mushacke, 1979)], it may prove feasible to test this hypothesis.

C. Movement by Inshore and Offshore Populations

Adult lobsters exhibit a range of movement patterns (nomadism, migration, homing; see Section V,D for definitions) that have consequences for the definition of population structure and fisheries management strategies. The annual scale of movement is related in part to the need to maximize annual thermal advantage (Pezzack and Duggan, 1986; Campbell, 1990). However, across the geographical range, this may lead to no pronounced seasonal depth-related redistribution, seasonal inshore–offshore return migrations, or offshore diffusion. These varied movement patterns may actually represent legacies of paleoenvironments along the North Atlantic seaboard (see Cobb, Chapter 7). While larval dispersal is an important mechanism potentially linking subpopulations (Ennis, 1986; Harding and Trites, 1988, 1989; Pezzack, 1989; Katz *et al.*, 1994), movements of adult lobsters may also facilitate genetic exchange (Fogarty, Chapter 6). Assessing the significance of linkages between inshore and offshore lobster populations from a fisheries management perspective is a multifaceted problem (Pezzack *et al.*, 1992; Harding *et al.*, 1993; Pringle and Burke, 1993; Miller, Chapter 5).

1. Scale of Movement of Inshore Lobsters

Our understanding of the scale of movement of adult lobsters comes largely from single mark–recapture studies undertaken in association with the commercial fishery, an endeavor with a certain capricious aspect, as noted in one of the first tagging studies (Bumpus, 1901). Early studies used tags that were lost at the molt, as the emphasis was on determining exploitation rates in the fishery rather than long-term movement (Stasko, 1980). The "sphyrion" tag, which is retained through one to several molts (Scarratt and Elson, 1965), was used extensively through the 1970s and 1980s. In spite of significant limitations (poor resolution of interannual variability in movement, incidental tag losses, and reliance on the fishery for recoveries), evidence from tagging studies led to the conclusion that inshore lobsters were relatively local in their distribution, moving <25 km seasonally (Fogarty *et al.*, 1980; Cobb and Wang, 1985; Miller *et al.*, 1989; early Canadian studies reviewed by Stasko, 1980; American studies reviewed by Krouse, 1980). However, tag–recapture data often demonstrate the intensity of fishing pressure on a lobster population, rather than the innate capacity for movement, as shown in studies in which 91% of recaptures by fishers are within 5 km of the release site (Northeast Utilities Service Company, 1993).

The impression of a seasonal inshore–offshore migration of lobsters is commonly held among inshore fishers, but may be due to environmental factors (e.g., temperature-related differences in catchability) and/or seasonal removals by an intensive fishery (Wilder and Murray, 1958). In the Gulf of Maine, lobsters engage in small-scale movements from shallow water into deeper water, apparently in response to strong winds and turbulence rather than the seasonal thermal regime (Cooper *et al.*, 1975). Seasonal movements in a relatively isolated lobster population in Bonavista Bay, Newfoundland, also take place over a restricted depth range; the mean depth of lobsters in the winter is 14 m, compared to ~9.5 m in the summer. Coastal physiography restricts along-shore movement, while the prevailing thermal regime in that area provides no clear advantage for pronounced inshore–offshore migration. Wind-induced turbulence in shallow water may also be implicated in the downslope movement (Ennis, 1984a). Adult inshore females seem to move to deeper water earlier in the fall than adult males (Campbell and Stasko, 1986; Robichaud and Campbell, 1991; Roddick and Miller, 1992). In Connecticut, long-term tag–release studies suggest that only a small percentage of the inshore lobsters migrate to deeper waters or offshore canyons. Of the 39% (21,136) of tagged lobsters (53,875) recaptured from 1978 to 1993, only 1117 (5%) were recovered at distances greater than 5 km from their release site. Of this 1117 lobsters, only several hundred were recaptured in the Race, a deepwater channel between Long Island and Block Island sounds, and only 24 lobsters were recaptured off the continental shelf (Northeast Utilities Service Company, 1993). Thus, migratory behavior of inshore lobsters in southern New England may be less common than was previously thought.

Instances of long-distance movement of inshore lobsters had been recorded from several areas by the 1980s and 1990s (Maine, Dow, 1974 and Fogarty *et al.*, 1980; Îles de la Madeleine, Bergeron, 1967 and Munro and Therriault, 1983; southwestern Nova Scotia, Stasko, 1980; Connecticut, Northeast Utilities Service Company, 1993). However, recaptures from large-scale releases of lobsters in the Bay of Fundy (Campbell, 1986; Campbell and Stasko, 1986) and southwestern Nova Scotia (Campbell and Stasko, 1985) revealed that many adult lobsters (16.4% of those recaptured) moved >92.6 km (50 nautical miles) from their release sites; the farthest distance traveled was 798 km for a male at liberty 3.5 years (Campbell and Stasko, 1986). These movements may lead to intermixing of lobsters within the Bay of Fundy, throughout the Gulf of Maine, and in the adjoining continental shelf and slope regions (Fig. 7). The fastest rate of movement recorded for lobsters tagged in the Bay of Fundy was 2.5 km/day for a 112-mm CL female lobster released and recaptured 328 km away after 130 days at liberty (Campbell and Stasko, 1986). Although lobsters were recaptured farther away from release locations with time, some were recaptured near release sites up to 4–6 years after tagging.

2. Scale of Movement of Offshore Lobsters

Initial tagging studies on offshore lobsters revealed the presence of well-defined migratory movements

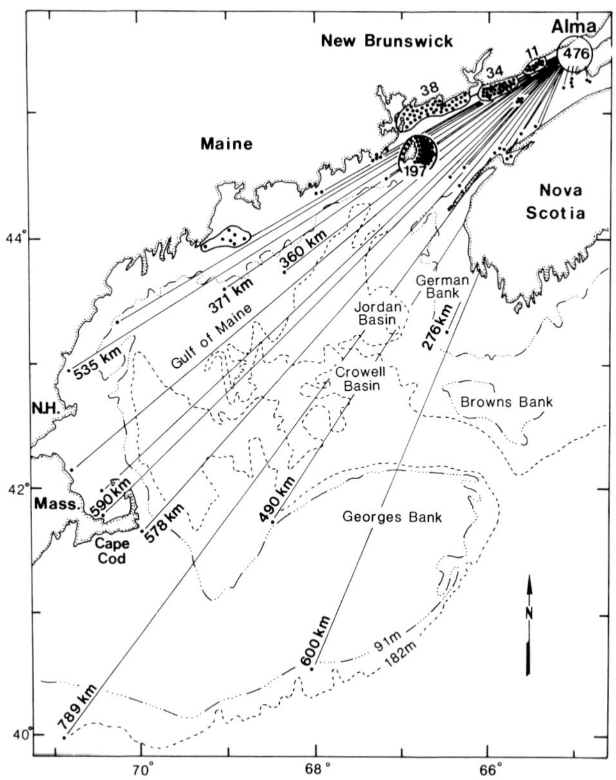

FIGURE 7 Recapture points and straight-line distances moved by tagged American lobsters released off Alma, New Brunswick, during 1979–1980. One dot represents one recapture, except where a group of dots has a numeral showing the number of recaptures in the grouping; 476 lobsters were recaptured within 30 km of the release site. (Reprinted from Campbell, A., and Stasko, A. B. (1986). Movements of lobsters (*Homarus americanus*) tagged in the Bay of Fundy, Canada. *Mar. Biol.* 92, 393–404, Fig. 5. Copyright © Springer-Verlag, used with permission.)

for a substantial portion of the tagged population. Probably 20%, and possibly 30–40%, of offshore lobsters annually migrate in directed shoalward movements in the spring and summer off southern New England (Cooper and Uzmann, 1971; Uzmann *et al.*, 1977a), complemented by an offshore migration in the fall and winter (Fogarty *et al.*, 1980). Lobsters from offshore portions of the Scotian Shelf and eastern Gulf of Maine undertake seasonal movements to shoal areas on Georges and Browns banks (Pezzack *et al.*, 1992). These movements from deep to shallow water may act to maintain lobsters within a temperature range of 8–14°C, allowing for more rapid growth than would be possible for lobsters either remaining inshore, where temperatures drop below 0°C in winter, or remaining in deep offshore habitats, where temperatures rarely rise above 12°C (Uzmann *et al.*, 1977a).

For offshore lobsters undergoing seasonally directed migrations, speed of movement over the ground is impressive. A maximum displacement of 345 km in 71 days (4.2 km/day) was recorded for one lobster, while 12 others made directed movements within 21–41 days of release ranging from 93 to 161 km (Uzmann *et al.*, 1977a). Noting some of the limitations in single release–recapture data, such as the inference of straight-line movement, and the possibility that lobsters arrive at recapture points earlier than reported, speeds over the ground of 7.4–9.3 km/day (4–5 nautical miles/day) are thought likely for migrant lobsters (Uzmann *et al.*, 1977a).

Lobsters occupying submarine offshore canyons do not appear to form discrete populations, as a significant east–west component to movement is revealed from single-point recaptures in American (Uzmann *et al.*, 1977a) and Canadian tagging programs (Pezzack and Duggan, 1987, 1988; Pezzack, 1987). However, multiple recaptures of some tagged lobsters (Pezzack and Duggan, 1986) and movement of displaced ovigerous lobsters (Duggan and Pezzack, 1988) suggest that a reappraisal of this conclusion may be warranted.

Some lobsters on the Scotian Shelf apparently undertake long-distance return migrations (>200 km) (Pezzack and Duggan, 1986). One individual (a 136-mm CL female) completed a round-trip movement of 211 km in 6 months, with a net displacement of 15 km. Another (a 121-mm CL female) made an estimated 236-km migration in the first year of release, again with a net displacement of only 15 km; this lobster was subsequently recaptured 2.5 years later <3 km from its initial release location. Such movement data reinforce earlier evidence (Saila and Flowers, 1968; Fogarty *et al.*, 1980) that lobsters captured offshore and displaced to inshore release sites returned to areas near their original offshore capture sites, indicative of homing behavior. Lobsters recaptured close to a release location, after 10 months or more at large, could have migrated substantial distances and returned. Similarly, recaptures considerable distances from a release site during another season do not necessarily indicate exploratory or removal migration (*sensu* Van der Steen, 1984, as cited by Pezzack and Duggan, 1986). The documentation of homing in lobsters from the Scotian Shelf suggests that even where extensive movement occurs, there may be relatively closed breeding populations if homing is well defined and adopted by a substantial portion of the adult population.

3. Movements of Ovigerous Lobsters

Historically, opinion has differed regarding the movement of ovigerous females. Prince (1899, as cited by Harding and Trites, 1989) wrote that off southwestern Nova Scotia "great schools pass the winter at depths of 40 or 50 fathoms (70 to 90 m); but during the warm summer months they move into shallow water, 2 to 10 fathoms (4 to 18 m), where the females ripen their eggs and hatch them out." Conversely, MacKay (1929) thought that late-stage ovigerous lobsters moved very little, and then only to procure food. However, late-stage ovigerous lobsters released in inshore waters off Cape Cod made substantial northerly movements [average distance, 28.2 km; mean velocity, 0.72 km/day (Morrissey, 1971)]. Seasonal migration of ovigerous lobsters into shallow lagoons off the Îles de la Madeleine is linked to water temperature variations (Munro and Therriault, 1983).

Offshore ovigerous females tend to remain offshore, moving into shallower water on offshore banks only during summer months (Uzmann *et al.*, 1977a; Pezzack *et al.*, 1992). In recognition of these seasonal movements, part of Browns Bank is closed to lobster fishing to protect broodstock (as noted by Miller, Chapter 5). Ovigerous lobsters captured in Veatch Canyon and released 218 km away in Narragansett Bay hatched their eggs in inshore waters, then moved back offshore (Saila and Flowers, 1968). Most of those displaced during late June from Georges Bank to coastal locations off southwestern Nova Scotia moved offshore >117 km toward the initial capture location (Duggan and Pezzack, 1988).

Such seasonal shallow–deep migrations may provide offshore lobsters with sufficiently high temperatures to meet the physiological demands of molting, mating, and egg extrusion (Cooper and Uzmann,

1971, 1980). These movements may also optimize egg development rates (Campbell, 1986; Talbot and Helluy, in Chapter 9, discuss rates of embryonic development). Ovigerous lobsters in the Grand Manan area undergo similar seasonal deep–shallow migrations that could serve to maintain the lobsters at the highest local water temperatures. Furthermore, it appears that some ovigerous lobsters return to the same coastal location in successive years. Other ovigerous females may move from overwintering sites in the Grand Manan basin northeast into the Bay of Fundy, west to the Maine coastline, or south to the shoal waters of the continental shelf (Campbell, 1986, 1990; P. Lawton and D. A. Robichaud, unpublished data).

D. *Habitat Types and Patterns of Abundance*

1. Inshore Habitats

In contrast to earlier life history phases, adult lobsters may be only seasonal members of shallow-water rocky inshore habitats. With the general expansion in types of habitat used, relaxation in predation pressure (Cooper and Uzmann, 1980; Steneck, 1989), and corresponding increase in scale of movement, habitat linkages for adult lobsters appear diffuse (see Section III,B). However, an organism that undertakes seasonal return migrations >50 km, and that exhibits localized homing behavior (Stewart, 1972; Ennis, 1984b; Karnofsky *et al.*, 1989b), probably relies on well-developed sensory cues to govern movement patterns and seasonal habitat utilization. Unfortunately, this is largely an unexplored area of research in clawed lobsters, in contrast to spiny lobsters (Herrnkind, 1983).

Lobsters above the minimum legal size are typically represented at low levels of abundance (<1 lobster per 100 m^2) in rocky inshore habitats subject to continued fishing pressure during the summer months (e.g., Stewart, 1972; Richards and Cobb, 1986). In locations where the fishery closes during the summer, adult lobsters increase in abundance in shallow coastal sites, but densities typically remain between 1 and 3 lobsters per 100 m^2 (Miller, 1989; Roddick and Miller, 1992).

The occupation of inshore mud habitats by adult lobsters has been noted in a number of areas (Thomas, 1968; Stewart, 1972; Cooper *et al.*, 1975; Cooper and Uzmann, 1980; Jarvis, 1989), with population densities probably <10% of that of inshore rocky habitats (P. Lawton, unpublished observations). However, these often expansive habitats are targeted seasonally by commercial fishers, and may support substantial lobster populations. A common perception is that adult lobsters migrate across these areas on their way to inshore rocky habitats, although the presence of occupied burrows suggests the possibility of resident populations (Thomas, 1968; Stewart, 1972).

The relative abundance of ovigerous lobsters can vary considerably in inshore waters over the summer months, even between adjacent areas, suggesting distinct physical habitat preferences (Campbell, 1990). Off northeastern Grand Manan, adult lobsters occupy several shallow coves between June and October. On the sand–clay bottom of one sheltered cove, the summer lobster population is heavily skewed toward females, and ovigerous females in particular (Campbell, 1990). Average densities of 2.33 ± 0.04 (SE) ovigerous lobsters per 100 m^2 were observed during September (Campbell, 1990). Female lobsters, and the few males that are present, excavate bowl-shaped depressions, with as little as 0.3 m between adjacent, occupied depressions (P. Lawton, unpublished observations). This aggregation of ovigerous lobsters contrasts with more typical summer residency patterns, in which ovigerous lobsters are dispersed among the lobster population occupying rocky inshore habitats (P. Lawton, unpublished observations). Similar bowl-shaped depressions have been described from other inshore areas (Stewart, 1972; Hudon, 1987; Jarvis, 1989), in offshore habitats (Cooper and Uzmann, 1980; Auster *et al.*, 1991), and under high-density holding conditions within commercial lobster pounds (McLeese and Wilder, 1964).

Canadian inshore waters known to support good commercial lobster catches exhibit wet weight biomass values clustered around 10 g/m^2 (Miller, 1985). Inshore sand substrates with overlying rocks in the Gulf of Maine may support a combined juvenile and adult biomass of up to 178 g/m^2 (Cooper and Uzmann, 1980), while inshore rocky bottoms (with abundant algal cover) off the Îles de la Madeleine evidently support a combined biomass up to 75 g/m^2 (Hudon and Lamarche, 1989). Off eastern Nova Scotia, a peak biomass of 30 g/m^2 is considered "extraordinarily high" (Roddick and Miller, 1992), yet biomass values for adult lobsters up to 570 g/m^2 have recently been recorded in summer spawning areas off Grand Manan (Lawton and Robichaud, 1992).

2. Offshore Habitats

There are severe logistic constraints to estimating patterns of abundance of offshore lobsters. Three survey techniques were evaluated by Uzmann *et al.* (1977b), working in the Veatch Canyon region of the outer continental shelf. Manned submersibles, a towed camera sled, and an otter trawl were deployed at selected sites, with subsequent sampling by trapping, to intercalibrate direct (submersible, sled) and indirect (trawling, traps) measures of abundance. The mean estimates of the standing stock of lobsters were comparable between the camera sled and submersible observations (1.41–1.46 lobsters per hectare), which were higher than the trawl-based estimate (0.24 lobsters per hectare). Average offshore densities may be only 0.1% of those in coastal areas, although densities are greater in certain offshore canyons (Cooper and Uzmann, 1977, 1980).

Submarine canyons are large geological features that represent unique ecosystems, largely because of their highly varied, three-dimensional habitats (Cooper *et al.*, 1987a,b). The species abundance and community structure of the megabenthic fauna are closely related to surface geology and sedimentary features, which, in turn, are related to the bottom gradient and currents. Cooper *et al.* (1987a,b) elaborated on an initial habitat classification for offshore areas (Cooper and Uzmann, 1977, 1980; see Section III,B), discriminating five habitat types.

Photographic documentation of these habitat types and their associated biota show lobsters to be among the 28 most commonly observed and photographed fish and invertebrates at Georges Bank and Lydonia, Oceanographer, and Veatch canyons from 1980 through 1984 (Cooper *et al.*, 1987a,b). Lobster abundance at Oceanographer and Lydonia canyons was 0–2 lobsters per hectare in habitat type I, 0–10 lobsters per hectare in habitat type II, and 5–1260 lobsters per hectare in habitat types III and IV combined. On the middle Atlantic and southern New England continental shelf, lobsters exhibit a clear microhabitat association with tilefish (*Lopholatilus chamaeleonticeps*), which are primary shelter excavators, responsible for extensive bioerosion of the shelf environment (Grimes *et al.*, 1987).

Pioneering attempts to measure the size of deep-water lobsters *in situ* and to estimate transect areal coverage from manned submersibles used relatively coarse techniques [e.g., by comparison with graduated bars (Cooper *et al.*, 1975)]. Contemporary studies in the offshore environment make use of acoustic positioning systems for monitoring the track of undersea vehicles and laser sizing scales for estimating individual animal size, transect path width, and hence, population density (Auster *et al.*, 1991; Babb *et al.*, 1993). Using these sophisticated techniques, a new research study on the linkages between lobsters and habitat in the offshore environment is presently under way (Steneck *et al.*, 1995).

E. Shelter-Related Behavior

1. Physical Attributes of Shelters

Physical attributes of shelter sites and hydrodynamic effects greatly influence shelter use by all size classes of juveniles and adolescent phase lobsters (see Section V,B), but this influence wanes in the adult phase. Adult lobsters occupy preexisting spaces under large boulders and crevices within exposed bedrock, often showing extensive periods of occupancy (Stewart, 1972; Ennis, 1984b; Karnofsky *et al.*, 1989a,b). However, they also exploit more ephemeral cover, such as attached kelp (Jarvis, 1989; Bologna and Steneck, 1993; Section VII,B), detached kelp, and other seaweed fronds (Stewart, 1972; P. Lawton, unpublished observations). Except during times of molting and mating, there appears to be only a residual affinity for cover among adults. However, some adult lobsters use shelter sites that provide no cover, merely a location from which to base daily activities. Adults occupying bowl-shaped depressions in summer spawning areas off Grand Manan are typically observed with sediment along the ventral margin of their claws, indicating active maintenance of these sites. The size of the depressions is related to lobster size (Campbell, 1990) and they can form impressive structures, up to 1 m in diameter and >0.3 m in depth, with a prominent lip at the substrate surface [these lobsters are typically >100-mm CL, with some large males exceeding 180-mm CL (Lawton and Robichaud, 1992)]. The functional significance of these well-maintained shelters is not yet fully understood, as the lobsters occupying this site create other, less well-defined, excavations in deeper-water locations, while foraging on ocean quahogs (*Arctica islandica*) (P. Lawton, unpublished observations). Some lobsters even take up residence in the depressions created within such beds. A similar pattern of microhabitat use and foraging behavior on ocean quahog beds has been described for a low-relief site on the outer continental shelf in the Middle Atlantic Bight (Auster *et al.*, 1991).

2. Shelter Competition

As with juvenile and adolescent lobsters, field observations on adult lobsters suggest that shelter competition occurs less frequently in the wild than in laboratory settings (Ennis, 1984b; Karnofsky *et al.*, 1989a,b). The ability to immigrate into or emigrate from an area probably plays an important role in lobster behavior. Thus, when lobster movements are restricted, which occurs even in large, naturalistic aquaria, increased aggression results and shelter competition is observed. During 2 months of observation in a large (180-m^2) pool, 447 shelter approaches were made and 50% of these resulted in evictions of the resident lobster (Karnofsky and Price, 1989a). When shelter is artificially provided in the field, similar evictions are seen, but only among animals larger than 60-mm CL (Steneck, 1989; R. S. Steneck, unpublished data). Because unoccupied shelters were always available in these trials (i.e., shelter was not limiting), these evictions probably serve to establish dominance rather than acquisition of the shelter (Karnofsky and Price, 1989a).

Interspecific competition for shelter space occurs principally with the crabs *Cancer borealis* and *C. irroratus*, which are sympatric with the lobster. These interactions involve several benthic phases and are markedly dependent on the relative sizes of the interacting individuals; they are therefore explored separately (see Section VII,A).

F. Foraging Behavior and Daily Activity Patterns

1. Natural Diet

The general feeding habits of the adult lobster were described somewhat theatrically by Herrick (1895), who observed that "it is very cautious and cunning, capturing its prey by stealth, and with weapons which it knows how to conceal. Lying hidden in a bunch of seaweed, in a crevice among the rocks or in its burrow in the mud, it waits until its victim is within reach of its claws, before striking." Adult lobsters certainly adopt ambush predation techniques, yet the accumulation of bivalve shell material around long-established shelter sites suggests that central place foraging also occurs (see also Weiss, 1970; Moody and Steneck, 1993). Food-caching behavior, which has been observed in the laboratory setting (Smith, 1976), is perhaps related to this foraging mode. Lobsters readily attack and consume crabs in the laboratory and have been observed in the wild at night actively feeding on, or carrying, the crabs *Cancer irroratus* and *Libinia emarginata* (Weiss, 1970). Adult lobsters are also capable of acting as "pursuers" (*sensu* Hughes, 1980), as they have been observed excavating large clams from sand or mud substrates. Observations such as these, together with the broad diet composition (Table 3) and documentation of nocturnal foraging movements (Weiss, 1970; Cooper and Uzmann, 1980; Section VI,F,4), indicate that ambush predation, while exploited by adult lobsters, may not be the predominant foraging mode, as originally speculated by Herrick (1895). Furthermore, *Homarus gammarus* adults (77- to 82-mm CL) seem to be capable of suspension feeding on coarse particles (600 μm) and may be able to utilize such food particles (in the form of marine snow or large epibenthic zooplankton) to maintain their nutritional state in periods of poor food supply or while overwintering (Loo *et al.*, 1993).

The principal components of the diet of adult lobsters are various crustaceans and mollusks (Table 3), with polychaetes and echinoderms increasing in relative importance in certain areas or times of the year (Hudon and Lamarche, 1989; Ojeda and Dearborn, 1991). The degree to which sea urchins form a significant component of the lobster diet has been the subject of intense scrutiny, related to the putative role of the lobster as a keystone predator (*sensu* Paine, 1969) in North Atlantic kelp bed ecosystems (Elner and Vadas, 1990; Section VII,B). Assessment of the relative importance of fish to the natural diet of adult lobsters is complicated by the presence of commercial fishing activity in most areas where diet studies have been conducted (e.g., Squires, 1970). Scavenging on dead prey, such as discarded fish bait, certainly occurs (Weiss, 1970); however, lobsters have also been observed capturing fish, using the cutter claw to close rapidly on the prey from an ambush predation situation (P. Lawton, unpublished observations). Notwithstanding these food preferences, lobsters have rather eclectic tastes (Scarratt, 1980), ingesting at times "pieces of plastic, wool, tea bags complete with cotton cover and tea leaves, and in one instance a small iron nail."

Certain aspects of the natural diet change over the year, as noted earlier for juvenile and adolescent lobsters (Section V,C,4). Changes in feeding activity have been correlated with seasonal temperatures in Bonavista Bay (Ennis, 1973). While the rise in feeding activity (as inferred from an index of stomach fullness) is gradual up to early summer, there is a rapid increase to peak feeding activity between June and July; feeding activity then remains high in September even as temperatures begin to fall; and females maintain a higher level of feeding activity than males, at least until mid-February. The latency in feeding activity is related to the need for physiological recovery from molting (for both sexes) and greater physiological demands on females because of gonadal development.

2. Selective Feeding

Lobsters predominantly use crushing tactics to open blue mussels (*Mytilus edulis*) and in this respect are similar to *Cancer borealis* (Moody and Steneck, 1993). Conversely, *C. irroratus* and *Carcinus maenas* resort to a wider variety of techniques, including those directed at the perimeter of mollusk shells (prying, edge chipping, boring, and entering while the shell is gaping). The reliance on crushing techniques when lobsters feed on mussels may not extend to other lobster–bivalve predatory interactions. In the Baie des Chaleurs, Quebec, adult lobsters feed on sea scallops (*Placopecten magellanicus*) and utilize perimeter tactics to open large scallop shells (Stokesbury and Himmelman, 1995). Examination of the shell damage inflicted on various sizes of scallops tethered along a transect line and documentation of predator abundance reveal a significant relationship between lobster density and the mortality of large scallops (70- to 90-mm shell height) (Stokesbury and Himmelman, 1995). The damage inflicted on scallops by *Homarus americanus* (and *C. irroratus*) was principally directed at the shell margin (76%), as opposed to direct crushing damage (24%). While perimeter-opening techniques have been previously described for lobsters feeding on scallops (Elner and Jamieson, 1979; Elner and Campbell, 1981), the general conclusion, now incorrect, it seems, was that scallops >70-mm shell height would be immune to lobster predation (Stokesbury and Himmelman, 1995). *In situ* observations of lobsters feeding on large ocean quahogs [>60-mm shell height (P. Lawton, unpublished observations)] also suggest that perimeter-opening techniques are more commonly used than was previously assumed. Clearly, information is required from a wider array of the predatory interactions in which adult lobsters are involved before generalizations on predominant foraging modes can be made. Quantitative information is almost exclusively restricted to lobster–bivalve and lobster–sea urchin interactions. Predation on soft-bodied prey, such as polychaetes and fish, is largely unknown, reflecting a general preoccupation in decapod crustacean foraging studies (Juanes, 1992; Lawton and Zimmer-Faust, 1992).

3. Fishery Effects

With the widespread introduction of escape vents on lobster traps (Miller, Chapter 5), it is now likely that most lobsters feed from traps before they are finally captured by the commercial fishery. In areas of intense fishing pressure, fishing bait may provide a significant trophic subsidy, supplementing the natural food resources available on lobster grounds.

Both far- and near-field aspects of the attraction of lobsters to baited traps provide insight into the natural foraging behavior of adult lobsters and highlight the often complex interactions between individual lobster sensory perception and social–behavioral interaction (Atema and Voigt, Chapter 13). One of the parameters used to assess the effectiveness of traps is the "catchability coefficient," also termed the "effective fishing area" (Miller, 1990). While this quantity (expressed as square meters fished per trap) is less than the true area of attraction, it gives some qualitative measure of the foraging range of lobsters (see Cobb, Chapter 7, for a more complete derivation of catchability terms). Decapod crustacean catchability coefficients on shallow bottoms with high relief are thought to be in the tens and hundreds of square meters, whereas values from deep flat habitats may range up to thousands of square meters (Miller, 1990).

Nonetheless, seminatural laboratory studies suggest that traps may be largely inefficient at capturing lobsters, with perhaps as low as 2% of trap interactions leading to capture (Karnofsky and Price, 1989b). Unfortunately, there are few direct observations from the natural environment that allow for calibration of trap efficiency (Auster, 1985; reviewed by Miller, 1990). Lobsters often attempt to remove bait through the trap laths, while some position themselves on the funnel entrance, feed on the bait, and then depart (Auster, 1985; Karnofsky and Price, 1989b). (Miller, in Chapter 5, discusses trap design.)

4. Daily Activities and Home Range

As with earlier benthic phases, adult lobsters are principally active nocturnally in inshore waters (Cooper and Uzmann, 1980; Ennis, 1984b). Offshore lobsters (Cooper and Uzmann, 1980) and inshore lobsters occupying deeper, turbid waters, such as those in western Long Island Sound (Stewart, 1972), may show periods of activity during the daytime. Off northeastern Grand Manan, adult lobsters have recently been observed actively foraging during daytime, around low-water slack periods (P. Lawton, unpublished observations). Thus, in areas subject to high near-bottom current speeds, it appears there may be greater synchrony of daily activity with tidal stage than with light level per se (see also Lund *et al.*, 1973; Howard and Nunny, 1983; Auster, 1985).

The range of foraging, or dispersal, on a given night generally does not exceed 300 m, but may possibly range up to 2 km (Cooper and Uzmann, 1977). In a shallow, sublittoral, rocky habitat in Bonavista Bay, nocturnal dispersal of resident lobsters is more restricted, typically <100 m (Ennis, 1984b). However, some lobsters move into and out of the study area over the summer, suggesting more extensive nocturnal movements for transient lobsters. Nocturnal activity in this population varies seasonally with temperature.

Underwater telemetry systems have not been widely applied in studying *Homarus americanus* activity and foraging range, as they have been for spiny lobsters (e.g., Phillips *et al.*, 1984; Jernakoff *et al.*, 1987) and crabs (Wolcott and Hines, 1989). This is surprising, as lobster researchers were among the early proponents of these techniques (Lund and Lockwood, 1970). The large size of adult lobsters has obvious advantages for external tag placement and surgical implantation of physiological sensors, yet the capacity for extensive movement poses logistic constraints that are only now being surmounted (O'Dor and Webber, 1991). Lobsters fitted with sonic tags and released on a featureless sand bottom in eastern Long Island Sound traveled farther during the first night following release than lobsters tagged and placed back on the bottom in shelters (Lund *et al.*, 1973). Thereafter, daily excursions of sonic-tagged lobsters were on the order of 30 m, although greater movements occurred where lobsters made a transit between areas of suitable habitat. Two sonic-tagged adult lobsters released in Bideford River (Maynard and Conan, 1984) and tracked over several days during November showed a pattern of directed movement, alternating with resting periods that may have been linked to possible searching behavior for burrow locations in which to overwinter (as described by Thomas, 1968) in the estuary (Maynard and Conan, 1984).

Recently, there has been a resurgence of interest in applying telemetry techniques to study lobster movement, specifically in relation to ovigerous lobster behavior (Jarvis, 1989; O'Dor and Webber, 1991). Fourteen ovigerous lobsters fitted with ultrasonic tags released in the Jeddore Harbor/Clam Bay area of eastern Nova Scotia showed maximum rates of movement of 190 m/hr (Jarvis, 1989). Resident behavior was more apparent in shallow nearshore areas with suitable habitat (e.g., boulder and bedrock reefs with algal cover), while transient behavior was observed over featureless mud and gravel bottoms. Subsequent studies in Jeddore Harbor tested a more sophisticated, computer-assisted tracking system in which location and environmental sensor data were relayed to shore from an offshore sonobuoy array, thereby providing continuous coverage over a 4-km^2 area (O'Dor and Webber, 1991). The total distance traveled by four ovigerous lobsters tracked for 5 days ranged from 1.5 to 15 km.

Telemetry systems will likely play an increasing role in focal animal studies, as environmental influences [e.g., temperature (O'Dor and Webber, 1991)] and internal motivational state [e.g., mandibular muscle contraction indicative of feeding activity (Wolcott and Hines, 1989)] can now be routinely monitored. However, the social context of behavior is not typically provided. Recent refinement of transmitter designs and data-logging systems offers the prospect of monitoring a number of lobsters concurrently, thereby expanding the utility of these approaches.

G. Predation Pressures

Annual natural mortality levels (excluding anthropogenic causes) between 2 and 8% are considered realistic for juvenile and adult lobster populations, whereas annual exploitation rates in the lobster fishery may exceed 90% (Fogarty, in Chapter 6, provides a comprehensive review of available mortality estimates). In addition to direct harvest pressure from the trap fishery, adult lobsters are susceptible to physical damage from groundfish trawling and scallop dragging, as well as entrapment in lost fishing gear (Smith and Howell, 1987; Cooper *et al.*, 1987b; Roddick and Miller, 1992). Historically, it was commonly believed that large predatory groundfish such as the cod (*Gadus morhua*) consume lobsters, particularly during the summer molting season. However, a recent review of food habit studies on 12 individual species and general taxonomic references reveal only anecdotal evidence for such predation (Bigelow and Schroeder, 1953; Scott and Scott, 1988; D. S. Pezzack, personal communication).

H. Mating

1. Sex Ratios in Populations

Population sampling typically shows a 1:1 sex ratio among lobsters in nature (Skud and Perkins, 1969; Krouse, 1973; Cooper *et al.*, 1975; Cooper and Uzmann, 1980), although there are some instances of male- and female-dominated local populations and seasonal assemblages (Briggs and Zawacki, 1974; Cooper *et al.*, 1975; Briggs and Mushacke, 1979; Karnofsky *et al.*, 1989a; Campbell, 1990, 1992; Howell and Watson, 1991). Competition between males could lead to such shifts in the sex ratio, as could differential susceptibility of the sexes to the trap fishery (Miller, 1990; Campbell, 1992). Alternatively, environmental factors such as salinity (Jury *et al.*, 1994a) may also influence the sex ratio in manners not yet completely understood. Seasonal changes in estuarine lobster populations can lead to sex ratios as high as 1:5 (females to males) and may be partially explained by differential behavioral responses to reductions in salinity (W. H. Howell and W. H. Watson, University of New Hampshire, personal communication). While all lobsters actively avoid low-salinity water, female lobsters appear to be more sensitive, initiating avoidance responses when salinity falls below 18 ± 1.74 ppt, whereas males do not initiate movements until salinity levels reach 11 ± 3.29 ppt (Jury *et al.*, 1994a).

2. Mate Selection Behavior

Adult females in naturalistic laboratory settings usually initiate and form pair bonds only with dominant males (Atema *et al.*, 1979; Atema, 1986; Cowan and Atema, 1990). Dominance in males is linked to large chela size, which probably has a reproductive display function as the claws are disproportionately large for the prey in natural diets (Table 3) (Elner and Campbell, 1981). Females stagger their molts to mate with the dominant male (Cowan and Atema, 1990; Cowan, 1992). There is some evidence that this mating system might occur in small, wild populations (Karnofsky *et al.*, 1989a), but more observations are needed. Atema and Voigt (Chapter 13) consider the details of mating behavior.

Females benefit from their cohabitations with males in that they are protected during their vulnerable soft-shelled period, suffering fewer injuries (Cowan, 1992), and they replace their shed spermatophores (Templeman, 1934, 1936b; Waddy and Aiken, 1990; Cowan, 1992). Since females will spawn unfertilized eggs if not inseminated (Templeman, 1936b; Waddy and Aiken, 1990; Cowan, 1992), replacement of the spermatozoa prevents the female from wasting energy during egg production. Males, too, may benefit from cohabitations by gaining paternity assurance (Atema, 1986), particularly as females are capable of being inseminated by more than one male (Nelson and Hedgecock, 1977).

However, if females molt in an area where males are scarce and do not mate within 48 hours of molting, they may remain receptive for up to 80 days (Snyder *et al.*, 1992) and can mate during intermolt (Dunham and Skinner-Jacobs, 1978; Waddy and Aiken, 1990). This alternative strategy ensures that females unmated at their molt will still be able to extrude fertilized eggs during intermolt. The mechanisms and frequency of intermolt mating in natural lobster assemblages are not known, since all of these matings have occurred in laboratory settings. Nonetheless, intermolt matings probably do occur in the wild, as adult, uninseminated, preovigerous females are common in some populations (Krouse, 1973; Ennis, 1980).

I. Social Behavior

1. Territoriality

In naturalistic laboratory aquaria, adult male lobsters appear to defend territories that can be larger than the area immediately surrounding their shelter. Dominant males defend several shelters simultaneously, evicting all lobsters that try to occupy them; males will only occupy one particular shelter while the others remain vacant (Karnofsky and Price, 1989a). If living singly, only the shelter itself is defended; however, during male–female cohabitations, male lobsters actively defend the areas around their shelters (Karnofsky and Price, 1989a). Presumably, defense of the mating area and eviction of other males from neighboring shelters demonstrate the suitability of the defender/evictor to the female.

Individual subdominant lobsters move frequently from shelter to shelter, possibly due to shelter evictions (Karnofsky and Price, 1989a). In the field, lobsters change shelters more frequently in the final weeks before molting, perhaps the result of increased activity during premolt (Tamm and Cobb, 1978) or to find a less conspicuous place in which to molt (Karnofsky *et al.*, 1989a). Overall, shelter fidelity seems to vary greatly among individuals, with some remaining in the same shelter for weeks to months and others changing shelters on a daily basis [as shown in field studies (Ennis, 1984b; Karnofsky *et al.*, 1989b) and in the laboratory mesocosm (Karnofsky and Price, 1989a)]. However, lobsters commonly return to a particular shelter after absences ranging from 2 to 222 days (Ennis, 1984b) and they are capable of homing back to such shelters after being displaced from their home area (Karnofsky *et al.*, 1989b).

2. Aggression and Dominance

Dominance hierarchies are generally established between males when the sex ratio is skewed in favor of females. If skewed toward males, agonistic encounters occur frequently, but no one male is able to establish dominance (Cowan, 1992). Dominant males win a high proportion of their encounters (90% or more) and occupy the best shelters for mating cohabitations. Subordinate males either do not occupy shelters or live within small shelters that are unsuitable for cohabitation (Cowan and Atema, 1990; Cowan, 1992). Dominance order can shift over time (Stein *et al.*, 1975), but dominant males in closed populations retain their status and gain mating access to most, if not all, of the molting females (Cowan and Atema, 1990). Female lobsters also establish dominance hierarchies, but these do not appear to be related to order of access to the dominant male (Cowan and Atema, 1990) and are less clear-cut than for the males. Generally speaking, larger lobsters win agonistic encounters and those that have previously won encounters win future ones as well. Females usually lose encounters against males (Scrivener, 1971).

VII. Community Role of Lobsters

Lobsters are conspicuous members of the benthic megafauna in coastal northwest Atlantic marine systems; however, their community role and interactions with various benthic fishes and crabs for shelter and trophic resources are still incompletely resolved. Recent studies in the rocky subtidal zone of the Gulf of Maine indicate that the lobster is one of a suite of mobile benthic predators (four species of decapod crustaceans, 23 fish species) that influence benthic community structure (Ojeda and Dearborn, 1990, 1991). Biotic interactions and the lobster's community role in offshore habitats remain enigmatic (Cooper and Uzmann, 1980; Cooper *et al.*, 1987b). However, considerable advances in understanding the interspecific interactions of juvenile and adult lobsters have been made, especially in two key areas.

A. *Interactions with Other Crustaceans*

Throughout their benthic life, lobsters coexist with several crustacean species, notably *Cancer borealis* and *C. irroratus.* The extent to which these two crab species are subject to harvesting pressure may play a role in determining the consequences of their interspecific interaction with lobsters (Wang, 1982; Cobb *et al.*, 1986; Cobb, Chapter 7). Independently, all three species prefer to occupy crevices or shelters under rocks. However, when shelters are limiting, both adolescent and adult lobsters displace these crabs from shelters (Cobb *et al.*, 1986). Even crabs larger than lobsters are unable to successfully defend shelters and are forced to switch to alternative microhabitats (Richards and Cobb, 1986; Richards, 1992). *Cancer borealis* is the primary competitor for shelter space in coastal New England (Wang, 1982; Cobb *et al.*, 1986), but in the Gulf of St. Lawrence, lobsters interact principally with *C. irroratus* (Hudon and Lamarche, 1989). Since lobsters do not bury themselves in sediment, while crabs do (Cobb *et al.*, 1986), competition with crabs may lead to an overall decline in shelter use by lobsters and a concomitant increase in predation rate upon lobsters (Richards and Cobb, 1986). *Cancer irroratus* does not experience the same degree of competitive interaction with *Homarus americanus*; rock crabs commonly burrow into sediment abutting rocks (Hudon and Lamarche, 1989) or directly into sand. Even when they occupy lobster shelters, aggressive interactions tend not to ensue (Ennis, 1984b).

In addition to the subtle yet consistent differences in the use of physical niches (*sensu* Caddy, 1986), coexistence of lobsters and crabs may also result from segregation along trophic niche axes (Hudon and Lamarche, 1989). As noted earlier (Sections V,C and VI,F), lobster diet seems to be consistent between habitats, implying selective feeding. The crab diet is more cosmopolitan and subject to change, both between habitats and in the presence of large lobsters (reviewed by Hudon and Lamarche, 1989). Attention to trophic partitioning between later benthic phase lobsters and crabs is needed, considering differences in prey composition, size, and handling.

Ecological interrelationships between crabs and juvenile lobsters (Barshaw and Lavalli, 1988; Sections IV,C and V,E) are more complex than in adolescents and adults (Cobb *et al.*, 1986). Ontogenetic changes in competitive or predatory roles may be related to the seasonal progression in size structure and density as new benthic recruitment occurs within lobster and crab populations (Hudon and Lamarche, 1989; Richards, 1992). While lobsters, including juveniles, feed on crabs (Sections V,C and VI,F), there is little evidence that crabs prey on lobsters (Hudon and Lamarche, 1989). However, depending on the timing of benthic settlement of crabs relative to postlarval lobsters, physical niche space may be preempted. Additionally, given the common physical constraints (relative claw size, small body size) and selection pressures (predation risk) faced by early benthic crabs and lobsters, there may be greater correspondence in "realized" foraging pattern (*sensu* Lawton and Zimmer-Faust, 1992) than exists for later benthic phases.

Unfortunately, few field studies have followed lobster and crab size frequencies and population densities concurrently (Wang, 1982; Hudon and Lamarche, 1989; Campbell, 1991). In shallow waters off the Îles de la Madeleine, Quebec, small lobsters (<25-mm CL) and small crabs (<25-mm CL) tend to remain hidden in gravel, sand, and the softer sediments between boulders. Larger rock crabs are found on all bottom types, while larger lobsters are most abundant on rocky bottoms, but absent from sand and rare on sand with eelgrass. Crabs generally outnumber lobsters, but have less biomass [similar results were achieved by Wang (1982), Campbell (1991), and K. L. Lavalli (unpublished data)].

The wider habitat range of small benthic crabs could result from differential habitat selection at settlement, migration following settlement, and/or heavier postsettlement mortality in certain habitats (Section IV,B). The spatial and temporal pattern of postlarval lobster settlement has not been adequately compared with that of the various decapod crustaceans that are sympatric along the lobster's inshore range.

B. *Interactions with Sea Urchins in Kelp Communities*

The putative role of lobsters as a keystone predator (*sensu* Paine, 1969) within northwestern Atlantic nearshore ecosystems, particularly kelp communities, has been vigorously debated since the 1970s (reviewed by Mann, 1985; Miller, 1985; Elner and Vadas, 1990; Vadas and Elner, 1992). The hypothesis was that lobsters, through predatory control of sea urchin populations, restricted the destructive grazing of kelp by sea urchins. However, under certain conditions (lowered predator abundance or enhanced sea urchin recruitment), predators would be "swamped" and the urchin population would be released from predatory constraints to form destructive feeding aggregations that reduce macroalgal abundance (Mann, 1985; Hagen and Mann, 1992, 1994). The crustose coralline community that replaces kelp is maintained because urchins are able to survive on benthic diatoms and algal detritus (Schiebling, 1986).

These coralline "barren" grounds were initially considered to be less productive than the kelp communities they replaced (Chapman, 1981), leading to the prospect of depressed lobster production. However, later assessments of invertebrate biomass suggest that food supply is not limiting for lobsters on barren grounds (Vadas and Elner, 1992). Although reduced by overharvesting in the late 1970s, lobster population size subsequently rebounded through the 1980s (Pezzack, 1992). In fact, the lobsters responsible for the increased landings in the early to mid-1980s would have spent their prerecruit years on these barrens (Miller, 1985; Elner and Campbell, 1991).

The association between lobsters and kelp is partly founded on the physical habitat structure provided by kelp; densities of various life phases are high within kelp (Sections IV,B, V,A, and VI,D). When kelp cover is modified experimentally in the field, lobsters are most frequently seen along edges of experimental patches and density corresponds positively with the perimeter–area relationship (Bologna and Steneck, 1993). These local density effects reveal the preference of lobsters for the prominent ecotone at the edge of kelp beds (Miller, 1985), which may provide shelter while allowing lobsters to remain vigilant against predators and competitors (Bologna and Steneck, 1993). However, it is unlikely that such preferences translate into an increase in biomass regionally; rather, they represent a local concentrating factor (Miller, 1985; Bologna and Steneck, 1993).

Consensus has developed that lobsters play a trophic role in kelp bed communities as one of a suite of predators [e.g., Atlantic wolffish (Hagen and Mann, 1992)] that feed on sea urchins, but not as a keystone predator (Elner and Vadas, 1990). However, functional relationships among lobsters, sea urchins, and kelp are still incompletely resolved and are certain to be subjected to further scrutiny.

VIII. Directions for Further Research

Elner and Campbell's (1991) admonition provides a useful springboard for further research: "Research thrusts seem to have changed opportunistically with ecological fads (Abrahamson *et al.*, 1989) rather than doggedly addressing fundamental questions until they were solved." The research program was parochial through to the 1970s and focused in various geographic locations on vital rates (growth, reproductive schedules, adult movement, etc.). It then divided further as specific ecological issues were addressed; however, this helped reinforce the close linkages that exist between lobster life history and environmental forcing factors. The program is now coalescing, as regionally focused research alliances address topics such as postlarval delivery to benthic recruitment sites, intercalibration of population sampling approaches, and recruitment prediction, as well as renewed research in the offshore environment. Nonetheless, several specific questions demand further attention.

Within the early benthic life history phases, information is lacking on settlement cues in the field and how these affect postlarval supply to settlement sites. Additionally, while postlarvae are known to exploit marginal habitats, their survival rates on such habitats are as yet unclear. A major avenue of research lies in the clearer definition of how density-dependent processes (e.g., predation and intra- and interspecific competition) act on the recently settled juveniles to further mold their demographics. Attention should also be focused on natural diet and foraging activity following settlement, particularly during the shelter-restricted and emergent juvenile phases of life history, with the eventual goal of resolving when dietary shifts occur and active out-of-shelter foraging begins.

Now that shelter-restricted, emergent, and vagile juvenile phases can be readily censused and are accessible for experimental manipulation, correlations of the numbers of lobsters surviving recruitment into the benthic environment with subsequent year classes and, eventually, with recruits into the fishery should be forthcoming. However, recruitment prediction must be founded on a firm grasp of lobster production dynamics. Thus, current debates on the carrying capacity of inshore lobster habitats have resurrected questions on the degree to which lobsters may be food-limited during their juvenile phases, or rather the degree to which there is scope for enhanced production of lobsters during favorable environmental conditions (Elner and Campbell, 1991).

Within the latter life history phases (adolescent and adult), there is a need to investigate further the effects of size at maturity on behavior, perhaps by comparing behavioral attributes of similar-sized lobsters from areas with different sizes at maturity. However, until we have better observations of behavioral interactions between individual lobsters in the field, we will not know how agonism and dominance, so common in laboratory studies, affect lobsters in natural populations. The apparent preharvest decline of maturing lobsters from high-density coastal habitats, hypothesized by Steneck (1989; R. S. Steneck, personal communiction) to be due to increased agonism, is a priority area for field corroboration.

Spatially explicit models of lobster movement should be amenable to integration with emerging landscape ecology theory (Wiens *et al.*, 1993) and harvesting models that view the lobster population (on a regional basis) as a metapopulation of interconnected subpopulations (as advocated by Fogarty, in Chapter 6). Full realization of these new opportunities for modeling lobster population dynamics requires further documentation of seasonal and interannual patterns of movement, for which traditional mark–recapture studies seem to be the only realistic approach. A priority area for technical development is a marking technique that would permit the multiple recapture and release of males and nonovigerous females. Unfortunately, technical considerations and cost implications of such a program, involving as it would the return of marketable lobsters to the wild by fishers, could prove difficult to surmount. Success will be predicated on establishing a rapport with and commitment from the fishing community and will only be feasible in specific localities, as has been the experience to date.

There are few long-term *in situ* studies of ecological interactions within well-defined seasonal assemblages of lobsters (Karnofsky *et al.*, 1989a,b; Ennis, 1984a,b). Advances in remote-sensing techniques, such as ultrasonic transmitters and *in situ* video monitoring, are providing novel approaches to capturing the social behavior, foraging, and reproductive ecology of mature lobsters in the wild. Additionally, underwater research conducted from manned submersibles and remotely operated vehicles has now made the transition from a primarily qualitative, descriptive field to quantitative science with calibrated sampling and sensing hardware and temporal and spatial replication (e.g., Auster *et al.*, 1991). With these new tools, more thorough descriptions of short-term movements and shelter residency patterns of mature lobsters can be obtained. Unfortunately, these approaches provide only high-density information on the activities of individuals, while missing the social context of such activities. Thus, there will be a continuing need for *in situ* observational studies on social interactions within free-ranging adult populations, similar to those undertaken by Karnofsky *et al.* (1989a,b) in a predominantly adolescent population.

Certain environments, such as estuaries (Jury *et al.*, 1994b) have been postulated to be "marginal" habitats that subdominant, or injured, lobsters can nonetheless exploit to advantage. These observations, together with the recent documentation of seasonal aggregations of large mature lobsters (primarily berried lobsters) in certain shallow coastal habitats (Campbell, 1990; Lawton and Robichaud, 1992) pose new scenarios under which traditional concepts of aggression and dominance in lobsters should be tested.

Individually focused research by lobster ecologists will remain crucial to the health of the general research program; however, the problems that require resolution demand multidisciplinary teams working across fishery management and political boundaries. While collaboration with geological and physical oceanographers has long been appreciated (e.g., Squires, 1970), the working relationship is now more intimate as sophisticated remote-sensing and modeling tools from these disciplines are brought to bear in the definition of lobster production systems (Steneck, 1989; Hudon, 1994). Lobster ecologists are utilizing the latest marine technological tools (remotely operated vehicles, submersibles, telemetry, etc.) as they study lobster ecology in the natural environment. With the caveat that the impacts these tools may have on lobster behavior must first be determined (e.g., Spanier *et al.*, 1994), they promise to extend our "window of opportunity" to capture the natural social behavior of free-ranging lobsters. In resolving the community role of lobsters in nearshore ecosystems, further collaborations with benthic and fish ecologists are indicated.

IX. Summary

A new life history scheme is presented for the American lobster (*Homarus americanus*) consisting of seven phases, in which the first three larval developmental stages are considered one (fully) pelagic phase. The postlarva (developmental stage IV) effects a transition from the pelagic to the benthic realm. The period from benthic settlement through adolescence spans a period of several years. While several cohorts may overlap in habitat use, different behaviors and physiology result in four distinct life history phases prior to functional maturity: shelter-restricted, emergent, and vagile phase juveniles and the adolescent phase, which is marked by the onset of physiological maturity. The final, adult phase of the life history is potentially the most prolonged; however, under contemporary exploitation, few lobsters reach the large body size and estimated longevity noted in the historical record.

The postlarva is well designed to effect the drastic change in habit from a pelagic to a benthic lifestyle. It is a strong swimmer that can detect thermal gradients—behaviors that allow it to dive to the bottom to seek suitable settlement sites. It has, at least in the laboratory, clear preferences for habitats that will provide it with the greatest protection against predators and the ability to use chemical cues to assess habitat quality and complexity. These preferences and the protective quality of the habitats are reflected in the areas where shelter-restricted phase juveniles are routinely found in the field: rocks on sand, cobble, and peat reefs. Immediate postsettlement behavior is cryptic, with the lobster remaining shelter-bound to avoid predators and exploiting shelter-based food supplies to avoid risk-prone out-of-shelter foraging bouts. Much of this sheltering behavior may be due to physiological constraints faced by the recently settled lobster, such as growth rate after settlement and allometric growth patterns of the claws and abdomen.

As the shelter-restricted phase juvenile grows, it must balance its dietary requirements from a shelter-based food supply with the increased growth rate possible from foraging outside of its shelter. At some point, dietary requirements will demand that the lobster begin short foraging bouts in the immediate area of its shelter; it then enters the emergent phase. As they grow further, the juveniles begin to outgrow their predators, become capable of protecting themselves via aggressive displays and claw weaponry, and can become more vagile. Physiological maturity may then be the next significant point in the lobster's life and may result in hormonal changes that drastically alter their behavior. Adolescent lobsters behave more like adult lobsters, shifting from shelter to shelter, displaying more aggression, exhibiting seasonal shifts in depth distribution, and participating in migratory movements. However, there is no evidence that these adolescent lobsters participate in mating behavior; thus, they are not functionally mature. Size at sexual maturity varies markedly over the geographical range of the lobster, yet is not comprehensively documented in some regions nor well linked to size at functional maturity. As size at functional maturity has important ramifications for the sustainability of local populations, and thus for management of the fishery, it requires further attention.

Geological history, present coastal physiography, seasonal thermal regimes, and possibly agonism interact to mold adult lobster movement patterns. Under this synthesis, both long-distance return migration and offshore diffusion are logical expressions of adult lobster movement. Shelter is less important for adult lobsters than for juvenile lobsters, except during periods of molting and mating. Then, agonism may play a significant role in determining sex ratios in seasonal assemblages and in determining spacing of individuals within a population. Movement patterns of ovigerous females appear to be related to the need to maximize annual thermal regimes and may result in an intermixing of subpopulations at the time of hatching.

Lobsters are conspicuous members of the benthic megafauna in coastal northwest North Atlantic marine systems in their later life history phases; however, their community role and interactions with various benthic fishes and crabs remain incompletely resolved. Rather than keystone predators in these systems, lobsters are members of a suite of large, mobile predators that likely exert significant influence on benthic community structure and function.

References

Able, K. W., Heck, K. L., Fahay, M. P., and Roman, C. T. (1988). Use of salt-marsh peat reefs by small juvenile lobsters on Cape Cod, Massachusetts. *Estuaries* 11, 83–86.

Abrahamson, W. G., Whitham, T. G., and Price, P. W. (1989). Fads in ecology. *BioScience* 39, 321–325.

Addison, J., and Fogarty, M. (1992). Juvenile lobster habitat limitation: What can landings tell us? *Lobster Newsl.*, 5(2), 10–12.

Aiken, D. E., and Waddy, S. L. (1980). Reproductive biology. *In* "The Biology and Management of Lobsters" (J. S. Cobb and B. F. Phillips, eds.), Vol. 1, pp. 215–276. Academic Press, New York.

Aiken, D. E., and Waddy, S. L. (1982). Cement gland development, ovary maturation, and reproductive cycles in the American lobster *Homarus americanus*. *J. Crust. Biol.* 2, 315–327.

Aiken, D. E., and Waddy, S. L. (1986). Environmental influences on recruitment of American lobsters (*Homarus americanus*): A perspective. *Can. J. Fish. Aquat. Sci.* 43, 2258–2270.

Atema, J. (1986). Review of sexual selection and chemical communication in the lobster, *Homarus americanus*. *Can. J. Fish. Aquat. Sci.* 43, 2283–2290.

Atema, J., and Cobb, J. S. (1980). Social behavior. *In* "The Biology and Management of Lobsters" (J. S. Cobb and B. F. Phillips, eds.), Vol. 1, pp. 409–450. Academic Press, New York.

Atema, J., Jacobson, S., Karnofsky, E., Oleszko-Szuts, S., and Stein, L. (1979). Pair formation in the lobster, *Homarus americanus*: Behavioral development, pheromones and mating. *Mar. Behav. Physiol.* 6, 277–296.

Atema, J., Leavitt, D. F., Barshaw, D. E., and Cuomo, M. C. (1982). Effects of drilling muds on the behavior of the American lobster, *Homarus americanus*, in water column and substrate exposures. *Can. J. Fish. Aquat. Sci.* 39, 675–690.

Atkinson, R. J. A., and Taylor, A. C. (1988). Physiological ecology of burrowing decapods. *Symp. Zool. Soc. London* 59, 201–226.

Auster, P. J. (1985). Factors affecting catch of American lobster, *Homarus americanus*, in baited traps. *Tech. Bull. Ser., Univ. Conn. Sea Grant Program* CT-SG-85-1, 1–46.

Auster, P. J., Malatesta, R. J., LaRosa, S. C., Cooper, R. A., and Stewart, L. L. (1991). Microhabitat utilization by the megafaunal assemblage at a low relief outer continental shelf site—Middle Atlantic Bight, USA. *J. Northw. Atl. Fish. Sci.* 11, 59–69.

Babb, I. G., Auster, P. J., Belt, W., Olivier, D., Langton, R., MacDonald, I. R., Popenoe, T., and Steneck, R. S. (1993). Dual use of a nuclear powered research submersible—the US-Navy's NR-1. *Mar. Tech. Soc. J.* 27(4), 39–48.

Bannister, R. C. A. (1993). Session 6: Restocking. *Lobster Newsl.*, 6(2), 6–7.

Bannister, R. C. A., Thompson, B. M., Addison, J. T., Lovewell, S. R. J., and Lawler, A. R. (1991). The 1990 Results from a Lobster Stock Enhancement Experiment on the East Coast of England. *Int. Counc. Explor. Sea* C. M. 1991/K:33.

Barshaw, D. E. (1988). "The Substrate-Related Behavior and Predator–Prey Interactions of the Early Juvenile Lobster, *Homarus americanus*," Ph.D. dissertation. Boston University, Boston, Massachusetts.

Barshaw, D. E. (1989). Growth and survival of post-larval lobsters, *Homarus americanus*, on a diet of plankton. *Fish. Bull.* 87, 366–370.

Barshaw, D. E., and Able, K. (1990). Tethering as a technique for assessing predation rates in different habitats: An evaluation using juvenile lobsters *Homarus americanus*. *Fish. Bull.* 88, 415–417.

Barshaw, D. E., and Bryant-Rich, D. R. (1988). A long-term study on the behavior and survival of early juvenile American lobster, *Homarus americanus*, in three naturalistic substrates: Eelgrass, mud, and rocks. *Fish. Bull.* 86, 789–796.

Barshaw, D. E., and Lavalli, K. L. (1988). Predation upon postlarval lobsters *Homarus americanus* by cunners *Tautogolabrus adspersus* and mud crabs *Neopanope sayi* on three different substrates: Eelgrass, mud, and rocks. *Mar. Ecol. Prog. Ser.* 48, 119–123.

Barshaw, D. E., Able, K., and Heck, K. L., Jr. (1994). Salt marsh peat reefs as protection for postlarval lobsters *Homarus americanus* from fish and crab predators: Comparisons with other substrates. *Mar. Ecol. Prog. Ser.* 106, 203–206.

Bergeron, J. (1967). Contribution à la biologie du homard (*Homarus americanus* M.-Edw.) des Îles-de-la-Madeleine. *Nat. Can. (Que.)* 94, 169–207.

Bernstein, B. B., and Campbell, A. (1983). Contribution to the Development of Methodology for Sampling and Tagging Small Juvenile Lobsters (*Homarus americanus*). *Can. Manuscr. Rep. Fish. Aquat. Sci.* 1741.

Berrill, M. (1974). The burrowing behavior of newly-settled lobsters, *Homarus vulgaris* (Crustacea–Decapoda). *J. Mar. Biol. Assoc. U.K.* 54, 797–801.

Berrill, M., and Stewart, R. (1973). Tunnel-digging in mud by newly-settled American lobsters, *Homarus americanus*. *J. Fish. Res. Board Can.* 30, 285–287.

Bigelow, H. B., and Schroeder, W. C. (1953). Fishes of the Gulf of Maine. *Fish. Bull.* 53, 1–577.

Bologna, P. A. X., and Steneck, R. S. (1993). Kelp beds as habitat for American lobster *Homarus americanus*. *Mar. Ecol. Prog. Ser.* 100, 127–134.

Boothroyd, F. A., and Ennis, G. P. (1992). Reproduction in American lobsters *Homarus americanus* transplanted northward to St. Michael's Bay, Labrador. *Fish. Bull.* 90, 659–667.

Bordner, C. E., and Conklin, D. E. (1981). Food consumption and growth of juvenile lobsters. *Aquaculture* 24, 285–300.

Botero, L., and Atema, J. (1982). Behavior and substrate selection during larval settling in the lobster, *Homarus americanus*. *J. Crust. Biol.* 2, 59–69.

Boudreau, B., Bourget, E., and Simard, Y. (1990). Benthic invertebrate larval response to substrate characteristics at settlement: Shelter preferences of the American lobster *Homarus americanus*. *Mar. Biol.* 106, 191–198.

Boudreau, B., Simard, Y., and Bourget, E. (1992). Influence of a thermocline on vertical distribution and settlement of post-larvae of the American lobster *Homarus americanus* Milne-Edwards. *J. Exp. Mar. Biol. Ecol.* 162, 35–49.

Boudreau, B., Bourget, E., and Simard, Y. (1993a). Effect of age, injury, and predator odors on settlement and shelter selection by lobster *Homarus americanus* postlarvae. *Mar. Ecol. Prog. Ser.* 93, 119–129.

Boudreau, B., Bourget, E., and Simard, Y. (1993b). Behavioural responses of competent lobster postlarvae to odor plumes. *Mar. Biol.* 117, 63–69.

Boulding, E. G. (1984). Crab-resistant features of shells of burrowing bivalves: Decreasing vulnerability by increasing handling time. *J. Exp. Mar. Biol. Ecol.* 76, 201–233.

Briggs, P. T., and Mushacke, F. M. (1979). The American lobster in western Long Island Sound. *N.Y. Fish Game J.* 26, 59–86.

Briggs, P. T., and Mushacke, F. M. (1980). The American lobster and the pot fishery in the inshore waters off the south shore of Long Island, New York. *N.Y. Fish Game J.* 27, 156–178.

Briggs, P. T., and Zawacki, C. S. (1974). American lobsters at artificial reefs in New York. *N.Y. Fish Game J.* 21, 73–77.

Bryden, C. G. (1973). "The Use of Shellfish Feces as a Food Source for Juvenile Lobsters in Aquaculture," M.A. thesis. Boston University, Boston, Massachusetts.

Bumpus, H. C. (1901). On the movements of certain lobsters liberated at Woods Hole. *Bull. U.S. Fish Comm.* 21, 225–230.

Burton, C. A. (1992). "Techniques of Lobster Stock Enhancement." Marine Farming Unit, Sea Fish Industry Authority, Ardtoe, Argyle, Scotland.

Caddy, J. F. (1986). Modelling stock-recruitment processes in Crustacea: Some practical and theoretical perspectives. *Can. J. Fish. Aquat. Sci.* 43, 2330–2344.

Caddy, J. F., and Stamatopoulos, C. (1990). Mapping growth and mortality rates of crevice-dwelling organisms onto a perforated surface: The relevance of 'cover' to the carrying capacity of natural and artificial habitats. *Estuarine, Coastal Shelf Sci.* 31, 87–106.

Campbell, A. (1983). Growth of tagged American lobsters, *Homarus americanus*, in the Bay of Fundy. *Can. J. Fish. Aquat. Sci.* 40, 1667–1675.

Campbell, A. (1986). Migratory movements of ovigerous lobsters, *Homarus americanus*, tagged off Grand Manan, eastern Canada. *Can. J. Fish. Aquat. Sci.* 43, 2197–2205.

Campbell, A. (1990). Aggregations of berried lobsters (*Homarus americanus*) in shallow waters off Grand Manan, eastern Canada. *Can. J. Fish. Aquat. Sci.* 47, 520–523.

Campbell, A. (1991). Juvenile *Homarus americanus* studies from McNutt Island, Nova Scotia. *J. Shellfish Res.* 10, 283–284. (abstract).

Campbell, A. (1992). Characteristics of the American lobster fishery of Grand Manan, New Brunswick, Canada. *North Am. J. Fish. Manage.* 12, 139–150.

Campbell, A., and Stasko, A. B. (1985). Movements of tagged American lobsters, *Homarus americanus*, off southwestern Nova Scotia. *Can. J. Fish. Aquat. Sci.* 42, 229–238.

Campbell, A., and Stasko, A. B. (1986). Movements of lobsters (*Homarus americanus*) tagged in the Bay of Fundy, Canada. *Mar. Biol.* 92, 393–404.

Carter, J. A., and Steele, D. H. (1982a). Attraction to and selection of prey by immature lobsters (*Homarus americanus*). *Can. J. Zool.* 60, 326–336.

Carter, J. A., and Steele, D. H. (1982b). Stomach contents of immature lobsters (*Homarus americanus*) from Placentia Bay, Newfoundland. *Can. J. Zool.* 60, 337–347.

Chapman, A. R. O. (1981). Stability of sea-urchin dominated barren grounds following destructive grazing of kelp in St. Margaret's Bay, eastern Canada. *Mar. Biol.* 62, 307–311.

Charmantier, G. (1989). Terminological tribulations. *Lobster Newsl.* 2(1), 5.

Charmantier, G., Charmantier-Daures, M., Bouaricha, N., Thuet, P., Aiken, D. E., and Trilles, J.-P. (1988). Ontogeny of osmoregulation and salinity tolerance in two decapod crustaceans: *Homarus americanus* and *Penaeus japonicus*. *Biol. Bull. (Woods Hole, Mass.)* 175, 102–110.

Charmantier, G., Charmantier-Daures, M., and Aiken, D. E. (1991). Metamorphosis in the lobster *Homarus* (Decapoda): A review. *J. Crust. Biol.* 11, 481–495.

Cobb, J. S. (1970). Effect of solitude on time between fourth and fifth larval molts in the American lobster (*Homarus americanus*). *J. Fish. Res. Board Can.* 27, 1653–1655.

Cobb, J. S. (1971). The shelter related behavior of the lobster, *Homarus americanus*. *Ecology* 52, 108–115.

Cobb, J. S. (1988). When is a larva a post larva? *Lobster Newsl.* 1(1), 8.

Cobb, J. S., and Tamm, G. R. (1975). Dominance status and molt order in lobsters (*Homarus americanus*). *Mar. Behav. Physiol.* 3, 119–124.

Cobb, J. S., and Wahle, R. A. (1994). Early life history and recruitment processes of clawed lobsters. *Crustaceana* 67, 1–25.

Cobb, J. S., and Wang, D. (1985). Fisheries biology of lobsters and crayfishes. *In* "The Biology of Crustacea" (D. E. Bliss, ed.), Vol. 10 of "Economic Aspects: Fisheries and Culture" (A. J. Provenzano, ed.), pp. 167–247. Academic Press, Orlando, Florida.

Cobb, J. S., Tamm, G. R., and Wang, D. (1982). Behavioral mechanisms influencing molt frequency in the American lobster, *Homarus americanus*. *J. Exp. Mar. Biol. Ecol.* 12, 185–200.

Cobb, J. S., Gulbransen, T., Phillips, B. F., Wang, D., and Syslo, M. (1983). Behavior and distribution of larval and early juvenile *Homarus americanus*. *Can. J. Fish. Aquat. Sci.* 40, 2184–2188.

Cobb, J. S., Wang, D., Richards, R. A., and Fogarty, M. J. (1986). Competition among lobsters and crabs and its possible effects in Narragansett Bay, Rhode Island. *Can. Spec. Publ. Fish. Aquat. Sci.* 92, 282–290.

Cobb, J. S., Wang, D., and Campbell, D. B. (1989). Timing of settlement by postlarval lobsters (*Homarus americanus*): Field and laboratory evidence. *J. Crust. Biol.* 9, 60–66.

Conan, G. Y. (1986). Summary of Session 5: Recruitment enhancement. *Can. J. Fish. Aquat. Sci.* 43, 2384–2388.

Conan, G. Y., Maynard, D. R., Merrien, A., and Guennegan, Y. (1984). Monitoring the Foraging Activity of Lobsters and Associated Species by Underwater Television. *Int. Counc. Explor. Sea* C.M. 1984/K:45.

Cooper, R. A. (1970). Retention of marks and their effects on growth, behavior, and migrations of the American lobster, *Homarus americanus*. *Trans. Am. Fish. Soc.* 99, 409–417.

Cooper, R. A., and Uzmann, J. R. (1971). Migrations and growth of deep-sea lobsters, *Homarus americanus*. *Science* 171, 288–290.

Cooper, R. A., and Uzmann, J. R. (1977). Ecology of juvenile and adult clawed lobsters, *Homarus americanus*, *Homarus gammarus*, and *Nephrops norvegicus*. *Div. Fish. Oceanogr. Circ. (Aust., C.S.I.R.O.)* 7, 187–208.

Cooper, R. A., and Uzmann, J. R. (1980). Ecology of juvenile and adult *Homarus*. *In* "The Biology and Management of Lobsters" (J. S. Cobb and B. F. Phillips, eds.), Vol. 2, pp. 97–142. Academic Press, New York.

Cooper, R. A., Clifford, R. A., and Newell, C. D. (1975). Seasonal abundance of the American lobster, *Homarus americanus*, in the Boothbay region of Maine. *Trans. Am. Fish. Soc.* 104, 669–674.

Cooper, R. A., Shepard, A., Valentine, P., Uzmann, J. R., and Hurlbert, A. (1987a). Pre and post drilling benchmarks and monitoring data of ocean floor fauna, habitats, and contaminant loads on Georges Bank and its submarine canyons. *NOAA Symp. Ser. Undersea Res.* 2(2), 17–48.

Cooper, R. A., Valentine, P., Uzmann, J. R., and Slater, R. A. (1987b). Submarine canyons. *In* "Georges Bank" (R. H. Backus, ed.), pp. 52–63. MIT Press, Cambridge, Massachusetts.

Costello, W. J., and Lang, F. (1979). Development of the dimorphic claw closer muscles of the lobster *Homarus americanus*. IV. Changes in functional morphology during growth. *Biol. Bull. (Woods Hole, Mass.)* 156, 179–195.

Cowan, D. F. (1992). "Pheromones, Molting, and Courtship Behavior in the American Lobster *Homarus americanus*," Ph.D dissertation. Boston University, Boston, Massachusetts.

Cowan, D. F., and Atema, J. (1990). Moult staggering and serial monogamy in American lobsters, *Homarus americanus*. *Anim. Behav.* 39, 1199–1206.

Crnkovic, D. (1968). Some observations regarding the burrows of juvenile *Nephrops norvegicus* (L.). *Rapp. Comm. Int. Mer Medit.* 19, 171–172.

Curry, J. D. (1988). Shell form and strength. *In* "The Mollusca" (E. R. Trueman and M. R. Clarke, eds.), Vol. 11, pp. 183–210. Academic Press, San Diego.

D'Agostino, A. (1980). Growth and color of juvenile lobsters (*Homarus americanus*) kept on diets of natural and artificial foodstuff. *Maine Sea Grant Publ. Tech. Rep.* 58, 41–48.

Dall, W. (1970). Osmoregulation in the lobster *Homarus americanus*. *J. Fish. Res. Board Can.* 27, 1123–1130.

Daniel, P. C., and Bayer, R. C. (1987a). Development of chemically mediated prey-search response in postlarval lobster (*Homarus americanus*) through feeding experience. *J. Chem. Ecol.* 13, 1217–1233.

Daniel, P. C., and Bayer, R. C. (1987b). Attraction of predatorily naive postlarval lobsters to extracts of metabolites of common prey: *Mytilus edulis*, *Mya arenaria*, *Cancer irroratus*, and *Asterias vulgaris*. *J. Chem. Ecol.* 13, 1201–1215.

Daniel, P. C., Bayer, R. C., and Chapman, S. (1985). Barnacle larvae (*Balanus* spp.) as a potential diet for juvenile lobsters (*Homarus americanus*). *Aquaculture* 46, 67–70.

de Lafontaine, Y., Lambert, T., Lilly, G. R., McKone, W. D., and Miller, R. J., eds. (1992). Juvenile Stages: The Missing Link in Fisheries Research. Report of a Workshop. *Can. Tech. Rep. Fish. Aquat. Sci.* 1890.

Delistraty, D. A., Carlberg, J. M., Van Olst, J. C., and Ford, R. F. (1977). Ammonia toxicity in cultured larvae of the American lobster (*Homarus americanus*). *Proc. Annu. Meet.—World Maric. Soc.* 8, 647–672.

Derby, C. D., and Atema, J. (1981). Selective improvement in responses to prey odors by the lobster, *Homarus americanus*, following feeding experience. *J. Chem. Ecol.* 7, 1073–1080.

Dow, R. L. (1974). American lobsters tagged by Maine commercial fishermen, 1957–59. *Fish. Bull.* 72, 622–623.

Duggan, D. R., and Pezzack, D. S. (1988). Movement of Offshore Lobster (*Homarus americanus*) Displaced to Coastal Areas of Nova Scotia. *Can. Atl. Fish. Sci. Advis. Comm. Res. Doc.* 88/67.

Dunham, P. J., and Skinner-Jacobs, P. (1978). Intermolt mating in the lobster *Homarus americanus*. *Mar. Behav. Physiol.* 5, 209–214.

Dybern, B. I. (1973). Lobster burrows in Swedish waters. *Helgol. Wiss. Meeresunters.* 24, 401–414.

Elner, R. W. (1980). Predation on the sea urchin (*Strongylocentrotus droebachiensis*) by the American lobster (*Homarus americanus*) and the rock crab (*Cancer irroratus*). *Can. Tech. Rep. Fish. Aquat. Sci.* 954, 48–65.

Elner, R. W., and Campbell, A. (1981). Force, function and mechanical advantage in the chelae of the American lobster *Homarus americanus* (Decapoda: Crustacea). *J. Zool.* 193, 269–286.

Elner, R. W., and Campbell, A. (1987). Natural diets of lobster *Homarus americanus* from barren ground and macroalgal habitats off southwestern Nova Scotia, Canada. *Mar. Ecol. Prog. Ser.* 37, 131–140.

Elner, R. W., and Campbell, A. (1991). Spatial and temporal patterns in recruitment for American lobster, *Homarus americanus*, in the northwestern Atlantic. *Mem. Queensl. Mus.* 31, 349–363.

Elner, R. W., and Hamet, S. L. (1984). The Effects of Ocean Dumping of Dredge Spoils onto Juvenile Lobster Habitat: A Field Evaluation. *Can. Tech. Rep. Fish. Aquat. Sci.* 1247.

Elner, R. W., and Jamieson, G. S. (1979). Predation of sea scallops, *Placopecten magellanicus*, by the rock crab, *Cancer irroratus*, and the American lobster, *Homarus americanus*. *J. Fish. Res. Board Can.* 36, 537–543.

Elner, R. W., and Lavoie, R. E. (1983). Predation on American oysters (*Crassostrea virginica*, Gmelin) by American lobsters (*Homarus americanus*, Milne-Edwards), rock crabs (*Cancer irroratus*, Say) and mud crabs (*Neopanope sayi*, Smith). *J. Shellfish Res.* 3, 129–134.

Elner, R. W., and Vadas, R. L. (1990). Inference in ecology: The sea urchin phenomenon in the north-west Atlantic. *Am. Nat.* 136, 108–125.

Emmel, V. E. (1908). The problem of feeding in artificial lobster culture. *Annu. Rep.—R. I. Comm. Inland Fish.* 38, 99–114.

Ennis, G. P. (1973). Food, feeding, and condition of lobsters, *Homarus americanus*, throughout the seasonal cycle in Bonavista Bay, Newfoundland. *J. Fish. Res. Board Can.* 30, 1905–1909.

Ennis, G. P. (1980). Size–maturity relationships and related observations in Newfoundland populations of the lobster (*Homarus americanus*). *Can. J. Fish. Aquat. Sci.* 37, 945–956.

Ennis, G. P. (1984a). Small-scale seasonal movements of the American lobster *Homarus americanus*. *Trans. Am. Fish. Soc.* 113, 336–338.

Ennis, G. P. (1984b). Territorial behavior of the American lobster *Homarus americanus*. *Trans. Am. Fish. Soc.* 113, 330–335.

Ennis, G. P. (1984c). Comparison of physiological and functional size–maturity relationships in two Newfoundland populations of lobsters *Homarus americanus*. *Fish. Bull.* 82, 244–249.

Ennis, G. P. (1984d). Tag–recapture validation of molt and egg extrusion predictions based upon pleopod examination in the American lobster, *Homarus americanus*. *Fish. Bull.* 82, 242–244.

Ennis, G. P. (1986). Stock definition, recruitment variability, and larval recruitment processes in the American lobster, *Homarus americanus*: A review. *Can. J. Fish. Aquat. Sci.* 43, 2072–2084.

Evans, P. D., and Mann, K. H. (1977). Selection of prey by American lobsters (*Homarus americanus*) when offered a choice between sea urchins and crabs. *J. Fish. Res. Board Can.* 34, 2203–2207.

Factor, J. R. (1989). Terminological tribulations. *Lobster Newsl.* 2(1), 5.

Fogarty, M. J., and Idoine, J. S. (1986). Recruitment dynamics in an American lobster (*Homarus americanus*) population. *Can. J. Fish. Aquat. Sci.* 43, 2368–2376.

Fogarty, M. J., Borden, D. V. D., and Russell, H. J. (1980). Movements of tagged American lobsters, *Homarus americanus*, off Rhode Island. *Fish. Bull.* 78, 771–780.

Good, L. K., Bayer, R. C., Gallagher, M. L., and Rittenburg, J. H. (1982). Amphipods as a potential diet for juveniles of the American lobster *Homarus americanus* (Milne Edwards). *J. Shellfish Res.* 2, 183–187.

Govind, C. K., and Lang, F. (1978). Development of the dimorphic claw closer muscle of the lobster *Homarus americanus*. III. Transformation to dimorphic muscles in juveniles. *Biol. Bull. (Woods Hole, Mass.)* 154, 55–67.

Govind, C. K., and Pearce, J. (1989). Critical period for determining claw asymmetry in developing lobsters. *J. Exp. Zool.* 249, 31–35.

Griffiths, C. L., and Seiderer, J. L. (1980). Rock-lobsters and mussels — limitations and preferences in a predator–prey interaction. *J. Exp. Mar. Biol. Ecol.* 44, 95–109.

Grimes, C. B., Able, K. W., Jones, R. S., Twichell, D. C., and Turner, S. C. (1987). Studies on tilefish fishery biology, ecology, and bioerosion on the middle Atlantic and southern New England continental shelf. *NOAA Symp. Ser. Undersea Res.* 2(2), 49–69.

Gunn, L. A. (1987). "The Feeding Ecology of American Lobster Larvae in Long Island Sound," M.S. thesis. Southern Connecticut State University, New Haven.

Hadley, P. B. (1908). The behavior of the larval and adolescent stages of the American lobster (*Homarus americanus*). *J. Comp. Neurol. Psychol.* 18, 199–301.

Hagen, N. T., and Mann, K. H. (1992). Functional response of the predators American lobster *Homarus americanus* (Milne-Edwards) and Atlantic wolffish *Anarhichas lupus* (L.) to increasing numbers of the green sea urchin *Strongylocentrotus droebachiensis* (Müller). *J. Exp. Mar. Biol. Ecol.* 159, 89–112.

Hagen, N. T., and Mann, K. H. (1994). Experimental analysis of factors influencing the aggregating behavior of the green sea urchin *Strongylocentrotus droebachiensis* (Müller). *J. Exp. Mar. Biol. Ecol.* 176, 107–126.

Harding, G. C. (1992). American Lobster (*Homarus americanus* Milne Edwards): A Discussion Paper on Their Environmental Requirements and the Known Anthropogenic Effects on Their Populations. *Can. Tech. Rep. Fish. Aquat. Sci.* 1887.

Harding, G. C., and Trites, R. W. (1988). Dispersal of *Homarus americanus* larvae in the Gulf of Maine from Browns Bank. *Can. J. Fish. Aquat. Sci.* 45, 416–425.

Harding, G. C., and Trites, R. W. (1989). A further elaboration on "Dispersal of *Homarus americanus* larvae in the Gulf of Maine from Browns Bank" in response to comments by D.S. Pezzack. *Can. J. Fish. Aquat. Sci.* 46, 1078–1082.

Harding, G. C., Drinkwater, K. F., and Vass, W. P. (1983). Factors influencing the size of American lobster (*Homarus americanus*) stocks along the Atlantic coast of Nova Scotia, Gulf of St. Lawrence, and Gulf of Maine: A new synthesis. *Can. J. Fish. Aquat. Sci.* 40, 168–184.

Harding, G. C., Kenchington, E., and Zeng, Z. (1993). Morphometrics of American lobster (*Homarus americanus*) larvae in relation to stock determination in the Maritimes, Canada. *Can. J. Fish. Aquat. Sci.* 50, 43–52.

Heck, K. L., Jr., Able, K. W., Fahay, M. P., and Roman, C. T. (1989). Fishes and decapod crustaceans of Cape Cod eelgrass meadows: Species composition, seasonal abundance patterns and comparisons with unvegetated substrates. *Estuaries* 12, 59–65.

Herrick, F. H. (1895). The American Lobster: A Study of Its Habits and Development. *Bull. U.S. Fish Comm.* 15, 1–252 + 54 plates.

Herrick, F. H. (1909). Natural history of the American lobster. *Bull. U.S. Bur. Fish.* 29, 149–408 + 20 plates.

Herrnkind, W. F. (1980). Spiny lobsters: Patterns of movement. *In* "The Biology and Management of Lobsters" (J. S. Cobb and B. F. Phillips, eds.), Vol. 1, pp. 349–407. Academic Press, New York.

Herrnkind, W. F. (1983). Movement patterns and orientation. In "The Biology of Crustacea" (D. E. Bliss, ed.), Vol. 7 of "Behavior and Ecology" (F. J. Vernberg and W. B. Vernberg, eds.), pp. 41–105. Academic Press, Orlando, Florida.

Howard, A. E. (1980). Substrate controls on the size composition of lobster (*Homarus gammarus*) populations. *J. Cons., Cons. Int. Explor. Mer* 39, 130–133.

Howard, A. E., and Bennett, D. B. (1979). The substrate preference and burrowing behavior of juvenile lobsters (*Homarus gammarus* (L.)). *J. Nat. Hist.* 13, 433–438.

Howard, A. E., and Nunny, R. S. (1983). Effects of near-bed current speeds on the distribution and behavior of the lobster, *Homarus gammarus* (L.). *J. Exp. Mar. Biol. Ecol.* 71, 27–42.

Howell, W. H., and Watson, W. H. (1991). Sex ratio differences between estuarine and coastal lobster populations. *J. Shellfish Res.* 10, 285.

Huber, R., and Kravitz, E. A. (1995). A quantitative analysis of agonistic behavior in juvenile American lobsters (*Homarus americanus* L.). *Brain Behav. Evol.* (in press).

Hudon, C. (1987). Ecology and growth of postlarval and juvenile lobster, *Homarus americanus*, off Îles de la Madeleine (Québec). *Can. J. Fish. Aquat. Sci.* 44, 1855–1869.

Hudon, C. (1994). Large-scale analysis of Atlantic Nova Scotia American lobster (*Homarus americanus*) landings with respect to habitat, temperature, and wind conditions. *Can. J. Fish. Aquat. Sci.* 51, 1308–1321.

Hudon, C., and Fradette, P. (1993). Wind-induced advection of larval decapods into Baie de Plaisance (Îles de la Madeleine, Québec). *Can. J. Fish. Aquat. Sci.* 50, 1422–1434.

Hudon, C., and Lamarche, G. (1989). Niche segregation between American lobster *Homarus americanus* and rock crab *Cancer irroratus*. *Mar. Ecol. Prog. Ser.* 52, 155–168.

Hudon, C., Fradette, P., and Legendre, P. (1986). La répartition horizontale et verticale des larves de homard (*Homarus americanus*) autour des îles de la Madeleine, golfe du Saint-Laurent. *Can. J. Fish. Aquat. Sci.* 43, 2164–2176.

Hughes, R. N. (1980). Optimal foraging in the marine context. *Oceanogr. Mar. Biol.* 18, 423–481.

Huntsman, A. G. (1923). Natural lobster breeding. *Bull. Biol. Board Can.* 5, 1–11.

Incze, L. S., and Wahle, R. A. (1991). Recruitment from pelagic to early benthic phase in lobsters *Homarus americanus*. *Mar. Ecol. Prog. Ser.* 79, 77–87.

Jacobson, S. (1977). "Agonistic Behavior, Dominance, and Territoriality in the American Lobster, *Homarus americanus*," Ph.D. dissertation. Boston University, Boston, Massachusetts.

Jacobson, S., and Atema, J. (1977). Social structure and agonistic behavior in juvenile lobsters (*Homarus americanus*): Effects of space and density. *Div. Fish. Oceanogr. Circ. (Aust., C.S.I.R.O.)* 7, 21–22.

Jarvis, C. (1989). "Movement Patterns of Late-Stage Ovigerous Female Lobsters (*Homarus americanus* Milne-Edwards) at Jeddore, Nova Scotia," M.S. thesis. Dalhousie University, Halifax, Nova Scotia, Canada.

Jatzke, P. (1970). The trichterkreisel, an *in situ* device for cultivation of marine animals in tidal currents. *Helgol. Wiss. Meeresunters.* 20, 685–690.

Jernakoff, P., Phillips, B. F., and Maller, R. A. (1987). A quantitative study of the nocturnal foraging distances of the western rock lobster *Panulirus cygnus* George. *J. Exp. Mar. Biol. Ecol.* 113, 9–21.

Johns, P. M., and Mann, K. H. (1987). An experimental investigation of juvenile lobster habitat preference and mortality among habitats of varying structural complexity. *J. Exp. Mar. Biol. Ecol.* 109, 275–285.

Juanes, F. (1992). Why do decapod crustaceans prefer small-sized molluscan prey? *Mar. Ecol. Prog. Ser.* 87, 239–249.

Juinio, M. A. R., and Cobb, J. S. (1992). Natural diet and feeding habits of the postlarval lobster *Homarus americanus*. *Mar. Ecol. Prog. Ser.* 85, 83–91.

Jury, S. H., Kinnison, M. T., Howell, W. H., and Watson, W. H., III (1994a). The behavior of lobsters in response to reduced salinity. *J. Exp. Mar. Biol. Ecol.* 180, 23–37.

Jury, S. H., Kinnison, M. T., Howell, W. H., and Watson, W. H., III (1994b). The effects of reduced salinity on lobster (*Homarus americanus* Milne-Edwards) metabolism: Implications for estuarine populations. *J. Exp. Mar. Biol. Ecol.* 176, 167–185.

Karnofsky, E. B., and Price, H. J. (1989a). Dominance, territoriality, and mating in the lobster, *Homarus americanus*: A mesocosm study. *Mar. Behav. Physiol.* 15, 101–121.

Karnofsky, E. B., and Price, H. J. (1989b). Behavioural response of the lobster *Homarus americanus* to traps. *Can. J. Fish. Aquat. Sci.* 46, 1625–1632.

Karnofsky, E. B., Atema, J., and Elgin, R. H. (1989a). Natural dynamics of population structure and habitat use of the lobster, *Homarus americanus*, in a shallow cove. *Biol. Bull. (Woods Hole, Mass.)* 176, 247–256.

Karnofsky, E. B., Atema, J., and Elgin, R. H. (1989b). Field observations of social behavior, shelter use, and foraging in the lobster, *Homarus americanus*. *Biol. Bull. (Woods Hole, Mass.)* 176, 239–246.

Katz, C. H., Cobb, J. S., and Spaulding, M. (1994). Larval behavior, hydrodynamic transport, and potential offshore-to-inshore recruitment in the American lobster *Homarus americanus*. *Mar. Ecol. Prog. Ser.* 103, 265–273.

Kirby, S. S. (1994). "The Distribution and Site Fidelity of Juvenile Lobsters, *Homarus americanus*, in the Intertidal Zone," B.A. honors thesis. Bates College, Lewiston, Maine.

Kittaka, J., Henocque, Y., Yamada, K., and Tabata, N. (1983). Experimental release of juvenile lobster at Koshiki Islands in South Japan. *Bull. Jpn. Soc. Sci. Fish.* 49, 1347–1354.

Krouse, J. S. (1973). Maturity, sex ratio, and size composition of the natural population of American lobster, *Homarus americanus*, along the Maine coast. *Fish. Bull.* 71, 165–173.

Krouse, J. S. (1980). Summary of lobster, *Homarus americanus*, tagging studies in American waters (1898–1978). *Can. Tech. Rep. Fish. Aquat. Sci.* 932, 135–140.

Krouse, J. S., and Nutting, G. E. (1990a). Effectiveness of the Australian western rock lobster tag for marking juvenile American lobsters along the coast of Maine. *Am. Fish. Soc. Symp.* 7, 94–100.

Krouse, J. S., and Nutting, G. E. (1990b). Evaluation of coded micro-wire tags inserted in legs of small juvenile American lobsters. *Am. Fish. Soc. Symp.* 7, 304–310.

Lang, F., Govind, C. K., Costello, W. J., and Greene, S. I. (1977). Developmental neuroethology: Changes in escape and defensive behavior during growth of the lobster. *Science* 197, 682–685.

Lavalli, K. L. (1991). Survival and growth of early-juvenile American lobsters, *Homarus americanus*, through their first season while fed diets of mesoplankton, microplankton and frozen brine shrimp. *Fish. Bull.* 89, 61–68.

Lavalli, K. L. (1992). "The Feeding Mechanisms of Post-larval and Early Juvenile American Lobsters (*Homarus americanus* Milne Edwards, 1837)," Ph.D. dissertation. Boston University, Boston, Massachusetts.

Lavalli, K. L., and Barshaw, D. E. (1986). Burrows protect postlarval lobsters *Homarus americanus* from predation by the non-burrowing cunner *Tautogolabrus adspersus* but not from the burrowing mud crab *Neopanope texani*. *Mar. Ecol. Prog. Ser.* 32, 13–16.

Lavalli, K. L., and Barshaw, D. E. (1989). Post-larval American lobsters (*Homarus americanus*) living in burrows may be suspension feeding. *Mar. Behav. Physiol.* 15, 255–264.

Lavalli, K. L., and Factor, J. R. (1992). Functional morphology of the mouthparts of juvenile lobsters, *Homarus americanus* (Decapoda: Nephropidae), and comparison with the larval stages. *J. Crust. Biol.* 12, 467–510.

Lavalli, K. L., Brayer, K., and Ayers, J. L. (1995). Shelter usage by juvenile American lobsters, *Homarus americanus*, while under predation risk: The role of prior experience with predators. (in preparation).

Lawton, P. (1985). Predatory interaction betweeen the brachyuran crab *Cancer pagurus* and decapod crustacean prey. *Mar. Ecol. Prog. Ser.* 52, 169–179.

Lawton, P. (1987). Diel activity and foraging behavior of juvenile American lobsters, *Homarus americanus*. *Can. J. Fish. Aquat. Sci.* 43, 1195–1205.

Lawton, P., and Robichaud, D. A. (1992). Lobster habitat ecology research in the Bay of Fundy. *In* "Science Review of the Bedford Institute of Oceanography, the Halifax Fisheries Research Laboratory, and the St. Andrews Biological Station, 1990 & 91," pp. 53–56. Department of Fisheries and Oceans, Dartmouth, Nova Scotia, Canada.

Lawton, P., and Zimmer-Faust, R. K. (1992). Context-sensitivity in crab foraging studies. *Am. Zool.* 32, 396–404.

Leavitt, D. F., Bayer, R. C., Gallagher, M. L., and Rittenburg, J. H. (1979). Dietary intake and nutritional characteristics in wild American lobsters (*Homarus americanus*). *J. Fish. Res. Board Can.* 36, 965–969.

Loo, L.-O., Baden, S. P., and Ulmestrand, M. (1993). Suspension feeding in adult *Nephrops norvegicus* (L.) and *Homarus gammarus* (L.) (Decapoda). *Neth. J. Sea Res.* 31, 291–297.

Lund, W. A., Jr., and Lockwood, R. C., Jr. (1970). Sonic tag for large decapod crustaceans. *J. Fish. Res. Board Can.* 27, 1147–1151.

Lund, W. A., Jr., Stewart, L. L., and Rathbun, C. J. (1973). Investigation on the Lobster. Final Report to Department of Commerce. *NOAA, NMFS, Comm. Fish. Res. Dev. Act, Conn., Proj.* 3-130-R.

MacKay, D. A. (1920). Report upon yearlings and two year old lobsters in Richmond Bay, P.E.I. for 1920. *Biol. Board Can., Manuscr. Rep. Biol. Stn.* 49, 1–13.

MacKay, D. A. (1929). Larval and postlarval lobsters. *Am. Nat.* 63, 160–170.

Mann, K. H. (1985). Invertebrate behaviour and the structure of marine benthic communities. *In* "Behavioural Ecology: Ecological Consequences of Adaptive Behaviour" (R. M. Sibley and R. H. Smith, eds.), pp. 227–246. Blackwell, London.

Maynard, D. R., and Conan, G. Y. (1984). Preliminary Monitoring of Short Term Movements for Lobsters (*Homarus americanus*) in the Bideford River, Prince Edward Island, Using Ultrasonic Telemetry. *Can. Atl. Fish. Sci. Advis. Comm. Res. Doc.* 84/99.

McKenzie, T. P. (1989). "Foraging Behavior of Juvenile American Lobsters, *Homarus americanus*, under Predation Risk," M.S. thesis. University of Rhode Island, Kingston.

McLeese, D. W. (1956). Effects of temperature, salinity and oxygen on the survival of the American lobster. *J. Fish. Res. Board Can.* 13, 247–272.

McLeese, D. W., and Wilder, D. G. (1964). Lobster Storage and Shipment. *Bull. Fish. Res. Board Can.* 147.

Miller, R. J. (1985). Seaweeds, sea urchins, and lobsters: A reappraisal. *Can. J. Fish. Aquat. Sci.* 42, 2061–2072.

Miller, R. J. (1989). Catchability of American lobsters (*Homarus americanus*) and rock crabs (*Cancer irroratus*) by traps. *Can. J. Fish. Aquat. Sci.* 46, 1652–1657.

Miller, R. J. (1990). Effectiveness of crab and lobster traps. *Can. J. Fish. Aquat. Sci.* 47, 1228–1251.

Miller, R. J., Mann, K. H., and Scarratt, D. J. (1971). Production potential of a seaweed–lobster community in eastern Canada. *J. Fish. Res. Board Can.* 28, 1733–1738.

Miller, R. J., Duggan, R. E., Robinson, D. G., and Zeng, Z. (1989). Growth and Movement of *Homarus americanus* on the Outer Coast of Nova Scotia. *Can. Tech. Rep. Fish. Aquat. Sci.* 1716.

Miller, R. J., Campbell, A., Sharp, G. J., and Hudon, C. (1992). Are small juvenile lobsters habitat limited? *Can. Tech. Rep. Fish. Aquat. Sci.* 1390, 55–61.

Moody, K. E., and Steneck, R. S. (1993). Mechanisms of predation among large decapod crustaceans of the Gulf of Maine coast: Functional vs. phylogenetic patterns. *J. Exp. Mar. Biol. Ecol.* 168, 111–124.

Moore, D. S., and Miller, R. J. (1983). Recovery of Macroalgae following Widespread Sea Urchin Mortality with a Description of the Nearshore Hard-Bottom Habitat: Eastern Shore and Cape Breton, Nova Scotia. *Can. Tech. Rep. Fish. Aquat. Sci.* 1230.

Moore, D. S., Miller, R. J., and Meade, L. D. (1986). Survey of Shallow Benthic Habitat: Eastern Shore and Cape Breton, Nova Scotia. *Can. Tech. Rep. Fish. Aquat. Sci.* 1546.

Morrissey, T. D. (1971). Movements of tagged American lobsters, *Homarus americanus*, liberated off Cape Cod, Massachusetts. *Trans. Am. Fish. Soc.* 1, 117–120.

Munro, J., and Therriault, J.-C. (1983). Migrations saisonnières du homard (*Homarus americanus*) entre la côte et les lagunes des Îles-de-la-Madeleine. *Can. J. Fish. Aquat. Sci.* 40, 905–918.

Nelson, K., and Hedgecock, D. (1977). Electrophoretic evidence of multiple paternity in the lobster *Homarus americanus* (Milne-Edwards). *Am. Nat.* 111, 361–365.

Northeast Utilities Service Company (1993). "Monitoring the Marine Environment of Long Island Sound at Millstone Nuclear Power Station, Waterford, Connecticut, 1993 Annual Report." Environmental Services Division, Aquatic Services Branch, Northeast Utilities Environmental Laboratory, Millstone, Connecticut.

O'Dor, R., and Webber, D. (1991). Ultrasonic telemetry. *Lobster Newsl.* 4(1), 1–2.

Ojeda, F. P. (1987). "Rocky Subtidal Community Structure in the Gulf of Maine: The Role of Mobile Predators," Ph.D. dissertation. University of Maine, Orono.

Ojeda, F. P., and Dearborn, J. H. (1990). Diversity, abundance, and spatial distribution of fishes and crustaceans in the rocky subtidal zone of the Gulf of Maine. *Fish. Bull.* 88, 403–410.

Ojeda, F. P., and Dearborn, J. H. (1991). Feeding ecology of benthic mobile predators: Experimental analyses of their influence in rocky subtidal communities of the Gulf of Maine. *J. Exp. Mar. Biol. Ecol.* 149, 13–44.

O'Neill, D. J., and Cobb, J. S. (1979). Some factors influencing the outcome of shelter competition in lobsters (*Homarus americanus*). *Mar. Behav. Physiol.* 6, 33–45.

Paine, R. T. (1969). A note on trophic complexity and community stability. *Am. Nat.* 103, 91–93.

Pezzack, D. S. (1987). Offshore Lobster Stocks in NAFO Subarea 4W: Potential for a New Fishery. *Can. Atl. Fish. Sci. Advis. Comm. Res. Doc.* 87/78.

Pezzack, D. S. (1989). Comments on "Dispersal of *Homarus americanus* larvae in the Gulf of Maine from Browns Bank" by G. C. Harding and R. W. Trites. *Can. J. Fish. Aquat. Sci.* 46, 1077–1078.

Pezzack, D. S. (1992). A review of lobster (*Homarus americanus*) landing trends in the Northwest Atlantic, 1947–86. *J. Northw. Atl. Fish. Sci.* 14, 115–127.

Pezzack, D. S, and Duggan, D. R. (1986). Evidence of migration and homing of lobsters (*Homarus americanus*) on the Scotian

Shelf. *Can. J. Fish. Aquat. Sci.* 43, 2206–2211.

Pezzack, D. S., and Duggan, D. R. (1987). Canadian Offshore Lobster Fishery, 1985–86, and an Assessment of the Potential for Future Increases in Catch. *Can. Atl. Fish. Sci. Advis. Comm. Res. Doc.* 87/79.

Pezzack, D. S., and Duggan, D. R. (1988). An Assessment of the Canadian Offshore Lobster Fishery (LFA 41) for 1986–87. *Can. Atl. Fish. Sci. Advis. Comm. Res. Doc.* 88/65.

Pezzack, D. S., and Duggan, D. R. (1989). Female Size–Maturity Relationships for Offshore Lobsters (*Homarus americanus*). *Can. Atl. Fish. Sci. Advis. Comm. Res. Doc.* 89/66.

Pezzack, D. S., Tremblay, M. J., Hudon, C., and Miller, R. J. (1992). The Inshore–Offshore Lobster Issue in Southwestern Nova Scotia. *Can. Manuscr. Rep. Fish. Aquat. Sci.* 2165.

Phillips, B. F., Joll, L. M., and Ramm, D. C. (1984). An electromagnetic tracking system for studying the movements of rock (spiny) lobsters. *J. Exp. Mar. Biol. Ecol.* 79, 9–18.

Pottle, R. A., and Elner, R. W. (1982). Substrate preference behavior of juvenile American lobsters, *Homarus americanus*, in gravel and silt–clay sediments. *Can. J. Fish. Aquat. Sci.* 39, 928–932.

Prince, E. E. (1899). Report of the Canadian lobster commission 1898. 31st *Annu. Rep. Dep. Mar. Fish.*, 1899 (Suppl. 1), 5–41.

Pringle, J. P., and Burke, D. L. (1993). The Canadian lobster fishery and its management, with emphasis on the Scotian Shelf and Gulf of Maine. *Can. Bull. Fish. Aquat. Sci.* 226, 91–122.

Reddin, D. (1973). "The Ecology and Feeding Habits of the American Lobster (*Homarus americanus* Milne-Edwards, 1837) in Newfoundland," M.S. thesis. Memorial University of Newfoundland, St. John's.

Richards, R. A. (1992). Habitat selection and predator avoidance: Ontogenetic shifts in habitat use by the Jonah crab *Cancer borealis* (Stimpson). *J. Exp. Mar. Biol. Ecol.* 156, 187–197.

Richards, R. A., and Cobb, J. S. (1986). Competition for shelter between lobsters (*Homarus americanus*) and Jonah crabs (*Cancer borealis*): Effects of relative size. *Can. J. Fish. Aquat. Sci.* 43, 2250–2255.

Roach, S. G. (1983). Survivorship, Growth, and Behavior of Juvenile Lobsters *Homarus americanus* Milne-Edwards in Controlled Environments in Nature. *N. S. Dep. Fish., Manuscr. Tech. Rep. Ser. Proj. Rep.* 83–02.

Robichaud, D. A., and Campbell, A. (1991). Annual and seasonal size-frequency changes of trap-caught lobsters (*Homarus americanus*) in the Bay of Fundy. *J. Northw. Atl. Fish. Sci.* 11, 29–37.

Roddick, D. L., and Miller, R. J. (1992). Spatial and temporal overlap of the American lobster (*Homarus americanus*) and sea scallop (*Placopecten magellanicus*) as related to the impact of inshore scallop dragging. *Can. J. Fish. Aquat. Sci.* 49, 1486–1492.

Saila, S. B., and Flowers, J. M. (1968). Movements and behavior of berried female lobsters displaced from offshore areas to Narragansett Bay, Rhode Island. *J. Cons., Cons. Int. Explor. Mer* 31, 342–351.

Sasaki, G. C., Capuzzo, J. M., and Biesiot, P. (1986). Nutritional and bioenergetic considerations in the development of the American lobster *Homarus americanus*. *Can. J. Fish. Aquat. Sci.* 43, 2311–2319.

Sastry, A. N., and Ehinger, R. E. (1980). Dominance hierarchies among communally held juvenile lobsters, *Homarus americanus*. *Mar. Behav. Physiol.* 7, 85–93.

Sastry, A. N., and Laczak, J. (1975). The ammonia tolerance of lobsters, *Homarus americanus*. *In* "Recent Advances in Lobster Aquaculture" (J. S. Cobb, ed.), Sea Grant Lobster Aquaculture Workshop Technical Session, April 1975. University of Rhode Island, Kingston (abstract).

Scarratt, D. J. (1968). An artificial reef for lobsters (*Homarus americanus*). *J. Fish. Res. Board Can.* 25, 2683–2690.

Scarratt, D. J. (1973). Abundance, survival, and vertical and diurnal distribution of lobster larvae in Northumberland Strait, 1962–1963, and their relationships with commercial stocks. *J. Fish. Res. Board Can.* 30, 1819–1824.

Scarratt, D. J. (1980). The food of lobsters. *Can. Tech. Rep. Fish. Aquat. Sci.* 954, 66–91.

Scarratt, D. J., and Elson, P. F. (1965). Preliminary trials of a tag for salmon and lobsters (*Homarus americanus*). *J. Fish. Res. Board Can.* 22, 421–423.

Schiebling, R. (1986). Increased macroalgal abundance following mass mortalities of sea urchins (*Strongylocentrotus droebachiensis*) along the Atlantic coast of Nova Scotia. *Oecologia* 68, 186–198.

Scott, W. B., and Scott, M. G. (1988). Atlantic Fishes of Canada. *Can. Bull. Fish. Aquat. Sci.* 219.

Scrivener, J. C. E. (1971). Agonistic Behavior of the American Lobster *Homarus americanus* (Milne-Edwards). *Tech. Rep.—Fish. Res. Board Can.* 235.

Sheehy, D. J. (1976). Utilization of artificial shelters by the American lobster (*Homarus americanus*). *J. Fish. Res. Board Can.* 33, 1613–1622.

Skud, B. E., and Perkins, H. C. (1969). Size Composition, Sex Ratio, and Size at Maturity. *U.S., Fish Wildl. Serv., Spec. Sci. Rep.—Fish.* 598.

Smith, E. M. (1976). "The Food Burial Behavior of the American Lobster (*Homarus americanus*)," M.S. thesis. University of Connecticut, Storrs.

Smith, E. M., and Howell, P. T. (1987). The effects of bottom trawling on American lobsters, *Homarus americanus*, in Long Island Sound. *Fish. Bull.* 85, 737–744.

Snyder, M. J., Ameyaw-Akumfi, C., and Chang, E. S. (1992). Mating behavior in vision-deprived American lobsters, *Homarus americanus*. *Mar. Behav. Physiol.* 21, 227–238.

Spanier, E., Cobb, J. S., and Clancy, M. (1994). Impacts of remotely operated vehicles (ROVs) on the behavior of marine animals: An example using American lobsters. *Mar. Ecol. Prog. Ser.* 104, 257–266.

Squires, H. J. (1970). Lobster (*Homarus americanus*) fishery and ecology in Port au Port Bay, Newfoundland, 1960–65. *Proc. Natl. Shellfish. Assoc.* 60, 22–39.

Squires, H. J. (1990). Decapod Crustacea of the Atlantic Coast of Canada. *Can. Bull. Fish. Aquat. Sci.* 221.

Squires, H. J., and Ennis, G. P. (1968). Lobsters from 95 fathoms deep off St. John's, Newfoundland. *J. Fish. Res. Board Can.* 25, 2723–2724.

Stasko, A. B. (1980). Tagging and lobster movements in Canada. *Can. Tech. Rep. Fish. Aquat. Sci.* 932, 141–150.

Stein, L., Jacobson, S., and Atema, J. (1975). Behavior of Lobsters (*Homarus americanus*) in a Semi-natural Environment at Ambient Temperatures and under Thermal Stress. *Woods Hole Oceanogr. Inst. Tech. Rep.* WHOI-75-48.

Steneck, R. S. (1989). The ecological ontogeny of lobsters: *In situ* studies with demographic implications. *In* "Life History of the American Lobster. Proceedings of a Workshop, November 29–30, 1989, Orono, Maine" (I. Kornfield, ed.), pp. 30–33. Lobster Institute, Orono, Maine.

Steneck, R. S., Pezzack, D. S., Fogarty, M. J., and Lawton, P. (1995). Manned submersibles as tools for assessing deep water lobster stocks. *Lobster Newsl.* (in preparation).

Stewart, L. L. (1972). "The Seasonal Movements, Population Dynamics, and Ecology of the Lobster, *Homarus americanus* off Ram Island, Connecticut," Ph.D. dissertation. University of Connecticut, Storrs.

Stokesbury, K. D. E., and Himmelman, J. H. (1995). Biological and physical variables associated with aggregations of the giant scallop *Placopecten magellanicus*. *Can. J. Fish. Aquat. Sci.* (in press).

Tamm, G. R., and Cobb, J. S. (1978). Behavior and the crustacean molt cycle: Changes in aggression of *Homarus americanus*. *Science* 200, 79–81.

Templeman, W. (1934). Mating in the American lobster. *Contrib. Can. Biol. Fish.* 8, 423–432.

Templeman, W. (1936a). Local differences in the life history of the lobster (*Homarus americanus*) on the coast of the Maritime Provinces of Canada. *J. Biol. Board Can.* 2, 41–87.

Templeman, W. (1936b). Further contributions to mating in the American lobster. *J. Biol. Board Can.* 2, 223–226.

Templeman, W. (1940). The Life History of the Lobster. *Newfoundland Dep. Nat. Resour., Res. Bull. (Fish.)* 15.

Templeman, W., and Tibbo, S. N. (1945). Lobster Investigations in Newfoundland 1938 to 1941. *Newfoundland Dep. Nat. Resour., Res. Bull. (Fish.)* 16.

Thomas, H. J. (1954). The oxygen uptake of the lobster (*Homarus vulgaris* Edw.). *J. Exp. Biol.* 31, 228–251.

Thomas, M. L. H. (1968). Overwintering of American lobsters, *Homarus americanus*, in burrows in Bideford River, Prince Edward Island. *J. Fish. Res. Board Can.* 25, 2725–2727.

Thomas, M. L. H., and White, G. N. (1969). Mass mortality of estuarine fauna at Bideford, P.E.I., associated with abnormally low salinities. *J. Fish. Res. Board Can.* 26, 701–704.

Thomas, M. L. H., Arnold, D. C., and Taylor, A. R. A. (1983). Rocky intertidal communities. *Can. Spec. Publ. Fish. Aquat. Sci.* 64, 35–73.

Uzmann, J. R., Cooper, R. A., and Pecci, K. J. (1977a). Migration and Dispersion of Tagged American Lobsters, *Homarus americanus*, on the Southern New England Continental Shelf. *NOAA Tech. Rep. NMFS SSRF*-705.

Uzmann, J. R., Cooper, R. A., Theroux, R. B., and Wigley, R. L. (1977b). Synoptic comparison of three sampling techniques for estimating abundance and distribution of selected megafauna: Submersible vs. camera sled vs. otter trawl. *Mar. Fish. Rev.* 39 (12), 11–19.

Vadas, R. L., Sr., and Elner, R. W. (1992). Plant–animal interactions in the north-west Atlantic. *Syst. Assoc. Spec. Vol.* 46, 33–60.

Van der Meeren, G. I. and Næss, H. (1991). Recatches of Marked Lobsters, *Homarus gammarus*, Released in 1988. *Int. Counc. Explor. Sea* C.M. 1991/K:7.

Van der Steen, W. J. (1984). Methodological aspects of migration and orientation in fish. *In* "Mechanisms of Migration in Fish" (J. D. McCleave, G. P. Arnold, J. D. Dodson, and W. H. Neil, eds.), pp. 421–444. Plenum, New York.

Van Olst, J. C., Carlberg, J. M., and Ford, R. F. (1975). Effects of substrate type and other factors on the growth, survival, and cannibalism of juvenile *Homarus americanus* in mass rearing systems. *Proc. World Maric. Soc.* 6, 261–274.

Vogel, S. (1974). Current-induced flow through the sponge, *Halichondria*. *Biol. Bull. (Woods Hole, Mass.)* 147, 443–456.

Vogel, S. (1977). Organisms that capture currents. *Sci. Am.* 239, 128–139.

Vogel, S., and Bretz, W. L. (1972). Interfacial organisms: Passive ventilation in the velocity gradients near surfaces. *Science* 175, 210–211.

Waddy, S. L., and Aiken, D. E. (1990). Intermolt insemination, an alternative mating strategy for the American lobster (*Homarus americanus*). *Can. J. Fish. Aquat. Sci.* 47, 2402–2406.

Wahle, R. A. (1990). "Recruitment, Habitat Selection, and the Impact of Predators on the Early Benthic Phase of the American Lobster (*Homarus americanus* Milne Edwards)," Ph.D. dissertation. University of Maine, Orono.

Wahle, R. A. (1991). Implications for lobster fishery enhancement from natural benthic recruitment, hatchery-reared 'blues,' and experimental cobbles. *J. Shellfish Res.* 10, 287 (abstract).

Wahle, R. A. (1992a). Substratum constraints on body size and the behavioral scope of shelter use in the American lobster. *J. Exp. Mar. Biol. Ecol.* 159, 59–75.

Wahle, R. A. (1992b). Body-size dependent anti-predator mechanisms of the American lobster. *Oikos* 65, 52–60.

Wahle, R. A. (1993). Recruitment to American lobster populations along an estuarine gradient. *Estuaries* 16, 731–738.

Wahle, R. A., and Incze, L. S. (1993). Pre- and post-settlement processes in the American lobster: Observations from an island in the larval stream. *Larval Ecol. Meet., State Univ. N. Y., Stony Brook, Sea Grant*, p. 45 (abstract).

Wahle, R. A., and Steneck, R. S. (1991). Recruitment habitats and nursery grounds of the American lobster *Homarus americanus*: A demographic bottleneck? *Mar. Ecol. Prog. Ser.* 69, 231–243.

Wahle, R. A., and Steneck, R. S. (1992). Habitat restrictions in early benthic life: Experiments on habitat selection and *in situ* predation with the American lobster. *J. Exp. Mar. Biol. Ecol.* 157, 91–114.

Wang, D. (1982). "The Behavioral Ecology of Competition among Three Decapod Species, the American Lobster, *Homarus americanus*, the Jonah Crab, *Cancer borealis*, and the Rock Crab, *Cancer irroratus*, in Rocky Habitats," Ph.D. dissertation. University of Rhode Island, Kingston.

Weiss, H. M. (1970). "The Diet and Feeding Behavior of the Lobster, *Homarus americanus*, in Long Island Sound," Ph.D dissertation. University of Connecticut, Storrs.

Wiens, J. A., Stenseth, N. C., Van Horne, B., and Ims, R. A. (1993). Ecological mechanisms and landscape ecology. *Oikos* 66, 369–380.

Wilder, D. G., and Murray, R. C. (1958). Do lobsters move offshore and onshore in the fall and the spring? *Fish. Res. Board Can., Prog. Rep. Atl. Coast Stn.* 69, 12–15.

Williams, L. W. (1907). The stomach of the lobster and the food of larval lobsters. *Annu. Rep.—R.I. Comm. Inland Fish.* 37, 153–180.

Witman, J. D., and Sebens, K. P. (1992). Regional variation in fish predation intensity: A historical perspective in the Gulf of Maine. *Oecologia* 90, 305–315.

Wolcott, T. G., and Hines, A. H. (1989). Ultrasonic biotelemetry of muscle activity from free-ranging marine animals: A new method for studying foraging by blue crabs (*Callinectes sapidus*). *Biol. Bull. (Woods Hole, Mass.)* 176, 50–56.

Wolff, T. (1978). Maximum size of lobsters (*Homarus*) (Decapoda, Nephropidae). *Crustaceana* 34, 1–14.

Young-Lai, W. W., Charmantier-Daures, M., and Charmantier, G. (1991). Effect of ammonia on survival and osmoregulation in different life stages of the lobster *Homarus americanus*. *Mar. Biol.* 110, 293–300.

Zeitlin-Hale, L., and Sastry, A. N. (1978). Effects of environmental manipulation on the locomotor activity and agonistic behavior of cultured juvenile American lobsters, *Homarus americanus*. *Mar. Biol.* 47, 369–379.

C H A P T E R

5

Fishery Regulations and Methods

ROBERT J. MILLER
Department of Fisheries and Oceans
Halifax, Nova Scotia, Canada B3J 2S7

I. Introduction

The North American lobster fishery is successful because *Homarus americanus* is a very adaptable species, it is a prized food for a broad spectrum of consumers, and fishing is controlled by a few strong regulations. The American lobster is found at commercial concentrations in habitats ranging from the intertidal to 600-m depth, from large boulders to eelgrass beds, from 25° to 50° N latitude, and from Boston Harbor to the most pristine coastal areas of Maine and Newfoundland. It is proudly served as a Mother's Day treat by families of modest means and as a gourmet delight in exclusive restaurants throughout the Western world. Regulations specify a precise minimum size, give some protection to brood stock, with a few exceptions limit the fishery to nondestructive traps, and are generally obeyed by the fishing industry.

Change has been a constant throughout the history of this fishery, although the major forces of change have been technology and competition rather than fisheries management. The ease with which such innovations as power haulers, depth sounders, synthetic materials, and loran have been accepted cannot help but make a fishery manager envious. Battles over optimum minimum size, seasons, and access to fishing grounds, on the other hand, have been waged for over a century with only incremental progress. Management changes require simultaneous acceptance by many participants, usually without tangible proof that the benefit will occur and with a delay before the benefit is realized. Adopting a technological change, however, is an individual decision that includes the option to revert to the old method; a demonstration can usually be arranged and the benefit is usually immediate. A challenge for the modern fishery manager is to emulate the benefits of technological innovation when introducing a management change.

II. Regulations

A. Changing Regulations

Accounts of the history of a fishery usually include the date when regulations were introduced and their purpose, but rarely discuss two other important issues: how management decisions were made and how well they were observed. These issues also do not receive the attention they deserve when a change in fishery management is contemplated. The following considerations are drawn from published anecdotes and personal experience.

Four effective models for making decisions to change lobster management are: small groups of fishers regulate their own activities; a small group and a management agency negotiate a division of management responsibilities; fisheries managers impose change on the industry with strong governmental enforcement until the change is accepted; and a small group of fishery and government representatives make decisions for the entire fishery with government and joint government–industry enforcement. The first two approaches require strong leadership and a methodical building of consensus, with the important constraint that the group be small. The third requires that the manager has ample enforcement support, is an effective communicator of intentions, and has the fortitude to stay the course until the change has been accepted. The fourth has the requirements of the third, but additionally the industry and government must support their representatives. All of these choices avoid the nearly impossible chore of convincing hundreds or thousands of fishers to agree to one solution (see Section IV).

In a fifth model the manager or a member of the industry convinces a large segment of the industry to democratically accept an advisable change. This is probably the most common and least successful approach. Democratic decision-making within a large and diverse group is very difficult because the interest groups simply cannot agree. These interests include the heavily indebted fishers who cannot afford to sacrifice a dollar today for two tomorrow, the fatalist fishers who believes that fish stocks are entirely beyond human control, rival fishing communities with a tradition of conflict, the government biologists and economists who are frustrated and impatient because their advice for change is being ignored, and the university scientists looking for the next research grant. Add the mutual mistrust of lobster fishers and buyers and it is no wonder that politicians, selected for their skill at following the electorate, and senior civil servants, selected for their willingness to follow the politicians, most often choose the status quo.

A regulation must be seen to be in the best interest of the individual fishers and buyers before it will be accepted. This best interest can be achieved as a result of penalties that are severe enough and certain enough to be convincing. Penalties can be administered by the courts or through peer pressure. A mere reminder of a misdeed by a colleague can be an effective form of peer pressure. Best interest is also achieved when the individual believes that personal gain will exceed personal sacrifice, for example, by returning small lobsters to the water today that will be recaptured as larger, more valuable lobsters tomorrow. In other words, a fisher will be personally rewarded by investing in a stock-enhancing measure. Fishery managers too often ignore individual best interest, seduced by the reasoning that fishers as a group will endorse a change because they will be better off on average. Convincing a group that a change will produce the promised benefits can be very difficult. And always, an individual can obtain a bigger share of the catch by ignoring constraints to which others adhere. The perception that a few are successfully flaunting the rules is enough to bring down the best-reasoned management regime.

B. Modern Canadian Regulations

All aspects of lobster fishery management in Canada are the responsibility of the federal Department of Fisheries and Oceans (DFO). The east coast is divided into four regions, each with responsibilities for all functions: Newfoundland, Quebec (northern Gulf of St. Lawrence), Gulf (southern Gulf of St. Lawrence), and Scotia–Fundy (Bay of Fundy and Nova Scotia, excluding the Gulf of St. Lawrence) (Fig. 1). Enforcement is carried out by fisheries officers stationed in literally hundreds of locations. They operate from DFO ships and aircraft and from shore. Records of licenses and landings are maintained by dedicated groups, but each group is assisted by fisheries officers in the field. Major changes in regulations must be approved by the federal cabinet, but existing regulations, such as lobster sizes, seasons, or trap limits, can be altered by the senior DFO manager

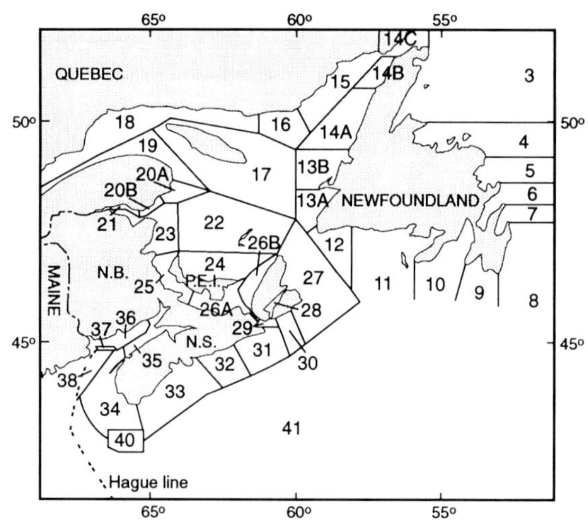

FIGURE 1 Lobster fishing areas (LFAs) in Canada. LFAs are administered by four regions: Newfoundland (LFAs 3–14C), Quebec (LFAs 15–22), Gulf (LFAs 23–26B), and Scotia–Fundy (LFAs 27–41).

in each region. Several committees made up of fishers and DFO officials (and sometimes other interest groups), each representing a section of coast, advise each regional manager. Most violators are tried in provincial courts, but the senior regional managers recently obtained authority to suspend licenses if, in their judgment, a violation has been committed. Each region has staff biologists and economists engaged in research, assessment, and monitoring. In the Scotia–Fundy region in 1987, the funds allotted to enforcement, administration, and research were 4, 1, and 2%, respectively, of lobster landed value (Pringle and Burke, 1993).

Regulations for the Canadian fishery are longstanding and most are effective. They enhance the yield in weight, protect the breeding stock, and control waste. Their social and economic benefits include targeting market demand, controlling the cost of fishing, and apportioning the catch among participants. They are generally well accepted by the fishery because they create a sense of resource ownership, are well understood by the fishing industry and enforcement officers, and allow fishers and managers to regulate the fishery differently in different areas. These regulations lack the complication of gear sectors and quota management. They could be improved with greater flexibility in the use of minimum legal size and management districts. Table 1 summarizes the following discussion by listing the principal regulations, where enforcement is carried out, a relative measure of the difficulty of enforcement, and the reasons for the regulations.

1. Lobster Fishing Areas

The provision for lobster fishing areas (LFAs) (Fig. 1) is a powerful regulation because it increases the effectiveness of other regulations. LFAs make it possible to adapt fishing seasons and minimum sizes to market demand and local conditions of weather and lobster life history. Equally important, they accommodate differences in opinion among groups of fishers and among fishery managers. For example, the trap limits in force in each LFA were based on votes by fishers. It would be difficult to justify the choices, ranging from 100 to 500 traps per license, on economic or conservation grounds. Further, LFAs provide the option to vary regulations on a small geographic scale. In the fishing industry, it is nearly impossible for a large or diverse group to agree on any issue. Introducing management changes on a small scale, as advocated by Tyler *et al.* (1982) and Bergh *et al.* (1990), increases the chance of industry acceptance.

The high-profile problems in lobster management during the last few years resulted from attempts to change regulations for large groups of fishers simultaneously. In 1989, the DFO championed an increase of 3 mm ($\frac{1}{8}$ inch) in the minimum legal size of lobsters over a large area to coincide with planned U.S. increases. After many meetings with industry, volumes of unfavorable press, and opposition by fishers in several parts of Nova Scotia, the DFO withdrew its support for the change. If given an option to accept or reject the change on an area-by-area basis, it seems likely that many LFAs would have adopted it.

TABLE 1 **Summary of Canadian Lobster Fishery Regulations**

Regulation	Enforcement	Difficulty	Benefits
Fishing areas (LFAs)	Fishers, sea	Easy	Permits regulations to be adjusted to weather, markets, size at sexual maturity, etc.; permits differences in opinion among fishers, managers, and scientists; prevents fleet movement to most productive areas; encourages sense of resource ownership
Season	DFO,[a] fishers, dock, sea	Easy	Matches landings to markets; avoids molting; reduces the capture of mature females; adaptable to weather conditions; controls cost of fishing
Minimum size	DFO, dock	Medium	Exact measure; good survival of discards; adjusts supply to markets; affects yield in weight; permits banking biomass and collecting "interest" and "capital gains"; affects eggs per recruit
Limited entry	DFO, fishers, dock	Easy	Increases earnings per license; eliminates (most) part-time and recreational fishers; creates sense of resource ownership
Type and amount of gear	DFO, dock, sea	Difficult	Reduces gear conflicts; controls cost of fishing; allocates equal effort to competing fishers
Ovigerous females	DFO, dock	Medium	Increases egg production

[a]DFO, Canada's Department of Fisheries and Oceans.

In another example, a 6-mm ($\frac{1}{4}$-inch) increase in minimum size was successfully introduced in western Nova Scotia (LFA 26B, Fig. 1) after extensive consultation (Maynard *et al.*, 1992). In 1990–1991, the DFO imposed the same change on a much larger area in the southern Gulf of St. Lawrence. A ground swell of opposition stopped the change in 1992, after one or two 1.5-mm ($\frac{1}{16}$-inch) increases. In a third example, fishers in LFA 33 (Fig. 1) expressed interest in 1984 in adopting trap escape gaps (also called vents); however, this was postponed until the DFO and industry could agree on a regulation for all of Atlantic Canada. Introduction was delayed until 1993, but even then pockets of industry resistance delayed enforcement in some areas.

Changes in regulations over large areas are now less necessary than before. A 1991 revision in the Canada Fisheries Act gave regional officers authority to vary a regulation in a designated area of any size. Thus, a new LFA could be created to include a group of fishers who agree to a more sound (economically or biologically) management regime. Canadian fisheries managers could take advantage of this provision to make improvements in any location where the industry is in agreement.

2. Limited Entry

No new lobster fishing licenses have been granted since 1968. Existing licenses may be transferred only to individuals employed in the fishery for at least 2–8 months per year (depending on the area) for at least 2 years, and whose other income, if any, is principally from fishing, farming, or forestry. A license may be used for fishing in only one LFA and licenses held in 1968 by part-time fishers are nontransferable. All sport fishing is prohibited.

Limiting entry obviously increases the average earnings per license and prevents an influx of fishers during times of high catch rates. In combination with LFA designations, limited entry also prevents the aggregation of fishing effort on the most productive grounds. These conditions encourage a sense of resource ownership, which in turn promotes the attitude that "I will take care of what is mine," and counters the "gold rush" approach, "If I don't grab as much as I can today, someone else will get it tomorrow." Fishers support and police this regulation in order to protect their share of the catch from outside competition.

3. Seasons

Fishing seasons match the fishery to markets and local conditions. With minor exceptions, Canadian seasons are closed from late July through October (Table 2), avoiding competition with U.S. landings during their peak. Late summer molting is also avoided. The spring season observed in the Gulf of St. Lawrence and Newfoundland misses winter ice and autumn storms, whereas winter seasons in western Nova Scotia take advantage of ice-free fishing grounds to supply lucrative Christmas and mid-winter markets. Most seasons also afford some protection to mature females, which are most catchable during the summer–autumn closed period (Knight, 1917; Robichaud and Campbell, 1991). There is good acceptance of seasons by fishers. Although many individuals may be involved in summer lobster poaching, and these are a great nuisance to enforcement personnel, the total out-of-season catch is likely an insignificant fraction of the legal catch.

4. Minimum Size

Atlantic Canada has six minimum size limits (Table 2). Two result from derailed attempts to increase the minimum, but the remainder are useful in supplying the market with a wide range of sizes and products: live, frozen whole, frozen meat, and canned meat (Barbara, 1991). Because of high exploitation, most lobsters do not survive past their first year at legal size (Miller *et al.*, 1987); a single minimum size would result in only a narrow range of sizes for the market.

Regulating size at first capture influences egg production (eggs per recruit) and weight yield (yield per recruit). A 10-mm ($\frac{3}{8}$-inch) increase in carapace length (CL) can easily provide a threefold increase in eggs (Campbell and Robinson, 1983). The Gulf of St. Lawrence, with the smallest minimum sizes, also has the lowest size at maturity. All inshore Canadian stocks are harvested below the size giving maximum yield per recruit (Ennis, 1980; Campbell and Robinson, 1983; Miller *et al.*, 1987); that is, if left to grow, the weight added annually by growth would exceed the weight lost from natural mortality. Maximum income per recruit is also remote. Wholesale prices paid for different lobster sizes (Barbara, 1991) show that the industry is oversupplying the market with small lobsters and undersupplying it with large lobsters.

If industry and fishery managers would accommodate change, the industry could profit by banking biomass ("capital"), and collecting "interest" and "capital gains." For example, during a time of high catches and low prices, fishers could choose to raise the minimum size by 6 mm ($\frac{1}{4}$ inch) and reduce their catch by about 25%. Each year thereafter that they fished at the new size they would collect "interest" in the form of higher yield per recruit. When prices

TABLE 2 **Canadian Regulations for Season, Minimum Size, and Trap Limits for Each Lobster Fishing Area (LFA)**

Region	LFA[a]	Open seasons	Minimum size (mm/inches)	Trap limit[b]
Newfoundland	3–11	April 12–July 15	81 / $3\frac{3}{16}$	100–300
	12	April 20–July 5	81 / $3\frac{3}{16}$	150
	13A	April 27–July 11	81 / $3\frac{3}{16}$	200
	13B	April 29–July 11	81 / $3\frac{3}{16}$	275
	14A	May 5–July 10	81 / $3\frac{3}{16}$	350
	14B	May 5–July 10	81 / $3\frac{3}{16}$	500
	14C	May 25–July 28	81 / $3\frac{3}{16}$	350
Quebec	15, 16, 18	May 20–July 31	76 / 3	300
	17	June 15–August 15	76 / 3	300
	19	May 10–July 27	76 / 3	300
	20A and B	May 1–July 17	76 / 3	300
	22	May 10–July 10	76 / 3	300
Gulf	23	May 7–July 6	67 / $2\frac{5}{8}$	375
	24	May 10–July 6	64 / $2\frac{1}{2}$	300
	25	August 10–October 11	67 / $2\frac{1}{2}$	250
	26A	May 5–July 5	65 / $2\frac{9}{16}$	300
	26B	May 5–July 5	70 / $2\frac{3}{4}$	300
Scotia–Fundy	27	May 15–July 15	70 / $2\frac{3}{4}$	275
	28, 29	May 9–July 9	81 / $3\frac{3}{16}$	275
	30	May 19–July 20	81 / $3\frac{3}{16}$	250
	31, 32	April 19–June 20	81 / $3\frac{3}{16}$	250
	33	Last Monday in November–May 31	81 / $3\frac{3}{16}$	250
	34	Last Monday in November–May 31	81 / $3\frac{3}{16}$	400
	35	October 15–December 31 and February 28–July 31	81 / $3\frac{3}{16}$	300
	36	2nd Tuesday in November–January 14 and March 31–June 29	81 / $3\frac{3}{16}$	300
	38	2nd Tuesday November–4th Thursday June	81 / $3\frac{3}{16}$	375
	40	Closed area, no fishery		
	41	Open all year	81 / $3\frac{3}{16}$	None

[a] See Fig. 1 for LFA locations.
[b] Part-time (class B) fishers have lower trap limits.

increased, they could choose to return to the original size, removing the "capital" on deposit, as well as "capital gains" in the form of a higher price per pound.

The temptation is great to sell undersized lobsters directly to consumers, making enforcement difficult. However, the benefit of the minimum size regulation is so large that it is worth diligent enforcement. Sizes can be measured precisely, and undersized lobsters returned to the sea have high survival rates.

5. Ovigerous Females

The prohibition against landing ovigerous females (those carrying external eggs) or removing their eggs substantially increases the number of eggs per recruit [threefold in the Bay of Fundy (Campbell, 1986)]. This regulation has strong support among lobster fishers. Biologists are unable to determine the number of eggs required to sustain a given level of recruitment to a fishery, but many are uneasy about the low fraction of females that survive the fishery to reach maturity.

6. Gear

Lobsters may be fished only by traps. Traps cannot exceed dimensions of 125 × 90 × 50 cm high or a number varying between 100 and 500. Lobsters taken as a bycatch in other fisheries may not be retained. Limiting the lobster fishery to one type of gear reduces conflict and avoids the problems of segregating fishing grounds and setting quotas for gear sectors. Traps are not destructive of habitat. Because they capture lobsters in good condition, there is minimum waste of the lobsters retained and a high survival rate among those returned to the sea. The one wasteful aspect of traps is that they continue to fish even if lost or abandoned ("ghost fishing"). Thus, traps are required to be equipped with a timed-release device that disables the trap after several months. Additionally, escape openings (vents) in each "parlor" allow undersized lobsters to escape and avoid being held in close proximity, which may result in injury.

Most fishers agree in principle with regulations on trap limit, trap size, and a trap-only fishery. When effective, these regulations tend to normalize effort among competing fishers and control the cost of fishing. Trap limits are administered by providing each fisher with a set of uniquely numbered tags each year. Any trap found on the fishing ground without a current year's tag attached may be destroyed. Any fisher in possession of an untagged trap may be prosecuted. This system is effective in areas where fishers support it by reporting to enforcement officers the location of untagged traps. However, where this support is lacking, trap limits are easily violated. This is probably the most expensive of all lobster regulations.

7. Excluded Regulations

Regulations that are not included are also important to the quality of a management regime. Unlike the lobster fishery, Canadian finfish fisheries have several fleet sectors based on gear type and boat size. In this management scheme, a total allowable catch (TAC) is set for a stock and divided among fleet sectors. In many cases this is further divided into individual transferable quotas (ITQs). The fleet sectoring provides the industry with a choice of fishing methods, and ITQs can increase the fleet's economic efficiency by reducing competition among vessels and aggregating quotas on fewer, more efficient vessels. However, these choices are bought at high costs for research to set TACs, lost credibility if the TACs are wrong, high administrative and political costs for allocating quotas among sectors and boats, and high enforcement costs (Copes, 1986; Sissenwine and Mace, 1992). This regime provides such a strong incentive to falsify catch statistics by misreporting catches and to high grade by discarding less desirable sizes and species that enforcement for a large fleet may be impossible at any reasonable cost. Conversely, it provides little incentive to invest in the resource so that it becomes more productive (Keen, 1988). Multiple fleet sectors competing for the same fish on the same grounds also reduce the sense of resource ownership. This regime could be applied to the lobster fishery by dividing it into fleets using trapping, trawling, and diving, each with large and small vessels, each assigned an ITQ. However, all of the above costs would apply. A large fleet of small boats landing easily transported, high-value product requiring no secondary processing could result in a nearly impossible enforcement problem.

C. History of Canadian Regulations

Most of this section refers to the outer coast of Nova Scotia (LFA 33, Fig. 1), but important regulations were introduced about the same time throughout Atlantic Canada. The sources of information are the bulletin by DeWolf (1974) and an unpublished manuscript by D. S. Moore of the DFO Halifax Laboratory (both of these draw extensively from Canada Sessional Papers, Ottawa); other sources include the work of Templeman (1941), Wilder (1954), and Gough (1991). The history of Canadian regulations (in LFA 33) is summarized in Fig. 2.

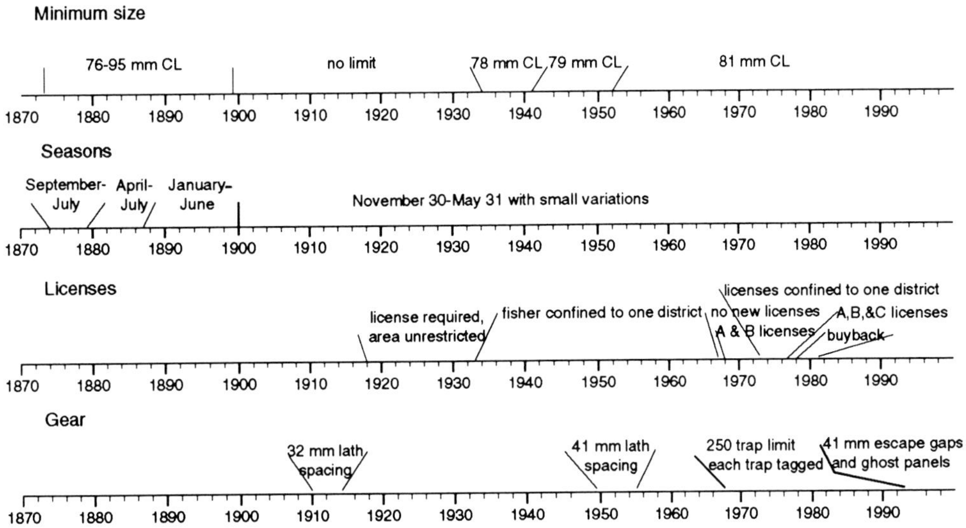

FIGURE 2 History of principal lobster fishing regulations in Canada's LFA 33 (Scotia–Fundy region). CL, Carapace length.

The Canada Fisheries Act was passed in 1868, 1 year after Canadian confederation, and still regulates fisheries on all marine and most anadromous/catadromous species. The first lobster regulations appeared in 1873 and set a minimum legal size and prohibited the retention of soft-shelled (recently molted) and ovigerous female lobsters. They were in response to a New Brunswick Fisheries Commissioner who cited evidence of overfishing scarcely 10 years after the commercial fishery began. The ovigerous female restriction is still in place. The next year, the restriction on soft shells was replaced by a summer closed season. In 1897 this was expanded to a 6-month closed season, similar to the June 1 to late November closure of today in LFA 33. The open season targeted the winter export of live lobsters to the United States. Avoiding the handling of ovigerous females was added later to the reasons for a summer closure (Knight, 1917).

An interesting variation on a seasonal closure was the complete 3-year closure of the Newfoundland fishery from 1925 to 1927. The landings were 264 mt (metric tons) in 1924, 2090 mt in 1928, and 1090 mt in 1930. Templeman (1941) interpreted this result as fast stock recovery and fast depletion by the fishery. The next landings peak occurred in 1936–37, perhaps representing the generation spawned by the large stock present at the end of the closed period.

Minimum sizes have varied greatly over time and still vary considerably among locations (Table 2), depending in large part on the mix between the small canner lobster and the live market lobster. LFAs 27–41 (and 3–14C in Newfoundland) (Fig. 1) have supplied the live market. It has been illegal for fishers to sell lobster parts since 1907.

Between 1887 and 1927 the federal government commissioned eight studies to resolve the cause of declining catches. Many recommendations were made and many regulations were added to the Fisheries Act, but each commissioner reported that important regulations were neither observed nor enforced. This continued to be a problem in some areas through the 1940s and 1950s (Wilder, 1954, 1965).

Lobster canneries have had a large impact on Canadian regulations. Because they could accept smaller and soft-shelled lobsters not suitable for the live trade, and sold a less perishable product not tied to an annual market cycle, they were less interested in season and size regulations. They preferred to operate their factories for a short time when supply was the greatest. In the 1880s canneries owned most of the boats and traps, hired fishers to fish for them, and tried unsuccessfully to obtain government leases on lobster grounds. The cash sales associated with the live lobster trade allowed fishers to become independent of cannery operators. As recently as 1993, cannery operators and workers in Prince Edward Island successfully resisted an increase in minimum size. It would have increased the cost of raw material and reduced the number of factory jobs because the larger lobsters would have been more valuable on the live market.

Lobster fishing districts were created in 1899 to make it easier to apply different regulations, such as season and size, to different areas. The number of districts has been increased several times to accommodate an increasing variety of regulations.

Traps were required to have 32-mm ($1\frac{1}{2}$-inch) lath spacing during 1910–1914 and 41-mm ($1\frac{5}{8}$-inch) spacing from 1949–1955; however, the regulations were dropped both times for lack of industry support. In 1993 a 44-mm ($1\frac{3}{4}$-inch) spacing was introduced to a chorus of familiar claims that the gaps let too many sublegal-sized lobsters in and too many of legal size out. Trap limits were introduced in 1968 and reduced in some areas in 1977, principally to control the cost of fishing. Fishing gear has been limited to traps for many decades, excluding hoop nets, all towed gears, and bycatch from other fisheries.

Lobster fishing licenses have been required since 1918. Since 1933, a license holder has been restricted to one district and, in an effort to control the flow of capital as well as labor, starting in 1945 the boat and traps could be used in only one district. The latter regulation was rescinded in 1959 because of difficulties with enforcement. With the exception of a few offshore licenses issued in the 1970s, no new lobster licenses have been issued since 1968.

Beginning in 1968 licenses were classified as A or B depending on the number of traps fished. B licenses could not be transferred between individuals and the number of traps could not be increased. From 1973, no license could be transferred between districts. In 1976, licenses were reclassified into A, B, and C categories, depending on the fisher's source of income. An A license was issued only to a fisher holding a lobster license continuously since 1968, or who received by transfer a license in effect in 1968, and who was not fully employed outside a resource industry in 1976. A B license was given to a license holder since 1968, whose income did not qualify for an A license; it was not transferable and allowed only about 30% as many traps (75 in LFA 33) as an A license. A C license was given to a fisher who received a license by transfer after 1968 and who was fully employed outside a resource industry. C licenses were withdrawn by 1978. A government-sponsored license-buyback program lasted from 1978–1981 and retrieved 22% of the licenses from the Maritime Provinces lobster fishery. The price paid was based on the previous 3 years of earnings, with a minimum and a maximum. Most licenses bought were inactive at the time, but as landings improved from a depressed state to record levels during the late 1980s, they surely would have reentered the fishery. From 1964 until 1990, the number of licenses in Nova Scotia, New Brunswick, and Prince Edward Island decreased by about 70%.

D. Modern U.S. Regulations

Administration of the lobster fishery in the United States is shared by 11 coastal states and the National Marine Fisheries Service (NMFS). The states have jurisdiction over the territorial sea, extending from shore to nominally 3 nautical miles from the headlands. The NMFS has jurisdiction over the Fisheries Conservation Zone, from the limits of the territorial sea to a line 200 miles offshore. The individual states and the NMFS grant licenses to fish their respective zones. Anyone licensed to fish for lobsters in federal waters must comply with federal regulations, regardless of where the lobsters are caught or landed. For example, V-notched lobsters may not be landed, whether or not a state prohibits their landing. If the state regulations where the lobsters are landed are more restrictive than federal regulations, the state regulations apply. For example, no maximum size limit exists in federal waters, but lobsters exceeding Maine's maximum limit may not be landed there. Many regulations are the same in state and federal waters. Regulations for the federal zone are set by the New England Fisheries Management Council in consultation with the Mid-Atlantic Fisheries Management Council. Each state is responsible for regulations within its territorial sea, but the Atlantic States Marine Fisheries Commission plays a coordinating role. Beginning in 1995, the Atlantic Coastal Fisheries Cooperative Management Act of 1993 will allow the federal government to suspend lobster fishing within a state's waters if that state does not comply with the management plan agreed on by the Atlantic States Marine Fisheries Commission.

Amendment 5 of the NMFS Fishery Management Plan for the American Lobster Fishery has recently been adopted (Anonymous, 1994) with the following salient points. Licenses for the federal zone were frozen for 5 years effective 1991. Overfishing was defined in this way: "The resource is recruitment overfished when, throughout its range, the fishing mortality rate (F), given the regulations in place at that time under the suite of regional management measures, results in a reduction in estimated egg production per recruit to 10 percent or less of non-fished populations ($F_{10\%}$)." Committees were appointed for each of four management areas (Gulf of Maine nearshore, southern New England nearshore, middle Atlantic nearshore, and offshore) to recommend stock-rebuilding programs to the New England Fisheries Management Council by early 1995.

The U.S. Coast Guard is responsible for at-sea enforcement of regulations in the federal zone, and individual states enforce regulations in coastal waters as well as on shore. Cases from the federal and state zones are tried in their respective court systems. Landings are recorded by both the NMFS and the states under a variety of shared responsibilities (New

England Fisheries Management Council, 1983; J. S. Krouse, Maine Department of Marine Resources, personal communication). Fisheries research, monitoring, and assessment are carried out to various degrees by biologists in each state and the NMFS.

The federal regulations protect ovigerous females and small lobsters and specify the type of fishing gear. Federal regulations also require that all lobster traps must be marked by a number assigned by the state or federal management agency. All traps must be equipped with one rectangular or two circular escape vents, to allow escape of undersized lobsters. All traps not constructed entirely of wood must be fitted with a timed-release (ghost) panel attached with material that will fail in about 1 year. This allows legal-sized lobsters to escape if a trap is lost or abandoned. Attaching the door of the trap with timed-release fasteners satisfies this requirement.

The federal minimum legal size is 82.5 mm ($3\frac{1}{4}$ inches), and only whole lobsters may be landed, although fishers without federal permits in New York and New Jersey are still allowed to land tails and claws in the shell.

To protect the female brood stock, ovigerous lobsters must be returned to the water immediately upon capture; it is illegal to possess ovigerous lobsters or females that have had their eggs forcibly removed. It is also illegal to possess a female with a V-notch cut in the right middle "flipper" (endopodite of the uropod) according to federal rules. (In Maine waters only, fishers voluntarily mark lobsters in this manner.)

Traps and otter trawling may be used for commercial fishing (except in Maine, New Hampshire, and Massachusetts, which prohibit trawling), but spears or other implements that would puncture the lobster shell may not. Recreational fishers may use traps or scuba; the daily bag limit is six. Principal regulations by jurisdiction are summarized in Table 3.

E. History of Maine Regulations

The history of Maine regulations is provided as an example of state management. This discussion is taken from an excellent report by Kelly (1992) and is summarized in Fig. 3. The first state lobster regulations in 1823 preserved lobster stocks for local fishers. By the end of the century, regulations were added to protect egg-bearing females and small lobsters and to require the marking of fishing gear.

Minimum size has been consistently regulated since 1874. Size limits were in force only part of the year until 1885, and smaller limits applied to lobsters sold to canneries than to the live trade until 1891. Beginning in 1919, all lobsters had to be landed whole, apparently to aid enforcement of the size regulations. The 78-mm ($3\frac{1}{16}$-inch) minimum CL intro-

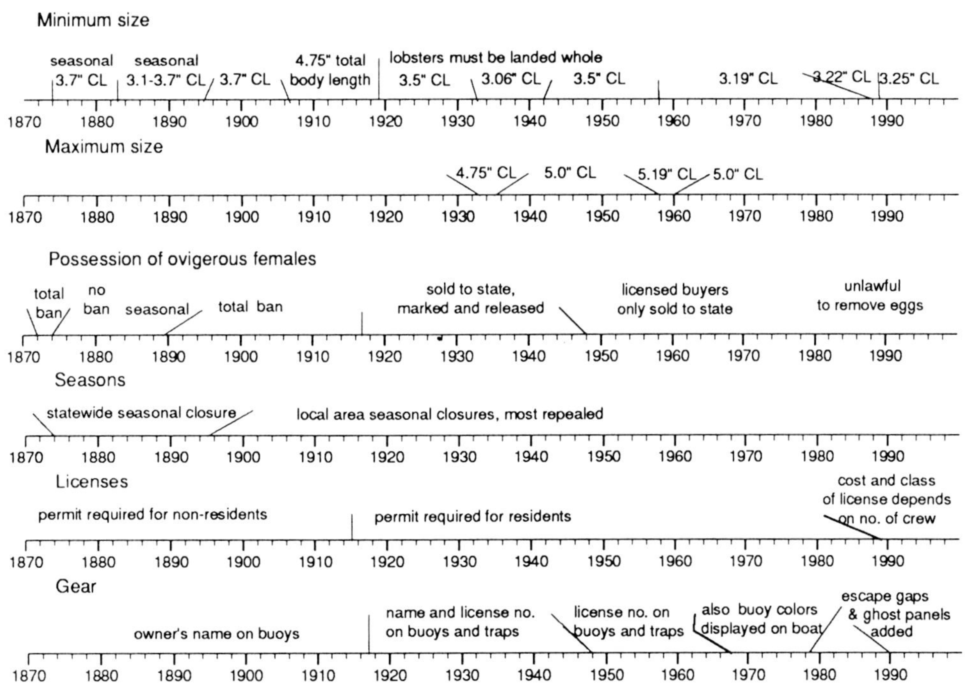

FIGURE 3 History of principal lobster fishing regulations in Maine. CL, Carapace length.

TABLE 3 **Summary of the Principal U.S. Lobster Fishery Regulations.**

Regulation	Jurisdiction							
	Federal	Maine	New Hampshire	Massachusetts	Rhode Island	Connecticut	New York	New Jersey
Minimum carapace length	———	———	$3\frac{1}{4}$ inches ———	———	———	———	———	$3\frac{3}{16}$ inches
Maximum carapace length		5 inches						
Ovigerous females protected	———	———	———	——— Yes ———	———	———	———	———
V-Notched females protected	———	——— Yes ———						
Trap escape gaps	——— $1\frac{7}{8} \times 6$ inches or 2 circles $2\frac{3}{8}$ inches ———	———	$1\frac{3}{4} \times 6$ inches or 2 circles $2\frac{1}{4}$ inches	———	——— $1\frac{7}{8} \times 6$ inches or 2 circles $2\frac{3}{8}$ inches ———	———	———	$1\frac{3}{4} \times 6$ inches or 2 circles $2\frac{1}{4}$ inches
Timed-release openings in traps	———	———	——— $3\frac{3}{4} \times 3\frac{3}{4}$ inches ———	———	———	———		
Commercial fishing gear (in addition to otter trawl and traps)	None	None	Scuba	Otter trawl	Otter trawl	Otter trawl	Otter trawl, scallop dredge	
Recreational fishing gear	Traps, scuba	None	5 Traps	10 Traps, scuba	5 Traps, scuba	10 Traps, scuba	5 Traps, scuba	Scuba
Trap limits		In small areas		800				
Landing lobster parts							In shell	In shell
Limited entry	Frozen at 1991 level for 5 years							

duced in 1933 has been slightly increased four times, leading to the current 82.5 mm ($3\frac{1}{4}$-inch) minimum. A maximum size was also introduced in 1933; this has varied only slightly from today's 127-mm (5-inch) CL limit.

To protect brood stock, selling ovigerous females was first prohibited in 1872, repealed in 1874, and replaced for part or all of the year from 1883 to 1917. Beginning in 1917, fishers could sell ovigerous females only to the state. A hole punched in the tail flipper before they were released marked these lobsters as property of the state which could not be legally retained if recaptured. Starting in 1948, only licensed lobster buyers could sell to the state lobsters that became ovigerous while in captivity, and the tail punch was replaced by a V-notch cut in the tail.

From 1874 through 1895 closed seasons applied to canning or to both canning and live markets, but there have been no statewide closures since then. In the early decades of the 20th century, seasons were applied in many local areas, but most have been repealed. Fishing has been prohibited at night (June through October) and on Sundays (June through August) since 1967.

Beginning in 1823, nonresidents of Maine were required to have a permit to fish lobsters. In 1917, 1-year residency was required; this was increased to 3 years in 1929, raised to 10 years in 1933, reduced again to 3 years in 1949, opened to any resident in 1975, and finally set at 6 months in 1979. Beginning in 1931, fishers were required to report their landings annually to the state. In 1991, licenses were divided into three classes based on whether the license holder fished alone, had one crew member, or had two or more crew members. The license fee increased with crew size. A lobster fund, financed by a portion of license fees since 1961, has been used to purchase ovigerous females from pound owners, to support lobster hatcheries, to fund lobster tagging studies, and for marketing.

Fishing gear has been regulated since 1885, when it was ruled that all trap buoys must bear the owner's name and that only the owner, or a designate, could haul traps. By 1917, all buoys, traps, boats, and lobster cars (for storage of live lobsters) were required to be marked with the owner's name and license number. The name requirement was dropped in 1948, but beginning in 1968 fishers were required to register their buoy color design and display it on their boat. From 1915 to 1953, it was unlawful to set traps on a trawl (groundline) without permission of the Commissioner of Fisheries. Starting in 1961, trawls were limited to a specified number of traps in some areas. Traps became the only legal fishing gear in 1961 and rectangular or circular vents were first required in 1979. Since 1990, traps made of synthetic material must be equipped with timed-release ghost panels.

F. U.S. versus Canadian Regulations

The United States and Canada each regulate lobster minimum size, protect ovigerous females, and require that traps be equipped with escape vents and timed-release (ghost fishing) devices.

The U.S. regulations afford greater protection to brood stock than do Canadian regulations. The United States has a higher minimum size, even in southern areas, which have a smaller size at maturity, thus allowing more of the populations to mature before being subjected to fishing. Part of the U.S. fishery also has a maximum size and tail notching to protect large and reproductively mature animals.

Canada's elaborate effort—with controls on seasons, number of licenses, number of traps, and fishing areas—has major social and economic consequences. Stabilizing who may fish and when and where they fish, and equalizing their fishing effort, reduce (somewhat) the competitive nature of the fishery and encourages a sense of common ownership. These controls also constrain the cost of fishing. Furthermore, a range of season and minimum sizes facilitates supplying different markets. Of course, these controls, largely absent in the United States, restrict freedom of choice.

All Canadian and most U.S. lobster fishers are allowed to fish with traps only. Both the retained and discarded catch from traps are usually healthy and traps do not alter the lobster habitat. Having only one gear sector avoids the inevitable between-sector conflicts over fishing grounds and animosities over share.

The presence of one management agency (DFO) in Canada versus several in the United States results in easier alteration of the management plan; however, regional differences in the quality of advice and enforcement remain.

The final major difference between the U.S. and Canadian management approaches is cost. In Canada's Scotia–Fundy region, total costs are about 7% of landed value versus an average of approximately half that amount in the major lobster-producing areas of the United States, including state and federal contributions. Of the several components of fisheries management—licensing, collating landing statistics, fishery monitoring, research, development, and enforcement of regulations—enforcement is by far the most expensive in both countries.

III. Nature of the Fishery

A. Canadian Offshore Fishery

The Canadian offshore fishery is much smaller than its U.S. counterpart offshore fishery in geographic extent, number of vessels, and catch (Fig. 4). It is confined to the northeastern edge of Georges Bank, the deepwater basins north of Georges Bank, and 100 km along the edge of the Scotian Shelf south and east of Browns Bank (Pezzack and Duggan, 1988), for a total area of about 10,000 km^2. The total allowable catch of 720 mt was taken by five vessels in 1993.

The Canadian offshore fishery also has a much shorter history (Pezzack *et al.*, 1992). Offshore lobster licenses were offered to swordfish fishers in 1972, following closure of that fishery due to high mercury levels in swordfish flesh. Offshore lobsters were not landed previously because trawlers were not permitted to sell their lobster bycatch. The inshore and offshore fleets were separated by a line 50 nautical miles from shore. In 1976, the offshore fleet was frozen at eight vessels in response to objections from inshore fishers. A trap limit and quota were imposed on the six vessels operating in Canadian waters, but not on the two boats sharing Georges Bank with the unlimited U.S. fishery. In 1979, a 5400-km^2 area of Browns Bank was closed to all lobster fishing because it had been identified as a lobster spawning area and a possible source of larvae to seed inshore grounds (Stasko, 1978). After the World Court set the United States–Canada boundary in the Gulf of Maine in 1984 (the "Hague Line"), each of the eight licenses was given a transferable quota of 90 mt. This 720-mt TAC has since been consolidated onto five boats. The 1000-trap limit was rescinded in 1992 because the boat quota was deemed to be a sufficient control of fishing mortality. The fishery has always been regulated by an 81-mm ($3\frac{3}{16}$-inch) minimum CL, a prohibition against retaining ovigerous females, and fishing with traps only.

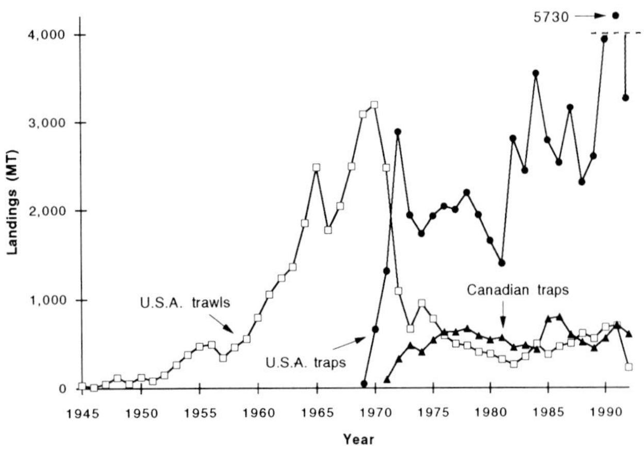

FIGURE 4 United States and Canadian catch of offshore lobsters, 1945–1992. MT, Metric tons.

Objections from inshore lobster fishers have had a large influence on regulation of the offshore fishery. A recent review (Pezzack *et al.*, 1992) addressed the question "Do catches offshore significantly reduce catches by the inshore fishery?" The reviewers concluded that lobsters taken by the offshore fleet would not migrate inshore to be taken during the December to May inshore season, that offshore ovigerous females would not move inshore to release larvae during the summer closed season, and that larvae released on Georges Bank or the Scotian Shelf east of Browns Bank would not seed inshore grounds. If indeed there is an effect from offshore removals, it is to reduce the drift of larvae inshore from the vicinity of Browns Bank. However, much of Browns Bank is closed to fishing, and even in the open areas the fishery has probably been too small to affect larval production. A consistent reduction in catch per unit effort and sizes of animals in the catch should accompany the fishing down of a virgin stock. Neither of these changes has occurred. (Fogarty, in Chapter 6, discusses the interaction between inshore and offshore populations; see Cobb, Chapter 7, on the interrelationship of ecology, behavior, and fisheries.)

The offshore fleet is most active from November through July. Summer fishing is constrained by conflicts with otter trawlers on Browns and Georges banks and lower survival in holding tanks. The fleet follows lobster migrations as shallow as 120 m in summer and as deep as 400 m in winter (Pezzack *et al.*, 1992).

Traps are fished in strings of 50–120. Each trap is about 1.2 m long with two opposite entrance ramps leading to a middle "kitchen" chamber. This is connected by ramps to "parlors" (also called "bedrooms") in each end of the trap. Construction material changed from nearly all wood to plastic-coated wire during 1990–1992. A long soak time of 7 days is more successful offshore than inshore, perhaps because of lower lobster density, a greater distance of attraction resulting from a larger foraging area for large lobsters, and odor trails that are easier to follow on a smoother bottom.

Four of the five vessels are converted steel side-draggers about 37 m long. All hold their catch in live wells with continuous renewal of surface water and aeration.

B. U.S. Offshore Fishery

In the United States, offshore lobsters have been taken as bycatch in a beam trawl fishery from 1891 and an otter trawl fishery since 1905 (Schroeder, 1959). A trawl fishery directed at lobsters began south of Massachusetts and Connecticut in either 1947 (McRae, 1960) or 1953 (Schroeder, 1959). It soon expanded along the continental slope edge as far as Corsair Canyon on the eastern end of Georges Bank and Norfolk Canyon off Virginia, ranging in depth from 100 to 500 m (Skud, 1969; Fogarty, Chapter 6) (Fig. 1).

In 1952–1953, several vessels working for the U.S. Bureau of Commercial Fisheries (now the NMFS) and exploring for commercial concentrations of finfish located lobsters between southern Georges Bank and Virginia (Schroeder, 1959). During 1955–1956, the same organization sponsored eight cruises of the R/V *Delaware* to explore for offshore lobster and red crab between Corsair and Hudson canyons (McRae, 1960). Schroeder described the trawlers of the 1950s as about 18 m in length; they held their catch in tanks supplied with pumped seawater and obtained a marketable catch averaging 500 kg per sea day. Landings peaked at approximately 3000 mt in 1969–1970 and decreased rapidly to about 500 mt by 1977, where it has remained (Fig. 4). The offshore fishery is now principally a winter fishery concentrated on the seaward edge of Georges Bank (Burns *et al.*, 1979).

The offshore trap fishery quickly overtook the trawl fishery. From 1969 to 1972, landings from traps increased from 52 to 2900 mt (Fig. 4) (Burns *et al.*, 1979), stimulated by the introduction of the hydraulic trap hauler (Fogarty *et al.*, 1982). Trapping had the advantage that lobsters sustain fewer injuries and mortalities. During the 1970s, the deepwater trap fishery spread throughout the Gulf of Maine and across the continental shelf between Cape Cod and New Jersey (Burns *et al.*, 1979; New England Fisheries Management Council, 1983). However, more than 80% of the landings have always been taken from the outer edge of the shelf from Georges Bank to southern New England (Burns *et al.*, 1979; Fogarty *et al.*, 1982). The Hague Line, created by the World Court decision in 1984, excluded U.S. vessels from Georges and Crowell basins located north of Georges Bank and Corsair Canyon on the ocean side of Georges Bank. These areas represented annual landings of at least 100 mt (Pezzack and Duggan, 1988).

By 1970, the fishing-up process was well along, as indicated by decreases in mean sizes (Skud, 1969) and catch per unit of effort (Fogarty *et al.*, 1982). Mean sizes of lobsters taken from Hudson and Veatch canyons, located south of New England, decreased by 20 and 40 mm ($\frac{3}{4}$ and $1\frac{1}{2}$ inches), respectively, between 1956 and 1966. From 1964 through 1970, catch per unit of effort from trawl surveys on Georges Bank and off southern New England decreased by more than 50%. An increase in landings during the 1980s (Northeast Fisheries Science Center, 1993) was presumably due to a pulse in recruitment (M. J. Fogarty, National Marine Fisheries Service, personal communication).

Inshore-style wooden traps were used initially, but by the early 1980s most boats had switched to metal because they withstood rough handling better, were not attacked by marine borers, and held their position better in strong currents (T. Angell, Rhode Island Department of Environmental Management, personal communication). In the 1990s, traps with two parlors in line have become popular (Fig. 7F). Fishers believe that over long soak times this design retains lobsters better than the single-parlor design (J. S. Krouse, Maine Department of Marine Resources, personal communication). During the 1970s, an offshore trap was expected to last less than 1 year, primarily because of interference by foreign trawlers. Much of this conflict has been resolved by lobster fishers reporting the position of their gear to the Coast Guard, which in turn broadcasts the locations.

A typical fishing enterprise in Rhode Island now uses a 20- to 28-m steel boat, fishes 2000 traps every 7–9 days, sets traps in strings of 40–70 spaced 30 m apart, and fishes year-round in depths of 120–360 m (T. Angell, Rhode Island Department of Environmental Management, personal communication).

The offshore lobster fishery has been regulated by federal statutes since the Magnuson Act claimed jurisdiction over a 200-mile zone in 1977. Regulations are the same as those in most states: a minimum CL, prohibition against landing ovigerous females and lobster parts, and a requirement for escape gaps in traps (New England Fisheries Management Council, 1983). There is no record of a directed foreign fishery. In 1974, the lobster was declared by the United States to be a creature of the continental shelf and reserved for the domestic fleet (New England Fisheries Management Council, 1983).

C. Landings and Markets

Early European explorers and immigrants to North America reported capturing and eating lobsters. Martin and Lipfert (1985) mention four visits of Englishmen to New England and Nova Scotia between 1602 and 1607 in which lobsters were described as being abundant and easily captured.

French settlers in Nova Scotia also ate lobsters (Lescarbot, 1609, as quoted in DeWolf, 1974; Denys, 1694). Herrick (1909) mentioned a feast in which lobsters were served to arriving colonists in 1623. Until the mid-1800s, they remained plentiful enough to be used as fish bait in the United States (Rathbun, 1887) and as garden fertilizer in Canada (DeWolf, 1974).

The early U.S. fishery was small and local, centered in Massachusetts until about 1840. It then developed rapidly to include the entire coastal distribution of the species by 1860 (Rathbun, 1887; Cobb, 1901). Although complete records are not available, the fishery peaked in Maine in 1889 (Cobb, 1901) and probably earlier in other states. Herrick (1909) estimated that it decreased by half between 1892 and 1906. Since 1928, the beginning of complete landings records, Maine landings have constituted 46–67% of the U.S. total. In the mid-1970s, Massachusetts began setting record catches annually, followed by upturns in all of the other important lobster-producing states (Fig. 5). In 1991, U.S. catches set a record of 29,000 mt with a landed value of U.S. $165 million (U.S. Department of Commerce, 1992).

The lobster fishery in the United States has traditionally served a domestic market for live lobsters. Most were transshipped live to Portland, Boston, or New York (Rathbun, 1887). Because of the propinquity of the fishery to the markets, the lobster canning industry was small and short-lived relative to the Canadian experience. All canneries were located in Maine during 1843–1895 and peaked at 23 in 1880 (Cobb, 1901). In 1991, the United States caught 50% of its total supply. Of this total, 18% were sold to growing export markets (U.S. Department of Commerce, 1992).

Most U.S. states allow recreational lobster fishing (with the notable exception of Maine). This represents about 1% of the total landings in New York (P. T. Briggs, New York Department of Environmental Conservation, personal communication) and about 5% each in Connecticut (Smith and Howell, 1987) and Massachusetts (B. T. Estrella, Massachusetts Department of Fisheries, Wildlife, and Environmental Law Enforcement, personal communication).

The Canadian commercial fishery began in about 1870 in New Brunswick and Nova Scotia (Fig. 6), the provinces closest to U.S. markets (DeWolf, 1974). It grew to include Prince Edward Island, Quebec, and Newfoundland by 1876 (Templeman, 1941; Bergeron, 1967; DeWolf, 1974). Canadian landings peaked in 1886, then began a 30-year decline. They continued at a low level, with minor upturns in the 1930s and 1950s, until the 1980s, when new historic highs were set in Canada, as in the United States. The value of Canadian landings in 1992 was U.S. $228 million (Department of Fisheries and Oceans, 1993). These very high landings extended from New Jersey to the Magdalen Islands in the Gulf of St. Lawrence. Only Newfoundland and a 300-km section of the outer coast of Nova Scotia (LFAs 28–32, Fig. 1) did not have near-record landings.

Canada, with a much smaller domestic market than the United States, has always exported most of

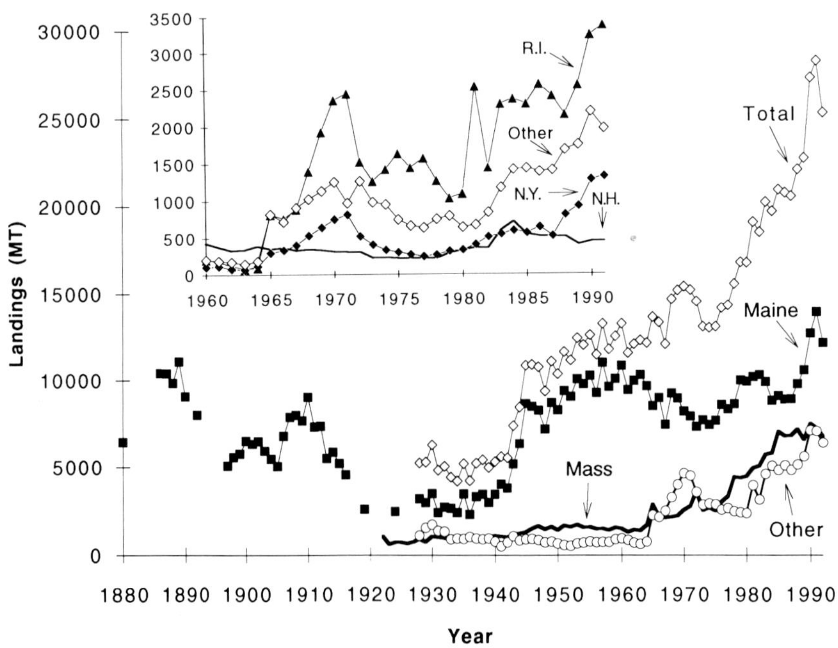

FIGURE 5 United States lobster landings through 1992, total and by state. MT, Metric tons.

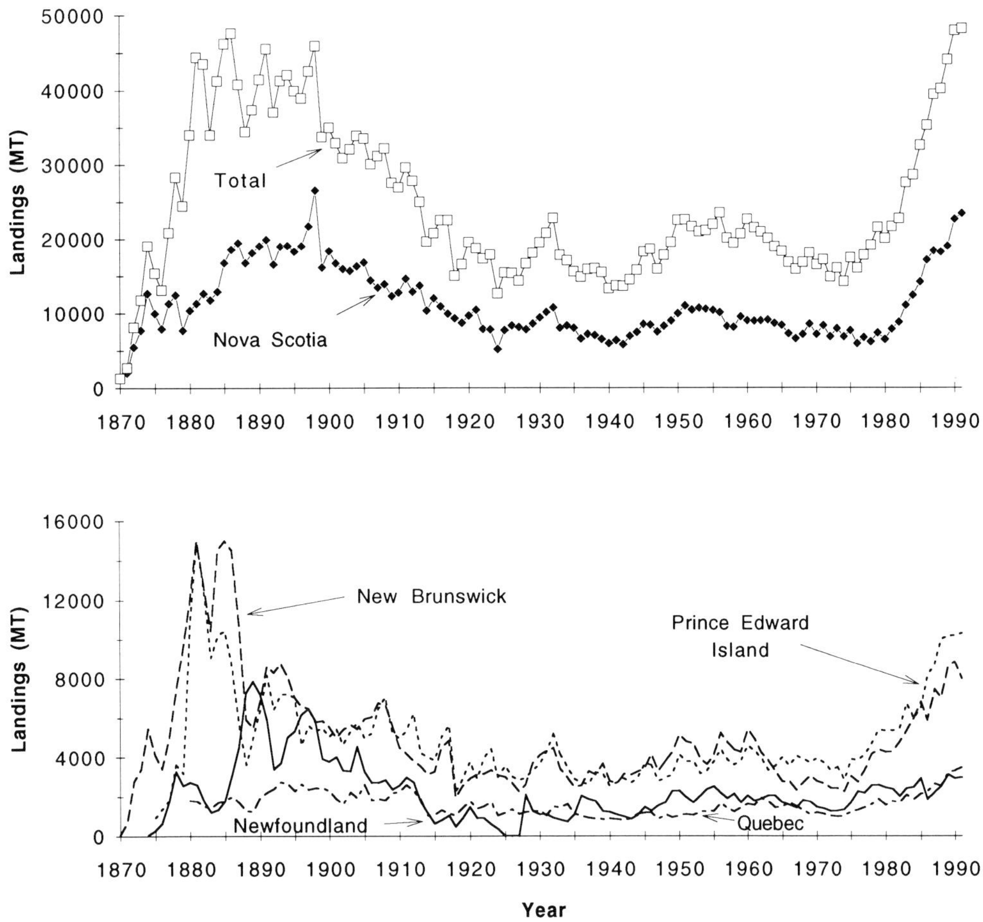

FIGURE 6 Canadian lobster landings, 1870–1992, total and by province. MT, Metric tons.

its catch. A U.S. market for Canadian lobsters developed in the 1860s, when U.S. landings no longer satisfied demand. By 1873, 64 canneries were in operation in Nova Scotia and New Brunswick. By the turn of the century, this number increased to about 1200 in all eastern provinces (including Newfoundland, which joined Canada in 1949) (DeWolf, 1974; Gough, 1991). Although a live market developed by 1881 in areas closest to the United States, processed meat has always been a large part of the Canadian market (Robinson, 1980). Only one half of the product was sold live by World War II, and today canned and frozen products still constitute 27% of the mix (Barbara, 1991). In 1990, the combined markets for live and processed Canadian products were 57% to the United States, 26% to Europe, 10% to Canada, and 7% to Japan and others (Barbara, 1991).

D. Gear Conflicts

Fixed gear, such as lobster traps, and towed gear, such as otter trawls or scallop dredges, cannot be fished on the same grounds at the same time. If conflicts are decided by force, towed-gear fishers usually win, as they destroy or displace fixed gear merely by towing through it. If conflicts are resolved through the political process, lobster fishers may win because of their superior numbers.

Lobster fishers are also concerned that towed gear destroys lobsters or unfavorably alters the habitat. Otter trawling has the greatest potential for capturing and damaging lobsters. Lobsters are most susceptible to injury soon after molting. In Long Island Sound, the mortality rate of trawl-captured lobsters was 6–21% during two molting periods, but 2% or less during the remainder of the year (Smith and Howell, 1987). In the directed trawl fishery for lobsters, as much as 30% of the catch may be damaged due to rough handling (New England Fisheries Management Council, 1983). The introduction of modern rock-hopper trawls has expanded the range of trawls to include a rougher bottom. A small commercial rock-hopper caught an average of 160 kg of lobster per 20-minute tow in shallow water near southwest

Nova Scotia (Simon and Campana, 1987). The possible damage of lobster habitat by trawls is part of the rationale for prohibiting their use in U.S. inshore waters (B. T. Estrella, Massachusetts Department of Fisheries, Wildlife, and Environmental Law Enforcement, personal communication).

Inshore dragging for scallop and Irish moss captures and injures few lobsters relative to the value of those fisheries. In the southern Gulf of St. Lawrence, $73 Canadian worth of lobsters were destroyed in the process of capturing $12,000 Canadian worth of scallops (Jamieson and Campbell, 1985). In coastal Nova Scotia, scallop and lobster co-occurred in only two of 52 locations of suspected conflict. In the worst of these two cases, only 1% of the lobster biomass would have been injured by intense scallop harvesting (Roddick and Miller, 1992). In northern Prince Edward Island, the value of lobsters destroyed by Irish moss harvesting was 18% of the value of the Irish moss harvest and 7% of the value of the lobster landings (Scarratt, 1973). In a nearby area, Irish moss dragging injured 1.3–2.7% as many lobsters as the lobster fishery took (Pringle and Sharp, 1980). Dragging for sea urchins and kelp are new activities for which the effect on lobsters has yet to be assessed. Although lobster fishers often express concern about habitat damage, there appears to be an absence of studies that have measured the effect of habitat alteration on lobster yield.

E. Fishing Technology

1. Traps

The evolution of lobster fishing gear in the 19th century is nicely summarized by Rathbun (1887). American lobsters were first captured using spears, gaffs, and dip nets in shallow water. Hoop nets came into general use around 1800. These were constructed of an iron hoop about 70 cm in diameter to which was attached a shallow net bag. Two wooden half-hoops crossed over the iron hoop at right angles. The bait and the rope to a surface buoy were attached to the intersection of the wooden hoops. Hoop nets were typically fished at night, when lobsters are most active. This gear was simple and inexpensive to make and it fished quickly because lobsters did not have to locate or pass through an entrance. However, lobsters could walk off the net as easily as they walked on, requiring the net to be hauled every 10–30 minutes.

The lath trap came into use by 1830. These were enclosures similar in shape to those used today. They were typically half-cylinders with rounded tops and flat bottoms and with a funnel-shaped entrance in one or both ends. The lath sides were nailed onto sturdy half-hoops made of tree branches. Funnels were usually made of twine, but sometimes of lath. Bait was impaled on a wooden or iron spear fixed vertically to the middle bottom of the trap. Stones or bricks were tied to the bottom of the trap for weight. Lobsters were removed through a leather-hinged door in the top of the trap.

The obvious advantage of the lath trap over the hoop net was that lobsters could not easily escape, allowing fishers to tend their gear once in 24 hours, work in the daytime, and set many more traps over a larger area of fishing ground. Lath traps were first set one or two to a buoy, but in some areas around 1860 this evolved to setting eight to 40 traps on a single groundline or "trawl" (not to be confused with a trawl net) buoyed at both ends. Buoying traps individually allowed each one to be placed strategically. However, this method was best carried out with two fishers in a boat, one to sail or row and the other to haul traps. Using trawls, once a fisher reached the first trap he could pull the boat along the groundline from trap to trap without concern as to steering.

The lath trap probably evolved from the Dutch *tejener* used for fishing European lobster since at least the early 1700s. The *tejener*, in turn, was developed from the eel trap. It was cylindrical, with sides made of roots or branches, and with a funnel-shaped entrance in each end.

It is not known when the modern parlor trap came into general use. Martin and Lipfert (1985) suggest about 1930 for the Maine coast, although a photo of a large number in northern Nova Scotia is included in a 1917 publication by Knight. These traps usually had the entrance heads on the sides rather than on the ends (Fig. 7), with the supposed advantage of preventing lobsters from falling out the end funnel when the trap was being hauled. More importantly, the addition of an internal compartment, the parlor, markedly reduced escape. After World War II, more durable nylon mesh replaced tarred natural fiber for use in trap entrance heads (Prudden, 1962). In some areas, traps were dipped in wood preservative to protect them from marine borers, following experiments by Dow and Baird (1953). Plastic-coated wire traps were first introduced to the U.S. offshore fishery in 1969 (Fig. 7C and F) and spread inshore soon after that (Martin and Lipfert, 1985). These were more recently adopted by the Canadians. About 85% of the traps in Maine (J. S. Krouse, Maine Department of Marine Resources, personal communication) and 65% of those in Nova Scotia (personal observation) were wire in 1993.

Escape gaps became mandatory in U.S. federal and

FIGURE 7 Several trap designs now in use, all shown with lids open: (A) six-sided trap with three entrances and no parlor, (B) lath trap with two entrances and one parlor, (C) wire trap with two entrances and one parlor, (D) wood-framed trap with mesh sides, two entrances, and one parlor, (E) combination wood–wire trap with two middle entrances and two end parlors, and (F) wire trap with steel frame, two end entrances, and two parlors in line.

most state jurisdictions by 1984, and ghost fishing provisions were required by 1992 (Table 3). Escape gaps were instituted for the third time in Canada in 1993 and ghost fishing provisions were required for the first time in the same year, except in the southern Gulf of St. Lawrence, which is still exempt.

2. Other Fishing Equipment

If a technical innovation makes fishing easier or more profitable, it is adopted sooner or later, although the rate of penetration varies from time to time and place to place. Variables affecting the level of acceptance are cost, economic state of the industry, and traditions of participants. For example, in the 1930s, when earnings declined more than operating costs, some fishers changed from gasoline-powered vessels to rowboats (Martin and Lipfert, 1985). The following discussion of technical innovation relies on information from Lunt (1981), Martin and Lipfert (1985), and W. Cross (a Halifax retailer of marine electronics, personal communication).

Engines began appearing in lobster boats in about 1900. These were one-cylinder gasoline engines of 1–5 hp. Engines increased a fisher's range, increased independence from weather, and required less attention than sailing or rowing. After World War I, four-cycle gasoline automobile engines were adapted to marine use. These had much more power per of unit of weight, a throttle, and a gear box with forward, reverse, and neutral. Because of the economic depression of the 1930s and the relatively high cost, these engines were slow to be adopted. In the 1970s, diesels began replacing gasoline engines. The purchase cost is much greater, but fuel consumption, reliability, and safety of operation are compensating advantages.

The added weight and power of engines dictated changes in hull design, from the rounded streamlines suited to sail power, to wider beams, wide and flat sterns, high bows, and heavier construction to absorb the vibration. Strong, low-maintenance, fiberglass hulls have been an alternative to wood since the 1970s. Boat design has been discussed in detail by more qualified observers (Rathbun, 1887; Lunt, 1981).

The multicylinder engine had enough power to drive a trap hauler. To raise a trap a fisher grabbed the buoy, turned the trap line around a capstan once or twice, and maintained tension on the line until the trap reached the surface. These haulers were first belt driven, then hydraulic. Since the late 1970s, they have been gradually replaced by hydraulic power blocks capable of hauling a trap so fast that it can be deformed by the water's resistance. They also grip the line automatically. The replacement of natural-fiber rope with stronger synthetic-fiber rope in the 1950s was a great boon to the fishery, if not to our scenic beaches.

The use of electronic aids has increased from none to a smorgasbord of high technology during the last 50 years. Radar has been available since World War II, but has not been adopted by all fishers because many can retain a shoreline map of their fishing area in their memory. However, a map of the sea bottom where lobsters are found requires a more difficult method of observation and a much finer resolution. Depth sounders have been improved many times, each giving more information on bottom relief than the previous generation. Paper recorders introduced in the 1940s gave way to black-and-white screen displays in the 1960s and to color displays using multiple frequencies in the late 1980s. New, forward-looking depth sounders permit a fisher to search to the side and front of a vessel for the best fishing bottom. Deca and loran A, using radio signals transmitted from shore-based stations, have helped fishers position themselves on nautical charts since the 1950s. The more precise loran C came into use in the early 1980s and the new and still-improving satellite-based global positioning system was introduced in the 1990s. Fishers have communicated boat to boat and with shore stations using FM or citizen band radios since the 1950s. Cellular phones are becoming the communication gadget of choice for the 1990s.

F. Fishing Success

The success of a lobster fisher in Canada depends primarily on location. Stories abound of fishers who could catch lobsters in a rain barrel and of others who could not catch one in a lobster pound. However, success is more likely determined by the section of coast being fished than by the individual's fishing skills (e.g., judgments about trap placement, choice of bait, and trap design). In areas of western Nova Scotia and western New Brunswick with similar fishing seasons and the same minimum lobster size, the mean annual earnings per license in the 1991–1992 season were Canadian \$16,000 and \$72,000 (LFAs 36 and 34, respectively). The top 10% of licenses in these same districts averaged earnings of \$38,000 and \$150,000. Clearly, location was a dominant factor.

In one community in LFA 34, records of annual landings for nine fishers for each of 3 years varied by only a factor of 2. Also, the lowest annual catch was higher than the highest annual catch among all 170 fishers in LFA 36. A combination of analysis of variance (ANOVA) and correlation analysis on the 27 records from LFA 34 shows that the variance in annual landings per fisher is explained 48% by number of trap hauls (also recorded), 32% by fisher (presumably skill), and 10% by year, with 10% due to other causes. Thus, in order of importance, factors contributing to a fisher's success were location, fishing effort, skill, and year.

The absence of limited entry and LFAs moderates

the importance of fishing location in the United States. There, fishers have more freedom to move into areas of high catch rates, diluting individual landings. However, some communities in both the United States and Canada exercise control over fishing effort on their grounds (see Section IV).

IV. Community-Based Management

Under community-based management, a fishing community collectively assumes at least part of the responsibility for managing a fishery resource. This arrangement changes a common property to a shared private property. This practice is found in many fisheries in many cultures worldwide (Berkes *et al.*, 1989). The community membership must be exclusive (i.e., clearly identified) to be effective and the borders of the fishing area must be well defined. Membership and enforcement of rules among members can be maintained through access to wharfage, access to the better fishing grounds, membership in a cooperative, or a range of social interactions. Maintaining the borders of community grounds can involve real or implied threats, destruction of traps, or other property damage (Acheson, 1988). Success of community management probably depends on a tradition of cooperation, quality leadership, and maintaining a small local membership. Government intervention has the potential to destroy cooperative arrangements by assuming responsibility for regulation, then destroy management by failing to provide adequate enforcement (Berkes *et al.*, 1989).

No discussion of lobster management is complete without mention of the Western Australian rock lobster fishery (see Brown and Phillips, 1994; Brown *et al.*, 1994). It includes a fleet of 700 vessels, it is well managed, and the management regime is adjusted often (Bowen and Hancock, 1989). It is comanaged by industry and government through an industry advisory committee that endorses every management measure before it is adopted (B. K. Bowen, Western Australian Department Fisheries, personal communication). The exceptional aspects of this arrangement are that (generally) the industry accepts the judgment of its representatives and (generally) the government accepts the judgment of its advisory committee. That is, industry factions do not often demonstrate against the committee decisions and government does not require industrywide votes. This exhibits a degree of maturity in the application of representative democracy not often seen in the management of North American lobster fisheries.

The lobster fishery is a good candidate for community management because both the lobster and the lobster fisher have small home ranges. Examples of successes in Maine (Acheson, 1988) include higher catch rates on grounds under the control of one group of fishers compared to rates on grounds shared among groups. Some groups have imposed seasons and trap limits not specified by regulations. A community in eastern Nova Scotia partitioned its grounds into several exclusive areas for individual members, plus a few common areas shared by all members. When the exclusive areas of two fishers began yielding above-average catches, the group voted to deny these members access to the common areas. It should be recognized, however, that these examples of success are the exception. As discussed in Section II, before government became involved in management, virgin lobster stocks were quickly fished out in both the United States and Canada. Over many decades and throughout much of the fishery, important conservation measures were ignored before government applied a heavy hand to enforcement.

The analysis and conclusion presented here are in agreement with those of Acheson (1988), Berkes *et al.* (1989), and Jentoft (1989), who suggested that cooperative arrangements between communities and management agencies (comanagement) may hold the most promise for effective change. Improvements in lobster fishery regulations have been slow since the 1960s, frustrating industry and government representatives alike. Even well-conceived proposals usually polarize positions among groups of fishers, or fishers and the management agency, then die from lack of consensus.

Fishery managers would do well to emulate the elements that lead fishers to accept technological change (Section I). These elements are individual decision-making, including the option to reverse the change, a demonstration of the benefits of a new technology before acceptance, and immediate feedback on the consequences of acceptance. Although none of these elements are completely attainable when changing fishery regulations, they can be approached. An agreement to change a regulation can include the condition that it can be reversed after an appropriate trial period. In order to demonstrate the consequences of a change, the result must be large enough to produce a measurable effect. If a change is first accepted by a small group of fishers, who are perhaps even paid for any inconvenience during a trial period, then their experience can be used as a demonstration for other groups. Immediate feedback on the consequences of change is usually not practical because biological populations take time to respond.

However, the management agency can act quickly to implement the change once agreed on. Also, during the trial period, agency staff can involve the industry in data collection to test the effect of change and keep fishers well informed of the results. This approach should reduce initial resistance to change and increase the chance of eventual acceptance. However, it would require a leader persuasive in both bureaucratic and industry circles—and one who is unusually persistent.

V. The Lobster Fishery in the Year 2020

Current trends suggest predictions about the nature of the lobster fishery in the next century. Lobster boats will be small, fast, and energy efficient. Fishers will all be equipped with precise positioning devices and many will have portable underwater video equipment to help them position traps on good lobster habitat. Traps will remain the method of capture, but will come in a greater variety than the "poverty crates" seen today. The most popular will be small, collapsible designs similar to those now in use in Japan. A design innovation will prevent escape through the side entrances. Lobster habitat will be enhanced by construction of artificial shelters as well as degraded by coastal construction and mineral extraction. Ownership of fishing enterprises will be concentrated in fewer hands and hired fishers will be more common. Lobster abundance will still fluctuate widely over time, and management will not radically change. Recreational and native fishers will take a larger share of the catch. A price ceiling will be set by the product from lobster aquaculture and markets will become more international.

VI. Summary

The fishery for the American lobster, *Homarus americanus*, has had uninterrupted success for over a century in both the United States and Canada. This chapter describes the past and present management regimes in both countries. Cornerstones of the present regimes are the protection of sublegal sized lobsters and the brood stock, the use of nondestructive fishing gear, and the proprietary interest of the fishers in their lobster stocks. Minimum size and brood stock protection were introduced early in the fishery, but were not well enforced until after World War II. Effective controls on fishing effort were added to Canadian management in the 1960s.

The weight of Canadian landings has been higher than that of U.S. landings since 1880, but the gap is narrowing from about 3:1 before 1945 to less than 2:1 since 1975. The offshore fisheries began in the 1940s in the United States and the 1970s in Canada, and have always composed only a small part of the total landings. U.S. production serves mostly a domestic market as a live lobster trade. Canadian production, on the other hand, is 90% exported, with a significant portion in canned or frozen lobsters.

Technological change has been constant throughout the history of the lobster fishery. Changes in boat design, power, traps, materials, and electronics have all been adopted when they made fishing easier or more profitable. The ease with which these technological changes have been accepted is in contrast to the industry's enormous resistance to management changes. Of several methods suggested for overcoming resistance, community-based management with a focus on small groups of fishers and working closely with these groups to demonstrate the effect of the proposed changes appears to hold the most promise.

Acknowledgments

J. S. Krouse, B. T. Estrella, M. J. Fogarty, and P. T. Briggs were generous in providing source material. D. S. Pezzack supplied the landings data for the fishers discussed in the section on fishing success, and R. Rodger performed the analysis. M. Comeau supplied most of the traps shown in Fig. 7. J. R. Factor and J. S. Krouse improved the readability and accuracy of the manuscript.

References

Acheson, J. M. (1988). "The Lobster Gangs of Maine." University Press of New England, Hanover, New Hampshire.

Anonymous (1994). American lobster fishery. *Fed. Regist.* 59(118), 31938–31955.

Barbara, R. M. (1991). Scotia–Fundy Live Lobster Price Trends: 1981–1990. *Can. Econ. Comm. Anal. Rep.* 109, 1–19.

Bergeron, J. (1967). La Pêche Commerciale du Homard (*Homarus americanus* Milne-Edwards) du Québec des Origines a Nos Jours. Québec Ministère, de l'Industrie et du Commerce, Station de Biologie Marine (Grande-Rivière). *Cah. Inf.* 42, 1–47.

Bergh, M. O., Pikitch, E. K., Skalski, J. R., and Wallace, J. R. (1990). Statistical design of comparative fishing experiments. *Fish. Res.* 9, 143–163.

Berkes, F., Feeny, D., McCay, B. J., and Acheson, J. M. (1989). The benefits of the commons. *Nature (London)* 340, 91–93.

Bowen, B. K., and Hancock, D. A. (1989). Effort limitation in the Australian rock lobster fisheries. *In* "Marine Invertebrate Fisheries: Their Assessment and Management" (J. F. Caddy, ed.), pp. 375–393. Wiley, New York.

Brown, R. S., and Phillips, B. F. (1994). The current status of Australia's rock lobster fisheries. *In* "Spiny Lobster Management" (B. F. Phillips, J. S. Cobb, and J. Kittaka, eds.), pp. 33–63. Blackwell, London.

Brown, R. S., Caputi, N., and Hall, N. G. (1994). Measurement of catch and fishing effort in the western rock lobster fishery. In "Spiny Lobster Management" (B. F. Phillips, J. S. Cobb, and J. Kittaka, eds.), pp. 262–284. Blackwell, London.

Burns, T. S., Clark, S. H., Anthony, V. C., and Essig, R. J. (1979).

Review and Assessment of the USA Offshore Lobster Fishery. *Int. Counc. Explor. Sea Res. Doc.* C.M. 1979/K:25, 1–31.

Campbell, A. (1986). Implications of size and sex regulations for the lobster fishery of the Bay of Fundy and southwestern Nova Scotia. *Can. Spec. Publ. Fish. Aquat. Sci.* 92, 126–132.

Campbell, A., and Robinson, D. G. (1983). Reproductive potential of three American lobster (*Homarus americanus*) stocks in the Canadian Maritimes. *Can. J. Fish. Aquat. Sci.* 40, 1958–1967.

Cobb, J. N. (1901). The lobster fishery of Maine. *Bull. U.S. Fish. Comm.* 19, 241–265.

Copes, P. (1986). A critical review of the individual quota as a device in fisheries management. *Land Econ.* 62, 279–291.

Department of Fisheries and Oceans (1993). "Canadian Fisheries Landings." Department of Fisheries and Oceans, Ottawa, Ontario, Canada.

Denys, N. (1694). "A Description and Natural History of the Coasts of North America (Acadia)" (W. F. Ganong, transl., 1908). Greenwood, New York.

DeWolf, A. G. (1974). The Lobster Fishery of the Maritime Provinces: Economic Effects of Regulations. *Bull., Fish. Res. Board Can.* 177, 1–59.

Dow, R. L., and Baird, F. T., Jr. (1953). Methods to Reduce Borer Damage to Lobster Traps. *Maine Sea Shore Fish. Tech. Bull.* 3, 1–15.

Ennis, G. P. (1980). Further Yield per Recruit Analysis for Newfoundland Lobsters. *Can. Atl. Fish. Sci. Advis. Comm. Res. Doc.* 80/39, 1–14.

Fogarty, M. J., Cooper, R. A., Uzmann, J. R., and Burns, T. (1982). Assessment of the USA Offshore American Lobster (*Homarus americanus*) Fishery. *Int. Counc. Explor. Sea Res. Doc.* C.M. 1982/K:14, 1–21.

Gough, J. (1991). Fisheries Management in Canada 1880–1910. *Can. Manuscr. Rep. Fish. Aquat. Sci.* 2105, 1–96.

Herrick, F. H. (1909). Natural history of the American lobster. *Bull. U.S. Bur. Fish.* 29, 149–408 + 20 plates.

Jamieson, G. S., and Campbell, A. (1985). Sea scallop fishing impact on American lobsters in the Gulf of St. Lawrence. *Fish. Bull.* 83, 575-586.

Jentoft, S. (1989). Fisheries co-management. *Mar. Policy* April 1989, 137–154.

Keen, E. A. (1988). "Ownership and Productivity of Marine Fishery Resources." McDonald and Woodward, Blacksburg, Virginia.

Kelly, K. H. (1992). A Summary of Maine Lobster Laws and Regulations, 1820–1992. *Maine Lobster Inf. Leafl.* 19, 1–22.

Knight, A. P. (1917). Official Report upon Lobster Conservation in Canada. *Annu. Rep. Fish., Br. Dep. Nav. Serv., 51st, Suppl.* 1916–1917, 1–48.

Lunt, C. R. K. (1981). The Origins and Early Evolution of the Maine Lobsterboat. *Maine Marit. Mus. Ann. Symp. Am. Marit. Hist., 9th* 1–16.

Martin, K. R., and Lipfert, N. R. (1985). "Lobstering and the Maine Coast." Maine Maritime Museum, Bath.

Maynard, D., Savoie, F., Landsburg, W., Roach, G., and Wade, E. (1992). The Cape Breton Experiment on Legal Minimum Lobster Size Increase: An Intermediate Report. *Can. Atl. Fish. Sci. Advis. Comm. Res. Doc.* 92/64, 1–47.

McRae, E. D., Jr. (1960). Lobster explorations on continental shelf and slope off northeast coast of the United States. *U.S. Comm. Fish. Rev.* 22(9), 1–7.

Miller, R. J., Moore, D. S., and Pringle, J. D. (1987). Overview of the Inshore Lobster Resources in the Scotia–Fundy Region. *Can. Atl. Fish. Sci. Advis. Comm. Res. Doc.* 87/85, 1–20.

New England Fisheries Management Council (1983). "American Lobster Fishery Management Plan." New England Fisheries Management Council, Saugus, Massachusetts.

Northeast Fisheries Science Center (1993). Report of the Sixteenth Northeast Regional Stock Assessment Workshop. *Northeast Fish. Sci. Center Ref. Doc.* 93-18. 1–116.

Pezzack, D. S., and Duggan, D. R. (1988). An Assessment of the Canadian Offshore Lobster Fishery (LFA 41) for 1986–87. *Can. Atl. Fish. Sci. Advis. Comm. Res. Doc.* 88/65, 1–21.

Pezzack, D. S., Tremblay, M. J., Hudon, C., and Miller, R. J. (1992). The Inshore-Offshore Lobster Issue in Southwestern Nova Scotia. *Can. Manuscr. Rep. Fish. Aquat. Sci.* 2165, 1–19.

Pringle, J. D., and Burke, D. L. (1993). The Canadian lobster fishery and its management, with emphasis on the Scotian Shelf and the Gulf of Maine. *Can. Bull. Fish. Aquat. Sci.* 226, 91–122.

Pringle, J. D., and Sharp, G. J. (1980). Multispecies resource management of economically important marine plant communities of eastern Canada. *Helgol. Wiss. Meeresunters.* 33, 711–720.

Prudden, T. M. (1962). "About Lobsters." Bond Wheelwright, Freeport, Maine.

Rathbun, R. (1887). The crab, lobster, crayfish, rock-lobster, shrimp, and prawn fisheries. *In* "The Fisheries and Fishery Industries of the United States" (G.B. Goode, ed.), Section V, Vol. II, pp. 629–810. U.S. Commission of Fish and Fisheries, Washington, D.C.

Robichaud, D. A., and Campbell, A. (1991). Annual and seasonal size-frequency changes of trap-caught lobsters (*Homarus americanus*) in the Bay of Fundy. *J. Northw. Atl. Fish. Sci.* 11, 29–37.

Robinson, D. G. (1980). History of the lobster fishery on the eastern shore of Nova Scotia. *Can. Tech. Rep. Fish. Aquat. Sci.* 94, 8–23.

Roddick, D. L., and Miller, R. J. (1992). Spatial and temporal overlap of the American lobster (*Homarus americanus*) and sea scallop (*Placopecten magellanicus*) as related to the impact of inshore scallop dragging. *Can. J. Fish. Aquat. Sci.* 49, 1486–1492.

Scarratt, D. J. (1973). The effects of raking Irish moss (*Chondrus crispus*) on lobsters in Prince Edward Island. *Helgol. Wiss. Meeresunters.* 24, 415–425.

Schroeder, W. C. (1959). The lobster, *Homarus americanus*, and the red crab, *Geryon quinquedens*, in the offshore waters of the western North Atlantic. *Deep-Sea Res.* 5, 266–282.

Simon, J. E., and Campana, S. E. (1987). Species composition and distribution in inshore waters of southern Nova Scotia: Results of exploratory trawl surveys. *Can Tech. Rep. Fish. Aquat. Sci.* 1582, 1–53.

Sissenwine, M. P., and Mace, P.M. (1992). ITQ's in New Zealand: The era of fixed quota in perpetuity. *Fish. Bull.* 90, 147–160.

Skud, R. E. (1969). The effect of fishing on size composition and sex ratio of offshore lobster stocks. *Fiskeridir. Skr., Ser. Havunders.* 15, 295–309.

Smith, E. M., and Howell, P. T. (1987). The effect of bottom trawling on American lobsters (*Homarus americanus*) in Long Island Sound. *Fish. Bull.* 85, 737–744.

Stasko, A. B. (1978). Inshore–offshore S.W. Nova Scotia Lobster Stock Interactions: A Hypothesis. *Can. Atl. Fish. Sci. Advis. Comm. Res. Doc.* 78/37, 1–10.

Templeman, W. (1941). The Newfoundland Lobster Fishery: An Account of Statistics, Methods and Important Laws. *Newfoundland, Dep. Nat. Resour., Newfoundland Gov. Res. Bull.* 11, 1–43.

Tyler, A. V., Gabriel, W. L., and Overholtz, W. T. (1982). Adaptive management based on structure of fish assemblages of northern shelves. *Can. Spec. Publ. Fish. Aquat. Sci.* 59, 149–156.

U.S. Department of Commerce (1992). Fisheries of the United States. *Curr. Fish. Stat.* 9100.

Wilder, D. G. (1954). The Lobster Fishery of the Southern Gulf of St. Lawrence. *Circ. Fish. Res. Board Can.* 24, 1–16.

Wilder, D. G. (1965). Lobster conservation in Canada. *Rapp. P.-V. Reun., Cons. Perm Int. Explor. Mer.* 156, 21–29.

CHAPTER

6

Populations, Fisheries, and Management

MICHAEL J. FOGARTY
National Oceanic and Atmospheric Administration
National Marine Fisheries Service
Northeast Fisheries Science Center
Woods Hole, Massachusetts 02543

I. Introduction

Variation in demographic characteristics, abundance, and yield of *Homarus americanus* populations is manifest on a broad range of temporal and spatial scales. Understanding the determinants of this variability and the implications for management is a primary objective of population dynamics studies of the American lobster. The response of lobster populations to exploitation depends on life history features and environmental conditions, both of which vary in space and time. Determining the resilience of these populations to harvesting is central to defining strategies for sustainable use of this valuable resource. American lobster populations have long supported intensive fisheries in the coastal regions of the northeastern United States and Canada. In many areas, the lobster fishery has defined the character of local fishing communities and regional economies are highly dependent on the viability of the resource. Issues related to the ecology and life history of lobsters are therefore deeply intertwined with socioeconomic factors in the development of fishery management policies.

Any attempt to understand the dynamics of an exploited population involves a synthesis of information on fundamental ecological and demographic processes, including growth, mortality, reproductive characteristics (sex ratios, maturation schedule, and fecundity), larval dispersal, recruitment, migration, the relative importance of density-dependent factors, and interspecific interactions. These basic elements are incorporated directly or implicitly in models to assess the implications of alternative management actions on yield and productivity of the resource. Aspects of crustacean biology require special consideration in model development. Age determination is not generally possible and size- or stage-based modeling approaches are generally necessary. This poses no analytical difficulties and may in some instances be preferable to more traditional, age-based methods of analysis. Other characteristics, such as sexual dimorphism in growth and the linkage between growth and reproductive cycles, also must be accommodated. Finally, the relationships between groups connected through migration and/or larval dispersal must be considered. This chapter considers aspects of the historical development of the fisheries, population biology, and the development of management models for populations of *Homarus americanus*.

II. History of the Fishery

The commercial lobster fishery developed in the United States and Canada during the early 1800s,

BIOLOGY OF THE LOBSTER *Homarus americanus*
Jan Robert Factor, Editor

although subsistence fisheries by native peoples and early settlers considerably predated commercial exploitation (DeWolf, 1974; Dow, 1980). The fisheries occur from Labrador to the mid-Atlantic region of the United States (Fig. 1). The primary mode of capture is baited traps, although bottom trawls are also used in certain areas and hoop nets and gaffs were used in the early phases of exploitation (Dow, 1980). Lobster landings have undergone dramatic fluctuations over the last century, with a general coherence in temporal patterns on a regional scale (Campbell and Mohn, 1983; Pezzack, 1992). Lobster populations have changed markedly in abundance and size structure since the inception of the commercial fishery. In the mid-1800s, the average market weight of lobsters in Massachusetts was reported to be about 4 lb (1.8 kg), but 10- to 12-pound (4.5- to 5.4-kg) lobsters were not uncommon (Gould, 1841). By the late 1800s in Maine, the average at market was about 3 lb (1.4 kg) and lobsters less than 10.5 inches (26.7 cm, equivalent to 0.8 kg) were rejected; they were abundant close inshore and could be gaffed at low tide (Goode, 1887). The early fishery stands in sharp contrast to current conditions. The size composition in the inshore fishery is now considerably smaller (mean size, approximately 0.75 kg) (Thomas, 1973; Cadrin and Estrella, 1993).

Reported American lobster landings were comparatively high during the latter 1800s, but decreased steadily through the early decades of this century (Fig. 2). The initial declines presumably reflected the depletion of easily accessible local stocks as exploitation rates increased. The sharply decreased landings elicited considerable concern for the future of the industry (G. W. Field, 1902; Herrick, 1906; Knight, 1916). The intervention of two world wars and the worldwide economic depression during the first half of this century subsequently affected lobster fisheries and markets (Dow, 1980), providing a respite for overexploited local stocks. Following World War II, landings increased dramatically as demand increased and technological changes became widespread, accompanied by changes in fishing strategy. In the early fishery, traps were typically set individually around rock outcrops or other favorable habitats and hauled after short soak times. However, with the advent of power equipment, gear was set in multiple-trap strings (trawls), soak times were increased, and the areas fished were expanded to include soft-bot-

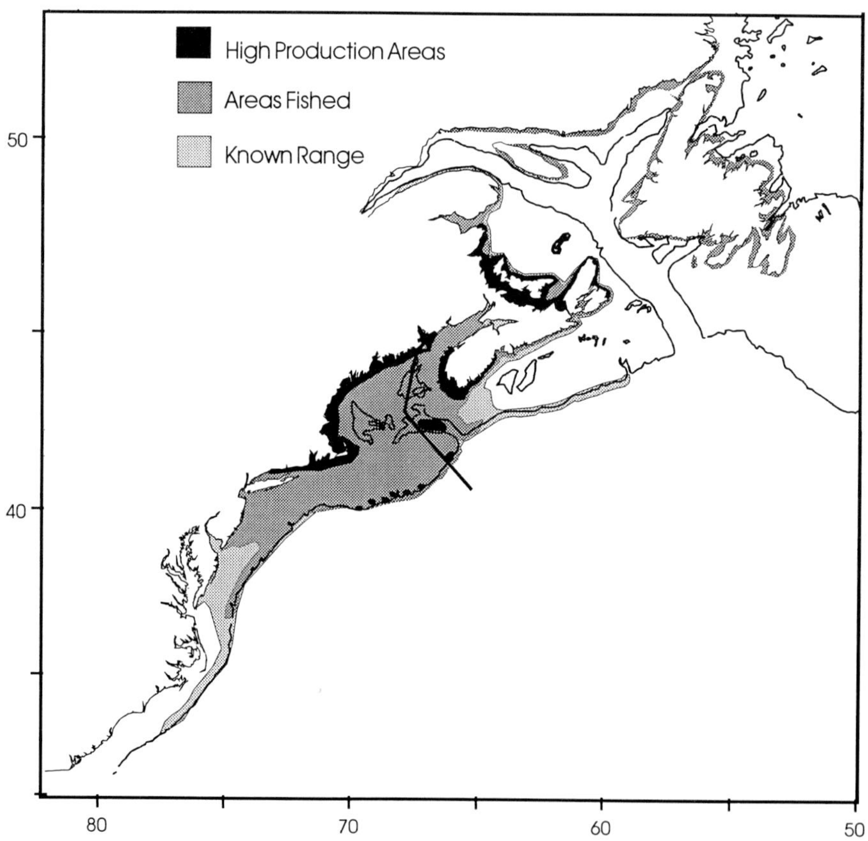

FIGURE 1 Distribution of American lobster populations and fisheries from Labrador to Cape Hatteras (Courtesy of D. S. Pezzack; modified from Pezzack, 1992).

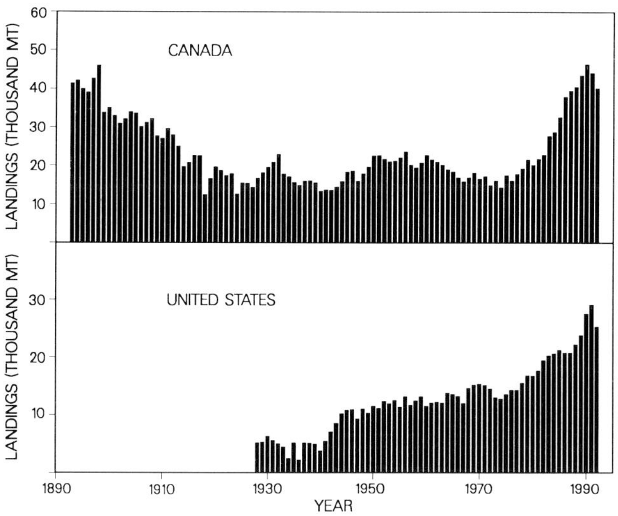

FIGURE 2 Lobster landings in Canada and the United States. Uninterrupted landings data are available from 1894 in Canada and from 1928 in the United States. MT, Metric tons. (Based on data from D. S. Pezzack, personal communication, and Anonymous, 1993.)

tom habitats. More recently, radar and satellite navigation systems have facilitated harvesting farther from shore. Advanced depth sounders provide detailed views of bottom relief and favorable lobster habitat. Miller (Chapter 5) provides further detail on the history of the fishery and changes in fishing methods.

The development of deepwater lobster fisheries on the outer continental shelf of the United States in submarine canyons and offshore banks during the 1950s (Schroeder, 1959) marked a new phase of fishery development. The fishery initially comprised solely trawlers, but by 1969 a trap fishery was initiated (Fogarty *et al.*, 1982) that soon dominated the U.S. offshore fishery. A Canadian fishery in the offshore regions of Browns and Georges Banks was initiated in 1971 (Pezzack, 1992). Changes occurred in the size structure of offshore lobsters under exploitation during the early phase of development of the U.S. offshore fishery and declining modal sizes have been attributed to harvesting effects (Skud and Perkins, 1969; Skud, 1969). These fisheries have potentially important implications for the resource as a whole.

Changes in landings are clearly related to changes in abundance. However, landings can also be affected by many factors other than abundance, including increases in the areal extent of the fishery, changes in the relative vulnerability to capture, and increases in fishing effort (Herrick, 1906; Addison and Fogarty, 1992, 1993). Important changes in these factors have occurred throughout the history of the fishery and it is not possible to infer changes in population size without further adjustment. For example, the reported number of Maine landings is now somewhat higher than that at the close of the last century (Fig. 3). However, the estimated number of traps being fished is currently higher by a factor of 20 (Fig. 3). The mean number of traps per boat doubled in Maine during the last two decades (K. Kelly, personal communication). Ennis *et al.* (1986) reported an increase in the number of traps and the number of trap hauls by factors of over 2 and 3, respectively, during the period 1971–1984 in Arnold's Cove, Newfoundland. Campbell (1989) documented an increase in the number of trap hauls per boat per day by a factor of 2 or more during two time periods (1944–1962 and 1985–1986). Temperature levels have also affected Maine lobster landings (Dow, 1969, 1977, 1978; Dow *et al.*, 1975), due to increased activity and vulnerability to capture at higher water temperatures (Fogarty, 1988). Changes in fishing effort and environmental conditions must therefore be considered in interpreting changes in landings. Expansion of the areal extent of local fisheries has been documented by Ennis *et al.* (1982), Campbell (1989), and Pringle and Burke (1993).

Fishery-independent measures of population size are not subject to these biases and recent increases in abundance have, in fact, been documented (Anonymous, 1993; Jeffries, 1994). Several hypotheses have been advanced to explain the recent increase in landings. It has been suggested that reduced predation levels due to the depletion of predators, such as Atlantic cod and other groundfish, have resulted in increased survival and recruitment (Elner and Campbell, 1991; Pezzack, 1992; see Section IV,D,1 on

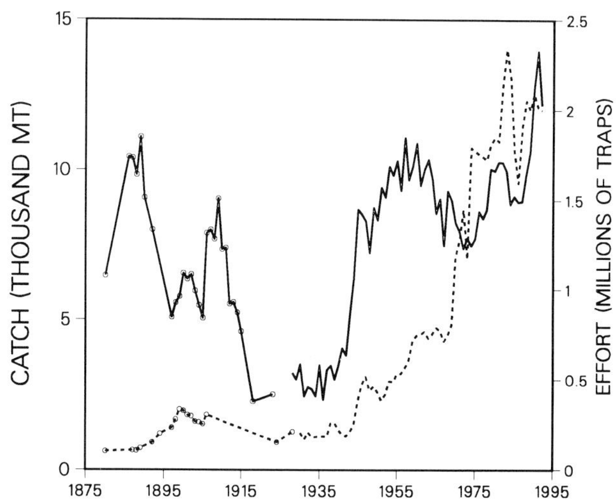

FIGURE 3 Landings (solid line) and effort (number of traps, dashed line) for the Maine fishery. Note that data prior to 1928 for landings and prior to 1930 for effort are not complete. MT, Metric tons.

natural mortality). Reduction in interspecific competition with flatfish has also been hypothesized (Jeffries, 1994). Reduced exploitation rates in Canada as a result of enhanced enforcement of existing regulations also may have contributed to increased yields (Elner and Campbell, 1991). The widespread use of escape vents in traps has reduced within-trap and discard mortality of sublegal-sized lobsters. In the United States, recent increases in the minimum legal size have also contributed to increased landings.

Other potential explanations involve changes in physical environmental conditions, such as water temperature, which may exert a broad-scale impact on survival. For example, increased landings in parts of Atlantic Canada during the last decade have been attributed to favorable water temperatures (Campbell *et al.*, 1991). Very small changes in the survival rate during the early life history stages can result in marked increases in recruitment levels. For example, a hypothetical change in survival from 1 to 2% during the larval stages would result in a doubling of the number of recruits to the fishery if all other factors remain constant. The increase in landings has focused attention on the need to understand factors underlying changes in the production of lobster populations.

American lobster fisheries have traditionally been managed using a combination of minimum legal size limits, protection of ovigerous females, and gear restrictions. In Canada, seasonal harvesting periods have also been regulated and limited entry and effort limitations have been imposed. A maximum size limit has been enacted in Maine. Miller (Chapter 5) provides a detailed review of existing regulations in various regions. The support for these measures among fishers is a key element in the success of any management plan (Acheson, 1975, 1988). The focus of this chapter is an evaluation of the implications of alternative management measures, such as changes in minimum size limits and fishing mortality rates.

III. Population Structure

Definition of population structure is essential to any analysis of the dynamics of exploited species. Here, a *population* is taken to be a relatively closed group of interacting individuals of a particular species. It is also useful to define a *stock* as a group of individuals of a species identified either as a true population or as a convenient management unit. Population differentiation within a species is often contingent on some degree of geographical isolation, although other mechanisms could also result in the formation of discrete populations. Attempts to define the stock structure of the American lobster have been based on several distinct approaches (Pezzack, 1987), including examination of genetic variability, consideration of patterns of migration and larval drift, analysis of regional differences in demographic characteristics, and comparison of trends in landings among regions. Of these methods, only the first can provide unambiguous evidence of true population differentiation. However, the other techniques can provide valuable supporting information on population structure and can be used directly to define stock units for management purposes. Because of the centrality of the population concept in both analysis and management, the evidence is reviewed in detail here.

Two general approaches to examining genetic variation among geographically separated groups of American lobsters have been employed to date: (1) application of allozyme electrophoresis to test for variation in nuclear gene frequencies and (2) studies of mitochondrial DNA (mtDNA). Electrophoretic studies have revealed low overall levels of genetic variation in *Homarus americanus*, complicating the attempt to define population structure. The generally low levels of genetic variability in *H. americanus* populations is consistent with observations in other lobster species (e.g., Smith *et al.*, 1980; Hedgecock *et al.*, 1982; Ovenden and Brasher, 1994).

Examination of esterase systems in cardiac tissue of lobsters collected in inshore Maine waters and from the outer continental shelf off New England demonstrated that polymorphic loci could be identified and used in population studies (Barlow and Ridgway, 1971). No significant differences in gene frequencies between inshore and offshore lobsters were found. It was shown that the electrophoretic activity of these esterase polymorphisms is invariant over molting and reproductive cycles. In contrast, serum proteins have been shown to vary with physiological state of the individual and therefore cannot be used in population discrimination studies without accounting for the molt or reproductive status of the lobster (Barlow and Ridgway, 1969). In an analysis of variations in gene frequencies, a larger number of loci documented low levels of genetic variation (3.8% heterozygous loci per individual) (Tracey *et al.*, 1975). Of 44 loci examined, only five showed sufficient variability to be useful in population separation studies. Gene frequency differences among locations were observed for only one enzyme (malic enzyme). Samples taken from the southern Gulf of St. Lawrence were nearly all homozygous for one allele, while those from two coastal locations in southern New England were nearly all homozygous for a second allele. In contrast, half of the individuals sampled on the outer continental shelf off New England were heterozygous for this locus. It was inferred that the Gulf of St. Lawrence, inshore New England, and offshore New England are characterized by discrete populations. Estimates of Nei's genetic distance (D) between the Canadian and two U.S. inshore locations (Woods Hole and Martha's Vineyard, Massachusetts) were 0.035 and 0.028, respectively (a value of 0 indicates complete identity). All other comparisons among locations showed values of $D < 0.01$, indicating negligible genetic distance. To place these results in context, note that the genetic distance between *Homarus americanus* and *H. gammarus* is estimated to be 0.11 (Hedgecock *et al.*, 1977). A comparison of gene frequencies at 30 loci in lobster samples collected from inshore Nova Scotia and an offshore location on the Scotian Shelf revealed no significant differences (Odense and Annand, 1978) and it was concluded that these groups were genetically identical.

Restriction enzyme analysis of mtDNA from the Gulf of St. Lawrence, the Gulf of Maine, and the offshore mid-Atlantic region of the United States indicates the existence of two principal lineages that diverged well before the last glaciation (Kornfield and Moran, 1990). Comparison of the relative frequency of each lineage indicates no significant differences among locations; however, an indication of possible separation is noted between lobsters from the Gulf of St. Lawrence and those from the inshore Gulf of Maine. Within the Gulf of Maine, there is no indication of discrete populations based on a comparison of the relative frequency of each clonal lineage among sites. Comparisons between American and European lobsters reveal clear divergence and a relatively rapid rate of mtDNA evolution, consistent with observations for other marine invertebrates.

The electrophoretic and mtDNA studies conducted to date indicate population differentiation between widely separated locations (the Gulf of St. Lawrence and coastal waters of New England), where it has been hypothesized that barriers to dispersal and interchange exist (Hedgecock, 1986). The evidence for genetic differences between inshore and offshore lobsters off New England and Nova Scotia appears to be equivocal. The primary studies conducted to date find no evidence for differentiation on finer spatial scales. However, in a reexamination of the results of Tracey *et al.* (1975), Burton and Feldman (1982; Burton, 1983) and Shaklee (1983) concluded that there is evidence for population differences within the Gulf of Maine. The number of lobsters examined in these studies has been relatively low (less than 300) and it appears that expanded studies using newer and more sensitive technologies would be useful.

Relatively low levels of interchange could be sufficient to maintain genetic homogeneity among groups. Cooper and Uzmann (1971), Uzmann *et al.* (1977), and Fogarty *et al.* (1980) demonstrated that intermixing occurs among inshore and offshore populations off the New England coast through migration. Seasonal onshore-offshore movements were documented for lobsters tagged on the outer continental shelf (Cooper and Uzmann, 1971; Uzmann *et al.*, 1977) and for those in a near–shore location (Fogarty *et al.*, 1980). Long-distance movements have also been documented by Saila and Flowers (1968), Dow (1974), Campbell (1986), Campbell and Stasko (1985, 1986), Campbell *et al.* (1984), Pezzack and Duggan (1986), and Pezzack (1987). Linkage among populations or subpopulations is also possible through larval drift (Cobb and Caddy, 1989; Cobb *et al.*, 1983, 1989; Cobb, Chapter 7). The potential for long-distance transport is substantial (Fogarty, 1983; see Cobb, Chapter 7, for caveats). Harding and Trites (1988) suggested that larvae advected from offshore regions off Nova Scotia could play an important role in recruitment to coastal areas. Pezzack (1989), however, concluded that this was unlikely. Katz *et al.* (1994) modeled surface drift in the southern New England region during the period of larval occurrence and indicated that surface water transport, coupled with directed swimming behavior, could result in larval recruitment from offshore to inshore areas. Despite the clear potential for interchange among some regions through migration and larval drift, it must be recognized that discrete populations can, in principle, be maintained through differential selection or other mechanisms (Hedgecock, 1986).

Regional differences in morphology and demographic characteristics have been identified. Saila and Flowers (1969) examined morphometric variation in lobsters and concluded that inshore and offshore populations in New England could be differentiated. Similar results for the Bay of Fundy, the Scotian Shelf, and the Gulf of Maine were reported by Campbell and Mohn (1982). Rogers *et al.* (1968) reported differences in mean size of stage I larvae between inshore and offshore southern New England sites. Uzmann (1970) and Campbell and Bratty (1986) noted that inshore and offshore lobsters could be characterized on the basis of parasitic infestation. Differences in growth and maturity patterns have been noted (reviewed in Russell, 1980a; Aiken, 1980; Aiken and Waddy, 1980, 1986; see also Sections IV,A and B). These phenotypic differences do require consideration in the analysis of lobster population dynamics, even in the absence of genetic variation.

Campbell and Mohn (1983) investigated affinities

in lobster groups based on analysis of patterns in landings. Three principal groupings in Canada and the northern United States were defined: (1) the Gulf of St. Lawrence, (2) the eastern coast of Nova Scotia, and (3) the Gulf of Maine. A recent evaluation of lobster stock status in the United States (Anonymous, 1993) also recognized the Gulf of Maine as a unit; an offshore group, including lobsters on Georges Bank to the mid-Atlantic region, was identified, as was a third group in the inshore region of southern New England. Linkage between the offshore group and the inshore southern new England region by migration and/or larval drift is probable. At least five major groupings of lobsters in Canada and the United States can therefore be identified for management purposes. This classification is consistent with known migratory patterns and broad-scale hydrographic patterns influencing linkage through larval transport. The current lack of clear evidence of finer-scale population differentiation, coupled with information suggesting interchange through migration and larval drift, suggests that a metapopulation model of interconnected subpopulations may be appropriate for some broad-scale areas (e.g., the Gulf of Maine). The significance of this representation for linked subpopulations subject to differential harvesting rates is explored in the following section.

IV. Population Dynamics and Vital Rates

Russell (1931) provided a convenient conceptual framework relating net production to the individual components of production for an exploited species. The change in biomass per unit of time (net production) can be viewed as the difference between increases due to recruitment and individual growth, and losses due to natural mortality (e.g., predation and disease) and exploitation

$$\frac{\Delta B}{\Delta t} = R + G - M_N - H, \tag{1}$$

where B is population biomass, t is time, R is the weight added through recruitment, G is the biomass increment due to individual growth, M_N is the biomass lost due to natural mortality, and H is the harvested biomass. A closed population is assumed in which immigration and emigration are negligible or, alternatively, in which the changes in biomass due to these factors balance. Here, *recruitment* refers to the number of individuals surviving to the age or size of vulnerability to the fishery. Recruitment represents the addition of new individuals to the population through reproduction and entails consideration of maturation, fecundity, and survival during the prerecruit phase. Individual growth rates of lobsters are quite labile and represent an important source of variation in production. Natural mortality is a function of conditions in the abiotic and biotic environments. Partitioning natural mortality into elements describing predation and all other sources of natural loss provides one approach to defining multispecies models of community dynamics. Harvesting is the only component of production that can be controlled; harvesting models examine the implications of varying control variables, such as the fishing mortality rate or minimum legal size limits. An understanding of these components of production is required in order to characterize the dynamics of an exploited population, and the elements of each are described in detail in the sections that follow. Descriptions of growth, maturation, and mortality parameters are first provided, followed by information on abundance and recruitment dynamics. The individual components of production are then combined in the development of management-oriented models.

A. *Growth*

Crustacean growth comprises two components: the probability of molting within a specified time frame (or its inverse, intermolt interval) and the size increase at molting (molt increment). Calcified structures with annual or other periodic markings are not retained through ecdysis and age cannot be directly determined. It is possible, however, to determine growth; estimates of molt probability and increment have been determined primarily by mark–recapture studies using tags that are retained through the molt (Scarratt and Elson, 1965; Cooper, 1970). Observations based on aquarium studies (Wilder, 1953; Hughes and Mathiessen, 1962; McLeese, 1972) have provided a basis for calibration of field estimates of growth. Two principal approaches have been taken in developing descriptors of lobster growth: (1) empirical models based on combining information on molt increment and molt probability, and (2) estimation of the parameters of the von Bertalanffy growth equation.

Molt increments have been documented in a number of mark–recapture studies. The relationship between premolt and postmolt carapace length, the so-called Hiatt (Kurata, 1962) or Gray–Newcombe (Botsford, 1985) growth diagram, has been widely reported in studies throughout the geographical range (Table 1). These studies demonstrate remarkably little variation in premolt–postmolt relationships among regions. The dominant component of the postmolt size is the premolt size and it is therefore not

TABLE 1 **Estimates of Parameters of the Hiatt Growth Diagram for the American Lobster, *Homarus americanus***

Location	Intercept	Slope	Sex	Reference
Bonavista Bay, Newfoundland	2.35	1.12	M	Ennis (1972)
	9.74	1.00	F	
Placentia Bay, Newfoundland	−2.35	1.18	M	Ennis (1980a)
	8.88	1.01	F	
Magdalen Islands	−0.27	1.15	M	Dubé (1986)
	9.67	1.00	F	
Egmont Bay, Prince Edward Island	8.73	1.05	M	Wilder (1963)
	10.52	1.00	F	
New Harbour, Nova Scotia	8.10	1.05	M	Miller *et al.* (1989)
	6.65	1.06	F	
Port Mouton, Nova Scotia	12.70	1.00	M	Miller *et al.* (1989)
	0.29	1.14	F	
Port Maitland, Nova Scotia	3.87	1.08	M	Campbell (1983b)
	5.59	1.06	F	
Bay of Fundy	10.10	1.04	M	Campbell (1983a)
	9.60	1.04	F[a]	
	18.01	0.95	F[b]	
Offshore Scotian Shelf	5.60	1.10	M	D. S. Pezzack (personal communication)
	8.20	1.04	F[a]	
	18.37	0.94	F[b]	
Kennebunkport, Maine	9.00	1.02	M and F	Krouse (1981)
Boothbay Harbor, Maine	9.45	1.24	M and F	
Jonesport, Maine	20.01	0.92	M and F	
Cape Cod Bay, Massachusetts	6.15	1.07	M	Lawton *et al.* (1984)[c]
	8.47	1.03	F	
Long Island Sound, New York	15.42	0.91	M	Northeast Utilities Service Company (1994)[c]
	14.58	0.91	F	
Western Long Island Sound	1.56	1.13	M	Briggs and Mushacke (1984)
	11.04	0.93	F	
Offshore New England	10.82	1.06	M	Fogarty (1986)
	16.26	0.97	F	

[a] <95-mm carapace length.
[b] 95-mm carapace length.
[c] Several regressions are provided in the reference. Relationship with the largest sample size is provided here.

surprising that a close relationship is indicated in this representation (Hartnoll, 1982; Botsford, 1985). Sexual dimorphism is clearly evident in the premolt–postmolt relationship. A change in this relationship at the size of maturity has been noted in some crustaceans. Hiatt (1948) proposed that conjoined linear models with an intersection near the size at maturity be used in this case (see also Somerton, 1980). Mauchline (1976) suggested that the apparent change at maturity was an artifact and that a hyperbola was a more appropriate functional form, although this argument was rejected by Botsford (1985). Easton and Misra (1988) proposed an alternative nonlinear form based on an allometric relationship between pre- and postmolt size. Evidence for a change at the size of maturity for American lobsters has been reported by Ennis (1972) and Campbell (1983a,b), but other studies have detected no obvious change.

The relationship between the percentage increase per molt and the premolt size has also been used to describe growth in crustaceans (Hartnoll, 1982). This representation is problematic, however, because of a confounding of the dependent and independent variables. The relationship between molt increment and premolt size would appear to provide the most informative representation of growth per molt (Botsford, 1985; Fogarty and Idoine, 1988), and it is recommended that this approach be generally adopted in preference to the premolt–postmolt relationship.

Molt probability has been most commonly estimated using the "anniversary" method of Hancock and Edwards (1967), based on tagging studies. In this technique, the proportion of lobsters having molted during a specified time period (usually 1 year ± some tolerance factor) is calculated for different size classes. Estimation of molt probability based on mark–recapture studies can be biased if tags are lost during molting; tag loss can be in the range of approximately 10–40% per year (Cooper, 1970; Russell, 1980b; Ennis, 1986a). Similarly, if mortality is associated with molting, the estimate of the proportion molting will be biased. The corrected estimate of the proportion molting adjusted for molt-related tag loss, is

$$\hat{\rho}_L = \frac{N_{m,L}(1-T)^{-1}}{N_{n,L} + N_{m,L}(1-T)^{-1}}, \tag{2}$$

where $N_{m,L}$ is the number of individuals in size class i that molted during a specified time period, $N_{n,L}$ is the number that did not molt in the time period, and T is the fraction of individuals that lose tags during the molt. Fogarty and Idoine (1988) provide a correction factor for the combined effects of molt-related tag loss and mortality.

Ennis *et al.* (1982, 1986) based molt probability estimates on examination of shell condition (see Ennis, 1977). This technique holds the potential for obtaining molt probability estimates based on larger sample sizes, with lower cost than tagging studies, and deserves broader application. It has proven exceedingly difficult to obtain a sufficient number of recaptures to define the molting probability at larger sizes (Fogarty and Idoine, 1988). Ennis *et al.* (1982, 1986) provide important temporal sequences of molt probability estimates and show that the annual molting probability of lobsters is directly related to water temperatures (cumulative degree days) in Comfort Cove and Arnold's Cove, Newfoundland. Ennis (1991) also documented a density-dependent effect on molt probability in Arnold's Cove. Campbell (1983b) also demonstrated that temperature strongly affects the annual molt probability of lobsters. Similar observations have been made in culture systems (Van Olst *et al.*, 1980; Hughes *et al.*, 1972). These results have important implications for understanding interannual variation in the production of lobsters.

Several empirical models of annual molt probability have been developed for *Homarus americanus*. Cooper and Uzmann (1980) and Campbell (1983a,b) employed linear and decaying exponential models, respectively, for molt probability as a function of premolt carapace length (Table 2). Molt probability was described using a probit analysis by Ennis *et al.* (1982, 1986; Ennis, 1980a), a second-degree polynomial by Dubé (1986), and a logistic function by Fogarty and Idoine (1988) and D. S. Pezzack (personal communication). A generalized logistic molt probability model that can accommodate additional variables (e.g., temperature) can be developed as

$$\rho_L = \left[1 + \exp\left(a + bL + \sum_{j=1}^{m} c_j X\right)\right]^{-1}, \tag{3}$$

where L is the premolt length, X is an auxiliary variable (e.g., temperature), and a, b, and c_j are coefficients. Quadratic or higher-order effects for premolt size and the auxiliary variables can, in principle, be specified in this formulation by including additional terms in the model. The logistic model is advantageous because it is both flexible and naturally constrained to reasonable probability values ($0 \le \rho_L \le 1$).

The estimates of molt probability per unit of time and molt increment can be combined to give an estimate of mean size as

$$\bar{L}_{t+\Delta t} = \bar{L}_t + \rho_L \Delta L, \tag{4}$$

where $\bar{L}_t$ is the mean length at time t, and ΔL is the molt increment for lobsters of size L. An illustration of an empirical growth model derived in this way for an offshore lobster stock is provided in Fig. 4, demonstrating the sexual dimorphism in growth rates. Idoine (1985) described a related approach based on a distributed-delay model for growth. Variations in molt increment and molt probability result in substantial differences in growth among individuals. As a result, individuals in a cohort (year class) will recruit to the fishery at different ages, resulting in a smoothing of recruitment variation.

Parameters of the von Bertalanffy growth equation have been estimated using two primary methods: (1) tag–recovery information and (2) estimation of growth from molt increment and molt probability data and fitting the von Bertalanffy equation to size-at-age data after first assigning an age to the initial size. The von Bertalanffy equation provides a useful descriptor of growth for comparative purposes (Cobb and Caddy, 1989). Further, the parameters of this growth model are used in the Beverton–Holt yield model and certain length-based mortality estimators. A summary of parameter estimates derived for

TABLE 2 **Molt Probability Models for Selected Populations of the American Lobster, *Homarus americanus*[a]**

Area	Sex	Model	Reference
Bay of Fundy	M	$\rho_L = 1.76e^{-0.0084L}$	Campbell (1983a)
	F	$\rho_L = 1.95e^{-0.0097L}$	
Port Maitland, Nova Scotia	M	$\rho_L = 14.40e^{-0.036L}$	Campbell (1983b)
	F	$\rho_L = 31.38e^{-0.066L}$	
	M	$\rho_L = 21.79e^{-0.041L}$	
	F	$\rho_L = 25.95e^{-0.045L}$	
Offshore southern New England	M	$\rho_L = 1.260 - 0.00489\ L$	Cooper and Uzmann (1980)
	F	$\rho_L = 1.285 - 0.00544\ L$	
	M	$\rho_L = (1 + e^{-6.886 + 0.052L})^{-1}$	Fogarty and Idoine (1988)
	F	$\rho_L = (1 + e^{-6.867 + 0.058L})^{-1}$	
Scotian Shelf	M	$\rho_L = (1 + e^{-9.896 + 0.082L})^{-1}$	D. S. Pezzack (personal communication)
	F[b]	$\rho_L = (1 + e^{-9.168 + 0.088L})^{-1}$	
	F[c]	$\rho_L = (1 + e^{-4.710 + 0.022L})^{-1}$	
Magdalen Islands north	M	$\rho_L = 2.155 - 0.015L - 0.00003L^2$	Dubé (1986)
	F	$\rho_L = -3.692 + 0.146L - 0.00115L^2$	
Magdalen Islands south	M	$\rho_L = -0.720 + 0.052L - 0.00041L^2$	
	F	$\rho_L = -9.019 - 0.287L - 0.00204L^2$	

[a] L, Carapace length (in millimeters); ρ_L, annual probability of molting.

[b] Nonovigerous females.

[c] Ovigerous females.

Homarus americanus populations is provided in Table 3. It should be noted, however, that von Bertalanffy parameter estimates derived from mark–recapture studies are sensitive to the size range represented. Further, the exact interpretation of the parameters is uncertain when the actual age of marked individuals is not known (Francis, 1988). The narrow size range available for tagging in heavily exploited coastal regions can result in biased estimates. Empirical models based on molt increment and frequency are preferable. Saila *et al.* (1979) advocated a similar approach for modeling rock lobster growth.

B. Maturity

Marked differences in lobster maturity have been noted on a regional basis. In particular, lobsters mature at a smaller size in relatively warm-water locations, such as the Gulf of St. Lawrence and the inshore bays and estuaries of the southern New England region (Aiken and Waddy, 1980, 1986; Van Engel, 1980; Waddy *et al.*, Chapter 10). Small size at maturity has also been noted in Newfoundland stocks (Ennis, 1980c). In contrast, lobsters in offshore regions off the New England coast, in the Gulf of Maine, and in the Bay of Fundy mature at larger sizes (Krouse, 1973; Campbell and Robinson, 1983; Fogarty and Idoine, 1988). The most broadly available measure of female maturity is the presence of external eggs, although other measures based on direct examination of the ovaries, morphometric considerations, and the development of cement glands on the pleopods have also been employed (Aiken and Waddy, 1980; Waddy *et al.*, Chapter 10). The focus in this section is female maturity, as indicated by the presence of external eggs on the abdomen, because of its widespread availability; further, certain fishery regulations are linked to this measure of maturity. The relationship between the proportion of mature female lobsters and size can be conveniently represented by the logistic function

$$m_L = \left[1 + \exp\left(c + dL\right)\right]^{-1}, \tag{5}$$

where m_L is the proportion of mature lobsters in size class L, and c and d are coefficients.

An illustration of the variation observed in maturation-at-size data for three regions off the New England coast is provided in Fig. 5. The inshore southern New England region is characterized by

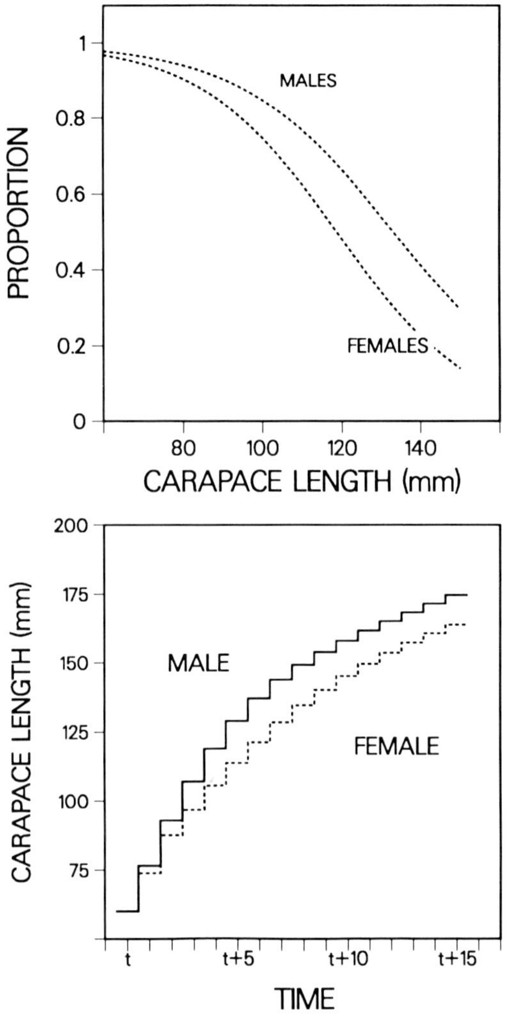

FIGURE 4 Estimates of (top) the proportion molting by sex as a function of premolt size and (bottom) mean size at successive annual time steps, starting at a size of 60-mm carapace length.

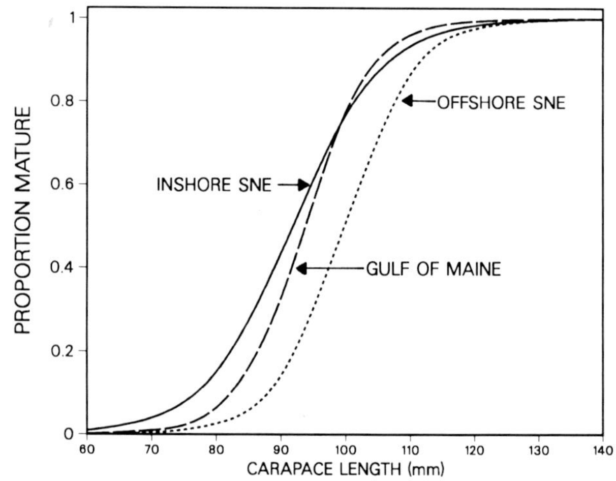

FIGURE 5 Estimates of the proportion of mature female lobsters at size for three regions off the New England coast based on the occurrence of ovigerous females. SNE, southern New England region. (Based on data from Anonymous, 1993.)

early maturation and relatively warm-water temperatures. In contrast, maturation occurs at larger sizes in the colder waters of the Gulf of Maine and offshore New England. Unfortunately, data are not available to assess whether differences in maturation rates have occurred naturally over time or whether the intensive fishery has exerted strong selective pressure for reduced size at maturity. No reduction in the size at onset of sexual maturity (indicated by the presence of external eggs) is evident in comparison to the size of the smallest ovigerous females in southern Massachusetts in samples separated by a century (Estrella and Cadrin, 1995). In contrast, reduction in the mean size at maturity as exploitation rates increase has been documented in spiny lobster populations (Polovina, 1989; Chubb, 1994).

It is clear that marked regional differences in maturation occur. The efficacy of minimum legal size limits as a conservation measure depends on the correspondence between the size at maturity and the minimum size limit. Clearly, the same size limit in different regions will vary in its effectiveness; in colder-water regions, the minimum legal size is often well below the mean size at female maturity.

C. Fecundity

The fecundity of the American lobster is low relative to many other crustacean species, ranging from tens of thousands of eggs to nearly 100,000 at the largest sizes. In contrast, many brachyurans produce clutches of over 1 million eggs, and spiny lobster fecundity is measured in hundreds of thousands of eggs. Ova size of the American lobster is large and represents a substantial energetic investment per individual egg relative to more fecund species. The eggs are incubated externally for an extensive period, approximately 9 months, depending on temperature (Aiken and Waddy, 1980), and are therefore afforded protection during a critical life history phase. The larvae are correspondingly large at the time of hatching, have larger energetic reserves, and are vulnerable to a more restricted range of predators than smaller planktonic organisms. These characteristics have the potential to reduce the variability in mortality rates during the early life history, and can result in lower levels of recruitment variability (Fogarty *et al.*, 1991; Fogarty, 1993) relative to species with smaller, more vulnerable progeny.

Regional differences in fecundity have been documented for American lobster groups (Table 4 and Fig.

TABLE 3 **Parameters of the von Bertalanffy Growth Equation Estimated for Selected Populations of the American Lobster, *Homarus americanus*[a]**

Area	Year	Sex	L_∞	k	t_0	Reference
Newfoundland		M	105	0.390	−0.80	Ennis (1980a)
		F	112	0.240	−0.69	
Bay of Fundy	1980	M	281	0.065	0.76	Campbell (1983a)
		F	207	0.089	0.42	
Port Maitland, Nova Scotia	1948–1949	M	165	0.110	−0.122	Campbell (1983b)
		F	103	0.389	2.064	
	1960–1961	M	119	0.242	1.247	
		F	109	0.288	1.303	
	1948–1967	M	118	0.229	0.963	
		F	109	0.288	1.303	
	1979–1980	M	135	0.193	1.224	
		F	108	0.372	2.262	
Maine		M	267	0.048	−0.773	Thomas (1973)
		F	241	0.087	−0.096	
Offshore New England		M	270	0.096	0.5	Cooper and Uzmann (1980)
		F	240	0.074	0.3	
Magdalen Islands		M	129.84	0.192	1.46	Dubé (1986)
		F	105.86	0.228	2.89	
Comfort Cove, Newfoundland		M	102.1	0.3701	3.6113	Ennis *et al.* (1982)
		F	99.0	0.3417	4.2504	

[a]The von Bertalanffy growth equation is: $l_t = L_\infty\{1 - \exp[-k(t - t_0)]\}$, where l_t is the size at time t_1, L_∞ is the asymptotic length, k is the rate at which the asymptote is approached, and t_0 is the age at zero length.

6) (Saila *et al.*, 1969; Ennis, 1981; Estrella and Cadrin, 1995). These differences can have important implications for considerations such as the total egg production of the population and egg production per recruit (see Section V,B). A comparison of lobster fecundity samples at one location separated by nearly a century shows no significant differences in fecundity-at-size data between contemporary samples from Massachusetts and earlier data collected and archived by V. Edwards of the U.S. Fish Commission in Woods Hole and subsequently published by F. H. Herrick (1909) in his classic monograph (Estrella and Cadrin, 1995). Temporal variation in annual egg production during an 11-year period has been documented in a Newfoundland lobster population and linked to interannual variation in female population size and molting frequency (Ennis, 1991). Estimated annual egg production varied by a factor of more than 10 during this period.

D. Mortality

Mortality rates can be partitioned into two components: natural mortality and fishing mortality. *Natural mortality* includes all mortality due to sources other than harvesting (or other anthropogenic effects). Harvesting is the principal anthropogenic factor considered here. However, recent improvements in water quality in coastal regions may be a factor in increased production.

Mortality rates can be expressed in terms of instantaneous rates or, equivalently, as the proportion of the population dying in a specified time interval. The instantaneous rate of total mortality (Z) is the sum of instantaneous rates of natural (M) and fishing (F) mortality ($Z = F + M$). The fraction of the population dying in a unit time period is $A = [1 - \exp(-Z)]$ and the fraction of the population removed by harvesting, the exploitation rate, is therefore $E = (F/Z)[1 - \exp(-Z)]$ (Ricker, 1975). In the following discussion, harvesting effects are expressed in terms of exploitation rates for ease of interpretation.

1. Natural Mortality

The dominant sources of natural mortality of lobsters include predation, disease, and extreme environmental conditions (e.g., storm-related damage and displacement). Predation on all phases of the life history (including eggs) has been documented. Infestation of lobster egg masses with an ovophagous

TABLE 4 **Estimates of Size–Fecundity Relationships for Selected Populations of the American Lobster, *Homarus americanus*[a]**

Area	*a*	*b*	Correction factor	Reference
Newfoundland				
Arnold's Cove	0.0045324	3.347062	1.011369	Ennis (1981)
Paradise	0.0126958	3.098418	1.014505	
Ship Harbor	0.4878665	2.318816	1.014745	
Boswarlos	0.0211542	2.938735	1.035498	
Northwestern coast	0.3982538	2.316141	1.040168	
Nova Scotia and New Brunswick				
Northumberland Strait	0.0000482	4.320195	1.038530	Campbell and Robinson (1983)
Eastern Nova Scotia	0.0000586	4.272524	1.041395	
Bay of Fundy	0.0031829	3.353501	1.032462	
New England				
Vineyard Sound	0.0005640	3.758048	1.01522	Herrick (1895)
Southern Gulf of Maine	0.0009198	3.58022	1.09886	Estrella and Cadrin (1995)
Outer Cape Cod	0.0127547	3.062789	1.03440	
Buzzard's Bay	0.0000764	4.17506	1.09672	
Vineyard Sound	0.0005640	3.72269	1.07265	

[a]The model is of the form: $f = aL_b$, where f is fecundity, L is carapace length (in millimeters), and a and b are coefficients. The correction factor is a bias correction term to adjust for logarithmic transformation. (Based on estimates provided by B. T. Estrella, Massachusetts Department of Fisheries, Wildlife, and Environmental Law Enforcement, personal communication.)

nemertean worm, *Pseudocarcinonemertes homari*, can result in potentially high egg mortality (Campbell and Bratty, 1986; Talbot and Helluy, Chapter 9).

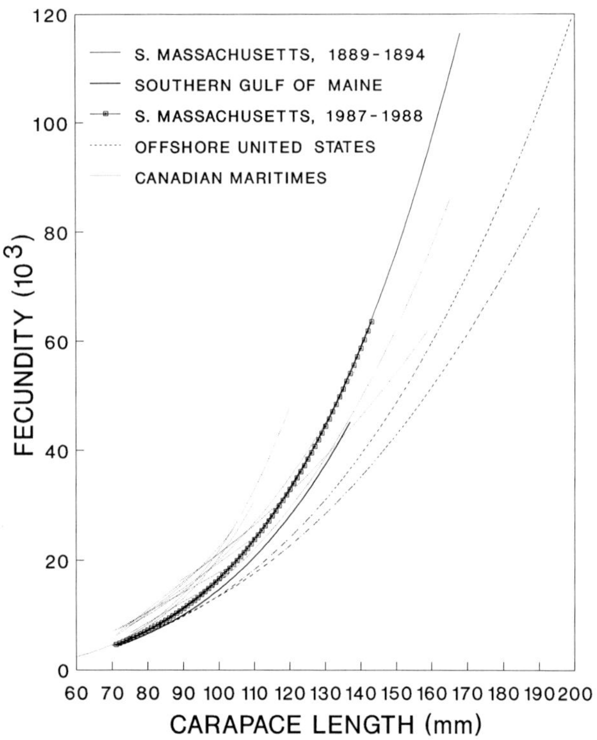

FIGURE 6 Fecundity as a function of size for lobster populations throughout the range. (After Estrella and Cadrin, 1995.)

Larval lobsters are subject to predation from sea birds (Mendall, 1934; Mills, 1957) and a broad spectrum of predatory fish (Herrick, 1895, 1909; I. A. Field, 1907; Ennis, Chapter 3). Preadult and adult lobsters apparently have fewer natural predators, although recently molted lobsters are presumably more vulnerable to predation (Herrick, 1909). Diet composition studies of fish populations off the northeastern United States indicate that Atlantic cod, smooth dogfish, spiny dogfish, little skate, thorny skate, white hake, red hake, and longhorn sculpin prey on lobsters (R. Rountree, personal communication). Of these, only the gadids are currently heavily exploited and depleted. Intraspecific predation has also been documented in natural populations (Elner and Campbell, 1987). Cannibalism is a potentially important regulatory mechanism in population dynamics and therefore deserves further study.

The role of disease and parasitism in natural lobster populations has not been extensively examined. A potentially fatal blood disease (gaffkemia) caused by the pathogen *Aerococcus viridans* var. *homari* has inflicted high mortality in lobsters in impoundments and other high-density holding conditions (Stewart, 1980), but its prevalence and effect on wild populations are not known. (See Waddy *et al.*, Chapter 10, and Martin and Hose, Chapter 17, for discussions of gaffkemia.) The role of density-dependent processes in factors such as transmission of disease and egg predation by nemerteans is potentially important as a population regulatory mechanism.

Quantitative estimates of mortality rates have been made for the larval stages and for juveniles and adults. Scarratt (1964) provided an estimate of average mortality of 98.9% between the first larval stage and the approximate midpoint of the fourth stage during 1949–1961 (Cobb, in Chapter 7, discusses potential biases). Thomas (1973) provided estimates of the instantaneous rate of natural mortality (M) ranging from 0.02 to 0.35 (equivalent to 2–30% per year) for juvenile and adult lobsters in Maine. Thomas considered the higher estimates to be unrealistic and suggested that intermediate estimates of $M \approx 0.08$ (8% per year) were more reasonable. Ennis (1979) described a technique for estimating natural mortality and provided an estimate of $M = 0.02$ (2% per year) in Newfoundland. Low natural mortality rates are consistent with the apparent longevity of this species.

2. Exploitation Rates

Estimated exploitation rates for *Homarus americanus* in coastal regions are typically high, reflecting an intense concentration of fishing effort in nearshore areas (Table 5). Both regional and temporal differences in exploitation rates are evident. Exploitation or fishing mortality rates have been primarily derived using two methods: (1) analysis of length composition data and (2) determination of the rate of return in tagging studies. The former method provides an estimate of the instantaneous rate of total mortality (Z), and the instantaneous fishing mortality rate (F) is determined as $F = Z - M$. It is possible to derive estimates of fishing mortality rates directly from mark–recapture studies. Mortality can be confounded with migration if lobsters undertake directional movements from areas of intensive fisheries to other regions, resulting in artificially high mortality rates. This factor cannot be discounted in many of the estimates of mortality that have been made to date. To counter this potential bias, the population must be sampled throughout its area of occurrence. Fishing mortality rates also differ with distance from shore. More dispersed offshore lobsters are subject to lower fishing rates and these lobsters may comprise a substantial fraction of the population in some regions. In the following discussion, instantaneous mortality rates have been converted to exploitation rates where the latter have not been specified. If an estimate of natural mortality has not been provided, an estimate of $M = 0.1$ was used to complete the conversion.

Paloheimo (1963) provided one of the earliest quantitative estimates of exploitation rates for lobsters based on mark–recapture studies. Estimated annual exploitation rates ranged from 41 to 83% at four sites in Nova Scotia and Prince Edward Island during the period 1947–1960 (Table 5). Ennis (cited in Campbell, 1980) provided exploitation estimates for five areas in Newfoundland in 1978, ranging from a low of 52 to 91% based on tagging studies. Estimated exploitation rates at Comfort Cove during 1975–1980 were 69.8–94.9% per year (Ennis *et al.*, 1982). Krouse (1981) reported extremely high mortality rates at three sites in Maine in tagging conducted during 1975; the corresponding exploitation rates (assuming $M = 0.1$) ranged from 98.3 to 98.8%. Krouse cited several possible factors leading to overestimates of mortality in this study.

Several methods are available for estimation of mortality based on size composition of the commercial fishery. The total mortality rate is given by

$$Z_L \Delta t_L = log_e\left(N_{L+1}/N_L\right), \tag{6}$$

where Z_L is the total mortality rate at size L, N_L is the number of individuals in size class L, and Δt_L is the time required to grow from size class L to class $L + 1$. Total mortality rates can therefore be estimated if an estimate of population numbers at size is available and growth rates can be determined. This technique measures the rate of decay in population numbers between successive size classes. Typically, a proxy for population size, such as catch per unit of effort, is used in these analyses. Total mortality estimates derived using this type of size-based analysis are summarized in Table 5. Caddy (cited in Campbell, 1980) estimated the total mortality rate for 11 locations in New Brunswick and Nova Scotia during 1977–1978 based on an analysis of size composition data. The exploitation rate estimates were low (29%) at locations that had recently experienced decreases in population and fishing effort, but ranged as high as 98% in western Northumberland Strait. Anthony (1980) summarized mortality estimates based on size composition analysis for inshore fisheries off the northeastern United States during 1972–1975; estimated exploitation rates were 82–91%. Earlier estimates from size composition data in Maine for the periods 1949–1953 and 1967–1970 were 81 and 85%, respectively. Cadrin and Estrella (1993) reported exploitation rates for Massachusetts on a statewide basis that ranged from 62 to 69% during 1981–1992; estimated exploitation rates were substantially higher in most regions of the Massachusetts coast, but were low at one location exploiting onshore migrants (see Table 5). Fogarty *et al.* (1982) provided mortality estimates for the offshore New England fishery on Georges Bank and in the southern New England region using length-based cohort analysis (Jones, 1974). Conser and Idoine (1992) derived an alternative approach to mortality estimation based on a

TABLE 5 **Estimates of the Percentage Exploitation Rate ($E \times 100$) and Instantaneous Rate of Total Mortality (Z) for Selected American Lobster Populations According to Method of Analysis (Tag Return or Size Composition Analysis)**

Year	Area	*E*	*Z*	Method	Reference
1953–1960	Fourchu, Nova Scotia	43–83		Tag	Paloheimo (1963)
1950–1960	Gabarus, Nova Scotia	57–71			
1947–1955	Port Maitland, Nova Scotia	41–56			
1951–1955	Tignish, Prince Edward Island	58–73			
1978	Arnold's Cove, Newfoundland	84		Tag	Ennis (in Campbell, 1980)
1978	Bonavista Bay, Newfoundland	69			
1978	Notre Dame Bay, Newfoundland	91			
1978	Port au Port, Newfoundland	64			
1978	Northwestern Newfoundland	52–86			
1978	Lismore, Nova Scotia		0.94–1.86	Size	Caddy (in Campbell, 1980)
1978	Grand Manan Island, New Brunswick		1.80–1.85		
1978	Abbots Harbour, Nova Scotia		1.68–1.85		
1978	Southeastern Browns Bank		1.31		
1978	Victoria Beach, Nova Scotia		0.46–1.17		
1977	Western Northumberland Strait		3.2–5.3		
1977	Central Northumberland Strait		1.4–4.0		
1977	Eastern Northumberland Strait		0.7–1.3		
1977	Prince Edward Island		1.6–2.6		
1977	Cape Breton Island, Nova Scotia		1.0–2.0		
1977	Southeastern Nova Scotia		0.5–0.9		
1949–1953	Maine		1.92	Size	Anthony (1980)
1967–1970	Maine		2.20		
1972–1974	New Hampshire		1.98		
1974–1975	Massachusetts		1.99–2.29		
1974–1975	Rhode Island		2.05	Tag	
1975	New York		1.94–2.80	Size	
1968–1971	Offshore New England		0.07–0.67	Tag	
1974–1975	Offshore Virginia		0.91–1.50	Size	
1975	Maine		6.36–8.73	Tag	Krouse (1981)
1978–1979	Long Island Sound, New York		1.47–2.38	Tag	Briggs and Mushacke (1984)
1975–1980	Comfort Cove, Newfoundland	69.8–94.9		Tag	Ennis *et al.* (1982)
1986	Massachusetts		0.47–2.11	Size	Cadrin and Estrella (1993)
1981–1986	Cape Anne, Massachusetts		1.32–1.52		
1981–1986	Beverly-Salem, Massachusetts		1.59–1.99		
1984–1986	Boston Harbor, Massachusetts		1.75–1.92		
1981–1986	Cape Cod Bay, Massachusetts		1.64 –2.07		
1981–1986	Outer Cape Cod, Massachusetts		0.52–0.57		
1981–1986	Buzzard's Bay, Massachusetts		2.21–2.97		
1985–1992	Gulf of Maine	38–47	0.50–0.67	Size	Anonymous (1993)
1985–1992	Georges Bank and south	24– 49	0.29–0.72		
1985–1992	Inshore southern New England–Long Island Sound	65–90	1.12–2.69		

modified DeLury analysis. A recent evaluation throughout the northeastern United States indicated increases in mortality rates during the last decade (Anonymous, 1993). This increase is consistent with known factors, such as a steady escalation in the number of traps, conversion to wire traps requiring less maintenance, and increases in trap efficiency with the use of escape vents.

E. Abundance

Lobster abundance has been measured at different life history stages in several long-term studies. The focus in this section concerns estimates of relative or absolute abundance measured over time to illustrate trends and levels of variation. Variation in population levels can occur in response to exploitation and changes in the natural production characteristics of the system. Populations or stocks at the extremes of the range can be expected to be most vulnerable to changes in environmental conditions. The most extensive time series of planktonic larval production and abundance was developed for the Northumberland Strait region (Scarratt, 1964, 1973). Production for the larvae (stages I–III) and the postlarva (stage IV) during 1949–1963 is depicted in Fig. 7. A generally declining trend in larval production was evident during this period and the variability in production increased with successive stages (Fogarty and Idoine, 1986). The population declined markedly shortly after the cessation of this sampling program.

The abundance of juvenile and adult lobsters has been monitored using three principal techniques: (1) mark–recapture studies, (2) monitoring trends in catch per unit of effort, and (3) research vessel trawl surveys. *In situ* measurements of density have also been made using scuba and other techniques, but will not be considered further here (reviewed by Cooper and Uzmann, 1980; Cobb and Wang, 1985; Lawton and Lavalli, Chapter 4). Paloheimo (1963) provided estimates of population size based on a modified Peterson mark–recapture technique for several sites in Nova Scotia and Prince Edward Island, after careful adjustment for changes in catchability due to interannual variation in temperature. Scarratt (1973) subsequently extended the series for one of these sites off Port Maitland, Nova Scotia (Fig. 7 provides the complete time series). These estimates suggest that the lobster population in this region fluctuated about a relatively stable level during the period 1949–1969. In contrast, Ennis *et al.* (1982) demonstrated a clear trend in abundance for a Newfoundland population at the northern end of the range using mark–recapture techniques; a doubling of population size was

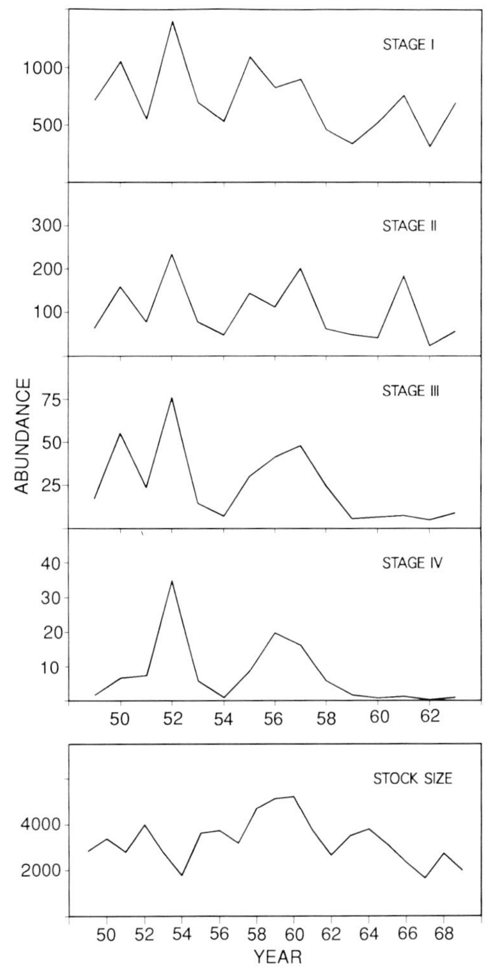

FIGURE 7 Trends in production for stages I–IV and stock size in Northumberland Strait. Note the difference in the time scale for larval production and stock size estimates.

documented during the period 1972–1980. A declining trend in standing stock of lobsters was noted in Arnold's Cove during 1976–1984, following a dramatic increase from 1971 to 1976 (Ennis *et al.*, 1986).

Research vessel surveys on the continental shelf from Cape Hatteras to the Gulf of Maine have been used to monitor trends in relative abundance for segments of the population vulnerable to trawl gear since 1963 (Fogarty *et al.*, 1982; Fogarty, 1986; Anonymous, 1993). Surveys of more limited geographical extent have also been conducted in several regions (Anonymous, 1993; Jeffries, 1994). Declines in relative biomass following the initial phase of exploitation in offshore waters were evident on Georges Bank and in the southern New England region. In contrast, a clear increase in relative abundance is evident for the Gulf of Maine during the last decade.

Trends in catch per unit of effort have also been monitored and time series of these estimates are available for several regions. Typically, the catch (number and/or weight) per trap haul or the catch per trap haul divided by soak (immersion) time is recorded. However, the catch rate in traps is temperature dependent (McLeese and Wilder, 1958) and is further subject to saturation effects (Caddy, 1977a; Skud, 1979; Fogarty and Borden, 1980; Krouse, 1989; Fogarty and Addison, 1995). Accordingly, catch per trap haul should be adjusted for temperature and trap saturation. A generalized model for the relationship between catch (C) and soak time (t) can be written as

$$C_t = \left\{\frac{a}{b}\left[1 - e^{-(1-m)bt}\right]\right\}^{1/(1-m)}, \tag{7}$$

where a is an entry parameter, b is an escapement rate, and m is a parameter determining the shape of the entry curve. The saturation level is given by the ratio of the capture to escapement or loss rates (a/b). The capture rate (adjusted for temperature and/or other factors) can provide the basis for an index of abundance after calibration studies to determine the relationship between entry rates and abundance levels (Fogarty and Addison, 1995). Clear evidence of a recent increase in catch per trap haul has been noted in the United States during the last decade (Anonymous, 1993).

F. Recruitment

Replenishment of a population through reproduction is an essential consideration in any evaluation of its dynamics. The relationship between the parental stock and the number of progeny surviving to some specified size or age (recruitment) is of particular interest. Populations are, of course, reduced by harvesting, and the implications of this reduction for recruitment determine the overall response to exploitation. If the stock can compensate in some way for lower population size, it is possible (in principle) to conduct a sustainable fishery. Conversely, a population with little or no compensatory capacity is highly vulnerable to overexploitation. Possible compensatory mechanisms include density-dependent growth, survival, maturation, and fecundity. Evidence for density-related effects have been obtained in culture systems (reviewed by Aiken and Waddy, 1978; Van Olst *et al.*, 1980) and in one natural lobster stock in Arnold's Cove (Ennis, 1991), where density-dependent molting frequency and egg production were observed. Density-dependent responses have been documented in spiny lobster populations (Chittleborough, 1970; Polovina, 1989; Chubb, 1994).

It has been suggested that competition for shelter sites is a potentially important regulatory mechanism (Caddy, 1986; Fogarty and Idoine, 1986). Shelter is clearly a critical component of lobster habitat (Cobb, 1971) and in its absence, lobsters are highly vulnerable to predation (Richards and Cobb, 1986; Wahle and Steneck, 1992). Wahle and Steneck (1991) suggested that cobble habitats, which supply appropriate shelter sites for early benthic phase lobsters, are a potentially limiting resource, although early postsettlement lobsters are found in other habitats (Able *et al.*, 1988; Hudon, 1987). Miller *et al.* (1992) suggested that the recent large-scale increase in landings, however, argues against habitat limitation (see also Miller, 1993). As noted earlier, however, landings cannot be taken as an index of abundance without adjustment for fishing effort and the areal extent of the fishery. Further, the size structure of the population has been radically altered in over a century of exploitation. Habitat or other resources may not be currently constraining, but may have been limiting for the virgin population (Addison and Fogarty, 1992).

Examination of the relationship between stock and recruitment is considerably complicated by environmentally induced fluctuations in survival during the early life history stages. Stock–recruitment relationships of marine populations are characterized by relatively high levels of stochastic variation (Fogarty *et al.*, 1991; Fogarty, 1993). A highly variable relationship between stock and recruitment is not unexpected in a fluctuating environment. For a closed population, the relevant question is whether the relationship is linear (density independent) or nonlinear (density dependent).

The relationship between stock and recruitment has not been directly examined for any population of *Homarus americanus*; however, the relationship between larval production and subsequent stock size and/or landings has been examined in detail in the southern Gulf of St. Lawrence (Scarratt, 1964, 1973; Harding *et al.*, 1983; Fogarty and Idoine, 1986). Scarratt (1964, 1973) related stage IV larval production and subsequent stock size in Northumberland Strait and concluded that no relationship was evident. However, Scarratt considered only linear relationships that were not constrained to pass through the origin. Fogarty and Idoine (1986) reexamined these data and concluded that an asymptotic model was more appropriate, indicating the possibility of a density-dependent interaction (Fig. 8). Other statistical tests for density dependence in the stock size estimates were consistent with these findings. In contrast, there was no evidence of density-dependent relationships between successive pelagic larval stages. These authors concluded that the apparent asymptotic form of the relationship between the post-

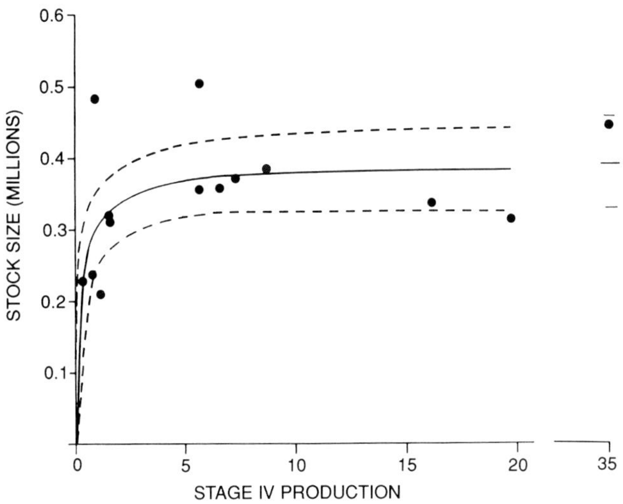

FIGURE 8 Relationship between stage IV production and resulting stock size in Northumberland Strait. (After Fogarty and Idoine, 1986.)

larval stage and subsequent stock size will be stabilizing. Further, the steep slope of the curve at the origin is consistent with strong resilience to exploitation. Pollock (1993) drew similar inferences for spiny lobster recruitment dynamics. Ennis (1986b) and Caddy (1986) argued that the most probable form of the stock recruitment relationship in lobsters is asymptotic, indicating the importance of intracohort competition for critical resources.

V. Population and Management Models

Traditional models of the dynamics of exploited populations treat each of the components of production in varying levels of detail and specificity. Two classes of management-oriented models (surplus production and yield-per-recruit models) are described in Section V,B. A stage-structured model that can integrate recruitment and yield-per-recruit models in a full-production model is also described. Surplus production models consider recruitment, growth, and natural mortality in aggregate. Yield- and egg production-per-recruit models consider growth, natural mortality, and fishing mortality, but do not consider recruitment explicitly. Each of these approaches offers different advantages in application to *Homarus americanus* populations and suffers from different limitations with respect to availability of necessary information (Saila and Marchesseault, 1980). The principal goal of these modeling approaches is to synthesize information on basic biological and demographic characteristics of the population and to examine the consequences of different management strategies such as changes in minimum legal size limits and fishing mortality rates.

A. Production Models

Extension of the Verhulst–Pearl logistic model to account for exploitation effects was developed by Hjort *et al.* (1933) and Graham (1935) and further refined by Schaefer (1954, 1957), Fox (1970), and Pella and Tomlinson (1969). Surplus production models have been extensively used in analysis of the Western Australian (Phillips and Brown, 1989; Hall and Brown, 1994) and New Zealand (Saila *et al.*, 1979; Booth and Breen, 1994) rock lobster fisheries. These models are attractive because of their analytical tractability and modest data requirements. This approach requires a time series of catch and standardized fishing effort data; catch per unit of effort is used as a proxy for population biomass. It is assumed that age/size or other structural features of the population are in equilibrium and that time delays in production processes can be ignored. It is further implicitly assumed that the population is homogeneously distributed in space and that environmental conditions remain constant. Clearly, these are overly simplistic assumptions, and approaches to address some of these limitations are described later in this section. The change in biomass is given by

$$\frac{dB_t}{dt} = f_1(B_t)B_t - qh_t^{\gamma}B_t , \tag{8}$$

and the change in yield (Y) is

$$\frac{dY}{dt} = qh_t^{\gamma}B_t , \tag{9}$$

where B_t is population biomass, $f_1(B_t)$ is a compensatory function representing the combined effects of recruitment, individual growth, and natural mortality, q is the constant of proportionality between fishing effort (h) and instantaneous fishing mortality, and γ is a parameter controlling the shape of relationship between yield and fishing effort. Virtually all analyses to date have set γ at 1 (but see Smith, 1980). The compensatory function is $f_1(B_t) = (\alpha_1 - \beta_1 B^{m-1})$, where α_1 and β_1 are density-independent and -dependent parameters, respectively, and m is the shape parameter controlling the degree of curvature. The Pella–Tomlinson function includes the Schaefer (m = 2) and Fox ($m \rightarrow 1$) models as special cases.

Simple production models have been applied to the American lobster fishery in an attempt to define optimal harvesting strategies. Jensen (1986) employed the Schaefer and Fox models in an analysis of the Maine lobster fishery and concluded that the Fox model (m = 1) provided a better description of the fishery dynamics than did the symmetric Schaefer model. Dow *et al.* (1975) and Smith (1980) based economic analyses of the Maine lobster fishery on a Schaefer-type model with modifications for stochastic influences. In all cases, it has been concluded that the level of effort in the Maine fishery greatly exceeded the level providing optimal biological and economic yield.

Several limitations affect the general utility of the production models applied to American lobster fisheries to date: (1) the principal measure of effort available (number of traps fished per year) is an extremely crude index of fishing intensity; (2) all existing analyses have been spatially limited and have not covered the entire population; (3) changes in environmental conditions, particularly temperature, appear to have affected catch rates over time; and (4) fishery regulations and the areal extent of the fishery have changed during the time period. To illustrate potential difficulties, the relationship between yield and effort in the inshore Maine fishery for the period 1928–1992 is depicted in Fig. 9. The superimposed curve represents an equilibrium production curve fit to the data up to 1980; it is clear that the more recent data (open circles) do not conform to the expected production relationship. Consideration of environmental conditions indicates an interactive effect between temperature and the number of traps fished, which must be taken into account (see inset, Fig. 9). Landings increased rapidly during the 1950s as effort increased, but this period was characterized by high water temperatures. Yield declined with increasing effort in the latter 1960s and early 1970s; however, water temperatures were relatively cool during this period. The recent increase in landings has occurred during a period of high effort and relatively high water temperatures. In addition, many factors have recently resulted in increases in efficiency in the fishery. Simple production models must therefore be used with extreme caution in defining sustainable yields and harvesting strategies. However, the empirical relationship of the number of traps fished, environmental factors, and fishery performance offers an instructive view of the interactions among these factors with important bioeconomic implications.

Modifications to production models to incorporate time delays (e.g., time lags between hatching and recruitment) have been developed specifically for

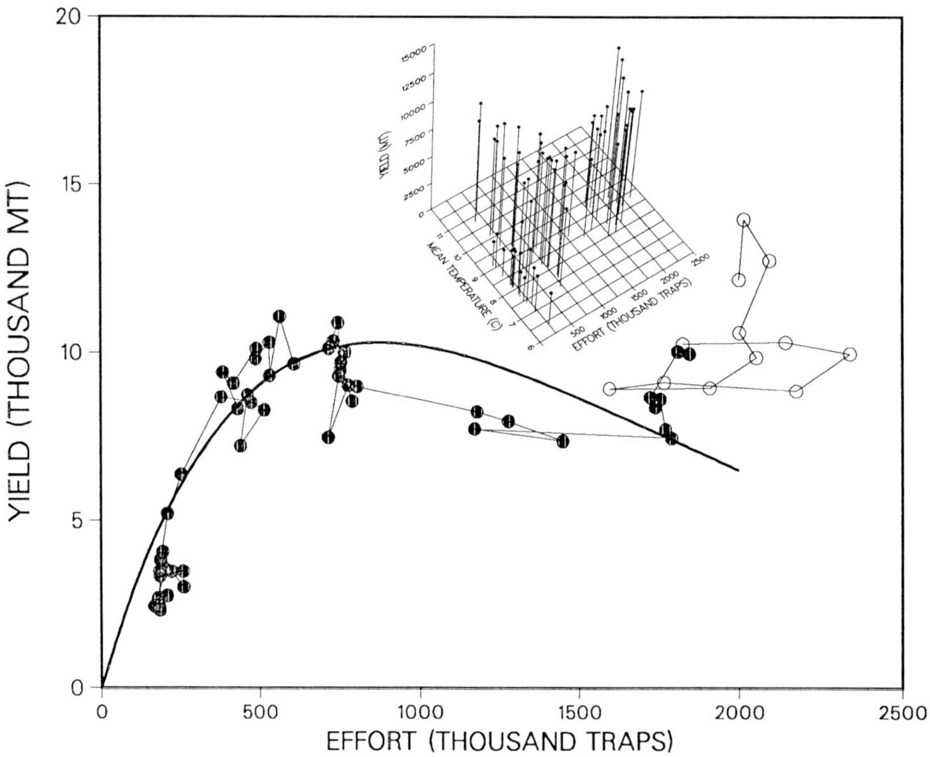

FIGURE 9 Relationship between yield and fishing effort (number of traps) for the Maine fishery during 1930–1992. Empirical observations for the period 1930–1980 (black circles) and 1981–1992 (white circles). The solid curve represents the equilibrium production model for the period 1930–1980. (Inset) The relationship between the number of traps, the mean annual water temperature, and yield. MT, Metric tons.

lobster populations (Marchesseault *et al.*, 1976; Fogarty and Murawski, 1986). The model can be written as

$$\frac{dB_t}{dt} = f_1(B_t)B_t + f_2(B_{t-r})B_{t-r} - qf_t^{\gamma}B_t, \quad (10)$$

where $f_1(B_t)$ is a function representing the effects of individual growth and natural mortality, $f_2(B_{t-r})$ is a recruitment function depending on the biomass at time $t - r$ (representing the time delay between spawning and recruitment), and all other terms are as previously explained. The function $f_1(B_t)$ can be defined as for Eq. (8). Similarly, the recruitment function can be represented as $f_2(B_{t-r}) = (\alpha_2 - \beta_2 B_{t-r})$. This specification imposes an implicit stage structure (prerecruits and recruits) in the model formulation.

Deriso (1980) introduced a delay-difference model that allows specification of the individual components of production. Fogarty and Murawski (1986) described an application of a simplified version of the Deriso model to lobster populations; the basic model formulation can be expressed as

$$B_t = B_{t-1}\exp\left(G - M - qf_{t-1}^{\gamma}\right) + f\left(B_{t-r}\right), \quad (11)$$

where G and M are instantaneous rates of individual growth and natural mortality, q is the catchability coefficient, f is fishing effort, and $f(B_{t-r})$ is the recruitment function. Again, an implicit stage structure is defined. As longer time series of detailed catch and effort data, which can be standardized for temperature effects and other factors, become available, application of delay-differential or delay-difference models can profitably be made. Delay-difference models have been used to advantage in analysis of the Western Australian rock lobster fishery (Hall and Brown, 1994, 1995).

B. Yield- and Egg Production-per-Recruit

In contrast to surplus production models, which provide a simplistic description of the dynamics of the entire population, yield- and egg production-per-recruit models track changes in biomass, yield, and reproductive output of a single cohort over its lifetime following recruitment to the fishery. However, yield and egg production models do incorporate detailed information on age- or size-specific growth, mortality, and maturation rates and can accommodate regulatory measures such as minimum and maximum size limits and protection of egg-bearing females, which generally cannot be directly addressed in production models.

Application of yield-per-recruit models, including the classical Beverton and Holt (1957) formulation and several models designed specifically for the American lobster, has been described (Thomas, 1973; Caddy, 1977b, 1979a; Ennis, 1980b; Fogarty, 1980; Campbell, 1985; Fogarty and Idoine, 1988). Despite differences in model structure and in the biological parameters describing different populations in various applications, results of these analyses share several common features. The fishing mortality rate at which yield per recruit is maximized under existing regulations (minimum legal size limits, etc.) is typically quite low (often less than $F = 0.3$ for males and somewhat higher for females). Estimated fishing mortality rates in most areas exceed F_{max}, in most cases considerably, suggesting that reductions in fishing mortality would have substantial benefits in terms of yield per recruit. Sexual dimorphism in growth rates and the linkage between growth and reproduction in females result in different yield-per-recruit patterns relative to males.

The size-based yield-per-recruit model described by Fogarty and Idoine (1988; Fogarty, 1986) can be written as

$$Y = \sum\sum \frac{p_{L,t}F}{p_{L,t}F + M_{L,t}}\left[1 - e^{-\left(p_{L,t}F + M_{L,t}\right)}\right]N_{L,t}\,\bar{w}_{L,t}, \quad (12)$$

where p_L is the proportion of the population vulnerable to exploitation at size L, $\bar{w}_{L,t}$ is the mean weight of an individual in size class L at time t, and all other terms are defined as before. The summations are taken over size classes and time. The growth component of the model (used to specify $\bar{w}_{L,t}$) allows variation introduced through differences in molting schedules and variability in the molt increment. The coefficient p_L can be used to designate minimum and maximum legal size limits, restricted harvesting of egg-bearing females, etc. Examples of yield analyses varying minimum legal size limits and fishing mortality rates for male and female offshore lobsters are illustrated in Fig. 10. Under the prevailing minimum legal size limits at the time of this study and at the estimated fishing mortality rates, clear benefits in yield per recruit are evident with increases in the minimum legal size and/or reduction of fishing mortality rates.

The expected lifetime reproductive output of a female is a function of growth and mortality rates, maturation schedule, and fecundity. Egg production per recruit is related to the concept of reproductive value used in demographic studies. Fogarty and Idoine (1988) described a size-based model:

$$E = \sum\sum m'_L f_L N_L, \quad (13)$$

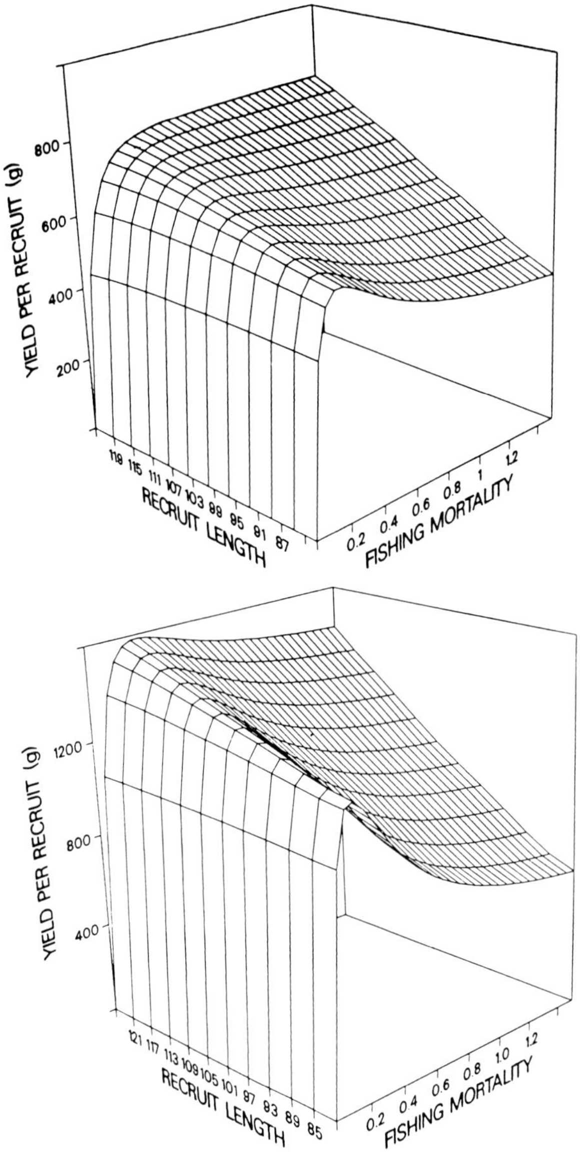

FIGURE 10 Yield per recruit for female (top) and male (bottom) offshore lobsters as a function of minimum legal size and fishing mortality rates. (After Fogarty and Idoine, 1988.)

where m'_L is the proportion of ovigerous females, f_L is the fecundity, and N_L is the number in size class L. Related formulations are described by Saila and Flowers (1965), Caddy (1977a, 1979b), Campbell and Robinson (1983), Campbell (1985), and Ennis (1985). An example of egg production per recruit at several different minimum legal size limits and for a range of fishing mortality rates for offshore lobsters is provided in Fig. 11.

C. Stage-Structured Model

Each of the biological parameters described above

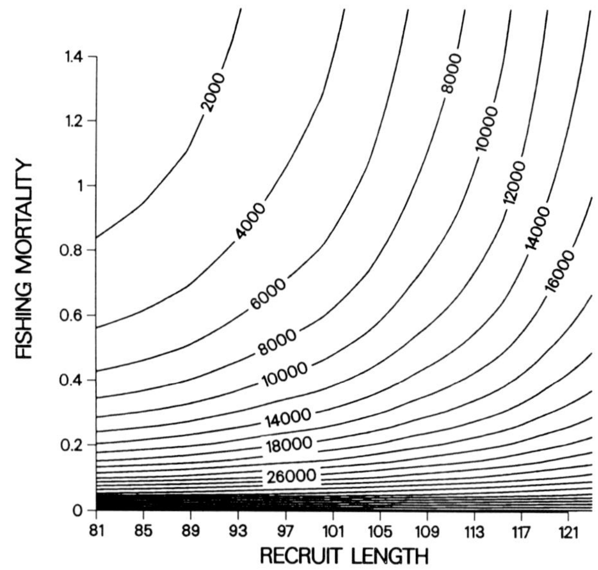

FIGURE 11 Egg production per recruit at several minimum legal size limits as a function of fishing mortality rates. (After Fogarty and Idoine, 1988.)

can be conveniently integrated in the context of a stage-structured population model (Caswell, 1989, provides a general overview). The basic form of the projection matrix is

$$\begin{bmatrix} N_{0,t+1} \\ N_{1,t+1} \\ N_{2,t+1} \\ \cdot \\ \cdot \\ \cdot \\ N_{n,t+1} \end{bmatrix} = \begin{bmatrix} a_{0,0} & s_0 m'_1 f_1 & s_0 m'_2 f_2 & \cdot & \cdot & \cdot & s_0 m'_n f_n \\ a_{1,0} & a_{1,1} & 0 & \cdot & \cdot & \cdot & 0 \\ a_{2,0} & a_{2,1} & a_{2,2} & \cdot & \cdot & \cdot & 0 \\ \cdot & \cdot & \cdot & \cdot & \cdot & \cdot & \cdot \\ \cdot & \cdot & \cdot & \cdot & \cdot & \cdot & \cdot \\ \cdot & \cdot & \cdot & \cdot & \cdot & \cdot & \cdot \\ a_{n,0} & \cdot & \cdot & \cdot & \cdot & a_{n,n-1} & a_{n,n} \end{bmatrix} \begin{bmatrix} N_{0,t} \\ N_{1,t} \\ N_{2,t} \\ \cdot \\ \cdot \\ \cdot \\ N_{n,t} \end{bmatrix}, \tag{14}$$

where the elements of the main diagonal represent the probabilities of surviving but remaining in the same size or stage class in the time interval, the elements of all subdiagonals represent the probability of surviving and growing into a larger category, m'_L is the proportion ovigerous in size class L, and f_L is the fertility of the Lth size class. Note that $a_{L,L} = (1 - \rho_L)s_L$, where ρ_L is given by Eq. (3) and $s_L = \exp(-M_L - p_L F)$, and the subdiagonal elements are functions of the survival fraction (s_L), the probability of molting (ρ_L), and the molt increment. The model is therefore defined in terms of the parameters used to specify vital rates in Section IV. It is possible to incorporate density-dependent processes in any of the elements of the projection matrix. For example, if the first stage represents the prerecruit phase and survival rates are density dependent (see Section IV,F), then

$$s_0 = \frac{s_0^*}{1 + kN}, \tag{15}$$

where s_0^* is the maximum survival rate, k is a density-dependent parameter, and N is some relevant measure of population size. A graphic representation of a simple, four-stage life history model incorporating a density-dependent survival component is shown in Fig. 12. This figure traces the trajectory of a cohort through successive life history stages (starting at point A). Note that the asymptotic relationship between stage IV production and recruitment [mediated by a survival function such as in Eq. (15)] is highly stabilizing. After two generations in this hypothetical case, the population has essentially reached an equilibrium (population is at point I after one generation and the nearby point Q after two generations). The yield is given by

$$\begin{bmatrix} Y_{0,t+1} \\ Y_{1,t+1} \\ Y_{2,t+1} \\ \cdot \\ \cdot \\ \cdot \\ Y_{n,t+1} \end{bmatrix} = \begin{bmatrix} E_0 & 0 & 0 & \cdot & \cdot & \cdot & 0 \\ 0 & E_1 & 0 & \cdot & \cdot & \cdot & 0 \\ 0 & 0 & E_2 & \cdot & \cdot & \cdot & 0 \\ \cdot & \cdot & \cdot & \cdot & \cdot & \cdot & \cdot \\ \cdot & \cdot & \cdot & \cdot & \cdot & \cdot & \cdot \\ \cdot & \cdot & \cdot & \cdot & \cdot & \cdot & \cdot \\ 0 & \cdot & \cdot & \cdot & \cdot & 0 & E_n \end{bmatrix} \begin{bmatrix} N_{0,t} \\ N_{1,t} \\ N_{2,t} \\ \cdot \\ \cdot \\ \cdot \\ N_{n,t} \end{bmatrix} \begin{bmatrix} \overline{w}_0 \\ \overline{w}_1 \\ \overline{w}_2 \\ \cdot \\ \cdot \\ \cdot \\ \overline{w}_n \end{bmatrix}, \tag{16}$$

where E_L is the exploitation rate for the Lth size class and w_L is the mean weight of the ith size class.

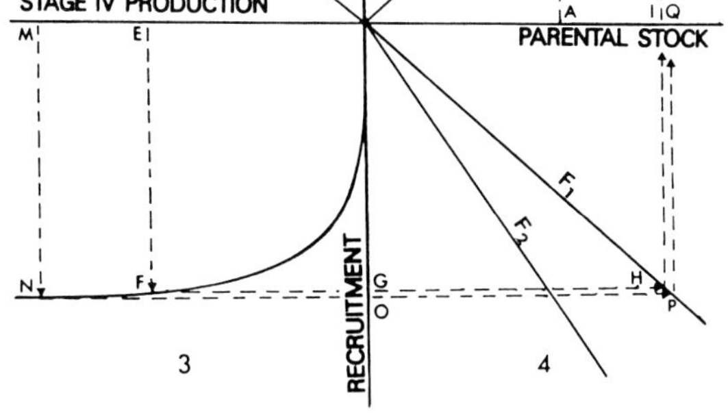

FIGURE 12 Graphic representation of a four-stage life history model with density-dependent survival between the fourth larval stage and recruitment to the fishery.

The matrix formulation can be viewed as including many of the other models described previously as special cases. For example, the delay-difference and delay-differential models can be formulated as a two-stage matrix model. The basic yield and egg production-per-recruit models can be emulated by specifying a constant number of recruits entering the projection model. It is further possible to specify a metapopulation model in the general context of this matrix formulation (see Caswell, 1989, pp. 50–53). In this case, additional terms specifying the rates of migration and/or dispersal by larval drift among subpopulations must be provided. A spatially explicit model that can accommodate these types of considerations has been very helpful in analysis of western rock lobster fishery dynamics (Walters *et al.*, 1993; Hall and Brown, 1994).

VI. Forecasting Models

The matrix formulation in the previous section can provide a convenient means of developing population projections (i.e., the time trajectory of population size given a specified set of demographic parameters). It is also useful to consider empirically based predictive models for providing catch or population forecasts. The development of empirical models for predicting lobster yields has received considerable attention (Dow, 1969, 1977, 1978; Flowers and Saila, 1972; Boudreault *et al.*, 1977; Orach-Meza and Saila, 1978; Harding *et al.*, 1983; Fogarty, 1988, 1989; Campbell *et al.*, 1991). In particular, the relationship between temperature and yield has been examined in detail using a variety of techniques. Landings data and environmental variables such as temperature often exhibit temporal correlation patterns. This autocorrelation complicates the evaluation of the statistical significance of relationships between yield and temperature. Development of time-series models based on the approach of Box and Jenkins (1976) for the construction of univariate and multivariate (transfer function) models has recently been explored to counter this problem and to assess the predictive capability of this class of models. The basic form for a univariate time-series model of lobster landings is

$$C_t = \Phi_1 C_{t-1} + \ldots \Phi_p C_{t-p} + a_t - \theta_1 a_{t-1} \ldots \theta_q a_{t-q} + \theta_o, \tag{17}$$

where C_t is the catch in year t, ϕ_i are autoregressive parameters reflecting the effect of past catch levels, a_t is a random "shock" or error term, θ_i are moving average parameters reflecting the effects of past ran-

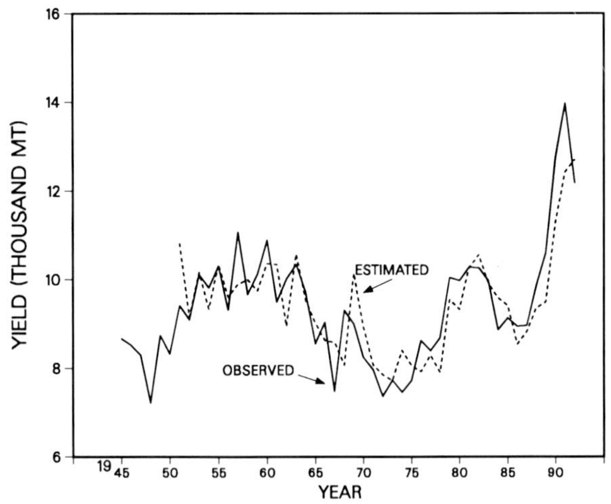

FIGURE 13 Observed and predicted yields based on a transfer function model incorporating the effects of temperature on yield. MT, Metric tons.

dom shocks, and θ_0 is an overall trend parameter. Boudreault *et al.* (1977) and Fogarty (1988) demonstrated that univariate time-series models can provide useful forecasts of landings. Fogarty (1988) developed univariate models for both annual and monthly landings. These models contain autoregressive terms at a lag of 1 year for the annual landings series and at lags of 1 and 12 months.

Consideration of additional "explanatory" variables can often improve the predictive capability of the model and provide insight into structural relationships. A bivariate model between lobster landings and water temperature can be written as

$$C_t - \delta_1 C_{t-1} \ldots - \delta_r C_{t-r} = \omega_o T_{t-b} - \omega_1 T_{t-b-1} \ldots - \omega_n T_{t-b-n} + \eta_t \ , \tag{18}$$

where C_t is the catch in year t, T_{t-b} is the catch in year $t - b$, δ_i and ω_i are model parameters, and η_t is an error term that can be modeled as an autoregressive integrated moving average process. Fogarty (1988) and Campbell *et al.* (1991) demonstrated that consideration of temperature effects improves the predictive capability of the models relative to corresponding univariate models. For a comparison of observed and predicted annual landings for the Maine lobster fishery including temperature at lags of zero, indicating an immediate temperature effect, and 6 years for the period 1945–1992, see Fig. 13). The immediate temperature effect is consistent with the known increase in catchability with increasing water temperature (McLeese and Wilder, 1958). The delayed temperature effect at a lag of 6 years may represent the importance of water temperature during the early life history. Caddy (1979b) postulated that in colder years, lobster larvae hatched at the end of season may not be able to complete development through the pelagic phase before winter; larval development rates are highly temperature dependent (reviewed by Fogarty, 1983). Faster development through the pelagic phase may also decrease the window of vulnerability to predators in the water column.

VII. Summary

Increases in landings of the American lobster, *Homarus americanus*, through the 1980s and early 1990s in both the United States and Canada have highlighted the need for greater understanding of the factors controlling production of this valuable resource. Regional variation is evident in critical demographic parameters, such as growth, maturation, fecundity, and mortality, and some populations may be more vulnerable to exploitation than others. Further, temporal variation in growth rates (annual fraction molting) is also pronounced, with important implications for production rates. An understanding of the stock–recruitment relationship is critical in determining the resilience of these populations to exploitation. The abundance and demographic structure of lobster populations, however, have been radically altered by over a century of exploitation, making a reconstruction of the recruitment dynamics over the full range of stock sizes impossible. Nevertheless, long-term monitoring of lobster populations at a number of critical life history phases must be undertaken if we are to understand the regulatory mechanisms controlling these populations. In general, American lobster populations have been relatively robust under exploitation. However, it is not known whether this apparent resilience is due to strong population feedback mechanisms, to the role of refugia of lightly exploited groups that provide a larval subsidy to more heavily exploited groups (Anthony and Caddy, 1980), or to some combination of both mechanisms (Fogarty and Idoine, 1986). Consideration of the recruitment dynamics of lobsters will therefore require detailed studies of potential compensatory mechanisms, such as density dependent growth, reproduction (maturity and fecundity), and survival at different life history stages, and the role of migration and larval dispersal in linking subpopulations. Although considerable attention has been devoted to studying the effects of abiotic environmental factors on lobster production and yield (Fogarty, 1989), the role of lobsters in ecosystem dynamics requires further research. While the implications of possible competitive interactions among lobsters and other decapods have been examined in the context of an exploited assemblage (Cobb *et al.*, 1986; Cobb and Caddy, 1989), far less is known about other ecological interactions, such as the role of predation on lobsters in controlling production.

The factors controlling the production rates of lobster populations are highly dynamic, presenting important challenges for the development of effective management strategies. To meet these challenges, we must understand the determinants of spatial and temporal changes in production processes and adjust management strategies accordingly.

Acknowledgments

I am grateful to Bob Miller for his comments on this chapter and to Jan Factor for many helpful editorial suggestions. Doug Pezzack and Bruce Estrella graciously provided graphic information and data. Joe Idoine has been instrumental in many aspects of my work and I am grateful for his help. I would particularly like to thank Stan Cobb and Saul Saila for the guidance they have given me over the years.

References

Able, K. W., Heck, K. L., Fahey, M. P., and Roman, C. T. (1988). Use of salt-marsh peat reefs by small juvenile lobsters on Cape Cod, Massachusetts. *Estuaries* 11, 83–86.

Acheson, J. M. (1975). Fisheries management and social context: The case of the Maine lobster fishery. *Trans. Am. Fish. Soc.* 104, 653–668.

Acheson, J. M. (1988). "The Lobster Gangs of Maine." University Press of New England, Hanover, New Hampshire.

Addison, J. T., and Fogarty, M. J. (1992). Juvenile lobster habitat limitation: What can landings tell us? *Lobster Newsl.* 5(2), 10–12.

Addison, J. T., and Fogarty, M. J. (1993). Lobster landings reprise. *Lobster Newsl.* 6(1), 10.

Aiken, D. E. (1980). Molting and growth. *In* "The Biology and Management of Lobsters" (J. S. Cobb and B. F. Phillips, eds.), Vol. 1, pp. 91–163. Academic Press, New York.

Aiken, D. E., and Waddy, S. L. (1978). Space, density, and growth of the lobster (*Homarus americanus*). *Proc. Annu. Meet.—World Maric. Soc.* 9, 461–467.

Aiken, D. E., and Waddy, S. L. (1980). Reproductive biology. *In* "The Biology and Management of Lobsters" (J. S. Cobb and B. F. Phillips, eds.), Vol. 1, pp. 215–276. Academic Press, New York.

Aiken, D. E., and Waddy, S. L. (1986). Environmental influences on recruitment of the American lobster, *Homarus americanus*. *Can. J. Fish. Aquat. Sci.* 43, 2258–2270.

Anonymous (1993). American lobster. *Northeast Fish. Sci. Cent. Ref. Doc.* 93-18, 75–107.

Anthony, V. C. (1980). Review of lobster mortality estimates in the United States. *Can. Tech. Rep. Fish. Aquat. Sci.* 932, 17–26.

Anthony, V. C., and Caddy, J. F., eds. (1980). Proceedings of the Canada–U.S. Workshop on Status of Assessment Science for N.W. Atlantic Lobster (*Homarus americanus*) Stocks (St. Andrews, N.B., Oct. 24–26, 1978). *Can. Tech. Rep. Fish. Aquat. Sci.* 932.

Barlow, J., and Ridgway, G. (1969). Changes in serum protein during the molt and reproductive cycles of the American lobster (*Homarus americanus*). *J. Fish. Res. Board Can.* 26, 2101–2109.

Barlow, J., and Ridgway, G. J. (1971). Polymorphisms of esterase isozymes in the American lobster (*Homarus americanus*). *J. Fish. Res. Board Can.* 28, 15–21.

Beverton, R. J. H., and Holt, S. J. (1957). On the Dynamics of Exploited Fish Populations. *Fish. Invest. (London), Ser.* 2.

Booth, J. S., and Breen, P. A. (1994). The New Zealand fishery for *Jasus edwardsii* and *J. verreauxi*. *In* "Spiny Lobster Management" (B. F. Phillips, J. S. Cobb, and J. Kittaka, eds.), pp. 64–75. Blackwell, London.

Botsford, L. W. (1985). Models of growth. *In* "Crustacean Issues" (F. R. Schram, ed.), Vol. 3, "Factors in Adult Growth" (A. Wenner, ed.), pp. 171–188. Balkema, Rotterdam, The Netherlands.

Boudreault, F. R., DuPont, J. N., and Sylvain, C. (1977). Modèles lineares de prediction des debarguements de homard aux Îles-de-la-Madeleine (Golfe du Saint-Laurent). *J. Fish. Res. Board Can.* 34, 379–383.

Box, G. E. P., and Jenkins, G. M. (1976). "Time Series Analysis, Forecasting and Control." Holden Day, San Francisco.

Briggs, P. T., and Mushacke, F. M. (1984). The American lobster in western Long Island Sound: Movement, growth and mortality. *N.Y. Fish Game J.* 31, 21–37.

Burton, R. S. (1983). Protein polymorphism and genetic differentiation of marine invertebrate populations. *Mar. Biol. Lett.* 4, 193–206.

Burton, R. S., and Feldman, M. W. (1982). Population genetics of coastal and estuarine invertebrates: Does larval behavior influence population structure? *In* "Estuarine Comparisons" (V. S. Kennedy, ed.), pp. 537–551. Academic Press, New York.

Caddy, J. F. (1977a). Some Considerations Underlying Definitions of Catchability and Fishing Effort in Shellfish Fisheries, and Their Relevance for Stock Assessment Purposes. *Int. Counc. Explor. Sea Doc.* C.M. 1977/K:18.

Caddy, J. F. (1977b). Approaches to a Simplified Yield-per-Recruit Model for Crustacea with Particular Reference to the American Lobster, *Homarus americanus*. *Fish. Mar. Serv. Manuscr. Rep. (Can.)* 1445.

Caddy, J. F. (1979a). Notes on a More Generalized Yield per Recruit Analysis for Crustaceans Using Size Specific Inputs. *Tech Rep.—Fish. Mar. Serv.) (Can.)* 1525.

Caddy, J. F. (1979b). The influence of variations in the seasonal temperature regime on survival of larval stages of the American lobster (*Homarus americanus*) in the southern Gulf of St. Lawrence. *Rapp. P.-V., Reun., Cons. Int. Explor. Mer* 175, 204–216.

Caddy, J. F. (1986). Modelling stock–recruitment processes in Crustacea: Some practical and theoretical perspectives. *Can. J. Fish. Aquat. Sci.* 43, 2330–2344.

Cadrin, S. X., and Estrella, B. T. (1993). Massachusetts Coastal Commercial Lobster Trap Sampling Program May–November, 1992. *Mass., Div. Mar. Fish., Publ.* 17405-39-200-.97-8/93-C.R.

Campbell, A. (1980). Review of mortality estimates of the lobster populations in the Canadian Maritimes. *Can. Tech. Rep. Fish. Aquat. Sci.* 932, 27–35.

Campbell, A. (1983a). Growth of the American lobster, *Homarus americanus*, in the Bay of Fundy. *Can. J. Fish. Aquat. Sci.* 40, 1667–1675.

Campbell, A. (1983b). Growth of Tagged Lobsters (*Homarus americanus*) off Port Maitland, Nova Scotia, 1948–80. *Can. Tech. Rep. Fish. Aquat. Sci.* 1232.

Campbell, A. (1985). Application of a yield and egg production model to the lobster fishery in the Bay of Fundy. *North Am. J. Fish. Manage.* 5, 91–104.

Campbell, A. (1986). Migratory movements of ovigerous lobsters, *Homarus americanus*, tagged off Grand Manan, eastern Canada. *Can. J. Fish. Aquat. Sci.* 43, 2197–2205.

Campbell, A. (1989). The lobster fishery of southwestern Nova Scotia and the Bay of Fundy. *In* "Marine Invertebrate Fisheries:

Their Assessment and Management" (J. F. Caddy, ed.), pp. 141–158. Wiley, New York.

Campbell, A., and Bratty, J. (1986). Egg loss from the American lobster, *Homarus americanus*, in relation to nemertean, *Pseudocarcinonemertes homari*, infestation. *Can. J. Fish. Aquat. Sci.* 43, 772–780.

Campbell, A., and Mohn, R. K. (1982). Quest for Lobster Stock Boundaries in the Canadian Maritimes. *Northwest Atl. Fish. Org. Res. Doc.* 82/IX/107 (Ser. N615:45).

Campbell, A., and Mohn, R. K. (1983). Definition of American lobster stocks for the Canadian Maritimes by analysis of fishery landings trends. *Trans. Am. Fish. Soc.* 112, 744–759.

Campbell, A., and Robinson, D. G. (1983). Reproductive potential of three American lobster (*Homarus americanus*) stocks in the Canadian Maritimes. *Can. J. Fish. Aquat. Sci.* 40, 1958–1967.

Campbell, A., and Stasko, A. B. (1985). Movements of tagged American lobsters, *Homarus americanus*, off southwestern Nova Scotia. *Can. J. Fish. Aquat. Sci.* 42, 229–238.

Campbell, A., and Stasko, A. B. (1986). Movements of lobsters *Homarus americanus* tagged in the Bay of Fundy, Canada. *Mar. Biol.* 92, 393–404.

Campbell, A., Graham, D. E., MacNichol, H. J., and Williamson, A.M. (1984). Movement of Tagged Lobsters Released on the Continental Shelf from Georges Bank to Baccaro Bank, 1971–73. *Can. Tech. Rep. Fish. Aquat. Sci.* 1288.

Campbell, A., Noakes, D. J., and Elner, R. W. (1991). Temperature and lobster, *Homarus americanus*, yield relationships. *Can. J. Fish. Aquat. Sci.* 48, 2073–2082.

Caswell, H. (1989). "Matrix Population Models." Sinauer, Sunderland, Massachusetts.

Chittleborough, R. G. (1970). Studies on recruitment in the Western Australian rock lobster *Panulirus longipes cygnus*. Density and natural mortality of juveniles. *Aust. J. Mar. Freshwater Res.* 21, 131–148.

Chubb, C. F. (1994). Reproductive biology: Issues for management. *In* "Spiny Lobster Management" (B. F. Phillips, J. S. Cobb, and J. Kittaka, eds.), pp. 181–212. Blackwell, London.

Cobb, J. S. (1971). The shelter-related behavior of the lobster, *Homarus americanus*. *Ecology* 52, 108–115.

Cobb, J. S., and Caddy, J. F. (1989). The population biology of decapods. *In* "Marine Invertebrate Fisheries: Their Assessment and Management" (J. F. Caddy, ed.), pp. 327–374. Wiley, New York.

Cobb, J. S., and Wang, D. (1985). Fisheries biology of lobsters and crayfish. *In* "The Biology of Crustacea" (D. E. Bliss, ed.), Vol. 10, "Fisheries and Aquaculture" (A. J. Provenzano, ed.), pp. 167–247. Academic Press, Orlando, Florida.

Cobb, J. S., Gulbranson, T., Phillips, B. F., Wang, D., and Syslo, M. (1983). Behavior and distribution of larvae and early juvenile *Homarus americanus*. *Can. J. Fish. Aquat. Sci.* 40, 2184–2188.

Cobb, J. S., Wang, D., Richards, R. A., and Fogarty, M. J. (1986). Competition among crabs and lobsters and its possible effects in Narragansett Bay, R.I. *Can. Spec. Publ. Fish. Aquat. Sci.* 92, 282–290.

Cobb, J. S., Wang, D., Campbell, D. B., and Rooney, P. (1989). Speed and direction of swimming by postlarvae of the American lobster. *Trans. Am. Fish. Soc.* 118, 82–86.

Conser, R., and Idoine, J. S. (1992). A Modified DeLury Model for Estimating Mortality Rates and Stock Sizes of America Lobster Populations. *Northeast Fish. Sci. Cent. Ref. Doc.* 14/7.

Cooper, R. A. (1970). Retention of marks and their effects on growth, behavior, and migrations of the American lobster, *Homarus americanus*. *Trans. Am. Fish. Soc.* 95, 239–247.

Cooper, R. A., and Uzmann, J. R. (1971). Migrations and growth of deep-sea lobsters, *Homarus americanus*. *Science* 171, 288–290.

Cooper, R. A., and Uzmann, J. R. (1980). Ecology of juvenile and adult *Homarus*. *In* "The Biology and Management of Lobsters" (J. S. Cobb and B. F. Phillips, eds.), Vol. 1, pp. 215–276. Academic Press, New York.

Deriso, R. B. (1980). Harvesting strategies and parameter estimates for an age-structured model. *Can. J. Fish. Aquat. Sci.* 37, 268–282.

DeWolf, G. A. (1974). The lobster fishery of the Maritime Provinces: Economic effects of regulations. *Bull., Fish. Res. Board Can.* 187, 1–59.

Dow, R. L. (1969). Cycles of geographical trends in seawater temperature on abundance of American lobster. *Science* 164, 1060–1063.

Dow, R. L. (1974). American lobsters tagged by Maine commercial fishermen, 1957-59. *Fish. Bull.* 72, 622–623.

Dow, R. L. (1977). Relationship of sea surface temperature to American and European lobster landings. *J. Cons., Cons. Int. Explor. Mer* 37, 186–191.

Dow, R. L. (1978). Effects of sea surface temperature cycles on landings of American, European, and Norway lobsters. *J. Cons., Cons. Int. Explor. Mer* 38, 271–272.

Dow, R. L. (1980). The clawed lobster fisheries. *In* "The Biology and Management of Lobsters" (J. S. Cobb and B. F. Phillips, eds.), Vol. 2, pp. 265–316. Academic Press, New York.

Dow, R. L., Bell, F. W., and Harriman, D. M. (1975). Bioeconomic Relationships for the Maine Lobster Fishery with Consideration of Alternative Management Schemes. *NOAA Tech. Rep., NMFS SSRF* 683, pp. 1–44.

Dubé, P. (1986). Croissance du Homard, (*Homarus americanus*) dans les Parties Nord and Sud del l'Archipel des Îles de la Madeleine. *Que. Min. Agric. Pech. Aliment., Dir. Gen. Peches Mar. Cah. Inf.* 117.

Easton, M. D. L., and Misra, R. K. (1988). Mathematical representation of crustacean growth. *J. Cons., Cons. Int. Explor. Mer* 45, 61–72.

Elner, R. W., and Campbell, A. (1987). Natural diets of lobster *Homarus americanus* from barren ground and macroalgal habitats off southwestern Nova Scotia, Canada. *Mar. Ecol. Prog. Ser.* 37, 131–140.

Elner, R. W., and Campbell, A. (1991). Spatial and temporal patterns in recruitment for American lobster, *Homarus americanus*, in the northwestern Atlantic. *Mem. Queensl. Mus.* 31, 349–363.

Ennis, G. P. (1972). Growth per molt of tagged lobsters (*Homarus americanus*) in Bonavista Bay, Newfoundland. *J. Fish. Res. Board Can.* 29, 143–148.

Ennis, G. P. (1977). Determination of shell condition in lobsters (*Homarus americanus*) by means of external macroscopic examination. *Proc. Natl. Shellfish. Assoc.* 67, 67–70.

Ennis, G. P. (1979). Estimates of abundance and recruitment to the standing stock for a Newfoundland population of the lobster, *Homarus americanus*, with a method of estimating its natural mortality. *Rapp. P.-V. Reun., Cons. Int. Explor. Mer* 175, 225–228.

Ennis, G. P. (1980a). Recent and current research on growth of lobsters in the wild. *Can. Tech. Rep. Fish. Aquat. Sci.* 932, 10–15.

Ennis, G. P. (1980b). Canadian efforts to assess yield per recruit in lobsters. *Can. Tech. Rep. Fish. Aquat. Sci.* 932, 45–49.

Ennis, G. P. (1980c). Size–maturity relationships and related observations in Newfoundland populations of the lobster (*Homarus americanus*). *Can. J. Fish. Aquat. Sci.* 37, 945–956.

Ennis, G. P. (1981). Fecundity of the American lobster, (*Homarus americanus*), in Newfoundland waters. *Fish. Bull.* 79, 1958–1967.

Ennis, G. P. (1985). An assessment of the impact of size limit and exploitation rate changes on egg production in a Newfoundland lobster population. *North Am. J. Fish. Manage.* 5, 86–90.

Ennis, G. P. (1986a). Sphyrion tag loss from the American lobster

(*Homarus americanus*). *Trans. Am. Fish. Soc.* 115, 914–917.

Ennis, G. P. (1986b). Stock definition, recruitment variability, and larval recruitment processes in the American lobster, *Homarus americanus*: A review. *Can. J. Fish. Aquat. Sci.* 43, 2072–2084.

Ennis, G. P. (1991). Annual variation in egg production in a Newfoundland population of the American lobster, *Homarus americanus*. *In* "Crustacean Issues" (F. R. Schram, ed.), Vol. 7, "Crustacean Egg Production" (A Wenner and A. Kuris, eds.), pp. 291–299. Balkema, Rotterdam, The Netherlands.

Ennis, G. P., Collins, P. W., and Dawe, G. (1982). Fisheries and Population Biology of Lobsters (*Homarus americanus*) at Comfort Cove, Newfoundland. *Can. Tech. Rep. Fish. Aquat. Sci.* 1116.

Ennis, G. P., Collins, P. W., Dawe, G., and Squires, W. R. (1986). Fisheries and Population Biology of Lobsters (*Homarus americanus*) at Arnold's Cove, Newfoundland. *Can. Tech. Rep. Fish. Aquat. Sci.* 1438.

Estrella, B. T., and Cadrin, S. X. (1995). Fecundity of the American lobster (*Homarus americanus*) in Massachusetts coastal waters. *ICES Mar. Sci. Symp.* (in press).

Field, G. W. (1902). The biological basis of legislation governing the lobster industry. *Science* 15, 612–616.

Field, I. A. (1907). Unutilized Fishes and Their Relation to the Fishing Industries. *Bull. U.S. Bur. Fish. Doc.* 622.

Flowers, J. M., and Saila, S. B. (1972). An analysis of temperature effects on the inshore lobster fishery. *J. Fish. Res. Board Can.* 29, 1221–1225.

Fogarty, M. J. (1980). Assessment of yield per recruit for the American lobster (*Homarus americanus*). *Can. Tech. Rep. Fish. Aquat. Sci.* 932, 38–44.

Fogarty, M. J. (1983). Distribution and relative abundance of American lobster, *Homarus americanus* larvae: A review. *NOAA, Tech. Rep. NMFS SSRF* 775, 3–8.

Fogarty, M. J. (1986). "Population Dynamics of the American Lobster, *Homarus americanus*," Ph.D. dissertation. University of Rhode Island, Kingston.

Fogarty, M. J. (1988). Time series models of the Maine lobster fishery: The effect of temperature. *Can J. Fish. Aquat. Sci.* 45, 1145–1153.

Fogarty, M. J. (1989). Forecasting yield and abundance of exploited invertebrate populations. *In* "Marine Invertebrate Fisheries: Their Assessment and Management" (J. F. Caddy, ed.), pp. 701–724. Wiley, New York.

Fogarty, M. J. (1993). Recruitment in randomly varying environments. *ICES J. Mar. Sci.* 50, 247–260.

Fogarty, M. J., and Addison, J. T. (1995). Modeling performance of individual traps: Entry, escapement, and soak time. *ICES J. Mar. Sci.* (in press).

Fogarty, M. J., and Borden, D. V. D. (1980). The effects of trap venting on gear selectivity in the inshore Rhode Island American lobster, *Homarus americanus* fishery. *Fish. Bull.* 77, 925–933.

Fogarty, M. J., and Idoine, J. S. (1986). Recruitment dynamics in an American lobster (*Homarus americanus*) population. *Can. J. Fish. Aquat. Sci.* 43, 2368–2376.

Fogarty, M. J., and Idoine, J. S. (1988). Application of a yield and egg production model based on size to an offshore American lobster population. *Trans. Am. Fish. Soc.* 117, 350–362.

Fogarty, M. J., and Murawski, S. A. (1986). Population dynamics and assessment of exploited invertebrate stocks. *Can. Spec. Publ. Fish. Aquat. Sci.* 92, 228–244.

Fogarty, M. J., Borden, D. V. D., and Russell, H. J. (1980). Movements of tagged American lobster, *Homarus americanus*, off Rhode Island. *Fish. Bull.* 78, 771–780.

Fogarty, M. J., Cooper, R. A., Uzmann, J. R., and Burns, T. (1982). Assessment of the USA Offshore American Lobster (*Homarus americanus*) Fishery. *Int. Counc. Explor. Sea Doc.* C.M. 1982/K:14.

Fogarty, M. J., Sissenwine, M. P., and Cohen, E. B. (1991). Recruitment variability and the dynamics of exploited marine populations. *Trends Ecol. Evol.* 6, 241–246.

Fox, W. W. (1970). An exponential yield model for optimizing exploited fish populations. *Trans. Am. Fish. Soc.* 99, 80–88.

Francis, R. I. C. C. (1988). Are growth parameters estimated from tagging and age-length data comparable? *Can. J. Fish. Aquat. Sci.* 45, 936–942.

Goode, G. B. (1887). "The Fisheries and Fishing Industries of the U.S." U.S. Government Printing Office, Washington, D.C.

Gould, A. A. (1841). "Report on the Invertebrata of Massachusetts Comprising the Mollusca, Crustacea, Annelida, and Radiata." Folsum, Wells, and Thurston, Cambridge, Massachusetts.

Graham, M. (1935). Modern theory of exploiting a fishery, an application to North Sea trawling. *J. Cons., Cons. Int Explor. Mer* 10, 264–274.

Hall, N. G., and Brown, R. S. (1994). Modelling for management—The western rock lobster fishery. *In* "Spiny Lobster Management" (B. F. Phillips, J. S. Cobb, and J. Kittaka, eds.), pp. 311–322. Blackwell, London.

Hall, N. G., and Brown, R. S. (1995). Delay-difference models for the western rock lobster (*Panulirus argus*) fishery of Western Australia. *ICES Mar. Sci. Symp.* (in press).

Hancock, D. A., and Edwards, E. (1967). Estimation of annual growth in the edible crab (*Cancer pagurus*). *J. Cons., Cons. Int. Explor. Mer* 31, 246–264.

Harding, G. C., and Trites, R. W. (1988). Dispersal of *Homarus americanus* larvae in the Gulf of Maine from Browns Bank. *Can. J. Fish. Aquat. Sci.* 45, 416–425.

Harding, G. C., Drinkwater, K. F., and Vass, W. P. (1983). Factors influencing the size of American lobster (*Homarus americanus*) stocks along the Atlantic coast of Nova Scotia, Gulf of St. Lawrence, and Gulf of Maine: A new synthesis. *Can. J. Fish. Aquat. Sci.* 40, 168–184.

Hartnoll, R. G. (1982). Growth. *In* "The Biology of Crustacea" (D. E. Bliss, ed.), Vol. 2, "Embryology, Morphology, and Genetics" (L. G. Abele, ed.), pp. 111–196. Academic Press, New York.

Hedgecock, D. (1986). Is gene flow from pelagic larval dispersal important in the adaption and evolution of marine invertebrates? *Bull. Mar. Sci.* 39, 550–564.

Hedgecock, D., Nelson, K., Simons, J., and Shleser, R. (1977). Genic similarity of American and European species of the lobster *Homarus*. *Biol. Bull. (Woods Hole, Mass.)* 152, 41–50.

Hedgecock, D., Tracey, M., and Nelson, K. (1982). Genetics. *In* "The Biology of Crustacea" (D. E. Bliss, ed.), Vol. 2, "Embryology, Morphology, and Genetics" (L. G. Abele, ed.), pp. 284–403. Academic Press, New York.

Herrick, F. H. (1895). The American lobster: A study of its habits and development. *Bull. U.S. Fish Comm.* 15, 1–252 + 54 plates.

Herrick, F. H. (1906). Effective protection for the lobster fishery. *Science* 23, 650–655.

Herrick, F. H. (1909). Natural history of the American lobster. *Bull. U.S. Bur. Fish.* 29, 149–408 + 20 plates.

Hiatt, R. W. (1948). The biology of the lined shore crab, *Pachygrapsus crassipes* Randall. *Pac. Sci.* 2, 135–213.

Hjort, J., Jahn, G., and Ottestad, P. (1933). The optimum catch. *Hvalradets Skr.* 7, 92–127.

Hudon, C. (1987). Ecology and growth of postlarval and juvenile lobster, *Homarus americanus*, off Îles de la Madeleine (Québec). *Can. J. Fish. Aquat. Sci.* 44, 1855–1869.

Hughes, J. T., and Mathiessen, G. C. (1962). Observations on the biology of the American lobster, *Homarus americanus*. *Limnol. Oceanogr.* 7, 414–421.

Hughes, J. T., Sullivan, J., and Schlesser, R. A. (1972). Enhancement of lobster growth. *Science* 177, 1110–1111.

Idoine, J. S. (1985). "A Distributed Delay Model: Growth of the American Lobster," M. S. thesis. University of Massachusetts, Amherst.

Jeffries, H. P. (1994). The impacts of warming climate on fish populations. *Maritimes* 37, 12–14.

Jensen, A. L. (1986). Assessment of the Maine lobster fishery with surplus production models. *North Am. J. Fish. Manage.* 6, 63–68.

Jones, R. (1974). Assessing the Long Term Effects of Changes in Fishing Effort and Mesh Size from Length Composition Data. *Int. Counc. Explor. Sea Doc.* C.M. 1974/G:33.

Katz, C. H., Cobb, J. S., and Spaulding, M. (1994). Larval behavior, hydrodynamic transport, and potential offshore recruitment in the American lobster, *Homarus americanus*. *Mar. Ecol. Prog. Ser.* 103, 265–273.

Knight, A. P. (1916). Lobster mating: A means of conserving the lobster industry. *Science* 44, 828–832.

Kornfield, I., and Moran, P. (1990). Genetics of population differentiation in lobsters. *In* "Life History of the American Lobster: Proceedings of a Workshop, November 29–30, 1989, Orono, Maine" (I. Kornfield, ed.), pp. 23–24 (abstract). Lobster Institute, Orono, Maine.

Krouse, J. S. (1973). Maturity, sex ratio, and size composition of the natural population of American lobster, *Homarus americanus*, along the Maine coast. *Fish. Bull.* 71, 165–173.

Krouse, J. S. (1981). Movement, Growth, and Mortality of American Lobsters along the Coast of Maine. *NOAA, Tech. Rep. NMFS SSRF* 747.

Krouse, J. S. (1989). Performance and selectivity of trap fisheries for crustaceans. *In* "Marine Invertebrate Fisheries: Their Assessment and Management" (J. F. Caddy, ed.), pp. 159–182. Wiley, New York.

Kurata, H. (1962). Studies of the age and growth of Crustacea. *Bull. Hokkaido Reg. Fish. Res. Lab.* 24, 1–114.

Lawton, R. P., Brady, P., Sheehan, C., Sides, W., Kouloheras, E., Brando, M., and Malkoski, V. (1984). Growth and movements of tagged lobsters released in western Cape Cod Bay, 1970–1977. *In* "Observations on the Ecology and Biology of Western Cape Cod Bay, Massachusetts" (J. D. Davis and D. Merriman, eds.), pp. 119–129. Springer-Verlag, New York.

Marchesseault, G. D., Saila, S. B., and Palm, W. J. (1976). Delayed recruitment models and their application to the American lobster (*Homarus americanus*) fishery. *J. Fish. Res. Board Can.* 33, 1779–1787.

Mauchline, J. (1976). The Hiatt growth diagram for Crustacea. *Mar. Biol.* 35, 79–84.

McLeese, D. W. (1972). Effects of several factors on the growth of the American lobster (*Homarus americanus*) in captivity. *J. Fish. Res. Board Can.* 29, 1725–1730.

McLeese, D. W., and Wilder, D. G. (1958). The activity and catchability of the lobster (*Homarus americanus*) in relation to temperature. *J. Fish. Res. Board Can.* 15, 1345–1354.

Mendall, H. L. (1934). The relationship of certain sea birds to the fishing industry of the State of Maine. *Bienn. Rep.—Comm. Sea Shore Fish., Maine, 8th* pp. 37–64.

Miller, R. J. (1993). More about habitat for juvenile *Homarus americanus*. *Lobster Newsl.* 6(1), 8–9.

Miller, R. J., Duggan, D. R., Robinson, D. G., and Zheng, Z. (1989). Growth and Movement of *Homarus americanus* on the Outer Coast of Nova Scotia. *Can. Tech. Rep. Fish. Aquat. Sci.* 1716.

Miller, R. J., Campbell, A., Sharp, G. J., and Hudon, C. (1992). Are small juvenile lobsters habitat limited? *Can. Tech. Rep. Fish. Aquat. Sci.* 44, 1855–1869.

Mills, D. H. (1957). Herring gulls and common terns as possible predators of lobster larvae. *J. Fish. Res. Board Can.* 14, 729–730.

Northeast Utilities Service Company (1994). Lobster studies. *In* "Monitoring the Marine Environment of Long Island Sound at Millstone Nuclear Power Station, Waterford, Connecticut. Annual Report 1993," pp. 9–34.

Odense, P. H., and Annand, C. (1978). Isozyume systems of an inshore and offshore lobster population. *Int Counc. Explor. Sea Doc. C.M.* 1978/K:15.

Orach-Meza, F. L., and Saila, S. B. (1978). Application of a polynomial distributed lag model to the Maine lobster fishery. *Trans. Am. Fish. Soc.* 107, 402–411.

Ovenden, J. R., and Brasher, D. J. (1994). Stock identity of the red (*Jasus edwardsii*) and green (*J. verreauxi*) rock lobsters inferred from mitochondrial DNA analysis. *In* "Spiny Lobster Management" (B. F. Phillips, J. S. Cobb, and J. Kittaka, eds.), pp. 230–249. Blackwell, London.

Paloheimo, J. E. (1963). Estimation of catchabilities and population sizes of lobsters. *J. Fish. Res. Board Can.* 20, 559–588.

Pella, J. J., and Tomlinson, P. K. (1969). A generalized stock production model. *Bull. Int. Am. Trop. Tuna Comm.* 13, 421–496.

Pezzack, D. S. (1987). Lobster (*Homarus americanus*) Stock Structure in the Gulf of Maine. *Int. Counc. Explor. Sea Doc.* C.M. 1987/K:17.

Pezzack, D. S. (1989). Comments on 'Dispersal of *Homarus americanus* larvae in the Gulf of Maine from Browns Bank' by G.C. Harding and R.W. Trites. *Can. J. Fish. Aquat. Sci.* 46, 1079–1083.

Pezzack, D. S. (1992). A review of lobster (*Homarus americanus*) landings and trends in the Northwest Atlantic, 1947–86. *J. Northwest Atl. Fish. Sci.* 14, 115–127.

Pezzack, D. S., and Duggan, D. R. (1986). Evidence of migration and homing of lobsters (*Homarus americanus*) on the Scotian Shelf. *Can. J. Fish. Aquat. Sci.* 43, 2206–2211.

Phillips, B. F., and Brown, R. S. (1989). The West Australian rock lobster fishery: Research for management. *In* "Marine Invertebrate Fisheries: Their Assessment and Management" (J. F. Caddy, ed.), pp. 159–182. Wiley, New York.

Pollock, D. E. (1993). Recruitment overfishing and resilience in spiny lobster populations. *ICES J. Mar. Sci.* 50, 9–14.

Polovina, J. (1989). Density-dependence in spiny lobster, *Panulirus marginatus*, in the Northwestern Hawaiian Islands. *Can. J. Fish. Aquat. Sci.* 46, 660–665.

Pringle, J. D., and Burke, D. L. (1993). The Canadian lobster fishery and its management, with emphasis on the Scotian Shelf and the Gulf of Maine. *Can. Bull. Fish. Aquat. Sci.* 226, 91–122.

Richards, R. A., and Cobb, J. S. (1986). Competition for shelter between lobsters (*Homarus americanus*) and Jonah crabs (*Cancer borealis*): Effects of relative size. *Can. J. Fish. Aquat. Sci.* 43, 2250–2255.

Ricker, W. E. (1975). Computation and Interpretation of Biological Statistics of Populations. *Bull., Fish. Res. Board Can.* 191, 1–381.

Rogers, B. A., Cobb, J. S., and Marshall, N. B. (1968). Size comparisons of inshore and offshore larvae of the lobster, *Homarus americanus*, off southern New England. *Proc. Nat. Shellfish Assoc.* 58, 78–81.

Russell, E. S. (1931). Some theoretical considerations on the 'overfishing' problem. *J. Cons., Cons. Int. Explor. Mer* 6, 1–20.

Russell, H. J., Jr. (1980a). The determination of growth rates for American lobsters. *Can. Tech. Rep. Fish. Aquat. Sci.* 932, 1–8.

Russell, H. J., Jr. (1980b). Analysis of double tagging experiments: An update. *Can J. Fish. Aquat. Sci.* 37, 114–116.

Saila, S. B., and Flowers, J. M. (1965). A simulation study of sex ratios and regulation effects with the American lobster (*Homarus americanus*). *Proc. Annu. Gulf Carrib. Fish. Inst.* 18, 66–78.

Saila, S. B., and Flowers, J. M. (1968). Movements and behavior of berried female lobsters displaced from offshore areas to

Narragansett Bay, Rhode Island. *J. Cons., Cons. Int, Explor. Mer* 31, 342–351.

Saila, S. B., and Flowers, J. M. (1969). Geographic morphometric variation in the American lobster. *Syst. Zool.* 18, 330–338.

Saila, S. B., and Marchesseault, G. (1980). Population dynamics of clawed lobsters. *In* "The Biology and Management of Lobsters" (J. S. Cobb and B. F. Phillips, eds.), Vol. 2, pp. 219–241. Academic Press, New York.

Saila, S. B., Flowers, J. M., and Hughes, J. T. (1969). Fecundity of the American lobster, *Homarus americanus*. *Trans. Am. Fish. Soc.* 98, 537–539.

Saila, S. B., Annala, J. H., McKoy, J. L., and Booth, J. D. (1979). Application of yield models to the New Zealand rock lobster fishery. *N.Z. J. Mar. Freshwater Res.* 13, 1–11.

Scarratt, D. J. (1964). Abundance and distribution of lobster larvae (*Homarus americanus*) in Northumberland Strait. *J. Fish. Res. Board Can.* 21, 661–680.

Scarratt, D. J. (1973). Abundance, survival, and vertical and diurnal distribution of lobster larvae in Northumberland Strait, 1962–63, and their relationship with commercial stocks. *J. Fish. Res. Board Can.* 30, 1819–1824.

Scarratt, D. J., and Elson, P. F. (1965). Preliminary trials of a tag for salmon and lobsters. *J. Fish. Res. Board Can.* 22, 421–423.

Schaefer, M. B. (1954). Some aspects of the dynamics of populations important to the management of the commercial fisheries. *Bull. Int. Am. Trop. Tuna Comm.* 1, 27–56.

Schaefer, M. B. (1957). A study of the dynamics of the fishery for yellowfin tuna in the eastern tropical Pacific Ocean. *Bull. Int. Am. Trop. Tuna Comm.* 2, 245–285.

Schroeder, W. C. (1959). The lobster, *Homarus americanus*, and the red crab, *Geryon quinquedens*, in the offshore waters of the western North Atlantic. *Deep-Sea Res.* 5, 266-282.

Shaklee, J. B. (1983). The utilization of isozymes as genetic markers in fisheries management and conservation. *Isozymes: Curr. Top. Biol. Med. Res.* 11, 214–247.

Skud, B. E. (1969). The effect of fishing on size composition and sex ratio of offshore lobster stocks. *Fiskeridir. Skr., Ser. Havunders.* 115, 295–309.

Skud, B. E. (1979). Soak Time and Catch per Pot in an Offshore Fishery for American Lobsters (*Homarus americanus*). *Rapp. P.-V. Reun., Cons. Int. Explor. Mer* 177.

Skud, B. E., and Perkins, H. C. (1969). Size Composition, Sex Ratio, and Size at Maturity of Offshore Northern Lobsters. *U.S. Fish. Wildl. Serv., Spec. Sci. Rep.—Fish.* 598.

Smith, J. B. (1980). Replenishable resource management under uncertainty: A reexamination of the U.S. northern lobster fishery. *J. Environ. Econ. Manage.* 7, 209–219.

Smith, P. J., McCoy, J. L., and Machin, P. J. (1980). Genetic variation in the rock lobsters *Jasus edwardsii* and *Jasus novaehollandiae*. *N.Z. J. Mar. Freshwater Res.* 14, 55–63.

Somerton, D. A. (1980). Fitting straight lines to Hiatt growth diagrams: A re-evaluation. *J. Cons., Cons. Int. Explor. Mer* 39, 15–19.

Stewart, J. E. (1980). Diseases. *In* "The Biology and Management of Lobsters" (J. S. Cobb and B. F. Phillips, eds.), Vol. 1, pp. 301–342. Academic Press, New York.

Thomas, J. T. (1973). An Analysis of the Commercial Lobster (*Homarus americanus*) Fishery along the Coast of Maine, August 1966 through December 1970. *NOAA, Tech. Rep. NMFS SSRF* 667.

Tracey, M. L., Nelson, K., Hedgecock, D., Schlesser, R. A., and Pressick, M. L. (1975). Biochemical genetics of lobsters: Genetic variation and structure of American lobster (*Homarus americanus*) populations. *J. Fish. Res. Board Can.* 32, 2091–2101.

Uzmann, J. R. (1970). Use of parasites in identifying lobster stocks. *Proc. Int. Conf. Parasitol., 2nd* 56, 12–20.

Uzmann, J. R., Cooper, R. A., and Pecci, K. J. (1977). Migration and Dispersion of Tagged American Lobsters, *Homarus americanus*, on the southern New England Continental Shelf. *NOAA, Tech. Rep., NMFS SSRF* 705.

Van Engel, W. A. (1980). Maturity and fecundity in the American lobster, *Homarus americanus*: A review. *Can. Tech. Rep. Fish. Aquat. Sci.* 932, 51–58.

Van Olst, J. C., Carlberg, J. M., and Hughes, J. T. (1980). Aquaculture. *In* "The Biology and Management of Lobsters" (J. S. Cobb and B. F. Phillips, eds.), Vol. 2, pp. 333–384. Academic Press, New York.

Wahle, R. A., and Steneck, R. S. (1991). Recruitment habitats and nursery grounds of the American lobster *Homarus americanus*: A demographic bottleneck? *Mar. Ecol. Prog. Ser.* 69, 231–243.

Wahle, R. A., and Steneck, R. S. (1992). Habitat restrictions in early benthic life: Experiments on habitat selection and in situ predation with the American lobster. *J. Exp. Mar. Biol. Ecol.* 157, 92–114.

Walters, C. J., Hall, N., Brown, R., and Chubb, C. (1993). Spatial model for the population dynamics and exploitation of the Western Australian rock lobster, *Panulirus cygnus*. *Can. J. Fish. Aquat. Sci.* 50, 1650–1662.

Wilder, D. G. (1953). The growth rate of the American lobster (*Homarus americanus*). *J. Fish. Res. Board Can.* 10, 371–412.

Wilder, D. G. (1963). Movements, growth, and survival of marked and tagged lobsters liberated in Egmont Bay, Prince Edward Island. *J. Fish. Res. Board Can.* 20, 305–318.

CHAPTER

7

Interface of Ecology, Behavior, and Fisheries

J. STANLEY COBB
Department of Zoology
University of Rhode Island
Kingston, Rhode Island 02881

I. Introduction

A commercial fishery for the American lobster has existed along the northeast coast of the United States and in the Canadian Maritime Provinces for approximately 150 years. *Homarus americanus* is among the most heavily exploited marine species in the world. In many parts of the range of the lobster, between 70 and 90% of all eligible animals are taken each year. Despite enormous fishing pressure, the fishery thrives; in fact, after many years of relatively constant catch and declining catch per unit of effort, the commercial catch increased steadily during the 1980s and 1990s. Much is known about the biology of *H. americanus*, yet this increase was not predicted, nor are the reasons for it understood. In order to understand how populations respond to the combination of exploitation and environmental change, a good understanding of survival, growth, and reproduction is needed. These are complex functions of physiological, behavioral, and ecological factors, all of which, in combination, result in population regulation. A quick review of the chapters in this book that cover these topics will convince the reader that we know a great deal about them with regard to lobsters. Much of the material in these chapters is important to fishery biologists.

The interaction among behavior, ecology, and fishery management is particularly complex. Fishers know and use the behavior of lobsters to trap them. Changes in the abiotic environment may result in altered growth rates or fecundity. Migrations cause shifts in population, raise issues about stock identity, and violate assumptions needed for the estimation of abundance. Many of these are implicitly understood in making management decisions, or explicitly included in mathematical population models. However, the simplification required results in ignoring some of the complexity. Conversely, some of the sophisticated stage-structured models now available may demand data that simply are not available. This chapter highlights some of the interesting and important interactions among lobster behavior, ecology, and management. It is structured in much the same way a fisheries biology review would be—with sections on abundance estimation, stock identity, growth, mortality, fecundity, and habitat limitation. Each section explores details of behavior and ecology that have important implications for management.

II. Estimation of Abundance

A. Sampling for Larvae

The abundance of larval stages is estimated by sampling with plankton nets. As in adults, behavior of the individuals being sampled markedly affects the results. Almost all sampling for larvae (stages I–III) and postlarvae (stage IV) is performed using surface-towed neuston nets that quantitatively sample the top 40–100 cm. A number of early reports (Smith, 1873; Herrick, 1909; Templeman, 1937; Scarratt, 1964) suggested that larval lobsters of all stages generally are found at or near the surface, although these reports also indicated that a few were discovered in subsurface tows. More recent evidence (Harding *et al.*, 1982, 1987; Nichols and Lovewell, 1987) suggests that a significant proportion of the larvae may be below the surface 0.5 m. Diel vertical migration in stage I larvae was demonstrated by Harding *et al.* (1987) to water 60 m deep. Postlarvae seem to be located at or near the surface at all times (Herrick, 1909; Harding *et al.*, 1987). Thus, a sampling design appropriate to stage and behavior is required if accurate estimates of absolute abundance are required. It is likely, however, that an index of larval abundance probably can be derived from carefully standardized neuston tows. If larval transport by currents is to be considered, the vertical distribution of all stages is very important, and not well enough understood.

Mortality estimates have been made for the larval stages (Lund and Stewart, 1970; Scarratt, 1973; Fogarty and Idoine, 1986). Such mortality estimates, as well as attempts to determine the relationship between the abundance of one stage and the abundance of the preceding stage, depend critically on standardized abundance estimates. Since each stage has a different duration, and thus is present in the plankton for a different length of time, the stages are not equally susceptible to capture. Scarratt (1964) adjusted density estimates for developmental time at the prevailing water temperature to provide standardized estimates of the production of each larval and the postlarval stage.

The postlarval stage settles to the benthic environment, making the transition from surface to bottom about midway through the postlarval period (Herrick, 1909; Scarratt, 1973; Cobb *et al.*, 1989a). Thus, the period of postlarval availability for capture by surface nets does not depend on temperature alone, but also on the behavioral decision to settle. Cobb *et al.* (1989a) showed that settlement behavior appears 2–6 days after metamorphosis in the laboratory. However, postlarvae in later molt cycle stages (D_0 and D_1) predominated in field samples, suggesting that settlement could be later and timing is likely to be quite variable in the field. In Maine, postlarvae captured near shore were found to be in late molt cycle stages (D_2 and D_3), suggesting an even later time of settlement, and certainly well past the estimate of Herrick (1909), Scarratt (1973), and Cobb *et al.* (1989a) of halfway through the stage.

Substrate choice during settlement may also affect the timing of the transition to the benthic environment, since postlarvae appear to be quite selective of bottom type (Hudon, 1987; Wahle and Steneck, 1991). Postlarvae in the laboratory may prefer dark shelters and alga-covered rock (Botero and Atema, 1982; Boudreau *et al.*, 1993). If these are not available, they return to the surface (Cobb *et al.*, 1983, 1989a). The relative proportion of suitable to unsuitable substrate thus may be a factor in determining the number of postlarvae remaining at the surface. Abrupt changes in water temperature in vertical cylinders in the laboratory cause postlarvae to interrupt their descent to the bottom, suggesting that thermoclines in the field may play a role in the distribution of larvae, limiting them to the warmer upper layers and discouraging settlement in deep, stratified water (Boudreau *et al.*, 1991, 1992; Hofe, 1993). This hypothesis has not been tested in the field and the very sharp thermoclines used in the laboratory do not replicate field conditions, so further testing is needed before building this behavior into distribution models.

Any behavioral factor that affects the timing of the descent to the bottom will also affect the temporal availability of the postlarvae to surface nets, and thus estimates of relative abundance. The time of settlement by postlarvae may differ among regions, years, or even time of year. Thus, a challenge to those who work with larval lobsters is to build behavioral factors into abundance estimates (starting, perhaps with Scarratt's adjustment, already discussed), drift patterns, and settlement sites. This has been started by Incze and Wahle (1991) in a model describing the coupling between abundance in pelagic and benthic environments, but considerable further refinement is needed.

B. Trapping Adults

Knowledge of the abundance of the population being fished is essential to the rational management of a fishery. In many instances, abundance estimates come from the fishery itself, by estimating catch per unit of effort. In lobster fisheries, this means reliance on trapping as a method for estimating abundance. Independent methods for abundance estimation, such as trawl or diver surveys, are limited—trawls cannot be towed through rocky areas and divers have depth restrictions and are expensive. Because lobsters seek refuge in shelters under rocks, remote methods, such as photography or video surveys with unmanned submersibles, may not be as reliable as other methods. Traps, then, are the most useful and the most used sampling method for providing an index of abundance. However, many factors affect the catch rate of lobster traps, including location, bait, trap design, temperature, and the presence of other animals (Miller, 1990). In contrast to trawl fisheries, in which the catch rate depends largely on the behavior of fishermen, catch rates in trap fisheries depend in great part on the behavior of the target species (Fogarty and Addison, personal communication).

In order for a lobster to be captured by a trap, it must find it, enter it, and remain within it. This has been stated more formally by Regier (personal communication, cited in Miller, 1990) as

$$P(C) = P(E)\, P(I)\, [1 - P_1(X) - P_2(X)] \qquad (1)$$

in which $P(E)$ is the probability of encountering a trap, $P(I)$ is the probability of entering once the trap is encountered, $P_1(X)$ is the probability of escape through the entrance, and $P_2(X)$ is the probability of escape through the escape gap. The discussion that follows focuses on the behavioral aspects of the first three components.

Lobsters encounter traps when actively foraging outside of their shelters. The probability of encounter depends on emergence from the shelter (a function of light intensity and probably hunger), walking rate, and response to the odor of bait in the trap. Lobsters are active nocturnally (Cobb, 1969; Lawton, 1987; Zeitlin-Hale and Sastry, 1978). In general, lobsters remain within shelters during the day and emerge only when light intensity falls below 2×10^{-2} $\mu W/cm^2$ (Weiss, 1970). In Long Island Sound, 80–90% of the lobsters were estimated to be out of shelter several hours after dusk (Weiss, 1970). Since emergence appears to be keyed to light intensity, it is likely that in deeper or murkier water, lobsters are active and foraging for greater proportions of the day (Cooper and Uzmann, 1980). In the Norway lobster (*Nephrops norvegicus*), the period of emergence corresponds to the same range of light intensities at different depths (Chapman, 1980). Thus, the duration of the foraging period, and presumably the probability of encountering a trap, vary with day length, depth, and water clarity. It is not clear, however, whether the total time spent foraging in a day is dependent on these factors or on the availability of prey items and hunger level.

Temperature affects the encounter rate. The walking rate increases approximately linearly between 2° and 10°C, is constant from 10 to 20°C, and increases again above 20°C (McLeese and Wilder, 1958). This suggests that catchability will decline as temperatures drop below 10°C, as the probability of the individual's encountering a trap declines simply because it is not walking as fast, therefore not covering as much ground. In addition, lobsters are much less likely to leave their shelters during the winter (Stewart, 1972). Feeding activity also is related to temperature, and thus a lower attraction to the odor of bait might be expected at decreased temperatures, but this has not been quantitatively demonstrated. However, the quantity of food in the gut varies seasonally, with the lowest values of the stomach fullness index occurring during the winter months (Ennis, 1973). This suggests that feeding activity also is lower in winter.

Catchability of crustaceans varies over the molt cycle. The lowest catchability is found around the time of ecdysis and lasts for 10–15% of the molt cycle (Miller, 1990). Significant changes in catchability over the molt cycle have been shown for the spiny lobster *Panulirus cygnus* (Morgan, 1974) and inferred for *P. argus* (Lipcius and Herrnkind, 1982). Feeding activity, as measured by stomach fullness, changes over the molt cycle in *Homarus americanus* (Weiss, 1970), presumably affecting the attraction to traps. Changes of serum protein in field-captured lobsters have been noted over the molt cycle, suggesting that a high level of feeding activity may be required in order to recover from the immediate postmolt low in condition (Ennis, 1973). Legal-sized lobsters molt once per year or less. Most areas are characterized by a spring molt, probably synchronized by low winter temperatures that slow or stop growth (Aiken and Waddy, 1986). This cyclical change in feeding motivation, superimposed on temperature effects on feeding and walking, may have to be accounted for in trying to estimate catchability by traps.

Behavioral observations in a large laboratory tank have shown that there is considerable individual variation in responsiveness to traps. Of 18 legal-sized lobsters observed in 20 trials, seven were never captured, and of those, six never entered a baited trap (Karnofsky and Price, 1989). The number of observed approaches to the traps ranged from one to 54. Some lobsters were seen to partially enter the trap and feed on the bait before leaving. These lobsters move past the ring opening, but hook their telson on the ring, enabling them to hang into the "kitchen" section of the pot and feed on the bait before pulling back out and leaving the trap. The individual variation in responsiveness may be due to many factors and suggests that caution must be used in making the assumption that all segments of the population are equally susceptible to trapping.

Sex and size are factors that appear to cause lobsters to respond to traps differently. Simultaneous diver and trap sampling showed that while the male–female ratio of large lobsters in the diver samples was 1:1, in the traps it was 3:1 (Ennis, 1978, and personal communication, cited in Miller, 1990). In a similar diver–trap comparison, 41% of the trap catches, but only 4% of the diver samples, were lobsters having greater than 80-mm carapace length (CL) (Miller, 1989). Ovigerous females often seem to trap at much lower rates than expected, perhaps as a result of differences in feeding activity (Templeman and Tibbo, 1945). This differential motivation results in differential susceptibility to the traps, making estimates of sex ratio and abundance of ovigerous females suspect. Caution must be used, however, to distinguish between true differences in abundance of males and females and differential trap susceptibility.

The response to bait is chemically mediated. Lobsters appear to be very sensitive to low-molecular-weight compounds (Derby, 1984; Daniel and Bayer, 1987) and walk upstream in the presence of prey odor (McLeese, 1973). They approach the trap from the downstream side, apparently orienting to an odor stream (Karnofsky and Price, 1989) and sometimes spending a considerable amount of time trying to reach the bait through the slats of the trap. The quantity and quality of the bait are important to attraction (Miller, 1990; Zimmer-Faust, 1989; Karnofsky and Price, 1989). Bait is affected by selection and treatment before placement in the trap and by the length of time the trap is in the water (soak time). The effectiveness of a trap as a sampling device depends in part on the nature of the bait. In addition, there may be some complex interactions between environmental conditions and behavioral state. Most of the research done so far has been performed in the laboratory and results have been difficult to replicate in the field (Zimmer-Faust, 1989, and personal communication). Creative field research is needed in order to understand how lobsters locate prey.

In the California spiny lobster (*Panulirus interruptus*), varied concentrations of chemical feeding stimulants cause different behavioral responses. High concentrations result in the initiation of locomotion and of grasping reflexes, but low levels only modify locomotor patterns of animals that are already aroused (Zimmer-Faust and Case, 1983; Zimmer-Faust, 1989). This suggests that in order for bait odors in the low concentrations probably found in the field to be stimulatory, lobsters must already be in an endogenous state of readiness. Entering this into the estimation of catchability may be difficult. Even more difficult may be the effects of physical factors, particularly water velocity and bottom rugosity. Most experimental work on odor detection has been done in laboratory flumes. In the field, current velocity may vary widely with stage of the tide and the effects of obstructions downstream of the bait are not known. Blue crabs (*Callinectes sapidus*) have a greater success of finding live prey in slow-flowing water than in fast-flowing or still water (Weissburg and Zimmer-Faust, 1993). This suggests that the magnitude and structure of turbulent eddies within the benthic boundary layer may affect the distribution of odor, and thus the ability of crustaceans to detect it. At greater velocities and greater bottom rugosities, the chemical signal may not be constant, making it more difficult to follow an odor trail. Even the orientation of the trap with respect to current direction is important. Crabs have a much lower entry rate to side-entrance traps if the entrance is not oriented properly with regard to the odor stream from the bait. The odor trail of bait led 65% of *Cancer productus* to enter the trap when the entrance was parallel to the current, but only 7% entered when the trap was turned 90° (Miller, 1980). Although the work cited here does not refer directly to *Homarus americanus*, the principles are general and probably apply. Attraction to the odor of the bait is influenced by factors affecting the internal motivation of the animal, as well as complex external factors that the fishery biologist wishing to estimate abundance may have neither control over nor knowledge of.

After remaining in the water for some time, traps become saturated, that is, they do not continue to capture animals at the same rate as when first placed. This is likely due to loss of bait, therefore lower attractiveness of the trap, and to the increased number of animals in the trap. Crab traps emptied at 2-hour intervals caught more crabs than did those emptied only at the end of the 12-hour fishing period, a clear indication of the effect of saturation (Miller, 1978). Traps stocked with either three or eight lobsters and fished for 24 hours caught many fewer lobsters than did unstocked traps (Richards *et al.*, 1983), suggesting that behavioral interactions among lobsters are partially the cause of trap saturation. Reduced entry, rather than increased escapement, is likely to be the cause of the saturation effect, since only about 10% of the stocked lobsters escaped from the traps. In a laboratory study, a lobster in the inner, "parlor" section of the trap never escaped, although some in the outer, "kitchen" section were able to find their way out (Karnofsky and Price, 1989). Reduced entry was important in this study; only 8% of the observed approaches to a trap resulted in entry, and of those lobsters that left the vicinity of the trap without entering it, 30% did so in response to an aggressive interaction with another lobster.

The results reported in studies of trapping behavior all suggest that of the three main components affecting catch rate (encounter, entry, and escape), encounter and entry are the most important. The encounter rate seems to depend on motivation of the lobster and dispersion of the odor plume from the bait, and may be quite variable among individuals. Entry may depend more on factors such as the presence of other lobsters, trap type, and trap orientation. The many factors affecting catch rates make the estimation of relative or absolute abundance difficult because of the variability and (worse) bias introduced to the data. Miller (1990) reviewed the use of catch per trap as an index of abundance and concluded that trapping clearly is the most convenient technique; despite its poor correlations with independent measures (e.g., diving), it remains the technique of choice because of the lack of suitable alternatives. Considerably more research is needed on behavioral and ecological factors affecting trap encounter and entry before traps can be a truly effective device for measuring lobster abundance.

III. Stock Identity

The fishery for the American lobster often is divided into *inshore* and *offshore* fisheries, recognizing that the individuals found in deep waters along the outer continental shelf, in the canyons, and off Browns and Georges Banks, may be different from those found in shallower, inshore waters (Skud and Perkins, 1969; Pezzack and Duggan, 1983). The question of whether the two are separate stocks, with reduced gene flow between them, is still open, and the evidence is equivocal (Cobb and Wang, 1985; Harding *et al.*, 1993). Gene flow among populations can be maintained by migration or dispersal and there is ample evidence for both (see Fogarty, Chapter 6).

Lobsters can move long distances, although much early evidence suggested that they did not. The early studies generally were short term and addressed immature animals (e.g., Templeman, 1935; Wilder, 1963). However, when Campbell and Stasko (1985) tagged both mature and immature lobsters, 70% of the immature lobsters recaptured were within 18 km of the release site, while only 40% of the mature ones were. Mature animals moved an average of 15.6 km, while immature lobsters moved 4.7 km. Nearly 20% of the mature animals were recaptured more than 93 km from the release site.

Approximately 20% of the lobsters tagged in offshore waters by Cooper and Uzmann (1971; Uzmann *et al.*, 1977) showed a springtime shoalward migration, and analysis of the depth of recapture suggested that there was a return migration to the offshore areas in the autumn. Fourteen lobsters were caught at inshore locations and many more were inshore of their release location. Lobsters captured in offshore waters, but transported to be tagged and released in inshore areas, moved offshore, with some of the individuals recaptured near their original sites of capture (Saila and Flowers, 1968; Fogarty *et al.*, 1980). Pezzack and Duggan (1986) provided evidence from multiple recaptures of tagged lobsters of long-distance (>200-km) return migrations in which some of the tagged animals returned to within 9–42 km of the original site of release.

Dispersal by the larval stage is facilitated by the interaction between passive drifting and behavior, which can modify the speed and/or direction of the drift. The three larval stages may last between 15 and 30 days, depending on water temperature, and the postlarval stage adds another 10–30 days at temperatures normally encountered (Templeman, 1936; MacKenzie, 1988). Settlement to the bottom probably occurs from the midpoint of the postlarval stage on (Scarratt, 1964; Cobb *et al.*, 1989a; Ennis, Chapter 3; Fogarty, Chapter 6). Thus, the potential for dispersal during the 25- to 60-day larval period is great. How might dispersal be modified by larval behavior?

Vertical migration by the larval stages may alter speed and direction of drift. Harding *et al.* (1987) showed that stage I larvae are found above 10 m at night and are most frequent between 10 and 30 m during the day over a part of Browns Bank where the water depth is 60 m. Too few stage II and III larvae were caught to distinguish between day and night distributions; however, a significant proportion was found below the surface. By contrast, studies made in inshore waters generally found stages I–III to be concentrated at the surface both day and night, although some always are found in subsurface tows (Scarratt, 1973; Fogarty and Lawton, 1983). There is no evidence for vertical migrations by postlarvae (Harding *et al.*, 1987, and studies cited therein). The ecological implications of vertical migration are twofold. Dispersal can be affected if subsurface water masses move at different directions or different speeds than the surface water. Growth rates may be slowed by spending up to half the day at depths where water temperatures are lower and prey are less abundant. Sampling for lobster larvae and postlarvae is usually done by towing plankton nets at the surface; clearly, accurate abundance estimates require that subsurface tows for larvae must also be made, at least in deep water (*Homarus americanus*, Harding *et al.*, 1987; *H. gammarus*, Nichols and Lovewell, 1987).

For many years, the dispersal of larval lobsters has been assumed to match the drift of surface water and thus to be discernible from the results of drift bottle surveys (e.g., Rogers *et al.*, 1968). However, the surface meter of water is much more affected by wind stress than is subsurface water. Thus, larvae found below the surface are likely to drift at different rates and in different directions from those at the surface. This possibility has not yet been explored in *Homarus americanus*, but is very likely to be a factor in the transport of the larvae of other species, for example, in the return of phyllosoma larvae of *Panulirus cygnus* to the coast of Australia (Phillips, 1981).

Lobster postlarvae are strong swimmers (Herrick, 1909; Ennis, 1986; Rooney and Cobb, 1991). It is striking that only during a brief part of the life cycle does sustained swimming in the horizontal plane appear to be a major part of the behavior of lobsters. Until recently, the adaptive significance of this swimming was not obvious. However, the observation that the swimming is directional and shoreward in southern New England waters (Cobb *et al.*, 1989b) suggests that the behavior may be an important component of dispersal during the pelagic period. The speed of swimming is rapid (averaging 18 cm/sec) and suggests that considerable displacement could be accomplished during a 10-day postlarval stage. (See Ennis, Chapter 3, for a detailed discussion of the potentials of directed swimming.) Early reports (Lund and Stewart, 1970; Rogers *et al.*, 1968) suggested that larvae hatched at the edge of the continental shelf in southern New England could not provide a recruitment subsidy to inshore areas because the surface residual drift was not sufficient to bring the larvae inshore within the duration of the larval period. Katz *et al.* (1994) reexamined this hypothesis using a model that integrated residual drift, wind-forced drift, directional swimming, and larval development time. The model showed that passive drift was not sufficient to provide larval subsidies to coastal waters, but when directional swimming was added to the model, the amount of time needed to reach the coast was within the developmental time at the prevailing 18–20°C temperatures. Larvae hatched on Browns Bank may be transported to Nova Scotia, the Bay of Fundy, and into the Gulf of Maine by passive drift alone (Harding and Trites, 1988); it is not known whether swimming by postlarvae is directional in this region.

The migratory behavior of adult lobsters and the directional swimming of postlarvae strongly suggest a connection between the offshore and inshore populations of *Homarus americanus*. These behaviors may have evolved during the Wisconsin regression, when glacier growth lowered sea level to as much as 130 m below the present level (Milliman and Emery, 1968; Emery *et al.*, 1988). At that time, the coastline probably was approximately at the current 100-m contour, near the shelf break, and thus near very deep water, unlike the broad expanse of shelf now present. Behavior of both adults and larvae that kept individuals near shore, and out of areas of deep, cold water, would have been adaptive. The present expression of this behavior, in the form of cross-shelf migration and directional postlarval swimming, has a predictable by-product: the maintenance of gene flow between offshore and inshore populations. It is only where coastline morphology and current patterns clearly suggest that there could be little connection that we should expect separate stocks (e.g., the Gulf of St. Lawrence and the remainder of coastal Canada, and perhaps north and south of Cape Cod). A recent analysis of larval morphology suggested separate stocks for lobster populations around the Canadian Maritimes (Harding *et al.*, 1993).

IV. Growth and Mortality

Growth and mortality estimates are incorporated into models used for stock assessment. In their simplest form, surplus production models use only catch and effort data, however, more sophisticated models incorporate terms for time-delayed recruitment. A delay-difference age-structured production model proposed by Fogarty and Murawoski (1986) employs explicit growth, mortality, and recruitment terms. Dynamic pool models require the estimation of a number of parameters, including natural and fishing mortality, age (or size) at recruitment, and growth. Fogarty (Chapter 6) reviews population dynamics models; some behavioral factors that affect growth and mortality are examined here.

A basic assumption in the estimation of mortality rates by analysis of size composition of the catch is that individuals of all size classes are equally vulnerable to the capture technique. As noted in Section II, this is not the case. Lobster traps are biased against the smaller and probably the largest individuals. Aggression between lobsters almost invariably results in the larger animal winning (Scrivener, 1971; O'Neill and Cobb, 1979; Atema and Cobb, 1980) and a dominant–subordinate relationship being established in laboratory settings. Field observations also suggest that larger animals displace smaller ones (O'Neill and Cobb, 1979; Karnofsky *et al.*, 1989). Lobsters leaving the vicinity of traps often do so in response to the approach of another lobster, and the submissive animal retreats or may be chased away (Karnofsky and Price, 1989). As a result, size-related dominance relationships may be a significant factor in the bias of trap data. Since it is the smaller animals that are excluded in higher proportion from the trap, and thus from the sample, estimation of mortality from size-based catch curves may not be accurate. Underrepresentation of smaller animals in a census would lead to an underestimation of mortality.

Growth rates are known to be size and temperature specific (Aiken and Waddy, 1986; see Waddy *et al.*, Chapter 10, for a discussion of the control of growth and reproduction). The interval between molts, and therefore the annual probability of molting, increases with increasing size. Increased temperature decreases the intermolt interval. Both of these have been incorporated into dynamic pool models for lobsters, allowing for the effects of seasonal changes in growth and mortality rates (Fogarty and Idoine, 1986; Fogarty and Murawoski, 1986). Lobsters of legal size molt about once a year, usually in the springtime in inshore waters or in July or August at the edge of the continental shelf. The synchrony of the inshore spring molt is a result of low winter temperatures, preventing the induction of premolt. As soon as the temperature rises above 5°C, all of the animals that became ready to enter premolt over the winter do so, and they molt 2 or 3 months later (Aiken and Waddy, 1986).

Growth rate is affected not only by temperature, but also by social conditions. When lobsters are held communally in the laboratory, average growth rates decrease (Stewart and Squires, 1968; Van Olst *et al.*, 1980). The lessened growth rate is due to the dominance relationships that develop among the animals (Cobb and Tamm, 1974, 1975); the dominant lobster has about the same growth rate as individually held lobsters, while the subordinates show an increased intermolt interval, resulting in a lower rate of growth. Subordinates feed less, even in the presence of an abundance of food, which is the probable cause of the delayed molting. Most of the delay occurs before the induction of premolt (in molt stage D_0), thus acting in a way similar to low winter temperatures (Cobb *et al.*, 1982). Because of differential growth rates induced by social interactions, the variability of size in lobsters raised communally is much greater than among individually reared lobsters. It is not known whether this phenomenon occurs under natural conditions, but aggressive encounters over shelter (O'Neill and Cobb, 1979; Karnofsky *et al.*, 1989) and bait in traps (K. Castro, University of Rhode Island, personal communication) have been observed, and the well-developed agonistic behavior in lobsters leads one to suspect that it probably is an important component of the lobster's behavioral repertoire. If smaller animals do not feed as much in the presence of larger ones, then individual growth rates may vary with the density of the lobster population in an area. Variance of growth rate also is likely to increase with population density, but this hypothesis has not been tested.

Molting inevitably is risky. The process is stressful; in the laboratory, more mortality is associated with the time of ecdysis than with any other part of the molt cycle (personal observations). Since smaller lobsters molt more frequently, natural mortality would be higher in juveniles than in adults and might be related to frequency of molting. Vulnerability to predators probably varies over the molt cycle as well. Immediately after molting, metabolites are in the water and the exoskeleton is soft, making the lobster more detectable, less able to defend itself, and easier to swallow, suggesting that predator-induced mortality is higher around the time of the molt. On the other hand, newly molted lobsters tend to remain in shelters more than intermolt lobsters (personal observations), which may afford a higher degree of protection from predators. Also, the form and intensity of the tail flip response used to escape from predators vary with stage of the molt cycle (Cromarty *et al.*, 1991). Despite the soft shell during postmolt, the total escape response carries these lobsters a greater distance than it does hardshelled animals. The number of tail flips and the duration of the response are greater in postmolt (stage B) animals than in intermolt or premolt lobsters (Cromarty *et al.*, 1991). This suggests that, at least with regard to the ability to move away from a predator quickly, the newly molted animal is at no greater risk (or perhaps even less risk) than a hard-shelled animal. Observations of lobster behavior in a shallow bay over a long period suggest that there is an increase in aggression before the molt (Karnofsky *et al.*, 1989). This may decrease the likelihood of encroachment during the molt by other lobsters resident in the area; presumably, the risk of injury or death during molting also decreases. No direct evidence exists concerning the relationship between molting and predator mortality, but this should be a high priority for further work (Cobb and Caddy, 1989).

Mortality due to predation is related to size and habitat. In large tanks, juvenile lobsters (49- to 59-mm CL) confined with predatory fish in "open" (sand or small cobble) habitats suffered a mortality rate of 25–31%, while mortality was only 2% in a habitat that afforded shelter (Cobb *et al.*, 1986; Richards, 1992). Of early benthic phase lobsters of three size classes tethered in the field, the smaller size classes suffered much higher mortality (Wahle and Steneck, 1992). Early benthic phase lobsters tethered in laboratory tanks with cobble bottom had survival rates twice as high as those tethered in shelterless tanks (Wahle and

Steneck, 1992). Size-dependent vulnerability to predators can be included in population dynamics models by providing a term for natural mortality that shows decreasing mortality with size or age. Recent advances in microwire tagging of postlarvae (M. J. James, University of Rhode Island, personal communication) and early benthic phase lobsters (Krouse and Nutting, 1990) may soon allow us to estimate the natural mortality of the youngest stages of lobsters.

Predators influence the dynamics of prey populations in many ways, not all of them direct. Individuals under risk of predation will change foraging, sheltering, and grouping behaviors (Krebs and Davies, 1993). In lobsters, the influence of predation risk and hunger was investigated by McKenzie (1989) in a laboratory setting. When a predator was present, separated by a clear partition, lobsters spent less time foraging and ingested up to two-thirds fewer calories than did those in the absence of a predator. Lobsters appear to trade off energetic considerations against risk of predation when foraging. Thus, in areas where predators are abundant, lobsters may not only suffer a higher rate of mortality, but also lower growth rates. These results also suggest that the probability of trap encounter will be lower when predators are abundant; thus, traps will fish differently in such locations.

Lobsters living in rocky areas compete with *Cancer* crabs for shelter and probably food (Cobb *et al.*, 1986; Hudon and Lamarche, 1989). In general, lobsters are the better competitors; they "win" most encounters with crabs and usually are able to displace them from the preferred habitat (Richards and Cobb, 1986; Richards, 1992). Nevertheless, in laboratory experiments on competition between lobsters and crabs with a predatory fish present, mortality rates were significantly higher for lobsters than for crabs (Richards and Cobb, 1986). This counterintuitive finding suggests that the loss of a shelter or failure to win in competition (or perhaps the lowering of vigilance while competing) has severe consequences for the lobster. Natural mortality, therefore, is likely to depend not only on the density of predators in an area, but the joint densities of predators and competitors.

V. Maturity and Fecundity

Physiological maturity, that is, the size at which male and female lobsters produce mature gametes, is often detected through changes in allometric growth of the claws in males and changes in setae on the pleopods in females (Aiken and Waddy, 1980; see Lawton and Lavalli, Chapter 4). The size at maturity varies greatly across the geographical range of *Homarus americanus*, depending on summer seawater temperature (Aiken and Waddy, 1986). The difference between physiological maturity and functional maturity is important, however, and the nature of the mating system in lobsters may significantly affect (and be affected by) management tactics.

Male lobsters as small as 40- to 45-mm CL may be capable of producing mature spermatozoa (Briggs and Mushacke, 1979), while females mature at sizes ranging from 70- to 120-mm CL (Aiken and Waddy, 1986). Small males may be incapable of mating with large females (Templeman, 1935), raising the question of whether functional maturity and physiological maturity in males are one and the same (Aiken and Waddy, 1980). Even if a male has mature sperm in its testes, it cannot be regarded as functionally mature unless it is able to attract a female and copulate with her.

Mating behavior has been described from observations in large, naturalistic, laboratory tanks (Atema, 1986; see Atema and Voigt, Chapter 13, on mating behavior). The dominant male (usually the largest) establishes a permanent shelter that is visited by females. Females check the shelters of all the males and make the initial choice for a mate, apparently mediated by chemical communication. Several days before she molts, the female moves into the male shelter. Copulation occurs shortly after she molts and the pair continues to cohabit for several more days. The period of cohabitation, which can be regarded as mate guarding by the male, may last from 1 to 3 weeks. In the laboratory, females mate almost exclusively with the dominant male (Cowan and Atema, 1990). This is likely to be true in nature as well, but observations in the field are very difficult to make. Male dominance, together with defense of a mating site, mate guarding, and female choice of partner, all suggest that lobsters exhibit the type of mating system called resource defense polygyny (Krebs and Davies, 1993). Male dominance may be directly correlated with mating success (Atema *et al.*, 1979; Atema and Voigt, Chapter 13). Since larger males dominate smaller ones, size is also a correlate of mating success, with the largest animal in an area performing most of the matings. This, together with the observation that females do not choose to mate with available but subdominant males, suggests that male fertilizations may be a limiting resource to females.

In populations that are heavily fished, the sex ratio of animals just over the minimum legal size is likely to be skewed toward females because egg-bearing females are protected from capture, and therefore have a lower mortality rate than males (Cobb and Wang, 1985). Because they grow to larger size, males tend to dominate the sex ratio in the larger size classes. Since fishing takes an enormous proportion of lobsters before they reach large size, we can suppose that there are relatively very few large males, at least as compared to an unfished population. This suggests that opportunities for larger females to mate may be particularly limited in those areas where fishing is intense and the majority of sexually mature animals are at or just above the legal size. This hypothesis may be difficult to test in the field.

An interesting hypothesis concerning molt delay by females waiting for dominant males to be "free" was suggested by Atema (1986) and Cowan and Atema (1990). Data from large laboratory tanks, in which several females and males are held together for long periods, suggest that the females may be able to prolong the time before molting in order to gain access to the dominant male for mating. If this is the case, a scarcity of males would not be as great a problem. Unfortunately, the data could not be shown to be significantly different from a random distribution (Hazlett, 1991; Cowan *et al.*, 1991), and this intriguing observation must be revisited for confirmation.

If mating opportunities are limited in heavily fished populations, a consequence might be that population fecundity is lower than would be expected based on the proportion of females in the population. Reproductive output may be lower than its potential because of the interaction between fishing pressure and the polygynous, mate-guarding nature of the reproductive biology of male lobsters. The implications of this biology on management decisions have not been explored, although similar suggestions have been made regarding snow crabs (Bailey and Elner, 1989) and Dungeness crabs (Smith and Jamieson, 1991), which also exhibit size-based dominance, mate guarding, and male polygamy.

VI. Habitat Limitation

In a provocative article, Caddy (1986) suggested the possibility of modeling the recruitment–mortality process as a function of the availability of physical "niche" (appropriately sized shelter), rather than the more usual factors of size and age of the lobster. Since size increase is discontinuous, the sudden size increase at molting may require changing shelters, or even habitat. An obligate crevice dweller such as the lobster will be faced with a declining availability of shelters as it grows. Three responses are possible: migration, competition for declining resources, or slower growth (Caddy, 1986). Susceptibility to predators declines with size (Wahle and Steneck, 1992), making dependence on shelter less critical as the lobster grows. Knowledge of the changes in behavior and habitat over the life cycle suggests that the response to habitat limitation may differ depending on the life history phase of the lobster (see the discussions of life history by Cobb and Wahle, 1994; Lawton and Lavalli, Chapter 4).

In the early benthic phase, lobsters are cryptic and found mainly in cobble or rock-on-sand substrates. The strong association with a particular type of habitat at this stage suggests that the availability of such habitat may limit population size (Wahle and Steneck, 1991). A "bottleneck" based on habitat availability has been reported in other crevice-dwelling crustaceans. The spiny lobster *Panulirus argus* strongly prefers the red alga *Laurencia* sp. as a settlement habitat. After several weeks, the small juveniles move into crevices found in rocks, sponges, and octocorals (Herrnkind and Butler, 1986). Both habitats have the potential to limit population size; however, a series of ingenious experiments showed that in the nursery areas of Florida Bay, it is the crevice habitat that acts as a bottleneck (Butler and Herrnkind, personal communication). Similarly, stomatopods that require close-fitting holes for residence are increasingly size limited as they grow, since the larger crevices are less abundant (Steger, 1987). Field experiments that demonstrate the presence of a bottleneck during the early life history of *Homarus americanus* have not yet been reported.

As lobsters grow, they become more vagile, less dependent on shelter, and less subject to predators. Above the size of maturity, lobsters may move considerable distances. While density-dependent processes are probably important at all times, very different factors associated with habitat may affect growth and mortality in each of the life history phases. Simple, stage-structured models of population dynamics have been developed (see Fogarty, Chapter 6) and can be extended to accommodate as many stages as are needed for adequate description. What is required now is better data from the field about how habitat type affects growth, mortality, and population size so that appropriate functions for each life history phase can be incorporated into the models. Such data are difficult to gather and are not the results of "trendy" science, so will be slow in coming.

The last 15 years have seen an enormous increase in the catch of American lobsters throughout the entire range of the species. In some Canadian areas, catches in the 1980s were three times the levels in the 1970s, and in some locations exceeded even the highest catches of the early history of the fishery. Several factors, probably in conjunction with one another, are likely to explain the increase. An increase in effort through greater efficiency and more gear may explain much of the increase in some areas, but not in others. Increased recruitment may have increased population size (Miller, 1993). A long-term warming trend may have increased growth and catchability. Finally, the heavy fishing on groundfish stocks may have decreased predation on prerecruits (Addison and Fogarty, 1992). However, it is unlikely that the physical, shelter-providing habitat has changed in any marked way, thus suggesting that habitat limitation, at least in Nova Scotia, may not have played a significant role at the relatively lower population densities during the 1970s (Miller, 1993).

VII. Directions for Further Research

Suggestions for further research are scattered throughout this chapter. More broadly, students of the lobster who are interested in its behavior or ecology might usefully take three approaches. First, plan field experiments. Wonderful hypotheses can be built on laboratory evidence of apparently compelling data (e.g., the settling time of postlarvae, Cobb *et al.*, 1989a) that simply are not borne out by field tests. Field work is time consuming and difficult, but results can be strong confirmation of laboratory evidence [e.g., Karnofsky *et al.* (1989) observed what appeared to be a male mating shelter, lending weight to Atema's laboratory analysis of mating behavior]. Finally, observations can be made in the field that are impossible in the laboratory. Denis Wang's and David Campbell's (personal communication) idle staring at the water's surface during a sampling break led to a description of the direction of postlarval swimming (Cobb *et al.*, 1989b), an idea that never would have come about in the laboratory alone.

Second, a careful consideration of the models developed by population biologists may reveal particularly fruitful avenues for research. Estimation of fishing and natural mortality is a critical component of most models, and the models are computationally sophisticated enough to handle changes in mortality rate between and within phases. Unfortunately, biological knowledge of how mortality changes or is controlled falls far short of the capacity of the models. On an even larger scale, the shape of the stock recruitment model postulated by Fogarty and Idoine (1986) for the lobster population in Northumberland Strait suggests that density-dependent population regulation is acting somewhere between the postlarval and adult phases. This is an important observation, but gives no insight into mechanisms. Knowledge of the mechanisms and timing of the limiting factors would facilitate protection or even enhancement of critical life history phases. Breaking down the life cycle into shorter phases and modeling the recruitment from one phase to the next (e.g., Paulik, 1973) seems to be the next appropriate step, and one that will give ethologists and ecologists much substrate for interesting experiments.

Finally, the application of new technology to behavioral and ecological problems is very likely to reap great rewards. Field observations are difficult, but remotely operated vehicles (Spanier *et al.*, 1994) hold promise, particularly when they are miniaturized and made less intrusive. Biochemical advances have allowed the estimation of recent growth and nutritional condition of individual postlarvae from the field (Juinio and Cobb, 1994). Understanding patterns of paternity has become important to understanding the complexity of mating systems in birds (Krebs and Davies, 1993). DNA fingerprinting might allow us to determine the extent of polygyny in populations of *Homarus americanus*.

VIII. Summary

The abundance of American lobsters (*Homarus americanus*) is estimated by sampling with nets (in the planktonic phases) or with traps (in the benthic phases). Both of these methods have inherent biases due to the behavior and physiology of the animals. Knowledge of daily movements, bottom-seeking behavior, and the duration of each larval stage is required for accurate estimation of larval abundance. Traps sample animals that first sense, then enter and remain within the trap. The chemosensory abilities and aggressive nature of lobsters, as well as walking rate and hunger levels (which vary with temperature and probably stage of the molt cycle), undoubtedly bias catches. Lobsters escape from traps at a fairly low rate once they have entered the "parlor" section; reduced entry, rather than escape, is probably the dominant mechanism affecting trap saturation.

Lobsters are capable of moving great distances. Dispersal appears to be concentrated in two life history phases. The larvae may travel long distances using currents and swimming, while after maturity, adults may show remarkably long migrations. Both of these suggest that there is likely to be little genetic differentiation among geographically separated segments of the population, except perhaps between the Gulf of St. Lawrence and the remainder of the range. The growth rate is affected not only by temperature and food availability, but also by social conditions, suggesting that estimates of the growth rate perhaps should include a coefficient that expresses population density (and, therefore, the probability of social interaction) in an area. Mortality rates are likely to be correlated with molt cycle stage, as well as predation, competition, and habitat type. Dominant male lobsters are polygynous and guard mates individually for 1–3 weeks before and after they molt and mate. The effects of this mating system on population fecundity have not been explored. Small lobsters are particularly dependent on shelter-providing habitat, but this requirement loosens progressively as the animal grows. Habitat may be limiting in the early benthic phase, but the mechanisms of the possible limitation have not been explored sufficiently.

References

Addison, J., and Fogarty, M. J. (1992). Juvenile lobster habitat limitation: What can landings tell us? *Lobster Newsl.* 5(2), 10–12.

Aiken, D. E., and Waddy, S. L. (1980). Reproductive biology. *In* "The Biology and Management of Lobsters" (J. S. Cobb and B. F. Phillips, eds.), Vol. 1, pp. 215–276. Academic Press, New York.

Aiken, D. E., and Waddy, S. L. (1986). Environmental influence on recruitment of the American lobster *Homarus americanus*: A perspective. *Can. J. Fish. Aquat. Sci.* 43, 2258–2270.

Atema, J. (1986). Review of sexual selection and chemical communication in the lobster, *Homarus americanus*. *Can. J. Fish. Aquat. Sci.* 43, 2283–2390.

Atema, J., and Cobb, J. S. (1980). Social behavior. *In* "The Biology and Management of Lobsters" (J. S. Cobb and B. F. Phillips, eds.), Vol. 1, pp. 409–450. Academic Press, New York.

Atema, J., Jacobson, S., Karnofsky, E., Oleszko-Szuts, S., and Stein, L. (1979). Pair formation in the lobster, *Homarus americanus*: Behavioral development, pheromones and mating. *Mar. Behav. Physiol.* 6, 277–296.

Bailey, R. F. J., and Elner, R. W. (1989). Northwest Atlantic snow crab fisheries: Lessons in research and management. *In* "Marine Invertebrate Fisheries: Their Assessment and Management" (J. F. Caddy, ed.), pp. 261–280. Wiley, New York.

Botero, L., and Atema, J. (1982). Behavior and substrate selection during larval settling in the lobster *Homarus americanus*. *J. Crust. Biol.* 2, 59–69.

Boudreau, B., Simard, Y., and Bourget, E. (1991). Behavioural responses of the planktonic stages of the American lobster *Homarus americanus* to thermal gradients, and ecological implications. *Mar. Ecol. Prog. Ser.* 76, 13–23.

Boudreau, B., Simard, Y., and Bourget, E. (1992). Influence of a thermocline on vertical distribution and settlement of post-larvae of the American lobster *Homarus americanus* Milne-Edwards. *J. Exp. Mar. Biol. Ecol.* 162, 35–49.

Boudreau, B., Bourget, E., and Simard, Y. (1993). Behavioural responses of competent lobster postlarvae to odor plumes. *Mar. Biol.* 117, 63–69.

Briggs, P. T., and Mushacke, F. M. (1979). The American lobster in western Long Island Sound. *N.Y. Fish Game J.* 26, 59–86.

Caddy, J. F. (1986). Modelling stock–recruitment processes in Crustacea: Some practical and theoretical perspectives. *Can. J. Fish. Aquat. Sci.* 43, 2330–2344.

Campbell, A., and Stasko, A. B. (1985). Movements of tagged American lobsters, *Homarus americanus*, off southwestern Nova Scotia. *Can. J. Fish. Aquat. Sci.* 42, 229–238.

Chapman, C. J. (1980). Ecology of juvenile and adult *Nephrops*. *In* "The Biology and Management of Lobsters" (J. S. Cobb and B. F. Phillips, eds.), Vol. 1, pp. 143–175. Academic Press, New York.

Cobb, J. S. (1969). "Activity, Growth, and Shelter Selection of the American Lobster," Ph.D. thesis. University of Rhode Island, Kingston.

Cobb, J. S., and Caddy, J. F. (1989). The population biology of decapods. *In* "Marine Invertebrate Fisheries: Their Assessment and Management" (J. F. Caddy, ed.), pp. 327–374. Wiley, New York.

Cobb, J. S., and Tamm, G. R. (1974). Social conditions increase intermolt period in juvenile lobsters. *J. Fish. Res. Board Can.* 32, 1941–1943.

Cobb, J. S., and Tamm, G. R. (1975). Dominance status and molt order in lobsters (*Homarus americanus*). *Mar. Behav. Physiol.* 3, 119–124.

Cobb, J. S., and Wahle, R. A. (1994). Early life history and recruitment processes of clawed lobsters. *Crustaceana* 67, 1–25.

Cobb, J. S., and Wang, D. (1985). Fisheries biology of lobsters and crayfishes. *In* "The Biology of Crustacea" (D. E. Bliss, ed.), Vol. 10, "Economic Aspects: Fisheries and Culture" (A. J. Provenzano, ed.), pp. 167–247. Academic Press, Orlando, Florida.

Cobb, J. S., Tamm, G. R., and Wang, D. (1982). Behavioral mechanisms influencing molt frequency in the American lobster, *Homarus americanus* Milne Edwards. *J. Exp. Mar. Biol. Ecol.* 62, 185–200.

Cobb, J. S., Gulbransen, T., Phillips, B. F., Wang, D., and Syslo, M. (1983). Behavior and distribution of larval and early juvenile *Homarus americanus*. *Can. J. Fish. Aquat. Sci.* 40, 2184–2188.

Cobb, J. S., Wang, D., Richards, R. A., and Fogarty, M. J. (1986). Competition among lobsters and crabs and its possible effects in Narragansett Bay, Rhode Island. *Can. Spec. Publ. Fish. Aquat. Sci.* 92, 282–290.

Cobb, J. S., Wang, D., and Campbell, D. B. (1989a). Timing of settlement by postlarval lobsters (*Homarus americanus*): Field and laboratory evidence. *J. Crust. Biol.* 9, 60–66.

Cobb, J. S., Wang, D., Campbell, D. B., and Rooney, P. (1989b). Speed and direction of swimming by postlarvae of the American lobster. *Trans. Am. Fish. Soc.* 118, 82–86.

Cooper, R. A., and Uzmann, J. R. (1971). Migrations and growth of deep-sea lobsters, *Homarus americanus*. *Science* 171, 288–290.

Cooper, R. A. and Uzmann, J. R. (1980). Ecology of juvenile and adult *Homarus*. *In* "The Biology and Management of Lobsters" (J. S. Cobb and B. F. Phillips, eds.), Vol. 2, pp. 97–142. Academic Press, New York.

Cowan, D. F., and Atema, J. (1990). Moult staggering and serial monogamy in American lobsters, *Homarus americanus*. *Anim. Behav.* 39, 1199–1206.

Cowan, D. F., Atema, J. A., and Solow, A. R. (1991). Moult staggering in the American lobster: A statistical analysis. *Anim. Behav.* 42, 863–864.

Cromarty, S. I., Cobb, J. S., and Kass-Simon, G. (1991). Behavioral analysis of the escape response in the juvenile lobster *Homarus americanus* over the molt cycle. *J. Exp. Biol.* 158, 565–581.

Daniel, P. C., and Bayer, R. C. (1987). Partial purification and characterization of post-larval lobster (*Homarus americanus*) feeding attractants from herring (*Clupea harengus*) tissue. *Mar. Behav. Physiol.* 13, 29–50.

Derby, C. D. (1984). Molecular weight fractions of natural foods that stimulate feeding in crustaceans, with data from the lobster *Homarus americanus*. *Mar. Behav. Physiol.* 10, 273–282.

Emery, K. O., Merrill, A. S., and Druffel, E. R. M. (1988). Changed late Quaternary marine environments on Atlantic continental shelf and upper slope. *Quat. Res. (N.Y.)* 30, 251–269.

Ennis, G. P. (1973). Food, feeding, and condition of lobsters (*Homarus americanus*) throughout the seasonal cycle in Bonavista Bay, Newfoundland. *J. Fish. Res. Board Can.* 30, 1905–1909.

Ennis, G. P. (1978). Growth curves for Newfoundland lobsters from data on molt increment and proportion molting. *Can. Atl. Fish. Sci. Adv. Comm. Res. Doc.* 78/29, 1–11.

Ennis, G. P. (1986). Swimming ability of larval American lobsters, *Homarus americanus*, in flowing water. *Can. J. Fish. Aquat. Sci.* 43, 2177–2183.

Fogarty, M. J., and Idoine, J. S. (1986). Recruitment dynamics in an American lobster (*Homarus americanus*) population. *Can. J. Fish. Aquat. Sci.* 43, 2368–2376.

Fogarty, M. J., and Lawton, R. (1983). An overview of larval American lobster, *Homarus americanus*, sampling programs in New England during 1974–79. *NOAA Tech. Rep., NMFS SSRF-775*, 9–14.

Fogarty, M. J., and Murawoski, S. A. (1986). Population dynamics and assessment of exploited invertebrate stocks. *Can. Spec. Publ. Fish. Aquat. Sci.* 92, 228–244.

Fogarty, M. J., Borden, D. V. D., and Russell, H. J. (1980). Movements of tagged American lobster *Homarus americanus*, off Rhode Island. *Fish. Bull.* 78, 771–780.

Harding, G. C., and Trites, R. W. (1988). Dispersal of *Homarus americanus* larvae in the Gulf of Maine. *Can. J. Fish. Aquat. Sci.* 45, 416–425.

Harding, G. C., Vass, W. P., and Drinkwater, K. F. (1982). Aspects of larval American lobster (*Homarus americanus*) ecology in St. Georges Bay, Nova Scotia. *Can. J. Fish. Aquat. Sci.* 39, 117–129.

Harding, G. C., Pringle, J. D., Vass, W. P., Pearre, S. J., and Smith, S. J. (1987). Vertical distribution and daily movements of larval lobsters *Homarus americanus* over Browns Bank, Nova Scotia. *Mar. Ecol. Prog. Ser.* 41, 29–41.

Harding, G. C., Kenchington, E., and Zheng, Z. (1993). Morphometrics of American lobster (*Homarus americanus*) larvae in relation to stock determinations in the Maritimes, Canada. *Can. J. Fish. Aquat. Sci.* 50, 43–52.

Hazlett, B. A. (1991). Moult staggering or are female lobsters moulting independently? *Anim. Behav.* 42, 862.

Herrick, F. H. (1909). Natural history of the American lobster. *Bull. U.S. Bur. Fish.* 29, 149–408 + 20 plates.

Herrnkind, W. F., and Butler, M. J., IV (1986). Factors regulating postlarval settlement and juvenile microhabitat use by spiny lobsters, *Panulirus argus*. *Mar. Ecol. Prog. Ser.* 34, 23–30.

Hofe, C. F. (1993). "The Effect of Thermoclines, Molt Stage and Postlarval Source on the Diving Behavior of Postlarval Lobster, *Homarus americanus*," M.S. thesis. University of Rhode Island, Kingston.

Hudon, C. (1987). Ecology and growth of larval and postlarval lobster, *Homarus americanus*, off Îsles de la Madeleine (Québec). *Can. J. Fish. Aquat. Sci.* 44, 1855–1869.

Hudon, C., and Lamarche, G. (1989). Niche segregation between the American lobster *Homarus americanus* and the rock crab *Cancer irroratus*. *Mar. Ecol. Prog. Ser.* 52, 155–168.

Incze, P. M., and Wahle, R. A. (1991). Recruitment from pelagic to early benthic phase in *Homarus americanus*. *Mar. Ecol. Prog. Ser.* 79, 77–87.

Juinio, M. A. R., and Cobb, J. S. (1994). Estimation of recent growth of field caught postlarval lobsters from RNA/DNA ratios. *Can. J. Fish. Aquat. Sci.* 51, 286–294.

Karnofsky, E. B., and Price, H. J. (1989). Behavioral response of the lobster *Homarus americanus* to traps. *Can. J. Fish. Aquat. Sci.* 46, 1625–1632.

Karnofsky, E. B., Atema, J., and Elgin, R. A. (1989). Field observations of social behavior, shelter use, and foraging in the lobster, *Homarus americanus*. *Biol. Bull. (Woods Hole, Mass.)* 176, 239–246.

Katz, C. H., Cobb, J. S., and Spaulding, M. (1994). Larval behavior, hydrodynamic transport, and potential offshore recruitment in the American lobster, *Homarus americanus*. *Mar. Ecol. Prog. Ser.* 103, 265–273.

Krebs, J. R., and Davies, N. B. (1993). "An Introduction to Behavioural Ecology." Blackwell, Oxford, England.

Krouse, J. S., and Nutting, G. E. (1990). Evaluation of coded microwire tags inserted in legs of small juvenile American lobsters. *Am. Fish. Soc. Symp.* 7, 304–310.

Lawton, P. (1987). Diel activity and foraging behavior of juvenile American lobsters, *Homarus americanus*. *Can. J. Fish. Aquat. Sci.* 44, 1195–1205.

Lipcius, R. N., and Herrnkind, W. F. (1982). Molt cycle alterations in behavior, feeding and diel rhythms of a decapod crustacean, the spiny lobster *Panulirus argus*. *Mar. Biol.* 68, 241–252.

Lund, W. A., and Stewart, L. L. (1970). Abundance and distribution of larval lobsters, *Homarus americanus*, off the coast of southern New England. *Proc. Natl. Shellfish. Assoc.* 60, 40–49.

MacKenzie, B. R. (1988). Assessment of temperature effects on interrelationships between stage durations, mortality, and growth in laboratory-reared *Homarus americanus* Milne Edwards larvae. *J. Exp. Mar. Biol. Ecol.* 116, 87–98.

McKenzie, T. P. (1989). "Foraging Behavior of Juvenile American Lobsters, *Homarus americanus*, under Predation Risk," M.S. thesis. University of Rhode Island, Kingston.

McLeese, D. W. (1973). Orientation of lobsters to odor. *J. Fish. Res. Board Can.* 30, 838–840.

McLeese, D. W., and Wilder, D. G. (1958). The activity and catchability of the lobster (*Homarus americanus*) in relation to temperature. *J. Fish. Res. Board Can.* 15, 1345–1354.

Miller, R. J. (1978). Saturation of crab traps: Reduced entry and escapement. *J. Cons., Cons. Int. Explor. Mer* 38, 338–345.

Miller, R. J. (1980). Design criteria for crab traps. *J. Cons., Cons. Int. Explor. Mer* 39, 140–147.

Miller, R. J. (1989). Catchability of American lobsters (*Homarus americanus*) and rock crabs (*Cancer irroratus*) by traps. *Can. J. Fish. Aquat. Sci.* 46, 1652–1657.

Miller, R. J. (1990). Effectiveness of crab and lobster traps. *Can. J. Fish. Aquat. Sci.* 47, 1228–1251.

Miller, R. J. (1993). More about habitat for juvenile *Homarus americanus*. *Lobster Newsl.* 6(1), 8–9.

Milliman, J. D., and Emery, K. O. (1968). Sea levels during the past 35,000 years. *Science* 162, 1121–1123.

Morgan, G. R. (1974). Aspects of the population dynamics of the western rock lobster, *Panulirus cygnus* George. I. Estimation of population density. *Aust. J. Mar. Freshwater Res.* 25, 235–248.

Nichols, J. H., and Lovewell, S. J. (1987). Lobster larvae (*Homarus gammarus* L.) investigations in Bridlington Bay. Can quantitative sampling be confined to the neuston layer? *J. Nat. Hist.* 21, 825–841.

O'Neill, D. J., and Cobb, J. S. (1979). Some factors influencing the outcome of shelter competition in lobsters (*Homarus americanus*). *Mar. Behav. Physiol.* 6, 33–45.

Paulik, G. J. (1973). Studies of the possible form of the stock–recruitment curve. *Rapp. P.-V. Reun., Cons. Int. Expl. Mer* 164, 302–315.

Pezzack, D. S., and Duggan, D. R. (1983). The Canadian Offshore Lobster (*Homarus americanus*) Fishery—1971 to 1982. *Int. Counc. Explor. Sea Shellfish. Comm.* C.M. 1983/K:34.

Pezzack, D. S., and Duggan, D. R. (1986). Evidence of migration and homing of lobsters (*Homarus americanus*) on the Scotian Shelf. *Can. J. Fish. Aquat. Sci.* 43, 2206–2211.

Phillips, B. F. (1981). The circulation of the southeastern Indian Ocean and the planktonic life of the western rock lobster. *Oceanogr. Mar. Biol.* 19, 11–39.

Richards, R. A. (1992). Habitat selection and predator avoidance: Ontogenetic shifts in habitat use by the Jonah crab *Cancer borealis* (Stimpson). *J. Exp. Mar. Biol. Ecol.* 156, 187–197.

Richards, R. A., and Cobb, J. S. (1986). Competition for shelter between lobsters (*Homarus americanus*) and Jonah crabs (*Cancer borealis*): Effects of relative size. *Can. J. Fish. Aquat. Sci.* 43, 2250–2255.

Richards, R. A., Cobb, J. S., and Fogarty, M. J. (1983). Effects of behavioral interactions on the catchability of American lobster (*Homarus americanus*) and two species of *Cancer* crabs. *Fish. Bull.* 81, 51–60.

Rogers, B. A., Cobb, J. S., and Marshall, N. (1968). Size comparison of inshore and offshore larvae of the lobster, *Homarus americanus*, off southern New England. *Proc. Natl. Shellfish. Assoc.* 58, 78–81.

Rooney, P., and Cobb, J. S. (1991). The effects of time of day, water temperature and water velocity on swimming by postlarvae of the American lobster, *Homarus americanus*. *Can. J. Fish. Aquat. Sci.* 48, 1944–1950.

Saila, S. B., and Flowers, J. M. (1968). Movements and behaviour of berried female lobsters displaced from offshore areas to Narragansett Bay, Rhode Island. *J. Cons., Cons. Int. Explor. Mer* 31, 342–351.

Scarratt, D. J. (1964). Abundances and distribution of lobster larvae (*Homarus americanus*) in Northumberland Strait. *J. Fish. Res. Board Can.* 21, 661–680.

Scarratt, D. J. (1973). Abundance, survival and vertical and diurnal distribution of lobster larvae in Northumberland Strait, 1952–1963, and their relationship with commercial stocks. *J. Fish. Res. Board Can.* 24, 1819–1824.

Scrivener, J. C. E. (1971). Agonistic Behavior of the American Lobster *Homarus americanus*. *Tech. Rep.—Fish. Res. Board Can.* 235, 1–128.

Skud, B. E., and Perkins, H. C. (1969). Size Composition, Sex Ratio, and Size at Maturity of Offshore Northern Lobsters. *U.S., Fish Wildl. Serv., Spec. Sci. Rep.—Fish.* 598.

Smith, S. I. (1873). The early stages of the American lobster *Homarus americanus* Milne-Edwards. *Trans. Conn. Acad. Arts Sci.* 2, 351–381.

Smith, B. D., and Jamieson, G. S. (1991). Possible consequences of intensive fishing for males on the mating opportunities of Dungeness crabs. *Trans. Am. Fish. Soc.* 120, 650–653.

Spanier, E., Cobb, J. S., and Clancy, M. (1994). Impacts of remotely operated vehicles (ROVs) on the behavior on marine animals: An example using American lobsters. *Mar. Ecol. Prog. Ser.* 104, 257–266.

Steger, R. (1987). Effects of refuges and recruitment on gonodactylid stomatopods, a guild of mobile prey. *Ecology* 68, 1520–1533.

Stewart, L. L. (1972). "The Seasonal Movements, Population Dynamics and Ecology of the Lobster, *Homarus americanus* (Milne-Edwards) off Ram Island, Connecticut," Ph.D. thesis. University of Connecticut, Storrs.

Stewart, J. E., and Squires, H. J. (1968). Adverse conditions as inhibitors of ecdysis in the lobster *Homarus americanus*. *J. Fish. Res. Board Can.* 25, 1763–1774.

Templeman, W. (1935). Local differences in the body proportions of the lobster, *Homarus americanus*. *J. Biol. Board Can.* 1, 213–226.

Templeman, W. (1936). The influence of temperature, salinity, light and food conditions on the survival and growth of the larvae of the lobster (*Homarus americanus*). *J. Biol. Board Can.* 2, 485–497.

Templeman, W. (1937). Habits and distribution of larval lobsters (*Homarus americanus*). *J. Biol. Board Can.* 3, 343–347.

Templeman, W., and Tibbo, S. N. (1945). Lobster Investigations in Newfoundland 1938–1941. *Newfoundland, Dep. Nat. Resour., Res. Bull. (Fish.)* 16, 1–98.

Uzmann, J. R., Cooper, R. A., and Pecci, K. J. (1977). Migration and Dispersion of Tagged American Lobsters, *Homarus americanus*, on the Southern New England Continental Shelf. *NOAA Tech. Rep. NMFS SSRF* 705.

Van Olst, J. C., Carlberg, J. M., and Hughes, J. T. (1980). Aquaculture. *In* "The Biology and Management of Lobsters" (J. S. Cobb and B. F. Phillips, eds.), Vol. 2, pp. 333–384. Academic Press, New York.

Wahle, R. A., and Steneck, R. S. (1991). Recruitment habitats and nursery grounds of the American lobster *Homarus americanus*: A demographic bottleneck? *Mar. Ecol. Prog. Ser.* 69, 231–243.

Wahle, R. A., and Steneck, R. S. (1992). Habitat restrictions in early benthic life: Experiments on habitat selection and in situ predation with the American lobster. *J. Exp. Mar. Biol. Ecol.* 157, 91–114.

Weiss, H. M. (1970). "The Diet and Feeding Behavior of the Lobster, *Homarus*," Ph.D. thesis. University of Connecticut, Storrs.

Weissburg, M. J., and Zimmer-Faust, R. K. (1993). Life and death in moving fluids: Hydrodynamic effects on chemosensory-mediated predation. *Ecology* 74, 1428–1443.

Wilder, D. G. (1963). Movements, growth and survival of marked and tagged lobsters liberated in Egmont Bay, Prince Edward Island. *J. Fish. Res. Board Can.* 20, 305–318.

Zeitlin-Hale, L., and Sastry, A. N. (1978). Effects of environmental manipulation on the locomotor activity and agonistic behavior of cultured juvenile American lobsters, *Homarus americanus*. *Mar. Biol.* 47, 369–379.

Zimmer-Faust, R. K. (1989). The relationship between chemoreception and foraging behavior in crustaceans. *Limnol. Oceanogr.* 34, 1367–1374.

Zimmer-Faust, R. K., and Case, J. F. (1983). A proposed dual role of odor in foraging by the California spiny lobster, *Panulirus interruptus* (Randall). *Biol. Bull. (Woods Hole, Mass.)* 164, 341–353.

CHAPTER

8

Aquaculture

D. E. AIKEN *and* S. L. WADDY
Biological Station
Department of Fisheries and Oceans
St. Andrews, New Brunswick, Canada E0G 2X0

I. Introduction

Twenty years ago people talked wistfully of lobster farms that would annually produce some 500 metric tons (mt) of 450-g "ocean chickens." The American lobster, *Homarus americanus*, with its strong international market appeal, high unit value, and well-understood life cycle, was considered the ideal seafood to test the ingenuity of science and attract the attention of entrepreneurs. Young lobsters had been hatched, reared, and released for stock enhancement purposes for over a century and the traditional storage and shipment industry had demonstrated that adults were reasonably resistant to disease and could survive long periods of confinement.

The lobster has always been a favorite of the general public, and in the early 1970s this popularity was utilized to promote the idea of lobster aquaculture. In succeeding years, lobster culture research prospered, but the intense media attention and generous funding created expectations among granting agencies, politicians, and entrepreneurs—expectations that could not be met by researchers who were attempting to domesticate a wild animal on an unrealistic schedule. Although the research effort was tremendously productive, the inflated expectations caused by media hype and the impatience of the various funding agencies made "failure" inevitable.

This early vision of lobster farming has faded, but *Homarus americanus* still has considerable potential as a cultured species. Two decades of research have closed the reproductive cycle, allowed seedstock to be produced on demand, and provided a variety of systems and strategies to grow lobsters to market size. Nevertheless, only one commercial facility is currently growing lobsters for sale, and this is on a very modest scale. The reasons for this, and the state of the science and technology required for commercial culture, are the subject of this chapter.

A. Modern Lobster Culture Research

Efforts to culture lobsters began in the mid-1800s (see Section I,B,1). Interest in intensive culture of *Homarus americanus* peaked approximately 100 years later, in the 1970s, when government funding agencies—led by the U.S. Sea Grant Program—gave priority to lobster culture research and lavished funds on lobster research projects. University and government programs existed from coast to coast in the United States. In Canada, research on nutrition, systems design, and rearing strategies was closely integrated with the key programs in the United States.

In 1974, the world's first integrated lobster culture facility was built at a federal government research institution at St. Andrews, New Brunswick, Canada. The facility was constructed to determine whether the cycle from egg to broodstock could be closed and whether the animal's performance under intensive cultivation could be improved through successive generations. Other research culture facilities existed, but none combined all stages from broodstock to marketable product in a continuous production system. The challenge was to integrate the different elements of the evolving culture technology into a year-round production system. A stable juvenile grow-out phase requires continuous, year-round production of postlarvae, which, in turn, requires a continuous and reliable supply of newly hatched larvae. This had never been accomplished with the American lobster.

Research and development on lobster culture was ready for pilot-scale evaluation by 1979, and in that year researchers from the major U.S. and Canadian programs prepared a "North American Lobster Culture Proposal" for funding by the Sea Grant Program. It proposed a pilot-scale facility that would incorporate the latest technology, address the remaining technical and biological problems, and obtain reliable data on capital and operating costs. The Sea Grant Program, perhaps responding to pressure from commercial interests, decided not to fund the pilot project, thereby transferring the development responsibility to the private sector. Several systems for intensive lobster culture were subsequently offered for sale by commercial interests in North America, but none ever produced lobsters on a commercial scale.

In retrospect, it is obvious the technology was not yet sufficiently advanced for direct transfer to the private sector. A commercial ration that would produce adequate growth and survival in intensive culture was lacking and our knowledge of broodstock management, water quality control, and rearing technology were still too primitive for commercial application. Ironically, the intensive promotion necessary to ensure continued research funding caused the private sector to move prematurely, resulting in the demise of productive university and government research programs that were essential for successful commercial application.

B. Types of Lobster Culture

Culture systems are classified as intensive, semi-intensive, or extensive, depending on the degree of husbandry involved. Extensive culture involves the least manipulation and complexity, and therefore the lowest capital, labor, and operating costs. However, extensive culture methods are not easily applied to lobsters because they are difficult to confine in the natural environment.

To most people, *lobster culture* suggests a farming system in which broodstock animals produce eggs that are hatched and reared to a marketable product, or *closed-cycle culture.* Lobster culture technology can also be utilized for *resource enhancement*, in which young are hatched, reared, and released to enhance natural stocks. Another variation is *product enhancement*, in which low-value animals taken from the fishery (soft-shelled, sublegal size, or suboptimum size) are maintained until their value increases. Other variations exist, including the production of scampi-sized lobsters, soft-shelled lobsters, and unusual color-morphs for the aquarium trade. All work to date indicates that most types of culture would be intensive, highly capitalized undertakings, but that semiintensive systems might be devised for product enhancement culture.

1. Resource Enhancement

The first lobster hatcheries were built in France more than 130 years ago to hatch and rear the European lobster (*Homarus gammarus*) and similar programs were subsequently developed in Norway (Herrick, 1909). The first successful attempt at hatching larvae of *Homarus americanus* was in 1885 at the newly opened laboratory of the U.S. Fish Commission in Woods Hole, Massachusetts (Rathbun, 1886). Lobster and cod were produced at the Dildo hatchery in Newfoundland as early as 1889 and Canada's first hatchery was built at Bay View, Nova Scotia, in 1891. By the early 1900s, there were lobster hatcheries throughout eastern Canada (Nova Scotia, New Brunswick, Prince Edward Island, and Quebec) and the northeastern United States (Maine, New Hampshire, Massachusetts, and Rhode Island). Between 1885 and 1903, 880 million first-stage lobster larvae were hatched and released along the eastern seaboard of the United States (Carlson, 1954). (See Factor, Chapter 1, and Ennis, Chapter 3, for discussions of the larvae and life history of *H. americanus*.)

The Canadian lobster hatcheries were closed (except for experimental work) after the 1917 season because of a critical report that claimed, with some justification, that there was no evidence that hatcheries had any beneficial effect on lobster populations (Knight, 1918). Hatcheries elsewhere continued for much longer. European lobsters were hatched in Norway until 1940 and in France until 1948. In the United States, hatcheries operated in Rhode Island, Connecticut, and Maine until the late 1940s and early 1950s.

It is important to realize that the early hatcheries used very crude technology. In Canada, for example, the eggs of ovigerous females destined for the cannery were scraped off with a spoon and held in

McDonald hatching jars. Hatching success under these conditions was abysmal because lobster eggs tend to become diseased and die after they are removed from the female. For example, the 27.6 million eggs held in two Nova Scotia hatcheries in 1917 produced fewer than 100,000 first-stage larvae (MacKay, 1929), a success rate of fewer than four larvae per 1000 eggs.

The technology for hatching eggs removed from egg-bearing lobsters was improved by Adolf Nielson, a Norwegian fisheries expert working for the Government of Newfoundland. In 1893, he developed and patented egg incubators that were placed directly into the sea. Nielson incubators were used at 50 sheltered locations along the coast of Newfoundland and were adopted by hatcheries in both North America and Europe (Roche, 1898; Carlson, 1954).

However, the release of hatchlings during the early larval stages has little measurable impact on natural production. Larvae drift in the plankton for several weeks and are subject to heavy mortality. Recognizing this, Herrick (1894) advocated rearing lobsters through the three larval stages and releasing them at the fourth stage, as they change to a benthic existence. The first successful procedure for rearing larvae to the fourth stage was developed by A. D. Mead of the Rhode Island Commission of Inland Fisheries (Mead, 1908). He held ovigerous females in partially submerged rafts until they released their larvae. As they hatched, the first-stage larvae were collected and transferred to separate rearing cages in which a mechanical device (a horizontal propeller) produced a constant circulation of water that kept both the larvae and their food in suspension (Fig. 1). Larvae were fed scrambled and pulverized hen's eggs, minced clams, cod liver, and shredded fish, beef liver, and crab meat. Coincidentally, during the years this hatchery was in operation the lobster catch in Rhode Island increased from less than 180 mt to over 726 mt, a fact regarded by some as proof that the program was successful (Carlson, 1954).

In the past 40 years, only one hatchery in North America has produced significant quantities of lobsters—the Massachusetts State Lobster Hatchery and Research Station in Vineyard Haven, Martha's Vineyard—which has been hatching larvae since 1951 and releases half a million stage IV lobsters (postlarvae) into Massachusetts coastal waters each year (Hughes, 1968). Two small hatcheries, one supported by fishers' groups and the other a private facility, are currently operating in Maine. In Europe, where wild stocks have declined significantly, the United Kingdom, Ireland, Norway, and France expended considerable effort in the last 15 years on stock enhancement (Bannister and Howard, 1991; Beard and Wickins, 1992; Grimsen *et al.*, 1987; Tveite, 1995).

With continued interest in lobster culture has come improved technology. The Massachusetts hatchery originally reared larvae in a 35-cm rearing box with rounded corners. This evolved into a properly engineered larval rearing tank (Fig. 2) that enabled 50–75% of the hatchlings to survive to the postlarval (fourth) stage (Carlson, 1954; Hughes *et al.*, 1974). The Massachusetts hatchery releases larvae in the fourth stage; however, this approach is not without critics, as it is difficult to demonstrate any beneficial effect on the wild population. No one has yet found a way to mark lobsters of this size for identification several years later, although carefully coordinated releases of distinctive color strains are a possibility that deserves further evaluation (Aiken and Waddy, 1989a).

A lobster is not committed to benthic existence until midway through the fourth stage (see Ennis, Chapter 3), so release during the early fourth stage still exposes the animal to heavy predation. The solution adopted in Europe was to rear the juvenile lobsters for up to 1 year before stocking them in refuges that offer the proper habitat. The advantage is vastly improved survival; the disadvantage is much greater cost. The long period of intensive cultivation prior to release raises the enhancement cost considerably. However, if minimum size regulations in the fishery permit survivors to reproduce and contribute their own progeny to the population, the return on investment would be compounded. In England, a release program conducted between 1983 and 1988, using stage XII, microtagged European lobsters, demonstrated that hatchery-reared stock can survive in the wild, grow to commercial size, and reproduce (Bannister and Howard, 1991).

2. Product Enhancement

In product enhancement, low-value lobsters taken in the traditional capture fishery (e.g., soft-shelled, about-to-molt, and suboptimum-sized lobsters) are maintained until their quality, and therefore their value, increases. Product enhancement is of interest in Canada because of the so-called "canner" fishery, in which lobsters that are too small for export to the lucrative New England market are canned or sold locally at prices well below those of "market" lobsters. A 74-mm carapace length (CL) canner lobster, purchased at bargain prices in the spring fishery, will molt to market size and be fully meated by the time the price reaches its peak in winter. However, there are biological and technical constraints to this

FIGURE 1 The floating laboratory of the Rhode Island Commission of Inland Fisheries at Wickford, Rhode Island. General view of a house boat with floats attached (top). The arrangement of the rearing cars, alleyways, and flotation barrels is visible on the right. Rearing cars are shown, with drive shaft and gears (bottom). The propeller is visible, submerged in the first car. (From Mead, 1908.)

approach. This type of culture requires a facility for holding lobsters individually, water temperatures of 15–20°C during the summer and autumn, and a good formulated food. Only males are used for this type of culture because females of this size tend to spawn rather than molt. Furthermore, capture and confinement are stressful, and stress increases susceptibility to diseases such as gaffkemia, which is fatal.

FIGURE 2 Planktonkreisel for rearing lobster larvae (so-called "Hughes' pot") developed at the Massachusetts State Lobster Hatchery and Research Station on Martha's Vineyard. (Courtesy of J. R. Factor.)

Stringent quarantine procedures are required for animals coming into the facility and vaccination may be required, increasing the cost (Keith *et al.*, 1988). Unfortunately, there are indications that the vaccine currently available also suppresses molting, which is counterproductive. Finally, there is the problem of food. Natural products are expensive to store and deliver, but there is no commercially available prepared ration of the required quality.

Although enhancement culture has been attempted several times in Canada (Fig. 3), problems specific to each site (e.g. inappropriate temperature, disease, low salinity, low oxygen levels, and poor growth) have made it impossible to evaluate the commercial viability of this type of culture (Waddy and Aiken, 1995). However, the concept of product enhancement for selected male lobsters is valid, provided the buy–sell price differential remains attractive and the technical problems can be managed. Current efforts to increase the minimum legal size in the fishery may make product enhancement culture impractical in the future (see Miller, Chapter 5, on fishery regulations).

3. Closed-Cycle Culture

In its ideal form, aquaculture utilizes domesticated broodstock to produce seedstock and ultimately adults from which new broodstock are selected. In the process, the successive generations are genetically modified to enhance desirable traits and suppress undesirable ones (Hedgecock *et al.*, 1976). Such selective breeding is the basic tenet of agriculture. It is much less common in aquaculture, however, in which the process of domestication has been replaced by selecting wild species that produce acceptable yield without generations of selection.

Wild seedstock of *Homarus americanus* can be difficult to obtain. Fishery regulations prohibit the possession or sale of egg-bearing females and lobster larvae cannot be harvested from the ocean like penaeid larvae. To farm the American lobster intensively, one must be able to produce the larvae and apply genetic selection, which means that reproductive biology must be understood and years must be devoted to rearing successive generations under intensive culture conditions. In addition, detailed biological knowledge is needed on growth processes, water quality requirements, behavioral and social require-

FIGURE 3 Trays of individually held lobsters in the semiintensive lagoon-based system in the Magdalen Islands, Quebec. (Reprinted with permission from Waddy and Aiken, 1995.)

ments, effects and nature of stress, causes and treatments regarding disease and pathology, and the influence of sex, age, and size on all of these. Once this information is available, holding systems can be designed to take these factors into account. (See the relevant chapters in this volume by Waddy *et al.*, Chapter 10; Atema and Voigt, Chapter 13; Martin and Hose, Chapter 17.)

Holding systems developed to date for intensive lobster culture have been complex, highly capitalized, land-based operations in which seawater is pumped through a building complex containing a variety of tanks designed for the different production phases (Van Olst *et al.*, 1980). The capital and operating costs of such systems are enormous compared to the simple impoundments used for most commercially viable aquaculture operations.

II. Culture Systems and Strategies

A. *Larval and Postlarval Rearing*

Because of the long history of larval rearing for stock enhancement (see Section I,B,1), the strategies and systems for rearing larval lobsters are the most refined in the culture cycle. Lobster larvae are hardy, are easy to culture, and have a short development time (9–18 days at 20°C). They grow and survive well on live adult *Artemia* brine shrimp, and even the frozen commercial product produces a survival rate of 35–65%. (See Conklin, Chapter 16, on nutrition and ration formulation.) Larvae can be hatched and reared year-round, although survival is best during the normal spring–summer hatching period (Eagles *et al.*, 1986). However, survival does appear to be related to the degree of temperature manipulation required to induce embryos to hatch, so the improved control we now have over the time of spawning (Waddy and Aiken, 1992; Waddy *et al.*, Chapter 10) may reduce the seasonal variation in larval survival.

The first larval molt occurs just after hatching, when the prelarva, still in the egg mass, molts to a first-stage larva and leaves the female to become planktonic. Hatching larvae are screened from the outflow of the maternal female's tank each evening and stocked in specially designed larval tanks (planktonkreisels) that disperse water in an upward rotation that keeps the larvae in uniform suspension, reducing injury and cannibalism (Figs. 2 and 4) (Hughes *et al.*, 1974). In commercial application, this basic design would be modified to reduce water consumption and improve the retention of small, live foods such as brine shrimp nauplii.

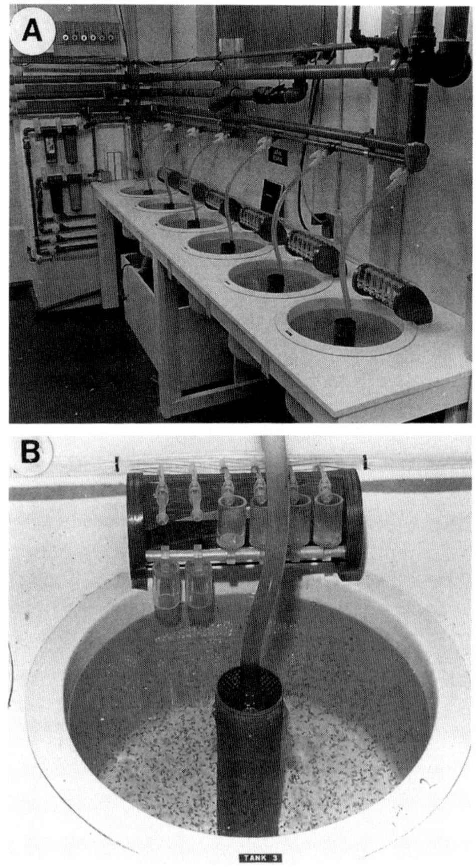

FIGURE 4 (A) Bank of planktonkreisels at the Biological Station, St. Andrews. (B) Larval pot in operation showing larvae and the automatic food dispenser. (Reprinted with permission from Waddy, 1988.)

Three additional molts occur before the lobster reaches the fourth stage and has the general appearance of an adult. Occasionally, larvae are found that are intermediate between the larval stages. These intermediate stages are often associated with unfavorable environmental conditions, such as the presence of pollutants or endocrine manipulation such as eyestalk ablation (Charmantier *et al.*, 1991).

Survival and growth of larvae depend on a variety of factors, including season, photoperiod, light intensity, temperature, salinity, water quality, stocking density, disease, type of rearing system, and quality and quantity of food. The importance of good water quality in larval rearing cannot be overemphasized. If disease-free eggs are hatched and good water quality is maintained, the likelihood of mortality from disease is remote. However, if water quality is poor, or untreated water is recirculated, bacterial and fungal fouling organisms will proliferate and mortality will be significant. A flow-through system using good-quality water is the best choice. If recirculation must be used, then effective treatment and sterilization systems are essential.

The time required to reach the fourth stage depends primarily on temperature and season of the year, varying from as little as 9 days to more than 2 months (Aiken, 1980). At low temperatures (<10°C), lobster larvae are unable to complete development and die before reaching the fourth stage (Templeman, 1936). Time to 50% metamorphosis varies dramatically with the time of year. At 20°C, development time varies between 9 and 18 days, being shortest in the winter and spring and longest in the summer and autumn (Waddy *et al.*, Chapter 10).

The effects of daylength and light intensity vary with season, and not all the relationships are well understood. In general, very long daylengths [light–dark (LD) 23:1] promote rapid development but reduce survival, whereas very short daylengths (LD 1:23) produce a high survival rate but extend development time. Larvae reared in continuous darkness develop rapidly and are almost twice the weight of larvae reared in normal daylengths, but survival is reduced (Eagles *et al.*, 1986). Until the relationships between daylength and larval development are more clearly understood, the use of near normal daylengths (LD 10:14 to LD 14:10) is recommended, as these produce reasonable survival and rate of development year-round. Less is known about the effects of light intensity, although low light intensity appears to enhance both survival and growth, so bright illumination should be avoided. (Ennis, in Chapter 3, discusses the effects of temperature and daylength on larval development.)

Food is very important. Larval lobsters require relatively large amounts of food, and inadequate feeding levels reduce survival and growth and result in a longer development time. Food is particularly critical to newly hatched larvae. Feeding should start immediately after hatching, as larvae may not recover if inadequate food is provided during the first larval stage, even if food is abundant later (Eagles *et al.*, 1986). However, all stages are responsive to the amount of food available and a suitable feeding schedule is essential, as excessive food also reduces survival and performance. An effective schedule for feeding frozen adult *Artemia* is 48 ml/day per 1000 larvae stocked, distributed among three daily feedings (Aiken and Waddy, 1989b). The feed may be provided in equal quantities each day throughout the development period or the amount can be gradually decreased over the larval period, with little reduction in survival (Eagles *et al.*, 1986).

When reared under culture conditions at 20°C, most lobster larvae molt to the fourth stage during the night. Larvae are even able to adjust the timing of the molt when the day–night cycle is reversed, so that the dark period is 12 hours out of phase with the normal cycle. Timing of the molt appears to be related to the onset of darkness (Aiken and Waddy, 1995).

The first larvae to reach the fourth stage are the largest (Eagles *et al.*, 1986) and in European lobsters these rapidly growing larvae also have the best survival (Galindo, 1985, cited in Lee and Wickins, 1992). Although it would appear that these large, rapidly growing larvae are the best ones to use for grow-out, it is not known whether they continue to perform better as juveniles.

Lobsters are still pelagic at the postlarval fourth stage and it is only during the latter half of the fourth stage that they become benthic (see Ennis, Chapter 3). The transition to benthic life is gradual, and during most of the fourth stage they exhibit both pelagic and benthic behavior, making them difficult to cope with in culture. Heavy mortality often occurs during the postlarval and early juvenile stages in culture and postlarvae are often reared individually for 8–10 weeks to minimize losses (Fig. 5) (Aiken and Waddy, 1989b). D'Abramo and Conklin (1985) suggested that postlarvae could be reared for about 4 months with minimal labor in individual perforated cubicles that would allow live brine shrimp to move throughout the unit.

B. *Juvenile and Adult Rearing*

1. Communal Rearing

Maximum survival is obtained when juvenile lobsters are held in individual containers where cannibalism and injuries due to agonistic interactions can be controlled, but individual cubicles are expensive to provide and daily feeding and maintenance are labor intensive. An economical alternative for small lobsters is communal rearing, as it reduces capital costs and requires little maintenance (Fig. 6). In such systems, juvenile lobsters are stocked and, except for feeding, ignored until harvest 2–3 months later.

Communal habitats for juvenile lobsters should provide a variety of niche sizes and utilize the three dimensions of the water column. The niches should also be accessible from two directions so that resident lobsters cannot be trapped and eaten by their tankmates. Studies with a variety of habitats, including mollusk shells, marine Astro Turf, PVC pipe sections, and corrugated fiberglass sheets, reveal that habitat configuration has a significant effect on survival and growth. Shells of the oyster (*Crassostrea virginica*), soft-shelled clam (*Mya arenaria*), and surf clam

FIGURE 5 Stage V lobster in a small individual container made of a vinyl downspout and perforated PVC. Cubicles are assembled into trays that are stacked vertically in deep tank wells. (Reprinted with permission from Waddy and Aiken, 1995.)

(*Spisula solidissima*), stacked randomly in tanks, have worked well and demonstrate principles that could be used to develop a commercial substrate.

Each habitat configuration supports a different size range of lobsters. Oyster shells support the widest range of lobster sizes (6- to 20-mm CL) due to the irregular surface of the shell, whereas the smooth valves of the soft-shelled and surf clams are effective habitats for lobsters over much narrower size ranges (5- to 11-mm CL for *Mya arenaria* and 15- to 24-mm CL for *Spisula solidissima*) (Aiken and Waddy, 1988). In addition, residency times and stocking densities vary with lobster size. In these mollusk shell habitats, lobsters of 5- to 10-mm CL are stocked at 100 per meter and harvested every 8 weeks, while lobsters 10-mm CL and larger are stocked at 50 per meter and harvested every 8 weeks. When lobsters reach 15-mm CL, the period of residency should be extended to 10–12 weeks.

Sibling lobsters in communal tanks exhibit a wide variation in growth rates, and after 2 months the largest lobster is often three times the size of the smallest (Fig. 7). When there is considerable variation in size, the larger lobsters dominate and grow well, while the small ones are intimidated, obtain less food, and suffer a greater incidence of injury and death. A well-designed habitat that provides refuge for a range of sizes will reduce this phenomenon, but harvesting and size grading at appropriate intervals have the greatest impact.

The effects of social factors on lobster growth can be seen by harvesting sibling lobsters from communal tanks, sorting them into groups of uniform size, and restocking them. Each of these "uniform" groups will display the original growth differential when subsequently harvested (Aiken and Waddy, 1988). Some that were previously "slow growers" become fast growers in a more favorable social situation, whereas others continue to grow slowly and should be considered "culls" and diverted to other markets, such as the aquarium or trinket trade.

Removal of the chelipeds (chelotomy) from lobsters smaller than 10-mm CL and of the dactyl (dactylotomy) from lobsters of 10- to 25-mm CL significantly increases survival in communal tanks (Figs. 8 and 9) (Aiken and Young-Lai, 1981). After 8 weeks in communal conditions, the survival in control tanks is only about 76%, whereas 91–94% of those without chelipeds or dactyls survive. Claws and dactyls regenerate to their original size in about three molts, so harvest times must be adjusted to molting frequency. Claw and dactyl removal has an insignificant effect on molt frequency and size increase when used on lobsters smaller than 25-mm CL; in fact, lobsters

FIGURE 6 Stacked cantilevered communal tanks for holding small juvenile lobsters. Habitat in this tank is randomly arranged valves of oyster shell (*Crassostrea virginica*). (Reprinted with permission from Waddy, 1988.)

without dactyls often have the highest growth rates. Although dactylotomy would appear to expose the animal to gaffkemia (see Section II,E), the technique has been used routinely in the St. Andrews laboratory for over 20 years without problems (D. E. Aiken and S. L. Waddy, unpublished observations).

Lobsters in communal tanks are usually fed frozen adult brine shrimp at the rate of 10 g/day for every 50 g of lobsters stocked in the tank. Once a lobster reaches 25- to 35-mm CL, it is too valuable to be held in a communal tank and should be transferred to an individual cubicle of appropriate size (Aiken and Waddy, 1978).

2. Individual Rearing

Large juveniles have sufficient value to justify the expense of individual rearing for the balance of the grow-out period. Next to water quality, the most important consideration in individual holding is container size. Molting frequency, size increase at molt, and survival are all reduced if the containers are too small, but the most dramatic effect is on size increase (Aiken and Waddy, 1978; Van Olst and Carlberg, 1978).

Circular, rectangular, and square containers have been used, but container shape seems to have no effect on growth rate (Shleser, 1974). The limiting factor is floor area, and the area required for unrestricted growth of wild lobsters has been estimated at $72 \times CL^2$. However, in a container area of only $55 \times CL^2$, the average increment of a 50-mm CL lobster is still 14%. Significant growth reduction does not occur until the container size is reduced to $15–18 \times CL^2$, at which the average increment may be only 10% (Aiken and Waddy, 1978; Van Olst *et al.*, 1980). Nelson and co-workers found a short-lived, density-dependent, growth-inhibiting, chemical factor that confounds the relationship between space and growth (Nelson and Hedgecock, 1983; Nelson *et al.*, 1983).

Lobsters have done well when kept in cubicles of 310 cm^2 for those of 25- to 39-mm CL and transferred to cubicles of 620 cm^2 when they reach 40-mm CL (Aiken and Waddy, 1989b). In this system, lobsters have grown to 63-mm CL at an average rate of 0.085 mm/day without culling, a rate sufficient to produce a 450-g animal in 30 months. Males grow slightly faster than females at all sizes (Fig. 10) and some individuals have grown as fast as 0.14 mm/day. Culling the smallest 10% increases the mean growth rate in males to 0.095 mm/day up to a size of approximately 55-mm CL (Waddy *et al.*, 1988). Above this size, even at a culling rate of 25%, the average growth rate declines to less than 0.08 mm/day. Much of this decline appears to be due to space restriction, but animal size and onset of maturity are also factors.

The ideal tank for rearing lobsters individually would be inexpensive to construct and operate and simple to maintain. It would be self-cleaning, use space in three dimensions, conserve water, and per-

FIGURE 7 Size variation of siblings reared in communal tanks. Strategies to reduce variation include periodic size grading and cheliped inactivation. (Reprinted with permission from Waddy and Aiken, 1995.)

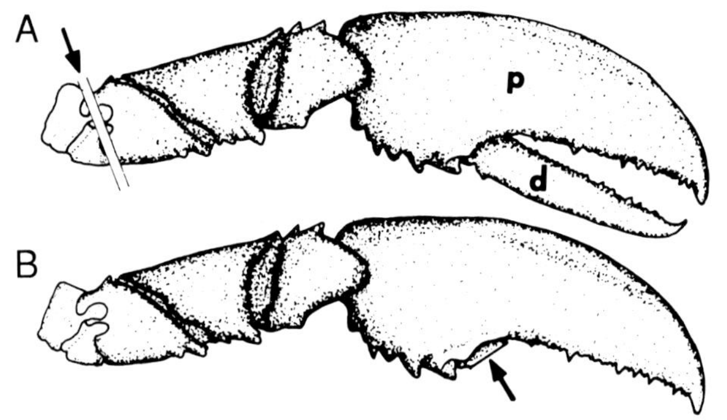

FIGURE 8 Methods of cheliped inactivation. Ventral surface of right cheliped, showing propus (p) and dactyl (d). (A) Cheliped ablation (chelotomy). (B) Dactyl ablation (dactylotomy). (From Aiken and Young-Lai, 1981.)

mit easy access to the livestock for inspection and feeding. So far, no one has successfully incorporated all of these features into a single design.

There are two basic designs: two-dimensional and three-dimensional tanks. The two-dimensional design holds lobsters in the horizontal plane and the tanks can be stacked to make use of vertical space. Cantilevered flushing tanks, cascade tanks, and "care-o-cells" fall into this category. Three-dimensional designs stack trays of cubicles vertically in the water column. Cylinders, wells, raceways, and silos are common variations, and upwelling, downwelling, or laminar flow may be used to move water past the animals (Van Olst *et al.*, 1980).

Two-dimensional designs are expensive to construct and operate, but simple to service. Three-dimensional designs are relatively inexpensive to construct and operate, but much more difficult to service. Of the two, the three-dimensional concept shows the most promise (Van Olst *et al.*, 1980), and recent designs have incorporated innovations such as buoyant, interlocking trays and mechanized tray handlers to facilitate inspection and feeding of lobsters (Figs. 11 and 12).

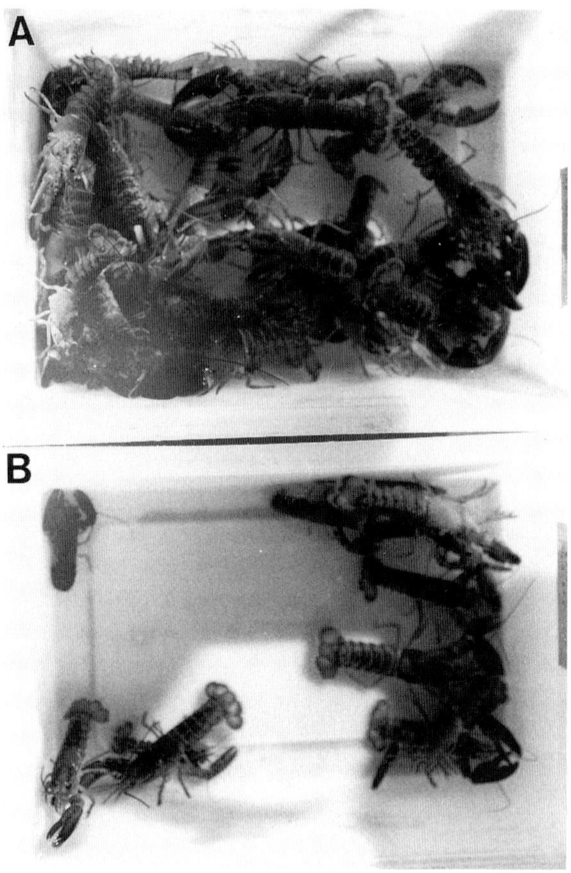

FIGURE 9 Effect of dactylotomy on survival of juvenile lobsters reared in an oyster or clamshell habitat. Initial stocking densities were identical but survival after 8 weeks in dactylotomized lobsters (A) may be more than twice that of intact lobsters (B). (Reprinted with permission from Waddy, 1988.)

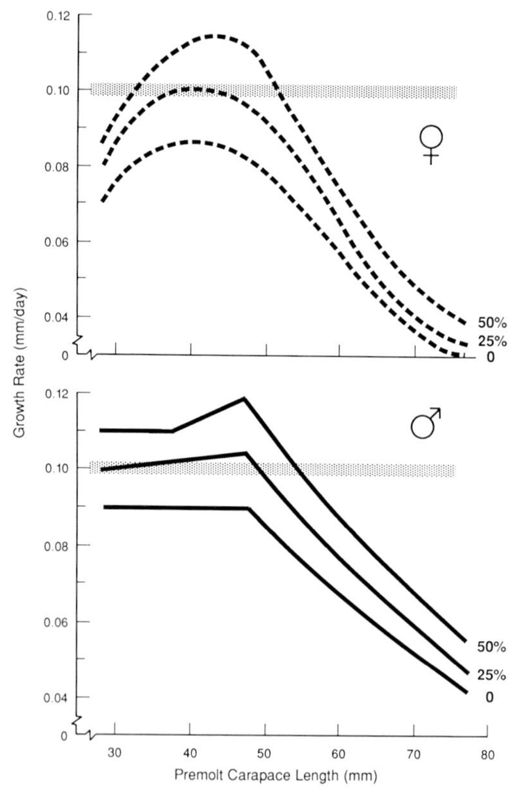

FIGURE 10 Growth rates of cultured lobsters. Even at the smallest sizes, males grow faster than females. Culling the smallest 10% increases the mean growth rate in males to 0.095 mm/day up to a size of approximately 55-mm carapace length (CL).

3. Growth Enhancement

Lobsters increase their size by molting and they must molt about 20 times to reach 500 g (the standard "market lobster" weighs approximately 1 lb, or 454 g). Under ideal conditions, a lobster grows about 15% in length and 40% in weight at each molt. The smallest lobsters molt every few days, while those that weigh several kilograms may molt only once in 4–5 years. Males grow faster than females, even at the smallest sizes, and as they approach maturity the difference is even greater (Waddy *et al.*, 1988).

Temperature is the primary regulator of growth, and molting frequency increases in proportion to temperature within the range of 6° to 24°C. The most rapid growth occurs when lobsters are held at 20–22°C year-round, with many juveniles growing at a rate of 0.1 mm/day or faster. At lower temperatures, grow-out time is extended and the normal, seasonal winter inhibition of molting may not be overcome (Aiken, 1980). However, even at 20°C, there are significant differences in the number of lobsters molting at various times of the year; molting incidence peaks in the spring and autumn and is lowest during the winter (Waddy and Aiken, 1989). In culture, the animals are able to perceive the natural photoperiod cycle and may have responded to it, so it is possible that this seasonal molting pattern could be eliminated by using an unchanging intermediate daylength (LD 12:12) (Waddy and Aiken, 1989).

Many other factors also affect growth, through either frequency of molting or size increase at molt (Aiken, 1980; see Waddy *et al.*, Chapter 10, on control of growth and reproduction). There are indications that photoperiod and light intensity are influential; juveniles reared in near-constant darkness eat more and grow faster (Bordner and Conklin, 1981), but few studies have involved photoperiod and none have elucidated its role in juvenile and adult lobsters. Quantity and quality of food are particularly important, not only for growth but also for reproduction and disease resistance. Many stress factors (e.g., social interaction, population density, perceived space, handling, habitat, and water quality) can have a negative influence on growth (Aiken and Waddy, 1986).

Many techniques have been used to accelerate growth rates in lobsters. The best known—bilateral eyestalk ablation—removes the source of the neuroendocrine hormones that control molting. Without these hormones, lobsters molt more frequently and increase more in size at each molt. However, eyestalk

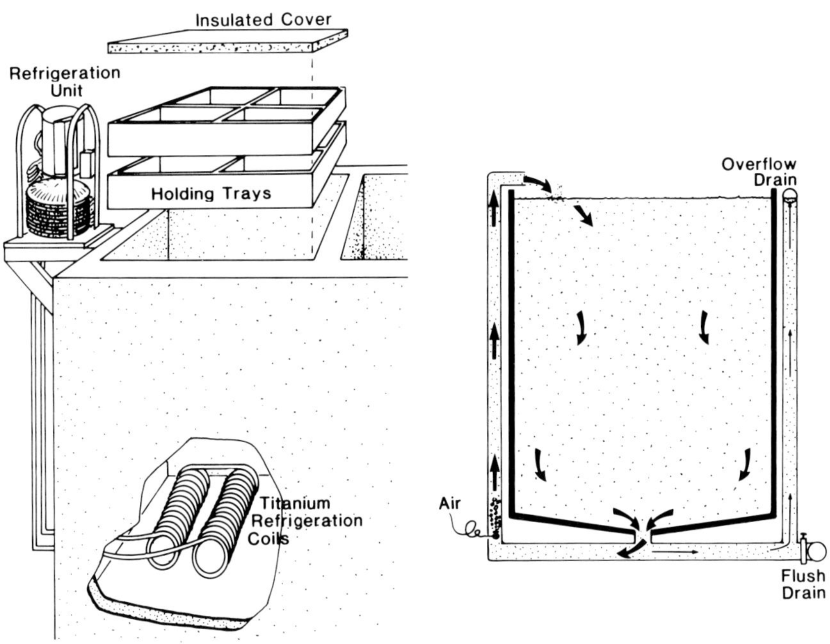

FIGURE 11 Deep tank system for holding lobsters at high density in vertically stacking trays. Water is circulated with airlift pumps and a small quantity of water is added continuously, with the amount dependent on temperature. Arrows indicate the direction and relative proportion of flow. Various configurations of compartment sizes are so that a range of lobster sizes can be held in the same system. For controlling spawning and development of embryos carried by ovigerous females, the tanks are insulated and a refrigeration system maintains the water at 2°C. (From Aiken and Waddy, 1985.)

FIGURE 12 Three-dimensional deep tank and tray of individually held lobsters at the Lobster Culture Facility in St. Andrews. (Reprinted with permission from Waddy, 1988.)

ablation produces a debilitating and frequently lethal hormonal imbalance. Meat yields are reduced (generally one-half that of an intact lobster), coloration becomes pale and unattractive, and the body becomes so malproportioned that molting is difficult. The eyestalk is the source of many neuroendocrine factors and removal of these neuroendocrine centers causes major physiological and metabolic instability. The animal becomes extremely sensitive to slight changes in temperature, salinity, and oxygen concentration, and heavy mortalities occur under conditions that would normally be tolerated. Also, postoperative mortality is very high in the spring, particularly among males, and few lobsters survive long enough to molt. There are indications that the problem of low meat yield can be alleviated with high-protein diets, but results are preliminary (Conklin and Chang, 1993).

Unilateral eyestalk ablation avoids many of the problems of bilateral ablation, but it is not nearly as effective in promoting growth. Unilaterally ablated lobsters are morphologically indistinguishable (except for the missing eye) from normal lobsters 2 years after ablation (Fig. 13). In a preliminary study, the growth of unilaterally ablated lobsters averaged 0.128 mm/day, well above the 0.096 mm/day achieved by the intact control lobsters, and after 1 year the ablated lobsters were 19% longer and 67% heavier (Peutz *et al.*, 1987). However, unilateral ablation must be reexamined, as other work has suggested that uniablated animals may not perform as well when reared communally (S. L. Waddy, unpublished data) and there may be no apparent effect on molting in large canner males (Coulombe and Motnikar, 1989).

Ecdysteroid (molting hormone) treatments to accelerate growth have been attempted, but they are costly and produce mixed results ranging from a protracted intermolt to an abnormal premolt culminating in death. The most active of the molting hormones is 20-hydroxyecdysone, but the dose is critical. An overdose will cause rapid, but abnormal, premolt development and death. However, the effective dose is difficult to determine because it varies dramatically with season, temperature, reproductive condition, and molt stage (Aiken, 1980; Aiken and Waddy, 1975). 20-Hydroxyecdysone seems to be involved more in the stimulation of cuticle synthesis than with the induction of premolt, and a dose that is adequate to overcome endogenous molt inhibition and induce premolt during a nonmolting period (autumn and winter) will often cause abnormal development. The lethal effects can usually be avoided by using lower doses preceded by eyestalk ablation, by using a slow-

FIGURE 13 Two-year-old lobster that underwent unilateral eyestalk ablation at the sixth stage. Note the normal body proportions and dark coloration. (Reprinted with permission from Waddy, 1988.)

release form of the hormone, or by pretreatment with ecdysone (Aiken, 1980). (Waddy *et al.*, in Chapter 10, provide details of the molt cycle.)

Other hormones may have potential for aquaculture, but results are preliminary. For example, larval and postlarval lobsters injected with human somatotropin grew 10–20% faster than controls, and the enhancing effect lasted for several molts (Charmantier *et al.*, 1989). Recombinant DNA techniques also have potential for promoting rapid growth, but considerably more research is required (Conklin and Chang, 1993).

C. Broodstock Management

1. Control of Spawning

A commercial culture facility requires a steady and reliable supply of larvae throughout the year. Fishery regulations in Canada and the United States prohibit the possession or sale of ovigerous lobsters and there is strong opposition to removing them from the fishery, so seed for culture must be obtained from females that spawn in captivity.

Under normal seawater temperatures, female lobsters spawn only once in 2 years and most—except the smallest and the largest—have an alternate-year molt and reproductive cycle in which each spawning is separated by a molt. Spawning usually occurs between June and September and the eggs, attached to the swimmerets, are carried for 9–11 months, during which the female aerates and cleans them of fouling organisms (Aiken and Waddy, 1980). (See Talbot and Helluy, Chapter 9, and Waddy *et al.*, Chapter 10, on reproduction.) To maintain a natural reproductive cycle with normal spawning frequency and reliable egg attachment, female lobsters in captivity must be maintained on a temperature–photoperiod cycle similar to that of the natural habitat. Water temperature must rise to at least 10–12°C during the summer and autumn and decline to 5°C or lower for several months in the winter. When these conditions are not maintained, reproductive failure is common. For example, when females are held in seawater temperatures >8–9°C year-round, spawning and molting cycles become desynchronized and spawning incidence is reduced (Waddy and Aiken, 1991).

Preovigerous lobsters required for seedstock can be selected from commercial landings in the autumn, winter, or spring by noting the degree of pleopod cement gland development (Aiken and Waddy, 1982). These mature females will spawn the following summer (>90%) when held under normal, seasonally changing seawater temperatures (in July and early August in St. Andrews, New Brunswick). To extend the spawning season through the late summer and autumn, the temperature is not allowed to increase in the spring. Females are held in very cold water (1–2°C) throughout the winter, spring, and summer and are transferred to a temperature of 10–12°C 3 weeks ahead of the required spawning time. About 50–60% of the females will spawn after transfer to a warmer temperature (Fig. 14). Spawning can be delayed by almost 5 months in this way (Waddy and Aiken, 1992).

It is also possible to induce precocious spawning. From early October through June, preovigerous females spawn within 1–4 months after being exposed to temperature and photoperiod manipulation. The time required to induce spawning varies with the season of the year and the temperature–daylength regime. The most efficient method is to obtain females from the wild in the winter or spring (January through May) and hold them under temperature and photoperiod conditions characteristic of the western North Atlantic (45° N) until 4–6 weeks prior to the desired spawning time. At that point, exposure to warm temperatures (10–15°C) and short photophases (8–13 hours of light) will produce rapid ovarian maturation and spawning. With these techniques, spawning is easily scheduled to occur as required from February through June (Waddy and Aiken, 1992).

Attempts to induce spawning in August and September, when females are newly molted, have not been successful. But shortly after the autumnal equinox (21 September), the ovary becomes competent to respond to warm temperatures if long daylengths (LD 16:8) are available (Waddy and Aiken, 1992). However, spawning induction in autumn takes 3–4 months instead of the 4–6 weeks required in the winter and spring, so there is little advantage to this approach.

2. Control of Hatching

The flexibility of the broodstock system can be further increased by manipulating the time of hatching and by using a variety of sizes of broodstock females. Females from different areas mature at different sizes and a wide range of female sizes is available (approximately 65- to 140-mm CL). A lobster farm producing 1 million 500-g lobsters per year would require between 200 and 800 ovigerous females, depending on the size of female used. These figures include 20% more lobsters than are actually required, in order to compensate for losses that might occur due to environmental manipulation during the year the females are spawning and brooding their eggs (Aiken and Waddy, 1985).

The egg development rate varies with temperature between 7° and 25°C, permitting considerable latitude in the control of hatching times (Perkins, 1972). (See Talbot and Helluy, Chapter 9, on embryonic development.) Females can be placed on a temperature schedule that will yield larvae for 15 months. Development can be easily retarded in newly spawned eggs, but it is difficult to retard development in eggs that are close to hatching. The latter will continue to develop even when held below a temperature of 5°C. Previously, the period from October through December was the most difficult in which to obtain larvae because maximum acceleration or retardation was required (Aiken and Waddy, 1985). However, with the control that is now possible over spawning time, this problem should be eliminated.

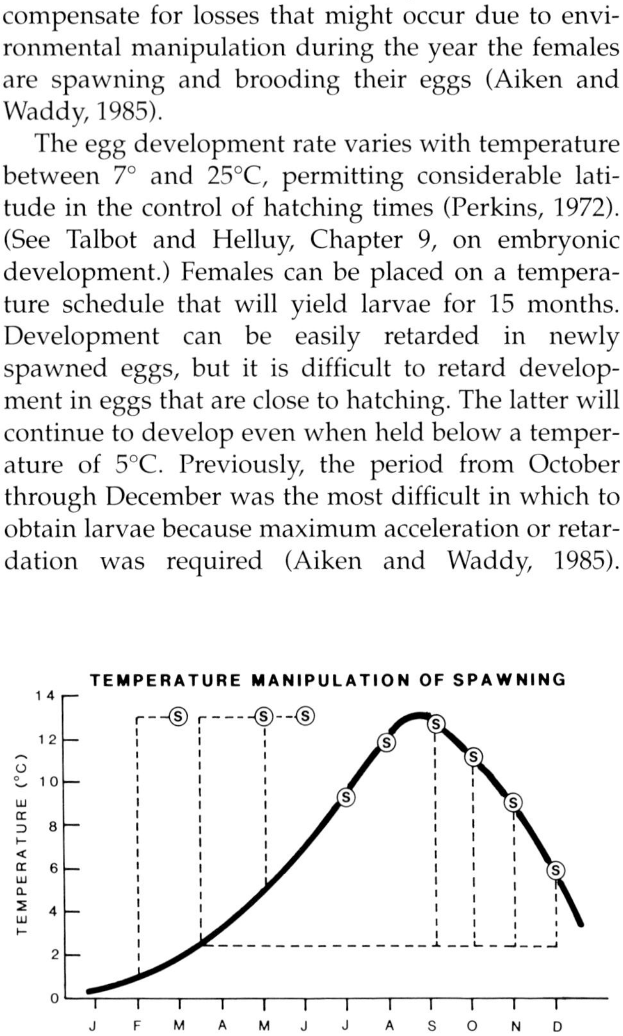

FIGURE 14 Control of spawning in the American lobster. Females normally spawning between late June and early August can be induced to spawn throughout the remainder of the year. To obtain spawning from February through early June, preovigerous lobsters are transferred to warm temperatures (10–15°C) and short daylengths (LD 8:16). Spawning occurs 4–6 weeks later. To delay spawning so that it occurs between late August and December, preovigerous females are maintained at a constant low temperature (2°C) throughout the spring, summer, and autumn. Spawning occurs 3–4 weeks after transfer to 10–12°C. (From Waddy and Aiken, 1992.)

Larvae must be produced for the culture system year-round to ensure a uniform flow of animals through the system. Larvae can be obtained throughout the year by manipulating the spawning cycle and controlling the hatching time. Once females have spawned, both acceleration and retardation of embryonic development by temperature manipulation can be employed to obtain hatching throughout the year.

3. Broodstock Facilities

Broodstock females must be provided with sufficient tank area and water depth to molt and spawn (Fig. 15). Those up to 100-mm CL can be accommodated in tanks 30–40 cm on each side and 25–40 cm deep (Aiken and Waddy, 1985). Spawning and successful egg attachment can occur as long as there is sufficient water to cover the lobster when she assumes the inverted spawning posture (see Talbot and Helluy, Chapter 9). Successful attachment has been achieved in only 14 cm of water (Aiken and Waddy, 1980), but a minimum of 25 cm is recommended.

Broodstock tanks should have glass or plexiglass front panels so that molting, mating, spawning, and hatching can be observed without handling or disturbing the lobsters. Water flowing through the tanks should move along the bottom, and the tanks should be self-cleaning and easily maintained (Aiken and Waddy, 1985). However, laminar flow along the bottom can cause stagnation and thermal stratification higher in the water column, conditions that could be detrimental if the water supply were interrupted. This problem can be eliminated by fitting each compartment with a simple airlift pump, an arrangement that also protects valuable broodstock in the event of an extended interruption in water supply.

The water supply system for maintaining broodstock must permit a reasonable degree of temperature control and manipulation, because year-round larval production requires that broodstock females be held in a variety of temperatures. For normal ovary development and spawning, the annual temperature must range from a low of 0–5°C to a high of 12–20°C. To retard embryonic development, the temperature must be held between 1° and 3°C. Maximum acceleration of embryonic development requires temperatures of 20–22°C.

Hatching may occur at any time of the day, but larvae are usually retained on the female until dusk and released by periods of pleopod fanning. Not all of the eggs on a female hatch on the same day; a female

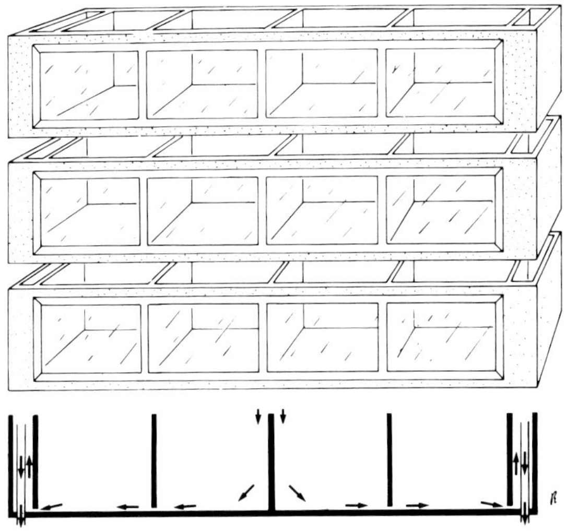

FIGURE 15 Brood tanks for holding ovigerous females. Various sizes and compartment configurations are used, but all incorporate the basic features of a clear acrylic front panel, laminar bottom flow (bottom of figure), and a scouring–flushing action activated by the removable standpipe. (From Aiken and Waddy, 1985.)

may release larvae every night for up to 3 weeks. The duration of the hatching period varies according to temperature, the number of eggs, and the degree of developmental acceleration or retardation the embryos have experienced. In a culture facility, hatchling larvae float to the surface and are carried by outflowing water to a collecting device, where they can be segregated for distribution to the planktonkreisels.

4. Egg Loss

To survive the months of incubation, embryos must attach securely to the pleopods; remain free from disease, predation, and exposure to water of poor quality; and have sufficient nutrient reserves. Eggs usually attach securely in wild females and in cultured females that have received proper environmental conditioning (Waddy and Aiken, 1991). However, entire egg masses can be lost during incubation. A high incidence of egg loss has been reported for both wild and cultured lobsters maintained in California and France (Bertran and Lorec, 1986; Hedgecock, 1983; Talbot and Harper, 1984; Talbot *et al.*, 1984) and cultured lobsters tend to lose more eggs than wild ones.

Studies indicate that the egg stalks of lobsters held at a consistently high temperature tend to be morphologically abnormal, causing poor attachment and retention of eggs. Although the cause of the problem is not known, environmental conditions, such as temperature, may be a factor (Talbot and Harper, 1984). Data from the St. Andrews laboratory also implicate temperature history as an important factor in egg attachment and hatching success; cultured females taken from constant 20°C and maintained for three or more consecutive spawnings in our natural water temperature cycles incubate and hatch orders of magnitude more eggs than cultured females that are spawning eggs for only the first or second time at seasonally changing temperatures (D. E. Aiken and S. L. Waddy, unpublished data).

Other negative influences on egg attachment and hatching success include nemertean predation, fungal and bacterial diseases, water quality, and excessive handling and crowding (Aiken and Waddy, 1986; Talbot and Helluy, Chapter 9).

5. Control of Mating

In order for a female to continue producing fertile eggs, she must be mated after each molt. Copulation usually occurs within hours of the female molt if a male is introduced to the female's tank. (See Talbot and Helluy, Chapter 9, and Atema and Voigt, Chapter 13, on reproductive and mating behavior.) The male deposits a spermatophore in the seminal receptacle of the female, where it remains until spawning (sperm viability is retained for several years). It was previously thought that unless mating occurred when the female molted, the opportunity for mating would be lost and the subsequent brood would be infertile. There are now two solutions to the problem: artificial insemination and intermolt mating.

Artificial insemination (AI) has considerable potential, as males in a culture facility can become impotent when held in isolation for long periods and sperm can then only be obtained with electroejaculation. However, results with AI have been disappointing. Although fertilization can be obtained, the eggs are often lost before hatching (Aiken *et al.*, 1984; Talbot *et al.*, 1986; Waddy and Aiken, 1985; Talbot and Helluy, Chapter 9). However, AI should be reexamined, as most of the work on cultured females was done before it was recognized that some factors cause problems of egg retention and that male lobsters have cycles of potency related to molt stage (Waddy and Aiken, 1990).

The second option—intermolt mating—has eliminated most problems of mating under confined conditions (Waddy and Aiken, 1990). Females that are not inseminated at their previous molt will mate prior to spawning if held with males. The most expedient way to ensure mating is to hold males and females together without shelter. Through a mechanism not well understood, mature but uninseminated females are attractive to males and copulation and insemination invariably occur. The only requirement for intermolt mating appears to be a potent male of suitable size (a male that is significantly smaller or larger has difficulty copulating with a female). Most females mated during intermolt retain their eggs as efficiently as females mated at molt.

6. Control of Reproduction

Progress in the development of cultured stocks and selection for desirable traits depends on the degree to which reproduction can be controlled. (See Waddy *et al.*, Chapter 10, for a discussion of factors controlling growth and reproduction.) The Massachusetts State Lobster Hatchery and Research Station is credited with the first success at closing the reproductive cycle of the American lobster (J. T. Hughes, personal communication). Researchers there found, as did others, that obtaining seed from wild or cultured stock is relatively easy if the seawater temperature approximates that of a nearshore lobster habitat. It is more of a challenge to produce eggs, sperm, and larvae from lobsters grown to maturity at temperatures of 20–22°C. Attempts to develop cultured broodstock have been hampered by the adverse effects of high temperature on both egg and sperm production. Females reared at 20°C attach few of the eggs they extrude and most of these are lost before embryonic development is completed (Waddy and Aiken, 1991). Cultured males often fail to produce spermatophores, or produce spermatophores without viable sperm (Aiken *et al.*, 1984). Although the reasons for the low fecundity of cultured lobsters are not known, the problem can be overcome by altering the temperature regime.

In order to environmentally condition cultured lobsters and improve egg production and attachment, as well as sperm and spermatophore production (Waddy and Aiken, 1991), animals are held from December through April at local winter seawater temperatures of 0–5°C. Spawning incidence, egg attachment, and male fecundity increase in proportion to the time spent at normal temperatures during the winter period. Unfortunately, 3–4 years (one or two spawning cycles) of normal summer–winter temperature cycles are required before most female lobsters spawn predictably; even more time is required (three or four spawning cycles) before egg attachment is equivalent to that in wild stock (Fig. 16).

D. Nutrition

Lobsters eat a variety of natural foods, including algae, fish, crustaceans, mollusks, and other invertebrates. The best growth rates in culture systems have been obtained with a varied diet of fresh and fresh-frozen marine fish, mollusks, crustaceans, and macroalgae. These foods produce normal coloration and good survival, growth, and reproduction. Unfortunately, fresh and frozen natural foods are expensive, difficult to obtain on a consistent basis, and impractical to store and feed in large-scale commercial culture. The storage problem alone would be overwhelming; it has been estimated that a facility producing 1 million market lobsters annually would require 24 mt of natural foods each day (Conklin *et al.*, 1983).

The solution is a prepared ration that would be attractive to the animal, resist fragmentation and leaching of water-soluble nutrients, produce good growth and survival, store and dispense easily, and not foul water systems. Researchers worked for almost two decades to develop such a ration for lobsters, but none evaluated so far can match the growth achieved with natural foods. Growth rates greater than 0.1 mm/day have been obtained with natural foods, compared with 0.05–0.07 mm/day with artificial diets (Conklin *et al.*, 1980; D'Abramo and Conklin, 1985). The lack of an adequate and cost-effective formulated diet is one of the major impediments to lobster culture today. (See Conklin, Chapter 16, on digestion, nutrition, and ration formulation.)

Brine shrimp (*Artemia* spp.) are an excellent food for lobsters from hatching through 6 months of age, producing better rates of survival and growth than other foods that have been used (e.g., Daniel *et al.*, 1985). However, as juvenile lobsters grow, brine shrimp no longer support maximum growth, pre-

FIGURE 16 Effect of environmental conditioning on egg attachment in cultured broodstock. (A) Good egg attachment after several years of temperature conditioning. (B) Few eggs are retained by a female after only 1 year at normal temperatures. (Reprinted with permission from Waddy, 1988.)

sumably because more energy is expended to capture small prey than is derived from eating it (D'Abramo and Conklin, 1985).

Lobsters are slow, intermittent feeders. Rapidly growing lobsters consume about 10% of their body weight each day, while adults consume only about 1% of their body weight (D'Abramo and Conklin, 1985). Environmental conditions affect food consumption and photoperiod may have a strong influence on feeding and food conversion; consumption increases as the length of the dark period increases. Interestingly, feeding twice daily does not significantly increase consumption (D'Abramo and Conklin, 1985).

E. Disease Management

Intensive culture has increased the incidence and diversity of disease in lobsters. Contributing factors are warm temperatures, stress, poor water quality, and inadequate nutrition. Fortunately, lobsters are hardy and when water quality and nutrition are good, disease is rarely a problem.

1. Egg and Larval Diseases

Eggs of broodstock females should be examined for three afflictions that can destroy eggs and spread throughout the system. Two are microbial diseases that can be transferred from egg to larva and are generally fatal unless controlled. Microbial epibionts are not usually a problem unless water quality is poor (Sindermann and Lightner, 1988). When flow-through seawater is used, disease outbreaks among eggs and larvae are rare.

Leucothrix mucor, a filamentous bacterium, is common in the marine environment and proliferates whenever nutrients are present in the water. *L. mucor* infection can slow embryonic development, kill the eggs or larvae, or contaminate larvae that hatch from the infected egg mass. A dip of the egg mass in malachite green or an iodophor such as Wescodyne will control this organism (Fisher *et al.*, 1978; Aiken and Waddy, 1985). The filaments of *L. mucor* interfere with larval respiration, feeding, and molting.

The phycomycetous fungus *Lagenidium* sp. can infect eggs but is a more serious problem in larvae. This fungus can destroy larval tissues within 24–48 hours and kill 90% of exposed larvae within 1 week. Infected eggs and larvae can be treated with malachite green. Live brine shrimp can also be infected and can serve as a source of infection for lobster larvae (Fisher *et al.*, 1978).

No treatment is consistently effective against these diseases. Antibiotics may destroy the function of biological filters and may actually enhance the growth of fungi or resistant strains of bacteria in the culture system. For example, streptomycin used to counteract *Leucothrix mucor* can enhance the growth of *Lagenidium* spp. (Fisher *et al.*, 1978).

The third problem is the nemertean *Pseudocarcinonemertes homari*, an egg predator that can destroy a lobster egg mass. This nemertean is so hardy, prolific, and destructive that great care should be taken to prevent its introduction to a culture facility. It is mobile and, because of sticky mucoid secretions, can attach to gloves, hands, brushes, and similar objects and transfer to other tanks and lobsters in the facility. Nemerteans are capable of rapid proliferation and ultimate destruction of all the eggs on a lobster (Aiken *et al.*, 1985). Talbot and Helluy (Chapter 9) provide additional discussion of the various sources of egg loss in *Homarus americanus*.

2. Juvenile and Adult Diseases

Diseases in juveniles are usually related to poor nutrition or water quality. Two fungi attack juveniles, the phycomycetous fungus *Haliphthoros milfordensis* and the imperfect fungus *Fusarium* sp. Both invade the tissues and cause melanization at the sites of infection. This usually occurs at the bases of the walking legs, the gills, and the interior of the branchial chamber, the latter causing the so-called "black gill" diseases.

Although fairly common in small lobsters, *Haliphthoros milfordensis* is usually not considered a problem for large lobsters (Sindermann and Lightner, 1988). However, mature males in the St. Andrews culture system have fallen prey to this fungus. Female broodstock have not been affected, possibly because of the better husbandry that they receive. The condition is difficult to detect because only the gills and the interior of the branchial chamber are affected. In small juveniles, *H. milfordensis* infections have been treated successfully with malachite green, but there is no known treatment for *Fusarium* sp. infections.

Lobsters in improperly designed facilities can be killed by the bacterial blood disease gaffkemia ("red tail"). Although gaffkemia is a serious problem in the storage and shipping industry, its incidence in a culture operation should be negligible as long as stress is minimized and care is taken to screen animals obtained from other facilities. A gaffkemia vaccine is available, but is best used prior to an outbreak (Keith *et al.*, 1988). Medicated feeds containing oxytetracycline have been used for both prevention and treatment of gaffkemia (Bayer and Daniel, 1987; Huang and Bayer, 1989). Its use has raised concerns, however, as it is available over the counter and lobsters can be marketed without inspection or determination of drug residues (J. E. Stewart and J. W. Cornick, Department of Fisheries and Oceans, Halifax, personal communication). Because of mounting concerns about the widespread prophylactic use of antibiotics in aquaculture and the months of withdrawal that may be required when animals are held at low temperatures (Armstrong, 1993; Brown, 1989), experts should be consulted for both diagnosis and treatment methods.

Shell disease is another condition that is usually related to poor water quality and is relatively common in lobsters held in tidal pounds. It is caused by chitinolytic bacteria and fungi that erode the shell, causing a pitted appearance (see Waddy *et al.*, Chapter 10). In severe cases, the shell can be penetrated, providing a route for secondary invaders. It is usually not a serious problem in lobster culture, and even when some animals are infected, it does not spread through the system if the lobsters are healthy and water quality is maintained. Iodophors appear to be effective in controlling the disease. It should not become a serious problem in a culture facility, and rapidly growing lobsters usually molt frequently enough to prevent the development of lesions deep enough to erode through the shell.

Occasional problems have been caused by a blood ciliate (*Mugardia* sp.), which destroys the blood cells of lobsters of all sizes (Aiken *et al.*, 1973; Shelburne and Bean, 1991). This ciliate appears to be a facultative parasite that is usually free-living in seawater. For some reason, it occasionally infects lobsters and can be present in wild lobsters brought into the culture facility. However, it is easily contained, has not affected cultured stock, and is usually a problem only at low temperatures (Aiken and Waddy, 1986).

A low level of mortality can occur from a disease of unknown origin that causes necrosis and melanization of the antennal glands ("kidneys"), destroying osmoregulatory function and causing death. This disease also occurs in impounded wild lobsters (Aiken and Waddy, 1986).

Occasional mortalities due to a yeast of the *Candida* type in the hepatopancreas have been recorded at the St. Andrews laboratory, and a *Vibrio* bacterium has been isolated from dead lobsters in California. *Vibrio* sp. is a facultative bacterium—a normal component of the marine bacterial flora that appears to infect lobsters only when they have been stressed (D'Abramo and Conklin, 1985).

F. Domestication

Strain improvement through selective breeding requires that the most desirable cultured animals become the broodstock for subsequent generations. Selective breeding was utilized at the St. Andrews lobster culture facility for almost 15 years. Part of the selection process involved the development of red, white, and blue "colormorphs"—distinctive color variants that occur naturally, but are rare (Fig. 17). Colored strains have also been developed at the Massachusetts State Lobster Hatchery and Research Station; at Fort Pond Bay, Montauk, New York; and at the Darling Marine Center, University of Maine, in Walpole. At Montauk, Anthony D'Agostino concentrated on the blue strain (Porterfield, 1982), whereas the orange–red strain has received the most attention at St. Andrews because it appears to be the most common, most easily recognized, and most inclined to breed true. At St. Andrews, this strain was carried to the third generation before the lobster culture facility was shut down.

Although the blue colormorphs are extremely attractive, both parents must be blue in order to produce blue offspring. Because of the rarity of these animals, inbreeding and loss of genetic variability could become a problem. The orange–red strain, on the other hand, will yield red progeny when only one parent is red.

Color variants have considerable potential as a cultivated product because they could be easily distinguished from the traditional wild product—an important consideration in those areas where fishery regulations prohibit the possession, importation, or sale of lobsters smaller than the minimum legal size established for the capture fishery. Distinctive color strains offer marketing opportunities (e.g., brand identity or distinction from the wild product) that are not available with normally colored animals, and the

FIGURE 17 Second- and third-generation blue- (left) and red-colored cultured juvenile lobsters reared at St. Andrews.

red is particularly attractive because many consumers consider red to be the normal color of a lobster.

III. Water Management

Most water supplies available to the culturist require some degree of conditioning, recirculation, and reconditioning. Water management is the manipulation of water quality to achieve conditions that approximate the ideal. This also requires the use of monitoring and control systems to ensure that established conditions are maintained or, if not maintained, at least reported to persons who can take corrective action.

A. Environmental Requirements

Homarus americanus can tolerate a wide range of environmental conditions. They survive well between 0° and 25°C, and if properly acclimated can withstand brief exposure to −1° to 31°C (McLeese, 1956). These lobsters are reasonably tolerant of low levels of dissolved oxygen and reduced salinity, although tolerance declines with increasing temperature (Table II in Van Olst *et al.*, 1980). (See McMahon, Chapter 18, on the physiology of *H. americanus*.)

When heated water is used, precautions must be taken to prevent gas supersaturation. When cold water is saturated with oxygen and nitrogen, it becomes supersaturated when warmed unless the excess gas is driven off. A lobster in supersaturated water develops gas bubbles in the hemolymph, which block blood flow through the gills and cause suffocation. Unless physical damage has occurred, the effects of supersaturation can be successfully reversed by transferring the animal to cold water.

Nitrogenous wastes can be a serious problem in intensive culture, especially when recirculation is used. Uneaten food, decomposing organic matter, and excretion from the lobsters and fouling organisms all contribute to the problem. Of the nitrogenous wastes, ammonia is the limiting factor, while nitrate and nitrite are much less toxic. The term *ammonia* usually refers to both un-ionized ammonia (NH_3) and its conjugate acid (NH_4^+), but un-ionized ammonia is the more toxic compound. In solution, they exist in equilibrium. Even sublethal exposure to ammonia is a problem, as functions such as osmoregulatory ability can be impaired and the detrimental effects can last for months (Young-Lai *et al.*, 1991). Alkaline conditions, high temperature, and low oxygen concentration exacerbate ammonia toxicity. First-stage larvae are the least tolerant of ammonia, and tolerance increases with developmental stage (Delistraty *et al.*, 1977; Young-Lai *et al.*, 1991). Ammonia excretion rates change with feeding levels, molt cycle, and size of the animal, making it difficult to determine minimal flow rates to avoid problems (Van Olst *et al.*, 1980).

B. Recirculation versus Flow-through

If water supply is unlimited and water temperature is optimum, flow-through (single pass) is the system of choice. Water is drawn in, pumped once through the system, and then discharged. If intake and discharge are properly isolated, a flow-through system provides minimum opportunity for disease or other problems to develop and requirements for water quality monitoring and treatment are therefore minimized.

If, however, the supply of incoming water is inadequate, or the temperature differs significantly from the requirements, then a portion of the water supply must be recirculated. The two combined—the incoming water and the recirculated water—should exceed by a comfortable margin the requirements for oxygen and nitrogenous wastes (Van Olst *et al.*, 1980). Likewise, if the water temperature required in the system is considerably different from that available from the source, recirculation can be used to ensure that heating or cooling costs are manageable. Because some of the system water is reused, some form of water treatment or reconditioning is necessary. In extreme cases, all of the water used in a facility might have to be recirculated, making the reconditioning capability a significant part of the operation.

C. Thermal Effluents

Lobsters held at water temperatures common to their native environment grow far too slowly to support commercial culture. Because the growth rate increases with temperature, a range of 20° to 22°C allows production of 0.5-kg lobsters in 2–3 years. In other words, 20–22°C water is essential to the economic viability of aquaculture. These temperatures, however, are difficult and expensive to maintain in northern temperate regions.

Conversely, low winter temperature is required for the proper synchronization of reproduction. With current knowledge, it is not possible to maintain broodstock at 20–22°C year-round and obtain the required level of successful spawning and egg attachment. Consequently, both low and high temperatures are necessary during the culture cycle. In Atlantic Canada and the northeastern United States, low water temperature is readily available, but high temperature is not. Thermal effluents are a potential source of elevated water temperature and two sources offer promise: power plants and heat pumps. Studies of the survival and growth of lobsters in power plant effluent have shown that the cooling effluent from power generation facilities can be safely used to grow lobsters for human consumption (reviewed by Van Olst *et al.*, 1980).

In certain tropical and subtropical areas, water at a consistently low temperature can be pumped from the ocean depths and combined with surface water to obtain an appropriate range of temperatures. At Hawaii's Natural Energy Laboratory, for example, water at a consistent 6°C from a depth of 600 m can be combined with surface water at 28°C to obtain the full range of culture temperatures.

Lobsters taken in the commercial fishery are often stored before shipment to market. Their survival during storage and subsequent shipment is greatly enhanced if they are held in very cold water (below 5°C). Large marine heat pumps used to chill seawater to these storage temperatures generate an effluent that is significantly warmer than the local seawater—which might be profitably used in an adjacent grow-out or product enhancement facility.

D. Environmental Monitoring

Water management involves a great many mechanical and electrical components, the malfunction of which can result in lethal consequences for lobsters. An environmental monitoring system is therefore essential to the successful operation of a culture facility. The complexity of the monitoring system depends on the nature of the water management system. If the water is simply pumped through the facility and discharged, it may be sufficient to monitor only a few basic parameters. If recirculation is employed, the monitoring requirements become much more complex. Whatever the system, it must be capable of alerting maintenance personnel whenever something goes wrong and it must be thoroughly tested at regular and frequent intervals. The ultimate failure is a failure in the system-monitoring device.

E. Backup Systems

Once the monitoring system detects a failure, the problem must be corrected immediately. This usually means repair or replacement of some component of the water distribution system, even though failures tend to occur in conditions that are unfavorable for repair or replacement. If the failure involves critical elements of the system—distribution pumps, main water lines, or main electrical supply—the result can be a major financial loss. One principle governs these situations: critical pumps, water lines, and power supplies must be duplicated to ensure that failures cause only brief interruptions.

IV. Economics and Marketing

A. Marketing Strategies

Homarus americanus is an attractive culture prospect because of its high value and well-established worldwide market. Demand for this luxury seafood appears to be almost insatiable. However, this demand is based on a minimum weight of 450–500 g, the size most commonly available from the traditional capture fishery. A 500-g lobster is not superior in quality to one that is half that size or smaller; in fact, the premium product at the present time is actually a larger lobster that weighs 1 kg or more. The 500-g "standard" is based on minimum legal size regulations in the fishery and is therefore the size to which the public is presently accustomed. The cultured product could be any size that the consumer would accept, and with consumers becoming familiar with such small crustaceans as the Norway lobster (scampi, *Nephrops norvegicus*), and freshwater prawns and crayfish, a smaller American lobster should be readily accepted.

The growth rate begins to decline at approximately 250 g and a disproportionate amount of time is required to double the weight to 500 g. Reducing the market size would have a considerable effect on aquaculture production costs, since space requirements, capital costs, feeding costs, inventory shrinkage, maintenance requirements, and grow-out time would be reduced (the latter, down to 14–18 months from the 27–30 months now required to produce a 450-g product) (Waddy, 1988).

There are other benefits to creating a market for small, farmed lobsters. In some areas, traditional fishers are concerned about competition from lobster farming, but there would be little direct competition between the two if different markets are targeted for noticeably different products. The minimum size of the wild product is being gradually increased to enhance natural production and increase the economic return from the fishery. A 250-g product from the culture industry is below the minimum size at maturity in most populations and is smaller than can be legally landed from the fishery. Thus, the cultivated product and the captured product could be directed toward different markets and would be complementary rather then competitive.

B. Novelty Products and Colormorphs

Cultured lobsters could be marketed at many sizes and in several ways. In a commercial culture system, many lobsters will be culled from production because of poor growth. Two markets exist for these small, culled lobsters: the saltwater aquarium trade and the tourist souvenir market. The color variants are particularly desirable in the aquarium market and would therefore command a much higher price than when sold as food (Fig. 18). Although a recent request to sell small (illegally short), colored lobsters for the aquarium market was denied by the State of Maine because of objections from commercial lobster fishers, this might be a difficult position to sustain against a determined marketing effort.

A variety of tourist products have been developed from small lobsters (e.g. jewelry and collages), and the cast shells are popular when dried, arranged, and painted. A few years ago, shells were selling for $2.00 Canadian each. Since a lobster can molt 20 times or more before reaching 500 g, there is potential for considerable revenue.

FIGURE 18 This red/normal bicolored lobster represents only one of the many bicolored, spotted, and mottled color patterns that occur in the American lobster. Small specimens of such unusual lobsters would be valuable in the saltwater aquarium trade. (Reprinted with permission from Waddy, 1988.)

V. Summary

Many of the biological and engineering problems of the culture of the American lobster, *Homarus americanus*, are under control, but problems remain in the areas of nutrition, food technology, disease control, and grow-out systems. Knowledge is adequate of both larval rearing and broodstock management to produce larvae year-round on a strict schedule; however, an inexpensive ration for larvae is needed. Communal culture is a viable option for reducing costs during the early and middle juvenile stages, but a suitable substrate must be designed. Current substrates are effective, but inconvenient to clean and harvest. New designs are also required for individual holding. Three-dimensional tanks utilizing stacked trays work well for large juveniles, but are costly.

Available evidence indicates that the concept of producing a 450- to 500-g product should be abandoned in favor of a more economical, 200- to 250-g animal. However, the cost of producing even this size will be prohibitive until innovative designs reduce the high cost of constructing individual holding systems and heating seawater to the optimal 20–22°C. In addition, an inexpensive and effective formulated diet is still required. Growth rates of lobsters reared on available formulated diets are only about one half or two thirds of those in lobsters fed natural foods.

Even though large-scale commercial lobster culture is not a reality, *Homarus americanus* has as much culture potential as currently popular species such as halibut. Improved techniques for stock enhancement, combined with renewed interest in parts of Europe, make enhancement a more attractive option than in years past. The remaining culture problems appear tractable, but the legacy of the 1970s has put commercial lobster aquaculture in limbo—and it is likely to remain there at least until the surprisingly productive wild fishery declines.

References

Aiken, D. E. (1980). Molting and growth. *In* "The Biology and Management of Lobsters" (J. S. Cobb and B. F. Phillips, eds.), Vol. 1, pp. 91–163. Academic Press, New York.

Aiken, D. E., and Waddy, S. L. (1975). Temperature increase can cause hyperecdysonism in American lobsters (*Homarus americanus*) injected with ecdysterone. *J. Fish. Res. Board Can.* 32, 1843–1847.

Aiken, D. E., and Waddy, S. L. (1978). Space, density and growth of lobster (*Homarus americanus*). *Proc. Annu. Meet.—World Maric. Soc.* 9, 461–467.

Aiken, D. E., and Waddy, S. L. (1980). Reproductive biology. *In* "The Biology and Management of Lobsters" (J. S. Cobb and B. F. Phillips, eds.), Vol. 1, pp. 275–276. Academic Press, New York.

Aiken, D. E., and Waddy, S. L. (1982). Cement gland development, ovary maturation and reproductive cycles in the American lobster, *Homarus americanus*. *J. Crust. Biol.* 2, 315–327.

Aiken, D. E., and Waddy, S. L. (1985). Production of seed stock for lobster culture. *Aquaculture* 44, 103–114.

Aiken, D. E., and Waddy, S. L. (1986). Environmental influence on recruitment of the American lobster (*Homarus americanus*). *Can. J. Fish. Aquat. Sci.* 43, 2258–2270.

Aiken, D. E., and Waddy, S. L. (1988). Strategies for maximizing growth of communally reared juvenile American lobster. *World Aquacult.* 19(3), 61–63.

Aiken, D. E., and Waddy, S. L. (1989a). Colormorphs as biological tags. *In* "Proceedings of a Workshop on Life History of the American Lobster" (I. Kornfield, ed.), p. 2. Lobster Institute, University of Maine, Orono.

Aiken, D. E., and Waddy, S. L. (1989b). Lobster culture. *In* "Cold-water Aquaculture in Atlantic Canada" (A D. Boghen, ed.), pp. 77–122. Canadian Institute for Research on Regional Development, Moncton, New Brunswick.

Aiken, D. E., and Waddy, S. L. (1995). Scotophase influence on the diel timing of the metamorphic molt in larval American lobsters, *Homarus americanus*. *Can. J. Fish. Aquat. Sci.* (in press).

Aiken, D. E., and Young-Lai, W. W. (1981). Dactylotomy, chelotomy and dactylostasis: Methods for enhancing survival and growth of small lobsters (*Homarus americanus*) in communal conditions. *Aquaculture* 22, 45–52.

Aiken, D. E., Sochasky, J. B., and Wells, P. G. (1973). Ciliate Infestation of the Blood of the Lobster *Homarus americanus*. *Int. Counc. Explor. Sea Doc.* C.M. 1973/K:46.

Aiken, D. E., Waddy, S. L., Moreland, K., and Polar, S. M. (1984). Electrically induced ejaculation and artificial insemination of the American lobster, *Homarus americanus*. *J. Crust. Biol.* 4, 519–527.

Aiken, D. E., Waddy, S. L., and Uhazy, L. S. (1985). Aspects of the biology of *Pseudocarcinonemertes homari* and its association with the American lobster, *Homarus americanus*. *Can. J. Fish. Aquat. Sci.* 42, 351–356.

Armstrong, R. (1993). Annual report of the Salmon Health Consortium. *Bull. Aquacult. Assoc. Can.* 93-3, 1–60.

Bannister, R. C. A., and Howard, A. E. (1991). A large-scale experiment to enhance a stock of lobster (*Homarus gammarus* L.) on the English east coast. *ICES Mar. Sci. Symp.* 192, 99–107.

Bayer, R. C., and Daniel, P. C. (1987). Safety and efficacy of oxytetracycline for control of gaffkemia in the American lobster (*Homarus americanus*). *Aquaculture* 5, 71–81.

Beard, T. W., and Wickins, J. F. (1992). Techniques for the Production of Juvenile Lobsters (*Homarus gammarus* (L.)). *Fish. Res. Tech. Rep. U.K. Dir. Fish. Res.*) 92.

Bertran, R., and Lorec, J. (1986). Observations sur le Cycle de Reproduction du Homard Européen (*Homarus gammarus*) en Captivité. *Int. Counc. Explor. Sea Doc.* C.M. 1986/K:12.

Bordner, C. E., and Conklin, D. E. (1981). Food consumption and growth of juvenile lobsters. *Aquaculture* 24, 285–300.

Brown, J. H. (1989). Antibiotics: Their use and abuse in aquaculture. *World Aquacult.* 20(2), 34–43.

Carlson, F. T. (1954). The American lobster fishery and possible applications of artificial propagation. *Yale Conserv. Stud.* 3, 3–7.

Charmantier, G., Charmantier-Daures, M., and Aiken, D. E. (1989). La somatotropine humaine stimule la croissance de jeunes américains, *Homarus americanus* (Crustacea, Decapoda). *C. R. Hebd. Seances Acad. Sci., Ser. III* 308, 21–26.

Charmantier, G., Charmantier-Daures, M., and Aiken, D. E. (1991). Metamorphosis in the lobster *Homarus* (Decapoda): A review. *J. Crust. Biol.* 11, 481–495.

Conklin, D. E., and Chang, E. S. (1993). Culture of juvenile lobsters (*Homarus americanus*). *In* "CRC Handbook of Mariculture" (J. P. McVey, ed.), 2nd ed., pp. 497–510. CRC Press, Boca Raton, Florida.

Conklin, D. E., D'Abramo, L. R., Bordner, C. E., and Baum, N. A. (1980). A successful purified diet for the culture of juvenile lobster: The effect of lecithin. *Aquaculture* 21, 243–249.

Conklin, D. E., D'Abramo, L. R., and Norman Boudreau, K. (1983). Lobster nutrition. *In* "CRC Handbook of Mariculture" (J. P. McVey, ed.), Vol. 1, pp. 413–423. CRC Press, Boca Raton, Florida.

Coulombe, F., and Motnikar, S. (1989). The acceleration of the premolt process by unilateral eyestalk ablation in the male "canner" lobster (76–81 mm CL) of the Gaspé Bay. *In* "Proceedings of a Workshop on Life History of the American Lobster" (I. Kornfield, ed.), p. 9. Lobster Institute, University of Maine, Orono.

D'Abramo, L. R., and Conklin, D. E. (1985). Lobster aquaculture. *In* "Crustacean and Mollusk Aquaculture in the United States" (J.V. Huner and E.E. Brown, eds.), pp. 159–201. AVI Publishing, Westport, Connecticut.

Daniel, P. C., Bayer, R. C., and Chapman, S. (1985). Barnacle larvae (*Balanus* spp.) as a potential diet for juvenile lobster (*Homarus americanus*). *Aquaculture* 46, 67–70.

Delistraty, D. A., Carlberg, J. M., Van Olst, J. C., and Ford, R. F. (1977). Ammonia toxicity in cultured larvae of the American lobster (*Homarus americanus*). *Proc. Annu. Meet.—World Maric. Soc.* 8, 647–672.

Eagles, M. D., Aiken, D. E., and Waddy, S. L. (1986). Influence of light and food on larval American lobsters (*Homarus americanus*). *Can. J. Fish. Aquat. Sci.* 43, 2303–2310.

Fisher, W. S., Nilson, E. H., Steenbergen, J. F., and Lightner, D. V. (1978). Microbial diseases of cultured lobsters: A review. *Aquaculture* 14, 115–140.

Galindo, C. G. (1985). "Direct and Indirect Effects of Natural Diets on Cultured Lobster (*Homarus gammarus* (L.)) Larvae and Their Subsequent Performance as Juveniles," M.S. thesis. University of Stirling, Scotland.

Grimsen, S., Jaques, R. N., Erenst, V., and Balchen, J. G. (1987). Aspects of automation in a lobster farming plant. *In* "Automation and Data Processing in Aquaculture" (J. G. Balchen, ed.), Proceedings of the IFAC Symposium, No. 9, Trondheim, Norway, August 18–21, 1986, pp. 221–224. International Federation of Automatic Control (IFAC).

Hedgecock, D. (1983). Maturation and spawning of the American lobsters, *Homarus americanus*. *In* "CRC Handbook of Mariculture" (J. P. McVey, ed.), Vol. 1, pp. 261–290. CRC Press, Boca Raton, Florida.

Hedgecock, D., Nelson, K., and Shleser, R. A. (1976). Growth differences among families of the lobster, *Homarus americanus*. *Proc. Annu. Meet.—World Maricul. Soc.* 7, 347–361.

Herrick, F. H. (1894). The habits and development of the lobster and their bearing upon its artificial propagation. *Bull. U.S. Fish Comm.* 1893, 75–86.

Herrick, F. H. (1909). Natural history of the American lobster. *Bull. U.S. Bur. Fish.* 29, 148–408 + 20 plates.

Huang, C., and Bayer, R. C. (1989). Gastrointestinal absorption of various antibacterial agents in the American lobster (*Homarus americanus*). *Prog. Fish-Cult.* 51, 95–97.

Hughes, J. T. (1968). Grow your own lobsters commercially. *Ocean Industry* 3(12), 46–49.

Hughes, J. T., Shleser, R. A., and Tchobanoglous, G. (1974). A rearing tank for lobster larvae and other aquatic species. *Prog. Fish-Cult.* 36, 129–133.

Keith, I., Paterson, W. D., Airdrie, D., and Boston, L. (1988). Commercial development of a vaccine to prevent gaffkemia in lobsters: Field and lab results. *Bull. Aquacult. Assoc. Can.* 88-2, 90.

Knight, A. P. (1918). "Official Report on Lobster Investigations, 1917, and Methods of Increasing the Lobster Supply in Canada." King's Printer, Ottawa.

Lee, D. O., and Wickins, J. F. (1992). "Crustacean Farming."

Halsted, New York.
MacKay, D. A. (1929). Larval and postlarval lobsters. *Am. Nat.* 63, 160–170.
McLeese, D. W. (1956). Effects of temperature, salinity, and oxygen on the survival of the American lobster. *J. Fish. Res. Board Can.* 13, 1949–1952.
Mead, A. D. (1908). A method of lobster culture. *Bull. U.S. Bur. Fish.* 28, 219–245.
Nelson, K., and Hedgecock, D. (1983). Size dependence of growth inhibition among juvenile lobsters (*Homarus*). *J. Exp. Mar. Biol. Ecol.* 66, 125–134.
Nelson, K., Hedgecock, D., Heyer, B., and Nunn, T. (1983). On the nature of short-range growth inhibition in juvenile lobsters (*Homarus*). *J. Exp. Mar. Biol. Ecol.* 72, 83–89.
Perkins, H. C. (1972). Development rates at various temperatures of embryos of *Homarus americanus*. *Fish. Bull.* 70, 95–99.
Peutz, A. V. H., Waddy, S. L., Aiken, D. E., and Young-Lai, W. W. (1987). Accelerated growth of juvenile American lobster induced by unilateral eyestalk ablation. *Bull. Aquacult. Assoc. Can.* 87-2, 28–29.
Porterfield, B. (1982). New York scientist produced "Montauk genetic blue lobsters." *Aquacult. Mag.* March/April, 36–41.
Rathbun, R. (1886). Notes on lobster culture. *Bull. U.S. Fish Comm.* 6, 17–32.
Roche, G. (1898). "La Culture des Mers en Europe." Bibliothèque Scientifique International, Paris.
Shelburne, S. W., and Bean, L. L. (1991). Mortalities of impounded and feral male lobsters, *Homarus americanus* H. Milne-Edwards, 1837, caused by the protozoan ciliate *Mugardia* (formerly *Anophrys =Paranophrys*), with initial prevalence data from ten locations along the Maine coast and one offshore area. *J. Shellfish Res.* 10, 315–326.
Shleser, R. A. (1974). Studies on the effects of feeding frequency and space on the growth of the American lobster (*Homarus americanus*). *Proc. Annu. Meet.—World Maric. Soc.* 5, 149–155.
Sindermann, C. J., and Lightner, D. V., eds. (1988). "Disease Diagnosis and Control in North American Marine Aquaculture," Dev. Aquacult. Fish. Sci. 17. Elsevier, Amsterdam.
Talbot, P., and Harper, R. (1984). Abnormal eyestalk morphology is correlated with clutch attrition in laboratory-maintained lobsters. *Biol. Bull. (Woods Hole, Mass.)* 166, 349–356.
Talbot, P., Thaler, C., and Wilson, P. (1984). Spawning, egg attachment and egg retention in captive lobsters (*Homarus americanus*). *Aquaculture* 37, 239–249.
Talbot, P., Thaler, C., and Wilson, P. (1986). Artificial insemination of the American lobster (*Homarus*). *Gamete Res.* 14, 25–31.
Templeman, W. (1936). The influence of temperature, salinity, light and food conditions on the survival and growth of the larvae of the lobster (*Homarus americanus*). *J. Biol. Board Can.* 2, 485–497.
Tveite, S. (1995). Survival on one year old artificially raised lobsters (*Homarus gammarus*) released in southern Norway. *ICES Mar. Sci. Symp.* (in press).
Van Olst, J. C., and Carlberg, J. M. (1978). The effects of container size and transparency on growth and survival of lobsters cultured individually. *Proc. Annu. Meet.—World Maric. Soc.* 10, 469–479.
Van Olst, J. C., Carlberg, J. M., and Hughes, J. T. (1980). Aquaculture. *In* "The Biology and Management of Lobsters" (J. S. Cobb and B. F. Phillips, eds.), Vol. 2, pp. 333–384. Academic Press, New York.
Waddy, S. L. (1988). Farming the homarid lobsters: State of the art. *World Aquacult.* 19(4), 63–71.
Waddy, S. L., and Aiken, D. E. (1985). Fertilization and egg retention in artificially inseminated female American lobsters, *Homarus americanus*. *Can. J. Fish. Aquat. Sci.* 42, 1954–1956.
Waddy, S. L., and Aiken, D. E. (1989). Seasonal molting patterns of juvenile American lobsters cultured at constant 20°C. *J. World Aquacult. Soc.* 20, 70A.
Waddy, S. L., and Aiken, D. E. (1990). Intermolt insemination, an alternative mating strategy for the American lobster (*Homarus americanus*). *Can. J. Fish. Aquat. Sci.* 47, 2402–2406.
Waddy, S. L., and Aiken, D. E. (1991). Egg production in the American lobster. *In* "Crustacean Issues" (F. R. Schram, ed.), Vol. 7, "Crustacean Egg Production" (A. Wenner and A. Kuris, eds.), pp. 267–290. Balkema Press, Rotterdam, The Netherlands.
Waddy, S. L., and Aiken, D. E. (1992). Environmental intervention in the reproductive process of the American lobster, *Homarus americanus*. *Invertebr. Reprod. Dev.* 22, 245–251.
Waddy, S. L., and Aiken, D. E. (1995). Lobster culture. *In* "Cold-water Aquaculture in Atlantic Canada" (A.D. Boghen, ed.), 2nd ed., pp. 145–188. Canadian Institute for Research on Regional Development, Moncton, New Brunswick.
Waddy, S. L., Aiken, D. E., and Eagles, M. D. (1988). Growth of cultured American lobsters and effect of culling on growth rates. *Bull. Aquacult. Assoc. Can.* 88-4, 66–68.
Young-Lai, W. W., Charmantier-Daures, M., and Charmantier, G. (1991). Effect of ammonia on survival and osmoregulation in different life stages of the lobster, *Homarus americanus*. *Mar. Biol.* 110, 293–300.

CHAPTER

9

Reproduction and Embryonic Development

P. TALBOT
Department of Biology
University of California at Riverside
Riverside, California 92521

SIMONE HELLUY
Department of Biological Sciences
Wellesley College
Wellesley, Massachusetts 02181

I. Introduction

The reproductive biology of *Homarus americanus* has been the subject of numerous investigations. Herrick's (1909) early monograph is a masterpiece rich in information and beautifully illustrated. Waterman's *The Physiology of Crustacea* (1960), Bliss' series *The Biology of Crustacea* (1982–1985), and Cobb and Phillips' *The Biology and Management of Lobsters* (1980) provide more recent information on lobster reproduction.

Several factors have contributed to advances in our understanding of the reproductive biology and embryology of the lobster. *Homarus americanus* is relatively easy to obtain in many areas of North America, adapts well to laboratory tanks, and can easily be maintained in captivity. It is economically important and much research has been stimulated by a desire to control its reproductive cycle and to develop an aquaculture technology suitable for completion of the life cycle of this species in captivity. Finally, *H. americanus* has attracted the attention of numerous researchers because certain features of its reproduction are unusual. For example, lobster sperm are nonflagellated and undergo a novel type of acrosome reaction, sometimes termed *explosion*, which has fascinated spermatologists for over 100 years. Research carried out during the past century has augmented our basic understanding of reproductive mechanisms and early development and has resulted in technological advancements to control reproduction, such as hybridization and artificial insemination.

Surprisingly, the embryonic development of *Homarus americanus* (reviewed by Aiken and Waddy, 1980) has attracted little attention compared to the reproduction and larval development of this species. (Factor, in Chapter 1, and Ennis, in Chapter 3, consider larval development.) The most extensive investigations date from the last century. The monograph by Bumpus (1891) is the only work focusing on early embryogenesis in *H. americanus*. Herrick (1895a, 1909) contributed to the study of ontogeny with the same care and attention to detail that he lavished on other aspects of the biology of lobsters, but without the help of modern techniques. More recent developmental studies have dealt primarily with the perihatching period (Davis, 1964; Ennis, 1973, 1975) and the prelarval embryonic molt cycle (Helluy and Beltz, 1991). Embryonic growth rates, osmotic regulation, biochemistry, bioenergetics, and development of the nervous system have also been explored. The purpose of this chapter is to summarize knowledge currently available on the reproduction and embryonic development of the lobster *H. americanus*.

BIOLOGY OF THE LOBSTER Homarus americanus
Jan Robert Factor, Editor

II. Sexual Differentiation

A. *Sexual Dimorphism*

Lobsters are dioecious. Externally, the main features distinguishing the sexes are the position of the gonopores and the form of the first pair of pleopods (Herrick, 1909). The gonopores of males open at the base of the coxae of the fifth pereiopods, while in females they are associated with the third pereiopods. In males, the first pair of pleopods is modified for spermatophore transfer, while in females it is reduced in size and vestigial. Pleopods of reproductively active females bear long, ovigerous setae for egg attachment.

At sexual maturity, the carapace is shorter (relative to body length) in females than males, and females develop a wider abdomen to facilitate the carriage of eggs (Templeman, 1944). The sternal spines on the abdomen of egg-bearing females are reduced to blunt knobs, apparently so eggs will not be punctured when the abdomen is flexed (Templeman, 1944). Templeman (1940a, 1944) and Aiken and Waddy (1980) provide additional information on dimorphism.

B. *Hermaphroditism*

Hermaphrodites are apparently very rare in the genus *Homarus* (early literature reviewed by Ridgewood, 1909), and both complete and incomplete hermaphrodites are known. The first complete homarid hermaphrodite was described by Nicholls in 1730 for *H. gammarus*. This lobster had external and internal female characteristics on the right side and male characteristics on the left side. This pattern of female features on the right and male features on the left has been observed in subsequent reports of *H. americanus* hermaphrodites (Halkett, 1919; Chace and Moore, 1959; Syslo and Hughes, 1981).

The most completely described specimen of a *Homarus americanus* hermaphrodite exhibited bilateral color differentiation as well as hermaphroditism (Chace and Moore, 1959). The body was skewed, probably due to differential growth of the male and female components. The external features were typical of the respective sexes, except that the seminal receptacle (thelycum) did not correspond exactly to the structure found in normal females. Internally, this lobster had a maturing ovary with eggs on the right and a functional testis producing normal sperm on the left. The gonads were connected to their respective gonopores by a normal oviduct and vas deferens. It is not known whether this animal was capable of successfully transferring or receiving sperm, since the intromittent organ and seminal receptacle were not complete.

Reproductive performance has been observed in a bicolored hermaphrodite of *Homarus americanus* (Syslo and Hughes, 1981). On two occasions this animal was placed with a normal, freshly molted female, but in neither case did it successfully transfer sperm. However, mating occurred when the gynandromorph molted and was placed with a normal male. After one such mating, the hermaphrodite extruded eggs that appeared normal and attached them to the pleopods on the right (female) side. Unfortunately, the eggs were aborted after 1 week and their development could not be followed.

C. *Sex Determination*

Most work on sexual determination in crustaceans has been limited to a few groups, for example, Branchiopoda, Copepoda, Isopoda, and Amphipoda (reviewed by Ginsburger-Vogel and Charniaux-Cotton, 1982; Legrand *et al.*, 1987). Little is known about this process in *Homarus americanus*.

III. Anatomy of the Female Reproductive System

The female reproductive tract consists of paired ovaries and oviducts and a receptacle for sperm storage. The gross anatomy of the tract has been described for a number of lobsters in addition to *Homarus americanus*, while microscopic studies, especially on the oviduct, are rare for all lobsters.

A. *Ovaries*

The ovaries of *Homarus americanus* are shaped like an elongated H, with the right and left lobes connected by a small cross-lobe at the level of the heart (Bumpus, 1891; Herrick, 1909). The ovaries lie in the dorsal cephalothorax. When mature, the posterior lobes extend into the abdomen and the anterior lobes extend to the level of the eyestalks. The ovaries undergo cyclic changes that culminate in ovulation and spawning of mature oocytes (see Section V,A).

The ovaries in lobsters are surrounded by a "wall" that comprises a thin outer epithelium, connective tissue, muscle, and blood vessels. Contraction of the ovarian wall is thought to be important in forcing ovulated oocytes out of the ovary and into the oviduct during spawning. The ovarian muscles of *Homarus americanus* resemble smooth muscle (Herrick, 1909). Ultrastructural studies show that they are indeed *not* striated; however, they are quite unlike typical smooth muscle (Talbot, 1981a; Howard, 1991). [Ovarian wall muscles of several palinurid lobsters have been characterized as striated (Silberbauer, 1971; Junio, 1987).] Ovarian muscles contain both actin and myosin. They are densely packed with microtubules that cross-react with monoclonal antibodies to tubulin and have been shown, using immunofluorescence microscopy, to be distributed throughout the cytoplasm (Fig. 1A) (Howard, 1991). Drugs that depolymerize specific cytoskeletal elements have been used to show that both actin filaments and microtubules are necessary to generate contraction in ovarian muscle strips (Howard, 1991). The ovarian muscles are apparently not innervated and the factors regulating their contraction *in vivo* are speculative. However, massive contraction of the ovarian wall can be stimulated *in vitro* in a dose-dependent fashion by octopamine, while 5-hydroxytryptamine only mildly stimulates contraction, and dopamine is ineffective (Howard and Talbot, 1992). Compounds that elevate intracellular cAMP [e.g., 3-isobutyl-1-methylxanthine (IBMX), forskolin, and dibutyryl-cAMP (dbcAMP)] inhibit the *in vitro* contraction of ovarian muscles. The opposing effects of octopamine and forskolin suggest that octopamine does not stimulate contraction through class 2 octopamine receptors, but may be acting on ovarian muscles through class 1 receptors or a novel type of octopamine receptor.

A germinal strand runs through the center of each lobe of the ovary, parallel to its long axis, and contains connective tissue and oocytes in various stages of oogenesis (Fig. 1B). More mature stages are located toward the periphery of the ovary, with fully mature oocytes lying adjacent to the wall (Kessel, 1968; Talbot, 1981a). Oocytes undergoing secondary vitellogenesis are surrounded by a single layer of follicle cells, which is quite thin and often difficult to see around mature oocytes. A lumen is present in the center of the ovary and mature oocytes are apparently ovulated into this cavity. After spawning, the ovary is greatly reduced in size and usually devoid of mature oocytes, although a small number of ovulated oocytes may be retained and resorbed after a successful spawn. The spawned ovary may be restored to its original size by injecting fluid into it through the oviduct (Bumpus, 1891). Herrick (1909) reported that "gland like organs" appear peripherally in ovaries shortly after spawning; these are shriveled by 5 weeks after spawning and their significance is not known.

Lobsters may resorb developing oocytes rather than maturing and spawning them (Allen, 1895; Herrick, 1909). This is generally considered abnormal and may be brought on by environmental or physiological stress, such as the transfer of wild females to laboratory conditions (Aiken and Waddy, 1985). During resorption, yolk proteins are rapidly passed into the hemolymph. This process is usually evident externally through the ventral abdominal surface, which takes on a green tint due to the presence of lipovitellin in the hemolymph. Resorption apparently also occurs in natural populations of *Homarus americanus* if the molting and reproductive cycles conflict (Ennis, 1984a). A female in middle or late premolt will absorb her ovary in preparation for molting.

B. Oviducts

The oviducts are paired, thin, transparent or white tubes that leave the ovary at its ventral surface (Herrick, 1909). They pass laterally and ventrally within the cephalothorax and join the gonopores at the base of the third walking legs. The distal portion of the oviduct is clear in immature females, but retains a yellow color in females that have spawned (Aiken and Waddy, 1980). Histological studies have shown the oviducts to comprise an outer layer of epithelium, connective tissue, muscle, basal lamina, and tall columnar epithelial cells (Herrick, 1909). The oviducal muscle is nonstriated in *Homarus americanus*, and is structurally similar to the microtubule-containing muscle surrounding the ovary (P. Talbot, unpublished data). [In contrast, the oviducal muscle of the palinurid lobster *Jasus lalandii* appears striated (Silberbauer, 1971).] The epithelium lining the oviducts is secretory and undergoes cyclic changes associated with development of the ovary and spawning (Herrick, 1909). Oviducal secretions are released prior to spawning and oocytes presumably pass through these secretions during spawning (Herrick, 1909); however, the effect of the secretions on oocytes is not known. The composition of oviducal secretions has not been examined in *H. americanus*, but may be similar to that of the crayfish, in which the secretory product is histochemically positive for acid and neutral mucopolysaccharides (Cheung, 1966).

C. Seminal Receptacle

The spermatophore is transferred to a blind pouch on the female's ventral surface (Bumpus, 1891). This pouch has been described using various terms,

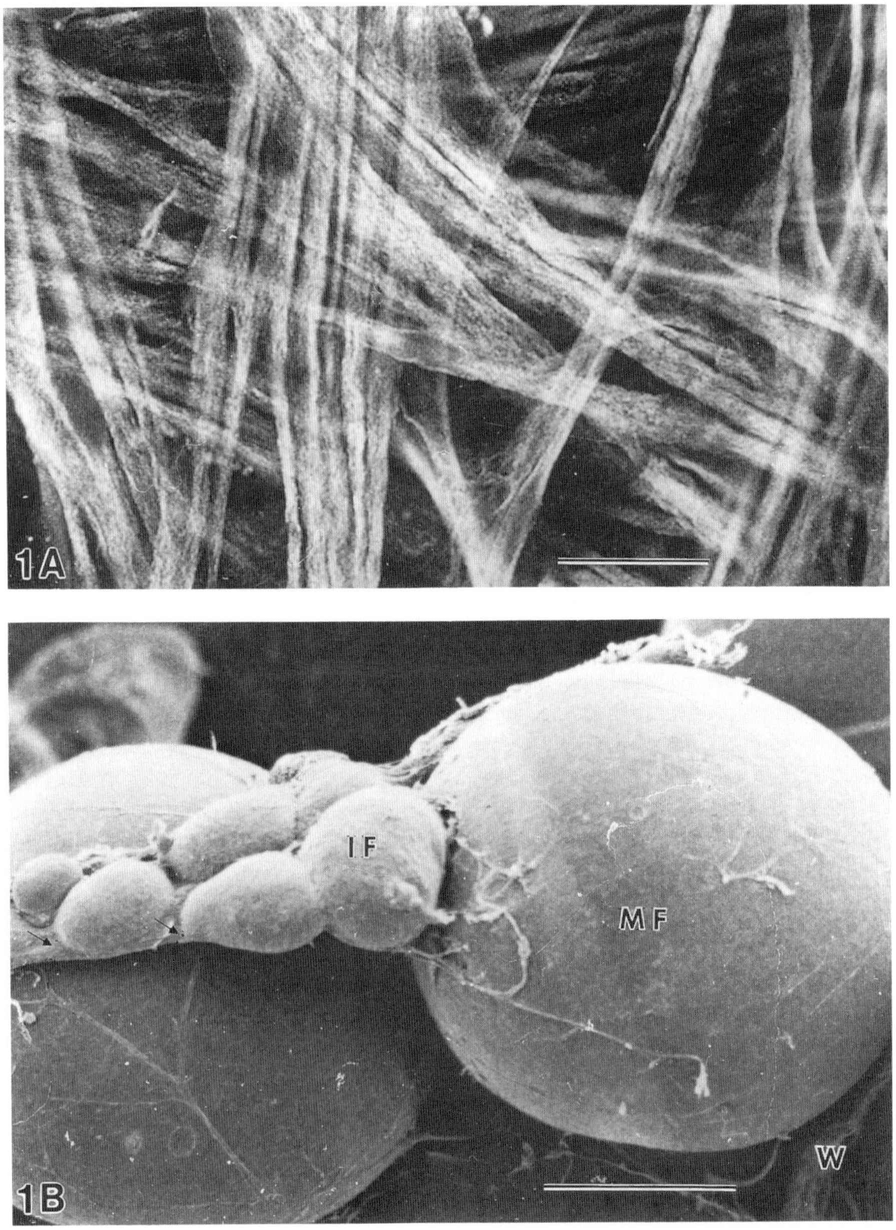

FIGURE 1 Features of the ovarian wall and follicles of *Homarus americanus*. (A) Confocal scanning laser micrograph of a lobster ovarian wall treated with a monoclonal antibody to tubulin and secondarily reacted with an anti-mouse immunoglobulin G conjugated to fluorescein isothiocyanate (FITC). Fluorescence indicates microtubules that pack these muscle cells. (Courtesy of David R. Howard.) (B) Scanning electron micrograph of mature (MF) and immature (IF) ovarian follicles. Mature follicles are located close to the wall (W) of the ovary. Scale bars: (A) approximately 10.0 μm; (B) 0.5 mm. (Reprinted from Talbot, 1981a, with permission of Academic Press.)

notably *thelycum* (Aiken and Waddy, 1980) and *seminal receptacle* (Bauer, 1986). After mating, the spermatophore is stored along the wall farthest from the opening of the receptacle (Fig. 2). The remainder of the receptacle in mated females is packed with a hard, brown substance.

IV. Anatomy and Histology of the Male Reproductive Tract

The male reproductive tract consists of paired testes and vasa deferentia. The androgenic gland, which is thought to regulate the expression of male

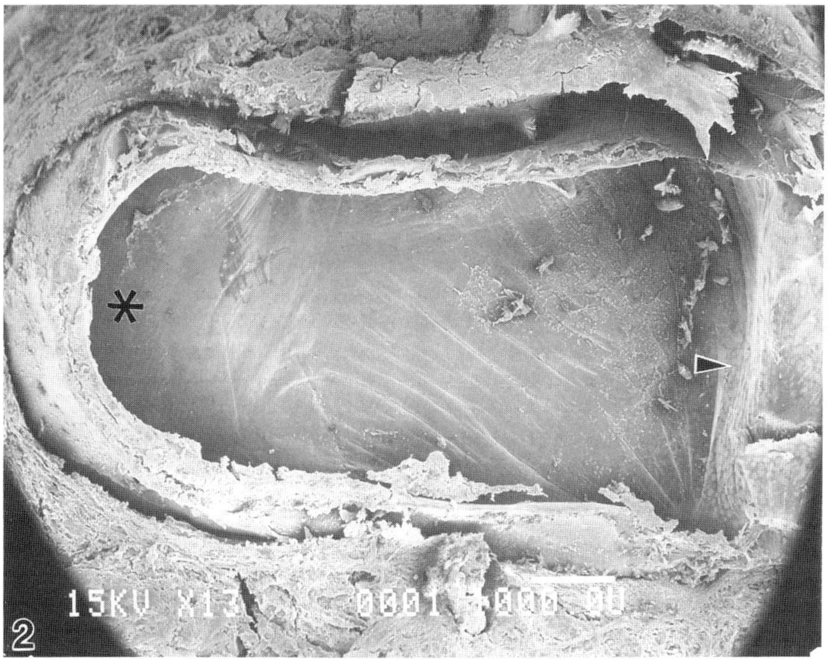

FIGURE 2 Scanning electron micrograph (SEM) of an empty seminal receptacle from *Homarus americanus*. The receptacle has been cut open longitudinally for the SEM. Arrowhead indicates the ventral opening of the receptacle. *, Position of spermatophore storage. Scale bar: 1.0 mm. (Courtesy of Tae Won Lee and Cynthia Alaso.)

sexual characteristics, is associated with the terminal portion of the vas deferens. The parts of the tract have been examined anatomically, histologically, and ultrastructurally in *Homarus americanus*, providing the basis for the following summary.

A. Testes

The two testes of *Homarus americanus* are connected by a short, medial segment of testicular tissue, giving them the shape of an elongated H (Herrick, 1909). They lie ventral to the pericardium and dorsal to the digestive tract in the cephalothorax. The posterior lobes are longer than the anterior lobes and extend into the abdomen, while the anterior lobes approach the cephalic ganglion. The testes are thin and change from translucent to white or opaque when sperm are being produced. They are held in place by connective tissue mesenteries.

The seminiferous epithelium has not been analyzed in detail at the cellular level in *Homarus americanus*; however, it probably resembles that of the nephropid red lobster, *Enoplometopus occidentalis*, which has been examined ultrastructurally (Haley, 1984, 1986). In addition to spermatogenic cells, the follicular epithelium contains somatic cells that envelop developing spermatocytes and spermatids and may be functionally equivalent to the Sertoli cells of mammalian testes.

The dual role of the mammalian testes in sperm and steroid hormone production is well established. Because testicular implants do not promote sexual differentiation in juvenile males, it is generally thought that the testes do not produce hormones in malacostracans, but that the androgenic gland is the main reproductive endocrine gland in males (Charniaux-Cotton and Payen, 1985). Several recent observations suggest that this matter needs further study. Haley (1986) identified a type of cell in the testes of *Enoplometopus occidentalis* that has structural features associated with steroid production. Gilgan and Idler (1967) showed that *Homarus americanus* testes have a 17β-hydrosteroid dehydrogenase capable of converting (4-C) androst-4-ene-3,17-dione to (4-C) testosterone. More recently, Burns *et al.* (1984a) demonstrated that *H. americanus* testes contain a steroid, 20-ketone reductase, which converts (C) progesterone to 20α-dihydroprogesterone. Although levels are low, testosterone has been identified in the serum and testes of *H. americanus* (Burns *et al.*, 1984b). These reports, although preliminary, suggest that metabolic pathways for the synthesis and catabolism of steroids exist in the testes of *H. americanus*, but further work is required to determine whether these steroids have reproductive functions.

B. Vas Deferens

The paired vasa deferentia of *Homarus americanus* (Fig. 3) extend laterally and ventrally from the middle region of the testes to the gonopores in the coxae at the base of the fifth walking legs (Herrick, 1909). The vasa deferentia function in secretion of the acellular components of the spermatophore, storage of the spermatophore, and extrusion of the spermatophore during mating. The overall appearance of the vasa deferentia is often quite similar in related species (*Nephrops norvegicus*, Farmer, 1974), but less similar in more distant groups (*Cherax albidus*, Talbot and Beach, 1989).

All regions of the vas deferens are organized histologically in a similar manner, although the relative amount of the different tissues may vary (Fig. 4). From the lumen to the periphery, the vas deferens comprises a single layer of epithelial cells, a thick basal lamina, connective tissue containing blood vessels, and striated muscle. The epithelial cells lining the vas deferens are covered apically with microvilli (Kooda-Cisco and Talbot, 1986), not cilia, as was originally reported (Fasten, 1917). Also, the basal surfaces of the epithelial cells are highly infolded (Kooda-Cisco and Talbot, 1986), a characteristic of cells involved in fluid transport (Diamond and Bossert, 1968); however, physiological studies have not yet been carried out to establish the significance of the infolding. The epithelial cells lining the lumen of the vas deferens are secretory throughout its length, although the composition of the secretory product varies in each region. The diameter of the vas deferens increases from the proximal to the distal end due to secretion of the spermatophore wall and an increase in the thickness of the muscle layer.

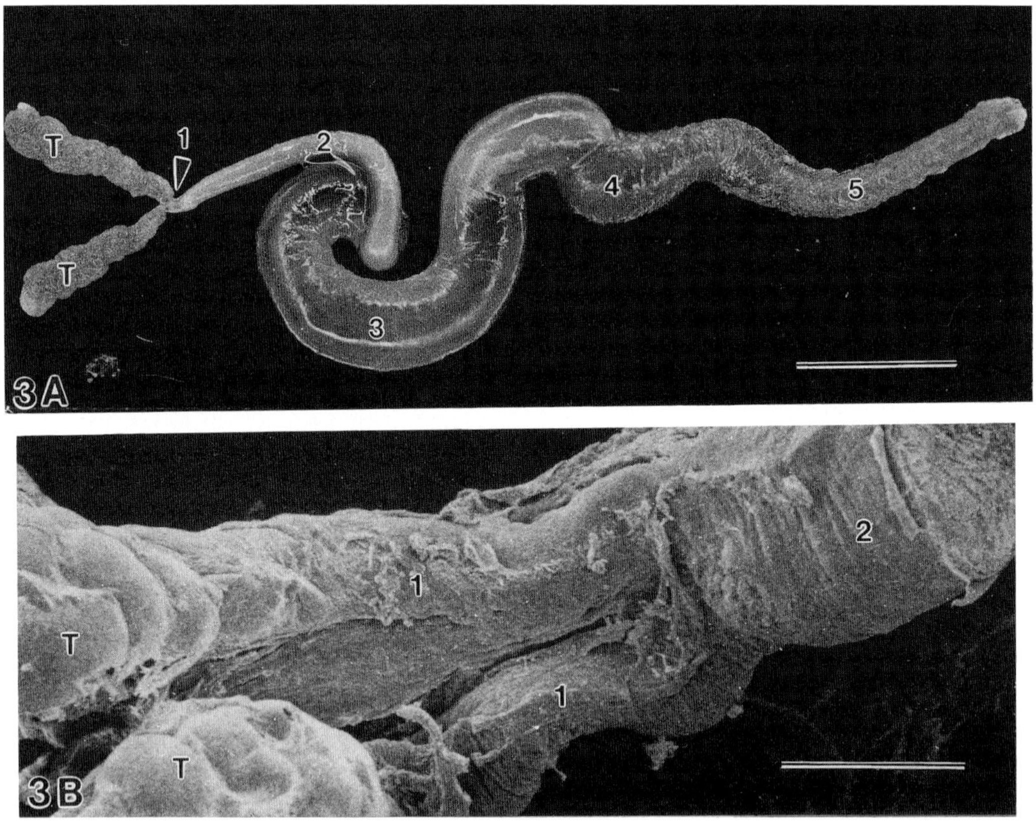

FIGURE 3 Overviews of vasa deferentia of *Homarus americanus*. (A) The vas deferens can be subdivided into five regions: the proximal vas deferens (PVD) (regions 1 and 2); the middle vas deferens (region 3); and the distal vas deferens, which includes the "bump" (regions 4 and 5). A small portion of the testes (T) is shown. (B) Region 1 (at higher magnification) leaves the testes (T) as two separate tubes that merge to form region 2 of the PVD. Scale bars: (A) 7.0 mm; (B) 1.0 mm. (From Kooda-Cisco, M., and Talbot, P. *J. Morphol.* 188. Copyright © 1986 Wiley-Liss. Reprinted with permission of Wiley-Liss, a division of John Wiley & Sons, Inc.)

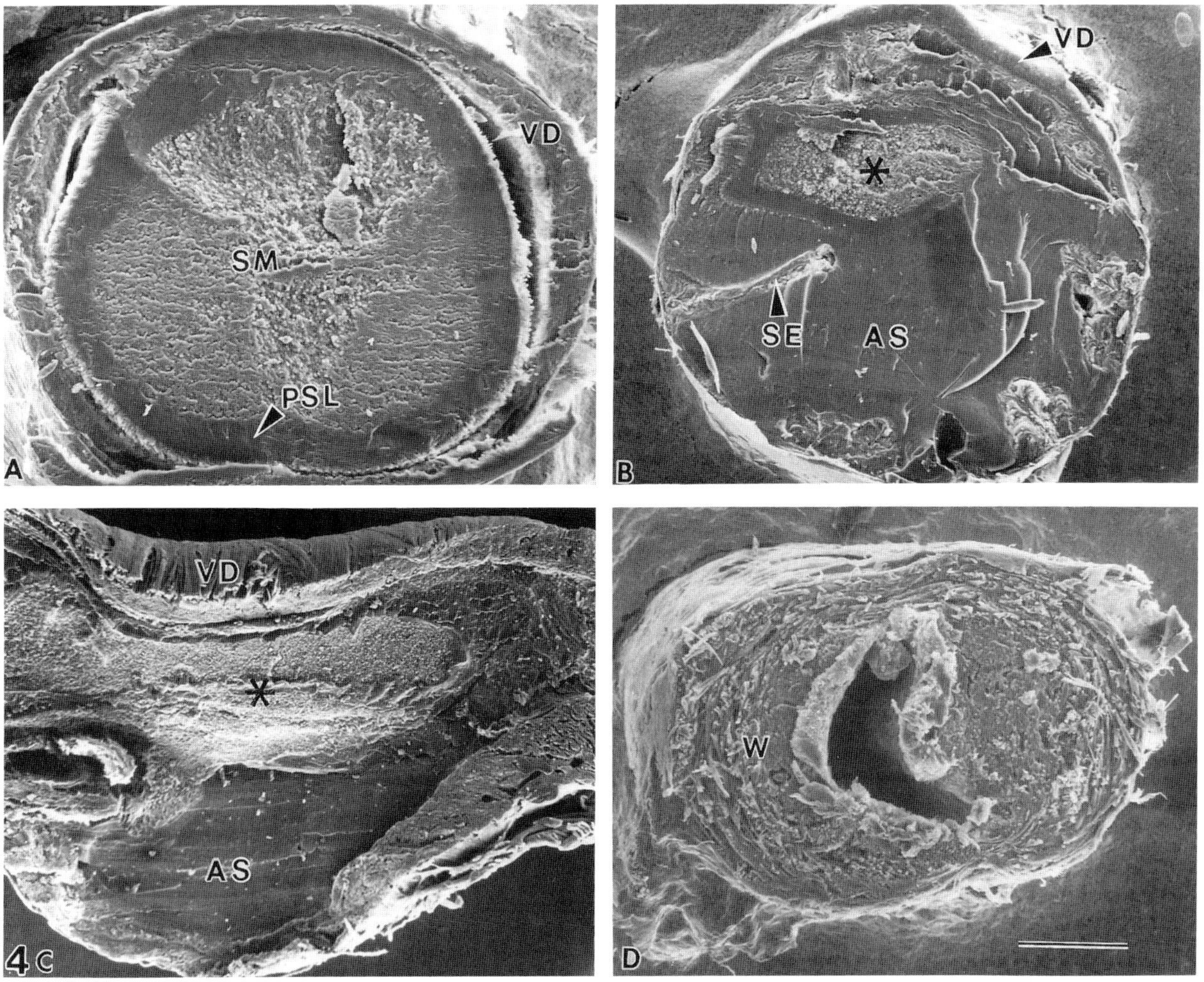

FIGURE 4 Scanning electron micrographs showing sections through the vas deferens of *Homarus americanus*. (A) Cross-section through the proximal vas deferens, segment 2 (see Fig. 3), showing the sperm mass (SM), primary spermatophore layer (PSL), and tissue of the vas deferens (VD). (B) Cross-section through the posterior part of the middle vas deferens (MVD), segment 3. The septum (SE) separating the sperm mass/primary spermatophore layer (*) from the acellular secretions (AS) of the MVD is incomplete; proximal to this, the septum forms a complete partition in the MVD. The tissue of the VD is relatively thin at this level. (C) Longitudinal section through the distal VD, segment 4. This is the region of the "bump," in which the spermatophore (*) and AS secreted in the MVD merge to form a complete spermatophore. AS are located in the bump. (D) Cross-section through the distal VD, segment 5. The lumen lacks a spermatophore and most of the wall (W) is formed of striated muscle. Scale bar in (D) represents 1.0 mm and applies to (A)–(D). (Courtesy of M. Kooda.)

The vasa deferentia of *Homarus americanus* have clear morphological subdivisions, regions that may themselves be subdivided into segments. We propose that the regions be called the proximal (segments 1 and 2), middle (segment 3), and distal (segments 4 and 5) vas deferens (Fig. 3).

1. Proximal Vas Deferens

The proximal vas deferens (PVD) is continuous with the collecting tubules of the testes and is relatively straight (Kooda-Cisco and Talbot, 1986). It is divisible into two subregions. Segment 1 consists of a small tube leaving each of the four lobes of the testes; on each side, the two tubes from segment 1 (one each from the anterior and posterior lobes) fuse to form a thicker tube, segment 2 (Fig. 3). Segment 1 functions mainly as a conduit for transferring sperm to segment 2. It is lined by a single layer of epithelium, which, in turn, is surrounded by a thick basal lamina, connective tissue, and circular striated muscle. The epithelium is secretory and may contribute to the sperm-supporting matrix, but it does not form the primary spermatophore layer. (Tubules similar to segment 1 of *Homarus americanus* have not been reported in other lobsters and crayfish, perhaps because they are small and difficult to find or because this feature is not characteristic of all species.) The main portion of the PVD, segment 2, has a similar structural organization; however, its epithelium secretes the primary spermatophore layer and portions of the secondary (intermediate) spermatophore layer (Fig. 4). The height of the epithelial cells varies, and this may be important in shaping the spermatophore (Kooda-Cisco and Talbot, 1986).

2. Middle Vas Deferens

The middle vas deferens (MVD) is wider than the PVD, is slightly curved, and retains the same histological organization. The MVD of nephropid lobsters generally exhibits two lumina; one contains the spermatophore and the other contains an acellular secretion thought to be transferred to the female at mating (Fig. 4). The epithelium is secretory and elaborates a major portion of the spermatophore wall. The MVD is also the primary site of spermatophore storage.

3. Distal Vas Deferens

The distal vas deferens (DVD), which has also been called the ejaculatory duct, is characterized by a thick coat of striated muscle (Fig. 4). Its main function appears to be extrusion of spermatophores during mating. Contraction of the muscular coat around the MVD moves its contents into the "bump" at the proximal end of the DVD (Figs. 3A and 4C). The acellular secretion and spermatophore of the MVD unite in the bump to form a complete spermatophore. A thick muscular sphincter in the DVD contracts to pinch off a spermatophore from the continuous spermatophoric mass within the MVD and the DVD. Contraction of muscle in the terminal part of the DVD forces the spermatophore through the gonopore. Spermatophores are not extruded during electrical stimulation unless the spermatophoric mass extends into the terminal vas deferens (Kooda-Cisco and Talbot, 1983). The epithelium lining the DVD is short columnar.

C. *Androgenic Gland*

The importance of the androgenic gland in promoting the differentiation of the primary, secondary, and behavioral sex characteristics of male malacostracans was first recognized by Charniaux-Cotton (1962). In *Homarus americanus*, this gland is located near the distal vas deferens, and dinitroblue tetrazolium stain has been used to help visualize it (Gilgan and Idler, 1967). Although the androgenic gland has been reported to synthesize testosterone in *H. americanus* (Gilgan and Idler, 1967), relatively little is known about this gland in lobsters.

V. Gametogenesis

A. *Oogenesis*

The ovaries of *Homarus americanus* normally mature on a biennial cycle and oocytes reach maturity during the summer (Herrick, 1909). Various criteria have been used to categorize ovarian development in *H. americanus*. According to the scheme proposed by Aiken and Waddy (1980), ovarian maturation can be grouped into six macroscopically identifiable stages. Stage 1 ovaries, which are immature, are white and contain oocytes <0.5 mm in diameter. As the ovary develops and stage 6 is reached, the color becomes dark green and the diameter of the oocytes increases to 1.4–1.6 mm. The color change results from deposition of pigment in the developing oocytes. A final stage, termed 6A, is used to describe the ovary following spawning or resorption. Spent ovaries are white or yellow and contain some dark-green residual oocytes. The cytological changes that occur during oogenesis have not yet been analyzed for *H. americanus*.

As maturation occurs, the ovaries extend into the abdomen, where they can be observed by "candling" (Hedgecock *et al.*, 1978), a valuable method for assessing ovarian development in broodstock females used for aquaculture. (See Aiken and Waddy, Chapter 8, on aquaculture.) Ovarian maturation can also be estimated by observing the stage of development of the pleopod tegumental glands, which cycle with the ovaries (Aiken and Waddy, 1982). The latter method is rapid, not traumatic to the lobster, and fairly accurate. (See Waddy *et al.*, Chapter 10, on molting.)

B. *Vitellogenesis*

The fertilized lobster egg contains a large amount of yolk necessary to sustain the embryo during its prolonged period of development. Most studies on oogenesis have focused on the deposition of yolk in developing oocytes. This process, vitellogenesis, occurs in two phases. In primary vitellogenesis, which usually lasts several months, yolk is deposited slowly and the oocyte undergoes a gradual increase in size. In secondary vitellogenesis, large amounts of yolk are deposited in oocytes relatively quickly and oocytes increase synchronously in diameter to their mature size. This is followed by ovulation. In lobsters and other macrurans, the major yolk protein of mature oocytes is called lipovitellin.

The source of yolk protein in *Homarus americanus* and other macrurans has been controversial for a number of years (Meusy, 1980; Dehn *et al.*, 1983). It was originally thought that the oocyte synthesizes proteinaceous yolk from amino acids circulating in the hemolymph. This concept was challenged when a female-specific protein was found in the hemolymph of the crab *Carcinus maenas* (Frentz, 1960), suggesting that decapods may synthesize at least some yolk protein at a site outside the ovary.

The idea that yolk protein originates in the ovary of *Homarus americanus* was first supported by ultrastructural observations (Kessel and Beams, 1963; Kessel, 1968). In developing oocytes, the endoplasmic reticulum actively synthesizes a proteinaceous product that was interpreted to be yolk precursor. However, it has since been shown that this product, which can be recognized by its distinctive ring shape, is, in fact, the precursor to one type of cortical granule that is released exocytotically at fertilization (Talbot and Goudeau, 1988; see Section X,E). Recent studies in which lobster ovaries were incubated *in vitro* in the presence of labeled amino acids are consistent with the endogenous synthesis of lipovitellin during primary vitellogenesis (Dehn *et al.*, 1983).

There is considerable evidence that lobster oocytes also incorporate exogenously produced yolk protein by means of endocytosis. The exogenous precursor to lipovitellin is called vitellogenin, and most evidence is consistent with its incorporation occurring during secondary vitellogenesis. Oocytes in secondary, but not primary, vitellogenesis take up horseradish peroxidase in coated pits and vesicles (Schade and Shivers, 1980). If females not undergoing secondary vitellogenesis are eyestalk ablated to induce vitellogenesis, a similar uptake of horseradish peroxidase occurs. Moreover, a female-specific protein, which is immunologically identical to egg lipovitellin, is present in the hemolymph of female lobsters during vitellogenesis, but not at times when the ovary is quiescent (Dehn *et al.*, 1983; Byard and Aiken, 1984). Females in secondary vitellogenesis have two immunologically related vitellogenins in their hemolymph, each of which corresponds to an ovarian lipovitellin (Nelson *et al.*, 1988a). The uptake of vitellogenin is inferred to occur by receptor-mediated endocytosis. A solid-phase binding assay for vitellogenin receptors in oocyte plasma membranes has been used to show increased vitellogenin binding when oocytes begin secondary vitellogenesis and decreased binding activity in older oocytes (Laverdure and Soyez, 1988).

The hepatopancreas (Wolin *et al.*, 1973), hemocytes (Kerr, 1968), and subepidermal adipose tissue (Tom *et al.*, 1987) have all been suggested as the extraovarian source of vitellogenin in other malacostracans. However, it has not been possible to confirm any of these tissues as the site of vitellogenin synthesis in *Homarus americanus* (Dehn *et al.*, 1983).

Although the literature in this area is complex, there seems to be emerging evidence that supports a dual origin for yolk protein. Oocytes appear to synthesize lipovitellin during primary vitellogenesis and incorporate exogenously synthesized vitellogenin by receptor-mediated endocytosis during secondary vitellogenesis.

C. Spermatogenesis

Spermatogenesis occurs in the seminiferous epithelium of the testes. Although detailed descriptive studies of spermatogenesis have been undertaken on other decapods, virtually nothing is known about this process in *Homarus americanus*. It is probable that the acrosome forms during spermiogenesis, as in *Nephrops norvegicus*, and originates from vesicles of rough endoplasmic reticulum that coalesce to form a large vesicle at one pole of the spermatid; this eventually elongates to become the acrosome, much of which is occupied by a periodic acid–Schiff (PAS)-positive material (Chevaillier, 1965).

D. Factors Affecting Gametogenesis

Both hormonal and exogenous factors regulate gametogenesis in *Homarus americanus*. Although genetic regulation is undoubtedly important, we know little about it in this group. Waddy *et al.* (Chapter 10) provide a discussion of the mechanisms that control reproduction.

1. Hormonal Control

Hormonal control of gametogenesis in *Homarus americanus* has been studied in some detail (reviewed by Adiyodi, 1985; Waddy *et al.*, Chapter 10). Gonadal activity appears to be regulated by both peptide and steroid hormones, although as yet no definitive functions have been established for the steroid hormones. The glands and hormones that may function in reproduction in lobsters include the following: (1) the X-organ–sinus gland complex of the eyestalk, which produces gonad-inhibiting hormone or vitellogenesis-inhibiting hormone in females; (2) the brain and the thoracic ganglia, which secrete gonad-stimulating hormone; (3) the Y-organ, which produces ecdysteroids that may function in vitellogenesis; (4) the gonads themselves, which have been reported to produce steroid hormones; (5) the androgenic gland of males, which may produce both steroid and peptide hormones; and (6) the mandibular organ, which synthesizes methyl farnesoate, an analog of juvenile hormone thought to function in reproduction.

2. Exogenous Control

Photoperiod and water temperature and their role in regulating oogenesis have been studied extensively in *Homarus americanus* (e.g., Nelson *et al.*, 1983, 1988a–c; Aiken and Waddy, 1989; Waddy and Aiken, 1992). From these recent studies, it seems probable that both water temperature and photoperiod can affect vitellogenesis and the time of egg extrusion. When animals are held at water temperatures characteristic of their natural environment, temperature regulates vitellogenesis and spawning. However, when females are held at warmer water temperatures, photoperiod overrides control by temperature. It would thus appear that water temperature may be the natural exogenous regulator of these events, but that photoperiod may be used to control spawning times in aquaculture facilities. However, the percentage of females that go on to spawn under photoperiod regulation is lower than that obtained under temperature regulation.

VI. Structure of the Gametes

A. *Mature Oocyte*

Oocytes develop in solid follicles in the ovary of *Homarus americanus* (Herrick, 1909). Follicles comprise a large, spherical oocyte surrounded by an acellular coat, a single layer of follicle cells, a thick, multilamellar basal lamina, and some connective tissue cells resembling fibroblasts (Fig. 5) (Talbot, 1981a; Schade and Shivers, 1980; Kessel, 1968). Close to the time of ovulation, oocytes are approximately 1.6 mm in diameter. In *H. gammarus*, the wet weight of each spawned egg is 3.69 mg (Pandian, 1970a,b), while the smaller egg of *H. americanus* weighs 2.2 mg. The total percentages of fat, protein, and carbohydrate in *H. gammarus* eggs are 43.8, 47.4, and 5.1%, respectively (Pandian, 1970a), but these percentages are not known for *H. americanus*.

Mature oocytes of *Homarus americanus* are recognized not only by their large size, but also by their dark-green color (Herrick, 1909). This color is due to the presence of a carotenoid pigment, astaxanthin (Goodwin, 1951; Cheesman *et al.*, 1967), which is noncovalently bound to a protein (Wallace *et al.*, 1967). Stern and Salomon (1937) named this lipoprotein complex *ovoverdin*, but later authors suggested that the major high-density lipoproteins found in mature crustacean eggs be referred to as *lipovitellins* (Wallace *et al.*, 1967; Nelson *et al.*, 1988a). Lipovitellins are the major yolk protein of crustacean oocytes. In *H. gammarus*, the astaxanthin has been found to be a mixture of three optical isomers (Ronneberg *et al.*, 1980). The molecular weight of lipovitellin has been estimated by several different procedures and is in the range of $3.0–3.8 \times 10^5$ (Wyckoff, 1937; Ceccaldi *et al.*, 1966; Wallace *et al.*, 1967). More than one type of lipovitellin has been reported in mature eggs of *H. americanus* (Nelson *et al.*, 1988b). The lipovitellin from *H. gammarus* is especially rich in glutamic acid and moderately rich in serine, aspartic acid, and leucine (Zagalsky, 1985). The biological significance of the carotenoid pigment associated with lipovitellin is not known. There are contradictory reports that it is *not* metabolized during embryonic development (Goodwin, 1951) and that it *is* metabolized (Ceccaldi, 1968).

Mature oocytes of *Homarus americanus* are packed with electron-dense yolk granules (Fig. 5) that are synthesized during vitellogenesis and are the presumed repositories of lipovitellin (Talbot and Goudeau, 1988). In addition, mature oocytes contain numerous cortical granules beneath their plasma membrane (Talbot and Goudeau, 1988). Based on differences in electron density, three types have been identified: high-density, low-density, and moderately dense vesicles. A fourth type, readily identifiable by its ring-shaped inclusions, has been called *ring vesicles* (Talbot and Goudeau, 1988). Because of their distinctive ultrastructural appearance, it has been possible to trace the formation of ring vesicles. They are synthesized on the rough endoplasmic reticulum, but are not processed through the Golgi apparatus (Kessel, 1968). The release of the cortical granules during fertilization is discussed in Section X,E.

The mature ovarian oocyte of *Homarus americanus* (Fig. 5) is surrounded by a single coat, originally called a *chorion* (Herrick, 1909; Talbot, 1981a,b) or *vitelline envelope* (Kessel, 1968), but which should be referred to as *envelope 1* to be consistent with the literature on other decapods. Envelope 1 can be subdivided into a thin, granular outer region (layer 1A), adjacent to the follicle cells, and a thicker inner zone (layer 1B), adjacent to the oocyte. Layer 1B comprises structures resembling bottlebrushes and small, elec-

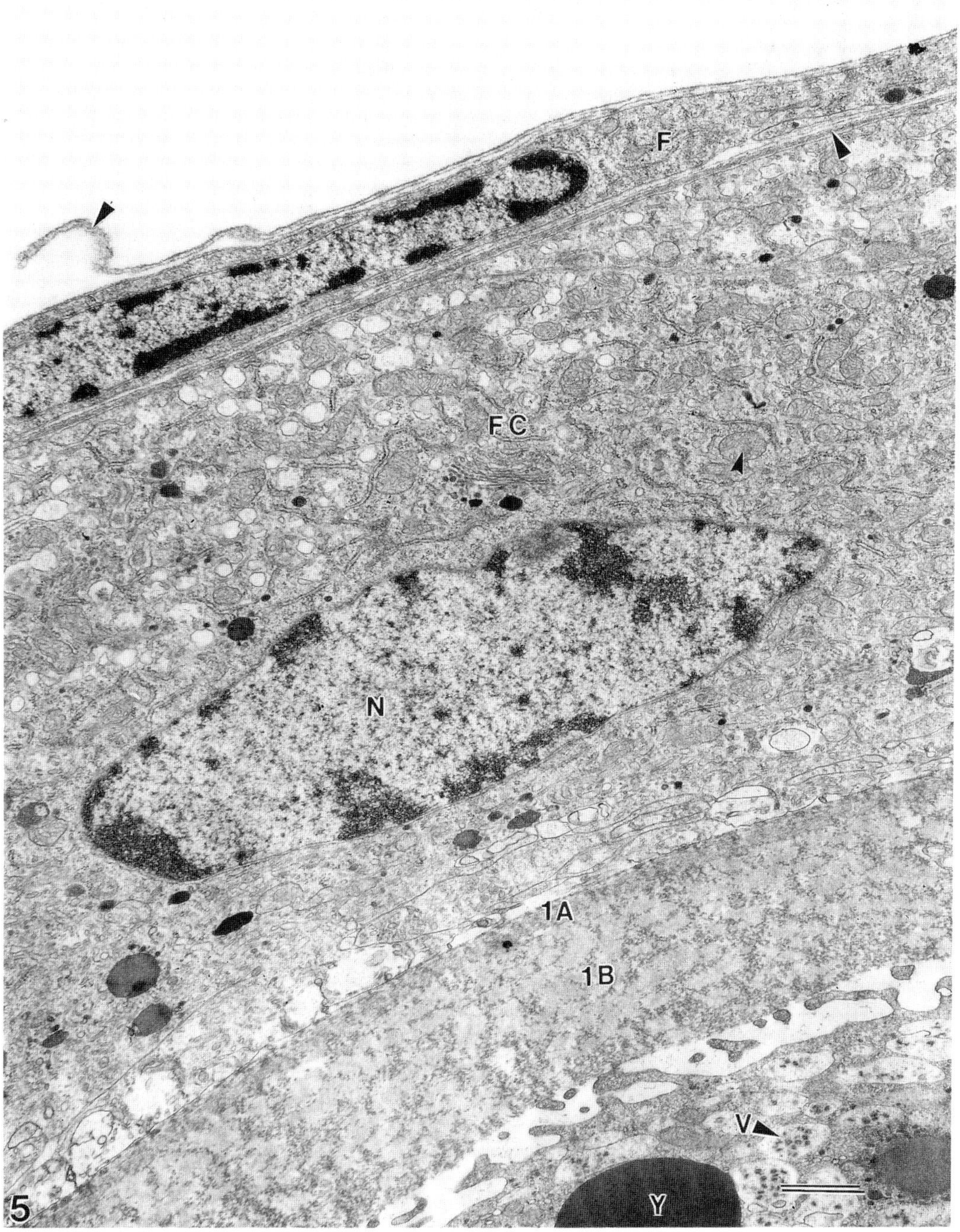

FIGURE 5 Transmission electron micrograph showing layers of a mature ovarian follicle of *Homarus americanus*. The cortex of the oocyte contains yolk (Y) and vesicles (V) with ring-shaped inclusions. (Three additional types of vesicles are not shown in this view.) Envelope 1, which surrounds the oocyte, is divided into outer (1A) and inner (1B) layers. A nucleus (N), a layer of follicle cells (FC), a basal lamina (arrow), fibroblast-like cells (F), and an outer basal lamina (arrow) are present outside envelope 1. Scale bar: 1.0 μm. (Reprinted from Talbot, 1981b, with permission of Academic Press.)

tron-dense granules, both of which have been identified in smooth-surfaced cisternae of the follicle cells that are presumed to synthesize them (Talbot, 1981a,b). The origin of layer 1A is not yet known. Envelope 1 is the only coat surrounding lobster oocytes at the time they are spawned (Cheung, 1966;

Goudeau *et al.*, 1987), and it must be penetrated by the fertilizing sperm.

B. Sperm

The sperm of *Homarus americanus* lack a flagellum and are nonmotile, except when undergoing the acrosome reaction. They exhibit the structural characteristics of other lobster and crayfish sperm, but are quite distinct from shrimp sperm. Probably because of their unorthodox appearance, lobster sperm have fascinated spermatologists for many years, and their structure has been studied extensively by both light and electron microscopy. Although numerous terms have been used to describe structures in decapod sperm, the terminology of Talbot and Chanmanon (1980a,b) is used here in order to be consistent with recent ultrastructural studies.

The sperm of *Homarus americanus* (Fig. 6) possess a cylindrical acrosome, a subacrosomal region, a "collar," and a nucleus with extensions called spikes (Fig. 6) that converge in the collar to form the arch complex (Talbot and Chanmanon, 1980b; Talbot, 1991a).

1. Acrosome

The cylindrical acrosome of *Homarus americanus* may represent the largest acrosome found in any animal species, and it is easily observed by light microscopy (Fig. 6). The acrosome is complex and shows a high degree of compartmentalization. It can be divided into an apical cap, inner acrosomal material, and outer acrosomal material. The apical cap can be subdivided into four discrete zones. Zone 1 in unreacted sperm is crystalline, but following the acrosome reaction, it contains numerous 18-nm-thick filaments. Zones 2 and 3 are crystalline and differ from each other in electron density. Zone 3 has been shown to contain an abundance of RCA_{120} (*Rincinus communis* agglutinin) binding sites, which may be involved in attaching sperm to envelope 1A of the oocyte during fertilization (Tsai and Talbot, 1994). Zone 4 is homogeneous and appears to be the only soluble part of the acrosome; it is not present in normally reacted sperm. Apical cap fractions have recently been isolated and on sodium dodecyl sulfate–polyacrylamide gel electrophoresis (SDS–PAGE) were found to contain eight major polypeptides ranging in molecular weight from 37 to 66 kDa and an additional seven bands between 16 and 29.5 kDa (Tsai and Talbot, 1994). The inner and outer acrosomal material have not yet been isolated and characterized biochemically. The complexity and size of the lobster acrosome may be related to the functional roles that have evolved for the decapod acrosome. In mature sperm, the acrosome is limited by a complete acrosomal membrane (Talbot and Chanmanon, 1980b), usually appears far more electron dense than the nucleus (which, in *H. americanus*, remains uncondensed), and can be identified histochemically by a positive PAS reaction.

2. Subacrosomal Region

The base of the acrosome is invaginated by subacrosomal material that lies outside the acrosomal membrane; it is neither PAS nor Feulgen positive (Talbot and Chanmanon, 1980b). The subacrosomal material has also been called the *percutor organ* (Pochon-Masson, 1968). The subacrosomal region is analogous in position to the perforatorium of flagellated sperm.

3. Collar

The collar is the region below the acrosome and the subacrosomal material. The nuclear envelope is not complete, and the contents of the collar are continuous with the chromatin (Talbot and Chanmanon, 1980b). While this is unusual in eukaryotic cells, an incomplete nuclear envelope has also been reported in the sperm of *Nephrops norvegicus* (Chevaillier and Maillet, 1965a,b) and natantians (Clark *et al.*, 1973). The collar contains two centrioles and the membrane lamellar complex, which comprises highly folded stacks of membrane that may contain mitochondria or degenerating mitochondria. The function of the membrane lamellar complex is not known.

4. Nucleus, Nuclear Envelope, and Spikes

Unlike nuclei in flagellated sperm, the nucleus of *Homarus americanus* sperm is uncondensed and usually appears granular or filamentous in electron micrographs (Fig. 6). It is Feulgen positive. The chromatin is confined to the main body of the nucleus and does not extend out into the spikes, which are compartments of the nucleus.

The nuclear envelope is usually closely apposed to the plasma membrane (Fig. 6); together, these two membranes have a pentalaminar appearance in electron micrographs, suggesting partial fusion between them. The nuclear envelope is absent at the anterior end of the sperm and the chromatin and the cytoplasm are in direct continuity. The plasma membrane–nuclear envelope complex is often highly infolded.

The spikes (or spines) are extensions of the nucleus and are lined by the nuclear envelope (Fig. 6). Nephropid lobsters have three spikes (other lobsters

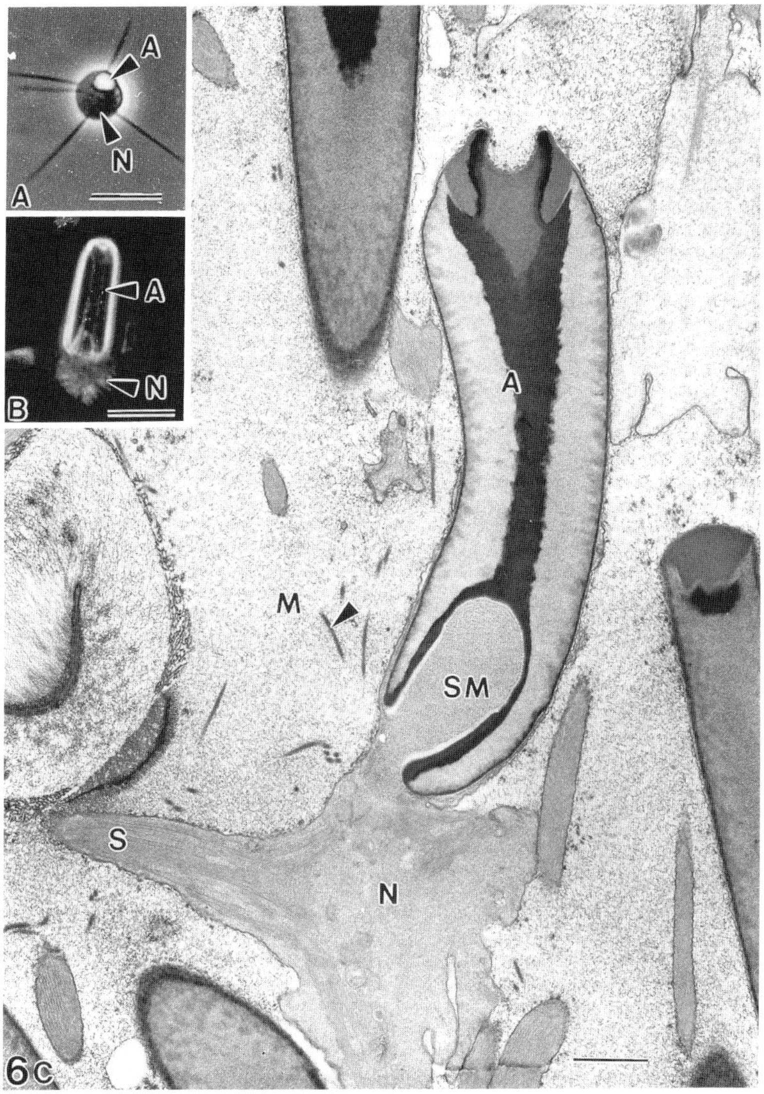

FIGURE 6 Sperm from (A) *Panulirus argus* and (B and C) *Homarus americanus*. A, Cylindrical acrosome; SM, subacrosomal material; N, nucleus; S, nuclear spikes; M, extracellular matrix; arrowhead, collagen-like fibers. Scale bars: (A) 10.0 μm; (B) 10.0 μm; (C) 1.0 μm. (Reprinted from Talbot and Chanmanon, 1980b, with permission of Academic Press.)

and crayfish may have as many as 15–20). The spikes contain microtubules that are not organized in the 9+2 pattern characteristic of the axonemes in flagellated sperm. The main function of the microtubules is apparently to give rigidity and form to the spikes, which do not beat like flagella. The spike microtubules of *Homarus americanus* sperm are not sensitive to 10^{-4} *M* colchicine (Talbot and Chanmanon, 1980b).

The microtubules in the spikes are surrounded by sheets of membrane that fluoresce bright blue when treated with the ionophore A23187 (Talbot and Chanmanon, 1980b). This "stain" has been used to show that the membrane–microtubule complex in each spike bifurcates in the collar and joins the adjacent spikes. The structure formed by the joining of these microtubules and membranes is called the arch complex (Talbot, 1991a).

5. Plasma Membrane

The plasma membrane is highly infolded (Fig. 6), creating a large surface area of membrane that is important in allowing the sperm volume to increase during the normal acrosome reaction (Talbot and Chanmanon, 1980a). The leading edge of the sperm possesses 15–16 small hooks or folds of the plasma membrane. Their function is unknown, but they are probably the first part of the sperm to contact the egg envelope during fertilization. The sperm of *Homarus americanus* have three surface antigens and share at least one antigen with *Pagurus pollicaris*, *Callinectes sapidus*, and *Libinia emarginata* sperm (Mowbray *et al.*, 1970).

6. Sperm Coats

Unlike some other species, sperm of *Homarus americanus* are packaged with their spikes extended and because of this are spaced fairly far apart in the male reproductive tract (Fig. 6) (Talbot and Chanmanon, 1980a). They also have no definitive coat or capsule. While in the male reproductive tract, sperm are suspended in an extracellular matrix that contains filamentous material (Talbot and Chanmanon, 1980b) and appears to be secreted by cells of the testes or the proximal vas deferens.

VII. Spermatophores

A. Structure and Formation

Spermatophores are packets of sperm that function foremost in sperm storage, transport, and protection. As Calman (1909) first recognized, macruran spermatophores are generally of the tubular type, in contrast to the pedunculate spermatophores of anomurans. Tubular is a misnomer, as these spermatophores are solid, not hollow; however, the term is widely used and will be retained here. The tubular spermatophores of macrurans vary in their complexity, with the simplest form being found in *Homarus americanus* (Kooda-Cisco and Talbot, 1982, 1986). Spermatophores have a central sperm mass surrounded by several acellular coats called the primary, secondary, and tertiary layers (Fig. 7). The sperm mass is produced in the testes and contains both the sperm and an extracellular supporting matrix, which is homogeneous at the ultrastructural level (Talbot and Chanmanon, 1980a,b) and often contains filaments that resemble collagen (Kooda-Cisco and Talbot, 1982). The supporting matrix is formed largely in the testes, although additions may be made in the most proximal region of the vas deferens (Kooda-Cisco and Talbot, 1986). The sperm mass is, in turn, surrounded by an acellular coat that has been given various names, but in recent literature has been described as "the primary spermatophore layer" (Kooda-Cisco and Talbot, 1982). This primary layer is secreted by the proximal region of the vas deferens and is moderately PAS positive. The sperm mass and the primary spermatophore layer remain straight and become surrounded by additional secretory material in the middle and/or distal vas deferens (Kooda-Cisco and Talbot, 1982, 1986). The acellular material outside the primary spermatophore layer is weakly PAS positive and in various species has been called the *homogeneous layer* (Matthews, 1954), the *eosinophilic layer* (Silberbauer, 1971), the *intermediate layer* (Kooda-Cisco and Talbot, 1982), the *matrix* (Martin *et al.*, 1987), and the *secondary spermatophore layer* (Talbot and Beach, 1989). The tertiary layer of the *H. americanus* spermatophore, also referred to as the *outer boundary layer* (Kooda-Cisco and Talbot, 1982), is made up of small filaments aligned parallel to each other. At the periphery of the spermatophore, these filaments lose their alignment and become loosely packed. In the future, it would simplify interpretations if the terms *primary*, *secondary*, and *tertiary spermatophore layers* were used to identify the acellular coats surrounding the sperm mass in *H. americanus* and other macrurans.

Apart from the PAS-positive nature of the primary spermatophore layer, little is known about the chemical composition of homarid spermatophores. In the spermatophores of other species, there is evidence for the presence of chondroitin sulfate, the absence of chitin, and the involvement of melanin production (reviewed by Subramoniam, 1990).

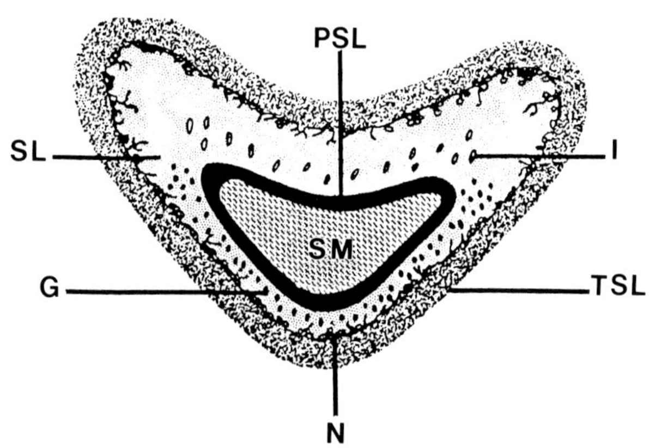

FIGURE 7 Schematic cross-section through a freshly extruded spermatophore from *Homarus americanus*. The sperm mass (SM) is surrounded by primary (PSL), secondary (SL), and tertiary (TSL) spermatophore layers. The SL contains granules (G), inclusions (I), and a network (N) of thin processes. Shown approximately 33 times the actual size. (From Kooda-Cisco, M., and Talbot, P. *J. Morphol.* 172. Copyright © 1982 Wiley-Liss. Reprinted with permission of Wiley-Liss, a division of John Wiley & Sons, Inc.)

B. Transfer to the Female

Because female lobsters do not spawn their eggs immediately after mating, it is important that sperm be transferred to them in a convenient, protective packet that can be stored until needed. The storage interval may last several years (Herrick, 1909; Waddy, 1989; see Section IX,B). Mating occurs with the ventral surfaces of the male and the female apposed. Spermatophores are extruded through the gonopores located at the base of the fifth walking leg by the contraction of striated muscle located around the distal vas deferens (Kooda-Cisco and Talbot, 1983). The transfer of spermatophores is aided by the modified first pair of pleopods of the male, which are rigid and contoured to hold the spermatophore.

Females store spermatophores internally, in a small pouch called the seminal receptacle (also called the thelycum or spermatheca), found in the midline on the ventral surface posterior to the gonopores (Fig. 2). When sperm are present in the seminal receptacle, they are found along the wall farthest from the opening; the spermatophore is held in this location by a large, brown plug that probably originates in segment 3 of the vas deferens and hardens after transfer to the female (M. Kooda and P. Talbot, unpublished data). While this hard plug is probably protective, it may not be essential for sperm survival, as females that are artificially inseminated lack this plug, yet they may fertilize eggs (Talbot *et al.*, 1986).

C. Release of Sperm

Students of lobster reproduction have not satisfactorily determined how sperm leave the seminal receptacle to fertilize eggs. The female does not release all sperm from the seminal receptacle during a single spawning. Herrick (1909) noted that "By pressing the lips of the spermatic receptacle of a female with internal eggs nearly ripe, I have observed the sperm in a thick grayish mass which gave up its cells freely to sea water," and Bumpus (1891) made similar observations. This suggests that, close to the time of spawning, sperm can exit through the ventral opening of the seminal receptacle and fertilization would then be external. In contrast, Farmer (1974) observed two very fine ducts leading from the seminal receptacle toward the oviduct in a single specimen of *Nephrops norvegicus*, suggesting that fertilization is internal. However, the ducts observed by Farmer have not been confirmed, and *Homarus americanus* oocytes that were not fertilized have been recovered from the oviducts by electrical stimulation (Talbot and Goudeau, 1988). The mechanism of sperm release from internally stored spermatophores of *H. americanus* and the precise site of fertilization are topics requiring further study.

D. Electrically Induced Extrusion

It is desirable, when working with spermatophores or sperm, to have a method of collection that does not necessitate sacrificing the male. Electrically induced extrusion of decapod spermatophores has been used in *Homarus americanus* (Kooda-Cisco and Talbot, 1983; Aiken *et al.*, 1984). A male is placed ventral side up and a pair of electrodes is positioned near the base of the fifth walking leg. Delivery of an electrical stimulus will generally induce spermatophore extrusion within several seconds. The procedure works only if a spermatophore is in the terminal vas deferens.

Sperm collected by this procedure are viable; they appear normal when examined with phase contrast and electron microscopy, they exclude trypan blue, and they are capable of undergoing normal acrosome reactions (Kooda-Cisco and Talbot, 1983). Moreover, spermatophores collected by electrical stimulation have been used successfully to artificially inseminate female lobsters, indicating that sperm collected by this method are fertile (Waddy and Aiken, 1985a; Talbot *et al.*, 1986).

Both freshly caught wild lobsters and those maintained or raised in captivity will extrude spermatophores when stimulated electrically (Aiken *et al.*, 1984; Talbot *et al.*, 1983). However, both the quality and quantity of spermatophores have been reported to decrease with time in captivity (Aiken *et al.*, 1984). Stressed males extrude poor-quality spermatophores, but they can recover and produce normal spermatophores after about 3 months in captivity (Talbot *et al.*, 1983). Viable sperm have been collected from *Homarus americanus* throughout the year with this procedure, indicating that there is not a seasonal production of sperm in this species (Kooda-Cisco and Talbot, 1983; P. Talbot, unpublished data). This is consistent with the finding that the vas deferens does not undergo regression or exhibit cyclic seasonal activity (Aiken and Waddy, 1986a).

Spermatophores collected by electrically induced extrusion have been used for morphological studies (Kooda-Cisco and Talbot, 1982), the creation of sperm banks (Ishida *et al.*, 1986), and artificial insemination of females (Aiken *et al.*, 1984; Talbot, 1984; Waddy and Aiken, 1985a; Talbot *et al.*, 1986). In addition, this technique has been used to collect spermatophores from laboratory-generated hybrids of *Homarus ameri-*

canus and *H. gammarus* and the reciprocal cross. Although the hybrid males extrude the components of the spermatophore wall, their spermatophores consistently lack sperm (Talbot *et al.*, 1983, and unpublished data). Thus, this procedure can be used to quickly cull infertile males from breeding programs.

E. Spermatophore Storage

The ability of female lobsters to store spermatophores for long intervals before using the sperm for fertilization has suggested the possibility of banking spermatophores for subsequent experimental manipulation or artificial insemination. In fact, the only macruran sperm to have been successfully stored to date are those of *Homarus americanus*. Spermatophores collected by electrical stimulation, placed under paraffin oil, and maintained at 4–7°C contained morphologically normal sperm capable of undergoing normal acrosome reactions after 289 days of storage (Ishida *et al.*, 1986). Thick-walled spermatophores store better than those with thin walls, and bacterial contamination of some samples results in degeneration of sperm. Females artificially inseminated with banked spermatophores have successfully fertilized eggs, indicating that banked sperm retain their fertility (Talbot *et al.*, 1986). This procedure is simple and may be readily adapted to aquaculture facilities or laboratories with breeding programs. Such methods for controlling reproduction are important in the domestication of any species. Development of these procedures will also aid in the domestication of other economically valuable crustaceans. (See Aiken and Waddy, Chapter 8, on aquaculture of *H. americanus*.)

F. Abnormal Spermatophores

In macrurans, unlike some other decapods, degeneration of the vas deferens and the spermatophore is rarely observed either in wild or captive specimens. Fresh, wild-caught specimens of *Homarus americanus* have infrequently exhibited darkening around the gonopore, and in such individuals spermatophores are not normally extruded upon electrical stimulation (Kooda-Cisco and Talbot, 1983). Occasionally, however, such males have released abnormal spermatophores that were hard and dark and lacked normal sperm (Talbot, 1984). In male lobsters with this problem, often only one gonopore is affected. The factors causing this problem have not yet been analyzed in lobsters; however, it seems to be relatively rare in both wild and captive males.

VIII. Mating

A. General Features

Information on the mating behavior of *Homarus americanus* has been obtained from both field and laboratory studies (Atema and Cobb, 1980; Atema, 1986). Atema and Voigt provide a detailed discussion of mating behavior in Chapter 13; a brief account follows. Males establish themselves in shelters and await the visit of premolt females. The dominant male in the area (in a naturalistic aquarium) is usually the largest and is the one selected by the premolt females. Pair bonding occurs between a dominant male and a premolt female, who approaches the male's shelter and ejects her gill current into his den. This current may contain a sex pheromone. As the female enters the male's shelter, he fans the water with his pleopods, and the pair engage in a "boxing" ritual, using their chelipeds. Boxing bouts are generally several seconds to minutes in length and the female may become quite violent just prior to molting. Pair formation is thought to be established by both the behavior of premolt females and a female sex pheromone (Atema *et al.*, 1979). One hour before she molts, the female engages in "knighting" behavior in which she faces the male and positions her claws on top of him. The purpose of this behavior is probably to convey a chemical signal to the male. Knighting may signal the male that molting is imminent (Atema, 1986). Mating follows molting by almost exactly 30 minutes, and copulation per se lasts about 8 (Atema and Engstrom, 1971) to 60 seconds (Templeman, 1940a). The newly molted female is thought to release a sex pheromone that suppresses aggression and induces courtship behavior (Atema and Engstrom, 1971). After mating, the female remains in the male's shelter for about 1 week. Presumably, the male protects the female and his investment in her offspring during the early postmolt period, when she is most vulnerable. A second female may begin cohabitation with a dominant male shortly after the first female leaves his shelter. In fact, premolt females stagger their molts so that they can each mate and cohabit with a dominant male (Cowan and Atema, 1990). This observation, which was made using large, naturalistic aquaria, suggests that female lobsters exert behavioral control over their molt cycles, enabling lobsters to use a mating system of serial monogamy.

B. Relationship between Mating and Molting

Mating is generally thought to take place within 48

hours of molting, while the female is still soft and the spermatophore can easily be internalized (Templeman, 1940a). However, there is evidence that mating may occur when the female is in intermolt and has a hard shell (Dunham and Skinner-Jacobs, 1978; Waddy and Aiken, 1990a,b). Under laboratory conditions, females may mate at any stage of the molt cycle. Intermolt mating occurs more frequently with uninseminated, preovigerous females than with uninseminated, nonpreovigerous females, and females appear to lose their receptivity to males once sperm are present in their seminal receptacle (Waddy and Aiken, 1990a,b). The remains of the spermatophore are shed with the exoskeleton at the time of molting (Templeman, 1936).

C. Relationship between Mating and Spawning

Females retain spermatophores for long periods before spawning and fertilization. Females often molt and mate one summer, then spawn eggs the following summer (Templeman, 1940a; Waddy and Aiken, 1986), although molting and spawning may occur in the same season (Ennis, 1984b). Spermatophores may be stored for as long as 2 years (Templeman, 1934). Large females (>120-mm carapace length) have been shown to molt, then undergo two successive spawns before molting again (Waddy and Aiken, 1986), suggesting that sperm from the initial mating may be used to fertilize eggs during more than one spawn (i.e., multiple fertilization) and may be stored for 3 years before they are used. After larvae hatch from a berried female, the seminal receptacle may still contain a rich supply of sperm (Templeman, 1936), suggesting that only a fraction of the total stored sperm is used in a single spawning. It is not yet understood how the female controls the partial release of stored sperm.

A female may spawn without carrying a spermatophore, in which case the eggs are not fertilized. In some populations of *Homarus americanus*, mature, uninseminated females are relatively common (Krouse, 1973; Ennis, 1980). While such females can spawn and fail to fertilize eggs, recent laboratory observations indicate that if these lobsters are kept in a communal tank, they will undergo intermolt mating (Waddy and Aiken, 1990a). If similar intermolt matings occur in wild populations, spawning without prior insemination may be very rare. The minimum size at which females are found to carry spermatophores and thus mate is 63- to 71-mm carapace length (reviewed by Van Engel, 1980). Spawning without prior mating has also been seen in palinurid lobsters.

D. Repeated Matings

More than one male may mate with a molted female (Templeman, 1934), and a particular male may mate with more than one molted female (Templeman, 1934; Atema, 1986). Electrophoretic data have determined that two different males may deposit sperm in a single female's seminal receptacle and that sperm from both males may be used to fertilize eggs in the subsequent spawning (Nelson and Hedgecock, 1977).

IX. Ovulation and Spawning

A. Ovulation

Ovulation is the escape of a mature oocyte from its follicle. Lobster oocytes develop in simple, solid follicles. At least three changes must occur in a follicle in order for ovulation to take place: the follicle cell processes must be retracted from the envelope surrounding the oocyte; a rupture site must form in the apex of the follicle; and the oocyte must pass through the rupture site and into the lumen of the ovary. *In vivo* ovulation has not been studied in lobsters because it is difficult to regulate precisely, and we do not yet have a means to predict the time of ovulation accurately. Nevertheless, *in vitro* experiments have shown that ovulation will occur when mature ovarian strips are incubated in solutions containing collagenase and that collagenase-induced ovulation can be blocked by collagenase inhibitors (Talbot, 1981c). During *in vitro* ovulation, the follicle wall contracts and causes extrusion of the oocyte (Talbot, 1991b). Follicular contraction is a common feature of ovulation in most animal phyla (reviewed by Schroeder and Talbot, 1985). In species having solid follicles, such as *Homarus americanus*, the oocyte is squeezed into a dumbbell shape as it passes through the rupture site. Contraction of the follicle wall is probably important in ensuring that oocytes escape through the rupture site. Follicle cells have recently been shown to contain numerous microfilaments that may bring about this contraction (H. Al Hajj, D. Howard, and P. Talbot, unpublished data). The factors that regulate the events associated with ovulation *in vivo* are not yet known. (See Waddy *et al.*, Chapter 10, on control of reproduction.)

B. Spawning

Spawning is the passage of oocytes from the ovary through the oviduct and gonopore to the exterior of the female. Females position themselves on their back during spawning (Fig. 8) and the abdomen is flexed forward to form a brood chamber that catches the eggs (Templeman, 1937; Herrick, 1909). As eggs exit the gonopore, they move over the ventral surface of the female (Scott, 1903), probably drawn toward the brood chamber by water currents created by the beating pleopods (Templeman, 1937). Upon reaching the brood chamber, the eggs become attached to the ovigerous setae. If a female is disturbed during spawning, the eggs may drop off and the clutch will not be completely formed.

Ovulation and spawning are not necessarily temporally coupled. Upon dissection, mature ovaries will often yield numerous free eggs that have been ovulated but not spawned (Herrick, 1891; Aiken and Waddy, 1980). It is not known how long lobsters hold eggs in the ovary before spawning occurs. Spawning is a delicate process that requires precise positioning of a female in an undisturbed state (Templeman, 1937). Initiation of a spawn may require that behavioral and environmental conditions be met. There would be an advantage, under such circumstances, to having oocytes already ovulated and ready for extrusion immediately upon assuming the spawning posture.

The mechanics of egg extrusion have not been investigated experimentally, but it is generally assumed that the muscles of the ovarian wall contract to force oocytes into the oviducts and eventually through the gonopore. This idea is supported by the observations that this muscle is capable of massive contraction (Howard and Talbot, 1992) and that the ovaries of females that have just spawned are very small compared to their size prior to spawning. The factor(s) that stimulates *in vivo* contraction of the ovarian musculature is not yet known. However, octopamine, a biogenic amine known to induce *in vivo* muscle contraction in lobsters (Kravitz, 1988), will induce strong contraction of *Homarus americanus* ovarian muscle *in vitro* (Howard and Talbot, 1992; see Section V,A).

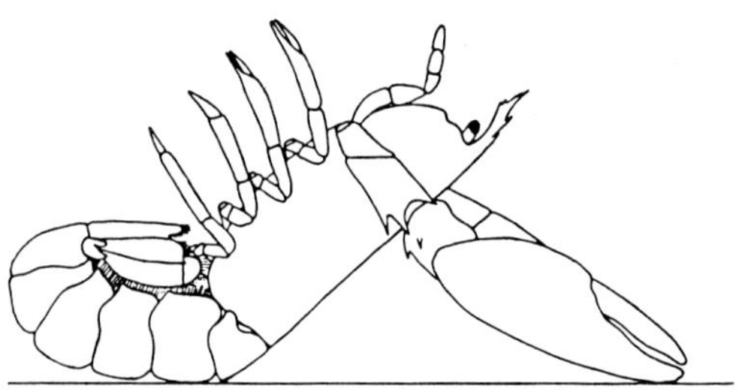

FIGURE 8 Spawning posture of *Homarus americanus*. The female is positioned ventral surface upright, and her tail folds upward to create a basket around the pleopods. (From Templeman, 1937.)

Females spawn their eggs in the spring or summer. Since the ovaries mature on a biennial cycle, most females spawn once in alternate summers. The exact time of spawning is often difficult to predict; however, there is a good correlation between the development of pleopod tegumental glands and egg extrusion (Aiken and Waddy, 1982; Ennis, 1984c). Most females spawn during July and August, but as many as 20–25% of the wild population may extrude eggs during other seasons (Herrick, 1894, 1895b). Water temperature can affect the time of spawning; for example, spawning occurs earlier in the warm water of the Gulf of St. Lawrence than in the cool water of the Bay of Fundy (Waddy and Aiken, 1990a).

In captivity, females held in unfavorable conditions may mature and spawn only a few hundred eggs (Knight, 1918). However, in laboratories where conditions are satisfactory, the number of eggs produced by females is equivalent to that expected in wild populations and can reach 10,000 (Talbot *et al.*, 1984). Fecundity is related to the size of the female, ranging from a few thousand eggs in a young animal to several tens of thousands in older individuals (Aiken and Waddy, 1980).

Ovulation and spawning are complex processes that are not yet well understood in lobsters. Much more work is required in order to fully comprehend the mechanisms underlying these events.

X. Fertilization

A. *Site of Fertilization*

Although the site of fertilization has not been established with certainty for *Homarus americanus*, most observations are consistent with external fertilization as the eggs pass over the seminal receptacle during spawning (reviewed by Aiken and Waddy, 1980; see Section VII,C).

B. *Binding of Sperm to Egg*

Little is known about the process of sperm binding to envelope 1A of the egg during normal spawning and fertilization in *Homarus americanus*; our understanding is based on *in vitro* studies. When *in vitro* inseminations are carried out, numerous sperm bind to envelope 1A, where they undergo the acrosome reaction (Talbot *et al.*, 1991a). Sperm that are not reacted are rarely observed on the vitelline envelope, suggesting that unreacted sperm readily detach from the oocyte surface or that once binding has occurred, sperm tend to rapidly undergo an acrosome reaction. Not all lobster sperm bind to envelope 1A with the proper orientation, and when poorly oriented sperm react, they fail to penetrate the envelope. Only sperm that bind by the front end of the acrosome appear to pass through the envelope during the acrosome reaction. In cross-insemination experiments, sperm from a variety of decapods were capable of attaching only to egg coats of the same species, with the exception of *H. americanus* sperm, which also bound weakly to *Callinectes sapidus*, *Cancer irroratus*, and *Ovalipes ocellatus* egg coats (Mowbray *et al.*, 1970).

C. Acrosome Reaction

The acrosome reaction is an exocytotic event that results in eversion of the acrosomal vesicle during fertilization; in lobsters, it is accompanied by formation of an acrosomal filament. Although acrosomes have been observed in all macruran species studied to date, the normal acrosome reaction has been well characterized only in *Homarus americanus*. This event fascinated many early microscopists (e.g., Herrick, 1909), who described the reaction as *sperm explosion*, *eversion*, or *devagination*. Barker and Austin (1963) first recognized that this explosive event in a decapod (*Emerita talpoida*) was equivalent to the acrosome reaction of flagellated sperm.

The morphological features of the normal acrosome reaction have since been determined for *Homarus americanus* at the ultrastructural level (Talbot and Chanmanon, 1980a). Frame-by-frame videotape analysis of ionophore-induced reactions have shown that the acrosome reaction occurs in four steps (Fig. 9) (Tsai and Talbot, 1993). During the first step, the apical cap expands and the remainder of the acrosomal cylinder swells in an apical-to-basal direction. In the second step, the apical half of the acrosome is gradually everted. During eversion, the inner and outer acrosomal material hydrate and in so doing move forward through the opening in the apical cap. Eversion of the apical part of the acrosome pulls sperm forward by about 12.7 μm. At the end of step 2, the apical cap contracts around the arch complex. During step 3, the subacrosomal material and the nucleus plus its associated spikes are ejected forward 3.9 μm into the cavity created by acrosomal eversion. Ejection is brought about by eversion of the basal half of the acrosomal cylinder. During step 4, the apical cap undergoes final contraction and causes a final forward movement of the nucleus and extrusion of the acrosomal filament, resulting in an additional advance of 1.1 μm. For sperm in suspension, the entire event takes about 1 second. As a result of the reaction, the sperm is turned inside out, a filament is formed from the subacrosomal material at the leading edge of the sperm, and the nucleus increases in volume by a factor of about 2.6. The contraction of the apical cap is an interesting feature of the lobster sperm acrosome reaction, as this contraction occurs in an extracellular compartment (Tsai and Talbot, 1993). The protein(s) responsible for apical cap contraction has not yet been identified, although actin appears not to be involved (Tsai and Talbot, 1994).

The acrosome reaction normally occurs as sperm pass through envelope 1 during fertilization (Talbot *et al.*, 1991b). An important established function of the reaction in *Homarus americanus* is to generate forward movement in this normally immotile cell. During the reaction, sperm are propelled forward a total of about 18 μm (Talbot and Chanmanon, 1980a; Tsai and Talbot, 1993), a distance sufficient to get the sperm through the 3- to 4-μm-thick envelope that surrounds the lobster egg. Presumably, sperm that react prema-

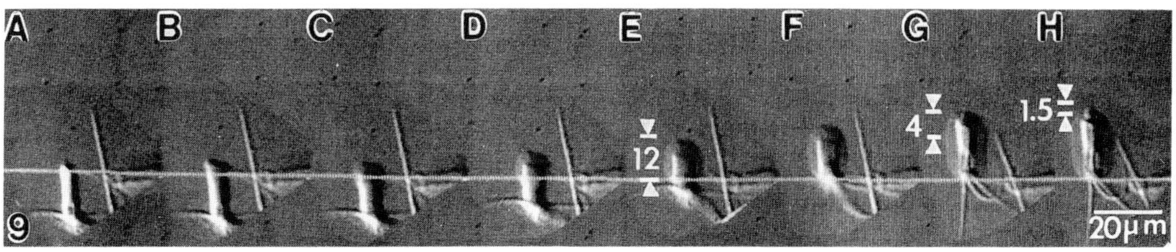

FIGURE 9 Sequential video images of a lobster sperm undergoing a normal acrosome reaction. The sperm in (A) is unreacted, and its apical cap is positioned on the white line. The remainder of the figure shows the relative forward movement of the sperm as it reacts. During step 2 (B–E) of the reaction, the acrosome begins everting and the sperm moves forward about 12 μm. During step 3 (F and G), the sperm advances another 4 μm. A final 1.5 μm is gained during step 4 (H). See Section X,C for a description of steps. Scale bar: 20.0 μm. (From Tsai, I., and Talbot, P. *Molec. Reprod. Develop.* 36. Copyright © 1993 Wiley-Liss. Reprinted with permission of Wiley-Liss, a division of John Wiley & Sons, Inc.)

turely lose their sole opportunity to move forward and thus are infertile. It is not yet known whether lobster sperm penetrate envelope 1 enzymatically or mechanically. It has been postulated in two crabs that the acrosome reaction exposes enzymes that aid in penetration of the egg coat (Brown, 1966; Hinsch, 1971). Electron micrographs of *H. americanus* sperm penetrating the vitelline envelope show no evidence of enzymatic digestion of the envelope (Talbot *et al.*, 1991b). The vitelline envelope is very soft at the time of fertilization, and the mechanical force produced by the reacting acrosome may be sufficient for penetration.

The acrosome reaction has been studied in *Homarus americanus* because of the ease with which it can be induced in this species. The acrosome is sufficiently fragile that mechanical agitation (Gesteira and Halcrow, 1988) or compression under a cover slip (P. Talbot, unpublished observations) will elicit a normal reaction. Of greater physiological significance, there is evidence that Ca^{2+} influx is important in initiating the acrosome reaction. Herrick (1909) observed that isotonic calcium chloride induces the reaction in lobster sperm. More recently, it has been shown that the calcium-transporting ionophore A23187 also induces the reaction (Talbot and Chanmanon, 1980a). Furthermore, the reaction fails to occur in the absence of extracellular calcium; cysteine, which may chelate calcium, is an inhibitor of the reaction (Gesteira and Halcrow, 1988).

D. Mechanism of Fertilization

The only macrurans in which fertilization has been directly observed are *Homarus gammarus* (Goudeau and Goudeau, 1986) and *H. americanus* (Talbot *et al.*, 1991a,b). Recently, a method has been developed for harvesting nearly mature oocytes from females close to natural spawning; when combined *in vitro* with sperm from the vas deferens or the seminal receptacle, they undergo polyspermic fertilization, but it is not known whether this is normal (Talbot *et al.*, 1991a,b). Unfortunately, methods are not yet available for the culture of embryos following *in vitro* fertilization.

Gamete interactions have been studied at the ultrastructural level using *Homarus americanus* oocytes undergoing *in vitro* fertilization (Talbot *et al.*, 1991b). All sperm found in the perivitelline space have undergone an acrosome reaction, providing further evidence that the reaction is necessary to propel sperm through envelope 1. Sperm appear to enter the oocyte by fusion of the membrane over the acrosomal filament with the oolemma (Fig. 10). A small fertilization cone forms at each site of sperm fusion. The acrosomal filament is still identifiable in the ooplasm 120 minutes after insemination; however, the microtubules of the spikes and the arch complex have depolymerized and the nuclear membrane has begun to fragment.

Using an *in vitro* system for fertilizing *Homarus gammarus* oocytes, several important consequences of fertilization have been observed (Goudeau and Goudeau, 1986). (Similar information is not available for *H. americanus*.) Meiosis, which is arrested in metaphase I at the time of ovulation, resumes and the first polar body is formed by 1 hour after fertilization. During final maturation, the oocyte plasma membrane becomes relatively more permeable to Cl^- and less permeable to K^+. After fertilization, the oocyte plasma membrane shows an increased permeability to K^+ and undergoes a fast, sustained hyperpolarization that corresponds to the fertilization potential. While similar increases in K^+ conductance have been observed in many other animals, the fertilization potential in most species is *hypo*polarizing. The significance of membrane hyperpolarization in reptantians has not yet been established, but may provide a mechanism for blocking polyspermy.

E. Cortical Reaction

Envelope 1 is the only coat surrounding oviducal oocytes or oocytes extruded through the gonopore in response to electrical stimulation. This coat swells in sea water even in the absence of fertilization. In fertilized oocytes of *Homarus americanus*, a complex cortical reaction results in deposition of a new coat—envelope 2—between envelope 1 and the oolemma (Fig. 11). During the cortical reaction, the four different types of cortical granules (see Section VII,A) are released in sequential order. The fate of the high- and low-density vesicles, which are released first, is not known, as their contents cannot be readily identified after release. The contents of the moderately dense and ring vesicles, which are released later, coalesce to form envelope 2. Envelope 1 later condenses and fuses with envelope 2 to form the fertilization envelope, which persists around the developing embryo

FIGURE 10 Transmission electron micrograph of a lobster sperm that has penetrated envelope 1 (1B) and fused with the oolemma of an oocyte. The acrosomal filament (F) appears unchanged in the ooplasm. The nuclear envelope (arrows) is fragmented and less compact. Egg cytoplasm (OC) surrounds the sperm nucleus (N) and forms a small fertilization cone on the surface of the oocyte. The remains of the acrosome (AM) are outside the oocyte in the perivitelline space (*). Scale bar: 5.0 μm. (Reprinted from Talbot *et al.*, 1991, with permission of Academic Press.)

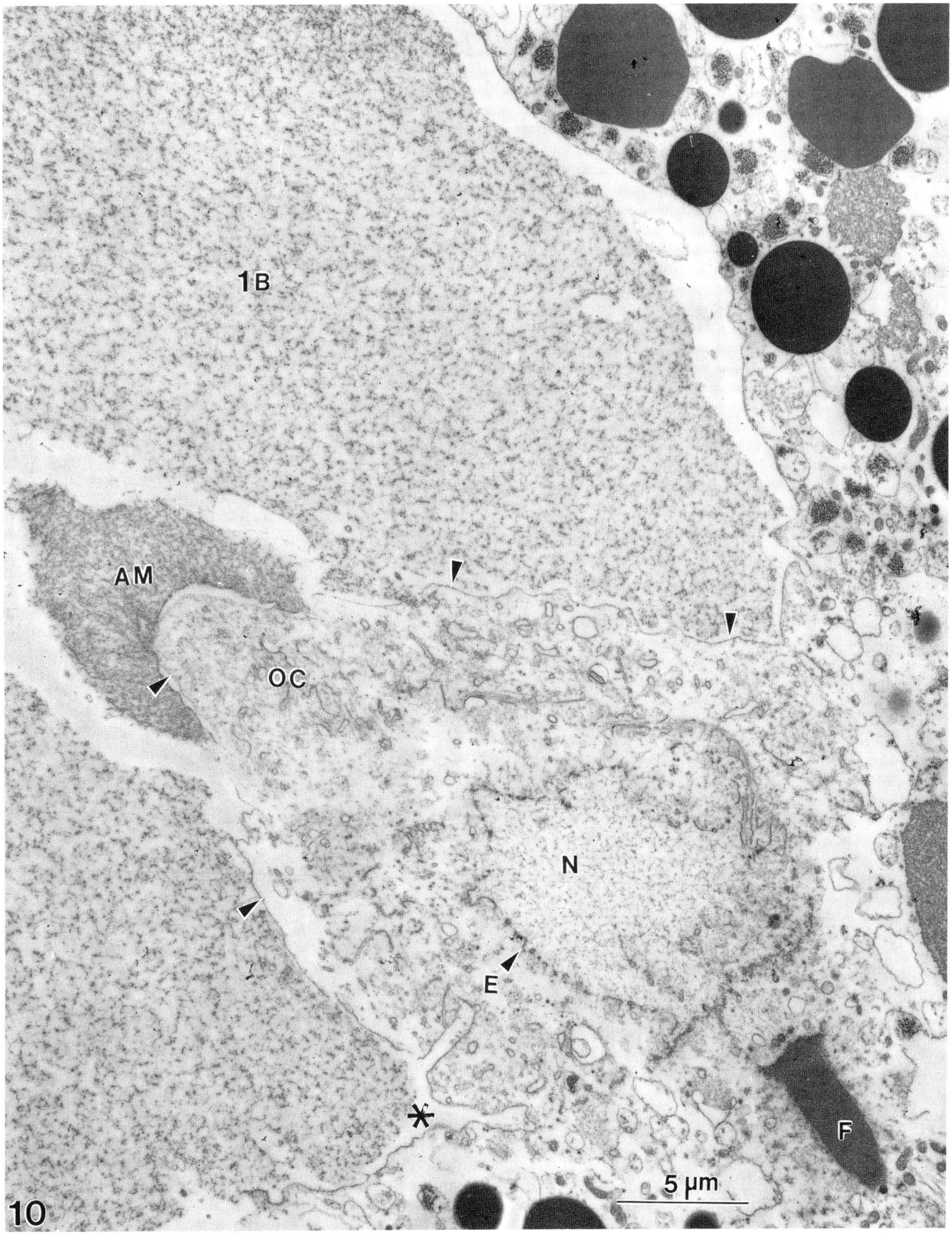
1B
AM
OC
N
E
F
5 μm
10

until the time of hatching. The time required to complete the cortical reaction is not yet known, but it may be quite long. Oocytes that are fertilized *in vitro* retain their low-density and ring vesicles up to 120 minutes after insemination (Talbot *et al.*, 1991b). The cortical reaction of *H. americanus* is more complex than that reported for any other decapod, but is similar to that of some crabs (e.g., *Carcinus maenas*, Goudeau and Becker, 1982).

F. Success of Fertilization

In natural spawns of *Homarus americanus*, the percentage of eggs that are fertilized and develop is not specifically known, but is thought to be generally high, since the greatest percentage of possible matings occurs with the male slightly larger than the female (Templeman, 1934). Insemination of mature females may be limited by the relative scarcity of larger males, which are exploited commercially (Dubé and Grondin, 1985).

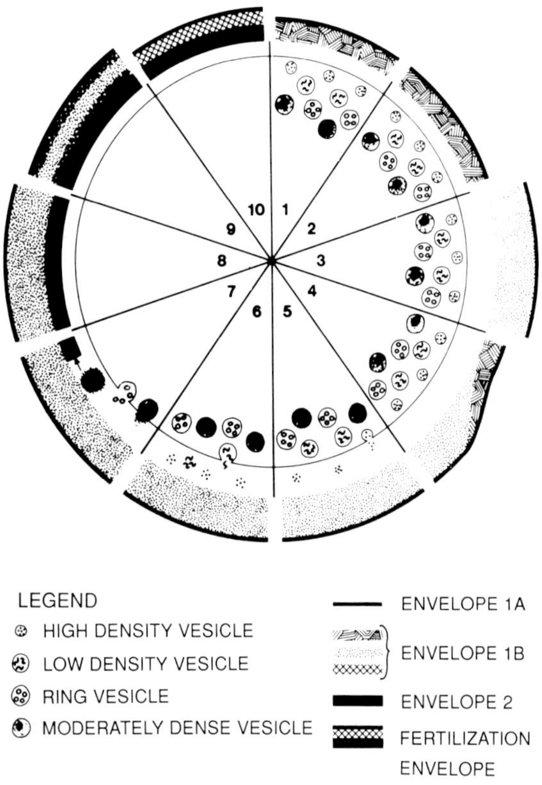

FIGURE 11 Schematic diagram showing the cortical reaction and formation of the fertilization envelope in *Homarus americanus*: (1) ovarian oocyte; (2) "dry" gonopore oocyte (i.e., oocyte extruded through the gonopore but fixed without exposure to sea water; (3) gonopore oocyte after 45 minutes in sea water; (4) oocyte fixed soon after onset of spawning; (5) oocyte fixed 1 hour after *in vitro* fertilization; (6–10) progressively later stages in the release of the cortical granules and formation of the fertilization envelope. (From Talbot, P., and Goudeau, M. *Gamete Res.* 19. Copyright © 1988 Wiley-Liss. Reprinted with permission of Wiley-Liss, a division of John Wiley & Sons, Inc.)

G. Hybridization

Interspecific fertilization is possible between closely related species of lobsters. Hybrids of *Homarus americanus* and *H. gammarus* have been produced in the laboratory (Hedgecock *et al.*, 1978). The sperm from these species are remarkably similar structurally to those of the wild type (Talbot and Chanmanon, 1980b; Pochon-Masson, 1968). Hybrid males are infertile, while females may produce fertile eggs if backcrossed to the wild stock (Hedgecock *et al.*, 1978).

XI. Egg Attachment and Loss

A. Egg Attachment

Although the precise site of fertilization is still conjectural for *Homarus americanus*, it is generally assumed that eggs have been fertilized by the time they reach the brood chamber. While in the brood chamber, eggs become attached to ovigerous setae and to other eggs by short stalks. These stalks were originally thought to be produced by secretions of the pleopod tegumental glands (Yonge, 1937; Lloyd and Yonge, 1940; Templeman, 1940a), which are known to cycle with the ovary (Herrick, 1909; Aiken and Waddy, 1982; Johnson and Talbot, 1987) and apparently release much of their content at the time of spawning. This idea has been challenged by several facts, for example: egg coats are present around decapod eggs (including those of *H. americanus*) while still in the ovary (Cheung, 1966); and electron microscopic analysis has shown that the attachment stalk is formed from the egg coat (envelope 1) (Goudeau *et al.*, 1987), which is secreted in the ovary by the follicle cells (Talbot, 1981a). Envelope 1 swells and becomes soft and sticky in sea water. After sticking to an ovigerous seta or another egg, the beating of the pleopods probably moves the egg, causing a portion of this coat to be drawn off the egg surface as the attachment stalk. The envelope and the stalk undergo condensation within about 24 hours of spawning, after which they are no longer sticky and the egg mass forms a cohesive clutch that is normally well anchored to the female. The egg coat and the attachment stalk are extremely durable, as they must withstand the prolonged brooding period characteristic of *H. americanus* (Branford, 1978; Templeman, 1940a,b).

Eggs of *Homarus americanus* may have more than one stalk, which may either connect to adjacent egg coats or wrap around an ovigerous seta (Goudeau *et al.*, 1987). Those stalks attached to ovigerous setae wrap around a single seta. The extensive interconnection that occurs between adjacent eggs probably contributes to the compact, cohesive nature of the egg mass.

After attachment is complete, females periodically fan their pleopods and groom their eggs with the walking legs. They may also remove unhealthy or dead eggs from the clutch.

Although it has been established that the pleopod tegumental glands (also called the cement glands) do not form the attachment stalks (Cheung, 1966; Goudeau *et al.*, 1987), their role in egg attachment has not yet been clarified. Their secretions may be involved in the condensation or hardening of the coat and the stalk material (Talbot, 1991b; Talbot and Demers, 1993). [Details of the structure and function of tegumental glands are reviewed by Talbot and Demers (1993).]

B. *Egg Loss*

It is not surprising that *Homarus americanus* females normally lose 30–50% of a clutch during the long brooding interval of 9–16 months (Perkins, 1971; Campbell and Brattey, 1986). Some females lose all of their clutch, although this may not occur very frequently (Ennis, 1984a). Although data are limited, there appears to be a general trend of greater egg loss in species having longer brooding periods.

The causes of natural egg loss are not well understood. The nemertean *Pseudocarcinonemertes homari* can reproduce in a lobster egg mass (Aiken *et al.*, 1985) and consume the contents of eggs (Waddy and Aiken, 1985b). However, this is not a factor in natural populations, unless a high density of nemerteans becomes established on a clutch during the early stages of brooding (Campbell and Brattey, 1986). This species can be destroyed by dipping infested egg masses into fresh water for 4 minutes, which kills the nemertean without harming the lobster or her eggs (Charmantier *et al.*, 1991). Other factors that could affect egg retention in natural populations include removal or eating of eggs by the female (Knight, 1918), fouling by epibionts (Miller and Fleming, 1983; Harper and Talbot, 1984), and cleaning of infested egg masses by the lobster (Aiken *et al.*, 1985). Capture of gravid females in traps with handling and release may also contribute to clutch attrition (Herrick, 1909).

In captive populations, egg retention is sometimes equivalent to that found in the wild (Waddy and Aiken, 1984); however, significant egg loss may also occur in captivity after both natural mating (Hedgecock *et al.*, 1978; Talbot *et al.*, 1984) and artificial insemination (Waddy and Aiken, 1985b). Poor egg retention in captivity can be caused by several factors. Captive females may spawn the normal number of eggs, but fail to attach them because they are disturbed or improperly balanced during spawning (Talbot *et al.*, 1984); fail to produce normal ovigerous setae, so that they are unable to attach a full complement of eggs (P. Talbot, unpublished data); or attach a full clutch of eggs, but lose them within 2–3 weeks (Talbot *et al.*, 1984) because of inadequate egg stalks (Talbot and Harper, 1984) that are probably the result of incomplete formation of envelope 1 during oocyte maturation. Other factors that may contribute to egg loss in captivity include reduced fertility of some males, nutritional or environmental stress, and aberrant cleaning behavior of the female (Hedgecock, 1983).

XII. Embryonic Development

A. *Terminology and Staging*

1. Terminology

As more than 70 terms have been used to refer to the various embryonic and larval stages of decapods (Gore, 1985), a brief review of the terms applying to *Homarus americanus* seems necessary. Early organogenesis results in a naupliar stage, often called the *egg nauplius*. The embryonic stage arising from the differentiation and growth of the postmandibular appendages in the embryo has been called *postnauplius* (Helluy and Beltz, 1990; Herrick, 1895a) and *metanauplius* (Shiino, 1988; Wear, 1974; Williamson, 1982). *Prelarva* and *prezoea* generally refer to the form that is released from the egg envelopes at eclosion (hatching) and rapidly molts into the first larval stage. The *metanauplius* and *prelarva* are equivalent in the sense that the form that starts developing after the naupliar stage is the same form that hatches after a protracted embryonic molt cycle (Helluy and Beltz, 1991). The first larval stage of *H. americanus* is sometimes called a *mysis larva* (Goudeau *et al.*, 1990; Shiino, 1988) because of its number of appendages, or a first *zoea* because it locomotes with the thoracic appendages (Williamson, 1982; Anderson, 1979, 1982).

Beyond the variety of terms, some facts and equivalences are clear. The first outcome of organogenesis is the egg nauplius. The growth of the postnaupliar appendages gives rise to the metanauplius (or postnauplius). Some brief molt cycles are likely to occur at the naupliar–metanaupliar transition. When about 12% of embryonic development is complete (E12%) (see Section XII,A,2), the last of these early molts initiates the prelarval (or prezoeal) embryonic molt cycle. The first larval stage (first zoea or first mysis) differentiates under the cuticle of the growing prelarva during the last two thirds of embryonic development. Eventually, the prelarva hatches and rapidly molts into the first larval stage (stage I). These developmental events are illustrated in Figs. 12–16.

2. Staging Schemes

In the last 100 years, various methods of identifying stages in the embryonic development of *Homarus americanus* have been designed. Bumpus (1891) used letters to characterize early developmental events (Fig. 12 and Table 1). Pandian (1970b) defined egg stages I–III based on size, color, and morphological characters visible on intact eggs. The color of the yolk, changing from dark green to green–brown and the amount of yolk consumed (Figs. 15 and 16) also give some indication of the extent of embryonic development.

Herrick (1895a) noticed that the size of the dark-brown compound lateral eye provides a convenient and accurate way of assessing embryonic growth. Using this feature, Perkins (1972) devised the first quantitative assessment, the eye index, which is the average of the length and the width (in micrometers) of the dark-screening pigment spot in the embryonic lateral eyes (Table 1). The eye index has been used subsequently in a number of investigations (Cole and Lang, 1980; Kirk and Govind, 1992; Sasaki, 1984; Sasaki *et al.*, 1986; Schuur *et al.*, 1976). The eye index at hatching can vary slightly, from 560 (Perkins, 1972) to 580 μm (Helluy and Beltz, 1991). In the percent staging system developed by Helluy and Beltz (1991), an eye index of 570 μm is used to represent 100% embryonic development (Table 1 and Fig. 17). Earlier embryonic stages are identified as lower percentages (based on eye index × 100 ÷ 570) and expressed, for example, as E50%. Prior to the appearance of lateral eye pigment (at about E13% with an eye index of 70 μm), the percent staging was based on time.

In addition, molt stages C, D_0, D_1, and D_{2-3} (developed by Drach, 1939; adapted to setal and epidermal changes by Aiken, 1973, 1980; Rao *et al.*, 1973; and Sasaki, 1984), based on setal and tegumentary changes in the telson, can be used to divide the prelarval embryonic molt cycle (Fig. 14 and Table 1).

The various staging schemes are all useful for different applications. In a field situation, the color of the egg and the amount of yolk consumed are practical criteria. The eye index is a precise way of assessing development, but only after eye pigment has appeared. Percent staging is convenient for studying embryos at regular intervals during development. Bumpus' (1891) system, using letters, is still the best tool for identifying early stages (before E10%), whereas the setal changes in the telson can indicate moments of biological relevance in the embryonic molt cycle, such as the D_0–D_1 transition, the possible time of embryonic molt and hatch induction.

Because of the usefulness of the various staging schemes, researchers have adopted one or another; no one scheme has emerged as superior in all respects or has become the standard. This presents problems for comparative purposes, however, as equating the different staging schemes is not always easy. For example, stage O of Bumpus (1891) appears equivalent to E10% of Helluy and Beltz (1991) (Figs. 12O and 13A). The proportion of cephalothorax and abdomen indicates that the last stage described by Bumpus (stage R) is about E35% (Figs. 12R and 13C and D). Notable stages in embryonic development and their equivalences in several of the staging schemes are presented in Table 1.

B. Early Embryonic Development

The newly fertilized egg is enclosed by two envelopes: a vitelline coat (envelope 1, also known as the external or outer egg envelope or the chorion) synthesized in the ovaries and a more internal coat originating from the complex cortical reaction after fertilization (envelope 2, or the internal or inner egg envelope) (see Sections VI,A and X,C). The yolk-laden egg of *Homarus americanus* measures approximately 1.4–1.6 mm in diameter (Pandian, 1970b; Sasaki *et al.*, 1986; Helluy and Beltz, 1991) and undergoes superficial cleavage. The nuclei, each surrounded by an amoeboid mass of protoplasm, divide within the yolk and approach the periphery (Bumpus, 1891). When two or three divisions have taken place, shallow furrows form at the animal pole (Fig. 12C). At the morula stage (Fig. 12D and E), all the nuclei are located at the periphery of the egg. The furrows between the blastomeres extend deep into the central mass of yolk, which remains undivided. On the third or fourth day, an elongated blastopore is forming (Fig. 12G). Gastrulation is followed by the formation of the naupliar stage that lies at the surface of the egg, dorsal side apposed to the yolk. The naupliar stage, a developmental hallmark of crustaceans, is characterized by the presence of a median eye and three pairs of appendages: the antennulae, antennae, and mandibles (Fig. 12M and N). At this stage, the antennulae are uniramous, the antennae are biramous, and the mandibles appear as two small, round protuberances. The posterior or postnaupliar region arises as a tubelike caudal papilla folded on the ventral side of the germ band. A growth zone located in the caudal papilla in front of the telson primordium generates most of the segments in an anteroposterior sequence. As in most malacostracans, the growth zone of *H. americanus* consists of an external ring of 19 ectoteloblasts and an internal ring of eight mesoteloblasts (Scholtz, 1992, 1993).

At about E10%, the first twitches in the two pairs of antennae occur upon dissection. It is likely that several molts occur at this stage within the egg envelopes. Both Bumpus (1891) and Herrick (1895a) mentioned the existence of embryonic molts in *Homarus americanus*. Bumpus noted that the cuticle lifts from the embryo in the region of the compound eye between his stages N and O and that a true ecdysis follows. Based on the number of membranes enclosing the embryo, Herrick presumed that at least three embryonic molts have occurred by the time the pigment appears in the lateral eyes. Using electron microscopy, five concentric envelopes can be seen beneath envelopes 1 and 2 of *H. gammarus*; they are secreted by ectodermal embryonic cells (Goudeau *et al.*, 1990). The concentrations of two major ecdysteroids studied at the time of secretion of three of the five embryonic envelopes fluctuate; high titers of 20-hydroxyecdysone and ponasterone A are found during the onset of envelope secretion and during the last phase of secretion, respectively (Goudeau *et al.*, 1990).

C. Prelarval Embryonic Molt Cycle

1. Setal and Tegumentary Changes in the Telson

The evidence for the existence of a protracted prelarval embryonic molt cycle is based on the setal and tegumentary changes occurring in the embryonic telson of *Homarus americanus* (Fig. 14 and Table 1) (Helluy and Beltz, 1991). These setal changes are parallel to those documented in the telson of larvae (Rao *et al.*, 1973; Sasaki, 1984) and in the pleopods of juveniles and adults (Aiken, 1973, 1980) during molt cycles. The subdivisions of the premolt stage, D_0, D_1, and D_{2-3}, and their distinguishing features are those described by Sasaki (1984). (See Waddy *et al.*, Chapter 10, on the molt cycle.)

At about E12%, the last of the early molts initiate the prelarval embryonic molt cycle. At this stage, the embryonic telson is provided with six setae on each side of the bilobed telson (expressed as "6+6" setae) (Fig. 14A). At E30%, the cuticle begins to separate from the underlying structures in the telson, signaling the start of stage D_0. During D_0, the setae of the first larval stage are forming proximally and medially in the bilobed telson under the prelarval cuticle. Between E80% and E90%, about 3–4 weeks before hatching, when the full complement of setae (15+15) of the first larval stage has been formed, several dramatic modifications take place in the embryo as it enters stage D_1. The egg, which has increased very gradually in size since extrusion to reach 1.7–1.8 mm (greatest axis), begins to expand more rapidly. A very active period begins in terms of rates of utilization of biochemical constituents (Sasaki *et al.*, 1986). The lateral spines of the telson become pointed, then start to retract along with the setae (Fig. 14). Stage D_{2-3} occurs between hatching and the prelarval molt (see Section XII,D), and the lateral spines are everted, presumably helping in the hatching/molting process. The prelarval cuticle, which begins to lift from the telson at E30%, possesses the imprint of the 6+6 setae that were present at the E12% molt; this same prelarval cuticle, identifiable by this imprint, is discarded at the prelarval molt, demonstrating that there is only one instar during this period (Fig. 14A and G). Whereas the setal changes occurring in the telson of the prelarva seem very similar to those occurring during the molt cycle of larval and juvenile lobsters (Aiken, 1973, 1980; Rao *et al.*, 1973; Sasaki, 1984), the cellular and biochemical changes in the epidermis and the cuticle must be somewhat different. In the growing prelarva, there is no fixed postecdysial volume as there is in postembryonic animals. Indeed, the cephalothoracic length of the embryo grows by a factor of about 4 from the early E12% molt to the hatch molt (Fig. 13). Presumably, there is no mineralization of the prelarval cuticle. In this respect, it has also been noted that the prezoeal cuticle is different from the exuvium of older lobsters (reviewed by Gore, 1985).

2. Developmental Landmarks

The most obvious developmental events that occur during the prelarval embryonic molt cycle (Fig. 17) include the following: pigment appears in the lateral eyes at E13%, the heart starts beating; refringent intestinal granules become visible in the intestine; and the red chromatophores appear, first internally near the brain, then along the edge of the carapace. At E30%, the dark-pigment crescents have filled and the eyes assume an almond shape; the buds of the endopods of each antennula are seen under the prelarval cuticle, representing one of the first signs of

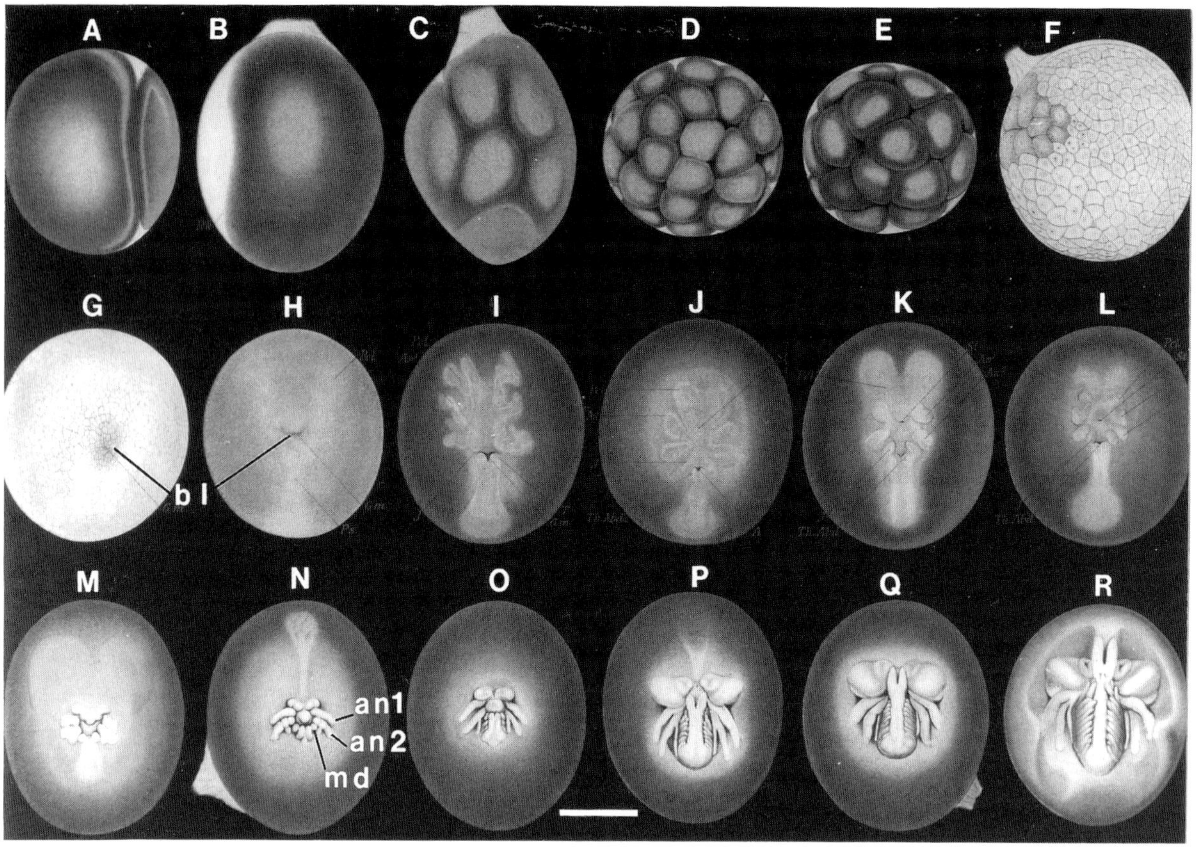

FIGURE 12 Stages in the early embryonic development of *Homarus americanus*. The drawings represent fresh eggs that have been treated with dilute nitric acid. The morula stage is viewed from the animal pole (D) and from the vegetative pole (E). The blastula is shown in (F). In (G), the egg is undergoing gastrulation, and the blastopore (bl) becomes visible. The first appearance of the germ band is represented in (H). The buds of antennulae (an1), antennae (an2), and mandibles (md) are labeled in the naupliar stage (N). Scale bar: 0.5 mm. (From Bumpus, H. C. *J. Morphol.* 5, plate XIV. Copyright © 1891 Wiley–Liss. Reprinted with permission of Wiley–Liss, a division of John Wiley & Sons, Inc.; labeling enhanced.)

the organogenesis of the first larval stage; and presumed sensory neurons and their axons are observable at the tip of the exopod of the antennulae, where a giant sensillum is differentiating. At E50%, the first tubules of the midgut gland (hepatopancreas) can be seen upon dissection at the junction of the pyloric stomach and midgut, just posterior to the two pairs of caeca containing the yolk. In embryos, the cephalothorax grows more slowly than the abdomen (Fig. 13), whereas the relationship is reversed in juveniles and adults (Lang *et al.*, 1977).

3. Significance of the Embryonic Molt Cycles

Envelopes equated to embryonic exuvia are found during embryonic life in several taxa of crustaceans, for example, amphipods (Graf, 1972), isopods (Goudeau, 1976), and decapods (Wear, 1974). Graf (1972) characterized the embryonic molt cycle of an amphipod on the basis of changes in the epidermis, the setae, and calcium storage; these changes parallel those occurring during juvenile and adult molt cycles. Embryonic molts are also known to occur in horseshoe crabs (Sekiguchi *et al.*, 1982). These aquatic chelicerates, as have many groups of insects, have

FIGURE 13 Ventral view of embryos of *Homarus americanus*. The embryos (unfixed) are dissected from the yolk at (A) E10%, (B) E20%, (C) E30%, (D) E40%, (E) E50%, (F) E60%, (G) E70%, (H) E80%, and (I) E90% embryonic development (see Section XII,A,2 for a description of the staging method). The line drawing in (J) represents a schematic embryonic prelarva, with the head up and the abdomen folded ventrally onto the thorax. te, Telson; mxp3, third maxilliped; me, median eye; le, lateral eye; ab, abdomen; an1, antennula; an2, antenna; ch, chromatophores; ig, intestinal granules; he, heart. Scale bar: 500 μm for (A)–(I). (Adapted from Helluy and Beltz, 1991, with permission of *Biological Bulletin*, Marine Biological Laboratory.)

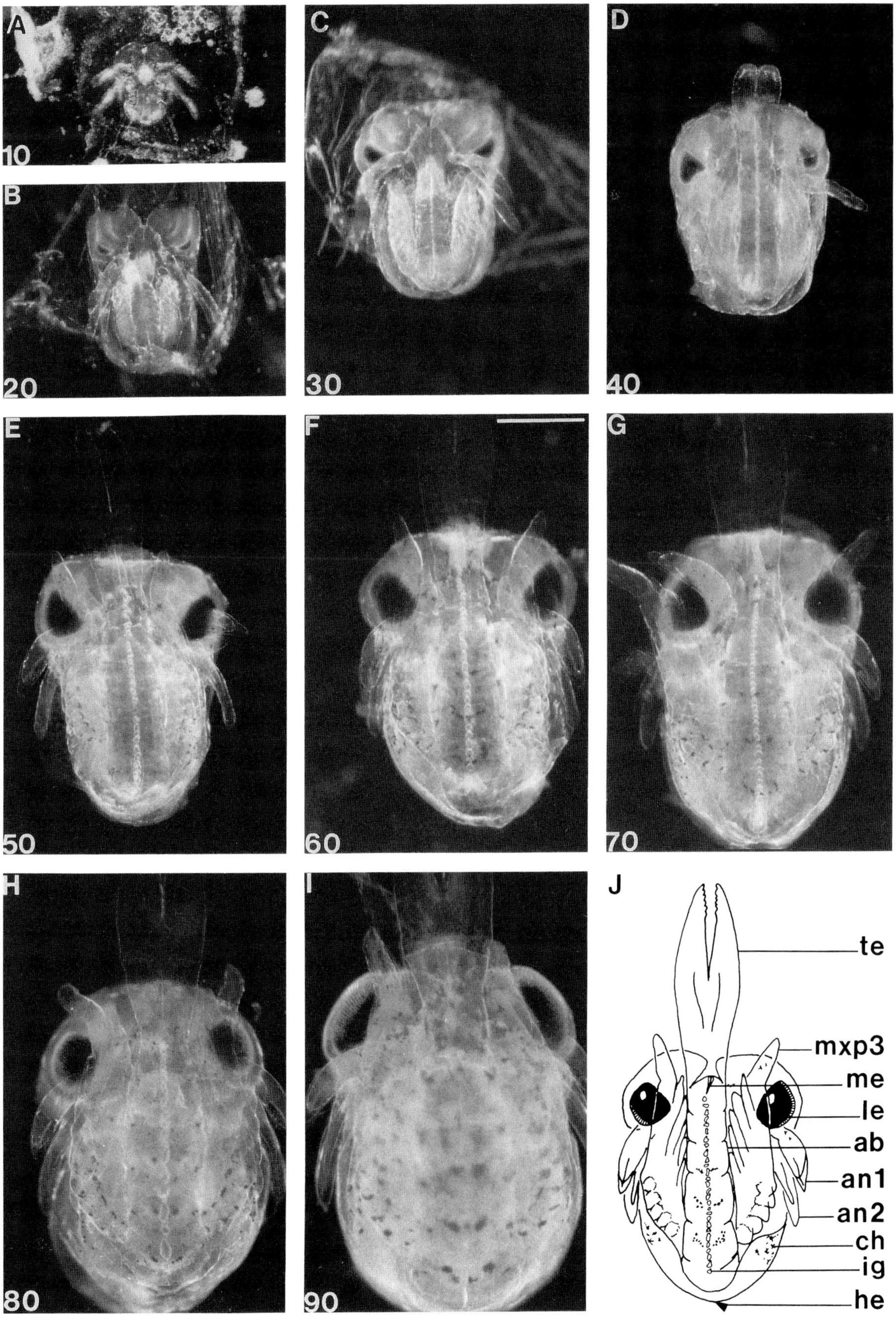
A
10
B
20
C
30
D
40
E
50
F
60
G
70
H
80
I
90
J
te
mxp3
me
le
ab
an1
an2
ch
ig
he

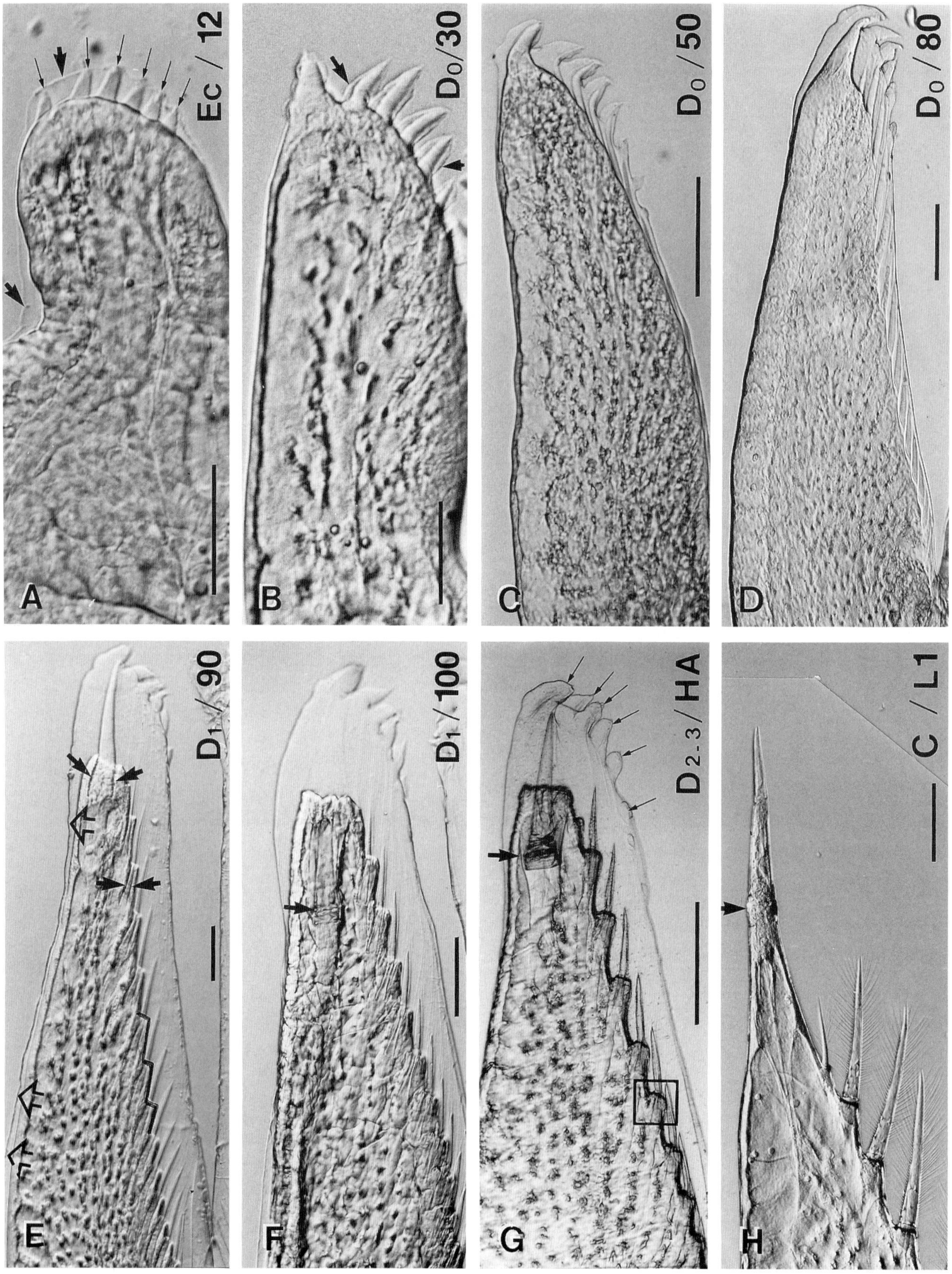
Ec / 12
Do/30
Do /50
Do /80
D1 / 90
D1 / 100
D2.3 / HA
C / L1
A
B
C
D
E
F
G
H

TABLE 1 **Equivalences among Various Systems for Staging Embryonic Development of *Homarus americanus*[a]**

(a) Percent staging	(b) Lettered stages	(c) Eye index (μm)	(d) Telson molt staging
0			
5	M		
10	O		
12			Embryonic exuvium separated from underlying structures; (6+6) prelarval setae
15		86	
20	P	114	
25	Q	142	
30		171	D_0—Prelarval cuticle begins to separate from under lying structures
35	R	200	D_0—First larval stage setae forming proximally and medially on each hemitelson under the prelarval cuticle
40		228	
45		256	
50		285	
55		314	
60		342	
65		370	
70		399	
75		428	
80		456	D_0—Full complement of setae present (15+15)
85		484	
90		513	D_1—Sharp lateral spines; spines and setae retracting; cuticle separated from the side of the telson
95		542	
100 (just prior to hatching)		570	D_1—Maximal retraction of spines and setae
Hatchling			D_{2-3}—Lateral spines extending; pronounced bulging of the epidermis around the setae

[a] (a) Percent staging system (E%) (Helluy and Beltz, 1991); (b) Bumpus' (1891) lettered stages; (c) Perkins' (1972) eye index; and (d) molt stages (developed by Drach, 1939; adapted by Aiken, 1973, 1980; Rao *et al.*, 1973; and Sasaki, 1984) of the prelarval embryonic molt cycle (Helluy and Beltz, 1991).

probably undergone embryonization (Spiridonov, 1992): formerly free larval stages, such as the prelarva, have become egg-bound during the course of evolution, a phenomenon already hypothesized by Wear (1974) for decapods. The fact that most developmental events occur early in the embryogenesis of *Homarus americanus* may reflect past functional requirements and the relegation to the egg of formerly independent stages, such as the nauplius and the prelarva.

FIGURE 14 Hemitelsons (unfixed) of *Homarus americanus* during embryonic development at (A) E12%, (B) E30%, (C) E50%, (D) E80%, (E) E90%, and (F) E100% development. (G) is a hatchling (HA) and (H) is a first-stage larva (L1). In all micrographs, the distal side is up and the medial side is to the right. The indications in the upper right corners refer to the stage of the embryonic molt cycle, followed by the percentage of embryonic development. Criteria for the various stages are described in Table 1 and features referred to are indicated by arrows. See Section XII,C,1 for a description of changes. Scale bars: (A) and (B) 50 μm; (C)–(H) 100 μm.

D. Hatching and Molt of Prelarva

At the time of hatching, the female assumes a characteristic posture, erect on the tip of her walking legs with her abdomen straightened out (Templeman, 1937; Davis, 1964; Ennis, 1975). Eggs ready to hatch display a light bluish hue ("blue embryo"), probably related to changes in hemolymph pigments. The stomach becomes dark blue; the remaining yolk,

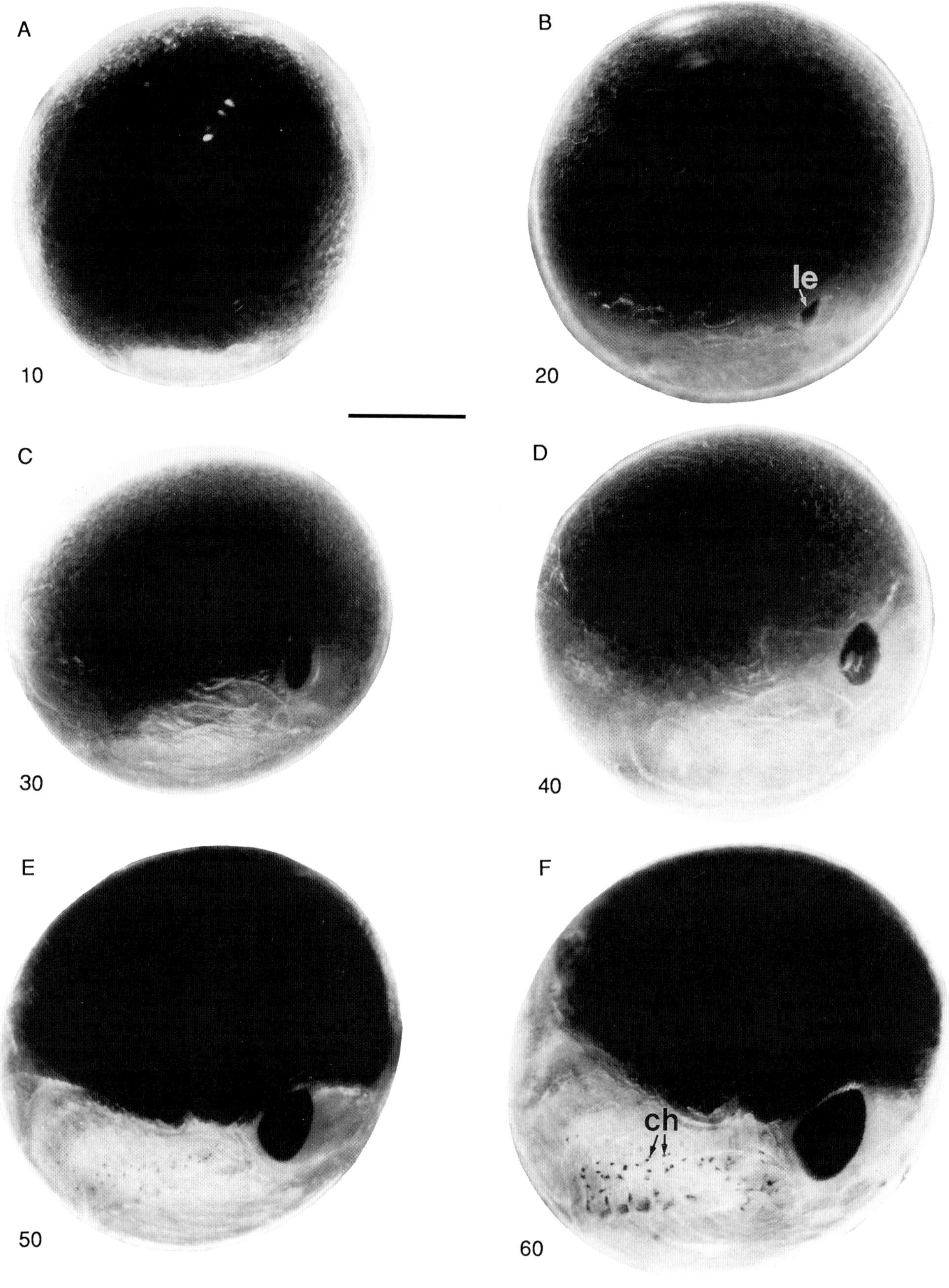
A
10
B
le
20
C
30
D
40
E
50
F
ch
60

bright yellow; and the digestive gland, light yellow. The red pigment is spreading in the chromatophores. At the time of hatching, the eggs have reached a size of 2.0 × 2.3 mm (Fig. 16).

Over the course of about 30 minutes, the outer egg envelope (envelope 1) bursts and is sloughed off from the egg completely. However, it remains attached, crumpled, both to the pleopod on one side and to the inner egg envelope (envelope 2) near the head of the prelarva on the other (Fig. 16C) (Davis, 1964). This stage is sometimes referred to as the *hatchling*. Thus, the prelarva normally remains hanging from the mother's swimmeret. Liberation from the inner egg envelope (Fig. 16D) and molt of the prelarva (Fig. 16E) is helped by the characteristic rapid movements of the swimmerets performed by the mother. The first larval stage is expelled in batches, generally at night (Ennis, 1975), ready for a planktonic life (Fig. 16F).

Hatching of all the individuals of a single brood (i.e., hatching period) can last from a few days (Hughes and Matthiessen, 1962; Pandian, 1970b) to more than 4 weeks (Ennis, 1975). The hatching period is shorter at higher temperatures. However, the thermal history of the clutch also influences the duration of the hatching period, which is longer in animals raised at continuously high temperatures (reviewed by Aiken and Waddy, 1986b).

In the ridgeback prawn, hatching is initiated by the penetration of the caudal spines through the hatching envelope (Kidd, 1991). In *Homarus americanus*, the sharp lateral spines of the telson, totally retracted in the so-called blue embryo, extend during hatching and molting (Fig. 14E–H) (Helluy and Beltz, 1991), but their exact role in the hatching/molting process is not clear.

E. Formation of Internal Organs

The embryonic development of the nervous system of *Homarus americanus* has given rise to a number of publications, whereas the ontogeny of other organs or systems has been largely ignored. The bulk of the "brain" (or supraesophageal ganglion) develops from three embryonic ganglia: the protocerebral primordium, associated with the paired lateral eyes and with the median eye; the deutocerebral primordium, associated with the antennulae and olfaction; and the tritocerebral primordium, associated with the antennae and mechanoreception. (See Beltz, Chapter 11, on the neurobiology of *H. americanus*.) The brain, which is initially longitudinally elongated, expands transversely as the olfactory and, later, the accessory neuropiles of the deutocerebrum enlarge. All the major tracts and neuropiles of the brain are established by E40%, with the notable exception of the accessory lobes (Beltz *et al.*, 1992), presumed integrative areas in the deutocerebrum. These lobes grow and differentiate later than the olfactory lobes, a situation unlike that occurring in the crayfish *Cherax destructor* (Helluy *et al.*, 1993).

In the developing embryo, the nervous system is more compact than in the adult. The paired lateral eye lobes are apposed to the protocerebrum. The eye stalks form only after hatching. The paired circumesophageal ganglia develop fused to the tritocerebrum (Allen, 1894). The six ganglia of the subesophageal ganglion and the five thoracic ganglia are fused into one mass, whereas the six abdominal ganglia are already separated by longitudinal connectives. Upon hatching, the circumesophageal ganglion, as well as the five thoracic ganglia, become individualized. In *Homarus gammarus*, the stomatogastric ganglion is established very early, at E20%, and displays at this time a single rhythmic motor pattern controlling movement of the entire stomodeum (Casanovas *et al.*, 1991). All nine neurons of the cardiac ganglion of *H. americanus* are present by the time the heart starts beating and the beats appear to be neurogenic from the beginning (Sherman and Burrage, 1979).

The neuromuscular system is also established early in the abdomen. The swimmerets appear only in larval stage II; however, the abdominal muscles responsible for the beating of the swimmerets are formed during embryonic life—that is, excitatory and inhibitory innervation of these muscles is functional before hatching (Kirk and Govind, 1992). Another abdominal neuromuscular system, the deep abdominal extensor muscles and their innervation, is functional as part of the tail flip reflex at least 6 months before hatching (Cole and Lang, 1980).

All of the neurotransmitters/neuromodulators assayed by immunocytochemistry are first detected in the central nervous system of *Homarus americanus* before midembryonic life (Beltz and Kravitz, 1987; Beltz *et al.*, 1990; Helluy and Beltz, 1990; Schneider *et al.*, 1993). However, the sequence and timing of

FIGURE 15 Intact eggs (unfixed) of *Homarus americanus* at (A) E10%, (B) E20%, (C) E30%, (D) E40%, (E) E50%, and (F) E60% embryonic development. The figures in the lower left corners refer to the percentage of development. In all photographs, the dorsal side is up, the head of the embryo is on the right, the abdomen is folded on the thorax, and the telson is on the right. At E10% development, the embryo is seen as a small halo at the bottom part of the egg. The eye pigment is visible in the lateral eyes (le) by E20%. The red chromatophores (ch), already present by E40%, are labeled at E60%. Scale bar: 500 μm for (A)–(F). (Adapted from Helluy and Beltz, 1991, with permission of *Biological Bulletin*, Marine Biological Laboratory.)

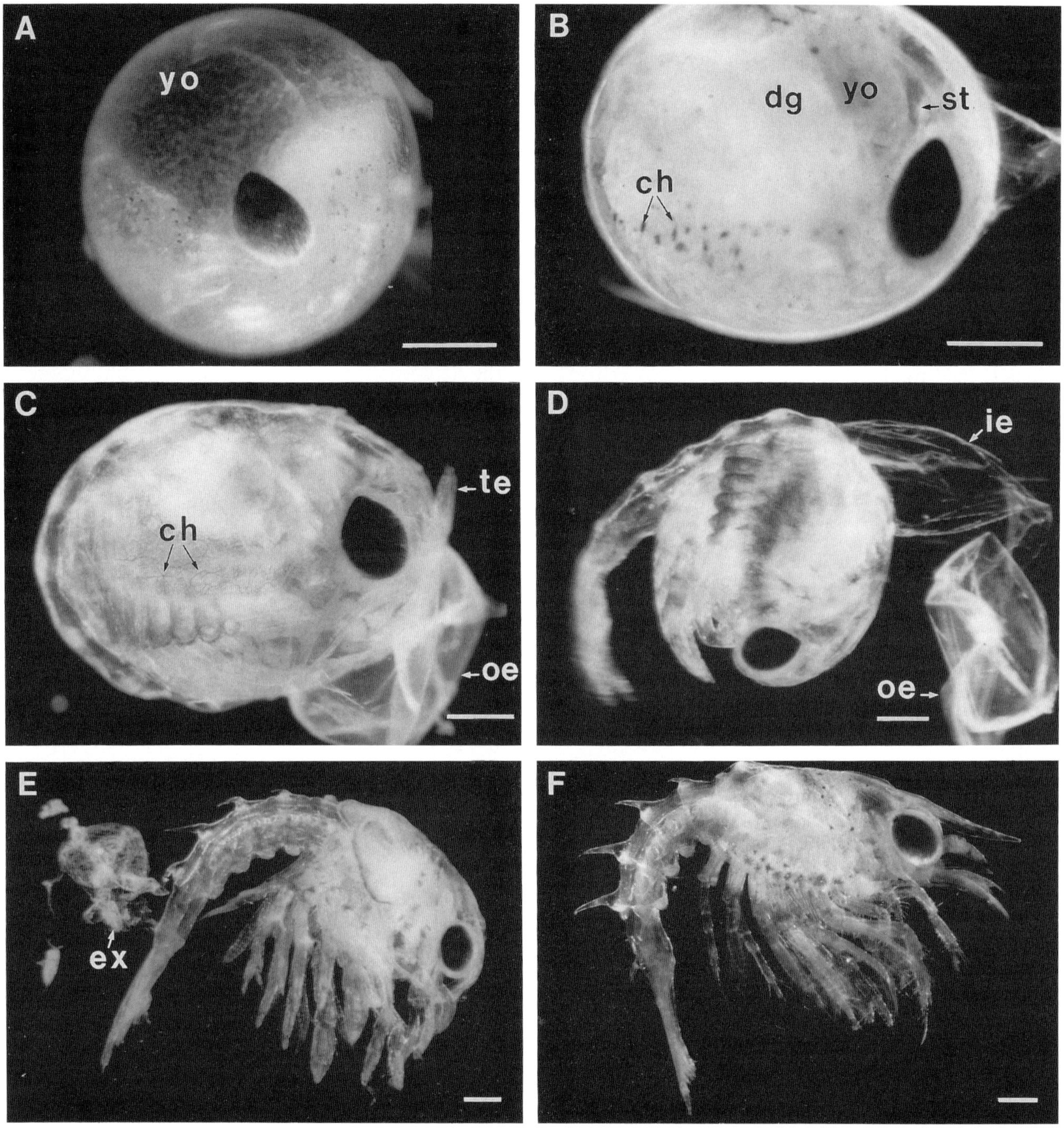

FIGURE 16 Perihatching development in *Homarus americanus*. In all photographs, specimens are unfixed, the dorsal side is up, and anterior is right. (A) E75% embryonic development. (B) Embryo just prior to hatching (E100%; "blue embryo"). (C) Hatchling; the outer egg envelope (oe) has burst, and the telson (te) is piercing the inner egg envelope; red pigment has spread in the star-shaped chromatophores (ch). (D) Prelarva with the inner (ie) and outer (oe) egg envelopes removed. (E) Early first larval stage; the exuvium (ex) of the prelarva has been sloughed. (F) Mature first larval stage; rostrum, abdominal spines, and other acuminate structures are now erect. dg, Digestive gland; st, stomach; yo, yolk. Scale bars: 500 μm. (Adapted and modified from Helluy and Beltz, 1991, with permission of *Biological Bulletin*, Marine Biological Laboratory.)

appearance of these different amines and peptides are very distinct (Beltz *et al.*, 1992).

F. Duration of Embryonic Development

Egg development requires approximately 10 months in New England waters (Bumpus, 1891; Herrick, 1895a; Hughes and Matthiessen, 1962) and 11.5–12 months in Canadian waters (Templeman, 1940b; reviewed by Aiken and Waddy, 1980, 1986b). First-stage larvae are released mostly in May and June in the southern part of the lobster's range, and from June to August in the northern part. Temperature has a strong influence on the developmental rates of *Homarus americanus* (Perkins, 1972; Templeman, 1940b). In a natural temperature regime, the embryonic growth curve is sigmoid, with an extended plateau during the winter months. Development stops or is barely discernible under 6°C, that is, from late November to early May in Maine (Perkins, 1972). In the northern part of the range, hatching can be delayed until autumn if the temperature remains at 2–3°C throughout the spring and summer (Aiken and Waddy, 1986b).

Under laboratory conditions, developmental rates of *Homarus americanus* can be manipulated by adjusting water temperatures (Table 2). Thus, hatches of larvae can be staggered throughout the year. Other environmental factors as well as genetic factors affect the length of embryonic development. Perkins (1972) mentioned that "lobster embryos develop differentially, under the same thermal conditions, depending on their age or extent of development when they are subjected to a given thermal environment." Even at a constant temperature, development may reach a temporary plateau in the latest part of stage D_0 (Fig. 17).

The following formulas from Perkins (1972),

$$Z_{ha-ti} = (W_{ha} - W_{ti})/Y \tag{1}$$

with

$$Y = -8.3151 + 2.6019X, \tag{2}$$

make it possible to calculate for *H. americanus* the time to hatching (*ha*) in weeks from time *ti* (Z_{ha-ti}), knowing the increase in eye index per week in micrometers (Y) at a given temperature between 5°C and 25°C (X), the eye index at the time of hatching (W_{ha}, approximately 560 μm), and the eye index at time *ti* (W_{ti}). In *H. gammarus*, the equation has nearly the same slope (2.5104), indicating a similar influence of temperature on embryonic development in the two species, and a slightly different intercept (–5.6240), reflecting differences in size (Charmantier and Mounet-Guillaume, 1992). The eye index of *H. gammarus* at hatching (600–620 μm, Richards and Wickens, 1979; 640–680 μm, Charmantier and

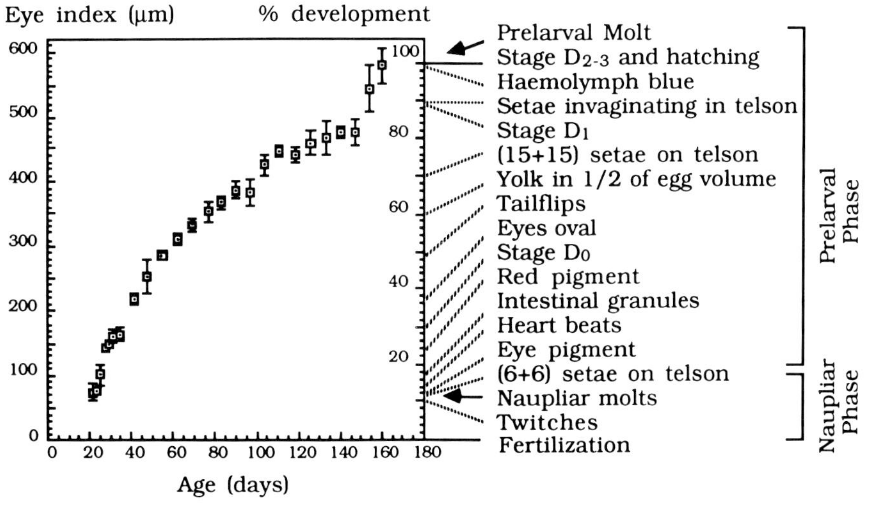

FIGURE 17 Age, eye index, and percent scale of embryonic development in a brood of *Homarus americanus* raised at 18°C. Developmental landmarks are indicated along a percent scale based on the eye index at hatching. Perkins' eye index (1972) is the mean of the length and the width of the screening pigment spot in the lateral eyes (see Section XII,A,2). Each data point represents the mean of five measurements of individuals of an experimental brood ± 1 SD. (Adapted and modified from Helluy and Beltz, 1991, with permission of *Biological Bulletin*, Marine Biological Laboratory.)

TABLE 2 **Duration of Development at Different Temperatures at Salinities near 31 ppt**[a]

Water temperature (°C)	Weeks from extrusion to onset of eye pigment	Weeks from onset of eye pigment to hatching	Weeks from extrusion to to hatching
5	40	120	160
10	9	30	39
15	5	18	23
20	4	12	16
25	3	9	12

[a]From Perkins (1972).

Mounet-Guillaume, 1992) is greater than that of *H. americanus* (560–580 μm, see Section XII,A,2). An approximately linear relationship between incubation period and the sum of monthly average temperatures was also reported for *H. gammarus* (Branford, 1978). Clearly, temperature plays a major role in determining the length of embryonic development in clawed lobsters. The influence of other environmental factors (e.g., photoperiod, salinity), as well as the importance of genetic factors on the regulation of embryonic life duration, have not been explored.

G. *Biochemistry and Physiology*

Between the early cleavage stage of *Homarus americanus* and an egg that is ready to hatch, water content increases from 56% to 87%, ash increases from 5.8% to 21.2%, and wet weight increases from 2.2 to 3.5 mg (Pandian, 1970b). Dry weight decreases throughout embryogenesis. Lipids provide most of the metabolic energy (Sasaki *et al.*, 1986). Energetically, hatching and molting are "expensive" processes for the prelarva. The energy content of the eggs, which decreases by about 19% over a few months, from 6.40 calories in an early cleavage stage to 5.17 calories in an egg ready to hatch, drops an additional 20% in a few hours, to 3.86 calories in a freshly hatched first-stage larva (Pandian, 1970b). Figures for *H. gammarus* follow similar trends (Pandian, 1970a), but the eggs of this species are larger (growing from 1.8 to 3.0 mm) with a higher energy content (10.49 calories per egg in the early cleavage stage; 8.6 calories before hatching). Upon hatching, the osmotic properties of the prelarva change. In dilute media, late egg-bound prelarvae are hyperosmotic, whereas hatchlings and free prelarvae are hyperosmoconformers (Charmantier and Aiken, 1987). The bursting of the relatively impermeable external egg envelope is responsible for the change in osmotic regulation at hatching. Temperature influences the rates of utilization of energy substrates during embryogenesis. Embryos raised at elevated temperatures have yolk remaining at the time of hatching (Sasaki *et al.*, 1986). In addition to temperature, other environmental and genetic factors contribute to the variability in the quantity and quality of reserves in the freshly hatched larvae (Anger *et al.*, 1985). Indeed, if conditions are not favorable, it is possible that the embryonic prelarva will continue to grow, and therefore to consume yolk at stage D_0 of the embryonic molt cycle without differentiating any further. In juvenile and adult lobsters, "molt induction" occurs at the D_0–D_1 transition (Aiken, 1980). (Waddy *et al.*, in Chapter 10, provide a detailed discussion of the molt cycle in *H. americanus*.) It seems likely that the D_0–D_1 transition in the prelarval embryonic molt cycle may also signal a point of no return. The hatching/molting events are probably proceeding ineluctably from there on. If the embryonic prelarva enters stage D_1 with a depleted storage of yolk due to a protracted stage D_0, it may lack the energy to complete the very demanding steps involved in the passage from a benthic embryonic life, attached to the swimmerets of the mother, to an independent planktonic existence.

XIII. Directions for Further Research

Our understanding of the reproduction and embryonic development of *Homarus americanus* is based on the classical studies carried out from the 1880s through the early years of this century. Subsequent descriptive work has been extensive and has clarified many aspects of the anatomy and histology of the reproductive organs and embryonic stages. Some understanding of hormonal regulation of reproduction has been acquired since the 1940s, although this area needs further development. For example, the androgenic gland has been extensively analyzed in anomurans, but little is known about its role in macrurans, including *H. americanus*. It is likely that more of the reproductive hormones will be isolated and characterized in the next decade and a finer understanding of the regulation of reproduction will be forthcoming.

Additional studies at the level of cellular and molecular biology are needed to clarify the exact events that occur during acrosome reaction and fertilization. A more complete understanding is also needed of the pleopod tegumental glands, their function, and factors regulating their secretory activity.

The study of embryonic development lags behind that of reproduction. A number of basic zoological investigations are still necessary to shed light on the early ontogeny of *Homarus americanus*. For example, there are no studies this century that focus on cleavage, germ layer formation, and gastrulation. Embryonic molts also deserve additional attention. Whereas the prelarval molt cycle has been characterized, embryonic molts occurring prior to the appearance of eye pigment remain to be documented. Among the more fruitful issues that might be explored are the coupling of hatching and molting of the prelarva and the endocrine control of these events. At the population level, it would be interesting to elucidate the relative importance of temperature, photoperiod, and other factors in determining hatching time and the health of the larval stages. Devising a reliable way of raising eggs detached from the maternal pleopods would also benefit both basic and applied research on the ontogeny of *Homarus americanus*.

XIV. Summary

Females and males of *Homarus americanus* have paired gonads that communicate with the oviducts and the vasa deferentia, respectively. Sperm are produced in the testes and then packaged into spermatophores during passage through the vas deferens. During mating, which often occurs after a molt but may occur during intermolt, spermatophores are transferred to a seminal receptacle on the female's ventral surface, where sperm may be stored for several years before being used to fertilize oocytes. Females ovulate oocytes on a biennial cycle. Freshly spawned oocytes are surrounded by an envelope that is formed in the ovary by the follicle cells. During spawning, oocytes pass over the seminal receptacle, sperm are extruded, and fertilization is thought to occur externally at this site. During fertilization, the normally nonmotile sperm bind to the envelope surrounding the oocyte and undergo a novel type of acrosome reaction that propels them forward through the egg envelope so that they may fuse with the oolemma. Fertilized oocytes move to the pleopods, where they attach to the ovigerous setae by an extension of the egg coat. The egg coat and the attachment stalk harden and remain around the developing oocytes until the time of hatching. The pleopod tegumental glands release a copious secretion at the time of spawning and may aid in hardening of the egg coat.

Most females of *Homarus americanus* spawn between June and September. The peak in spawning intensity changes with the geographic area along the North American coast and depends on the thermal history of the population, among other factors. Embryos develop attached to the swimmerets of the mother for about 10 months in New England waters and 12 months in Canadian waters. The rate of development is strongly dependent on temperature, to the extent that hatching can be staggered throughout the year by maintaining broods in different thermal environments. The egg undergoes superficial cleavage and the central mass of yolk remains undivided. The naupliar stage (i.e., egg nauplius) occurs early in embryonic life. Most of the embryonic period is devoted to the growth of the prelarva. The first-stage larva differentiates under the cuticle of the prelarva. This differentiation is apparent in the telson, which undergoes setal and tegumentary changes similar to those observed in the telson of larvae and in the pleopods of juvenile lobsters during a molt cycle. The prelarval molt cycle ends just after hatching by the liberation of the first-stage larva. The hatching/molting process is helped by the evagination of the lateral spines of the telson.

Systems for staging embryonic development rely on morphological descriptions for the early stages and on a morphometric criterion, the Perkins eye index, once pigment has appeared in the lateral eyes. A percent staging system based on the eye index has also been developed. Reliable staging schemes have made possible some detailed studies of the remarkably lengthy embryonic life of *Homarus americanus*. In particular, the bioenergetics and osmotic properties of the eggs and the development of the embryonic heart and nervous system have been investigated.

References

Adiyodi, R. G. (1985). Reproduction and its control. *In* "The Biology of Crustacea" (D. E. Bliss, ed.), Vol. 9, "Integument, Pigments, and Hormonal Processes" (L. H. Mantel, ed.) pp. 147–215. Academic Press, Orlando, Florida.

Aiken, D. E. (1973). Proecdysis, setal development, and molt prediction in the American lobster (*Homarus americanus*). *J. Fish. Res. Board Can.* 30, 1337–1344.

Aiken, D. E. (1980). Molting and growth. *In* "The Biology and Management of Lobsters" (J. S. Cobb and B. F. Phillips, eds.), Vol. 1, pp. 91–163. Academic Press, New York.

Aiken, D. E., and Waddy, S. L. (1980). Reproductive biology. *In* "The Biology and Management of Lobsters" (J. S. Cobb and B. F. Phillips, eds.), Vol. 1, pp. 215–276. Academic Press, New York.

Aiken, D. E., and Waddy, S. L. (1982). Cement gland development, ovary maturation and reproductive cycles in the American lobster *Homarus americanus*. *J. Crust. Biol.* 2, 315–327.

Aiken, D. E., and Waddy, S. L. (1985). Production of seed stock for lobster culture. *Aquaculture* 44, 103–114.

Aiken, D. E., and Waddy, S. L. (1986a). Growth of the vasa deferentia of mature *Homarus americanus*: Conflicting results from field and laboratory studies. *Can. J. Fish. Aquat. Sci.* 43, 1453–1457.

Aiken, D. E., and Waddy, S. L. (1986b). Environmental influence on recruitment of the American lobster, *Homarus americanus*: A perspective. *Can. J. Fish. Aquat. Sci.* 43, 2258–2270.

Aiken, D. E., and Waddy, S. L. (1989). Interaction of temperature and photoperiod in the regulation of spawning by American lobsters (*Homarus americanus*). *Can. J. Fish. Aquat. Sci.* 46, 145–148.

Aiken, D. E., Waddy, S. L., Moreland, K., and Polar, S. M. (1984). Electrically induced ejaculation and artificial insemination of the American lobster *Homarus americanus*. *J. Crust. Biol.* 4, 519–527.

Aiken, D. E., Waddy, S. L., and Uhazy, L. S. (1985). Aspects of the biology of *Pseudocarcinonemertes homari* and its association with the American lobster *Homarus americanus*. *Can. J. Fish. Aquat. Sci.* 42, 351–356.

Allen, E. J. (1894). Studies on the nervous system of Crustacea. I.—Some nerve elements of the embryonic lobster. *Q. J. Microsc. Sci.* 36, 461–482.

Allen, E. J. (1895). The reproduction of the lobster. *J. Mar. Biol. Assoc. U.K.* 4, 60–69.

Anderson, D. T. (1979). Embryos, fate maps, and the phylogeny of arthropods. *In* "Arthropod Phylogeny" (A.P. Gupta, ed.), pp. 59–105. Van Nostrand Reinhold, Princeton, New Jersey.

Anderson, D. T. (1982). Embryology. *In* "The Biology of Crustacea" (D. E. Bliss, ed.), Vol. 2, "Embryology, Morphology, and Genetics" (L. G. Abele, ed.) pp. 1–41. Academic Press, New York.

Anger, K., Storch, V., Anger, V., and Capuzzo, J. M. (1985). Effects of starvation on moult cycle and hepatopancreas of stage I lobster (*Homarus americanus*) larvae. *Helgol. Wiss. Meeresunters.* 39, 107–116.

Atema, J. (1986). Review of sexual selection and chemical communication in the lobster *Homarus americanus*. *Can. J. Fish. Aquat. Sci.* 43, 2283–2390.

Atema, J., and Cobb, J. S. (1980). Social behavior of lobsters. *In* "The Biology and Management of Lobsters" (J. S. Cobb and B. F. Phillips, eds.), Vol. 1, pp. 409–450. Academic Press, New York.

Atema, J., and Engstrom, D. (1971). Sex pheromone in the lobster *Homarus americanus*. *Nature (London)* 232, 261–263.

Atema, J., Jacobson, E., Karnofsky, E., Oleszko-Szuts, S., and Stein, L. (1979). Pair formation in the lobster, *Homarus americanus*: Behavioral development, pheromones and mating. *Mar. Behav. Physiol.* 6, 277–296.

Barker, K. R., and Austin, C. R. (1963). Sperm morphology of *Emerita talpoida*. *Biol. Bull. (Woods Hole, Mass.)* 125, 361–362.

Bauer, R. T. (1986). Phylogenetic trends in sperm transfer and storage complexity in decapod crustaceans. *J. Crust. Biol.* 6, 313–325.

Beltz, B. S., and Kravitz, E. A. (1987). Physiological identification, morphological analysis, and development of identified serotonin–proctolin containing neurons in the lobster ventral nerve cord. *J. Neurosci.* 7, 533–546.

Beltz, B. S., Pontes, M. S., Helluy, S. M., and Kravitz, E. A. (1990). Patterns of appearance of serotonin and proctolin immunoreactivities in the developing nervous system of the American lobster. *J. Neurobiol.* 21, 521–542.

Beltz, B. S., Helluy, S., Ruchhoeft, M. L., and Gammill, L. S. (1992). Aspects of the embryology and neural development of the American lobster. *J. Exp. Zool.* 261, 288–297.

Bliss, D. E., ed. (1982–1985). "The Biology of Crustacea," Vols. 1–10. Academic Press, New York.

Branford, J. R. (1978). Incubation period for the lobster Homarus gammarus at various temperatures. *Mar. Biol.* 47, 363–368.

Brown, G. G. (1966). Ultrastructural studies of sperm morphology and sperm–egg interaction in the decapod. *J. Ultrastruct. Res.* 14, 425–440.

Bumpus, H. C. (1891). The embryology of the American lobster. *J. Morphol.* 5, 215–262.

Burns, B. G., Sangalang, G. B., Freeman, H. C., and McMenemy, M. (1984a). Bioconversion of steroids by the testes of the American lobster, *Homarus americanus*, *in vitro*. *Gen. Comp. Endocrinol.* 54, 422–428.

Burns, B. G., Sangalang, G. B., Freeman, H. C., and McMenemy, M. (1984b). Isolation and identification of testosterone from the serum and testes of the American lobster (*Homarus americanus*). *Gen. Comp. Endocrinol.* 54, 429–432.

Byard, E. H., and Aiken, D. E. (1984). The relationship between molting, reproduction, and a hemolymph female-specific protein in the lobster, *Homarus americanus*. *Comp. Biochem. Physiol. A* 77A, 749–757.

Calman, W. T. (1909). Crustacea. *In* "A Treatise on Zoology" (R. Lankester VII, ed.). Black, London.

Campbell, A., and Brattey, J. (1986). Egg loss from the American lobster, *Homarus americanus*, in relation to nemertean, *Pseudocarcinonemertes homari*, infestation. *Can. J. Fish. Aquat. Sci.* 43, 772–780.

Casanovas, B., Cournil, I., Meyrand, P., and Moulins, M. (1991). Development of neuronal networks in the lobster stomatogastric nervous system. *Soc. Neurosci. Abstr.* 17, 939.

Ceccaldi, H. J. (1968). "Recherches sur la Biologie des Associations entre Protéines et Caroténoides chez les Crustacés Décapodes; Aspects Métaboliques et Moléculaires," Ph.D. thesis. Université d'Aix-Marseille, France.

Ceccaldi, H. J., Cheesman, D. F., and Zagalsky, P. F. (1966). Quelques propriétés et caractéristiques de l'ovoverdine. *C. R. Seances Soc. Biol. Ses Fil.* 160, 587.

Chace, F. A., and Moore, G. M. (1959). A bicolored gynandromorph of the lobster, *Homarus americanus*. *Biol. Bull. (Woods Hole, Mass.)* 116, 226–231.

Charmantier, G., and Aiken, D. E. (1987). Osmotic regulation in late embryos and prelarvae of the American lobster *Homarus americanus* H. Milne-Edwards, 1837 (Crustacea, Decapoda). *J. Exp. Mar. Biol. Ecol.* 109, 101–108.

Charmantier, G., and Mounet-Guillaume, R. (1992). Temperature-specific rates of embryonic development of the European lobster *Homarus gammarus* (L.). *J. Exp. Mar. Biol. Ecol.* 160, 61–66.

Charmantier, G., Charmantier, M., Waddy, S. L., and Aiken, D. E. (1991). Salinity tolerance and osmoregulation in the nemertean *Pseudocarcinonemertes homari*, an egg predator of American lobster, *Homarus americanus*. *Can. J. Fish. Aquat. Sci.* 48, 209–214.

Charniaux-Cotton, H. (1962). Androgenic gland of crustaceans. *Gen. Comp. Endocrinol., Suppl.* 1, 241–247.

Charniaux-Cotton, H., and Payen, G. (1985). Sexual differentiation. *In* "The Biology of Crustacea" (D. E. Bliss, ed.), Vol. 9, "Integument, Pigments, and Hormonal Processes" (D. E. Bliss and L. H. Mantel, eds.), pp. 217–299. Academic Press, Orlando, Florida.

Cheesman, D. F., Lee, W. L., and Zagalsky, P. F. (1967). Carotenoproteins in invertebrates. *Biol. Rev. Cambridge Philos. Soc.* 42, 131–160.

Cheung, T. S. (1966). The development of egg membranes and egg attachment in the shore crab, *Carcinus maenas*, and some related decapods. *J. Mar. Biol. Assoc. U.K.* 46, 373–400.

Chevaillier, P. (1965). Formation de la capsule du spermatozoïde de la langoustine, *Nephrops norvegicus* L: Évolution des polysaccharides au cours de la spermiogénèse. *C. R. Hebd. Seances Acad. Sci.* 261, 3225–3228.

Chevaillier, P., and Maillet, P. L. (1965a). Structure et ultrastructure du spermatozoïde de la langoustine. *J. Microsc. (Paris)* 4, 128.

Chevaillier, P., and Maillet, P. L. (1965b). Structure fine et constitution cytochimique du spermatozoïde de la langoustine *Nephrops norvegicus* (Crustacea, Decapoda). *J. Microsc. (Paris)* 4, 679–700.

Clark, W. H., Jr., Talbot, P., Neal, R. A., Mock, C. R., and Salser, B. R. (1973). In vitro fertilization with non-motile spermatozoa of the brown shrimp *Penaeus aztecus*. *Mar. Biol.* 22, 353–354.

Cobb, J. S., and Phillips, B. E., eds. (1980). "The Biology and Management of Lobsters," Vols. 1–2. Academic Press, New York.

Cole, J. J., and Lang, F. (1980). Spontaneous and evoked postsynaptic potentials in an embryonic neuromuscular system of the lobster, *Homarus americanus*. *J. Neurobiol.* 11, 459–470.

Cowan, D., and Atema, J. (1990). Moult staggering and serial monogamy in American lobsters, *Homarus americanus*. *Anim. Behav.* 39, 1199–1206.

Davis, C. C. (1964). A study of the hatching process in aquatic invertebrates. XIII. Events of eclosion in the American lobster, *Homarus americanus* Milne-Edwards (Astacura, Homaridae). *Am. Midl. Nat.* 72, 203–210.

Dehn, P. F., Aiken, D. E., and Waddy, S. L. (1983). Aspects of vitellogenesis in the lobster *Homarus americanus*. *Can. Tech. Rep. Fish. Aquat. Sci.* 1161, 1–24.

Diamond, J. M., and Bossert, W. H. (1968). Functional consequences of ultrastructural geometry in backwards fluid transporting epithelia. *J. Cell Biol.* 37, 694–702.

Drach, P. (1939). Mue et cycle d'intermue chez les crustacés décapodes. *Ann. Inst. Oceanogr. (Paris)* 19, 103–392.

Dubé, P., and Grondin, P. (1985). Sexual maturity of the female lobster *Homarus americanus* on the Magdalen Islands, Canada. *Que. Minist. Agric. Pech. Aliment., Dir. Rech. Sci. Tech. Cah. Inf.* 10, 1–41.

Dunham, P. J., and Skinner-Jacobs, D. (1978). Intermolt mating in the lobster (*Homarus americanus*). *Mar. Behav. Physiol.* 5, 209–214.

Ennis, G. P. (1973). Endogenous rhythmicity associated with larval hatching in the lobster *Homarus gammarus*. *J. Mar. Biol. Assoc. U.K.* 53, 531–538.

Ennis, G. P. (1975). Observations on hatching and larval release in the lobster *Homarus americanus*. *J. Fish. Res. Board Can.* 32, 2210–2213.

Ennis, G. P. (1980). Size-maturity relationships and related observations in Newfoundland, Canada, populations of the lobster (*Homarus americanus*). *Can. J. Fish. Aquat. Sci.* 37, 945–956.

Ennis, G. P. (1984a). Comparison of physiological and functional size–maturity relationships in two Newfoundland populations of lobsters *Homarus americanus*. *Fish. Bull.* 82, 244–249.

Ennis, G. P. (1984b). Incidence of molting and spawning in the same season in female lobsters *Homarus americanus*. *Fish. Bull.* 82, 529–530.

Ennis, G. P. (1984c). Tag–recapture validation of molt and egg extrusion prediction based upon pleopod examination in the American lobster *Homarus americanus*. *Fish. Bull.* 82, 242–244.

Farmer, A. S. (1974). Reproduction in *Nephrops norvegicus* (Decapoda: Nephropidae). *J. Zool.* 174, 161–183.

Fasten, N. (1917). Male reproductive organs of decapoda, with special reference to Puget Sound forms. *Puget Sound Mar. Stn. Publ.* 26, 285–307.

Frentz, R. (1960). Contribution à l'Etude Biochimique du Milieu Intérieur de *Carcinus maenas* Linn. *Bull. Soc. Sci. Nancy* 19, 1–176.

Gesteira, T., and Halcrow, K. (1988). Influence of some external factors on the acrosome reaction in the spermatozoa of *Homarus americanus* Milne-Edwards, 1837. *J. Crust. Biol.* 8, 317–321.

Gilgan, M. W., and Idler, P. R. (1967). The conversion of androstenedione to testosterone by some lobster (*Homarus americanus*) (Milne-Edwards) tissues. *Gen. Comp. Endrocrinol.* 7, 319–324.

Ginsburger-Vogel, T., and Charniaux-Cotton, H. (1982). Sex determination. *In* "The Biology of Crustacea" (D. E. Bliss, ed.), Vol. 2, "Embryology, Morphology, and Genetics" (L. G. Abele, ed.), pp. 257–281. Academic Press, New York.

Goodwin, T. W. (1951). Carotenoid metabolism during development of lobster eggs. *Nature (London)* 7, 559.

Gore, R. H. (1985). Molting and growth in decapod larvae. *In* "Crustacean Issues" (F. R. Schram, ed.), Vol. 2, "Larval Growth" (A. D. Wenner, ed.) pp. 1–65. Balkema, Rotterdam, The Netherlands.

Goudeau, M. (1976). Secretion of embryonic envelopes and embryonic molting cycles in *Hemioniscus balani* Buchholtz, Isopoda, Epicaridea. *J. Morphol.* 148, 427–451.

Goudeau, H., and Goudeau, M. (1986). Electrical and morphological response of the lobster egg to fertilization. *Dev. Biol.* 114, 325–335.

Goudeau, M., and Becker, J. (1982). Fertilization in a crab. II. Cytological aspects of the cortical reaction and fertilization envelope elaboration. *Tissue Cell* 14, 273–282.

Goudeau, M., Talbot, P., and Harper, R. (1987). Mechanism of egg attachment stalk formation in the lobster *Homarus*. *Gamete Res.* 18, 279–290.

Goudeau, M., Lachaise, F., Carpentier, G., and Goxe, B. (1990). High titers of ecdysteroids are associated with the secretory process of embryonic envelopes in the European lobster. *Tissue Cell* 22, 269–281.

Graf, F. (1972). Stockage de calcium et formation des soies chez l'embryon d'*Orchestia* (Crustacé, Amphipode, Talitridé). Notion d'intermue embryonnaire. *C. R. Hebd. Seances Acad. Sci. Ser. D* 275, 1669–1672.

Haley, S. R. (1984). Spermatogenesis and spermatophore production in the Hawaiian red lobster *Enoplometopus occidentalis* (Randall) (Crustacea, Nephropidae). *J. Morphol.* 180, 181–193.

Haley, S. R. (1986). Ultrastructure of spermatogenesis in the Hawaiian red lobster *Enoplometopus occidentalis*. *J. Morphol.* 190, 81–92.

Halkett, A. (1919). An hermaphrodite lobster. *Can. Field-Nat.* 33, 40.

Harper, R., and Talbot, P. (1984). Analysis of the epibiotic bacteria of lobster (*Homarus*) eggs and their influence on the loss of eggs from the pleopods. *Aquaculture* 36, 9–26.

Hedgecock, D. (1983). Maturation and spawning of the American lobster, *Homarus americanus*. *In* "Crustacean Aquaculture" (J. P. McVey, ed.), Vol. 1, pp. 261–270. CRC Press, Boca Raton, Florida.

Hedgecock, D., Moffett, W. L., Borgeson, W., and Nelson, K. (1978). Progress and problems in lobster broodstock development. *Proc. Annu. Meet.—World Maric. Soc.* 9, 497–506.

Helluy, S., and Beltz, B. S. (1990). Stages in the embryonic development of the American lobster *Homarus americanus* with special emphasis on its nervous system. *In* "Frontiers in Crustacean Neurobiology" (K. Wiese, W. D. Krenz, J. Tautz, H. Reichert, and B. Mulloney, eds.), pp. 530–536. Birkhäuser, Basel, Switzerland.

Helluy, S., and Beltz, B. S. (1991). Embryonic development of the American lobster *Homarus americanus*: Quantitative staging and characterization of an embryonic molt cycle. *Biol. Bull. (Woods Hole, Mass.)* 180, 355–371.

Helluy, S., Sandeman, R., Beltz, B., and Sandeman, D. (1993). Comparative brain ontogeny of the crayfish and clawed lobster: Implications of direct and larval development. *J. Comp. Neurol.* 335, 343–354.

Herrick, F. H. (1891). The reproductive organs and early stages of development of the American lobster. *Johns Hopkins Univ. Circ.* 88, 98–100.

Herrick, F. H. (1894). The reproduction of the lobster. *Zool. Anz.* 17, 289–292.

Herrick, F. H. (1895a). The American lobster: A study of its habits and development. *Bull. U.S. Fish Comm.* 15, 1–252 + 54 plates.

Herrick, F. H. (1895b). The reproduction of the lobster. *Zool. Anz.* 18, 226–228.

Herrick, F. H. (1909). Natural history of the American lobster. *Bull. U.S. Bur. Fish.* 29, 149–408 + 20 plates.

Hinsch, G. W. (1971). Penetration of the oocyte envelope by spermatozoa in the spider crab. *J. Ultrastruct. Res.* 35, 86–97.

Howard, D. (1991). "Ovarian Muscle of the Lobster, *Homarus americanus*: Functions of Microtubules and Actin Filaments in Contraction," Ph.D. dissertation. Department of Biology, University of California, Riverside.

Howard, D., and Talbot, P. (1992). In vitro contraction of lobster (*Homarus*) ovarian muscle: Methods for assaying contraction and effects of biogenic amines. *J. Exp. Zool.* 263, 365–366.

Hughes, J. T., and Matthiessen, G. C. (1962). Observations on the biology of the American lobster, *Homarus americanus*. *Limnol. Oceanogr.* 7, 414–421,

Ishida, T., Talbot, P., and Kooda-Cisco, M. (1986). Technique for the long-term storage of lobster (*Homarus*) spermatophores. *Gamete Res.* 14, 183–195.

Johnson, B., and Talbot, P. (1987). Ultrastructural analysis of the pleopod tegumental glands in male and female lobster, *Homarus americanus*. *J. Crust. Biol.* 7, 288–301.

Junio, M. A. R. (1987). Some aspects of the reproduction of *Panulirus penicillatus* (Decapoda: Palinuridae). *Bull. Mar. Sci.* 41, 242–252.

Kerr, M. S. (1968). Protein synthesis by hemocytes of *Callinectes sapidus*: A study of in vitro incorporation of 14C-leucin. *J. Cell Biol.* 39, 72a–73a.

Kessel, R. G. (1968). Mechanisms of protein yolk synthesis and deposition in crustacean oocytes. *Z. Zellforsch. Mikrosk. Anat.* 89, 17–38.

Kessel, R. G., and Beams, H. W. (1963). Electron microscope studies on developing oocytes of the lobster, *Homarus*, with special reference to the method of yolk formation. *J. Cell Biol.* 19, A87.

Kidd, R. J. (1991). Development of embryonic and naupliar setae and spines and their role in hatching in the penaeid *Sicyonia ingentis*: A light and electron microscopic study. *J. Crust. Biol.* 11, 40–55.

Kirk, M. D., and Govind, C. K. (1992). Early innervation of abdominal swimmeret muscles in developing lobsters. *J. Exp. Zool.* 261, 298–309.

Knight, A. P. (1918). Lobster investigations at Long Beach Pond. *N.S. Contrib. Can. Biol.* 53–71.

Kooda-Cisco, M., and Talbot, P. (1982). A structural analysis of the freshly extruded spermatophore from the lobster, *Homarus americanus*. *J. Morphol.* 172, 193–207.

Kooda-Cisco, M., and Talbot, P. (1983). A technique for electrically stimulating extrusion of spermatophores from the lobster, *Homarus americanus*. *Aquaculture* 30, 221–227.

Kooda-Cisco, M., and Talbot, P. (1986). Ultrastructure and role of the lobster vas deferens in spermatophore formation: The proximal segment. *J. Morphol.* 188, 91–103.

Kravitz, E. A. (1988). Hormonal control of behavior: Amines and the biasing of behavioral output in lobsters. *Science* 241, 1775–1781.

Krouse, J. S. (1973). Maturity, sex ratio and size composition of a natural population of American lobster, *Homarus americanus*, along the Maine coast. *Fish. Bull.* 71, 165–173.

Lang, F., Govind, C. K., Costello, W. J., and Greene, S. I. (1977). Developmental neuroethology: Changes in escape and defensive behavior during growth of the lobster. *Science* 197, 682–684.

Laverdure, A.-M., and Soyez, D. (1988). Vitellogenin receptors from lobster oocyte membrane: Solubilization and characterization by a solid phase binding assay. *Int. J. Invertebr. Reprod. Dev.* 13, 251–266.

Legrand, J. J., Legrand-Hamelin, E., and Juchault, P. (1987). Sex determination in Crustacea. *Biol. Rev. Cambridge Philos. Soc.* 62, 439–470.

Lloyd, A. J., and Yonge, C. M. (1940). Correlation between egg-carrying setae and cement glands in decapod Crustacea. *Nature (London)* 146, 334.

Martin, G. G., Herzig, C., and Narimatsu, G. (1987). Fine structure and histochemistry of the freshly extruded and hardened spermatophore of the spiny lobster *Panulirus interruptus*. *J. Morphol.* 192, 237–246.

Matthews, D. C. (1954). The origin and development of the spermatophoric mass of a nephropsid lobster, *Enoplometopus occidentalis* (Randall). *Pac. Sci.* 8, 115–120.

Meusy, J. J. (1980). Vitellogenin, the extraovarian precursor of the protein yolk in Crustacea: A review. *Reprod. Nutr. Dev.* 20, 1–21.

Miller, J., and Fleming, L. (1983). Fungi associated with an infestation of *Pseudocarcinonemertes homari* in *Homarus americanus*. *Trans. Br. Mycol. Soc.* 80, 9–12.

Mowbray, R., Brown, G. B., and Metz, C. B. (1970). Cytological and immunological investigations of sperm–egg interactions on selected decapods (Crustacea) and *Limulus polyphemus* L. (Merostomata). *Biol. Bull. (Woods Hole, Mass.)* 139, 313–320.

Nelson, K., and Hedgecock, D. (1977). Electrophoretic evidence of multiple paternity in lobster *Homarus americanus* (Milne-Edwards). *Am. Nat.* 111, 361–365.

Nelson, K., Hedgecock, D., and Borgeson, W. (1983). Photoperiodic and ecdysial control of vitellogenesis in lobsters *Homarus* (Decapoda: Nephropidae). *Can. J. Fish. Aquat. Sci.* 40, 940–947.

Nelson, K., Heyer, B., Johnson, E., Hedgecock, D., and Chang, E. S. (1988a). Photoperiod-induced changes in hemolymph vitellogenins in female lobster (*Homarus americanus*). *Comp. Biochem. Physiol. B* 90B, 809–821.

Nelson, K., Hedgecock, D., and Borgeson, W. (1988b). Effects of reproduction upon molting and growth in female American lobster (*Homarus americanus*). *Can. J. Fish. Aquat. Sci.* 45, 805–821.

Nelson, K., Hedgecock, D., and Borgeson, W. (1988c). Factors influencing egg extrusion in the American lobster (*Homarus ameri-*

canus). *Can. J. Fish. Aquat. Sci.* 45, 797–804.

Nicholls, F. (1730). An account of the hermaphrodite lobster presented to the Royal Society on Thursday, May the 7th, by Mr. Fisher of Newgate-Market, examined and dissected, pursuant to an order of the Society. *Philos. Trans. R. Soc. London* 36, 290–294.

Pandian, T. J. (1970a). Ecophysiological studies on the developing eggs and embryos of the European lobster *Homarus gammarus*. *Mar. Biol.* 5, 154–167.

Pandian, T. J. (1970b). Yolk utilization and hatching in the Canadian lobster *Homarus americanus*. *Mar. Biol.* 7, 249–254.

Perkins, H. C. (1971). Egg loss during incubation from offshore northern lobsters (Decapoda: Homaridae). *Fish. Bull.* 69, 451–453.

Perkins, H. C. (1972). Developmental rates at various temperatures of embryos of the northern lobster (*Homarus americanus* Milne-Edwards). *Fish. Bull.* 70, 95–99.

Pochon-Masson, J. (1968). L'ultrastructure des spermatozoïdes chez les crustacés décapodes avant et au cours de leur dévagination expérimentale. II. Macroures. Discussion et conclusions. *Ann. Sci. Nat.* 10, 367–454.

Rao, K. R., Fingerman, S. W., and Fingerman, M. (1973). Effects of exogenous ecdysones on the molt cycles of fourth and fifth stage American lobsters, *Homarus americanus*. *Comp. Biochem. Physiol. A* 44A, 1105–1120.

Richards, P. R., and Wickens, J. F. (1979). Lobster Culture Research. *Lab. Leafl. MAFF Dir. Fish. Res. Lowestoft* 47, 1–33.

Ridgewood, W. G. (1909). A case of abnormal oviducts in *Homarus vulgaris*. *Ann. Mag. Nat. Hist.* 8(3), 1–7.

Ronneberg, H., Renstrom, B., Aareskjold, K., and Liaaen-Jensen, S. (1980). Natural occurrence of enantiomeric and meso-astaxanthin. 1. Ex lobster eggs (*Homarus gammarus*). *Helv. Chim. Acta* 63, 711–715.

Sasaki, G. C. (1984). "Biochemical Changes Associated with Embryonic and Larval Development in the American Lobster *Homarus americanus* Milne-Edwards," Ph.D. thesis, Woods Hole Oceanogr. Inst. Tech. Rep. WHOI-84-8. Massachusetts Institute of Technology, Cambridge/Woods Hole Oceanographic Institution, Woods Hole, Massachusetts.

Sasaki, G. C., Capuzzo, J. M., and Biesiot, P. (1986). Nutritional and bioenergetic considerations in the development of the American lobster *Homarus americanus*. *Can. J. Fish. Aquat. Sci.* 43, 2311–2319.

Schade, M. L., and Shivers, R. R. (1980). Structural modulation of the surface and cytoplasm of oocytes during vitellogenesis in the lobster, *Homarus americanus*. An electron microscope–protein tracer study. *J. Morphol.* 163, 13–26.

Schneider, H., Peckol, E., Budhiraja, P., Beltz, B., and Kravitz, E. (1993). Different patterns of appearance of octopamine- and serotonin-immunoreactive neurons in the developing lobster CNS. *Soc. Neurosci. Abstr.* 19, 1273.

Scholtz, G. (1992). Cell lineage studies in the crayfish *Cherax destructor* (Crustacea, Decapoda): Germ band formation, segmentation, and early neurogenesis. *Roux's Arch. Dev. Biol.* 202, 36–48.

Scholtz, G. (1993). Teloblasts in decapod embryos: An embryonic character reveals the monophyletic origin of freshwater crayfishes (Crustacea, Decapoda). *Zool. Anz.* 230, 45–54.

Schroeder, P., and Talbot, P. (1985). Ovulation in the animal kingdom: A review with emphasis on the role of contractile elements. *Gamete Res.* 11, 191–221.

Schuur, A., Fisher, W. S., Van Olst, J. C., Carlberg, J., Hughes, J. T., Shleser, R. A., and Ford, R. F. (1976). Hatchery methods for the production of juvenile lobsters (*Homarus americanus*). *Inst. Mar. Res., Calif.* 48, 1–20.

Scott, A. (1903). On the spawning of the common lobster. *Trans. Liverpool Biol. Soc.* 17, 106–113.

Sekiguchi, K. S., Yamamichi, Y., and Costlow, J. D. (1982). Horseshoe crab developmental studies I. Normal embryonic development of *Limulus polyphemus* compared with *Tachypleus tridentatus*. *In* "Physiology and Biology of Horseshoe Crabs: Studies on Normal and Environmentally Stressed Animals" (J. Bonaventura, C. Bonaventura, and S. Tesh, eds.), pp. 53–73. Liss, New York.

Sherman, R. G., and Burrage, T. G. (1979). Cellular development of the American lobster heart. *Am. Zool.* 19, 29–38.

Shiino, S. M. (1988). Crustacea. *In* "Invertebrate Embryology" (M. Kumé and K. Dan, eds.), pp. 333–388. Garland, New York. (Reprint of the 1968 edition, NOLIT, Belgrade, Yugoslavia, which is a reprint of the original 1957 Japanese edition, Bai Fu Kan Press, Tokyo.)

Silberbauer, B. I. (1971). The biology of the South African rock lobster *Jasus lalandii* (H. Milne-Edwards) 2. The reproductive organs, mating and fertilization. *S. Afr., Div. Sea Fish., Invest. Rep.* 93, 1–46.

Spiridonov, V. A. (1992). Review of the book by A.L. Tichomirova: Ontogeny transformation as a mechanism of insect evolution. *Arthropoda Sel.* 1, 91–93.

Stern, K. G., and Salomon, K. (1937). Ovoverdin, a pigment chemically related to visual purple. *Science* 86, 310.

Subramoniam, T. (1990). Chemical composition of spermatophores in decapod crustaceans. *In* "Crustacean Sexual Biology" (R. T. Bauer and J. W. Martin, eds.), pp. 308–321. Columbia University Press, New York.

Syslo, M., and Hughes, J. (1981). Reproductive behavior of a bicolored gynandromorphic American lobster. *Prog. Fish-Cult.* 43, 214–216.

Talbot, P. (1981a). The ovary of the lobster, *Homarus americanus*. I. Architecture of the mature ovary. *J. Ultrastruct. Res.* 76, 235–248.

Talbot, P. (1981b). The ovary of the lobster, *Homarus americanus*. II. Structure of the mature follicle and origin of the chorion. *J. Ultrastruct. Res.* 76, 249–262.

Talbot, P. (1981c). Collagenase solutions induce in vitro ovulation in lobsters (*Homarus americanus*). *J. Exp. Zool.* 216, 181–185.

Talbot, P. (1984). Problems and progress in controlling reproductive biology in the American lobster (*Homarus*). *In* "Advances in Invertebrate Reproduction" (W. Engels, ed.), pp. 473–480. Elsevier, New York.

Talbot, P. (1991a). Compartmentalization in the acrosome. *In* "Comparative Spermatology—20 Years After" (B. Baccetti, ed.), Serono Symp. Publ., 75, pp. 255–258. Raven, New York.

Talbot, P. (1991b). Ovulation, attachment, and retention of lobster eggs. *In* "Crustacean Issues" (F. R. Schram, ed.), Vol. 7, "Crustacean Egg Production" (A. Wenner and A. Kuris, eds.), pp. 9–18. Balkema, Rotterdam, The Netherlands.

Talbot, P., and Beach, D. (1989). Role of the vas deferens in the formation of the spermatophore of the crayfish (*Cherax*). *J. Crust. Biol.* 9, 9–24.

Talbot, P., and Chanmanon, P. (1980a). Morphological features of the acrosome reaction of lobster (*Homarus*) sperm and the role of the reaction in generating forward sperm movement. *J. Ultrastruct. Res.* 70, 287–297.

Talbot, P., and Chanmanon, P. (1980b). The structure of sperm from the lobster, *Homarus americanus*. *J. Ultrastruct. Res.* 70, 275–286.

Talbot, P., and Demers, D. (1993). Tegumental glands of Crustacea. *In* "The Crustacean Integument: Morphology and Biochemistry" (M. N. Horst and J. A. Freeman, eds.), pp. 151–192. CRC Press, Boca Raton, Florida.

Talbot, P., and Goudeau, M. (1988). A complex cortical reaction leads to formation of the fertilization envelope in the lobster

Homarus. Gamete Res. 19, 1–18.

Talbot, P., and Harper, R. (1984). Abnormal egg stalk morphology is correlated with clutch attrition in laboratory-maintained lobsters (*Homarus*). *Biol. Bull. (Woods Hole, Mass.)* 166, 349–356.

Talbot, P., Hedgecock, D., Borgeson, W., Wilson, P., and Thaler, C. (1983). Examination of spermatophore production by laboratory-maintained lobsters (*Homarus*). *J. World Maric. Soc.* 14, 271–278.

Talbot, P., Thaler, C., and Wilson, P. (1984). Spawning, egg attachment and egg retention in captive lobsters (*Homarus americanus*). *Aquaculture* 37, 239–249.

Talbot, P., Thaler, C., and Wilson, P. (1986). Artificial insemination of the American lobster (*Homarus*). *Gamete Res.* 14, 25–31.

Talbot, P., Poolsanguan, W., Poolsanguan, B., and Al Hajj, H. (1991a). In vitro fertilization of lobster oocytes. *J. Exp. Zool.* 258, 104–112.

Talbot, P., Poolsanguan, B., Al Hajj, H., and Poolsanguan, W. (1991b). Gamete interactions during *in vitro* fertilization of lobster (*Homarus americanus*) oocytes. *J. Struct. Biol.* 106, 125–134.

Templeman, W. (1934). Mating in the American lobster. *Contrib. Can. Biol. Fish.* 8, 423–432.

Templeman, W. (1936). Further contributions to mating in the American lobster. *J. Biol. Board Can.* 2, 223–226.

Templeman, W. (1937). Egg-laying and hatching postures and habits of the American lobster (*Homarus americanus*). *J. Biol. Board Can.* 2, 339–342.

Templeman, W. (1940a). The Life History of the Lobster (*Homarus americanus*). *Newfoundland, Dep. Nat. Resour., Res. Bull. (Fish.)* 15, 1–42.

Templeman, W. (1940b). Embryonic developmental rates and egg-laying of Canadian lobsters. *J. Fish. Res. Board Can.* 5, 71–83.

Templeman, W. (1944). Sexual dimorphism in the lobster (*Homarus americanus*). *J. Fish. Res. Board Can.* 6, 228–232.

Tom, M., Goren, M., and Ovadia, M. (1987). Localization of the vitellin and its possible precursors in various organs of *Parapenaeus longirostris* (Crustacea, Decapoda, Penaeidae). *Int. J. Invertebr. Reprod. Dev.* 12, 1–12.

Tsai, I., and Talbot, P. (1993). Video microscopic analysis of ionophore induced acrosome reactions of lobster (*Homarus americanus*) sperm. *Molec. Reprod. Develop.* 36, 454–461.

Tsai, I., and Talbot, P. (1994). Ultrastructural and biochemical analysis of the apical cap, a contractile element in the lobster sperm acrosome. *J. Exp. Zool.* 268, 186–199.

Van Engel, W. A. (1980). Maturity and fecundity in the American lobster *Homarus americanus*. A review. *Can. Tech. Rep. Fish. Aquat. Sci.* 932, 51–58.

Waddy, S. L. (1989). Farming the homarid lobsters: State of the art. *World Aquacult.* 19(4), 1–9.

Waddy, S. L., and Aiken, D. E. (1984). Brood stock management for year-round production of larvae for culture of the American lobster. *Can. Tech. Rep. Fish. Aquat. Sci.* 1272, 1–14.

Waddy, S. L., and Aiken, D. E. (1985a). Fertilization and egg retention in artificially inseminated female American lobsters *Homarus americanus*. *Can. J. Fish. Aquat. Sci.* 42, 1954–1956.

Waddy, S. L., and Aiken, D. E. (1985b). Immunofluorescent localization of American lobster *Homarus americanus* egg yolk protein in the alimentary tract of the nemertean *Pseudocarcinonemertes homari*. *Can. J. Fish. Aquat. Sci.* 42, 357–359.

Waddy, S. L., and Aiken, D. E. (1986). Multiple fertilization and consecutive spawning in large American lobsters *Homarus americanus*. *Can. J. Fish. Aquat. Sci.* 43, 2291–2294.

Waddy, S. L., and Aiken, D. E. (1990a). Mating and insemination in the American lobster, *Homarus americanus*. *In* "Crustacean Sexual Behavior" (R. T. Bauer and J. W. Martin, eds.), pp. 126–144. Columbia University Press, New York.

Waddy, S. L., and Aiken, D. E. (1990b). Intermolt insemination, an alternative mating strategy for the American lobster (*Homarus americanus*). *Can. J. Fish. Aquat. Sci.* 47, 2402–2406.

Waddy, S. L., and Aiken, D. E. (1992). Seasonal variation in spawning by preovigerous American lobster (*Homarus americanus*) in response to temperature and photoperiod manipulation. *Can. J. Fish. Aquat. Sci.* 49, 1114–1117.

Wallace, R. A., Walker, S. L., and Hauschka, P. V. (1967). Crustacean lipovitellin. Isolation and characterization of the major high-density lipoprotein from the eggs of decapods. *Biochemistry* 6, 1582–1590.

Waterman, T. H., ed. (1960). "The Physiology of Crustacea," Vols. 1–2. Academic Press, New York.

Wear, R. G. (1974). Incubation in British decapod Crustacea, and the effects of temperature on the rate and success of embryonic development. *J. Mar. Biol. Assoc. U.K.* 54, 745–762.

Williamson, D. I. (1982). Larval morphology and diversity. *In* "The Biology of Crustacea" (D. E. Bliss, ed.), Vol. 2, "Embryology, Morphology, and Genetics" (L. G. Abele, ed.), pp. 43–110. Academic Press, New York.

Wolin, E. H., Laufer, H., and Albertini, D. F. (1973). Uptake of the yolk protein lipovitellin by developing crayfish oocytes. *Dev. Biol.* 35, 160–170.

Wyckoff, R. W. G. (1937). The sedimentation constant of ovoverdin. *Science* 86, 311–312.

Yonge, C. M. (1937). The nature and significance of the membrane surrounding the developing eggs of *Homarus vulgaris* and other decapods. *Proc. Zool. Soc. London, Ser. A* 107, 499–517.

Zagalsky, P. F. (1985). A study of the astaxanthin–lipovitellin ovoverdin isolated from the ovaries of the lobster *Homarus gammarus*. *Comp. Biochem. Physiol. B* 80B, 589–598.

CHAPTER

10

Control of Growth and Reproduction

S. L. WADDY *and* D. E. AIKEN
Biological Station
Department of Fisheries and Oceans
St. Andrews, New Brunswick, Canada E0G 2X0

D. P. V. DE KLEIJN
Department of Animal Physiology
University of Nijmegen
Toernooiveld, 6525 ED Nijmegen, The Netherlands

I. Introduction

Descriptions of growth and reproductive mechanisms in *Homarus americanus* date to the late 1800s, when Francis Hobart Herrick wrote the first of two impressive monographs on the natural history of the American lobster (Herrick, 1895, 1909). Forty years later, Wilfred Templeman conducted extensive field and laboratory studies on lobster behavior, development, growth, and reproduction. He was the first to define maturity indices for both male and female lobsters and he suggested several mechanisms involved in the environmental control of lobster biology, including the role of temperature in regulating size at maturity and seasonal timing of spawning, hatching, and molting (Templeman, 1936a, 1940a, 1940b). The work of both of these individuals has withstood the test of time and, even today, they remain major contributors to the field.

Interest in intensive lobster culture peaked during the 1970s. As attempts were made to grow lobsters under intensive conditions, unlike the extensive systems used for shrimp and crayfish culture, the need for biological information became acute and the field moved very quickly from the descriptive to the experimental. It became important to define the factors regulating development, growth, and reproduction so they could be used to control biological cycles. Thus, a generation of behavioral, physiological, and endocrinological studies was spawned in both the United States and Canada leading to hundreds of publications by researchers and graduate students from Nova Scotia to California.

Much of this work was summarized in reviews published in 1980 (Aiken, 1980; Aiken and Waddy, 1980; Van Olst *et al.*, 1980). At that time, the mechanisms controlling molting and growth were much better understood than those controlling reproduction. This has changed. Concepts of reproductive mechanisms and the factors that regulate them have been altered dramatically. Mechanisms such as intermolt mating, multiple extrusion, and multiple fertilization, unheard of in 1980, are now well accepted. Simple field techniques for assessing maturity in both males and females, cement gland staging and crusher propus index (CPI), have been developed and refined. Some of the most significant advances have been in the field of environmental control of reproduction, which was summarized in just seven lines in 1980 (Aiken and Waddy, 1980). Conflicting results on the role of photoperiod and temperature in the regulation of spawning cycles that have been resolved, resulting in the identification of an intriguing temperature–photoperiod facultative control mechanism and the development of strategies that enable spawning to be induced throughout the year. Suggestions of similarities between the environmentally regulated mechanisms which control molting and reproduction continue to surface as more information is obtained about seasonal influences on biological responses. It is now known that dramatic changes in reproductive physiology occur at the autumnal equinox and winter solstice and there are suggestions that molting physiology may be similarly affected.

The past 15 years have also brought impressive advances in endocrinology. Studies have revealed the existence of a family of novel multifunctional eyestalk peptides that includes the crustacean hyperglycemic hormone (CHH), molt inhibiting hormone (MIH), and gonad inhibiting hormone (GIH). These hormones have diverse functions and are important in both growth and reproduction. Data indicate that CHH may play a role in the regulation of both molting and spawning. Even though the function of the mandibular organ remains an enigma in *Homarus americanus*, research is continuing. It is interesting that the role of methyl farnesoate (MF, produced by the mandibular organ) in ovarian maturation appears quite different from other species, and it may be important in the regulation of premolt ecdysteroid levels.

It is appropriate that this book on the biology of the American lobster is published 100 years after Herrick produced his first tome on the subject. As a result of this century of study, more is probably known about the control of growth and reproduction of *Homarus americanus* than of any other commercially important marine invertebrate.

II. Molting and Growth

A. *The Molt Cycle*

In the historical sense, "molt" includes all of the physiological, morphological, biochemical, and behavioral changes involved in preparation for and recovery from ecdysis. Terms such as molt cycle and molting physiology recognize this broader connotation. However, there is a well-entrenched tendency to use molt in synonymy with ecdysis and terms such as premolt, intermolt, and postmolt are derived from this colloquial use. The more precise terms are proecdysis, metecdysis, and postecdysis, respectively.

Lobsters spend much of their lives preparing for or recovering from ecdysis. Premolt induction and preparation for ecdysis are under the control of the endocrine system, which in turn responds to internal (nutritional state, health, etc.) and external (temperature, day length, etc.) cues. Premolt activities include limb regeneration, resorption and storage of cuticular components, deposition of new cuticle, histolysis of somatic muscle, selective water and ion absorption, and shifts in biochemical pathways. These culminate in ecdysis, the shedding of the exoskeleton.

After ecdysis, the lobster assumes its new length and volume by actively absorbing water. The epidermis regresses, additional cuticle is secreted, mineralization occurs, water is replaced by new tissue, and metabolic reserves are replenished. The lobster may then move directly into preparation for the next molt (diecdysis) or pause for an extended period (anecdysis) (Carlisle and Dohrn, 1953).

1. The Integument

The lobster integument consists of a basement membrane, cellular epidermis, and cuticle (Fig. 1). The cuticle is composed of four distinct layers that are secreted by the epidermis during the molt cycle. A variety of terms have been used for these layers, but the nomenclature most widely used is that of Richards (1951): epicuticle, exocuticle, endocuticle, and membranous layer.

The epicuticle is very thin and is composed of proteins, lipids, and calcium salts, but not chitin, and the exocuticle is a calcified matrix of chitin and protein. These two layers form the soft shell of a newly molted lobster (pre-exuvial layers). These layers harden after ecdysis by tanning and calcification. The endocuticle is also a calcified matrix of chitin and protein that in *Homarus americanus* consists of two parts: an outer, rigid, calcified, thick lamina portion characterized by vertical striations; and an inner, flexible, thin lamina portion without obvious vertical striations. Endocuticle synthesis and calcification begins after molt and continues through stage C_3. The membranous layer also contains protein and chitin, but is not calcified. It is the innermost layer of the cuticle and its completion signals the onset of intermolt (Aiken, 1980).

The cuticle is usually thought of as a nonliving shell protecting the living tissue beneath. However, cytoplasmic extensions of the epidermal cells penetrate to the epicuticle through the pore canals, suggesting that the cuticle is actually a complex living tissue (Fig. 2). The pore canals comprise about 20% of the volume of the cuticle and 50–90 canals emanate from each epidermal cell, leaving no part of the cuticle more than 25 μm from epidermal cytoplasm (Halcrow, 1976). The pore canals are in contact with the epidermis until the cuticle separates from the epidermis (apolysis), suggesting that the cuticle is able to receive cell products throughout stages A to C_4.

2. Stages of the Molt Cycle

In progressing from one molt to the next, a lobster passes through a series of well-defined morphological and physiological states. Herrick (1909) identified three stages: *preparing to molt*, the growth of the new

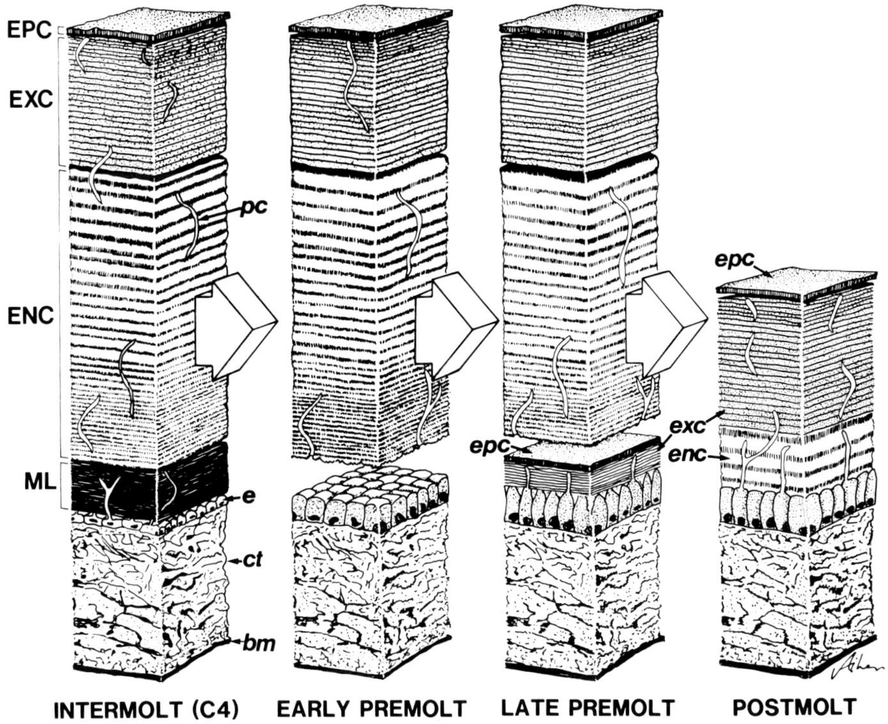

FIGURE 1 Changes in the structure of the integument during the molt cycle. EPC, Epicuticle; EXC, exocuticle; ENC, endocuticle; ML, membranous layer (the lower case labels identify equivalent layers of the new cuticle); e, epidermis; ct, connective tissue; bm, basement membrane; pc, pore canals. (From Aiken, D. E., and Waddy, S. L. 1992. The growth process in crayfish. *Rev. Aquat. Sci.* 6, 335–381. Reprinted by permission of CRC Press, Boca Raton, Florida.)

shell under the old; *shedding*, the shedding of the outgrown shell and a sudden expansion in size; and *recently molted*, the gradual hardening of the new shell. Fishermen of the time used such descriptive terms as shedder, black shell, and crack shell for late premolt lobsters, and shadow, rubber shell, paper shell, and buckle shell for lobsters that had recently molted (Herrick, 1909).

Drach (1939) was the first to devise a staging system that recognized the morphological and physiological changes associated with molting. He divided the intermolt cycle into four periods (A to D) with numerous subdivisions (Drach, 1939, 1944; Drach and Tchernigovtzeff, 1967). Knowles and Carlisle (1956) added stage E (ecdysis), making the designation "molt cycle" more appropriate. Postmolt and its synonym, postecdysis, encompasses stages A, B, and C_1 through C_3. Intermolt, or metecdysis, usually refers to stage C_4 (but see Drach and Tchernigovtzeff, 1967). Premolt and proecdysis are alternate terms for stage D. Drach's system has been adapted in a number of ways for use with lobsters (see Aiken, 1973, 1980, for references).

a. Stage E In *Homarus americanus*, where premolt can span months and the final preparation for ecdysis may extend over days or even weeks, it is essential that normal mobility, agility, and awareness of the environment be maintained, even while neuromuscular control is being transferred from the old cuticle to the new one underneath. This transition is accomplished so efficiently that a lobster is capable of maintaining coordinated activity up to and including the passive phase of ecdysis. Only after the old carapace has split away from the abdomen is the lobster immobilized and defenseless. Lobsters normally spend the vulnerable period in the safety of a shelter.

Ecdysis includes a passive and an active phase. In the passive phase (often classified as stage D_4), water is ingested, absorbed, and redistributed within the body (Mykles, 1980). The carapace, because of increased hydrostatic pressure, lifts from the decalcified ecdysial sutures above the bases of the legs and the thoracoabdominal membrane bulges outward at the juncture of the elevated carapace and the abdomen (Fig. 3). The lobster is still fully mobile and may become agitated if disturbed. If conditions for molt are unfavorable, the passive phase may be prolonged for several hours (Aiken, 1980).

The active phase of ecdysis starts with the rupture of the thoracoabdominal membrane, lasts 10–20 minutes, and is unresponsive to external stimuli. When

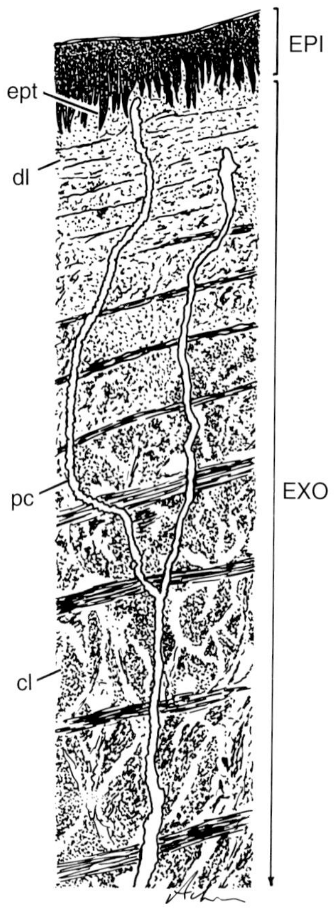

FIGURE 2 Relationship between the pore canals of the exocuticle and the "tailings" of the epicuticle of *Uca pugilator*. cl, Continuous lamella; dl, discontinuous lamella; ept, epicuticular tailing; pc, pore canal; EPI, epicuticle; EXO, exocuticle. (Drawn from micrographs in Green and Neff, 1972; from Aiken, D. E., and Waddy, S. L. 1992. The growth process in crayfish. *Riv. Aquat. Sci.* 6, 335–381. Reprinted by permission of CRC Press, Boca Raton, Florida.)

the membrane ruptures, the lobster loses mobility and rolls onto its side. The old carapace lifts clear along the posterior margin and pivots forward, allowing the head to be raised and the cephalic appendages to be withdrawn. This is followed by withdrawal of the chelipeds and other thoracic appendages, and finally the abdomen.

The large muscles of the chelipeds create a special problem because their diameter can be several times that of the upper leg segments through which they must pass. To facilitate withdrawal, the upper portion of the cheliped merus is decalcified until the cuticle is soft and transparent, and the cheliped muscle protein is degraded until there is a loss of 30–60% of the tissue mass (Mykles, 1992). Lubrication required for withdrawal is provided by a molting fluid. Although claw loss occasionally occurs during ecdysis, the success rate is surprisingly high.

b. Stages A and B Stage A occupies only 2% of the molt cycle (Drach, 1944) and begins as soon as the lobster is free of the exuvia. The newly emerged animal is flaccid and wrinkled, but ingestion and absorption of water expands the animal to its new volume, some 50% greater than before ecdysis. Growth is determined by the amount of epidermal cell division that occurs immediately before the onset of cuticle synthesis at molt stage D_1''' (Cheng and Chang, 1993, 1994) and the degree to which the new cuticle is stretched when water is taken up after ecdysis. The ripples that develop as the epidermal cells divide allow expansion after molting (Aiken, 1973).

Equally interesting is the mechanism for ensuring that strategic parts of the new cuticle are hardened immediately after ecdysis so that feeding can begin. During premolt, calcium carbonate is deposited in dome-shaped concretions (gastroliths) on the lateral walls of the stomach. When the cuticular lining of the stomach is shed at ecdysis, the gastroliths drop into the stomach where they disintegrate and are used for immediate hardening of the tips of the chelipeds and the cutting and tearing edges of the mandibles and maxillipeds. At this point, the "paper-shell" lobster is able to begin foraging (unlike other hard parts, the teeth of the gastric mill remain hard during ecdysis). The lobster then consumes its exuvia, except for the tips of the large chelate appendages, thereby recovering additional organic and inorganic matter. (See Lavalli and Factor, Chapter 14, and Factor, Chapter 15, on the mandibles and gastric mill.)

The first endocuticle lamella appears within 24–48 hours of ecdysis and marks the onset of stage B. Additional lamellae form on succeeding days, but there is little correlation between the number of postmolt days and the number of lamellae, and little consistency in numbers of lamellae from one anatomical region to another (Aiken, 1980). Although the endocuticle in the gastrolith forms in stage A in some decapods (Travis, 1960), deposition begins in stage D_3 in *Homarus americanus* and is completed in stage B (Aiken, 1980). The pleopods in stages A and B are characterized by rows of large cells that give them a striated appearance. Stage B occupies about 8% of the molt cycle and ends when postmolt changes in the pre-exuvial layers are complete (Drach, 1939, 1944).

c. Stage C Stage C begins when the chemical changes in the pre-exuvial layers are complete, but shell rigidity is a more convenient index (Aiken, 1973). In what has become the accepted molt cycle classification (Passano, 1960), stage C is divided into four substages and stage C_3 is the final stage of postmolt. The integument has attained maximum rigidity by the end of this stage, although some areas may

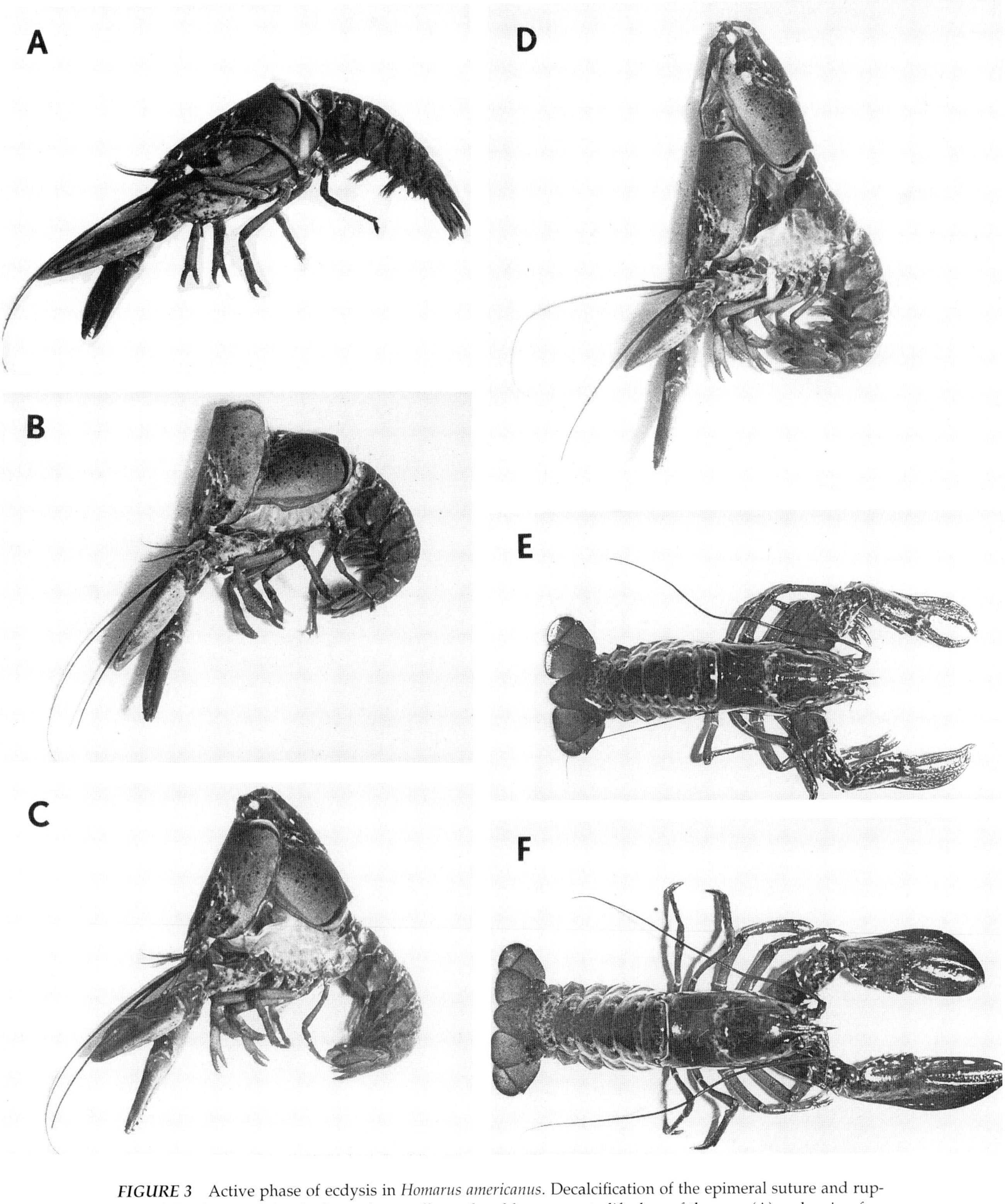

FIGURE 3 Active phase of ecdysis in *Homarus americanus*. Decalcification of the epimeral suture and rupture of the thoracoabdominal membrane allows the old carapace to lift clear of the new (A) and swing forward (B). Head appendages are withdrawn first (C), then the thoracic appendages, and finally the abdomen (D). Time from A to D is approximately 15 minutes. A linear increase of 9% was achieved after only 15 minutes postmolt, even though the chelipeds remain dehydrated (E). This continued to an increase of 16% in length and 55% in weight after four hours (F). (From Aiken, 1980, with permission.)

retain a degree of flexibility. Stage C_4 represents "intermolt," the protracted period after the cuticle is completed and the time when organic reserves are accumulating. Final ovary maturation and spawning occurs in C_4 in most crustaceans (Passano, 1960). In some species, stage C occupies as much as 65% of the molt cycle (Drach, 1944).

Although stage C can be arbitrarily divided into four substages in *Homarus americanus*, on the basis of rigidity, it is difficult to do so histologically because of the variation in onset, rate, and degree of formation of the different cuticle layers. The problem can be avoided if one anatomical location is selected as the standard, but this reduces the criteria available for staging. In the anterior carapace, for example, it is only possible to identify the onset of stage C_2 (start of thin lamina endocuticle formation) and stage C_4 (membranous layer present). Even when a variety of histological criteria are included, there is no meaningful way stage C can be subdivided more than three times. Donahue (1954) used only three substages, but because C_4 has become synonymous with intermolt, it is confusing to have intermolt as C_3 in *H. americanus*, but C_4 in other decapods. For this reason, C_2 and C_3 are combined here (as stage C_{2-3}) and stage C_4 represents intermolt.

Stage C_1. The endocuticle begins to form on the gastrolith disc in stage C_1 and the thin lamina portion of the endocuticle appears in the merus of the cheliped, but there is no thin lamina endocuticle in the carapace. In a typical 450-g lobster held at a temperature of 10°–15°C, stage C commences about 14–18 days after ecdysis (Aiken, 1980).

Stage C_{2-3}. Stage C_2 is characterized by the deposition of the thin lamina endocuticle in the anterior carapace and begins about 25–30 days after ecdysis in a 450-g lobster. By stage C_3, this portion of the endocuticle is present in all carapace sections. The gastrolith endocuticle continues to develop through stages C_2 and C_3.

Stage C_4. Stage C_4 commences with the deposition of the membranous layer. Drach (1939) distinguished between onset of membranous layer formation (C_4') and completion of the membranous layer (C_4), which he identified with a "broken cuticle test." However, this test is unreliable in *Homarus americanus* as the inner endocuticle is flexible and holds together when broken, even before the membranous layer has formed. In a 450-g lobster at 10°–15°C, the membranous layer is present after about 55 days postmolt. The gastrolith endocuticle is also completed at this time and the pleopods are "normal," that is, no large cells are visible and the epidermis is closely applied to the cuticle at the pleopod tip (Aiken, 1973, 1980).

d. Stage D Stage D is premolt or proecdysis, the physiologically significant period during which the lobster prepares for the next molt. Stage D is divided into substages D_0 through D_4 and is further subdivided by the addition of superscript characters (D_1', D_1'', D_1''') (Charniaux-Cotton, 1957; Drach, 1939, 1944; Drach and Tchernigovtzeff, 1967).

In the lobster, there is a gradual transition from intermolt to premolt with the characteristics of one blending with those of the other over weeks or even months. A developmental plateau (analogous to anecdysis) is common following apolysis, but before epidermal retraction is completed. Even in the winter, lobsters pass from C_4 into D_0, but do not proceed beyond the initial stages of epidermal retraction until the water warms in the spring. In this sense, D_0 is more of a resting stage than is C_4 and should be considered a broad transitional phase when lobsters display the characteristics of both intermolt and premolt. Although a lobster in stage D_0 is not clearly in either intermolt (C) or premolt (D), D_0 is a valid stage with well-defined characteristics. The dramatic and irreversible changes characteristic of premolt begin with the transition to stage D_1. It is interesting that Drach (1939) originally included the phenomena of D_0 in stage C_4 and started proecdysis with stage D_1.

Each subdivision of stage D is characterized by a distinct change in the morphology of the integument (Drach and Tchernigovtzeff, 1967):

D_0, retraction of the epidermis from the cuticle (apolysis);
D_1, invagination of the setal matrices;
D_2, secretion of the pre-exuvial layers;
D_3, resorption along epimeral sutures;
D_4, opening of the epimeral sutures.

Stage D_0. The most reliable indicator of the onset of stage D_0 is epidermal retraction (apolysis) in the tips of flattened appendages such as the pleopods. Onset of gastrolith formation or limb bud growth are occasionally used, but are too variable to be reliable indicators (Aiken, 1973). Apolysis is preceded by the appearance of an amber zone at the tip of the pleopod that is associated with a high incidence of gastrolith development. Its presence is a morphological indication of the onset of stage D (pleopod stage 1.0, $D^{1.0}$, $=D_0'$). At this substage, there is considerable variation in the degree of gastrolith development, but little evidence of apolysis in the general cuticle (Aiken, 1973).

By pleopod stage $D^{1.5}$ ($=D_0''$), the amber zone has condensed and there is clear area between the retracting epidermis and the chromatophores of the pleopod tip. When the epidermal line is clearly retracted from the pleopod tip, but still in contact along the lateral margins, the lobster is in pleopod stage $D^{2.0}$ ($=D_0''$); if epidermal retraction also includes the lateral margins of the pleopod, the stage is $D^{2.5}$ ($=D_0'''$). At this stage, the epidermis in the general integument begins to show signs of activity, indicating imminent transition into the irreversible premolt development of stage D_1.

Stage D_1. The first indication of stage D_1 is the scalloping of the epidermis at the tip of the pleopod, indicating the onset of setal invagination and stage $D^{3.0}$ ($=D_1'$). At this point, apolysis and epidermal activity are evident throughout the general integument. In pleopod stage $D^{3.5}$ ($=D_1''$), the epidermis has formed distinctive papillae around the invaginating setae. An "amorphous layer," a nonlaminar structureless band of varying thickness that stains light blue with Mallory's triple stain, appears between the epidermis and the cuticle at about stage $D^{3.0}$, becomes more widespread in $D^{3.5}$, and has the gross appearance of a thin membrane by $D^{4.0}$ ($=D_1'''$). It appears to be a modification of the membranous layer and thin lamina portion of the endocuticle (Herrick, 1909; Aiken, 1980).

In pleopod stage $D^{4.0}$, the shafts of the invaginated setae are obvious and new epicuticle can be seen on the epidermis of the dorsal carapace, although it may not be present on the pleopod epidermis and other locations. The new epicuticle is formed during stage $D^{4.0}$ and the new exocuticle is formed at stage $D^{4.5}$ ($=D_2'$) (Aiken, 1980).

Stage D_2. By pleopod stage $D^{4.5}$, new epicuticle is present on the pleopod epidermis and new exocuticle is present in the gastric region of the carapace. By stage $D^{5.0}$ ($=D_2''$), the exocuticle is well developed throughout the general integument and is developing on the pleopod epidermis. Exocuticle development occurs very rapidly at this point.

Stage D_3. In *Homarus americanus*, stage D_3 is the last useful substage of premolt, as D_4 (encompassing water absorption and opening of the ecdysial sutures) is transitory and more properly belongs in stage E. In stage D_3 (pleopod stage $D^{5.5}$), the pre-exuvial cuticle is well developed. Exocuticle development on the gastrolith is of particular interest in the American lobster, because it occurs just before the molt, late in stage D_3, instead of after the molt as in other decapods (Aiken, 1980).

3. Molt Staging

Two types of molt staging systems are used for lobsters: those based on shell hardness and those based on changes in developing setae (see Aiken, 1980, for references). Accurate molt prediction can only be achieved with pleopod staging (Fig. 4). Among the external characteristics, only decalcification of the merus and the lower carapace margin are reliable. Both of these occur at stage D_3 and indicate the molt will occur in less than a week at temperatures of 12°–15°C (Aiken, 1973).

a. Setal Staging Because of the variation in onset and rate of completion of different premolt events (Table 1), reliability in molt staging is achieved by basing subdivisions on changes in a single developmental system—setal development. Premolt staging with setal development is accurate and convenient and the American lobster is one of the best decapods for setogenic molt staging; the eight major pleopods each have a pair of flattened, transparent blades, making 16 successive samples possible (Aiken, 1973). This fact, plus the lobster's slow progression through the molt stages, permits setogenic changes to be followed in individuals.

It must be remembered, however, that setal staging and molt staging are different. Drach's (1939) intermolt stages are based on changes in the integument, but onset of a given integumentary process varies from one area of the body to another. Because of this, setal changes in the American lobster were correlated with histological evidence before stages were assigned (Aiken, 1980).

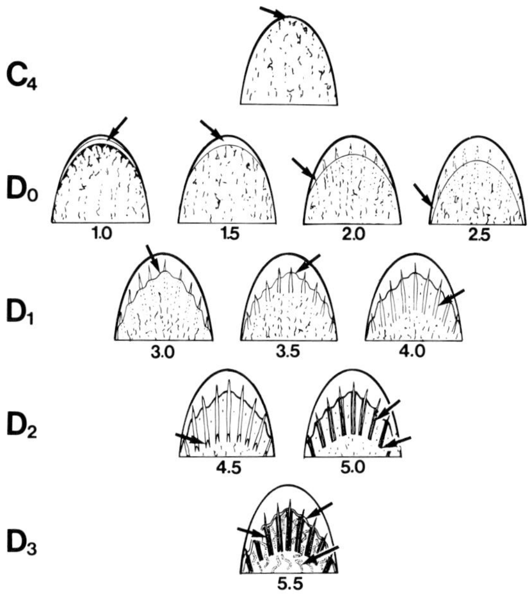

FIGURE 4 Epidermal and setal changes in the pleopods of American lobster from stage C_4 to D_3. Pleopod stage (1.0–5.5) and Drach's molt stage (D_0–D_3) are indicated. Criteria for each stage are given in Table 2. (From Aiken, D. E., and Waddy, S. L. 1992. The growth process in crayfish. *Rev. Aquat. Sci* 6, 335–381. Reprinted by permission of CRC Press, Boca Raton, Florida.)

TABLE 1 Time of Occurrence of Selected Premolt Events[a] in Various Regions of *Homarus americanus*

Molt stage:	D_0'	D_0''		D_0'''	D_1'	D_1''	D_1'''	D_2'	D_2''	D_3
Pleopod stage:	1.0	1.5	2.0	2.5	3.0	3.5	4.0	4.5	5.0	5.5
Pleopod										
Ep	X									
Apo		X	X	X						
ML			X	X	X					
Seto					X	X	X	X	X	X
Epic								X	X	X
Exoc									X	X
Carapace										
Ep				X						
Apo					X	X				
ML					X	X	X			
Epic							X	X	X	X
Exoc								X	X	X
Merus										
Apo					X	X				
ML					X	X	X			
Epic							X	X	X	X
Exoc								X	X	X
Gastrolith										
Ep	X	X								
Calc		X	X	X	X	X	X	X		
Epic									X	X
Exoc										X

[a]Apo, Apolysis; Calc, calcium deposition; Ep, epidermal activity; Epic, new cuticle present; Exoc, new exocuticle present; ML, membranous layer degeneration; Seto, setal development.

The first morphological indication of premolt is the retraction of the epidermis and setal cones from the cuticle. Externally, this can be seen along the transparent, flattened margins of the cuticle, especially the tips of the pleopods, branchial epipodites, uropods, and telson. When retraction is complete, a papilla forms around the base of each retracted seta and a new seta develops through epidermal growth and invagination (Fig. 5). Barbules, the new setal hairs, begin developing as invagination continues. These barbules and the setal shafts thicken and become more pronounced as the new epicuticle and pigmented layers are formed. This series of changes forms the basis for staging premolt lobsters (Table 2; Aiken, 1973, 1980).

b. External Criteria External characteristics are not very reliable as indicators of molt stage, but at times they are the only criteria available. A postmolt (stage A) lobster is uniformly soft except for the mouthparts, tips of the chelipeds, and similar essential parts. Lobsters harden progressively through stages B and C. The rate of calcification—and apparent molt stage—can be altered by diet, temperature, and other factors.

A lobster is considered to be in stage B or C_1 if the anteriodorsal region of the carapace can be depressed by light finger pressure; in stage C_2 if that area is rigid, but the dorsolateral portion of the posterior carapace can be depressed; in stage C_3 if only the ventrolateral carapace can be depressed; and in stage C_4 or D if all these areas are hard (Fig. 6). To distinguish stage C_4 from D where pleopod examination is not possible, the criteria in Fig. 7 are useful. Of these, a soft lateral margin of the carapace and a dark, soft, merus are the most reliable indicators of impending molt, and both indicate stage D_3. A pink and mottled ventral abdomen is a reliable sign of premolt, but its absence does not mean the animal is not in premolt.

B. Control of Molting and Growth

1. Larvae

a. Growth and Development Most available data on the control of larval growth and development have come from laboratory studies and the results have demonstrated principles useful in interpreting field data. Frequency of molting and size increase at molt are influenced by many factors including food, light, temperature, salinity, genetics, disease, mutilation, social conditions, water quality, and seasonal factors. These factors often act synergistically, making

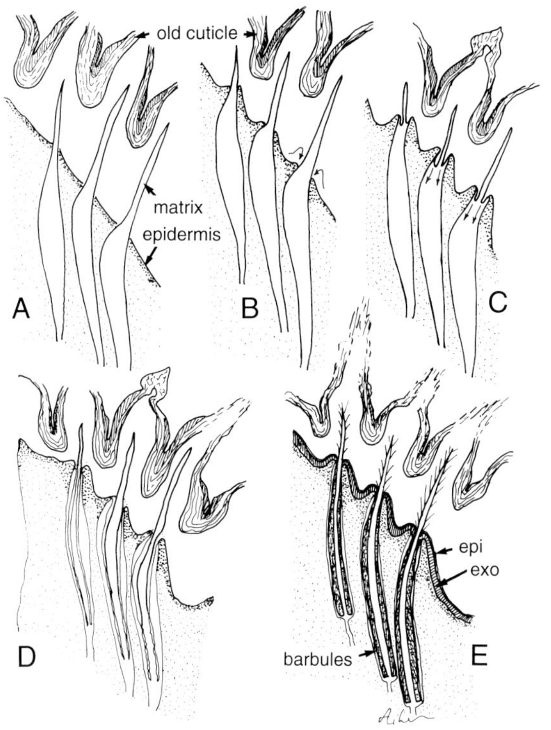

FIGURE 5 Steps in the formation of new setae in the tips of flattened appendages. (A) Epidermis retracts from cuticle (stage D_0). (B and C) Epidermis invaginates at the site of the new setae (stage D_1). (D) Invaginating epidermis visible as new setal shafts. (E) New barbules apparent on well-developed setal shafts (stage D_2). epi, Epicuticle; exo, exocuticle. (From Aiken, 1973, with permission.)

individual effects difficult to predict. (See Factor, Chapter 1, for a description of larval development, and Ennis, Chapter 3, on larval ecology.)

Larval development in the American lobster cannot be completed at low temperature, so it is critical that hatching occur well before the water temperature declines in the autumn. The temperature limit for larval development is about 10°C, below which survival is often poor. Temperature limits for development vary geographically in many decapods, but it is not known whether such variation occurs in *Homarus americanus* (Sastry and Vargo, 1977). Although time to stage IV is generally considered to be about 10 days at 22°–24°C and 2 months at 10°C (Hughes and Matthiessen, 1962; Huntsman, 1923; Templeman 1936a,c; cf. MacKenzie, 1988), development times vary seasonally. At 20°C, time for 50% of sibling larvae to reach stage IV varies from 11.8 to 18.5 days (Fig. 8). Similarly, time to stage IV is reported to be 65 days at 12°C (MacKenzie, 1988), while at St. Andrews development at similar temperatures was completed in only 36 days. Surprisingly, the most rapid development occurs in the spring, before the normal hatching season, and the slowest development occurs in the late summer and autumn (Waddy and Aiken, 1986). There are also seasonal differences in survival. Larvae hatching at the time of normal larval release generally survive better than those hatched in other seasons, and larvae hatching early in the season survive better than those hatching later

TABLE 2 **Criteria for Staging Proecdysis in *Homarus americanus***

Molt stage	Pleopod stage	Description
C_4	0	Epidermis closely applied to cuticular nodes at tip of pleopod; no amber zone or epidermal retraction at pleopod tip
D_0'	1.0	First indication of apolysis—amber or double-bordered region forms at the pleopod tip; chromatophores often show signs of reorganization, but there is no epidermal retraction from the cuticle
D_0''	1.5	Epidermis retracting from the terminal cuticular nodes; may have double-bordered appearance
D_0''	2.0	Epidermal line clearly formed and retracting from lateral cuticular nodes
D_0'''	2.5	Maximum epidermal retraction—not touching any lateral cuticular nodes
D_1'	3.0	Invagination papillae form at site of future setae; epidermis becomes scalloped
D_1''	3.5	Invagination papillae clearly formed, but shafts of new setae not well defined
D_1'''	4.0	Shafts of developing setae visible but proximal ends not clearly defined; shafts now invaginated to maximum length
D_2'	4.5	Shafts visible full length, but proximal ends are bifurcate instead of blunt; barbules becoming visible on setal shafts
D_2''	5.0	Shafts of developing setae thick, proximal ends blunt
D_3	5.5	Shafts of setae very thick and dark, proximal ends blunt; classify as D_3'' if fold or ripples are visible in cuticle on upper surface of pleopod

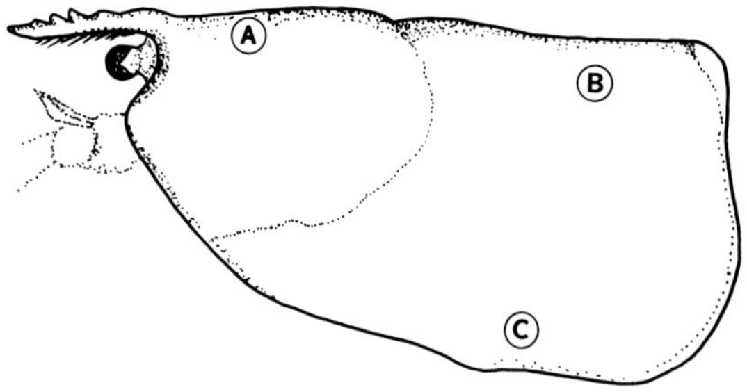

FIGURE 6 Shell rigidity as a guide to molt staging in *Homarus americanus*. If carapace in the vicinity of (A) can be easily depressed by finger pressure—stage B or C_1. If (A) is rigid but (B) can be depressed—stage C_2. If (B) is rigid but (C) can be depressed—stage C_3. If (A), (B), and (C) are all rigid—stage C_4 or D. To separate C_4 from D, refer to criteria in Fig. 7. (From Aiken, 1980, with permission.)

(Aiken and Waddy, 1986a; Sastry and Vargo, 1977).

The literature contains conflicting reports on the effect of environmental conditions on larval size. Larvae reared at 15°–18°C have been reported to be larger than those reared at lower or higher temperature, but larval size could not be correlated with geographic origin (MacKenzie, 1988). In the Northumberland Strait, larvae from the cooler, eastern end are reportedly larger than those from the warmer, western end (Templeman, 1936c). Hadley (1906) felt that higher rearing temperatures produce larger larvae. However, comparisons are difficult as many factors other than temperature influence larval size. Stage I larvae collected from Georges and Browns Banks and along the continental shelf have been found to be significantly larger than those from inshore sites on the Nova Scotia and Rhode Island coasts (Harding *et al.*, 1993; Rogers *et al.*, 1968), while no correlation could be found between larval size and geographic location in the Magdalen Islands (Hudon and Fradette, 1988). A decrease in the size of wild stage III larvae and stage IV postlarvae as the season progresses has been noted (Hudon and Fradette, 1988) and larval growth rates are reported to be higher in June than in July, even though temperatures are not as favorable and no differences in zooplankton biomass could be identified (Juinio and Cobb, 1994). Even groups of sibling larvae can be significantly different in size. The first larvae to hatch from a brood are usually the largest, as are the first to reach stage IV (Aiken *et al.*, 1981a, 1982; MacKenzie, 1988). Larvae continue to increase in size for at least 36 hours after they are released from the maternal female, and size differences are found in stage I siblings in different molt stages and in those fed different amounts of food (S. L. Waddy and W. W. Young-Lai, unpublished data).

Size and developmental differences between broods are often due to seasonal differences in growth rates, while differences within a brood are thought to be related to the decline in energy content that occurs in larvae hatching on successive nights. The n3:n6 fatty acid ratio of triacylglyceride lipids (TAG) in the first larvae to hatch from a brood is 38% higher than in larvae hatching later (Wickins *et al.*, 1995). High TAG levels are thought to confer survival advantages by increasing the ability to withstand adverse conditions, such as low salinity, pollution, starvation, or high temperature. The temperature experienced during embryogenesis also influences energy reserves; larvae hatching from eggs incubated at warmer temperatures have greater reserves than larvae hatched from eggs held for long periods at colder temperatures (Sasaki *et al.*, 1986).

Larvae require relatively more food than adults and they may not recover from inadequate quantities of food during the immediate post-hatching period, even if food is abundant later. Those with access to more food generally develop faster, grow larger, and survive better (Eagles *et al.*, 1986; Templeman, 1936b). It appears, however, that once sufficient metabolic reserves have been acquired, molting occurs regardless of food availability (Cobb and Wahle, 1994). Lobster larvae may be similar to other decapod larvae that have both obligatory and facultative feeding periods—although food is very important in molt stages A, B, and C, sufficient energy reserves are accumulated by stage D_0 for completion of premolt and ecdysis, independent of food availability (reviewed by Anger, 1991).

Providing food over a longer period each day can have a positive effect: larval development is more rapid when food is available 17 hours each day, compared to feeding the same amount over an 8-hour period (Eagles *et al.*, 1984). Larvae fed during the dark phase of the photoperiod cycle survive better than those fed during the light phase, but larval size and rate of development are unaffected (Eagles *et al.*, 1984). Excess food can slow development and reduce survival; larvae fed four times a day can take longer to reach stage IV than larvae fed three times a day if too much food is provided. Results from experiments using fragmented or carotenoid-deficient frozen adult *Artemia* as food support the principle that poor food quality increases development time, decreases growth and survival, and results in a higher incidence of claw loss and molting difficulties (Eagles *et al.*, 1984, 1986).

Larval responses to day length and light intensity vary with the season: in spring and early summer the most rapid development occurs at very short day lengths (LD 1:23, 1 hour of light and 23 hours of dark

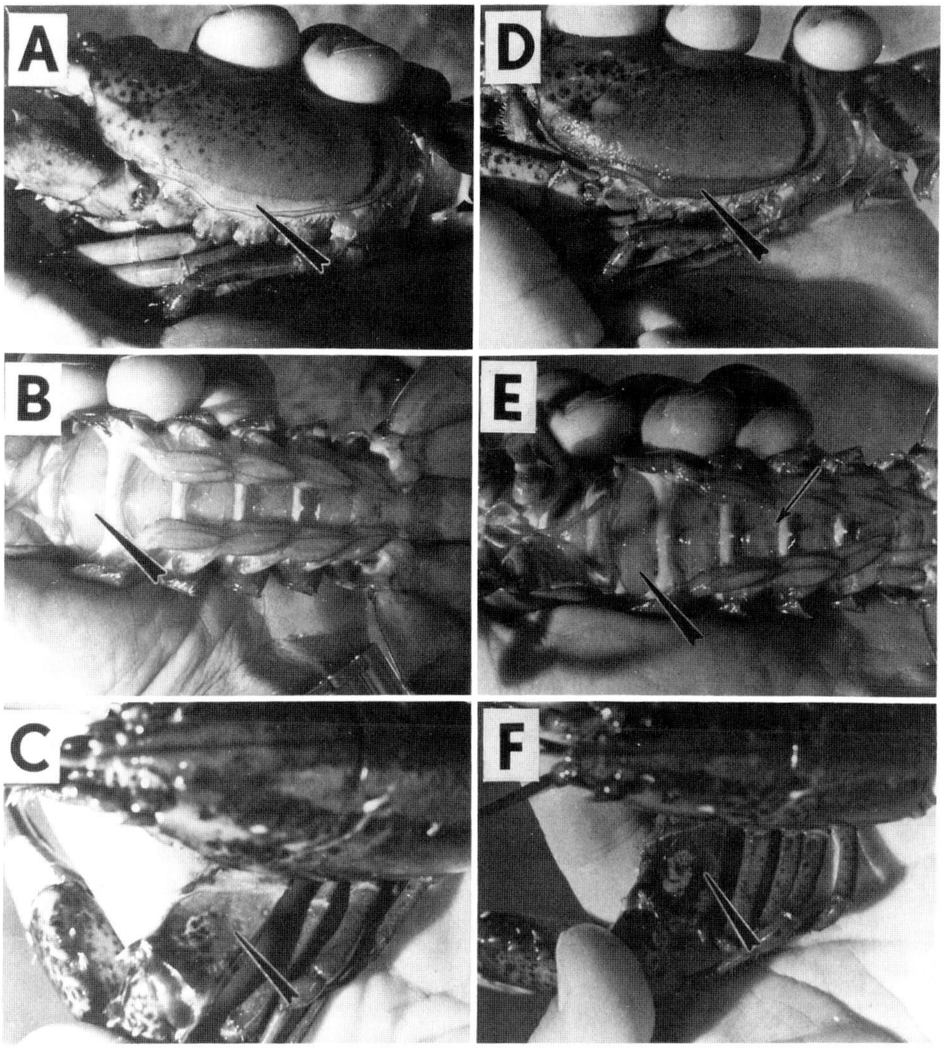

FIGURE 7 Changes in external appearance of *Homarus americanus* from intermolt (A–C) to late premolt (D–F). The lower margin of the carapace is firm and light colored during stage C (A), but becomes progressively darker during premolt (D), and turns soft within 3 days of ecdysis. Similar changes occur in the merus of the cheliped, which is firm and pale during stage C and most of premolt (C), but becomes dark about a week, and very soft 2–3 days, before ecdysis (F). The clear white ventral abdomen of an intermolt lobster (B) becomes pink during premolt and may develop dark mottling (arrow in E) as much as 2 weeks before ecdysis. (From Aiken, 1973, with permission.)

per day), whereas in September very long day lengths (LD 23:1) are most favorable for growth. There are also seasonal trends in the effect of day length on larval size (Aiken *et al.*, 1982). Bright lights inhibit larval growth (Hadley, 1906; Templeman, 1936c); larvae are heavier when reared under low light intensity and those grown in continuous darkness (LD 24:0) are almost twice as heavy as those reared in normal light cycles (Eagles *et al.*, 1986; Templeman, 1936b).

Human impacts on the environment can affect larval growth and development. The construction of a causeway across the Strait of Canso is believed by

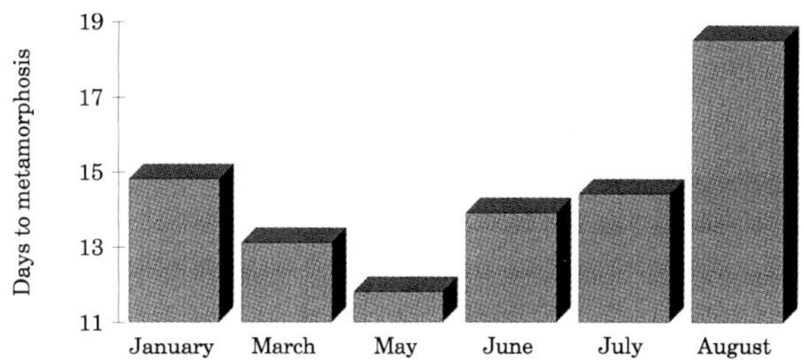

FIGURE 8 Seasonal difference in development rate (days from hatch to fourth stage) of larval American lobsters reared at 20°C.

some to have had an indirect effect on larvae by reducing the genetic mixing of lobsters from the Gulf of St. Lawrence and the eastern shore of Nova Scotia. Poor agricultural practices, quarrying, dredging, and construction in nearshore areas have been suggested as causes of siltation and turbidity, negatively affecting growth and development. Heated and unheated effluents from power plants, municipalities, and industrial facilities can alter thermal patterns and have local effects on survival, growth, development, and metabolic rates. Some larval stages are more sensitive than others; stage II larvae, for example, have a lower upper temperature tolerance than other larval stages (Aiken and Waddy, 1986a; see Sastry and Vargo, 1977, for references).

Larvae are 14–1000 times more sensitive to contaminants than adults, and stage I larvae are usually the most sensitive of the larval stages (Connor, 1972). Sublethal effects include reduced growth and regeneration, increased incidence of malformation, delayed development and inhibition of molting, changes in metabolism and energetics, respiratory stress, and increased incidence of intermediate larval stages (Ennis, Chapter 3; reviewed by Sastry and Pechenik, 1977).

b. Intermediate Stages Templeman (1936b) described a larval form intermediate between stages III and IV that occurs when environmental conditions are stressful, such as low temperature, low salinity, or inadequate food. Exposure to pollutants such as crude oil or K^+ brine also induces a significant number of intermediates, as does eyestalk ablation (Charmantier *et al.*, 1988). Although rare under favorable conditions, intermediates are occasionally found in the wild (Charmantier and Aiken, 1987).

Intermediate stages retain functional appendages from the previous stage and they can molt to either the next normal stage or to another intermediate, prolonging the time required to reach metamorphosis and in some cases increasing the time spent in surface waters and exposure to factors that also negatively affect survival (Charmantier *et al.*, 1988). Two types of intermediate larvae can follow stage III, designated IV′ and IVa (Fig. 9). Stage IVa is an additional larval stage, intermediate between stages III and IV but more like stage III in morphology, behavior, and physiology. This stage always molts to another intermediate stage, metamorphosis is spread over three molts, and final modifications of telson shape and exopodite length are deferred until stage VI. Stage IV′ larvae molt to a normal stage V and are morphologically similar to stage IV lobsters, except they retain the larval characters of long setae-bearing exopodites on the pereiopods, abdominal dorsal spines, and, to a lesser extent, posterolateral spines on the telson (Charmantier and Aiken, 1987).

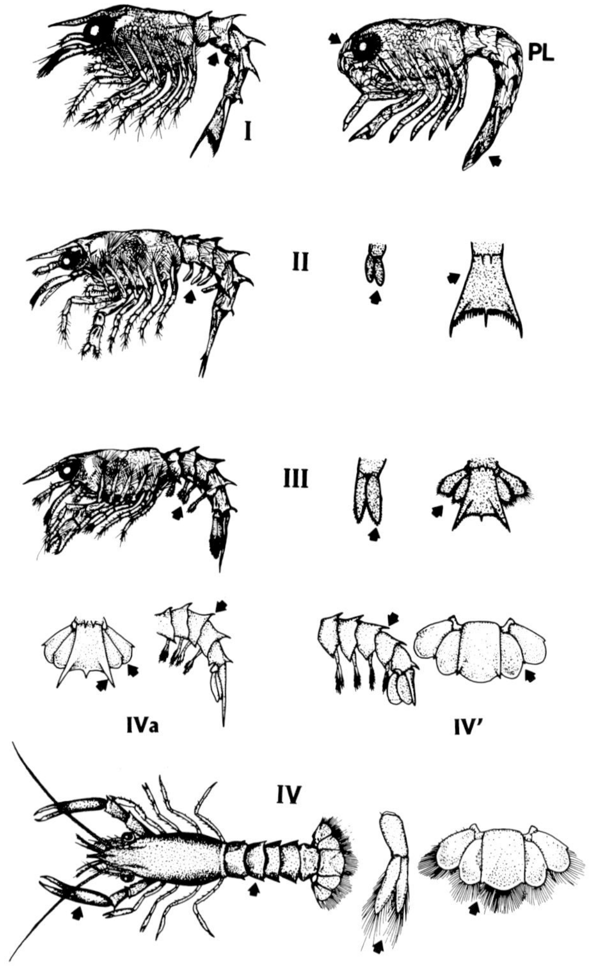

FIGURE 9 Normal and intermediate stages of larval American lobsters. Stages PL (prelarva), I, II, and III are normal larval forms. Stages designated IVa and IV′ are intermediate stages: stage IVa is intermediate between stages III and IV and always molts to another intermediate stage; stage IV′ molts to a normal stage V. Arrows point to distinguishing features of each stage. (Modified from Aiken and Waddy, 1986a, with permission.)

c. Endocrine Control Metamorphosis occurs at the molt from the third to the fourth stage, when many of the morphological, anatomical, ethological, and physiological characteristics are significantly altered (see Factor, Chapter 1). Lesser modifications are made during the molt from the fourth to the fifth stage (Charmantier *et al.*, 1991). Endocrine control of the metamorphic molt is assumed to be similar to that in juvenile and adult lobsters (see Section II,B,4), as responses to eyestalk ablation and ecdysteroid treatment are similar: eyestalk ablation induces premolt, the peak of hemolymph ecdysteroid concentration occurs in the middle of each larval stage and occurs earlier in eyestalk-ablated larvae than in intact larvae, and injection of sinus-gland extracts delays the molt and decreases ecdysteroid titers (Charmantier *et al.*, 1988; Snyder and Chang, 1986a).

Molting and morphogenesis are thought to be independent processes that are synchronized under normal conditions (McConaugha, 1985). There may be additional factors influencing metamorphosis, which, when disturbed by the stress of adverse environmental conditions or eyestalk removal, result in intermediate stages. Larvae may need to reach a critical size before metamorphosis can be completed, and, when malnourished, may molt to intermediate stages prior to attaining the required size (Snyder and Chang, 1986b).

Eyestalk ablation causes heavy mortality in larvae. When ablation occurs early in stage I, none survive. When the eyestalks are removed in the middle of larval stage II, all resulting stage III larvae molt to intermediate stages. Different intermediates form depending on the time of ablation: if ablation occurs well before the critical point (molt stage D_1'''), stage III larvae molt to stage IVa and metamorphosis is delayed by one or two molts; if ablation occurs near the critical point, larvae molt to stage IV′, retain some larval morphology, but undergo physiological metamorphosis; and the incidence of intermediates decreases if ablation is performed late in stage II. No intermediates develop when the eyestalks are removed in stage III. The intimate association between molt stage at the time of ablation and incidence and type of intermediate suggests that eyestalk neuroendocrine tissue is directly involved in the control of metamorphosis (Charmantier *et al.*, 1988, 1991).

Although juvenile hormones (JH) are important in the control of insect metamorphosis, there is little information on crustacean juvenile factors or the effects of exogenous JH on lobster larvae. Immersion of lobster larvae in a solution of JH III, one of several JH isomers that is a homolog of JH I and JH II, results in only a slight delay in the timing of the molt and alters the relative lengths of the carapace, antennae, and chelae (Hertz and Chang, 1986). JH I has no effect on the duration of the larval period and produces few intermediate stages (Charmantier *et al.*, 1991).

d. Molting Rhythms When lobster larvae are reared at 20°C under controlled laboratory conditions, the metamorphic molt occurs in a well-defined rhythm that peaks during the scotophase (the dark phase of the photoperiod cycle) (Fig. 10). Even when the light–dark cycle is reversed after the molt to the first stage, the third-stage larvae will adjust and molt in the dark phase (Aiken and Waddy, 1995). It appears that the metamorphic molt from third to fourth stage is triggered by the onset of the scotophase, and that some degree of entrainment is necessary. Under continuous illumination, the metamorphic molt occurs randomly throughout the day, but if a group of larvae are entrained to a light–dark cycle for a few days, they will molt predominantly in the scotophase (S. L. Waddy, unpublished data).

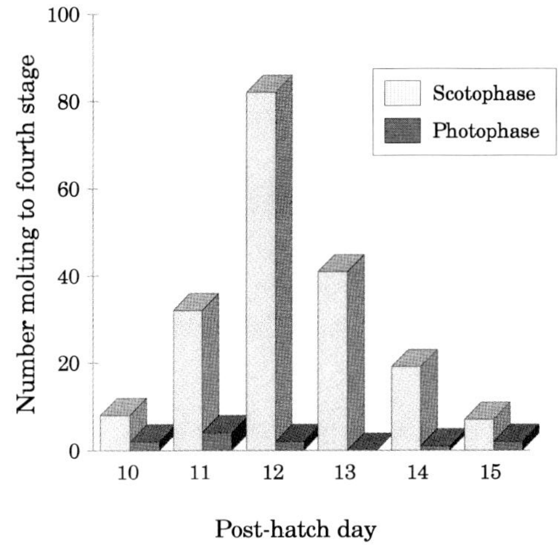

FIGURE 10 Diel timing of the molt in third-stage larvae American lobsters reared under controlled temperature and photoperiod conditions.

It is not known whether other larval stages exhibit similar molting rhythms, but over 80% of postmetamorphic stage IV lobsters molt to stage V during the day (Tamm and Cobb, 1976). This suggests that a change in timing of the molt may occur at metamorphosis, perhaps related to the change in phototaxis that occurs at this stage (Hadley, 1908).

2. Postlarvae

Stage IV postlarval lobsters are selective of bottom type and if substrates are not suitable they return to the surface and delay settlement for up to 2 weeks (Botero and Atema, 1982; Cobb, 1970, 1995; see Ennis, Chapter 3, and Lawton and Lavalli, Chapter 4). Although it was initially thought that stage IV lobsters could delay molting when exposed to unsuitable substrates, it was subsequently shown this was the result of social interaction, not substrate type (Cobb, 1968, 1970).

It has been estimated that in the warm, shallow inshore areas of the Magdalen Islands the majority of newly settled larvae spend the winter at stage VIII (approximately 14.5-mm carapace length (CL); Hudon, 1987). In the cooler waters of coastal Maine, new recruits are thought to reach a size of only 10-mm CL or less during their first summer, overwinter in stage VII, and grow to 11- to 16-mm CL during their second year (Incze and Wahle, 1991). At St. Andrews, under inshore temperature regimes that do not exceed 13°C, larvae reach stage IV between late August and mid-October and newly settled postlarvae grow rapidly until the temperature drops to 5°C in early December. The postlarvae display wide variations in growth rate and a small proportion reach 12- to 16-mm CL before winter, demonstrating that rapid growth is possible at relatively low temperatures (S. L. Waddy, unpublished data). Although it is not known how these growth rates compare with wild postlarvae in that area, growth rates of wild larvae always exceed those obtained in captivity in studies conducted in Rhode Island (Juinio and Cobb, 1994).

There continue to be suggestions that larval development and benthic recruitment are limited to areas where temperatures exceed 12°–15°C and that inadequate temperature limits postlarval settlement in the Bay of Fundy and some areas of coastal Maine (Cobb and Wahle, 1994; Huntsman, 1923; Wahle and Steneck, 1991). This hypothesis is weakened, however, by results from St. Andrews and the Fundy Isles region of the Bay of Fundy, where summer seawater temperatures do not exceed 13°C. High survival and settlement rates have been obtained in the laboratory and significant numbers of newly settled recruits have been found in field studies (Lawton and Robichaud, 1992; Waddy and Aiken, 1994; Lawton and Lavalli, Chapter 4).

3. Juveniles and Adults

a. Temperature Temperature is the most pervasive influence on lobster growth. Elevated temperatures accelerate metabolic processes and growth rate is proportional to temperature within the range of approximately 8° to 25°C (Aiken, 1977), although the highest temperatures may adversely affect linear increase at molt (Carlberg and Van Olst, 1976). For this reason, lobsters from relatively contained inshore areas, such as the southern Gulf of St. Lawrence and western Long Island Sound, grow faster than lobsters from colder areas, such as southwest Nova Scotia, the Bay of Fundy, and the Gulf of Maine. However, rapidly growing lobsters also mature earlier, which has a negative effect on growth, particularly in females, and can distort the perception of growth rate. The impression given by von Bertalanffy growth equations is that lobsters from colder areas grow faster (reviewed by Ennis, 1986).

In areas of relatively high summer temperature there are usually two molting peaks each year, one in the spring and another in the autumn. In colder areas, such as the Bay of Fundy, molting occurs in a single peak in late summer (Templeman, 1936a). Molting occurs 2–3 weeks later in colder years than in warmer ones and is delayed by about 1 week for each degree of reduction in mean summer water temperature (Templeman, 1936a, 1940b). Near the northern limits of the range, the proportion of lobsters of a given size that molt each year varies dramatically from year to year and in particularly cold years molting incidence can decrease by almost 50% (Ennis, 1983a).

When the water temperature drops to approximately 5°C, molt induction is blocked. Intermolt lobsters can be prevented from molting for at least 2 years by maintaining the temperature at ≤5°C. Lobsters held at this temperature progress slowly through the molt cycle to stage D_0, but no further progress is made until the temperature increases. The few lobsters that reach premolt (stage D_1) before the temperature declines in the autumn continue slowly through premolt and complete ecdysis at temperatures as low as 0°C. The effect of low winter temperature on progression through the molt cycle is largely responsible for the synchrony of the spring molt in nearshore American lobsters (Aiken and Waddy, 1986a).

Offshore lobsters undergo seasonally directed movements and may be more able to optimize their temperature regimes throughout the year (Cooper and Uzmann, 1980). The uniform temperatures in offshore areas tend to desynchronize molting (Waddy and Aiken, 1995a) and lobsters in these areas probably molt throughout a greater portion of the year and grow considerably faster than their nearshore counterparts, which endure several months below the minimum temperature for growth. It is thought that offshore lobsters recruit to the fishery after 4–5 years, instead of the 7–12 years required by lobsters in nearshore waters (Cooper and Uzmann, 1980).

Length increase at molt for a given size of lobster is relatively consistent from year to year for lobsters in a given locality (Conan, 1978; Ennis, 1971, 1972; but compare Campbell, 1983), but can vary significantly between areas (see Ennis, 1972, 1986, for references; cf. Wilder, 1953; Conan, 1978), being slightly greater in the warm waters of the southern Gulf of St. Lawrence than in the cooler waters of the Bay of Fundy (Wilder, 1953).

Thermal shock may have some inductive effects, especially for lobsters reared for long periods at constant temperature (Aiken, 1973). It is known that a rapid temperature increase alters the response to injected 20-hydroxyecdysone (Aiken and Waddy, 1975a), but the potential use of an abrupt temperature change to induce molting has not been adequately examined.

b. Season Molting occurs almost exclusively during seasons of optimum temperature (spring, summer, and autumn) and seasonally changing environmental cues schedule the molt and ensure completion before conditions become unfavorable. For many years it was assumed that lobsters molted in response to favorable temperature, irrespective of the time of year, but this is not the case. While lobsters exposed to temperatures of 10°C or higher in the spring quickly enter premolt and progress to ecdysis, lobsters held at the same temperature beginning in the autumn experience prolonged disruptive effects and do not molt, do not enter premolt, and do not even molt the following summer (Aiken and Waddy, 1976).

Temperatures of 15°–20°C disengage or override the seasonal molt inhibition (Aiken, 1980) and molting may be induced at any time of the year. Similarly, lobsters reared from the egg at a constant 20°C do not enter a distinct winter refractory period, although seasonal differences in molting incidence persist (Fig. 11). A multiyear study of 2283 molts of cultured lobsters shows that molting incidence peaks in the spring, decreases to a significantly lower rate in the summer, increases to a second, smaller peak in the autumn, and falls to its lowest level in the winter (Waddy and Aiken, 1989), suggesting that seasonal factors influence molting at all temperatures.

The circannual cycle of sensitivity to molt induction is mediated by the neurosecretory centers of the eyestalk. Eyestalk ablation eliminates the refractory period and returns summer and winter lobsters to a

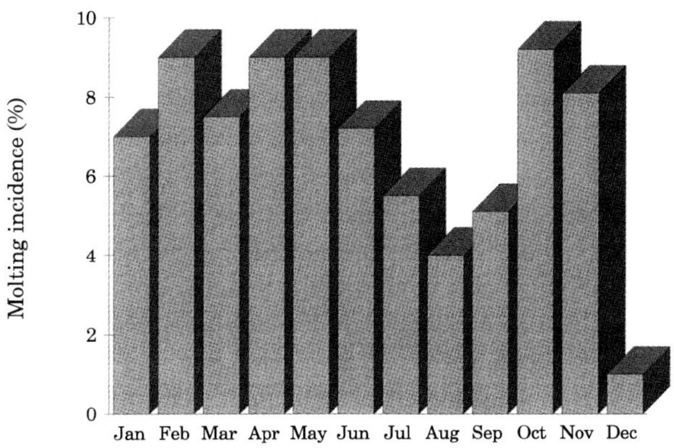

FIGURE 11 Seasonal differences in molting incidence in cultured American lobsters reared at 20°C.

common physiological baseline. The refractory period can be reinstated with eyestalk implants (Aiken and Waddy, 1976). The response of lobsters to injected 20-hydroxyecdysone also varies seasonally and much higher doses are required for premolt induction in the autumn than in the spring (Aiken and Waddy, 1975b).

c. Light and Photoperiod The effects of light are more specific than those of temperature and can be expressed either directly or indirectly. Direct effects, such as photoperiodism, operate through the central nervous system to control physiological processes. Indirect effects, such as those of light intensity, often influence growth through a secondary mechanism such as behavior or activity levels. Although the molt cycle of stage III larvae reared at 20°C is synchronized by photoperiod (Aiken and Waddy, 1995), there is little evidence that photoperiodism directly controls either molting or growth in juveniles and adults. There are strong indications, however, that light influences related processes.

Bright light depresses lobster activity and, since an inactive animal obtains less food, growth suffers. Hyperactivity caused by constant darkness can also reduce growth as it can increase the frequency of antagonistic behavior and the level of stress and injury (Aiken, 1980). Juvenile lobsters molt predominantly during the day (Tamm and Cobb, 1976) and food consumption and growth of small juveniles is enhanced by a short photoperiod (Bordner and Conklin, 1981). In larger juveniles, long photoperiods increase molt frequency and molt increment at 10°C, but not at 15° or 20°C (Aiken and Waddy, 1976), suggesting that response to day length may vary with both life history stage and temperature.

d. Density and Space Population density is an important factor influencing growth. The effect can be due to social interactions, lack of shelter or suitable habitat, or to a deterioration in environmental conditions such as oxygen levels or food availability. Crowding increases the frequency of social interactions at all stages and has a negative influence on survival, frequency of molting, and size increase at molt (Aiken and Waddy, 1978; Van Olst *et al.*, 1980). Although most of the available information on the effect of density on growth has come from laboratory studies, population density has been reported to affect molting rates in a wild population at Arnold's Cove in Newfoundland. The variation in the percentage of pre-recruit and recruit lobsters that molt each summer is thought to be due to changes in population density, with a lower incidence of molting at higher abundance (Ennis, 1991).

Physical barriers impose the same restrictions on growth as does crowding (Van Olst *et al.*, 1980). Lobsters stocked at equivalent density in either communal or individual holding perceive the same space relationship, even though the "apparent" space in the communal tank is much greater (Aiken and Waddy, 1978). Growth rates vary in proportion to the space available to each lobster, irrespective of the shape or depth of the cubicle (Sastry *et al.*, 1975; Sastry and French, 1977; Shleser, 1974, 1975). Although it has been estimated that an area of $75 \times (CL)^2$ is required for totally unrestricted growth (15% increase in CL; Van Olst *et al.*, 1980), length increases in 50- to 60-mm CL lobsters average 14% in an area of $55 \times (CL)^2$, and 10% in an area of only $18 \times (CL)^2$ (Aiken and Waddy, 1978).

The response of lobsters to changes in available space varies with molt stage (Templeman, 1948). Increasing the available space has a positive effect as long as the transfer to larger cubicles is made before molt stage D_1''', when formation of the new cuticle begins. Transfer to smaller cubicles negatively affects growth regardless of when the transfer occurs; however, the reduction in growth is smaller if the lobsters are moved after stage D_1''' (Cheng and Chang, 1994).

Relationships between available space and growth rates may be confounded by a short-lived, density-dependent, growth-inhibiting chemical factor (Nelson *et al.*, 1980, 1983). Juvenile lobsters grown in compartments immediately downstream from somewhat older individuals were 40% lighter in weight than lobsters farther downstream, suggesting that growth suppression may be an indirect effect caused by behavioral modifications that reduce mobility and food intake. In light of this, the influence of space restriction on growth might profitably be reexamined with regard to water flow and the location of animals in the tanks.

e. Habitat Juvenile lobsters have distinct substrate preferences, but adapt to a variety of habitats. In controlled environments in nature, no differences in growth rate are found among postlarval lobsters living on rock, vegetation, or mud substrates. The lobsters not only grow at similar rates, but all display the characteristic wide range of growth rates (Roach, 1983). (See Lawton and Lavalli, Chapter 4, on juvenile habitats.)

However, habitat type can affect growth and both bottom cover and shelter are influential. Shelters increase carrying capacity as lobsters use them to segregate themselves (Van Olst *et al.*, 1980). Juvenile lobsters grown in tanks at 20°C without substrate suffer heavy mortality and after 4 months only a single large lobster will remain (Aiken and Waddy, 1988). Laboratory studies have demonstrated some important principles on the effects of substrate type and niche size. Mortality is reduced if the habitat provides refuge for a range of lobster sizes. The irregular margins and surfaces of an oyster-shell substrate (*Crassostrea virginica*) provide a variety of niche sizes and shapes and support a wide size range of lobsters from stage IV to about 15-mm CL. The smaller niche sizes provided by a substrate of soft-shell clam shells (*Mya arenaria*) protect lobsters up to 11-mm CL, but larger individuals are captured and eaten by their tank mates. The large and symmetrical valves of the surf clam (*Spisula solidissima*) provide good cover for juveniles 8-mm CL and larger, but are inferior to oysters or soft-shell clams for smaller juveniles (Aiken and Waddy, 1988). These are probably indirect effects, resulting from varying social pressure and nutrition associated with these habitats, whereas the effects of space restriction appear to be more direct.

f. Nutrition and Food Food and nutritional state regulate both the duration of the molt cycle and growth at molt (Aiken 1980; Castell and Budson, 1974). Although lobsters are able to withstand long periods of starvation, they do not store appreciable amounts of energy and require a nearly constant supply of nutrients to achieve maximum growth (Bordner and Conklin, 1981). Rapidly growing juvenile lobsters consume about 10% of their body weight each day, while adults consume only about 1% of their weight (D'Abramo and Conklin, 1985).

There is a direct relationship between food consumption and growth in juvenile lobsters. Reducing available food by 29% results in a growth reduction of 31% (Bordner and Conklin, 1981). Decreasing the level of protein in the diet can lead to weight loss and decreased incidence of molting (Capuzzo and Lancaster, 1979; Castell and Budson, 1974). There is an inverse relationship between serum protein levels and intermolt time and a linear relationship between serum protein and weight gain at molt (Castell and Budson, 1974).

Environmental conditions affect food consumption. Consumption increases as temperatures increase and decreases with exposure to light. Lobsters maintained in near-constant darkness eat more and grow faster. Growth is more rapid when lobsters are fed 7 days per week rather than 5, but feeding twice per day does not increase growth (Bordner and Conklin, 1981). Feeding to excess in culture can decrease growth because of toxic effects from the decomposition of unconsumed food (Bartley *et al.*, 1980).

When nutrition is poor, cultured lobsters are prone to claw loss, limb deformities, death at molt, and poor survival following molt (Bowser and Rosemark, 1981). Small, rapidly growing lobsters are more prone to diet-related molting difficulties than larger ones, and the so-called "molt-death syndrome" can be prevented by supplementing the diet with fresh, live food (Ali and Wickins, 1994). When starved, the molt cycle is arrested in stage C (Juinio *et al.*, 1992) and time to 50% mortality varies inversely with temperature. Younger, smaller lobsters are the least able to withstand long periods without food; postlarvae survive for 12–29 days when starved, while small juveniles live for 47 to 114 days at temperatures of 10°–25°C (Bartley *et al.*, 1980; Juinio *et al.*, 1992).

g. Behavior and Social Conditions Behavioral and social pressures influence molting and growth. The presence of other lobsters causes molt delay and when lobsters are held in pairs a dominant-subordinate relationship develops (see Atema and Voigt, Chapter 13). Dominants control access to food and shelter, are the first to molt, and have growth rates similar to lobsters held individually. In subordinates, stage D_0 is prolonged, which can extend the length of intermolt up to 70%. Even in the presence of an abundance of food, subordinates do not feed as readily or rapidly and do not hoard food as often, so the slower growth rates are probably due to lower food intake (Cobb *et al.*, 1982; Cobb and Tamm, 1974, 1975). Although dominants are seldom killed or injured, subordinates are susceptible to injury and death at molt (Sastry and Ehringer, 1980).

The growth rate of an individual is thus modified by the relative size of other lobsters, and sibling lobsters reared in communal tanks display wider variations in growth rates than lobsters raised individually—after only 2 months some lobsters are three to five times the size of others. The behavioral suppression of growth can be reduced somewhat by removing or immobilizing the chelipeds (Aiken and Waddy, 1978; Aiken and Young-Lai, 1979, 1981; Cobb and Tamm, 1977). When lobsters are harvested and sorted into narrow size categories, new hierarchies develop as previously subordinate animals assume new status. The slowest-growing lobsters immediately display the same growth pattern as the original population (Fig. 12). Among the fastest-growing lobsters, some continue to grow rapidly while others grow so slowly they are bypassed by the fastest-growing individuals from the initially slow-growing group (Aiken and Waddy, 1988).

Other behavioral influences have been suggested. Lobsters are thought to have slower growth rates in the presence of predators, as they spend more time in their shelters and less time foraging and feeding (Cobb and Wahle, 1994; MacKenzie, 1988). The timing of molt in mature females has been reported to depend upon the availability of dominant males (Cowan and Atema, 1990), but this requires confirmation.

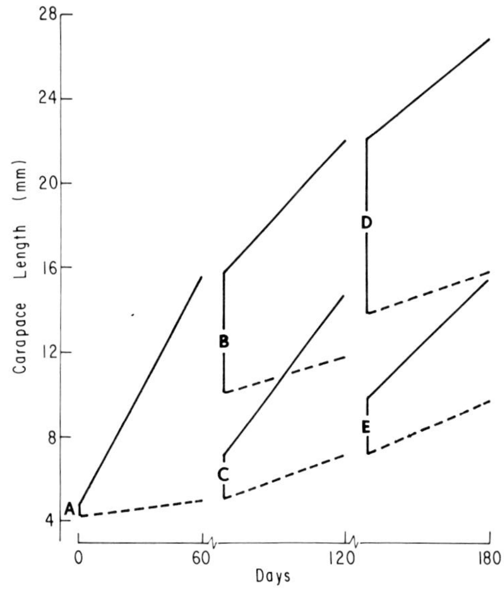

FIGURE 12 Effect of size grouping on growth rate of fourth stage American lobsters stocked in communal tanks. Solid lines indicate the growth rate of the fastest growing lobsters; broken lines, the growth of the slowest animals. At harvest, Group A was divided into large and small animals and restocked as B and C, respectively. At the second harvest, they were restocked as D and E. Each group, whether consisting of fast- or slow-growing animals, produced a pattern in which both fast- and slow-growing animals emerged. (From Aiken and Waddy, 1988, with permission.)

h. Size and Sex Growth rate declines with increasing size (Templeman, 1940b). Small juveniles molt every month or so in the summer, while lobsters that weigh several kilograms molt once in 4 or 5 years (Aiken, 1977).

Growth rate also varies with sex. Males grow faster than females even at the smallest sizes (Waddy *et al.*, 1988), but the difference becomes more pronounced after sexual maturity because the female must divert energy to egg production and delay molting to brood her eggs (Templeman, 1933). Mature males also weigh more than females of comparable carapace length because of allometric increase in cheliped volume. However, direct comparison of male and female growth based on carapace length is confounded by the differential growth associated with gender. "Growth" has both a linear and a volumetric (weight) component. For a given total length, males have a relatively longer carapace length (Templeman, 1940b) and heavier chelipeds, whereas females have a greater body volume.

i. Anthropogenic Effects Although there is little information on the sublethal effects of pollutants on lobsters, studies of other species demonstrate that molting and growth in crustaceans can be adversely affected by the presence of aromatic hydrocarbons, polychlorinated biphenyls (PCBs), ammonia, chlorophenols, dithiocarbamates, phosphorous, methylmercury, cadmium, zinc, crude and refined oils, and a variety of pesticides. In addition, a synergism exists between pollutants and natural environmental stressors (Fingerman, 1985). Some contaminants decrease growth at molt or reduce molting incidence, while others interfere with feeding behavior, regeneration, cuticle formation, or ecdysis (reviewed by Aiken and Waddy, 1986a).

Industrial discharges, such as bleached kraft mill effluent, potash brine, and drilling muds, have been identified as causes of reduced survival, growth, and regeneration, changes in ATPase activity, increased oxygen consumption, increased susceptibility to disease, thinning of shells, morphological abnormalities, inability to molt, and changes in behavior in *Homarus americanus*. Human impacts on the environment, such as the construction of causeways, poor agricultural practices, dredging, and drilling for oil, can produce environmental changes that directly affect growth or produce indirect effects by destroying or altering habitat (reviewed by Aiken and Waddy, 1986a).

j. Miscellaneous Any stressful situation can delay or inhibit ecdysis and even handling is thought to be inhibitory (Stewart and Squires, 1968). Tagging or marking may cause a molt delay or reduced molt increment under certain conditions (Cooper, 1970). Boring holes in the tail to mark lobsters, for example, inhibits molting temporarily (Stewart and Squires, 1968), but once the animal has recovered, size increase at molt is thought to be unaffected. Diver-collected data from newly molted lobsters still in their burrows suggests that data obtained from internally tagged lobsters is reasonably representative of the growth of untagged lobsters (Cooper, 1970; Ennis, 1972).

Lobsters in deeper, darker waters are thought to be more active and spend more time foraging than lobsters in shallower waters, which may affect growth (Cooper and Uzmann, 1980). Both reduced dissolved oxygen and trauma can affect growth in other decapods, and probably also in the American lobster (Chittleborough, 1974). An intense fishery and large catches of sublegal and ovigerous lobsters can decrease molt frequency and reduce linear growth increments (Aiken and Waddy, 1986a). Scallop dredges, Irish moss rakes, and a variety of trawl gear undoubtedly have a negative impact on growth as well (Scarratt, 1973; Roddick and Miller, 1992).

4. Endocrine Control

The basic paradigm of decapod molt regulation is that a molt-inhibiting hormone MIH from the lobster X organ–sinus gland complex of the eyestalk suppresses the biosynthetic activity of the molting glands (the Y organs). Environmental factors are assumed to act through the central nervous system to control the synthesis and release of MIH. Once released from this inhibition, the Y organs (molting glands) synthesize and secrete ecdysteroids, initiating the suite of biochemical and physiological changes that culminate in ecdysis. Thus, molting is often presented as a relatively straightforward system involving two antagonistic hormones, one a molt-inhibiting neuropeptide, the other a molt-promoting steroid. In reality, crustacean molting physiology is a profoundly complex system about which much is known, but little is clearly understood.

Studies of juvenile and adult *Homarus americanus* have revealed the existence of a novel eyestalk neuropeptide family of multifunctional peptides that includes crustacean hyperglycemic hormone CHH, molt inhibiting hormone MIH, and gonad inhibiting hormone GIH (Kegel *et al.*, 1991; Keller, 1992). This family of hormones is thought to be important in both growth and reproduction. Six eyestalk hormones from this family have been separated with high performance liquid chromatography (HPLC) techniques: two isoforms of gonad inhibiting hormone; two isoforms of the most potent crustacean hyperglycemic hormone, CHH-A; and two isoforms of a second crustacean hyperglycemic hormone, CHH-B, only one of which has significant hyperglycemic activity. At least one of the CHH-B isoforms has gonad stimulating hormone (GSH) activity and it has been suggested that a CHH peptide has molt-inhibiting activity. If this seems confusing, it is because these hormones were named many years ago when their existence was speculative and before they were known to exist in different isoforms with varying activities. The diverse functions of these hormones—regulation of blood glucose, molting, and vitellogenesis—are thought to be due to the variety in the C-terminal region (Webster, 1991). Studies on the expression of each of these hormones during the molt and reproductive cycle should reveal their role in molt and reproductive control in *H. americanus*.

a. Molt Inhibition The neuropeptide inhibitor MIH is thought to be the primary regulator of crustacean molting glands. It is produced and secreted by the X organ, a group of neurosecretory neurons, and is stored in the sinus gland, a neurohemal organ in the eyestalks. There is rapid elevation in the concentration of circulating molting hormones following X organ–sinus gland removal, a decrease in ecdysteroid titers in eyestalk-ablated animals after injection of eyestalk extracts, and an inhibition of ecdysteroid secretion when Y organs are cultured with eyestalk extracts (reviewed by Aiken and Waddy, 1992).

MIH appears to regulate molting by controlling cholesterol uptake and metabolism (Spaziani and Wang, 1993). There is also evidence that MIH can act by increasing the rate of hormone metabolism or excretion, or by modifying target tissue response to circulating ecdysteroids. The neurotransmitter 5-hydroxytryptamine mediates MIH release in a reversible dose-dependent manner and MIH release appears to be regulated primarily by serotonergic neurons and secondarily by ecdysteroid feedback and environmental stimuli. MIH secretion is diminished or overridden during premolt when ecdysteroid production is high (reviewed by Aiken and Waddy, 1992).

The amino acid sequence of a lobster MIH (Bruce and Chang, 1984; Chang *et al.*, 1990) is reported to be very similar (96% identity) to lobster CHH-B and almost identical to CHH-A, the most potent of the crustacean hyperglycemic hormones (Tensen *et al.*, 1991). It is not surprising, therefore, that lobster MIH reportedly has significant CHH activity (Chang *et al.*, 1993). A putative MIH has also been isolated from the green crab (*Carcinus maenas*) using a bioassay for the inhibition of ecdysteroid synthesis in Y organs *in vitro*. This MIH has only a 28% identity with crab CHH, but has a 65% identity with lobster GIH (Webster and Keller, 1986; Webster, 1991). However, its function as a molt inhibitor *in vivo* has yet to be demonstrated (Keller, 1992).

The similarity among MIH, GIH, and CHH is intriguing, as are suggestions that GIH and CHH may also have molt-inhibiting functions (de Kleijn, 1995). Although MIH and GIH are specific to the eyestalk and thus eliminated by eyestalk ablation, CHH—although predominantly in the eyestalk—occurs throughout the central nervous system (de Kleijn *et al.*, 1994a). Despite the reported similarity between lobster MIH and CHH (Chang *et al.*, 1990), the two hormones are produced by different neurosecretory cells. *In situ* hybridization studies using MIH-encoding cDNA have confirmed the results of immunocytochemical studies showing that MIH does not colocalize with the crustacean hyperglycemic hormone (Dircksen *et al.*, 1988; Klein *et al.*, 1993).

Further complicating our understanding of molt inhibition are suggestions that decapods are able to regulate their ecdysteroid levels in a normal manner even when there is no eyestalk tissue (Chang, 1984), that methyl farnesoate stimulates ecdysteroid secretion (Tamone and Chang, 1993), and that CHH inhibits ecdysteroid production (Webster, 1991; Yasuda *et al.*, 1994). There are also reports of nonpeptide ecdysteroid inhibitors with MIH activity in the X organ–sinus gland complex: 3-hydroxy-L-kynurenine (3-OH-K) and xanthurenic acid (XA); 3-OH-K is converted to XA at the Y organ. Results suggest that 3-OH-K is released into the hemolymph, accumulates on the surface of the Y organ, and is converted into xanthurenic acid, which suppresses ecdysteroid synthesis (Naya *et al.*, 1988, 1989; Kleinholz, 1985).

b. Molt Induction and Molt Accelerating Hormone "Molt induction" in the American lobster occurs with the transition from stage D_0 to D_1 (see Section II,A,2). After this point, premolt events proceed irreversibly at a rate that is dependent on animal size and temperature. The weight of current opinion favors passive molt induction with premolt being induced and sustained by the absence, or relative insignificance, of MIH (Aiken and Waddy, 1992). However, metabolic changes of early premolt are mediated by substances from something other than the eyestalk or molting gland (Bollenbacher *et al.*, 1972), and there is considerable historical evidence for central nervous system enhancement of Y-organ activity, thus the postulation of a molt-accelerating hormone (reviewed by Vernet, 1976). It has been recently reported that MF produced by the mandibular organs is a positive regulator of the Y organs (Chang *et al.*, 1993; Tamone and Chang, 1993), suggesting that MIH and MF may both be important in the regulation of premolt ecdysteroid levels.

c. Molting Gland The molting glands (called Y organs in decapods) are endocrine structures of ectodermal origin, located bilaterally beneath the epidermis of the anterolateral carapace (Fig. 13), and are neither vascularized nor innervated. They are maintained and their products are removed by superfusion of the circulating hemolymph. In *Homarus americanus*, the Y organs are an integral part of the hypodermis and much less conspicuous than those in crabs, which are detached from the hypodermis (Sochasky *et al.*, 1972). The physiological changes of premolt are induced and regulated by secretions of ecdysteroids from the molting glands. In contrast to

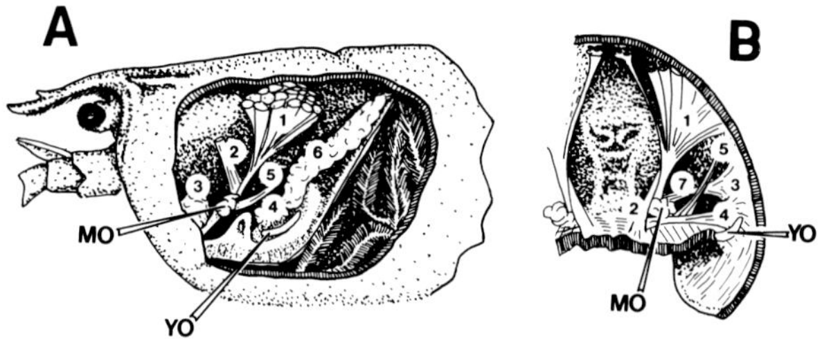

FIGURE 13 Location of mandibular organs and Y organs in American lobsters. Cephalothorax of adult lobster viewed from lateral (A) and posterior (B) aspects showing position of mandibular organ (MO) and Y organ (YO). Muscles and tendons: (1) maxillularly abductor coxopoditis; (2) posterior mandibular adductor; (3) anterior dorsoventral; (4) lateral mandibular adductor; (5) major mandibular abductor; (6) posterior dorsoventral; (7) transverse mandibular tendon. (From Sochasky *et al.*, 1972.)

insect systems, relatively little is known about the synthesis of ecdysteroids in crustaceans. In at least some decapods, ecdysone is found mainly within the mitochondria of the molting gland; in insects, however, ecdysone occurs predominantly in the smooth endoplasmic reticulum, suggesting substantial differences in ecdysone synthesis (Birkenbeil and Eckert, 1983).

The molting gland is only one of several known sources of molting hormones in insects, but the Y organ is generally thought to be the principal source in decapods, even though there are reports of normal ecdysteroid levels in crabs without Y organs. Ecdysteroids have been identified in the ovaries and eyestalks of crabs, but the ovarian ecdysteroids are contained for use during embryogenesis and the eyestalk ecdysteroids do not, as yet, have a demonstrated function (reviewed by Aiken and Waddy, 1992).

*d. **Molting Hormones*** Molting hormone is a general term for several polyhydroxylated steroid hormones (ecdysteroids) that induce premolt and the physiological and biochemical changes that lead to molt. The major ecdysteroids in decapods are ecdysone, 3-dehydroxyecdysone, 20-hydroxyecdysone, and ponasterone A (Naya and Ikeda, 1993; Rudolph and Spaziani, 1992). Ecdysteroid production is complex and the types and relative quantities of ecdysteroids synthesized vary among species (reviewed by Rudolph and Spaziani, 1992).

In early premolt, the Y organs rapidly convert sterol precursor and secrete ecdysteroids that circulate in the hemolymph in free form and are then taken up by the target tissues. Ecdysteroid titers follow a complex pattern that allows precise control over the various molting processes. Small transient peaks occur and ecdysteroid titers often fluctuate daily in *Homarus americanus*. Despite the complexity, concentrations of molting hormones in the hemolymph follow a predictable pattern throughout the molt cycle. They are barely detectable immediately after ecdysis and minor peaks occur in stage C. There is usually a small increase at stage D_0, coinciding with apolysis and initial gastrolith deposition, but the developmental processes of D_0 seem to require relatively small quantities of ecdysteroids. The major ecdysteroid peak occurs during stages D_1''' and D_2', when the pre-exuvial cuticle is being formed. In stage D_3', when the old cuticle is being resorbed, hemolymph ecdysteroid titers drop rapidly to the low levels characteristic of stage A (reviewed by Aiken and Waddy, 1992) and the decrease is associated with an increase in the titer of highly polar metabolites (Snyder and Chang, 1991a).

Ecdysone and 20-hydroxyecdysone have different roles in integumental morphogenesis in insects: ecdysone brings about changes in epidermal cells prior to apolysis, whereas 20-hydroxyecdysone induces apolysis and stimulates cuticle deposition. Their titers during the crustacean molt cycle are consistent with their roles: ecdysone is the primary ecdysteroid prior to apolysis, but 20-hydroxyecdysone predominates when the cuticle is being deposited. Specific events are assumed to result from a delicate balance between the two hormones (Walgraeve and Verhaert, 1988; Porcheron *et al.*, 1984) and the variation in type and quantity of ecdysteroids produced in crustaceans suggests there is a high degree of regulatory sophistication and coordination. At least three distinct levels of control have been identified: rate of synthesis, type of ecdysteroid metabolism, and rate of ecdysteroid inactivation and excretion (Chang, 1989).

Significant alterations in the pathways of ecdysteroid metabolism occur during the molt cycle. Ecdysone produced by the Y organ is hydroxylated to several different products, but the 20-hydroxyecdysone metabolic pathway is the only one of significance during premolt (McCarthy, 1982). In the American lobster, highly polar metabolites are the major circulating ecdysteroids at all stages except in mid- to late premolt (Snyder and Chang, 1991a). The rate of hydroxylation is thought to vary during the molt cycle, as there is a threefold increase in ecdysone-20-hydroxylase activity in crab testes following eyestalk ablation (Chang, 1989).

There are molt-stage-specific differences in the rate of turnover of ecdysone metabolites (McCarthy, 1982), which involves both polar and apolar pathways in *Homarus americanus* (Snyder and Chang, 1991b). Ecdysteroids are eliminated by forming highly polar products in organs such as the nonvitellogenic ovary, hindgut, and epidermis; polar conjugates loaded into vitellogenic ovaries are potential sources of ecdysteroids for developing embryos (Spindler *et al.*, 1987). Inactivation mechanisms are capable of effecting dramatic rate changes, producing a 10-fold increase in the rate of ecdysteroid elimination as the animal approaches molt (McCarthy, 1982). In the American lobster, ecdysteroids are thought to be eliminated primarily by excretion of ecdysteroid metabolites in the urine. Much higher levels of highly polar conjugates are excreted during stage D_2 than during other molt stages (Snyder and Chang, 1991c).

Although Y-organ secretory regulation is ascribed primarily to the X organ–sinus gland complex (Chang, 1985), it is recognized that other regulatory mechanisms must exist. The relative competence with which the new integument is formed in eyestalkless animals certainly confirms this, as does a report that methyl farnesoate stimulates ecdysteroid production (Tamone and Chang, 1993).

e. Crustacean Hyperglycemic Hormone There is accumulating evidence that crustacean hyperglycemic hormones (CHH) play a role in the regulation of molting. Unlike MIH, hyperglycemic hormones are produced at sites other than the eyestalk and CHH mRNAs have been found throughout the central nervous system of *Homarus americanus* (de Kleijn *et al.*, 1994a). The prepro crustacean hyperglycemic hormones (preproCHHs) of crab, crayfish, and the American lobster have been cloned and sequenced (Weidemann *et al.*, 1989; de Kleijn *et al.*, 1994a, 1995) and a high identity has been found between crab and crayfish preprohormones and lobster preproCHH-A and -B (Fig. 14). Advanced HPLC, microsequencing, and molecular biology techniques have revealed the presence of four lobster hyperglycemic hormones: two immunoreactive groups (CHH-A and CHH-B), each occurring as two peptide isoforms (Kegel *et al.*, 1989, 1991; Tensen *et al.*, 1991; Soyez *et al.*, 1990). Isoforms within each group have identical amino acid sequences, weight, and isoelectric points, and seem to originate from a posttranslational event that changes the third L-phenylalanine to D-phenylalanine (Soyez *et al.*, 1994). Interestingly, the two isoforms of lobster CHH-A, the most potent of the hyperglycemic hormones, have different effects on the time course of the elevation of glucose levels in the hemolymph (Soyez *et al.*, 1994), suggesting that the different isoforms have different functions and possibly different targets. Stereoinversion of CHH

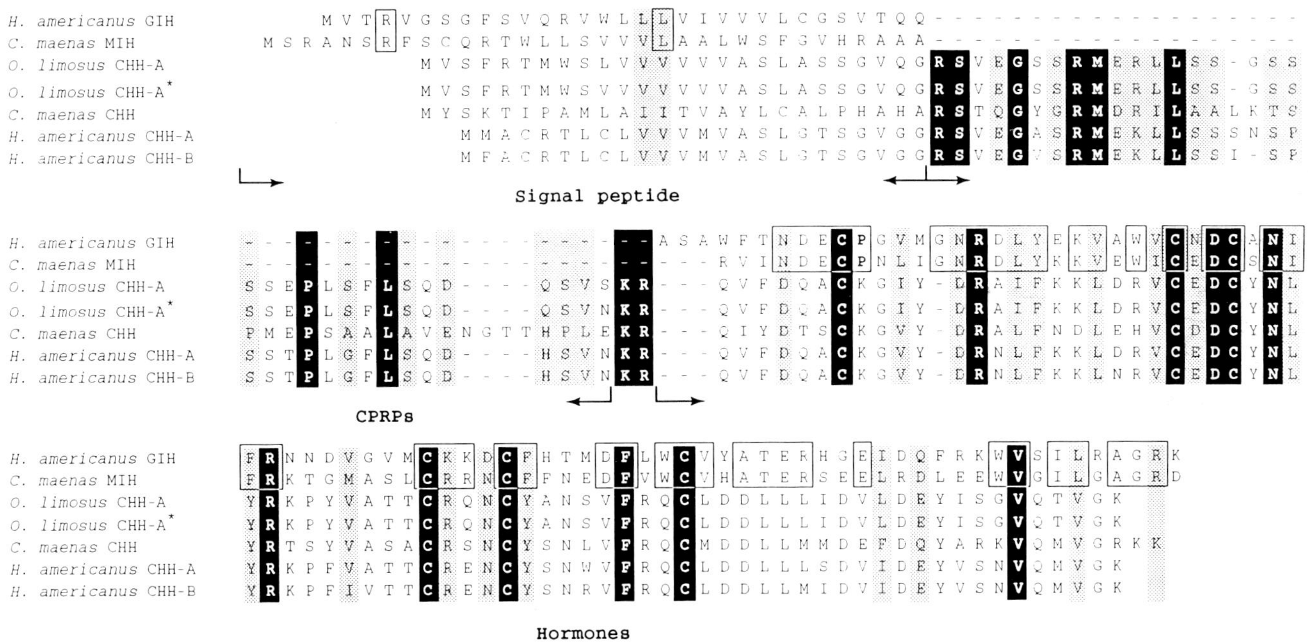

FIGURE 14 Alignment of the preprohormones of *Homarus americanus* GIH (de Kleijn *et al.*, 1994c), *Carcinus maenas* (Klein *et al.*, 1993) and *Orconectes limosus* CHH-A, *Orconectes limosus* CHH-A* (de Kleijn *et al.*, 1994a), *Carcinus maenas* CHH (Weidemann *et al.*, 1989), *Homarus americanus* CHH-A and *Homarus americanus* CHH-B (de Kleijn *et al.*, 1994b). Sets of identical amino acid residues and conservative substitutions between all preprohormones are indicated in black and gray, respectively. Sets of identical amino acids in preproGIH and preproMIH are boxed. (From de Kleijn *et al.*, 1994a.)

leads to the alteration of hormone function and a CHH with a D configuration may serve as MIH in *Procambarus clarkii* (Yasuda *et al.*, 1994). These findings, along with earlier results from eyestalk ablation experiments (Aiken and Waddy, 1980; Gomez and Nayar, 1965; Otsu, 1963) and bioassays (Tensen *et al.*, 1989) showing the presence of both molt-inhibiting and gonad-stimulating factors in the nervous system, suggest that CHH plays a significant role in the inhibition of molting. Because CHH and MIH have multiple biological activities (e.g., lobster MIH has both MIH and CHH activity, while crayfish CHH has both CHH and MIH activities), the designation of any one of these peptides as the molt-inhibiting hormone or as the crustacean hyperglycemic hormone is premature.

f. Exuviation Factor The titer of molting hormones normally decreases before ecdysis, but if exogenous ecdysteroids are used to maintain the titer at abnormally high levels, molting may not occur. One possible explanation of this phenomenon that has received attention is an exuviation factor proposed as a crustacean analog of the lepidopteran eclosion hormone (EH). The concept was developed from work on amphipods and was later extended to shrimp and crabs (reviewed by Aiken and Waddy, 1992). However, peptide extracts from brains and thoracic ganglia of premolt and postmolt crabs have failed to reveal eclosion hormone activity (Cameron, 1989) and it is thought that the American lobster does not use an EH-like factor to initiate ecdysis (Cheng and Chang, 1991). There are indications that events in late premolt are negatively modulated by ecdysteroids and that time of ecdysis is controlled through regulation of the rate of decline of ecdysteroid titers (Cheng and Chang, 1991).

g. Mandibular Organs and Juvenile Hormones The mandibular organs of *Homarus americanus* have been a functional enigma since their discovery and subsequent distinction from the Y organs (Fig. 13) (Sochasky *et al.*, 1972). While implants and injections of homogenates of mandibular organs reduce intermolt times and accelerate the transition from stage C_4 to D_0 in some decapods (Taketomi *et al.*, 1989; Yudin *et al.*, 1980), American lobsters deprived of their mandibular organs molt normally (Byard *et al.*, 1975), suggesting little involvement in the control of the molt cycle.

Because of the similarities in molting regulation between crustaceans and insects, it has long been thought that an analog to juvenile hormone may be present in crustaceans. Insect JH and related compounds have biological activity in crustaceans and extracts of crustacean tissues have been reported to have JH activity in insects (reviewed by Chang, 1993, and Laufer *et al.*, 1993). However, exposure to synthetic analogs of JH result in only minor morphological abnormalities in newly metamorphosed larvae (Charmantier *et al.*, 1988) and slight reductions in rate of development to the fourth stage (Hertz and Chang, 1986). Although current thinking is that JH is probably not present in crustaceans, the mandibular organs do synthesize a terpenoid hormone—methyl farnesoate, an unepoxidated precursor of JH III—that is thought to be homologous to insect juvenile hormone (reviewed by Laufer *et al.*, 1993).

MF production is negatively regulated by eyestalk factors (Laufer *et al.*, 1986; Tsukimura and Borst, 1992). Eyestalk ablation causes hypertrophy of the mandibular organ (Byard *et al.*, 1975) and produces dramatic increases in MF levels in the hemolymph, while injections of sinus-gland extracts decrease MF to undetectable levels (Borst *et al.*, 1994). MF stimulates ecdysteroid production in the crab Y organs *in vitro* and a hemolymph binding protein for MF has been identified that has variable activity over the molt cycle, suggesting that MF may be a positive regulator of the Y organs (Chang *et al.*, 1993; Tamone and Chang, 1993).

5. Manipulation of Molting and Growth

a. Exogenous Hormones Attempts to control molting in *Homarus americanus* with exogenous molting hormones have produced a range of effects, from a protracted intermolt to an accelerated premolt culminating in death at ecdysis (Aiken and Waddy, 1975a,b). Of the ecdysteroids used, 20-hydroxyecdysone is the most active. Dose is critical, however, as an overdose causes rapid, but fatal, premolt development and a low dose is ineffective in inducing premolt. 20-Hydroxyecdysone is an integumentary hormone, not the inducer of premolt, so doses adequate to overcome endogenous molt inhibition often result in abnormalities in the sequence of physiological and morphological events leading to ecdysis. Although treated animals undergo accelerated premolt, they often die while attempting to molt as the new shell is formed before apolysis, setogenesis, and gastrolith and limb-bud development are completed. Response to exogenous ecdysteroids is mediated by eyestalk tissue and can be upset by eyestalk implants. Eyestalk removal followed by sequential 20-hydroxyecdysone injections usually results in successful completion of an accelerated molt, suggesting that a separate factor is necessary for the vital processes of early premolt. The lethal effects can also be avoided by using a slow-release form of the hormone, by administering smaller, more frequent doses, or by pretreating with ecdysone (Aiken and Waddy, 1975b, 1976; Gilgan *et al.*, 1977; Gilgan and Zinck, 1975). However, the optimum dose is difficult to determine because it varies with season, reproductive state, molt stage, and temperature.

The threshold dose of 20-hydroxyecdysone injected into 50- to 60-mm CL lobsters held at 10°C is 0.5 μg per gram of body weight in the spring and 1.5 μg in the autumn (Aiken and Waddy, 1975b; Gilgan and Zinck, 1975). However, temperature exerts a profound effect on the response: a dose that is ineffective at 10°C produces a rapid and fatal molt at 17°C (Aiken and Waddy, 1975a). Lobsters are also more sensitive to ecdysteroids if they have progressed into premolt (stage D_1) (Gilgan and Farquharson, 1977) and mortality is more common when lobsters are injected with high doses in the spring than when they are treated in the autumn.

Preliminary results suggest that other hormones may also have the potential for enhancing growth. Larval and postlarval lobsters injected with human somatotropin grew 10–20% faster than controls and the enhancing effect lasted for several molts (Charmantier *et al.*, 1989).

b. Multiple Autotomy and Regeneration Although multiple leg autotomy induces premolt in many crustaceans (Aiken and Waddy, 1992), there is little acceleratory effect in the American lobster (Aiken and Young-Lai, 1979, 1981). Emmel's thorough study (1906) of regeneration in small juveniles demonstrated conclusively that regeneration can actually retard the growth rate by reducing the size increase at molt and extending the intermolt period. Regeneration often reduces the molt increment in proportion to the magnitude of regenerative effort, due to energy partitioning between molting and regeneration (Aiken and Waddy, 1992). A critical stage for regeneration is D_1''', when deposition of the new cuticle begins. If chelipeds are autotomized before this stage, regeneration proceeds; otherwise, regeneration is delayed until the following molt. Molt intervals are decreased or increased, depending on the stage at which autotomy occurs. Molt intervals are shortened when the chelipeds are removed at the beginning of the molt cycle and are lengthened when autotomy occurs close to the critical stage (Cheng and Chang, 1994).

c. Eyestalk Ablation Eyestalk ablation is the best-known technique for accelerating the growth of crustaceans. The relationship between eyestalk removal and accelerated molting and growth in *Homarus americanus* was established many years ago (Sochasky *et al.*, 1973). Although there have been conflicting reports of erratic responses, these are due mainly to temperature effects and reproductive interference. Eyestalk removal accelerates whichever process—growth or reproduction—is dominant at the time of the operation. When the eyestalks with their neurosecretory complex are removed, the processes leading to molt are induced and rapidly completed. Eyestalk removal eliminates most nonreproductive blocks on molt induction, including seasonal inhibition, and allows premolt to proceed at the physiological rate determined by temperature and to a lesser extent by nutrition (Aiken, 1977). Without the eyestalk hormones, lobsters molt more frequently and increase more in size at each molt. For this reason, eyestalk ablation is often suggested as a way to accelerate growth in lobsters.

Lobsters without eyestalks have shorter intermolt periods than those with eyestalks, but the rate of progression through the premolt stages is no greater than that of the fastest intact lobsters. Weight gains of 75% are common in ablated animals, gains of over 100% are not unusual, and weight increases in males are significantly greater than in females. Temperature has a dramatic effect on the response to eyestalk ablation: at 15°C, 100% of "canner" lobsters ablated in the autumn molt within 4 months, while molting occurs in only 9% of those held at seasonally decreasing temperatures (10°–0°C) (Aiken *et al.*, 1977; Aiken, 1980).

Although it has generally been assumed that increased growth in eyestalk-ablated lobsters is due to an excess influx of water at ecdysis, it was recently demonstrated that the growth-promoting effect is mainly a result of the increased surface area of epicuticle produced by each epidermal cell (Cheng and Chang, 1993). The growth-promoting effects are dramatically reduced if eyestalk removal does not occur until premolt is underway, and there is no growth enhancement if the eyestalks are removed at or after molt stage D_1''' (Cheng and Chang, 1994).

The nutritional value of the diet plays a significant role in the growth of eyestalk-ablated animals and is more critical to the survival of ablated animals than to intact controls. For example, a low-protein diet that has little effect on intact lobsters, reduces growth and survival in ablated ones (Mauviot and Castell, 1976). The abrupt physiological changes and accelerated growth may also lead to unique nutritional demands, as eyestalk factors affect digestion and are involved in the control of protein metabolism (D'Abramo and Conklin, 1985).

Natural eyestalk ablation does occur, either genetically or accidentally during the larval stages, and these so-called "blind Homers" grow several times faster than their sighted siblings (Van Olst *et al.*, 1980). Regardless of how eyestalk ablation occurs, long-term survival is poor because the loss of hormones involved in such processes as limb regeneration, gonad development, cuticle deposition, metabolism, osmoregulation, and reproduction causes hormonal imbalance and metabolic disturbance. Eyestalk-ablated lobsters are particularly sensitive to stress and often die when subjected to changes in salinity, oxygen concentration, or temperature that are sublethal to intact lobsters (Waddy and Aiken, 1995b). Eyestalk ablation causes a disproportionate increase in volume at molt, producing bloating and abnormal body proportions, particularly in the chelipeds (Aiken, 1980; Cheng and Chang, 1994), resulting in frequent claw loss and death at molt. Ablated lobsters are also uncoordinated and pale in color (Koshio *et al.*, 1992). Growth is so rapid there is insufficient time for increase of body tissues; after several molts, meat yields in eyestalk-ablated lobsters may be only half those of intact lobsters (Castell *et al.*, 1977). These effects make eyestalk ablation unattractive as a method for enhancing growth.

Preliminary studies indicate that the removal of only one eyestalk eliminates the problems inherent in bilateral ablation, while still providing some degree of growth enhancement. When one eyestalk is removed in stage VI, early juvenile lobsters grow significantly faster than intact controls and survive almost as well. In a 1-year study, intact lobsters grew at a rate of 0.096 mm/day, while ablated ones grew at a rate of 0.128 mm/day (Peutz *et al.*, 1987). Further work is needed, however, as unilaterally ablated lobsters may not perform as well as intact animals when reared communally (S. L. Waddy, unpublished observation) and an attempt to enhance the growth of "canner" lobsters with unilateral eyestalk ablation did not have the expected molt-enhancing effect (Coulombe and Motnikar, 1989).

III. Maturation and Reproduction

A. Maturation

1. Determining Maturity

Estimates of size at maturity are an important management tool and can provide clues to the origin and homogeneity of lobsters in a management area. In addition, seasonal variations in size at maturity provide evidence of migrants into an area and the size at maturity of the migrants may provide clues about their origin. Maturity in female lobsters has traditionally been estimated from the sizes of ovigerous lobsters in the population, but this assessment is difficult because reproductive patterns vary, only half the mature females are ovigerous in any one year, and ovigerous females may not be adequately sampled by traditional procedures. It is desirable, therefore, to improve maturity determinations by including other criteria. The detection of subtle differences in size at maturity requires that estimates must be carefully based on large numbers of animals. Errors arise when techniques are improperly applied or interpreted. For example, the presence of stored spermatozoa in the seminal receptacle is an unreliable indicator of maturity, and allometric changes in the width of the female abdomen can be difficult to interpret because they occur gradually over several molts.

The most accurate methods require sacrificing females to determine the degree of ovarian development (size of oocytes, weight and color of the ovary) and the color of the oviducts. Oviduct color is particularly useful as it can distinguish between females that have never spawned and those that have spawned but are not carrying eggs at the time of the sample (Aiken and Waddy, 1979; see Talbot and Helluy, Chapter 9, on the reproductive system). Estimates of maturity based on cement gland development offer the best combination of speed, reliability, and ease of application. Cement glands on the pleopods of female lobsters become progressively more engorged as ovarian maturation proceeds (Fig. 15). Although the exact function of these glands is still unknown, there is a good correlation between their degree of development and ovary maturation (Aiken and Waddy, 1982). By examining the cement glands it is possible to distinguish between immature and mature females, and among the various stages of preovigerous females. This method is most accurate in the 1 or 2 months preceding the onset of the spawning season, as interpretation is more difficult at other times of the year. The cement gland method used alone is 96% as accurate as estimates based on all other available criteria, including ovary staging (S. L. Waddy, unpublished observations).

Maturity is not as easy to assess in males as in females. The presence of spermatozoa in the testes or vasa deferentia is misleading as there is little correlation between the presence of spermatozoa and the ability to mate (Aiken and Waddy, 1979). Lawton and Lavalli (Chapter 4) draw a distinction in life history phases between adolescent lobsters, which are marked by physiological maturity, and adults, which actually engage in mating. The most reliable indicator of maturity, other than mating capability, is the rela-

FIGURE 15 Cement glands of female American lobsters. Stages 1 to 4 are illustrated with whole mounts of pleopod endopodites (upper) and enlarged view of lateral or medial region (lower) showing individual cement glands (cg). (From Aiken and Waddy, 1982, with permission.)

tive size of the propus of the crusher claw (Aiken and Waddy, 1989). The cheliped propodite index (CPI = CPV/CL^3, where CPV, crusher propus volume, = crusher length × width × thickness, with measurements to the nearest 0.01 cm) is well correlated with the ability to mate and shows an inflection at maturity when plotted against carapace length. CPI values greater than 22–24 are indicative of maturity (Fig. 16), regardless of geographic origin and size at maturity (Aiken and Waddy, 1989, and unpublished data).

2. Control of Maturation

There are wide geographical variations in the size at first spawning. The only factor consistently associated with variation in size at maturity is summer seawater temperature. Templeman (1936a) first recognized that high summer water temperature induces maturation at a small size and young age, while low summer temperature delays maturity and allows growth to a larger size before spawning. In wild populations, initial spawning size varies from as small as 55-mm CL in Long Island Sound to 110-mm CL in the Bay of Fundy. Among females in Canadian waters, size at 50% maturity varies from 72-mm CL in Northumberland Strait to 102-mm CL in the Bay of Fundy (Fig. 17) (S. L. Waddy and D. E. Aiken, unpublished data; cf. Campbell and Robinson, 1983).

The effect of temperature on maturation is apparent even within localized areas: lobsters at the warmer, western end of Northumberland Strait mature at a smaller size than those at the cooler, eastern end; lobsters in western Long Island Sound mature at a smaller size than those at the open eastern end; and lobsters from Long Island Sound mature at a smaller size than those along the Atlantic coastline of the south shore of Long Island (Briggs and Mushacke, 1980; Smith, 1977; Templeman, 1936a). One of the best examples of the relationship between size at maturity and seawater temperature was provided by Estrella and McKiernan (1989) in coastal Massachusetts waters where size at 50% maturity varied from 76-mm CL in Buzzard's Bay to 89- to 97-mm CL in nearshore Gulf of Maine waters (Fig. 18).

Long-term laboratory studies have verified the role of temperature. Juvenile lobsters mature at approximately the same size when reared at a com-

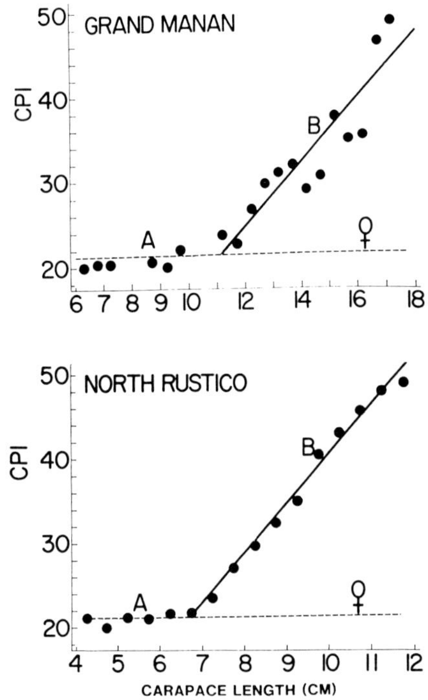

FIGURE 16 Allometric growth of the crusher cheliped of male American lobsters as indicated by the plot of crusher propodite index (CPI) against carapace length (CL). Intersect of mature male regression line (B) on female regression line (A) indicates size at onset of maturity of male lobsters from Grand Manan in the Bay of Fundy and North Rustico in the southern Gulf of St. Lawrence. (From Aiken and Waddy, 1989, with permission.)

mon temperature, even if they come from parental stocks with widely differing sizes at maturity (Waddy and Aiken, 1991). The smallest recorded size at maturity (52-mm CL, which is probably close to the physiological minimum) occurred in a small proportion of

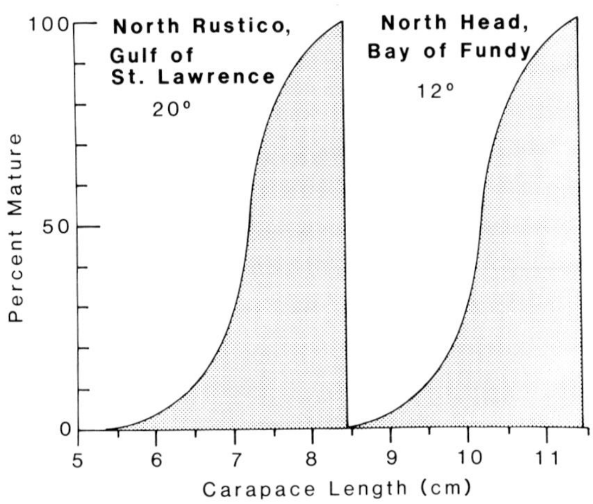

FIGURE 17 Size at maturity in female American lobsters from areas with different maximum summer seawater temperatures. [Reprinted from Waddy and Aiken, 1991. Crustacean egg production. (Adrian Wenner and Armand Kuris, eds.), Crustacean Issues, 7. A. A. Balkema, Rotterdam.]

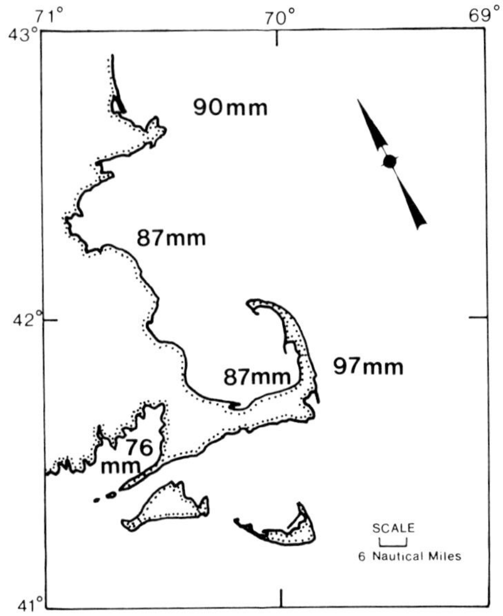

FIGURE 18 Relationship between size at maturity and geographic location in coastal Massachusetts waters. Size at maturity in these areas is well correlated with maximum summer seawater temperature. (Data from Estrella and McKiernan, 1989.)

cultured lobsters raised at a high year-round temperature of 20°C (Waddy and Aiken, 1991).

The relationship between temperature and size at maturity is not uncontested, however. Size at maturity in female lobsters along the eastern shore of Nova Scotia is reported to vary from 89- to 101-mm CL between years and is not well correlated with summer seawater temperature (Miller and Watson, 1991). This may indicate that lobsters are moving into the sampled area, confounding the results. In the Gulf of St. Lawrence and western Long Island Sound, lobsters mature at a small size and are found in relatively high density, suggesting a relationship between population density and size at maturity. This hypothesis is weakened, however, by observations from the southern end of Grand Manan and southwest Nova Scotia, where maturity occurs at a large size in spite of high density. It has also been suggested that the duration of the intermolt period, and thus the ability to incubate eggs, might determine size at maturity (Nelson, 1986), and that intensive fishing effort exerts genetic pressure for maturation at a smaller size (Fogarty, Chapter 6). Nevertheless, the only factor consistently associated with small size at maturity is high summer seawater temperature.

B. *Control of Reproductive Cycles*

1. Reproductive Cycles and Spawning

a. Size and Age The reproductive cycle of mature female American lobsters smaller than 120-

mm CL typically takes 2 years. The female molts and mates one summer, spawns the following summer, and carries the eggs on her pleopods until the third summer, when they hatch. One to three months after the eggs hatch, the female molts again, mates to replenish her sperm supply, and the cycle is repeated. During mating, the male deposits a spermatophore in the seminal receptacle, where it remains until the eggs are extruded and the sperm are required for fertilization. In most inshore areas, spawning is synchronized within the population and occurs well before the molting peak. A molt always occurs between successive spawnings and successive molts are always separated by spawning and egg incubation (Aiken and Waddy, 1980; Waddy and Aiken, 1991; see also Atema and Voigt, Chapter 13, and Talbot and Helluy, Chapter 9). The spawning and egg hatching periods usually overlap, so that in early summer females carrying eggs spawned the previous year are found at the same time as females carrying newly spawned eggs (Templeman, 1936a).

There are several variations to the typical pattern that are related to the size of the female and the temperature regime. Female lobsters are classified as Adult-I in the year they extrude eggs for the first time. This reproductive class is unusual in that ecdysis and egg extrusion can occur in the same summer, resulting in two reproductive patterns: Adult-Ia, the more common and typical pattern, in which ecdysis and spawning occur in alternate years; and Adult-Ib, an alternate pattern, in which both events occur in the same summer (Aiken and Waddy, 1982). The incidence of the Adult-Ib cycle depends on summer seawater temperature; it is common in the warm waters of the Northumberland Strait and the southern Gulf of St. Lawrence, occurring in about 19% of Adult-I females (Aiken and Waddy, 1982). Spawning in this area occurs as two peaks: hard-shelled Adult-Ia females and subsequent reproductive classes spawn in late-June and July, while newly molted Adult-Ib females spawn 4–6 weeks later (Templeman, 1940b; Aiken and Waddy, 1982). In Newfoundland, the incidence of the Adult-Ib cycle varies from 0 to 38% between locations and years (Ennis, 1980), presumably because of the pronounced variability in summer seawater temperature in the area. The Adult-Ib cycle is rare along the cool Atlantic coast of Nova Scotia, the Bay of Fundy, and the Gulf of Maine (Aiken and Waddy, 1979; Templeman, 1940b).

In most areas, it is unusual for molting and spawning to occur in the same summer after the Adult-I year (Aiken and Waddy, 1982), although exceptions occur in Long Island Sound and offshore areas (Briggs and McGroaty, 1985; Pezzack, 1991). The incidence is related to annual temperature regimes. Laboratory studies have shown that the phenomenon can be induced in sizes and stocks of lobsters where it does not normally occur by altering the temperature cycle (Waddy and Aiken, 1991). The ability to molt and spawn in the same season might allow females <120-mm CL in Long Island Sound to spawn in consecutive years, a possibility that receives some support from the high incidence of ovigerous females (68–80%) in some size classes (Briggs and McGroaty, 1985).

Reproductive cycles also change with increasing size. Females molt less frequently at larger sizes and they use this as an opportunity to increase their fecundity by spawning twice between molts (Fig. 19). This phenomenon is termed *consecutive spawning* and occurs in two forms: *successive-year*, spawning in two consecutive summers without an intervening molt; and *alternate-year*, spawning in alternate summers without an intervening molt (Waddy and Aiken, 1986). Lobsters are usually able to fertilize two broods with sperm from a single insemination (*multiple fertilization*), but if necessary they can obtain additional sperm from an *intermolt mating* (Waddy and Aiken, 1990a).

Consecutive spawning occurs in both warm- and cold-water areas and begins when females reach 120-mm CL. In general, females on a 3-year molt cycle spawn in successive years, while those on a 4-year molt cycle spawn in alternate years. The highest incidence of spawning occurs in the 120- to 150-mm CL size group. Over an extended period of time, consecutive spawners produce many more broods than smaller females, making their relative fecundity greater than often assumed. Even the largest lobsters (200-mm CL) continue to spawn as often as their smaller counterparts and the viability of the eggs and larvae produced by large lobsters is no different from that of smaller, younger lobsters (Waddy and Aiken, 1986). Moreover, large lobsters produce relatively more eggs than smaller ones (Aiken and Waddy, 1979) and reportedly produce eggs with a higher energy content per gram of egg, which should increase the ability of the larvae to survive adverse conditions (Attard and Hudon, 1987).

b. Stage of the Molt Cycle In order to maximize egg production, it is important that the maternal female not molt while incubating eggs, but yet molt soon after the eggs hatch. If the molt occurs too early the eggs will be lost, while if the molt is delayed, the subsequent egg extrusion will also be delayed, perhaps by as much as a year. Although coordination of the molting and spawning cycles is critical to repro-

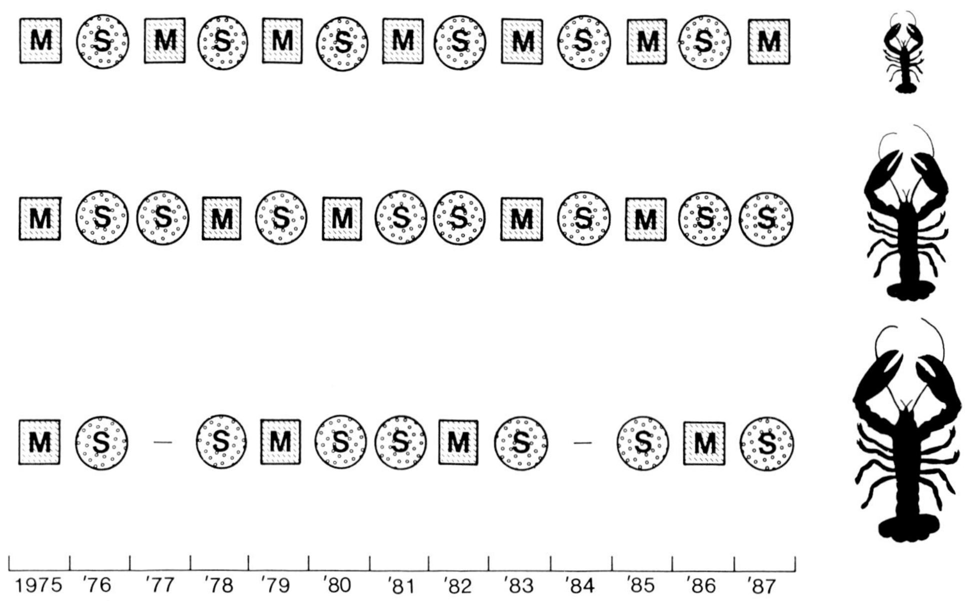

FIGURE 19 Molt and spawning cycles of female American lobsters. Molt is indicated by squares and spawning by circles. Upper row, cycle characteristic of female lobsters <120-mm CL; middle row, successive-year spawning cycles of lobsters >120-mm CL; and lower row, alternate-year spawning cycles of lobsters >120-mm CL. (From Waddy and Aiken, 1986, with permission.)

ductive success, little is known about the controlling mechanisms. There are indications that spawning would occur annually, rather than biennially, if not for the constraints imposed by the molt cycle. The coordination between the two cycles appears to be associated with the eyestalk organs, as females without eyestalks often spawn during premolt (Aiken *et al.*, 1977). Under normal conditions, vitellogenesis only occurs in molt stages B, C, and D_0. If the molt cycle progresses into premolt (D_1), vitellogenesis is inhibited and the ovarian yolk may be resorbed from the oocytes so that the eggs are not lost at molt. Except for Adult-Ib females that spawn early in stage C, lobsters are usually in stage C_4 or D_0 at the time of spawning and the molt cycle progresses at a normal rate so that ovigerous females are often in mid- to late premolt when their eggs hatch (Waddy and Aiken, 1991).

*c. **Insemination*** Contrary to common belief, spawning is independent of insemination. There have been suggestions that female lobsters may resorb their oocytes if they have not mated (Ennis, 1984), but laboratory studies involving hundreds of females have shown there is no relationship between ovarian development and insemination. In fact, lobsters held for many years in isolation spawn on the same schedule as their inseminated counterparts (S. L. Waddy, unpublished observation). Dubé and Grodin (1985) felt the relative scarcity of larger males in the Magdalen Islands limits the fertilization success of large females and this is reflected in the low percentage of large berried females in the area. Similar concern has been expressed over the differential removal of large males from the Bay of Fundy and other fisheries (Campbell, 1992; Cobb, Chapter 7). However, recent information allays these concerns somewhat as it appears that unless males are rare, females will have sufficient opportunity to mate (Waddy and Aiken, 1990a).

*d. **Distribution and Movements*** Seasonal movements of mature lobsters are thought to be related to reproduction, but little is known about migrations and biological events. As the temperature declines in the autumn, lobsters move to deeper water and ovigerous females tend to move sooner, farther, and deeper than males or immature females (Campbell, 1984, 1986; Ennis, 1983b, 1984; Herrick, 1909), suggesting a direct link with egg and larval production. Offshore lobsters winter in the canyons of the continental slope, but move to shallower and warmer portions of the shelf and coastal areas for the summer, where they molt and reproduce. Large, mature females concentrate in certain coastal areas, returning to the same location year after year (Campbell and Pezzack, 1986). Seasonal aggregations of ovigerous females have been found in shallow waters off Grand Manan (Campbell, 1990) and it is thought that other egg-bearing lobsters may move from overwintering sites in the Grand Manan basin northeast into the Bay of Fundy, west to the Maine coastline, or south to the shoal waters of the continental shelf (Lawton and Lavalli, Chapter 4). These movements are thought to maintain lobsters within a temperature range of 8°–14°C, to congregate lobsters to facilitate breeding, to provide optimal conditions for larval release, to synchronize molting and spawning, and to optimize exposure to temperatures suitable for ovarian maturation and embryonic development (Campbell, 1986; Cooper and Uzmann, 1980).

The relative contribution of these migratory females to certain downstream areas may therefore be greater (and possibly more important) than their contribution to the population as a whole. Such a mechanism has been implied for the Grand Manan fishery (Campbell and Duggan, 1980) and there are other areas where comparable situations exist (Dadswell, 1979). It is estimated that up to 77% of eggs produced in the Bay of Fundy, southwest Nova Scotia, and the offshore areas of Georges Bank and Browns Bank are from females from offshore areas (Campbell and Pezzack, 1986). It has been shown that larvae hatched on Browns Bank can be transported to southwest Nova Scotia and into the Bay of Fundy and Gulf of Maine by passive drift alone (Harding and Trites, 1988). In view of the spawning frequency of large lobsters, the concentrations of large lobsters in certain offshore and inshore areas, and the recent fishing effort directed at these large size classes, it would be prudent to reassess the contribution that these large lobsters make to specific stocks and to lobster populations in general.

e. Environmental Factors Wilfred Templeman (1936a, 1940b) recognized over 50 years ago that seawater temperature cycles are the dominant regulator of egg production in *Homarus americanus*. Reproduction occurs within a narrower range of environmental conditions than those suitable for growth or survival, and the temperature requirements for reproduction influence latitudinal distribution. Temperature is the major factor controlling size at maturity, oocyte maturation, incidence, timing, and synchronization of spawning, success of egg attachment and incubation, and time of hatching. Other environmental influences, such as day length, season, nutrition, social conditions, stress, pollutants, and parasites, are also important and often interact synergistically.

i. Temperature Spawning in most inshore areas occurs between June and October and peaks occur about a month earlier in the warmer waters of the southern Gulf of St. Lawrence than in the colder waters of the Bay of Fundy (Templeman, 1940a). Spawning within a geographical area is usually more synchronized than molting and occurs before the peak of the molting season. With the exception of the Adult-Ib cycle, 80% of female lobsters from a given area spawn over a 3-week period when held in the laboratory (Waddy and Aiken, 1991).

Temperature has a direct effect on yolk synthesis and ovarian maturation and is largely responsible for spawning synchrony. During the winter, preovigerous females have low levels of vitellogenin (the yolk protein) in their hemolymph, but when temperature is increased to 13°–14°C in the spring, vitellogenin levels increase to >5000 μg/ml within a few weeks (Fig. 20) (Tsukimura *et al.*, 1993). Just as with the molt cycle, however, there is a seasonal inhibition to spawning and a temperature increase alone is often not sufficient to induce ovarian maturation and spawning.

Maturation and spawning in nearshore stocks requires an extended period of winter temperatures below 8°C, and perhaps below 5°C, followed by an increase to a temperature of 10°C or higher in the summer. The incidence of spawning is reduced if either summer or winter temperatures are inappropriate (Waddy and Aiken, 1995a). While low summer temperature inhibits egg production, low winter temperatures are essential. Ovarian development in nearshore stocks requires a cold period near the time of the winter solstice. If temperatures remain above 6°–8°C, the molt and spawning cycles become uncoupled and responses of preovigerous females are variable: some females spawn, some molt and then spawn, some molt twice without spawning, and a few spawn twice without molting. Interestingly, these temperature-induced variations to the normal alternate-year molting and reproductive cycle have been reported in offshore lobsters that do not experience the same degree of seasonal temperature variation as

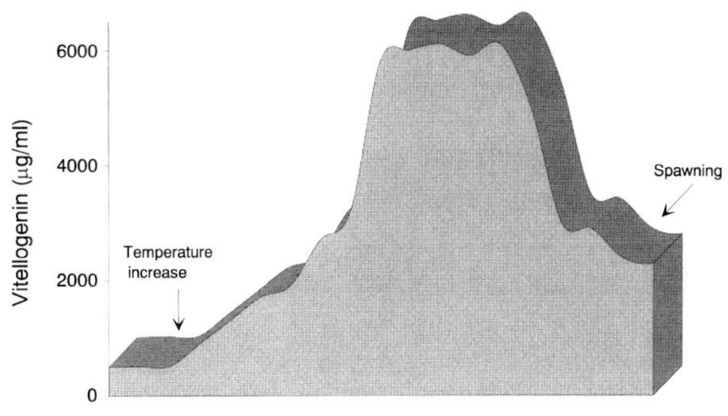

FIGURE 20 Hemolymph vitellogenin profile in preovigerous female lobster after induction of vitellogenesis and spawning with temperature and photoperiod manipulation. (B. Tsukimura, S. Waddy, and D. Borst, unpublished data.)

inshore lobsters (Pezzack, 1991).

Only a few weeks of low temperature at the appropriate time of year appear sufficient to maintain the spawning cycle. Most females are competent to resume vitellogenesis and spawning by late January and the extended winter period in most nearshore areas forces the females into a "waiting period" until the appropriate temperature is available. Lobsters respond rapidly to increasing spring temperatures and can be induced to spawn as early as February or as late as December, simply by altering the timing of the temperature increase (Waddy and Aiken, 1992). Studies also suggest that preovigerous females require at least 4–6 weeks of temperatures of 10°C or higher in the spring or summer for spawning to occur (Waddy and Aiken, 1992). Temperatures during the two previous summers are also important, but we know little about the temperature requirements of the first phase of ovarian development.

If the spring increase in seawater temperature is delayed, spawning will also be delayed. If temperatures remain low, spawning will not occur at all (Waddy and Aiken, 1991). Like progression through the molt cycle, ovarian development is suspended when seawater temperature drops below 5°C. At Comfort Cove, Newfoundland, water temperatures in mid-July vary from 4° to 18°C between years (Ennis, 1982). Although little is known about reproductive cycles in this area, the effect of temperature on the variable proportion of lobsters that molt each year (22–73% in the 81- to 90–mm size range) may explain the variation in the proportion of ovigerous females (2–30%) from year to year (Ennis, 1980, 1983a). The effect of low summer temperature has been noted in other areas of Newfoundland as well. Lobsters in two arms of a Newfoundland bay spawn on different cycles: in the warmer arm they spawn biennially, while in the colder arm they spawn every third year (Squires *et al.*, 1971). Lobsters at the northern limit of their range off the northeast coast of Newfoundland only spawn every fourth year (Ennis, 1971) and lobsters transplanted beyond their range to Labrador spawn infrequently (Boothroyd and Ennis, 1992).

A significant number of mature females in Newfoundland waters abort ovarian development and resorb the yolk from the mature oocytes (Ennis, 1984), which may be related to the relatively low summer water temperature in the area. In the laboratory, 50% or more of females resorb their oocyte vitellin when low summer temperatures delay spawning beyond the normal season (Waddy and Aiken, 1995a). A similar phenomenon has been reported in northern populations of the crayfish *Astacus astacus* (Huner and Lindqvist, 1987).

At the southern limit of the range and in offshore areas, the opposite situation prevails. Winter water temperature must decline below 5°–8°C for proper synchronization of molting and reproduction. Following a winter of relatively warm temperatures, there is a proportional increase in the incidence of molting and oocyte resorption (Waddy and Aiken, 1995a; Hedgecock *et al.*, 1978), and there is a reduction in the incidence of spawning in direct proportion to the degree of temperature increase (Fig. 21). For the same reasons, cultured females held at a constant 20°C rarely spawn because of the effect of uniform temperatures on the molting and spawning cycles. Even when the interval between molts is sufficient to allow spawning and egg incubation, the precise phasing required between somatic and reproductive events seldom occurs. However, when cultured lobsters are transferred to conditions with a normal period of low winter temperature, spawning incidence increases dramatically and after several years will approach that of wild females (Fig. 22) (Waddy, 1988).

Despite the potential for conflict between the molt and reproductive cycles, 95–99% of female lobsters spawn on schedule in the laboratory when maintained at temperature and photoperiod regimes characteristic of nearshore lobster habitats and it is rare for lobsters to molt while carrying eggs (Waddy and Aiken, 1991). The molting and spawning cycles are somehow synchronized by temperature so as to minimize conflict and an extended period of low winter temperature is an essential part of the regulation.

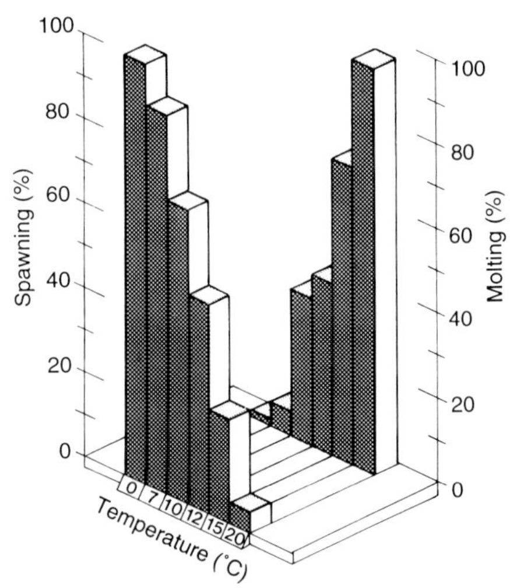

FIGURE 21 Effect of various winter seawater temperatures on the relative incidence of molting or spawning the following summer. Spawning incidence is inversely correlated with increasing winter temperature. (From Aiken and Waddy, 1986, with permission.)

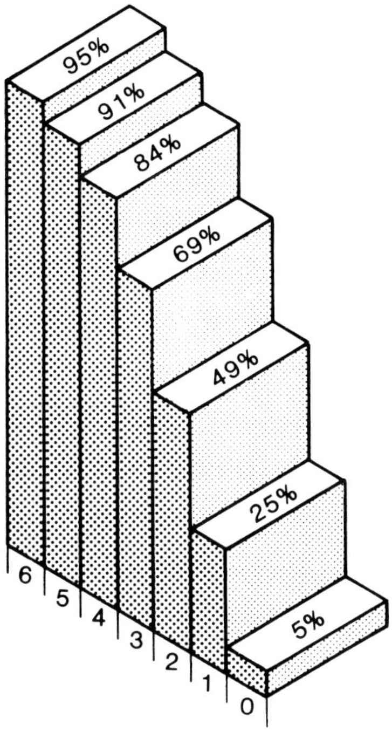

FIGURE 22 Incidence of spawning in cultured females reared and held at 20°C compared to cultured females transferred from 20° to local seasonally changing seawater temperature that varies from 0° to 12°C annually. Incidence of spawning (vertical axis) increases with number of years spent in local temperatures (horizontal axis); after 6 years, spawning occurs with the same frequency as in wild stock. [Reprinted from Waddy and Aiken, 1991. Crustacean egg production. (Adrian Wenner and Armand Kuris, eds.), Crustacean Issues, 7. A. A. Balkema, Rotterdam.]

ii. Photoperiod and season Photoperiod appears to have little influence on the regulation of spawning in nearshore populations. Even when preovigerous lobsters are exposed to extreme photoperiod cycles such as short (LD 1:23 or 8:16) or decreasing spring day lengths, spawning time and incidence are unaffected (Waddy and Aiken, 1992). If seasonal temperature cues are not available, however, photoperiod becomes critically important.

At the autumnal equinox (21 September), only 10% of preovigerous lobsters spawn in response to a 13°C temperature increase, and then only if photophases are long (LD 16:8). By early November, 95–100% of preovigerous females respond to a combined temperature and photophase increase (13°C, LD 16:8) and spawn 90–120 days later. A similar response occurs throughout the remainder of the winter and spring; regardless of when warm temperatures and long day lengths begin, spawning occurs synchronously 3 to 4 months later (Fig. 23).

Shifting the onset of the temperature increase by only a few days at the critical time of year can produce entirely different responses to day length. Exposure to warm temperatures (13°C) and short

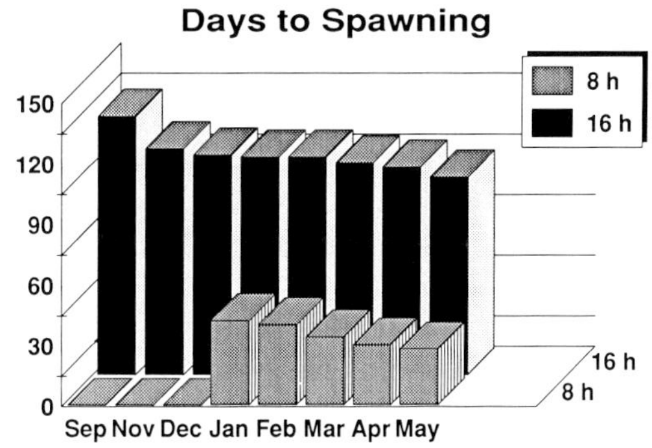

FIGURE 23 Seasonal variation in the number of days required to induce spawning in groups of preovigerous lobsters exposed to elevated temperature (13°C) beginning at various times between the autumnal equinox and the following spring. Females in long photophases (LD 16:8) spawn within 90–120 days after the temperature and photophase change regardless of the time of year. Females held in short photophases beginning on 21 September, 7 November, or 21 December do not spawn. The identical conditions beginning on 8 January or later induces spawning in 26–42 days, and the shortest development times are in those groups receiving the longest period of normal winter conditions prior to environmental manipulations. (From Waddy and Aiken, 1992.)

photophases (LD 8:16), beginning between the autumnal equinox and the winter solstice, inhibits vitellogenesis; lobsters are prevented from spawning as long as these conditions are maintained (Waddy and Aiken, 1992). These females can be induced to spawn with a shift to long day lengths (LD 16:8; Nelson, 1986), but only with a 50% success rate, indicating conditions that are not very favorable for ovarian maturation and spawning. In contrast, exposure to the previously inhibitory conditions of warm temperatures and short day lengths (13°C, LD 8:16), beginning at any time from early January through May, rapidly induces vitellogenesis; spawning occurs within 26–42 days. The shortest development time (less than 4 weeks) and highest spawning incidence (100%) are associated with the longest periods of winter conditioning prior to environmental manipulation (Fig. 24) (Waddy and Aiken, 1992). It is significant that the response time is so much longer under long day lengths (LD 16:8) than short (LD 8:16): 90–120 days vs. 26–42 days, respectively, but as yet there is no explanation for this (Waddy and Aiken, 1992). Aiken and Waddy (Chapter 8) discuss techniques for manipulating spawning in preovigerous females.

The most intriguing aspect of the response of preovigerous lobsters to day length is the facultative nature of the control mechanism: under normal conditions with strong seasonal temperature cues, the system is regulated by temperature without regard for day length; under constant temperature conditions, ovarian development and spawning have very

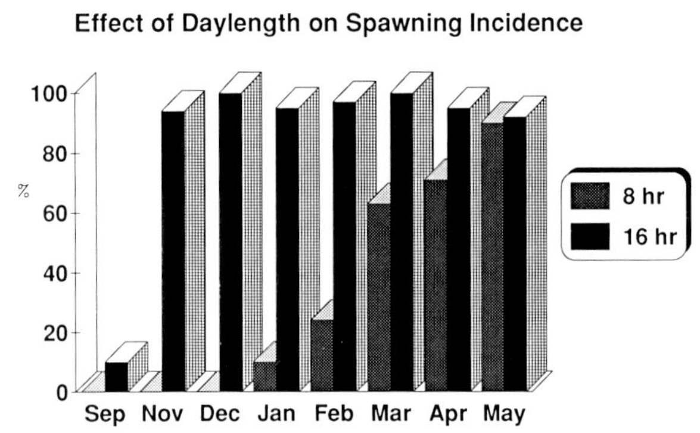

FIGURE 24 Seasonal variation in the incidence of spawning in preovigerous females induced to spawn with environmental manipulation. Few females spawn (10%) when exposed to elevated temperature and long days (LD 16:8) beginning at the autumnal equinox, but when the same conditions begin between November and May, most females (>90%) respond by spawning. In contrast, females held at the same temperature and short photophases (LD 8:16) beginning between the autumnal equinox and the winter solstice do not spawn. Response to elevated temperature and short day lengths after the winter solstice increases and is related to the length of time that the females were conditioned under normal winter temperature and photoperiod. (From Waddy and Aiken, 1992.)

specific day length requirements. In addition, response to temperature and photoperiod are profoundly influenced by real calendar time and significant alterations in reproductive physiology occur at the autumnal equinox and winter solstice (Waddy and Aiken, 1992). It is presumed that underlying this is a free-running, circannual biological clock. The fact that lobsters spawn in response to elevated temperatures only if they have received the appropriate environmental cues is consistent with a need for flexibility in the timing of spawning in wild populations. An early increase in seawater temperatures in one year is not predictive of springs in subsequent years; thus, it would not be advantageous to permanently reset the clock.

iii. Nutrition Little is known about the nutritional requirements for ovarian development and spawning, but reproductive success can be negatively influenced by shortages of food and deficiencies in dietary lipid or protein. Yolk reserves are the sole source of nutrients for the developing embryo, so essential fatty acids and total lipids are very important to successful embryonic development. The reported effects of nutritional deficiencies include oocyte resorption, reduced fecundity, and low hatching success (Castell and Budson, 1974; Castell and Kean, 1986; Harrison, 1990).

iv. Stress, trapping, handling, and confinement Handling and shipping in the 2–4 weeks prior to spawning can cause the degeneration of the oocytes and resorption of vitellin. In areas where lobsters mature at a small size and experience heavy fishing pressure during spring and summer, those lobsters that mature below the minimum legal size may be susceptible to oocyte resorption as a result of repeated capture and release. Oocyte resorption, resulting in "black lobsters," is also associated with unfavorable holding conditions, such as impoundment, or exposure to stressful environmental conditions (Aiken and Waddy, 1985).

v. Miscellaneous There is little information on the effect of contaminants on egg production in crustaceans. Studies on the crayfish, *Procambarus clarkii*, have shown that exposure to heavy metals can reduce fecundity and hatching success (Naqvi and Howell, 1993) and lobsters may suffer negative effects as well. Correlations have been reported between population density and egg production in Arnold's Cove, Newfoundland. As standing stock increases, fewer females molt and more females spawn (Ennis, 1991).

2. Egg Attachment, Embryogenesis, and Hatching

The spawned eggs are carried beneath the abdomen of the female for about a year, during which time the female routinely aerates them and cleans them of fouling organisms. Fanning and grooming with the tips of the walking legs is critical to the health and successful incubation of the eggs. Adverse conditions that are easily tolerated by the maternal female may quickly kill the embryos she carries. To survive the 10–12 months of normal incubation in the wild, an embryo must have sufficient organic reserves, be securely attached to the pleopods, and remain free from disease, predation, and exposure to water of poor quality. Although egg attachment is usually secure, entire egg masses can be lost during incubation. Natural attrition in an offshore population has been estimated at 36% (Perkins, 1971), while egg loss in inshore lobsters from western Newfoundland in the southern Gulf of St. Lawrence is consistently between 15 and 27% (Savoie and Maynard, 1991). Egg loss is considerably higher in some areas: 14–15% of ovigerous females carry fewer than 150 embryos in areas of Newfoundland and Long Island Sound (Ennis, 1984; Ennis and Collins, 1983; Smith, 1977). It may be significant that these areas represent environmental conditions near the extremes of the lobster's range. In contrast, only 1% of ovigerous females in the Bay of Fundy lose their eggs before they hatch (Campbell and Pezzack, 1986). Such differences in attrition rates may account for the reported geographic variations in fecundity (see Ennis, 1981). It is interesting that comparisons of size-fecundity relationships in one location over a 100-year span indicated that no change had occurred (see Fogarty, Chapter 6).

a. Temperature and Salinity Embryonic development is regulated primarily by temperature (Perkins 1972; Templeman, 1940a), so hatching occurs earlier in areas of high summer water temperature and earlier in warmer years (Templeman, 1936a). Embryogenesis requires 9–12 months in most populations, but can be reduced to as little as 3–4 months by holding ovigerous females at 20°–22°C (Aiken and Waddy, 1985; Perkins, 1972). Similarly, hatching can be delayed for as long as 6 months beyond the normal season if the increase in summer temperature is delayed. (See Talbot and Helluy, Chapter 9, on embryonic development.)

The time required from onset to completion of hatching of a brood can extend from a few days to 3 weeks or more (Ennis, 1975). The duration of the hatching period is determined primarily by temperature and in years (or areas) with high summer temperatures the hatching period will be shorter than in colder areas. Exposure to low winter seawater temperatures tends to synchronize development within a brood. Broods held at low temperatures for many months will hatch rapidly when the water is warmed, whereas those held at 20°C for several months will have a prolonged hatching period (Waddy and Aiken, 1995b). Temperature can also affect the lipid conversion efficiency in embryos and, as a result, the quantity of yolk remaining at hatching and the chance for survival (Sasaki *et al.*, 1986). However, even in a relatively small area, such as the southern Gulf of St. Lawrence, hatching time within the population may extend over a couple of months due in large part to the variation in time of spawning between Adult-Ia and -Ib females the previous year (Aiken and Waddy, 1982; Templeman, 1936b; see Section III,B,1).

In general, the relationship between temperature and the rate of embryonic development is linear from 5° to 25°C, while below 5°C development proceeds slowly. However, the effect of temperature is modified somewhat by the degree of egg development; older or more advanced embryos develop at a slightly slower rate than younger embryos (Perkins, 1972). Aiken and Waddy (Chapter 8) provide techniques for control of egg development and time of hatching.

During winter, lobster eggs become "dormant" and the duration of the dormant period depends upon the stage of the eggs when the temperature declines. Egg development is suspended at a specific point by low temperature; the oldest eggs reach that point sooner and thus are dormant for a longer period (Perkins, 1972). Low temperature, therefore, synchronizes embryo development in much the same way that it synchronizes the molt cycle of juveniles and adults (see Section II,B,3).

There is strong evidence that environmental factors, particularly temperature and salinity, are involved in egg attachment and retention. *Homarus gammarus* broodstock exposed to low salinity suffer substantial egg loss (Wickins *et al.*, 1995) and there is a high incidence of egg loss in areas where temperatures are marginal for egg production. Both wild and cultured lobsters held in uniformly warm temperatures in California and France retain few eggs (Bertran and Lorec, 1986; Hedgecock, 1983; Talbot *et al.*, 1984), while cultured females transferred to seasonally changing seawater temperatures (0°–12°C) incubate and hatch orders of magnitude more eggs than cultured females at 20°C (Waddy, 1988). These results implicate temperature history as an important factor in the success of egg incubation. The factors that influence chorion synthesis and egg-stalk formation are not well understood, but environmental conditions may be important, e.g., the morphology of the egg stalk is often inferior in cultured lobsters (Talbot and Harper, 1984). Consistently low temperatures may also have a negative impact on egg attachment, as lobsters spawning in Labrador waters often carry few eggs, probably because the incubation period may extend through a second winter (Boothroyd and Ennis, 1992).

b. Disease Fouling by both filamentous and nonfilamentous microorganisms on the external surfaces of egg membranes causes anoxia. Although the effect of the most common epibiont, *Leucothrix mucor*, on wild populations is unknown, it can prolong embryonic development by 6 weeks or more under laboratory conditions (Aiken and Waddy, 1985). The phycomycetous fungus, *Lagenidium*, can also infect lobster eggs and prevent hatching (Fisher *et al.*, 1978). The nemertean *Pseudocarcinonemertes homari* (Fig. 25) created considerable problems in the Bay of Fundy fishery in the late 1970s and early 1980s (Aiken *et al.*, 1985; Waddy and Aiken, 1985a). At one point, as many as 74% of the ovigerous lobsters in the area were infected (Campbell and Brattey, 1986), but it disappeared as unexpectedly as it appeared and has not been reported in more than a decade. (See Martin and Hose, Chapter 17, and Talbot and Helluy, Chapter 9, on diseases of the eggs.)

c. Miscellaneous Severe loss of embryos can occur when ovigerous females are captured in traps

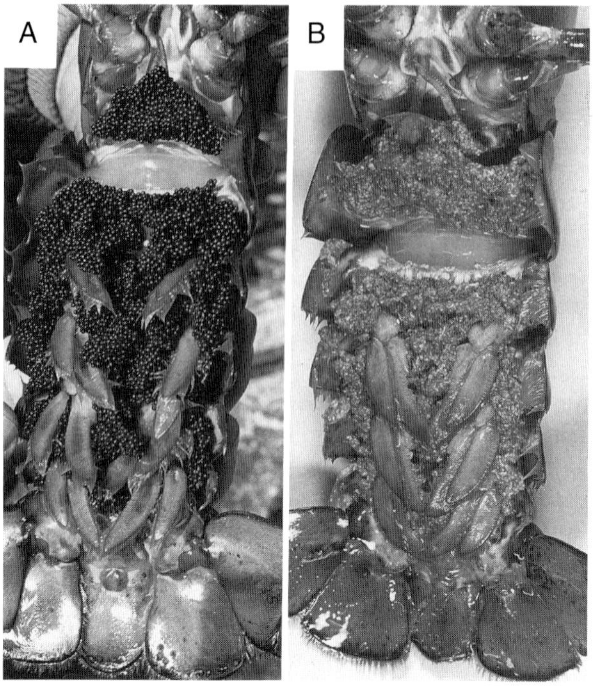

FIGURE 25 Massive infestation of a lobster egg mass by the nemertean *Pseudocarcinonemertes homari*. (A) Normal egg mass, free of nemerteans; (B) heavily infested egg mass with more than 14,000 individuals counted. (From Aiken *et al.*, 1985, with permission.)

or trawls, taken onboard ship, and then thrown back. The female compounds the problem by flapping her tail, dislodging even more eggs. This effect is more pronounced immediately after spawning or late in development when the eggs are less securely attached (Aiken and Waddy, 1986a; Herrick, 1909).

Nelson *et al.* (1983) postulated that a relationship exists between incubation success and the length of time between molting and subsequent egg extrusion. However, in areas such as the southern Gulf of St. Lawrence and Northumberland Strait, it is common for females to spawn within a few weeks of molting (the Adult-Ib cycle, see Section III,B,1) (Aiken and Waddy, 1980, 1982; Templeman, 1940a), with egg retention rates comparable to those of females that spawn a year after molting (S. L. Waddy, unpublished observation).

C. Control of Mating Behavior, Insemination, and Fertilization

1. Mating Behavior

a. Female Molt and Reproductive Stage Templeman (1934) felt that most matings occur within 24 hours of the female molt and found an inverse relationship between time following the female molt and incidence of successful mating. Subsequent studies appeared to confirm Templeman's observations, and matings after molt stage A or B were considered unusual (Aiken and Waddy, 1980; Hughes and Matthiessen, 1962; Atema, 1986; Atema and Cobb, 1980). However, recent studies have shown there are no limitations to mating and insemination because of molt stage. Although receptivity peaks immediately after molting, mating can occur at any time in the female molt cycle (Dunham and Skinner-Jacobs, 1978; Waddy and Aiken, 1990a).

Female reproductive stage is an important factor in the interest shown by males. Males are much more active than females (Fig. 26) and are able to recognize and distinguish between immature and mature females and between inseminated and uninseminated females (Waddy and Aiken, 1990a). Preovigerous, uninseminated females broadcast behavioral and possibly chemical signals to alert male lobsters to their lack of insemination: females become more active and remain outside their shelters for longer periods of time each night. These signals become pronounced as the time of spawning approaches. Once insemination occurs, female activity levels drop dramatically. Inseminated females are usually no longer receptive and males rarely attempt to mate with a female carrying stored spermatophores. Even with ample opportunity, only 5–10% of females are inseminated twice (Snyder *et al.*, 1992; Waddy and Aiken, 1990a). (See Atema and Voigt, 1995, Chapter 13, on mating behavior.)

In most areas, preovigerous lobsters not carrying spermatophores are unusual, while females with immature (stage 1) ovaries are rarely inseminated (Fig. 27). Of several hundred preovigerous females examined from the southern Gulf of St. Lawrence,

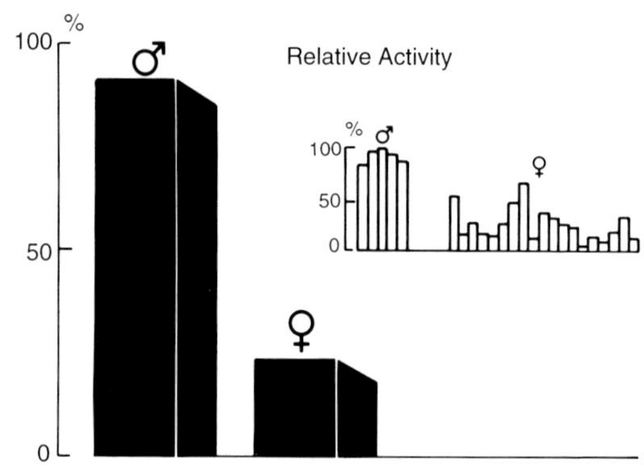

FIGURE 26 Relative activity of mature male ($N = 5$) and female ($N = 19$) lobsters held communally with excess shelter during June and July. The large histograms indicate the percentage of lobsters that leave their shelters during each 24-hour period. Upper right: percentage of days during June and July that individual lobsters left their shelters. [From Waddy and Aiken, 1990a. *In* "Crustacean Sexual Biology" (Raymond T. Bauer and Joel W. Martin, eds.). Copyright © 1990 by Columbia University Press. Reprinted with permission of the publisher.]

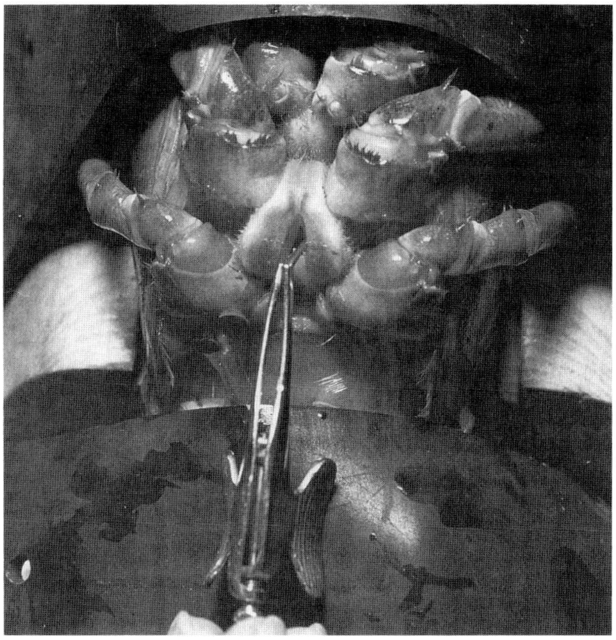

FIGURE 27 Examining a female American lobster to determine whether or not she has been inseminated. Barraquer forceps are used to open the seminal receptacle so the "sperm plug" can be seen. (From Waddy and Aiken, 1990b, with permission.)

Northumberland Strait, southwest Nova Scotia, and the Bay of Fundy, virtually all were carrying spermatophores (Waddy and Aiken, 1990a). Although mature females without sperm are reportedly common in some stocks (Krouse, 1973; Ennis, 1980), these studies were done before criteria for assessing maturity were well defined, so the samples may have included females that would have molted again prior to spawning.

It has been suggested that behavioral controls over the female molt cycle, possibly acting through pheromones, enable mature females to delay molting if the male they have selected as a mate is unavailable (Cowan and Atema, 1990). This requires verification, however, as the females were not monitored for progression through the molt cycle and the data could not be shown to be different from a random distribution (Hazlett, 1991; Cowan *et al.*, 1991).

b. Male Maturity, Potency, and Virility There is little information on the reproductive capacity of male lobsters. Early studies suffered from a lack of criteria for assessing maturity and it is probable that many included immature males. In the Bay of Fundy and southern Gulf of St. Lawrence, males reach the adult phase and become functionally mature (i.e., actually engage in mating) at about the same size as females; this occurs several molts after they reach the adolescent phase and become physiologically mature (i.e., first produce sperm) (Aiken and Waddy, 1980, 1989; see Section III,A,1, and Lawton and Lavalli, Chapter 4). Even when males are known to be functionally mature, it can be difficult to predict whether or not a male will mate with a particular female (Waddy and Aiken, 1990a).

Males are often capable of mating with two females in one day, but their recovery time is variable, as is potency among individuals (Hughes and Matthiessen, 1962; Waddy and Aiken, 1990a). In one laboratory study, a male lobster mated with three females within a 21-hour period and over 11 days mated with a total of nine females, all of which subsequently produced fertilized eggs. Since this male mated with every female available to him, his capability may have been even greater. Such behavior may be quite variable, as several males demonstrated similar capability while others did not (Waddy and Aiken, 1990b).

Relative size of the male and female also influences mating success. Small males may be incapable of mating with significantly larger females, because they lack the physical strength to subdue the female (Hughes and Matthiessen, 1962).

c. Male Molt Stage Although male lobsters can mate in all stages from buckle-shell (early stage C) through intermolt and into premolt (Templeman, 1934), there is a strong influence of molt stage on male potency and frequency of mating. Mature males from the Bay of Fundy molt only once every 2 or 3 years and frequency of mating in individuals varies dramatically between years (Fig. 28). Most mating activity is confined to the years between molts and little mating activity occurs in the same year as the

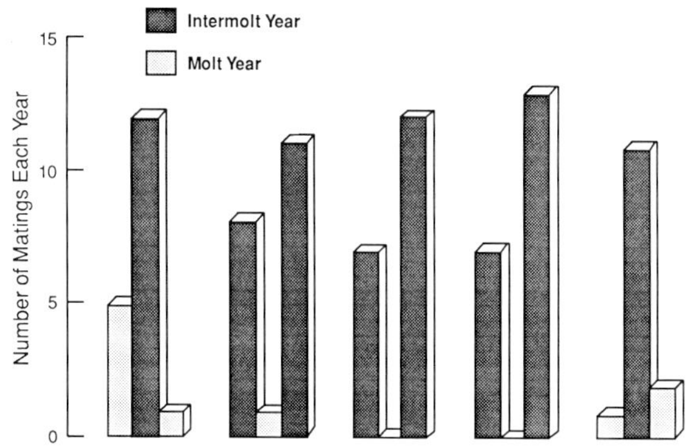

FIGURE 28 Variation in mating incidence in male lobsters held in a communal tank with both newly molted and preovigerous females. Mating activity of five males was monitored over three successive years and related to molt stage of the males. [From Waddy and Aiken, 1990a. *In* "Crustacean Sexual Biology" (Raymond T. Bauer and Joel W. Martin, eds.). Copyright © 1990 by Columbia University Press. Reprinted with permission of the publisher.]

molt (Waddy and Aiken, 1990a). Before the effect of molt stage was recognized, loss of interest in mating was attributed to social and environmental effects (Waddy and Aiken, 1985b). The relationship between molting and reproductive capability requires further investigation, as mature male lobsters in some localities molt each year and are obviously able to mate.

d. Environmental Effects Mating occurs throughout the year, even when seawater temperature is only 0°–2°C. There is no evidence of seasonal changes in the weight of the vasa deferentia and no indication of seasonal differences in sperm production or mating capability (Aiken and Waddy, 1986b; Waddy and Aiken, 1990a,b). Although mating incidence is low (28%) during the winter if lobsters are provided with shelter, mating occurs readily (70%) if lobsters are held without shelters, suggesting that incidence of mating during the winter is related more to the effect of temperature on activity than on sperm or spermatophore production (Waddy and Aiken, 1990a).

Despite the fact that temperature does not seem to be a major factor regulating the reproductive cycle of male lobsters, long-term exposure to a constant elevated temperature has deleterious effects on sperm and spermatophore production; males hatched and reared to maturity at 20°C produce small spermatophores that contain little or no sperm. This problem can be alleviated by transferring the males to a seasonally fluctuating temperature regime when they are between 40- to 50-mm CL, well below the size at maturity. When environmentally conditioned from this size on, cultured males breed almost as well as wild males by the time they reach 65- to 75-mm CL (Waddy and Aiken, 1990a).

Habitat and shelter affect the diel timing of mating in intermolt females; when lobsters are held without shelter, all matings occur during the day, but when shelters are provided, 90% of the matings occur at night (Waddy and Aiken, 1990b).

2. Insemination and Fertilization

a. Multiple Insemination Once inseminated, females are usually no longer receptive. If the spermatophore transferred is small, however, the female will mate a second time, often with the same male. Newly molted females have been reported to mate with as many as 10 males, but such behavior is rare (Templeman, 1934) and it has not been determined how many of the males in such cases actually transfer a spermatophore. Although lobsters usually do not mate a second time as the male discourages other suitors (Templeman, 1934; Snyder *et al.*, 1992; Waddy and Aiken, 1990a), genetic evidence has confirmed that multiple paternity occurs in wild lobsters (Nelson and Hedgecock, 1977).

A second type of multiple insemination occurs in jumbo lobsters that spawn twice within a molt cycle. Reinsemination is necessary for some of these females to fulfill their reproductive potential, as their sperm supply can be exhausted after the first spawning. Such females readily mate a second time, 2 or 3 years after the first mating. Although reinsemination is difficult to confirm in nature, all wild females from the Bay of Fundy that have been examined at the appropriate size and reproductive condition were carrying sperm; this seems unlikely without reinsemination (Waddy and Aiken, 1986, 1990a).

b. Sperm Viability Lobster sperm retain their viability for long periods in the seminal receptacle. (See Talbot and Helluy, Chapter 9, on the reproductive system.) Templeman (1934) concluded that sperm remain viable for at least 2 years and later studies extended the viability to 3 years. There seems to be little deterioration in the quality of stored sperm; no differences in fertilization or hatching success have been detected between successive broods fertilized from a single insemination (Waddy and Aiken, 1986).

D. Endocrine Control

Reproduction and molting are entwined in *Homarus americanus*, and, consequently, so are their respective hormone systems. Once the eggs are extruded, molting must be avoided to prevent the attached eggs from being lost with the cast shell, and yet it is advantageous for lobsters to molt (and mate) as soon as their eggs hatch. For nearshore stocks of the American lobster, the changes in the seasons are extreme and frequently result in conditions that are either very favorable or very unfavorable for reproduction and survival of offspring. Fortunately, these seasonal changes are accompanied by equally dramatic changes in photoperiod and temperature, and these are utilized to synchronize reproduction to appropriate seasons. The central nervous system channels this information into a usable form and induces appropriate organismic responses by way of the endocrine system and the complex of growth and reproductive hormones.

Gonadal maturation in crustaceans is generally thought to be regulated by two antagonistic neurohormones: one a GIH from the sinus gland (Panouse, 1943, 1947) and the other a GSH found in the brain and thoracic ganglion (Eastman-Reks and Fingerman, 1984; Otsu, 1963). In reality, regulation of reproduction is more complicated: CHH and androgenic hormone are also involved in reproductive control, and MF is thought to have a role as well. (Compare with Section II,B,4, on endocrine control of molting.) It is still unclear how these various hormones interact to control reproduction, but recent progress in the structural and functional characterization of several of these hormones is providing new information on reproductive mechanisms and the factors that control them.

1. Inhibition of Gonad Growth

The existence of a GIH in the eyestalk of crustaceans was first inferred from eyestalk ablation studies (Panouse, 1943, 1947). This neurohormone (also called VIH, vitellogenesis inhibiting hormone) inhibits ovarian synthesis of yolk proteins *in vitro* (Fingerman, 1987; Quackenbush and Keeley, 1988) and it has been postulated that lobster GIH inhibits the internalization of vitellogenin (Van Herp, 1993). GIH has been isolated from *Homarus americanus* and sequenced (Soyez *et al.*, 1991) and is found in the eyestalks of larvae as well as in mature males and females (de Kleijn *et al.*, 1992; Rotllant *et al.*, 1993). Removal of the eyestalks, and presumably the source of GIH, in American lobsters in the autumn accelerates ovarian maturation and spawning and causes a highly significant increase in the size and weight of the vasa deferentia (Aiken *et al.*, 1977, 1981b).

Lobster GIH mRNA is only found in the eyestalks, indicating that GIH is not produced in other parts of the nervous system (de Kleijn *et al.*, 1994b). Lobster GIH occurs as two isoforms, only one that has GIH activity (Soyez *et al.*, 1991). Work using cDNA cloning has shown that GIH contains 78 amino acids (instead of the 77 originally reported by Soyez *et al.*, 1991) and there is a potential amidation site at the C terminus (de Kleijn *et al.*, 1994b). Lobster preproGIH has a much higher degree of identity with crab preproMIH than with lobster, crab, or crayfish preproCHHs, suggesting that GIH and MIH form a group separate from CHH in the CHH/MIH/GIH neuropeptide family (de Kleijn *et al.*, 1994b).

In situ hybridization studies have revealed partial colocalization of GIH and CHH mRNAs, demonstrating that the hormones are produced by the same cells in the eyestalk (de Kleijn *et al.*, 1992). Although crab GIH is not colocalized with CHH, the close preprohormone identity of lobster and crab GIH and MIH and the expression of GIH only in the medulla terminalis (MT) could suggest that GIH and MIH are hormones with similar function(s). They may directly or indirectly regulate the production and/or release of other hormones involved in reproduction and molting, analogous to vertebrate gonadotropic releasing hormones.

The expression, storage, and release of GIH has been followed through various stages of the reproductive cycle in female American lobsters. Levels of GIH mRNA in the eyestalk are significantly lower immediately after spawning (just before vitellogenesis resumes), than at other stages in the ovarian cycle. Hemolymph levels of GIH are significantly lower during secondary vitellogenesis than they are at other stages in the reproductive cycle. Levels increase significantly just before spawning, when vitellogenin levels are declining, and they remain high after spawning and during primary vitellogenesis (de Kleijn, 1995).

2. Stimulation of Gonad Growth

The discovery that thoracic or supraesophageal ganglion implants induce ovarian growth in crabs (*Paratelphusa hydrodromous*, Gomez and Nayar, 1965; *Potamon dehaani*, Otsu, 1963) led to speculation that a GSH exists in the central nervous system of crustaceans. Injection of brain extracts from *Homarus americanus* induced ovarian maturation in a shrimp (*Penaeus vannamei*, Yano and Wyban, 1992) and suggested that a brain hormone (GSH-releasing hormone, GSH-RH) stimulates the release of GSH from the thoracic ganglion.

The hypothesis that CHH may have a role in reproduction was suggested by the results of a heterologous bioassay for oocyte growth of HPLC fractions of *Homarus americanus* sinus-gland peptides—one or both of the isoforms of CHH-B has GSH activity (Tensen *et al.*, 1989). Lobster prepro-CHH-A and -B have been cloned and expression studies have shown that mRNA of both these hormones is present throughout the nervous system (de Kleijn *et al.*, 1994b). Studies on the expression, storage, and release of CHH during the reproductive cycle of female lobsters have produced intriguing results (de Kleijn, 1995). Levels of CHH-B mRNA are low immediately after spawning, but increase to significantly higher levels when vitellogenesis resumes shortly after spawning. Levels of CHH-B mRNA during secondary vitellogenesis are significantly lower than during primary vitellogenesis, but increase to their highest levels just before spawning. CHH-A mRNA levels are also highest during primary vitellogenesis, but unlike CHH-B mRNA, do not increase prior to spawning. Hemolymph CHH-A and CHH-B levels are significantly higher just before spawning than at other times in the reproductive cycle. These results suggest that CHH may be involved in oocyte maturation (de Kleijn, 1995).

3. Androgenic Gland and Androgenic Hormone

The androgenic gland (AG) of decapods, first described by Charniaux-Cotton (1956), produces androgenic hormone (AH) which regulates spermatogenic activity in the testes and is responsible for the development and maintenance of the secondary sexual characteristics (reviewed by Charniaux-Cotton and Payen, 1988). The chemical nature of androgenic hormone is still uncertain and it is thought that several compounds may be produced by the androgenic gland (Laufer and Landau, 1991). In the isopod *Armadillium vulgare*, androgenic hormone is a protein (Katakura *et al.*, 1975; Juchault *et al.*, 1978) that exists as two similar peptides with different amino acid structures (Martin *et al.*, 1990), whose physiological function and structure is still unknown. In *Carcinus maenas*, the androgenic gland produces farnesylacetone (Fingerman, 1987), which is similar in structure to methyl farnesoate (Laufer and Landau, 1991). Androgenic gland activity is thought to be regulated by an eyestalk hormone as eyestalk ablation produces hyperactive androgenic glands in several species, including the American lobster. In crabs, the ganglionic mass seems to play a role in androgenic gland development as well (Charniaux-Cotton and Payen, 1988). Little is known about the androgenic gland of *Homarus americanus*.

4. Ecdysteroids

Ecdysteroids have not been as extensively studied in crustacean reproduction as in insects and little work has been done to develop a model of the role of ecdysteroids in lobster reproduction (Chang, 1993). Ecdysteroids have been identified in the follicle cells and oocytes of several species (Chang, 1993; Laufer *et al.*, 1993) and correlations between vitellogenesis and hemolymph ecdysteroid titers have been reported (Chaix and de Reggi, 1982; Lachaise *et al.*, 1981). In some species, however, evidence suggests that ecdysteroids are not involved in vitellogenesis (Meusy and Payen, 1988). In the crab *Cancer anthonyi*, hemolymph levels of ecdysteroids gradually decline during ovarian maturation and increase when the attached eggs are being brooded (Chang, 1993). The effects of ecdysteroid injections and *in vitro* incubations have been both inhibitory and stimulatory (Adiyodi, 1985; Chang, 1985). Administration of ecdysteroids just after ecdysis inhibits the onset of secondary vitellogenesis in amphipods and shrimp (Tourir and Charniaux-Cotton, 1974; Blanchet *et al.*, 1975), while ovary protein synthesis in early vitellogenesis is stimulated *in vitro* by ecdysone and 20-hydroxyecdysone in the isopod *Porcelio dilatatus* (Gohar and Souty, 1984).

Even less is known about the involvement of ecdysteroids in male reproduction. Several studies have suggested that ecdysteroids stimulate gonad growth in males (Laufer *et al.*, 1993). Although it is not clear whether the hormone is responsible for the associated morphological and biochemical changes (reviewed by Adiyodi, 1985), *in vitro* studies have demonstrated some direct effects of ecdysteroids on male gonads (Matlock and Dornfield, 1982; Brody and Chang, 1989).

5. Vertebrate-like Hormones

The presence of vertebrate-like hormones, including pregnenolone, progesterone, and 17β-estradiol, is well documented in crustaceans, but little is known of their role. Progesterone and 17β-estradiol have been found in the mandibular organ, kidney, hepatopancreas, ovary, and hemolymph of *Homarus americanus*, and levels of these hormones vary over the course of the reproductive cycle. There are suggestions that progesterone may be a precursor of 17β-estradiol in invertebrates, just as it is in vertebrates (Couch *et al.*, 1987; Quinitio *et al.*, 1991). Progesterone has been reported to stimulate ovarian development and induce spawning in shrimps and prawns. Hemolymph progesterone increases at the onset of vitellogenesis in *Pandalus kessleri* and decreases during the peak of vitellogenesis, while 17β-estradiol concentrations increase during vitellogenesis and decrease after spawning (Quinitio *et al.*, 1991).

The highest concentrations of estradiol in *Homarus americanus* have been found in the mandibular organs. There seems to be a correlation between the level of estradiol and ovarian development, as estradiol is found only in lobsters with maturing ovaries (Couch *et al.*, 1987). Attempts to alter vitellogenin synthesis in female lobsters by treating them with 17β-estradiol, progesterone, or 17α-hydroxyprogesterone prior to and during induced vitellogenesis have not been successful (Tsukimura *et al.*, 1993).

There are reports of other vertebrate hormones affecting crustacean reproduction as well. Human chorionic gonadotropin (HCG), follicle stimulating hormone (FSH), and luteinizing hormone (LH) have been reported to stimulate vitellogenesis synthesis in isopods and shrimp (reviewed by Laufer and Landau, 1991). Although it is generally thought that crustacean testes do not produce reproductive hormones, the testes of *Homarus americanus* have many of the enzyme systems necessary for steroidogenesis (Burns *et al.*, 1984a), and testosterone has been isolated from both the hemolymph and testes (Burns *et al.*, 1984b).

6. Effect of Eyestalk Ablation

Female American lobsters respond to eyestalk ablation, as do other decapods, and either molting or reproduction, or both, can be accelerated. Eyestalk ablation has the greatest effect on final ovarian maturation and spawning and the effects are modified by temperature. At 15°C in the autumn and winter, eyestalk ablation induces vitellogenesis in all mature females, leading to oviposition in some and massive resorption of the oocyte vitellin in others (Aiken *et al.*, 1981b). Eyestalk-ablated females often spawn in mid-premolt and molt a few weeks later, suggesting that eyestalk factors are necessary for proper synchronization of the molt and reproductive cycles (Aiken *et al.*, 1977). Although eyestalk-ablated pubertal females at 15°C convert immature stage 1 ovaries to stage 2, the effect may be due to temperature alone (Aiken *et al.*, 1981b).

Vas deferens development in *Homarus americanus* parallels ovarian response to eyestalk ablation, but there does not appear to be a comparable effect on the testes (Aiken *et al.*, 1981b). Male lobsters have a seasonal sensitivity to eyestalk ablation and often die shortly after ablation. It is rare for males that are eyestalk ablated in April to survive long enough to molt, while those ablated in November survive well. This seasonal sensitivity is puzzling, as it does not occur in females (Aiken and Waddy, unpublished observations).

7. Mandibular Organs and Methyl Farnesoate

Studies on the mandibular organs (MO) of the American lobster began over 20 years ago when ultrastructural and other evidence suggested that the MO were analogous to the insect corpus allatum and involved in the reproductive cycle (Byard *et al.*, 1975; Sochasky *et al.*, 1972). Interestingly, the MO of male lobsters increase dramatically in size after maturity, while the female MO do not (Fig. 29).

The mandibular organs of lobsters and other decapods secrete methyl farnesoate (MF) and farnesoic acid (FA) (Laufer *et al.*, 1987b; Tobe *et al.*, 1989). MF is the precursor of the insect juvenile hormone JH III, so it has been thought that the MO may be producing either a crustacean juvenile hormone or a prohormone that is converted to some JH-active compound in peripheral tissues (Laufer *et al.*, 1987a,b; Borst *et al.*, 1987; Tobe *et al.*, 1989). There are strong indications that the MO of *Homarus americanus* produce other hormones as well; they convert pregnenolone to progesterone *in vitro* (Borst *et al.*, 1994) and they contain significant amounts of progesterone (Couch *et al.*, 1987).

Studies in other species have provided considerable evidence that MO are important in reproduction in both males and females. High levels of MF have been found in vitellogenic females and reproductively active male crabs, while MF levels are low in immature and nonvitellogenic females (reviewed by Laufer *et al.*, 1993, 1994). Implants of active MO induce ovarian growth in immature spider crabs (*Libinia emarginata*, Hinsch, 1980), and MF induces small but significant increases in hemolymph levels of vitellogenin when injected into eyestalkless crabs (Vogel and Borst, 1989). There is a strong correlation between gonad maturity and production of MF in male *L. emarginata* (Sagi *et al.*, 1993), suggesting that MF acts similarly to JH in insects—terminating reproductive diapause and stimulating reproduction.

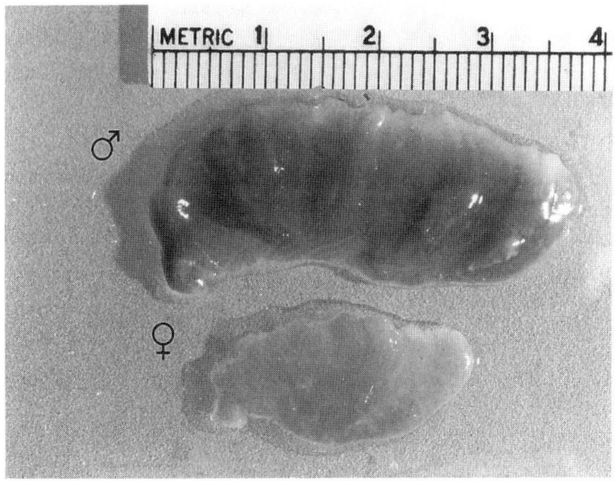

FIGURE 29 The mandibular organs of both male and female lobsters are multilobate, pale- to dark-green, foliaceous structures that can approach 3 g in weight in large males.

Despite the data supporting the involvement of the MO in the regulation of reproduction in both sexes of crustaceans, results on *Homarus americanus* continue to be confusing and recent studies question the role of MF in secondary vitellogenesis. Female American lobsters deprived of their MO maintain normal spawning rhythms (S. L. Waddy and D. E. Aiken, unpublished observations) and MF levels in preovigerous lobsters do not follow the patterns found in other species. MF levels in the hemolymph are high (4–10 ng/ml) in preovigerous females during the winter, when there is no ovarian activity and serum vitellogenin levels are low. When vitellogenesis is induced in the spring, MF drops to undetectable levels (Fig. 30) (Tsukimura *et al.*, 1992). These results seem very different from those in other species showing that mandibular organ activity is positively correlated with reproductive stage and that vitellogenic females have high levels of MF in the hemolymph (Laufer *et al.*, 1986; Sagi *et al.*, 1993; Vogel and

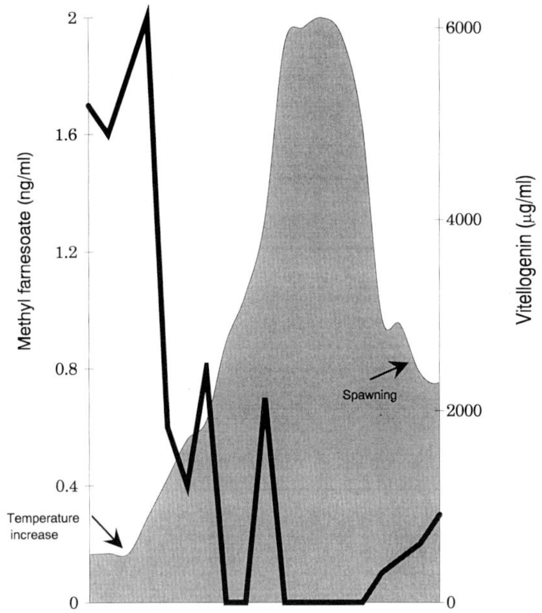

FIGURE 30 Hemolymph titers of methyl farnesoate and vitellogenin in female American lobsters induced to spawn in the spring with environmental manipulation. (B. Tsukimura, S. Waddy, and D. Borst, unpublished data.)

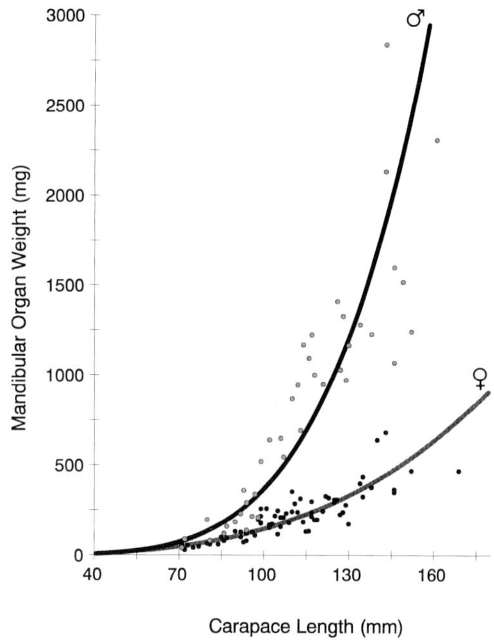

FIGURE 31 Relative size of the mandibular organs in male and female American lobsters from the Bay of Fundy. (D. E. Aiken, unpublished data.)

Borst, 1989).

MO in *Homarus americanus* appear to be more complex than those in other decapods. It has been speculated that lobster MO result from the fusion of two tissues, one which synthesizes MF and the other which produces other products (Borst *et al.*, 1994). Lobster MO, unlike those in other species, have two distinct morphological regions: a fan-folded region along one side of the gland and a smooth unfolded region comprising the remainder of the gland (Fig. 31). Almost all MF synthesis takes place in the fan-folded region (Borst *et al.*, 1994). These results suggest that lobster MO are quite different from those in other crustaceans, which may help to explain some of the puzzling results that have been obtained.

The MO of *Homarus americanus* appear to be regulated by eyestalk factors (Tsukimura and Borst, 1992). Serotonin may function as a neuroregulator of MF synthesis in *Libinia emarginata* (Homola *et al.*, 1989). The red pigment concentrating hormone (RPCH) is able to stimulate MF synthesis in the crayfish *Procambarus clarkii*, while the pigment dispersing hormone (PDH) inhibits MF synthesis (Rao *et al.*, 1985).

IV. Directions for Further Research

Biological cycles play an important role in population dynamics and in the responses of lobsters to exploitation and environmental change. In most cases, the regulatory mechanisms for these cycles are poorly understood, especially at the population level. Although considerable information is available on the responses of individuals under controlled conditions, much less is known about the responses of wild stocks, which live in a variety of environments and are, for the most part, free to move elsewhere when conditions are unsuitable. The following areas of investigation deserve further attention.

1. The waters around Newfoundland are much colder in the spring than those in the Gulf of Maine and water temperatures in both of these locations are more extreme than those in the deep canyons off Georges Bank. However, lobsters grow and reproduce in all of these locations, raising the following questions. How do their biological control mechanisms compensate for the differences in temperature? Are the inductive thermal phenomena cumulative (as in degree-day relationships), threshold, or some combination of the two? Are they independent of, or integrated with, photoperiodic input?

2. Long-distance migrations and seasonal shoaling are well known in lobsters and assumed to be important in reproductive success. Mature females have been shown to move to specific areas for spawning or hatching and to return to the same areas year after year. How are these directed movements induced and synchronized with molting and reproductive

requirements?

3. There have been suggestions that some of the areas that attract concentrations of ovigerous females have thermal regimes which would make it difficult for hatched larvae to complete metamorphosis before the onset of winter. Are these larvae lost to the population, or do they benefit from survival and distribution mechanisms about which we have no knowledge? If the ovigerous females were transferred to a distant location, would their larvae be released at a similar time and survive at a similar rate, or are the larvae from a given area genetically "programmed" to take advantage of the thermal and oceanographic conditions of a specific geographic area?

4. There is the very interesting question of seasonal influence on biological responses. Studies have shown that real time has a profound influence on biological responses to environmental stimuli. Larval development under specific temperature and photoperiod regimes varies dramatically with the time of year; the response of lobsters to molt induction changes dramatically in early spring and again in the autumn, as does the response to eyestalk ablation and ecdysteroid treatment. Also, the photoperiod and temperature requirements for spawning induction change dramatically at the autumnal equinox and the winter solstice. Almost nothing is known about the regulatory mechanisms involved and how time of year and biological rhythms interact to produce these effects.

5. There are questions about the hormonal processes that drive the biological mechanisms. New techniques are providing the tools to characterize and quantitatively measure a variety of peptide hormones in *Homarus americanus* at the mRNA, peptide storage, and hemolymph levels. Molecular biology is making it possible to produce large amounts of recombinant crustacean peptide hormones in eukaryotic systems for studies on the mode of action and target sites, as well as for studies of the structure and functional relationships of these peptides. *H. americanus* is an ideal crustacean for these studies because of the degree of control over the molt and reproductive cycles that is possible. What is the nature of the relationship between GIH, MIH, and CHH? What is the role of CHH in regulating ovarian maturation and spawning? Do CHH and GIH have molt-inhibiting functions? How do these neurohormones interact to coordinate the alternate-year molt and reproductive cycles?

6. Questions on the mandibular organs have puzzled researchers for over 20 years. Results on the role of MF in final ovarian maturation in *Homarus americanus* seem inconsistent with those of other crustaceans. What is the explanation for this? Recent suggestions of a role for MF in control of the Y organs should be pursued. No work has been done on the role of the mandibular organs in male lobsters, despite the fact that the weight of the organs increases dramatically near the time of maturity and can exceed 2 g in the largest males, several times that in females of similar length. Is the function of this organ in males different from females?

V. Summary

Growth and reproduction in *Homarus americanus* have been studied for more than 100 years and during that time an impressive quantity of information has been produced on the subject. Experimental biology has been the key to recent advances in the understanding of the factors that control growth and reproductive functions.

The integument of *Homarus americanus* consists of a cellular epidermis overlain by four distinct layers of the cuticle. In preparing for molt, the lobster undergoes complex biochemical and morphological changes, leaving muscle and nerve connections attached. The epidermis separates from the old cuticle and begins secreting enzymes that dissolve the inner layers and remove minerals for storage. Two of the layers of the cuticle are formed before the molt and underlie the exoskeleton during premolt. When the latter is cast off at the molt, the premolt cuticle becomes the new exoskeleton, complete with muscle and nerve attachments and direct communication with the cells of the epidermis. The epidermis then synthesizes the final two layers and mineralizes the new cuticle. The minerals that were stored during premolt are used to harden critical parts of appendages, such as the mouthparts, tips of the chelipeds, walking legs, and other components that enable the postmolt lobster to begin foraging.

The molt cycle has been subdivided into a series of stages based on these integumentary and physiological changes, during which the body is first expanded in volume (stage A) and then calcified (stages B and C). The end of stage C is characterized by completion of new tissue growth, complete calcification of the cuticle, and storage of organic reserves. This, and subsequent stages leading to the molt, can be readily identified by examination of setal development in flattened, transparent structures such as the pleopods and uropods.

Growth in lobsters is affected by temperature, light, nutrition, season, social interaction, habitat and population density, state of reproductive development, health, genetic potential, animal size, and a variety of miscellaneous factors. These regulatory parameters often act synergistically and some, such as photoperiod, appear to be thermally coupled. Molting occurs almost exclusively during seasons of optimum temperature and there are strong seasonal cycles of molting competence that can be uncoupled by eyestalk removal or temperatures of 15°–20°C. Although lobsters are able to withstand long periods of starvation, they require a nearly constant supply of nutrients to achieve maximum growth. Population density is also an important factor and physical barriers impose the same restrictions as crowding. Habitat type also affects growth and both bottom cover and shelter are important. Male lobsters generally grow faster than females, especially after maturity, and with advancing age there is a decrease in the percent length increase at molt and an increase in the time between molts.

Larval growth and development are influenced by similar factors and many of these act synergistically, making it difficult to predict effects. For example, although larval development is controlled predominantly by temperature, responses vary seasonally. There are conflicting reports on the effect of temperature on larval size; comparisons are complicated by the fact that even groups of siblings can be significantly different in size. Larvae require relatively more food than adults and may have obligatory and facultative feeding periods. Metamorphosis occurs at the molt from the third to the fourth stage in a well-defined rhythm that peaks during the scotophase at 20°C.

Male and female lobsters become functionally mature at similar sizes and there is wide geographic variation in the size at maturity because of differences in summer seawater temperature. Male reproductive biology is not as well understood as that of females and maturity is more difficult to define as males produce mature spermatozoa at a smaller size (physiological maturity) than when they begin to mate. There also appear to be cycles in mating competence. Female maturity is more easily identified and a variety of criteria for assessing maturity have been developed.

The reproductive cycle of female lobsters typically takes two years, but there are several exceptions related to female size and seawater temperature. Large females (≥120-mm CL) often spawn twice between molts, usually fertilizing both broods with sperm from a single insemination. A proportion of Adult-I females from some areas molt and spawn in the same season. Variations are also common at the northern and southern extremes of the range, where temperatures are marginal for reproduction, and in offshore areas, where seasonal variations in temperature are minimal.

The effect of temperature on reproduction influences latitudinal distribution, as temperature is the major factor determining size at maturity, incidence, timing and synchronization of spawning, success of egg attachment and incubation, and time of hatching. Photoperiod seems to have little influence on the regulation of spawning in nearshore populations, but laboratory studies have demonstrated an intriguing facultative response of preovigerous females to day length when seasonal temperature cues are not available. Responses to temperature and photoperiod are also profoundly influenced by real calendar time and significant alterations in reproductive physiology occur at the autumnal equinox and the winter solstice. Other environmental factors such as nutrition, social conditions, stress, pollutants, and parasites can also influence egg production.

Adverse conditions that are easily tolerated by the maternal female may kill the attached embryos. There is strong evidence that environmental factors, such as temperature history and salinity, are involved in incubation success. Bacterial and fungal diseases, nemerteans, and repeated handling during fishing operations can all cause egg loss. Embryonic development is regulated primarily by temperature and requires 9–12 months in most populations.

Most matings occur within 24 hours of the female molt, but there are no limitations to mating and insemination imposed by the female molt stage. Mating can occur throughout the year and males are able to recognize and distinguish between immature and mature females and between inseminated and uninseminated females. Male lobsters mate from early stage C into premolt, but there is a strong influence of male molt stage on potency and frequency of mating.

The complex processes involved in both molting and reproduction are controlled by the endocrine system. Environmental information is translated into hormonal stimuli, which are orchestrated to produce the appropriate effect on the animal. Regulation of both molting and spawning are often presented as relatively straightforward systems involving two antagonistic hormones. However, molting and reproductive physiology are profoundly complex and it is difficult to provide an overall scheme of their regulation. Recent studies have revealed that molt inhibiting hormone (MIH), gonad inhibiting hormone (GIH), and crustacean hyperglycemic hormones (CHH-A and -B) comprise a novel family of neuropeptides with diverse functions. There is accumulating evidence that CHH plays a role in the regulation of both molting and spawning and at least one of the CHH isoforms has gonad stimulating hormone (GSH) activity and may be involved in oocyte maturation. Unlike MIH and GIH, hyperglycemic hormones are produced at sites other than the eyestalk and CHH mRNAs have been found throughout the nervous system. The role of the mandibular organs is still unclear and some of the results seem inconsistent with other studies; MF levels in preovigerous *Homarus americanus* decline and drop to undetectable levels during secondary vitellogenesis, while in other species positive correlations between MF levels and ovarian maturation have been reported.

References

Adiyodi, R. G. (1985). Reproduction and its control. *In* "The Biology of Crustacea" (D. E. Bliss, ed.), Vol. 9, "Integument, Pigments and Hormonal Processes" (L. H. Mantel, ed.), pp. 147–215. Academic Press, New York.

Aiken, D. E. (1973). Proecdysis, setal development and molt prediction in the American lobster (*Homarus americanus*). *J. Fish. Res. Board Can.* 30, 1337–1344.

Aiken, D. E. (1977). Molting and growth in decapod crustaceans with particular reference to the lobster *Homarus americanus*. *Div. Fish. Oceanogr. Circ. (Aust., CSIRO)* No. 7, pp. 41–73.

Aiken, D. E. (1980). Molting and growth. *In* "The Biology and Management of Lobsters" (J. S. Cobb and B. F. Phillips, eds.), Vol. 1, pp. 91–162. Academic Press, New York.

Aiken, D. E., and Waddy, S. L. (1975a). Temperature increase can cause hyperecdysonism in American lobsters (*Homarus americanus*) injected with ecdysterone. *J. Fish. Res. Board Can.* 32, 1843–1847.

Aiken, D. E., and Waddy, S. L. (1975b). Induction and inhibition of molting in *Homarus*. *In* "Recent Advances in Lobster Aquaculture" (J. S. Cobb, ed.), p. 13. Sea Grant Lobster Aquaculture Workshop, University of Rhode Island, Kingston.

Aiken, D. E., and Waddy, S. L. (1976). Controlling growth and reproduction in the American lobster. *Proc. Annu. Meet.—World Maric. Soc.* 7, 415–430.

Aiken, D. E., and Waddy, S. L. (1978). Space, density and growth of lobster (*Homarus americanus*). *Proc. Annu. Meet.—World Maric. Soc.* 9, 461–467.

Aiken, D. E., and Waddy, S. L. (1979). Maturity and reproduction in the American Lobster. *Can. Tech. Rep. Fish. Aquat. Sci.* 932, 59–71.

Aiken, D. E., and Waddy, S. L. (1980). Reproductive biology. *In* "The Biology and Management of Lobsters" (J. S. Cobb and B. F. Phillips, eds.), Vol. 1, pp. 275–76. Academic Press, New York.

Aiken, D. E., and Waddy, S. L. (1982). Cement gland development, ovary maturation and reproductive cycles in the American lobster, *Homarus americanus*. *J. Crust. Biol.* 2, 315–327.

Aiken, D. E., and Waddy, S. L. (1985). Production of seed stock for lobster culture. *Aquaculture* 44, 103–114.

Aiken, D. E., and Waddy, S. L. (1986a). Environmental influence on recruitment of the American lobster (*Homarus americanus*). *Can. J. Fish. Aquat. Sci.* 43, 2258–2270.

Aiken, D. E., and Waddy, S. L. (1986b). Growth of the vasa deferentia of mature *Homarus americanus*: Conflicting results from field and laboratory studies. *Can. J. Fish. Aquat. Sci.* 43, 1453–1457.

Aiken, D. E., and Waddy, S. L. (1988). Strategies for maximizing growth of communally reared juvenile American lobsters. *World Aquaculture* 19(3), 61–63.

Aiken, D. E., and Waddy, S. L. (1989). Allometric growth and onset of maturity in male lobsters (*Homarus americanus*): The crusher propodite index. *J. Shellfish Res.* 8, 7–11.

Aiken, D. E., and Waddy, S. L. (1992). The growth process in crayfish. *Rev. Aquat. Sci.* 6, 335–381.

Aiken, D. E., and Waddy, S. L. (1995). Scotophase influence on the diel timing of the metamorphic molt in larval American lobsters, *Homarus americanus*. *Can. J. Fish. Aquat. Sci.* (in press).

Aiken, D. E., and Young-Lai, W. W. (1979). Cheliped ablation and immobilization: Methods for improving survival and growth of juvenile American lobsters in communal culture tanks. *Proc. Annu. Meet.—World Maric. Soc.* 10, 159–161.

Aiken, D. E., and Young-Lai, W. W. (1981). Dactylotomy, chelotomy and dactylostasis: Methods for enhancing survival and growth of small lobsters (*Homarus americanus*) in communal conditions. *Aquaculture* 22, 45–52.

Aiken, D. E., Waddy, S. L., Castell, J. D., and Covey, J. F. (1977). Effect of sex, temperature, diet and holding conditions on growth of intact and eyestalk ablated lobsters (*Homarus americanus*). *Div. Fish. Oceanogr. Circ. (Aust., CSIRO)* No. 7, pp. 13–14 (abstract).

Aiken, D. E., Martin, D. J., Meisner, D. J., and Sochasky, J. B. (1981a). Influence of photoperiod on survival and growth of larval American lobsters (*Homarus americanus*). *J. World Maric. Soc.* 12, 225–230.

Aiken, D. E., Waddy, S. L., and Castell, J. D. (1981b). Effect of temperature and eyestalk ablation on the reproductive organs of the lobster *Homaraus americanus*. *Dev. Endocrinol.* 11, 330.

Aiken, D. E., Rowe, W. J., Martin-Robichaud, D. J., and Waddy, S. L. (1982). Seasonal factors in the effect of photoperiod on survival and growth of larval American lobsters (*Homarus americanus*). *J. World Maric. Soc.* 13, 287–293.

Aiken, D. E., Waddy, S. L., and Uhazy, L. S. (1985). Aspects of the biology of *Pseudocarcinonemertes homari* and its association with the American lobster, *Homarus americanus*. *Can. J. Fish. Aquat. Sci.* 42, 351–356.

Ali, Y. O., and Wickins, J. F. (1994). The use of fresh food supplements to ameliorate moulting difficulties in lobsters, *Homarus gammarus* (L.), destined for release to the sea. *Aquacult. Fish. Manage.* 25, 483–496.

Anger, K. (1991). Developmental changes in the bioenergetics of decapod larvae. *Mem. Queensl. Mus.* 31, 289–308.

Atema, J. (1986). Review of sexual selection and chemical communication in the lobster, *Homarus americanus*. *Can. J. Fish. Aquat. Sci.* 43, 2283–2390.

Atema, J., and Cobb, J. S. (1980). Social behavior. *In* "The Biology and Management of Lobsters" (J. S. Cobb and B. F. Phillips, eds.), Vol. 1., pp. 409–450. Academic Press, New York.

Attard, J., and Hudon, C. (1987). Embryonic development and energetic investment in egg production in relation to size of female lobster (*Homarus americanus*). *Can. J. Fish. Aquat. Sci.* 44, 1157–1164.

Bartley, D. M., Carlberg, J. M., Van Olst, J. C., and Ford, R. F. (1980). Growth and conversion efficiency of juvenile American lobster (*Homarus americanus*) in relation to temperature and feeding level. *Proc. Annu. Meet.—World Maric. Soc.* 11, 355–368.

Bertran, R., and Lorec, J. (1986). Observations sur le Cycle de Reproduction du Homard Européen (*Homarus americanus*) en captivité. *Int. Counc. Explor. Mer.* C.M.1986/K:12.

Birkenbeil, H., and Eckert, M. (1983). Immuno-electronmicroscope evidence of ecdysteroids in Y-organ of *Orconectes limosus* (Crustacea, Decapoda). *Cell Tissue Res.* 233, 223–230.

Blanchet, M. F., Junera, H., and Meusy, J.-J. (1975). Molting and vitellogenesis in *Orchestia gammarella* Pallas (Crustacea, Amphipoda): Study of female-specific protein synthesis after ecdysterone administration. *Experientia* 31, 865–867.

Bollenbacher, W. F., Borst, D. W., and O'Connor, J. D. (1972). Endocrine regulation of lipid synthesis in decapod crustaceans. *Am. Zool.* 12, 381–384.

Boothroyd, F. A., and Ennis, G. P. (1992). Reproduction in American lobster *Homarus americanus* transplanted northward to St. Michael's Bay, Labrador. *Fish. Bull.* 90, 659–667.

Bordner, C. E., and Conklin, D. E. (1981). Food consumption and growth of juvenile lobsters. *Aquaculture* 24, 285–300.

Borst, D. W., Laufer, H., Landau, M., Chang, E. S., Hertz, W. A., Baker, F. C., and Schooley, D. A. (1987). Methyl farnesoate and its role in crustacean reproduction and development. *Insect Biochem.* 17, 1123–1128.

Borst, D. W., Tsukimura, B., Laufer, H., and Couch, E. F. (1994). Regional differences in methyl farnesoate production by the lobster mandibular organ. *Biol. Bull. (Woods Hole, Mass.)* 186, 9–15.

Botero, L., and Atema, J. (1982). Behavior and substrate selection during larval settling in the lobster *Homarus americanus*. *J. Crust. Biol.* 2, 59–69.

Bowser, P. R., and Rosemark, R. (1981). Mortalities of cultured lobsters, *Homarus*, associated with a molt death syndrome. *Aquaculture* 23, 11–18.

Briggs, P. T., and McGroaty, W. M. (1985). Departures from the normal molting and reproductive cycle of female lobsters from Long Island Sound, New York. *N.Y. Fish Game J.* 32, 211–212.

Briggs, P. T., and Mushacke, F. M. (1980). The American lobster and the pot fishery in the inshore waters off the south shore of Long Island, New York. *N.Y. Fish Game J.* 27, 156–178.

Brody, M. D., and Chang, E. S. (1989). Ecdysteroid and juvenile hormone induced protein synthesis in crayfish testis cell culture. *Am. Zool.* 29, 62A.

Bruce, M. J., and Chang, E. S. (1984). Demonstration of a molt-inhibiting hormone from the sinus gland of the lobster, *Homarus americanus*. *Comp. Biochem. Physiol. A* 79A, 421–424.

Burns, B. G., Sangalang, G. B., Freeman, H. C., and McMenemy, M. (1984a). Bioconversion of steroids by the testes of the American lobster, *Homarus americanus, in vitro*. *Gen. Comp. Endocrinol.* 54, 422–428.

Burns, B. G., Sangalang, G. B., Freeman, H. C., and McMenemy, M. (1984b). Isolation and identification of testosterone from the serum and testes of the American lobster (*Homarus americanus*). *Gen. Comp. Endocrinol.* 54, 429–432.

Byard, E. H., Shivers, R. R., and Aiken, D. E. (1975). The mandibular organ of the lobster, *Homarus americanus*. *Cell Tissue Res.* 162, 13–22.

Cameron, J. N. (1989). Post-moult calcification in the blue crab, *Callinectes sapidus*: Timing and mechanism. *J. Exp. Biol.* 143, 285–304.

Campbell, A. (1983). Growth of tagged lobsters (*Homarus americanus*) off Port Maitland, Nova Scotia, 1946–80. *Can. Tech. Rep. Fish. Aquat. Sci.* 1232, 1–10.

Campbell, A. (1984). Aspects of lobster biology and fishery in the upper reaches of the Bay of Fundy. *Can. Tech. Rep. Fish. Aquat. Sci.* 1256, 469–488.

Campbell, A. (1986). Migratory movements of ovigerous lobster, *Homarus americanus*, tagged off Grand Manan, eastern Canada. *Can. J. Fish. Aquat. Sci.* 43, 2197–2205.

Campbell, A. (1990). Aggregations of berried lobsters (*Homarus americanus*) in shallow waters off Grand Manan, eastern Canada. *Can. J. Fish. Aquat. Sci.* 47, 520–523.

Campbell, A. (1992). Characteristics of the American lobster fishery of Grand Manan, New Brunswick, Canada. *North Am. J. Fish. Manage.* 12, 139–150.

Campbell, A., and Brattey, J. (1986). Egg loss from the American lobster, *Homarus americanus*, in relation to the nemertean, *Pseudocarcinonemertes homari*, infestation. *Can. J. Fish. Aquat. Sci.* 43, 772–780.

Campbell, A., and Duggan, D. R. (1980). Review of the Grand Manan lobster fishery with an analysis of recent catch and effort trends. *Can. Tech. Rep. Fish. Aquat. Sci.* 997, 1–20.

Campbell, A., and Pezzack, D. (1986). Relative egg production and abundance of berried lobsters, *Homarus americanus*, in the Bay of Fundy and off southwestern Nova Scotia. *Can. J. Fish. Aquat. Sci.* 43, 2190–2196.

Campbell, A., and Robinson, D. G. (1983). Reproductive potential of three American lobster (*Homarus americanus*) stocks in the Canadian Maritimes. *Can. J. Fish. Aquat. Sci.* 40, 1958-1967.

Capuzzo, J. M., and Lancaster, B. A. (1979). The Effects of Diet on the Growth Energetics of Postlarval Lobsters (*Homarus americanus*). *Woods Hole Oceanogr. Inst. Tech. Rep.* WHOI-79-55.

Carlberg, J., and Van Olst, J. C. (1976). Brine shrimp consumption by the larval stages of the American lobster (*Homarus americanus*) in relation to food density and water temperature. *Proc. Annu. Meet.—World Maric. Soc.* 7, 12–22.

Carlisle, D. B., and Dohrn, P. F. R. (1953). Studies on *Lysmata seticaudata* Risso (Crustacea Decapoda). II. Experimental evidence of a growth and moult-accelerating factor obtainable from eyestalk. *Pubbl. Stn. Zool. Napoli* 24, 69–83.

Castell, J., and Kean, J. (1986). Evaluation of the role of nutrition in lobster recruitment. *Can. J. Fish. Aquat. Sci.* 43, 2320–3227.

Castell, J. D., and Budson, S. D. (1974). Lobster nutrition: The effect on *Homarus americanus* of dietary protein levels. *J. Fish. Res. Board Can.* 31, 1363–1370.

Castell, J. D., Covey, J. F., Aiken, D. E., and Waddy, S. L. (1977). The potential for eyestalk ablation as a technique for accelerating growth of lobsters (*Homarus americanus*) for commercial culture. *Proc. Ann. Meet. — World Maric. Soc.* 10, 895–914.

Chaix, J.-C., and de Reggi, M. (1982). Ecdysteroid levels during ovarian development and embryogenesis in the spider crab *Acanthonyx lunulatus*. *Gen. Comp. Endocrinol.* 47, 7–14.

Chang E. S. (1984). Ecdysteroids in Crustacea: Role in reproduction, molting and larval development. *In* "Advances in Invertebrate Reproduction" (W. Engels, W. H. Clark, Jr., A. Fischer, P. J. W. Olive, and D. F. Went, eds.), Vol. 2, pp. 223–230. Elsevier, Amsterdam.

Chang, E. S. (1985). Hormonal control of molting in decapod Crustacea. *Am. Zool.* 25, 179–185.

Chang, E. S. (1989). Endocrine regulation of molting in Crustacea. *Rev. Aquat. Sci.* 1, 131–157.

Chang, E. S. (1993). Comparative endocrinology of molting and reproduction: Insects and crustaceans. *Annu. Rev. Endocrinol.* 38, 161–180.

Chang E. S., Prestwich, G. D., and Bruce, M. J. (1990). Amino acid sequence of a peptide with both molt-inhibiting and hyperglycemic activities in the lobster, *Homarus americanus*. *Biochem. Biophys. Res. Commun.* 171, 818–826.

Chang, E. S., Bruce, M. J., and Tamone, S. L. (1993). Regulation of crustacean molting: A multi-hormonal system. *Am. Zool.* 33, 324–329.

Charmantier, G., and Aiken, D. E. (1987). Intermediate larval and postlarval stages of *Homarus americanus* H. Milne Edwards, 1837 (Crustacea: Decapoda). *J. Crust. Biol.* 7, 525–535.

Charmantier, G., Charmantier-Daures, M., and Aiken, D. E. (1988). Larval development and metamorphosis of the American lobster *Homarus americanus* (Crustacea: Decapoda): Effect of eyestalk ablation and juvenile hormone injection. *Gen. Comp.*

Endocrinol. 70, 319–333.

Charmantier, G., Charmantier-Daures, M., and Aiken, D. E. (1989). La somatotropine humaine stimule la croissance de jeunes américains, *Homarus americanus* (Crustacea, Decapoda). *C.R. Acad. Sci. Paris* 308, Sér. III, 21–26.

Charmantier, G., Charmantier-Daures, M., and Aiken, D. E. (1991). Metamorphosis in the lobster *Homarus* (Decapoda): A review. *J. Crust. Biol.* 11, 481–495.

Charniaux-Cotton, H. (1956). Determinisme endocrinien caractères sexuels d'*Orchestia gammarella* (Crustace Amphipode). *Ann. Sci. Nat., Zool.* II 18, 305–310.

Charniaux-Cotton, H. (1957). Croissance, régénération et determinisme endocrinien de caractères sexuels d'*Orchestia gammarella* (Pallas). *Ann. Sci. Nat., Zool. Biol. Anim., Ser. 11* 19, 411–559.

Charniaux-Cotton, H., and Payen, G. (1988). Crustacean reproduction. *In* "Endocrinology of Selected Invertebrate Types" (H. Laufer and R. G. H. Downer, eds.), pp. 279–304. Liss, New York.

Cheng, J.-H., and Chang, E. S. (1991). Ecdysteroid treatment delays ecdysis in the lobster, *Homarus americanus*. *Biol. Bull. (Woods Hole, Mass.)* 181, 169–174.

Cheng, J.-H., and Chang, E. S. (1993). Determinants of postmolt size in the lobster *Homarus americanus*. III. Scute density. *Invertebr. Reprod. Dev.* 24, 169–178.

Cheng, J.-H., and Chang, E. S. (1994). Determinants of postmolt size in the American lobster (*Homarus americanus*). 1. D_1^3 is the critical stage. *Can. J. Fish. Aquat. Sci.* 50, 2106–2111.

Chittleborough, R. G. (1974). Review of prospects for rearing rock lobsters. *Aust. Fish.* 33, 1–5.

Cobb, J. S. (1968). Delay of moult by the larvae of *Homarus americanus*. *J. Fish. Res. Board Can.* 25, 2251–2253.

Cobb, J. S. (1970). Effect of solitude on time between fourth and fifth stage larval moults in the American lobster (*Homarus americanus*). *J. Fish. Res. Board Can.* 27, 1653–1655.

Cobb, J. S., and Tamm, G. R. (1974). Social conditions increase intermolt period in juvenile lobster. *J. Fish. Res. Board Can.* 32, 1941–1943.

Cobb, J. S., and Tamm, G. R. (1975). Dominance status and molt order in lobsters *Homarus americanus*. *Mar. Behav. Physiol.* 3, 119–124.

Cobb, J. S., and Tamm, G. R. (1977). Social status and molt order in juvenile American lobsters (*Homarus americanus*). *Div. Fish. Oceanogr. Circ. (Aust., CSIRO)* No. 7, p. 25 (abstract).

Cobb, J. S., and Wahle, R. A. (1994). Early life history and recruitment processes of clawed lobsters. *Crustaceana* 67, 1–25.

Cobb, J. S., Tamm, G. R., and Wang, D. (1982). Behavioral mechanisms influence molt frequency in the American lobster, *Homarus americanus* Milne Edwards. *J. Exp. Mar. Biol. Ecol.* 62, 185–200.

Conan, G. Y. (1978). Comparison between Growth at Molt of the European and American Lobsters in Different Geographical Locations According to Sex and State of Maturity. *Int. Counc. Explor. Sea Doc.* C. M. 1978/K:18.

Connor, P. M. (1972). Acute toxicity of heavy metals to some marine larvae. *Mar. Pollut. Bull.* 3, 190–192.

Cooper, R. A. (1970). Retention of marks and their effects on growth, behavior and migrations of the American lobster, *Homarus americanus*. *Trans. Am. Fish. Soc.* 99, 409–417.

Cooper, R. A., and Uzmann, J. R. (1980). Ecology of juvenile and adult *Homarus*. *In* "The Biology and Management of Lobsters" (J. S. Cobb and B. F. Phillips, eds.), Vol. 2, pp. 97–178. Academic Press, New York.

Couch, E., Hagino, N., and Lee, J. W. (1987). Changes in estradiol and progesterone immunoreactivity in tissues of the lobster, *Homarus americanus*, with developing and immature ovaries. *Comp. Biochem. Physiol. A* 87A, 765–770.

Coulombe, F., and Motnikar, S. (1989). The acceleration of the premolt process by unilateral eyestalk ablation in the male "canner" lobster (76–81 mm CL) of the Gaspé Bay. *In* "Life History of the American Lobster" (I. Kornfield, ed.), p. 9. Lobster Institute, University of Maine, Orono.

Cowan, D. F., and Atema, J. (1990). Molt staging and serial monogamy in American lobster *Homarus americanus*. *Anim. Behav.* 39, 1199–1206.

Cowan, D. F., Atema, J., and Solow, A. (1991). Moult staggering in the American lobster: A statistical analysis. *Anim. Behav.* 42, 863–864.

D'Abramo, L. R., and Conklin, D. E. (1985). Lobster aquaculture. *In* "Crustacean and Mollusk Aquaculture in the United States" (J. V. Huner and E. E. Brown, eds.), pp. 159–201. AVI Publishing, Westport.

Dadswell, M. J. (1979). A review of the decline in lobster (*Homarus americanus*) landings in Chedabucto Bay between 1956 and 1977 with a hypothesis for a possible effect by the Canso Causeway on the recruitment mechanism of eastern Nova Scotia stocks. *Can. Fish. Mar. Serv., Tech. Rep.* 834, 113–144.

de Kleijn, D. P. V. (1995). "Cloning, Sequence Analysis and Expression of the Hyperglycemic Neurohormone Family in Crustaceans," Ph.D. thesis. University of Nijmegan, The Netherlands.

de Kleijn, D. P. V., Coenen, T., Laverdure, A. M., Tensen, C. P., and Van Herp, F. (1992). Localization of mRNAs encoding the crustacean hyperglycemic hormone (CHH) and gonad inhibiting hormone (GIH) in the X-organ sinus gland complex of the lobster *Homarus americanus*. *Neuroscience* 51, 121–128.

de Kleijn, D. P. V., Janssen, K. P. C., Martens, G. J. M., and Van Herp, F. (1994a). Cloning and expression of two crustacean hyperglycemic hormone mRNAs in the eyestalk of the crayfish *Orconectes limosus*. *Eur. J. Biochem.* 224, 623–629.

de Kleijn, D. P. V., Sleutels, F. J. G. T., Martens, G. J. M., and Van Herp, F. (1994b). Cloning and expression of mRNA encoding prepro gonad inhibiting hormone (GIH) in the lobster *Homarus americanus*. *FEBS Lett.* 353, 255–258.

de Kleijn, D. P. V., De Leeuw, E. P. H., Van Den Berg, M. C., Martens, G. J. M., and Van Herp, F. (1995). Cloning and expression of two mRNAs encoding structurally different crustacean hyperglycemic hormone precursors in the lobster *Homarus americanus*. *Biochim. Biophys. Acta* 1260, 62–66.

Dircksen, H., Webster, S., and Keller, R. (1988). Immunocytochemical demonstration of the neurosecretory systems containing putative moult inhibiting hormone and hyperglycemic hormone in the eyestalk of brachyuran crustaceans. *Cell Tissue Res.* 251, 3–12.

Donahue, J. K. (1954). Studies on ecdysis in the American lobster (*Homarus americanus*). 3. A method for differentiating stages of the intermolt cycle. *Maine Dep. Sea Shore Fish. Res. Bull.* 20, 1–4.

Drach, P. (1939). Mue et cycle d'intermue chez les Crustacé Décapodes. *Ann. Inst. Océanogr. (Paris)* [N.S.] 18, 103–391.

Drach, P. (1944). Etude préliminaire sur le cycle d'intermue et son conditionnement hormonal chez *Leander serratus* (Pennant). *Bull. Biol. Fr. Belg.* 78, 40–62.

Drach, P., and Tchernigovtzeff, C. (1967). Sur le méthode de détermination des stades d'intermue et son application générale aux Crustacés. *Vie Milieu (Ser. A. Biol. Mar.)* 18, 595–610.

Dubé, P., and Grodin, P. (1985). Maturité Sexuelle du Homard (*Homarus americanus*) Femelle aux Iles-de-la-Madeleine. *Can. Atl. Fish. Sci. Advis. Comm., Res. Doc.* 85/85.

Dunham, P. W., and Skinner-Jacobs, D. (1978). Intermolt mating in the lobster (*Homarus americanus*). *Mar. Behav. Physiol.* 5, 209–214.

Eagles, M. D., Aiken, D. E., and Waddy, S. L. (1984). Effect of food quality and feeding schedule on survival, growth and development of larval American lobsters fed frozen adult brine shrimp. *Proc. Annu. Meet.—World Maric. Soc.* 15, 142–143.

Eagles, M. D., Aiken, D. E., and Waddy, S. L. (1986). Influence of light and food on larval American lobsters, *Homarus americanus*. *Can. J. Fish. Aquat. Sci.* 43, 2303–2310.

Eastman-Reks, S., and Fingerman, M. (1984). Effects of neuroendocrine tissue and cyclic AMP on ovarian growth in vivo in the fiddler crab, *Uca pugilator*. *Comp. Biochem. Physiol. A* 79A, 679–684.

Emmel, V. E. (1906). The relation of regeneration to the molting process of the lobster. *R.I. Comm. Inland Fish., Annu. Rep.* 36, 258–313.

Ennis, G. P. (1971). Lobster (*Homarus americanus*) fishery and biology in Bonavista Bay, Newfoundland 1966–70. *Fish. Res. Board Can., Tech. Rep.* 289, 1–46.

Ennis, G. P. (1972). Growth per moult of tagged lobsters (*Homarus americanus*) in Bonavista Bay, Newfoundland. *J. Fish. Res. Board Can.* 29, 143–148.

Ennis, G. P. (1975). Observations on hatching and larval release in the lobster *Homarus americanus*. *J. Fish. Res. Board Can.* 32, 2210–2213.

Ennis, G. P. (1980). Size-maturity relationships and related observations in Newfoundland populations of the lobster (*Homarus americanus*). *Can. J. Fish. Aquat. Sci.* 37, 945–956.

Ennis, G. P. (1981). Fecundity of the American lobster, *Homarus americanus*, in Newfoundland Waters. *Fish. Bull.* 79, 796–800.

Ennis, G. P. (1982). Fisheries and population biology of lobsters (*Homarus americanus*) at Comfort Cove, Newfoundland. *Can. Tech. Rep. Fish. Aquat. Sci.* 1116, 1–45.

Ennis, G. P. (1983a). Variation in Annual Growth in Two Newfoundland Lobster (*Homarus americanus*) Populations in Relation to Temperature Conditions. *Int. Counc. Explor. Sea Doc.* C.M. 1983/K:24.

Ennis, G. P. (1983b). Observations on the behaviour and activity of lobsters, *Homarus americanus*, in nature. *Can. Tech. Rep. Fish. Aquat. Sci.* 1165, 1–26.

Ennis, G. P. (1984). Comparison of physiological and functional size-maturity relationships in two Newfoundland populations of lobsters, *Homarus americanus*. *Fish. Bull.* 82, 244–249.

Ennis, G. P. (1986). Sphryion tag loss from the American lobster, *Homarus americanus*. *Trans. Am. Fish. Soc.* 115, 914–917.

Ennis, G. P. (1991). Annual variation in egg production in a Newfoundland population of the American lobster, *Homarus americanus*. *In* "Crustacean Issues" (F. R. Schram, ed.), Vol. 7, "Crustacean Egg Production" (A. Wenner and A. Kuris, eds.), pp. 291–299. Balkema, Rotterdam, The Netherlands.

Ennis, G. P., and Collins, P. W. (1983). A computer program to assess egg production per recruit in a lobster (*Homarus americanus*) population. *Can. Tech. Rep. Fish. Aquat. Sci.* 1198, 1–15.

Estrella, B. T., and McKiernan, D. J. (1989). Catch-per-unit-effort and biological parameters from the Massachusetts coastal lobster (*Homarus americanus*) resource: Description and trends. *U.S. Dept. Commer., NOAA Tech. Rep. NMFS SSRF* 81.

Fingerman, M. (1987). Endocrine mechanisms in crustaceans. *J. Crust. Biol.* 7, 1–24.

Fingerman, S. (1985). Non-metal environmental pollutants and growth. *In* "Crustacean Issues" (F. R. Schram, ed.), Vol. 3, "Factors in Adult Growth" (A. M. Wenner, ed.), pp. 219–234. Balkema, Rotterdam, The Netherlands.

Fisher, W. S., Nilson, E. H., Steenbergen, J. F., and Lightner, D. V. (1978). Microbial diseases of cultured lobsters: A review. *Aquaculture* 14, 115–140.

Gilgan, M. W., and Farquharson, T. E. (1977). A change in the sensitivity of adult male lobsters (*Homarus americanus*) to ecdysterone on changing from intermolt to active premolt development. *Comp. Biochem. Physiol. A* 58A, 29–32.

Gilgan, M. W., and Zinck, M. E. (1975). Response of the adult lobster (*Homarus americanus*) to graded and multiple doses of ecdysterone. *Comp. Biochem. Physiol. A* 56A, 43–49.

Gilgan, M. W., Farquharson, T. E., and Burns, B. G. (1977). The effect of α-ecdysone, ecdysterone and inokosterone treatment separately or in combination, on premolt development and molting in adult male lobsters (*Homarus americanus*). *Comp. Biochem. Physiol. A* 56A, 43–49.

Gohar, M., and Souty, C. (1984). Action temporelle d'ecdysteroids sur la synthese proteique ovarienne *in vitro* chez le Crustacea Isopoda terrestre *Porcellio dilatatus* (Brandt). *Reprod. Nutr. Dev.* 24, 137–145.

Gomez, R., and Nayar, K. K. (1965). Certain endocrine influences in the reproduction of the crab *Parathelphusa hydrodromous*. *Zool. Jahrb., Abt. Allg. Zool. Physiol. Tiere* 71, 694–701.

Green, J. P., and Neff, M. R. (1972). A survey of fine structure of the integument of the fiddler crab. *Tissue Cell* 4, 137–171.

Hadley, P. B. (1906). Observations on some influences of light upon the larval and early adolescent stages of *Homarus americanus*: Preliminary report. *R.I. Comm. Inland Fish., Annu. Rep.* 36, 237–257.

Hadley, P. B. (1908). The behavior of the larvae and adolescent stages of the American lobster. *J. Comp. Neurol. Psychol.* 19, 199–302.

Halcrow, K. (1976). The fine structure of the carapace integument of *Daphnia magna* Straus (Crustacea, Branchiopoda). *Cell Tissue Res.* 169, 267–276.

Harding, G., Kenchington, E., and Zheng, Z. (1993). Morphometrics of American lobster (*Homarus americanus*) larvae in relation to stock determination in the Maritimes, Canada. *Can. J. Fish. Aquat. Sci.* 50, 43–52.

Harding, G. C., and Trites, R. W. (1988). Dispersal of *Homarus americanus* larvae in the Gulf of Maine from Browns Bank. *Can. J. Fish. Aquat. Sci.* 45, 416–425.

Harrison, K. (1990). The role of nutrition in maturation, reproduction and embryonic development of decapod crustaceans: A review. *J. Shellfish Res.* 9, 1–28.

Hazlett, B. A. (1991). Moult staggering or are female lobsters moulting independently? *Anim. Behav.* 42, 462.

Hedgecock, D. (1983). Maturation and spawning in the American lobster, *Homarus americanus*. *In* "CRC Handbook of Mariculture" (J. P. McVey, ed.), Vol. 1, pp. 261–270. CRC Press, Boca Raton, Florida.

Hedgecock, D., Moffet, W. L., Borgeson, W., and Nelson, K. (1978). Progress and problems in lobster broodstock development. *Proc. Annu. Meet.—World Maric. Soc.* 9, 947–506.

Herrick, F. H. (1895). The American lobster: A study of its habits and development. *Bull. U.S. Fish. Comm.* 15, 1–252 + 54 plates.

Herrick, F. H. (1909). Natural history of the American lobster. *Bull. Bur. Fish.* 29, 149–408 + 20 plates.

Hertz, W. A., and Chang, E. S. (1986). Juvenile hormone effects on metamorphosis of lobster larvae. *Int. J. Invertebr. Reprod. Dev.* 10, 71–77.

Hinsch, G. W. (1980). Effect of mandibular organ implants upon the spider crab ovary. *Trans. Am. Microsc. Soc.* 99, 317–322.

Homola, E., Landau, M., and Laufer, H. (1989). An *in vitro* bioassay for the regulation of methyl farnesoate synthesis in disaggregated mandibular organ cells from *Libinia emarginata* and the effect of serotonin. *Biol. Bull. (Woods Hole, Mass.)* 176, 69 (abstract).

Hudon, C. (1987). Ecology and growth of postlarval and juvenile lobster, *Homarus americanus*, off Îles de la Madeleine (Québec). *Can. J. Fish. Aquat. Sci.* 44, 1855–1869.

Hudon, C., and Fradette, P. (1988). Planktonic growth of larval lobster (*Homarus americanus*) off Îles de la Madeleine (Québec). *Can. J. Fish. Aquat. Sci.* 45, 868–878.

Hughes, J. T., and Matthiessen, G. C. (1962). Observations on the biology of the American lobster, *Homarus americanus*. *Limnol. Oceanogr.* 7, 414–421.

Huner, J. V., and Lindqvist, O. V. (1987). Response of a slow-growing noble crayfish, *Astacus astacus* L., population from a pond in central Finland to several years of exploitation. *Freshwater Crayfish* 7, 81–89.

Huntsman, A. G. (1923). Natural lobster breeding. *Bull. Biol. Board Can.* 5, 3–11.

Incze, L. S., and Wahle, R. A. (1991). Recruitment from pelagic to early benthic phase in lobsters *Homarus americanus*. *Mar. Ecol. Prog. Ser.* 79, 77–87.

Juchault, R., Maissiat, J., and Legrand, J.-J. (1978). Caracterisation chimique d'une substance ayant les effets biologiques de l'hormone androgene chez le crustacé isopode terrestre *Armadillidium vulgare* Latreille. *C.R. Acad. Sci. Paris, Ser. D* 286, 73–76.

Juinio, M. A. R., and Cobb, J. S. (1994). Estimation of recent growth of field-caught postlarvae American lobsters, *Homarus americanus*. *Can. J. Fish. Aquat. Sci.* 51, 286–294.

Juinio, M. A. R., Cobb, J. S., Bengston, D., and Johnson, M. (1992). Changes in nucleic acids over the molt cycle in relation to food variability and temperature in *Homarus americanus* postlarvae. *Mar. Biol.* 114, 1–10.

Katakura, Y., Fujimaki, Y., and Uno, K. (1975). Partial purification and characterization of androgenic gland hormone from the isopod crustacean, *Armadillidium vulgare*. *Annot. Zool. Jpn.* 48, 203–209.

Kegel, G., Reichwein, B., Weese, S., Gaus, G., Peter-Katalinic, J., and Keller, R. (1989). Amino acid sequence of the crustacean hyperglycemic hormone (CHH) from the shore crab, *Carcinus maenas*. *FEBS Lett.* 255, 10–14.

Kegel, G., Reichwein, B., Tensen, C. P., and Keller, R. (1991). Amino acid sequence of crustacean hyperglycemic hormone (CHH) from the crayfish *Orconectes limosus*. Emergence of a novel neuropeptide family. *Peptides (N.Y.)* 12, 909–913.

Keller, R. (1992). Crustacean neuropeptides: Structures, functions and comparative aspects. *Experientia* 48, 439–448.

Klein, J. M., de Kleijn, D. P. V., Hunemeyer, G., Keller, R., and Weidemann, W. (1993). Demonstration of the cellular expression of genes for moult inhibiting hormone and crustacean hyperglycemic hormone in the eyestalk of the shore crab *Carcinus maenas*. *Cell Tissue Res.* 274, 515–519.

Kleinholz, L. H. (1985). Biochemistry of crustacean hormones. *In* "The Biology of Crustacea" (D. E. Bliss, ed.), Vol. 9, "Integument, Pigments and Hormonal Processes" (L. H. Mantel, ed.), pp. 463–522. Academic Press, New York.

Knowles, F. G. W., and Carlisle, D. B. (1956). Endocrine control in the Crustacea. *Biol. Rev. Cambridge Philos. Soc.* 31, 396–473.

Koshio, S., Castell, J. D., and O'Dor, R. K. (1992). The effect of different dietary energy levels in crab-protein-based diets on digestibility, oxygen consumption, and ammonia excretion of bilaterally eyestalk-ablated and intact juvenile lobsters. *Aquaculture* 108, 285–297.

Krouse, J. S. (1973). Maturity, sex ratio, and size composition of the natural population of American lobster, *Homarus americanus*, along the Maine coast. *Fish. Bull.* 71, 165–173.

Lachaise, F., Goudeau, M., Hetru, C., Kappler, C., Hoffman, J. A. (1981). Ecdysteroids and ovarian development in the shore crab, *Carcinus maenas*. *Hoppe Seyler's Z. Physiol. Chem.* 362, 521–529.

Laufer, H., and Landau, M. (1991). Endocrine control of reproduction in shrimp and other Crustacea. *In* "Frontiers in Shrimp Research" (P. DeLoach, W. J. Dougherty, and M. A. Davidson, eds.), pp. 65–81. Elsevier, Amsterdam.

Laufer, H., Landau, M., Borst, D., and Homola, E. (1986). The synthesis and regulation of methyl farnesoate, a new juvenile hormone for crustacean reproduction. *In* "Advances in Invertebrate Reproduction" (M. Porchet, J.-C. Andries, and A. Dhainaut, eds.), pp. 135–143. Elsevier, Amsterdam.

Laufer, H., Borst, D., Baker, F. C., Carrasco, C., Sinkus, M., Reuter, C. C., Tsai, L. W., and Schooley, D. A. (1987a). Identification of a juvenile hormone-like compound in a crustacean. *Science* 235, 202–205.

Laufer, H., Landau, M., Homola, E., and Borst, D. W. (1987b). Methyl farnesoate: Its site of synthesis and regulation of secretion in a juvenile crustacean. *Insect Biochem.* 17, 1129–1131.

Laufer, H., Ahl, J. S. B., and Sagi A. (1993). The role of juvenile hormones in crustacean reproduction. *Am. Zool.* 33, 365–374.

Laufer, H., Sagi, A., and Ahl, J. S. B. (1994). Alternate mating strategies of polymorphic males of *Libinia emarginata* appear to depend on methyl farnesoate. *Invertebr. Reprod. Dev.* 26, 41–44.

Lawton, P., and Robichaud, D. A. (1992). Lobster habitat ecology research in the Bay of Fundy. *In* "Science Review," pp. 53–56. Department of Fisheries and Oceans, Scotia-Fundy Region, Dartmouth.

MacKenzie, B. (1988). Assessment of temperature effects on interrelationships between stage durations, mortality, and growth in laboratory-reared *Homarus americanus*. *J. Exp. Mar. Biol. Ecol.* 116, 87–98.

Martin, G., Juchault, P., Sorokine, O., and Van Dorsselaer, A. (1990). Purification and characterization of androgenic hormone from the terrestrial isopod *Armadillidium vulgare* Latr. (Crustacea, Oniscidea). *Gen. Comp. Endocrinol.* 80, 349–354.

Matlock, D. B., and Dornfield, E. J. (1982). The effect of crustecdysone on DNA synthesis in polyploid somatic cells of an isopod. *Comp. Biochem. Physiol. B* 73B, 603–605.

Mauviot, J. C., and Castell, J. D. (1976). Molt- and growth-enhancing effects of bilateral eyestalk ablation on juvenile and adult American lobsters, *Homarus americanus*. *J. Fish. Res. Board Canada* 33, 1922–1929.

McCarthy, J. F. (1982). Ecdysone metabolism in premolt land crabs (*Gecarcinus lateralis*). *Gen. Comp. Endocrinol.* 47, 323–332.

McConaugha, J. R. (1985). Nutrition and larval growth. *In* "Crustacean Issues" (F. R. Schram, ed.), Vol. 2, "Larval Growth" (A. M. Wenner, ed.), pp. 127–154. Balkema, Rotterdam, The Netherlands.

Meusy, J.-J., and Payen, G. (1988). Female reproduction in malacostracan Crustacea. *Zool. Sci.* 5, 217–265.

Miller, R. J., and Watson, F. L. (1991). Changes in lobster size at maturity among years and locations. *J. Shellfish Res.* 10, 286–287 (abstract).

Mykles, D. L. (1980). The mechanism of fluid absorption at ecdysis in the American lobster, *Homarus americanus*. *J. Exp. Biol.* 84, 89–101.

Mykles, D. L. (1992). Getting out of a tight squeeze—enzymatic regulation of claw muscle atrophy in molting. *Am. Zool.* 32, 485–494.

Naqvi, S. M., and Howell, R. D. (1993). Toxicity of cadmium and lead to juvenile red swamp crayfish, *Procambarus clarkii*, and effects on fecundity of adults. *Bull. Environ. Contam. Toxicol.* 51, 303–308.

Naya, Y., and Ikeda, M. (1993). Comprehensive survey of endogenous regulators of crustacean molting. *Pure Appl. Chem.* 65, 1265–1270.

Naya, Y., Kishida, K., Sugiyama, M., Murata, M., Miki, W., Ohnishi, M., and Nakanishi, K. (1988). Endogenous inhibitor of ecdysone synthesis in crabs. *Experientia* 44, 50–52.

Naya, Y., Ohnishi, M., Ikeda, M., Miki, W., and Nakanishi, K. (1989). What is molt inhibiting hormone? The role of an ecdysteroidogenesis inhibitor in the crustacean molting cycle. *Proc. Natl. Acad. Sci. U.S.A.* 68, 6826–6829.

Nelson, K. (1986). Photoperiod and reproduction in lobsters (*Homarus*). *Am. Zool.* 26, 447–457.

Nelson, K., and Hedgecock, D. (1977). Electrophoretic evidence of multiple paternity in the lobster *Homarus americanus* (Milne-Edwards). *Am. Nat.* 111, 361–365.

Nelson, K., Hedgecock, D., Borgeson, W., Johnson, E., Daggett, R., and Aronstein, D. (1980). Density-dependent growth inhibition in lobsters, *Homarus* (Decapoda, Nephropidae). *Biol. Bull. (Woods Hole, Mass.)* 159, 162–176.

Nelson, K., Hedgecock, D., Heyer, B., and Nunn, T. (1983). On the nature of short-range growth inhibition in juvenile lobsters (*Homarus*). *J. Exp. Mar. Biol. Ecol.* 72, 83–89.

Otsu, T. (1963). Bihormonal control of sexual cycle in the freshwater crab *Potamon dehaani*. *Embryologia* 8, 1–20.

Panouse, J. B. (1943). Influence de l'ablation, de pédoncle oculaire sur la croissance de l'ovaire chez la crevette *Leander serratus*. *C.R. Hebd. Seances Acad. Sci.* 217, 553–555.

Panouse, J. B. (1947). The sinus gland and maturation of genital products in shrimp. *Bull. Biol. Fr. Belg. Suppl.* 33, 160–163; Fish. Res. Board Canada Transl. Ser. No. 2797.

Passano, L. M. (1960). Molting and its control. *In* "The Physiology of Crustacea" (T. H. Waterman, ed.), Vol. 1, pp. 473–536. Academic Press, New York.

Perkins, H. (1971). Egg loss during incubation from offshore northern lobsters (Decapoda: Homaridae). *Fish. Bull.* 69, 452–453.

Perkins, H. (1972). Developmental rates at various temperatures of embryos of the northern lobster (*Homarus americanus* Milne-Edwards). *Fish. Bull.* 70, 95–99.

Peutz, A. V. H., Waddy, S. L., Aiken, D. E., and Young-Lai, W. W. (1987). Accelerated growth of juvenile American lobster induced by unilateral eyestalk ablation. *Bull. Aquacult. Assoc. Can.* 87-2, 28–29.

Pezzack, D. (1991). Growth patterns of female offshore lobsters. *J. Shellfish Res.* 10, 299 (abstract).

Porcheron, P., Caruelle, J.-P., Baehr, J.-C., and Cassier, P. (1984). Ecdysteroids and integuments in locusts. *In* "Biosynthesis. Metabolism and Mode of Action of Invertebrate Hormones" (J. Hoffman and M. Porchet, eds.), pp. 234–244. Springer-Verlag, Berlin.

Quackenbush, L. S., and Keeley, L. L. (1988). Regulation of vitellogenesis in the fiddler crab, *Uca pugilator*. *Biol. Bull. (Woods Hole, Mass.)* 175, 321–331.

Quinitio, E. T., Yamauchi, K., Hara, A., and Fuji, A. (1991). Profiles of progesterone- and estradiol-like substances in the hemolymph of female *Pandalus kessleri* during an annual reproductive cycle. *Gen. Comp. Endocrinol.* 81, 343–348.

Rao, K. R., Riehm, J. P., Zahnow, C. A., Kleinholz, L. H., Tarr, G. E., Johnson, L., Norton, S., Landau, M., Semmes, O. J., Sattelberg, R. M., Jorenby, W. H., and Hitz, M. F. (1985). Characterization of a pigment dispersing hormone in the eyestalks of the fiddler crab, *Uca pugilator*. *Proc. Nat. Acad. Sci. U.S.A.* 82, 5319–5322.

Richards, A. G. (1951). "The Integument of Arthropods." University of Minnesota Press, Minneapolis.

Roach, S. G. (1983). Survivorship, Growth and Behavior of Juvenile Lobsters *Homarus americanus* in Controlled Environments in Nature. *N. S. Dep. Fish., Manusc. Tech. Rep. Ser., Proj. Rep.* 83–02.

Roddick, D. L., and Miller, R. J. (1992). Spatial and temporal overlap of the American lobster (*Homarus americanus*) and the sea scallop (*Placopecten magellanicus*) as related to the impact of inshore scallop dragging. *Can. J. Fish. Aquat. Sci.* 49, 1486–1492.

Rogers, B. A., Cobb, J. S., and Marshall, N. (1968). Size comparisons of inshore and offshore larvae of the lobster, *Homarus americanus*, off southern New England. *Proc. Natl. Shellfish. Assoc.* 58, 78–81.

Rotllant, G., de Kleijn, D. P. V., Charmantier-Daures, M., Charmantier, G., and Van Herp, F. (1993). Localization of crustacean hyperglycemic hormone (CHH) and gonad inhibiting hormone (GIH) in the eyestalk of *Homarus gammarus* larvae by immunocytochemistry and *in situ* hybridization. *Cell Tissue Res.* 271, 507–512.

Rudolph, P. H., and Spaziani, E. (1992). Formation of ecdysteroids by Y-organs of the crab, *Menippe mercenaria*. II. Incorporation of cholesterol into 7-dehydrocholesterol and secretion products *in vitro*. *Gen. Comp. Endocrinol.* 88, 235–242.

Sagi, A., Homola, E., and Laufer, H. (1993). Distinct reproductive types of male spider crabs *Libinia emarginata* differ in circulating and synthesizing methyl farnesoate. *Biol. Bull. (Woods Hole, Mass.)* 185, 168–173.

Sasaki, G. C., Capuzzo, J. M., and Biesiot, P. (1986). Nutritional and bioenergetic considerations in the development of the American lobster, *Homarus americanus*. *Can. J. Fish. Aquat. Sci.* 43, 2311–2319.

Sastry, A. N., and Ehringer, R. S. (1980). Dominance hierarchies among communally held juvenile lobster, *Homarus americanus*. *Mar. Behav. Physiol.* 7, 85–93.

Sastry, A. N., and French, D. P. (1977). Growth of American lobster, *Homarus americanus* Milne-Edwards under controlled conditions. *Div. Fish. Oceanogr. Circ. (Aust., CSIRO)* No. 7, p. 11 (abstract).

Sastry, A. N., and Pechenik, J. A. (1977). A review of ecology, physiology and behavior of lobster larvae (*Homarus americanus* and *H. gammarus*). *Div. Fish. Oceanogr. Circ. (Aust., CSIRO)* 7, 159–173.

Sastry, A. N., and Vargo, S. L. (1977). Variations in the physiological responses of crustacean larvae to temperature. *In* "Physiological Responses of Marine Biota to Pollutants" (F. J. Vernberg, A. Calabrese, F. P. Thurberg, and W. B. Vernberg, eds.), pp. 401–422. Academic Press, New York.

Sastry, A. N., Wilcox, D., and Laczak, J. (1975). Effects of space on growth and survival of lobsters, *Homarus americanus*. *In* "Recent Advances in Lobster Aquaculture" (J. S. Cobb, ed.), p. 36. Sea Grant Lobster Aquaculture Workshop, University of Rhode Island, Kingston.

Savoie, F., and Maynard, D. (1991). Egg production of lobster (*Homarus americanus*) in the Gulf of St. Lawrence, Canada. *J. Shellfish Res.* 10, 299 (abstract).

Scarratt, D. J. (1973). The effect of raking Irish moss (*Chondrus crispis*) on lobsters in Prince Edward Island. *Helgol. Wiss. Meeresunters.* 24, 415–425.

Shleser, R. A. (1974). Studies on the effects of feeding frequency and space on the growth of the American lobster (*Homarus americanus*). *Proc. Annu. Meet.—World Maric. Soc.* 5, 149–155.

Shleser, R. A. (1975). Studies on the effects of feeding frequency and space on the growth of the American lobster, *Homarus americanus*. *In* "Recent Advances in Lobster Aquaculture" (J. S. Cobb, ed.), p. 18. Sea Grant Lobster Aquaculture Workshop, University of Rhode Island, Kingston.

Smith, E. M. (1977). "Some Aspects of Catch/Effort, Biology, and the Economics of the Long Island Sound Lobster Fishery during 1976." Final Report, USDC, NMFS, Connecticut, Project No. 3-253-R-1.

Snyder, M. J., and Chang, E. S. (1986a). Effects of sinus gland extracts on larval molting and ecdysteroid titers of the American lobster, *Homarus americanus*. *Biol. Bull. (Woods Hole, Mass.)* 170, 244–254.

Snyder, M. J., and Chang, E. S. (1986b). Effects of eyestalk ablation on larval molting rates and morphological development of the American lobster, *Homarus americanus*. *Biol. Bull. (Woods Hole, Mass.)* 170, 232–243.

Snyder, M. J., and Chang, E. S. (1991a). Ecdysteroids in relation to the molt cycle of the American lobster, *Homarus americanus*. I. Hemolymph titers and metabolites. *Gen. Comp. Endocrinol.* 81, 133–145.

Snyder, M. J ., and Chang, E. S. (1991b). Metabolism and excretion of injected [^{3}H]-ecdysone by female lobsters, *Homarus americanus*. *Biol. Bull. (Woods Hole, Mass.)* 180, 475–484.

Snyder, M. J., and Chang, E. S. (1991c). Ecdysteroids in relation to the molt cycle of the American lobster, *Homarus americanus*. II. Excretion of metabolites. *Gen. Comp. Endocrinol.* 83, 118–131.

Snyder, M. J., Ameyaw-Akumfi, C., and Chang, E. S. (1992). Mating behavior in vision-deprived American lobsters, *Homarus americanus*. *Mar. Behav. Phyiol.* 21, 227–238.

Sochasky, J. B., Aiken, D. E., and Watson, N. H. F. (1972). Y-organ, molting gland, and mandibular organ: A problem in decapod Crustacea. *Can. J. Zool.* 50, 993–997.

Sochasky, J. B., Aiken, D. E., and McLeese, D. W. (1973). Does eyestalk ablation accelerate molting in the lobster *Homarus americanus*? *J. Fish. Res. Board Can.* 30, 1600–1603.

Soyez, D., Noel, P. Y., Van Deijnen, J. E., Martin, M., Morel, A., and Payen, G. G. (1990). Neuropeptides from the sinus gland in the lobster *Homarus americanus*: Characterization of hyperglycemic peptides. *Gen. Comp. Endocrinol.* 79, 261–274.

Soyez, D., Le Caer, J. P., Noel, P. Y., and Rossier, J. (1991). Primary structure of two isoforms of the vitellogenesis inhibiting hormone from the lobster *Homarus americanus*. *Neuropeptides* 20, 25–32.

Soyez, D., Van Herp, F., Rossier, J., Le Caer, J. P., Tensen, C., and Lafont, R. (1994). Evidence for a conformational polymorphism of invertebrate neurohormones. *J. Biol. Chem.* 269, 18295–18298.

Spaziani, E., and Wang, W. L. (1993). Biosynthesis of ecdysteroid hormones by crustacean Y-organs: Conversion of cholesterol to 7-dehydrocholesterol is suppressed by a steroid 5 α-reductase inhibitor. *Mol. Cell. Endocrinol.* 95, 111–114.

Spindler, K. D., Van Wormhoudt, A., Sellow, D., and Spindler-Barth, M. (1987). Ecdysteroid levels during embryogenesis in the shrimp, *Palaemon serratus* (Crustacea Decapoda): Quantitative and qualitative changes. *Gen. Comp. Endocrinol.* 66, 116–122.

Squires, H. J., Tucker, G. E., and Ennis, G. P. (1971). Lobsters (*Homarus americanus*) in Bay of Islands, Newfoundland, 1963–65. *Fish. Res. Board Can., Tech. Rep.* 1151, 1–58.

Stewart, J. E., and Squires, H. J. (1968). Adverse conditions as inhibitors of ecdysis in the lobster *Homarus americanus*. *J. Fish. Res. Board Can.* 25, 1763–1774.

Taketomi, Y., Motona, M., and Miyawaki, M. (1989). The biological function of the mandibular organ of decapod Crustacea. *Cell Biol. Int. Rep.* 13, 463–469.

Talbot, P., and Harper, R. (1984). Abnormal eyestalk morphology is correlated with clutch attrition in laboratory-maintained lobsters. *Biol. Bull. (Woods Hole, Mass.)* 166, 349–356.

Talbot, P., Thaler, C., and Wilson, P. (1984). Spawning, egg attachment and egg retention in captive lobsters (*Homarus americanus*). *Aquaculture* 37, 239–249.

Tamm, G. R., and Cobb, J. S. (1976). Diel ecdysis rhythms in juvenile lobsters *Homarus americanus*. *J. Fish. Res. Board Can.* 33, 819–821.

Tamone, S. L., and Chang, E. S. (1993). Methyl farnesoate stimulates ecdysteroid secretion from crab Y-organs *in vitro*. *Gen. Comp. Endocrinol.* 89, 425–432.

Templeman, W. (1933). Female lobsters handicapped in growth by spawning. *Biol. Board Can., Prog. Rep. Atl. Biol. Stn.* 6, 5–6.

Templeman, W. (1934). Mating in the American lobster. *Contrib. Can. Biol. Fish.* 8, 423–432.

Templeman, W. (1936a). Local differences in the life history of the lobster (*Homarus americanus*) on the coast of the Maritime provinces of Canada. *J. Biol. Board Can.* 2, 41–87.

Templeman, W. (1936b). Fourth stage larvae of *Homarus americanus* intermediate in form between normal third and fourth stages. *J. Biol. Board Can.* 2, 349–354.

Templeman, W. (1936c). The influence of temperature, salinity, light and food conditions on the survival and growth of the larvae of the lobster (*Homarus americanus*). *J. Biol. Board Can.* 2, 485–497.

Templeman, W. (1940a). Embryonic developmental rates and egg-laying of Canadian lobsters. *J. Fish. Res. Board Can.* 5, 71–83.

Templeman, W. (1940b). The life history of the lobster. *Newfoundland Gov. Serv. Bull.* 15, 1–42.

Templeman, W. (1948). Growth per moult in the American lobster. *Bull. Newfoundland Gov. Lab.* 18, 26–48.

Tensen, C. P., Janssen, K. P. C., and Van Herp, F. (1989). Isolation, characterization and physiological specificity of the crustacean hyperglycemic factors from the sinus gland of the lobster *Homarus americanus* (Milne-Edwards). *Invertebr. Reprod. Dev.* 16, 155–164.

Tensen, C. P., de Kleijn, D. P. V., and Van Herp, F. (1991). Cloning and sequence analysis of cDNA encoding two crustacean hyperglycemic hormones from the lobster *Homarus americanus*. *Eur. J. Biochem.* 200, 103–106.

Tobe, S. S., Young, D. A., Khoo, H. W., and Baker, F. C. (1989). Farnesoic acid as a major product of release from crustacean mandibular organs *in vitro*. *J. Exp. Zool.* 249, 165–171.

Tourir, A., and Charniaux-Cotton, H. (1974). Influence de l'introduction d'ecdystérone sur l'exuviation et le démarrage de la vitellogènese chez la crevette *Lysmata seticaudata* Risso. *C.R. Acad. Sci. Paris, Ser. D* 278, 119–122.

Travis, D. F. (1960). Matrix and mineral deposition in skeletal structures of the decapod Crustacea. *In* "Calcification in Biological Systems" (R. F. Sognnaes, ed.), pp. 57–116. Publ. No. 64, Am. Assoc. Adv. Sci., Washington, D.C.

Tsukimura, B., and Borst, D. W. (1992). Regulation of methyl farnesoate in the hemolymphe and mandibular organ of the lobster *Homarus americanus*. *Gen. Comp. Endocrinol.* 86, 297–303.

Tsukimura, B., Waddy, S., Vogel, J. M., and Borst, D. W. (1992). The regulation of vitellogenesis in the lobster, *Homarus americanus*. *Am. Zool.* 32, 28A (abstract).

Tsukimura, B., Waddy, S., Burow, C. W., and Borst, D. W. (1993). The characterization and regulation of lobster vitellogenin. *Am. Zool.* 33, 122A (abstract).

Van Herp, F. (1993). Importance of reproductive biological knowledge for application of biotechnology in aquaculture: Crustaceans as an example. *J. Mar. Biotechnol.* 1, 17–20.

Van Olst, J. C., Carlberg, J. M., and Hughes, J. T. (1980). Aquaculture. *In* "The Biology and Management of Lobsters" (J. S. Cobb and B. F. Phillips, eds.), Vol. 2, pp. 333–384. Academic Press, New York.

Vernet, G. (1976). Données actuelles sur le déterminisme de la mue chez les Crustacés. *Année Biol.* 15 (3–4), 155–188.

Vogel, J. M., and Borst, D. (1989). Spider crab yolk protein: Molecular characterization and the effects of methyl farnesoate (MF) on its hemolymph levels. *Am. Zool.* 29, 49A (abstract).

Waddy, S. L. (1988). Farming the homarid lobsters: State of the art. *World Aquacult.* 19(1), 63–71.

Waddy, S. L., and Aiken, D. E. (1985a). Immunofluorescent localization of American lobster egg yolk protein in the alimentary tract of the nemertean *Pseudocarcinonemertes homari*. *Can. J. Fish. Aquat. Sci.* 1272, 1–14.

Waddy, S. L., and Aiken, D. E. (1985b). Fertilization and egg retention in artificially inseminated female American lobsters, *Homarus americanus*. *Can. J. Fish. Aquat. Sci.* 42, 1954–1956.

Waddy, S. L., and Aiken, D. E. (1986). Multiple fertilization and consecutive spawning in large American lobsters, *Homarus americanus*. *Can. J. Fish. Aquat. Sci.* 43, 2291–2294.

Waddy, S. L., and Aiken, D. E. (1989). Seasonal molting patterns of juvenile American lobsters cultured at constant 20°C. *J. World Aquacult. Soc.* 20, 78A (abstract).

Waddy, S. L., and Aiken, D. E. (1990a). Mating and insemination in the American lobster, *Homarus americanus*. *In* "Crustacean Sexual Biology" (R. T. Bauer and J. W. Martin, eds.), pp. 126–144. Columbia Univ. Press, New York.

Waddy, S. L., and Aiken, D. E. (1990b). Intermolt insemination, an alternative mating strategy for the American lobster (*Homarus americanus*). *Can. J. Fish. Aquat. Sci.* 47, 2402–2406.

Waddy, S. L., and Aiken, D. E. (1991). Egg production in the American lobster. *In* "Crustacean Issues" (F. R. Schram, ed.), Vol. 7, "Crustacean Egg Production" (A. Wenner and A. Kuris, eds.), pp. 267–299. Balkema, Rotterdam, The Netherlands.

Waddy, S. L., and Aiken, D. E. (1992). Environmental intervention in the reproductive process of the American lobster, *Homarus americanus*. *Invertebr. Reprod. Dev.* 22, 245–251.

Waddy, S. L., and Aiken, D. E. (1995a). Temperature regulation of reproduction in female American lobsters, *Homarus americanus*. *ICES Mar. Sci. Symp.* 199 (in press).

Waddy, S. L., and Aiken, D. E. (1995b). Culture of the American lobster, *Homarus americanus*. *In* "Cold-water Aquaculture in Atlantic Canada" (A. D. Boghen, ed.), pp. 145–188. The Canadian Institute for Research on Regional Development, Moncton.

Waddy, S. L., Aiken, D. E., and Eagles, M. D. (1988). Growth of cultured American lobsters and effect of culling on growth rates. *Bull. Aquacult. Assoc. Can.* 88(4), 66–68.

Wahle, R. A., and Steneck, R. S. (1991). Recruitment habitats and nursery grounds of the American lobster *Homarus americanus*: A demographic bottleneck. *Mar. Ecol. Prog. Ser.* 69, 231–243.

Walgraeve, H. R. M. A., and Verhaert, P. D. E. M. (1988). Presence and function of ecdysteroids in invertebrates. *In* "ISI Atlas of Science: Animal and Plant Sciences" Vol. 1, pp. 164–172. Institute of Scientific Information, Philadelphia.

Webster, S. G. (1991). Amino acid sequence of putative moult-inhibiting hormone from the crab *Carcinus maenas*. *Proc. R. Soc. London* B 244, 247–252.

Webster, S. G., and Keller, R. (1986). Purification, characterization and amino acid composition of the putative moult-inhibiting hormone (MIH) of *Carcinus maenas* (Crustacea, Decapoda). *J. Comp. Physiol. B* 156B, 617–624.

Weidemann, W., Gromoll, J., and Keller, R. (1989). Cloning and sequence analysis of cDNA for precursor of a crustacean hyperglycemic hormone. *FEBS Lett.* 257, 31–34.

Wickins, J. F., Beard, T. W., and Child, A. R. (1995). Maximizing lobster (*Homarus gammarus*) egg and larva viability. *Aquacult. Fish. Manage.* (in press).

Wilder, D. G. (1953). The growth rate of the American lobster (*Homarus americanus*). *J. Fish. Res. Board Can.* 10, 371–412.

Yano, I., and Wyban, J. A. (1992). Induced ovarian maturation of *Penaeus vannamei* by injection of lobster brain extract. *Bull. Natl. Inst. Aquacult.* 21, 1–7.

Yasuda, A., Yasuda, Y., Fujita, T., and Naya, Y. (1994). Characterization of crustacean hyperglycemic hormone from the crayfish (*Procambarus clarkii*): Multiplicity of molecular forms by stereoinversion and diversion functions. *Gen. Comp. Endocrinol.* 95, 387–398.

Yudin, A. I., Diener, R. A., Clark, W. H., Jr., and Chang, E. S. (1980). Mandibular gland of the blue crab, *Callinectes sapidus*. *Biol. Bull. (Woods Hole, Mass.)* 159, 760–762.

C H A P T E R

11

Neurobiology and Neuroendocrinology

BARBARA BELTZ
Department of Biological Sciences
Wellesley College
Wellesley, Massachusetts 02181

...Hail to thee most noble of crustaceans!
Worthy of Homaric ode,
Behemoth of the briny deep,
Traveler, of a wet, sandy road....

...Nerve cells here, the muscles there,
The hormones all around;
Big cells thou art, and pluckable,
Over and over can't be found.

Transmitters peptides and the like
For studies fundamental.
O joyous day! Callooh! Callay!
The data're transcendental.

Excerpt from "The Rime of the Ancient Scientist"
—E. A. Kravitz (1991)

I. Introduction

Very few nervous systems offer the many advantages of that of the American lobster for studying the cellular and integrative properties of neurons, and even fewer nervous systems inspire poetic tributes from those who study their function! *Homarus americanus* offers a rare blend of virtues for neurobiologists: a relatively simple and accessible nervous system with a segmental organization and large identifiable neurons, a fairly well-understood peripheral innervation pattern, a complex set of motor functions and social behaviors, and a host of neurotransmitters and hormones that regulate defined behaviors. Several lines of research have taken advantage of these benefits, ranging from the neural control of behavior to the development and molecular biology of the nervous system.

There is remarkable uniformity in how nerve cells function throughout the animal kingdom. Most neurons that have been examined in the lobster, like those of mammals, lower vertebrates, and other invertebrates, have resting potentials in the range of −60 to −80 mV, created largely by differences in the potassium ion concentration on the inside and outside of the cell. In lobsters, conventional ions (e.g., sodium, potassium, and chloride) are also responsible for active and passive conduction along the membrane, as well as for synaptic activity.

Likewise, the categories of neurotransmitters utilized by vertebrate and invertebrate nervous systems have been largely conserved. Acetylcholine, γ-aminobutyric acid (GABA), glutamate, serotonin, and dopamine are among the many transmitters that function in the lobster, just as they do in vertebrate systems. Although certain peptide transmitters may be unique to the lobster, or to crustaceans as a group, many of these peptides are closely related to peptide transmitters that are found in the vertebrates.

While there is surprising uniformity in ionic mechanisms and chemistry across species, there is great diversity in how nervous systems are organized. The degree of centralization and cephalization in nervous systems varies a great deal. In the lobster, there is moderate cephalization. The anterior ganglia (supra- and subesophageal ganglia) that receive sensory information and control feeding structures are enlarged. In contrast to the diffuse, ladderlike nervous system of flatworms, a well-defined ventral nerve cord with a prominent "brain" at the anterior end occurs in all of the arthropods. The nervous system exhibits fairly extensive fusion of ganglia relative to invertebrates such as the echinoderms and the annelids. Therefore, the central nervous system of the lobster is less uniformly segmented than the annelid system, but more highly segmented than related crustaceans such as the brachyurans. Each segmental ganglion in the lobster is primarily concerned with coordinating actions in that segment, so there is a high degree of local control of function.

The lobster nervous system has been utilized extensively for studies of motor circuits, sensory systems, and chemical coordination because of its relatively simple organization compared to the vertebrate nervous system and because it has retained many of the same ionic and chemical features that are important in the vertebrates. Therefore, neural mechanisms elucidated in the lobster may very well be relevant to higher organisms as well. This "simple systems" approach has been a powerful stimulus for studies of invertebrate neurons; the many neural investigations of the lobster serve as elegant examples of the strength of this approach.

II. Anatomical and Histological Organization of the Nervous System

A. *The Central Nervous System*

The organization of the nervous system of *Homarus americanus* is representative of decapods in general. Fifteen midline ganglia span the length of the animal (Fig. 1). Paired ganglia are joined to each other across the midline via commissures and to adjacent pairs of ganglia by longitudinal connectives to form the ventral nerve cord. Nerves projecting to peripheral targets originate from each ganglion. Cell bodies form a rind around each ganglion, with most of the somata concentrated on the ventral surface (Figs. 1 and 2). In general, interneurons, which serve as communication links between other neurons, are restricted in their entirety to the central nervous system. Motoneurons, which control muscle contractility, have their somata and dendritic arbors in the central nervous system, while their axons project to the periphery to innervate target muscles. The sensory neurons, with their associated accessory structures, are generally located at the periphery, and their axons travel to the central nervous system to contribute to the ganglionic neuropil, which is the region of dense synapses found in each ganglion. In addition to nerve and glial cells, which make up the primary tissue of the central nervous system, there are also connective and vascular tissues. The most prevalent form of connective tissue, the neural sheath, or lamella, forms an external covering around the ganglia, connectives, and nerves. The sheath surrounding some of the ganglia contains varicose neural endings and is also thought to function as a neurohormonal organ (Kobierski *et al.*, 1987).

The brain, or supraesophageal ganglion, is the anteriormost region of the central nervous system and lies between the eyes, just dorsal to the esophagus. This ganglion represents the evolutionary fusion of several segments in the head and is characterized by clusters of cell bodies and associated neuropils distributed among the protocerebrum, deutocerebrum, and tritocerebrum (Figs. 1 and 3). The identity and location of cell clusters, neuropils, and tracts follow a basic decapod plan (Sandeman *et al.*, 1992, 1993). The cell clusters are composed primarily of motoneurons and interneurons, although a few sensory neurons have centrally located somata. The brain is generally concerned with primary processing of visual, chemosensory, and mechanosensory inputs, as well as higher-order processing and integration of these modalities. The different neuropils of the cerebral ganglion may be broadly classified into those that are structured and unstructured. In structured neuropils, there is a repeated or geometric order (as in the deutocerebral olfactory and accessory neuropils; see Fig. 3B) that is lacking in unstructured neuropils (e.g., the hemiellipsoid body and central body of the protocerebrum) (Sandeman *et al.*, 1993). Unstructured neuropils are composed of a tangle of axon profiles mainly characterized by synaptic contacts, with presynaptic specializations (e.g., dense bars or ribbons) and synaptic vesicles (Govind, 1992).

The brain, which is the only major ganglion in the dorsal half of the body, is joined to the ventral nerve cord via connectives that emerge from the posterior margin of the brain (Fig. 1). Along these connectives next to the esophagus reside the circumesophageal (or commissural) ganglia, which are relatively small, paired ganglia whose neurons are known to be involved in neurohormonal functions and in the control of the stomatogastric ganglion, located on the stomach. The nerve cord lies ventral to the circumesophageal ganglia and contains the subesophageal ganglion, five thoracic ganglia, and six abdominal ganglia (Fig. 1). The subesophageal ganglion lies just below the mouth and contains neurons innervating

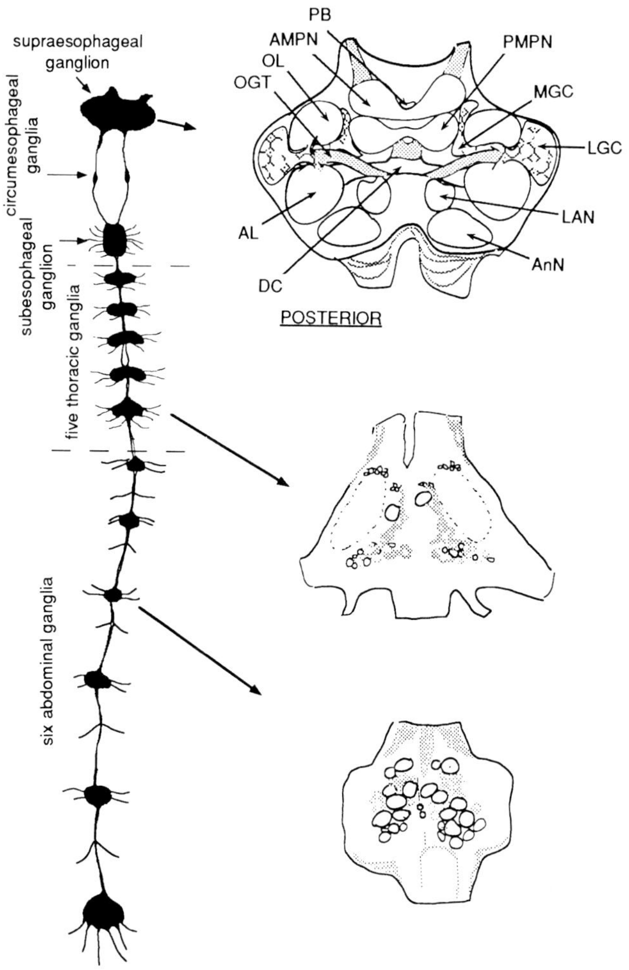

FIGURE 1 (Left) Fifteen midline ganglia compose the central nervous system: the supraesophageal ganglion (the brain), two paired circumesophageal ganglia, the subesophageal ganglion, five thoracic ganglia, and six abdominal ganglia. The brain receives primary visual information, olfactory information via the first pair of antennae (antennulae) and mechanosensory information via the second pair of antennae. Circumesophageal and subesophageal ganglia are involved in neural control of the gills and stomach, among many functions. The thoracic and abdominal ganglia are devoted largely to local control of functions within, or related to, the segment where each ganglion resides. (Right) The brain (compressed in the dorsal ventral plane), ventral surface of the fifth thoracic ganglion, and ventral surface of the third abdominal ganglion are drawn schematically, illustrating the positions of some neuropil regions and cell body clusters. AL, Accessory lobe (or neuropil); AMPN, anterior medial protocerebral neuropil; AnN, antenna II neuropil; DC, deutocerebral commissure; LAN, lateral antenna I neuropil; LGC, lateral globuli cells (cell cluster 10); MGC, medial globuli cells (cell clusters 9 and 11); OGT, olfactory globular tract; OL, olfactory lobe (or neuropil); PB, protocerebral bridge; PMPN, posterior medial protocerebral neuropil. [Terminology is according to Sandeman *et al.* (1992).]

the mouthparts and gills, as well as neurons controlling other local functions. The thoracic and abdominal ganglia generally innervate the body wall, musculature, gills, and appendages relevant to the segment within which they reside.

B. Sensory Systems

Sensory neurons in *Homarus americanus*, as in other decapods, are elaborated into a variety of receptors sensitive to the internal and external environments. In general, their somata are located at the periphery, their dendrites are associated with an accessory structure, and their axons travel to the central nervous system to contribute to the ganglionic neuropil. A few have centrally located somata. Among the sensory systems are photoreceptors, chemoreceptors, mechanoreceptors (e.g., myochordotonal organs, muscle receptor organs, and statocysts), and bimodal receptors (e.g., funnel canal organs, setae, and cuticular cylinders) that are sensitive to both mechanical and chemical stimuli (Derby, 1982; microanatomy is described by Govind, 1992). Sensory systems are discussed at length by Atema and Voigt (Chapter 13).

1. Photoreceptors

Lobsters possess paired, lateral compound eyes, as do most adult arthropods, and are highly visual organisms. Relatively few data specific to the eye of *Homarus americanus* are available. Most of what we know about the function of compound eyes has been obtained from studies of insects, and it is generally assumed that the lobster eye follows the basic organizational and functional rules of other compound eyes (Ache and Macmillan, 1980). Evidence also exists for central neural photosensitivity in some crustaceans, but these caudal photoreceptors have not been found in *H. americanus* (Wilkens and Larimer, 1976). Earlier reviews provide a basic introduction to the crustacean compound eye (Waterman, 1961; Bullock and Horridge, 1965).

2. Chemoreceptors

Innervated sensilla on the external cuticle provide the lobster with information about its external chemical environment. Most of these take the form of hairlike setae localized primarily on the first antennae (antennules), the pereiopod dactyls, and the mouthparts (Ache, 1982; Derby, 1982). Aesthetasc sensilla on the lateral filament of the bifid antennule are the primary sites of distance chemoreception, while leg receptors are utilized at close range to locate and recognize food (Derby and Atema, 1982a,b).

3. Mechanoreceptors

Three major groups of mechanoreceptors are distinguishable based on structural considerations (Bush and Laverack, 1982). (1) Internal mechanoreceptors

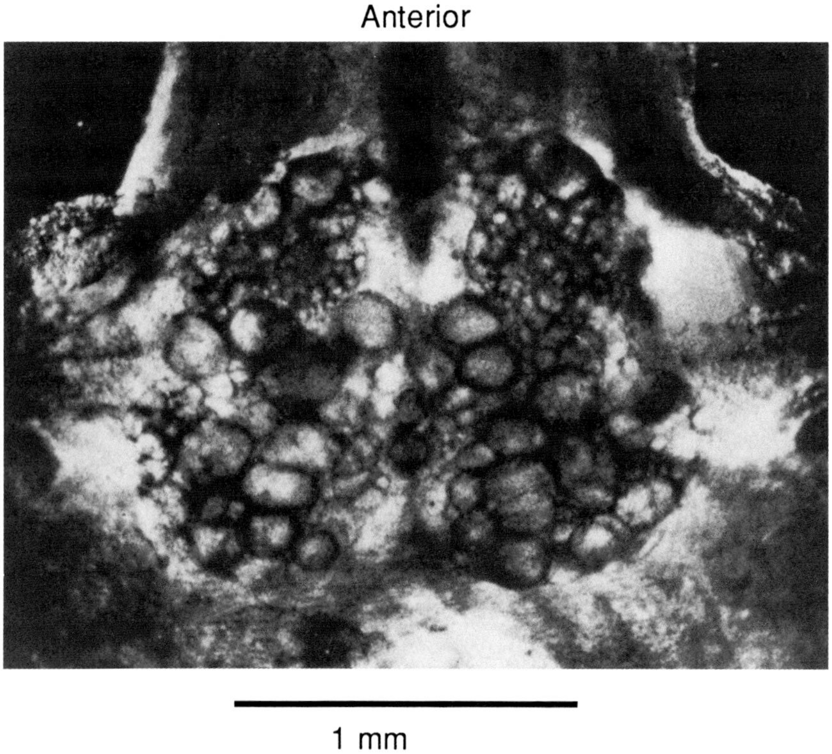

FIGURE 2 The ventral surface of the third abdominal ganglion, showing the appearance of the living cell bodies under the dissecting microscope. (Reprinted from Otsuka *et al.*, 1967, with permission.)

lie wholly within the body and are primarily proprioceptors that monitor movement, posture, and equilibrium. (2) Cuticular receptors insert on the cuticle and monitor stress. (3) Supracuticular receptors are structures in which the dendrites pass through the cuticle and attach to external projections or supracuticular structures.

Statocysts (Cohen, 1955, 1960) and muscle receptor organs (MROs) (Wiersma *et al.*, 1953; Kuffler, 1958) are among the most well-studied mechanoreceptors in *Homarus americanus*. Statocyst receptors are located in the hollow basal segment of each antennule and provide information about spatial equilibrium. MROs are internal mechanoreceptors that consist of a sensory cell that is functionally linked to an accessory structure in the form of a specialized muscle (Fig. 4). This type of receptor is found in the thorax and the dorsal abdomen, at the thoracic–coxal joints, and in the mandible. The ultrastructure of muscle receptor organs of *H. americanus* have been described, as well as the ultrastructural characteristics of synapses correlated with their function (Nadol and de Lorenzo, 1968, 1969). MROs are discussed further by Govind (Chapter 12) and by Atema and Voigt (Chapter 13); the chemical modulation of MROs is addressed in Section III,F.

C. The Neuromuscular Junction

Crustacean muscles typically receive a dual innervation from two types of motoneurons (Wiersma and Ripley, 1952; Govind, Chapter 12). The excitatory neuron, using glutamate as its transmitter (Kravitz *et al.*, 1970), depolarizes the muscle membrane and promotes contraction. The inhibitory neuron, using GABA (Otsuka *et al.*, 1966), tends to hyperpolarize the muscle and prevents contraction. The final tension generated by the muscle therefore reflects the combined inputs of both the excitor and inhibitor signals.

One of the most intensively studied neuromuscular systems in the lobster is the dactyl opener muscle of the walking leg. In addition to the conventional inhibitory and excitatory neural input, an additional layer of complexity in the function of this synapse is added by circulating neuromodulators that act presynaptically and/or postsynaptically to modify the release of transmitter or resting tension of the muscle. The amines serotonin, octopamine, and dopamine (Kravitz *et al.*, 1980) and three peptides, proctolin (Kravitz *et al.*, 1980; Schwarz *et al.*, 1980), a lobster FMRFamide-like compound with the amino acid sequence TNRNFLRFamide (also called lobster pep-

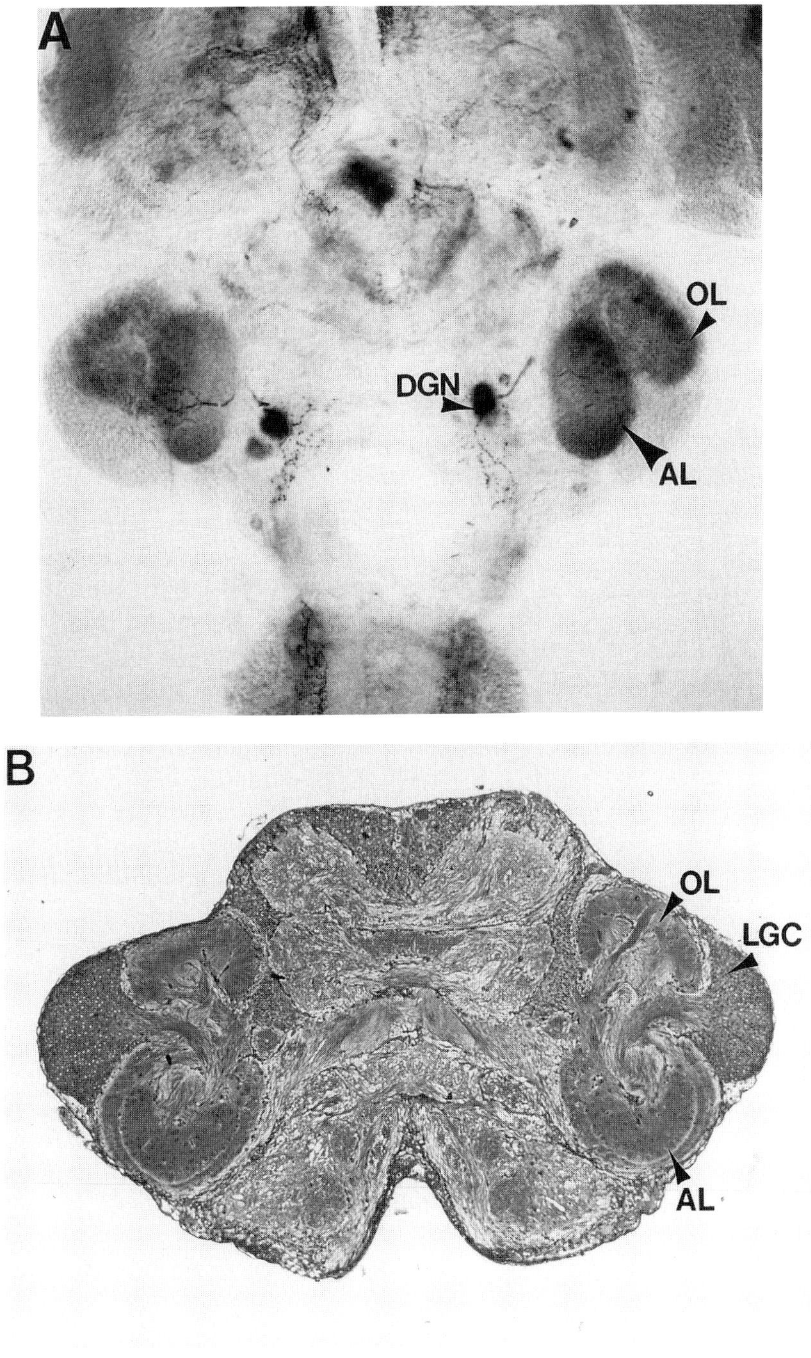

FIGURE 3 (A) The brain of a lobster embryo at the time of hatching, stained for serotonin using immunocytochemical methods. Among the immunoreactive areas are three prominent bilateral deutocerebral structures: the olfactory lobes (OL), accessory lobes (AL) and deutocerebral giant neurons (DGN). Each DGN innervates the ipsilateral OL and AL and provides the primary serotoninergic input to these regions. Serotonin is believed to function as a modulator in these regions, as well as serving as a morphogen during embryonic life. (See Beltz *et al.*, 1993.) (B) A 4-μm horizontal section through a juvenile lobster brain illustrates the relative positions of several neuropil and cell body regions. The glomerular organization of the OL and AL neuropils is also evident. The lateral globuli cells (LGC) are the largest group of neurons projecting from the OLs and ALs.

tide F_1 in Goy *et al.*, 1987b; Trimmer *et al.*, 1987; Worden *et al.*, 1995), and peptide G_1 (Goy *et al.*, 1987a), appear to be involved as neuromodulators at the dactyl neuromuscular junction (Fig. 5). Muscular control in the dactyl, as well as at other sites such as cardiac and visceral muscles, therefore is likely to reflect the finely tuned orchestration of many chemical transmitters and modulators (Worden *et al.*, 1995; Kravitz, 1988). Together, these substances are capable of changing the ways that lobster muscles are used for periods ranging from minutes up to 1 day, likely in response to specific physiological needs.

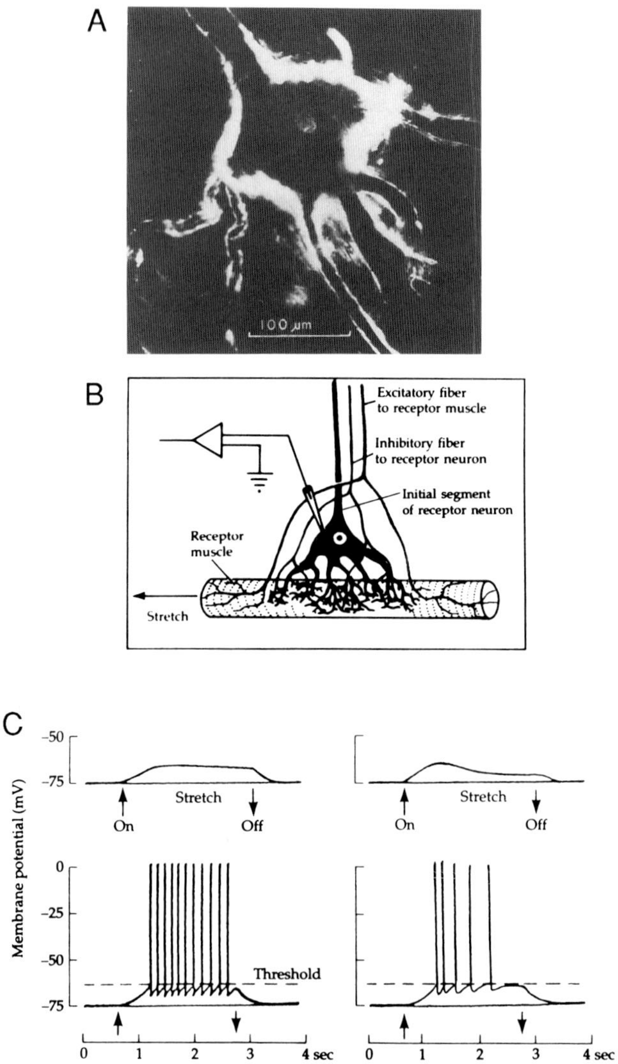

FIGURE 4 (A) Living stretch receptor neuron viewed with dark-field illumination. Distal portions of six dendrites insert into the receptor muscle, which is not visible. (B) Relationship between the stretch receptor neuron and muscle, indicating the method of intracellular recording. Excitatory fiber to the muscle produces contraction; inhibitory fiber innervates the neuron. Two additional inhibitory fibers are not shown. (C) Responses of stretch receptor neurons to increases in muscle length, recorded intracellularly as indicated in (B). In a slowly adapting receptor (left), a weak stretch for about 2 seconds produces a subthreshold receptor potential that persists throughout the stretch (upper record). With a stronger stretch, a larger receptor potential sets up a series of action potentials (lower record). In a rapidly adapting receptor (right), the receptor potential is not maintained (upper record), and during the large stretch the action potential frequency declines (lower record). (Reprinted from Nicholls *et al.*, 1992, with permission; originally adapted from Eyzaguirre and Kuffler, 1955, *The Journal of General Physiology*, 39, 87–119, used with copyright permission of the Rockefeller University Press.)

Ion channel activity in lobster skeletal muscle membranes has been described (Worden *et al.*, 1993, 1995). A number of channels are active in the absence of added calcium, metabolites, or cytoplasmic factors, and can be distinguished from each other on the basis of single channel conductance, degree of selectivity for potassium and sodium, and gating properties. Further understanding of muscle membrane channels is necessary in order to establish the cellular mechanisms by which the variety of neurohormones regulates muscle contractility.

D. Neurohormonal Organs

The actions of central neurons and peripherally located sensory organs are extended by an elaborate network of neurohormonal organs that release chemical substances directly into the hemolymph. Once circulating, every tissue in the body is a potential target for the action of these compounds. Serotonin, octopamine, dopamine, and a variety of peptides are

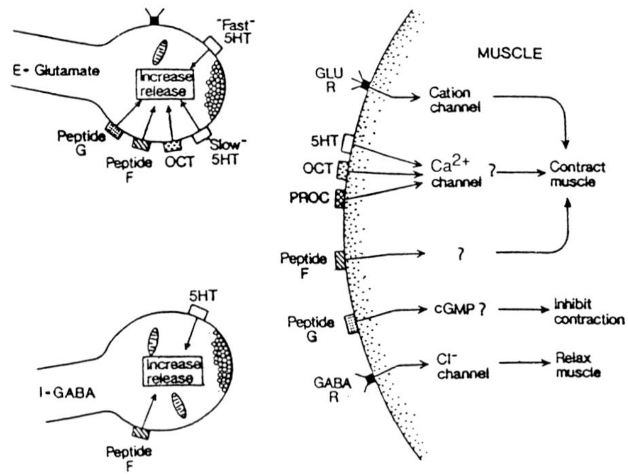

FIGURE 5 Schematic drawing of the sites of actions of modulators. A variety of amine and peptide substances exert short- and long-term actions on both pre- and postsynaptic elements of the neuromuscular apparatus. (Left) Excitatory (glutamate) and inhibitory (GABA) endings are illustrated, as well as the known effects of these substances. (Right) A postsynaptic muscle target is represented. GABA R, GABA receptor; GLU R, glutamate receptor; OCT, octopamine; PROC, proctolin; 5HT, serotonin. (Courtesy of E. A. Kravitz.)

known to be released from several neurohormonal sites. The effects of individual neurohormones have been explored in the lobster dactyl opener muscle preparation (Section II,C), as well as in the postural neuromuscular system (Sections III,C and D).

Homarus americanus contains the full complement of typical crustacean neurosecretory organs (Beltz, 1988); however, only the sinus gland–X-organ complex, pericardial organs, thoracic second roots, and neural sheath release zones have been investigated in any detail. The Y-organ and the postcommissural organ, although present, have not been the subject of studies in *H. americanus*.

Crustacean neurohormonal organs generally contain no neuronal cell bodies; they are instead composed of the swollen axon endings (varicosities) of neurons located in various regions, including neurons in the central ganglia. These organs are usually located within a hemolymph sinus; therefore, the circulation is able to carry the released contents of the varicosities to a variety of distant target organs. Varicosities range in size from 2 to 30 μm in diameter and are generally 1–5 μm from the blood space, separated from it by a limiting membrane and connective tissue fibers.

Of the crustacean neurohormonal regions, the sinus gland located in the eyestalk is the most fully characterized, although most studies have focused on species other than *Homarus americanus*. The varicose terminals that constitute the sinus gland originate primarily from cell bodies located in the X-organ of the eyestalk, although a minor component may come from other cell groups in the eyestalk, brain, and thoracic ganglia. The X-organ–sinus gland system, which is the major neuroendocrine regulatory center in crustaceans, is structurally and functionally analogous to the vertebrate hypothalamic–neurohypophyseal system, as well as to the corpus cardiacum of insects (Cooke and Sullivan, 1982; Gabe, 1966).

Ultrastructurally, the neurohormonal cells of vertebrates and invertebrates are remarkably similar. Like other active secretory cells, the cell bodies of neurohormonal cells generally contain well-defined Golgi complexes and abundant endoplasmic reticulum. The major distinguishing feature of peptidergic neurohormonal cells is the presence of large (100- to 300-nm-diameter) electron-dense granules having a homogeneous matrix bounded by a smooth enclosing membrane (Beltz, 1988). In contrast, terminals known to contain amines are generally filled with smaller (40- to 100-nm) clear vesicles (Livingstone *et al.*, 1981).

The mechanisms of excitation–secretion coupling in neurosecretory cells are analogous to those of conventional neurons that release transmitter in close proximity to their postsynaptic target. Neurohormonal cells can receive and integrate signals from other neurons and propagate action potentials. The electrical activity in neurohormonal cells regulates hormone secretion. Aminergic and peptidergic compounds known to be released from neurohormonal regions in the lobster are discussed further in Section III,F.

III. Chemistry of the Nervous System

A. *Excitation and Inhibition: Glutamate and GABA*

Historically, the first neuroactive substances discovered in lobster neurons were GABA and glutamate, which were isolated from identified motoneurons by chemical analysis (Otsuka *et al.*, 1967). There is little doubt that GABA is the inhibitory transmitter at the lobster neuromuscular junction; it is 100 times more concentrated in inhibitory than in excitatory axons. Glutamate is thought to act as an excitatory transmitter at the neuromuscular junction.

B. *Sensory Transmission: Acetylcholine*

Acetylcholine is the primary transmitter in sensory neurons in the lobster (Barker *et al.*, 1972). Experimentally, several different peripheral sensory structures incorporate radioactive choline and convert it to acetylcholine. Isolated sensory axons contain at least 500 times as much choline acetyltransferase per centimeter of axon as in efferent excitatory and inhibitory fibers. Furthermore, curare and atropine, which block transmission at cholinergic synapses, successfully block an identified sensory synapse in the abdominal ganglion, also indicating the cholinergic nature of sensory transmission. Data suggest, however, that acetylcholine is not working alone in some sensory neurons. Acetylcholine can act as a cotransmitter with serotonin or with the peptide proctolin in some stretch-sensitive neurons (see Section III,E).

C. *Amines*

Serotonin, octopamine, dopamine, and histamine are all present in the nervous system of *Homarus americanus* and are known to exert a variety of physiological and behavioral effects (Battelle and Kravitz, 1978; Kravitz *et al.*, 1980, 1985; Livingstone *et al.*, 1980; Beltz and Kravitz, 1983; Claiborne and Selverston, 1984; Mulloney and Hall, 1991). As in vertebrate systems, aminergic neurons are relatively few in number, but the projections of individual neurons are widespread throughout the nervous system (serotonin, Beltz and Kravitz, 1983; octopamine, Schneider *et al.*, 1993a; dopamine, Cournil *et al.*, 1994; histamine, Mulloney and Hall, 1991). The far-reaching projections of these cells make them perfect candidates for a modulatory role in the nervous system, setting the "gain" of other neural systems.

In the dactyl opener neuromuscular preparation, serotonin acts on excitatory and inhibitory nerve endings to facilitate transmitter release and on muscle fibers to produce a contracture and induce the appearance of calcium action potentials (Glusman and Kravitz, 1982). In total, just over 100 neurons contain serotonin-like immunoreactivity in the central nervous system, and the functions of a few of these neurons have been examined (Beltz and Kravitz, 1983; Ma *et al.*, 1992). Studies of embryonic and larval development show that the first appearance of serotonin is very early, by 10% embryonic development (E10%); the full complement of adult neurons is present and immunoreactive by midembryonic life (Beltz *et al.*, 1990; see Talbot and Helluy, Chapter 9).

Octopamine, like serotonin, acts directly on muscle fibers to produce a contracture and induce the appearance of calcium action potentials (Battelle and Kravitz, 1978; Kravitz *et al.*, 1980), as well as increasing the strength of nerve-evoked contractions. Exoskeletal muscle, hemolymph, and heart muscle respond to octopamine with increases in cAMP. The addition of octopamine to lobster hemolymph increases the rate and changes the nature of the clotting reaction, effects that fit logically with the involvement of octopamine in defensive postural displays. Octopamine, like serotonin, increases the rate and amplitude of the heart beat.

Octopamine is released from varicosities in the thoracic second roots, and the cell bodies of origin of these fibers are located in the central nervous system (Livingstone *et al.*, 1981; Schneider *et al.*, 1993a). The peripheral neuronal somata in the thoracic second roots do not contain octopamine, as was previously thought (Kravitz *et al.*, 1976). The entire system of octopamine neurons in the central ganglia has been mapped, with octopamine localized to 86 neurons in the brain and ventral nerve cord (Schneider *et al.*, 1993a). All of the octopamine neurons are paired and located along the ganglionic midline. Of the 86 neurons, 28 have been identified as neurosecretory and 26 as intersegmental interneurons. Developmental immunocytochemical mapping has also been completed (Schneider *et al.*, 1993b) and, in contrast to the early appearance of serotoninergic neurons, the adult complement of octopaminergic neurons is not immunoreactive until early larval life. These differences suggest that regulation of the onset of octopamine and serotonin syntheses in each neuronal type is distinct, although the factors involved in such regulation are not known.

Dopamine has not been studied extensively, as serotonin and octopamine have been. Dopamine also acts on muscle fibers directly; in contrast to the actions of serotonin and octopamine, however, dopamine relaxes muscle baseline tension (Battelle and Kravitz, 1978; Kravitz *et al.*, 1980). Immunocytochemical mapping studies for dopamine have been completed for *Homarus gammarus* (Cournil *et al.*, 1994), and these studies are likely to be relevant to dopamine localization in *H. americanus*. To date, neuronal labeling patterns for several substances (e.g., serotonin, octopamine, and GABA) have proven to be identical between these two species, and their fundamental neural organization is believed to be the same. In *H. gammarus*, approximately 100 neuronal somata stain for the catecholamine or its synthetic enzyme in the brain and ventral nerve cord; every segmental ganglion contains at least one pair of labeled neurons; and several of the labeled neurons are anatomically identifiable. In these studies, no cell bodies are labeled in the stomatogastric ganglion, but fibers and terminals in the neuropil are stained.

Histamine has been proposed as a transmitter in the nervous system of *Homarus americanus* based on a combination of biochemical, pharmacological, and immunohistochemical evidence (Orona *et al.*, 1990; Mulloney and Hall, 1991). As with the other amines, the pattern of neurons labeling immunocytochemically for histamine is remarkably sparse. No labeling occurs in peripheral nerves (with the exception of the nerves of the stomatogastric system), suggesting that the histamine neurons function as interneurons.

D. Peptides

The pentapeptide proctolin (Schwarz *et al.*, 1984; Siwicki and Bishop, 1986) and several distinct FLRFamide molecules (Trimmer *et al.*, 1987; Kobierski *et al.*, 1987) have been localized, isolated, and characterized in *Homarus americanus*. In addition, immunoreactivities for small cardioactive peptide B (SCP_b) (Arbiser and Beltz, 1991; Section III,E) and red pigment-concentrating hormone (B. S. Beltz, unpublished results) have also been examined in the lobster.

1. Proctolin

Proctolin-like immunoreactivity is found in nearly every portion of the nervous system; over 1000 immunoreactive neurons have been found in the brain and ventral nerve cord, and immunoreactive fibers are located in the connectives of the ventral nerve cord and in many of the nerve roots that emerge from the cord (Siwicki and Bishop, 1986). The greatest amounts of proctolin are found in the pericardial organs, where immunocytochemistry reveals a dense collection of labeled varicosities (Schwarz *et al.*, 1984; Siwicki *et al.*, 1985). Proctolin-like material is released from the pericardial organs when they are depolarized in the presence of calcium, suggesting that proctolin functions, at least in part, as a neurohormone (Schwarz *et al.*, 1980, 1984).

2. FMRFamide-like Peptides

FMRFamide-like peptides have been isolated and characterized from the neural tissues of *Homarus americanus* (Kobierski *et al.*, 1987; Trimmer *et al.*, 1987). FMRFamide-like immunoreactivity is found in low levels throughout the ventral nerve cord and the brain, and in much higher amounts in the pericardial organs. Approximately 350 neuronal somata label with anti-FMRFamide antibodies, as well as distinct neuropil regions, neuronal fiber tracts, and varicose endings. The connective tissue sheath surrounding the ventral nerve cord, thoracic second roots, and pericardial organs contains dense plexuses of varicose fibers. Ultrastructural examination shows varicosities filled with large, dense granules and small, clear vesicles that are located a few micrometers from the surface of the tissues, with no obvious target in their immediate vicinity. These findings, as well as the observation of low levels of FMRFamide immunoreactivity in the hemolymph, suggest a neurohormonal function for the FMRFamide-like molecules. In addition, the localization of immunoreactivity to particular somata, fiber tracts, and neuropil regions suggests possible functional roles for the FMRFamide-like peptides in (1) integration of visual and olfactory information, (2) function of the anterior and posterior gut, and (3) control of exoskeletal and cardiac muscles (Kobierski *et al.*, 1987).

Gel filtration and high-pressure liquid chromatographies have established that little or no authentic FMRFamide is present in *Homarus americanus*, but rather several closely related FMRFamide-*like* peptides. Of the major immunoreactive components, two peptides have been purified and identified as the octapeptides SDRNFLRFamide and TNRNFLRFamide (Trimmer *et al.*, 1987). The FMRFamide immunoreactivity in neural structures is, in fact, preadsorbed with a synthetic TNRNFLRFamide, which is the most abundant FMRFamide-like molecule in *H. americanus* (Arbiser and Beltz, 1991).

The TNRNFLRFamide molecule has also been tested in *Homarus americanus* for physiological actions on three different muscle preparations: exoskeletal, cardiac, and visceral (Worden *et al.*, 1995). This peptide causes long lasting enhancement of contractility in each target tissue, acts at nanomolar concentrations, and is three to four orders of magnitude more potent than authentic FMRFamide. These physiological findings support the suggestion that TNRNFLRFamide is a neurohormone with widespread myogenic actions throughout peripheral tissues. Several unusual features of the responses to this peptide have been noted at the neuromuscular junction, for example: (1) the preparation shows pronounced hysteresis when subjected to consecutive peptide treatments; and (2) the peptide potentiates the effects of other stimuli that induce contraction in this preparation (serotonin or elevated potassium). The molecular mechanisms by which the peptide acts are not yet known, but do not seem to involve changes in cAMP, cGMP, or arachidonate metabolism (Goy *et al.*, 1987a; Worden *et al.*, 1995).

E. Transmitter Coexistence

As in many other invertebrate as well as vertebrate systems, amines and peptides coexist with other neuroactive compounds in lobster neurons (Siwicki *et al.*, 1987; Katz *et al.*, 1989). For instance, proctolin coexists with serotonin, dopamine, and acetylcholine in identified neurons (Siwicki *et al.*, 1987). It is interesting to note that homologous neurons are identifiable in other closely related species and that the transmitter molecules are often distinct in those animals. For instance, the large, identifiable, fifth thoracic and first abdominal serotoninergic neurons (Beltz and Kravitz, 1983) are believed to be involved in postural regulation (Ma *et al.*, 1992). In *Homarus americanus*, proctolin coexists with serotonin in these neurons, but in the crayfish (*Procambarus clarkii*), only serotonin is present (Siwicki and Bishop, 1986). Therefore, although there are many similarities between systems of identified neurons in related species, there may also be distinct features of neuronal chemistry. The significance of these differences is not understood.

Acetylcholine coexists with other compounds in identified sensory cells. In the gastropyloric (GPR) cells, stretch-sensitive muscle receptors in the stomatogastric system, serotonin and acetylcholine act as cotransmitters (Beltz *et al.*, 1984; Katz *et al.*, 1989; Katz and Harris-Warrick, 1989, 1991). Although most of the studies on these cells are from other crustacean species, the GPR cells have also been identified in *Homarus americanus* (Katz, 1991).

The pentapeptide proctolin also colocalizes with acetylcholine in identified sensory neurons. Several proctolinergic/cholinergic fibers terminate in sensory dendrites of the oval organ mechanoreceptor in the scaphognathite (Pasztor, 1979; Pasztor and Bush, 1982; Siwicki *et al.*, 1987). The oval organ is a relatively simple stretch receptor in the second maxilla at the base of the scaphognathite ("gill bailer"). It lacks a receptor muscle and has no centrifugal control (Pasztor, 1979; Pasztor and Bush, 1982; Pasztor *et al.*, 1988). It consists of three large sensory neurons whose numerous peripheral dendrites are supported by a cone-shaped array of connective tissue strands (Fig. 6A). These span the second maxilla at the base of the scaphognathite and are stretched rhythmically as the beating of the scaphognathite drives water through the gill system of the lobster (Fig. 6B). The neurons have thick afferent fibers (20- to 40-μm diameter) and central cell bodies situated in the subesophageal ganglion, in a cluster of ventilatory motoneurons (Pasztor and Bush, 1987). Two of the three sensory neurons contain the peptide proctolin in addition to acetylcholine (Siwicki *et al.*, 1987; Pasztor *et al.*, 1988). Repetitive stretch stimulation of the oval organ results in the release of proctolin from the oval organ *in vitro* (Fig. 6). Proctolin is known to excite the primary sensory afferents (Pasztor *et al.*, 1985; Pasztor and Bush, 1987) and it has therefore been proposed that endogenous proctolin may self-modulate the sensory transduction mechanism of the oval organ sensory terminals (Pasztor *et al.*, 1988; Pasztor and Bush, 1989).

Several studies have reported the possible colocalization of SCP-like and FMRFamide-like peptides in invertebrate neurons, including the crab *Cancer irroratus* (Calloway *et al.*, 1987). In *Homarus americanus*, however, using antibodies against the molluscan SCP_b in conjunction with anti-FMRFamide antibodies (Kobierski *et al.*, 1987), virtually all of the SCP_b-immunoreactive neurons (about 60 cells) are also labeled by the FMRFamide antibodies (Arbiser and Beltz, 1991). These studies reveal that SCP_b and FMRFamide immunoreactivities are both successfully preadsorbed with the most abundant FMRFamide-like molecule in *Homarus americanus*, the octapeptide TNRNFLRFamide. Met-enkephalin-Arg-Phe-amide (YGGFMRFamide), an extended opioid peptide containing the FMRFamide sequence, also preadsorbs SCP_b- and FMRFamide-like immunoreactivities, while YGGFMRF has no effect on the staining properties of these antibodies. These results suggest that the SCP_b antibody can bind to extended forms of FMRFamide-like molecules, and that anti-SCP_b and anti-FMRFamide antibodies may be colabeling one or more FMRFamide-like molecules. It is unlikely, therefore, that the colabeling of lobster neurons with these two antibodies indicates the coexistence of two distinct peptides (Arbiser and Beltz, 1991).

F. Neurohormones

Amines and some peptides often serve a dual function in being released from centrally located synapses, as well as from neurosecretory organs. In the postural neuromuscular system (see Section VI), serotonin and octopamine appear to exert both central (synaptic) and peripheral (hormonal) effects. Neurohormonal influences of amines and peptides

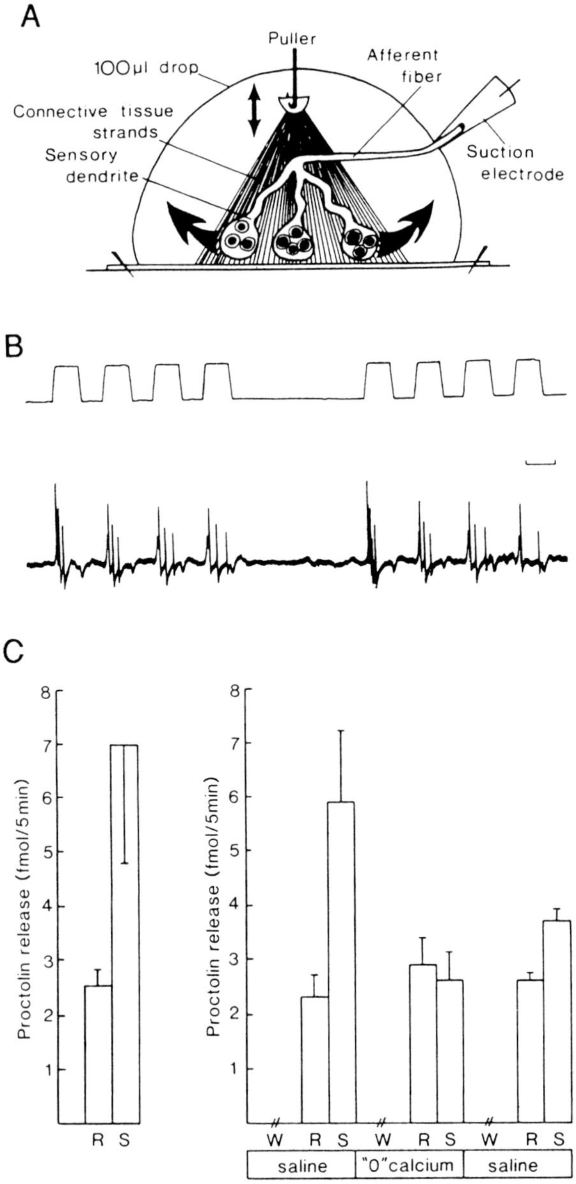

are also apparent in the abdominal MROs, where the sensitivity of these neurons to passive stretch is enhanced by serotonin and proctolin (Pasztor and Macmillan, 1990) and the receptor muscles themselves are responsive to serotonin, proctolin, and TNRNFLRFamide (Pasztor and Golas, 1993). Excitatory junction potential amplitude and nerve-evoked tension development are enhanced by all three neurohormones. The findings of Pasztor and Golas (1993) suggest that circulating neurohormones may act at multiple loci in the MRO system in a concerted and hormone-specific manner to alter the flow of proprioceptive feedback. Studies conducted on several types of proprioceptors in both lobsters and crayfish suggest that sensitivity to aminergic and peptidergic neurohormones is a general property of these neurons, but that the specific type of response (excitatory or inhibitory) to the substance may vary from one species to another (Pasztor and Macmillan, 1990).

In addition to substances that serve as both "transmitters" and "hormones" in lobsters, there are molecules that appear to be confined to the neuroendocrine system and function only as circulating hormones. The neuroendocrine X-organ–sinus gland complex located in the eyestalks of *Homarus americanus* and other decapods elaborates a variety of

FIGURE 6 (A) Isolated preparation of an oval organ from *Homarus americanus* (not to scale) mounted in a drop of saline to collect substances released from sensory nerve endings during stretch stimulation. Only one of three afferent fibers is shown. Each has an extensive dendritic arborization with varicosities and bulbous endings containing dense-cored vesicles. Vertical pulls applied to the connective tissue strands stretch the sensory dendrites, resulting in depolarization of the dendrites and afferent discharge in the sensory nerve. (B) Representative extracellular recording of oval organ responses during a period of repetitive stretch stimulation. (Upper line) Monitor of pull stimuli showing the groups of four 1-second pulls at 0.5 Hz interspersed with rest periods of 5 seconds. (Lower line) Bursts of sensory impulses from three fibers from the cut end of the sensory nerve. Note that pull amplitude and time parameters elicit spiking responses from all three sensory afferents. Scale bar: 1 second. (C) Stretch-induced release of proctolin from an isolated stretch receptor. (Left histogram) Release of proctolin from a relaxed organ (R) and from one organ repetitively stretched (S). Significantly higher ($P < 0.05$) release of proctolin occurred in response to the stretch. Data are expressed as the mean ± SE from six oval organs. (Right histogram) In the first pair of collections, in the presence of calcium, the amount of proctolin released by stretch (S) was significantly higher ($P < 0.05$) than that released during the same period from a relaxed organ (R). After 45 minutes of wash (W) in calcium-free saline ("O" calcium), stretch stimulation resulted in no more proctolin release than that which leaked out during 5 minutes of relaxation. Reintroduction of normal calcium levels reversed the effect, although the amount of proctolin released was then reduced. Data are expressed as the mean ± SE from three oval organs. All histograms represent release during a 5-minute period. (Reprinted from Pasztor *et al.*, 1988, with permission.)

neurohormones involved in the control of almost all known physiological processes: pigment migration, osmoregulation, glycemia, molting, growth and regeneration, reproduction, and hormones acting on nerves and muscles (reviewed by Beltz, 1988; Webster and Keller, 1988). Other neurohormonal regions, such as the pericardial organs or the mandibular organs, may also release some of these neuroendocrine compounds.

Several sinus gland hormones may be related. Crustacean hyperglycemic hormones (CHHs) (Keller *et al.*, 1985; Kleinholz and Keller, 1979; Tensen *et al.*, 1991a,b), molt-inhibiting hormone (MIH) (Chang, 1985; Skinner, 1985; Chang *et al.*, 1990), vitellogenesis-inhibiting hormone (VIH) (Soyez *et al.*, 1987, 1991; Meusy et al., 1987), and peptide G_1 (Goy et al., 1987a; Pavloff and Goy, 1990) are all related to each other structurally and, to some degree, functionally. There appear to be four CHH activities, two peptide G activities, one (possibly two) MIH activities, and two VIH isoforms (Soyez *et al.*, 1991). Of these, one of the CHH molecules (Tensen *et al.*, 1991b), MIH (Chang et al., 1990), and peptide G_1 (Pavloff and Goy, 1990) appear to have identical sequences. VIH also has a high degree of structural similarity to these molecules (Soyez *et al.*, 1991). However, because each laboratory working in this area uses different purification procedures, and not all of the peptides in the families have been sequenced, it is not yet possible to unambiguously identify to what degree these peptide families are related. It is clear that on the basis of sequence alone, there is at least one peptide in *Homarus americanus* that has variously been called CHH, MIH, and peptide G_1. A single sinus gland hormone, therefore, may have many physiological activities and further molecular studies must be completed before it will be possible to understand the structural and functional relationships of these compounds.

IV. Molecular Basis of Chemical Action

A. *Amines and Peptides*

Many potential targets of amines and peptides have been identified in *Homarus americanus*, including the exoskeletal muscles (e.g., the dactyl opener muscle described previously), chromatophores, hemolymph, heart, stomatogastric system, cardiac ganglia, and ganglia of the ventral nerve cord. What is the molecular basis of hormone action on these effectors? Are there unifying principles underlying the actions of amines and peptides? In other vertebrate and invertebrate systems, cyclic nucleotides mediate many transmitter- and hormone-regulated physiological responses and have been implicated as second messengers of amine and peptide actions on crustacean targets. In *H. americanus*, the known molecular actions of serotonin and octopamine are reviewed below.

Studies with the dactyl opener muscle of the walking leg have shown that serotonin, octopamine, and proctolin have similar actions on the fibers of the opener muscle: they induce long lasting contractures that are accompanied by little or no change in membrane potential or input resistance of the muscle fibers, and they produce action potentials in previously quiescent muscle fibers (Glusman and Kravitz, 1982; Kravitz *et al.*, 1985). The effect of serotonin on the muscle fibers has been studied using voltage–clamp techniques; no change is seen in an outward voltage-sensitive potassium conductance, but a large increase occurs in an inward current, probably carried by calcium ions (Glusman and Kravitz, 1982; Kravitz *et al.*, 1985). It is not clear whether this increase in inward current results from a direct action of serotonin on calcium channels or from an indirect action, for example, on calcium-activated potassium channels. Nor has it been shown whether this increased inward current can account for the appearance of the serotonin-induced action potentials. Serotonin also acts directly on excitatory and inhibitory nerve endings in this preparation to facilitate transmitter release. When serotonin is washed out of the bathing medium, there is a biphasic decay of the enhanced synaptic response: one component decays with a time constant of several minutes, while the other requires more than 1 hour to return to control levels (Goy *et al.*, 1984; Kravitz *et al.*, 1980). Pharmacological experiments suggest that different serotonin receptors mediate these two components.

Octopamine has much smaller presynaptic actions on excitatory nerve terminals in this preparation (Beltz and Kravitz, 1986; see Breen and Atwood, 1983, on crayfish), and proctolin has no measurable presynaptic effect. The multiplicity of synaptic effects seen with these different neuroactive chemicals hints at the potential complexity of modulatory influences. It is very clear, even in this relatively simple neuromuscular system, that the combined actions of many substances, acting via pre- and/or postsynaptic effects, will orchestrate the ultimate behavioral output (Fig. 5).

B. Cyclic Nucleotides

Most transmitters and modulators affect their postsynaptic targets by binding to a membrane receptor. This binding triggers a burst of biochemical activity within the cell. But how can a chemical signal received on the *outside* of a cell initiate metabolic changes on the *inside* of the cell? The solution is a "second messenger," an intracellular signal produced by the cytoplasmic surface of the plasma membrane in response to extracellular binding of the first messenger, i.e., the transmitter or modulator. Examples of well-known messengers in animal systems are cAMP, cGMP, and inositol triphosphate. Without such intermediate molecular signals, the message transmitted between neurons and their targets would not be relayed.

Serotonin causes increases in cAMP in heart and exoskeletal muscles (Battelle and Kravitz, 1978) and in the cardiac ganglion (Lemos and Berlind, 1980), while octopamine stimulates cAMP production in hemocytes and cardiac muscle (Battelle and Kravitz, 1978) of *Homarus americanus*. Incubation of intact opener muscle preparations with serotonin leads to large increases, and with octopamine to small increases, in cAMP levels. Proctolin, however, causes no detectable increase in cAMP levels (Goy *et al.*, 1984). Serotonin also causes phosphorylation of a 29-kDa protein and octopamine can cause phosphorylation of the same protein only in the presence of isobutylmethylxanthine (IBMX, a phosphodiesterase inhibitor), while proctolin never causes phosphorylation of this protein. None of these substances alter cGMP levels. The ability of these three compounds to increase cAMP levels closely parallels their ability to enhance phosphorylation of the 29-kDa protein. Other agents that increase intracellular cAMP levels also lead to phosphorylation of this protein; incubation of tissue homogenates with cAMP or cGMP in the presence of ATP causes phosphorylation of a protein with the same biochemical characteristics. The parallel between changes in cAMP levels and phosphorylation of the 29-kDa protein suggests that the two are related (Goy *et al.*, 1984; Kravitz *et al.*, 1985). Indeed, in intact tissues, agents that mimic the action of serotonin on cAMP levels (e.g., 8-bromo-cAMP, isobutylmethylxanthine, and forskolin) cause phosphorylation of the 29-kDa protein, but fail to mimic the postsynaptic actions of serotonin. These agents facilitate transmitter release, although none of them has as large an effect as that of serotonin. Several lines of evidence now indicate that the postsynaptic increase in muscle tone occurs independently of cAMP and that while the cyclic nucleotide does play a role in the facilitation of transmitter release by serotonin, there is also a cAMP-independent component to this facilitation (Goy and Kravitz, 1989).

Several observations suggest that cGMP may also

regulate some aspects of neuromuscular physiology or metabolism in *Homarus americanus*: (1) lobster muscle is one of the richest known sources of cGMP-dependent protein kinase; (2) neuromuscular preparations contain several phosphoproteins whose state of phosphorylation is affected by cGMP more effectively than by cAMP; and (3) guanylate cyclase and phosphodiesterase are active in this tissue. A polypeptide from the sinus gland in the eyestalk of the lobster is known to induce cGMP accumulation in lobster neuromuscular preparations (Goy and Kravitz, 1989). This molecule, termed peptide G_1, has now been isolated, characterized, and shown to be identical to MIH and one of the CHH isoforms isolated from *H. americanus* (see Section III,F; Pavloff and Goy, 1990; Chang *et al.*, 1990; Tensen *et al.*, 1991c). A second molecule isolated from the sinus gland that stimulates cGMP metabolism, called peptide G_2, appears to be the same as another CHH isoform. Collectively, therefore, this group of molecules is known to be involved in molting, carbohydrate metabolism, and cGMP metabolism. (See Waddy *et al.*, Chapter 10, on the control of molting, growth, and reproduction.)

V. Neural Regulation of Peripheral Targets

Many important concepts about the neuronal organization of motor systems have evolved from studies of the control of movement and coordination of crustacean appendages. In addition to the intrinsic interest in understanding how simple rhythmic behaviors are generated, limb movements are important because they play a major role in a variety of behaviors, such as walking, feeding, and elaborate agonistic and courtship displays. In *Homarus americanus*, the neural control of walking (MacMillan, 1975; Ayers and Davis, 1977a,b, 1978) and swimmeret beating (Davis, 1973) have been explored.

Movements of crustacean appendages involve coordinated sequences of contraction of several muscles. Within an appendage, individual muscles may participate in different types of movement, and muscles that work together during one type of movement may be antagonistic during another type of movement. Some motor actions, such as walking, require that the motor outputs to different segments be synchronized, so that the actions of the appendages are coordinated. The role of sensory feedback in some of these rhythmic behaviors has also been examined (Davis, 1969a,b, 1973). Limb movements, intrasegmental neural control, and coordination between limbs and segments in *Homarus americanus* are described below.

Four levels of control are involved with rhythmic movements of appendages: (1) the neurons in each segment that generate intrasegmental motor programs, (2) "command" neurons controlling intersegmental coordination, (3) sensory modulation, and (4) the target musculature in the limb. These four control systems work in concert to produce a coordinated movement.

A. Locomotion

Interest in the walking system has focused on the extent to which walking is centrally programmed versus generated by a reflexive response to sensory stimuli from the environment (MacMillan, 1975; Ayers and Davis, 1977a,b). Locomotory rhythms are sensitive to proprioceptive feedback, so the role of sensory feedback in controlling or modifying the central motor pattern has been closely examined.

Lobsters walk equally well in any direction; it is therefore presumed that the participating motor units must be capable of tremendous flexibility. How can the same motor units be employed to produce different variations of walking patterns? Additionally, since limb movements play a major role in a variety of complex behaviors, such as agonistic and courtship displays, the means by which movements are produced and modified are of exemplary importance to our understanding of neural integration and behavior.

The cycle of stepping in *Homarus americanus* consists of a power stroke, during which the limb bears the weight and provides propulsive force, followed by a return stroke, when the limb is elevated and returned to its original position to initiate the subsequent power stroke (Evoy and Ayers, 1982). In each walking leg, skeletal muscles exert force on a mechanically constrained system of levers formed by the apodemes and joints of the segments. The movements of an individual joint are brought about by a pair of antagonistic muscles. In some joints, however, a muscle may be made up of several heads that insert at different angles; the parts of the muscle serve different functions in locomotion and posture. Successive joints along a particular appendage operate at an angle of approximately 90° to one another and most movements result from various combinations of planar displacements of several joints (Evoy and Ayers, 1982). For instance, the major movements of the third walking legs occur around three joints and involve eight muscles (Ayers and Davis, 1977a,b).

Two approaches have been used to examine limb movements and locomotion in *Homarus americanus*. Macmillan (1975), studying unrestrained lobsters, found that variation in stepping frequency results in proportionate variation in both power and return strokes. Ayers and Davis (1977a,b, 1978), experimenting with tethered lobsters on a motor-driven treadmill, found that the duration of the return stroke remains constant and that changes in stepping frequency result from variation in the duration of the power stroke, regardless of the direction of walking. Two explanations of these contrasting results are offered: (1) the walking movements analyzed occurred over two distinct walking speeds which may account for some differences in the results; and (2) by analogy to studies of crabs (Evoy and Fourtner, 1973), it may well be that different motor programs are used in loaded (i.e., tethered, walking on a treadmill) and in unrestrained walking.

On the basis of timing criteria, motoneurons driving the walking legs of the lobster can be divided into three functional classes (Ayers and Davis, 1977a,b). (1) Some neurons discharge only during the return stroke. The duration of firing bursts in these neurons remains constant during variations in stepping period, independent of the direction of walking. One example of this group of neurons is the motoneurons of the anterior elevator muscle of the coxobasal (C–B) joint. (2) The second class of motoneurons, such as those of the anterior depressor muscle of the C–B joint, discharge only during the power stroke. The burst duration in these neurons varies directly with the stepping period; therefore, one can attribute all variations in stepping frequency to variation in the duration of power stroke bursts (Ayers and Davis, 1977a), except during the highest speeds of walking (Macmillan, 1975). During walking, these two classes of neurons (those responsible for the power and return strokes) typically discharge in alternating reciprocal bursts that drive opposing muscles (Ayers and Davis, 1977a). (3) The remaining "bifunctional" motor units can discharge during either the power or return stroke (Ayers and Davis, 1977a; Ayers and Clarac, 1978). They have a variety of discharge patterns ranging from tonic to burst discharges in synchrony with either the power or return motoneurons. With bursting of these bifunctional neurons, the burst duration varies with the motoneuronal groups with which they are synergistic (Ayers and Davis, 1977a). Examination of intra- and intersegmental coordination has shown the predominant mode of coordination to be the alternating tetrapod rhythm, in which phase relationships remain constant as stepping frequency varies.

One of the basic issues related to locomotion is the relative contributions of the central nervous system and peripheral feedback to the control of walking. The fundamental criterion proving the existence of a central oscillator (and therefore central control) is that the central nervous system must generate a replica of the behaviorally significant motor output ("motor program") in the complete absence of sensory feedback. The existence of a central oscillator has been difficult to demonstrate in the walking system of *Homarus americanus* for the following reasons. (1) Many of the motor nerves to the walking legs are mixed, containing both sensory and motor fibers, making it extremely difficult to eliminate sensory feedback. (2) Very few pharmacological agents are available that disable muscles while allowing central oscillators to run, as has been demonstrated in the vertebrates (Grillner, 1975). (3) The thoracic nerve cord has been intolerant to surgical manipulation. As a result of these factors, there has been no clear demonstration of a "walking motor program" from the isolated chain of thoracic ganglia in the absence of sensory feedback.

Sensory modulation of two types is clearly important in the normal expression and variation in walking (Davis and Ayers, 1972; Ayers and Davis, 1977b): exteroceptive inputs, such as optokinetic stimuli, provide information about the animal's orientation, direction, and rate of motion; and movements of the leg joints are monitored by a variety of proprioceptors responsive to angular change, joint position, muscular tension, and cuticular deformation.

B. *Swimmeret Movement and Intersegmental Coordination*

Swimmerets are versatile abdominal appendages that are involved in tail flips, in swimming behavior in postlarvae (stage IV), and in egg ventilation in gravid females. Swimmerets beat in a metachronal rhythm. They contain 26 muscle bundles that are morphologically identifiable, but can be grouped into 12 functional muscles based on their innervation patterns and their roles in movement of the swimmerets (Fig. 7) (Davis, 1968; Evoy and Ayers, 1982). As with the stepping movements of the walking legs (Section V,A), the swimming movements of the abdominal swimmerets consist of alternating power and return strokes (Fig. 7). Electromyographic recordings from these muscles during swimmeret beating allow separation of the muscles into power and return stroke groups that approximate reciprocity in their discharge, although there is overlap in the motor bursts. Other associated movements include the opening or curling of the rami of the swimmerets during the power stroke, or the closing and uncurling during the return stroke to limit resistance through the water.

Swimmeret movement is of greater amplitude during more rapid beating (Davis, 1971). Variation in amplitude is produced by increases in the number of active motor neurons as the frequency of beating increases. These motoneurons are recruited in order of increasing size and neuromuscular effect. Individual motoneurons also increase their discharge frequency as the beat frequency increases.

Unlike the walking system, in which central oscillatory control has not been unequivocally proven, the swimmeret rhythm of motoneuronal activity (motor program) can be produced by an isolated abdominal *hemi*ganglion. The motor program consists of alternating discharges in the motor neurons to the power and return stroke musculature. Although the rhythm-generating mechanism has not been completely defined, it appears that at least some of the swimmeret motoneurons are part of the pattern generator (Heitler, 1978; Heitler and Pearson, 1980).

Interneuronal elements ("command fibers") are known to control both the posture and rhythmic movements of the appendages (Evoy and Kennedy, 1967; Davis and Kennedy, 1972a–c). The idea of command neurons dates back to the finding that electrical stimulation of single neurons elicits recognizable behavior (Wiersma, 1938). The term *command* was first applied to single cells by Wiersma (1962), but it gained most acceptance when applied to single nerve cells capable of eliciting the coordinated movements of swimmerets in crayfish (Wiersma and Ikeda, 1964).

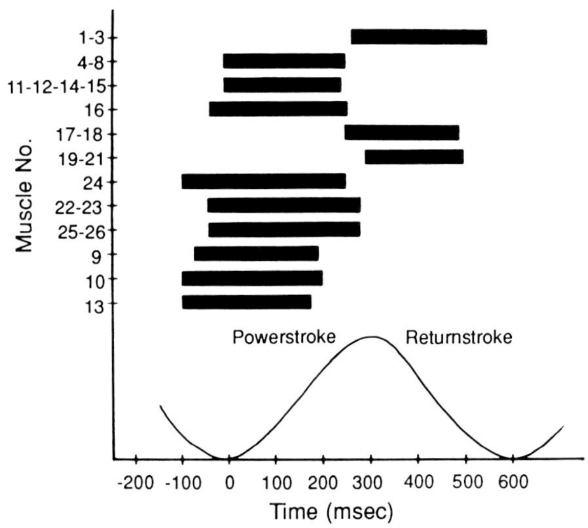

FIGURE 7 The sequential pattern of electrical activity in all of the swimmeret muscles during a single, representative cycle of swimmeret beating. Bars correspond to the active periods of the muscles indicated on the ordinate. Twelve functional muscle groupings are indicated. The pattern was reconstructed by combining numerous individual records. Muscle activity reflects the pattern of activity of motoneurons innervating each respective muscle. (From Davis, W. J., The neuromuscular basis of lobster swimmeret beating, *J. Exp. Zool.* 168, 363–378. Copyright © 1968 Wiley–Liss, Inc.; reprinted with permission of John Wiley & Sons, Inc.)

The swimmeret system of *Homarus americanus* is controlled in part by roughly five pairs of command fibers, which, when stimulated, produce a part of the swimmeret rhythm. However, no single command element can evoke the full range of swimmeret outputs seen in behaving animals, suggesting that the command fibers are used in concert rather than as individuals (Davis, 1973). Coordinating interneurons are also critically important to the maintenance of metachronal rhythm in the swimmeret system; such interneurons often function via transmission of an "efference copy" to the next segmental ganglion (Stein, 1974, 1977).

The "autonomous" development of central pattern generators has also been examined in the swimmeret system, in which the existence of a central motor program is so clear. In these studies, Davis and Davis (1973) attempted to extirpate the presumptive swimmeret appendages prior to differentiation in newly hatched larvae. Following extirpation, they tested for the presence of swimmeret motoneurons and for normal patterns of rhythmic motor discharge and reflexes. Both motoneuronal interconnectivity and reflexive responsiveness were normal following extirpation. In a more recent study that duplicates the goals of Davis and Davis (1973), Kirk and Govind (1992) recognized that the swimmeret muscles are innervated much earlier than was previously thought. Some functional innervation of swimmeret muscles is present even at the end of embryonic life before the external swimmerets are formed. Kirk and Govind (1992) suggest that such early innervation of the swimmeret muscles may be influential in establishing the central circuitry for pattern generation in this system and that these circuits may not emerge autonomously, as was previously suggested by Davis and Davis (1973).

C. The Stomatogastric System

The stomatogastric nervous system of decapods controls the foregut, which contains the apparatus used to macerate and filter food (Fig. 8A). The muscles of the foregut are striated skeletal muscles that are innervated by motor neurons whose cell bodies are located in the stomatogastric ganglion, which contains only 30 neurons. Because of the small number of neurons involved in the production of the behavior and because of the accessibility of the system, the neural connectivity pattern for the entire ganglion has been determined. In this system, the neural basis of the foregut "behaviors" can be studied at the cellular level. The gastric mill, which consists of a set of teeth that macerate the food, and the pylorus, which filters and separates the food before it reaches the midgut and hindgut, are controlled by neural circuits located within the stomatogastric ganglion. The esophagus and cardiac sac, which temporarily stores and mixes the food, are controlled by neurons located in the stomatogastric ganglion and three other ganglia. The circuits that control these various regions and functions work independently of each other and can produce rhythmic movements that differ in their periods of oscillation. Many of these rhythmic oscillations are triggered either by neuromodulatory substances released by axons projecting from other parts of the nervous system, or by hormones released into the blood. The phasing of oscillations can also be altered by a variety of neurohormonal substances (Fig. 8B). Finally, a single circuit can be "reconfigured" by chemicals that alter the synaptic strengths and cellular properties of individual neurons in the group. Therefore, the composition of circuits is not static and neurons can participate in more than one stomatogastric "behavior."

While the individual neurons involved in these circuits have been carefully defined and are similar in a variety of crustacean species (see Selverston and Moulins, 1987), very little of the pioneering work has been done in *Homarus americanus*. For the most part, studies have been initiated in either the spiny lobster or the crab, and the American lobster has been used for the purpose of chemical, functional, and evolutionary comparisons (Beltz *et al.*, 1984; Katz, 1991; Marder, 1987; Harris-Warrick and Marder, 1991). (See Factor, Chapter 15, for the detailed anatomy of the ossicles, muscles, and nerves of the stomatogastric system.)

VI. The "Integrated" Lobster: Central Neurons, Circulating Neurohormones, Postural Regulation, and Behavior

...Collect from my nerves;
Find all my cells;
Inject hormones by the score.
I'll behave for thee, and fight, not flee,
To please thee even more....

Excerpt from "The Rime of the Ancient Scientist"
—E. A. Kravitz (1991)

The ultimate goal of neural research is to understand the neural basis of behavior. What part of the behavior is programmed into the central nervous system? What signals can elicit the behavior? How are the sensory, motor, and hormonal systems controlled so that they work in concert to produce behavior? For the past 15 years, a substantial effort has been made to dissect aggressive behaviors in the lobster to the level of single neurons, specific transmitters, and identified hormones (Kravitz, 1988).

Serotonin and octopamine, when injected into freely moving lobsters, generate stable and stereotyped postures that resemble those seen in dominant and subordinate lobsters, respectively (Fig. 9) (Livingstone *et al.*, 1980). Serotonin injection causes lobsters to stand high on their walking legs, with their claws open in front of them and their abdomens loosely tucked downward; octopamine injection causes animals to stand low to the substrate, with their claws and walking legs extended and pointing forward and their abdomens hyperextended upward. These opposing postures result from serotonin-induced contraction of the postural flexor muscles and octopamine-induced contraction of the postural extensors. Such actions could result from opposing actions of these amines on peripheral skeletal muscles, or from their differential action on the neurons within the ventral nerve cord that innervate these muscles.

In addition to the excitatory (glutaminergic) and inhibitory (GABAergic) innervation of postural muscles, amines are known to influence lobster skeletal muscles in dramatic ways (see Section III,C; Glusman and Kravitz, 1982; Grundfest and Reuben, 1961). However, peripheral effects on the muscles cannot explain the opposing postures. Serotonin increases transmitter release from both excitatory and inhibitory nerve terminals and causes muscles to undergo a longlasting voltage- and Ca^{2+}-sensitive contracture and to generate Ca^{2+} action potentials. Octopamine also increases release from excitatory nerves, but its effects are smaller. Like serotonin, octopamine causes contractures and action potentials to appear in muscle fibers. When flexor and extensor muscle pairs are examined, both serotonin and octopamine facilitate transmitter release and enhance contractility in both types of muscle. Therefore, the peripheral effects of the amines, while serving to "prime" the muscles to respond more vigorously, cannot explain the opposing behavioral effects that are seen.

An alternative possibility is that the contrasting effects of serotonin and octopamine on posture are generated by central neurons. When serotonin or octopamine is applied to a dissected ventral nerve

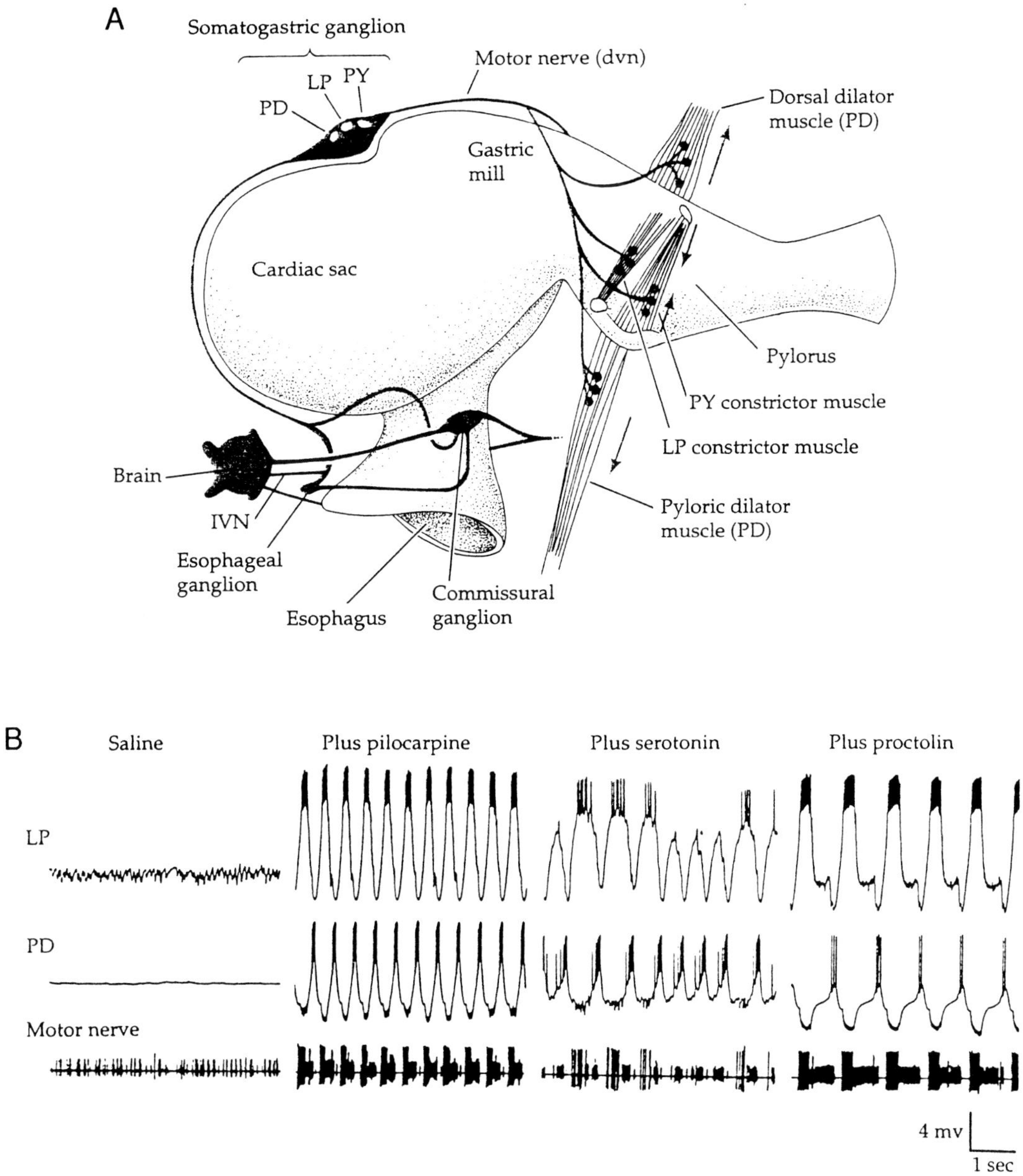

FIGURE 8 (A) The stomatogastric nervous system as it is positioned relative to the stomach. The complete stomatogastric nervous system includes four ganglia: the stomatogastric ganglion on the dorsal surface of the stomach, the esophageal ganglion on the anterior surface of the esophagus, and the paired commissural ganglia lateral to the esophagus. The ganglia and the nerves that connect them are shown in black. Muscles that move the esophagus, cardiac sac, gastric mill, and pylorus are controlled by the stomatogastric nervous system. The position of the two sets of muscles that dilate and constrict the pylorus are shown as bundles of parallel lines. Contraction of the dilator muscles opens the pylorus, whereas contraction of the constrictor muscles narrows it. The dilator muscles are innervated by the two pyloric dilator (PD) motoneurons, the anterior constrictor muscles by the lateral pyloric (LP) neuron, and the more posterior constrictor muscles by the pyloric (PY) neurons. IVN, Inferior ventricular nerve. (B) Modulatory agents produce different forms of the pyloric rhythm. Each panel shows intracellular recording from the LP neuron and the PD neuron, respectively, and the bottom trace is an extracellular recording from the motor nerve. In control saline, the pyloric rhythm is not active. Bath application of pilocarpine (a muscarinic agonist), serotonin, and proctolin turns on the pyloric rhythm. However, the form of the rhythm in the presence of each modulator is different. These recordings were made in a crab preparation; however, the general types of modulatory changes seen are typical of stomatogastric neurons in other crustacean species. (Reprinted from Kennedy and Marder, 1992, with permission.)

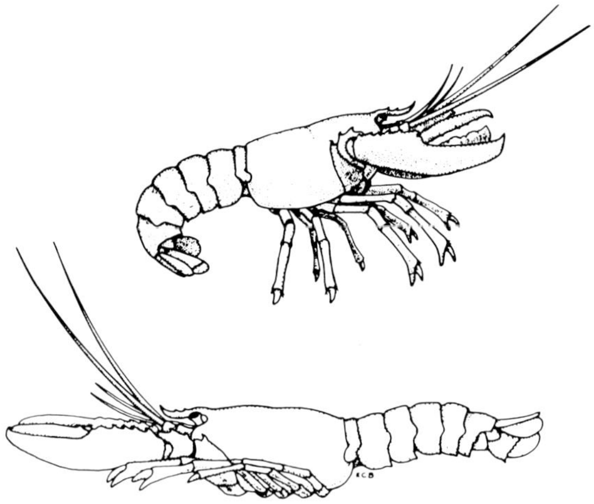

FIGURE 9 Postures generated when serotonin (top) or octopamine (bottom) is injected into freely moving lobsters. These flexed and extended poses can last for several hours following injection. (From Beltz, B. S., Crustacean neurohormones. *In* "Endocrinology of Selected Invertebrate Types" (H. Laufer and R. G. H. Downer, eds.), Vol. 2, pp. 235–258. Copyright © 1988 Wiley–Liss, Inc.; reprinted with permission of John Wiley & Sons, Inc.)

cord preparation, each amine evokes a distinct motor program (Livingstone *et al.*, 1980; Harris-Warrick and Kravitz, 1984). Serotonin increases the firing rates of excitatory motoneurons to the postural (slow) flexor muscles and inhibitory motoneurons to the postural extensor muscles, while simultaneously reducing the firing rates of excitatory motoneurons to the slow extensor muscles and inhibitory motoneurons to the slow flexors (Fig. 10). These combined effects on central neurons result in a motor output that directs a flexion posture. Octopamine causes the opposite pattern, thereby directing the read-out of an extension program (Fig. 10). The total effect of the amines is to cause the activation of opposing central motor programs while simultaneously priming the peripheral muscles to respond.

The actions of the amines on central neurons are

	SLOW FLEXORS		SLOW EXTENSORS	
	E	I	E	I
SEROTONIN	↑	↓	↓	↑
OCTOPAMINE	↓	↑	↑	↓

FIGURE 10 Actions of serotonin and octopamine on the firing of excitatory and inhibitory motoneurons innervating postural (slow) flexor and extensor muscles. Arrows indicate increase (↑) and decrease (↓) in activity of postural excitor (E) and inhibitor (I) motoneurons. (Reprinted from Kravitz *et al.*, 1983, with permission.)

reminiscent of the effects of "command neuron" stimulation described many years ago (Wiersma and Ikeda, 1964; Evoy and Kennedy, 1967), raising the possibility that the amines are somehow interacting with the command networks to activate the opposing postures (Harris-Warrick and Kravitz, 1984). The most direct way to test this hypothesis is to activate specific aminergic neurons in the ventral nerve cord to see whether firing these neurons reproduces the effects of injected or bath-applied amines. Serotonin and octopamine have been localized to central neurons, using immunocytochemical methods (Beltz and Kravitz, 1983; Schneider *et al.*, 1993a); some of the serotonin-containing neurons have been identified electrophysiologically (Beltz and Kravitz, 1987); and specific paired neurons in the first abdominal ganglion have been tested for effects on the flexor and extensor command fiber system (Ma *et al.*, 1992). These neurons appear to act as "gain-setters," rather than as control elements, modulating the interaction between command inputs and motoneuron outputs (Ma *et al.*, 1992). When flexor commands are activated, the rate of firing of the serotoninergic cells increases; when extensor commands are activated, the rate of firing of the serotoninergic cells decreases. Activation of the serotoninergic cell by the command neuron increases the rate of firing of certain neurons and decreases the rate in others. Comparable studies have not yet been carried out on octopamine neurons.

VII. Directions for Further Research

Since postural displays are an integral part of the expression of dominant and subordinate positions in the social hierarchy, recent studies have attempted to relate the function of the amine neurons to aggressive behaviors in *Homarus americanus*. One approach is to set up behavioral hierarchies in laboratory-reared juvenile lobsters, so that clear dominance is established by an individual lobster (R. Huber and E. A. Kravitz, personal communication), allowing future studies to assess the aminergic state of the dominant and subordinant animals. Several important questions can be posed. What are the circulating levels of serotonin and octopamine (and their metabolic products) in the behaviorally biased animals? What are the properties of the identified amine-containing neurons in the two behaviorally distinct groups? How do dominant animals communicate to other lobsters their "dominant" state? Could this also be related to amine levels?

Another approach is to examine possible sensory inputs that can activate the amine neurons involved in posture. One hint at the complexity of these inputs comes from the demonstration that sexually dimorphic mechanosensory sensilla on the second swimmerets are linked to the postural neuronal system (Page, 1985; Kotak and Page, 1986). Serotoninergic neurons involved in posture also receive sensory feedback from the telson (Weiger and Kravitz, 1994).

While a complete understanding of the neural basis of aggressive behavior in the lobster is still on the horizon, dramatic progress has clearly been made toward cellular explanations for sophisticated, socially relevant behaviors. This is a powerful example of how the nervous system of *Homarus americanus* can be utilized to help us understand basic neuronal properties, transmitter effects and mechanisms, and the coordinated action of central, neurohormonal, and sensory neurons in producing complex behaviors.

The lobster nervous system is also being utilized for developmental studies (Beltz *et al.*, 1993; see Talbot and Helluy, Chapter 9, on embryonic development). In a variety of vertebrate and invertebrate systems, aminergic transmitters have been proposed as morphogens—architects of the nervous system during differentiation. In the lobster, the serotoninergic dorsal giant neuron in the brain is the primary source of serotonin in the olfactory and accessory lobes (Fig. 3A). Serotonin depletion during embryonic life results in a dramatic slowing of the growth rate of the olfactory and accessory lobes, while the growth of the antenna II neuropil (Fig. 1), which receives no serotoninergic input during midembryonic life, is unaffected by serotonin depletion (Benton *et al.*, 1994). These data have led to the conclusion that serotonin, most likely from the dorsal giant neurons, may be directing the growth and/or differentiation of the olfactory and accessory lobes in the lobster brain. The cellular and molecular bases of these serotonin depletion effects are being examined, and should yield interesting insights into how amines can shape developing neurons.

On the theme of the cellular properties of neurons and their targets, the 21st century should bring a clearer understanding of how lobster neurons translate chemical signals into action as neurobiologists address the following questions: How do muscle and neuronal membrane channels function? How do neuronal circuits and circulating chemicals fine-tune behavioral outputs? Can natural behaviors be linked with specific identified neurons and chemical changes in specific neural circuits? Can these mechanisms be related to higher organisms on the phylogenetic scale? Although many of these issues are already being addressed, there is still a great deal of challenge ahead to understand how specific targets and behaviors are controlled at the level of individual neurons.

VIII. Summary

Because the American lobster, *Homarus americanus*, offers a relatively simple nervous system combined with a fairly sophisticated behavioral repertoire, it has been heavily utilized for studies of neuronal control systems. The dual innervation of muscles by separate excitor and inhibitor axons allows for finely tuned peripheral control over muscular contraction. But it is also clear that the motoneurons and their specific muscular targets involved in walking, swimmeret beating, digestion, or postural regulation are only a small part of the total story. The impact of sensory input on these systems varies and in some cases the role of sensory neurons in the circuit is not even known. The complexity of chemical effects is very clear, although the subtleties of hormonal action have yet to be worked out.

A host of amines and peptides, whose actions are overlaid on the excitatory transmitter glutamate and inhibitory transmitter GABA, have the ability to dramatically alter neural circuits that generate behavior. These chemical compounds include serotonin and dopamine, which are also well known as modulators in vertebrate systems, as well as distinctive peptides such as proctolin and several FLRFamide molecules that appear to be found only in invertebrates, and perhaps only in certain arthropod groups. Chemical aspects of relatively simple behaviors have been most carefully examined in the postural system of *Homarus americanus*, as well as in the stomatogastric system of related crustaceans. For some of the neuroactive compounds, the molecular functions have been at least partially defined in terms of sites of action and types of second-messenger involvement.

Taken together, the host of neurobiological studies conducted in the lobster ranges from the molecular analysis of channels, receptors, and messengers to the other end of the spectrum—the behaviors that are produced. It is this final synthesis of studies at many levels, conducted by numbers of research groups utilizing a wide variety of methods, that has given the lobster its laudable position as a model system for integrated ethological, chemical, physiological, and anatomical studies.

References

Ache, B. W. (1982). Chemoreception and thermoreception. *In* "The Biology of Crustacea" (D. E. Bliss, ed.), Vol. 3, "Neurobiology: Structure and Function" (H. L. Atwood and D. C. Sandeman, eds.), pp. 369–398. Academic Press, New York.

Ache, B. W., and Macmillan, D. L. (1980). Neurobiology. *In* "The Biology and Management of Lobsters" (J. S. Cobb and B. F. Phillips, eds.), Vol. 1, pp. 165–213. Academic Press, New York.

Arbiser, Z. K., and Beltz, B. S. (1991). SCP_b- and FMRFamide-like immunoreactivities in lobster neurons: Colocalization of distinct peptides or colabeling of the same peptide(s)? *J. Comp. Neurol.* 306, 417–424.

Ayers, J. L., and Clarac, F. (1978). Neuromuscular strategies underlying different behavioral acts in a multifunctional crustacean leg joint. *J. Comp. Physiol. A* 128A, 81–129.

Ayers, J. L., and Davis, W. J. (1977a). Neuronal control of locomotion in the lobster, *Homarus americanus*. I. Motor programs for forward and backward walking. *J. Comp. Physiol. A* 115A, 1–27.

Ayers, J. L., and Davis, W. J. (1977b). Neuronal control of locomotion in the lobster *Homarus americanus*. II. Types of walking leg reflexes. *J. Comp. Physiol. A* 115A, 29–46.

Ayers, J. L., and Davis, W. J. (1978). Neuronal control of locomotion in the lobster, *Homarus americanus*. III. Dynamic organization of walking leg reflexes. *J. Comp. Physiol. A* 123A, 289–298.

Barker, D. L., Herbert, E., Hildebrand, J. G., and Kravitz, E. A. (1972). Acetylcholine and lobster sensory neurones. *J. Physiol. (London)* 226, 205–229.

Battelle, B.-A., and Kravitz, E. A. (1978). Targets of octopamine in the lobster: Cyclic nucleotide changes and physiological effects in haemolymph, heart, and exoskeletal muscle. *J. Pharmacol. Exp. Ther.* 205, 438–448.

Beltz, B. S. (1988). Crustacean neurohormones. *In* "Endocrinology of Selected Invertebrate Types" (H. Laufer and R. G. H. Downer, eds.), Vol. 2, pp. 235–258. Wiley–Liss, New York.

Beltz, B. S., and Kravitz, E. A. (1983). Mapping of serotonin-like immunoreactivity in the lobster nervous system. *J. Neurosci.* 3, 585–602.

Beltz, B. S., and Kravitz, E. A. (1986). Aminergic and peptidergic neuromodulation in Crustacea. *J. Exp. Biol.* 124, 115–141.

Beltz, B. S., and Kravitz, E. A. (1987). Physiological identification, morphological analysis, and development of identified serotonin–proctolin containing neurons in the lobster ventral nerve cord. *J. Neurosci.* 7, 533–546.

Beltz, B. S., Eisen, J. S., Flamm, R., Harris-Warrick, R. M., Hooper, S. L., and Marder, E. (1984). Serotonergic innervation and modulation of the stomatogastric ganglion of three decapod crustaceans (*Panulirus interruptus, Homarus americanus* and *Cancer irroratus*). *J. Exp. Biol.* 109, 35–54.

Beltz, B. S., Pontes, M. C., Helluy, S. M., and Kravitz, E. A. (1990). Patterns of appearance of serotonin and proctolin immunoreactivities in the developing nervous system of the American lobster. *J. Neurobiol.* 21, 521–542.

Beltz, B. S., Helluy, S. M., Ruchhoeft, M. L., and Gammill, L. S. (1993). Aspects of the embryology and neural development of the American lobster. *J. Exp. Zool.* 261, 288–297.

Benton, J., Helluy, S., Huber, R., and Beltz, B. (1994). Serotonin depletion by 5,7 dihydroxytryptamine alters deutocerebral development in the lobster, *Homarus americanus*. *Soc. Neurosci. Abstr.* 20, 775.

Breen, C., and Atwood, H. L. (1983). Octopamine—a neurohormone with presynaptic activity-dependent effects at crayfish neuromuscular junctions. *Nature (London)* 303, 716–718.

Bullock, T. H., and Horridge, G. A. (1965). "Structure and Function of the Nervous Systems of Invertebrates," Vol. 2. Freeman, San Francisco.

Bush, B. M. H., and Laverack, M. S. (1982). Mechanoreception. *In* "The Biology of Crustacea" (D. E . Bliss, ed.), Vol. 3, "Neurobiology: Structure and Function" (H. L. Atwood and D. C. Sandeman, eds.), pp. 399–468. Academic Press, New York.

Calloway, J. C., Masinovsky, B., and Graubhard, K. (1987). Colocalization of SCP_b-like and FMRFamide-like immunoreactivities in the crustacean nervous system. *Brain Res.* 405, 295–304.

Chang, E. S. (1985). Hormonal control of molting in decapod Crustacea. *Am. Zool.* 25, 179–185.

Chang, E. S., Prestwich, G. D., and Bruce, M. J., (1990). Amino acid sequence of a peptide with both molt-inhibiting and hyperglycemic activities in the lobster, *Homarus americanus*. *Biochem. Biophys. Res. Commun.* 171, 818–826.

Claiborne, B. J., and Selverston, A. I. (1984). Histamine as a neurotransmitter in the stomatogastric nervous system of the spiny lobster. *J. Neurosci.* 4, 708–721.

Cohen, M. J. (1955). The function of receptors in the statocyst of the lobster, *Homarus americanus*. *J. Physiol. (London)* 130, 9–34.

Cohen, M. J. (1960). The response patterns of single receptors in the crustacean statocyst. *Proc. R. Soc. London, Ser. B* 152, 30–49.

Cooke, I. M., and Sullivan, R. (1982). Hormones and Neurosecretion. *In* "The Biology of Crustacea" (D. E. Bliss, ed.), Vol. 3, "Neurobiology: Structure and Function" (H. L. Atwood and D. C. Sandeman, eds.), pp. 205–290. Academic Press, New York.

Cournil, I., Helluy, S. M., and Beltz, B. S. (1994). Dopamine in the lobster *Homarus gammarus*: I. Comparative analysis of dopamine and tyrosine hydroxylase immunoreactivities in the nervous system of the juvenile. *J. Comp. Neurol.* 341, 1–15.

Davis, W. J. (1968). The neuromuscular basis of lobster swimmeret beating. *J. Exp. Zool.* 168, 363–378.

Davis, W. J. (1969a). Reflex organization in the swimmeret system of the lobster. I. Intrasegmental reflexes. *J. Exp. Biol.* 51, 547–563.

Davis, W. J. (1969b). Reflex organization in the swimmeret system of the lobster. II. Reflex dynamics. *J. Exp. Biol.* 51, 565–573.

Davis, W. J. (1971). Functional significance of motoneuron size and soma position in the swimmeret system of the lobster. *J. Neurophysiol.* 34, 274–288.

Davis, W. J. (1973). Neuronal organization and ontogeny in the lobster swimmeret system. *In* "Control of Posture and Locomotion" (R. B. Stein, K. G. Pearson, R. G. Smith, and J. B. Redford, eds.), pp. 437–455. Plenum, New York.

Davis, W. J., and Ayers, J. L. (1972). Locomotion: Control by positive-feedback optokinetic responses. *Science* 177, 183–185.

Davis, W. J., and Davis, K. B. (1973). Ontogeny of a simple motor system: Role of the periphery in the development of central nervous circuitry. *Am. Zool.* 13, 409–425.

Davis, W. J., and Kennedy, D. (1972a). Command neurons controlling swimmeret movements in the lobster. I. Types of effects on motor neurons. *J. Neurophysiol.* 35, 1–12.

Davis, W. J., and Kennedy, D. (1972b). Command neurons controlling swimmeret movements in the lobster. II. Interaction of effects on motor neurons. *J. Neurophysiol.* 35, 13–19.

Davis, W. J., and Kennedy, D. (1972c). Command neurons controlling swimmeret movements in the lobster. III. Temporal relationships among bursts in different motor neurons. *J. Neurophysiol.* 35, 20–29.

Derby, C. D. (1982). Structure and function of cuticular sensilla of the lobster *Homarus americanus*. *J. Crust. Biol.* 2, 1–21.

Derby, C. D., and Atema, J. (1982a). The function of chemo- and mechanoreceptors in lobster (*Homarus americanus*) feeding behavior. *J. Exp. Biol.* 98, 317–327.

Derby, C. D., and Atema, J. (1982b). Narrow-spectrum chemoreceptor cells in the walking legs of the lobster *Homarus americanus*: Taste specialists. *J. Comp. Physiol. A* 146A, 181–189.

Evoy, W. H., and Ayers, J. (1982). Locomotion and control of limb movements. *In* "The Biology of Crustacea" (D. E. Bliss, ed.), Vol.

3, "Neurobiology: Structure and Function" (H. L. Atwood and D. C. Sandeman, eds.), pp. 62–105. Academic Press, New York.

Evoy, W. H., and Fourtner, C. R. (1973). Nervous control of walking in the crab *Cardisoma granhumi*. III. Proprioceptive influences on intra- and intersegmental coordination. *J. Comp. Physiol. A* 83A, 303–318.

Evoy, W. H., and Kennedy, D. (1967). The central nervous organization underlying control of antagonistic muscles in the crayfish. I. Types of command fibers. *J. Exp. Zool.* 165, 223–238.

Eyzaguirre, C., and Kuffler, S. W. (1955). Processes of excitation in the dendrites and in the soma of single isolated sensory nerve cells of the lobster and crayfish. *J. Gen. Physiol.* 39, 87–119.

Gabe, M. (1966). "Neurosecretion." Pergamon, Oxford.

Glusman, S., and Kravitz, E. A. (1982). The action of serotonin on excitatory nerve terminals in lobster neuromuscular preparations. *J. Physiol. (London)* 325, 223–241.

Govind, C. K. (1992). Nervous System. *In* "Microscopic Anatomy of Invertebrates" (F. W. Harrison, ed.), Vol. 10, "Decapod Crustacea" (F. W. Harrison and A. G. Humes, eds.), pp. 395–438. Wiley–Liss, New York.

Goy, M. F., and Kravitz, E. A. (1989). Cyclic AMP only partially mediates the actions of serotonin at lobster neuromuscular junctions. *J. Neurosci.* 9, 369–379.

Goy, M. F., Schwarz, T. L., and Kravitz, E. A. (1984). Serotonin-induced protein phosphorylation in a lobster neuromuscular preparation. *J. Neurosci.* 4, 611–626.

Goy, M. F., Mandelbrot, D. A., and York, C. M. (1987a). Identification and characterization of a polypeptide from a lobster neurosecretory gland that induces cyclic GMP accumulation in lobster neuromuscular preparations. *J. Neurochem.* 48, 954–966.

Goy, M. F., Trimmer, B. A., and Kravitz, E. A. (1987b). Peptide F1: A myoactive lobster peptide related to FMRFamide. *Soc. Neurosci. Abstr.* 13, 1257.

Grillner, S. (1975). Locomotion in vertebrates: Central mechanisms and reflex interaction. *Physiol. Rev.* 55, 247–304.

Grundfest, H., and Reuben, J. P. (1961). Neuromuscular synaptic activity in lobster. *In* "Nervous Inhibition" (E. Florey, ed.), pp. 92–104. Pergamon, New York.

Harris-Warrick, R. M., and Kravitz, E. A. (1984). Cellular mechanisms for modulation of posture by serotonin and octopamine in the lobster. *J. Neurosci.* 4, 1976–1993.

Harris-Warrick, R. M., and Marder, E. (1991). Modulation of neural networks for behavior. *Annu. Rev. Neurosci.* 14, 39–57.

Heitler, W. J. (1978). Coupled motor neurons are part of the crayfish swimmeret central oscillator. *Nature (London)* 275, 231–234.

Heitler, W. J., and Pearson, K. G. (1980). Non-spiking interactions and local interneurones in the central pattern generator of the crayfish swimmeret system. *Brain Res.* 187, 206–211.

Katz, P. S. (1991). Neuromodulation and the evolution of a simple motor system. *Semin. Neurosci.* 3, 379–389.

Katz, P. S., and Harris-Warrick, R. M. (1989). Serotonergic/cholinergic muscle receptor cells in the crab stomatogastric nervous system. II. Rapid nicotinic and prolonged modulatory effects on neurons in the stomatogastric ganglion. *J. Neurophysiol.* 62, 571–581.

Katz, P. S., and Harris-Warrick, R. M. (1991). Recruitment of crab gastric mill neurons into the pyloric motor pattern by mechanosensory afferent stimulation. *J. Neurophysiol.* 65, 1442–1451.

Katz, P. S., Eigg, M. H., and Harris-Warrick, R. M. (1989). Serotonergic/cholinergic muscle receptor cells in the crab stomatogastric nervous system. I. Identification and characterization of the gastropyloric receptor cells. *J. Neurophysiol.* 62, 558–570.

Keller, R., Jaros, P. P., and Kegel, G. (1985). Crustacean hyperglycemic neuropeptides. *Am. Zool.* 25, 207–221.

Kennedy, M. B., and Marder, E. (1992). Cellular and molecular mechanisms of neuronal plasticity. *In* "An Introduction to Molecular Neurobiology" (Z. W. Hall, ed.), pp. 463–495. Sinauer, Sunderland, Massachusetts.

Kirk, M. D., and Govind, C. K. (1992). Early innervation of abdominal swimmeret muscles in developing lobsters. *J. Exp. Zool.* 261, 298–309.

Kleinholz, L. H., and Keller, R. (1979). Endocrine regulation in Crustacea. *In* "Hormones and Evolution" (E. J. W. Barrington, ed.), Vol. 1, pp. 159–213. Academic Press, New York.

Kobierski, L. A., Beltz, B. S., Trimmer, B. A., and Kravitz, E. A. (1987). FMRFamidelike peptides of *Homarus americanus*: Distribution, immunocytochemical mapping, and ultrastructural localization in terminal varicosities. *J. Comp. Neurol.* 266, 1–15.

Kotak, V. C., and Page, C. H. (1986). Sexually dimorphic mechanosensitive swimmeret sensilla affect abdominal posture in the lobster. *J. Neurobiol.* 17, 421–429.

Kravitz, E. A. (1988). Hormonal control of behavior: Amines and the biasing of behavioral output in lobsters. *Science* 241, 1775–1781.

Kravitz, E. A. (1991). The rime of the ancient scientist. *Biol. Bull. (Woods Hole, Mass.)* 180, 329–331.

Kravitz, E. A., Slater, C. R., Takahashi, K., Bownds, M. O., and Grossfeld, R. M. (1970). Excitatory transmission in invertebrates: Glutamate as a potential neuromuscular transmitter compound. *In* "Excitatory Synaptic Transmission" (P. Anderson and J. K. S. Jansen, eds.), pp. 85–93. Universitetsforlaget, Oslo.

Kravitz, E. A., Battelle, B.-A., Evans, P. D., Talamo, B. R., and Wallace, B. G. (1976). Octopamine neurons in lobsters. *Soc. Neurosci. Symp.* 1, 67–81.

Kravitz, E. A., Glusman, B. S., Harris-Warrick, R. M., Livingstone, M. S., Schwarz, T., and Goy, M. F. (1980). Amines and a peptide as neurohormones in lobsters: Actions on neuromuscular preparations and preliminary behavioral studies. *J. Exp. Biol.* 89, 159–175.

Kravitz, E. A., Beltz, B. S., Glusman, S., Goy, M. F., Harris-Warrick, R. M., Johnston, M. F., Livingstone, M. S., Schwarz, T. L., and Siwicki, K. K. (1983). Neurohormones and lobsters: Biochemistry to behavior. *Trends NeuroSci.* 6, 346–349.

Kravitz, E. A., Beltz, B., Glusman, S., Goy, M., Harris-Warrick, R., Johnston, M., Livingstone, M., Schwarz, T., and Siwicki, K. K. (1985). The well-modulated lobster: The roles of serotonin, octopamine, and proctolin in the lobster nervous system. In "Model Neural Networks and Behavior" (A. Selverston, ed.), pp. 339–360. Plenum, New York.

Kuffler, S. W. (1958). Synaptic inhibitory mechanisms, properties of dendrites and problems of excitation in isolated sensory nerve cells. *Exp. Cell Res., Suppl.* 5, 493–519.

Lemos, J. R., and Berlind, A. (1980). Cyclic adenosine monophosphate mediation of peptide neurohormone effects on the lobster cardiac ganglion. *J. Exp. Biol.* 90, 307–326.

Livingstone, M. S., Harris-Warrick, R. M., and Kravitz, E. A. (1980). Serotonin and octopamine produce opposite postures in lobster. *Science* 208, 76–79.

Livingstone, M. S., Schaeffer, S. F., and Kravitz, E. A. (1981). Biochemistry and ultrastructure of serotonergic nerve endings in the lobster: Serotonin and octopamine are contained in different nerve endings. *J. Neurobiol.* 12, 27–51.

Ma, P. M., Beltz, B. S., and Kravitz, E. A. (1992). Serotonin-containing neurons in lobsters: Their role as gain-setters in postural control mechanisms. *J. Neurophysiol.* 68, 36–54.

Macmillan, D. L. (1975). A physiological analysis of walking in the American lobster, *Homarus americanus*. *Philos. Trans. R. Soc. London, Ser. B* 270, 1–59.

Marder, E. (1987). Neurotransmitters and neuromodulators. *In* "The Crustacean Stomatogastric System" (A. I. Selverston and M. Moulins, eds.), pp. 263–300. Springer-Verlag, Berlin.

Meusy, J.-J., Martin, G., Soyez, D., Van Deijnen, J. E., and Gallo, J.-M. (1987). Immunochemical and immunocytochemical studies of the crustacean vitellogenesis-inhibiting hormone. *Gen. Comp. Endocrinol.* 55, 208–216.

Mulloney, B., and Hall, W. M. (1991). Neurons with histaminelike immunoreactivity in the segmental and stomatogastric nervous systems of the crayfish *Pacifastacus leniusculus* and the lobster *Homarus americanus*. *Cell Tissue Res.* 266, 197–207.

Nadol, J. B., Jr., and de Lorenzo, A. J. D. (1968). Observations on the abdominal stretch receptor and the fine structure of associated axo-dendritic synapses and neuromuscular junctions in *Homarus*. *J. Comp. Neurol.* 132, 419–443.

Nadol, J. B., Jr., and de Lorenzo, A. J. D. (1969). Observations on the organization of dendritic processes and receptor terminations in the abdominal muscle receptor organ of *Homarus*. *J. Comp. Neurol.* 137, 19–58.

Nicholls, J. G., Martin, A. R., and Wallace, B. G. (1992). "From Neuron to Brain," Chap. 14. Sinauer, Sunderland, Massachusetts.

Orona, E., Battelle, B.-A., and Ache, B. W. (1990). Immunohistochemical and biochemical evidence for the putative inhibitory neurotransmitters histamine and GABA in lobster olfactory lobes. *J. Comp. Neurol.* 294, 633–646.

Otsuka, M., Iverson, L. L., Hall, Z. W., and Kravitz, E. A. (1966). Release of gamma-aminobutyric acid from inhibitory nerves of lobster. *Proc. Natl. Acad. Sci. U.S.A.* 56, 1110–1115.

Otsuka, M., Kravitz, E. A., and Potter, D. D. (1967). Physiological and chemical architecture of a lobster ganglion with particular reference to gamma-aminobutyrate and glutamate. *J. Neurophysiol.* 30, 725–752.

Page, C. H. (1985). Sexual dimorphism of pleopod motor neurons in lobster. *Brain Res.* 339, 154–157.

Pasztor, V. M. (1979). The ultrastructure of the oval organ, a mechanoreceptor in the second maxilla of decapod crustacea. *Zoomorphologie* 193, 171–191.

Pasztor, V. M., and Bush, B. M. H. (1982). Impulse-coded and analog signaling in single mechanoreceptor neurons. *Science* 215, 1635–1637.

Pasztor, V. M., and Bush, B. M. H. (1987). Peripheral modulation of mechano-sensitivity in primary afferent neurons. *Nature (London)* 326, 793–795.

Pasztor, V. M., and Bush, B. M. H. (1989). Primary afferent responses of a crustacean mechanoreceptor are modulated by proctolin, octopamine, and serotonin. *J. Neurobiol.* 20, 234–254.

Pasztor, V. M., and Golas, L. B. (1993). The modulatory effects of serotonin, neuropeptide F1 and proctolin on the receptor muscle of the lobster abdominal stretch receptor and their exoskeletal homologues. *J. Exp. Biol.* 174, 363–374.

Pasztor, V. M., and Macmillan, D. L. (1990). The actions of proctolin, octopamine and serotonin on crustacean proprioceptors show species and neuron specificity. *J. Exp. Biol.* 152, 485–504.

Pasztor, V. M., Katz, R., Weitzner, S., and Bush, B. M. H. (1985). Physiological actions of proctolin in the ventilatory system of crayfish and lobster. *Soc. Neurosci. Abstr.* 11, 711.

Pasztor, V. M., Lange, A. B., and Orchard, I. (1988). Stretch-induced release of proctolin from the dendrites of a lobster sense organ. *Brain Res.* 458, 199–203.

Pavloff, M. S., and Goy, M. F. (1990). Purification and chemical characterization of peptide G1, an invertebrate neuropeptide that stimulates cyclic GMP metabolism. *J. Neurochem.* 55, 788–797.

Sandeman, D., Sandeman, R., Derby, C., and Schmidt, M. (1992). Morphology of the brain of crayfish, crabs, and spiny lobsters: A common nomenclature for homologous structures. *Biol. Bull. (Woods Hole, Mass.)* 183, 304–326.

Sandeman, D. C., Scholtz, G., and Sandeman, R. E. (1993). Brain evolution in decapod crustacea. *J. Exp. Zool.* 265, 112–133.

Schneider, H., Trimmer, B. A., Rapus, J., Echert, M., Valentine, D., and Kravitz, E. A. (1993a). Mapping octopamine-immunoreactive neurons in the central nervous system of the lobster. *J. Comp. Neurol.* 329, 129–142.

Schneider, H., Peckol, E., Budhiraja, P., Beltz, B. S., and Kravitz, E. A. (1993b). Different patterns of appearance of octopamine- and serotonin-immunoreactive neurons in the developing lobster CNS. *Soc. Neurosci. Abstr.* 19, 1273.

Schwarz, T. L., Harris-Warrick, R. M., Glusman, S., and Kravitz, E.A. (1980). A peptide action in a lobster neuromuscular preparation. *J. Neurobiol.* 11, 623–628.

Schwarz, T. L., Lee, G. M.-H., Siwicki, K. K., Standaert, D. G., and Kravitz, E.A. (1984). Proctolin in the lobster: The distribution, release, and chemical characterization of a likely neurohormone. *J. Neurosci.* 4, 1300–1311.

Selverston, A. I., and Moulins, M., eds. (1987). "The Crustacean Stomatogastric System." Springer-Verlag, Berlin.

Siwicki, K. K., and Bishop, C. A. (1986). Mapping of proctolin-like immunoreactivity in the nervous systems of lobster and crayfish. *J. Comp. Neurol.* 243, 435–453.

Siwicki, K. K., Beltz, B. S., Schwarz, T. L., and Kravitz, E. A. (1985). Proctolin in the lobster nervous system. *Peptides (N.Y.)* 6(Suppl. 3), 393–402.

Siwicki, K. K., Beltz, B. S., and Kravitz, E. A. (1987). Proctolin in identified serotonergic, dopaminergic, and cholinergic neurons in the lobster, *Homarus americanus*. *J. Neurosci.* 7, 522–532.

Skinner, D. M. (1985). Interacting factors in the control of the crustacean molt cycle. *Am. Zool.* 25, 275–284.

Soyez, D., Van Deijnen, J. E., and Martin, M. (1987). Isolation and characterization of a vitellogenesis-inhibiting factor from sinus glands of the lobster, *Homarus americanus*. *J. Exp. Zool.* 244, 479–484.

Soyez, D., Le Caer, J. P., Noel, P. Y., and Rossier, J. (1991). Primary structure of two isoforms of the vitellogenesis-inhibiting hormone from the lobster *Homarus americanus*. *Neuropeptides* 20, 25–32.

Stein, P. S. G. (1974). Neural control of interappendage phase during locomotion. *Am. Zool.* 14, 1003–1016.

Stein, P. S. G. (1977). A comparative approach to the neural control of locomotion. *In* "Identified Neurons and Behavior of Arthropods" (G. Hoyle, ed.), pp. 227–329. Plenum, New York.

Tensen, C. P., de Kleijn, D. P. V., and Van Herp, F. (1991a). Cloning and sequence analysis of cDNA encoding two crustacean hyperglycemic hormones from the lobster *Homarus americanus*. *Eur. J. Biochem.* 200, 103–106.

Tensen, C. P., Janssen, K. P. C., Soyez, D., and Van Herp, F. (1991b). Comparative characterization of hyperglycemic neuropeptides from the lobster *Homarus americanus*. *Peptides (N.Y.)* 12, 241–249.

Tensen, C. P., Verhoeven, A. H. M., Gaus, G., Janssen, K. P. C., Keller, R., and Van Herp, F. (1991c). Isolation and amino acid sequence of crustacean hyperglycemic hormone precursor-related peptides. *Peptides (N.Y.)* 12, 673–681.

Trimmer, B. A., Kobierski, L. A., and Kravitz, E. A. (1987). Purification and characterization of FMRFamidelike immunoreactive substances from the lobster nervous system: Isolation and sequence analysis of two closely related peptides. *J. Comp. Neurol.* 266, 16–26.

Waterman, T. H. (1961). Light sensitivity and vision. *In* "The Physiology of Crustacea" (T. H. Waterman, ed.), Vol. 2, pp. 1–64. Academic Press, New York.

Webster, S. G., and Keller, R. (1988). Physiology and biochemistry of crustacean neurohormonal peptides. *In* "Neurohormones in Invertebrates" (M. C. Thorndyke and G. J. Goldsworthy, eds.), pp. 173–196. Cambridge University Press, New York.

Weiger, W. A., and Kravitz, E. A. (1994). Serotonergic neurons in the lobster: Excitatory inputs from the tail fan. *Soc. Neurosci. Abstr.* 20, 816.

Wiersma, C. A. G. (1938). Function of giant fibers of the central nervous system of the crayfish. *Proc. Soc. Exp. Biol. Med.* 38, 661–662.

Wiersma, C. A. G. (1962). The organization of the arthropod central nervous system. *Am. Zool.* 2, 67–78.

Wiersma, C. A. G., and Ikeda, K. (1964). Interneurons commanding swimmeret movements in the crayfish, *Procambarus clarkii* (Girard). *Comp. Biochem. Physiol.* 12, 509–525.

Wiersma, C. A. G., and Ripley, S. H. (1952). Innervation patterns of crustacean limbs. *Physiol. Comp. Oecol.* 2, 391–405.

Wiersma, C. A. G., Furshpan, E., and Florey, E. (1953). Physiological and pharmacological observations on muscle receptor organs of the crayfish, *Cambarus clarkii* Girard. *J. Exp. Biol.* 30, 136–150.

Wilkens, L. A., and Larimer, J. L. (1976). Photosensitivity in the sixth abdominal ganglion of decapod crustaceans: A comparative study. *J. Comp. Physiol. A* 106A, 69–75.

Worden, M. K., Rahamimoff, R., and Kravitz, E. A. (1993). Ion channel activity in lobster skeletal muscle membranes. *J. Exp. Biol.* 182, 113–130.

Worden, M. K., Kravitz, E. A., and Goy, M. F. (1995). Peptide F_1, an N-terminally extended analogue of FMRFamide, enhances contractile activity in multiple target tissues in lobster. *J. Exp. Biol.* 198, 97–108.

CHAPTER

12

Muscles and Their Innervation

C. K. GOVIND
Life Sciences Division
Scarborough College
University of Toronto
Scarborough, Ontario, Canada M1C 1A4

I. Introduction

Mention the word *lobster* and it tends to evoke in most people a vision of delicate fleshy meat to be picked from cavities and crevices of a hard-shelled animal. This culinary delight in the form of muscles forms the most abundant soft tissue in the animal and dictates the importance of lobsters in our market economy. To the lobster, however, muscles bring about movements and constitute the principal motor components of behavior. The behavior of *Homarus americanus* is varied and complex, ranging from gross movements, as in escape tail flips, to fine movements, as in courtship and mating (see Atema and Voigt, Chapter 13). This requires that muscles vary in their fiber composition and in the level of control exerted by their innervation. Understanding the adaptations that permit this wide contractile repertoire provides the conceptual framework for this review of muscles and their innervation.

Our interest in lobsters goes beyond mere culinary aspects to a profound respect for their apparent ability not to age (Govind, 1992a). The American lobster undergoes 20–25 molts from the time of hatching to sexual maturity over a 5- to 7-year span (Hughes *et al.*, 1972). This primary development encompasses three larval stages, a single postlarval or metamorphic stage, and several juvenile stages over which time the lobster grows in weight from <10 mg to 500 g. Growth continues throughout life, as lobsters lack a terminal molt, and 15- 20-kg lobsters are not uncommon (Herrick, 1895). Such large lobsters are estimated to be 50–70 years old, based on the facts that molting frequency declines with age and that body weight at each molt is not doubled (Cooper and Uzmann, 1980). Despite this long life span, lobsters show few conventional signs of aging. For example, some motor reflexes, such as claw closing, are as rapid in old animals as in young adults. Consequently, development and growth of muscles and their innervation are topics of considerable interest and are also considered here.

II. Muscles

A. Organization

Located entirely within the hard shell, or exoskeleton, muscles serve primarily to bring about movements of the exoskeleton. Hence, they are organized into functional groups in the head, thorax, and abdomen and in various appendages of the three body segments. In such groupings, the muscles usually operate as antagonistic pairs, for example, extensor/flexor, levator/depressor, and promotor/remotor. As well, muscles in the heart are responsible for the movement of blood and in the alimentary canal for the movement of digesta.

Lobster muscle has a basic structure similar to that of striated muscle in other crustaceans (Atwood, 1967; Chapple, 1982). The two major structural proteins, myosin and actin, occur as filaments grouped into bundles of various diameters termed myofibrils. Within the myofibrils, the filaments occur in a highly organized linear manner that repeats itself and is symbolized by the sarcomere (Fig. 1A, and B). In longitudinal view, the sarcomere is delimited by adjacent Z-lines. The thin actin filaments project from the Z-line toward the middle of the sarcomere and surround the centrally located thick myosin filaments (Fig. 1C, and D). This gives rise to a central dark band (A-band) with the overlapping actin and myosin filaments and on either side a light band (I-band) with only actin filaments. The overlapping arrangement of thin and thick filaments in the A-band shows each thick filament surrounded by several thin filaments. The middle of the A-band, where only myosin filaments prevail, is not as dark as the rest of the A-band and is referred to as the H-band. In the middle of the H-band can be seen a dark M-line where the myosin filaments attach to each other.

Surrounding each myofibril in the form of a collar is a fenestrated system of tubules forming the sarcoplasmic reticulum. At points along the myofibril are found elements of the transverse tubular system, closely juxtaposed to the sarcoplasmic reticulum and forming diads or triads. The myofibrils are congregated into a fiber that also contains granular sarcoplasm, mitochondria, and nuclei, as well as a delimiting sheath of connective tissue. Muscle fibers, however, subdivide and branch extensively and are electrically connected, forming functional units of several fibers.

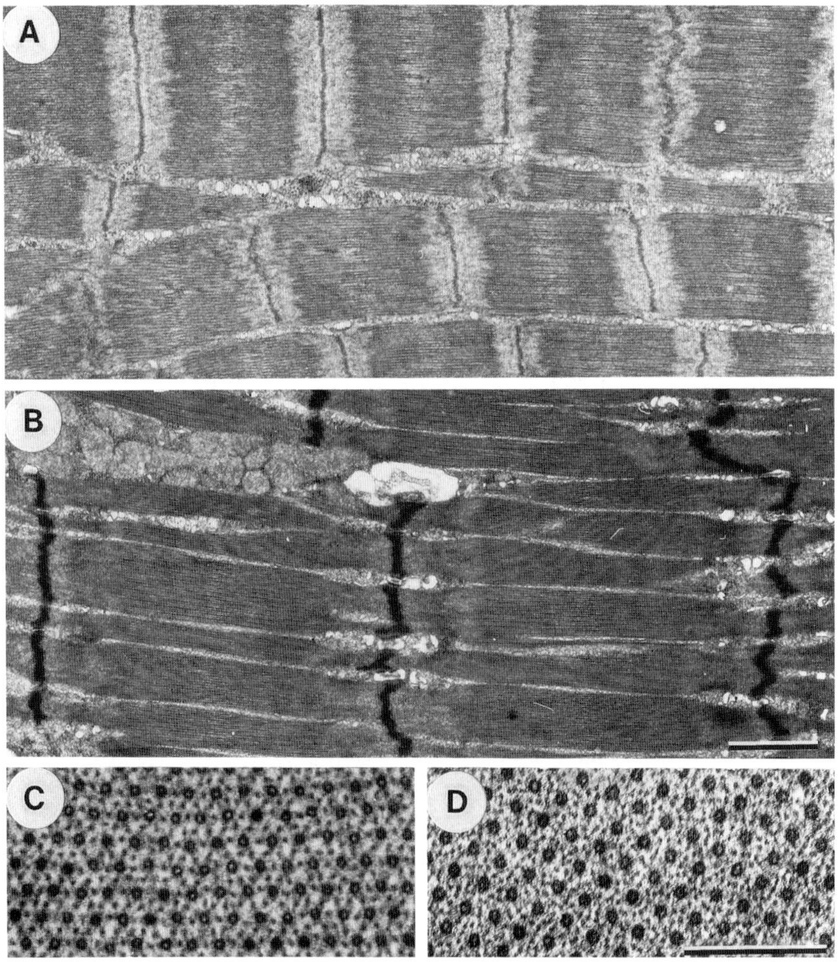

FIGURE 1 Fine structure of fast and slow myofibrils from the antennal remotor muscle. Fast myofibrils have short sarcomeres, narrow, straight Z-lines (A), and a low thin-to-thick filament ratio (C), while slow myofibrils have long sarcomeres, thick, wavy Z-lines (B), and a high thin-to-thick filament ratio (D). Scale bars: (A and B) 2 μm; (C and D) 0.2 μm. (From Bevengut *et al.*, 1993, with permission.)

They are also well vascularized (Govind and Guchardi, 1986).

B. Differentiation of Fiber Types

Historically, crustacean muscle fibers are divided into two broad categories of fast (phasic) and slow (tonic) fibers based on a number of criteria (Atwood, 1973; Govind and Atwood, 1982). Muscles in the lobster *Homarus americanus* fall into these two broad categories and have been used to define the categories. Thus, in terms of contractile properties, fast fibers respond to a standard depolarizing pulse with a quick rise to peak tension and the tension is maintained for the duration of the stimulus (Fig. 2) (Jahromi and Atwood, 1971; Costello and Govind, 1983). Slow fibers, on the other hand, show a continual increase in tension for the duration of the stimulus. Intermediate fibers display both types of contractile responses: an initial rapid rise followed by a slower rise. Usually, the tension developed by single fibers is closely coupled to the degree of depolarization, although in some fibers the increase in tension is disproportionate to the increased level of depolarization. Where such depolarization results in a regenerative response (action potential), a rapid, large contractile response is seen. Fibers also vary in their ability to maintain tension; some are able to maintain the initial level of tension for the entire duration of the stimulus, while others show a decrease, indicating fatigue.

Crustacean muscles display a wide range in sarcomere length, from 2 to 20 μm, a diversity unparalleled in the animal kingdom. Insect muscles show a narrower range, while most vertebrate muscles have a limited range of 2–4 μm. Other things being equal, a fiber with more sarcomeres per unit of length will contract more rapidly than one with fewer sarcomeres (Jahromi and Atwood, 1969; Josephson, 1975). Hence, sarcomere length has been a reliable indicator of the contractile speed of a muscle fiber and, following the vertebrate idiom, crustacean muscle has generally been classified as being fast (phasic) or slow (tonic). Fast fibers have shorter sarcomere lengths than slow fibers and the qualifying length for each type has varied, with fast fibers ranging from 2 to 6 μm and slow, from 6 to 20 μm.

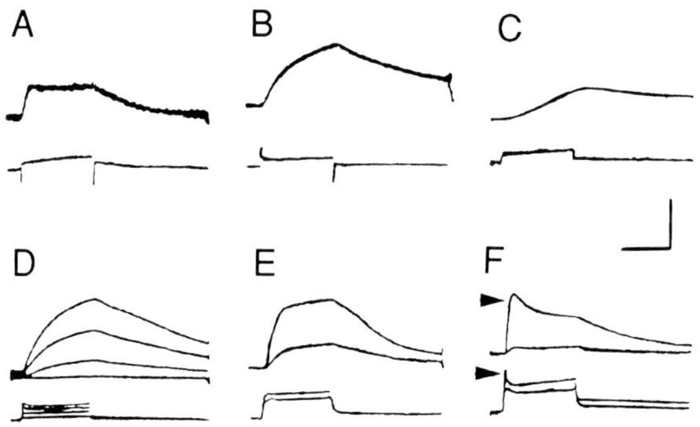

FIGURE 2 Contractile responses of single claw closer muscle fibers (upper traces) to depolarizing pulses (lower traces) demonstrating (A) fast, (B) intermediate, and (C) slow types. Development of tension is closely coupled to the level of depolarization (D) in some fibers, but not in others (E), including those displaying a regenerative response [arrowheads in (F)]. Scale bars: vertical, 5 mg in upper traces of A–E, 10 mg in F; horizontal, 400 msec. (From Costello, W. J., and Govind, C. K. *J. Exp. Zool.* 227, 381–393. Copyright © 1983 Wiley-Liss, Inc. Reprinted by permission of Wiley-Liss, Inc., a subsidiary of John Wiley & Sons, Inc.)

Lobster muscles show an equally wide range in sarcomere lengths from 2 to 15 μm, broadly divisible into the fast (<6-μm) and slow (>6-μm) types listed above (Fig. 1) (Lang *et al.*, 1977a). However, the range in sarcomere length within each type, particularly within the slow type, suggests the existence of subtypes (Govind *et al.*, 1981). Differentiation into fast and slow types is most convincingly seen in that the former has six thin filaments surrounding a thick filament, while the latter has almost twice as many, with 10–13 thin filaments encircling a single thick one (Bevengut *et al.*, 1993).

Each fiber type has a characteristic assemblage of structural and regulatory proteins (Mykles, 1985a). Analysis of myosin, a structural protein, from lobster fast and slow muscle by biochemical techniques reveals that the light-chain isoforms are similar, if not identical, between fast and slow fibers. At least two distinct isoforms of the myosin heavy chain are seen, however, and these regulate the differences in ATPase activity between fast and slow fibers: 11 peptides are unique to fast-muscle myosin and seven are unique to slow-muscle myosin (Li and Mykles, 1990). Analysis of regulatory proteins reveals that fast fibers (F) possess two isoforms of paramyosin (P_1 and P_2), while slow fibers contain only one (P_2) (Fig. 3) (Mykles, 1985b, 1988). Slow fibers, however, differentiate into twitch (S_1) and tonic (S_2) types based on the fact that S_2 fibers contain troponin T_1 (a high-molecular-weight troponin) in addition to the usual lower-molecular-weight variants of troponin (T_2, T_3) found in both S_1 and S_2 fibers. Both types of slow fibers (S_1 and S_2) occur in the cutter closer muscle, while only the tonic type occurs in the superficial abdominal muscles (S_f) (Fig. 3).

Histochemical detection of two enzymes, myofibrillar ATPase and NADH-diaphorase, reveal distinct fast and slow fibers. For example, the deep extensor and flexor muscles in the abdomen, which are uniformly fast, stain much more intensely for ATPase

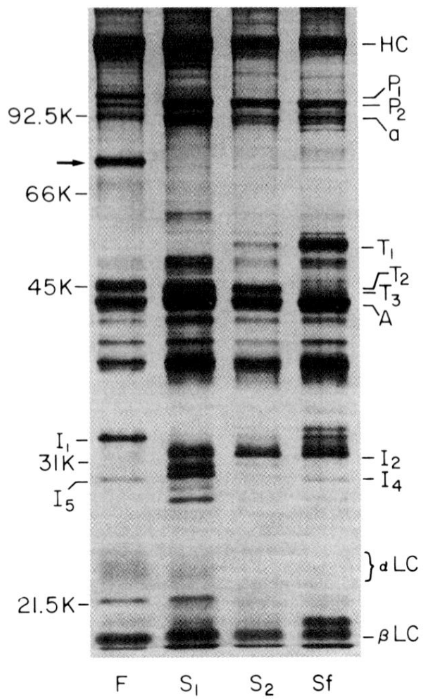

FIGURE 3 Myofibrillar proteins in fast (F) muscle and two types of slow muscle, twitch (S_1) and tonic (S_2 and S_f), revealed in sodium dodecyl sulfate-polyacrylamide gels of glycerinated fibers from the cutter closer muscle (F, S_1, and S_2) and the abdominal superficial flexor muscle (S_f). Molecular weight standards are indicated on the left and identified proteins are mostly on the right (However, I_1 and I_5 are on left). Common proteins include heavy (HC) and light αLC, βLC) chain myosin, actin (A) and alpha-actinin (a). Fast fibers have a higher concentration of troponin isoform I_1 and exclusively paramyosin 1 and an unknown 75-kDa protein. Slow tonic fibers (S_2, S_f) have exclusively a troponinT isoform T_1 and troponinI isoform I_2 as the major form. (From Mykles, D. L. *J. Exp. Zool.* 245, 232–243. Copyright © 1988 Wiley-Liss, Inc. Reprinted by permission of Wiley-Liss, Inc., a division of John Wiley & Sons, Inc.)

than their antagonists, the superficial extensors and flexors, which are uniformly slow (Fig. 4) (Ogonowski and Lang, 1979). Conversely, the abdominal slow muscles stain much more intensely for NADH-diaphorase than their fast counterparts, denoting a higher oxidative capacity for slow fibers. This correlation between high ATPase and low NADH-diaphorase for fast fibers and the reverse for slow fibers applies for most muscles in the lobster, although some fast muscles have a high oxidative capacity. For example, muscles at the base of the abdomen, controlling the swimmerets, have the highest oxidative capacity (Fig. 4), perhaps because these muscles are employed in rapid, prolonged swimming and burrowing behavior in the early juvenile stages. Among the slow fibers, differences in staining intensity for these two enzymes further subdivide them into twitch and tonic types (Kent and Govind, 1981; Mykles, 1985b).

C. Fiber Composition of Muscles

The typing of lobster muscle fibers into fast and slow, based on the parameters listed, allows us to analyze the fiber composition of various muscles (Table 1). Muscles vary in fiber composition; some are homogeneously fast or slow, while others are mixed. The abdominal extensor and flexor systems are examples of homogeneous muscles. They are subdivided into exclusively fast (deep) and exclusively slow (superficial) muscles, which respectively elicit rapid tail flips or slow postural movements. A mixed muscle, such as the closer in the cutter claw, with its predominantly fast-fiber population and a small ventral slow band, elicits both rapid and slow closing movements. Also in mixed muscles, fast and slow fibers usually occur in separate bundles and are not interspersed with each other in a mosaic pattern, as in vertebrate muscle.

The relative proportions of myofibrils to other muscle constituents may also vary, depending on the role of the muscle. In most muscles, myofibrils form the major component (60–80%), with mitochondria and sarcoplasmic reticulum forming much smaller components (10%) (Fig. 5). However, in continuously active muscles such as the larval exopodites (Govind *et al.*, 1988), myofibrils and mitochondria form approximately equal volumes (30–40%), while in very fast-acting muscles such as the antennal remotors, which cause high-frequency (200-Hz) vibration of the antenna (Mendelson, 1969), the sarcoplasmic reticulum is highly elaborate and forms 60% of the volume, compared to 30% for the myofibrils.

D. Development

Development of muscle encompasses not only the initial appearance of myofibrils (myogenesis), but their subsequent divergence into fiber types as well. These two aspects have been examined in various muscles in *Homarus americanus*.

1. Myogenesis

Myogenesis takes place at different times during primary development of the lobster, depending on when the muscle becomes functional during ontogeny. For instance, muscles within the abdomen undergo myogenesis early in embryonic development (Govind *et al.*, 1988) and are functional at the time of hatching, whereas those in the abdominal swimmerets form in late embryos and larval forms (Kirk and Govind, 1992) and become functional only in early juveniles. In larvae, the thoracic exopodites are the principal locomotory organs (Govind *et al.*, 1988).

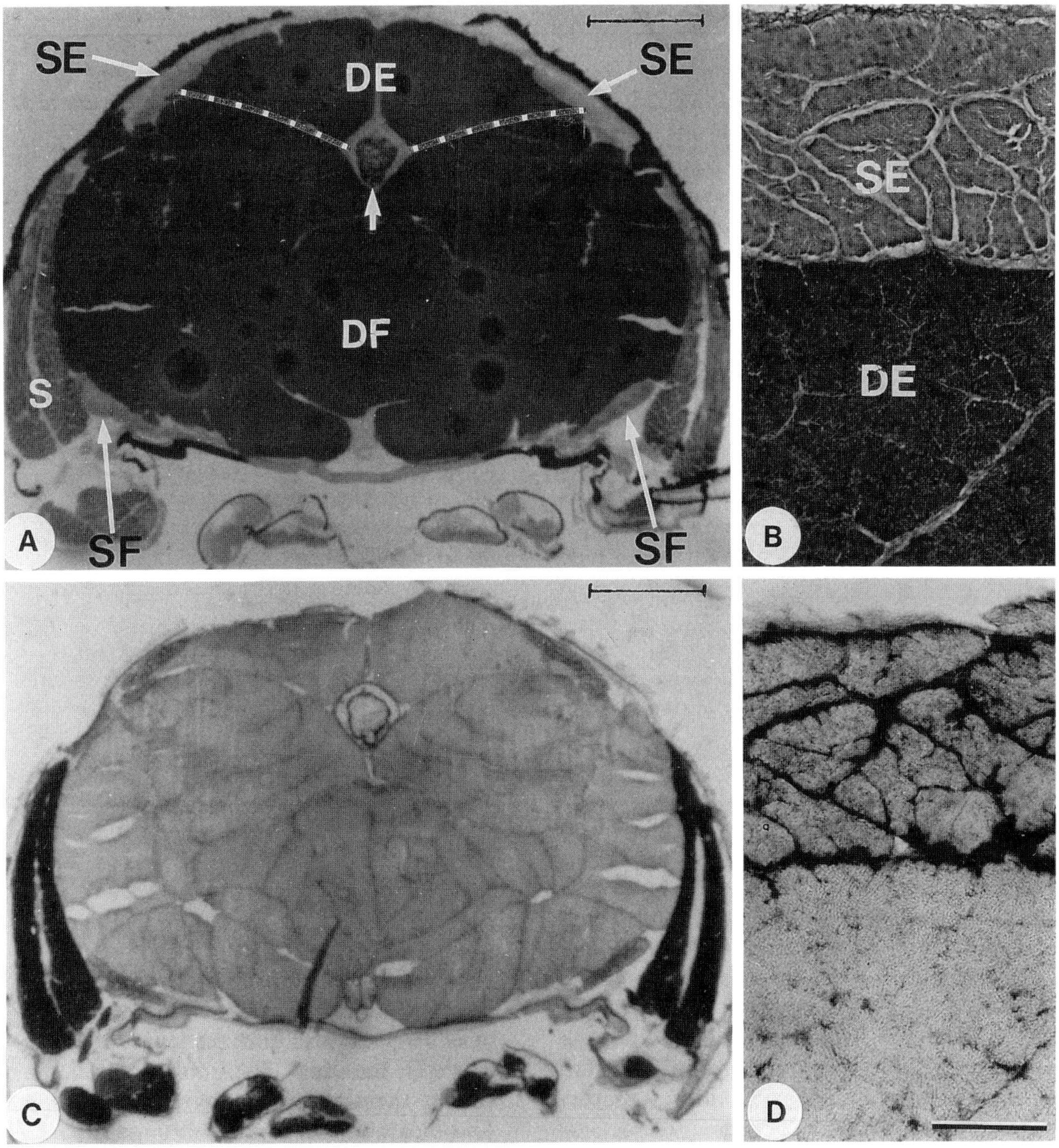

FIGURE 4 Enzymatic profiles of fast and slow muscles histochemically treated for (A and B) myofibrillar ATPase and (C and D) NADH-diaphorase activity in frozen cross-sections through the abdomen. Staining intensity for ATPase denotes higher specific activity for the fast, deep extensor and flexor (DE and DF) muscles than for their slow, superficial (SE and SF) counterparts. Staining intensity for NADH-diaphorase indicates relative oxidative capacity; hence, the slow (SE and SF) muscles have a higher oxidative capacity than their fast counterparts (DE and DF), while the swimmeret muscles (S) show the highest oxidative capacity. Fast–slow differences are shown at higher magnification for the extensor muscles (B and D). The small arrow indicates the alimentary canal. Scale bars: (A and C) 1 mm; (B and D, which are to the same scale) 50 μm. (From Ogonowski, M. M., and Lang, F. *J. Exp. Zool.* 207, 143–151. Copyright © 1979 Wiley-Liss, Inc. Reprinted by permission of Wiley-Liss, Inc., a subsidiary of John Wiley & Sons, Inc.)

Structural observations of myogenesis are, however, similar in muscles with differing developmental time tables.

Presumptive myotubes (syncytia of muscle cells) are recognized by the appearance of a tissue mass containing patches of myofilaments and many prominent nuclei with diffuse chromatin. Next, the myofilaments are organized into their longitudinal arrays and the characteristic latticework of thin filaments surrounding thick-filament forms. At this stage, how-

TABLE 1 **Percentage of Composition of Fast and Slow Fibers in Lobster Muscles**

	Muscle		
	Fast	Slow	Reference
Head			
Antennal remotor	85	15	Bevengut *et al.* (1993)
Antennal promotor	30	70	Bevengut *et al.* (1993)
Thoracic limb			
Claw:			
Cutter closer	80–90	10–20	Govind and Pearce (1988)
Crusher closer	0	100	Jahromi and Atwood (1971), Lang *et al.* (1977a), Ogonowski et al. (1980)
Opener	0	100	Govind *et al.* (1981)
Chelate walking leg			
Closer	40	60	Mearow and Govind (1986)
Opener	0	100	Mearow and Govind (1986)
Rotator	0	100	Sherman and Atwood (1971)
Extensor	0	100	El-Haj *et al.* (1984)
Accessory flexor	0	100	Govind *et al.* (1978)
Nonchelate walking leg			
Closer	0	100	Mearow and Govind (1986)
Opener	0	100	Mearow and Govind (1986)
Exopodite return stroke (larvae only)	100	0	Govind *et al.* (1988)
Abdomen			
Deep extensor	100	0	Jahromi and Atwood (1969)
Superficial extensor	0	100	Jahromi and Atwood (1969)
Stomach			
Gastric mill, GM8	0	100	Katz et al. (1993)
Gastric mill, GM9	100	0	Katz *et al.* (1993)

ever, neither Z-lines nor I-bands are visible, suggesting that the arrangement of thick and thin filaments is the initial event in the formation of sarcomeres. The first signs of innervation are visible at this time in the form of neuromuscular terminals with clear synaptic vesicles and occasional synaptic contacts with the muscle sarcoplasm. Subsequently, these myofilament patches become distinguishable as myofibrils characterized by sarcomeres with well-defined Z-lines, A-bands, and I-bands.

These early sarcomeres are considerably shorter in length (1.5–3 μm) than their mature counterparts (3–15 μm), although, even at this stage, sarcomeres of the putative slow fibers (3 μm) are twice as long as their fast counterparts (Govind *et al.*, 1974). Presumably, fast and slow phenotypes in lobster muscle are established at the time of assembly of the sarcomeres and may be genetically predetermined. Following their initial assembly, sarcomeres gradually elongate to their final length. During this process, the putative slow sarcomeres tend to elongate at a much greater rate than the putative fast sarcomeres, with the result that fast sarcomeres display a two- to threefold increase in length from their initial size (1.5 to 3–4 μm), whereas slow fibers show a three- to fivefold increase (3 to 9–15 μm).

2. Differentiation of Fiber Types

Fibers differentiate into fast and slow types when myofibrils first organize into sarcomeres in those muscles that are homogenous for each type, for example, the deep (fast) and superficial (slow) muscles in the abdominal extensor and flexor systems. This indicates that fiber phenotype in these muscles is genetically determined.

In contrast, differentiation of fiber types is more plastic in the closer muscle of the paired claws, in which there is a mixture of fast and slow fibers early in development (Lang *et al.*, 1977b). In the three larval stages, the paired claws and their closer muscles are symmetrical; both claws are slender and their closer muscles have a central band of putative fast fibers sandwiched by putative slow fibers. From this condition of bilateral symmetry, the paired claws and their closer muscles gradually develop into an asymmetrical major (crusher) and minor (cutter) claw (Fig. 6) (Costello and Govind, 1984; Govind and Lang, 1978; Ogonowski *et al.*, 1980). The crusher is a stout, molar-toothed, slow-acting claw with a closer muscle composed entirely of slow fibers, while the cutter is a

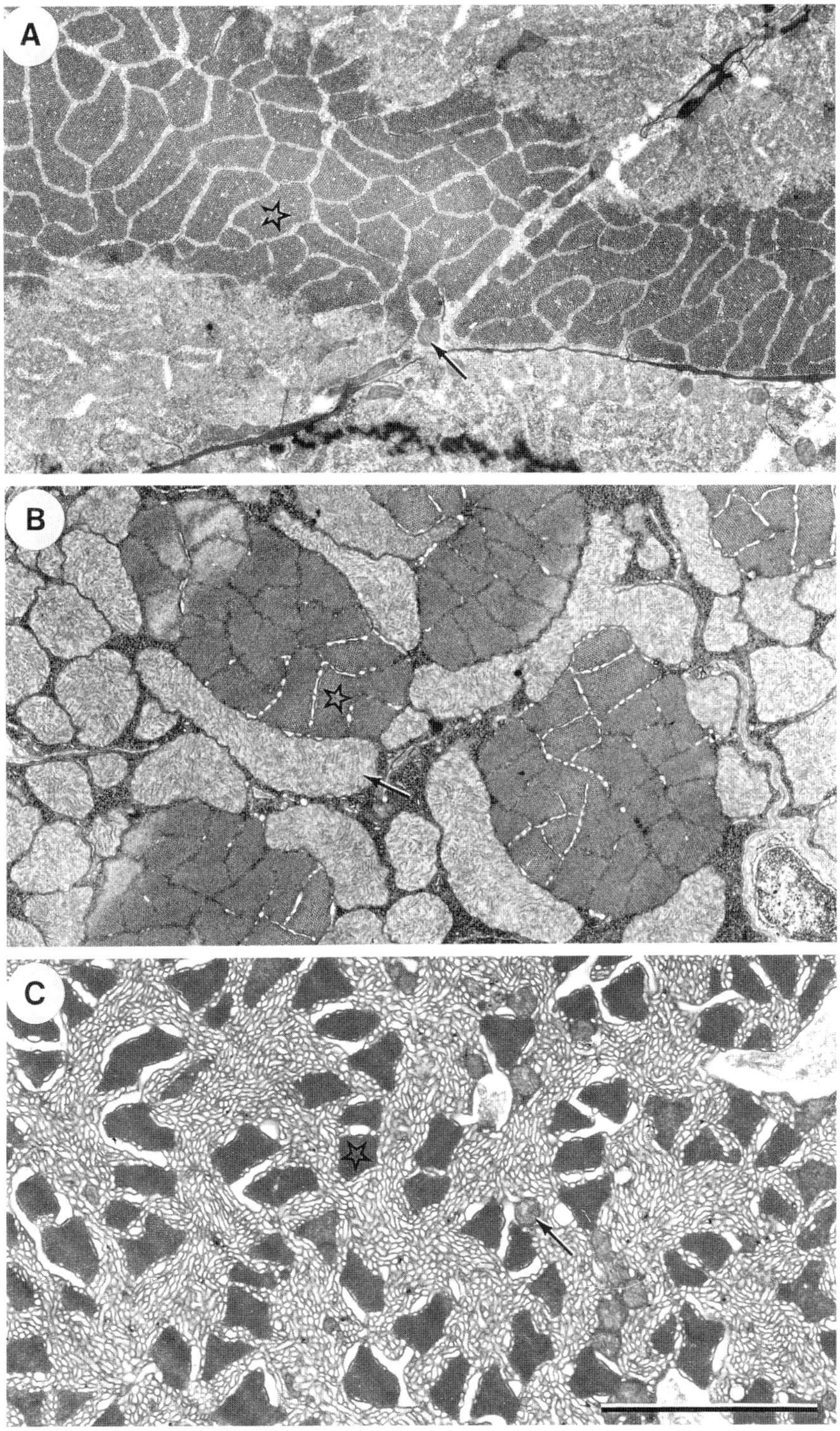

FIGURE 5 Specialization of lobster muscle fibers. (A) Most fibers have myofibrils (star) as the major component, with few scattered mitochondria (arrow) and a thin layer of sarcoplasmic reticulum surrounding the myofibrils. (B) Continuously active fibers, as in the larval exopodite remotor muscle, have myofibrils and mitochondria in about equal volumes. (C) Fibers adapted for high-speed activity, such as in the antennal remotor muscle, have sarcoplasmic reticulum as the major component. Scale bar: 5 μm. (A and C from Bevengut *et al.*, 1993, with permission; B from Govind *et al.*, 1988, with permission from Springer-Verlag.)

slender, incisor-toothed, fast-acting claw with a closer muscle composed of 80–90% fast fibers and a small, ventral band of slow fibers (Govind and Pearce, 1988).

Claw type is determined in the fourth (postlarval) and fifth (first juvenile) stages (Emmel, 1908; Govind and Pearce, 1989a). Once determined during this critical period, it is fixed for life, as subsequent loss of one or both claws results in the regeneration of the same type of claw (Kent *et al.*, 1989). Since the crusher or cutter claw occurs with equal probability on the right or left side of the animal, claw laterality is not genetically specified but is determined during the critical juvenile period by extrinsic factors (Lang *et al.*, 1978). Thus, reflex activity of the claws generates mechanoreceptive and primarily proprioceptive input that presumably lateralizes the claw ganglion; the side with the greater input becomes the crusher side, while the contralateral side becomes the cutter side (Govind and Pearce, 1986). The process may be likened to a teeter-totter, with either the right or left claw being used more often and lateralizing that side of the ganglion into a crusher side; the other side automatically becomes a cutter side and at the same time is precluded from ever becoming a crusher side (Govind and Pearce, 1992). This explains why almost all wild lobsters have bilaterally asymmetrical claws, with a 50:50 ratio of the crusher being on the right or left side. Occasionally, lobsters with paired cutter claws appear. This is explained by a level of sensory input from the developing paired claws that is either similar or subthreshold, keeping the teeter-totter on an even keel and failing to lateralize the ganglion or claws. On the other hand, unless the teeter-totter is broken, both sides of the ganglion are unlikely to become crusher sides, explaining why lobsters with paired crusher claws rarely occur in the wild (Herrick, 1895). Even in these rare lobsters with paired crusher claws, however, only one is a true crusher displaying characteristically slow-closing behavior and a closer muscle with 100% slow fibers. The other claw mimics a cutter in its rapid closing behavior and a closer muscle with 40% fast fibers (Govind and Lang, 1979).

Thus, bilateral asymmetry of the paired claws is initially determined in the claw ganglion; it is subsequently expressed, via unknown pathways, at the periphery, in terms of claw morphology, fiber composition of the closer muscle, and subsequent behavior (Govind, 1992b). In the closer muscle, the putative fast fibers are transformed to slow ones during differentiation of the crusher claw, while some of the putative slow fibers are transformed to the fast type during differentiation of the cutter claw (Govind and Kent, 1982).

E. Growth

During primary development, muscles acquire their individual characters, reflected primarily in their fiber type composition. Characteristics of sarcomere length and the actin–myosin ratio, as well as enzymatic and protein profiles, are established early in juvenile development. Throughout the rest of this stage of development, to maturity and beyond, muscles enlarge in longitudinal and cross-sectional area. Since lobsters lack a terminal molt, growth continues throughout life.

Lobster muscle fibers are complex, multibranched structures, with the branches cytoplasmically connected to each other and to adjacent fibers. This makes it difficult to define discrete fibers and therefore difficult to assess whether fiber number increases during cross-sectional growth of a muscle. A central, difficult issue is whether growth entails the addition of new fibers. In some muscles (e.g., the limb accessory flexor) the fiber number is predicted to remain constant during cross-sectional growth (Govind *et al.*, 1977), while in others (e.g., the claw closer) it is believed to increase (Jahromi and Atwood, 1971). A less contentious finding is the enlargement of existing fibers via the enlargement of individual myofibrils, a process that takes place over the intermolt by the addition of actin and myosin filaments (El-Haj *et al.*, 1984). Just before the molt, the enlarged myofibrils split into smaller units and these hypertrophy during the intermolt.

Growth in length is by the addition of sarcomeres in series, a process that takes place a few hours after the animal has molted (El-Haj *et al.*, 1984; Govind *et al.*, 1977). Immediately after the molt, the integument is enlarged by water uptake and the consequent stretching of the muscle serves to trigger the addition of sarcomeres of a fairly constant size. Such addition probably takes place at the exoskeletal ends of the fibers, based on the presence of slightly shorter sar-

FIGURE 6 Development of the paired homologous closer muscles in lobster claws as viewed in frozen cross-sections of the claw stained for myofibrillar ATPase activity. The paired muscles are symmetrical in the third (larval) and fourth (postlarval) stages, each comprised of a central, dark-staining band of fast fibers sandwiched by light-staining slow fibers. In subsequent juvenile stages (fifth and seventh), most of the fibers transform to fast (except for a small ventral slow band) in the putative cutter muscle and to exclusively slow in the putative crusher muscle. Scale bar: 1 mm. (From Ogonowski *et al. J. Exp. Zool.* 213, 359–367. Copyright © 1980 Wiley-Liss, Inc. Reprinted by permission of John Wiley & Sons, Inc.)

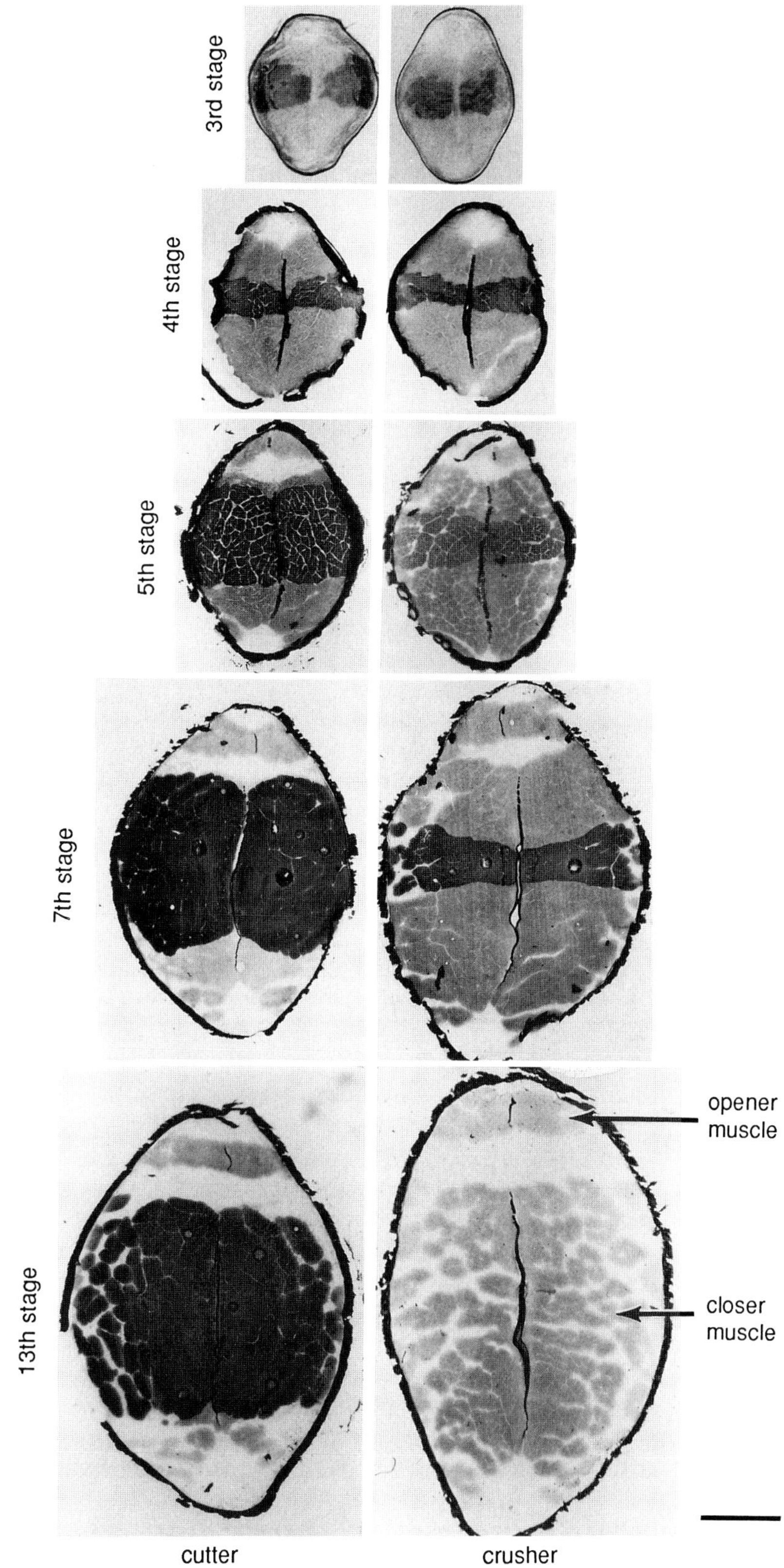
3rd stage
4th stage
5th stage
7th stage
13th stage
opener muscle
closer muscle
cutter
crusher

comeres in these positions. The serial addition of sarcomeres persists throughout life, as fibers continue to lengthen and grow in size with each molt.

F. Regeneration

In juvenile lobsters, the loss of the paired claws at their autotomy plane triggers regeneration of new claws. These resemble their predecessors in external morphology and in the fiber composition of the closer muscle not only after one or two (Kent *et al.*, 1989) but even after four (Govind *et al.*, 1991) successive cycles of limb loss and regeneration. The ability to regenerate entire muscles with unerring fidelity occasionally deviates with the regeneration of a mosaic pattern, wherein slow fibers are interspersed within a fast-fiber band.

III. Motor Innervation

A. Organization

Motoneurons to lobster muscles are located in the ganglia. There, a relatively large soma gives rise to a single primary neurite, from which emerge higher-order branches, including dendrites, to form the neuropilar segment (Fig. 7A) (Davis, 1970; Hill and Govind, 1983). At its distal end, the neurite leaves the ganglion as an axon that travels to the target muscle. The motor axon branches extensively over individual muscle fibers; distributed along these branches are synaptic nerve terminals. There are two types of motoneurons: excitatory neurons, which, via the neurotransmitter glutamate, depolarize the muscle membrane and elicit contraction; and inhibitory neurons, which, via the transmitter γ-aminobuyric acid (GABA), stabilize or hyperpolarize the muscle membrane and prevent contraction (Atwood, 1976; Otsuka *et al.*, 1967).

The morphology of motor innervation of lobster muscles is known from examination of several muscles, although one in particular—the accessory flexor muscle in the walking limb—has been most intensively studied (Govind *et al.*, 1978; Govind and Chiang, 1979). This muscle consists of two heads joined by a tendon and innervated by an excitor axon and an inhibitor axon (Table 2). This account arises largely from our understanding of this muscle, as representative of lobster neuromuscular synapses. In gross morphology, innervation of lobster muscle conforms to the typical crustacean pattern: higher-order branches of the primary motor axon ramify as thin, filament-like branchlets randomly punctuated along their length with varicosities (Fig. 7B).

The axonal branches are wrapped with glial tissue, which becomes thinner in the finer branches and is often absent, thus allowing contact with the muscle membrane (Fig. 8). Synaptic contacts occur in these

TABLE 2 **Number and Types of Motoneurons to Lobster Muscles**[a]

Muscle	Excitor	Inhibitor	Reference
Head			
Antennal remotor	2 (f)	1 (c)	Mendelson (1969)
Antennal extensor	3 (?)	None	Sigvardt (1977)
Antennal flexor	2 (?)	None	Sigvardt (1977)
Thoracic limb			
Closer	2 (f and s)	1 (c)	Wiersma (1955)
Opener	1 (s)*	2 (c and p)	Wiens (1990)
Stretcher	1 (s)*	2 (c and p)	Wiersma (1961)
Bender	2 (s?)	1 (c)	Wiersma (1961)
Rotator	1 (s)†	1 (c)	Sherman and Atwood (1971), Wiens and Govind (1990)
Flexor	4 (f to s)†	1 (c)	Wiersma (1961), Wiens *et al.* (1991)
Accessory flexor	1 (s)	1 (c)	Wiersma (1961)
Abdomen			
Deep extensor	5 (f)	1 (c)	Govind *et al.* (1985)
Swimmerets	1–6 (?)	None	Davis (1968), Kirk and Govind (1992)
Stomach			
Gastric mill, GM8	1 (?)‡	None	Katz *et al.* (1993)
Gastric mill, GM9	1 (?)‡	None	Katz *et al.* (1993)

[a]Excitor motoneurons include fast (f) and slow (s) types and inhibitor neurons include a single neuron to all limb muscles, the common (c) and private or specific (p) neurons. *†‡ Three cases in which a single excitor motoneuron is shared between two muscles.

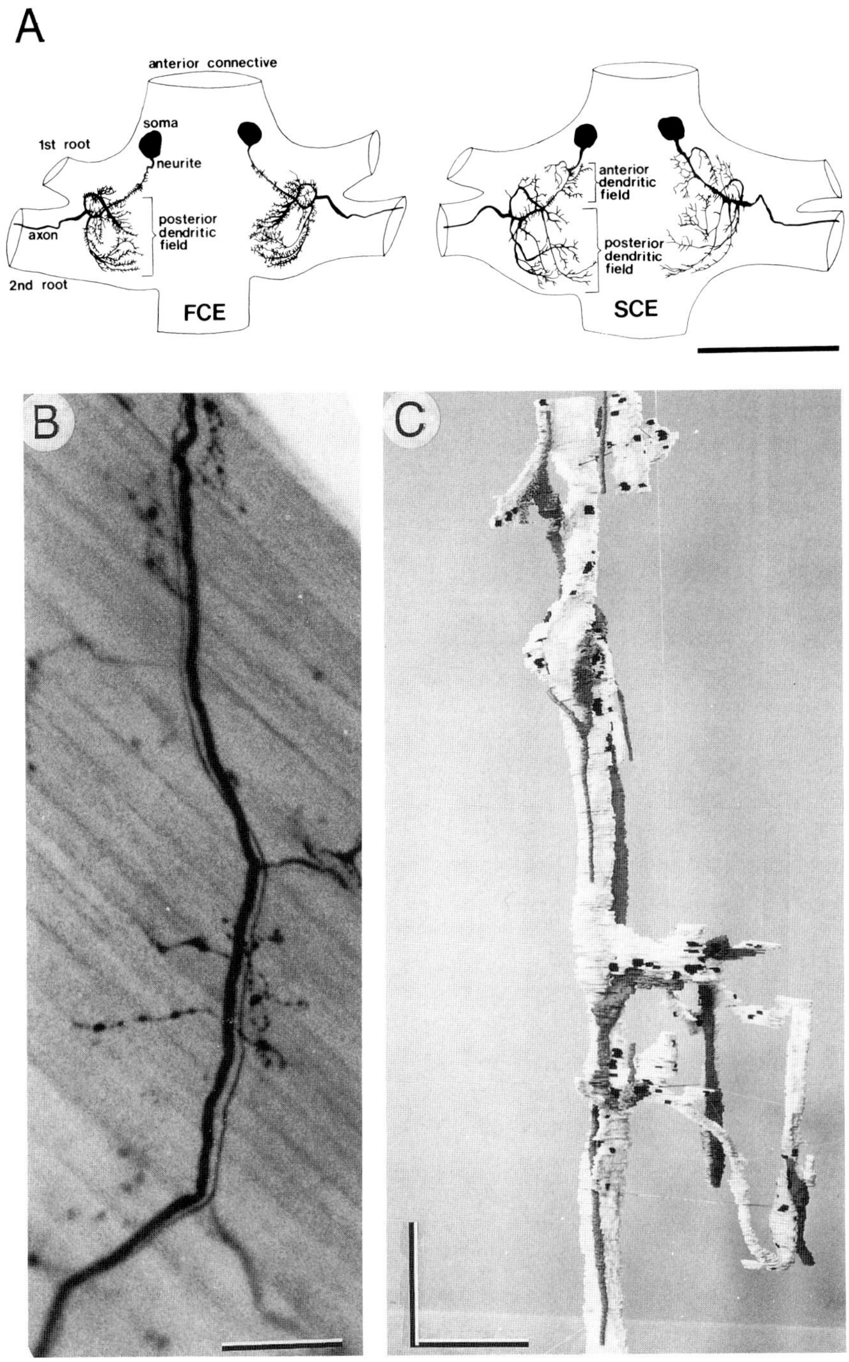

FIGURE 7 Motor innervation of lobster muscle. (A) Morphology of the fast (FCE) and slow (SCE) claw closer motoneurons on the cutter (right side) and crusher (left side) of the ganglion; the SCE has a more elaborate dendritic field than the FCE. (B) Methylene blue staining of excitor (thick) and inhibitor (thin) axons to the accessory flexor muscle showing serially arranged varicosities on the higher-order branches. (C) Three-dimensional reconstruction from serial thin sections of excitor (white) and inhibitor (gray) axon branches to a larval accessory flexor muscle; note the intricate branching pattern and the predominance of the excitor axon, with its larger diameter and higher density of synapses (black areas) compared to the inhibitor axon. Scale bars: (A) 1 mm; (B) 100 μm; (C) 5 μm. (A from Hill and Govind, 1983, with permission from Springer-Verlag; B and C from Govind, C. K., and Pearce, J. *J. Morphol.* 199, 197–205. Copyright © 1989 Wiley-Liss, Inc. Reprinted by permission of John Wiley & Sons, Inc.)

regions, hence their designation as nerve terminals. Internally, the axonal branches are supported by neurotubules, which also decrease in number in the finer branches and are absent in the nerve terminals. Mitochondria occur in the branches and terminals as do other organelles, such as clear and dense-core vesicles.

Dense populations of clear synaptic vesicles, however, are confined to nerve terminals. These vesicles assume a characteristic shape when treated with aldehyde fixatives: round for excitor axons and elliptical for inhibitor axons (Fig. 8B) (Tisdale and Nakajima, 1976). The defining feature of nerve terminals are synaptic contacts, which are recognized by densely stained pres- and postsynaptic membranes, aligned in a very regular manner, and separated by a synaptic gap (Fig. 9D). Often there is a filamentous substructure within the synaptic gap. Synapses tend to be delimited by glial fingers that insinuate themselves in the muscle granular sarcoplasm.

Dense bars are found on the presynaptic membrane; these are T-bar–shaped structures with synaptic vesicles aligned on either side under the transverse arms and in contact with the presynaptic membrane (Fig. 9C and D) (Govind *et al.*, 1980). This arrangement gives the impression that dense bars are docking sites for synaptic vesicles poised for release. This impression is confirmed with the observation of omega-shaped figures representing exocytosis, in which vesicles have fused with the presynaptic membrane along the dense bar. Clearly, transmitter release occurs preferentially along these dense bars or active zones.

The postsynaptic membrane, aside from being more densely stained, appears relatively unspecialized. Occasionally, the muscle sarcoplasm immediately adjacent to this membrane has a fine granular appearance and stains more intensely than the surrounding tissue, denoting the postsynaptic receptor area.

Fracturing the synaptic membrane along its lipid bilayer into two leaflets, the protoplasmic or (P-face) leaflet and the external or (E-face) leaflet, reveals a characteristic organization of the intramembrane particles for the excitor (Fig. 9A and B) (Pearce *et al.*, 1986) and inhibitor synapses (Walrond *et al.*, 1990). For the excitor nerve terminal, the P-face of the presynaptic membrane reveals synapses as elevated plateaus separated by shallow depressions representing an unspecialized terminal membrane. Each plateau has a single active zone in the form of a cluster of large intramembrane particles on a narrow, ridgelike elevation. At the base of the ridge are often found small, circular depressions, presumed to be synaptic vesicles captured in the process of exocytosis. The E-face of the postsynaptic membrane shows tightly packed clusters of particles, presumably representing the receptor proteins for the neurotransmitter (Fig. 10A). These clusters occur in circular and oval patches that correspond in size and position to the P-face plateaus of the presynaptic membrane.

Inhibitory synapses, which are larger than their excitatory counterparts, also appear in the presynaptic P-face membrane as slightly raised plateaus with several, scattered active zones, each as a small cluster of large particles. In the postsynaptic membrane, the putative neurotransmitter receptors display particles in both leaflets: on the P-face, large particles are arranged in parallel rows with complementary furrows in the E-face; on the E-face, shallow particles occur in doublet form (Fig. 10B). This contrasts sharply with excitatory synapses, in which the receptor particles occur only on the postsynaptic E-face and are not arranged with the strict regularity of the inhibitory particles (Fig. 10A).

The preceding description of the morphology of motor innervation derived primarily from skeletal muscles, although stomach muscles show a similar organization (Katz *et al.*, 1993).

B. Differentiation of Innervation

Lobster muscles function so exquisitely with so few motoneurons largely because their innervation is highly differentiated (Atwood, 1976; Govind and Atwood, 1982). Not only are the motoneurons differentiated into excitor and inhibitor types, but the excitor axons themselves are differentiated into fast and slow axons. Moreover, innervation provided by a single motor axon is nonuniform amongst individual muscle fibers and even along a single muscle fiber. These various levels of differentiation allow a single motoneuron or a small number of them to extend the functional range of the muscle, particularly since the degree of contraction of a fiber is dependent on the degree to which it is depolarized (Atwood, 1973, 1976). Differentiation at various levels of the motor synapses is briefly considered in the following sections.

1. Excitatory and Inhibitory Synapses

Each motor axon provides many small nerve terminals over a muscle fiber (i.e., multiterminal innervation). These correspondingly give rise to distributed small synaptic junctional potentials, which, for an excitor axon, activate the contractile apparatus and,

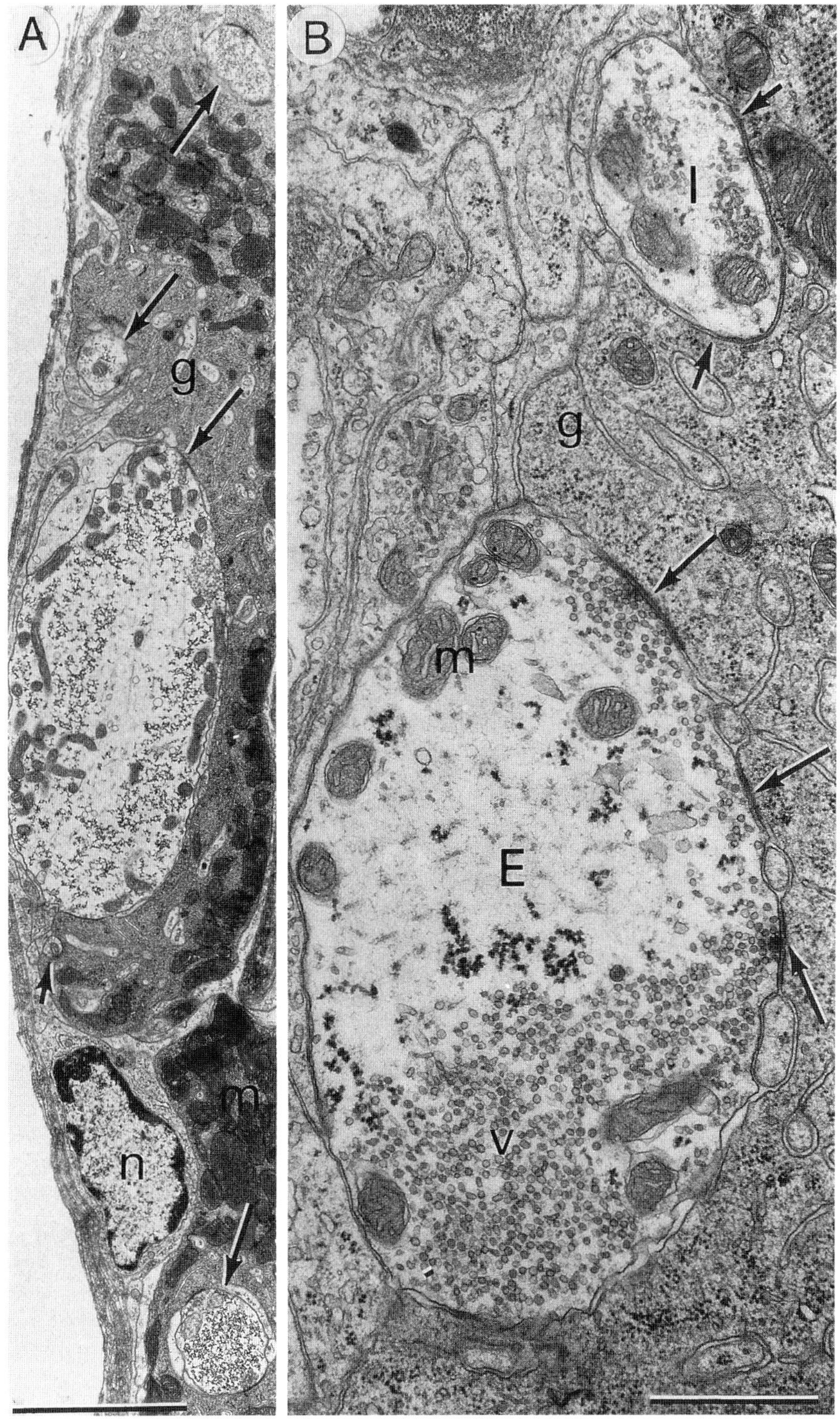

FIGURE 8 Fine structure of synaptic nerve terminals to the distal accessory flexor muscle. (A) Several different-sized excitor nerve terminals (long arrows) and a very small inhibitor terminal (short arrow) located in a bed of granular sarcoplasm (g) at the surface of a muscle fiber. The large excitor terminal is prominently populated by mitochondria and glycogen granules. m, Mitochondria; n, nucleus. (B) The excitor nerve terminal (E), recognized by its population of round, clear synaptic vesicles (v), shows three synaptic contacts (long arrows), each delimited by glial fingers. The inhibitor terminal (I) with elliptical, clear synaptic vesicles shows a single, characteristically long synaptic contact (between short arrows). Scale bars: (A) 4 μm; (B) 1 μm.

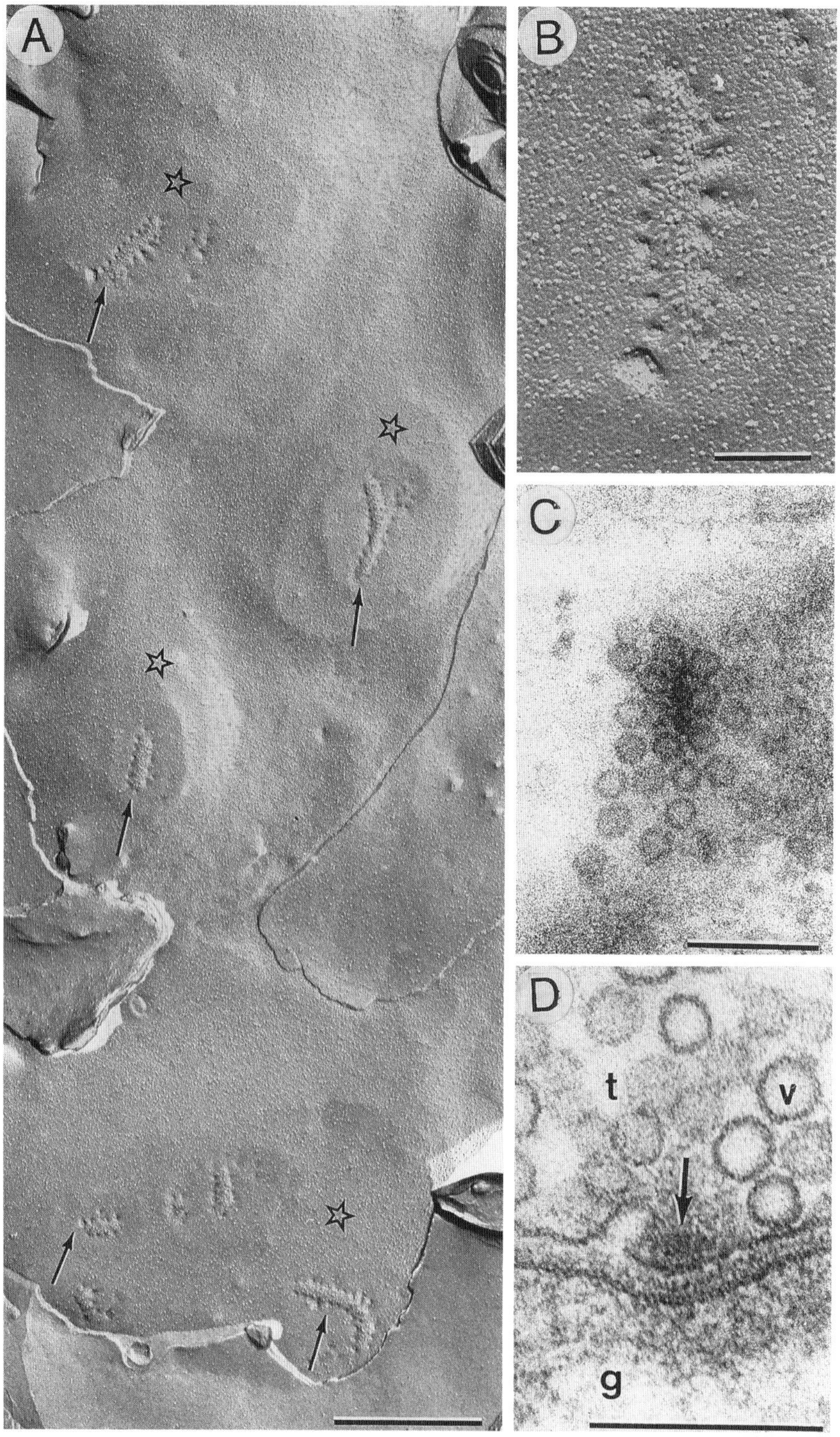

for an inhibitor axon, inhibit the contractile apparatus (Atwood, 1976). In addition to these conventional neuromuscular synapses provided by each type of motor axon, the inhibitor also makes synaptic contact with the excitor axon. This axoaxonal, inhibitor-to-excitor contact is the morphological correlate of presynaptic inhibition by which a suitably timed impulse in the inhibitor axon can, by shunting of the membrane potential of the excitor terminal, effectively reduce transmitter release at the excitatory neuromuscular terminals. Presynaptic inhibition is recorded in the claw opener muscle, where the common inhibitor axon forms axoaxonal synapses with the excitor nerve terminals (Kass-Simon and Govind, 1989).

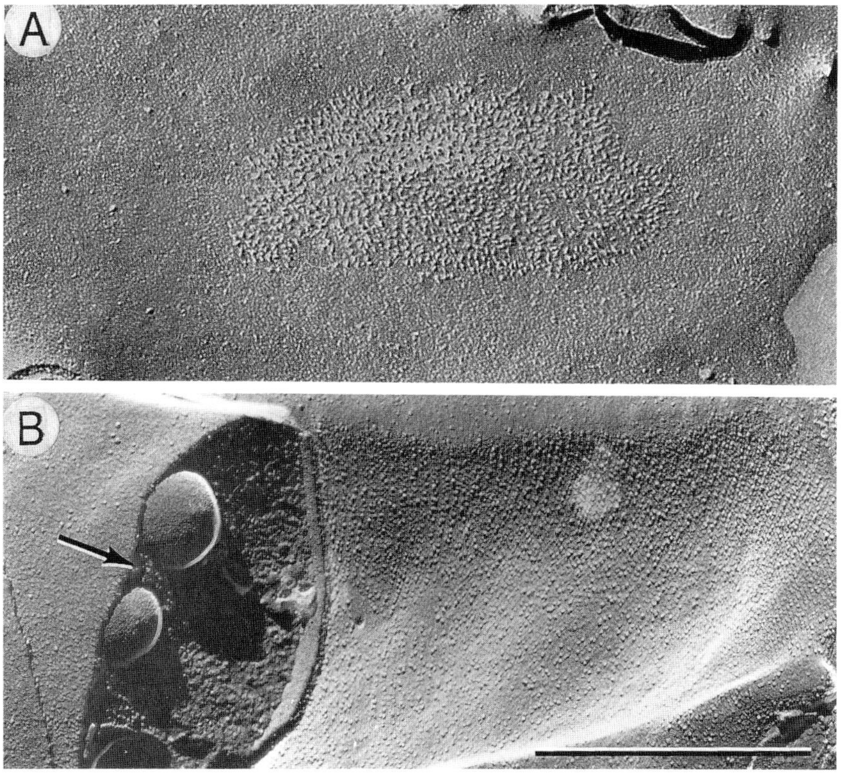

FIGURE 10 Organization of receptor sites on the postsynaptic (muscle) membrane for excitor (A) and inhibitor (B) synapses in the distal accessory flexor muscle (freeze–fracture preparations). Excitor synapses are characterized by an aggregation of large particles on the E-face, while inhibitor synapses show parallel rows of large particles on the P-face. The fracture plane also passes through the inhibitor axon profile (arrow). Scale bar: 1 μm. (From Walrond *et al.*, 1990, with permission from Springer-Verlag.)

2. Fast and Slow Synapses

Excitor neuromuscular synapses are differentiated into fast and slow in keeping with motoneuron differentiation (Govind and Lang, 1981; Hill and Govind, 1983), but also display intermediate forms between these two extremes. The two extreme types of neuromuscular innervation are best represented in the claw closer muscle, which receives two excitors and a common inhibitor axon (Fig. 11A–C) (Govind and Lang, 1974; Costello *et al.*, 1981). The fast synapses generate initially large junctional potentials that facilitate little or even depress with repetitive stimuli. Slow synapses generate initially small junctional potentials which facilitate strongly. Moreover, slow synapses are more fatigue resistant and show better recovery than their fast counterparts. The synapses are also morphologically distinct (Hill and Govind, 1981). Fast axon innervation is relatively simple, with small-diameter nerve terminals populated by few synaptic vesicles, a single long synaptic contact, and little, if any, postsynaptic apparatus. In contrast, slow axon innervation is more complex, with a wide range in size of nerve terminals, each with many synaptic vesicles, several short synapses, and an elaborate postsynaptic apparatus.

Excitor neuromuscular synapses intermediate in character between the fast and slow types are seen in the limb flexor muscle, which receives four excitor axons and the common inhibitor axon. Two of these

FIGURE 9 Organization of excitor synapses and their active zone in the distal accessory flexor muscle. (A) Broad expanse of the P-face of the nerve terminal membrane showing several synapses (stars) as plateaus, each with one or more active zones (arrows) seen as a ridge demarcated by circular, shallow depressions (freeze–fracture preparation). (B) P-face view of the active zone showing ridgelike elevation with large particles (putative calcium channels) bordered by circular, shallow depressions representing synaptic vesicles (freeze–fracture preparation). (C) Grazing thin section through a dense bar (active zone) with clear synaptic vesicles docked around its periphery. (D) Cross-section through synaptic contact showing densely stained pre- and postsynaptic membranes of nerve terminal (t) and muscle granular sarcoplasm (g). The nerve terminal has synaptic vesicles (v), with one in exocytosis adjacent to the dense bar (arrow). Scale bars: (A) 1μm; B–D, 0.2 μm. (From Pearce *et al.* *J. Neurocytol.* 15, 241–252. Copyright © 1986 Chapman & Hall. Reprinted with permission.)

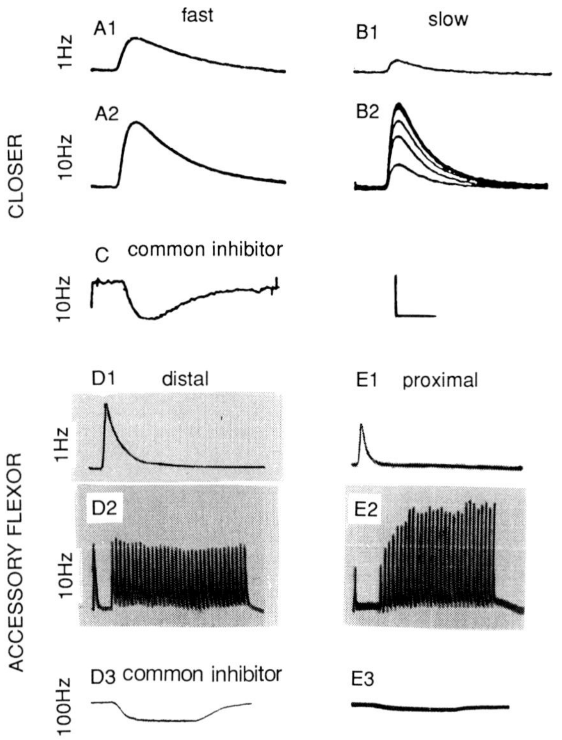

FIGURE 11 Representative junctional potentials from the cutter claw closer muscle (upper three panels) in response to stimulation of fast and slow excitor axons and the common inhibitor axon at different frequencies. Responses to stimulation of the single excitor and inhibitor axon in different regions of the distal accessory flexor muscle (lower three panels). Vertical scale bar: (C) 0.2 mV; (A2, B2, E1, and E2) 4 mV; (A1, B1, D3, and E3) 5 mV; (D1 and D2) 20 mV. Horizontal scale bar: (C) 10 msec; (A and B) 20 msec; (D1 and E1) 40 msec; (D3 and E3) 250 msec; 1 second (D2 and E2). (Upper three panels from Costello *et al.*, 1981; lower three panels from Meiss and Govind, 1979, all with permission from the Company of Biologists Ltd.)

excitors provide innervation reminiscent of fast and slow types, while the other two excitor axons show intermediate characteristics (Wiens *et al.*, 1991).

3. Regional Distribution of Synapses

a. Along Single Muscle Fibers Because muscle fibers are innervated via a multiplicity of nerve terminals that are not uniform in number or size, strength of innervation may also be nonuniform over the length of single muscle fibers. Excitatory synapses at the two ends of single muscle fibers in the distal accessory flexor muscle differ significantly in performance and morphology (Meiss and Govind, 1980). Those at the tendon end release more transmitter, facilitate weakly, and have many more and larger dense bars or active zones than their counterparts at the opposite, exoskeletal end of the fiber. Such regional variability in single muscle fibers may serve to extend their contractile range and ensure smooth, graded contractions.

b. Between Muscle Fibers A single motor axon gives rise to synapses that differ on individual muscle fibers. This fiber-to-fiber variation has a regional pattern best seen in the distal accessory flexor muscle in the thoracic limbs. This muscle, which receives a single excitor and the common inhibitor, is a thin, flag-shaped muscle divisible into five bundles. The two end bundles show marked differentiation in neuromuscular synapses of both axons (Fig. 11D and E) (Meiss and Govind, 1979; Walrond *et al.*, 1990). Excitatory synapses on the distal fiber bundle generate much larger junctional potentials and have a higher quantal output of transmitter than those on the proximal fiber bundle. Despite these differences in transmitter output, the size and number of synapses, as well as the number of dense bars or active zones per synapse, are similar between fibers of the proximal and distal bundles. The major difference is in the size of the active zone; distal ones are much longer and contain more intramembrane particles (Govind and Meiss, 1979; Walrond *et al.*, 1993). Since these particles are putative calcium channels and since calcium is required for transmitter release, the greater number of active-zone particles in the distal synapses most likely results in higher quantal transmitter release.

Firing of the common inhibitor axon generates larger junctional potentials in the distal fibers than in proximal fibers; hence, the distal fibers are much more strongly inhibited than their proximal counterparts (Walrond *et al.*, 1990). The size of active zones and their number per synapse are, however, similar between fibers of the proximal and distal bundles (Govind and Pearce, 1989b). However, there is a higher density of similar-sized varicosities of the inhibitor axon on distal muscle fibers compared to proximal fibers (Walrond *et al.*, 1993). Since most synaptic contacts formed by the motor axon occur in such varicosities, the density of synapses is higher on distal fibers than on proximal ones. This accounts for the much stronger inhibition of the distal muscle fibers.

The structural basis for the higher transmitter release on distal fibers compared to that of their proximal counterparts in the accessory flexor muscle is unique for each of the two axons. For the excitor axon, differences between these two regions are regulated by the size of the active zone, while, for the inhibitor axon, differences in synapse number regulate regional disparities. These different strategies, resulting in regional specialization of neuromuscular synapses, provide fine control of muscle contraction.

c. Between Muscles When synaptic differentiation from a single motor axon occurs in two functionally separate muscles, its role becomes more evident

(Katz *et al.*, 1993). A single excitor axon innervates two gastric mill muscles and produces in one (GM8) initially small junctional potentials that summate and facilitate strongly, while in the other (GM9), initially large junctional potentials that depress. With these synaptic patterns, the GM8 muscle shows a slow, gradually increasing contraction that is maintained for the duration of the stimulation, while the GM9 muscle shows an initial, rapid, near-maximal contraction that subsides to a lower level. In the intact foregut, these contractile patterns are functionally adaptive. A burst of firing in the motoneuron immediately contracts GM9, momentarily closing a valve and then reopening it as the muscle relaxes. Meanwhile, the slow, gradual build-up of contraction in GM8 ensures that movements of the lateral tooth occur as the valve is reopened. The contractile behavior of these two functionally divergent muscles is therefore orchestrated by differences in their neuromuscular synapses and not by differences in the motor firing pattern.

C. Innervation of Muscles

A listing of the innervation of lobster muscles (Table 2) reveals that the simplest innervation pattern is that of a single excitor axon in some stomach muscles (e.g., GM8 and GM9), a pattern suited to the relatively stereotypic activity of these muscles. The different roles of these two muscles are achieved via frequency-dependent modulation of the neuromuscular synapses. More complex innervation patterns are seen with multiple axons and these may be exclusively excitor axons, as in the antennal extensor and flexor muscles used in rapid antennal whipping during social interactions (Atema and Voigt, Chapter 13), or a mixture of excitors and inhibitors, as in the remaining muscles listed in Table 2. In muscles with inhibitory innervation, a common inhibitor is invariably present and its role may be to accentuate the activity of fast fibers by preferentially inhibiting slow ones (Wiens, 1989). A separate private or specific inhibitor to the stretcher and opener muscles permits these two widely separated and functionally divergent muscles to operate via a common excitor axon.

D. Development

Embryonic events in the formation of neuromuscular synapses are not known. However, small, vesicle-filled nerve terminals with well-defined synaptic contacts and dense bars are present for both excitatory and inhibitory axons while sarcomeres are still forming in the myotubules (Lang, 1977; Kirk and Govind, 1992). Subsequent steps in the elaboration of innervation are known in the distal accessory flexor muscle (Fig. 12) (Govind and Pearce, 1981, 1982). The single excitor axon traversing the muscle provides the initial innervation together with its five primary branches. Most of the rest of the innervation arises from these five primary branches by a gradual proliferation of higher-order branches that keep pace with the growth in muscle mass. As new synapses are being added to the ever more distal axonal branches, the main axon and the primary branches become increasingly ensheathed by glial and connective tissues and lose their synaptic contacts, resulting in the continual remodeling of multiterminal innervation. Moreover, the occurrence of synapses at putative sprouting sites, such as at branch points and at the very ends of branches, implicates them in the sprouting process (Fig. 13A).

Elaboration of neuromuscular synapses consists of an increase in number and size during primary development for both excitor (Govind *et al.*, 1982) and inhibitor (Govind and Pearce, 1989b) axons in the distal accessory flexor muscle. As synapses enlarge, they acquire more dense bars and become perforated; the latter feature may functionally subdivide synapses. The larval excitatory synapses in the distal accessory

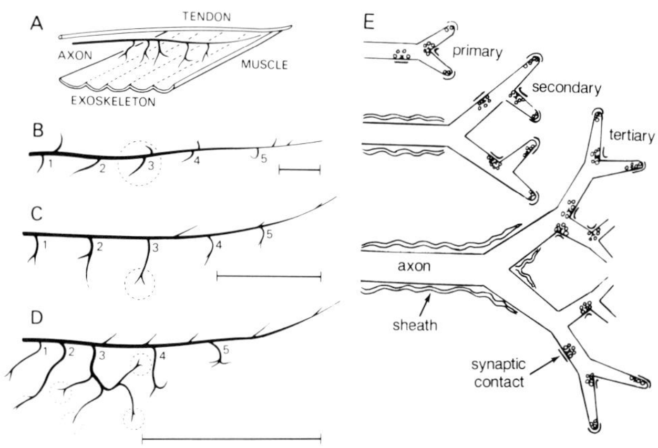

FIGURE 12 (A) Pattern of innervation of the mature distal accessory flexor muscle by five primary branches of the single excitor axon and development of this pattern in a (B) larval, (C) early juvenile, and (D) adult lobster via progressive elaboration of its five primary branches and a concurrent shifting of its synaptic terminals (encircled for branch 3) to more distal sites. (E) Schematic drawing depicting development and growth of the excitor axon to the accessory flexor muscle via sprouting of branches from existing synaptic terminals; note the loss of synapses in the more proximal branches and their generation in the more distal branches. Scale bars: (B) 0.01 mm; (C) 1 mm; (D) 10 mm. (A, B, C, and D from Govind, C. K., and Pearce, J. 1981, *Science* © AAAS; E from Govind and Walrond, 1989.)

flexor muscle are undifferentiated, but gradually during juvenile development acquire their phenotype as high- and low-output synapses based on structural features (Govind *et al.*, 1982). Similarly, neuromuscular synapses of the fast excitor axon are physiologically similar in the degree of facilitation between the paired claws in the earliest juvenile stage and diverge in later stages with greater facilitation in the cutter than the crusher claw (Lnenicka *et al.*, 1988).

E. Growth

Growth beyond sexual maturity for the single excitor axon to the limb accessory flexor muscle has been deduced by comparing adults ranging in weight from 0.3 to 10 kg (DeRosa and Govind, 1978; Pearce *et al.*, 1985). Muscle and its excitatory innervation grow in tandem, and as muscle fibers lengthen there is an almost linear increase in quantal transmitter output. Thus, axonal branches proliferate, increasing the number of nerve terminals, synapses, and active zones. As well, synapses hypertrophy and show perforations, particularly adjacent to active zones, providing a ready source of nonsynaptic axolemma. As with primary development, the putative sprouting sites at the very ends of axonal branches are usually synaptic terminals, suggesting that these are growth points (Fig. 13B).

In addition to growth-associated changes, the excitatory synapses show short-term plasticity. Activation of the motor axon for 1–2 hours results in increased transmitter output accompanied by an increase in the number of active zones (Chiang and Govind, 1986). Conversely, inactivation of terminals for several days leads to a decrease in transmitter output associated with a decrease in the number of synapses, but not in the number of their active zones (Chiang and Govind, 1984).

Innervation to lobster muscles displays considerable plasticity, adjusting the transmitter output and underlying morphology to meet short-term needs as well as the long-term demands associated with growth (Govind and Walrond, 1989). Indeed, the ability for continuing growth may permit the lobster to maintain healthy reflexes and, in this respect, defy conventional aging (Govind, 1992a).

IV. Sensory Innervation

A. Muscle Receptor Organ

Situated on either side of the midline in each abdominal segment is a pair of stretch-sensitive neurons with their dendrites embedded in a central tendinous region of a receptor muscle strand, hence the name muscle receptor organ (MRO). Stretching of the dendritic branches gives rise to generator potentials that spread over the soma and initiate action potentials in the sensory axon (Nadol and de Lorenzo, 1968, 1969). Each MRO is composed of a fast and a slow sensory neuron, each specialized for monitoring powerful, rapid tail flips and slow, postural adjustments, respectively. Correspondingly, mitochondrial content of the soma is higher in the slow cell than in the fast cell (Mayes and Govind, 1989).

The sensory neurons receive input from several inhibitory axons via axoaxonal synapses on the soma and the dendrites, while the receptor muscle receives input from both excitor and inhibitor neurons. Consequently, excitatory and inhibitory neuromuscular synapses, as well as several types of inhibitory output axoaxonal synapses, characterize the MRO (Schaeffer, 1984).

B. Chordotonal Organ

Movements of a joint are monitored by a chordotonal organ that consists of a strand of connective tissue attached proximally to a muscle tendon and distally to the cuticle of the next segment (Bush and Laverack, 1982). The chordotonal organ spanning the propus–dactyl joint in the claw has several hundred bipolar sensory neurons embedded in the connective tissue strand (Cooper and Govind, 1991). The distal dendrites terminate in the connective tissue in specialized scolopidia, while the proximal axons travel in a nerve bundle to the ganglion. The nerve contains many small and a few large axon profiles, suggesting that the organ has a majority of position-sensitive cells and a few movement-sensitive cells. This organ has twice as many cells in the crusher claw compared to its counterpart in the cutter claw, reflecting an asymmetry in the sensory system consistent with the asymmetry in muscles, motoneurons, and behavior between these two claws (Govind, 1992b).

V. Directions for Further Research

Mechanisms that regulate the diversity of fiber types within the basic framework of striated muscle in *Homarus americanus* are ripe for exploration, with the advent of cellular and molecular techniques, and should ultimately lead to uncovering genomic control of muscle differentiation. In particular, transformation of fiber type from fast to slow and vice versa

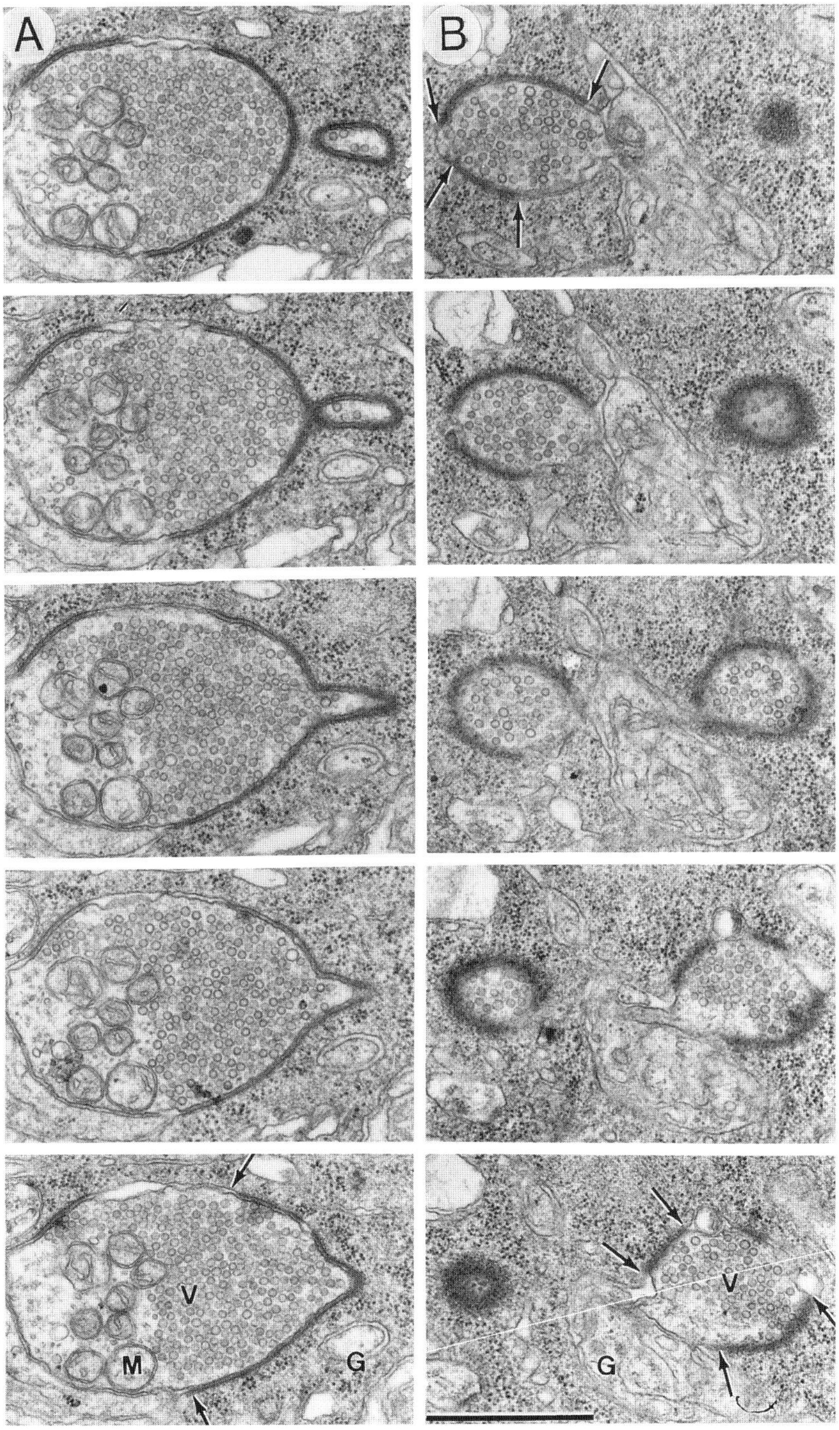

FIGURE 13 Electron micrographs of serial thin sections from the distal accessory flexor muscle of excitatory nerve terminals (recognized by clear spherical vesicles) showing (A) sprouting of a small terminal from an existing synaptic contact (between arrows) and (B) nerve terminals with synaptic contacts (between arrows) that continue to the end. M, Mitochondria; V, vesicles; G, muscle granular sarcoplasm. Scale bar: 1 μm. (A from Pearce *et al.*, 1985, with permission from Elsevier Science Publishers BV; B from Govind, C. K., and Pearce, J. 1981, *Science* © AAAS.)

during development of the bilaterally asymmetric claw closer muscles promises to provide insights into the regulation of muscle plasticity. Along similar lines, studies of neuromuscular synapses arising from single identifiable motoneurons should reveal epigenetic influences regulating the number and size of synaptic contacts, as it is unlikely that there is enough genetic material to specify these properties of multiterminal innervation. Moreover, since lobsters have a long life span (>50 years), punctuated by molts during which growth occurs, changes in the muscle and its innervation may be studied over an individual molt (short-term plasticity) as well as over an entire life span (long-term plasticity). In the latter case, very large and old lobsters display claw-closing reflexes that occur as rapidly as in young adults. Study of long-term neuromuscular plasticity will address how synapses appear to bypass conventional aging.

VI. Summary

The muscle of *Homarus americanus*, conventionally constructed of actin and myosin filaments linearly arranged into sarcomeres, is unique in an unusually large range of sarcomere lengths from 2 to 20 μm. Since contractile speed is dependent on the number of sarcomeres in series, the diversity in sarcomere length reflects an equally wide range in contractile properties. This wide range in fiber types is grouped into two broad categories, fast and slow, based on the number of actin filaments surrounding a myosin filament: typically six for the fast category and 10–12 for the slow category. Individual muscles may be purely fast or slow or mixed, in which case fast- and slow-fiber types are segregated into bundles. Fiber type composition of a muscle is usually specified early in development, although in the paired claw closer muscles an initial mix of fast and slow fibers differentiates during juvenile development into predominantly fast ones in the cutter claw and purely the slow type in the crusher claw. These same muscles continue to grow in mass, in keeping with the allometric growth of the paired claws, and consequently maintain rapid closing reflexes in large (>10-kg) and long-lived (>50 years) lobsters.

Lobster muscle is innervated by excitor motoneurons, which elicit contractions, as well as by inhibitor motoneurons, which prevent contractions by acting on the muscle itself (postsynaptic inhibition) or on the excitor motor nerve terminals (presynaptic inhibition). These motor axons provide a multiplicity of synaptic terminals distributed over a muscle fiber and chemical transmission at these synapses generates small changes in muscle membrane potential that lead to graded contractions. Although few motoneurons (one to six) innervate a muscle, differentiation of neuromuscular synapses between motor axons (fast and slow synapses) and within a motor axon (high- and low-output synapses) permits fine control of muscle activation such that a single excitor motoneuron may operate two functionally divergent muscles. Motor innervation is remodeled during development and growth, with synapse loss in the more proximal branches and proliferation in the more distal branches. Specialization within a basic arthropod blueprint for muscle and its innervation allow lobster muscle to execute the behavioral repertoire characteristic of *Homarus americanus*.

References

Atwood, H. L. (1967). Crustacean neuromuscular systems. *Am. Zool.* 7, 527–551.

Atwood, H. L. (1973). An attempt to account for the diversity of crustacean muscles. Am. Zool. 13, 357–378.

Atwood, H. L. (1976). Organization and synaptic physiology of crustacean neuromuscular systems. *Prog. Neurobiol. (Oxford)* 7, 291–391.

Bevengut, M., McTeague, J. M., and Govind, C. K. (1993). Fiber composition of antennal muscles in lobster (*Homarus americanus*) and crayfish (*Procambarus clarkii*). *J. Crust. Biol.* 13, 256–267.

Bush, B. M. H., and Laverack, M. S. (1982). Mechanoreception. *In* "The Biology of Crustacea" (D. E. Bliss, ed.), Vol. 3, "Neurobiology: Structure and Function" (H. L. Atwood and D. C. Sandeman, eds.), pp. 399–468. Academic Press, New York.

Chapple, W. D. (1982). Muscle. *In* "The Biology of Crustacea" (D. E. Bliss, ed.), Vol. 3, "Neurobiology: Structure and Function" (H. L. Atwood and D. C. Sandeman, eds.), pp. 151–184. Academic Press, New York.

Chiang, R. G., and Govind, C. K. (1984). Decrease in transmitter output and synaptic ultrastructure at lobster neuromuscular terminals with decentralization. *Brain Res.* 299, 265–279.

Chiang, R. G., and Govind, C. K. (1986). Reorganization of synaptic ultrastructure at facilitated lobster neuromuscular terminals. *J. Neurocytol.* 15, 63-74.

Cooper, R. L., and Govind, C. K. (1991). Axon composition of the proprioceptive PD nerve during growth and regeneration of lobster claws. *J. Exp. Zool.* 260, 181–193.

Cooper, R. A., and Uzmann, J. R. (1980). Ecology of juvenile and adult lobsters. *In* "The Biology and Management of Lobsters" (J. S. Cobb and B. F. Phillips, eds.), Vol. 2, pp. 97–142. Academic Press, New York.

Costello, W. J., and Govind, C. K. (1983). Contractile responses of single fibers in lobster claw closer muscles: Correlation with structure, histochemistry and innervation. *J. Exp. Zool.* 227, 381–393.

Costello, W. J., and Govind, C. K. (1984). Contractile proteins of fast and slow fibers during differentiation of lobster claw muscle. *Dev. Biol.* 104, 434–440.

Costello, W. J., Hill, R. H., and Lang, F. (1981). Innervation patterns of fast and slow motor neurones during development of a lobster neuromuscular system. *J. Exp. Biol.* 91, 271–284.

Davis, W. J. (1968). The neuromuscular basis of lobster swimmeret beating. *J. Exp. Zool.* 16, 363–378.

Davis, W. J. (1970). Motoneuron morphology and synaptic contacts; determination by intracellular dye injection. *Science* 168, 1358–1360.

DeRosa, R. A., and Govind, C. K. (1978). Transmitter output increases in an identifiable lobster motoneurone with growth of its muscle fibres. *Nature (London)* 273, 676–678.

El-Haj, A., Govind, C. K., and Houlihan, D. F. (1984). Growth of lobster leg muscle fibers over intermolt and molt. *J. Crust. Biol.* 4, 536–545.

Emmel, V. (1908). The experimental control of asymmetry at different stages in the development of the lobster. *J. Exp. Zool.* 5, 471–484.

Govind, C. K. (1992a). Age-related remodeling of lobster neuromuscular terminals. *Exp. Gerontol.* 27, 63–74.

Govind, C. K. (1992b). Claw asymmetry in lobsters: Case study in developmental neuroethology. *J. Neurobiol.* 23, 1423–1445.

Govind, C. K., and Atwood, H. L. (1982). Organization of neuromuscular systems. *In* "The Biology of Crustacea" (D. E. Bliss, ed.), Vol. 3, "Neurobiology: Structure and Function" (H. L. Atwood and D. C. Sandeman, eds.), pp. 63–103. Academic Press, New York.

Govind, C. K., and Chiang, R. G. (1979). Correlation between presynaptic dense bodies and transmitter output at lobster neuromuscular terminals. *Brain Res.* 161, 377–388.

Govind, C. K., and Guchardi, J. (1986). Vascularization of a lobster muscle. *J. Morphol.* 188, 327–333.

Govind, C. K., and Kent, K. S. (1982). Transformation of fast fibres to slow prevented by lack of activity in developing lobster muscle. *Nature (London)* 298, 755–757.

Govind, C. K., and Lang, F. (1974). Neuromuscular analysis of closing in the dimorphic claws of the lobster, *Homarus americanus*. *J. Exp. Zool.* 190, 281–288.

Govind, C. K., and Lang, F. (1978). Development of the dimorphic claw closer muscles of the lobster *Homarus americanus*. III. Transformation to dimorphic muscles in juveniles. *Biol. Bull. (Woods Hole, Mass.)* 154, 55–67.

Govind, C. K., and Lang, F. (1979). Physiological asymmetry in the bilateral crusher claws of a lobster. *J. Exp. Zool.* 207, 27–32.

Govind, C. K., and Lang, F. (1981). Physiological identification and asymmetry of lobster claw motoneurons. *J. Exp. Biol.* 94, 329–339.

Govind, C. K., and Meiss, D. E. (1979). Quantitative comparison of low- and high-output neuromuscular synapses from a motoneuron of the lobster *Homarus americanus*. *Cell Tissue Res.* 198, 455–463.

Govind, C. K., and Pearce, J. (1981). Remodeling of multiterminal innervation by nerve terminal sprouting in an identifiable lobster motoneuron. *Science* 212, 1522–1524.

Govind, C. K., and Pearce, J. (1982). Proliferation and relocation of developing lobster neuromuscular synapses. *Dev. Biol.* 90, 67–78.

Govind, C. K., and Pearce, J. (1986). Differential reflex activity determines claw and closer muscle asymmetry in developing lobsters. *Science* 233, 354–356.

Govind, C. K., and Pearce, J. (1988). Independent development of bilaterally homologous closer muscles in lobster claws. *Biol. Bull. (Woods Hole, Mass.)* 175, 430–433.

Govind, C. K., and Pearce, J. (1989a). Critical period for determining claw asymmetry in juvenile lobsters. *J. Exp. Zool.* 249, 31–35.

Govind, C. K., and Pearce, J. (1989b). Growth of inhibitory innervation in a lobster muscle. *J. Morphol.* 199, 197–205.

Govind, C. K., and Pearce, J. (1992). Mechanoreceptors and minimal reflex activity determining claw laterality in developing lobsters. *J. Exp. Biol.* 171, 149–162.

Govind, C. K., and Walrond, J. P. (1989). Structural plasticity at crustacean neuromuscular synapses. *J. Neurobiol.* 20, 409–421.

Govind, C. K., Atwood, H. L., and Lang, F. (1974). Sarcomere length increases in developing crustacean muscle. *J. Exp. Zool.* 189, 395–400.

Govind, C. K., She, J., and Lang, F. (1977). Lengthening of lobster muscle fibers by two age-dependent mechanisms. *Experientia* 33, 35–36.

Govind, C. K., Meiss, D. E., She, J., and Yap-Chung, E. (1978). Fiber composition of the distal accessory flexor muscle in several decapod crustaceans. *J. Morphol.* 157, 151–160.

Govind, C. K., DeRosa, R. A., and Pearce, J. (1980). Presynaptic dense bars at neuromuscular synapses of the lobster, *Homarus americanus*. *Cell Tissue Res.* 207, 81–88.

Govind, C. K., Stephens, P. J., and Trinkaus-Randall, V. (1981). Differences in motor output and fiber composition of the opener muscle in lobster dimorphic claws. *J. Exp. Zool.* 218, 363–370.

Govind, C. K., Meiss, D. E., and Pearce, J. (1982). Differentiation of identifiable lobster neuromuscular synapses during development. *J. Neurocytol.* 11, 235–247.

Govind, C. K., Stephens, P. J., and Eisen, J. (1985). Polyneuronal innervation of an adult and embryonic lobster muscle. *J. Embryol. Exp. Morphol.* 87, 13–26.

Govind, C. K., Kirk, M. D., and Pearce, J. (1988). A highly active neuromuscular system in developing lobsters with programmed obsolescence. *J. Comp. Neurol.* 272, 437–449.

Govind, C. K., Gee, C., and Pearce, J. (1991). Retarded and mosaic phenotype in regenerated claw closer muscles of juvenile lobsters. *Biol. Bull. (Woods Hole, Mass.)* 180, 28–33.

Herrick, F. H. (1895). The American lobster: A study of its habits and development. *Bull. U.S. Fish Comm.* 15, 1–252.

Hill, R. H., and Govind, C. K. (1981). Comparison of fast and slow synaptic terminals in lobster muscle. *Cell Tissue Res.* 221, 303–310.

Hill, R. H., and Govind, C. K. (1983). Fast and slow motoneurons with unique forms and activity patterns in lobster claws. *J. Comp. Neurol.* 218, 327–333.

Hughes, J. T., Sullivan, J. J., and Schloser, R. (1972). Enhancement of lobster growth. *Science* 177, 1110–1111.

Jahromi, S. S., and Atwood, H. L. (1969). Correlation of structure, speed of contraction and total tension in fast and slow abdominal muscles of the lobster (*Homarus americanus*). *J. Exp. Zool.* 171, 25–38.

Jahromi, S. S., and Atwood, H. L. (1971). Structural and contractile properties of lobster leg muscle fibers. *J. Exp. Zool.* 176, 475–486.

Josephson, R. K. (1975). Extensive and intensive factors determining the performance of striated muscles. *J. Exp. Zool.* 194, 135–154.

Kass-Simon, G., and Govind, C. K. (1989). Excitatory and inhibitory interactions in the opener muscle of lobster claws. *J. Crust. Biol.* 9, 1–8.

Katz, P. S., Kirk, M. D., and Govind, C. K. (1993). Facilitation and depression at different branches of the same motor axon: Evidence for presynaptic differences in release. *J. Neurosci.* 13, 3075–3089.

Kent, K. S., and Govind, C. K. (1981). Two types of tonic fibers in lobster muscle based on enzyme histochemistry. *J. Exp. Zool.* 215, 113–116.

Kent, K. S., Pearce, J., Gee, C., and Govind, C. K. (1989). Regenerative fidelity in the paired claw closer muscles of lobsters. *Can. J. Zool.* 67, 1573–1577.

Kirk, M. D., and Govind, C. K. (1992). Early innervation of abdominal swimmeret muscles in developing lobsters. *J. Exp. Zool.* 261, 298–309.

Lang, F. (1977). Synaptic and septate neuromuscular junctions in embryonic lobster muscle. *Nature (London)* 268, 458–460.

Lang, F., Costello, W. J., and Govind, C. K. (1977a). Development of the dimorphic claw closer muscles of the lobster, Homarus americanus. I. Distribution of muscle fiber types in adults. *Biol. Bull. (Woods Hole, Mass.)* 152, 75–83.

Lang, F., Govind, C. K., and She, J. (1977b). Development of the dimorphic claw closer muscles of the lobster Homarus americanus. II. Distribution of muscle fiber types in larval forms. *Biol. Bull. (Woods Hole, Mass.)* 152, 382–391.

Lang, F., Govind, C. K., and Costello, W. J. (1978). Experimental transformation of muscle fiber properties in lobster. *Science* 201, 1037–1039.

Li, Y., and Mykles, D. M. (1990). Analysis of myosins from lobster muscle: Fast and slow isozymes differ in heavy chain composition. *J. Exp. Zool.* 255, 163–170.

Lnenicka, G. A., Blundon, J. A., and Govind, C. K. (1988). Early experience influences the development of bilateral asymmetry in a lobster motoneuron. *Dev. Biol.* 129, 84–90.

Mayes, J. I., and Govind, C. K. (1989). Higher mitochondrial density in slow versus fast lobster sensory neurons. *Neurosci. Lett.* 102, 87–90.

Mearow, K. M., and Govind, C. K. (1986). Neuromuscular properties in the serially homologous lobster limbs. *J. Exp. Zool.* 239, 197–204.

Meiss, D. E., and Govind, C. K. (1979). Regional differentiation of neuromuscular synapses in a lobster receptor muscle. *J. Exp. Biol.* 79, 99–114.

Meiss, D. E., and Govind, C. K. (1980). Heterogeneity of excitatory synapses at the ends of single muscle fibers in lobster, *Homarus americanus*. *J. Neurobiol.* 11, 381–395.

Mendelson, M. (1969). Electrical and mechanical characteristics of a very fast lobster muscle. *J. Cell Biol.* 42, 548–563.

Mykles, D. M. (1985a). Heterogeneity of myofibrillar proteins in lobster fast and slow muscles: Variants of troponin, paramyosin, and myosin light chains comprise four different assemblages. *J. Exp. Zool.* 234, 23–32.

Mykles, D. M. (1985b). Multiple variants of myofibrillar proteins in single fibers of lobster claw muscles: Evidence for two types of slow fibers in the cutter closer muscle. *Biol. Bull. (Woods Hole, Mass.)* 169, 476–483.

Mykles, D. M. (1988). Histochemical and biochemical characterization of two slow fiber types in decapod crustacean muscles. *J. Exp. Zool.* 245, 232–243.

Nadol, J. B., Jr., and de Lorenzo, A. J. D. (1968). Observations on the abdominal stretch receptor and the fine structure of associated axo-dendritic synapses and neuromuscular junctions in *Homarus*. *J. Comp. Neurol.* 132, 419–443.

Nadol, J. B., Jr., and de Lorenzo, A. J. D. (1969). Observations on the organization of dendritic processes and receptor terminations in the abdominal muscle receptor organ of *Homarus*. *J. Comp. Neurol.* 137, 19–58.

Ogonowski, M. M., and Lang, F. (1979). Histochemical evidence for enzyme differences in crustacean fast and slow muscle. *J. Exp. Zool.* 207, 143–151.

Ogonowski, M. M., Lang, F., and Govind, C. K. (1980). Histochemistry of lobster claw closer muscles during development. *J. Exp. Zool.* 213, 359–367.

Otsuka, M., Kravitz, E. A., and Potter, D. D. (1967). Physiological and chemical architecture of a lobster ganglion with particular reference to gamma-aminobutyrate and glutamate. *J. Neurophysiol.* 30, 725–752.

Pearce, J., Govind, C. K., and Meiss, D. E. (1985). Growth-related features of lobster neuromuscular terminals. *Dev. Brain Res.* 21, 215–228.

Pearce, J., Govind, C. K., and Shivers, R. R. (1986). Intramembranous organization of lobster excitatory neuromuscular synapses. *J. Neurocytol.* 15, 241–252.

Schaeffer, S. F. (1984). Intramembrane organization of synapses in the lobster stretch receptor organ. *J. Neurocytol.* 13, 351–368.

Sherman, R. G., and Atwood, H. L. (1971). Structure and neuromuscular physiology of a newly discovered muscle in the walking legs of the lobster, *Homarus americanus*. *J. Exp. Zool.* 176, 461–474.

Sigvardt, K. A. (1977). Sensory–motor interactions in antennal reflexes of the American lobster. *J. Comp. Physiol.* 118, 195–214.

Tisdale, A. D., and Nakajima, Y. (1976). Fine structure of synaptic vesicles in two types of nerve terminals in crayfish stretch receptor organs: Influence of fixation methods. *J. Comp. Neurol.* 165, 369–386.

Walrond, J. P., Wiens, T. J., and Govind, C. K. (1990). Inhibitory innervation of a lobster muscle. *Cell Tissue Res.* 260, 421–429.

Walrond, J. P., Govind, C. K., and Huestis, S. E. (1993). Two structural adaptations for regulating transmitter release at lobster neuromuscular synapses. *J. Neurosci.* 13, 4831–4845.

Wiens, T. J. (1989). Common and specific inhibition in leg muscles of decapods: Sharpened distinctions. *J. Neurobiol.* 20, 458–469.

Wiens, T. J. (1990). The inhibitory innervation of the walking leg of the lobster *Homarus americanus*. *J. Comp. Physiol. A* 167A, 43–50.

Wiens, T. J., and Govind, C. K. (1990). The innervation of the lobster limb rotator muscle: Physiology and fine structure. *J. Exp. Zool.* 253, 252–262.

Wiens, T. J., Pearce, J., and Govind, C. K. (1991). Neuromuscular properties of the quintuply innervated flexor muscle in lobster limbs. *Can. J. Zool.* 69, 477–488.

Wiersma, C. A. G. (1955). An analysis of the functional differences between the contraction of the adductor muscles in the thoracic legs of the lobster *Homarus vulgaris* L. *Arch. Neerl. Zool.* 11, 1–13.

Wiersma, C. A. G. (1961). The neuromuscular system. *In* "The Physiology of Crustacea" (T. H. Waterman, ed.), Vol. 2, pp. 191–240. Academic Press, New York.

C H A P T E R

13

Behavior and Sensory Biology

JELLE ATEMA *and* **RAINER VOIGT**
Boston University Marine Program
Marine Biological Laboratory
Woods Hole, Massachusetts 02543

I. Introduction

The study of animal behavior concerns itself with the daily decisions made by individuals, with the mechanisms involved in these decisions, and with the consequences of such decisions. The decisions are not implied to be conscious; they are our human view of an animal's choice to do or not to do—to turn left or right, to enter or not to enter, to flee or fight, etc. Behavioral decisions take place continuously, generally on a time scale of 0.1–10 seconds, and are often based on sensory information playing on memory and motivational state. To know the sensory input that is actually used by animals for behavioral decisions, we need to understand the animal's sensory *Umwelt* (von Uexküll, 1921). Such understanding results from knowledge of the sensory stimuli that are physically present in the animal's specific environment and the many layers of sensory filters used by that animal to generate the signals on which decisions depend.

The behavior of an individual contributes significantly to its survival, growth, and reproductive success, and thus fitness in an evolutionary sense. In an ecological sense, the sum of the daily behavioral decisions made by individuals allows the population to adapt to a constantly changing environment. Consequently, sequences of behavioral decisions on the time scale of seconds modulate the large-scale effects of food availability, predation, and disease (weeks to years), and the even larger-scale effects of climate (years to centuries) on populations. The behavior of individual animals is therefore an important component of any ecological model attempting to predict population size and stability.

Interacting with behavior in determining the fate of organisms and populations are their genetically determined life history strategies. Basic strategies of lobsters include length of embryonic, larval, juvenile, adolescent, and adult stages; seasonal timing of reproduction; number of reproductive cycles; clutch size; growth; and longevity. Complete behavioral studies, therefore, include an understanding of developmental stages that generally have different capabilities and constraints, including behavior and sensory biology. As the lobster grows, its sensory and motoric worlds change.

Finally, the study of behavior can provide some insight into an animal's ability to "predict the future," as measured by the accuracy and range of its forecasting. At one extreme, simple escape responses of planktonic animals are performed in milliseconds, anticipating the predator from the bow wave typical of its approach. At the other extreme, humans modulate their behavioral responses to thousands of other individual acquaintances, remembering their faces and voices for a century with obvious consequences for social organization. Somewhere along the continuum, lobsters modulate their behavioral responses to at least one other individual, remembering the smell of its urine for 1 week, again with social consequences. This memory probably includes several individuals and shelter locations at any time. Clearly, sensory capability, memory, and the ability to combine them play significant roles in the behavior and survival of animals.

This chapter considers the behavior and sensory biology of the American lobster, *Homarus americanus*. The existing data base is focused on late juveniles and adults and is based primarily on laboratory work. Fortunately, research in the last 15 years has begun to document behavior of larval, postlarval, and early juvenile stages, and field research is beginning to make significant contributions to our knowledge of behavioral ecology of the lobster. The boundaries of this chapter fuse easily with the chapters on ecology (Ennis, Chapter 3; Lawton and Lavalli, Chapter 4), muscle physiology (Govind, Chapter 12), and neurobiology (Beltz, Chapter 11). Here, "behavior" includes those data that are obtained by observational or manipulative–experimental study of the motor patterns and movements of individual animals, but not indirect methods such as tag–recapture studies. "Sensory biology" includes primarily the immediate behavioral or neurophysiological responses to sensory stimuli. These animals are emerging as one of the primary models for our understanding of chemical signals and chemoreceptive signal transduction and processing. Far less is known about their other senses.

Information about adults and adolescents is followed, where possible, by a consideration of earlier life history stages following the criteria of Lawton and Lavalli (Chapter 4). Previous reviews of lobster behavior can be found in the work of Dunham (1978), Atema and Cobb (1980), Salmon (1983), Atema (1986), and Waddy and Aiken (1991).

II. Behavior

A. Information Currents

1. Lobster–Generated Currents and Their Role in Behavior

To understand the behavior of *Homarus americanus*, one must appreciate the importance of the mostly invisible, lobster-generated currents that serve locomotion, feeding, and information exchange, as well as the metabolic functions of breathing and waste removal (Fig. 1). Juvenile and adult *H. americanus* utilize three current–generating mechanisms that can operate separately or in combination, all of which are implicated in chemical communication: gill current, exopodite fan current, and pleopod current.

The scaphognathites inside the gill chambers generate a powerful gill current that jets forward from bilateral "nozzles." This current reaches distances of up to seven body lengths in adults (Atema, 1985) and velocities of 3 cm/sec near the nozzle. It is usually a bilateral current and it carries the gill metabolites. Adult lobsters under summer temperatures rarely cease producing this ventilating current; in winter, the current stops for episodes of several seconds, presum-

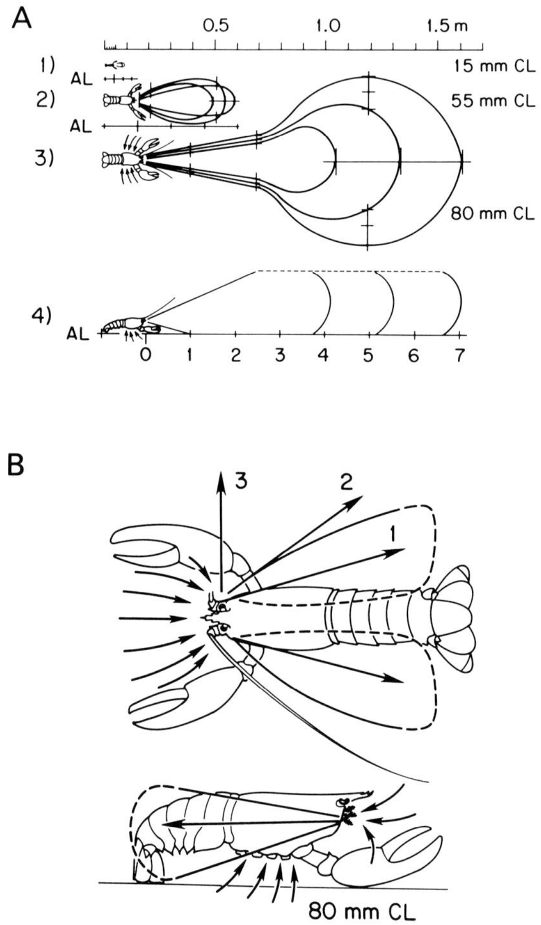

FIGURE 1 Information currents of *Homarus americanus*. (A) Gill currents with mean and standard deviations; top view of three different-sized animals (15-, 55-, and 80-mm CL) and (4) side view of (3) or adult animal (broken line indicates that vertical expansion of plume was limited by vertical stratification of water). Arrows indicate water uptake into gill chamber. AL, Animal length from rostrum to telson to provide the relative scale of current for the three animal sizes; CL, carapace length from eye socket to posterior carapace margin, the standard measurement for lobsters. (B) Exopodite "fan" current. Direction 1 is commonly observed (bilateral outflow); directions 2 and 3 occur occasionally as a result of partial or unilateral fanning. Small arrows show region and direction of inflow water flow drawn toward the lobster; "wings" are areas of turbulent directional current. (From Atema, 1985.)

ably reflecting lower metabolic demands. Into this current, urine can be released from bilateral bladders through small, ventrally directed excretory pores (nephropores) at the base of the antennae. Lobsters appear to release into the urine the products of a pair of small glands located along the ureter, some 100 μm inside the nephropore (Bushmann and Atema, 1993). The glands with their ducts, surrounding muscle tissue, and the valve of the nephropore all appear designed to give the animal control over chemicals released into the gill current, that is, no urine, urine alone, or urine plus glandular products. The glands are composed of rosettes of cells, similar to the rosettes of tegumental glands (Yonge, 1932). However, their size, highly organized structure, and histological staining patterns are different from adjacent tegumental rosettes (Babu *et al.*, 1985; Brunet *et al.*, 1991) and suggest a novel function (Bushmann and Atema, 1994).

A second current, the exopodite fan current, exerts further control over signaling. It appears that the exopodite of the first maxilliped can be positioned directly in front of the gill chamber outflow nozzle, thus deflecting and redirecting forward water flow (T. Breithaupt, unpublished data). The large feathery exopodites of the second and third maxillipeds then fan the deflected water backward, while drawing in a slow flow of water from around the head within a radius of about the length of the antennules (Atema, 1985). The lateral filament of each antennule flicks (Shepheard, 1974; Berg *et al.*, 1992) and thus samples odor within this area. The exopodite fan current can be bilateral or unilateral. The water displaced by these outgoing currents draws in incoming currents carrying chemical signals from the environment that can be sampled by the antennular chemoreceptors.

The third and most powerful current is the pleopod current, which draws water from below the lobster and moves it posteriorly (Atema, 1985). Typically, the lobster raises its tail and beats its pleopods. This current is sufficiently powerful in adults to assist forward motion and rock climbing. Its use in sensory biology is seen in the "advertising" of cohabiting adult males (Cowan and Atema, 1990; see Section II,D). During copulation, both partners beat their pleopods extensively without obvious function.

2. Developmental Stages of Currents

Larval currents are generated by different appendages and serve locomotion and feeding (see Ennis, Chapter 3; Lavalli and Factor, Chapter 14). Postlarval currents are similar to juvenile and adult currents. Settling stage IV lobsters use the pleopod current to swim extensively in search of a suitable settling substrate. Once settled, they use this major current to blow away sand and debris during shelter construction (Botero and Atema, 1982; Atema *et al.*, 1982) and to ventilate their burrows and capture suspended food particles (Lavalli and Barshaw, 1989; see Lawton and Lavalli, Chapter 4). The gill current is weak, extending about one body length (2 cm) in front of the animal (Fig. 1) (Atema, 1985). The exopodite current has not been studied.

The three larval stages of *Homarus americanus* possess feathery exopodites on the outer branches of the thoracic limbs (Herrick, 1895, 1909; Factor, Chapter 1; Lavalli and Factor, Chapter 14). These exopodites provide lift and propulsive power for swimming (Neil *et al.*, 1976). With each larval molt, the thoracic exopodites are reduced in size, until, in stage IV postlarval lobsters, they become small and nonfunctional (Hadley, 1908; Herrick, 1909). At this stage, however, the swimmerets (pleopods) become functional and are used for swimming prior to settlement (Davis, 1974). As the lobster develops, the pleopod current loses its swimming function and assumes a function in agonistic pushing, while continuing to serve an assisting role in rock climbing and general forward motion, in suspension feeding, and presumably in obtaining chemical information. Eventually, in adults, the pleopod current loses its role in suspension feeding and, in breeding females, is used in assisting the metabolism of the egg mass. Behavioral observation suggests that the gill current becomes more and more informational as social behavior develops, although this has not been studied experimentally.

Although a locomotory function is quite possible, the function of the exopodite fan current is probably mostly informational in adults, that is, shifting the gill current from a forward flow, sending chemical signals, to a rearward flow, obtaining chemical signals at the antennules. As carriers of chemical signals, these currents play important roles in dominance and courtship. In all postlarval stages, one may assume a waste removal function for the gill current and possibly also the pleopod current. Larval currents appear to be different (L. Farley and J. Atema, unpublished observations).

B. Dominance

1. Definitions and Function

When two lobsters meet, they may fight or avoid one another, depending on a number of behavioral, physiological, and morphological conditions (see Section IV) that determine their level of *aggression*.

Winning and losing determine *dominance* within the social group. An animal is dominant when others defer to it: the more dominant, the less challenged. Dominance is *established* by behavioral displays and, frequently, by physical fights; it is *maintained* by displays and, rarely, by fights. As displays carry symbolic information about real physical action, chemical signals can be considered displays as much as visual and acoustic signals. Dominance results from the memory of agonistic experience and serves to secure access to shelter and courtship (see Sections II, C and D). Shelters and mating opportunities are limiting resources. Food, in contrast, does not appear to be a limiting resource. There is no evidence that lobsters use dominance to secure food, although a larger lobster occasionally may chase away another animal from a piece of food. Whereas shelters are defended to some degree, there is still no compelling evidence that lobsters defend the perimeter of an area that might be called a *territory*, comparable to what is commonly seen in all classes of vertebrates.

2. Description of Behavior Units and Fight Sequences

The details of agonistic behavior are known predominantly from observations of forced encounters. When two size-matched lobsters are put into a small tank, a "boxing match" results in which aggression can escalate to the point of physical damage (Fig. 2). Animals that may have chosen to avoid one another in the field are in constant, unavoidable proximity. Agonistic encounters in adult *Homarus americanus* follow a few basic patterns that were first described extensively by Scrivener (1971), who summarized the following four fight sequences (italicized behavior units are described in Table 1): (1) the pathway of established winners—*antenna point, approach, meral spread, follow, rush, scissor, meral spread, antenna point*; (2) the first pathway of established losers—*antenna point, back, abdomen flex, back, antenna point*; (3) the less frequently used alternative pathway of losers—*antenna point, walk, run, walk, antenna point*; and (4) the long complex pathway of mutual aggression between unfamiliar and roughly equally matched opponents—*antenna point, approach, meral spread, push, meral spread, antenna point*.

Scrivener's (1971) account has been generally confirmed by subsequent studies (Atema and Engstrom, 1971; Todd *et al.*, 1972; Stein *et al.*, 1975; Jacobson, 1977; Karavanich and Atema, 1991; Huber and Kravitz, 1995). Aggressive behavior escalates in stages from *approach*, to display such as *meral spread*, to physical contact in *push* and *claw lock*, and finally to potentially damaging behavior such as *claw rip*. The opponent can break off into *avoid, run away*, or *tail flip* at any stage.

During *claw lock*, curiously absent in Scrivener's account, lobsters carefully engage the crusher claws in a type of handshake. This "arm wrestling" contest may last from a few seconds to over 1 minute and may include several sharp pulls by either animal. Eventually, one attempts to withdraw and the other releases its grip. In intermolt animals, *claw lock* does not generally cause damage and often settles the test of strength. Handedness refers to the side of the crusher claw, which is randomly distributed in the population (see Govind, Chapter 12). It is remarkable to see that both "right-handed" and "left-handed" lobsters generally use their heavily armed crusher claws, so that in *claw locks* between a right- and a left-handed lobster, one crusher grabs the outside of the other's crusher.

If *claw lock* does not settle the match, then *claw snapping* and *claw ripping* may develop, which can cause real physical damage (Fig. 2). Encounters in large, naturalistic aquaria have resulted in a claw dactyl being snapped off, the tip of the claw propodus being broken off, various puncture wounds being inflicted on claws and body, damaged antennae and antennules, and autotomized claws (J. Atema *et al.*, unpublished observations). Such unrestrained attacks occur typically at the final stages of fights and usually only between unfamiliar and closely matched opponents.

After the first few encounters establish dominance between a pair of lobsters, subsequent agonistic interactions are mostly simple approach–avoid sequences without physical contact (Scrivener, 1971; Karavanich and Atema, 1991). Similar results have been obtained in more natural conditions for groups of five lobsters observed for a 9-month period in large aquaria (Stein *et al.*, 1975): 71% of the observed encounters consisted of simple approach–avoid behavior; 21% were characterized as *attack* and *flee*; and only 8% involved prolonged fights and *claw lock*. Most of the high-intensity aggressive encounters take place soon after the introduction of lobsters into the aquarium, followed by months of approach–avoid encounters (Stein *et al.*, 1975; Atema *et al.*, 1979a; Karnofsky and Price, 1989; Cowan and Atema, 1990). Only molting and mating sometimes induce elevated levels of aggression among long-term aquarium residents (J. Atema *et al.*, unpublished data).

An interesting ritualized agonistic behavior is *knock-on-the-door*, in which the (established?) dominant animal approaches and *taps* or *boxes* at the subordinate's claw blocking the shelter entrance. The dominant then steps aside and allows the subordinate to leave its shelter, whereupon the dominant lob-

TABLE 1 **Terms Used to Describe Agonistic Encounters between Juvenile and Adult Lobsters**[a]

Scrivener (1971)	Atema and Engstrom (1971)	Stein *et al.* (1975)	Huber and Kravitz (1995)	J. Atema *et al.* (unpublished observations)	Brief description
Submissive					
Backing	—	—	Retreating	1. Walking backward, slow	—
Walking away	—	—	Retreating	Walking away, slow	Slow equals <1 body length/5 sec
—	—	—	Retreating	Turning away	Turn so rostrum points away from opponent
Sideways	—	—	—	—	Lateral walk away from opponent
Backing	—	Retreating	Retreating	2. Walking backward, fast	—
Running away	Fleeing	Fleeing	Retreating	Walking away, fast	Fast equals >1 body length/5 sec
Jumping	—	—	—	Jumping	Quick, upward-directed tail flip
Abdomen flexing	Tail flipping	Tail flipping	Tail flipping	3. Tail flipping	Fast, backward escape (with claws and legs extended)
	Submissive posture	—	—	Crouching	Crouch (in corner), claws extended
Aggressive					
—	—	Facing off	—	1. Facing	Facing each other, within 1 body length, no touching
Approaching	Frontal approaching	Advancing	Approaching	Approaching, slow	Walking forward, toward opponent
—	—	—	—	Turning toward	Turn directed toward opponent
Following	—	Following	—	Following, slow	Walking toward opponent while opponent moves away
Rushing	Chasing	Chasing	Lunging	Approaching, fast	—
—	Chasing	—	—	Following, fast	—
—	—	Defensive posturing	—	2. Claw open	Either or both claws completely open, usually seizer
—	—	En garde	—	Claw forward	One claw, usually seizer, extended directly forward
—	—	—	Body up	High on legs	Body raised high off substrate
Meral spreading	—	Meral spreading	Claw up	Meral spreading	Both claws extended up and out from body
Antenna pointing	Antenna searching	—	—	3. Antenna pointing	One or both antennae directed (forward) toward opponent
—	—	—	—	Antenna touching	One or both antennae continuously touching opponent
Antenna whipping	—	Antenna feeling	Antennae tapping, whipping	Antenna whipping	Lashing of opponent with antenna(e) in sweeping motion
—	—	—	Claw touching open, closed	Claw touching	Continuous touching of opponent with (closed) claw
Pushing	Pushing	Pushing	Pushing	Claw pushing	Continuous pressing of claw(s) on opponent's body
Boxing	—	Jabbing	Claw striking	Claw boxing	Back- or forehanded striking motion toward opponent
—	—	Lunging	—	Claw lunging	Thrust claw(s) forward
Scissoring	—	—	—	Claw scissoring	Both claws rapidly crossing in front of body from meral spread
—	—	Swatting	—	—	Partial scissoring
—	Claw locking, grasping	Claw locking	Claw grasping	4. Claw locking	Clamping of claws onto opponent's claw(s) or body
—	—	Snapping	Claw grasping	5. Claw snapping	Rapid opening and closing of (seizer) claw (toward opponent)
—	—	Ripping	Claw ripping	Claw ripping	(Rapid) grasp and pull with either claw
Neutral					
—	—	—	—	0. Tail under	Abdomen flexed under cephalothorax
—	—	—	—	Walking	—
Rapid turning	—	—	Turning	Rapid turning	Turning not related to opponent
—	—	—	Antennae up	Antenna up	Antenna pointing directly up
—	—	—	—	Pleopod beating	Fanning movement of pleopod appendages
—	—	—	—	Antennule flicking	Rapid downstroke of lateral filament
—	—	—	—	Exopodite fanning	Beating of exopodites of maxillipeds
—	—	—	—	Separating	>1 bo9dy length apart

[a]Numbers indicate ranks of aggression. Other, comparable descriptions are found in the work of Snyder *et al.* (1992, 1993).

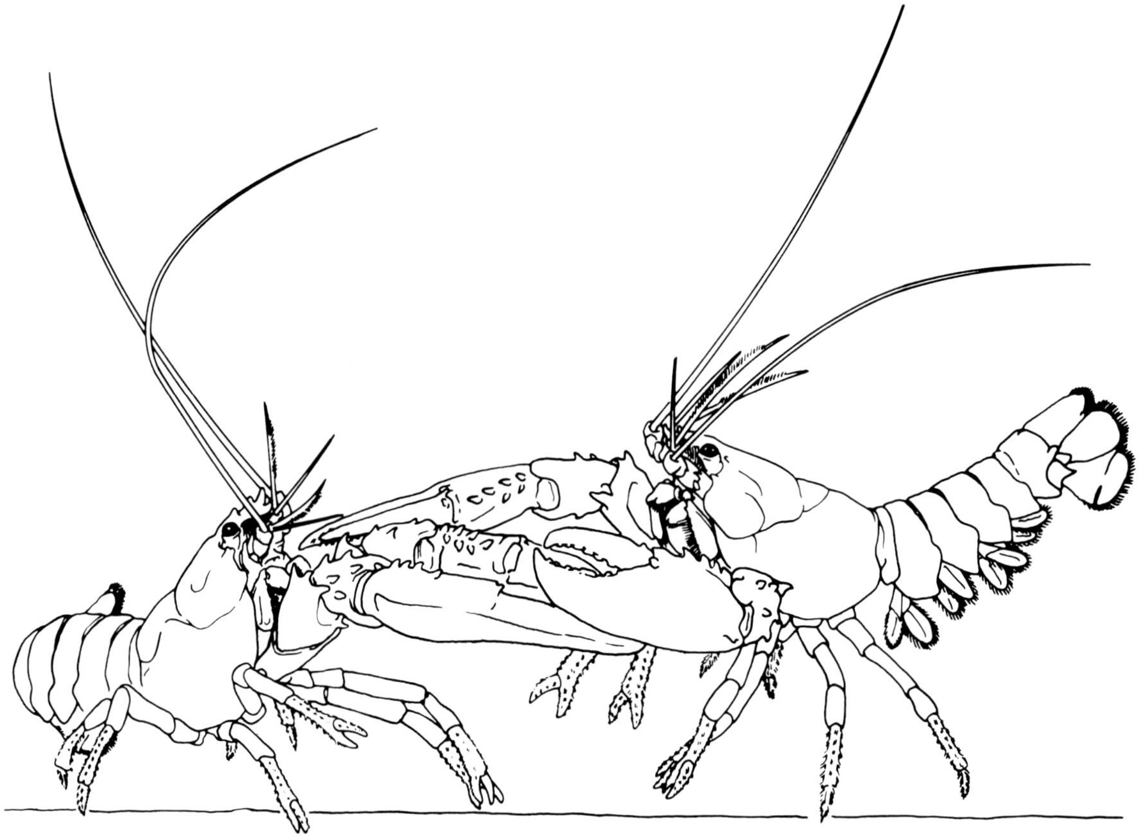

FIGURE 2 Agonistic encounter between mature lobsters when first placed in a small enclosure. Under "boxing match" conditions, lobsters display aggressive behavior; they attack each other with thrusts and jabs of their claws and often lock claws in a push–pull test of strength. Under conditions of ample space, as in nature, such fights probably occur infrequently, because smaller and subordinate animals avoid approaching superiors. (Reproduced from Atema, 1977b, with permission.)

ster sometimes enters and remains inside for a brief period (minutes or less); the original resident may return (Jacobson, 1977; O'Neill and Cobb, 1979; J. Atema, unpublished observations in both the laboratory and the field). Finally, during and after encounters [but not at rest (E. B. Karnofsky, unpublished data)] dominants tend to carry their body higher than subordinates do (Snyder *et al.*, 1993; J. Atema and R. Voigt, unpublished observations). This *high on legs* stance is a common crustacean display of aggression (Schöne, 1968) and may carry symbolic value. However, in adult lobsters this behavior is not predictable; both winners and losers may exhibit it at times (J. Atema and R. Voigt, unpublished observations). Early juveniles show it more systematically (Huber and Kravitz, 1995).

3. Agonistic Behavior in Field versus Laboratory Conditions

In laboratory confinement, lobsters of all postlarval phases and both sexes are aggressive toward each other as long as they are in close proximity and are not too disparate in size. However, only 67 interactions were observed during 333 hours of observations in a shallow field habitat (mostly during the peak activity period of 1–3 hours after sunset); most of these were among the 30 residents and only four involved transient visitors to the area (Karnofsky *et al.*, 1989a). None of the interactions included physical contact; most were shelter related; five included *scissoring*, *lunge*, and *snap*; and 13 resulted in *tail flip*. Although aggressive interactions are rarely observed, wounds of presumably lobster-inflicted origin, such as puncture wounds on the claws and cut antennae, are documented regularly in field-caught animals (Scarratt, 1971). Similarly, lobsters missing one or both claws are quite frequently seen in the field, both in shallow (1-m) and deep (200-m) water (Karnofsky *et al.*, 1989a; J. Atema and J. S. Cobb, unpublished observations). The causes of claw loss may include conspecific fights as well as attacks by predators and handling by fishermen. Newly molted lobsters seem to autotomize claws very easily. Thus, indirect evidence suggests that serious fights occasionally take place in nature, as well as in large aquaria, after which a dominance order is established in which losers avoid winners. Shallow-water habitats may serve as a refuge for wounded adult males escaping from dominant males in deeper water (Karnofsky *et al.*, 1989b; Moriyasu, 1984).

4. Factors Influencing Aggression

Several factors influence the outcome of agonistic encounters, roughly in order of importance: size, molt stage, sex, egg-bearing, and the memory of prior experience (based in part on urine cues).

a. Body Size, Claw Size, and Claw Loss Careful measurements of body and claw size differences show that in adult lobsters, larger animals with larger claws have a greater chance of winning an encounter (Scrivener, 1971). The probability of winning approached 90% when one animal is about 5% larger in carapace length (CL) (i.e., 5 mm larger for animals 84- to 114-mm CL) or has about a 15% larger claw index, a combination of length and circumference (20 cm^2 larger for claw indices of 88–258 cm^2). Animals missing one or both claws are unlikely to challenge an opponent and flee readily from an encounter (O'Neill and Cobb, 1979).

b. Modulation with the Molt Cycle The clearest documentation of the effects of molt stage on aggression comes from carefully controlled studies in early (normally cryptic) stage juveniles (Tamm and Cobb, 1978). Since comparable work in adults has not been performed, the juvenile data are discussed here rather than in the developmental section. For stage X and XI juveniles (approximately 1 year old) in a boxing match situation, winning and losing are significantly correlated with molt stage. Lobsters in early to middle proecdysis (stages D_1 and D_2 of Aiken, 1973) are dominant over intermolt animals (stage C), whereas late proecdysis (D_3) and postecdysis (A or B) animals are subordinate to those in intermolt (Tamm and Cobb, 1978).

The decrease in aggression as the delicate molting process approaches (D_3) and following molting (A and B), when the soft-shelled animal is vulnerable, is not surprising. However, the sudden rise in aggression in midproecdysis is remarkable, and speculation as to its significance is tempting. When juveniles are kept in pairs in small containers, the dominant animal loses its status for a few days following molting, but regains it afterward (Cobb and Tamm, 1974, 1975). The sudden rise in aggression before entering the vulnerable late pro- and postecdysis states may thus instill a "fear" in opponents (see Section II,B,4,e), which carries over into the molting period, when a subordinate could—but rarely does—kill his former dominant. Indeed, dominants invariably kill a molting subordinate in these close quarters, while surviving their own molting (Cobb and Tamm, 1975).

In adults, there are marked increases in aggressive behavior during the premolt, almost up until the moment of molting, and a sharp decline immediately following molting (Atema *et al.*, 1979a). The rise in premolt aggression may also serve to secure suitable shelter, including alternative shelters. In the field, juvenile and adult lobsters generally hold one shelter at a time, but in the weeks before molting the number of shelters regularly occupied increased to about two (Karnofsky *et al.*, 1989a). The majority of interactions involves resident animals that molt subsequently, possibly reflecting the well-known premolt activity peak (Tamm and Cobb, 1978; Karnofsky *et al.*, 1989a; J. Atema and D. F. Cowan, unpublished observations). In a naturalistic aquarium group of five females and two males, it took 2 months for a postmolt male to gain codominant status with a smaller male (Cowan and Atema, 1990). This may not be surprising given the time it takes for a newly molted adult to fill the larger shell with muscle. It is not known how (or even if) molt cycle hormones (perhaps ecdysone and methyl farnesoate) modulate the agonistic motivation [perhaps through serotonin and octopamine (see Beltz, Chapter 11)].

c. Sexual Dimorphism In many species, males and females are under different selection pressures. At maturity, lobsters begin to show increasing sexual dimorphism: males grow faster and develop increasingly larger claws, whereas females develop broader abdomens (Templeman, 1935; McLeese and Wilder, 1964; Lang *et al.*, 1977; Waddy and Aiken, 1991). Female abdomen broadening has obvious fitness value: it increases the number of eggs that can be carried on the underside of the abdomen, attached to the pleopods. It is not known whether a broad tail conveys other advantages, such as in female attractiveness or mating success.

Sexual dimorphism in claw size appears to be linked to male dominance. Natural diet studies (Elner and Campbell, 1981) show that male claws are disproportionately large for the prey consumed. They are not needed for feeding and may even be a liability in escape [the tail flip escape becomes increasingly ineffective in large males (Lang *et al.*, 1977; J. Atema *et al.*, unpublished observations)]. Thus, the male investment in large claws starting at maturity may be primarily an adaptation to establish dominance. Indeed, adult males tend to win fights with females of equal CL (Scrivener, 1971). A small difference in claw size is highly predictive for the outcome of a staged fight, which often involves *claw lock*. One could speculate that large claws serve as a mate selection signal for females, but it remains to be shown that females can assess male claw size during courtship. A similar argument has been made for crayfish: claw size confers selective advantage to males in display behaviors and in physical combat (Stein, 1976).

Males grow faster than females, who molt less frequently and put energy into egg production and brood care (Templeman, 1935). Such a growth differential could give adult males an additional advantage over females in terms of agonistic experience. Since adult males are larger, they accumulate more winning experiences than females of comparable age. Males are more active than females: every night males spend hours exploring and checking shelters, whereas females can spend days without leaving their shelters (Waddy and Aiken, 1991).

Dominance appears to function primarily as a means to gain preferential access to shelters. One must assume that shelters serve males and females equally in protection against predation. However, adult males must secure shelters that are also suitable for cohabitation if they are to attract females and succeed reproductively. Therefore, dominance may be more important for males than for females. Theoretically, intrasexual competition and differences in male and female priorities should favor some separation of male and female dominance orders: male dominance for establishing a mating shelter and female dominance for preferential access to the dominant male. Male dominance has been directly correlated with mating success (Atema *et al.*, 1979a; J. Atema and D. F. Cowan, unpublished data). Female dominance, however, is not correlated with first access to the dominant male (Cowan and Atema, 1990); perhaps it functions to allow a dominant female to take up shelter near the dominant male.

A size-based dominance order, in which males dominate nearly equal-sized females, would accommodate the separate male and female reproductive interests. The only field evidence of sex and size influences on dominance comes from an analysis of observed changes in shelter occupancy, in which replacement is assumed to be the result of evictions. Generally, larger animals successfully replace smaller ones, except that larger females do not replace smaller males (Karnofsky *et al.*, 1989a). Brief, staged evictions in the field were generally not successful (O'Neill and Cobb, 1979) probably because no proper social context was established.

d. Behavior of Ovigerous Females The behavior of ovigerous, or egg-bearing, females (also called "eggers" and "berried" females) stands out: they keep their abdomens tightly curved and become asocial, secretive, and defensive. When approached, they back up and raise up with claws wide open and held close to the body. From this position they *claw snap* and *lunge* quickly. This behavior has not been studied experimentally, but holds real promise for the study of hormonal regulation of behavior.

e. Memory of Prior Agonistic Experience and Isolation In most animals studied, including Crustacea, the outcome of previous encounters influences subsequent encounters (Rubenstein and Hazlett, 1974). Some form of memory of previous agonistic encounters results in a dominance relationship among animals in an interacting group.

In boxing matches between two size- and sex-matched adult lobsters, an initially small difference in aggressiveness develops into a clear winner–loser distinction within a few minutes, resulting in a permanent dominant–subordinate situation. Both animals also carry their winning or losing experience with them in subsequent fights against other opponents. Accounting for effects of body size and claw size, experienced winners win, and losers lose, significantly more encounters than predicted by chance (Scrivener, 1971). For adult males during a second encounter, after 24 hours of separation in different communal holding tanks, the previous loser does not challenge his known former opponent (Karavanich and Atema, 1991). The loser will challenge and can defeat an unknown opponent, however, even if the new opponent is the recent winner of another fight. The memory of a specific opponent lasts about 1 week when the two contestants are housed in isolation. Blocking or catheterizing nephropores and lesioning antennules demonstrate that this memory is based on chemical signals in urine mediated by olfaction (Karavanich and Atema, 1991). The composition of this urine and possible pheromones is not known. Aggressive animals release urine during a fight, but not when strongly disturbed by a lobster-sized plate moved through the water (Breithaupt and Atema, 1993; P. Bushmann and J. Atema, unpublished data). As soon as an animal has lost the fight, as evidenced by avoid and flee behavior, it stops releasing urine; the winner continues to deliver urine (Breithaupt *et al.*, 1994). All this implies that urine release by aggressive animals is dependent on social context and is not merely a function of the metabolic demands of increased activity, such as fleeing from an object. Thus, urine is used to influence the memory of the fight opponent. Memory of former opponents should play an important role in the maintenance of dominance hierarchies. It is not clear whether urine is also used during a fight as a "weapon" to influence the outcome. It should be mentioned that urine release also increases in the hours (Lindstrom, 1991; Breithaupt and Atema, 1993) or days (Snyder and Chang, 1991) following feeding.

In the context of prior experience, it is important to discuss the effect of isolation, that is, no recent social experience. When male lobsters are placed together after 14 days of individual isolation, they show a dramatic (95%) decrease in aggressiveness over the first 3 days (Hoffman *et al.*, 1975). The aggression-reducing effects of communal housing do not result solely from immediate subordination to a despot, since these effects persist in subsequent boxing matches with strangers. Boxing matches between previously isolated animals show frequent frontal attacks, while matched animals previously held communally partially ignore and even back into one another (Dunham, 1972). Physical contact is not necessary to reduce aggressive motivation—lobsters that are physically separated, but in visual and chemical contact, show the same reduction in aggressiveness (Hoffman *et al.*, 1975). Thus, communal housing, perhaps through habituation, causes a general suppression of aggressiveness. Although the crowding of communal housing probably does not reflect natural conditions, these general effects would interact with specific suppression of aggressiveness through urine-based memory of a specific opponent (Karavanich and Atema, 1991, 1993). Field studies show frequent shelter visits and suggest that residents of an area "know" each other (Karnofsky *et al.*, 1989a). If so, lobsters would recognize several individuals and their shelter locations at any one time.

5. Development of Agonistic Behavior and Dominance

Larval lobsters do not appear to have agonistic encounters, although they may grab and eat each other when held in high density. Upon metamorphosis into the postlarva (stage IV), they initiate agonistic interactions but their encounters are usually very brief approach–avoid bouts. The behavior units are essentially the same as those used by adults, but their agonistic behavior repertoire is not complete until the end of stage VII (J. Mitchell, personal communication), when differentiation in crusher and seizer claws first becomes visible (Herrick, 1907; Govind, Chapter 12).

Naive juvenile lobsters demonstrate a more stereotyped agonistic behavior (Huber and Kravitz, 1995) and exhibit six common behavioral patterns: meral spread, wrestling, *do-si-do* (the animals touch both claw tips in *meral spread* and push each other back and forth), *retreat*, *antenna tap*, and *strike/rip*. Furthermore, these behaviors occur in a temporal sequence, resulting in an increase in fight intensity during confrontations. A typical fight starts with intensive threat displays upon first contact, continues with periods of ritualized aggression and restrained use of the claws, and terminates in a brief session of unrestrained combat.

Changes in the relative size of body parts, especially the claws and the tail (Templeman, 1935), have an important influence on the readiness to fight or flee. Allometric growth is paralleled by changes in the threshold and conduction velocity of the giant nerve fiber system that is responsible for the *tail flip* response. Thus, larger lobsters have relatively smaller tails and higher *tail flip* thresholds than smaller ones (Lang *et al.*, 1977). Small juveniles have relatively large tails and use pleopod-powered forward swimming as an escape (Cobb and Tamm, 1975), whereas larger animals can no longer use their pleopods for swimming. These morphological changes affect the frequency with which various behavior units occur. To a lesser degree, these changes continue to take place during the lobster's natural life span. The escape response and its neural control also change over the molt cycle (Cromarty *et al.*, 1991).

As in adults, boxing matches between stage V–XI juveniles (about 12-mm CL) show that first bouts in a match determine the outcome; a 5% weight difference results in the dominance of the heavier lobster (Cobb and Tamm, 1974, 1975). In juveniles, in which claw size is not yet dimorphic between the sexes (Waddy and Aiken, 1991), males should not have an advantage in fights unless there is already a sex-linked difference in aggression. Indeed, in 50-mm CL juveniles, aggression and the probability of winning appear not to be sex linked, because either sex may emerge equally as the despotic alpha animal in sexually heterogeneous groups (Jacobson, 1977). With size differences on the order of 2-mm CL, the largest animal is dominant in most cases (90%). (One assumes that the dominant lobster establishes its position by winning fights, although this has not been measured.)

C. Shelter (and Territoriality)

1. Critical Resource

Juvenile and adult lobsters live alone in close-fitting shelters where they spend most of their time; however, this general rule has many interesting exceptions. Perhaps in evolutionary competition with fish, lobsters and many other large invertebrates that did not develop chemical defenses were driven into a shelter-dependent existence.

In shallow water, lobsters generally emerge from their shelters about 1 hour after sunset and show greatest activity in the following 2 hours, after which they gradually return to their own shelter or a nearby alternate shelter (Karnofsky *et al.*, 1989a; similar observations reported by Weiss, 1970; Cooper and Uzmann, 1980; Ennis, 1984a,b). Some animals are resident in one shelter for up to 9 months including overwintering, others move among different shelters in a general area, and yet others are transient (Karnofsky *et al.*, 1989a). Similar results have been obtained in the field (Stewart, 1972) and in naturalistic settings (Waddy and Aiken, 1991). Preliminary work shows that animals caught near their shelter and released 35 m away can return home in 2 hours, using well-directed initial return paths (Karnofsky *et al.*, 1989a). However, systematic homing research has not been done for this species. Direct field observations (Karnofsky *et al.*, 1989a) also show that lobsters can quickly locate a number of different, sometimes hidden, shelters when chased, whether running forward or tail flipping backward on or off the ground. Observation showed a cohabiting, presumably dominant, male running from his shelter directly to another shelter 5 m away to evict a large male, after which he returned directly home. Similar observations have been made regularly in naturalistic aquaria. Lobsters seem to "know" their physical environment. It has been suggested that they use their activity period primarily to forage for information, not food: to update their knowledge of the physical (and social) environment (Karnofsky *et al.*, 1989a). Observations from submersibles show single lobsters in shelters in submarine canyons at depths of 400–600 m (Cooper and Uzmann, 1971). Artificial reefs attract lobsters (Scarratt, 1968; Briggs and Zawacki, 1974), suggesting that shelter is a limiting resource. These direct field observations, although sparse and anecdotal, provide indirect evidence for the importance of shelters.

Under certain circumstances, lobsters share shelters. The best known situation is during courtship, when one female at a time shares the dominant male's shelter, each for up to 3 weeks (Cowan and Atema, 1990; see also Section II,C,2). The dominant male's shelter is exceptionally large to accommodate two animals, whereas lobsters generally prefer well-fitting shelters as demonstrated experimentally for early juveniles (Cobb, 1971). In rare cases, particularly in winter (Thomas, 1968), more than one lobster has been observed in a single shelter. This may result from the fact that activity stops almost entirely at temperatures below 2°C. In naturalistic aquaria, in summer or winter, multiple occupancy has not been observed outside courtship (J. Atema *et al.*, unpublished observations).

2. Shelter Exchange, Construction, and Housekeeping

Lobsters seem to prefer ready-to-use shelters. Laboratory and field studies show that they also may prefer two entrances: a main entrance and an escape door (Cobb, 1971; Karnofsky *et al.*, 1989a). They shop around for a perfect fit and may engage in shelter exchange, much like hermit crabs exchange shells. Hermit crabs "rap" their shell against the shell of the animal they try to evict in aggressive or negotiating encounters (Hazlett, 1987). Lobsters do not rap, but *knock on the door* (see Section II,B,2; O'Neill and Cobb, 1979; Atema and Cobb, 1980).

When fitting shelters are not available, lobsters construct their own, *digging, bulldozing, fanning*, and carrying rocks and other material. They dig tunnels under eel grass beds, often using a depression or rock as a starting point. Sandy clay and peat are other possible substrates. If no suitable substrate is available, they dig pits or squeeze between solid objects, mostly rocks.

Lobsters engage in extensive housekeeping, fanning debris and silt out, pushing sand and rocks, and modifying the entrances. They may block the entrances for up to 2 weeks with rocks or other objects (Karnofsky *et al.*, 1989a), probably in preparation for molting [as seen in the laboratory (J. Atema *et al.*, unpublished observations)].

3. Shelter Use: Molting and Sex Differences

In the field, each lobster generally uses one shelter at a time, sometimes for several months (Karnofsky *et al.*, 1989a). If a lobster moves to a different shelter, it is for a length of time. There are considerable differences in shelter fidelity (Karnofsky *et al.*, 1989a; Ennis, 1980). However, in the 2 months preceding molting, lobsters gradually increase the number of shelters regularly occupied to 1.9 during the last 2-week period preceding molting; one third of premolt animals spend time in three or more shelters during that time (Karnofsky *et al.*, 1989a). Increased shelter use corresponds with increased premolt activity (Tamm and Cobb, 1978; J. Atema, unpublished observations) and may serve to discourage other lobsters from inhabiting the area where a lobster is going to molt, thus buying time for the just-molted lobster before being evicted. Despite these preparatory efforts of controlling, and perhaps decoying, an area and barricading shelters, other lobsters often break through the barricade and evict the newly molted animal (J. Atema and R. Voigt, unpublished observations). In naturalistic aquaria, lobsters are generally not approached closely during and following molting, even outside shelters; one may speculate that some chemical protection operates. Lobsters may buy time by molting in early morning, when other lobsters do not move about, thus avoiding eviction until the next night (Cowan, 1992).

In naturalistic aquaria, where shelter evictions can be observed directly, dominant males can evict all others and do so, in some cases regularly. This has also been observed in the field, although the number of observations is limited. Larger animals are usually successful in evicting smaller ones, except larger females, who rarely evict smaller males. Males approach mostly male shelters, whereas females approach shelters of both sexes equally (Karnofsky *et al.*, 1989a).

The "mating" shelter of the dominant male is the focus of social activities in naturalistic aquaria (Cowan and Atema, 1990; Atema, 1986) and in the field (Karnofsky *et al.*, 1989a). All residents visit this shelter frequently, particularly premolt females and smaller animals of both sexes. Females appear to be waiting their turn for cohabitation (Cowan and Atema, 1990). Small animals may derive protection from the dominant male's presence, since he tends to drive off larger competitors but ignores much smaller animals. The curious association of a small or recently molted animal with the dominant male's shelter has been noted regularly (J. Atema *et al.*, unpublished observations).

4. Development

It is suspected that planktonic larval stages reduce predation by crypticity. Visual crypticity appears to be achieved by their bluish transparency, which matches their pelagic environment in the upper euphotic zone. Mechanical crypticity would require reduced movements and matching necessary movements to the flow patterns prevalent in their specific environment. Chemical crypticity would require storing waste products in the gut and the bladders, occasionally releasing them and leaving them behind as decoys (Atema, 1995). If continuous chemical output is necessary, it would have to match ambient chemistry to reduce the signal-to-background ratio. The alternative of chemical toxicity or repellency does not appear to be used, since larval lobsters are eaten voraciously by fish and crabs. These theoretically compelling adaptations to reduce predation await experimental evidence and provide an important context for postlarval behavior.

Upon metamorphosis into stage IV (the postlarva), the lobster becomes a brownish-pigmented, strong swimmer in search of a benthic habitat suitable for settling (Ennis, 1975; Cobb *et al.*, 1983). This brief nectonic stage may rely mostly on fast maneuvering for predator escape. Once a substrate has been found—perhaps after several dives to the bottom (Cobb, 1971; Cobb *et al.*, 1983)—the animal becomes a cryptic, shelter-restricted, benthic juvenile living in already existing rock or peat/eelgrass shelter, or constructing a tunnel with two or more entrances in muddy substrates. Experimental evidence, mostly from laboratory studies and supplemented by field collection, is discussed by Lawton and Lavalli (Chapter 4). The behavior patterns of shelter construction in mud are immediately effective in naive animals. They include *digging* down with claws, front walking legs, and third maxillipeds; *bulldozing* forward using the same appendages spread out into a basket, while forward power is provided by rear walking legs and pleopod current; and *pleopod fanning*, during which the lobster stands still with raised tail blowing silt and mud rearward. Solid surfaces such as walls and rocks are often used as starting points (Botero and Atema, 1982; Atema *et al.*, 1982). Behavioral mechanisms against predation seem rather ineffectual (J. Atema and R. Voigt, unpublished observations), as the main escape, *tail flip*, does not work in tunnel systems; face-to-face interactions between lobsters or between lobsters and crabs usually end with the larger animal eating the smaller one. This lack of behavioral defense places great weight on the shelter itself as an antipredator device. Rock and mud shelters have different physical attributes and provide different effectiveness against fish and crustacean predators (see Lawton and Lavalli, Chapter 4).

At about 25-mm CL, juveniles begin to leave their shelters for brief periods (J. Atema and R. Voigt, unpublished observations). As they grow larger, this behavior is more commonly seen in nocturnal forays (Karnofsky *et al.*, 1989a). Shelter remains critical throughout their lives, as animals of all sizes are mostly found associated with a shelter.

D. Courtship

1. Behavior Patterns and Chemical Signals

Under the most natural study conditions reported so far, courtship starts with male dominance. Naturalistic aquarium studies (Atema *et al.*, 1979a; Cowan and Atema, 1990; Karnofsky and Price, 1989) supported by limited field observations (Karnofsky *et al.*, 1989b) show that the dominant male occupies a shelter sufficiently large for two animals. This mating shelter becomes a focus of social interactions, including frequent visits by adult, premolt females. Visiting females stand still at its entrance for many seconds. They alternate between exopodite fanning and not fanning. When not fanning, they blow their gill current into the male shelter (see Section II,A). At the time of female visits, the male stands inside and often moves away from the female entrance, flicking his antennules (i.e., "sniffing"), fanning his exopodites (thus drawing water toward his antennules and redirecting his gill current backward), and occasionally fanning his pleopods. Together, these behaviors, both male and female, result in female odor reaching the male (Fig. 3).

Generally, lobsters do not accept other lobsters into their shelters (see Section II,C). However, in simplified choice tests (Bushmann and Atema, 1993), mature males accept mature females regardless of female molt state, but not males who either are prevented from entering or evict the male resident. In a naturalistic environment with two males and five females, the dominant male accepts only premolt females and cohabits with them each for a few days to weeks, with a mean duration of 12 days in naturalistic aquaria (Cowan and Atema, 1990). The behavioral context is probably provided by chemical signals. Male and female sex pheromones appear to be evident, but have not been identified chemically (Atema and Engstrom, 1971; McLeese *et al.*, 1977; Atema and Cowan, 1986; Cowan, 1991; reviewed by Atema and Cobb, 1980; Atema, 1986). Urine, and possibly special glandular products in the urine, may be involved to help the male identify the visiting female as an adult and premolt female lobster. Both males and females possess an active nephropore gland at all molt stages (Bushmann and Atema, 1993). Thus, if this gland is involved in setting the courtship context, its product or the perception of that product would have to be sexually dimorphic.

Urine is the major route of ecdysteroid elimination. Urine metabolites include conjugated forms of various ecdysteroids, primarily 20, 26-dihydroxyecdysone and 20-hydroxyecdysonic acid, and also 20-hydroxyecdysone, ecdysone, and ponasterone (Snyder and Chang, 1991). The latter three tested did not cause behavioral responses in adult males (Gagosian and Atema, 1973), although proper behavioral context was not provided.

The female molts sometime during cohabitation, and mating follows after 30 minutes (Templeman, 1934; Atema *et al.*, 1979a; Cowan and Atema, 1990). Cohabitation is usually intermittent at first and near the end, with several days during which the female comes and goes (Atema, 1986; Cowan and Atema, 1990). In the days just before and particularly after the female molts, she does not leave the shelter. The

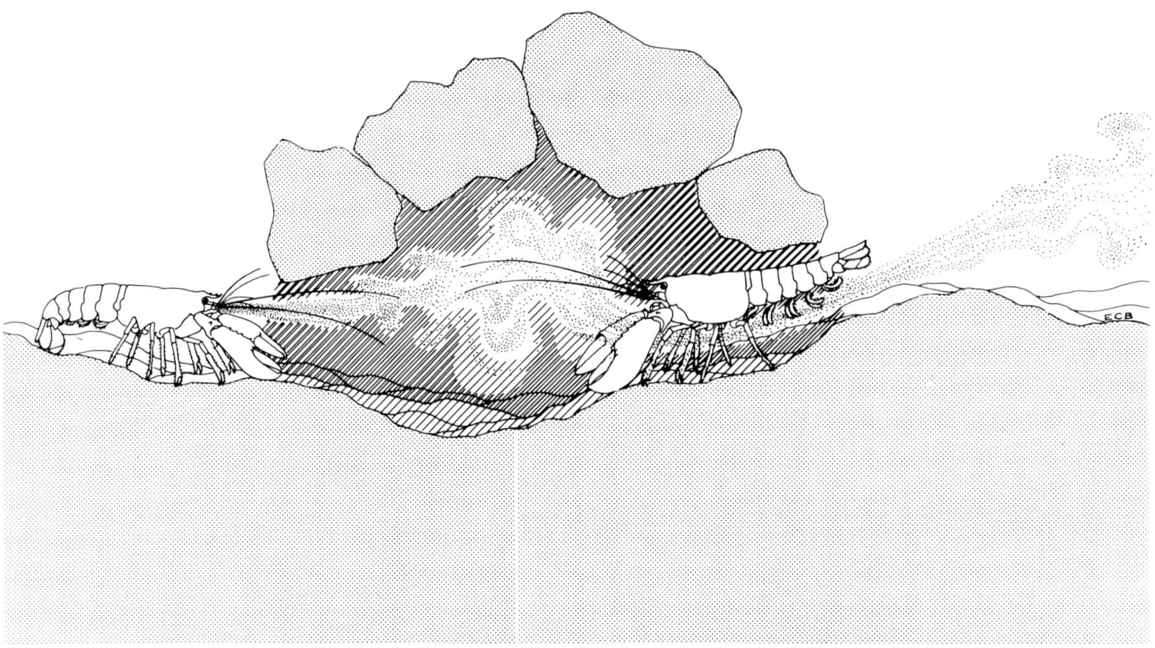

FIGURE 3 Female courting male with pheromone plume. The female (left) injects her gill current into the male's shelter; he responds by retreating to the opposite entrance and fanning his pleopods and exopodites. Water flows from female to male and into the environment. (Reproduced from Atema, 1986, with permission; drawing by Ellen Chase Botkin.)

male leaves rarely, usually for short bouts to chase and evict other lobsters. Over the premolt cohabitation period, male pleopod fanning increases, reaches a maximum during the day of female molting, and wanes in the postmolt cohabitation days. Fanning males often stand at one of their shelter entrances with their abdomen raised high and slightly outside the entrance (Fig. 3). This results in a strong current running through and out of his shelter, which carries with it all male and female metabolites (including possible pheromones) released by the cohabiting pair. Male fanning can be seen as advertising, since it is positively correlated with visits of other lobsters to the shelter, including premolt females (Cowan and Atema, 1990).

2. Mating Behavior

Both postmolt and intermolt mating have been described. Results from long-term, naturalistic aquarium observations in which ample shelters, including

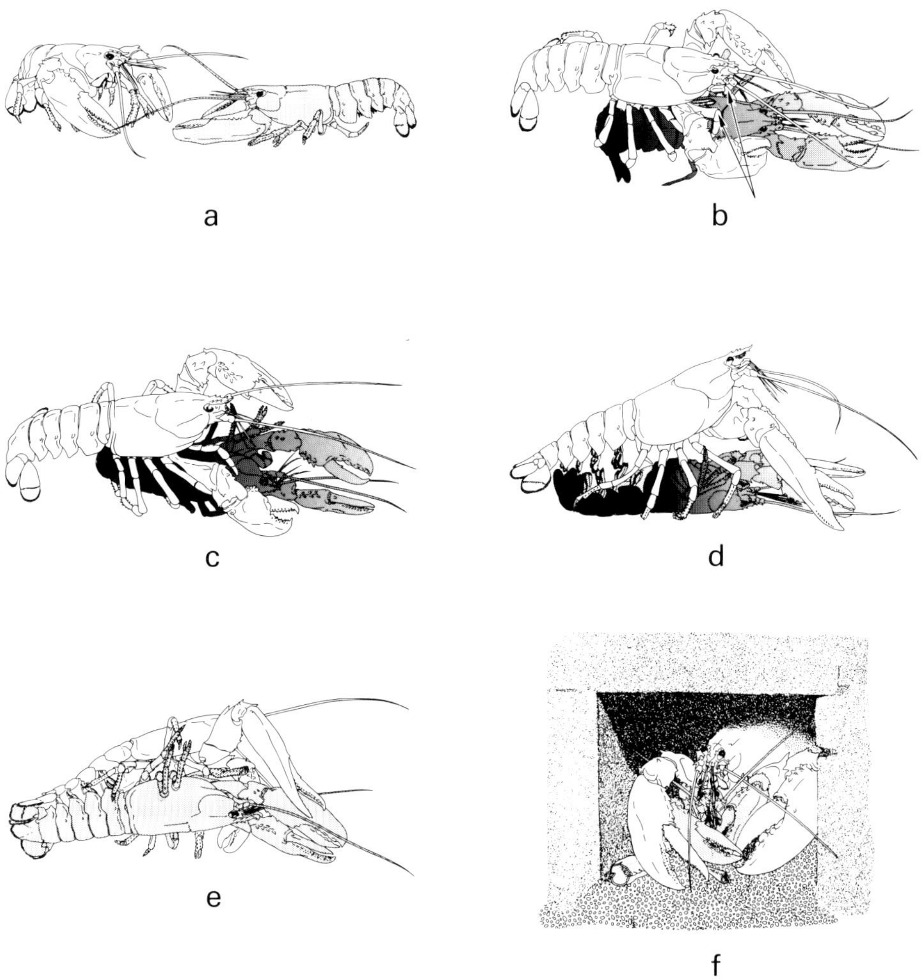

FIGURE 4 Pair formation and mating behavior of *Homarus americanus*. Female is shaded for identification. See text for details. (a) Recently molted female (right) cannot stand on legs; male is *high on legs* with claws closed and down. Frequent mutual antenna touching. (b) Male mounts female. Note the use of his third maxillipeds (arrows) to turn her over; male displays intense antenna touching and antennule flicking and touching. Male's chelipeds are closed and used for support. (c) Male turns female over with his maxillipeds and first two pairs of walking legs. Female remains passive, with claws stretched [seconds after (b)]. (d) Female on her back, stretches abdomen, with pleopod fanning. Male moves slightly further forward, extends gonopods for insertion into female sperm receptacle, with pleopod fanning [seconds after (c)]. (e) Copulation. Male deposits spermatophore. (f) Postmating cohabitation. Male eats female molt shell; note claw remainders near shelter entrance. Female remains in male's shelter for several days postmolt. (Reproduced with permission from Atema *et al.*, 1979a, *Mar. Behav. Physiol.*, Vol. 6, pp. 277–296. © 1979 by Gordon and Breach Science Publishers.)

large mating shelters, were available, show that almost all mating is postmolt, taking place during cohabitation in the male shelter after the female molts (Atema *et al.*, 1979a; reviewed by Atema and Cobb, 1980). Just prior to molting, the female faces the male and places her claws on the male's anterior carapace and claws in a *knighting* gesture. This curious behavior may be accompanied by urine release as it places the nephropores almost directly onto the male's antennules. This is seen only very rarely during the remainder of the cohabitation period. Shortly thereafter, the female lifts her left third walking leg and rolls over on her side to start the ecdysis process, which, for recently mature animals, takes about 10 minutes at summer temperatures and much longer in winter (up to 1 hour has been observed).

A half-hour after the female completes her molt, the male, who until then has stood by quietly, begins to approach and mount the female. He climbs on her back starting at the abdomen and begins to turn her over with his first two pairs of walking legs and his third maxillipeds, appendages richly provided with chemo- and mechanoreceptive sensilla. The female unfolds her abdomen and stretches her claws forward to allow the turning to take place easily. At this stage, she can resist and even tail flip away. The male supports himself with his closed claws on either side of the female.

The male then inserts his gonopods—the modified first pair of pleopods (swimmerets)—into the female seminal receptacle and deposits his spermatophore with a few thrusting movements. Actual copulation lasts only a few seconds, after which the female rights herself with a tail flip out from under the male. The male usually starts extensive grooming. Cohabitation continues, during which the male begins to eat most of the female molt shell, starting with the softest parts. In these circumstances, the subsequent matings that have been seen in confinement (Snyder *et al.*, 1992) have not been observed.

Intermolt mating behavior shows a few differences with this postmolt mating behavior (Waddy and Aiken, 1991). Continuous video observations have been made in large tanks without shelters large enough to hold two animals. A female without adequate sperm supply is recognized by one of the males (perhaps the dominant?), who could either subdue her forcefully (but not always successfully) or engage in a brief face-off, after which the female would turn around, inducing the male to mount and turn her over for copulation. The latter occurs mostly in prespawning, noninseminated females. Premolt animals and prespawning females tend to increase their activity (Karnofsky *et al.*, 1989a; Waddy and Aiken, 1991) and thus increase their chances of encountering a suitable mate either for cohabitation or for intermolt matings. The function of intermolt matings may be to allow females to replenish sperm they either failed to receive in the postmolt period or lost as a result of previous fertilization and spawning. The latter applies only to older females who spawn each year but molt only every 2 or 3 years (Waddy and Aiken, 1991; Snyder *et al.*, 1992). No limitations to mating and insemination are found because of molt stage, although there is a peak in female sexual receptivity immediately after molting. Once inseminated, female receptivity ceases and males are no longer attracted (Waddy and Aiken, 1991), although repeated mating has been reported when pairs are placed in small (three body lengths) aquaria (Snyder *et al.*, 1992). In such confinement males have been seen to mount and overturn other males (J. Atema and P. Bushmann, unpublished observations). Mating outside shelter, even when mating-sized shelters are available in a naturalistic aquarium, has been seen only once, when a new female was introduced (C. Karavanich, unpublished observations). Waddy and Aiken (1986, 1990, 1991) have cited regular cases of intermolt mating *outside* shelter when mating shelters are not available. The frequency and relative importance of postmolt and intermolt matings under natural conditions remain to be determined. Females appear far more discriminating than males in courtship and exert greater control over mating.

3. Serial Polygamy

Nearly all information concerning cohabitation comes from laboratory observations in which certain sex ratios are imposed and immigration and emigration are restricted. Adult female molt staggering is seen in such closed systems—one female at a time cohabits with the dominant male and ignores the subdominant animal (Cowan and Atema, 1990; Cowan *et al.*, 1991). The mechanism causing molt staggering is currently unknown; pheromones released from both the cohabiting male and female are implicated (Atema, 1986; Cowan, 1991). The male advertising current may not only attract females (releaser pheromones, Wilson and Bossert, 1963), but also contain cues (primer pheromones, Wilson and Bossert, 1963) that suppress female molting. As soon as a new female enters, she would be released from this inhibition and continue rapidly toward molting. Such a hypothesized mechanism (Atema, 1986) would result in the observed molt staggering. Hazlett (1991) questioned the evidence for molt staggering.

If a mating system of serial polygamy occurs in natural populations, one or two males in an area could monopolize the sexually mature and receptive females (Atema, 1986; Cowan and Atema, 1990). There is some evidence that such a mating system might exist in small, natural populations (Karnofsky *et al.*, 1989a), but this topic has not been well studied in the field. Immigration and emigration of both males and females could have significant effects on the dominance relationships of males and thus on the mating system itself.

4. Female Choice

Studies in large aquaria (Atema *et al.*, 1979a; Karnofsky and Price, 1989; Cowan and Atema, 1990), supplemented by rare observations in the field (Karnofsky *et al.*, 1989a), suggest that female lobsters choose the dominant male and initiate cohabitation in his shelter. The length of premolt cohabitation is greater when more females are present (Cowan and Atema, 1990) and may be a response to female competition for access to the dominant male. In controlled laboratory tests, females at different stages of the molt cycle have entered male-occupied shelters (P. Bushmann and J. Atema, unpublished observations). Since females may be able to enter at any time and respond to intrasexual competition, they may control the duration of premolt cohabitation. Usually, the female molts after a few days (range, 0–14 days). On rare occasions, a female has been seen to enter a male-occupied shelter and, without molting, mate with him and leave (P. Bushmann and J. Atema, unpublished data).

In female choice experiments in a large Y-shaped maze, females are attracted to the odor of a male. Both a urine cue and some other odor from the male are necessary to inform the female of his presence, without providing information concerning male dominance status or size (P. Bushmann and J. Atema, unpublished data). Females follow this odor mixture to its source, a male-occupied mating shelter. At the shelter entrance, females stop and appear to inspect the male, probably chemically. The decision to enter seems to be based on both information from the male and the female's own internal state. At the time of female visits, the resident male releases large amounts of urine (P. Bushmann and J. Atema, unpublished data).

Urine cues described in cohabiting females, just prior to molting and mating (see Section II,D,2), also appear to facilitate entering the male shelter by lowering male aggression (cf. Atema and Engstrom, 1971; Snyder *et al.*, 1993). It may convey other information to the male (e.g., species, sex, or receptivity). Females often enter showing the premolt *knighting* behavior, claws outstretched over the head of the male, an ideal behavior for delivering a critical urine cue to the male antennules (Atema, 1986). Both the decision to enter a male shelter and the decision to molt would seem to be of great importance for female mating success and survival. This, *knighting* may be more important than realized thus far. Whereas urine may give important distance cues and reduces aggression, catheterization shows that it is not necessary for successful mating in small aquaria (Snyder *et al.*, 1993).

Females are capable of entering male shelters at all molt stages, but they show the greatest overall interest in entering during late premolt. Early premolt animals, which tend to be more aggressive, can also enter, but their behavior is less consistent and they seem equally interested in empty shelters (Bushmann and Atema, 1993). At this early stage of molt preparation, it may be that female lobsters actively look for suitable shelter, including male-occupied shelters. Subsequent development of the reproductive state may then influence female behavior to select a dominant male. The presence of a sperm plug has been shown to reduce female receptivity (Waddy and Aiken, 1986), but it may not prevent her from seeking male shelter.

Information from the male influences a female's decision to enter his shelter. Females try to enter less often if the male urine cue is missing (Bushmann and Atema, 1993). Artificial release of male urine is not sufficient, however, suggesting either that males release a specific substance into the urine when they notice a female nearby or that the male releases urine in a particular temporal pattern to signal the female. When presented in a Y-maze with a single unknown male, females more often try to enter the shelter when the occupant is dominant, suggesting that they can detect dominance status. When presented with a choice of two males, however, females demonstrate significant preference for one only if he is both dominant and larger than the other (Bushmann and Atema, 1993). Familiarity with the male may also be important in the entering decision. Field and laboratory work have described females approaching sheltered males several times before entering. This repeated evaluation of the male would diminish the risk of male "cheating" with hollow displays. Thus, females evaluate and discriminate among males.

These results imply that females, even if they are not immediately ready for mating, have the ability to enter shelters and may possess the necessary chemical cues at all times. Not all females can enter, however, and the sheltered males react to entering females in many ways, from acceptance to vigorous rejection. Clearly, males can detect differences in females and determine the success of female entering. Thus, premolt females enter for molting and mating in a cohabitation of 1–2 weeks, intermolt females can enter for mating alone, and some females may use their chemical cue as a means of obtaining shelter. It is not known whether males can protect themselves from the latter form of cheating. A variety of chemical signals is likely exchanged between males and females.

5. Male versus Female Benefits of Cohabitation

The biological function of the cohabitation period may be primarily to assure the female protection during her most vulnerable postmolt days. It appears that the duration of the cohabitation period provides the male with a compromise between two alternatives: mating with more females, if each cohabitation period were shorter or absent; and protecting his mate, the key to his biological success, for even longer periods. The 1- to 2-week "compromise" may be sufficient for the female to harden her new shell and still allow the male sequential access to other females. The lobster's pair bond consists of pre- and postcopulatory guarding. Since females are most receptive for 1 or 2 days postmolt (Templeman, 1934, 1936; Hughes and Matthiesen, 1962; Aiken and Waddy, 1980) and are progressively less likely to mate with other males in the week after they molt, guarding may serve not only to prevent predation on the female, but also to prevent her access to other males (Atema, 1986). To the male, paternity assurance is important because females are capable of being inseminated by more than one male (Nelson and Hedgecock, 1977). Laboratory experiments with isolated females have shown that older and larger females, who typically molt only every second or third year, can fertilize each yearly batch of eggs from one postmolt insemination, up to 3 years in a row (Waddy and Aiken, 1986, 1991). This would be a very strong enforcer of postcopulatory guarding, because the male would ensure his contribution to not only one but several year classes of progeny by mating with one such female, unless she engages in intermolt matings to replenish her sperm supply. Theoretically, postcopulatory guarding benefits not only the male, but also the female, because predation on the postmolt female would affect the female's lifetime reproductive success more than the male's. Postcopulatory guarding, then, appears to benefit both the male and the female.

In contrast, the precopulatory cohabitation period may benefit only the female: it allows her to monopolize the dominant male and prevent him from mating with other females. If this were true, one would expect that dominant females have priority access to the dominant male and have longer premolt cohabitations. Surprisingly, this is not the case. One would also expect that female competition results in increased duration of premolt cohabitation, and this has been confirmed (Atema, 1986).

6. Intermolt Mating and Sperm Storage

Intermolt matings are clearly documented in the laboratory (Dunham and Skinner-Jacobs, 1978; Waddy and Aiken, 1990; J. Atema and R. Voigt, unpublished observations), where females remain receptive for up to 80 days (Snyder *et al.*, 1992). Males seem capable of distinguishing between females with mature ovaries and females already inseminated. They rarely attempt to mate with a previously inseminated female, unless that female has used much or all of her stored spermatozoa (Waddy and Aiken, 1990). Such matings may occur in the wild, as mature, noninseminated, preovigerous females are common in some populations (Krouse, 1973; Ennis, 1980). Intermolt mating may therefore be an alternative reproductive strategy to ensure that females unmated at their molt will still be able to extrude fertilized eggs during intermolt. It has been argued that such matings are necessary in the field because male and female lobsters would not encounter each other sufficiently frequently to ensure cohabitations at the time of female molting (Waddy and Aiken, 1991). If so, the discovery of sperm storage (Waddy and Aiken, 1990) would reduce the necessity for intermolt matings as sperm can be stored and used for several years to fertilize successive annual batches of eggs. Furthermore, the only field evidence available suggests that lobsters live more in social groups than was believed earlier (Karnofsky *et al.*, 1989b). Finally, molt- and courtship-related migrations into and out of social groups are likely, but have not been documented.

7. Courtship and Dominance in the Context of Life History Strategy

This section summarizes the complex (and still incomplete) data of the previous sections. Courtship must fit into the life history of the lobster together with many other physiological, behavioral, and environmental requirements, all of which interact directly or indirectly.

The larvae of *Homarus americanus* develop for 9–11 months to an advanced developmental stage (the prezoea) inside the egg and under maternal care. Larvae are released in June, when the water temperature begins to approach its yearly high, presumably when larval food is abundant, allowing them to grow rapidly. To avoid losing the brood or the new sperm, adult female molting must be timed after larval release and before mating; by tying mating to molting, the female gains the maximum possible yearly interval in which to carry eggs and still molt. After reaching a certain size, females begin to molt less frequently, thus making it possible to produce a new clutch of eggs as soon as the previous one has hatched. This necessitates sperm storage and multiyear fertilization from one insemination and/or additional yearly intermolt matings.

Female size is directly proportional to reproductive success; a broader abdomen allows females to raise larger broods. Thus, males should ideally compete for larger females, particularly those large enough to have multiple broods from one fertilization. To become fertilized, the female selects a dominant male who can protect her from competing females and predation during her soft-shelled postmolt state. A long cohabitation period, in which she can monopolize the dominant male, would enhance her competitive position with other females. This necessitates behavioral/pheromonal control over the female molt cycle when several females are competing. As in most polygamous mating systems, female lobsters control mate selection, presumably because female investment in yolk-filled eggs and extended brood care is greater than the male's one-time delivery of a spermatophore. This may turn out to be a simplification.

However, males have costs too. Adult males must establish a mating shelter and advertise their presence, dominance, and, when possible, proven mating success to females. They do this by fanning their odor into the environment and particularly the odor of their shelter during cohabitation. Since a good shelter is essential to their mating success, they must be dominant over males as well as females because both sexes need shelters. The male's enlarged claws may be both real and symbolic adaptations to accomplish this. To dominate the most desirable large females, the male must be larger or at least of comparable size. Indeed, males grow faster than females. There may be additional physiological reasons for this: females put energy into egg production; males, into body growth. To mate with as many females as possible, males may control female molt cycles in situations of severe female competition for one male. The male should attempt to keep the cohabitation as brief as possible, but not so brief that the female could still find another male to mate with or be overly vulnerable to predation. The length of cohabitation may thus be a compromise between the male and female interests.

Cohabiting females mate only very briefly. It is surprising, given our current, still incomplete knowledge of lobster behavior, that in small tanks they are behaviorally and physiologically receptive for up to 80 days after molting and longer (Waddy and Aiken, 1991; Snyder *et al.*, 1992). However, Templeman (1934) observed that females begin to show resistance to courtship after 12 hours. It appears that holding conditions are a critical variable. Brief courtship, repeated mating, prolonged receptivity, and intermolt matings appear under conditions of relative isolation and crowding. Prolonged courtship, cohabitation, single mating, and female molt staggering are reported in large, complex aquaria and (with scant evidence) in the field.

E. Chemotaxis

For many of the described behaviors, lobsters localize important odor sources, such as shelters, other animals, and food, over relatively short distances of a few animal body lengths. A persistent question is whether chemical signals can also be used in long–distance migrations, and if so, how? For this we need to know how much directional information can be extracted from a widely dispersed odor plume. Spatial odor gradients useful for source localization can be determined with appropriate sampling.

Recent experiments on lobster chemotactic orientation have used high resolution (30-μm, 5-msec) sensors to determine the distribution of odor concentration patches and filaments in aquatic odor plumes. These sensors are size-scaled to the 30-μm-diameter aesthetasc sensilla of the lobster antennule, the organ critical for efficient chemotaxis (Devine and Atema, 1982). With these sensors, spatial gradients in small odor plumes can be determined from a series of local measurements of odor patches, similar to lobster chemical sampling (Moore and Atema, 1991). Lobster localization behavior is consistent with the hypothesis that they use local odor patches to guide them through a turbulently dispersed odor field toward the source of odor release (Moore *et al.*, 1991b). That lobsters use patch information is plausible given the following temporal properties of their receptor cells: quick stimulus integration (0.2 second) and adaptation (~1 second); moderately quick to very slow recovery (10–30 seconds); a maximum flicker fusion frequency of 4–5 Hz in antennules, corresponding to the maximum observed flick rate of antennular lateral filaments (4–5 Hz); and continued exposure to a fixed odor concentration, usually resulting in complete adaptation within a few seconds (Gomez *et al.*, 1994; Gomez and Atema, 1994; Voigt and Atema, 1988, 1990; Leonard *et al.*, 1994). These properties allow receptor cells to respond best to the edges of arriving odor patches.

In these initial experiments, the localization distances are small, for example, a 15-cm CL lobster moving toward a constantly emitting source 230 cm away in a flume—some 15 body lengths. This scale of movement is far removed from the extremes seen in 200-km migrations of 20-cm CL animals (Cooper and Uzmann, 1971)—10^6 body lengths! However, the lobster is providing the first sensory information on underwater chemotaxis, with sensors scaled to the size and temporal resolution of biological sensors. Parallel experiments with robots designed after lobster chemoreception (Consi *et al.*, 1993) hold promise to determine the limits of open ocean chemotaxis. It is theoretically obvious that the ability of different species to sample odor distributions varies with animal size and locomotion capabilities (Atema, 1985).

Long–distance odor signals and the turbulent gill and pleopod plumes are useful to lobsters on the large and small scale, respectively. Spatial gradient information may be used to locate inhabited shelters (P. Bushmann and J. Atema, unpublished data) at the space scale of their daily social environment, some 10 m (Karnofsky *et al.*, 1989b). Interactions near a shelter (within 0.1 m) do not require spatial information. Locating food sources beyond 10 m may be possible, but would depend on the odor concentration of the source. This remains to be studied.

F. Pollution

Lobsters inhabit coastal waters and the continental shelf. Pollution due to human activities is a constant threat to these environments. Two major sources are oil or petroleum, and drilling and dredge spoils. Exposure may result in acute toxicity or in sublethal effects, which cause behavioral and physiological changes that affect the fitness of the animal in its ecological niche. While sublethal exposure may not result in visible signs of locomotor difficulties, interference with the chemical communication systems may cause problems in feeding (detection and ingestion, Atema and Stein, 1974), in finding a mate, and in escaping from a predator (Atema, 1980).

1. Petroleum

Exposure to low concentrations (0.1 and 1 ppm of No. 2 fuel oil) causes no apparent behavioral effects in adults. At high concentrations (10 ppm of No. 2 fuel oil, microliter amounts of kerosene in 100 liters of seawater), petroleum is acutely toxic within hours. Fuel oil causes noticeable short-term effects on lobster behavior at 3 ppm, but not at 1 ppm. It appears that the branched-cyclic fraction of kerosene is responsible for acute toxicity (Atema, 1976; Atema *et al.*, 1979b). The branched-cyclic and straight-chain aliphatic fractions depress general activity and feeding behavior for several days, even after only a brief exposure (minutes to hours). However, the polar aromatic fraction causes increased activity and attraction, a behavioral change utilized by fishermen who use kerosene-soaked bricks as bait (Atema, 1977b, 1980). At close range, such as inside a lobster trap, this fraction causes repulsion. In neurophysiological experiments, chemoreceptor cells of the lateral antennule respond to petroleum, and the addition of petroleum changes the response pattern to mussel extracts (Atema, 1980). Lobster larvae have been proposed as test organisms to detect acute toxicity in the marine environment (Olla *et al.*, 1980; Wells, 1976). Lobster larvae are temporarily paralyzed by dispersed petroleum products; they swim weakly and may be easier prey (Wells and Sprague, 1976).

2. Drilling Mud and Dredge Spoil

The toxicity of different drilling muds varies from lethal to adults to apparently harmless to postlarvae. In postlarval lobsters, toxicity is apparent in feeding and molting delays, severe delays in shelter construction, increased walking and swimming, unprovoked tail flipping, and lethargy (Atema *et al.*, 1982). In addition to toxicity, the physical properties of drilling mud (which are designed to sink and seal and to coat the drilling hole) make it impossible for settling-stage lobsters to get into the substrate with as little as a 1-mm layer of drilling mud covering the bottom. This exposes them to a large variety of benthic predators. In neurophysiological recording, drilling mud changes the response of leg chemoreceptors to food odors, usually causing a marked decrease in response magnitude and a change in the temporal response pattern (Derby and Atema, 1981a). Although behavioral observations and neurophysiological recording can complement each other as pollution detection assays, a causal relationship between chemoreceptor interference and behavioral deficit remains to be shown.

III. Sensory Biology

The sensory *Umwelt* of *Homarus americanus* is composed of many stimuli, but the systematic study of the sensory filter properties this animal uses to make behavioral decisions has only just begun. It is clear that chemical and mechanical stimuli are of great importance. Lobsters are emerging as one of the primary models for our understanding of the spatial–temporal structure of chemical signals and chemoreceptive signal transduction and processing. Mechanoreception has been studied more extensively in crayfish, but some work exists for *H. americanus*. Both chemo- and mechanoreception are distributed senses found with different degrees of specialization across the entire integument. Vision is the only other sensory modality in which some information exists for lobsters. This sense is located in two stalked lateral eyes and the associated "retinal" brain areas of the distal eyestalks. The role of vision is still a question of casual observation and inference. Other senses have not been studied. This section considers primarily the immediate behavioral and neurophysiological responses to sensory stimuli.

A. Behavioral Functions of Appendages

All cephalic, thoracic, and abdominal appendages of lobsters appear to be specialized for a great variety of behavioral functions, from locomotion, to grooming, to sensing. Peripheral specialization of appendages seems to be a general crustacean characteristic, not limited to lobsters. Some of the best studied sensory organs are shown in Fig. 5. Table 2 provides a summary of appendage functions, with emphasis on sensory functions and associated references. The appendages gain their special functions in part through appendage-specific arrays of different setal types. Setae are cuticular "hairs" described in detail by Factor (1978) and Lavalli and Factor (1992, Chapter 14). Sensilla are setae with sensory function and thus are innervated. Three important chemosensory sensilla are shown in Fig. 6.

B. Chemoreception

Typically, chemoreception has been divided into at least two different categories, smell and taste, and the criteria for this decision in Crustacea have been debated (e.g., Atema, 1977a, 1980; Schmidt and Ache, 1992). However, it may be useful to consider chemoreception across the phyla originally as a distributed sense covering the entire integument of an

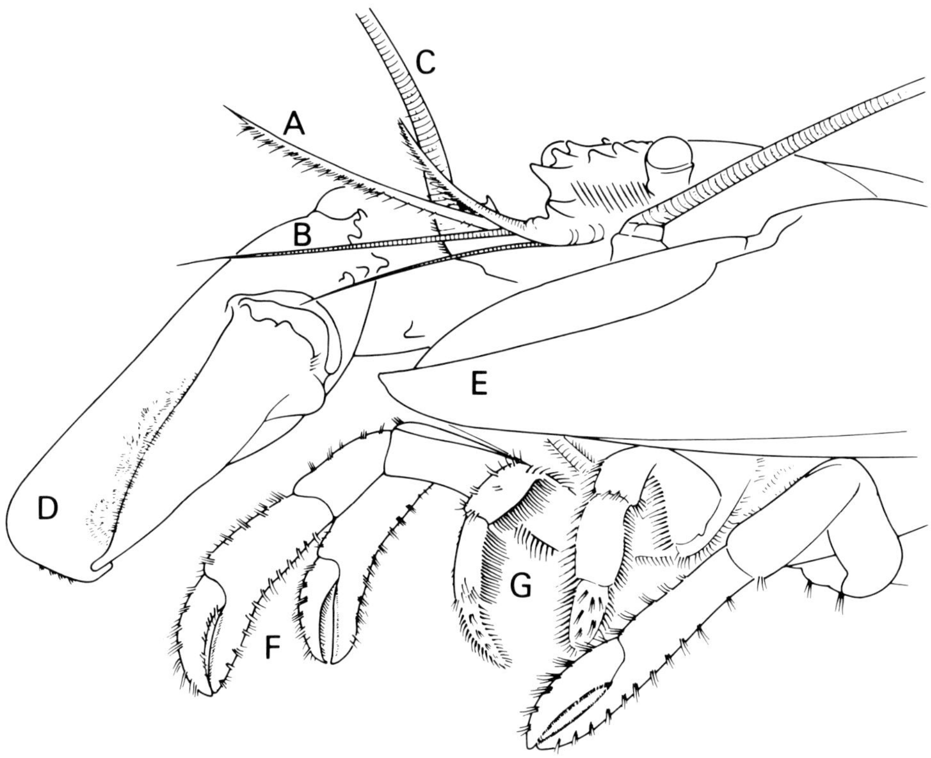

FIGURE 5 Schematic representation of the cephalothorax of *Homarus americanus* and its appendages. (A) Lateral flagellum of the antennule (first antenna); (B) medial flagellum of the antennule; (C) second antenna; (D) crusher claw (first pereiopod); (E) seizer claw (first pereiopod); (F) chelate walking legs (second and third pereiopods); and (G) third maxillipeds.

TABLE 2 **Behavioral Function of Appendages of Juvenile through Adult Lobsters (with Emphasis on Sensing and Communication)**

Appendage	Function	Behavior	Neurophysiology
Antennules (=first antennae)		Atema (1977a), Carter and Steele (1982)	
Lateral flagella	Olfaction for chemotaxis	McLeese (1970, 1973), Devine and Atema (1982), Moore et al. (1991b)	Levandowsky and Hodgson (1965), Ache (1972), Shepheard (1974), Johnson and Atema (1983), Derby *et al.* (1984), Johnson *et al.* (1985, 1989), Atema *et al.* (1989), Weinstein (1991), Gomez and Atema (1994), Gomez *et al.* (1992, 1994)
	Social information: dominance, courtship	Atema (1977a), Karavanich and Atema (1991)	
Medial flagella	Unknown, despite excellent chemoreceptivity		Tierney *et al.* (1988)
Basal segments	Statocyst organ for equilibrium		Cohen (1955, 1960)
Antennas (=second antennae)			
Flagella	Unknown, despite excellent chemoreceptivity; assumed to serve touch, and some chemoreception; used socially (agonistic, courtship); see "Antenna whipping" (Table 1)	Solon and Cobb (1980a)	Solon and Cobb (1980b), Voigt and Atema (1992)
Bases	Nephropore with gland	Bushmann and Atema (1994)	
Mandibles			
Endopodites	Grasping/ripping food		
Exopodites	Directing food	Lavalli and Factor (Chapter 14)	
First maxillae			
Endopodites	Food handling	Lavalli and Factor (Chapter 14)	
Exopodites	Currents	Lavalli and Factor (Chapter 14)	
Second maxillae			
Endopodites	Food handling	Lavalli and Factor (Chapter 14)	
Exopodites/epipodites	=Scaphognathite driving gill current (see Section II,A)	Lavalli and Factor (Chapter 14)	
First maxillipeds			
Endopodites	Feeding (pushing food into mouth)	Lavalli and Factor (Chapter 14)	
Exopodites	Fan for redirecting gill current (Section II,A)		
Second maxillipeds			
Endopodites	Feeding (pushing food into mouth)	Lavalli and Factor (Chapter 14)	
Exopodites	Redirecting gill current (Section II,A)		

continues

TABLE 2 Continued

Third maxillipeds			
Endopodites	Food recognition/acceptance, capture, ripping; grooming of antennules and antennae; bulldozing; in male, turning over female during molting	Derby and Atema (1982b), Atema and Engstrom (1971), Lavalli and Factor (Chapter 14)	Corotto *et al.* (1992)
Exopodites	Fan for redirecting gill current (Section II,A)		
Great chelipeds (=first pereiopods)	Defense; aggressive display (particularly in males); feeding	Solon and Cobb (1980a), Waddy and Aiken (1991), Lavalli and Factor (Chapter 14)	Solon and Cobb (1980b), Solon and Kass-Simon (1981)
Crusher	Crushing food (e.g., mollusk, crustacean); claw locking	See text	Derby (1982)
Seizer	Ripping food; catching prey; agonistic encounters	See text	
Walking legs			
Second pereiopods, chelate endopodites	Walking; grasping; feeding	Borroni *et al.* (1986), Lavalli and Factor (Chapter 14)	Derby (1982), Derby and Atema (1982a,c), Johnson *et al.* (1984), Borroni and Atema (1988, 1989), Voigt and Atema (1990)
	Grooming of eyestalks, major chelae, rostrum		Derby and Atema (1982b,c), Johnson *et al.* (1984), Bayha *et al.* (1993)
	Chemical food recognition	Atema (1977a), Borroni *et al.*, (1986)	
	Assist chemotaxis	Devine and Atema (1982)	
Third pereiopods, chelate endopodites	Same as second pereiopod		
Fourth pereiopods, nonchelate endopodites	Walking		
Fifth pereiopods, nonchelate endopodites	Walking; grooming abdomen; chemoreception		Bayha *et al.* (1993)
Pleopods	Generating current (see Section II,A); in male, first pair modified for sperm deposition; in female, all pairs develop long dark setae for egg attachment	Davis (1974)	
	Chemo- and mechanoreceptivity		Kotak and Page (1986), Killian and Page (1992a,b)
Uropods	Tail flip escape; chemo- and mechanoreceptivity; in female, cover brooding eggs		
Telson	Same as uropods		

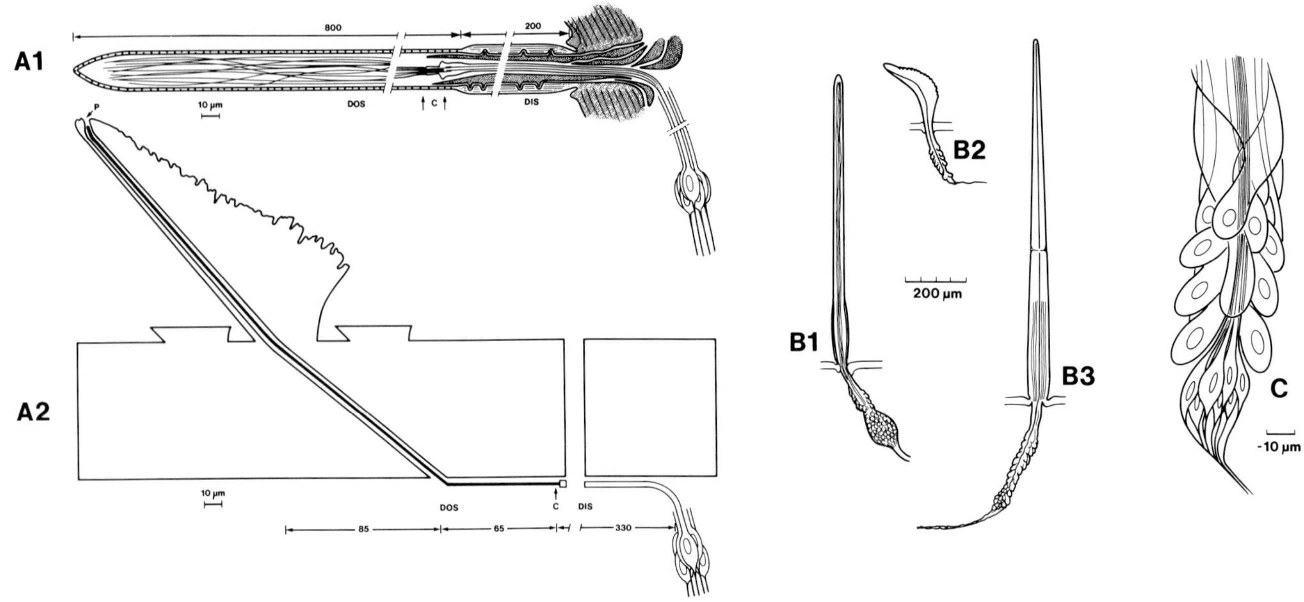

FIGURE 6 Schematic representation of major chemoreceptive sensilla of *Homarus americanus* and their receptor cells. (A) Innervation of two major sensilla. (A1) Aesthetasc sensillum on the lateral antennule. The entire sensillum is about 1 mm long and 30 μm in diameter. The somata of a few of the approximately 400 receptor cells are shown. Their long dendrites innervate the sensillum. Dendrites (one is shown greatly enlarged) sprout cilia, which soon branch and extend the length of the thin-walled distal part of the sensillum. Chemical stimuli are believed to reach dendrites by diffusing through the spongy, thin-walled cuticle (Oleszko-Szuts and Atema, 1977). (A2) Fringed setae on the leg of the crayfish *Austropotamobius torrentium* (after Altner *et al.*, 1983), closely resemble the tooth or "hedgehog" sensillum of lobster legs. Note again the peripheral location of the somata and the bipolar nature of these modified ciliary cells, with their dendrites extending in one direction into the lumen of the sensillum and their axons beginning their course toward the central nervous system. Chemical stimuli are believed to enter through the subapical pore. (Reproduced from Derby and Atema, 1988, with permission, and unpublished data.) (B) Relative dimensions of lobster chemoreceptive sensilla: (B1) smooth sensillum (30 neurons); (B2) tooth sensillum (eight neurons); (B3) aesthetasc sensillum (400 neurons). (C) Reconstruction of innervation of the tooth sensillum based on light and electron microscopic examination of serial sections. Many sheath cells wrap around the bundle of dendrites well into the sensillar lumen; the neural somata are exposed and easily seen at the proximal side, where an axon bundle leaves the cell cluster (J. Atema, P. F. Borroni, K. Hammar, M. S. Laverack, unpublished observations). Distances and scale are in micrometers. DOS, Outer (ciliary) segment of the sensory dendrite; DIS, inner segment of the sensory dendrite; C, ciliary junction; P, pore near the tip of the setae.

animal. Specializations take place in which specific organs condense into sensory clusters with associated innervation and brain development. In lobsters, the most specialized organ is the lateral flagellum of the antennules (Figs. 5 and 7), where crustacean-specific aesthetasc sensilla contain on the order of 10^5 olfactory neurons that all project into the deutocerebral olfactory lobe. This well-defined brain region is made up of olfaction-specific glomeruli in which chemical signals are processed in a manner similar to vertebrates and insects. (See Beltz, Chapter 11, on the nervous system.)

Other chemosensory "condensations" are found on the lobster antennules (=first antennae), antennae (=second antennae), maxillipeds, and pereiopods (Fig. 5). (See Factor, Chapter 1, on anatomy, and Lavalli and Factor, Chapter 14, on the feeding appendages.) Some of these sensory clusters have been shown to have typically gustatory functions; others remain unknown. These nonolfactory chemosensory clusters (organs) are generally organized through sensilla with both chemo- and mechanoreceptor neurons, making for a close association between these two sensory modalities.

1. Behavioral Responses to Chemical Stimuli

Behavioral studies have identified stimulatory amino acids and other components of food extracts in crustacean feeding behavior (reviewed by Carr, 1988); in *Homarus americanus*, the stimulatory effectiveness

FIGURE 7 Scanning electron micrograph of the lateral antennule of *Homarus americanus*. Dense rows of delicate aesthetasc hairs are surrounded by sturdy guard hairs. Scale bar: 100 μm.

of several food extracts and their constituents, as well as tank water from aquaria containing prey or conspecifics, has been described (Levandowsky and Hodgson, 1965; Atema and Engstrom, 1971; McLeese, 1970, 1973, 1974; Fuzessery and Childress, 1975; Evans and Mann, 1977; Derby and Atema, 1981b; Carter and Steele, 1982; Daniel and Bayer, 1987a–c; Bushmann and Atema, 1993, 1994). Chemotaxis toward food odors and conspecific body odors is described in Section II,E. Natural extracts of cod, shrimp, and lobster muscle are more stimulatory than any of the single compounds or simple artificial mixtures. A synthetic blend of 21 amino acids as found in mussel extract (Mackie, 1973) elicits a full behavioral response when presented to the anterior walking legs (Borroni *et al.*, 1986). A proportional blend of six of these amino acids hardly stimulates the legs behaviorally, yet, when tested individually, its components cause over 90% of the total electrophysiological response (Johnson *et al.*, 1984). Subtle differences in mixtures can have important behavioral consequences (Borroni *et al.*, 1986). Lobsters can discriminate between the body odors of two mussel species (*Mytilus edulis* and *Geukensia demissus*) and show significantly improved detection and localization behavior for the species odor to which they have been exposed for a few weeks (Derby and Atema, 1981b).

2. Physiological Responses of Chemoreceptor Cells: Spectral and Temporal Tuning

The task of sensory receptor cells is to interface efficiently with the physical environment and provide the brain with information that can be processed for adaptive behavior. Two key elements of this information are stimulus identification and stimulus localization. Important constraints on receptor cells are the amount and speed of information processing, sensitivity, and noise rejection. Indeed, beyond truly random noise, sensory systems determine what is signal and what is biological noise. Receptor cells act as filters with both spectral and temporal properties. Analogous to acoustic and visual senses, one may consider chemoreceptor cells for their spectral and temporal properties. The chemical spectrum is composed of the vast number of chemical compounds that exist in the environment. Each compound stimulates to various degrees some of the hundreds of thousands of relatively narrowly tuned receptor cells. Thus, each compound or mixture elicits a characteristic spatial pattern of excited receptor cells. The spectral tuning of a chemoreceptor organ results from the organ-specific assemblages of different receptor cells. Cell processes, such as receptor desensitization, second messenger release, and ion channel dynamics, give each receptor cell its temporal properties. These temporal filter properties vary from cell to cell and form a substrate for the discrimination of rapid concentration fluctuations, such as those found in odor plumes. *Homarus americanus* has contributed considerably to our understanding of such chemoreception principles.

a. Spectral Tuning and Mixture Discrimination The spectral tuning properties of several appendages have been determined by measuring the sensitivity of individual receptor cells to various single compounds. At biologically relevant concentrations, single receptor cells often respond to only one or a few stimuli, that is, they are narrowly tuned (Fig. 8). Effective stimuli include amino acids and amines, but not sugars and alcohols. This is a common feature of chemoreceptors of marine decapod crustaceans. Furthermore, chemosensitivity toward other compounds has been established (secondary plant compounds, Derby *et al.*, 1984; collagens, Johnson *et al.*, 1988; oligopeptides, Merrill, 1992; fractions of fuel oil, Atema, 1980, Derby and Atema, 1981a), often with remarkably strong and selective responses by a few receptor cells. Narrow tuning may allow the identification of key stimuli without (complete) cross-adaptation by chemicals in the background (Borroni and Atema, 1988; Derby and Atema, 1988; Atema *et al.*, 1989). Most chemoreceptor studies of the various lobster species, including *Homarus americanus*, have focused on lateral antennules (Ache, 1972; Shepheard, 1974; Johnson and Atema, 1983; Weinstein *et al.*, 1989) and anterior walking legs (Derby and Atema, 1982a; Johnson *et al.*, 1984; Bayha *et al.*, 1993), but a few have investigated the chemosensory properties of the medial antennules (Tierney *et al.*, 1988), maxillipeds

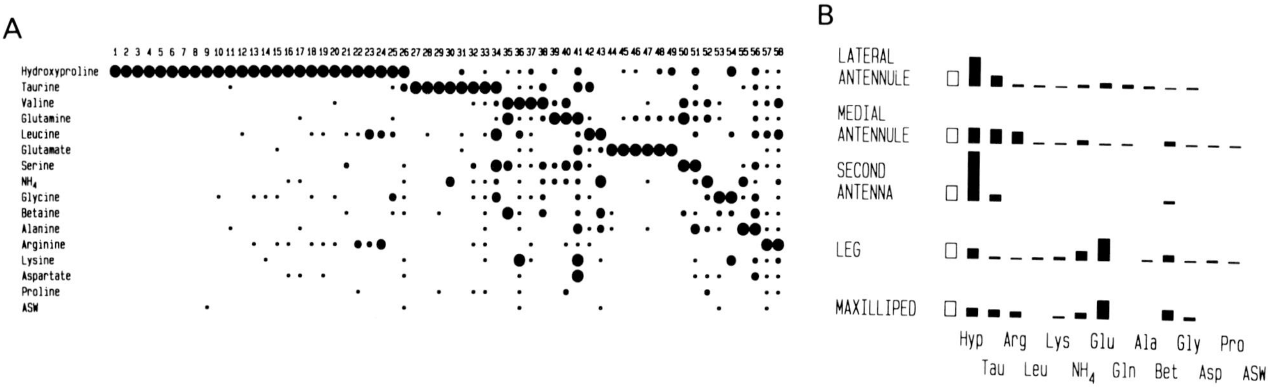

FIGURE 8 Spectral tuning of chemoreceptor cells and organs. (A) Response spectra of 58 chemoreceptor cells of the lateral antennule of *Homarus americanus*. Responses of each cell are normalized to their best response. The smallest dots indicate less than 20% of the maximum response; the next largest dots indicate response magnitudes of 20–40%, 40–60%, 60–80%, and 80–100%, respectively. Cells are grouped by their best stimulus and within each group are ordered by increasing tuning breadth based on their H-metric value (Smith and Travers, 1979). Stimuli are ordered by their overall efficacy: starting with 26 hydroxyproline-best cells, followed by eight taurine-best cells, etc. (After Weinstein, 1991.) (B) Response spectra of five chemoreceptive organs of *Homarus americanus*, expressed as relative numbers of "best" cell types. Data were reanalyzed for a comparison of responses to the 13 compounds tested in all five organs. The relative number of best cells from the antenna (n = 47, after Voigt and Atema, 1992), lateral antennule (n = 88, after Johnson and Atema, 1983; Weinstein, 1991), medial antennule (n = 53, after Tierney *et al.*, 1988), walking legs (n = 66, after Johnson *et al.*, 1984), and maxillipeds (n = 42, after Corotto *et al.*, 1992). ASW, Artificial sea water. Open scale bars on the left represent 25%. (After Voigt and Atema, 1992, reprinted with permission.)

(Corotto *et al.*, 1992), and antennae (Derby, 1982; Voigt and Atema, 1992). Among the five chemoreceptor organs so far investigated in *H. americanus*, only the second antennae are dominated by one narrowly tuned cell population: hydroxyproline-best cells make up 85% of the population. By comparison, hydroxyproline cells account for 46% of all cells tested in the lateral antennules, 26% in the medial antennules, 13% in the maxillipeds, and 16% in the walking legs (Fig. 8). The organ-level tuning is similar within the cephalic appendages (lateral and medial antennules and the antennae), which differ from the thoracic appendages (walking legs and maxillipeds). Each organ represents a differently tuned spectral filter that allows the animal to extract different information from its chemical environment. Resemblance in spectral tuning could indicate a functional overlap of chemoreceptor organs, and the tuning differences between antennal and leg organs may reflect the behavioral separation of olfaction and taste (Atema, 1977a). Some compounds are most likely the preferred long-distance signals (olfactory) because of their longevity in seawater and their signal-to-background ratio, while other compounds could allow a refined evaluation of food quality near the source of release and/or upon contact (gustatory); behavioral evidence for this notion is still sparse. The great variety of differently tuned chemoreceptor cells in each of the five chemoreceptor organs points to the idea that these receptor cells are given their tuning properties by different blends of receptor sites. These sites are themselves distributed in different proportions in the various chemoreceptor organs, thus giving the organs their specific "view" on the chemical world (Atema *et al.*, 1989).

In its natural environment, *Homarus americanus* must extract important chemical cues from a fluctuating chemical background. Mopper and Lindroth (1982) found spatially averaged coastal marine concentrations of amino acids in the submicromolar range; data on naturally occurring concentrations in microodor patches are lacking (Atema, 1988; Manahan, 1990). Background adaptation represents the change in responsiveness of a cell exposed to a constant background concentration of chemical stimulus. This effect becomes evident in seconds and does not fully recover for at least 30 seconds after the cell leaves that background (Gomez and Atema, 1994). Self-adaptation, in which stimulus and background are identical, results in a shift in the stimulus–response function and is thought to extend the cell's momentary working range, from two log steps of stimulus concentration to five or more log steps (Borroni and Atema, 1988). Cross-adaptation, in which stimulus and background are different, results in a change of slope in stimulus–response functions (Borroni and Atema, 1989). Individual receptor cells show a wide variety of effects under experimental conditions (Atema *et al.*, 1989; Johnson *et al.*, 1989).

The simultaneous presentation of two or more stimuli can result in mixture suppression or enhancement of response in comparison to the response to the cell's best single compound (Johnson *et al.*, 1985, 1989; Merrill, 1992). Also here, individual cells show the wide variety of effects necessary for a discriminating nose (Atema *et al.*, 1989).

It has been proposed that the brain encodes stimulus quality as an across-neuron pattern, whereas stimulus intensity might be coded by the response magnitude of a receptor cell population. However, recent experiments suggest that intensity may also be encoded by an across-neuron pattern (Johnson *et al.*, 1991; Merrill *et al.*, 1994). Furthermore, a population of unreliable single receptor cells can encode stimulus intensity reliably using either code (Merrill *et al.*, 1994).

b. Temporal Tuning and Signal Processing for Chemotaxis The patchy nature of underwater odor plumes appears to a chemoreceptor organ as a series of bursts of odor in varying strength and duration. These pulse patterns reflect the spatial–temporal distribution of odor patches, which contains directional information that can be used for orientation and distance estimates (Atema, 1985, 1988; Moore and Atema, 1991; Moore *et al.*, 1991b). The lateral antennules and associated brain areas, in addition to extracting information regarding quality and intensity of the odor source described above, may also extract information regarding the spatial distribution of odor concentration based on temporal analysis. This information would be useful for chemotaxis (see Section II,E).

The antennular chemoreceptor cells innervate thick tufts of aesthetasc hairs (Derby, 1982), which create thick boundary layers of water and thus a physical barrier to efficient odor access. To overcome this, antennules flick: the fast downward stroke of an antennule (100 msec) removes water, and therefore odor, caught between the aesthetasc hairs and replaces it with a new odor sample. During the slower upward stroke, odor concentration in the tuft changes only slightly due to diffusion and some wash-out (Moore *et al.*, 1991a). Antennular flicking increases stimulus access (Moore *et al.*, 1991a) and therefore enhances the response magnitude (Schmitt and Ache, 1979); the maximal flicking rate is 4–5 Hz (Berg *et al.*, 1992; Leonard *et al.*, 1994). This odor-sampling strategy is similar to vertebrate sniffing. The response fusion frequency (i.e., the limit of pulse frequency resolution, in which responses to individual stimuli begin to fuse) of antennular hydroxyproline cells is 4–5 Hz, perhaps not coincidentally (Gomez *et al.*, 1994).

Cumulative adaptation, as a measure of recovery from the effects of prior stimulation, reflects the disadaptation rate following adaptation to a series of stimulus pulses. This recovery has been measured directly in antennular hydroxyproline and taurine cells and follows an exponential time course, reaching full recovery in 30 seconds, with a range of 20–50 seconds for individual cells (Voigt and Atema, 1990; Gomez and Atema, 1994). To avoid response reduction due to prior stimulation, most studies used interstimulus intervals of 1 minute or more (Merrill *et al.*, 1994; Voigt and Atema, 1990).

Surprisingly, in leg chemoreceptors of *Homarus americanus*, interstimulus intervals of 5, 10, or 20 seconds do not affect the degree of cumulative adaptation of the glutamate-sensitive cell population (Voigt and Atema, 1990). This suggests that cumulative adaptation results from a change in the cell's state, rather than a graduated, stimulus-determined process. As always, individual cells reveal great diversity in the time course of cumulative adaptation, regardless of the interstimulus interval. Legs are probably not involved in detailed temporal analysis, although they play a minor role in chemotaxis (Devine and Atema, 1982).

Receptor cells integrate several adaptation processes. Some of the temporal dynamics of this process have been modeled (Mountain and Atema, 1993) based on a series of intracellular transduction mechanisms (McClintock and Ache, 1989). The model was tested successfully, with published results on second messenger kinetics and *in situ* receptor cell responses to pulsed stimuli in various backgrounds.

C. Mechanoreception

1. Cuticular Mechanoreception

Similar to chemoreceptor condensations, one finds functional clusters of mechanoreceptors that, in various phyla, take on the role of distinct organs with specialized functions such as equilibrium and motion control (e.g., statocysts or semicircular canals), hearing, hydrodynamic flow analysis (e.g., fish lateral line organs), various proprioreceptors, and touch. A general tactile sensitivity remains distributed across the integument. In lobsters, obvious specializations are the statocyst and the rows and clusters of sensilla on the two antennae, distal portions of maxillipeds, pereiopods, and telson. There are undoubtedly many more to be found, but even the few mentioned have not been well studied in lobsters. For example, one would expect to find the functional equivalent of the fish lateral line in the form of hydrodynamically sensitive directional sensilla over the lobster's body. Such investigations have been productive in crayfish (Breithaupt and Tautz, 1990).

In addition to chemical stimuli, mechanical (hydrodynamic and tactile) stimuli can be of great importance in detecting and locating prey, predators, and conspecifics. Contact mechanoreception of substrate vibrations and noncontact mechanoreception of waterborne particle movement are examples of mechanical stimuli detected by lobsters. Certain organs of *Homarus gammarus* respond to water currents and water-propagated pressure waves (Laverack, 1962a,b). The information currents of *Homarus americanus* (Fig. 1) (Atema, 1985; McPhie and Atema, 1984) may carry hydrodynamic information as well as chemical cues. This section considers supracuticular (external) mechanoreceptors. Internal mechanoreceptors (muscle receptor organs, apodeme receptors, chordotonal organs, innervated strands, and nerve chord receptors) are reviewed by Bush and Laverack (1982) and Govind (Chapter 12). The predominantly chemoreceptive antennules also have mechanoreceptor innervation, which may play a role in monitoring flicking behavior.

The second antenna in the Decapoda has been viewed as a major mechanoreceptive organ, both physiologically and behaviorally. Its mechanoreceptors have been classified as touch receptors and uni- and bidirectional water movement detectors [*Homarus gammarus* (=*H. vulgaris*), Laverack, 1962a,b; Tazaki, 1977; *H. americanus*, Derby, 1982; reviewed by Bush and Laverack, 1982). In *H. gammarus*, two types of smooth hairs are stimulated directly either by water flow or by contact (Tazaki, 1977). One type of receptor shows directional sensitivity and is sensitive to position and movement; a second type is classified as a phasic movement detector. In addition, feathered hairs are stimulated by antennal movements—that is, receptors in these hairs respond when the flagellum is bent; each feathered hair responds to bending of the joint it spans. The best frequency resolution is about 100 Hz, a frequency also produced by the "sonic" muscle (see Section III,C). These receptors are described in detail in crayfish (Tautz *et al.*, 1981; Zeil *et al.*, 1985), in which the flagellum feathered hair system is thought to detect moving objects in the crayfish's immediate environment. These hair types are also found on the flagellum of the first antenna.

There are four classes of cuticular hair organs on the great chelae (seizer and crusher claws) of *Homarus americanus* (Solon and Cobb, 1980a). Type I and type II organs each possess a single long sensillum, 30–60 μm and 70–130 μm long, respectively. Type III organs are toroid bumps, 20–30 μm in diameter, with a small tuft of fibers projecting from the center. Type IV organs are small, conical hairs, 1 μm long. While type IV organs are uniformly distributed, the others are not. Type I organs are similar in size and appearance to receptors on the antennal flagellum of *H. gammarus* (Tazaki, 1977). Their articulation and the depression in which they are located are similar to the hydrodynamic receptors of *Palinurus vulgaris* (Vedel and Clarac, 1976). Types I and II may be mechanoreceptors; their asymmetry and consistent orientation may reflect directional sensitivity. The role of types III and

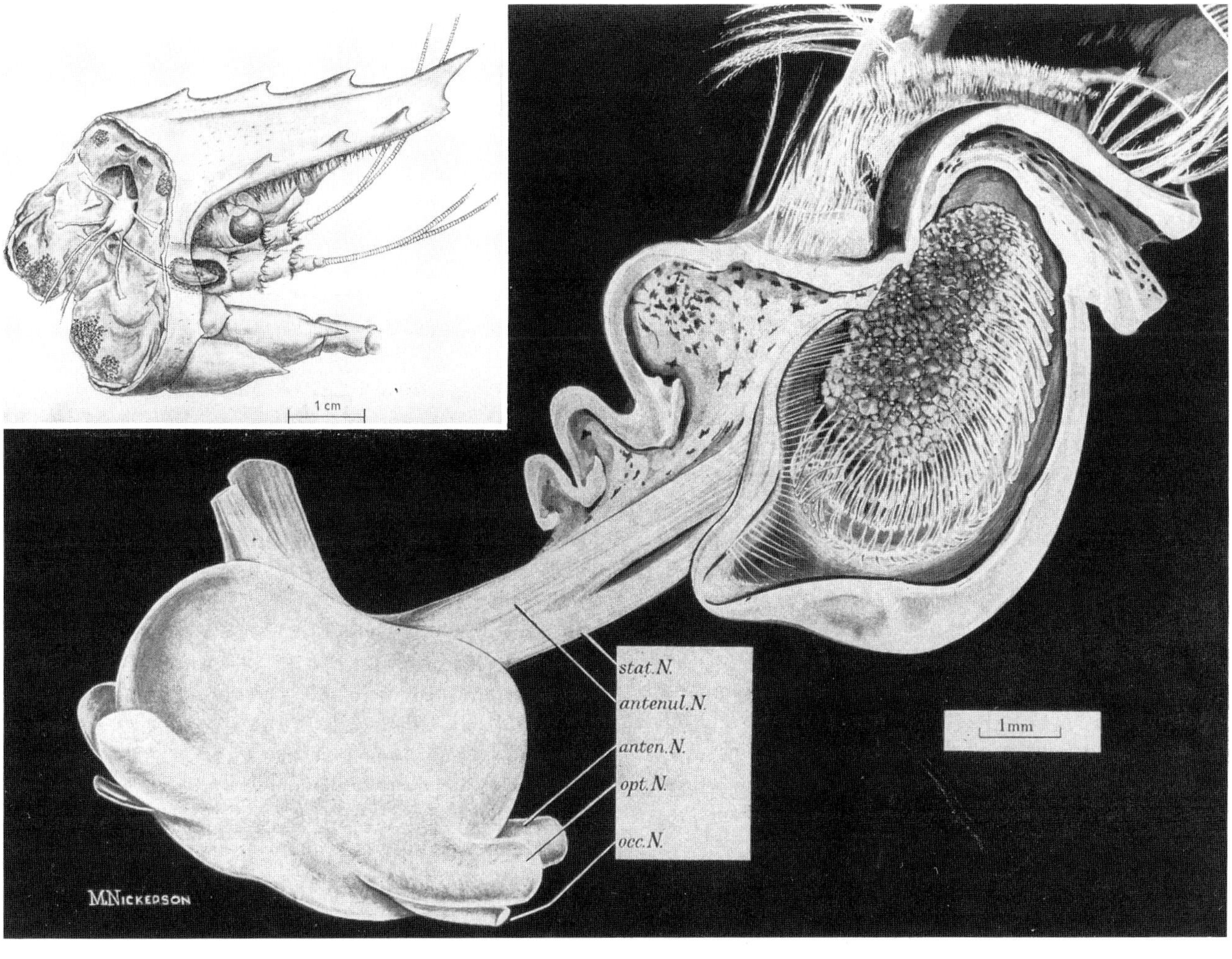

FIGURE 9 (Inset) The anterior cephalothoracic region of *Homarus americanus* with the exposed statocyst (in the basal segment of the right antennule). (Reproduced from Cohen, 1955, with permission). The right statocyst and its nervous connection to the supraesophageal ganglion (dorsal view). The ganglion is tipped posteriorly to expose the anterior surface. The statocyst nerve (*stat.N.*) on the dorsolateral aspect of the antennular nerve (*antennul.N.*) is seen as a distinct bundle. The shell of the antennular segment has been removed to expose the cyst lying in the lumen of the segment. The dorsal wall of the cyst has been removed and the compact statolith composed of cemented sand grains is seen on the floor of the cyst. The sensory hairs on the cyst floor are arranged in a crescent shape, with the inner three rows contacting the statolith while the irregular outer row projects freely into the fluid. The fine thread hairs are projecting horizontally into the cyst fluid from the medial cyst wall. *anten.N.*, Antennal nerve; *opt.N.*, optic nerve; *occ.N.*, oculomotor nerve. (Reproduced from M. J. Cohen, 1960. The response pattern of single receptors in the crustacean statocyst. *Proc. R. Soc. London, Ser. B*, vol. 152, pp. 30–49, with permission of The Royal Society.)

IV is not clear. Perhaps these receptors detect flow over the claws.

2. Sound Production and Perception

Homarus americanus produces a growl or rasplike sound (100–130 Hz, 100- to 500-msec duration, 16-dB maximum sound pressure) internally by contraction of a small, modified sonic muscle, which is part of the remotor muscle of the coxapodite of the second antennae (Fish, 1966; Mendelson, 1969). No sounds have been recorded either acoustically or via electromyograms of the sonic muscle during agonistic or courtship encounters in naturalistic aquaria (G. Pollock and J. Atema, unpublished data). There have been repeated and credible anecdotal reports of sound production from lobster-occupied shelters heard by divers. Laboratory recordings have confirmed this for single lobsters resting in shelter (A. Ludlow, unpublished observation). Sound generation by appendage stridulation, as described in Palinuridae (Moulton, 1957, 1958), has not been established. Thus, the biological significance of sound production in *H. americanus* is not known, although it has been suggested as a defense mechanism (Fish, 1966). If hearing is defined as the reception of the pressure component of sound, then it does not occur in Crustacea (Breithaupt and Tautz, 1990). However, low-frequency hydrodynamic vibrations that accompany all sound production can be detected at biologically relevant distances, for instance, by hair-fan organs (Offutt, 1970).

3. Statocyst

As in all decapod crustaceans, the statocysts of *Homarus americanus* are associated with geo-orientation and serve primarily as an equilibrium organ (Cohen, 1955). A statocyst lies in the basal segment of each of the two antennules (Fig. 9). The statocyst has been described as "an ectodermal sac, fluid-filled and lined by hairs which are in contact with a relatively dense mass, the statolith" (Cohen, 1955; reviewed by Cohen and Dijkgraaf, 1961). Four to five hundred chitinous hairs are distributed in four rows on each organ floor; they form a crescent-shaped elevation, called the sensory cushion (Cohen, 1955). Each statolith is composed of many sand grains cemented together and flexibly attached to the cyst floor. The hairs of the most peripheral row (row 1) on the sensory cushion probably never come into contact with the statolith. The hairs of row 2 are regularly arranged and project from the sensory cushion into the statolith mass at an angle of approximately 40°; only their tips come into contact with the statolith and seem to be fastened to the statolith. The hairs of rows 3 and 4 are spaced farther apart and seem to be fastened between sand grains. A row of approximately 70 long hairs (thread hairs, Prentiss, 1901) on the most posterior aspect of the medial wall seems to float freely in the lumen and does not come into contact with the statolith. These hairs respond to the slightest fluid motion. The relatively constant relationship between any single hair and the statolith suggests that a given stimulus will always affect an individual receptor in the same manner (Cohen, 1955).

Each hair is innervated by only one receptor neuron. In physiological experiments, four different types of receptor cells have been identified (Cohen, 1955, 1960). One static position receptor indicates the absolute position about the transverse axis. A second static position receptor responds to the direction of movement in addition to the static response. Acceleration receptors respond only to movement about any horizontal or vertical axis, not to position. Vibration receptors respond to high-intensity and low-frequency vibrations of the substratum, but not to air- or waterborne vibrations.

D. Vision

Underwater, light has a more restricted angular distribution, a generally lower intensity, and a narrower wavelength range than in typical terrestrial environments (Jerlov, 1976). Contrast underwater is reduced (Lythgoe, 1988), placing a premium on adaptations to enhance sensitivity and perception of contrast and motion. Camouflage in the open ocean is an excellent example of tricks used to reduce contrast, which tell us much about the visual perception of potential visual predators (Saidel, 1988).

1. Behavioral Responses

It has been suggested that vision plays an important role in several behaviors, but experimental evidence is sparse. *Antenna pointing* initiated by visual cues, as described in crayfish (Zeil *et al.*, 1985), has been reported anecdotally in *Homarus americanus*, probably as part of predator avoidance. Motion detection also plays a role in feeding behavior: lobsters respond only to moving crabs or sea urchins, but not to the sight of stationary specimens (Hirtle and Mann, 1978). Meral spread, an important part of agonistic behavior (see Section II,B,2), may be, in part, a visual display. Blindfolded lobsters appear to rely more on mechano- and chemoreception in agonistic encounters than do seeing lobsters (Kaplan *et al.*, 1993; Snyder *et al.*, 1993).

2. Morphology

The eyes, on two movable stalks, are lateral compound eyes that follow the basic organization of other compound eyes (reviewed by Waterman, 1961; Cronin, 1986). Stalked eyes supply a broader field of view and increased binocular spread. Eyestalk movements are used to protect the eyes from mechanical damage, not to track visual targets (J. Atema and R. Voigt, unpublished observations). *Homarus americanus* has a reflecting superposition eye. Parker's (1890) detailed description of the ommatidium in *H. americanus* (Fig. 10) has been confirmed by Rutherford and Horridge (1965) in *H. vulgaris*. In the adult *H. americanus*, each eye is composed of 13,500 ommatidia (Fig. 10) (Parker, 1890). Two corneagenous cells form an external corneal cuticular facet. Four pairs of cells form the corneal hypodermis. Four groups of cone cells form a long crystalline tract. At the proximal end of the crystalline tract, seven retinula cells, 250 μm long, are grouped around a central rhabdom, 125 μm long. The rhabdoms lie very close together, with little space between them for retinula cell cytoplasma. A rudimentary eighth retinula cell lies near the distal end of the rhabdom. The retinula cell axons pass through the basement membrane. Dark pigments are found in the retinula cells and the distal pigment cells. The distal pigment is contained in two cells around each, enclosing it in a black screen. Each distal pigment cell has one proximal process extending between the ommatidia and terminating just below the basement membrane. A white reflecting pigment is found in accessory pigment cells, which extend from between the rhabdoms through the basement membrane to the underlying lamina ganglionaris (first optic ganglion). One accessory cell fills the spaces between about 10 ommatidia. Each rhabdom is spindleshaped [125 μm long and 25 × 25 μm at the widest point in *H. gammarus* (Rutherford and Horridge, 1965)] and composed of microvilli that extend from the seven surrounding retinula cells. The characteristic feature is that the microvillar tubules are arranged in flat, saddle-shaped plates, each about 2 μm thick, which lie superimposed on each other in a vertical stack along the optical axis, with the tubules of each plate at right angles to the tubules of the adjacent plates (*H. gammarus*, Rutherford and Horridge, 1965). While the large size of the rhabdom increases the probability of

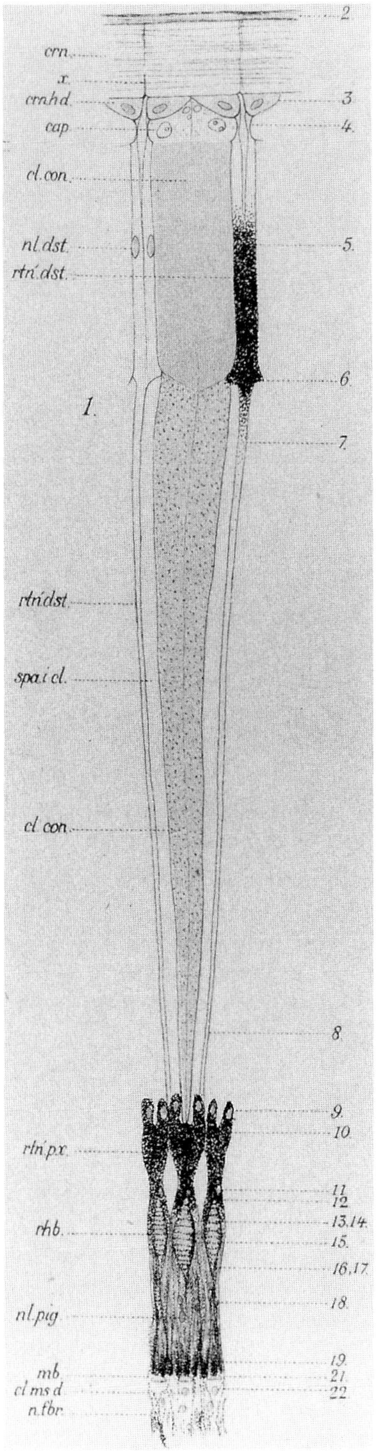

FIGURE 10 Longitudinal section of an ommatidium of the eye of *Homarus americanus*. The distal retinula on the right side of the cone contains its natural pigment; that on the left side has been depigmented. (Numbers on the right refer to drawings in the original paper which are not included here.) *cap.*, Protoplasmic cap of the cone cell; *cl.con.*, cone cell; *cl.msd.*, mesodermic cell; *crn.*, corneal cuticula; *crn.hd.*, corneal hypodermis; *mb.*, basement membrane; *n.fbr.*, nerve fiber; *nl.dst.*, nucleus of distal retinula; *nl.pig.*, nucleus of accessory pigmentcell; *rhb.*, rhabdom; *rtn'.dst.*, distal retinula; *rtn'.px.*, proximal retinula; *spa.icl.*, intercellular space of retina; *X*, position of the band that limits the corneal facet. (From Parker, 1890.)

photon capture, it seems a matter of chance which of the seven retinula cells will be excited, that is, in an apparent loss of possible directional sensitivity (Kuiper, 1962). The crossed arrangement of tubules suggests sensitivity to the plane of polarization of light (Rutherford and Horridge, 1965).

3. Physiology

The visual pigment has been identified as rhodopsin with maximum absorbance at λ_{max} 515 nm, both in the rhabdom (microspectrophotometry, Goldsmith and Bruno, 1973; Bruno *et al.*, 1977) and in digitonin extract (Wald and Hubbard, 1957); a stable metarhodopsin has a maximum absorbance at 490 nm (Wald and Burg, 1957). The λ_{max} of both species is 15 nm shorter than the peak of the spectral sensitivity function of the electroretinogram (520–525 μm) (Kennedy and Bruno, 1961; Wald, 1968). The spectral shift can be accounted for quantitatively by the screening effect of the red–brown granules of accessory pigment (Goldsmith, 1978); a 30- to 35-nm red shift in crayfish is due to screening pigments. The retinal reflecting pigment, which does not undergo photomechanical changes, occurs as a compact layer distal to the fenestrated basement membrane, as well as in substantial deposits proximal to this membrane (Kleinholz, 1959). It includes five substances, among them uric acid, xanthine, hypoxanthine, xanthopterin, and another (unidentified) pterine. Regeneration of the visual pigment is temperature dependent (Bruno *et al.*, 1977; Barnes and Goldsmith, 1977a,b). The proportion of rhodopsin to metarhodopsin affects only the sensitivity, that is, the probability of photon catch.

True color discrimination is only possible if an animal has at least two receptor types with distinct but overlapping spectral ranges. *Homarus americanus* probably has a single receptor type and, therefore, no true color vision. Its eyes are highly effective for photon capture and its overall efficacy has inspired a novel design for an orbital X-ray telescope (Hartline, 1980). Caudal photoreceptors, as described in other crustaceans, have not been demonstrated in *H. americanus* (Wilkens and Larimer, 1976). Thus, lobster eyes are designed to operate efficiently at low light levels and the vision is monochromatic. Information on spatial resolution is not available, but it is probably poor.

IV. Directions for Further Research

In the past decade, lobsters have become important oceanographic and biomedical models for understanding underwater chemical signals, chemoreceptive sensory transduction and sampling, and chemotaxis in turbulent environments. The next decade(s) will probably see further integration of multidisciplinary research approaches focused on chemoreception in this one species, including biochemical and biophysical transduction, molecular genetics of receptors, computational neural models, robotics, central nervous system processing, and integration with other sensory input, particularly hydrodynamic mechanoreceptors.

Lobsters are also important as biomedical models linking internal hormonal to external pheromonal regulation of behavior. This work is currently focused on urine, in which interesting chemical signals are seen and suspected, but not yet linked to behavior. Neuromodulators such as serotonin and octopamine may regulate behavioral postures and aggressive motivation, and may be linked to ecdysteroids and molt-inhibiting hormones that regulate the molt cycle. Metabolites of these compounds may be used as pheromones in urine or other excretory pathways to regulate dominance and courtship. In addition, the nephropore gland has great potential as a pheromone center. Research identifying these natural products is becoming more promising now that chemical methods and behavioral understanding have advanced significantly in the past decade.

Behavioral decisions play an important role in population dynamics. The study of lobster behavior in the field is lagging greatly behind laboratory analyses. It will be very important to devote time and effort to describing the natural behavioral ecology of lobsters to understand the role of seasonal migrations, possible hatching sites related to ocean currents, mating areas, intermolt versus postmolt mating, residency patterns, and food-foraging patterns. Now that laboratory work has generated a number of plausible hypotheses and well-defined questions, field research should become better focused and more promising, despite its intrinsic difficulty.

A wide-open area for research lies in larval and early juvenile lobster biology. Larval behavior and sensory biology remain nearly unknown, particularly in natural conditions. The postlarva (stage IV) has received much attention, but many questions remain regarding its natural behavior and sensory capabilities. Of particular interest for this settling stage are mechanisms of substrate recognition and predator avoidance. Subsequent juvenile stages that live in burrows have been studied, but additional understanding of their relationship to food, predators, and each other over the molt cycle and in the field is needed. This research will have important implications for the preservation of nursery habitat, the regulation of fishing practices, and mariculture.

Overall, our understanding of the behavioral and sensory biology of lobsters has improved dramatically in the last 25 years. These improvements are mostly in specific areas, leaving great unknowns in other areas expected to be important for lobsters. These include mechanoreception, vision, natural feeding behavior, migration, and larval and early juvenile behavioral and sensory biology.

V. Summary

Adult lobsters (*Homarus americanus*) are excellent models for the study of marine animal behavior and sensory biology. Their obvious dependence on chemical signals and shelters and aggressive behavior in the laboratory have led to initial focus on dominance and related courtship behavior and their regulation by chemical signals. Many other behaviors are important, but less studied.

Chemical signals are transported in part via urine and lobster-generated water currents. Three major currents are used frequently in all postlarval stages: gill current, pleopod current, and exopodite fan current. These currents link the animal to its environment, sending and receiving chemical information. Since the nephropore gland releases products into the urine, it is a potential source of pheromones.

Aggression is common when lobsters first meet. Fights escalate from approach to ripping with claws. This leads to dominance, which is maintained by frequent encounters in which the loser remembers the individual winner to which it lost. Small differences in body and claw size determine who wins. Sexual dimorphism in claw size starts at maturity, giving males an advantage over females. Generally, males are more aggressive, more active, and more dominant than females. Postmolt animals and those missing a claw are less aggressive. Ovigerous females behave defensively and become solitary. Isolation enhances aggression. Aggression appears at metamorphosis, develops with subsequent molts, and is stereotyped in naive animals. Experiences and memory greatly modulate aggressive behavior. Urine is released in large quantities by fight opponents; as soon as the loser becomes obvious and retreats, it also stops urine release.

Shelters are critical resources from the day of settling in stage IV. In the initial 2 years, lobsters do not leave their shelter system. At about 25-mm CL they start nocturnal excursions outside the shelter, regularly checking each other's shelters. Evictions are common; however, some adult animals reside in the same shelter for at least 9 months. Males can evict larger females, but the reverse is not seen. Dominant males set up a mating shelter sufficiently large for two and attract females for week-long cohabitation and mating.

Courtship is largely controlled by the female. They can enter the male shelters at any time during the molt cycle, but some females are vigorously rejected. One brief mating takes place generally 30 minutes after the female completes the molt, but in certain circumstances, intermolt matings occur without or with brief cohabitation. This appears to be an alternative strategy to allow the female to replenish sperm, which can be stored for 3 years and used to fertilize several batches of eggs.

Lobsters have many pairs of appendages, all of which are chemoreceptive and probably mechanoreceptive. The major chemoreceptor organs are the antennules, the dactyl and propus of the walking legs, and third maxillipeds. Commonly found receptor cells respond to amino acids and amines, but not to carbohydrates. Some chemoreceptor cells respond strongly to unusual stimuli, such as plant secondary compounds. Whereas receptor cells respond best to single compounds, lobster behavior depends on very specific blends, in which subtle differences can be discriminated.

Chemoreceptor cells show preferential responses to brief stimuli, such as the patches of odor that make up a turbulently dispersed odor plume. Odor sources can be located and temporal–spatial analysis of patches may lead the animal to the source.

Mechanoreception has not been well studied. Although a sonic muscle in the antennal base and a specific sound have been described, no behavioral significance is known. A typical decapod crustacean statocyst lies in the basal segment of the antennule, but hydrodynamic responses have not been described.

Vision does not appear to be of great importance. Several behaviors are barely affected by blindfolding. The eyes are monochromatic and designed for low-light, low-resolution performance, presumably predator-shadow detection at night or in deep water.

Acknowledgments

We thank Drs. Jenny Basil, Thomas Breithaupt, Paul Bushmann, Stanley Cobb, George Gomez, Christy Karavanich, and Lisa Karnofsky for collaboration, discussion, unpublished observations, and critical review. This chapter was written with support by grants from NSF (BNS 84-13661, BNS 84-11969, BNS 85-12585, BNS 88-12952, IBN-9212650) and NIH (2P01NS25915-04A1) to J. A.

References

Ache, B. W. (1972). Amino acid receptors in the antennules of *Homarus americanus*. *Comp. Biochem. Physiol. A* 42A, 807–811.

Aiken, D. E. (1973). Proecdysis, setal development and molt prediction in the American lobster (*Homarus americanus*). *J. Fish. Res. Board Can.* 30, 1337–1344.

Aiken, D. E., and Waddy, S. L. (1980). Reproductive biology. *In* "The Biology and Management of Lobsters" (J. S. Cobb and B. F. Phillips, eds.), Vol. 1, pp. 215–276. Academic Press, New York.

Altner, I., Hatt, H., and Altner, H. (1983). Structural properties of bimodal chemo- and mechanosensitive setae on the pereiopod chelae of the crayfish, *Austropotamobius torrentium*. *Cell Tissue Res.* 228, 357–374.

Atema, J. (1976). Sublethal effects of petroleum fractions on the behavior of the lobster, *Homarus americanus*, and the mud snail, *Nassarius obsoletus*. *In* "Estuarine Processes" (M. Wiley, ed.), Vol. 1, pp. 302–312. Academic Press, New York.

Atema, J. (1977a). Functional separation of smell and taste in fish and Crustacea. *In* "Olfaction and Taste" (J. Le Magnen and P. MacLeod, eds.), Vol. 6, pp. 165–174. Information Retrieval, London.

Atema, J. (1977b). The effects of oil on lobsters. *Oceanus* 20, 67–73.

Atema, J. (1980). Smelling and tasting underwater. *Oceanus* 23, 4–18.

Atema, J. (1985). Chemoreception in the sea: Adaptations of chemoreceptors and behavior to aquatic stimulus conditions. *Soc. Exp. Biol. Symp.* 39, 387–423.

Atema, J. (1986). Review of sexual selection and chemical communication in the lobster, *Homarus americanus*. *Can. J. Fish. Aquat. Sci.* 43, 2283–2290.

Atema, J. (1988). Chemoreceptor cells in aquatic invertebrates: Peripheral mechanisms of chemical signal processing in decapod crustacea. *In* "Sensory Biology of Aquatic Animals" (J. Atema, A. N. Popper, R. R. Fay, and W. N. Tavolga, eds.), pp. 29–56. Springer–Verlag, New York.

Atema, J. (1995). Chemical signals in lobster (*Homarus americanus*) behavior. Proceedings of the 4th International Workshop on Lobster Behavior. *Crustaceana* (in press).

Atema, J., and Cobb, J. S. (1980). Social behavior. *In* "The Biology and Management of Lobsters" (J. S. Cobb and B. F. Phillips, eds.), Vol. 1, pp. 409–450. Academic Press, New York.

Atema, J., and Cowan, D. F. (1986). Sex identifying urine and molt signals in the lobster *Homarus americanus*. *J. Chem. Ecol.* 12, 2065–2080.

Atema, J., and Engstrom, D. G. (1971). Sex pheromone in the lobster, *Homarus americanus*. *Nature (London)* 232, 261–263.

Atema, J., and Stein, L. S. (1974). Effects of crude oil on the feeding behaviour of the lobster *Homarus americanus*. *Environ. Pollut.* 6, 77–86.

Atema, J., Jacobson, S., Karnofsky, E. B., Oleszko-Szuts, S., and Stein, L. (1979a). Pair formation in the lobster, *Homarus americanus*: Behavioral development, pheromones and mating. *Mar. Behav. Physiol.* 6, 277–296.

Atema, J., Karnofsky, E. B., and Oleszko-Szuts, S. (1979b). Lobster behavior and chemoreception: Sublethal effects of number 2 fuel oil. *In* "Advances in Marine Environmental Research" (F. S. Jacoff, ed.), pp. 122–134. Environmental Research Laboratory, Office of Research and Development, Environmental Protection Agency, Narragansett, Rhode Island.

Atema, J., Leavitt, D. F., Barshaw, D. E., and Cuomo, M. C. (1982). Effects of drilling muds on behavior of the American lobster, *Homarus americanus*, in water column and substrate exposures. *Can. J. Fish. Aquat. Sci.* 39, 675–690.

Atema, J., Borroni, P. F., Johnson, B. R., Voigt, R., and Handrich, L.S. (1989). Adaptation and mixture interactions in chemoreceptor cells: Mechanisms for diversity and contrast enhancement. *In* "Perception of Complex Smells and Tastes" (D. L. Laing, W. Cain, R. McBride, and B. W. Ache, eds.), pp. 83–100. Academic Press, San Diego.

Babu, D. E., Hanumantha Rao, K., Shyamasundari, K., and Uma Devi, D. V. (1985). Histochemistry of the cuticle of the crab *Menippe rumphii* (Fabricius) (Crustacea: Brachyura) in relation to moulting. *J. Exp. Mar. Biol. Ecol.* 88, 129–144.

Barnes, S. N., and Goldsmith, T. H. (1977a). Dark adaptation of the lobster. *Biol. Bull. (Woods Hole, Mass.)* 145, 423–424.

Barnes, S. N., and Goldsmith, T. H. (1977b). Dark adaptation, sensitivity and rhodopsin level in the eye of the lobster, *Homarus*. *J. Comp. Physiol. A* 120A, 143–159.

Bayha, K. M., Voigt, R., and Atema, J. (1993). A comparison of the tuning properties of chemoreceptor cells in the first and fourth walking legs of female American lobsters. *Biol. Bull. (Woods Hole, Mass.)* 185, 316–317.

Berg, K., Voigt, R., and Atema, J. (1992). Flicking in the lobster *Homarus americanus*: Recordings from electrodes implanted in antennular segments. *Biol. Bull. (Woods Hole, Mass.)* 183, 377–378.

Borroni, P. F., and Atema, J. (1988). Adaptation in chemoreceptor cells: I. Self-adapting backgrounds determine threshold and cause parallel shift of response function. *J. Comp. Physiol. A* 164A, 67–74.

Borroni, P. F., and Atema, J. (1989). Adaptation in chemoreceptor cells: II. The effects of cross-adapting backgrounds depend on spectral tuning. *J. Comp. Physiol. A* 165A, 669–677.

Borroni, P. F., Handrich, L. S., and Atema, J. (1986). The role of narrowly tuned taste cell populations in lobster (*Homarus americanus*) feeding behavior. *Behav. Neurosci.* 100, 206–212.

Botero, L., and Atema, J. (1982). Behavior and substrate selection during larval settling in the lobster *Homarus americanus*. *J. Crust. Biol.* 2, 59–69.

Breithaupt, T., and Atema, J. (1993). Evidence for the use of urine signals in agonistic interactions of the American lobster. *Biol. Bull. (Woods Hole, Mass.)* 185, 318.

Breithaupt, T., Karavanich, C., and Atema, J. (1994). Urine as a chemical signal in lobster dominance recognition. *Chem. Senses* 19, 446–447.

Breithaupt, T., and Tautz, H. (1990). The sensitivity of crayfish mechanoreceptors to hydrodynamics and acoustic stimuli. *In* "Frontiers in Crustacean Neurobiology" (T. Wiese, ed.), pp. 114–120. Birkhäuser, Boston.

Briggs, P. T., and Zawacki, C. S. (1974). American lobsters at artificial reefs in New York. *N. Y. Fish Game J.* 21, 73–77.

Brunet, M., Cuoc, C., Arnaud, J., and Mazza, J. (1991). Tegumental glands in a copepod *Hemidiaptomus ingens*: Structural, ultrastructural and cytochemical aspects. *Tissue Cell* 23, 733–743.

Bruno, M. S., Barnes, S. N., and Goldsmith, T. H. (1977). The visual pigment and visual cycle of the lobster, *Homarus*. *J. Comp. Physiol. A* 120A, 123–142.

Bush, B. M. H., and Laverack, M. S. (1982). Mechanoreception. *In* "The Biology of Crustacea" (D. E. Bliss, ed.), Vol. 3, "Neurobiology: Structure and Function" (H. L. Atwood and D. C. Sandeman, eds.), pp. 399–468. Academic Press, New York.

Bushmann, P., and Atema, J. (1993). A novel tegumental gland in the nephropore of the lobster, *Homarus americanus*: A site for the production of chemical signals? *Biol. Bull. (Woods Hole, Mass.)* 185, 319–320.

Bushmann, P., and Atema, J. (1994). Nephropore glands in the lobster, *Homarus americanus*: A source of urine–carried chemical signals? *Chem. Senses* 19, 448.

Carr, W. E. S. (1988). The molecular nature of chemical stimuli in the aquatic environment. *In* "Sensory Biology of Aquatic Animals" (J. Atema, A. N. Popper, R. R. Fay, and W. N. Tavolga, eds.), pp. 3–29. Springer–Verlag, New York.

Carter, J. A., and Steele, D. H. (1982). Attraction to and selection of prey by immature lobsters (*Homarus americanus*). *Can. J. Zool.* 60, 326–336.

Cobb, J. S. (1971). The shelter–related behavior of the American lobster, *Homarus americanus*. *Ecology* 52, 108–115.

Cobb, J. S., and Tamm, G. R. (1974). Social conditions increase intermolt period in juvenile lobsters. *J. Fish. Res. Board Can.* 32, 1941–1943.

Cobb, J. S., and Tamm, G. R. (1975). Dominance status and molt order in lobster (*Homarus americanus*). *Mar. Behav. Physiol.* 3, 119–124.

Cobb, J. S., Gulbransen, T., Phillips, B. F., Wand, D., and Sislo, M. (1983). Behavior and distribution of larval and early juvenile *Homarus americanus*. *Can. J. Fish. Aquat. Sci.* 40, 2184–2188.

Cohen, M. J. (1955). The function of receptors in the statocyst of the lobster, *Homarus americanus*. *J. Physiol. (London)* 130, 9–34.

Cohen, M. J. (1960). The response pattern of single receptors in the crustacean statocyst. *Proc. R. Soc. London, Ser. B* 152, 30–49.

Cohen, M. J., and Dijkgraaf, S. (1961). Mechanoreception. *In* "The Physiology of Crustacea" (T. H. Waterman, ed.), Vol. 2, pp. 65–108. Academic Press, New York.

Consi, T. R., Goudey, C. A., and Atema, J. (1993). A small benthic robot for studies in marine animal chemosensing. *Soc. Neurosci. Abstr.* 19, 1617.

Cooper, R. A., and Uzmann, J. R. (1971). Migrations and growth of deep–sea lobsters, *Homarus americanus*. *Science* 171, 288–290.

Cooper, R. A., and Uzmann, J. R. (1980). Ecology of juvenile and adult *Homarus*. *In* "The Biology and Management of Lobsters" (J. S. Cobb and B. F. Phillips, eds.), Vol. 2, pp. 97–142. Academic Press, New York.

Corotto, F., Voigt, R., and Atema, J. (1992). Spectral tuning of chemoreceptor cells on the third maxilliped of the lobster, *Homarus americanus*. *Biol. Bull. (Woods Hole, Mass.)* 183, 456–462.

Cowan, D. F. (1991). The role of olfaction in courtship behavior of the American lobster *Homarus americanus*. *Biol. Bull. (Woods Hole, Mass.)* 181, 402–407.

Cowan, D. F. (1992). "Pheromones and Courtship Behavior in the American Lobster *Homarus americanus*," Ph.D. thesis. Boston University, Boston.

Cowan, D. F., and Atema, J. (1990). Moult staggering and serial monogamy in American lobsters, *Homarus americanus*. *Anim. Behav.* 39, 1199–1206.

Cowan, D. F., Atema, J., and Solow, A. R. (1991). Moult staggering in the American lobster: A statistical analysis. *Anim. Behav.* 42, 863–864.

Cromarty, S. I., Cobb, J. S., and Kass-Simon, G. (1991). Behavioral analysis of escape response in the juvenile lobster *Homarus americanus* over the molt cycle. *J. Exp. Biol.* 158, 565–581.

Cronin, T. W. (1986). Optical design and evolutionary adaptation in crustacean compound eyes. *J. Crust. Biol.* 6, 1–23.

Daniel, P. C., and Bayer, R. C. (1987a). Attraction of predatorily naive postlarval lobsters to extracts of metabolites of common prey: *Mytilus edulis, Mya arenaria, Cancer irroratus,* and *Asterias vulgaris*. *J. Chem. Ecol.* 13, 1201–1216.

Daniel, P. C., and Bayer, R. C. (1987b). Development of chemically mediated prey–search response in postlarval lobsters (*Homarus americanus*) through feeding experience. *J. Chem. Ecol.* 13, 1217–1233.

Daniel, P. C., and Bayer, R. C. (1987c). Partial purification and characterization of post–larval lobster (*Homarus americanus*) feeding attractants from herring (*Clupea harengus*) tissue. *Mar. Behav. Physiol.* 13, 29–50.

Davis, W. J. (1974). Neuronal organization and ontogeny in the lobster swimmeret system. *In* "Control and Locomotion" (R. B. Stein, K. B. Pearson, R. S. Smith, and J. B. Redford, eds.), pp. 437–455. Plenum, New York.

Derby, C. D. (1982). Structure and function of cuticular sensilla of the lobster *Homarus americanus*. *J. Crust. Biol.* 2, 1–21.

Derby, C. D., and Atema, J. (1981a). Influence of drilling muds on the primary chemosensory neurons in walking legs of the lobster *Homarus americanus*. *Can. J. Fish. Aquat. Sci.* 38, 268–274.

Derby, C. D., and Atema, J. (1981b). Selective improvement in responses to prey odors by the lobster *Homarus americanus*, following feeding experience. *J. Chem. Ecol.* 7, 1073–1080.

Derby, C. D., and Atema, J. (1982a). Chemosensitivity of walking legs of the lobster, *Homarus americanus*: Response spectrum and thresholds. *J. Exp. Biol.* 98, 303–315.

Derby, C. D., and Atema, J. (1982b). Function of chemo– and mechanoreceptors in lobster (*Homarus americanus*) feeding behaviour. *J. Exp. Biol.* 98, 317–327.

Derby, C. D., and Atema, J. (1982c). Narrow–spectrum chemoreceptor cells in the walking legs of the lobster *Homarus americanus*: Taste specialists. *J. Comp. Physiol. A* 146A, 181–189.

Derby, C. D., and Atema, J. (1988). Chemoreceptor cells in aquatic invertebrates: Peripheral mechanisms of chemical signal processing in decapod crustaceans. *In* "Sensory Biology of Aquatic Animals" (J. Atema, A. N. Popper, R. R. Fay, and W. N. Tavolga, eds.), pp. 365–385. Springer–Verlag, New York.

Derby, C. D., Reilly, P. M., and Atema, J. (1984). Chemosensitivity of the lobster *Homarus americanus* to secondary plant compounds: Unused receptor capabilities. *J. Chem. Ecol.* 10, 879–892.

Devine, D. V., and Atema, J. (1982). Function of chemoreceptor organs in spatial orientation of the lobster, *Homarus americanus*: Differences and overlap. *Biol. Bull. (Woods Hole, Mass.)* 163, 144–153.

Dunham, P. J. (1972). Some effects of group housing upon the aggressive behavior of the lobster, *Homarus americanus*. *J. Fish. Res. Board Can.* 29, 598–601.

Dunham, P. J. (1978). Sex pheromones in Crustacea. *Biol. Rev. Cambridge Philos. Soc.* 53, 555–583.

Dunham, P. J., and Skinner–Jacobs, D. (1978). Intermolt mating in the lobster (*Homarus americanus*). *Mar. Behav. Physiol.* 5, 209–214.

Elner, R. W., and Campbell, A. (1981). Force, function and mechanical advantage in the chelae of the American lobster *Homarus americanus* (Decapoda: Crustacea). *J. Zool.* 193, 269–286.

Ennis, G. P. (1975). Behavioral responses to changes in hydrostatic pressure and light during larval development of the lobster *Homarus americanus*. *J. Fish. Res. Board Can.* 32, 271–281.

Ennis, G. P. (1980). Size–maturity relationships and related observations in Newfoundland populations of the lobster (*Homarus americanus*). *Can. J. Fish Aquat. Sci.* 37, 945–956.

Ennis, G. P. (1984a). Small-scale seasonal movements of the American lobster *Homarus americanus*. *Trans. Am. Fish. Soc.* 113, 336–338.

Ennis, G. P. (1984b). Territorial behavior of the American lobster *Homarus americanus*. *Trans. Am. Fish. Soc.* 113, 330–335.

Evans, P. D., and Mann, K. H. (1977). Selection of prey by American lobsters (*Homarus americanus*) when offered a choice between sea urchins and crabs. *J. Fish. Res. Board Can.* 34, 2203–2207.

Factor, J. R. (1978). Morphology of the mouthparts of larval lobsters, *Homarus americanus* (Decapoda: Nephropidae), with special emphasis on their setae. *Biol. Bull. (Woods Hole, Mass.)* 154, 383–408.

Fish, J. F. (1966). Sound production in the American lobster, *Homarus americanus* H. Milne Edwards (Decapoda, Reptantia) notes and news. *Crustaceana* 11, 105–106.

Fuzessery, Z. M., and Childress, J. (1975). Comparative chemosensitivity to amino acids and their role in feeding activity of bathypelagic and littoral crustaceans. *Biol. Bull. (Woods Hole, Mass.)* 149, 522–538.

Gagosian, R. B., and Atema, J. (1973). Behavioral responses of male lobsters to ecdysone metabolites. *Mar. Behav. Physiol.* 2, 115–120.

Goldsmith, T. H. (1978). The effects of screening pigments on the spectral sensitivity of some Crustacea with scotopic (superposition) eyes. *Vision Res.* 18, 475–482.

Goldsmith, T. H., and Bruno, M. S. (1973). Behavior of rhodopsin and metarhodopsin in isolated rhabdoms of crabs and lobster. *In* "Biochemistry and Physiology of Visual Pigments" (H. Langer, ed.), pp. 147–153. Springer-Verlag, New York.

Gomez, G., and Atema, J. (1994). Stimulus integration time of lobster olfactory receptor neurons. *Chem. Senses* 19, 476–477.

Gomez, G., Voigt, R., and Atema, J. (1992). High resolution measurement and control of chemical stimuli in the lateral antennule of the lobster *Homarus americanus*. *Biol. Bull. (Woods Hole, Mass.)* 183, 353–354.

Gomez, G., Voigt, R., and Atema, J. (1994). Frequency filter properties of lobster chemoreceptor cells determined with high–resolution stimulus measurement. *J. Comp. Physiol. A* 174A, 803–811.

Hadley, P. B. (1908). The behavior of larval and adolescent stages of the American lobster (*Homarus americanus*). *J. Comp. Neurol. Psychol.* 18, 199–302.

Hartline, B. K. (1980). Lobster-eye x-ray telescope envisioned. *Science* 207, 47.

Hazlett, B. A. (1987). Information transfer during shell exchange in the hermit crab *Clibanarius antillensis*. *Anim. Behav.* 35, 218–226.

Hazlett, B. A. (1991). Moult staggering or are female lobsters moulting independently? *Anim. Behav.* 42, 862.

Herrick, F. H. (1895). The American Lobster: A Study of Its Habits and Development. *Bull. U.S. Fish Comm.* 15, 1–252 + 54 plates.

Herrick, F. H. (1907). Symmetry in big claws of the lobster. *Science* 25, 275–277.

Herrick, F. H. (1909). Natural history of the American lobster. *Bull. U.S. Bur. Fish.* 29, 149–408 + 20 plates.

Hirtle, R. W. M., and Mann, K. H. (1978). Distance chemoreception and vision in the selection of prey by American lobster (*Homarus americanus*). *J. Fish. Res. Board Can.* 35, 1006–1008.

Hoffman, R. B., Dunham, P. J., and Kelly, P. V. (1975). Effects of water temperature and housing conditions upon the aggressive behavior of the lobster *Homarus americanus*. *J. Fish. Res. Board Can.* 32, 713–717.

Huber, R., and Kravitz, E. A. (1995). A quantitative analysis of agonistic behavior in juvenile American lobsters (*Homarus americanus* L.). *Brain Beh. Evol.* (in press).

Hughes, J. T., and Matthiesen, G. C. (1962). Observations on the biology of the American lobster, *Homarus americanus*. *Limnol. Oceanogr.* 7, 414–421.

Jacobson, S. (1977). "Agonistic Behavior, Dominance and Territoriality in the American Lobster, *Homarus americanus*," Ph.D thesis. Boston University, Boston.

Jerlov, N. G. (1976). "Marine Optics." Elsevier, New York.

Johnson, B. R., and Atema, J. (1983). Narrow-spectrum chemoreceptor cells in the antennules of the American lobster, *Homarus americanus*. *Neurosci. Lett.* 41, 145–150.

Johnson, B. R., Voigt, R., Borroni, P. F., and Atema, J. (1984). Response properties of lobster chemoreceptors: Tuning of primary taste neurons in walking legs. *J. Comp. Physiol. A* 155A, 593–604.

Johnson, B. R., Borroni, P. F., and Atema, J. (1985). Mixture effects in primary olfactory and gustatory receptor cells from the lobster. *Chem. Senses* 10, 367–373.

Johnson, B. R., Merrill, C. L., Ogle, R. C., and Atema, J. (1988). Response properties of lobster chemoreceptors: Tuning of olfactory neurons sensitive to hydroxy–proline. *J. Comp. Physiol. A* 162A, 201–211.

Johnson, B. R., Voigt, R., and Atema, J. (1989). Response properties of lobster chemoreceptor cells: Response modulation by stimulus mixtures. *Physiol. Zool.* 62, 559–579.

Johnson, B. R., Voigt, R., Merrill, C. L., and Atema, J. (1991). Across–fiber patterns may contain a sensory code for stimulus intensity. *Brain Res. Bull.* 26, 327–331.

Kaplan, L. J., Lowrance, C., Basil, J., and Atema, J. (1993). The role of chemical and visual cues in agonistic interactions of the American lobster. *Biol. Bull. (Woods Hole, Mass.)* 185, 320–321.

Karavanich, C., and Atema, J. (1991). Role of olfaction in recognition of dominance in the American lobster, *Homarus americanus*. *Biol. Bull. (Woods Hole, Mass.)* 181, 359–360.

Karavanich, C., and Atema, J. (1993). Agonistic encounters in the American lobster, *Homarus americanus*: Do they remember their opponents? *Biol. Bull. (Woods Hole, Mass.)* 185, 321–322.

Karnofsky, E. B., and Price, H. J. (1989). Dominance, territoriality and mating in the lobster, *Homarus americanus*: A mesocosm study. *Mar. Behav. Physiol.* 15, 101–121.

Karnofsky, E. B., Atema, J., and Elgin, R. H. (1989a). Field observations of social behavior, shelter use and foraging in the lobster, *Homarus americanus*. *Biol. Bull. (Woods Hole, Mass.)* 176, 239–246.

Karnofsky, E. B., Atema, J., and Elgin, R. H. (1989b). Natural dynamics of population structure and habitat use of the lobster, *Homarus americanus*, in a shallow cove. *Biol. Bull. (Woods Hole, Mass.)* 176, 247–256.

Kennedy, D., and Bruno, M. S. (1961). The spectral sensitivity of crayfish and lobster vision. *J. Gen. Physiol.* 44, 1089–1102.

Killian, K. A., and Page, C. H. (1992a). Mechanosensory afferents innervating the swimmerets of the lobster. I. Afferents activated by cuticular deformation. *J. Comp. Physiol. A* 170A, 491–500.

Killian, K. A., and Page, C. H. (1992b). Mechanosensory afferents innervating the swimmerets of the lobster. II. Afferents activated by hair deflection. *J. Comp.Physiol. A* 170A, 501–508.

Kleinholz, L. H. (1959). Purines and pteridines from the reflecting pigment of the arthropod retina. *Biol. Bull. (Woods Hole, Mass.)* 116, 125–135.

Kotak, V. C., and Page, C. H. (1986). Sexually dimorphic mechanosensitive swimmeret sensilla affect abdominal posture in the lobster. *J. Neurobiol.* 17, 421–429.

Krouse, J. S. (1973). Maturity, sex ratio, and size composition of the natural population of the American lobster, *Homarus americanus*, along the Maine coast. *Fish. Bull.* 71, 165–173.

Kuiper, J. W. (1962). Optics of the compound eye. *Soc. Exp. Biol. Symp.* 16, 58–71.

Lang, F., Govind, C. K., Costello, W. J., and Greene, S. I. (1977). Developmental neuroethology: Changes in escape and defensive behavior during growth of the lobster. *Science* 197, 682–685.

Lavalli, K. L., and Barshaw, D. E. (1989). Post-larval American lobsters (*Homarus americanus*) living in burrows may be suspension feeding. *Mar. Behav. Physiol.* 15, 255–264.

Lavalli, K. L., and Factor, J. R. (1992). Functional morphology of the mouthparts of juvenile lobsters, *Homarus americanus* (Decapoda: Nephropidae), and comparison with the larval stages. *J. Crust.*

Biol. 12, 487–510.

Laverack, M. S. (1962a). Responses of cuticular sense organs of the lobster, *Homarus vulgaris* (Crustacea). I. Hair–peg organs as water current receptor. *Comp. Biochem. Physiol.* 5, 319–325.

Laverack, M. S. (1962b). Responses of cuticular sense organs of the lobster, *Homarus vulgaris* (Crustacea). II. Hair–fan organs as pressure receptors. *Comp. Biochem. Physiol.* 6, 137–245.

Leonard, A. E., Voigt, R., and Atema, J. (1994). Lobster orientation in turbulent odor plumes: Electrical recording of bilateral olfactory sampling (antennular "flicking"). *Biol. Bull. (Woods Hole, Mass.)* 187, 273.

Levandowsky, M., and Hodgson, E. S. (1965). Amino acid and amine receptors of lobsters. *Comp. Biochem. Physiol.* 16, 159–161.

Lindstrom, D. P. (1991). "Crustacean Sex Pheromones: A New Technique for Collection of Urine from the American Lobster, *Homarus americanus*," M.S. thesis. Boston University, Boston.

Lythgoe, J. N. (1988). Light and vision in the aquatic environment. *In* "Sensory Biology of Aquatic Animals" (J. Atema, A. N. Popper, R. R. Fay, and W. N. Tavolga, eds.), pp. 57–82. Springer–Verlag, New York.

Mackie, A. M. (1973). The chemical basis of food detection in the lobster *Homarus gammarus*. *Mar. Biol.* 21, 103–108.

Manahan, D. T. (1990). Adaptations by invertebrate larvae for nutrient acquisition from seawater. *Am. Zool.* 30, 147–160.

McClintock, T. S., and Ache, B.W. (1989). Hyperpolarizing receptor potentials in lobster olfactory receptor cells: Implications for transduction and mixture suppression. *Chem. Senses* 14, 637–647.

McLeese, D. W. (1970). Detection of dissolved substances by the American lobster (*Homarus americanus*) and olfactory attraction between lobsters. *J. Fish. Res. Board Can.* 27, 1371–1378.

McLeese, D. W. (1973). Orientation of lobsters (*Homarus americanus*) to odor. *J. Fish. Res. Board Can.* 30, 838–840.

McLeese, D. W. (1974). Olfactory responses of lobsters (*Homarus americanus*) to solutions from prey species and to seawater extracts and chemical fractions of fish muscle and effects of antennule ablation. *Mar. Behav. Physiol.* 2, 237–249.

McLeese, D. W., and Wilder, D. G. (1964). Lobster Storage and Shipment. *Bull., Fish. Res. Board Can.* 147, 1–69.

McLeese, D. W., Spraggins, R. L., Bose, A. K., and Pramanik, B. N. (1977). Chemical and behavioral studies of the sex attractant of the lobster (*Homarus americanus*). *Mar. Behav. Physiol.* 4, 219–232.

McPhie, D., and Atema, J. (1984). Chemical communication in lobsters: Information currents. *Biol. Bull. (Woods Hole, Mass.)* 167, 510–511.

Mendelson, M. (1969). Electrical and mechanical characteristics of a very fast lobster muscle. *J. Cell Biol.* 42, 548–563.

Merrill, C. L. (1992). "Reliability and Specificity of Hydroxy–L–proline Sensitive Chemoreceptor Cells," Ph.D. thesis. Boston University, Boston.

Merrill, C. L., Voigt, R., and Atema, J. (1994). Reliability of chemoreceptor cell response. Intensity coding by pattern and response magnitude. *J. Comp. Physiol. A* 175A, 95–105.

Moore, P. A., and Atema, J. (1991). Spatial information contained in three–dimensional fine structure of an aquatic odor plume. *Biol. Bull. (Woods Hole, Mass.)* 181, 408–418.

Moore, P. A., Atema, J., and Gerhardt, G. A. (1991a). Fluid dynamics and microscale odor movement in the chemosensory appendages of the lobster, *Homarus americanus*. *Chem. Senses* 16, 663–674.

Moore, P. A., Scholz, N., and Atema, J. (1991b). Chemical orientation of lobsters, *Homarus americanus*, in turbulent odor plumes. *J. Chem. Ecol.* 17, 1293–1307.

Mopper, K., and Lindroth, P. (1982). Diel and depth variations in dissolved free amino acids and ammonium in the Baltic Sea determined by shipboard HPLC analysis. *Limnol. Oceanogr.* 27, 336–347.

Moriyasu, M. (1984). Lobster claw loss phenomena in Bideford River estuary, Prince Edward Island, Canada. *Int. Counc. Explor. Sea Doc.* C.M. 1984/K:41, 1–21.

Moulton, J. M. (1957). Sound production in the spiny lobster *Panulirus argus* (Latreille). *Anat. Rec.* 132, 286–295.

Moulton, J. M. (1958). Age changes in stridulation and the stridulatory apparatus of the spiny lobster, *Panulirus argus* (Latreille). *Anat. Rec.* 132, 480.

Mountain, D., and Atema, J. (1993). A kinetic model of chemoreceptor cell function: Transduction, adaptation and phase locking. *Soc. Neurosci. Abstr.* 19, 121.

Neil, D. M., MacMillan, D. L., Robertson, R. M., and Laverack, M. S. (1976). The structure and function of thoracic exopodites in the larvae of the lobster, *Homarus gammarus* (L.). *Philos. Trans. R. Soc. London, Ser. B* 274, 53–68.

Nelson, K., and Hedgecock, D. (1977). Electrophoretic evidence of multiple paternity in the lobster, *Homarus americanus* (Milne–Edwards). *Am. Nat.* 111, 361–365.

Offutt, G. C. (1970). Acoustic stimulus perception by the American lobster *Homarus americanus* (Decapoda). *Experientia* 26, 1276–1278.

Oleszko–Szuts, S., and Atema, J. (1977). Chemical senses of the lobster *Homarus americanus*: Fine structure of aesthetasc hairs. *Div. Fish. Oceanogr. Circ. (Aust., C.S.I.R.O.)* 7, 16–17.

Olla, B. L., Atema, J., Forward, R., Kittredge, J., Livingston, R. J., McLeese, D. W., Miller, D. C., Vernberg, W. B., Wells, P. G., and Wilson, K. (1980). The role of behavior in marine pollution monitoring. *Rapp. P.-V. Reun., Cons. Int. Explor. Mer* 179, 174–181.

O'Neill, D. J., and Cobb, J. S. (1979). Some factors influencing the outcome of shelter competition in lobsters (*Homarus americanus*). *Mar. Behav. Physiol.* 6, 33–45.

Parker, G. H. (1890). The histology and development of the eye of the lobster. *Bull. Mus. Comp. Zool.* 20, 1–60.

Prentiss, C. W. (1901). The otocyst of decapod Crustacea. *Bull. Mus. Comp. Zool.* 36, 167–254.

Rubenstein, D. I., and Hazlett, B. A. (1974). Examination of agonistic behavior of the crayfish *Orconectes virilis* by character analysis. *Behaviour* 50, 194–216.

Rutherford, D. J., and Horridge, G. A. (1965). The rhabdom of the lobster eye. *Q. J. Microsc. Sci.* 106, 119–130.

Saidel, W. M. (1988). How to be unseen: An essay in obscurity. *In* "Sensory Biology of Aquatic Animals" (J. Atema, A. N. Popper, R.R. Fay, and W.N. Tavolga, eds.), pp. 487–514. Springer-Verlag, New York.

Salmon, M. (1983). Courtship, mating systems, and sexual selection in decapods. *In* "Studies in Adaptation: The Behavior of Higher Crustaceans" (S. Rebach and D. Dunham, eds.), pp. 143–169. Wiley, New York.

Scarratt, J. C. E. (1968). An artificial reef for lobsters (*Homarus americanus*). *J. Fish. Res. Board Can.* 25, 2683–2690.

Scarratt, J. C. E. (1971). Claw loss and other wounds in commercially caught lobsters (*Homarus americanus*). *J. Fish. Res. Board Can.* 30, 1370–1373.

Schmidt, M., and Ache, B. W. (1992). Antennular projections to the midbrain of the spiny lobster. II. Sensory innervation of the olfactory lobe. *J. Comp. Neurol.* 318, 291–303.

Schmitt, B. C., and Ache, B. W. (1979). Olfaction: Responses of a decapod crustacean are enhanced by flicking. *Science* 205, 204–206.

Schöne, H. (1968). Agonistic and sexual display in aquatic and semiterrestrial Brachyura crabs. *Am. Zool.* 8, 641–654.

Scrivener, J. C. (1971). Agonistic Behavior of the American Lobster, *Homarus americanus* (Milne Edwards). *Tech. Rep.—Fish. Res.*

Board Can. 235, 1–128.

Shepheard, P. (1974). Chemoreception in the antennule of the lobster, *Homarus americanus*. *Mar. Behav. Physiol.* 2, 261–273.

Smith, D. V., and Travers, J. B. (1979). A metric for the breadth of tuning of gustatory neurons. *Chem. Senses Flavour* 4, 215–229.

Snyder, M. J., and Chang, E. S. (1991). Ecdysteroids in relation to the molt cycle of the American lobster *Homarus americanus*. II. Excretion of metabolites. *Gen. Comp. Endocrinol.* 83, 118–131.

Snyder, M. J., Ameyaw-Akumfi, C., and Chang, E. S. (1992). Mating behavior in vision-deprived American lobster, *Homarus americanus*. *Mar. Behav. Physiol.* 21, 227–228.

Snyder, M. J., Ameyaw-Akumfi, C., and Chang, E. S. (1993). Sex recognition and the role of urinary cues in the lobster, *Homarus americanus*. *Mar. Behav. Physiol.* 24, 101–116.

Solon, M. H., and Cobb, J. S. (1980a). The external morphology and distribution of cuticular hair organs on the claws of the American lobster, *Homarus americanus* (Milne–Edwards). *J. Exp. Mar. Biol. Ecol.* 48, 205–215.

Solon, M. H., and Cobb, J. S. (1980b). Antennae whipping behavior in the American lobster *Homarus americanus* (Milne–Edwards). *J. Exp. Mar. Biol. Ecol.* 48, 217–224.

Solon, M. H., and Kass–Simon, G. (1981). Mechanosensory activity of hair organs on the chelae of *Homarus americanus*. *Comp. Biochem. Physiol. A* 68A, 217–223.

Stein, R. A. (1976). Sexual dimorphism in crayfish chelae: Functional significance linked to reproductive activities. *Can. J. Zool.* 54, 220–227.

Stein, L., Jacobson, S., and Atema, J. (1975). Behavior of Lobsters (*Homarus americanus*) in a Semi–natural Environment at Ambient Temperatures and under Thermal Stress. *Woods Hole Oceanogr. Inst., Tech. Rep.* 75–48, 1–49.

Stewart, L. L. (1972). "The Seasonal Movements, Population Dynamics and Ecology of the Lobster, *Homarus americanus* (Milne–Edwards), off Ram Island, Connecticut," Ph.D. thesis. University of Connecticut, Storrs.

Tamm, G. R., and Cobb, J. S. (1978). Behavior and the crustacean molt cycle: Changes in aggression of *Homarus americanus*. *Science* 200, 79–81.

Tautz, J., Masters, W. M., Aicher, B., and Markl, H. (1981). A new type of water vibration receptor on the crayfish *Astacus leptodactylus* antenna. 1. Sensory physiology. *J. Comp. Physiol. A* 144A, 533–542.

Tazaki, K. (1977). Nervous responses from mechanosensory hairs on the antennal flagellum in the lobster *Homarus gammarus*. *Mar. Behav. Physiol.* 5, 1–18.

Templeman, W. (1934). Mating in the American lobster. *Contrib. Can. Biol. Fish.* 8, 423–432.

Templeman, W. (1935). Local differences in the body proportions of the lobster *Homarus americanus*. *J. Biol. Board Can.* 1, 213–226.

Templeman, W. (1936). Further contributions to mating in the American lobster. *J. Biol. Board Can.* 2, 223–226.

Thomas, M. L. H. (1968). Overwintering of American lobster, *Homarus americanus*, in burrows in Bideford River, Prince Edward Island. *J. Fish. Res. Board Can.* 25, 2525–2527.

Tierney, A. J., Voigt, R., and Atema, J. (1988). Response properties of chemoreceptors from the medial antennule of the lobster *Homarus americanus*. *Biol. Bull. (Woods Hole, Mass.)* 174, 364–372.

Todd, J. H., Atema, J., and Boylan, D. B. (1972). Chemical communication in the sea. *Mar. Technol. Soc. J.* 6, 54–56.

Vedel, J. P., and Clarac, F. (1976). Hydrodynamic sensitivity by cuticular organs in the rock lobster *Palinurus vulgaris*. Morphological and physiological aspects. *Mar. Behav. Physiol.* 3, 235–251.

Voigt, R., and Atema, J. (1988). Effects of stimulus profile on temporal response profile and disadaptation time course. *Chem. Senses* 13, 742.

Voigt, R., and Atema, J. (1990). Adaptation in chemoreceptor cells: III. Effects of cumulative adaptation. *J. Comp. Physiol. A* 166A, 865–874.

Voigt, R., and Atema, J. (1992). Tuning of chemoreceptor cells of the second antenna of the American lobster (*Homarus americanus*) with a comparison of four of its other chemoreceptor organs. *J. Comp. Physiol. A* 171A, 673–683.

von Uexküll, J. (1921). "Umwelt und Innenwelt der Tierre," 2nd ed. Springer–Verlag, Berlin.

Waddy, S. L., and Aiken, D. E. (1986). Multiple fertilization and consecutive spawning in large American lobsters, *Homarus americanus*. *Can. J. Fish. Aquat. Sci.* 43, 2291–2294.

Waddy, S. L., and Aiken, D. E. (1990). Intermolt insemination in the American lobster, *Homarus americanus*. *Can. J. Fish. Aquat. Sci.* 47, 2402–2406.

Waddy, S. L., and Aiken, D. E. (1991). Mating and insemination in the American lobster, *Homarus americanus*. *In* "Crustacean Sexual Biology" (R. T. Bauer and J. W. Martin, eds.), pp. 126–144. Columbia University Press, New York.

Wald, G. (1968). Oscillations of potential in the electroretinogram of the lobster. *J. Gen. Physiol.* 51, 261–271.

Wald, G., and Burg, S. P. (1957). The vitamin A of the lobster. *J. Gen. Physiol.* 40, 609–625.

Wald, G., and Hubbard, R. (1957). Visual pigment of a decapod crustacean: The lobster. *Nature (London)* 180, 278.

Waterman, T. H. (1961). Light and vision. *In* "The Physiology of Crustacea" (T. H. Waterman, ed.), Vol. 2, pp. 1–64. Academic Press, New York.

Weinstein, A. (1991). "Tuning and Mixture Response Characteristics of Chemoreceptor Cells in the American Lobster, *Homarus americanus*," M.A. thesis. Boston University, Boston.

Weinstein, A., Voigt, R., and Atema, J. (1989). Spectral tuning to amino acids and mixture effects on antennular chemoreceptor cells in the lobster, *Homarus americanus*. *Biol. Bull. (Woods Hole, Mass.)* 177, 330.

Weiss, H. M. (1970). "The Diet and Feeding Behavior of the Lobster, *Homarus americanus*, in Long Island Sound," Ph.D. thesis. University of Connecticut, Storrs.

Wells, P. G. (1976). Lobster and other decapod crustacean larvae as test organisms in marine toxicity bioassays. *In* "Proceedings of the 1975 Aquatic Toxicity Workshop, November 4–5, 1975," pp. 266–290. Ontario Ministry of the Environment, Toronto.

Wells, P. G., and Sprague, J. B. (1976). Effects of crude oil on the American lobster (*Homarus americanus*) larvae in the laboratory. *J. Fish. Res. Board Can.* 33, 1604–1614.

Wilkens, L. A., and Larimer, J. L. (1976). Photosensitivity in the sixth abdominal ganglion of decapod crustaceans: A comparative study. *J. Comp. Physiol. A* 106A, 69–75.

Wilson, E. O., and Bossert, W. H. (1963). Chemical communication among animals. *Recent Prog. Horm. Res.* 19, 673–716.

Yonge, C. M. (1932). On the nature and permeability of chitin. I. The chitin lining the foregut of decapod crustacea and the function of the tegumental glands. *Proc. R. Soc. London, Ser. B* 111, 298–329.

Zeil, J., Sandeman, R., and Sandeman, D. (1985). Tactile localization: The function of active antennal movements in the crayfish *Cherax destructor*. *J. Comp. Physiol. A* 157A, 607–617.

C H A P T E R

14

The Feeding Appendages

KARI L. LAVALLI
Marine Science Center
Northeastern University
Nahant, Massachusetts 01908

JAN ROBERT FACTOR
Division of Natural Sciences
Purchase College
State University of New York
Purchase, New York 10577

I. Introduction

The feeding apparatus of the American lobster, *Homarus americanus*, comprises both external and internal structures involved in the acquisition, manipulation, ingestion, and digestion of the food. These structures include the pereiopods (the great chelae and four pairs of walking legs), mouthparts, foregut (esophagus, cardiac stomach with gastric mill, and pyloric stomach), midgut, and hindgut.

The external feeding appendages—the mouthparts and the pereiopods—are intimately involved in the several phases of feeding behavior (Maynard and Sallee, 1970). The first phase, initiation of feeding behavior, includes movements designed to bring these appendages into contact with food items and is stimulated by the chemosensory setae of the antennules. The pereiopods and the mouthparts locate food via their own chemo- and mechanosensory setae and then grasp, manipulate, and transport it to the mouth. The mouthparts alone are involved in the actual process of ingestion, constantly assessing food quality via their numerous chemoreceptive setae.

Once swallowed, the food collects in the cardiac sac via peristaltic waves of the esophagus and is prepared for grinding in the gastric mill—a gizzardlike structure (see Factor, Chapter 15, on the internal components of the feeding system). The gastric mill is better suited to the maceration of food than are either the external, toothed third maxillipeds or the mandibles. Thus, the mouthparts are used to assess food quality and to rip and rend food only grossly before ingestion. In younger lobsters, particularly the postlarvae and shelter-restricted juveniles, mouthparts may also be used to effect capture of particulate matter suspended in water surrounding the lobster. Similarly, the pereiopods are used primarily to effect capture of food; the great chelae crush any covering within which the food may lie and pass the flesh to the mouthparts. Because of the allometric growth pattern of these claws (see Govind, Chapter 12), however, they are less important than the other pereiopods or the mouthparts in food capture and manipulation for larvae, postlarvae, and shelter-restricted juveniles.

Structure, development, and function of the external feeding appendages are considered in several stages representative of the life history phases of *Homarus americanus* (see Lawton and Lavalli, Chapter 4, on these life history phases). Information on adult mouthparts is preliminary and incomplete, but is included where available.

II. Mouthparts

A. Location, Generalized Structure, and Orientation

The cephalothorax of *Homarus americanus* is formed by the fusion of five cephalic and eight thoracic segments, all of which bear paired appendages (see Factor, Chapter 1, on external anatomy of the lobster). The first and second antennae of the first two cephalic segments play an important sensory role in feeding (see Atema and Voigt, Chapter 13). The third through fifth cephalic segments bear the mandibles, first maxillae (or maxillules), and second maxillae (or maxillae), respectively. The eight thoracic segments bear the first, second, and third maxillipeds and five pairs of pereiopods (or walking legs), the first three of which are chelate (Phillips *et al.*, 1980; Schram, 1986).

Mouthparts of the lobster follow the plan of the typical crustacean limb, consisting of a protopodite from which branches may arise distally and proximally (Fig. 1). The protopodite is typically composed of two segments: the coxa (proximal) and the basis (distal), each of which may project specialized, medial endites for handling food. The distal lateral branch (the exopodite) comprises two portions: a basal region and a distal flagellum with many "segments" formed by annulations of the cuticle. At the time of metamorphosis, exopodites are lost from all thoracic appendages except the maxillipeds. The distal medial branch (the endopodite) is usually segmented and consists of two segments in the first and second maxillae and first maxillipeds. In the second and third maxillipeds, the endopodite is segmented (distal to proximal) into the dactyl (=dactylus, or dactylopodite), propus (=propodus, or propodite), carpus (=carpopodite), merus (=meropodite), and ischium (=ischiopodite). Proximally, the protopodite may bear an epipodite. A podobranch, if present, extends from the base of the coxa. Only the second maxillae possess a scaphognathite (or gill bailer), which is thought to represent the fusion of the exopodite and the epipodite (McLaughlin, 1982; Schram, 1986; Lavalli and Factor, 1992).

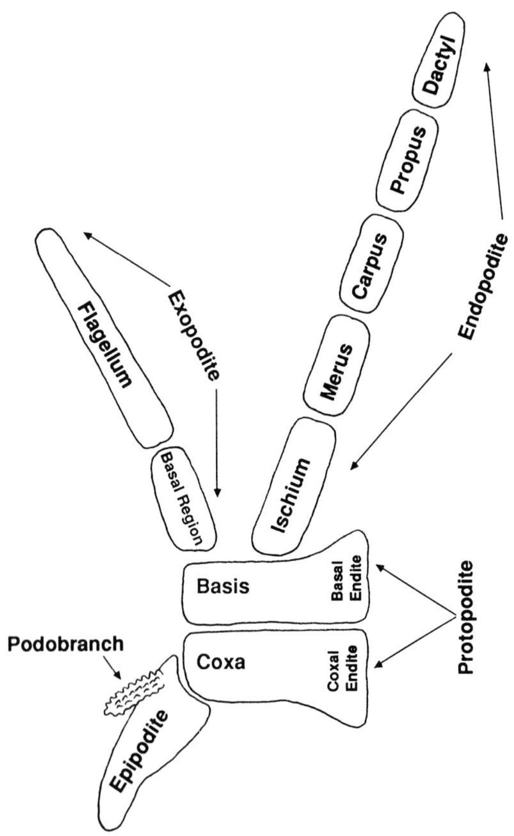

FIGURE 1 General plan of the mouthparts. (Redrawn from Factor, 1977, with permission.)

The mouth is marked anteriorly by a labrum (upper lip) and posteriorly by a pair of paragnaths (lower lips). The first five pairs of mouthpart appendages (mandibles through second maxillipeds) are flattened and layered over the mouth. The sixth pair (third maxillipeds) is not flattened, but extends anteriorly to act as a grasping structure. Since the mouthparts lie roughly in the frontal plane and do not have the same attitude as most appendages of the lobster, and because they are flattened and layered, the terms *inner* and *outer* are more useful than *dorsal* and *ventral* for describing features toward and away from the mouth, respectively. *Medial* and *lateral* (toward and away from the midline, or sagittal plane, respectively) and *proximal* (=*basal*) and *distal* (toward the base and toward the tip, respectively) are also used to describe and compare the positions of the mouthparts and the structures found on them.

B. Setae of the Mouthparts

1. Types of Setae

The classification of setae is based largely on their external morphology, particularly the nature and distribution of the setules. Thomas' (1970) scheme for naming the setae of crayfish provides the basis of the system used here, as modified by Factor (1978) and Lavalli and Factor (1992) for *Homarus americanus*.

The types of setae found on the mouthparts of *Homarus americanus* are described in Table 1 and illustrated in Fig. 2 for all stages thus far studied in detail: larval stages I-III, postlarval stage IV, shelter-restricted juvenile stage VI, and emergent juvenile stage XII. (Factor, in Chapter 1, and Lawton and Lavalli, in Chapter 4, discuss the life history of *H. americanus*.) The setae have been arranged into 13 categories; the variation within categories is sometimes sufficient to warrant subdivision into several types (designated by a letter for the general category and a number for the specific variation).

TABLE 1 **Description of Types of Setae Found on the Mouthparts of *Homarus americanus*[a]**

Setal category	Type	Description
Plumose	A	Two rows of long, fine, ribbonlike setules along most of the shaft; setules densely or sparsely arranged; rows of setules always opposite each other, forming an angle of 180°; annulations of the shaft may be present
Pappose	B1	Long, fine setules loosely arranged about the shaft in a random manner; terminal pore may be present
	B2	Similar to B1, but bearing more setules along the shaft
	B3	Similar to B2, but with fine denticulations at the tip
Plumodenticulate	C1	Sparse, random setules proximally; finer, denser setules distally
	C2	Sparse setules proximally; finer, denser setules distally; regions may be separated by bulbous swelling of the shaft
	C3	No setules proximally; fine, densely packed setules distally; bulbous swelling may be present midway along the shaft
	C4	Long, sparse setules proximally (identical to those of types C1 and C2); shorter, coarser setules distally
Serrate	D1	Large, distinct, toothlike setules distally arranged in two rows, forming an angle of less than 180°
	D2	Scalelike setules opposite two rows of toothlike setules; scalelike setules arranged sparsely, in overlapping rows of three, or densely and randomly packed
	D3	Row of shorter, finer setules opposite larger, toothlike setules
Triserrate	E	Three rows of serrate setules, all approximately equal in length
Serrulate	F1	Short, fine, peglike, distal setules arranged in two rows, forming an angle of less than 180°; similar to serrate (D1) setae, but with shorter, finer setules
	F2	Shaft thicker than that in F1, with a narrower lumen and a bulbous base; visible subterminal pore
	F3	Similar to F2, but bearing scalelike setules on the opposite side of the shaft
	F4	Scalelike setules on the opposite side of the shaft from short, fine, peglike setules
Triserrulate	G	Short, fine, peglike setules arranged in three rows distally; differ from triserrate (E) setae by having finer setules, and from serrulate (F1) setae by the presence of three rows of setules rather than two
Cuspidate	H1	Long, conical, toothlike; stout with thick walls and relatively narrow lumina; lack setules
	H2	Similar to H1, but bearing sparsely arranged, fine, short setules on the shaft
Simple	I	May be long and thin or short; generally bear no setules whatsoever; may have a bulb midway along the length of the shaft; some apparently simple setae exhibit a scalelike texture (or hair lines suggestive of borders of scales closely pressed to the shaft)
Hamate	J	Small, short setae shaped like hooks; lack setules
Grooved	K	Two to five ridges along one side of the distal portion of the shaft with scalelike setules on the opposite side; setules sparsely or densely arranged
Multiscaled	L	Scalelike setules, densely arranged around 270° of the distal portion of the shaft

[a]See Fig. 2 for illustrations of setal types. [Based on data from Factor (1978) and Lavalli and Factor (1992).]

2. Functions of Setae

All adult mouthparts bear chemo- and mechanosensory setae (Shelton and Laverack, 1970; Derby, 1982; Corotto *et al.*, 1992). While the walking legs may be the initial gustatory appendages, locating the food and passing it to the mouthparts, the mouthparts provide the final check of food quality before ingestion (Corotto *et al.*, 1992), since food that has been accepted by the walking legs may be later rejected after handling by the mouthparts (Derby *et al.*, 1984). Chemoreceptive setae on the surfaces of mouthparts are extremely important because lobsters with deafferented maxillipeds drop their food (Derby and Atema, 1982). The chemoreceptors of the third maxillipeds are broadly tuned, which may be necessary for their evaluative function (Corotto *et al.*, 1992). (See Atema and Voigt, Chapter 13, on sensory biology.)

The functions of only a few types of setae have been inferred from their surface and/or internal structure. Surface structure alone, as observed by scanning electron microscopy, does not demonstrate neurological function. Transmission electron microscopy is needed to observe any innervating neurons and conclusive determination of function

A
PLUMOSE
A
A
B1
B2
B3
PAPPOSE
B1
B1
B3
B3
C1
C2
C3
C4
PLUMODENTICULATE
C1
C3
C3
E
TRISERRATE
E
G
TRISERRULATE
G

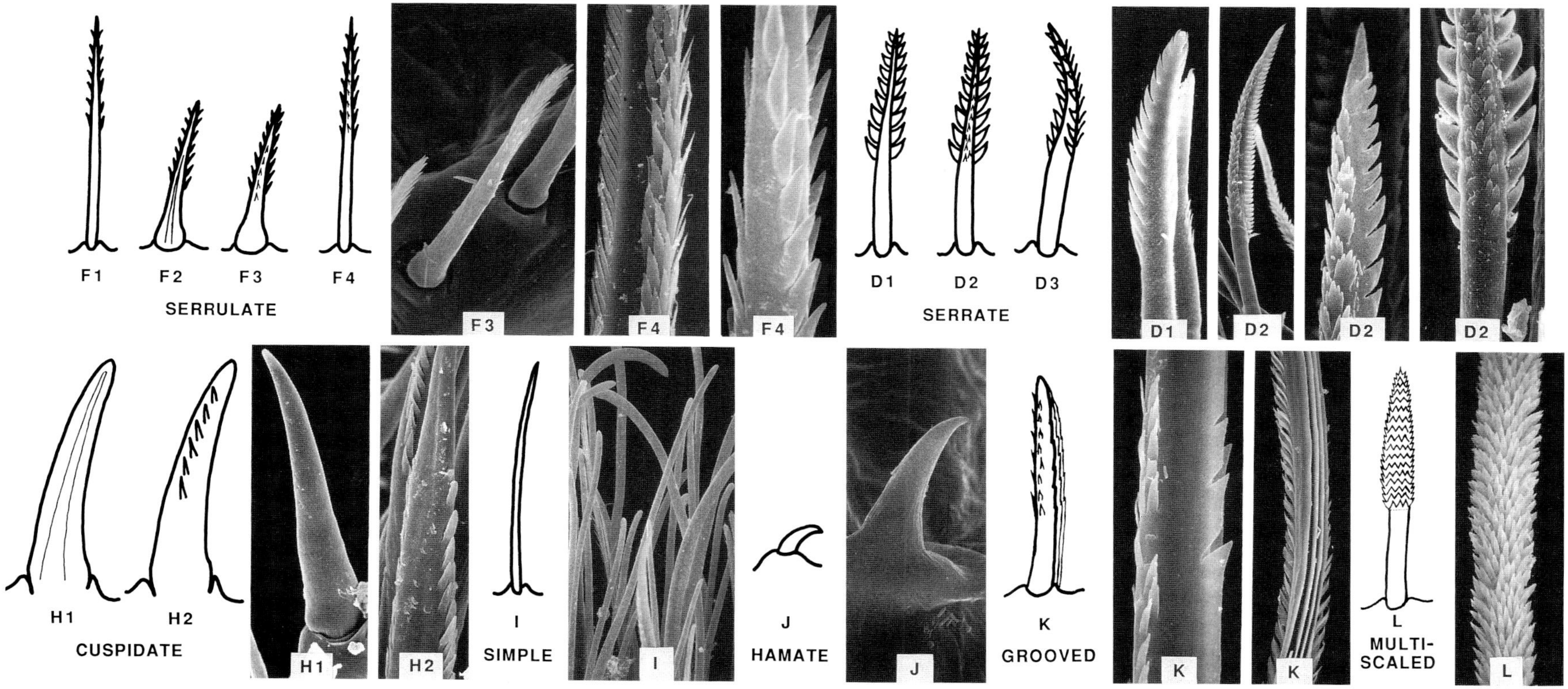

FIGURE 2 Types of setae found on the mouthparts of larval (stages I–III), postlarval (stage IV), shelter-restricted juvenile (stage VI), and emergent phase juvenile (stage XII) lobsters, *Homarus americanus*. Typical examples of setal types are illustrated with both line drawings and scanning electron micrographs, which are not to the same scale. See Table 1 for descriptions of setal types. (Modified from Factor, 1978; reprinted from Lavalli and Factor, 1992, with permission.)

requires neurological experimentation.

Only several types of setae thus far have been shown conclusively to be chemo- and/or mechanosensitive. In *Homarus gammarus*, isolated serrate setae are responsive to chemical stimuli and the serrations may increase the surface area of the receptor (Shelton and Laverack, 1970). Serrate setae are both chemo- and mechanosensory in *H. americanus* (Derby, 1982). The toothlike serrations of serrate setae form a comb-like, scraping structure adapted for grooming, as has been suggested for *H. americanus*, *Nephrops norvegicus*, and other decapods (Herrick, 1909; Roberts, 1968; Farmer, 1974; Bauer, 1975). The functions of grooming and chemo- and mechanoreception need not be mutually exclusive (Bauer, 1975; Factor, 1978; Felgenhauer and Schram, 1979).

Multiscaled setae have been shown to be involved in grooming (Bauer, 1987) and "squamous" setae (which are similar in form to multiscaled setae) are both chemo- and mechanosensory (Derby, 1982). Similarly, "smooth" (or simple) setae are both chemo- and mechanosensory (Derby, 1982).

The featherlike plumose setae have been assigned several functions. The short plumose setae on the scaphognathite (gill bailer) may aid the production of aerating water currents by acting like gaskets, sealing the space between the edge of the scaphognathite and the branchial chamber (Farmer, 1974; Factor, 1978). They may also filter particles from the water entering the branchial chamber (Farmer, 1974). A similar function could be ascribed to the fanlike plumose setae at the base of the endopodite of the first maxillae. The long plumose setae on the beating exopodites of the maxillipeds extend the surface area of the distal flagella and help create water currents, which serve to remove debris from the mouth region (Factor, 1978) or to adjust the sensing environment so that setae may better detect chemical signals (*sensu* information currents of McPhie and Atema, 1984; Atema, 1985; Atema and Voigt, Chapter 13). The ability of plumose setae to create currents may be related to both the density of the setae and the spacing of the long setules arising from the shaft. The short plumose setae on the outer surfaces of the merus and ischium of the third maxillipeds may be involved in mechanoreception; nearly identical in appearance to hydrodynamic receptors of *Palinurus vulgaris* (Vedel and Clarac, 1976), they would be ideal for determining the position of zooplankton swimming close to the mouth region in *Homarus americanus*.

The long apical hairs of the scaphognathite (triserrate in *Homarus americanus*) may act to seal the gap near the limb bases and the inner wall of the branchiostegite, and pappose setae may be used to clean adjacent parts of the appendages from which they arise (Farmer, 1974). Plumodenticulate setae presumably might perform the same function as the pappose setae. The stout cuspidate setae on the second maxillipeds and first maxillae seem well adapted for adding purchase when grasping and passing food toward the mouth (Factor, 1978) and for jabbing and piercing small particles of food (Lavalli, 1992; Lavalli and Trager, 1995).

3. Development of Setal Types

The presence of so many different types of setae is perplexing, and while 13 distinct categories are found on the mouthparts of *Homarus americanus*, it is difficult to imagine 13 unique functions. Some types of setae may be developmental stages of other, more common types. Farmer (1974) suggested that all setal types of *Nephrops norvegicus* arise from simple setae; those that develop setules become pappose, plumose, or plumodenticulate setae, while those that develop grooves become either serrate or grooved setae. Observations of *H. americanus* mouthparts offer several examples that lend some support to this developmental explanation of setal diversity. Under high magnification, the ridges between grooves on grooved setae can often be seen to be broken up by notches that create toothlike structures (Fig. 2, type K), perhaps due to elongation of the ridges. It is at least conceivable that these structures might become the typical peg- or toothlike setules of serrulate or serrate setae. Furthermore, the serrulate setae may be an intermediate type that could conceivably develop into the stouter serrate setae. Similarly, pappose setae may develop into plumodenticulate setae with the addition of denticulations at the tip. Early stages of *H. americanus* bear cuspidate setae with setules (type H2) on the first maxillae and second maxillipeds, but later stages possess only bare cuspidate setae (type H1) in the same locations, lending support to Farmer's (1974) contention that the scales and setules found on setae may also exhibit developmental patterns.

A variety of interesting changes in setae at particular locations occurs from the postlarva (stage IV) through the juvenile (stages VI and XII) life history phases (Factor, 1978; Lavalli and Factor, 1992), further suggesting the possibility of a developmental sequence in *Homarus americanus*. On the first maxillipeds, some of the plumodenticulate (C3) setae of the medial edge of the basal endite in stage IV are replaced by serrulate (F4) setae by stage VI and by serrate (D2) setae by stage XII. The row of setae on the outer surface of the basal endite shows the same pattern. Some of the pappose (B1) setae found along the distal edge of the basal endite, the medial edge of the endopodite, and the medial edge of the coxal endite in stage IV are replaced by plumodenticulate (C1/C4) setae in stages VI and XII. The B3 type of pappose setae may be an intermediate form between the pappose and plumodenticulate types, as it is clearly pappose along the proximal portion of the shaft but bears distal denticulations typical of the plumodenticulate type.

Similar patterns can be found on the other mouthparts. For example, the second maxilliped bears a row of plumodenticulate (C1/C4 and C3) setae on the merus (and clusters elsewhere) in stage VI, which are replaced by serrulate (F4) setae in stage XII.

Jacques (1989) and Watling (1989) provided alternative views of setal classification and development and Felgenhauer (1992) reviewed cuticular structures among the decapods.

C. Structure and Development of the Mouthparts

An overview and summary of the morphology of the mouthparts of *Homarus americanus* are presented below. The gross structural changes of the mouthparts from stage I larvae to the adult are illustrated here with drawings (Figs. 3, 6, 9, 11, and 13), which also demonstrate changes in the density, location, and relative length of setae. [More detailed drawings and maps of setal distribution are presented by Factor (1978) and Lavalli and Factor (1992), on which this summary is based; information on adult mouthparts is preliminary and incomplete and is based on the work of Chen (1992), for mandibles and first and second maxillae, and Neilsen (1993) for second and third maxillipeds; Smith (1873) provides additional information.] Table 2 provides a summary of the distribution of setal types and other prominent structures, and the changes for each mouthpart, segment by segment, for stages I–IV, VI, XII, and the adult. In addition, scanning electron micrographs illustrate interesting details of the mouthparts.

1. Mandibles

The mandibles comprise a large gnathal lobe and a mandibular palp (Fig. 3). The gnathal lobe is mostly an incisor process, with a molar process in some stages (Fig. 3F, H, J, and L). The mandibles are asymmetrical, with a single major tooth on the incisor process of the right side that fits between two major teeth on the left side. Herrick (1909) described the mandibles as being similar in form to "hinged double doors" that swing toward and away from the midline. The mandibular muscles act to pivot the mandibles on their hinges. The setose, three-segmented mandibular palp helps to direct food into the mouth (Herrick, 1909).

Of all the mouthparts, the most dramatic developmental transformation occurs in the mandibles. In stage I–III larvae, the teeth along the medial cutting edge of the incisor process of the gnathal lobes are thin and sharp (Fig. 4A–L). By stage IV, they have been transformed into more substantial, broader, duller teeth (Fig. 5A–D). In all of the larval stages (I–III), the right mandible bears a single major tooth with two smaller, associated teeth at the anterior end of the cutting edge (Figs. 4A, D, E, H, I, and L and 5A and D). In the larval stages, the associated teeth are proximal to the major tooth (Figs. 3A–C and 4A–L). In stage IV, there is a single major tooth with a distal protuberance (Figs. 3E and 5A–D). By stage VI, the single major tooth of the right mandible has become larger still and more blunt, with a slight indentation at its tip (Fig. 5G and J). In the adult, the single major tooth is represented by a point near the anterior end of the otherwise toothless cutting edge (Fig. 3K and L). The teeth along the medial cutting edge undergo a reduction in number. There are 10–13 teeth in the three larval stages, five or six teeth in stage IV, four or five teeth in stage VI, five to seven indistinct teeth in stage XII, and in the adult the teeth have coalesced into a continuous cutting edge (Fig. 3), giving the shield-shaped mandible typical of many adult decapods.

The left mandible undergoes similar changes. The larval and postlarval stages bear two major teeth with two smaller, associated teeth at the anterior end of the cutting edge (Figs. 4B, C, F, G, J, and K and 5B and C) and only one associated tooth in stages VI, XII, and the adult (Figs. 3G, I, and K and 5H, I, L, and M). By stage VI, the two major teeth have increased in size, with one becoming flat-edged and the other retaining its pointed tip (Fig. 5I). In the adult, the two major teeth are represented by two points near the anterior end of the otherwise toothless cutting edge (Fig. 3K and L). The teeth along the cutting edge of the left mandible also undergo a reduction in number. The three larval stages bear 10–13 teeth, the postlarvae six teeth, and stage VI five teeth (Fig. 3A–H). By stage XII, the separations between the four remaining teeth are barely noticeable and in the adult, the teeth have coalesced into a continuous cutting edge (Fig. 3I–L).

At the basal end of the cutting edge, on the inner surface, a dense field (or "pad") of small setae in stages I–III (Fig. 4M) is transformed in stage IV into a recessed molar process bearing one denticle and cuticular texturing (Fig. 5E and F). In stage VI, the left molar process bears three denticles and the right bears

TABLE 2 **Developmental Changes in the Mouthparts of Larval (Stages I–III), Postlarval (Stage IV), Juvenile (Stages VI and XII), and Adult Lobsters, *Homarus americanus*[a]**

Mouthpart	Stage I	Stage II	Stage III	Stage IV	Stage VI	Stage XII	Adult
Mandibles							
Palp: first segment	No setae	No setae	No setae	Serrulate setae, outer surface	Row of serrulate setae, outer surface	Row of plumodenticulate and serrate setae, outer surface	Row of plumodenticulate, serrulate, and triserrate setae, outer surface
Palp: second segment	No setae	No setae	No setae	Pappose and simple setae	Plumodenticulate and simple setae, lateral edge; plumodenticulate and serrulate setae, medial edge	Plumodenticulate setae, lateral edge; serrate setae near first-segment articulation; plumodenticulate and serrate setae, medial edge	Serrulate, serrate, and simple setae, lateral edge; triserrate, serrulate, and serrate setae, outer surface; serrulate setae, inner surface
Palp: third segment	2 pappose setae	2 plumodenticulate setae	10 plumodenticulate setae	Plumodenticulate and serrulate setae	Plumodenticulate and serrulate setae along edges; serrulate setae, inner surface	Plumodenticulate and serrulate setae along edges; plumodenticulate setae, outer surface; serrulate, serrate, and plumodenticulate setae, inner surface	Pappose, plumodenticulate, triserrate, serrulate, and simple setae along edges
Gnathal lobe	10–13 sharp teeth along cutting edge; anterior end, right: 1 major tooth, 2 associated teeth; anterior end, left: 2 major teeth, 2 associated teeth; inner basal end: pad of setae	10–14 sharp teeth along cutting edge; anterior end, right: 1 major tooth, 2 associated teeth; anterior end, left: 2 major teeth, 2 associated teeth; inner basal end: pad of setae	10–12 sharp teeth along cutting edge; anterior end, right: 1 major tooth, 2 associated teeth; anterior end, left: 2 major teeth, 2 associated teeth; inner basal end: pad of setae	5 or 6 dull teeth along cutting edge; anterior end, right: 1 major tooth, 1 associated distal protuberance; anterior end, left: 2 major teeth, 1 associated distal protuberance; inner basal end: molar process with simple setae	4 or 5 dull teeth along cutting edge; anterior end, right: 1 major tooth, 1 associated distal protuberance; anterior end, left: 2 major teeth, 1 associated distal protuberance; inner basal end: molar process; serrulate setae, outer surface	5–7 indistinct teeth along cutting edge, beginning to coalesce; anterior end, right: 1 major tooth, 1 associated distal protuberance; anterior end, left: 2 major teeth, 1 associated distal protuberance; inner basal end: molar process with plumodenticulate setae, distal end; serrate and serrulate setae, outer surface; plumodenticulate and serrate setae near first palp segment	Basal teeth coalesce; right, remnant of 1 major tooth; left, remnant of 2 major teeth on otherwise toothless cutting edge; inner basal end: molar process; field of serrulate setae, outer surface

First Maxillae							
Basal endite	Row of cuspidate setae (H2) along medial edge	Row of cuspidate setae (H2) along medial edge	Row of cuspidate setae (H2) along medial edge	3 rows of cuspidate setae (H1) along medial edge; row of serrulate setae near and parallel with medial edge, inner surface; pappose, plumodenticulate, and serrulate setae, inner surface; row of serrulate setae near and parallel with medial edge, outer surface	3 rows of cuspidate setae (H1) along medial edge; row of serrulate setae, near and parallel with medial edge, inner surface; plumodenticulate and serrulate setae, inner surface; row of plumodenticulate and serrulate setae near and parallel with medial edge, outer surface	5 rows of cuspidate setae (H1) along medial edge; serrulate setae, proximal end of medial edge; plumodenticulate and serrulate setae, lateral edge; serrate, serrulate, and plumodenticulate setae, near and parallel with medial edge, inner surface; serrulate setae, outer surface	5 rows of cuspidate setae (H1) along medial edge; serrulate and serrate setae, proximal end of medial edge; serrulate setae, inner and outer surfaces
Coxal endite	No information	No information	No information	Serrulate setae, distal medial edge; pappose setae, proximal medial edge; row of serrulate setae, near and parallel with medial edge, outer surface	Serrulate and plumodenticulate setae, distal medial edge; pappose setae, proximal medial edge; plumodenticulate setae, distal edge near basal endite; serrulate and plumodenticulate setae, inner surface; serrulate setae, outer surface	Serrulate and plumodenticulate setae, distal medial edge; pappose and serrulate setae, proximal medial edge; plumodenticulate and serrulate setae, distal edge near basal endite; serrulate setae, inner surface; serrulate and plumodenticulate setae, outer surface	Cuspidate setae, distal medial edge; serrulate and simple setae, proximal medial edge; plumodenticulate and serrulate setae, distal edge near basal endite; serrulate setae, inner and outer surfaces
Endopodite	1 segment; 3–5 pappose and plumodenticulate setae, tip	1 segment; 3–5 pappose and plumodenticulate setae, tip	1 segment; 3–5 pappose and plumodenticulate setae, tip; 3 pappose setae, base	2 segments; 1 pappose and 1 simple setae, tip of second segment; 1 pappose and 2 spinelike setae, medial edge of first segment; clump of pappose setae, base of first segment	2 segments; 3 serrulate setae, tip of second segment; serrate setae, lateral edge of first segment; serrulate setae, medial edge of first segment; fanlike clump of plumose setae, base of first segment	2 segments; 3 serrate setae, tip of second segment; pappose setae along edges of second segment; serrate setae, medial edge of first segment; serrate and serrulate setae, lateral edge of first segment; fanlike clump of plumose setae, base of first segment	2 segments; serrate and simple setae, tip of second segment; serrate setae, medial edge of first segment; serrate and serrulate setae, lateral edge of first segment; fanlike clump of plumose and pappose setae, base of first segment

continues

TABLE 2 Continued

Mouthpart	Stage I	Stage II	Stage III	Stage IV	Stage VI	Stage XII	Adult
Second Maxillae							
Distal and proximal basal endites	No information	No information	No information	Simple, serrulate, and pappose setae, medial edges; plumodenticulate setae, outer surface, proximal endite	Simple, serrulate, and plumodenticulate setae, medial edges; pappose setae, distal end of distal endite; plumodenticulate setae, outer surfaces; simple and plumodenticulate setae, inner surface	Simple, serrulate, and plumodenticulate setae, medial edges; plumodenticulate setae, distal end of distal endite; serrulate setae, outer and inner surfaces	Simple, cuspidate, and plumodenticulate setae, medial edges; serrulate setae, outer surface; simple, serrate, and serrulate setae, inner surface
Distal and proximal coxal endites	No information	No information	No information	Pappose, simple, serrulate, and plumodenticulate setae, medial edges	Simple and serrulate setae, medial edges; serrulate setae, distal and proximal edges of proximal endite; plumodenticulate and pappose setae along proximal end of proximal endite; plumodenticulate setae, outer and inner surfaces	Serrulate, simple, plumodenticulate, and cuspidate setae, medial edges; pappose setae, proximal edge of proximal endite; plumodenticulate setae, base of proximal endite; serrulate setae, inner and outer surfaces	Serrulate, simple, plumodenticulate, and cuspidate setae, medial edges; serrulate setae, outer surface; simple, serrate, and serrulate setae, inner surface
Endopodite	Pappose setae, tip	No information	No information	Plumodenticulate setae, tip; pappose setae, medial edge; plumose setae, base of lateral edge	Serrate setae, tip; serrulate setae, distal medial edge; plumose setae, base of lateral edge	Serrate setae, tip; serrulate setae, distal medial and lateral edges; plumose setae, base of lateral edge; serrulate setae, near base of inner surface	Serrate setae, tip; serrulate setae, medial edge; pappose setae, base of lateral edge; serrate setae, inner and outer surfaces

Scaphognathite	No information	No information	No information	Plumose setae fringe all edges	Plumose setae fringe all edges; triserrate setae, posterior tip	Plumose setae fringe all edges; triserrate setae, posterior tip	Plumose setae fringe all edges; hamate setae, inner and outer surfaces; pappose setae, inner and outer surfaces of anterior portion; serrulate setae, posterior tip
First maxillipeds							
Basal endite	No information	No information	No information	Plumodenticulate setae, medial edge; row of plumodenticulate setae parallel to medial edge of outer surface; pappose setae, lateral edge; serrulate setae, inner surface	Plumodenticulate and serrulate setae, medial edge; plumodenticulate setae, distal edge; serrulate and plumodenticulate setae parallel to medial edge, outer surface; plumodenticulate setae, base of outer surface; serrulate setae, inner surface	Serrate, serrulate, and plumodenticulate setae, medial edge; plumodenticulate setae, distal edge; row of serrate setae parallel to medial edge, outer surface; serrate setae, inner surface	No information
Coxal endite	Pappose setae distally	No information	Pappose setae distally	Serrulate setae, inner surface	Plumodenticulate and pappose setae, medial edge; plumodenticulate and serrulate setae, outer surface; serrulate setae, inner surface	Plumodenticulate and pappose setae, medial edge; serrulate setae, outer and inner surfaces	No information
Endopodite	No information	No information	No information	2 segments; row of plumose setae, inner surface; row of pappose setae, medial and lateral edges	Serrate setae, tip of terminal segment; row of plumose setae, lateral edges; pappose setae, medial edges; serrate setae near base of terminal segment	Serrate setae, tip of terminal segment; row of plumose setae, lateral edges of basal and terminal segments; serrate and plumodenticulate setae, medial edge of terminal segment	No information

continues

TABLE 2 Continued

Mouthpart	Stage I	Stage II	Stage III	Stage IV	Stage VI	Stage XII	Adult
Exopodite	No flagellum; plumose setae, lateral edge	No information	3 flagellar segments	7 flagellar segments; plumose setae, lateral and medial edges and tip of flagellum; serrulate setae, distal medial edge	9 flagellar segments; plumose setae at tip and lateral edge of flagellum and basal segment; plumodenticulate setae, medial edge of basal segment; serrate setae, inner surface of basal segment	18 flagellar segments; plumose setae, tip of flagellum and lateral edge of basal segment; serrate setae, medial edge of basal segment and inner and outer surfaces	No information
Epipodite	No hamate setae	No hamate setae	Row of hamate setae, lateral edge	Hamate setae, inner surface and lateral edge; serrulate setae, outer surface	Hamate setae, inner surface; plumodenticulate setae at base of inner surface; plumodenticulate and serrulate setae, outer surface	Hamate setae, inner surface; plumodenticulate setae at base and lateral edge, inner surface; plumodenticulate, serrulate, and serrate, outer surface	No information
Second maxillipeds							
Basis	No information	No information	No information	Plumodenticulate setae, medial edge; pappose setae, inner surface	Plumodenticulate and serrulate setae and 2 spines, medial edge; plumodenticulate setae, inner and outer surfaces	Serrulate setae and 3 spines, medial edge; plumodenticulate setae, outer surface; serrate, simple, and plumodenticulate setae, inner surface	Plumodenticulate, serrate, and grooved setae and 3 spines, medial edge; plumodenticulate setae, inner and outer surfaces
Coxa	No information	No information	No information	Plumodenticulate setae, medial edge; pappose and plumodenticulate setae, inner and outer surfaces	Plumodenticulate setae, medial edge and inner and outer surfaces	Plumodenticulate setae, medial edge and inner and outer surfaces; blunt spine, inner surface	Plumodenticulate setae, medial edge and inner surface; blunt spine, inner surface

Endopodite: dactyl	1 cuspidate (H2) seta, tip; serrulate setae	No information	1 cuspidate (H1), and several stout setae with fine setules (unnamed type)	2 cuspidate (H1) setae; serrate, serrulate, and simple setae along edges	2 or 3 cuspidate (H1), serrate, plumodenticulate, and grooved setae along edges; plumodenticulate setae, outer surface; plumodenticulate and serrate setae, inner surface	6 or 7 cuspidate (H1), serrulate, serrate, and grooved setae along edges; serrulate, serrate, and grooved setae, inner surface; pappose setae, outer surface	9 cuspidate (H1), serrulate, and grooved setae along edges and inner and outer surfaces; cuspidate, serrate, and serrulate setae, inner surface
Endopodite: propus	No information	No information	No information	Serrate, simple, and serrulate setae, edges and outer surface	Plumodenticulate, serrulate, and serrate setae, lateral edge; serrate setae, medial edge; plumodenticulate setae, distal edge; plumodenticulate setae, inner surface	Serrulate and serrate setae, lateral edge; serrate setae, medial edge; simple setae, outer surface; serrate, simple, serrulate, and grooved setae, inner surface	Serrate and serrulate setae, lateral edge; serrate, serrulate, and plumodenticulate setae, medial edge; serrulate setae, distal edge of outer surface; serrate, serrulate, and simple setae, lateral edge of inner surface; serrulate setae, medial edge of inner surface
Endopodite: carpus	Cluster of 3 setae	No information	No information	Serrulate and simple setae, inner surface	Serrulate setae, distal edge; serrate setae, outer surface; plumodenticulate setae, inner surface	Serrulate setae, lateral edge; serrulate and serrate setae, distal edge of inner surface; serrulate and simple setae, inner surface	Serrate and serrulate setae, distal lateral edge of outer surface; serrate setae, proximal lateral edge of outer surface; serrate and serrulate setae, medial distal edge of outer surface; serrate and serrulate setae, inner surface

continues

TABLE 2 *Continued*

Mouthpart	Stage I	Stage II	Stage III	Stage IV	Stage VI	Stage XII	Adult
Endopodite: merus	Serrate, serrulate, and triserrulate setae, medial edge	No information	No information	Serrulate and triserrulate setae, medial edge; serrate setae, distal medial edge; plumodenticulate setae, proximal medial edge	Serrate and plumodenticulate setae, distal medial edge; plumodenticulate, serrulate, and grooved setae, medial edge; plumodenticulate setae and spine, outer surface; spine, inner surface	Serrate setae, distal medial edge; serrulate, grooved, and multiscaled setae, medial edge; row of serrulate setae parallel to lateral edge on outer surface; row of serrulate and serrate setae parallel to medial edge on inner surface; serrulate setae, distal inner surface; blunt spines, both surfaces	Serrate, serrulate, and plumodenticulate setae, medial edge; row of serrate and serrulate setae parallel to lateral edge on outer surface; row of serrate, serrulate, grooved, and plumodenticulate setae parallel to medial edge on inner surface; serrate setae, distal lateral edge of inner surface; blunt spines, both surfaces
Endopodite: ischium	No information	No information	No information	Serrulate and plumodenticulate setae, medial edge	Plumodenticulate, serrate, and serrulate setae, medial edge; plumodenticulate setae, inner and outer surfaces; spine, distal edge	Serrulate, serrate, and grooved setae, medial edge; spine, distal end; serrulate and plumodenticulate setae, outer surface; plumodenticulate setae, inner surface	Plumodenticulate, serrate, and grooved setae, medial edge; spine, distal end; plumodenticulate and grooved setae, outer surface; plumodenticulate setae, inner surface
Exopodite	Plumose setae, tip of flagellum; indistinct separation between flagellum and basal segment	No information	6 flagellar segments; plumose setae, tip	12 flagellar segments; plumose setae, tip; plumodenticulate setae, basal segment	13 or 14 flagellar segments; plumose setae, tip; plumodenticulate, serrate, and serrulate setae, edges of basal segment; plumodenticulate setae, outer surface of basal segment	20–22 flagellar segments; plumose setae, tip; serrate and serrulate setae, lateral and distal medial edges and inner and outer surfaces of basal segment; plumodenticulate and simple setae, proximal medial edge of basal segment	24 flagellar segments; plumose setae, tip; serrate and serrulate setae, lateral edge and outer surface of basal segment

Epipodite	No hamate or serrulate setae present	No information	No information	Serrulate and hamate setae, inner surface and edges	Serrulate setae, outer surface; hamate and serrulate setae, inner surface and edges; plumodenticulate setae, base	Hamate setae, inner surface; serrate, multiscaled, serrulate, and plumodenticulate setae, outer surface; plumodenticulate setae, base; serrulate setae, base of podobranch	Hamate and serrulate setae, inner surface; serrate and multiscaled setae, outer surface; plumodenticulate setae, base; serrate, grooved, and plumodenticulate setae, base of podobranch
Third maxillipeds							
Basis	Serrulate setae	No information	No information	Serrate and triserrulate setae, medial edge	Serrulate and grooved setae and spine, medial edge; plumodenticulate setae, outer surface; serrate and plumodenticulate setae and spine, inner surface	Serrulate and grooved setae and spine, medial edge; plumodenticulate setae, outer surface; serrulate and simple setae and 5 spines, inner surface	Serrulate, grooved, and simple setae, medial edge; plumodenticulate setae and large pits, outer surface; 5 spines, inner surface
Coxa	Serrulate setae	No information	No information	Plumodenticulate setae, both surfaces	Plumodenticulate setae, lateral edge and both surfaces; serrulate setae and spine, medial edge	Plumodenticulate setae, lateral and medial edges and both surfaces; 2 spines and serrulate and grooved setae, medial edge	Plumodenticulate setae, lateral edge and both surfaces; 2 spines and serrulate and grooved setae, medial edge
Endopodite: dactyl	Simple setae, tip; serrulate setae, inner and outer edges	No information	No information	Serrate, simple, and triserrate setae, inner and outer edges	Serrate, serrulate, and grooved setae, lateral, medial, and distal edges; some setae on both surfaces	Serrate, serrulate, and grooved setae, lateral and medial edges and both surfaces; plumodenticulate setae, inner surface; papposе setae, outer surface	Serrate and simple setae, lateral and medial edges and both surfaces; simple and grooved setae, distal edge

continues

TABLE 2 Continued

Mouthpart	Stage I	Stage II	Stage III	Stage IV	Stage VI	Stage XII	Adult
Endopodite: propus	Triserrate, serrulate, and serrate setae, inner and outer edges	No information	No information	Serrate, simple, and triserrate setae, inner and outer edges	Serrate and serrulate setae, medial edge; serrulate, grooved, and pappose setae, lateral edge; simple and pappose setae, outer surface; serrate setae, inner surface	Serrate and grooved setae, medial edge; pappose and grooved setae, lateral edge ; serrate, grooved, and plumodenticulate setae, inner surface; simple and pappose setae, outer surface	Simple and serrate setae, medial and lateral edges; simple setae, inner and outer surfaces
Endopodite: carpus	Serrate, serrulate, and triserrate setae, inner medial edge	No information	No information	Serrate, triserrate, and triserrulate setae, inner medial edge	Serrate and serrulate setae, medial surface and inner medial edge; simple and serrate setae, outer and inner lateral surfaces; pappose, serrulate, and grooved setae, lateral edge; spine, outer medial edge	Grooved, serrate, serrulate, and plumodenticulate setae, inner medial edge; spine, outer medial edge; simple and serrate setae, inner lateral surface; simple and plumose setae, outer lateral surface; pappose, grooved, simple, serrate, and serrulate setae, lateral edge	Simple, serrate, and grooved setae, inner medial edge and surface; spine, outer medial edge; simple, serrate, and grooved setae, inner and outer lateral surfaces
Endopodite: merus	Serrate, serrulate, and triserrate setae, inner and outer medial edges	No information	No information	Triserrulate and serrate setae, inner and outer medial edges; spine, distal end of lateral edge	Serrate, grooved, and plumodenticulate setae, inner medial edge and medial surface; simple setae, lateral edge and inner lateral surface; row of simple setae, outer lateral surface; row of plumose setae parallel to lateral edge; 4 or 5 spines, outer medial edge; spine, distal lateral edge	Grooved, multi-scaled, and serrulate setae, medial surface; serrate setae, inner medial edge; plumose setae, outer medial edge; simple setae, lateral edge and both lateral surfaces; plumose setae, outer lateral surface; 5 spines, outer medial edge; spine, distal lateral and inner medial edges	Serrate, simple, grooved, and plumose setae, medial surface; simple setae and spine, inner medial edge; serrate, simple, grooved, and pappose setae, outer medial edge; plumose setae, outer lateral surface; 8 spines, outer medial edge; spine, distal lateral edge

Endopodite: ischium	2–4 teeth, inner medial edge	6 teeth, inner medial edge	6 teeth, inner medial edge	14 teeth, inner medial edge; serrate and tri-serrulate setae, outer medial edge; simple setae parallel to teeth, inner lateral surface; spine, distal end of lateral edge	16 or 17 teeth, inner medial edge; 4 or 5 spines, outer medial edge; plumodenticulate and grooved setae, medial surface; row of serrulate setae parallel to teeth,392 392inner lateral surface; plumose setae, distal end near lateral edge, inner lateral surface; row of plumose setae parallel to lateral edge on outer lateral surface; simple setae, outer lateral surface	18 teeth, inner medial edge; 7 spines, outer medial edge; serrate, multiscaled, serrulate, and grooved setae, medial surface; serrulate setae parallel to teeth; plumose setae, inner lateral surface; simple and plumose setae, outer lateral surface and edge	15 teeth, inner medial edge; 7 spines, outer medial edge; simple, serrate, and grooved setae, medial surface; simple, grooved, and pappose setae, parallel to lateral edge; large pits, outer lateral surface; simple and pappose setae, lateral edge
Exopodite	8 flagellar segments, each with a pair of plumose setae	No information	No information	12 flagellar segments; plumose setae at tip and on each segment	13 or 14 flagellar segments; plumose setae at tip and on 7 distal segments; serrate and plumodenticulate setae, both surfaces of basal segment	29 flagellar segments; plumose setae at tip and on each segment; serrate and plumodenticulate setae, both surfaces of basal segment	29 flagellar segments; plumose setae at tip and on each segment; simple, serrate, plumodenticulate, and pappose setae, both surfaces of basal segment
Epipodite	Serrulate setae, no hamate setae	Hamate setae	Hamate setae	Hamate, serrulate, and plumdenticulate setae, outer edge	Hamate setae, edges and inner surface; serrate, plumodenticulate, and serrulate setae, outer surface; plumodenticulate	Hamate setae, edges and inner surface; serrate, serrulate, and triserrulate setae, outer surface; plumodenticulate setae, base	Hamate and serrate setae, inner surface; plumodenticulate, serrulate, serrate, and multiscaled setae, outer surface

[a]Data on larvae, postlarvae, and juveniles are based on the work of Factor (1978) and Lavalli and Factor (1992); preliminary and incomplete data on adults are based on the work of Chen (1992) and Neilsen (1993).

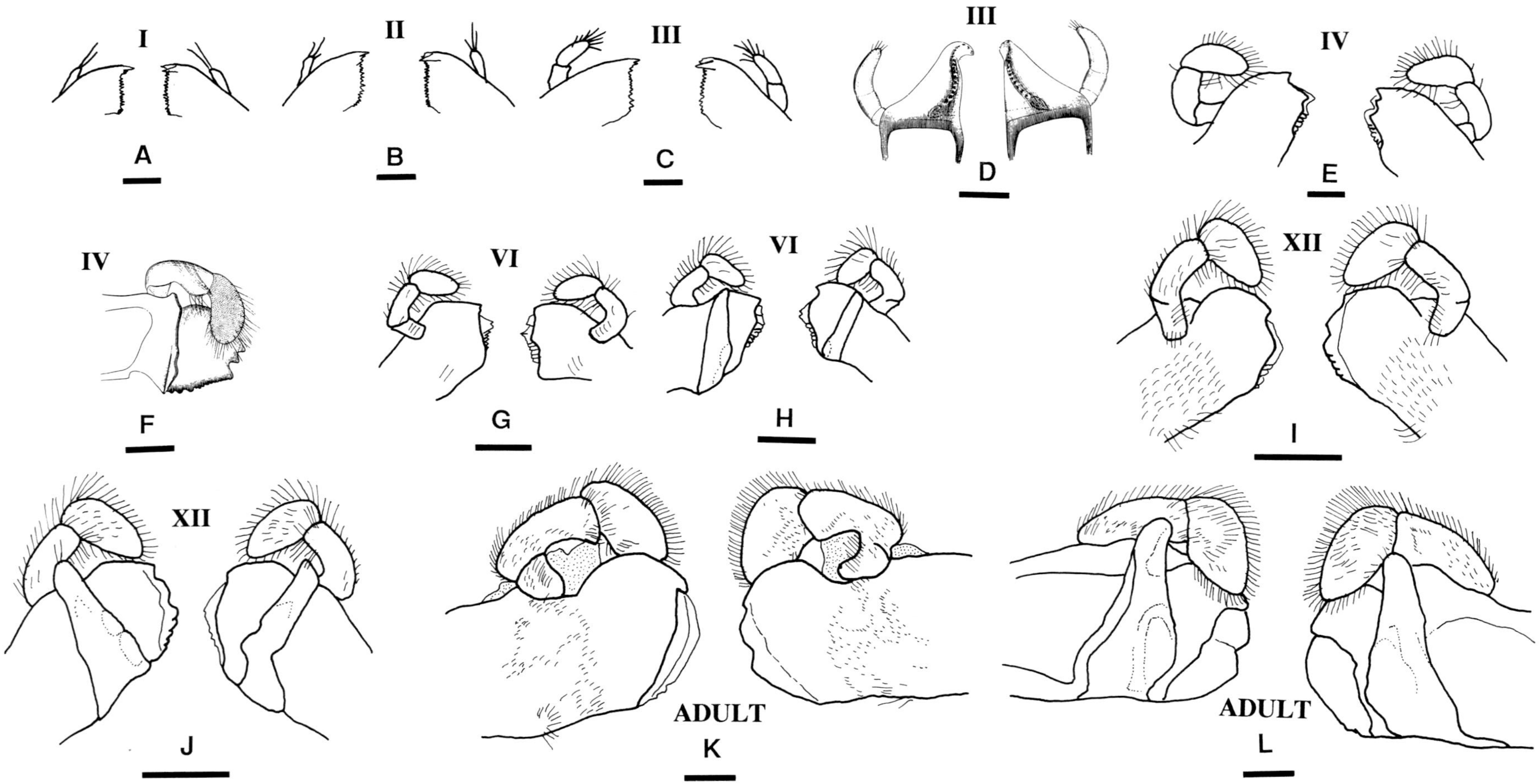

FIGURE 3 Mandible structure in larval, postlarval, juvenile, and adult lobsters, *Homarus americanus*. Line drawings of stages I–III, IV, VI, XII, and adult. Right and left outer surfaces: (A) stage I; (B) stage II; (C) stage III; (E) stage IV; (G) stage VI; (I) stage XII; (K) adult. Left and right inner surfaces: (D) stage III; (F) stage IV; (H) stage VI; (J) stage XII; (L) adult. Scale bars: (A–C and E) 0.25 mm; (D, F, K, and L) 0.3 mm; (G–J) 1.0 mm. [(D and F) modified from Smith, 1873; (A–C and E) redrawn from Factor, 1978, and (G–J) reprinted from Lavalli and Factor, 1992, with permission.]

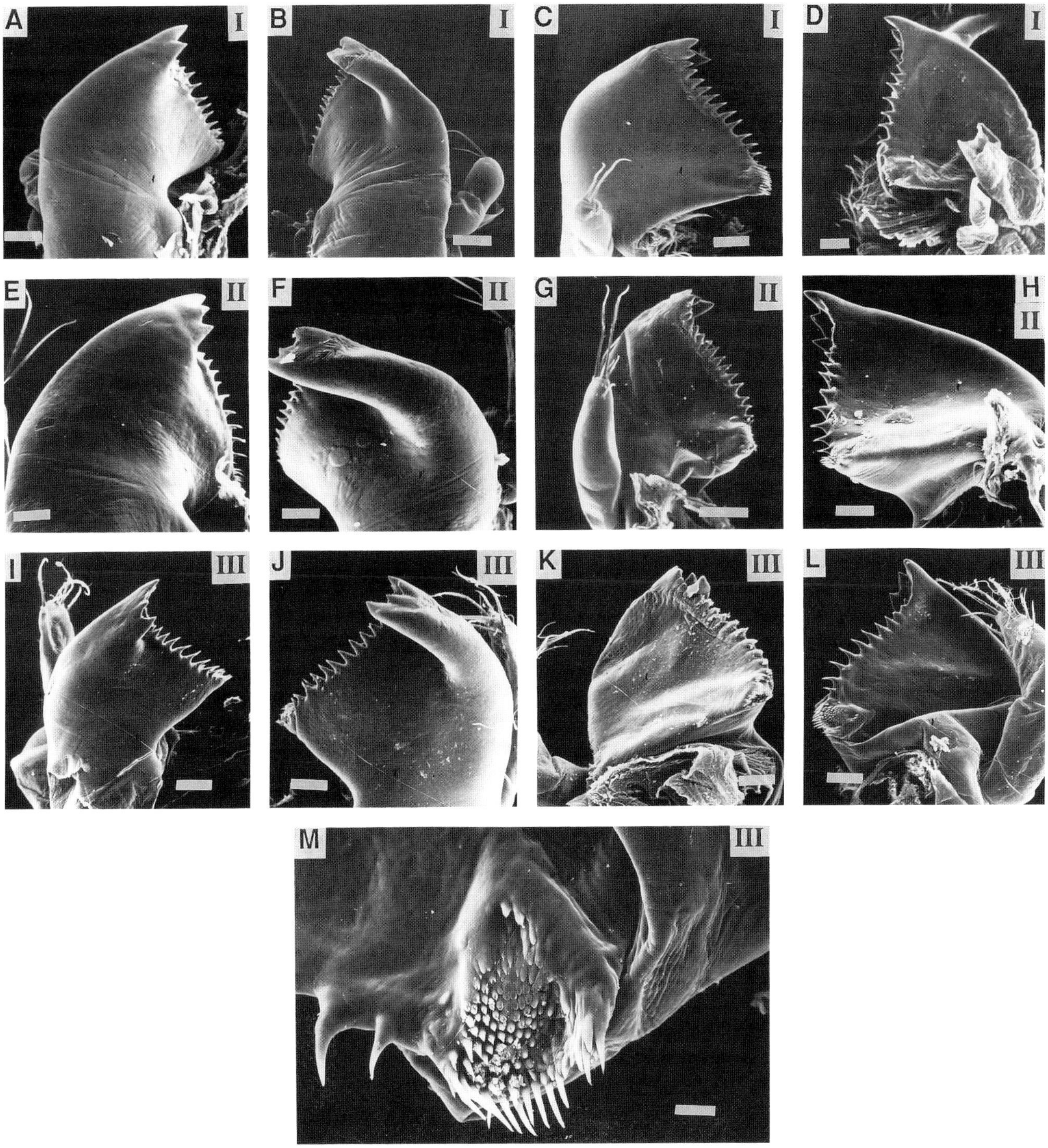

FIGURE 4 Mandibles of *Homarus americanus*. Scanning electron micrographs of the gnathal lobe of stages I–III. Stage I, outer surface: (A) right; (B) left. Stage I, inner surface: (C) left; (D) right. Stage II, outer surface: (E) right; (F) left. Stage II, inner surface: (G) left; (H) right. Stage III, outer surface: (I) right; (J) left. Stage III, inner surface: (K) left; (L) right; (M) setal pad at the proximal end of the cutting surface. Scale bars: (A) 0.078 mm; (B) 0.067 mm; (C) 0.075 mm; (D) 0.05 mm; (E) 0.073 mm; (F) 0.071 mm; (G) 0.1 mm; (H) 0.07 mm; (I) 0.105 mm; (J) 0.083 mm; (K) 0.088 mm; (L) 0.09 mm; (M) 0.017 mm.

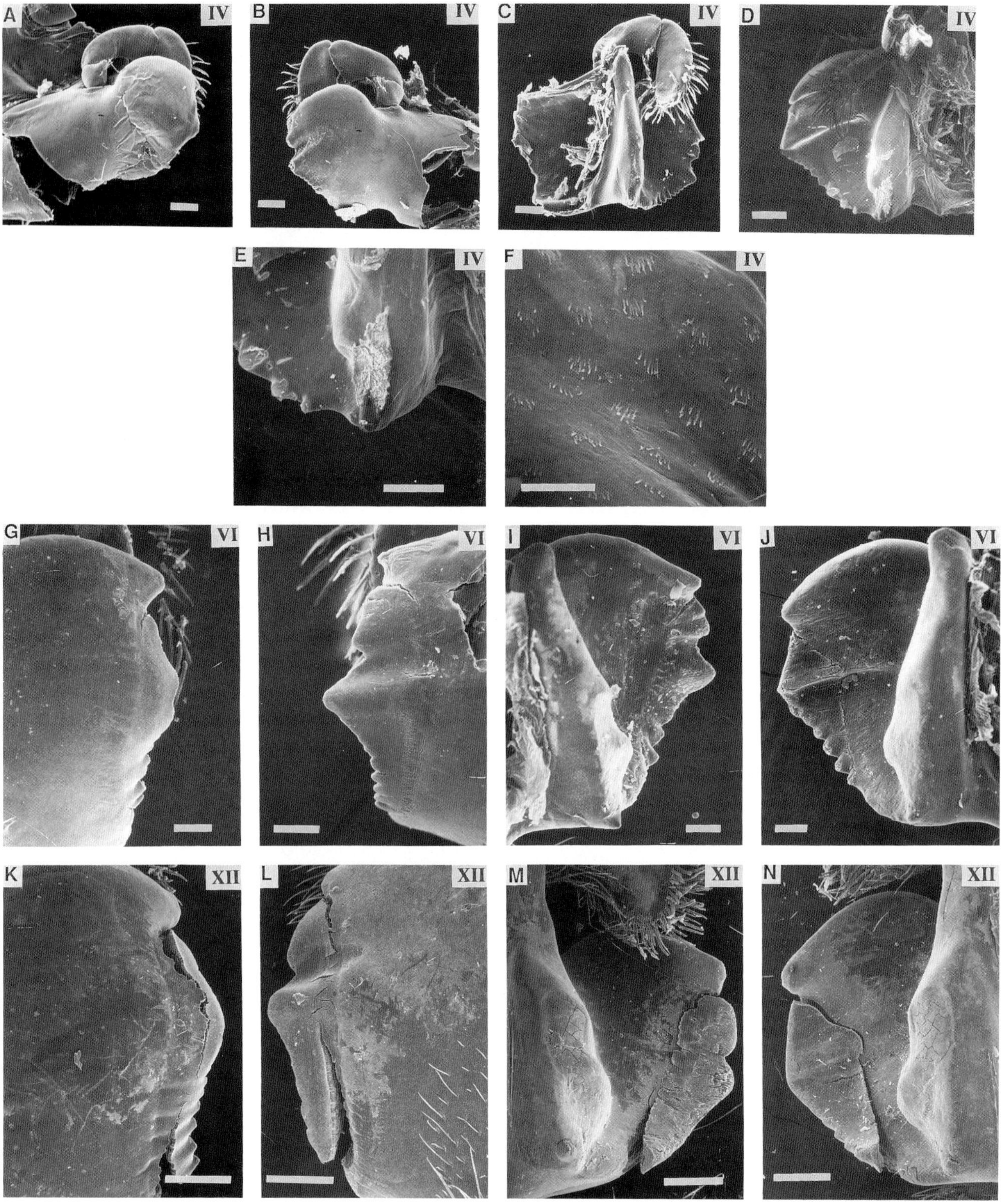

FIGURE 5 Mandibles of *Homarus americanus*. Scanning electron micrographs of stages IV, VI, and XII. Stage IV, outer surface: (A) right; (B) left. Stage IV, inner surface: (C) left; (D) right; (E) molar process; (F) cuticular texturing on molar process. Stage VI, outer surface: (G) right; (H) left. Stage VI, inner surface: (I) left; (J) right. Stage XII, outer surface: (K) right; (L) left. Stage XII, inner surface: (M) left; (N) right. Scale bars: (A) 0.15 mm; (B and C) 0.13 mm; (D and G–J) 0.1 mm; (E and K–N) 0.05 mm; (F) 0.01 mm. [(G–N) reprinted from Lavalli and Factor, 1992, with permission.]

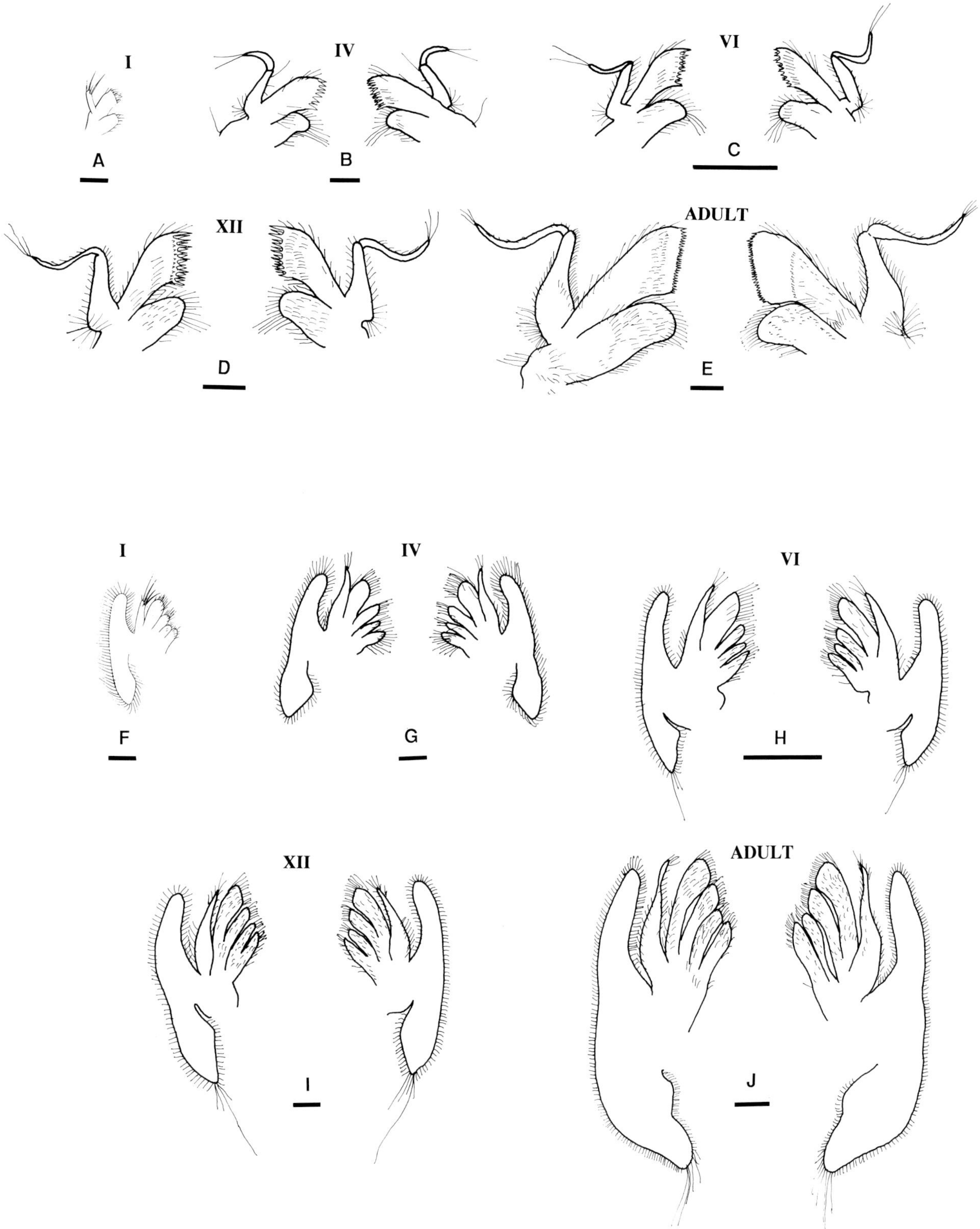

FIGURE 6 First (A–E) and second (F–J) maxillae of *Homarus americanus*. Line drawings of stages I, IV, VI, XII, and adult. Left inner and outer surfaces: (B) stage IV; (C) stage VI; (D) stage XII; (G) stage IV; (H) stage VI. Right inner surface: (A) stage I; (F) stage I. Right outer and inner surfaces: (I) stage XII; (E) adult; (J) adult. Scale bars: (A and F) 0.1 mm; (B and G) 0.25 mm; (C, D, H, and I) 1.0 mm; (E and J) 4.0 mm. [(A and F) modified from Smith, 1873; (B and G) redrawn from Factor, 1978, and (C, D, H, and I) reprinted from Lavalli and Factor, 1992, with permission.]

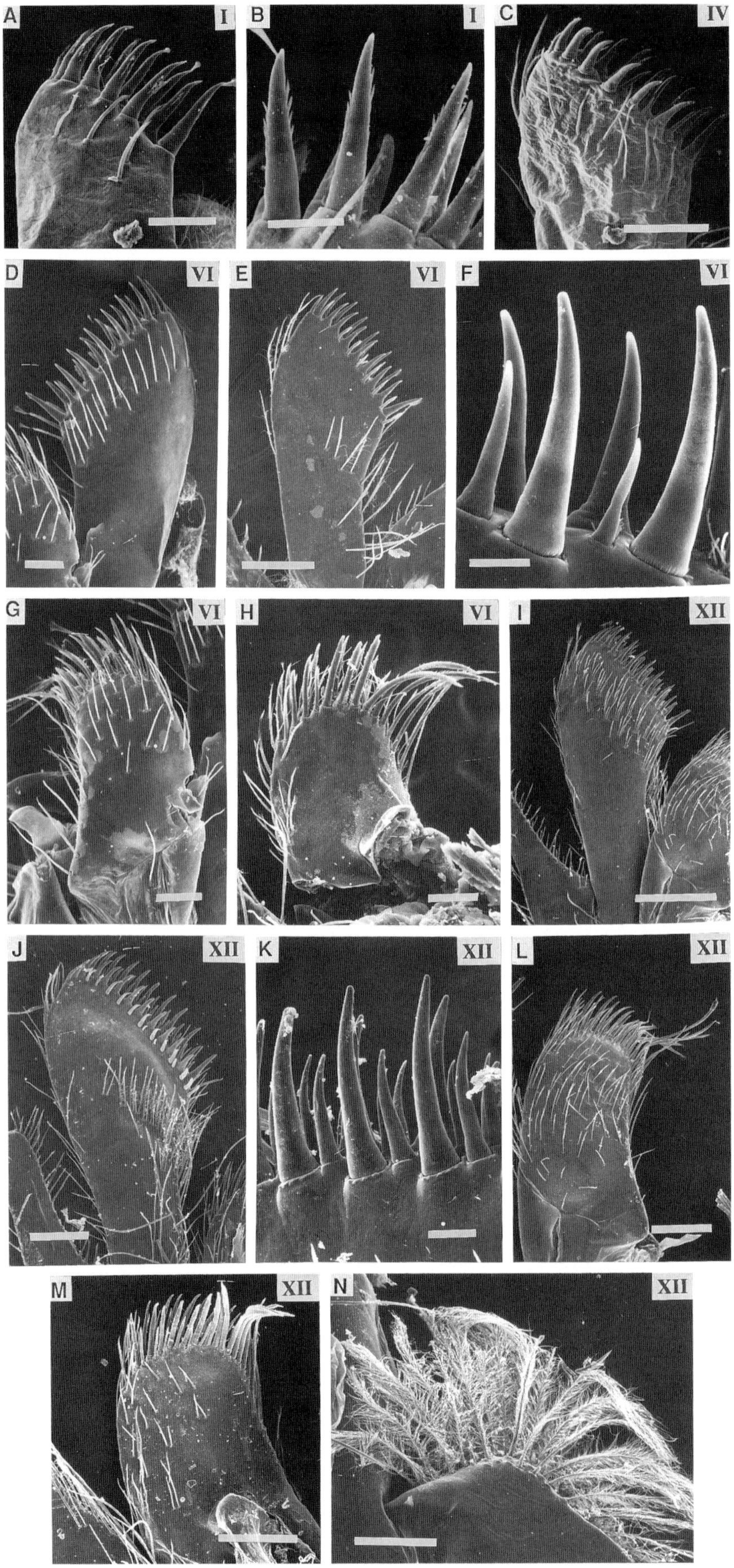

A I
B I
C IV
D VI
E VI
F VI
G VI
H VI
I XII
J XII
K XII
L XII
M XII
N XII

one or two (Fig. 5I and J). By stage XII, the molar process has increased significantly in size, with both right and left processes composed of three denticles and setae at the distal end of the process (Fig. 5M and N).

The mandibular palp also changes dramatically. The articulations between the three segments become more pronounced throughout the larval stages and there is an overall increase in the number of setae on all edges and surfaces.

2. First Maxillae

The first maxillae comprise a protopodite, with coxal and basal endites, and a segmented endopodite (Fig. 6A–E). The endites are thin, platelike structures with setose medial edges.

The overall structure of the first maxillae changes very little from the larvae to the adult. Those changes that do occur include an increase in the number of setae along all edges and on both surfaces, with the most interesting changes occurring on the basal endite and endopodite. The medial edge of the basal endite of the larval stages differs from that of all subsequent stages because its cuspidate setae are of the H2 type and thus possess rows of fine setules (Fig. 7A and B); these setules are lost by the postlarval stage and type H2 are replaced with type H1 cuspidate setae.

Stage I larvae bear only two rows of cuspidate setae along the medial edge (Fig. 7A), while stage IV postlarvae and stage VI juveniles bear three rows (Fig. 7C, D, and F), and stage XII juveniles and adults bear five rows (Fig. 7J and K). The coxal endite bears less stout types of setae (see Table 2 for a description and Figs. 6B–E and 7G, H, L, and M) until the adult, in which cuspidate setae first appear. Only a single segment is present in the endopodite of the larval stages, but this increases to two segments by stage IV (Fig. 6A and B). No setae are present at the base of the endopodite in the first two larval stages; however, three setae are present in stage III and a clump of pappose setae is present in stage IV. By stage VI, these setae have changed to a clump of plumose setae and by stage XII this clump has expanded into a fanlike structure (Figs. 6A–E and 7N), which persists in the adult and may act as a seal for the branchial chamber.

FIGURE 7 First maxillae of *Homarus americanus*. Scanning electron micrographs of stages I, IV, VI, and XII. Basal endite: (A) right outer surface, medial edge, stage I; (B) cuspidate (H2) setae, medial edge, stage I; (C) right outer surface, medial edge, stage IV; (D) left outer surface, stage VI; (E) right inner surface, stage VI; (F) cuspidate (H1) setae, medial edge, stage VI; (I) right outer surface, stage XII; (J) left inner surface, stage XII; (K) cuspidate (H1) setae, medial edge, stage XII. Coxal endite: (G) left outer surface, stage VI; (H) right inner surface, stage VI; (L) left outer surface, stage XII; (M) right inner surface, stage XII. Endopodite: (N) fanlike clump of plumose (A) setae, base, stage XII. Scale bars: (A) 0.05 mm; (B and F) 0.025 mm; (C, D, G, H, and K) 0.1 mm; (E and N) 0.25 mm; (I) 1.0 mm; (J, L, and M) 0.5 mm. [(A–C) reprinted from Factor, 1977, and (D–N) reprinted from Lavalli and Factor, 1992, with permission.]

3. Second Maxillae

The second maxillae comprise a protopodite with thin, platelike, bilobed coxal and basal endites, an endopodite, and a scaphognathite (Fig. 6F–J). The medial margins of the endites are heavily setose, as is the entire margin of the scaphognathite. The scaphognathite (gill bailer) extends into the branchiostegal chamber.

Fewer changes occur in the development of the second maxillae, perhaps due to the need for a functional scaphognathite for respiratory purposes in all stages (Factor, 1978). The medial edges of both distal and proximal lobes of the basal and coxal endites are heavily setose (Figs. 6F–J and 8A–I) in all stages, although the trend is an increasing number of setae. The inner and outer surfaces also become more setose from the larval stages to the adult. Other than an increase in the number of setae and changes in some types, little structural change occurs on the endopodite. Similarly, no structural change occurs on the scaphognathite, other than an increase in the number of plumose setae fringing its edges and the development of apical hairs at stage IV, which persist through the adult.

4. First Maxillipeds

The first maxillipeds are similar to the first and second maxillae in that the protopodite bears platelike coxal and basal endites with setose medial margins (Fig. 9). They also have a segmented endopodite, an exopodite, and an epipodite.

The first maxillipeds undergo few changes except for an increase in size, number of setae, and segmentation of the exopodite (Fig. 9A–F). The endopodite remains two-segmented throughout the larval, postlarval, and juvenile stages (Fig. 9A–F). Its edges are fringed with setae from stage IV on, although the surfaces remain bare throughout stage XII (Fig. 10B, E, and H). The basis bears setae on its medial edge and both surfaces in all stages examined (Fig. 10A, C, D, F, and G), as does the coxa (Fig. 9B–F), except for stage I larvae, in which the coxa is almost bare. The exopodite is not divided into two regions in stage I larvae and bears plumose setae only on its lateral edge (Fig. 9A). By stage III, a distal flagellum with three segments is present and by stage IV, the distal flagellum possesses

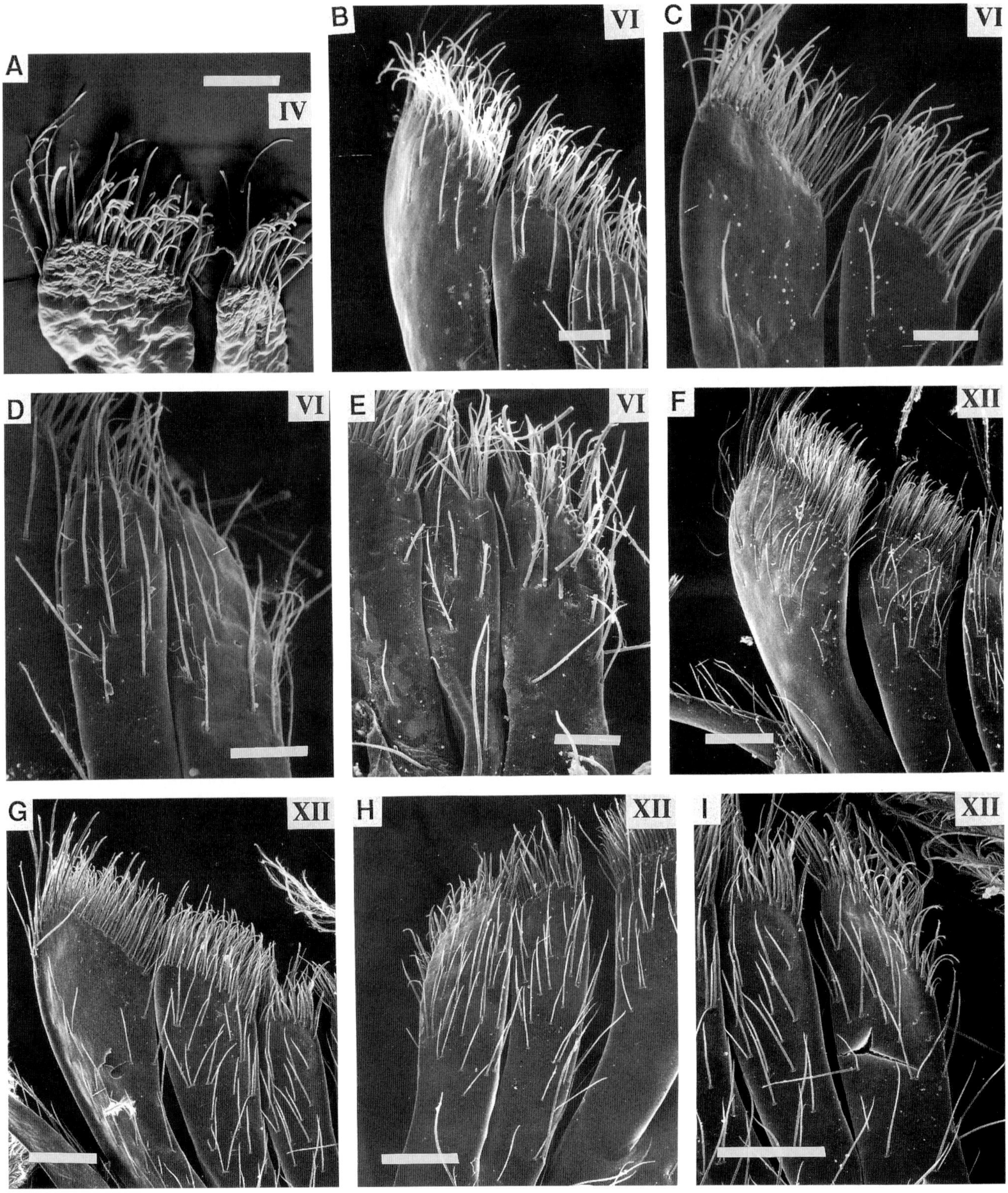

FIGURE 8 Second maxillae of *Homarus americanus*. Scanning electron micrographs of stages IV, VI, and XII. Basal endite, distal and proximal lobes: (A) right outer surface, stage IV; (B) right outer surface, stage VI; (C) left inner surface, stage VI; (F) right outer surface, stage XII; (G) left inner surface, stage XII. Coxal endite, distal and proximal lobes: (D) right outer surface, stage VI; (E) left inner surface, stage VI; (H) left outer surface, stage XII; (I) left inner surface, stage XII. Scale bars: (A–E) 0.1 mm; (F–I) 0.5 mm. [(A) reprinted from Factor, 1977, and (B–I) reprinted from Lavalli and Factor, 1992, with permission.]

seven segments (Fig. 9B and C). The number of segments increases to nine in stage VI (Fig. 10E) and 18 in stage XII (Fig. 10H).

The epipodite is a flattened structure that aids the adjacent scaphognathite of the second maxillae in directing water flow through the gill chamber (Factor, 1978). It bears scattered setae on both of its surfaces and hamate setae are located on the inner surface in all stages (Fig. 9A–F).

5. Second Maxillipeds

The second maxillipeds comprise a two-segmented protopodite bearing a five-segmented endopodite, an exopodite, an epipodite, and a single podobranch (Fig. 11). The endopodite is divided into the dactyl, propus, carpus, merus, and ischium (distal to proximal). The ischium and the merus project anteriorly and the terms *distal* and *proximal* and *medial* and *lateral* apply as they have for the other mouthparts. The carpus, propus, and dactyl, however, turn medially to give the second maxillipeds the shape of an inverted L. For these segments, the *lateral* edge actually faces anteriorly and the *medial* edge faces posteriorly; *distal* faces medially (toward the midline) and *proximal* faces laterally (away from the midline). The medially facing edges of these endopodite segments are heavily setose.

The endopodal segments of the second maxilliped undergo some significant structural changes, other than just an increase in the number of setae borne on edges and surfaces. The dactyl bears only a single cuspidate seta of type H2 in stage I (Fig. 12A), which loses its setules in stage III to become a type H1 cuspidate seta. Three cuspidate setae are present at stage VI, six to seven at stage XII, and nine in the adult (Fig. 12D, E, and H). These setae are important in processing captured food items (see Section II,D). Other setae are present on the dactyl, which, like those found on the propus and the carpus, merely increase in number. After stage IV, distal spines appear on the outer and inner surfaces of the merus (Fig. 12C, F, G, J, and K), the medial edges of the basis, and the inner surface of the coxa.

The exopodite is divided into two segments at stage I, with only four or five setae at its tip (Fig. 11). At stage III, a distal flagellum with six segments is present. By stage IV, the distal flagellum is divided into 12 segments; this number increases to 13 or 14 in stage VI , to 20 or 21 in stage XII, and to 24 in the adult. After stage I, the epipodite bears setae on both surfaces, with a clump of setae appearing at the distal end in stage VI. The podobranch, which is a simple lobe in stage I, gains secondary lobes in every stage examined.

6. Third Maxillipeds

The third maxillipeds are not flattened against the body, as are the other mouthparts, but extend anteriorly to act as grasping structures. They comprise a protopodite bearing a five-segmented endopodite, an exopodite, an epipodite, and one podobranch of the trichobranch architecture (Fig. 13). As in the second maxillipeds, the endopodite is divided into the dactyl, propus, carpus, merus, and ischium (distal to proximal), and the dactyl, propus, and carpus are similarly curved toward the midline. The carpus, merus, and ischium have a triangular cross-sectional shape with three edges—inner medial, outer medial, and lateral—which define three surfaces—medial, outer lateral, and inner lateral (Fig. 14). The endopodal segments are very setose and the ischium bears ischial teeth (=crista dentata) along its inner medial edge (Schram, 1986; Lavalli and Factor, 1992).

The developmental trend in the third maxillipeds is an increase in the number of setae and teeth (Fig. 13A–E). Unlike other segments, which gain diversity in setal types with development to the adult stage, the dactyl, propus, and carpus undergo a reduction in the types, but not the numbers, of setae (Fig. 15C–F and J–M). A spine appears on the outer medial edge of the carpus and the distal end of the lateral edge of the merus after stage IV. In addition, four or five spines arise along the outer medial edge of the merus at stage VI; these increase in number to seven in stage XII (Fig. 15G and N) and eight by the adult stage. Ischial teeth (crista dentata) are present on the inner medial edge of the ischium in all stages (Fig. 13B–E): two to four rudimentary teeth in stage I, six in stages II and III, 14 in stage IV (Fig. 15A), 16 or 17 in stage IV (Fig. 15H), 18 in stage XII (Fig. 15O), and 15 or 16 in the adult. A row of setae lies parallel to the ischial teeth on the inner lateral surface, beginning in stage IV (Fig. 15A, H, and O). The outer medial edge of the ischium also bears spines: four or five in stages IV and VI (Fig. 15B and I) and seven in stage XII (Fig. 15P) and the adult. Teeth are also present on the basis: one on the inner surface in stage VI and five from stage XII onward (Fig. 13C–E). A flattened spine is present on the medial edge of the coxa at stage VI, whereas two are present after stage XII (Fig. 13C–E).

The flagellum of the exopodite is composed of eight segments in stage I, 12 in stage IV, 13 or 14 in stage VI, and 29 in stage XII and the adult. After stage II, the epipodite bears setae on both surfaces, and, like the second maxillipeds, the podobranch gains additional filaments and secondary lobes throughout its development (Fig. 13).

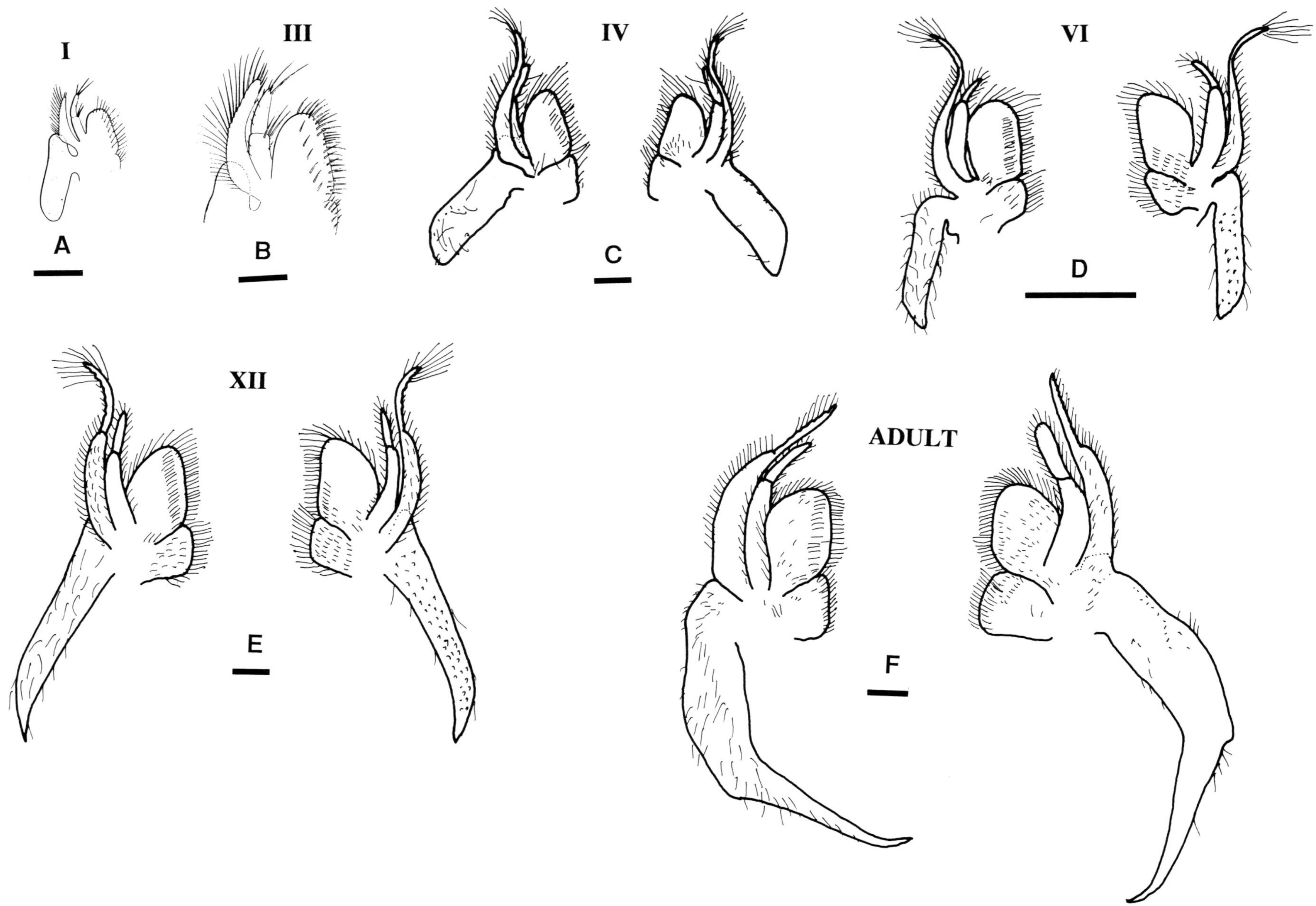
I
III
IV
VI
A
B
C
D
XII
ADULT
E
F

D. Function of the Mouthparts

Larval and early postlarval (presettlement) lobsters are planktonic and feed on other planktonic organisms that are also suspended in surface waters (Ennis, in Chapter 3, and Lawton and Lavalli, in Chapter 4, discuss the food preferences of larvae and pelagic postlarvae). They are apparently nondiscriminatory in their choice of food (Barnes, 1906; Herrick, 1909) and presumably feed on anything they can seize and swallow (Herrick, 1909). Incessant swimming activity of the larvae may be necessary to bring them into contact with suspended food (Mead and Williams, 1902). Thus, the larvae and postlarvae have been considered raptorial feeders that can pursue prey items with fairly directed swimming movements when necessary (Herrick, 1909). Analyses of gut contents indicate that their diet can include diatoms, copepods, crustacean larvae, gastropod larvae, floating fish eggs, and insect pieces (Herrick, 1895, 1909; Williams, 1907; Harding *et al.*, 1983; Gunn, 1987; Juinio and Cobb, 1992).

Factor (1978) suggested the possibility that larvae might filter-feed on small suspended particles, but this suggestion was never investigated. Of all the life history phases, however, the larvae bear the least setose mouthparts, suggesting that suspension feeding might not be as likely as in the more setose postlarvae and the shelter-restricted and emergent phase juveniles. Videographic analysis of tethered larvae suggests that while the current created by the thoracic exopodites can redirect particles flowing past the lobster, only small particles such as diatoms circulate and travel within this current (K. L. Lavalli and E. Nesti, unpublished data); these may be purposefully captured or ingested incidentally, possibly due to their accumulation on appendages (Williams, 1907). Larger items, such as adult copepods and crustacean nauplii, are unaffected by the thoracic exopodite current and are likely to be captured when they accidentally strike the legs or mouthparts or when the lobsters see and pursue them (K. L. Lavalli, personal observations). Such copepods may be superabundant in warm, inshore waters during the summer months when the larvae hatch (Harding *et al.*, 1983; Gunn, 1987). It is unlikely that the chelae of the first pereiopods are used in their capture, as they

FIGURE 9 First maxillipeds of *Homarus americanus*. Line drawings of stages I, III, IV, VI, XII, and adult. Right inner surface: (A) stage I; (B) stage III. Right outer and inner surfaces: (C) stage IV; (D) stage VI; (E) stage XII; (F) adult. Scale bars: (A and B) 0.3 mm; (C) 0.25 mm; (D and E) 1.0 mm; (F) 5.0 mm. [(A and B) modified from Smith, 1873; (C) redrawn from Factor, 1978, and (D and E) reprinted from Lavalli and Factor, 1992, with permission.]

seem to be incapable of prehensile activity until stage IV (Govind, Chapter 12), however, the pereiopods and the third maxillipeds are clearly capable of wrapping around and holding a prey item while it is processed by the mouthparts (K. L. Lavalli, personal observations).

Stage IV postlarvae and settled, stage V (shelter-restricted phase) juveniles can remove particles in the range of 70–1000 μm from the water (Lavalli and Barshaw, 1989) and produce two currents important in capturing these particles: the exopodite current of the maxillipeds and the pleopod current of the abdomen (Fig. 16) (Lavalli, 1992; Lavalli and Trager, 1995). The exopodite current, most commonly used in capturing zooplankton, is produced by beating of the distal flagella of the exopodites, which are fringed by long plumose setae. Analysis of high-speed videography indicates that only the exopodites of the second and third maxillipeds are capable of beating back and forth in an arc; in the first maxilliped, the distal flagellum merely vibrates because the epipodite extends over its basal portion, thereby preventing the beating motion. The exopodites on one side are capable of beating in unison, in opposition, or singly. Likewise, a pair of left and right exopodites can beat simultaneously or singly. This flexibility in beating pattern allows the creation of varying flow fields around the head and mouth region, as well as patterns along the cephalothorax (Fig. 16A). These currents can be used to draw particles into the mouth region or to eject them from the area altogether.

The pleopods are biramous, fringed with long, numerous, plumose setae, and can beat with a metachronal rhythm. Vigorous fanning creates the pleopod current that moves a large volume of water past the lobster, from front to back, both laterally and ventrally (Fig. 16B). Gentle fanning results in a redirection of the exopodite current, so that water flows underneath the lobster. The pleopod current brings a volume of water toward the animal, which the exopodite current can then circulate around the head region for capture of zooplankton (Lavalli, 1992; Lavalli and Trager, 1995).

Postlarvae and shelter-restricted phase juveniles use both raptorial and suspension feeding techniques to capture zooplankton (Lavalli, 1992; Lavalli and Trager, 1995) at a time of transition in their life history. Lobsters can survive on planktonic diets through stage IX (Barshaw, 1989; Lavalli, 1991), the stage at which they are thought to overwinter in their first year (Hudon, 1987). Their claws are not capable of the same crushing movements as the adult. Furthermore, the setae found on the mouthpart segments at stages IV, VI, and XII are much longer, rela-

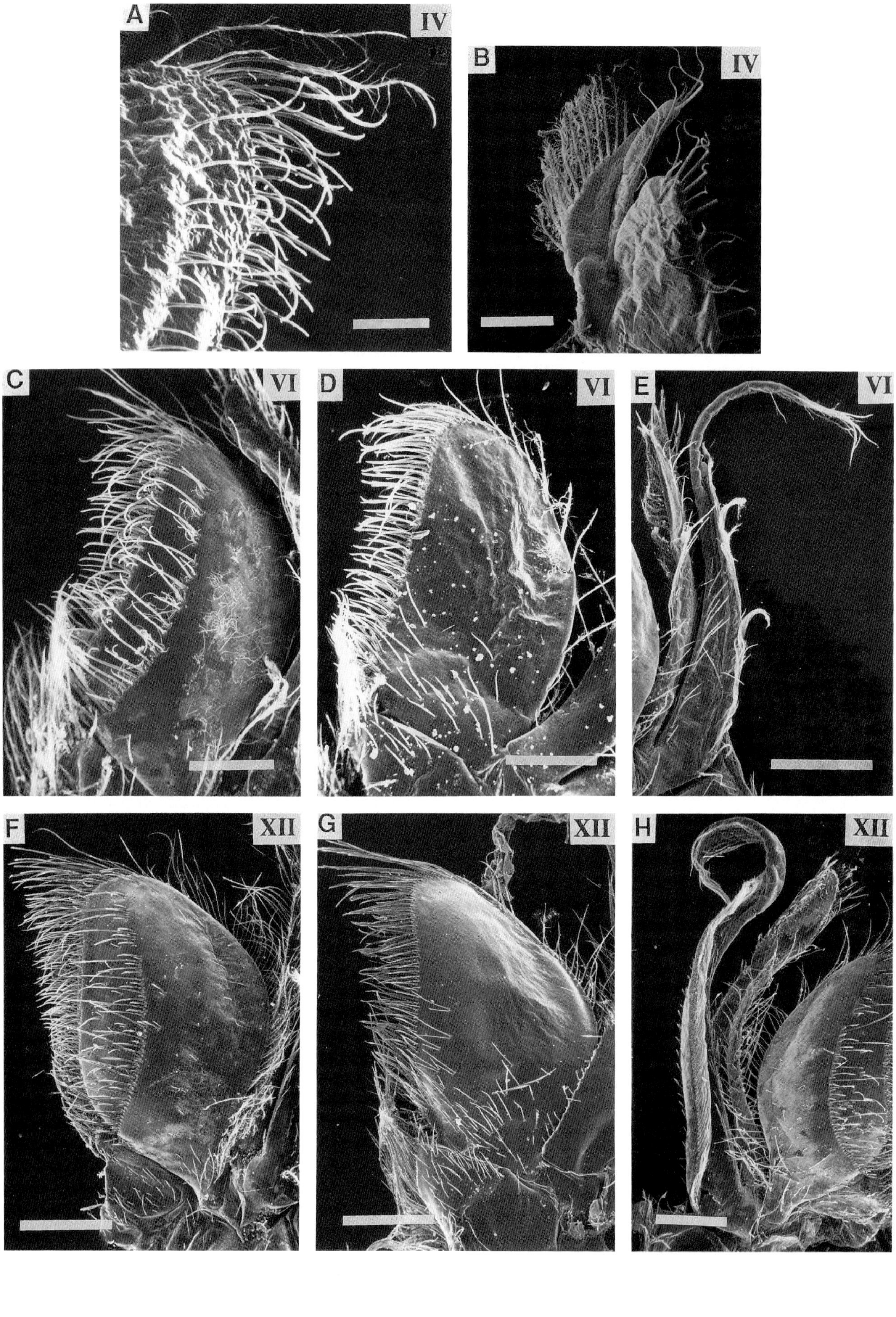
A
IV
B
IV
C
VI
D
VI
E
VI
F
XII
G
XII
H
XII

tive to overall size of the mouthparts, than those found on adult mouthparts, suggesting that suspension feeding is more important for the small juveniles than for later life history phases. However, suspension feeding may be used by adults in periods of poor food supply (*Homarus gammarus*, Loo *et al.*, 1993).

The specific capture techniques of stage IV–VI lobsters vary with the size of the prey item. Small zooplankters are most often captured by the third maxillipeds after being drawn toward the mouth region via the exopodite current (or, less frequently, the pleopod current). By waving both the third maxillipeds and the walking legs, the zooplankters are positioned to lie above the dactyls of the second maxillipeds and between the merus–ischium joint of the third maxillipeds. The three distal segments of the third maxillipeds (the dactyl, propus, and carpus) then sweep down over the prey, trapping it behind the mesh of setae, and push it toward the dactyls of the second maxillipeds (Fig. 17). The distal segments may sweep downward when prey strike a particular segment or are positioned elsewhere between the third maxillipeds; however, the zooplankton typically escape if the third maxillipeds unfurl. The walking legs often curl in toward the third maxillipeds during a downward sweep, presumably in case they are needed to contain the prey. Captures also result when a zooplankter touches the dactyls of the second maxillipeds, which quickly move together and push it toward the mouth. The third maxillipeds do not sweep downward in these cases.

The second maxillipeds appear to determine whether or not ingestion occurs. Once the captured item contacts the dactyls, the distal segments (the dactyl, propus, and carpus) can push accepted items inward toward the mouth or can pivot in a wide arc at the carpus–merus joint, flicking rejected material into the exopodite current (Fig. 18).

There seems to be a gradual, rather than abrupt, shift in feeding strategies from the planktonic, raptorial-feeding larvae and early postlarvae, to the benthic, raptorial- and suspension-feeding late postlarvae and shelter-restricted juveniles, and finally to the actively foraging emergent and vagile phase juveniles, adolescents, and adults. Suspension-feeding abilities may be reduced at the same time that the great chelae become capable of crushing hard foods. However, the capability of adult *Homarus gammarus* to feed on 600-μm particles (Loo *et al.*, 1993) raises the possibility that suspension feeding may persist into later life history phases in *H. americanus*.

FIGURE 10 First maxillipeds of *Homarus americanus*. Scanning electron micrographs of stages IV, VI, and XII. Basal endite (coxal endite may be visible at bottom left): (A) medial edge, outer surface, stage IV; (C) left outer surface, stage VI; (D) right inner surface, stage VI; (F) left outer surface, stage XII; (G) left inner surface, stage XII. Endopodite and exopodite: (B) stage IV; (E) left outer surface, stage VI; (H) right outer surface, stage XII. Scale bars: (A and B) 0.1 mm; (C and D) 0.25 mm; (E) 0.5 mm; (F–H) 1.0 mm. [(A) reprinted from Factor, 1977, and (B–H) reprinted from Lavalli and Factor, 1992, with permission.]

Understanding of the function of the larval mouthparts in processing whatever food is captured remains incomplete. For example, the larval mandibles of the Thalassinidea, Palinura, and Astacidea (including *Homarus americanus*) lack the very pronounced molar process that is typical of most other groups of decapods and is capable of truly masticating, grinding action in larvae that lack a gastric mill (Factor, 1989). Instead, the larval mandibles of *H. americanus* have a "pad" of setae (Fig. 4M), of uncertain function, where the postmetamorphic molar process will be located. Since the masticatory gastric mill does not develop until metamorphosis in *H. americanus* (Factor, 1981; see Factor, Chapter 15), the means of masticating food in larvae is not clear, although pads of setae in the cardiac stomach may function similarly to the gastric mill (Lavalli and Ayers, 1994). It is also possible that the pad of setae in the position of the molar process may contribute to mandibular food processing. The coordinated development of the mandibles and the third maxillipeds, however, emphasizes the coordinated manner in which they function and appears important in enabling the mechanism of food handling that is typical of the later juveniles and adults (Factor, 1978).

Surprisingly, few formal observations have been made of the feeding movements of adult mouthparts. The oldest is from Herrick (1909, p. 188), who placed the lobster upside down and stimulated the mouthparts with clam juice:

> . . . the plates of the first pair of maxillae come together over the lower posterior half of the mandibles. The movements of the masticatory parts of the second maxillae are synchronous with the beating of the scaphognathite. These leaf-like plates project somewhat obliquely over the convex surfaces of the jaws [mandibles], and are directed inward and slightly upward. The large plates of the first maxillipeds work up and down and at the same time inward toward the middle line, describing an ellipse. The second pair of maxillipeds move alternately or together, inward and outward, with slight up-and-down movement. The large [third] maxillipeds move together, the toothed margins [ischial teeth] meeting like the jaws of a nutcracker, while the three terminal joints are bent inward and somewhat downward, as in the case of the second maxillipeds, so as to meet on the middle line below and hold the food up to the mouth.

Herrick thought that "stiff hairs" (actually cuspidate setae) on the first maxillae and second maxillipeds might process softer food, but concluded that

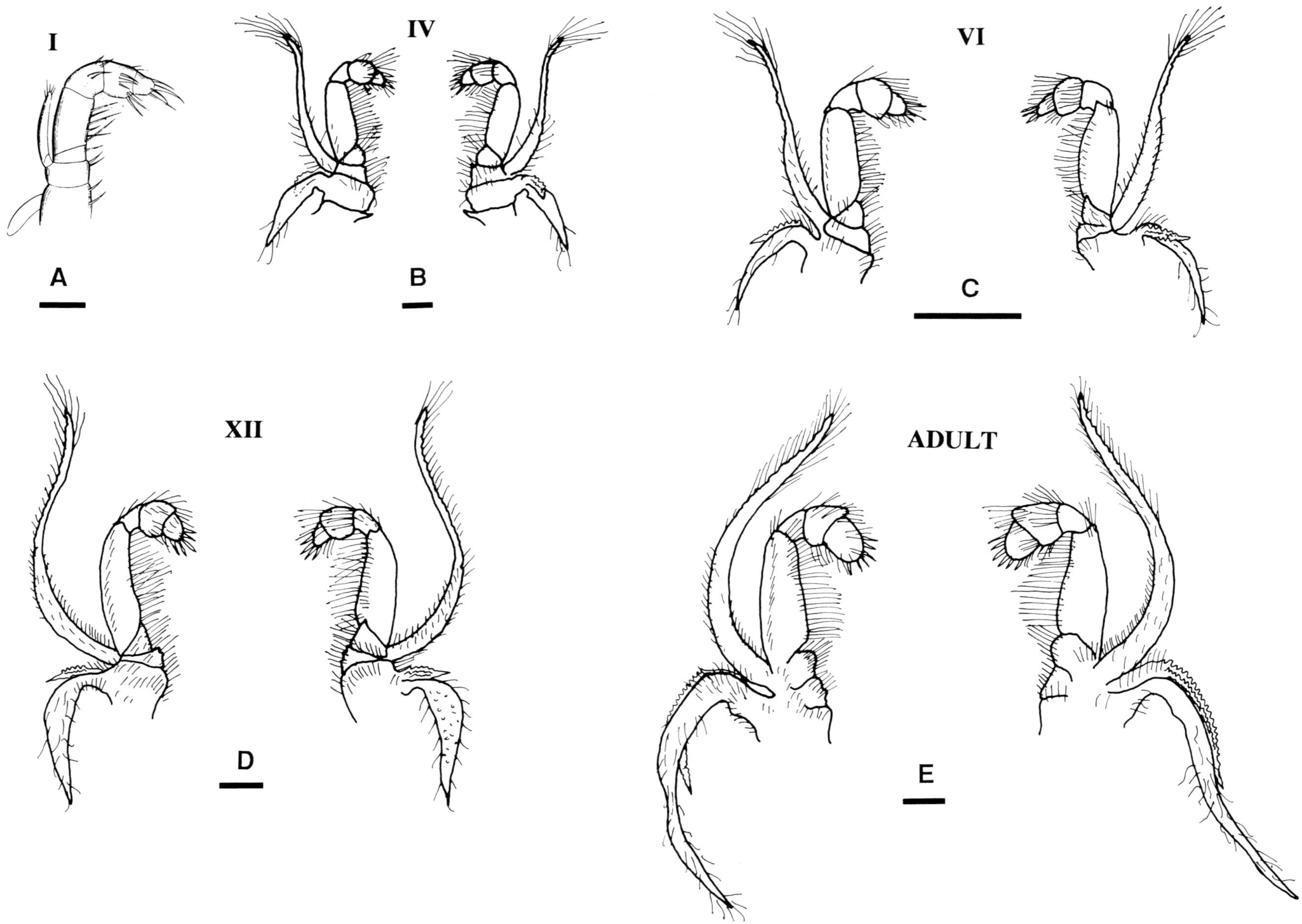

FIGURE 11 Second maxillipeds of *Homarus americanus*. Line drawings of stages I, IV, VI, XII, and adult. Right inner surface: (A) stage I. Right outer and inner surfaces: (B) stage IV; (C) stage VI; (D) stage XII; (E) adult. Scale bars: (A) 0.3 mm; (B) 0.25 mm; (C and D) 1.0 mm; (E) 4.0 mm. [(A) modified from Smith, 1873; (B) redrawn from Factor, 1978, and (C and D) reprinted from Lavalli and Factor, 1992, with permission.]

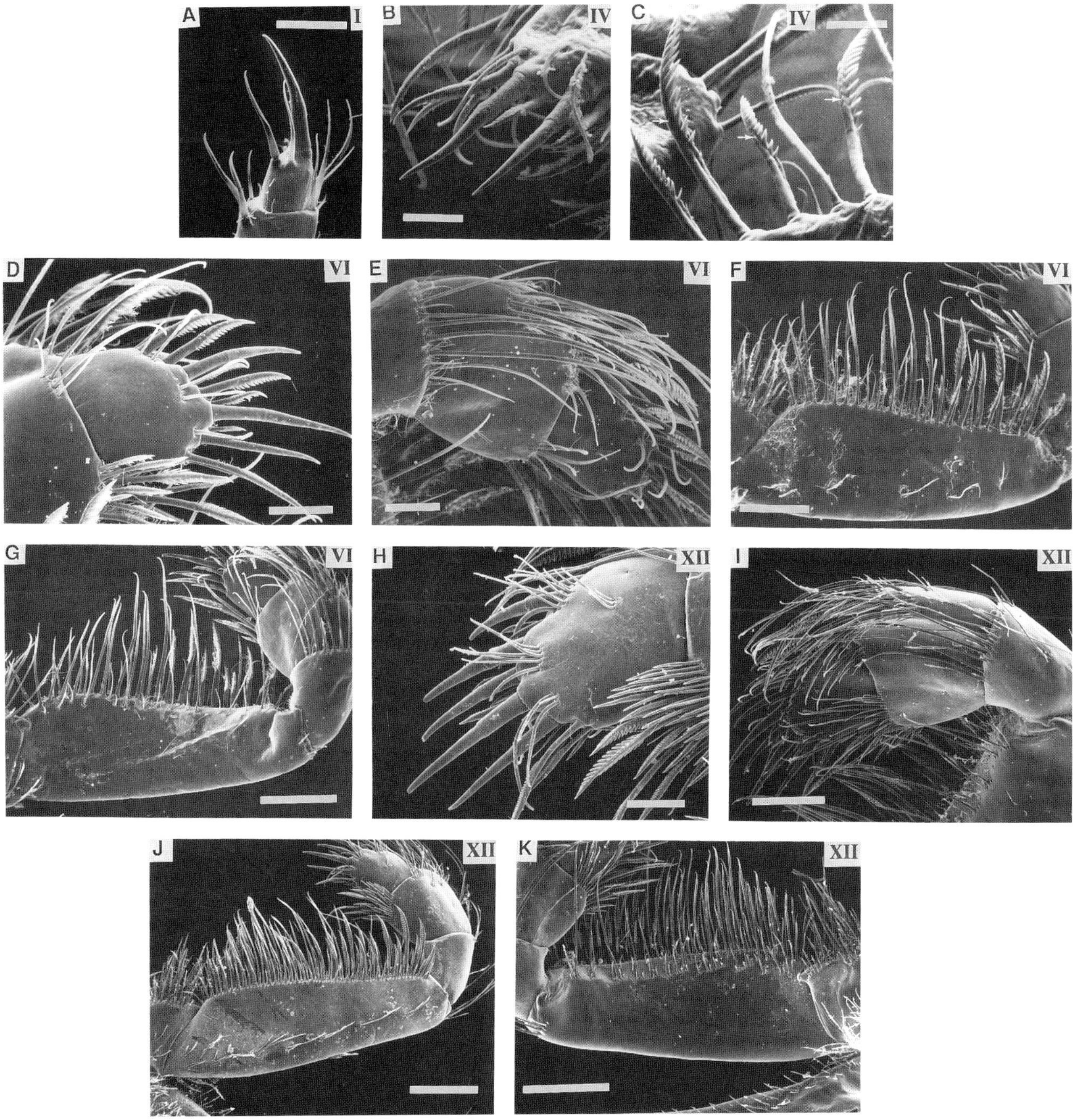

FIGURE 12 Second maxillipeds of *Homarus americanus*. Scanning electron micrographs of stages I–III, IV, VI, XII. Dactyl: (A) stage I (propus also visible); (B) stage IV; (D) right outer surface, stage VI; (E) left inner surface, stage VI; (H) left outer surface, stage XII; (I) right inner surface, stage XII. Merus: (C) medial edge, stage IV; (F) left outer surface, stage VI; (G) right inner surface, stage VI; (J) left outer surface, stage XII; (K) right inner surface, stage XII. Scale bars: (A) 0.1 mm; (B, C, and I) 0.5 mm; (D and E) 0.1 mm; (F–H) 0.25 mm; (J and K) 1.0 mm. [(B) reprinted from Factor, 1977, (A and C) reprinted from Factor, 1978, and (D–K) reprinted from Lavalli and Factor, 1992, with permission.]

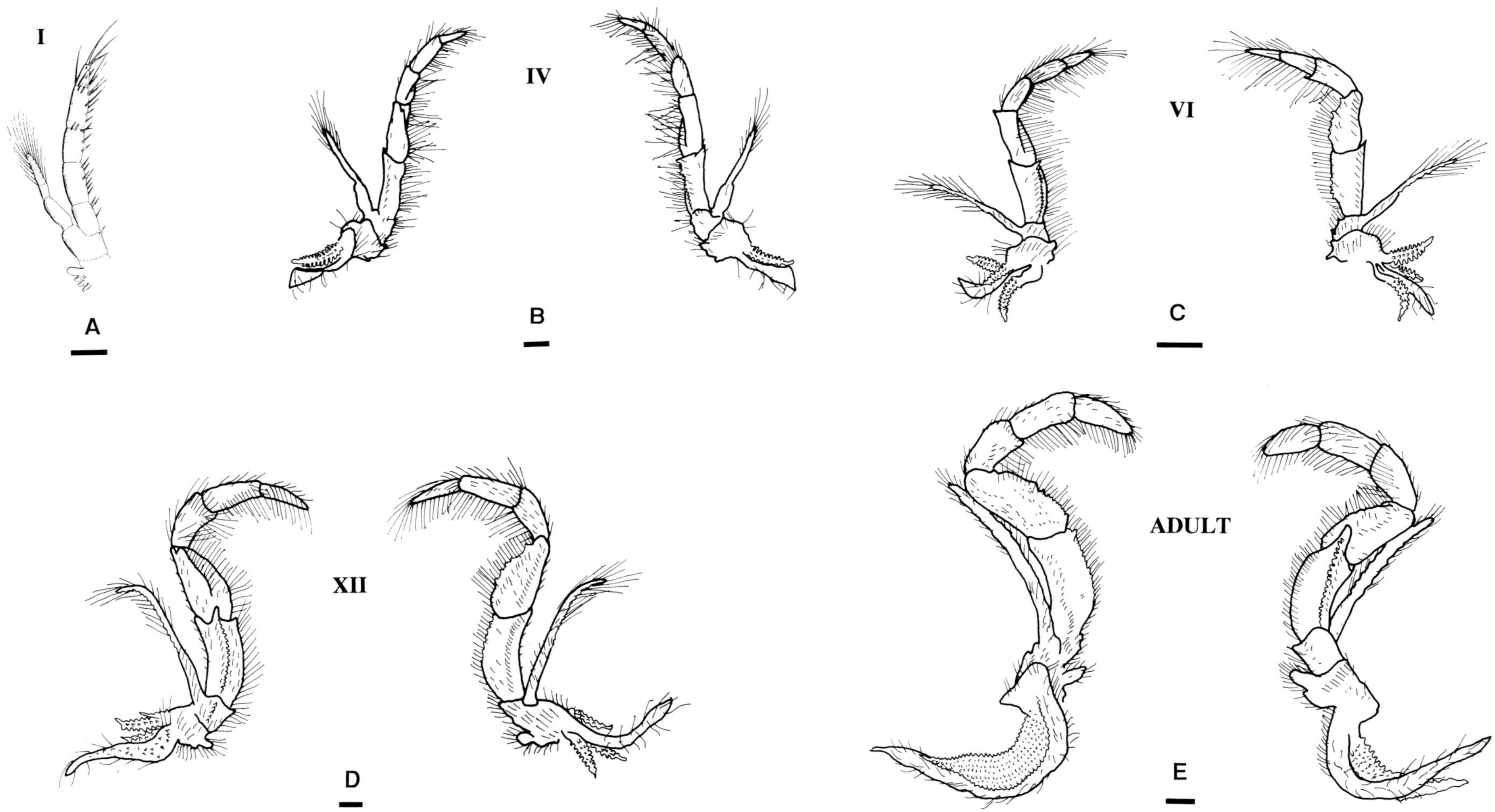

FIGURE 13 Third maxillipeds of *Homarus americanus*. Line drawings of stages I, IV, VI, XII, and adult. Right outer surface: (A) stage I. Right outer and left inner surfaces: (B) stage IV. Left inner and outer surfaces: (C) stage VI; (D) stage XII. Right outer and inner surfaces: (E) adult. Scale bars: (A) 0.4 mm; (B) 0.25 mm; (C and D) 1.0 mm; (E) 4.0 mm. [(A) modified from Smith, 1873; (B) redrawn from Factor, 1978, and (C and D) reprinted from Lavalli and Factor, 1992, with permission.]

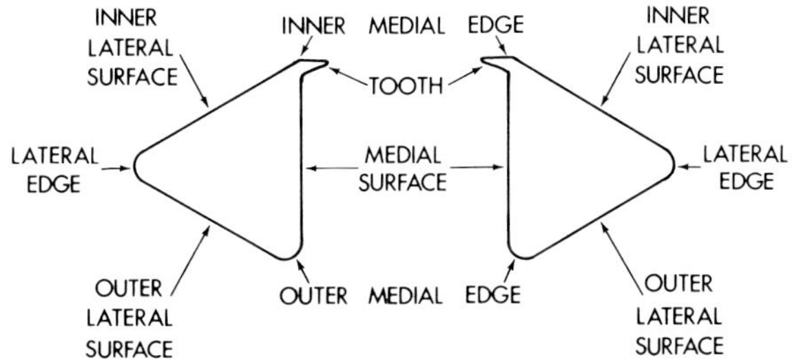

FIGURE 14 Diagrammatic representation of a cross-section through the paired ischia of the third maxillipeds. The section illustrates the various surfaces and edges of the ischium and the position of the rows of ischial teeth. (Reprinted from Factor, 1977, with permission.)

only the third maxillipeds and the mandibles are important in the reduction of food prior to ingestion (Herrick, 1909).

Observations of adult *Homarus gammarus* provide another description of feeding (Barker and Gibson, 1977). The three distal segments of the third maxilliped (the dactyl, propus, and carpus) are raised, food is placed between the two ischial segments, and the distal segments fold over the food to press it against the ischial segments. As the third maxillipeds move the food toward the mandibles, the first and second maxillipeds and the first and second maxillae move aside laterally. The mandibles then grip the food at one end, while the ischial teeth of the third maxillipeds grasp the food at the other end. Downward movement of the third maxillipeds through an arc of ~90° causes the food to be stretched and torn. The second maxillipeds push the food down, while the first maxillipeds move medially to keep the food centered. The first and second maxillae then move the food toward the mouth. During this process, the mandibular palps move up and down, positioning the food and pushing it into the mouth. Barker and Gibson (1977) viewed the role of the mandibles to grasp food during the stretching and tearing accomplished by the third maxillipeds, rather than to masticate food. Factor (1978) observed similar use of the mandibles and third maxillipeds in adult *H. americanus*.

The mouthparts are also used to manipulate and examine hard food items, such as mussels and snails. The prey is picked up by the chelate walking legs, the third maxillipeds, or both, and moved toward the mandibles, which may try to close down on the item. Hard items are often rotated repeatedly by the third maxillipeds (Derby and Atema, 1982). The item may be rejected or passed to the chelae for crushing, then returned to the third maxillipeds for further processing and examination (Lee, 1994). The mandibles then bite down on one end of the food, while the third maxillipeds pull and tear the other end or remove flesh from shell fragments (Elner and Lavoie, 1983; Lee, 1994). This behavior has been referred to as *eat* (Derby and Atema, 1982) and *rip* (Lee, 1994). If the chemoreceptors of the third maxillipeds are experimentally rendered nonfunctional, food is often dropped prior to crushing or, if crushed, not consumed. Thus, the mouthparts seem important as taste organs (via contact chemoreception of setae), providing information that influences ingestion (Derby and Atema, 1982).

The movements of the mandibles and second maxillipeds of adults are similar to those of postlarvae and juveniles. However, adult third maxillipeds are used to process food, while postlarval and juvenile third maxillipeds are used to capture prey. Furthermore, while the setae are still extremely important for food assessment, they may become less important in the capture and retention of prey in the adult. These structural and behavioral changes in the mouthparts accompany the dramatic change in diet from small, planktonic organisms to soft and hard, large, benthic organisms that require crushing and tearing of flesh.

III. Walking Legs and Claws

A. Generalized Structure and Orientation

The remaining five thoracic appendages, the pereiopods, are similar to the maxillipeds in that they consist of a two-segmented protopodite (coxapodite and basipodite) from which branches arise distally and proximally (Fig. 19). Larval pereiopods are biramous and bear a distal lateral branch, the exopodite (used for swimming), and a distal medial branch, the endopodite (used for manipulating food). The endopodite is segmented (distal to proximal) into the dactyl, propus, carpus, merus, and ischium.

The first three pairs of pereiopods are chelate from stage I and the first pair is destined to become the large claws (great chelae) of the adult. While they are symmetrical in the larvae, postlarvae, and shelter-restricted phase juveniles (through stage V or sometimes VI), they develop asymmetrically into crusher and cutter claws during the first year (see Govind, Chapter 12). The coxae of the third and fifth pereiopods bear the genital pores in the female and the male, respectively (Schram, 1986; Talbot and Helluy, in Chapter 9, discuss reproduction).

B. Setae of the Walking Legs and Claws

1. Types of Setae

The setae of the pereiopods of *Homarus americanus* are described in Table 3 and illustrated in Fig. 20.

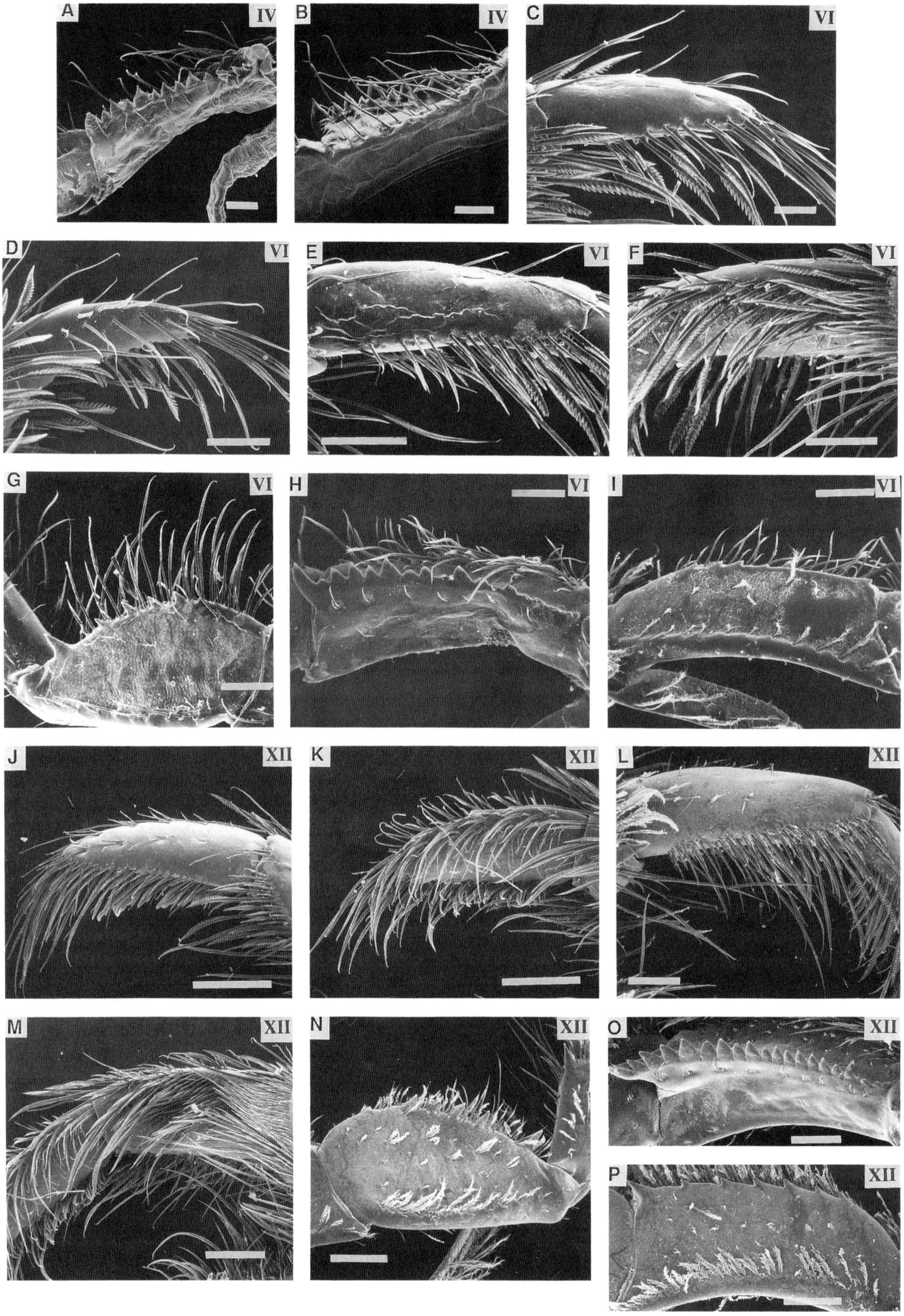
A
IV
B
IV
C
VI
D
VI
E
VI
F
VI
G
VI
H
VI
I
VI
J
XII
K
XII
L
XII
M
XII
N
XII
O
XII
P
XII

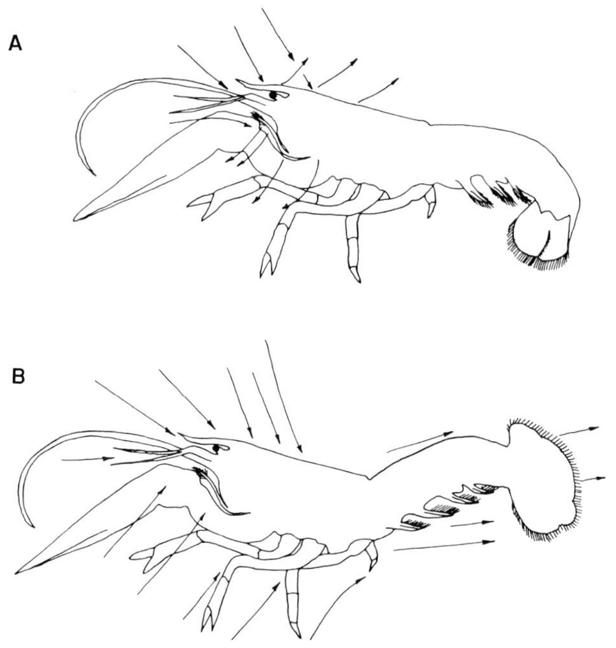

FIGURE 16 Flow fields of exopodite and pleopod currents of stage IV–V *Homarus americanus*. (A) Exopodite current. Arrows indicate water flow toward and under the head and out along the sides. (B) Pleopod current (with exopodite current). Arrows indicate water flow from front to back, laterally and ventrally. (Reprinted from Lavalli, 1992, with permission.)

Some types are similar or identical to setae found on the mouthparts.

2. Functions of Setae

Smooth, squamous, and serrate setae (Fig. 2, types I, L, and D, respectively; Fig. 20) are chemically and mechanically sensitive. Tufts of these setae on the dactyls of the pereiopods are purportedly mechanosensory and rows are chemosensory in *Homarus gammarus* (Shelton and Laverack, 1968, 1970). As mechanoreceptors, they are sensitive to both direct touch and water vibrations. As contact chemoreceptors, the smooth and squamous setae help locate food while the lobster is probing the substrate; however, they are also distance chemoreceptors (Derby, 1982; Derby and Atema, 1982). The presence of serrate setae on the last two pairs of walking legs suggests a grooming function. In ovigerous females, this combined grooming and chemosensory function may be important for distinguishing eggs from fouling material (Derby, 1982).

Similarly, hedgehog hairs (also called fringed setae; Figs. 20F and 24) are both contact chemo- and mechanoreceptors. As mechanoreceptors, however, they are sensitive only to bending and direct touch, not water vibrations. They may also be involved in grooming and grasping, as well as chemically sensing food (Derby, 1982, 1989).

In contrast, the peg sensilla (type I–IV hair organs of *Homarus americanus*, Solon and Cobb, 1980; hair fans of *H. gammarus*, Laverack, 1962a,b, 1963) seem to be only mechanoreceptive. Their external morphology varies greatly from species to species, but all lie within pits and are dually innervated to permit bidirectional sensitivity to water currents, pressure waves, or direct touch (Laverack, 1962a,b, 1963; Derby, 1982). The type I hair organ lies within an asymmetrical depression with a small bump at one end (Fig. 20G). The proximal bump and shaft insertion (Solon and Cobb, 1980) imply a preferential directional sensitivity. The type II hair organ lacks the bump (Fig. 20D). Type I responds tonically to stimuli, while type II responds phasically (Solon and Kass-Simon, 1981). The distribution of these organs and the orientation of the setae is similar to the hydrodynamic receptors of *Palinurus vulgaris* (Vedel and Clarac, 1976), suggesting near-field vibration detection. Type III and IV hair organs (Fig. 20D and I) have a less obvious role. Clusters of type IV hair organs are called scutes and correspond to the superficial cuticular limit built by each epidermal cell, and the elaborations within scutes are called scutelles. Since they apparently lack innervation, they may represent cuticular sculpturing rather than sensory organs (Derby, 1982).

Cuticular articulated peg (CAP) organs (Fig. 20H) are located only in the joints and are external stretch receptors. When the joint flexes, the joint membrane stimulates the CAP organ (Oakley and Macmillan, 1980; this function has been demonstrated only in *Jasus novaehollandiae*). Similarly, funnel canal organs typically provide information about cuticular strain and are the functional equivalents of campaniform sensilla in insects. They are not easily visualized by scanning electron microscopy and are obvious only through histological sectioning. Funnel canal organs have been found on the tips of the walking legs of *Homarus gammarus* (Shelton and Laverack, 1968); they

FIGURE 15 Third maxillipeds of *Homarus americanus*. Scanning electron micrographs of stages IV, VI, and XII. Ischium: (A) inner medial edge, stage IV; (B) outer medial edge, stage IV; (H) left inner surface, stage VI; (I) left outer surface, stage VI; (O) left inner surface, stage XII; (P) right outer surface, stage XII. Merus: (G) right outer surface, stage VI; (N) left outer surface, stage XII. Propus: (E) right outer surface, stage VI; (F) right inner surface, stage VI; (L) right outer surface, stage XII; (M) right inner surface, stage XII. Dactyl: (C) right outer surface, stage VI; (D) left inner surface, stage VI; (J) left outer surface, stage XII; (K) right inner surface, stage XII. Scale bars: (A–C) 0.1 mm; (D–I) 0.25 mm; (J–P) 1.0 mm. [(B) reprinted from Factor, 1977, (A) reprinted from Factor, 1978, and (C–P) reprinted from Lavalli and Factor, 1992, with permission.]

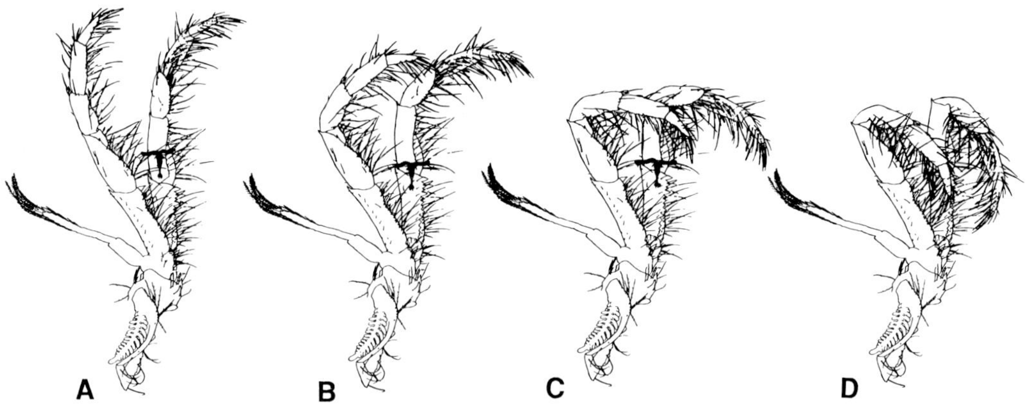

FIGURE 17 The capture of a copepod by the third maxillipeds (stages IV–VI). (A) normal position prior to capture sequence; (B and C) downward sweep of the three distal segments; (D) final position, trapping the copepod behind the mesh of setae. (Reprinted from Lavalli, 1992, with permission.)

may also be present in *H. americanus*, but may not be readily visible because they lie in sunken pits (Derby, 1982).

C. Structure and Development of the Pereiopods

Few studies have been concerned with developmental changes of the pereiopods of *Homarus americanus*. The account that follows is based on studies by Smith (1873, on gross morphological changes in larvae), Solon and Cobb (1980, on distribution of the peg sensilla in juveniles and subadults), Herrick (1909, on the great chelae), and Derby (1982, on setae of adults).

In the larvae, the first pereiopods are oriented such that they hang downward relative to the body of the animal. Thus, the dactyl and the propus have medial and lateral surfaces, as well as anterior- and posterior-facing edges. The dactyl is the movable finger of the chela and opens anteriorly, while the distal (or digital) portion of the propus is the fixed finger (Fig. 21).

The left and right first pereiopods of stage I larvae are symmetrical, not much longer than the third maxillipeds, and imperfectly subcheliform and have no power of prehension (Fig. 22A). The endopodite is somewhat stouter than that in the second and third pereiopods, but not much longer. All segments, except the coxa, are fringed with slender spines along their medial edge. Only the carpus, propus, and dactyl bear slender spines along all edges. The propus has two portions: the basal portion is as long as the merus and ends abruptly at the dactyl articulation; the distal portion is shorter than the dactyl and tapers rapidly to a short, spinelike tip (Fig. 22B). The posterior-facing edge ("inferior side" of Smith, 1873) bears several pairs of serrate setae; one serrate seta is found on either side of the base. Two or three teeth appear along the articulating edge. The dactyl also tapers rapidly to a tip and terminates in a spine nearly as long as the segment itself (Fig. 22A and B). Both of its edges bear several small, slender spines (Fig. 22B). The exopodite of all pereiopods is similar in

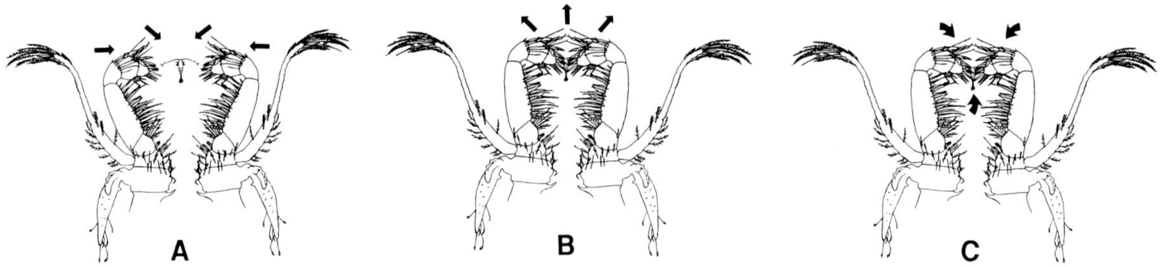

FIGURE 18 Rejection and acceptance by the second maxillipeds (stages IV–VI). (A) Normal position of the second maxillipeds prior to movement. (B) Rejection: upward and outward movements of the three distal segments, flicking the rejected item into the exopodite current. (C) Acceptance: downward and inward movements of the three distal segments, pushing the particle toward the mouth. (Modified from Lavalli, 1992, with permission.)

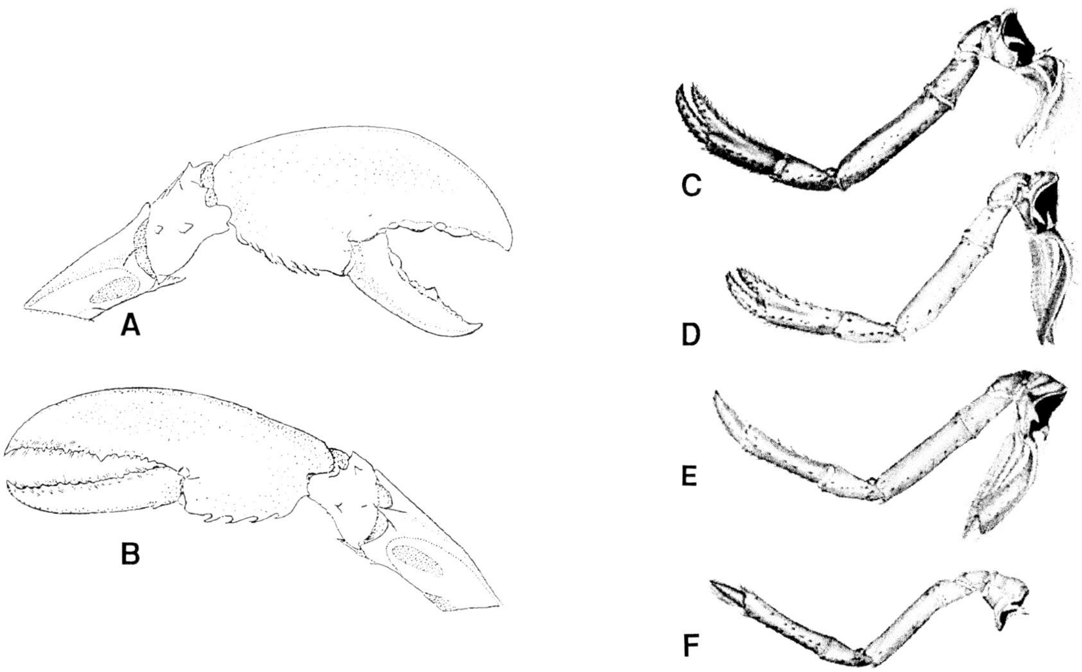

FIGURE 19 General plan of adult pereiopods. (A) First pereiopod, crusher claw; (B) first pereiopod, cutter claw; (C) second pereiopod; (D) third pereiopod; (E) fourth pereiopod; (F) fifth pereiopod. (Reprinted from Herrick, 1909.)

TABLE 3 **Description of Types of Setae Found on the Pereiopods of *Homarus americanus*[a]**

Setal category	Description
Smooth	Shaft surface smooth, although there may be annulations or some grooves at the base (Fig. 20A and B); identical to "simple" setae (type I, Fig. 2) of mouthparts
Squamous	Similar to smooth setae, but the distal portion of the shaft bears scales (Fig. 20C); similar to "multiscaled" setae (type L, Fig. 2) of mouthparts
Serrate	Distal portion of the shaft bears two lateral rows of tooth-shaped setules (Fig. 21K; identical to type D1, Fig. 2) or a smooth ridge along the back of the shaft (Fig. 20J); can also have scalelike setules on the back of the shaft (Fig. 20L and M; identical to type D2, Fig. 2)
Hedgehog hairs	Stout, conical hairs with numerous spines extending from one surface (Fig. 21F); resemble cuspidate setae when viewed from one side; identical to "squat" hairs in *Homarus gammarus* (Shelton and Laverack, 1970)
Peg sensilla	Type I hair organ: long, erect peg with fine rami near the distal end; lies within an asymmetrical depression, often with a single knob at one end of the depression (Fig. 21G)
	Type II hair organ: short, central rod from which several fine branches extend in the same direction as the shaft; projects from the joint in the center of the depression (Fig. 21D and E)
	Type III hair organ: small tuft of fine fibers projecting from the central depression within the elevated mound
	Type IV hair organ: conical hairs arranged in circular clusters scattered across the cuticular surface in a gridlike pattern
	Hair fans: small, globular base covered by hairs arranged like petals on a flower bud; only described for *H. gammarus* (Laverack, 1962a,b, 1963; Shelton and Laverack, 1970)
Cuticular articulated peg (CAP) organs	Groups of simple, peg-shaped hairs (Fig. 21H); located only within joints
Funnel canal organs	Small, knob-shaped hairs; sunken into pits

[a]Based on the work of Solon and Cobb (1980) and Derby (1982).

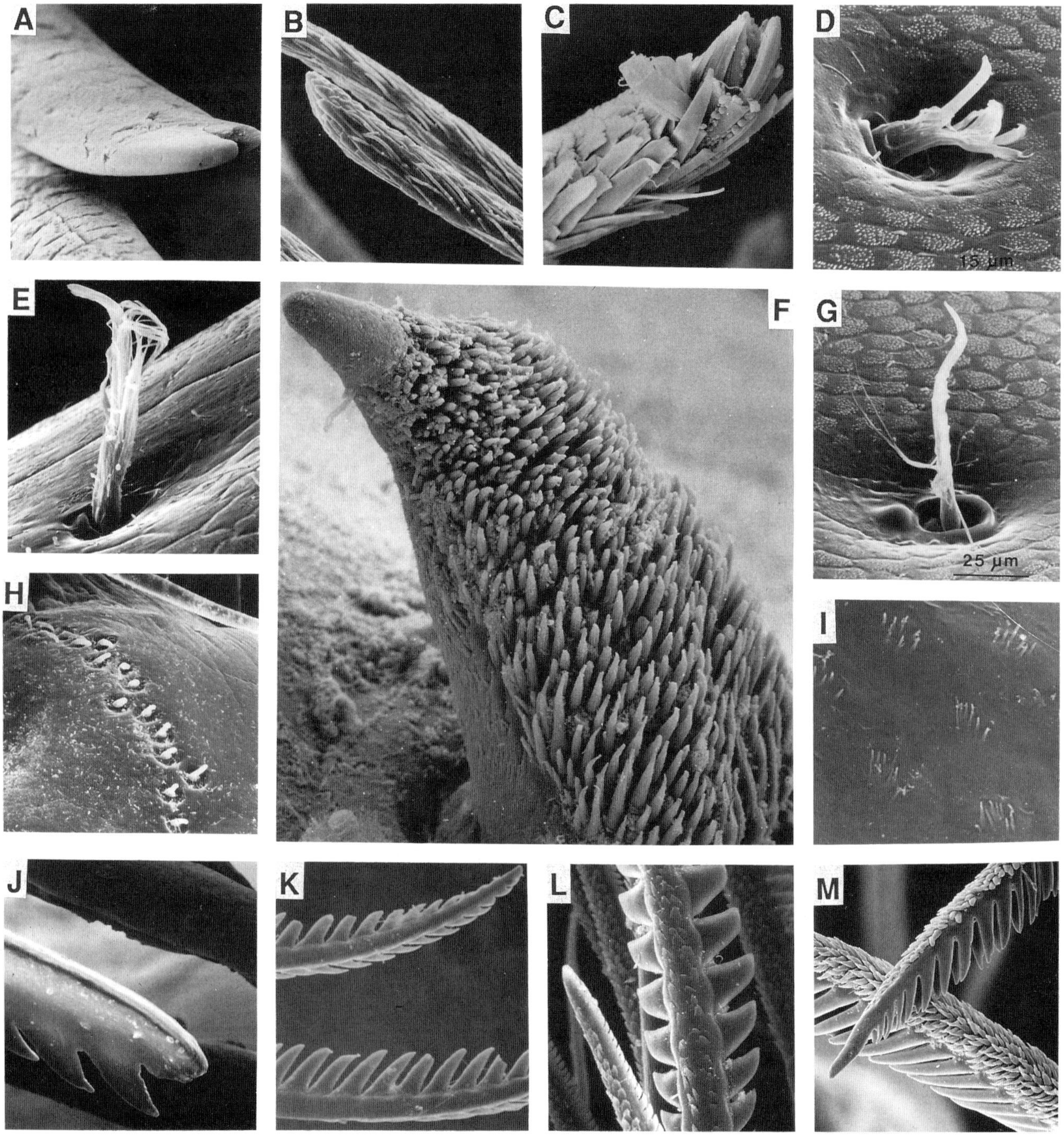

FIGURE 20 Setae of the pereiopods of *Homarus americanus*. Scanning electron micrographs of the adult. (A) Smooth setae, dactyl, fourth pereiopod; (B) squamous setae, dactyl, fifth pereiopod; (C) setules, squamous setae; (D) type II peg sensillum, propus, crusher claw (type IV peg sensilla on the cuticular surface); (E) type II peg sensillum, propus, fifth pereiopod; (F) hedgehog hair, inner edge of chela, second pereiopod; (G) type I peg sensillum, dactyl, crusher claw (note the knob near the rim of the depression); (H) cuticular articulated peg organ, propus–carpus joint, second pereiopod; (I) cuticular sculpturing reminiscent of a scutelle (type IV hair organ); (J) serrate setae with smooth ridge on back of shaft, propus, fifth pereiopod; (K) serrate setae with smooth back of shaft, dactyl, third maxilliped; (L) scales on serrate setae; (M) serrate setae with scales on the back of the shaft, propus, fifth pereiopod. (Modified from Derby, 1982, with permission.)

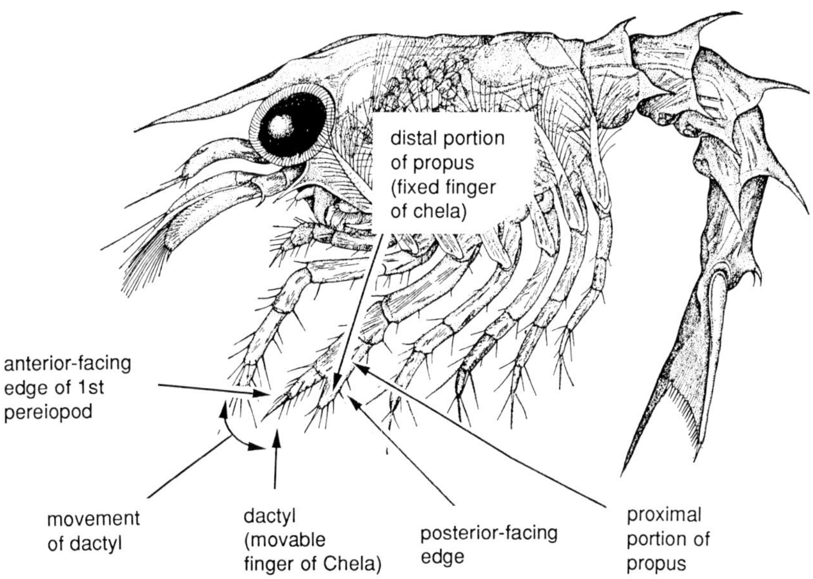

FIGURE 21 Stage I larva showing the orientation of the pereiopods (resting position). (Modified from Herrick, 1909.)

form to that of the maxillipeds, but the flagellum is slightly longer and is composed of 10 "segments" that bear long plumose hairs; this does not change during the larval stages.

The second and third pereiopods are identical and nearly as long as the first, but more slender (Fig. 22D, E, and G). The propus is not much stouter than the carpus and bears fewer spines; the dactyl here is more slender than that of the first pereiopods and is terminated by a longer spine (Fig. 22D). The fourth and fifth pereiopods are styliform (nonchelate and pointed) and more slender than the second or third pereiopods (Fig. 22C). The propus bears long, serrate spines on the medial edge near the dactyl. The dactyl terminates in a long, serrate spine nearly twice as long as the segment itself. The endopodite of the fifth pereiopod is even more slender than that of the fourth. The propus is proportionately longer and the dactyl terminates in a slender, serrate spine more than twice as long as the segment itself (Fig. 22C).

By stage II, all of the pereiopods have increased proportionately in size and the spines borne on their segments have decreased in length (Fig. 22E, F, and I). The first pereiopods are truly cheliform, with the proximal and distal portions of the propus nearly equal in length (Fig. 22F). The distal portion has an incurved tip that terminates in a short, slender "nail" (perhaps a cuspidate seta) and three to five teeth along the edge articulating with the dactyl (Herrick, 1909). The dactyl is slightly longer than the distal portion of the propus and seemingly has some power of prehension (Fig. 22F). The second and third pereiopods are alike (Fig. 22E); the spines of the propus are shorter and the distal portion is elongated and incurved and terminates in a short nail (perhaps a cuspidate seta). The dactyl is more slender, longer than the distal portion of the propus, and terminates in a serrate spine considerably shorter than that in stage I. The fourth and fifth pereiopods bear shorter spines on the propus and the dactyl (Fig. 22I).

In stage III, the first pereiopods are much longer and stouter than the other pereiopods and differ conspicuously (Fig. 22G and H). The propus is broad and stout with a convex upper (anterior-facing) edge and a nearly straight lower (posterior-facing) edge. The distal portion of the propus is nearly two-thirds longer than the proximal portion and tapers to an obtuse tip. The dactyl curves downward toward the tip. The exopodite is smaller and bears shorter plumose hairs. The endopodites of the second and third pereiopods are now truly cheliform (Fig. 22H); the propus bears only short spines, but still retains the serrate setae located on either side of the base of the dactyl. The dactyl projects slightly beyond the propus, is toothed along the inner edge, and terminates in a slender tip. The endopodites of the fourth and fifth pereiopods are of the same length and similar to the second and third pereiopods; the spines are shorter on the propus and the dactyl (Fig. 22I).

Upon metamorphosis to the postlarva (stage IV), the exopodites are lost, appearing only as stumps that disappear completely after stage V (Fig. 22J). The first

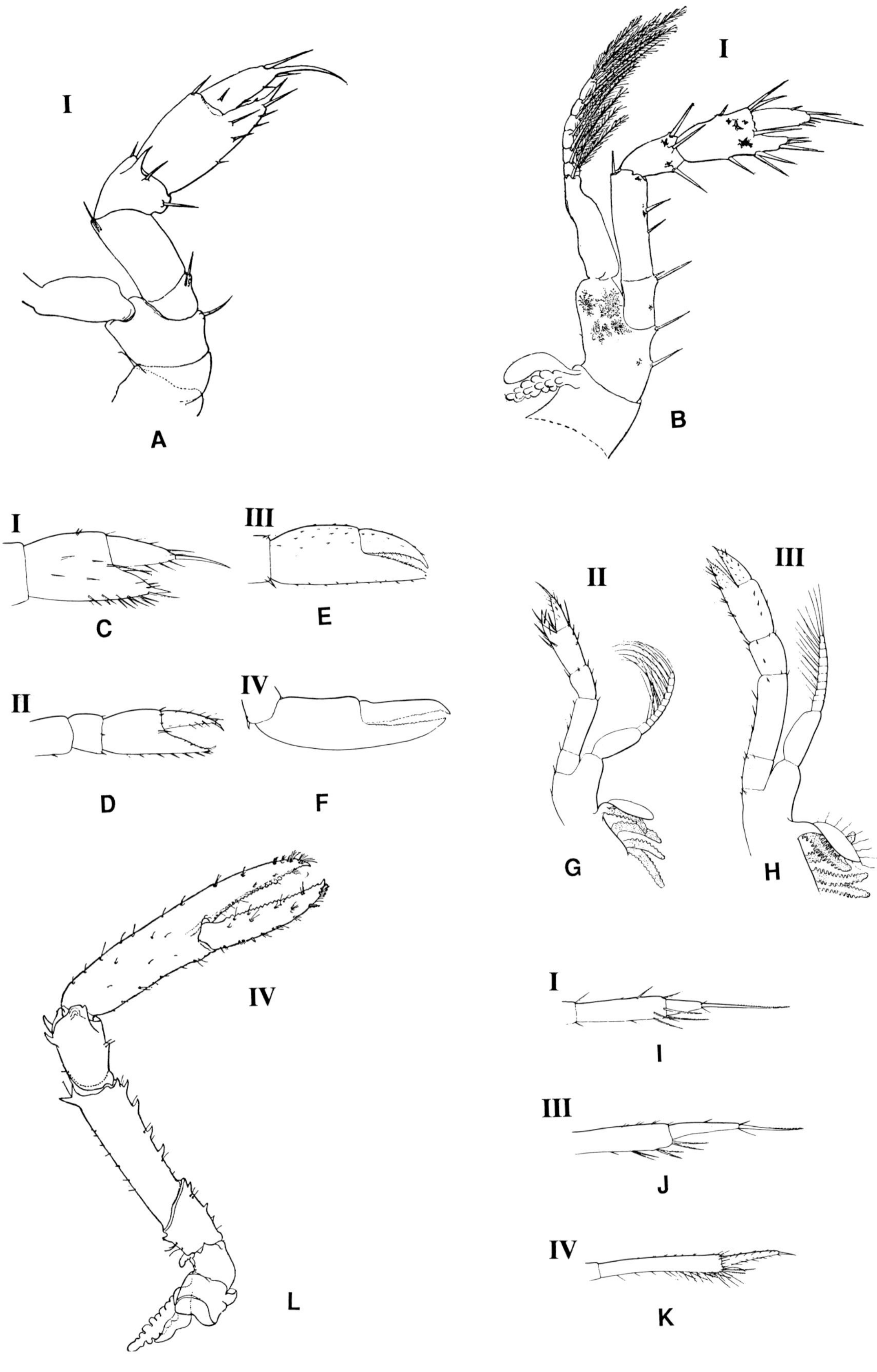
I
A
I
B
I
C
III
E
II
D
IV
F
II
G
III
H
IV
L
I
I
III
J
IV
K

pereiopods become very conspicuous and bear long, slender, prehensile chelae (Fig. 22J and K). Unlike the first pereiopods of the larvae, in which the chela opens anteriorly, the first pereiopods of the postlarvae are rotated so that the dactyl of the chelae opens medially, in a nearly horizontal plane (Fig. 23), giving the orientation of all subsequent stages. This rotation is due to a torsion of the carpus (Herrick, 1909). The left and right appendages rotate in opposite directions so that both dactyls open medially. The dactyl and the propus of the fifth pereiopods are also rotated, but in a direction opposite that in the first pereiopods. This torsion and the placement of setae allow the fifth pereiopods to act as grooming instruments for the abdominal pleopods and may be especially important to females brooding eggs (Herrick, 1909). The second through fourth pereiopods do not undergo torsion.

In the adult, smooth and squamous setae are found on the great chelae and second through fifth pereiopods. They border the cutting edges of both crusher and cutter claws, but are less numerous on the crusher. Typically, they occur in rows and tufts on the dactyl and the propus and as tufts on most other segments of all pereiopods (Fig. 24A). Smooth setae are more common than squamous setae. Serrate setae are located on the fourth and fifth pereiopods, where they occur as three tufts on the propus along the propus–dactyl joint, and as three rows only on the dactyl of the fifth pereiopods (Fig. 24B). Hedgehog hairs occur in rows along the cutting edges of the chelae of the second and third pereiopods and on the fifth pereiopods near the propus–dactyl joint. Type II peg sensilla are widely distributed over the ventral and dorsal surfaces of the great chelae (Solon and Cobb, 1980) and on all segments of the second through fifth pereiopods, except the dactyl (Derby, 1982). Type I peg sensilla are found only on the great chelae (Derby, 1982). Type III peg sensilla are also found on the ventral and dorsal surfaces of the great chelae, but their distribution on the other pereiopods has not yet been examined (Solon and Cobb, 1980). CAP organs are found on the distal side of all joints of all pereiopods (Derby, 1982).

FIGURE 22 Pereiopods of *Homarus americanus*. Line drawings of stages I–III and IV. First pereiopod: (A) left, stage I, showing the anteriorly opening dactyl before torsion; (L) right, stage IV, showing the loss of the exopodite and torsion of the carpus. Dactyl and propus of the first pereiopod, showing the position of the setae and spines: (C) stage I; (D) stage II; (E) stage III; (F) stage IV. Second pereiopod, before loss of the exopodite: (B) left, stage I; (G) stage II; (H) stage III. Styliform pereiopods, distal end: (I) stage I; (J) stage III; (K) stage IV. (Reprinted from Smith, 1873, and Herrick, 1909.)

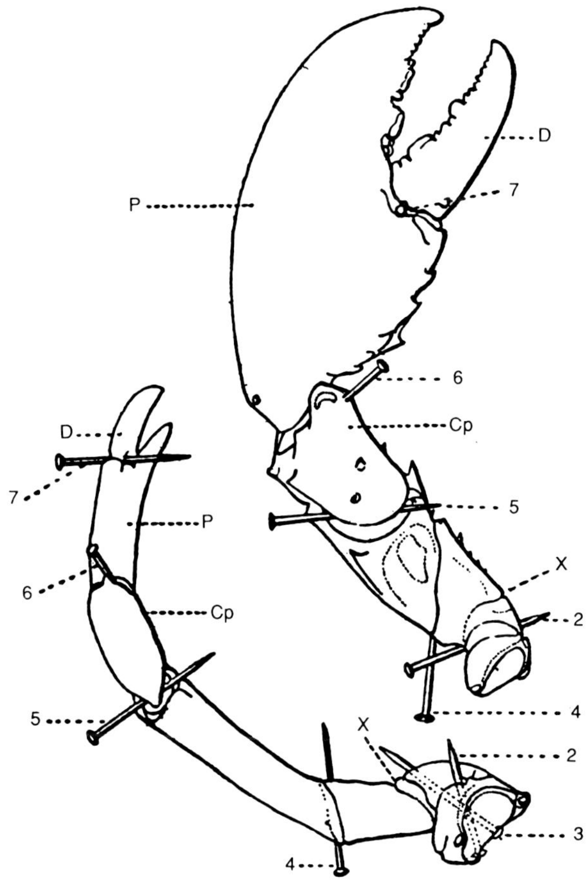

FIGURE 23 First pereiopod (crusher claw) and third chelate pereiopod of adult *Homarus americanus*, with pins inserted in the axes of articulation. Comparison of the first (rotated) and third (unrotated) pereiopods illustrates the effect of torsion. Cp, Carpus; D, dactyl; P, propus; X, ischium-basis joint. (From Herrick, 1909.)

D. Function of the Pereiopods

Although the first pereiopods (great chelae) of larval lobsters seem incapable of prehensile action (Smith, 1873; Herrick, 1909), the second and third pereiopods and third maxillipeds are capable of gripping zooplankton prey (K. L. Lavalli, unpublished observations). The feeding mechanisms of the larvae, however, have not been examined in detail. In contrast, presettlement postlarvae are well-directed swimmers (Ennis, 1986; Cobb *et al.*, 1989; Ennis, Chapter 3) and have been observed to pick at patches of foam and particulate surface matter with their chelipeds (Cobb *et al.*, 1983).

Settled postlarvae and shelter-restricted phase juveniles use both mouthparts and walking legs to capture zooplankton, with the size of the zooplankter determining the mode of capture (Lavalli, 1992; Lavalli and Trager, 1995). With small zooplankton, "walking leg assists" typically occur when a zooplankter strikes a particular leg. If extended, the

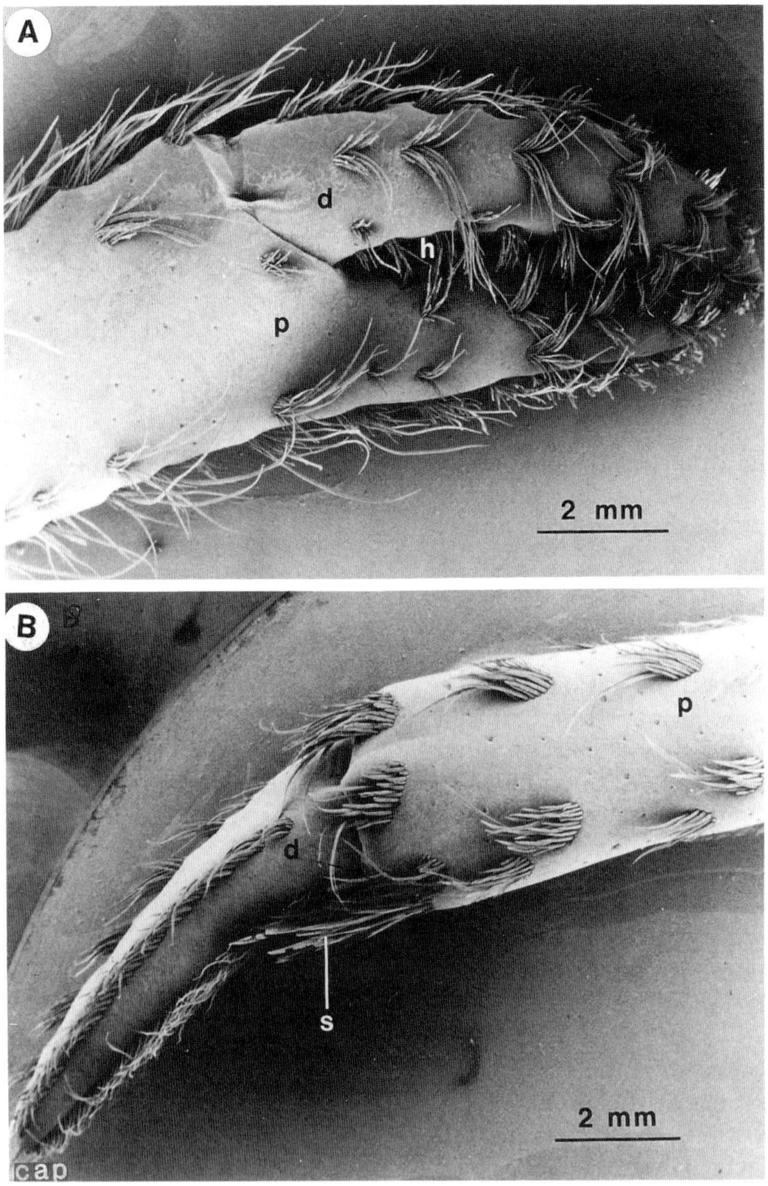

FIGURE 24 Pereiopods of *Homarus americanus* dactyl. Scanning electron micrographs of adult. (A) Second pereiopod, chelate; (B) fourth pereiopod, styliform. Note the tufts and rows of long setae, serrate setae (s), hedgehog hairs (h) along the inner cutting edges of the chela, epicuticular cap (cap) at the tip of the dactyl, and small depressions containing peg sensilla on the propus (p), but not on the dactyl (d). (Reprinted from Derby, 1982, with permission.)

walking leg curls and guides the prey toward the mouth. Some zooplankters initiate escape maneuvers at this point. If the walking leg is already near the mouth, it merely pushes the prey directly into the mouth region. Once delivered, the mouthparts manipulate the prey as described in Section II,D.

Direct captures by the chelate walking legs are most common with larger zooplankters and occur only when zooplankters touch or move near the chelae (Lavalli, 1992; Lavalli and Trager, 1995), which then extend and grasp the prey. The prey is delivered to the second maxillipeds and the third maxillipeds sweep downward to entrap the prey within the setal mesh. The walking leg then withdraws and the third maxillipeds remain furled for several moments.

Surprisingly, captures effected by the great chelae are extremely rare and have been observed only when large copepods are positioned immediately beside or in front of the claw (Lavalli, 1992; Lavalli and Trager, 1995), which then clamps down on it.

After capture, the cheliped moves above the dactyls of the third maxillipeds, which scrape at the copepod, attempting to pull it into the mouth area, and the chelate second and third pereiopods attempt to pull the copepod from the claw. Finally, using both left and right chelate walking legs, the copepod is delivered to the second maxillipeds, where it is handled as described in Section II,D.

Adult lobsters typically search for food while walking high on their legs to probe the substrate (Derby and Atema, 1982). Upon contact with a potential food item, the lobster may probe further, grasp it with the second and third pereiopods, and transfer it to the mouthparts for ingestion. Alternatively, the lobster may lunge over the food, lower its body, and grab the food item with the second and third pereiopods and third maxillipeds. The food is then rotated by the walking legs and third maxillipeds while the other mouthparts move rapidly over it, presumably examining it further (Derby and Atema, 1982). If the food requires crushing, it is passed forward by the third maxillipeds and chelate walking legs (Lee, 1994), which hold it in position for the crusher claw (Derby and Atema, 1982; Lee, 1994). The crushing action compresses opposite valves of such items as mussels, clams, and oysters. The cutter claw frequently will cover the crusher during this process to help hold food in place or to subdue mobile prey such as crabs (Derby and Atema, 1982; Lee, 1994). Food is then passed back to the mouthparts and reexamined (Lee, 1994). If the crushing attempt is unsuccessful, the food is repositioned for another attempt.

If the chemoreceptors of the walking legs are deafferented, the grasping response is disrupted and lobsters will not pick up food. Without their stimulation, the initial phases in food handling cannot be elicited (Derby and Atema, 1982).

The pereiopods develop into effective grasping and crushing structures for the capture and manipulation of food. The number of setae on the pereiopods increases developmentally to assist in their sensory functions. Concomitant with these morphological changes is a shift in habitat and behavior, all of which prepares *Homarus americanus* for the changes it encounters as it moves from the pelagic to the benthic environment and from a cryptic, benthic existence to a more vagile, benthic lifestyle.

IV. Directions for Further Research

The developmental changes in the mouthparts of *Homarus americanus*, from larva to adult, have been reasonably well studied, although data on the adult are incomplete. Our understanding permits a knowledgeable discussion of the functional morphology of these appendages as it relates to the ecology of the various life history phases. However, some important functional details of the larval mouthparts remain unclear, for example, the use of the mandibles in processing food. The kind of developmental details known for the mouthparts is lacking for the pereiopods. While much is known about the internal changes in the great chelae of postlarvae and juveniles, only some of the more obvious changes in the external morphology of the pereiopods have been examined. As the pereiopods bear some of the better studied chemo- and mechanoreceptors, details of developmental changes in the distribution of the setae are needed.

Even though the sensory abilities of the pereiopods have been well studied, those of the mouthparts need further work. The specific functional significance of the many types of mouthpart setae requires further explanation. In addition, the roles of the first maxillae and the first and second maxillipeds require further elucidation; they may play a greater role in feeding than simply to pass food along to the mandibles and the mouth.

Finally, videographic studies of larval feeding are notably lacking, as are studies of the capture of food in currents by postlarvae and shelter-restricted phase juveniles. As these life history phases of the lobster are exposed to variable flow environments, they may be more adaptive in their feeding strategies than is currently known.

V. Summary

The development of the great chelae and walking legs parallels that of the mouthparts in structural complexity. As *Homarus americanus* passes through the larval stages to become a postlarva, dramatic changes occur in the structure and complexity of these feeding appendages. The mouthparts become heavily setose and gain more teeth and spines along their edges. The fine teeth of the mandibles develop into the heavily cuticularized, toothless cutting edge of the adult and the pad of setae is replaced by the molar process. The pereiopods also become heavily setose. The first and fifth pereiopods undergo torsion, rotating the plane in which they operate in subsequent stages. Torsion of the first pereiopods results in opposing claws that operate in a horizontal plane; torsion of the fifth pereiopods results in an appendage that can extend backward to groom the abdominal appendages.

These morphological changes occur at the time of a dramatic habitat change—from the plankton to the benthos—and suggest that the lobster undergoes a dramatic change in diet. However, postlarvae and shelter-restricted phase juveniles are also capable of feeding on plankton (as are the larvae), as well as on small benthic organisms. This appears to be a versatile intermediate feeding strategy. The feeding modes used during these life history phases are dominated by the mouthparts and the walking legs, rather than the great chelae. The slow development of the musculature and mechanical advantage in the cutter and crusher claws suggests that the postlarvae and shelter-restricted phase juveniles may have diets that differ from those of the more vagile juveniles, adolescents, and adults. As the great chelae become fully functional in crushing small molluscs and other items, the mouthparts become less important in the capture of food, but remain important in the assessment and processing of food. These changes are reflected in a shortening of the setae on the mouthparts, which are no longer needed to entrap food items. The transition in feeding habits and food as lobsters move from a planktonic to a benthic life appears to be more gradual than abrupt, and is reflected in the developmental changes of the feeding appendages.

Acknowledgment

Dr. Charles Derby provided valuable comments and suggestions.

References

Atema, J. (1985). Chemoreception in the sea: Adaptations of chemoreceptors and behavior to aquatic stimulus conditions. *Soc. Exp. Biol. Symp.* 39, 387–423.

Barker, P. L., and Gibson, R. (1977). Observations on the feeding mechanism, structure of the gut, and digestive physiology of the European lobster, *Homarus gammarus* (L.) (Decapoda: Nephropidae). *J. Exp. Mar. Biol. Ecol.* 26, 297–324.

Barnes, E. W. (1906). Methods of protecting and propagating the lobster, with a brief outline of its natural history. *Annu. Rep.—R.I. Comm. Inland Fish.* 36, 119–152.

Barshaw, D. E. (1989). Growth and survival of post-larval lobsters, *Homarus americanus*, on a diet of plankton. *Fish. Bull.* 87, 366–370.

Bauer, R. T. (1975). Grooming behavior and morphology of the caridean shrimp *Pandalus danae* Stimpson (Decapoda: Natantia: Pandalidae). *Zool. J. Linn. Soc.* 56, 45–71.

Bauer, R. T. (1987). Stomatopod grooming behavior: Functional morphology and amputation experiments in *Gonodactylus oerstedii*. *J. Crust. Biol.* 7, 414–432.

Chen, C.-L. (1992). "Scanning Electron Microscopy of Setae of Adult American Lobster Mouthparts, *Homarus americanus*," baccalaureate thesis. State University of New York at Purchase.

Cobb, J. S., Gulbransen, T., Phillips, B. F., Wang, D., and Syslo, M. (1983). Behavior and distribution of larval and early juvenile *Homarus americanus*. *Can. J. Fish. Aquat. Sci.* 40, 2184–2188.

Cobb, J. S., Wang, D., Campbell, D. B., and Rooney, P. (1989). Speed and direction of swimming by postlarvae of the American lobster. *Trans. Am. Fish. Soc.* 118, 82–86.

Corotto, F., Voigt, R., and Atema, J. (1992). Spectral tuning of chemoreceptor cells of the third maxilliped of the lobster, *Homarus americanus*. *Biol. Bull. (Woods Hole, Mass.)* 183, 456–462.

Derby, C. D. (1982). Structure and function of cuticular sensilla of the lobster *Homarus americanus*. *J. Crust. Biol.* 2, 1–21.

Derby, C. D. (1989). Physiology of sensory neurons in morphologically identified cuticular sensilla of crustaceans. *In* "Crustacean Issues" (F. R. Schram, ed.), Vol. 6, "Functional Morphology of Feeding and Grooming in Crustacea" (B. E. Felgenhauer, L. Watling, and A. B. Thistle (eds), pp. 27–47. Balkema, Rotterdam, the Netherlands.

Derby, C. D., and Atema, J. (1982). The function of chemo- and mechanoreceptors in lobster (*Homarus americanus*) feeding behavior. *J. Exp. Biol.* 98, 317–327.

Derby, C. D., Reilley, P. M., and Atema, J. (1984). Chemosensitivity of the lobster *Homarus americanus* to secondary plant compounds: Unused receptor capabilities. *J. Chem. Ecol.* 10, 879–892.

Elner, R. W., and Lavoie, R. E. (1983). Predation on American oysters (*Crassostrea virginica* [Gmelin]) by American lobsters (*Homarus americanus* Milne-Edwards), rock crabs (*Cancer irroratus* Say), and mud crabs (*Neopanope sayi* [Smith]). *J. Shellfish Res.* 3(2), 129–134.

Ennis, G. P. (1986). Swimming ability of larval American lobsters, *Homarus americanus*, in flowing water. *Can. J. Fish. Aquat. Sci.* 43, 2072–2084.

Factor, J. R. (1977). "Morphology of the Mouthparts of Larval Lobsters, *Homarus americanus* (Decapoda: Nephropidae), with Special Emphasis on Their Setae," M.S. thesis. Cornell University, Ithaca, New York.

Factor, J. R. (1978). Morphology of the mouthparts of larval lobsters, *Homarus americanus* (Decapoda: Nephropidae), with special emphasis on their setae. *Biol. Bull. (Woods Hole, Mass.)* 154, 383–408.

Factor, J. R. (1981). Development and metamorphosis of the digestive system of larval lobsters, *Homarus americanus* (Decapoda: Nephropidae). *J. Morphol.* 169, 225–242.

Factor, J. R. (1989). Development of the feeding apparatus in decapod crustaceans. *In* "Crustacean Issues" (F. R. Schram, ed.), Vol. 6, "Functional Morphology of Feeding and Grooming in Crustacea" (B. E. Felgenhauer, L. Watling, and A. B. Thistle (eds.), pp. 185–203. Balkema, Rotterdam, the Netherlands.

Farmer, A. S. (1974). The functional morphology of the mouthparts and pereiopods of *Nephrops norvegicus* (L.) (Decapoda: Nephropidae). *J. Nat. Hist.* 8, 121–142.

Felgenhauer, B. E. (1992). External anatomy and integumentary structures. *In* "Microscopic Anatomy of Invertebrates" (F. W. Harrison, ed.), Vol. 10, "Decapod Crustacea" (F. W. Harrison and A. G. Humes, eds.), pp. 7–43. Wiley–Liss, New York.

Felgenhauer, B. E., and Schram, F. R. (1979). The functional morphology of the grooming appendages of *Palaemonetes kadiakensis* Rathbun, 1902. *Fieldiana, (Zool.)* 73, 1–17.

Gunn, L. A. (1987). "The Feeding Ecology of American Lobster Larvae in Long Island Sound," M.S. thesis. Southern Connecticut State University, New Haven.

Harding, G. C., Vass, W. P., and Drinkwater, K. F. (1983). Factors influencing the size of American lobster (*Homarus americanus*) stocks along the Atlantic coast of Nova Scotia, Gulf of St. Lawrence, and the Gulf of Maine: A new synthesis. *Can. J. Fish. Aquat. Sci.* 40, 168–184.

Herrick, F. H. (1895). The American lobster: A study of its habits and development. *Bull. U.S. Fish Comm.* 15, 1–252 + 54 plates.

Herrick, F. H. (1909). Natural history of the American lobster. *Bull. U.S. Bur. Fish.* 29, 149–408 + 20 plates.

Hudon, C. (1987). Ecology and growth of postlarval and juvenile lobster, *Homarus americanus*, off Îles de la Madeleine (Quebec). *Can. J. Fish. Aquat. Sci.* 44, 1855–1869.

Jacques, F. (1989). The setal system of crustaceans: Types of setae, groupings, and functional morphology. *In* "Crustacean Issues" (F. R. Schram, ed.), Vol. 6, "Functional Morphology of Feeding and Grooming in Crustacea" (B. E. Felgenhauer, L. Watling, and A. B. Thistle (eds.), pp. 1–13. Balkema, Rotterdam, the Netherlands.

Juinio, M. A. R., and Cobb, J. S. (1992). Natural diet and feeding habits of the postlarval lobster *Homarus americanus*. *Mar. Ecol. Prog. Ser.* 85, 83–91.

Lavalli, K. L. (1991). Survival and growth of early juvenile American lobsters *Homarus americanus* through their first season while fed diets of mesoplankton, microplankton, and frozen brine shrimp. *Fish. Bull.* 89, 61–68.

Lavalli, K. L. (1992). "The Feeding Mechanisms of Post-larval and Early Juvenile American Lobsters (*Homarus americanus* Milne Edwards, 1837)," Ph.D. dissertation. Boston University, Boston.

Lavalli, K. and Ayers, J. (1994). Development of stomatogastric mediated motor programs in the American lobster. *Soc. Neurosci. Abstracts* 20, 1596.

Lavalli, K. L., and Barshaw, D. E. (1989). Post-larval American lobsters (*Homarus americanus*) living in burrows may be suspension feeding. *Mar. Behav. Physiol.* 15, 255–264.

Lavalli, K. L., and Factor, J. R. (1992). Functional morphology of the mouthparts of juvenile lobsters, *Homarus americanus* (Decapoda: Nephropidae), and comparison with the larval stages. *J. Crust. Biol.* 12, 467–510.

Lavalli, K. L., and Trager, G. C. (1995). Suspension feeding in post-larval and early juvenile American lobsters, *Homarus americanus*. (in preparation.)

Laverack, M. S. (1962a). Responses of cuticular sense organs of the lobster, *Homarus vulgaris* (Crustacea). I. Hair-peg organs as water current receptors. *Comp. Biochem. Physiol.* 5, 319–325.

Laverack, M. S. (1962b). Responses of cuticular sense organs of the lobster, *Homarus vulgaris* (Crustacea). II. Hair-fan organs as pressure detectors. *Comp. Biochem. Physiol.* 6, 137–145.

Laverack, M. S. (1963). Responses of cuticular sense organs of the lobster, *Homarus vulgaris* (Crustacea). III. Activity invoked in sense organs of the carapace. *Comp. Biochem. Physiol.* 10, 261–272.

Lee, K. (1994). "Interactions between the Feeding and Stomatogastric Motor Programs in the American Lobster, *Homarus americanus*," Ph.D. dissertation. Northeastern University, Boston.

Loo, L.-O., Baden, S. P., and Ulmestrand, M. (1993). Suspension feeding in adult *Nephrops norvegicus* (L.) and *Homarus gammarus* (L.) (Decapoda). *Neth. J. Sea Res.* 31, 291–297.

Maynard, D. M., and Sallee, A. (1970). Disturbance of feeding behavior in the spiny lobster, *Panulirus argus*, following bilateral ablation of the medulla terminalis. *Z. Vergl. Physiol.* 66, 123–140.

McLaughlin, P. A. (1982). Comparative morphology of crustacean appendages. *In* "The Biology of Crustacea" (D. E. Bliss, ed.), Vol. 2, "Embryology, Morphology, and Genetics" (L. G. Abele, ed.), pp. 197–256. Academic Press, New York.

McPhie, D., and Atema, J. (1984). Chemical communication in lobsters: Information currents. *Biol. Bull. (Woods Hole, Mass.)* 167, 510–511 (abstract).

Mead, A. D., and Williams, L. W. (1902). Habits and growth of the lobster, and experiments in lobster-culture. *Annu. Rep.—R.I. Comm. Inland Fish.* 32, 57–79.

Neilsen, S. (1993). "Scanning Electron Microscopy of Adult Lobster Mouthparts," baccalaureate thesis. State University of New York at Purchase.

Oakley, J. K., and Macmillan, D. L. (1980). The morphology and physiology of CAP organs in *Jasus novaehollandiae* (Crustacea, Decapoda, Reptantia, Macrura). *Mar. Behav. Physiol.* 7, 233–247.

Phillips, B. F., Cobb, J. S., and George, R. W. (1980). General biology. *In* "The Biology and Management of Lobsters" (J. S. Cobb and B. F. Phillips, eds.), Vol. 1, pp. 1–82. Academic Press, New York.

Roberts, M. H., Jr. (1968). Functional morphology of mouthparts of hermit crabs, *Pagurus longicarpus* and *Pagurus pollicaris*. *Chesapeake Sci.* 9, 9–20.

Schram, F. R. (1986). "Crustacea." Oxford University Press, New York.

Shelton, R. G., and Laverack, M. S. (1968). Observations on a redescribed crustacean cuticular sense organ. *Comp. Biochem. Physiol.* 25, 1049–1059.

Shelton, R. G., and Laverack, M. S. (1970). Receptor hair structure and function in the lobster *Homarus gammarus* (L.). *J. Exp. Mar. Biol. Ecol.* 4, 201–210.

Smith, S. I. (1873). The early stages of the American lobster (*Homarus americanus*). *Trans. Conn. Acad. Arts Sci.* 2(2), 351–381.

Solon, M. H., and Cobb, J. S. (1980). The external morphology and distribution of cuticular hair organs on the claws of the American lobsters, *Homarus americanus* (Milne-Edwards). *J. Exp. Mar. Biol. Ecol.* 48, 205–215.

Solon, M. H., and Kass-Simon, G. (1981). Mechanosensory activity of hair organs on the chelae of *Homarus americanus*. *Comp. Biochem. Physiol. A* 68A, 217–223.

Thomas, W. J. (1970). The setae of *Austropotamobius pallipes* (Crustacea: Astacidae). *J. Zool.* 160, 91–112.

Vedel, J. P., and Clarac, F. (1976). Hydrodynamic sensitivity by cuticular organs in the rock lobster *Palinurus vulgaris*. Morphological and physiological aspects. *Mar. Behav. Physiol.* 3, 235–251.

Watling, L. (1989). A classification system for crustacean setae based on the homology concept. *In* "Crustacean Issues" (F. R. Schram, ed.), Vol. 6, "Functional Morphology of Feeding and Grooming in Crustacea" (B. E. Felgenhauer, L. Watling, and A. B. Thistle, eds.) pp. 15–26. Balkema, Rotterdam, the Netherlands.

Williams, L. W. (1907). The stomach of the lobster and the food of larval lobsters. *Annu. Rep.—R.I. Comm. Inland Fish.* 37, 153–180.

CHAPTER

15

The Digestive System

JAN ROBERT FACTOR
Division of Natural Sciences
Purchase College
State University of New York
Purchase, New York 10577

I. Introduction

The organization of the digestive system is generally similar throughout the Decapoda. [Icely and Nott (1992) and Felgenhauer (1992) consider the decapod digestive system.] The complex digestive system of *Homarus americanus* is divided into three major regions based primarily on developmental considerations, recognized as foregut, midgut, and hindgut (Table 1). Each region is elaborated into several organs. Figure 1 illustrates the general anatomy of the digestive system.

The foregut is lined with a thin layer of cuticle and comprises the mouth, the esophagus (a relatively short, muscular tube that connects the mouth and the stomach), the cardiac stomach (including the masticating mechanism known as the gastric mill), and the pyloric stomach (which serves as a sorting and filtering device).

The midgut comprises the intestine, a gently curved tube, and several associated organs: the digestive gland (=hepatopancreas, or midgut gland), the anterior midgut caeca, and the posterior midgut caecum.

The hindgut, like the foregut, is lined with cuticle and forms a muscular rectum that leads to the anus.

The division of the digestive system into three regions is based primarily on considerations of embryonic development (see Section VII,A). The organs of the foregut and the hindgut, being ectodermal derivatives, are lined with cuticle (Herrick, 1909), while the organs of the midgut have no such cuticular lining. Although the presence of a cuticular lining is developmentally determined, it has considerable functional significance. It is the elaborations of this cuticle, for example, that form the grinding and sorting structures of the foregut. The foregut and hindgut cuticles are continuous with the cuticle that forms the exoskeleton, and they are shed through the mouth and the anus at each molt.

II. Foregut

A. Mouth

The mouth lies between and behind the mandibles, on the anteroventral surface of the head. The anterior margin of the mouth is formed by the labrum (=upper lip), a fleshy, median lobe. A pair of ridges constitutes the sides of the mouth; they converge posteriorly to produce the metastoma (=lower lip) (Williams, 1907), with extended paragnaths. These four structures are continuous with the four ridges of the esophagus.

B. Esophagus

1. Gross Anatomy

The esophagus of *Homarus americanus* is a relatively short canal connecting the mouth and the anteroventral wall of the cardiac stomach. The thick, muscular wall of the esophagus is formed into four longitudinal ridges that protrude into the esophageal lumen: the labral ridge, the most prominent of the four, is a thick-

TABLE 1 **Regions and Organs of the Digestive System of *Homarus americanus*[a]**

Region		
Foregut	Midgut	Hindgut
Mouth	Intestine (=midgut trunk, midgut proper)	Rectum (=hindgut)
Esophagus	Digestive gland (=hepatopancreas, midgut gland)	Anus
Cardiac stomach	Anterior midgut caeca	
Pyloric stomach	Posterior midgut caecum	

[a]Only reasonable synomyms currently in use are listed; this is not an exhaustive list of all terms applied to these organs.

ening of the anterior wall and represents an inward extension of the labrum (upper lip) of the mouth; two lateral ridges are less prominent; and a metastomal ridge is a thickening of the posterior wall.

2. Arrangement of Tissues

Like the other regions of the digestive system, the esophagus consists of two tissue layers: the lining epithelium and the underlying connective tissue. The epithelium is simple and columnar. It is covered by a thin cuticle that consists of the layers typical of the cuticle in other parts of the lobster. (Waddy *et al.*, in Chapter 10, provide a detailed account of the integument.) The epithelium is underlain by a basement membrane of simple construction; it is a typical basement membrane, not at all similar to the complex basement membranes found in the organs of the midgut (Sections III,A and B). [The term *basement membrane* is used throughout this chapter to refer to the extracellular layer that underlies and supports many epithelia, and which is also called the *basal lamina*; Factor (1981a) reviews terms and definitions.]

The four longitudinal ridges of the esophagus are composed of both epithelium and connective tissue. Substantial elements of the connective tissue extend into the ridges, just underneath the folded layer of epithelial cells.

The connective tissue of the esophagus can be thought of as a matrix of collagen fibers in which the other subepithelial elements are embedded. Discrete bundles of circular muscles lie near the periphery of the esophagus and bundles of longitudinal muscles are centrally located, mostly in the longitudinal ridges. This arrangement of central longitudinal and peripheral circular muscles is similar to the arrangement of muscles in the rectum (Section IV,A), but is the reverse of the situation found in the midgut (Section III,A). A system of hemal sinuses (i.e., blood spaces) is embedded in the connective tissue matrix and supplies the tissue with hemolymph (i.e., blood).

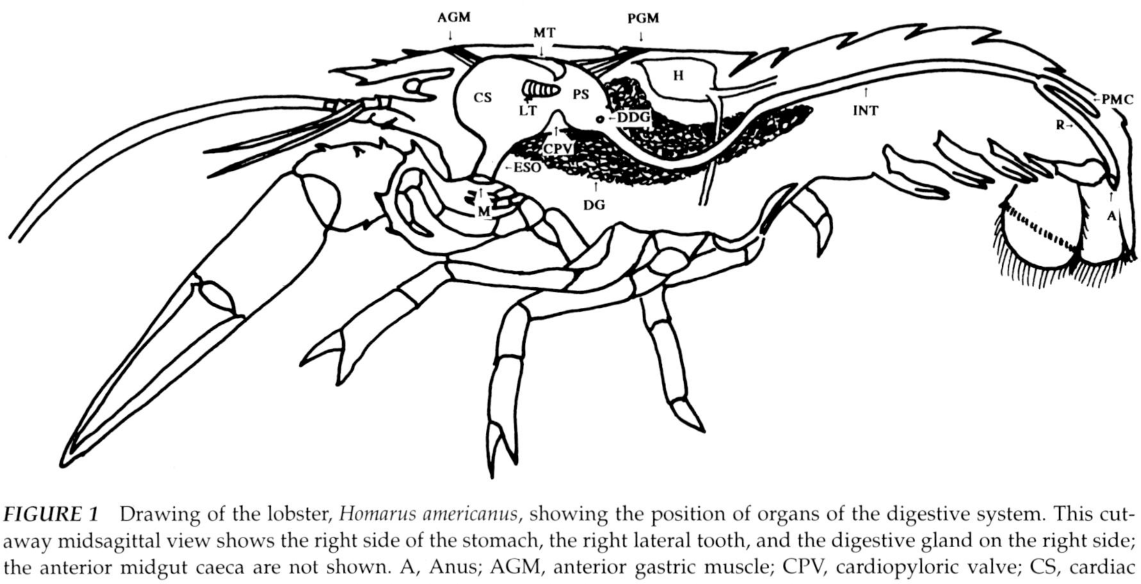

FIGURE 1 Drawing of the lobster, *Homarus americanus*, showing the position of organs of the digestive system. This cutaway midsagittal view shows the right side of the stomach, the right lateral tooth, and the digestive gland on the right side; the anterior midgut caeca are not shown. A, Anus; AGM, anterior gastric muscle; CPV, cardiopyloric valve; CS, cardiac stomach; DDG, duct of the digestive gland; DG, digestive gland; ESO, esophagus; H, heart; INT, intestine; LT, lateral tooth; M, mouth; MT, medial tooth; PGM, posterior gastric muscle; PMC, posterior midgut caecum; PS, pyloric stomach; R, rectum. [Based on Herrick's (1909) Plate XXXIII.]

As part of the open circulatory system of the lobster, these spaces are not blood vessels; blood bathes the tissues directly without intervening capillary walls. The connective tissue also contains a variety of cell types: fibroblasts, which have long cellular processes and are presumably responsible for secreting the fibrous elements of the matrix; granulocytes, which contain either acidophilic or basophilic granules that are electron dense, and whose function and possible relationship to the granulocytic hemocytes are not known; and circulating hemocytes (i.e., blood cells) in the hemal sinuses. (Martin and Hose, in Chapter 17, consider the circulatory system and the types of circulating hemocytes.)

The longitudinal ridges also contain abundant tegumental glands. They are generally organized into clusters and lie embedded in the connective tissue, just beneath the epithelium. Structurally, these esophageal tegumental glands are virtually identical to those found in the rectum and are discussed in Section V.

C. Cardiac and Pyloric Stomachs

The cardiac stomach is a spacious, saclike chamber that receives food from the esophagus. Food is prevented from reentering the esophagus by a single medial and a pair of lateral esophageal valves, which guard the opening (Williams, 1907). In the cardiac stomach, the cuticle provides the lining of the sac; in addition, elaborations of the cuticle produce a complex series of ossicles forming the plates and teeth of the gastric mill. Digestion begins in the cardiac stomach, where enzymes from the digestive gland are mixed with food (see Conklin, Chapter 16, on digestive physiology). The pyloric stomach is a complex filtering apparatus whose primary elements are also formed as ossicular elaborations of the cuticle. These ossicles of the cardiac and pyloric stomachs, together with their associated muscles and nerves, constitute the stomatogastric system, whose function is the mastication of ingested food into fine particles in the cardiac gastric mill and the sorting of these particles in the cardiac sac and the pyloric filter.

1. Wall of the Cardiac Stomach

The wall of the cardiac stomach includes an epithelium and a connective tissue layer. The epithelium is a single layer of columnar cells that produces the cuticular lining apically and is underlain by a typical basement membrane. The basic arrangement of tissues is similar to that of the esophagus, but the wall of the cardiac stomach is generally much thinner and the connective tissue is simpler than in the esophageal wall. The lining cuticle and the musculature, however, are much more complex where they form the series of ossicles, teeth, and muscles of the gastric mill. Yonge (1932) and Barker and Gibson (1977) reported that the cardiac stomach of *Homarus gammarus* has no tegumental glands.

Three elaborations of the wall of the cardiac stomach are especially noteworthy. (1) The paired gastroliths are formed at the gastrolith fields, which are large, oval, specialized epithelia associated with the anterolateral walls of the cardiac stomach. (2) The main hematopoietic tissues are two specialized areas that are only loosely associated with the dorsal surface of the cardiac stomach (see Martin and Hose, Chapter 17). (3) The chitinous lining of the cardiac stomach is elaborated into a series of more or less calcified plates and teeth that form the mechanism of the gastric mill, as well as fields of setae that help direct the flow of food.

2. Gastric Mill

Much of the interest in the decapod digestive system has centered on the gastric mill. Our understanding of the gastric mill relies on a series of historically important works, including Aristotle's reference to the teeth in the stomach of a crab, the classic work of Milne Edwards (1834, 1840), who originated many of the terms in use today for parts of the gastric mill, and the remarkable paper by Mocquard (1883), who first concluded that the gastric mill of all the stalked-eyed Crustacea follows a single structural plan in which homologies are recognizable. The seminal series by Patwardhan (1935a–f) compared 50 species of decapods, but there have been many other descriptive and comparative studies of a wide variety of decapods. The gastric mills of other nephropidean lobsters have been described by Yonge (1924) and Patwardhan (1935d). Studies of the gastric mill of *Homarus americanus* are few in number and include those of Williams (1907) and Maynard and Dando (1974).

The most obvious features of the gastric mill are the single medial tooth, which is an elaboration of the urocardiac ossicle; the paired lateral teeth, formed by the zygocardiac ossicles; and the paired accessory lateral teeth lying just ventral to the lateral teeth. Setal pads are associated with these teeth. The medial tooth hangs down into the cardiac stomach from the dorsal wall and is anteriorly directed. The lateral teeth are elongate ossicles that lie along an anteroposterior axis in the lateral walls of the cardiac stomach. Several prominent, posteriorly directed cusps protrude from each lateral tooth. The mechanism that allows these teeth to perform their grinding function is complex, however, and is described below.

3. Anatomy of the Stomatogastric System

The most detailed account of the anatomy of the cardiac and pyloric stomachs of *Homarus americanus*—the ossicles, muscles, and nerves of the stomatogastric system—is presented by Maynard and Dando (1974). Their terminology is modified from Mocquard (1883) and their study forms the basis of the following account (supplemented by Williams, 1907; Claiborne and Ayers, 1987). Although much work on the neurobiology of the stomatogastric system has been carried out since 1974, neurobiologists have generally concentrated their most extensive efforts on several spiny lobsters (Palinuridae; see Katz and Tazaki, 1992), which are now better understood in this respect than *H. americanus*. [Wales (1982), Wiens (1982), and the chapters in the books edited by Selverston and Moulins (1987) and Harris-Warrick *et al.* (1992) provide comprehensive treatments of the crustacean stomatogastric system.]

a. Ossicles The ossicles of the cardiac and pyloric stomachs are generally paired and often calcified. The ossicles form not only the grinding teeth of the gastric mill and the filter of the pyloric stomach, but also the supporting structures in the wall of the cardiac and pyloric stomachs and the hinges and articulations that allow the movement and mechanical advantage of the entire stomatogastric system. Claiborne and Ayers (1987) list four important functions for the ossicles: (1) semirigid supporting elements for the stomach wall; (2) a complex array of levers, joints, and fulcrums on which the muscles operate; (3) attachment points for the stomatogastric muscles that operate the stomach; and (4) cutting and grinding surfaces of the gastric mill (hard ridges in some ossicles). The ossicles of the stomatogastric system have been divided into seven categories (Maynard and Dando, 1974): (1) ossicles of the cardiac gastric mill (I–VII); (2) lateral supporting cardiac ossicles (VIII–XV); (3) ossicles of the cardiopyloric valve (XVI–XVIII); (4) supporting ossicles of the dorsal pyloric stomach (XIX–XXI); (5) supporting ossicles of the ventral pylorus and ampullae (=gland filters) (XXII–XXVII); (6) supraampullary supporting ossicles (XXVIII–XXX); and (7) supporting ossicles of the lateral pylorus (XXXI–XXXIII). The stomatogastric ossicles are listed in Table 2 and illustrated in Fig. 2.

b. Muscles The muscles of the cardiac and pyloric stomachs are generally paired and are typically composed of parallel striated fibers running between points of origin and insertion on skeletal elements. Some spread over the surface of the flexible cuticle of the stomach and may include branched elements forming a loose network of fibers. Extrinsic muscles originate on the inner surface of the thoracic

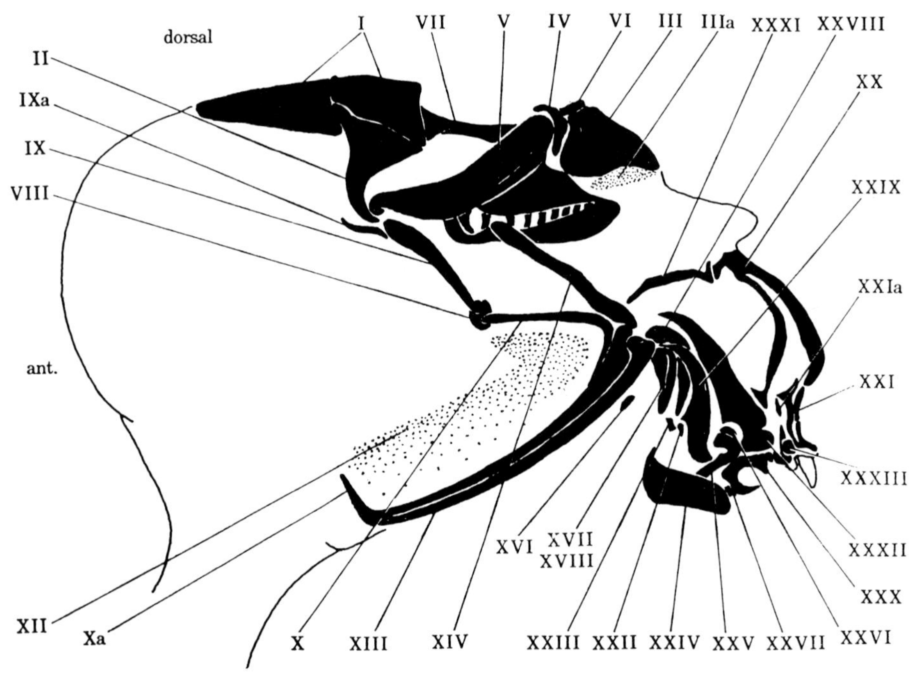

FIGURE 2 Ossicles of the stomatogastric system of *Homarus americanus*. Lateral view of the foregut (anterior toward the left, dorsal uppermost). Key to abbreviations in Table 2. The arrangement of ossicles VI and VII, which form the medial tooth, is more clearly seen in Fig. 6. (Reprinted from Maynard and Dando, 1974, with permission.)

exoskeleton and insert on stomach ossicles or cuticle; they support the stomach within the thoracic cavity and act as dilators. Most of the extrinsic dilators run anteriorly or anterolaterally. Intrinsic muscles have both attachments on the stomach wall. The muscles of the stomatogastric system have been divided into five categories (Maynard and Dando, 1974): (1) muscles of the cardiac stomach (c.); (2) gastric mill complex (g.m.); (3) ventral cardiac muscles (c.v.); (4) cardiopyloric valve (cp.v.); and (5) pyloric muscles (p.). The stomatogastric muscles are listed in Table 3 and illustrated in Fig. 3.

c. Nerves The stomatogastric ganglia include the commissural ganglia (paired), the esophageal ganglion (unpaired), and the stomatogastric ganglion (unpaired; located within the ophthalmic artery): the commissural and esophageal ganglia are connected via the paired, transverse inferior and superior esophageal nerves; the stomatogastric ganglion is connected to other ganglia by the stomatogastric nerve; the esophageal ganglion is connected to the brain by the inferior ventricular nerve (Claiborne and Ayers, 1987).

While most of the stomatogastric nerves are paired, there may be considerable variation in the branching patterns of the left and right pairs. Stomatogastric nerves that innervate the wall and muscles of the stomach and the esophagus have been divided into four categories (Maynard and Dando, 1974): (1) motor fibers that originate in the commissural and subesophageal ganglia (and possibly the median esophageal ganglion) of the ventral nerve cord and innervate the muscles of the esophagus (and perhaps some anterior cardiac muscles); (2) primarily sensory nerves that originate in sensory cells of the stomach and the esophagus and pass to the ventral ganglia; (3) a small group of mostly neurosecretory fibers in the esophagus and the lower stomach; and (4) motor fibers that originate in the stomatogastric ganglion and pass to the stomach muscles, as well as nerves that link the stomatogastric ganglion with the esophageal ganglion and the rest of the central nervous system. Maynard and Dando (1974), who called these categories "classes" of nerves, cautioned that nerves may carry more than one kind of fiber and nerves from more than one group may anastomose. Not all nerves, therefore, fall clearly into one of the four groups, and the classification should be considered a loose one. The stomatogastric nerves are listed in Table 4 and illustrated in Figs. 4 and 5.

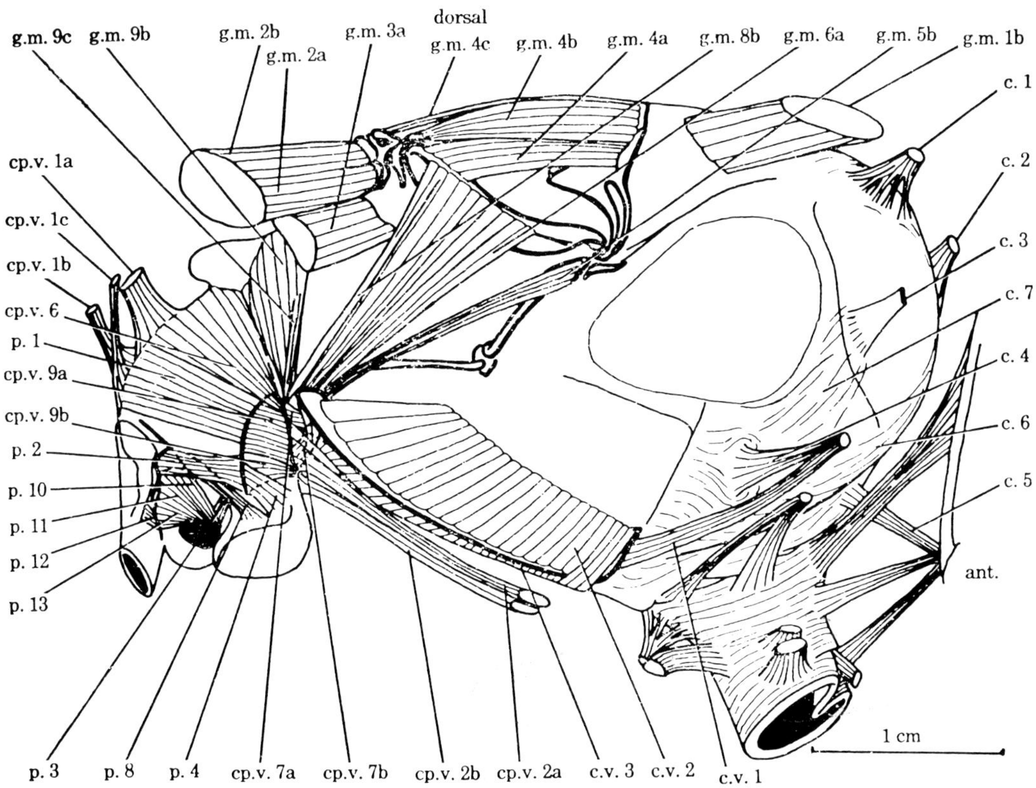

FIGURE 3 Muscles of the stomatogastric system of *Homarus americanus*. Lateral view of the foregut (anterior toward the right, dorsal uppermost). Key to abbreviations in Table 3. (Reprinted from Maynard and Dando (1974) *Philos. Trans. R. Soc. London Ser. B*, 268, 161–220, with permission from The Royal Society.)

TABLE 2 **Ossicles of the Stomatogastric System of *Homarus americanus*[a]**

Group 1. Cardiac gastric mill

I	Mesocardiac ossicle (single)
II	Pterocardiac ossicle
III	Pyloric ossicle (single, sometimes paired)
IIIa	Plate, ventral to the posterolateral border of the pyloric ossicle
IV	Exopyloric ossicle
V	Zygocardiac ossicle (dentate)
VI	Propyloric ossicle (single, dentate)
VII	Urocardiac ossicle (single, dentate)

Generally well calcified and paired; I and III are especially well developed; IIIa is a lightly calcified, oval plate that often occurs on the posterolateral border of III.

Group 2. Lateral supporting cardiac ossicles

VIII	Pectineal ossicle (dentate on the inner side)
IX	Prepectineal ossicle
IXa	Accessory prepectineal ossicle
X	Postpectineal ossicle
Xa	"Quill" of the postpectineal ossicle
XI	Anterior lateral cardiac plate (not illustrated)
XII	Posterior lateral cardiac plate
XIII	Inferior lateral cardiac ossicle
XIV	Lateral cardiopyloric ossicle

Well developed and heavily calcified (especially XIII); IXa is a small, accessory ossicle that extends anteriorly from IX and ventrally from II; X often appears separated from the "quill" segment (Xa); XII is generally uncalcified.

Group 3. Ossicles of the cardiopyloric valve

The cardiopyloric valve (cp.v.) is a ventral invagination at the cardiopyloric border. It partially separates the cardiac and pyloric stomachs, limiting the movement of food particles to the pyloric stomach. The lateral ossicles (XVIII) include both the auricle and the bar of Mocquard (1883) and have been confused with the anterior supraampullary ossicle (XXVIII) or with the preampullary ossicle (XXII) (Cochran, 1935).

cp.v.	Cardiopyloric valve
XVI	Anterior ossicle of the cardiopyloric valve (single) (=posterior inferior cardiac ossicle)
XVII	Posterior ossicle of the cardiopyloric valve (single) (=preanterior inferior pyloric ossicle)
XVIII	Lateral ossicle of the cardiopyloric valve (=cardiopyloric auricular ossicle)

XVI is reduced to a small, angular, calcified plate, connected by ligaments to the auricular complex; XVII is heavily calcified in the midline of the cp.v.; on each side, calcified bars extend anteriorly along the inner wall of the cp.v. and form the base of the expanded, flaplike XVIII, which extends outward from XVII.

Group 4. Supporting ossicles of the dorsal pyloric stomach

XIX	Anterior mesopyloric ossicle (lacking)
XX	Posterior mesopyloric ossicle (usually single)
XXI	Uropyloric ossicle (uncertain)
XXIa	Infrauropyloric fragment
up.v.	Uropyloric fold

A heavily calcified fragment on either side articulates with XXXI and represents part or all of XX; it is continuous with a thickened, often calcified fold that extends parallel to the dorsal ridge; XXIa is a small, calcified strip beneath the anterior half of the uropyloric fold and may be regarded either as a portion of the uropyloric complex or as an isolated extension of one of the pleuro-ossicles.

Group 5. Supporting ossicles of the ventral pylorus and ampullae

XXII	Preampullary ossicle
XXIII	Anterior inferior pyloric ossicle (single)
XXIV	Interior ampullary ossicle
XXV	Ampullary roof ossicle, lower portion
XXVI	Ampullary roof ossicle, upper portion
XXVII	Posterior inferior pyloric ossicle (single)

XXII is poorly calcified and located posterolaterally to XXIII, which is reduced to a small, median, slightly calcified oval; XXVII is strongly calcified; Mocquard (1883) did not recognize two portions of the ampullary roof (XXV and XXVI) and placed these ossicles in group 6.

continues

TABLE 2 Continued

Group 6.	**Supraampullary supporting ossicles**
XXVIII	Anterior supraampullary ossicle
XXIX	Middle supraampullary ossicle
XXX	Posterior supraampullary ossicle
Neither XXVIII nor XXIX is very strongly calcified; XXIX is substantial and provides extended sites for ventral attachments of lateral pyloric muscles.	
Group 7.	**Supporting ossicles of the lateral pylorus**
XXXI	Anterior pleuropyloric ossicle
XXXII	Middle pleuropyloric ossicle
XXXIII	Posterior pleuropyloric ossicle
XXXI is a calcified bar articulating with XX, terminating in a chitinous fold anterior to XXVIII; XXXII is a flattened, calcified plate; XXXIII has one well-calcified part under the uropyloric fold and one lightly calcified part forming a lateral projection posterior to the fold.	

[a]Compare with Fig. 2. (Based on information from Maynard and Dando, 1974.)

4. Operation of the Gastric Mill

The operation of the gastric mill results from the complex array of ossicles and muscles, whose movement is coordinated by the stomatogastric nervous system. It is a mechanical complex of ossicular articulation and leverage. Although the following may be a somewhat oversimplified account, it provides some understanding of the main engine driving the gastric mill.

The fundamental movement of the medial tooth is controlled primarily by the anterior and posterior gastric muscles. The anterior gastric muscle (g.m.1b) extends from its origin on the anterior, dorsal wall of the carapace to the mesocardiac ossicle (I; the "cardiac ossicle" of Williams, 1907), which forms the dor-

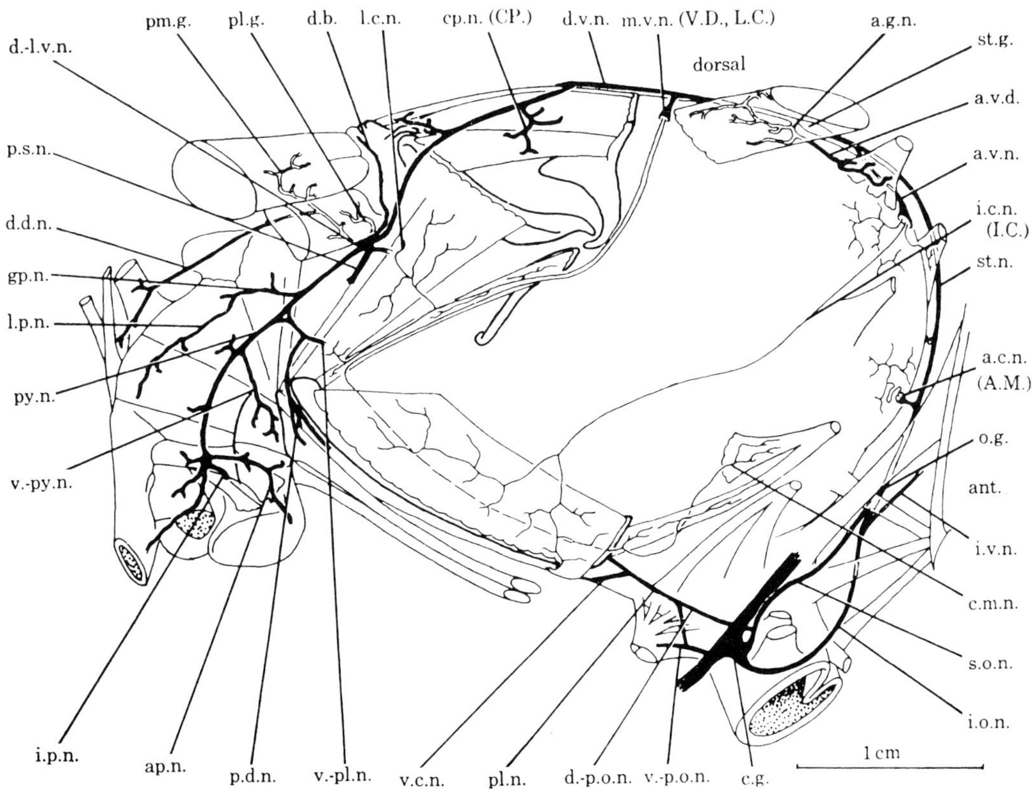

FIGURE 4 Nerves of the stomatogastric system of *Homarus americanus*. Lateral view of the foregut (anterior toward the right, dorsal uppermost). Key to abbreviations in Table 4. (Reprinted from Maynard and Dando (1974) *Philos. Trans. R. Soc. London Ser. B*, 268, 161–220, with permission from The Royal Society.)

TABLE 3 **Muscles of the Stomatogastric System of *Homarus americanus*[a]**

Muscle	Name	Origin	Insertion
Group 1. Cardiac stomach			
Extrinsic muscles			
c.1		Procephalic apophysis	sw: Anterior–dorsal wall
c.2	Anterior superior ventricle dilator {200}	Ventrolateral edge of eye cup	sw: Anterior wall of stomach, ventral to c.1
c.3	[Straplike ligament with few contractile elements]	Near eye cup	sw: Anterolateral cardium, in front of gastrolith field
c.4	Lateral anterior ventricle dilator {202}	Inner wall of thorax, with c.v.1	sw: Below gastrolith field
c.5	Anterior inferior ventricle dilator {201}	Midline of epistome, above labrum	sw: Anteriorly, above esophageal junction
Intrinsic muscles			
c.6	Anterior mesial cardiac muscle {213}	Mesocardiac ossicle (I), near insertion of g.m.1	sw: Either side of esophagus, along ventrolateral region of stomach (ossicle XI)
c.7	{214}	At insertion of c.3	sw: Terminating bundles of c.6 (ossicle XII)
Group 2. Gastric mill complex			
Extrinsic muscles			
g.m.1a	[Lacking]		
g.m.1b	Anterior gastric muscle {197}	Procephalic apophysis, behind rostrum	Anterior plate, mesocardiac ossicle (I)
g.m.2a	Mesial posterior gastric muscle	Apodeme, inner side of medial dorsal thoracic carapace near cervical groove	Pyloric ossicle (III)
g.m.2b	Mesial posterior gastric muscle {198}	Lateral to above apodeme, just anterior to cervical groove	Pyloric ossicle (III)
g.m.3a	Lateral posterior gastric muscle	Dorsal thoracic carapace	Zygocardiac ossicle (V)
Intrinsic muscles			
g.m.4a–c	Cardiopyloric muscle [forms thin sheet spanning dorsal region between anterior and posterior arches of gastric mill]		
g.m.4a	Cardiopyloric muscle	Pterocardiac ossicle (II)	Zygocardiac ossicle (V)
g.m.4b	Cardiopyloric muscle {210b}	Mesocardiac ossicle (I)	Exopyloric ossicle (IV)
g.m.4c	Cardiopyloric muscle {210a}	Mesocardiac ossicle (I)	Propyloric ossicle (VI)
g.m.5a	[Lacking]		
g.m.5b	Lateral interior cardiac muscle	Inferior lateral cardiac ossicle (XIII)	Prepectineal ossicle (IX)
g.m.6a	Lateral interior cardiac muscle {211b}	Narrow sheet on inferior lateral cardiac ossicle (XIII)	Zygocardiac ossicle (V)
g.m.6b	[Lacking]		
g.m.7	[Lacking]		
g.m.8a	[Lacking]		
g.m.8b	Lateral interior cardiac muscle {211e}	Lateral ossicle of cardiopyloric valve (XVIII)	Zygocardiac ossicle (V)
g.m.9a–c	Lateral interior cardiac muscle [forms sheet of muscle connecting posterior arch of gastric mill with anterior pyloric and cardiopyloric ossicles]		
g.m.9a	Lateral interior cardiac muscle {211f}	Anterior pleuropyloric ossicle (XXXI)	pyloric ossicle (III)
g.m.9b	Lateral interior cardiac muscle	Supraampullary ossicle (XXIX)	pyloric ossicle (III)
g.m.9c	Lateral interior cardiac muscle	Lateral ossicle of pyloric valve (XVIII)	pyloric ossicle (III)

Group 3. Ventral cardiac muscles			
Extrinsic muscles			
c.v.1	Lateral posterior ventricular dilator	Lateral carapace, with c.4	"Quill" of postpectineal ossicle (Xa)
Intrinsic muscles			
c.v.2	Lateral interior cardiac muscle {211a}	Posterior lateral cardiac plate (XII)	Inferior lateral cardiac ossicle (XIII)
c.v.3	Posterior inferior cardiac muscle {212}	Posterior inferior cardiac ossicle (XVI)	Inferior lateral cardiac ossicle (XIII)
Group 4. Cardiopyloric valve			
Extrinsic muscles			
cp.v.1a-c	Anterior dorsal pyloric dilator		
cp.v.1a	Anterior dorsal pyloric dilator {204a}	Apodeme, inner side of medial thoracic carapace, ventral to g.m.2a	Posterior mesopyloric ossicle (XX), anterior portion
cp.v.1b	Anterior dorsal pyloric dilator	Apodeme, inner side of medial thoracic carapace, ventral to g.m.2a	Posterior mesopyloric ossicle (XX), posterior portion
cp.v.1c	Anterior dorsal pyloric dilator	Strand of muscle and tissue connecting posterior end of above apodeme with pericardial septum	Posterior mesopyloric ossicle (XX), posterior portion
cp.v.2a	External inferior pyloric dilator {205a}	Endophragmal skeleton, at back of esophagus	Lateral ossicle of cardiopyloric valve (XVIII)
cp.v.2b	Internal inferior pyloric dilator {205b}	Inner margin of mandible	Anterior inferior pyloric ossicle (XXIII)
Intrinsic muscles			
cp.v.3a	[Lacking]		
cp.v.3b	Pyloric muscle {216k}	Anterior pleuropyloric ossicle (XXXI)	Posterior mesopyloric ossicle (XX), anterior region
cp.v.4a,b	[Lacking]		
cp.v.5	[Lacking]		
cp.v.6	Pyloric muscle {216a,b}	Posterior mesopyloric ossicle (XX), anterolateral edge	Lateral ossicle of cardiopyloric valve (XVIII)
cp.v.7a	Pyloric muscle {216n}	Inferior lateral cardiac ossicle (XIII), inner posterior edge	Lateral ossicle of cardiopyloric valve (XVIII), anterior lateral edge
cp.v.7b	Pyloric muscle	Inferior lateral cardiac ossicle (XIII), inner posterior edge, medial to cp.v.7a	Near midline of cardiopyloric valve ossicle complex, closer to XVII
cp.v.8	[Lacking]		
cp.v.9a,b	[Muscle sheet spanning lateral ossicle of cardiopyloric valve (XVIII) and middle supraampullary ossicle (XXIX)]		
cp.v.9a		Supraampullary ossicle (XXIX), anterior lateral edge	Lateral ossicle of cardiopyloric valve (XVIII)
cp.v.9b		Supraampullary ossicle (XXIX), posterior and medial to cp.v.9a	Preampullary ossicle (XXII)
cp.v.10	Pyloric muscle	Inferior ampullary ossicle (XXIV), anterior medial region	Preampullary ossicle (XXII), ventral edge
cp.v.11	[Lacking]		
Group 5. Pyloric muscles			
p.1	Pyloric muscle {216c}	Posterior mesopyloric ossicle (XX), dorsolateral edge	Middle supraampullary ossicle (XXIX), anterior half
p.2	Pyloric muscle {216d}	Uropyloric ossicle (XXI), anterolateral edge	Middle supraampullary ossicle (XXIX), posterior end
p.3	[Continuous with p.7]	Middle supraampullary ossicle (XXIX), posterior medial edge	Posterior supraampullary ossicle (XXX)
p.4	Pyloric muscle {216j}	Middle supraampullary ossicle (XXIX), posterior end	Middle supraampullary ossicle (XXIV)
p.5	[Lacking]		
p.6a		Middle pleuropyloric ossicle (XXXII)	Under uropyloric fold, near infrauropyloric fragment (XXIa)

continues

TABLE 3 *Continued*

Muscle	Name	Origin	Insertion
p.6b		Middle pleuropyloric ossicle (XXXII)	Middle supraampullary ossicle (XXIX), inner edge
p.7	Pyloric muscle {216f} [continuous with p.3]	Middle pleuropyloric ossicle (XXXII)	Posterior supraampullary ossicle (XXX), inner margin
p.8	Pyloric muscle {216e}	Dorsolateral ampulla	Uropyloric fold, beneath p.2
p.9		Inferior ampullary ossicle (XXIV), at tip of ampulla	Posterior supraampullary ossicle (XXX), ventral part
p.10	Pyloric muscle {216g}	Uropyloric ossicle (XXI), anterolateral edge	Posterior supraampullary ossicle (XXX)
p.11	[Continuous with p.10]	Uropyloric ossicle (XXI), lateral edge, just posterior to p.10	Posterior supraampullary ossicle (XXX)
p.12		Surface of gut and caecum, just above uropyloric ossicle (XXI)	Earlike projection of posterior pleuorpyloric ossicle (XXXIII)
p.13	Pyloric muscle {216h}	Earlike projection of posterior pleuorpyloric ossicle (XXXIII)	Posterior supraampullary ossicle (XXX), dorsolateral edge
p.14	[lacking]		

[a]Nomenclature in the "Muscle" column is that of Maynard and Dando (1974); muscles without listed names are not specifically stated; numbers in braces in the "Name" column indicate the muscle identification scheme of Cochran (1935), where available. sw, Insertion on the stomach wall. Compare with Fig. 3. (Based on information from Maynard and Dando, 1974.)

sal wall of the stomach just anterior to the medial tooth. The posterior gastric muscle (g.m.2a,b,3a) connects the posterior, dorsal wall of the carapace and the pyloric ossicle (III), which forms the dorsal wall of the stomach posterior to the medial tooth. The mesocardiac and pyloric ossicles are connected to each other via the urocardiac ossicle anteriorly and the prepyloric ossicle posteriorly. The sequence of muscles and skeletal elements, then, involved in the movement of the medial tooth is, from anterior to posterior: anterior dorsal wall of carapace, anterior gastric muscle (g.m.1b), mesocardiac ossicle (I), urocardiac ossicle (VII) descending into the stomach, prepyloric ossicle (VI) ascending to the dorsal wall, pyloric ossicle (III), posterior gastric muscle (g.m.2a,b,3a), and posterior dorsal wall of carapace. The urocardiac and prepyloric ossicles, both of which have dentate portions, together form the medial tooth. Contraction of the anterior and posterior gastric muscles causes the mesocardiac and pyloric ossicles to move anteriorly and posteriorly, respectively. The movement of these two ossicles affects the leverage of the system in such a way as to cause the medial tooth to move downward and forward (ventrally and anteriorly). This dilating, extrinsic muscle system (anterior and posterior gastric muscles) is opposed by the intrinsic cardiopyloric muscle (g.m.4a–c), which forms a sheet spanning the dorsal anterior and posterior arches of the gastric mill and causes the medial tooth to move posteriorly and dorsally (Fig. 3).

The medial tooth is thus drawn across the lateral teeth, which have been moved together (toward the midline of the cardiac stomach) by the lateral interior cardiac muscle (g.m.5b,6a). This causes the mastication of food among the three major teeth of the gastric mill (i.e., between the dentate surfaces of each of the lateral teeth and the dentate medial tooth). The intrinsic muscles of the lateral teeth are opposed by the extrinsic lateral posterior gastric muscle (g.m.3a), which moves the lateral teeth apart. The relationship of the ossicles and the major muscle bundles is illustrated in Fig. 6, which is helpful in the difficult task of visualizing the action of muscles and the movement of ossicles. (The arrangement of ossicles VI and VII, which form the medial tooth, is more clearly seen in Fig. 6 than in Fig. 2.)

Endoscopic studies of spiny lobsters have shown a flexibility of gastric mill movements (reviewed by Turrigiano and Heinzel, 1992). Movements of the medial or lateral teeth alone have been observed. In addition, there are several modes of masticating movements, most notably the "squeeze," "cut and grind," and "cut and squeeze" modes. The squeeze mode involves the movement of the lateral teeth

TABLE 4 **Nerves of the Stomatogastric System of *Homarus americanus*[a]**

Group 1. Motor fibers to oesophageal[b] muscles

These have not been examined in detail.

Group 2. Sensory fibers

d.-p.o.n.	Dorsal posterior oesophageal nerve
l.c.n.	Lateral cardiac nerve
l.v.n.	Lateral ventricular nerve
m.v.n.	Median ventricular nerve
pl.n.	Posterolateral nerve
p.s.n.	Posterior stomach nerve
s.o.n.	Superior oesophageal nerve
v.c.n.	Ventral cardiac nerve
v.-l.v.n.	Ventral branch of l.v.n.
v.-pl.n.	Ventral branch of pl.n.
v.-p.o.n.	Ventral posterior oesophageal nerve

Group 3. Neurosecretory fibers

This group includes fine fibers that originate from fibers in the core of the nerves involved. The principal nerves include the ventral portion of the stomatogastric nerve (just dorsal to its junction with the superior oesophageal nerves), the ventral portion of the inferior ventricular nerve, the superior and inferior oesophageal nerves, and the central trunk containing the oesophageal ganglion.

Group 4. Stomatogastric ganglion system

A.M.	Anterior median neuron
CP.	Cardiopyloric neuron
I.C.	Inferior cardiac neuron
L.C.	Lateral cardiac neuron
P.D.	Pyloric dilator neurons
V.D.	Ventricular dilator neuron
a.c.n.	Anterior cardiac nerves
a.g.n.	Anterior gastric nerve
ap.n.	Ampullary nerve
a.v.d.	Anterior ventricular dilator nerve
a.v.n.	Anterior ventricular nerves
c.g.	Commissural ganglia
c.m.n.	Cardiac branch (cardiac muscle nerve)
cp.n.	Cardiopyloric nerve (branch of l.v.n.)
d.b.	Dorsal branch of p.s.n.
d.d.n.	Dorsal dilator branch of d.-l.v.n.
d.-l.v.n.	Dorsal branch of l.v.n.
d.v.n.	Dorsal ventricular nerve
gp.n.	Gastropyloric nerve
i.c.n.	Inferior cardiac nerve
i.o.n.	Inferior oesophageal nerves
i.p.n.	Internal pyloric branch of py.n.
i.v.n.	Inferior ventricular nerve
l.c.n.	Lateral ventricular nerve
l.p.n.	Lateral pyloric nerve
l.v.n.	Lateral ventricular nerve
m.v.n.	Median ventricular nerve
o.g.	Oesophageal ganglion
p.d.n.	Pyloric dilator nerve
pl.g.	Posterolateral gastric branch of d.-l.v.n.
pm.g.	Posteromedian gastric branch of d.-l.v.n.
p.s.n.	Posterior stomach nerve
py.n.	Pyloric nerve
st.g.	Stomatogastric ganglion
st.n.	Stomatogastric nerve
s.o.n.	Superior oesophageal nerves
v.-l.v.n.	Ventral branch of l.v.n.
v.-py.n.	Ventral pyloric nerve

[a]Compare with Fig. 4. (Based on information from Maynard and Dando, 1974.)

[b]The spelling *oesophageal* of Maynard and Dando (1974) is retained in this table, so that names will match abbreviations; *esophageal* is used elsewhere in this chapter.

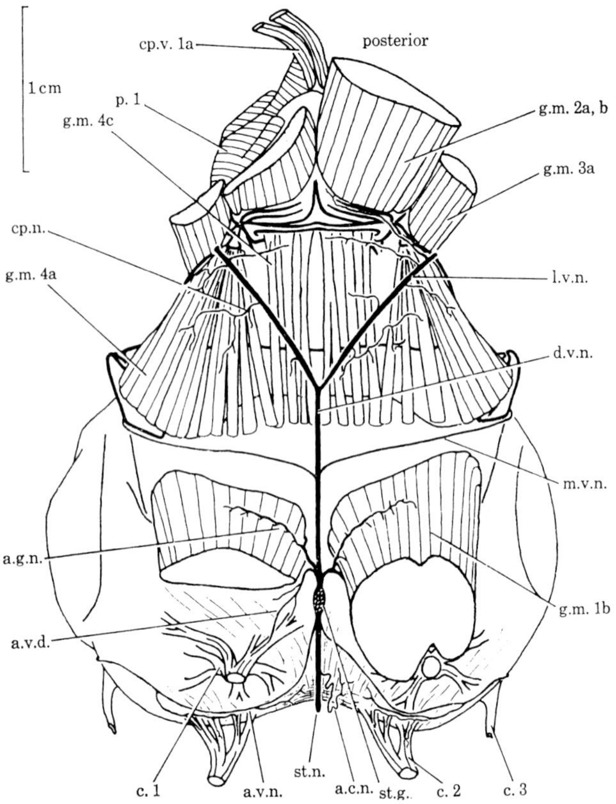

FIGURE 5 Nerves of the stomatogastric system of *Homarus americanus*. Dorsal view of the foregut (posterior uppermost). Key to abbreviations in Table 4. (Reprinted from Maynard and Dando (1974) *Philos. Trans. R. Soc. London Ser. B*, 268, 161–220, with permission from The Royal Society.)

toward the medial tooth, causing a convergence of the three teeth that squeezes food among them, followed by a return to the resting position. In the cut and grind mode, the mastication results from the forward movement of the medial tooth and backward movement of the lateral teeth, causing the cusps of the lateral teeth to grind against the surface of the medial tooth. In the cut and squeeze mode, the lateral teeth converge in a "cut," then open, and finally close again to meet the medial tooth in a "squeeze." Other variations of movement have also been observed in spiny lobsters, but less frequently. Movements of the teeth of the gastric mill of *Homarus americanus* have yet to be described in such detail.

5. Cardiopyloric Valve

The cardiac and pyloric portions of the stomach are separated primarily by the cardiopyloric valve, which limits and controls the movement of food from the cardiac to the pyloric chamber. It is a high ridge that protrudes dorsally from the floor of the stomach along the midline. In order to enter the pyloric stomach, food particles must pass over the top of the valve or around the valve in lateral channels.

6. Wall of the Pyloric Stomach

The arrangement of tissues in the wall of the pyloric stomach is similar to that in the wall of the cardiac stomach. The pyloric ossicles and other elabo-

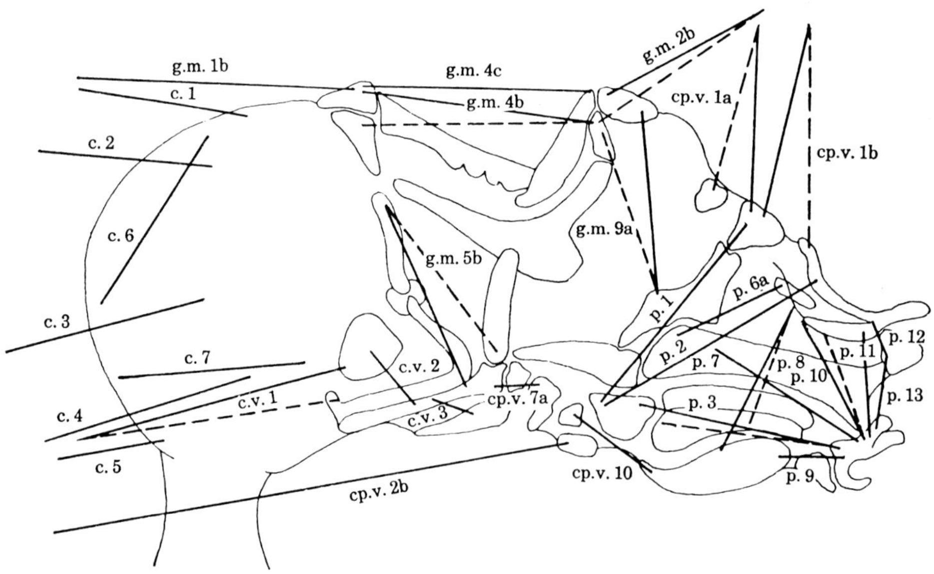

FIGURE 6 Diagram of the muscle bundles common to three decapod species (*Homarus americanus*, *Callinectes sapidus*, and *Panulirus argus*), showing the relationship of the ossicles and the major muscle bundles. Bundles are indicated by solid lines connecting idealized decapod ossicles (not specifically those of *H. americanus*); broken lines indicate minor variations among species. Key to abbreviations in Table 3. (Reprinted from Maynard and Dando (1974) *Philos. Trans. R. Soc. London Ser. B*, 268, 161–220, with permission from The Royal Society.)

rations of the lining cuticle form the pyloric filter apparatus.

7. Operation of the Pyloric Filter

The pyloric stomach is sometimes considered to contain a more dorsal, muscular "press" (a muscular wall that surrounds the middle pyloric canal) and a more ventral "filter" (Yonge, 1945). The press accepts material from the cardiac stomach and squeezes it through the more dorsal aspects of the pyloric filtering apparatus. The particles that pass then enter the more ventral gland filters, which are responsible for final filtration and act as gatekeepers for the digestive gland, which would become clogged by larger particles. Only the finest particles (1.0 μm or less) pass through the gland filters into the digestive gland (Bayer *et al.*, 1979). Yonge (1945) suggested that material to be passed into the intestine is drained of fluid and compacted by the pyloric press.

It appears that less information is available about the functioning of the pyloric portion of the stomatogastric system than about the cardiac portion, and very little of that is specific to *Homarus americanus*. Based on other decapods, Claiborne and Ayers (1987) suggest the following sequence: (1) dilation of the pyloric chamber by contraction of the anterior dorsal and external inferior pyloric dilators (cp.v.1,2) and opening of the cardiopyloric valve by the lateral posterior ventricular dilator (c.v.1); (2) constriction of the anterior region of the pyloric chamber (cp.v.4,5, which are both lacking in *H. americanus*, and p.11) and closing of the cardiopyloric valve by the external inferior pyloric dilators (c.v.2); and (3) constriction of the posterior region of the pyloric chamber by the pyloric muscles (p.2–14). They caution, however, that this scheme is probably too simple, considering the complex anatomy of the pylorus. Johnson and Hooper (1992) state that it is not understood how the contractions of these muscles result in the sorting of food particles.

D. Movement of Food through the Foregut

Food may move through the stomach along several complex paths, determined largely by the size of the particles and directed by channels, setae, valves, and filters. The following account is based primarily on Williams (1907) and Yonge (1924).

The anterior portion of the cardiac stomach, often called the cardiac sac, receives food from the esophagus, which enters ventrally near the anterior end. The muscular contractions of the wall of the cardiac stomach, together with the masticating movements of the gastric mill, serve to circulate food around the chamber, bring it into contact with the grinding teeth, and mix it with enzymes from the digestive gland. The movement of food is aided by setae that protrude from the walls of the cardiac stomach and are inclined in the direction of the proper flow of material. Setae may also serve as filters that allow only appropriately small particles to enter the ventral channels that convey food toward the pyloric stomach.

The cardiac stomach is separated from the pyloric stomach by the cardiopyloric valve, a high ridge between the two chambers that is responsible for retention of large food particles in the cardiac stomach. Particles too large to pass over the cardiopyloric valve are directed dorsally and are processed again by the gastric mill. After mastication by the gastric mill is complete, suitably small particles of food may pass laterally around the cardiopyloric valve via the lower cardiac canals (the "ventral channels" of Yonge, 1924), into the lower pyloric canals of the gland filters of the pyloric stomach (the "lateral pouches" of Williams, 1907; the "lateral pyloric pouches" of Herrick, 1909; and the "ampullae" of Maynard and Dando, 1974). Particles that can pass this final filter move through ducts into the digestive gland.

Some particles may pass over the cardiopyloric valve along the midline and enter the dorsal portion of the pyloric stomach (the dorsal, or upper, pyloric canal). Particles small enough to percolate down past the chitinous ridge and through the sieves of the pyloric stomach may then enter the gland filters, where final sifting occurs. These paired, ventrolateral chambers have the finest of all the sieves and allow only the smallest particles to enter the digestive gland (1.0 μm) (Bayer *et al.*, 1979), where final digestion and absorption occur. Particles of intermediate size, which are too large to pass through the gland filters, are routed posteriorly into the intestine, where they are mixed with waste material from the digestive gland (Conklin, 1980).

III. Midgut

The intestine is a tube that connects the foregut to the hindgut. Associated with the intestine are several diverticular organs of greater or lesser complexity. The largest and most complex of these organs, located at the anterior end of the midgut, is the digestive gland (=hepatopancreas, or midgut gland). Also associated with the intestine are the paired anterior midgut caeca, located at the anterior end of the intestine, and the posterior midgut caecum, near the junction of the intestine and the rectum. Based on ultrastructural and physiological studies, the epithelia of the intestine, midgut caeca, and hindgut appear to be involved in ion and water transport (Mykles, 1979).

A. *Intestine*

1. Gross Anatomy

The intestine of *Homarus americanus* is a gently curved, tubular organ of simple construction at the gross level (Fig. 1). It is approximately circular in cross-sectional shape (Fig. 7) and comprises an epithelium and a connective tissue, separated by a complex basement membrane. The intestine begins in the cephalothorax at the posterior margin of the pyloric stomach, where it is first directed ventrally, then gently curves dorsally, and finally levels off as it continues posteriorly. It continues through the posterior portion of the cephalothorax and most of the abdomen and ends in the sixth abdominal segment, where it joins the hindgut. A peritrophic membrane, lining the lumen and presumably secreted by the intestinal epithelium, has been reported in adult *H. gammarus* (Barker and Gibson, 1977) and larval (stage

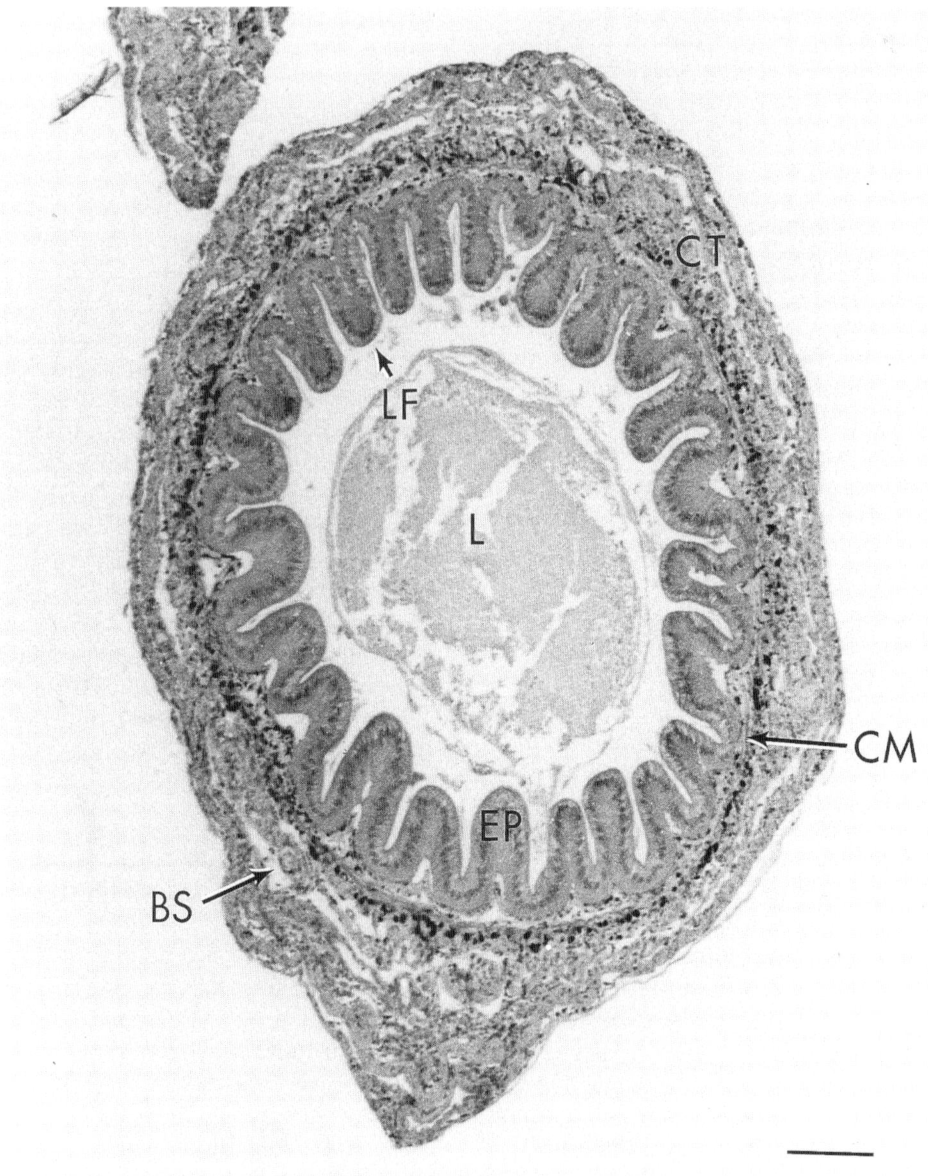

FIGURE 7 Cross-sectional view of the intestine. BS, Blood (hemal) sinus; CM, circular muscle; CT, connective tissue; EP, epithelium; L, lumen; LF, longitudinal fold (ridge). Photomacrograph of paraffin section stained with hematoxylin and eosin. Scale bar: 0.2 mm.

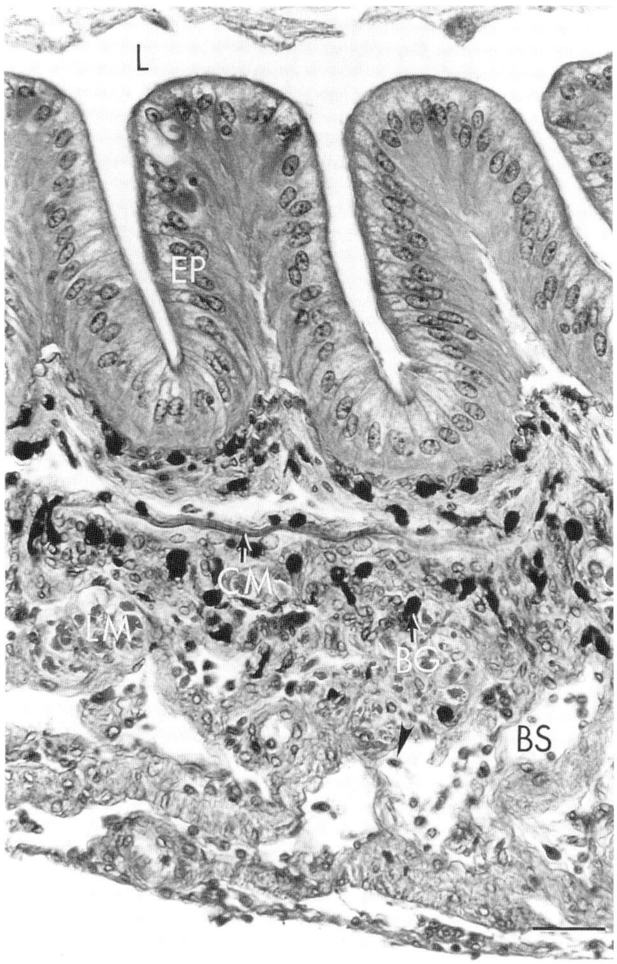

FIGURE 8 Cross-sectional view of the wall of the intestine. BG, Basophilic granulocyte; BS, blood (hemal) sinus; CM, circular muscle; CT, connective tissue; EP, epithelium; L, lumen; LF, longitudinal fold (ridge); LM, longitudinal muscle. Photomicrograph of paraffin section stained with hematoxylin and eosin. Scale bar: 40 µm.

I) *H. americanus* (Factor, 1981b).

2. Arrangement of Tissues

The wall of the intestine consists of two tissue layers: epithelium and connective tissue. The simple columnar epithelium is gently folded to produce low longitudinal ridges or folds (Figs. 7 and 8). As with the epithelia of all organs of the midgut, it does not produce a cuticular lining. The apical cell membrane forms a brush border of microvilli; adjacent lateral membranes are joined apically by septate junctions; the lateral membrane is generally straight, but interdigitates with lateral membranes of adjacent cells basally (Mykles, 1979). Basal cells are interspersed among the bases of the epithelial cells (Johnson, 1980).

The epithelium is underlain by a basement membrane (Fig. 9) that is much more complex structurally than those underlying the esophageal and rectal epithelia. This complex intestinal basement membrane varies from approximately 0.2 to 0.4 µm thick and has three layers (Factor, 1981a). The central layer is finely granular and extremely uniform; it is coated on both sides with thin, fibrous, electron-dense outer layers. Branching channels pierce the basement membrane. Processes of the basement membrane proper lie primarily on the connective tissue side. Analysis of serial sections shows that the larger processes are often continuous with the basement membrane proper; they consist of the same finely granular, central material, surrounded by the same fibrous, electron-dense material (Factor, 1981a; Bryant, 1992; E. A. Bryant and J. R. Factor, unpublished observations). The processes form a network that extends into the loose connective tissue, which may serve to anchor the basement membrane to the underlying connective tissue. Strands of the fibrous outer layer interconnect the processes and bridge the channels in the basement membrane. The biochemical constituents of the intestinal basement membrane have not been determined, nor is the functional significance of this complex structure known.

The connective tissue of the intestine is a matrix of collagen fibers in which the other subepithelial elements are embedded (Figs. 8 and 10). Bundles of longitudinal muscles lie near the periphery of the intestine and thinner aggregates of circular muscles are centrally located, some just beneath the epithelium (Fig. 8). The matrix is permeated by the system of irregular hemal sinuses (=blood spaces) in which circulating hemocytes can be seen. Fibroblasts and granulocytes are embedded in the collagenous matrix. The electron-dense granules in acidophilic granulocytes are larger than those in basophilic granulocytes (Fig. 10).

B. Digestive Gland

The term *digestive gland* is used in this chapter for the organ often called "hepatopancreas" or "midgut gland" and sometimes called "liver" or "digestive diverticula"; van Weel (1974) and Gibson and Barker (1979) provide discussions of the appropriateness of the various terms applied to this organ. The digestive gland is involved in diverse metabolic activities. It is primarily responsible for the synthesis and secretion of digestive enzymes, the final digestion of food, and absorption of nutrients. It is also involved in excretion, the molt cycle, storage of inorganic reserves, lipid and carbohydrate metabolism, and the storage of organic reserves, primarily lipids (Barker and Gibson, 1977; reviewed by Gibson and Barker, 1979).

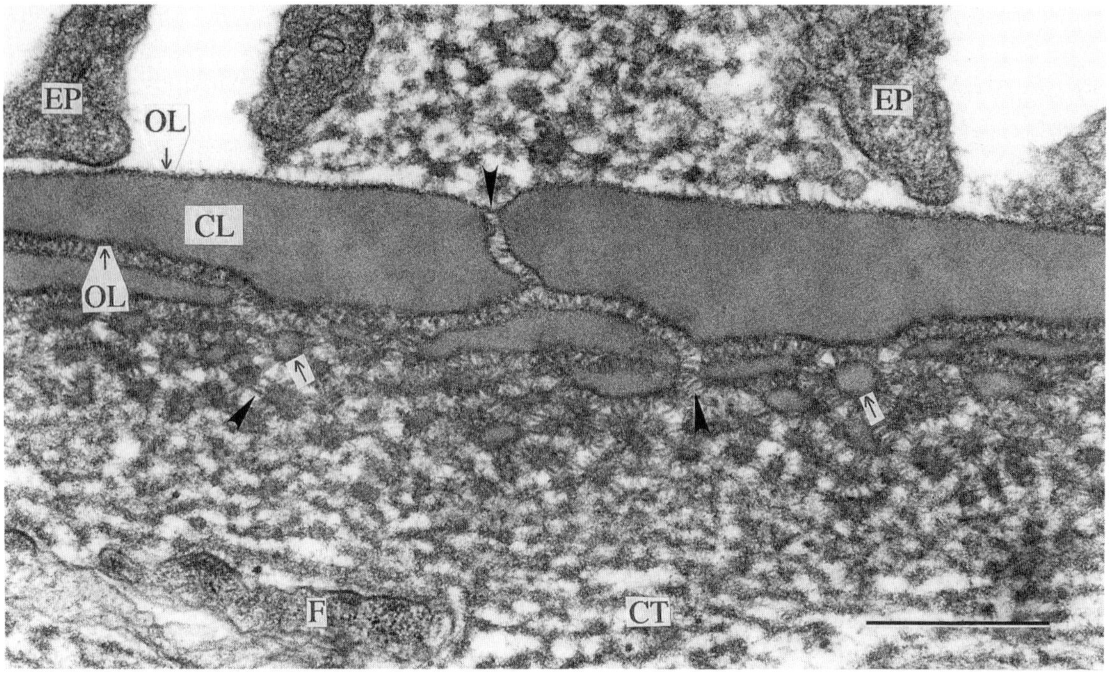

FIGURE 9 Transmission electron micrograph of a complex intestinal basement membrane. CL, Central layer of basement membrane; CT, connective tissue; EP, epithelial cell; F, fibroblast; OL, outer layer of basement membrane. Arrows indicate processes of the basement membrane; arrowheads, fibrous strands that interconnect processes and bridge channels. Scale bar: 0.5 μm. (From Factor, 1981a, *J. Morphol.* with permission of Wiley–Liss. Copyright © 1981 Alan R. Liss, Inc.)

1. Gross Anatomy

The digestive gland of *Homarus americanus* is actually a pair of large, complex glands, each with its own duct, lying primarily in the cephalothorax. Each gland, left and right, is divided into three lobes: anterior, posterior, and dorsal. They extend along either side of the cardiac stomach anteriorly, into the anterior portion of the abdomen posteriorly, and fill much of the hemocoelic space in the cephalothorax. Each lobe comprises a series of blindly ending tubules, called digestive tubules or tubules of the digestive gland (Figs. 11a–c and 12). The lumina of these digestive tubules run together into a major duct for each of the three lobes. The three ducts of the lobes then coalesce into the single duct for the digestive gland on each side of the animal. These two ducts communicate with the gut proper near the junction of the pyloric stomach and the intestine (there are conflicting reports of the precise location). The digestive gland is invested in an outer sheath of connective tissue.

2. Arrangement of Tissues

The tubules of the digestive gland are approximately 100 μm in diameter and vary in length. They have a rather simple construction (Figs. 12 and 13). Starting at the lumen and progressing peripherally, each digestive tubule includes a digestive epithelium, basement membrane, contractile cells, and tunica propria. The nonepithelial elements can be considered part of the extensive network of connective tissue in this organ.

The digestive epithelium of the tubules has been shown to comprise four cell types in *Homarus gammarus*: E cells (embryonic cells), R cells (resorptive cells), F cells (fibrillar cells), and B cells (blisterlike cells) (Barker and Gibson, 1977). Conklin (Chapter 16) discusses the roles of these cells in digestion. In addition, basal cells with indistinct cytoplasmic granules are inconspicuous and only found uncommonly among the bases of the epithelial cells (Johnson, 1980).

The digestive epithelium is underlain by a complex, two-layered basement membrane approximately 0.3 μm thick (Figs. 11f and 14) (Factor and Naar, 1985; Bryant, 1992; E. A. Bryant and J. R. Factor, unpublished observations). The layer adjacent to the epithelium contains a network of interconnecting fibers (Fig.

FIGURE 10 Connective tissue of the wall of the intestine. AG, acidophilic granulocyte; BG, basophilic granulocyte; BL, basal lamina (=basement membrane); BS, blood (hemal) sinus; CM, circular muscle; EP, epithelium; F, fibroblast; HA, hemocyte with acidophilic granules (=acidophilic granulocyte); HB, hemocyte with basophilic granules (=basophilic granulocyte); L, lumen; LF, longitudinal fold (ridge); LM, longitudinal muscle. Scale bar: 5.0 μm.

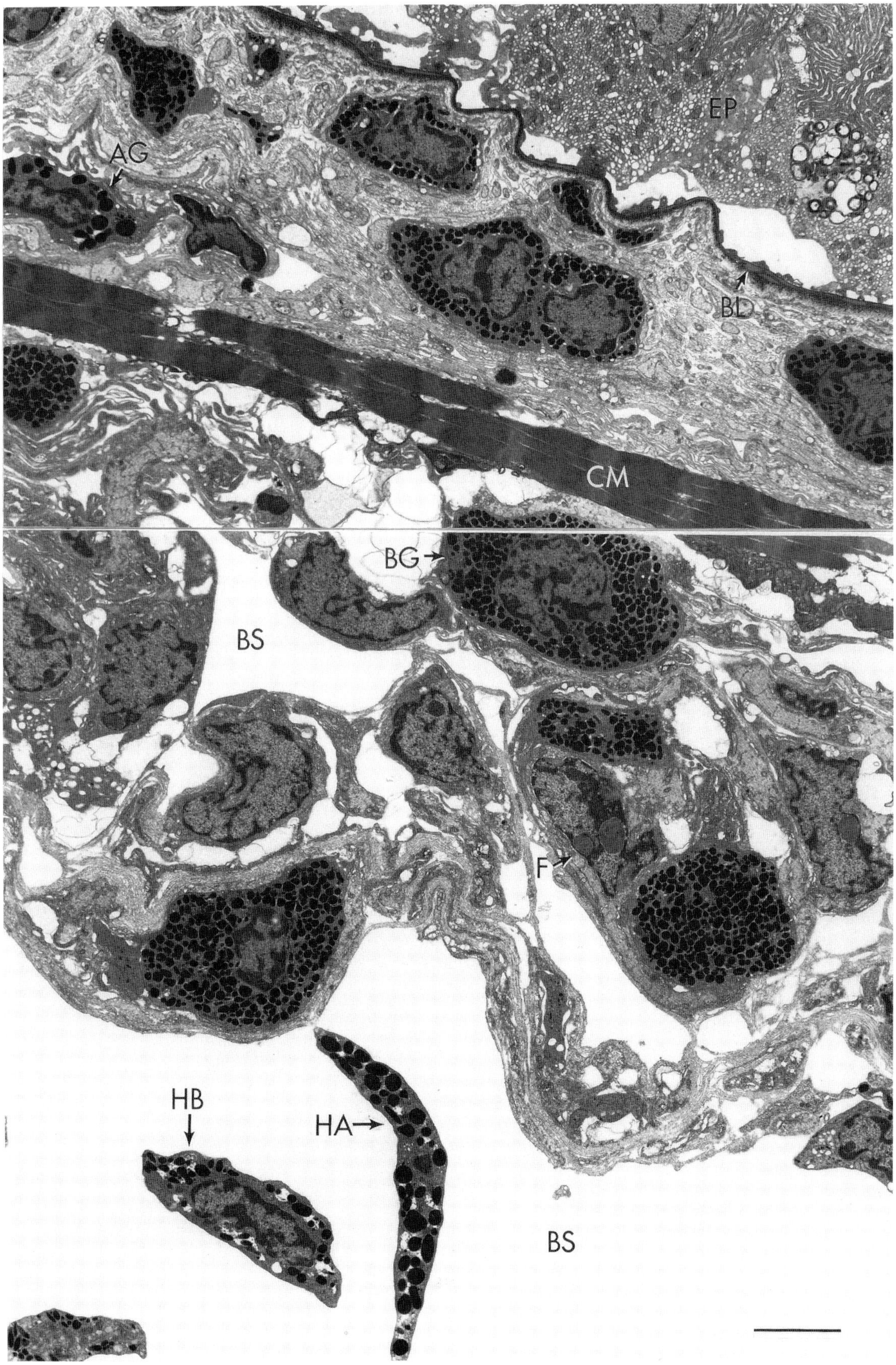
EP
AG
BL
CM
BG
BS
F
HB
HA
BS

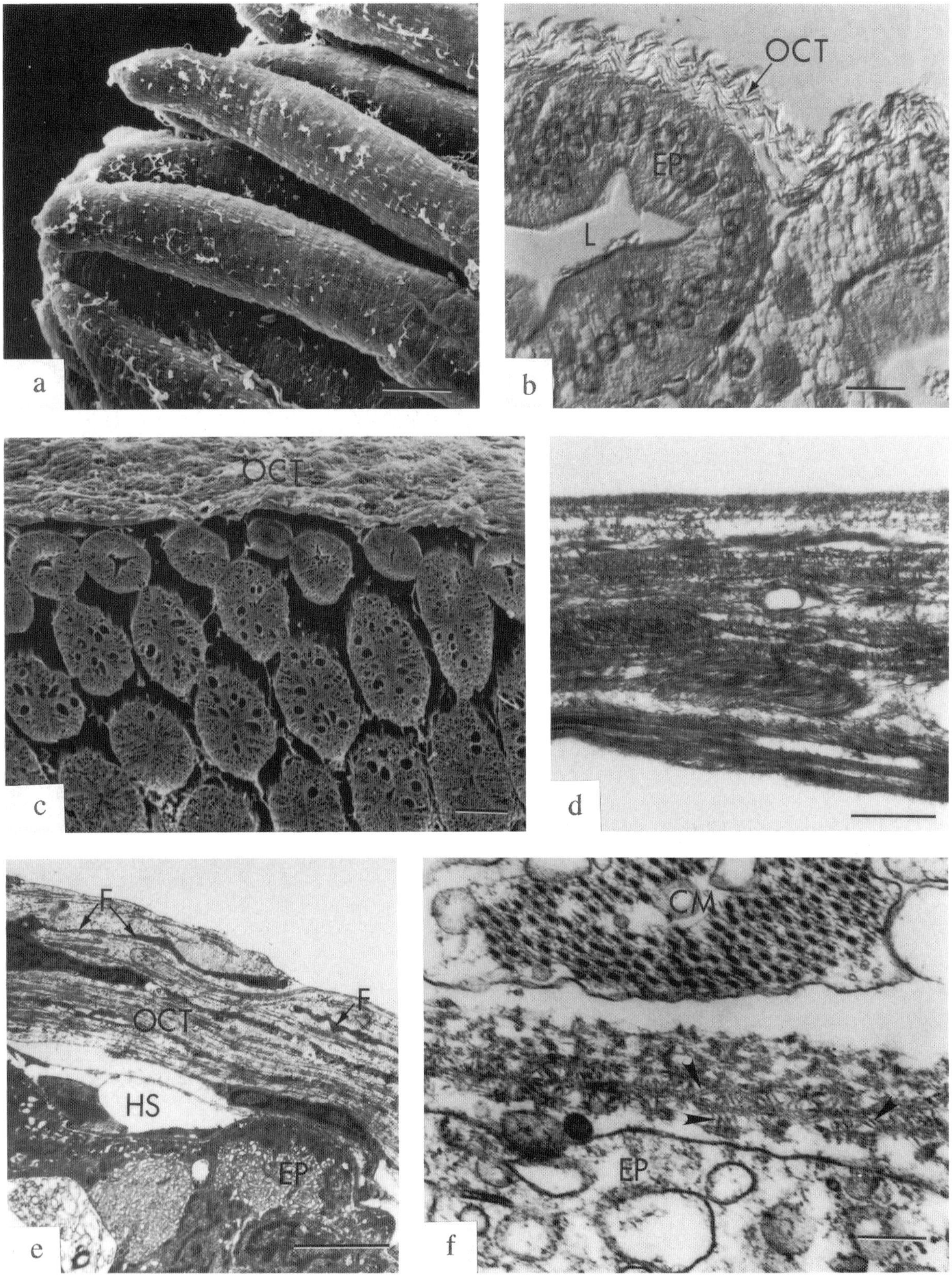
OCT
EP
L
OCT
F
F
OCT
HS
EP
CM
EP
a
b
c
d
e
f

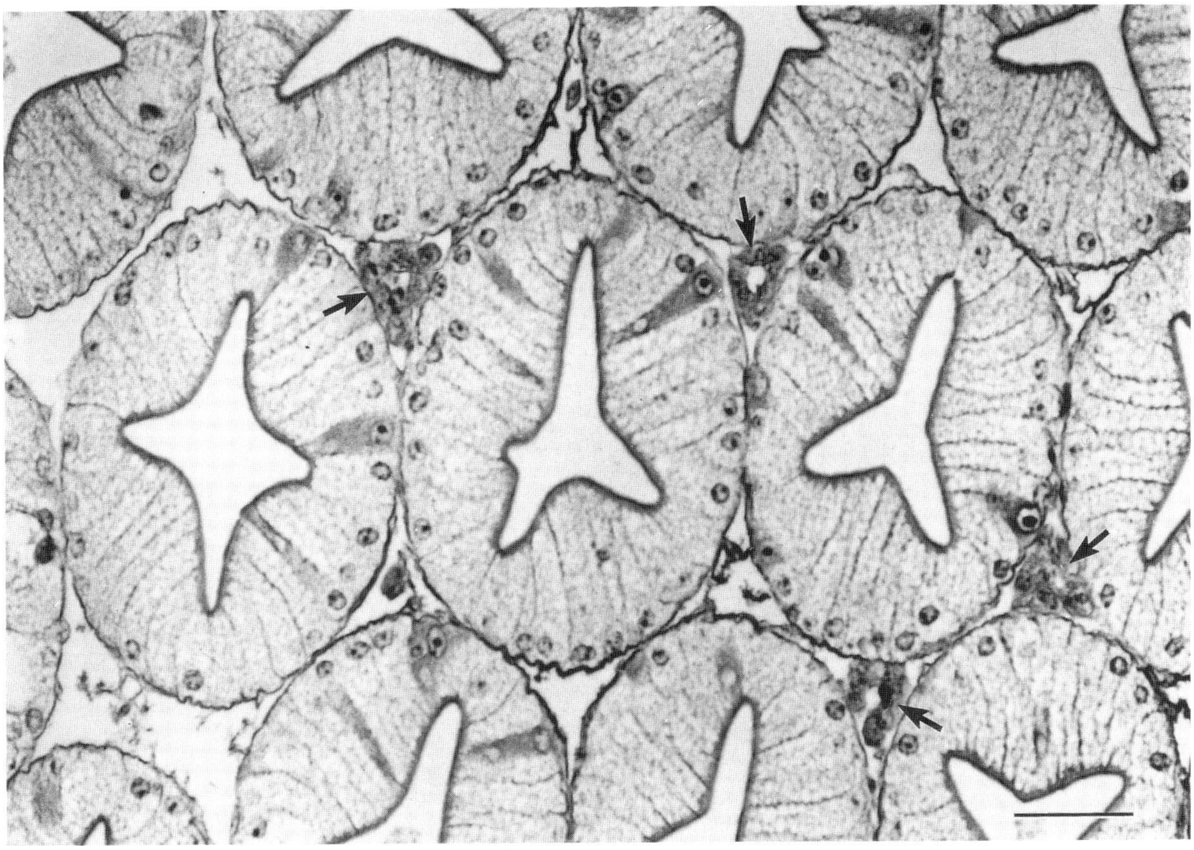

FIGURE 12 Cross-sectional view of the tubules of the digestive gland. Arrows indicate terminal hepatic arterioles. Photomicrograph of paraffin section stained with Mallory's triple stain. Scale bar: 50 μm. (From Factor and Naar, 1990, with © permission of Wiley–Liss, Inc.)

11f). At lower magnifications, the interconnecting fibers are not obvious and the epithelial side of the basement membrane simply appears to be more electron dense than the peripheral side (Fig. 14a and b).

The digestive gland has one continuous network of connective tissue in which the tubules are embedded and suspended and which forms the outer limiting layer of the organ (Factor and Naar, 1985). This tissue serves a typical connective tissue function—supporting the functional units of the organ, the digestive tubules. The layer surrounding each tubule—indeed, all of the space among tubules from the epithelium of one tubule to the epithelium of the next—can be viewed as a modified form of connective tissue, analogous to connective tissues in other regions of the digestive system. It is similar to other digestive connective tissues in position (peripheral to the epithelium), in function (support of the organ), and in components (including blood vessels, hemal sinuses, and muscular elements). It differs, however, in that the interior spaces do not have an extensive fibrous (collagenous) matrix, the hemal sinuses are more extensive, and the contractile cells do not form typical bands of muscles, but instead have cell processes containing myofilaments.

Contractile cells containing bundles of myofilaments lie between the basement membrane and the tunica propria (Fig. 14). Each cell body sends out both circular and longitudinal processes that contain bundles of circular and longitudinal myofilaments,

FIGURE 11 (a) Scanning electron micrograph of several tubules dissected from the digestive gland. (b) Tubules of the digestive gland and the outer connective tissue layer (OCT) that invests the organ. EP, Epithelium; L, lumen of the tubule. Photomicrograph of transverse section of tubules stained with phosphotungstic acid hematoxylin. Nomarski differential interference contrast optics. (c) Outer connective tissue layer (OCT). Scanning electron micrograph of a partially sectioned organ. (d) Transmission electron micrograph of the outer connective tissue layer (OCT). (e) Transmission electron micrograph of the outer connective tissue layer (OCT) illustrating fibroblasts (F), epithelial cells (EP), and hemal sinuses (HS). (f) Transmission electron micrograph of the basement membrane underlying the digestive epithelium (EP) of a tubule. Note the network of interconnecting fibers (arrowheads) in the epithelial half of the basement membrane and a bundle of circular myofilaments of a contractile cell (CM). Scale bars: (a) 100.0 μm; (b) 20.0 μm; (c) 100.0 μm; (d) 1.0 μm; (e) 3.0 μm; (f) 0.2 μm. (From Factor and Naar, 1985, with permission of Wiley–Liss. © Alan R. Liss, Inc.)

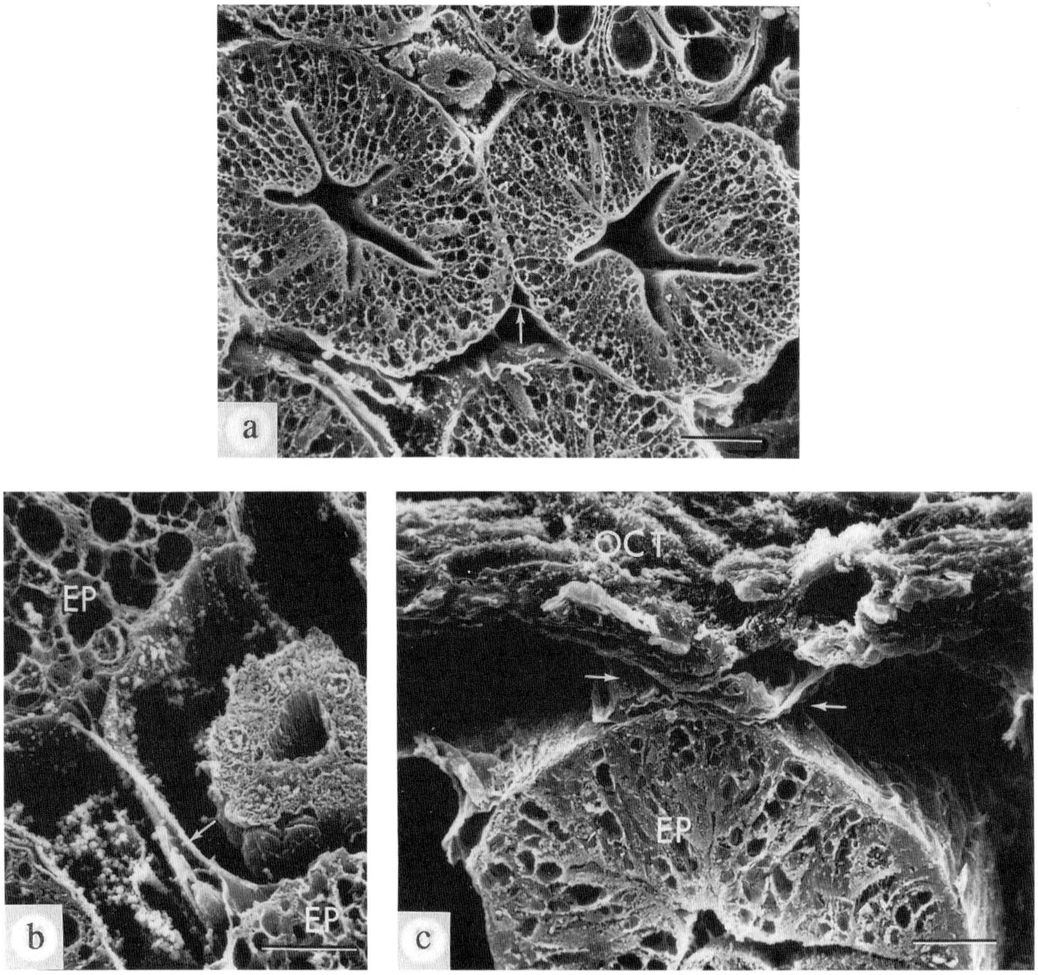

FIGURE 13 (a) Cross-sectional view of the tubules of the digestive gland. (b) Continuous tunica propria (arrow) between two tubules. White arrow indicates tunica propria, continuous from one tubule to the next. (c) A tubule adjacent to the outer connective tissue layer (OCT) that limits the digestive gland, showing the confluence between the tunica propria of the tubule and the outer layer (arrows). EP, epithelium of digestive tubule. Scanning electron micrographs of a partially sectioned organ. Scale bars: (a) 50.0 μm; (b and c) 20.0 μm. (From Factor and Naar, 1985, with permission of Wiley–Liss, Inc. © Alan R Liss, Inc.)

respectively. Both of these myofilaments can be seen in the same cell in fortuitous planes of section (Fig. 14d and e). In such areas, the thicker, circular bundles are peripheral to the less substantial, central, longitudinal bundles. Processes of the contractile cells form a network surrounding each tubule (the "muscle net" of Leavitt and Bayer, 1982; Factor and Naar, 1985). The circular myofilaments are interconnected by longitudinal myofilaments of smaller diameter (Fig. 14f). Depending on the level of any particular section, the space between the basement membrane and the tunica propria may appear to be populated by a large number of cells and processes (Fig. 14a), or it may seem to contain only an occasional cell process (Fig. 14b). Not much of the surface of the tubules is cov-

FIGURE 14 Peripheral connective tissue layer (PCT) of a tubule of the digestive gland. The space between the basement membrane and the tunica propria is well populated by contractile cell processes in some planes of section (a), and sparsely populated in others (b). Cell body and nucleus of a contractile cell (c). Bundles of both circular and longitudinal myofilaments in the same contractile cell, illustrated in cross-section (d) and longitudinal section (e). "Muscle net" of circular and longitudinal processes of contractile cells; arrows indicate the edge of the disrupted tunica propria. BM, Basement membrane; CM, circular myofilaments of the contractile cell; CP, cell process of the contractile cell; EP, epithelial cell (basal portion); LM, longitudinal myofilaments of the contractile cell; LP, longitudinal process of the contractile cell; TP, tunica propria. (a–e) Transmission electron micrographs; (f) scanning electron micrograph. Scale bars: (a and b) 1.0 μm; (c) 3.0 μm; (d and e) 0.2 μm; (f) 20 μm. (From Factor and Naar, 1985, with permission of Wiley-Liss. © Alan R. Liss, Inc.)

ered by contractile cell processes (Fig. 14f). Coordinated contraction of the myofilaments of the network causes the peristaltic movements of the tubules that is responsible for the transport of the luminal contents (Leavitt and Bayer, 1982).

The tunica propria (Yonge, 1924) is the thin, extra-

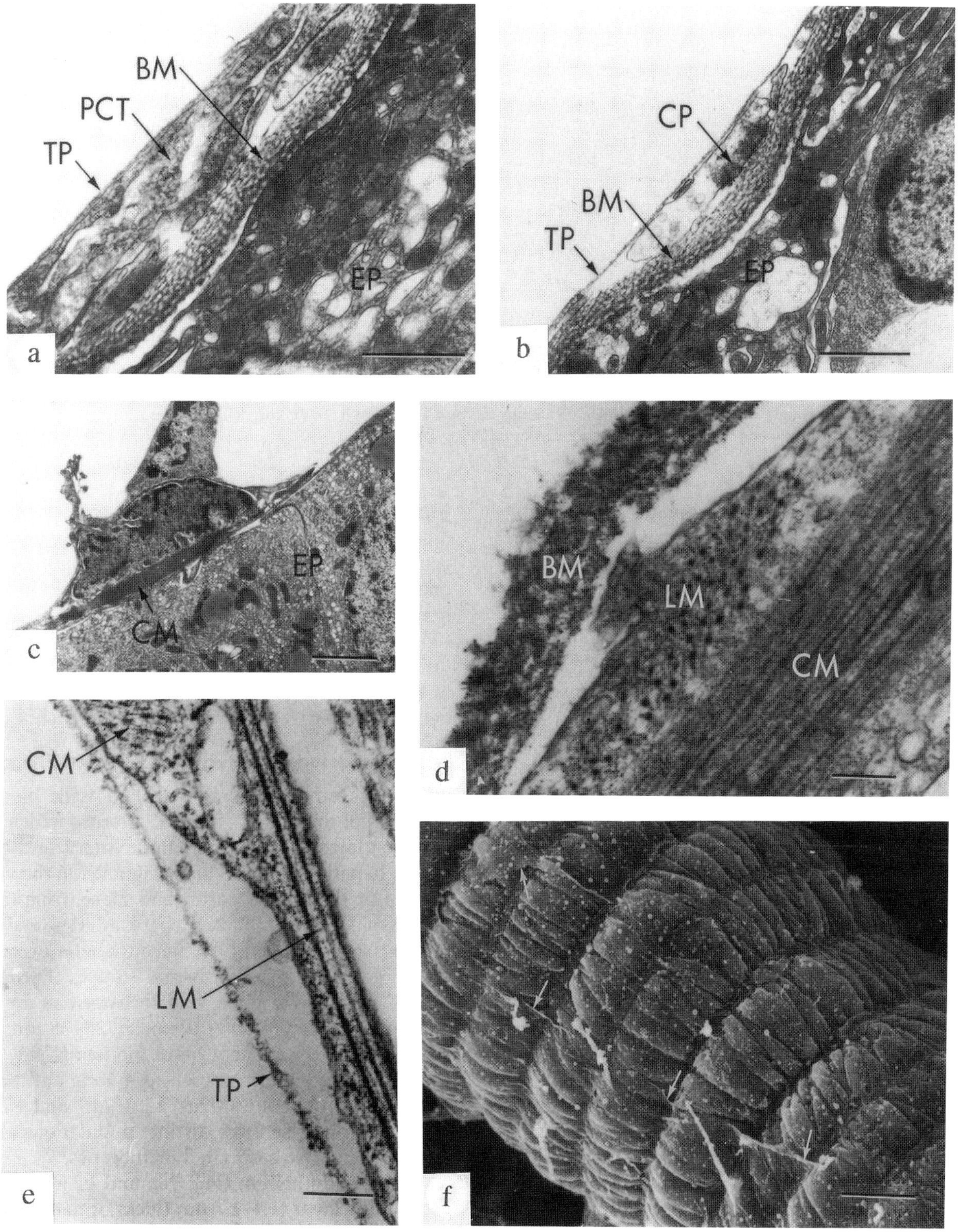

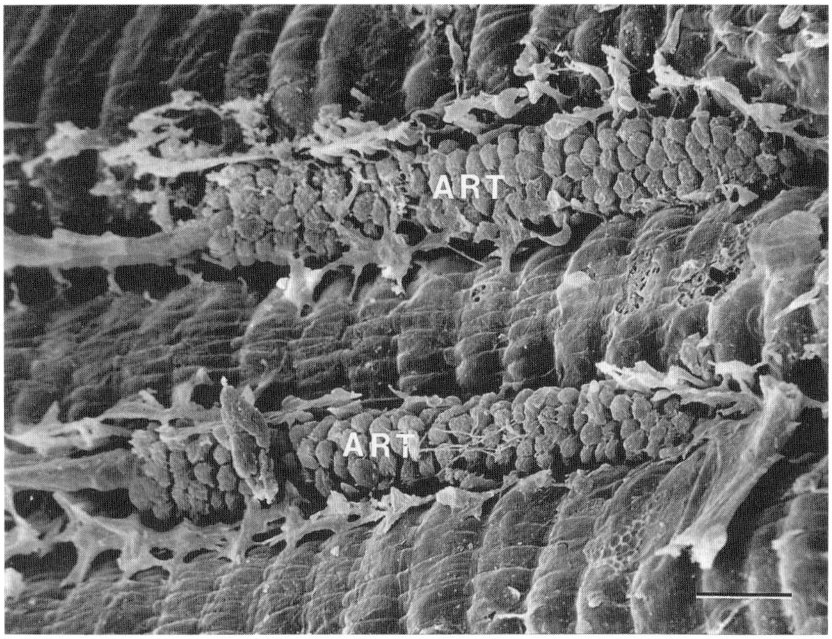

FIGURE 15 Longitudinal view of two terminal hepatic arterioles (ART) interspersed among three digestive tubules. Scanning electron micrograph. Scale bar: 50 μm. (From Factor and Naar, 1990, with © permission of Wiley–Liss, Inc.)

cellular, membranelike layer forming the outermost boundary of the tubules. In transmission electron micrographs, it appears as a uniform, apparently collagenous layer approximately 0.05 μm thick (Factor and Naar, 1985). In areas between contractile cell processes, the tunica propria lies directly against the basement membrane with no intervening structures or space; where cell processes are present, however, the tunica propria extends outward to cover them. The disrupted tunica propria can be seen when dissected tubules are viewed with the scanning electron microscope (Fig. 14f).

Much of the volume of the connective tissue of the digestive gland is occupied by hemal sinuses of the open circulatory system (Fig. 12), which can often be seen to contain circulating hemocytes and plasma. A variety of cells can be found among the tubules but outside the hemal sinuses, in places where tubules abut, including several types of granulocytes and cells that resemble fibroblasts. These are typical of the cells embedded in the connective tissue of other parts of the gut, for example, the intestine and the rectum. The nonepithelial cells of the digestive gland lie between the basement membrane and the tunica propria (as do the contractile cells). The tunica propria extends away from one tubule, covers the cells between tubules, and is continuous with the tunica propria of the adjacent tubule; it is continuous from one tubule to the next (Fig. 13) and is apparently disrupted when tubules are separated. There is no sharp distinction or boundary, then, between the peripheral connective tissue layer surrounding the epithelium of one tubule and that of the next; they are confluent. All hemal spaces are lined by an extracellular layer. The tunica propria can be considered to serve as such a lining in the digestive gland. In summary, the tubules of the digestive gland are suspended in an extensive network or matrix of loose connective tissue.

3. Terminal Hepatic Arterioles and Fixed Phagocytes

The digestive gland is supplied with blood by branches of the paired antennal arteries, which originate just lateral to the ophthalmic artery, and by the paired hepatic arteries, which originate on the ventral surface of the heart (Martin and Hose, Chapter 17). Numerous branches of the hepatic arteries, ending in terminal hepatic arterioles, permeate the organ and discharge blood into the hemal sinuses (Factor and Naar, 1990). The terminal hepatic arterioles are elongate tubes of tissue (outer diameter, 28–38 μm; inner diameter, 14–17 μm) that lie in the hemal sinuses of the digestive gland, interspersed among and parallel to the digestive tubules (Figs. 12, 13, 15, and 16). The arterioles comprise three distinct populations of cells (Fig. 16a) and two associated membranes.

1. The endothelium (Fig. 16a and b) is an almost continuous layer (1.4–1.7 μm thick) of flattened cells, with an often fimbriate inner surface, that forms the

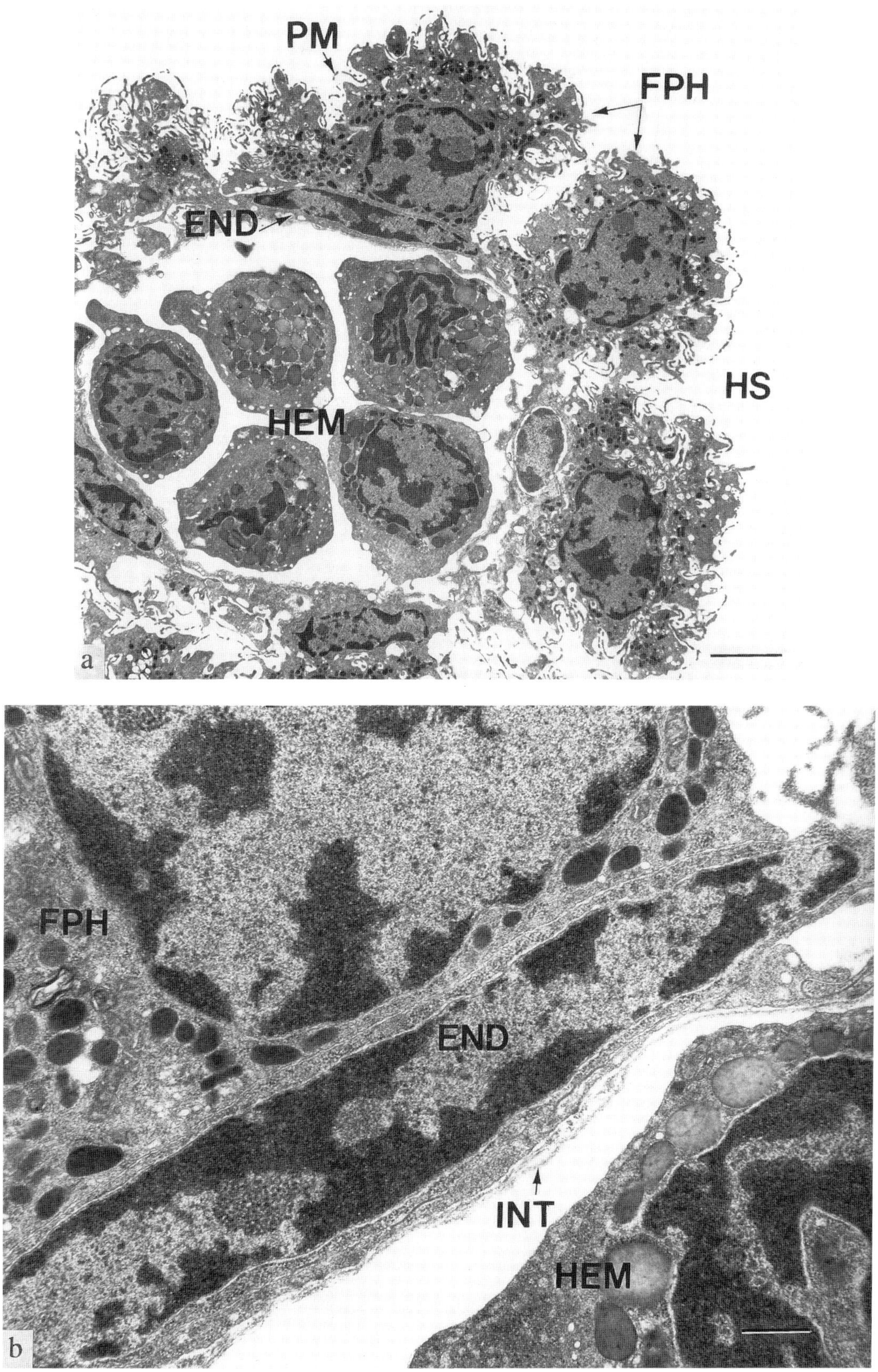

FIGURE 16 (a) Overview of a single terminal hepatic arteriole, cross-section. (b) Relationship of fixed phagocyte, endothelial cell, and intima. END, Endothelium; FPH, fixed phagocytes; HEM, circulating hemocytes in the lumen; HS, hemal sinus; INT, intima of endothelium; PM, perforated membrane. Transmission electron micrographs. Scale bars: (a) 3.0 μm; (b) 0.5 μm. (From Factor and Naar, 1990, with © permission of Wiley–Liss, Inc.)

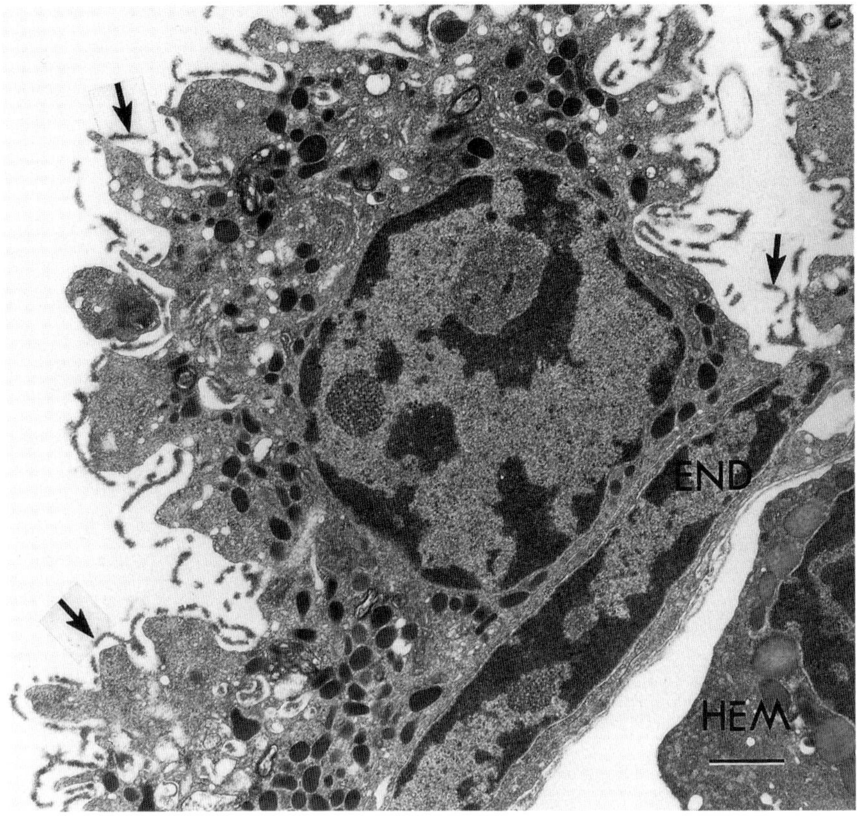

FIGURE 17 Transmission electron micrograph of a fixed phagocyte. Arrows indicate the perforated membrane; END, endothelium; HEM, circulating hemocyte in the lumen of the arteriole. Scale bar: 1.0 µm. (From Factor and Naar, 1990, with © permission of Wiley–Liss, Inc.)

wall of the arteriole. An acellular, endothelial intima (20–30 nm thick) on the inner (central) surface of the endothelial cells lines the lumen. There is a possibility of pores in the endothelium, but conclusive evidence is lacking.

2. Circulating hemocytes (Fig. 16) are present in the lumen of the arterioles. (Martin and Hose, in Chapter 17, discuss the various types of hemocytes.)

3. Fixed phagocytes (Figs. 16a and 17) are intimately associated with and firmly attached to the peripheral surfaces of the endothelial cells. They do not form a typical epithelial layer, but are attached by their basal surface only, with most of the lateral and apical surface area exposed to the hemolymph. Fixed phagocytes (7.0–9.5 µm across) contain a large, central-to-basal nucleus. The cytoplasm contains numerous electron-dense granules, roughly circular to irregular in shape (170- to 300-nm diameter). [These granules may be scarce or absent in quiescent cells and more numerous in activated cells (Johnson, 1987).] The exposed lateral and apical surfaces have a series of pronounced cell processes (i.e., ridges) that serve to increase the cell surface area (Fig. 17). The outer cell surface—indeed, the entire outer surface of the arteriole—is covered with a perforated membrane (70–100 nm thick; the "interrupted layer" of Johnson, 1980) that resembles a basement membrane but contains numerous pores (140- to 150-nm diameter, spaced 150–350 nm apart) (Figs. 17 and 18a and b). [The pores may be few or absent in quiescent cells (Johnson, 1987).] The perforated membrane is usually separated from the cell surface membrane of the fixed phagocytes by the pericellular space, yet points of contact are common. Cell processes sometimes extend through the pores.

The primary role of the terminal hepatic arterioles is to distribute blood to the extensive network of hemal sinuses of this very large organ. The flow of blood is presumably important in transferring the products of digestion and stored nutrients from the digestive tubules to the rest of the animal, as well as in supporting the metabolic activity of the tubules (the production of digestive enzymes and the final digestion of food) by supplying oxygen and removing cellular waste.

Along with circulating hemocytes and podocytes of the gills, the fixed phagocytes of the digestive gland also play an important role in cellular defense mecha-

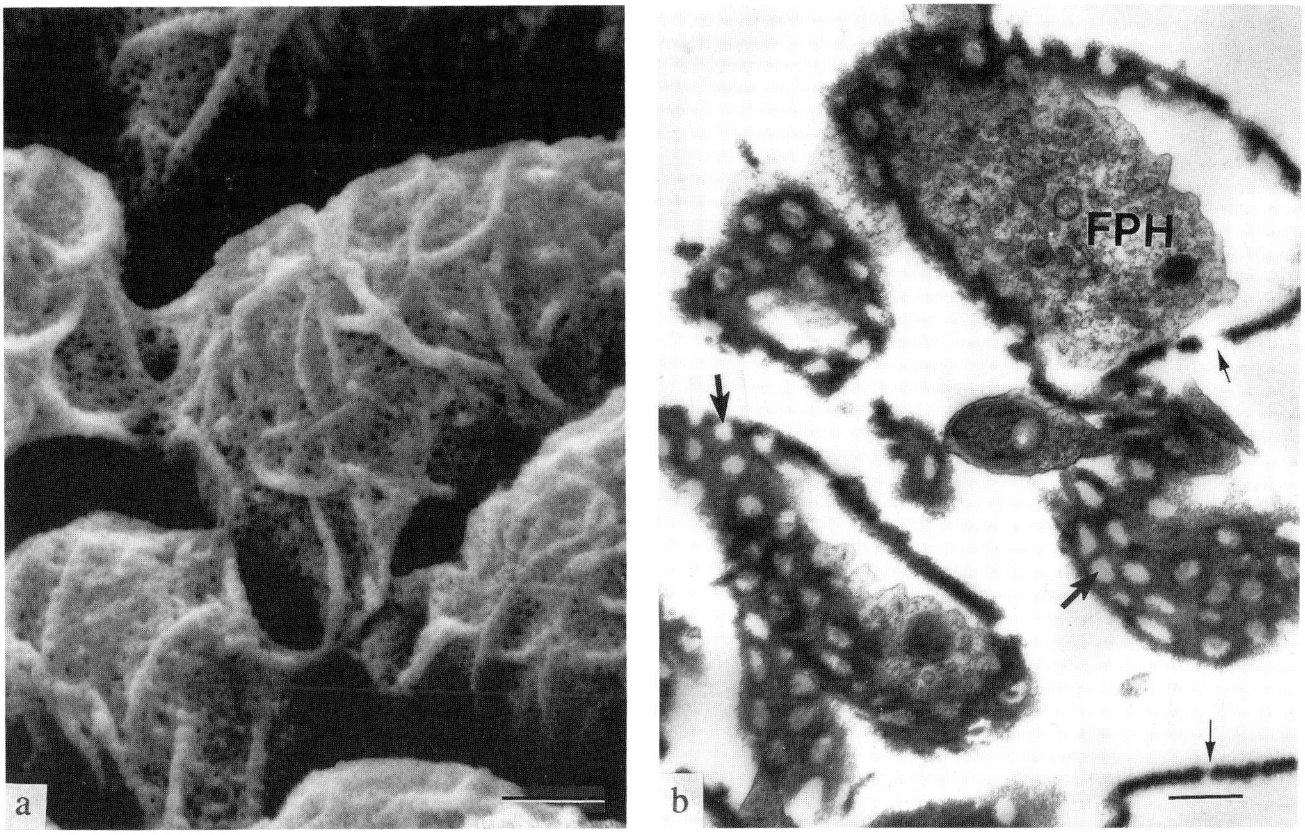

FIGURE 18 (a) Scanning electron micrograph showing the surface view of fixed phagocytes covered by the perforated membrane. Pronounced ridges of cell surface and numerous pores in the perforated membrane are visible. (b) Transmission electron micrograph of the perforated membrane. Pores are shown in face view (large arrows) with the plane of section approximately parallel to the membrane; pores are visible in transverse view (small arrows) with the plane of section approximately perpendicular to the membrane. FPH, Fixed phagocyte. Scale bars: (a) 2.0 μm; (b) 0.5 μm. (From Factor and Naar, 1990, with © permission of Wiley–Liss, Inc.)

nisms by removing foreign material from the blood (Figs. 19 and 20) (Johnson *et al.*, 1981; Johnson, 1987; Factor and Beekman, 1990; Martin and Hose, Chapter 17). Because they lie in the hemal sinuses, they are bathed in circulating blood as it percolates through the digestive gland. The irregular layer of fixed phagocytes and the cell processes on their surfaces increase the surface area and may break up boundary layers by causing turbulent flow of the blood; therefore, they serve to increase contact with circulating foreign particles. The role of the perforated membrane is not clear. Experimental studies (reviewed by Factor and Beekman, 1990) suggest that it does not act as a selective barrier (filter), but may be a very fluid layer that allows particles of varying sizes (even larger than the pores) to pass into the pericellular space. It may also act to sequester particles within the pericellular space when the fixed phagocytes are overwhelmed by large quantities of particles (Johnson, 1987). Phagocytosis occurs at the surface of the fixed phagocytes. When considered together, the fixed phagocytes constitute an important phagocytic organ, located within the digestive gland, which Cuénot (1903, 1905) recognized as *l'organe phagocytaire*. The location of important phagocytic components, which are usually thought of as elements of the circulatory system, within the limits of the digestive gland highlights the multifunctional nature of this organ.

4. Outer Layer of Connective Tissue

The entire digestive gland is invested in a substantial outer layer of connective tissue (Fig. 11b and c), approximately 4.5 μm thick. This surrounding connective tissue contains a variety of characteristic elements, including hemal sinuses, circulating hemocytes, granulocytes, and fibroblasts, all embedded in a collagenous matrix.

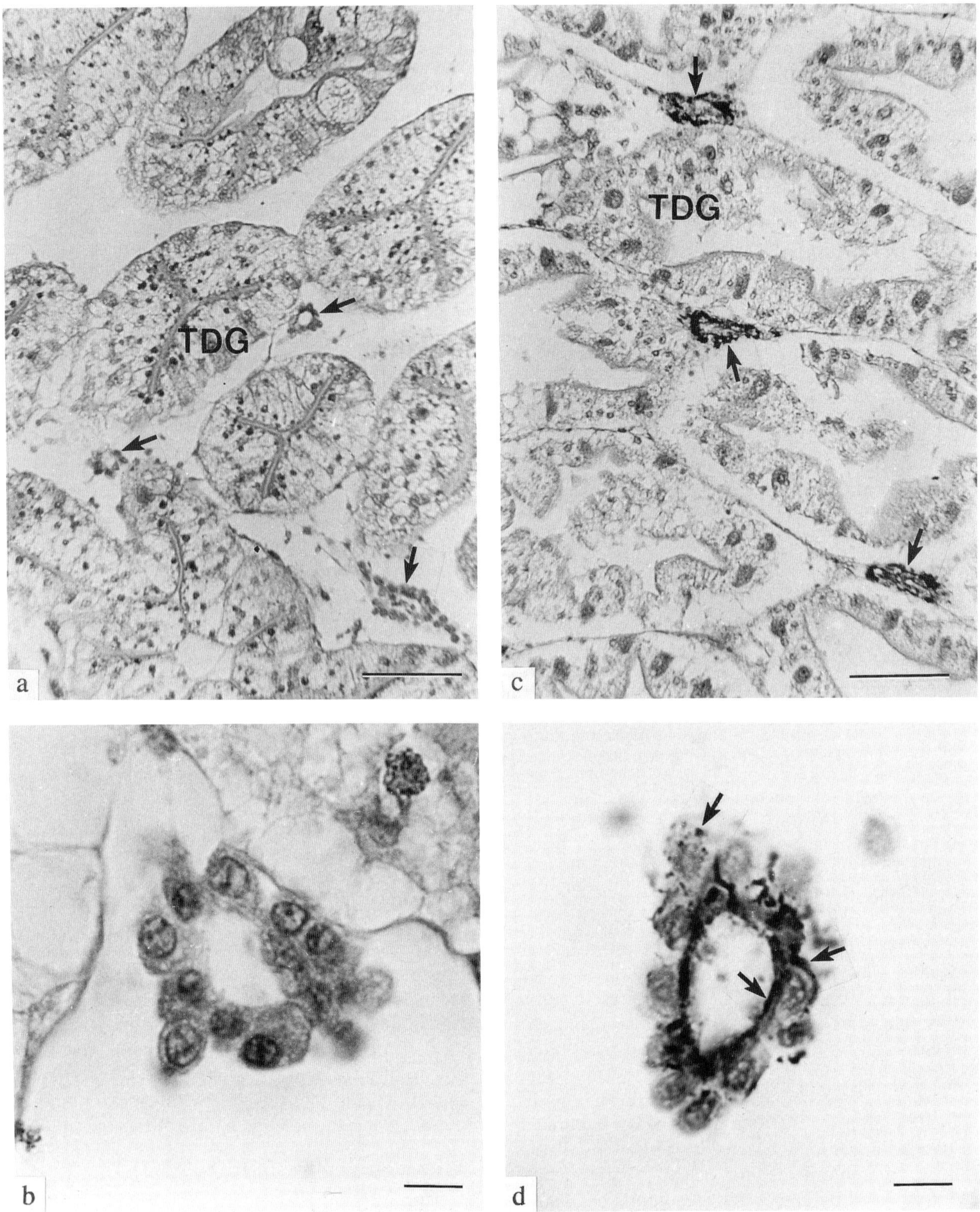

FIGURE 19 (a) Terminal hepatic arterioles [arrows in (a) and (c)] among tubules of the digestive gland (TDG). Sections of the digestive gland from an uninjected animal (a and b) and from a lobster injected with carbon particles (c and d), which have been removed from circulation and concentrated in fixed phagocytes [arrows in (d)]. Photomicrographs of paraffin sections. Scale bars: 100 μm. (From Factor and Beekman, 1990, with © permission of Wiley–Liss, Inc.)

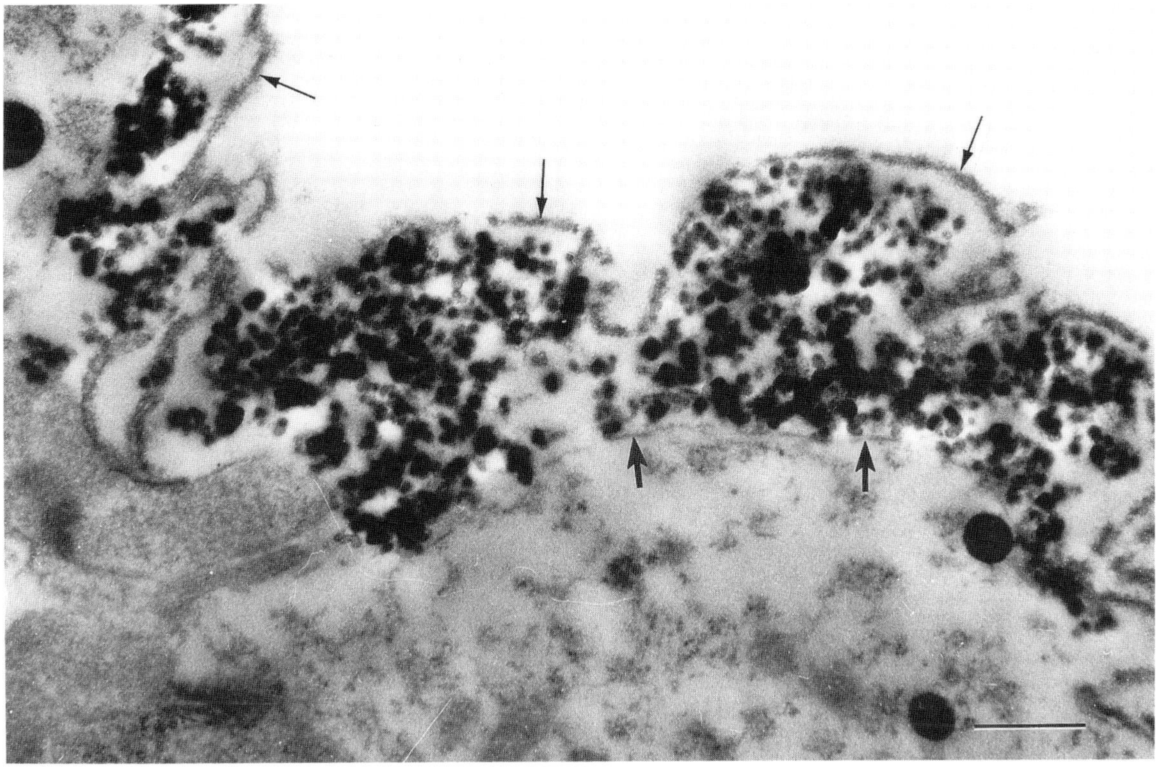

FIGURE 20 Transmission electron micrograph of injected carbon particles that have been concentrated in pericellular space between the cell membrane (large arrows) of a fixed phagocyte and the perforated membrane (small arrows). Scale bar: 1.0 μm. (From Factor and Beekman, 1990, used with © permission of Wiley–Liss, Inc.)

C. Anterior Midgut Caeca

1. Gross Anatomy

The variety of references to the anterior midgut caeca, using a variety of names, may be confusing to the casual reader. Herrick (1909, p. 250) reported "a small intestinal caecum, which extends forward over the dorsal wall of the stomach . . . " in *Homarus americanus*. In *H. gammarus*, Barker and Gibson (1977) reported an anterior midgut diverticulum that originates from the dorsal wall of the intestine and extends forward over the posterior pyloric stomach. Mykles (1979, p. 206) states that "*Homarus* lacks a pair of AMC [anterior midgut caeca] but has a short PMC [posterior midgut caecum]." He adds, however, that a bilobed, flattened outpocketing of the intestine is present in adults in the location of the anterior midgut caecum and that this structure is undoubtedly the equivalent of the structures reported by Herrick and by Barker and Gibson (D. L. Mykles, personal communication). The term *pyloric caeca* has been erroneously applied to this organ (reviewed by Smith, 1978).

In larval stages I–III and the postlarva (stage IV) of *Homarus americanus*, the paired anterior midgut caeca arise from a common point along the dorsal midline of the anterior end of the intestine. This common opening bifurcates immediately, sending one caecum forward on either side of the pyloric stomach (Figs. 30d–f, 31f and g, 32c, and 35) (Factor, 1981b). The anterior midgut caeca appear to be better developed in larvae and the postlarva than in the adult (D. L. Mykles, personal communication), suggesting the possibility of a reduction during development.

2. Arrangement of Tissues

The histological structure of the anterior midgut caeca of *Homarus americanus* has not been studied in detail. In *H. gammarus*, both anterior and posterior midgut caeca are histologically similar to the intestine, but lack its longitudinal musculature and possess considerably infolded walls (Barker and Gibson, 1977).

D. Posterior Midgut Caecum

1. Gross Anatomy

The posterior caecum arises dorsally from the posteriormost end of the intestine, in the sixth

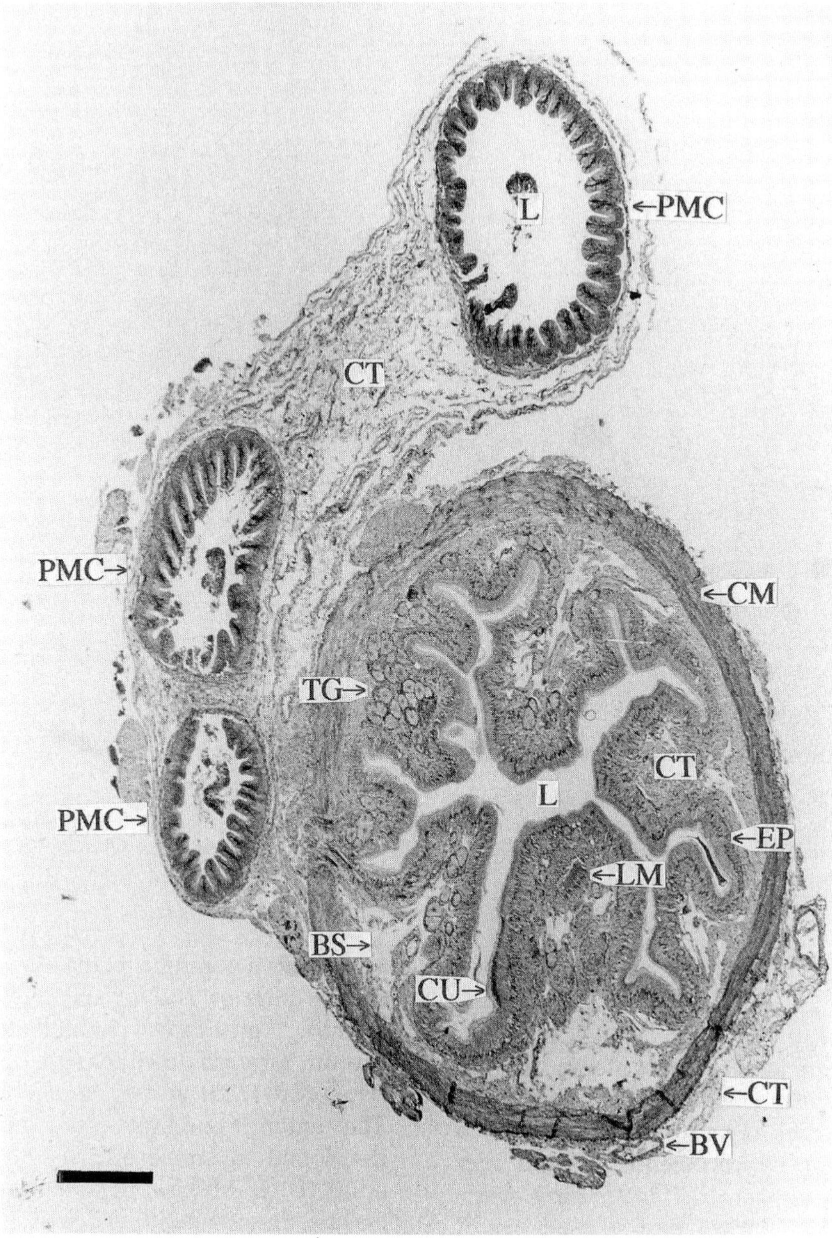

FIGURE 21 Cross-sectional view of the rectum. The posterior midgut caecum passes through the plane of section three times. BS, Blood sinus (=hemal sinus); BV, blood vessel; CM, circular muscle; CT, connective tissue; CU, cuticle; EP, epithelium; L, lumen; LM, longitudinal muscle; PMC, posterior midgut caecum; TG, tegumental glands. Photomacrograph of a paraffin section stained with Mallory's triple stain. Scale bar: 0.4 mm.

abdominal segment, just before the intestine joins the hindgut (Fig. 1). It contains several bends, extends posteriorly, and lies directly above the rectum (Fig. 21), which may be why it has sometimes been called, incorrectly, the "hindgut caecum" or "rectal caecum" (see Section VI).

2. Arrangement of Tissues

The histological structure of the posterior midgut caecum of *Homarus americanus* has not been studied in detail. Preliminary observations (Fig. 21) indicate an arrangement of tissues similar to that of the intestine and a complex basement membrane similar to that described for the tubules of the digestive gland (Factor, unpublished observations). The epithelial cells have a brush border of microvilli, are joined apically by septate junctions, contain pleomorphic and spherical vesicles, and adjacent lateral cell membranes interdigitate basally; the ultrastructure suggests an ion and water transport function for this epithelium (Mykles, 1979).

IV. Hindgut

A. Rectum

1. Gross Anatomy

The hindgut of *Homarus americanus* constitutes a muscular rectum that begins in the sixth abdominal segment and continues to the anus (Fig. 1). The thick, muscular wall of the rectum is formed into six longitudinal ridges which protrude so far that they almost completely occlude the rectal lumen when it is not filled with digesta (Fig. 21). No diverticula (or caeca) are associated with the hindgut (see Smith, 1978).

2. Arrangement of Tissues

Like the other regions of the digestive system, the hindgut consists of an epithelium and a connective tissue layer (Fig. 22). The simple columnar epithelium is covered by a thin cuticle that consists of the layers typical of the cuticle in other parts of the lobster. (See Waddy *et al.*, Chapter 10, on the cuticle.) The epithelium is underlain by a typical basement membrane of simple construction; it is of the same type found underlying the esophageal epithelium (Section II,A), but is not at all similar to the complex basement membranes found in the organs of the midgut (Sections III,A and B). The apical membrane of

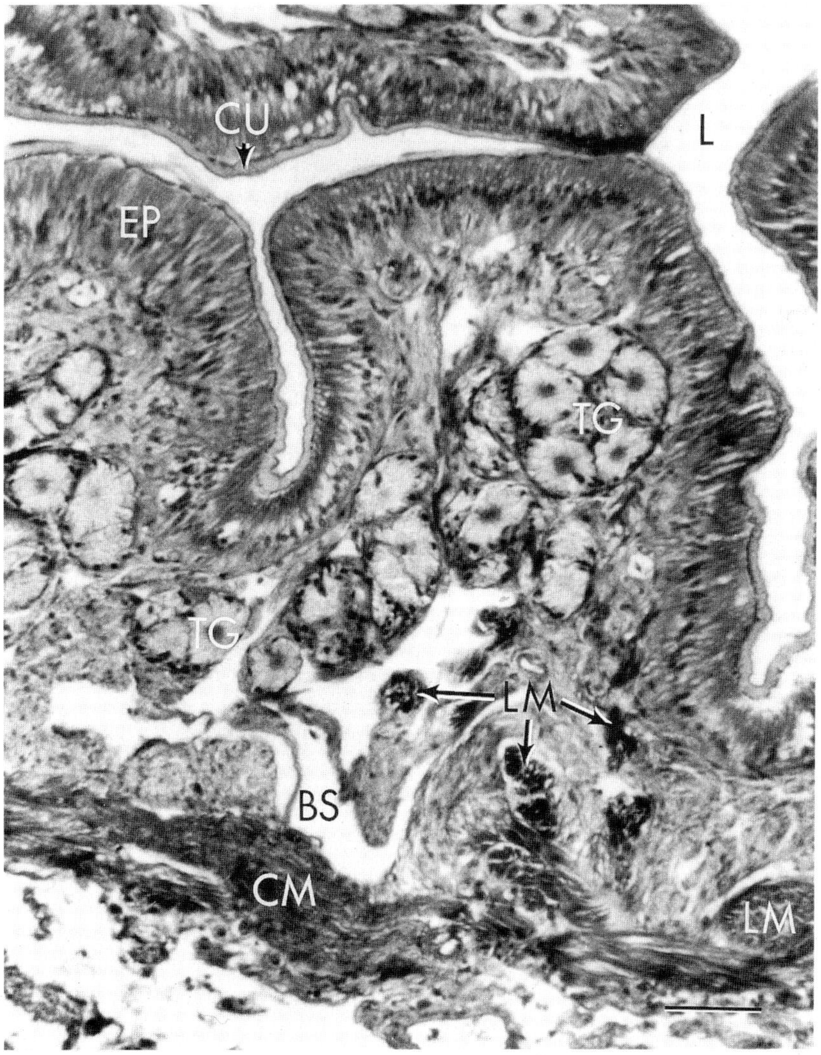

FIGURE 22 Cross-sectional view of the wall of the rectum. BS, Blood sinus (=hemal sinus); CM, circular muscle; CU, cuticle; EP, epithelium; L, lumen; LM, longitudinal muscle; TG, tegumental glands. Photomicrograph of a paraffin section stained with Mallory's triple stain. Scale bar: 0.1 mm.

epithelial cells is occasionally infolded, particularly near the septate junctions; adjacent lateral cell membranes in the junctional region are interdigitated; bundles of microtubules extend the entire length of the cell; and the basal cell membrane is plicated, especially in the longitudinal ridges (Mykles, 1979). The hindgut epithelium may be involved in ion and water transport (Mykles, 1979). The role of the hindgut in nutrition has yet to be defined (Conklin, Chapter 16).

The six longitudinal ridges of the rectum are made of both epithelium and connective tissue. Substantial elements of the connective tissue extend into the ridges, just beneath the folded layer of epithelial cells (Fig. 22). The connective tissue layer of the adult hindgut is approximately 1.0 mm thick when measured to the tip of a ridge, and 0.33 mm thick at its thinnest point between ridges. These hindgut ridges differ from the epithelial folds present in the midgut in their larger size and significant connective tissue component.

The connective tissue of the hindgut is a fibrous matrix in which the other subepithelial elements are embedded. It is permeated by the system of irregular hemal sinuses containing circulating hemocytes. Fibroblasts and both acidophilic and basophilic granulocytes are embedded in the collagenous matrix. The well-developed circular muscles lie near the periphery of the rectum and bundles of longitudinal muscles are centrally located, mostly in the longitudinal ridges. This arrangement of central longitudinal and peripheral circular muscles is the reverse of the situation found in the midgut (Section III,A), but similar to the arrangement of muscles in the esophagus (Section II,B).

The longitudinal ridges also contain numerous tegumental glands. They are generally organized into clusters and lie just beneath the epithelium, embedded in the connective tissue. Structurally, these rectal tegumental glands are virtually identical to those found in the esophagus and are discussed in Section V of this chapter. Yonge (1924) described a "glandular swelling" formed by a mass of tegumental glands at the anterior end of the hindgut of *Nephrops norvegicus*.

B. Anus

The anus vents on the ventral surface of the telson (Fig. 1). Herrick (1909) reported that it is closed by a sphincter muscle in *Homarus americanus*. *Homarus gammarus* (Barker and Gibson, 1977) and *Nephrops norvegicus* (Yonge, 1924) both have controlling radial muscles; in the latter, they attach to the cuticle lining the lumen, extend through the connective tissue, and terminate on the dorsal and ventral exoskeleton.

V. Tegumental Glands of Esophagus and Rectum

Virtually identical tegumental glands have been described in both the esophagus and the rectum of *Homarus americanus* (Factor *et al.*, 1995). They lie embedded in the loose connective tissue beneath the cuticle-lined epithelia of these organs (Fig. 22). Each tegumental gland has a roughly spherical structure, comprises three types of cells (secretory, central, and canal), and is surrounded by a thin connective tissue layer (Fig. 23).

The secretory cells are tall, columnar cells, radially arranged around the center of each tegumental gland (Fig. 24). They contain abundant Golgi bodies basally (peripherally), an extensive network of rough endoplasmic reticulum, and numerous membrane-bound secretory granules that virtually fill the apical (central) portion of most cells. Most granules are irregular in shape with loose, flocculent, relatively electron transparent contents, while others are ovoid in section with electron-dense contents. Neither esophageal nor rectal tegumental glands have the honeycomb granules found in pleopod tegumental glands (Johnson and Talbot, 1987).

The central lobe of the bilobed central cell fills the space at the very center of the spherical gland (Figs. 23 and 24). The cytoplasm contains a system of ductules, each lined with a tube of cytoplasm that represents the interdigitating process of a secretory cell (Fig. 25). These extensions of secretory cell cytoplasm contain longitudinal cytoskeletal filaments. The lining extends only so far and forms an incomplete ring or is absent altogether in some planes of section. The peripheral lobe, the main portion of the central cell, contains the large, elongate nucleus (Figs. 23 and 26). It lies between the canal cell and adjacent secretory cells, in a radial position, and it partially surrounds the canal cell.

The canal cell is long, narrow, and tube shaped (Figs. 23 and 26) and encloses an extracellular space that forms the lumen of the main duct.

The collecting and transporting canal system of the tegumental glands is formed by all three cell types. The tubelike, apical processes of the secretory cells protrude into the central cell and deliver secretory product. The central lobe of the central cell gathers secreted material in ductules, which collect into a central space, and delivers it to the canal cell. The canal cell forms a duct that accepts material from the central cell and transports it out of the tegumental gland,

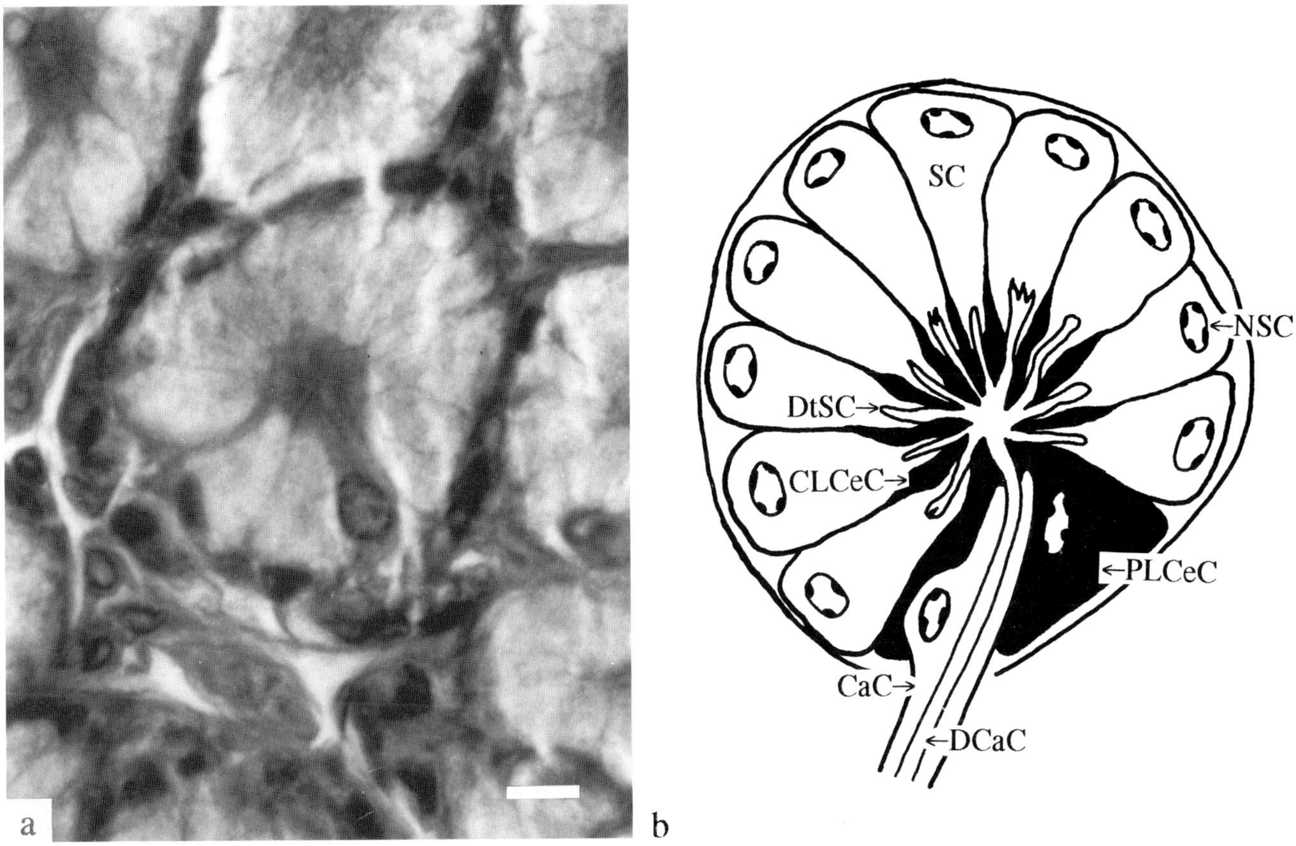

FIGURE 23 Tegumental glands of the esophagus and the rectum. (a) Photomicrograph of an esophageal tegumental gland; paraffin section stained with hematoxylin and eosin. Scale bar: 10.0 μm. (b) Diagram of the generalized tegumental gland illustrating major features. (From Factor *et al*., 1995.) **Abbreviations for Figs. 23–27:** CaC, Canal cell; CeC, central cell; CLCeC, central lobe of the central cell; CySC, cytoplasmic extension of the secretory cell, forming the ductule; DCaC, duct of the canal cell; DtSC, ductule of the secretory cell; LDt, lumen of the ductule of the secretory cell; NCaC, nucleus of the canal cell; NCeC, nucleus of the central cell; NSC, nucleus of the secretory cell; PLCeC, peripheral lobe of the central cell; SC, secretory cell.

through the connective tissue, and toward the cuticular surface. The luminal surface of the cuticle in both the esophagus and the rectum has pores (Fig. 27) that are presumably the termini of individual ducts; however, this has not been firmly established. The lumen of all the canals is apparently extracellular space.

Each tegumental gland is surrounded by a thin, investing layer that resembles, at low magnification, a simple basement membrane, but is actually a thin connective tissue layer that consists of a fibrous matrix and a basement membranelike layer that encloses the outer surface. It lacks contractile elements.

The function of the esophageal and rectal tegumental glands of *Homarus americanus* is not known. Various functions have been suggested (reviewed by Yonge, 1932), including the possibilities that they secrete a lubricant, that they may participate in the production of the thin cuticle that lines these two regions of the gut, or that the esophageal tegumental glands may secrete an enzyme that begins digestion. Their secretory product, however, has yet to be determined in *H. americanus*. In *H. gammarus*, these glands have been shown to contain acid muco-polysaccharide, acid phosphatase, and ATPase, which play no role in digestion; Barker and Gibson (1977) suggested that they are the sites of mucus production, as there are no other obvious candidates, and that they produce lubricant to facilitate the passage of food in both the esophagus and the rectum and to bind the feces in the rectum. Yonge (1924, 1932) also suggested that the glands have nothing to do with digestion, but may contribute to the secretion or preservation of the chitinous lining.

Structurally, the esophageal and rectal tegumental glands are virtually identical to each other and are very similar to the pleopod (=cement glands) (Johnson and Talbot, 1987) and antennal (Arsenault et

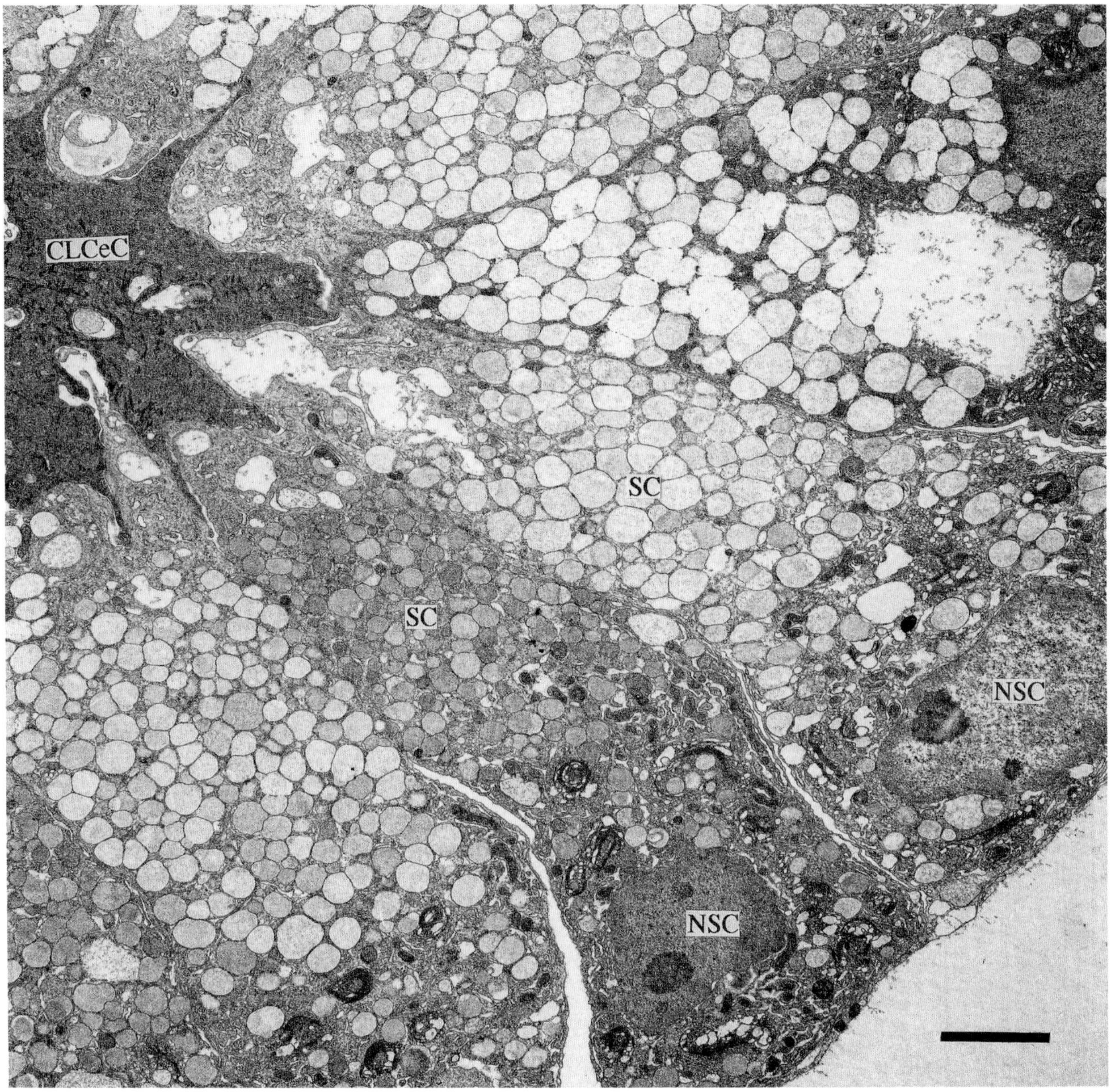

FIGURE 24 Overview of the tegumental gland ultrastructure from the center to the periphery. Secretory cells are radially arranged around the central lobe of the central cell and contain numerous secretory granules. See Fig. 23 for abbreviations. Transmission electron micrograph of the rectal tegumental gland. Scale bar: 5.0 µm. (From Factor *et al.*, 1995.)

al., 1979) tegumental glands. [Talbot and Demers (1993) reviewed tegumental glands; see Waddy *et al.* (Chapter 10) on the integument.]

VI. Midgut–Hindgut Transition

Homarus americanus has a very long midgut and a short hindgut, which begins in the sixth abdominal segment. By comparison, this is the reverse of the situation found in freshwater crayfish, which have a very short midgut, with most of the length of the gut in the abdomen being hindgut (Huxley, 1880).

Historically, there has been considerable confusion among carcinologists about the extent of the midgut and the hindgut. In brachyurans, the midgut was often mistaken for the hindgut and the posterior midgut caecum was frequently mistaken for a diverticulum of the hindgut and called "hindgut caecum"

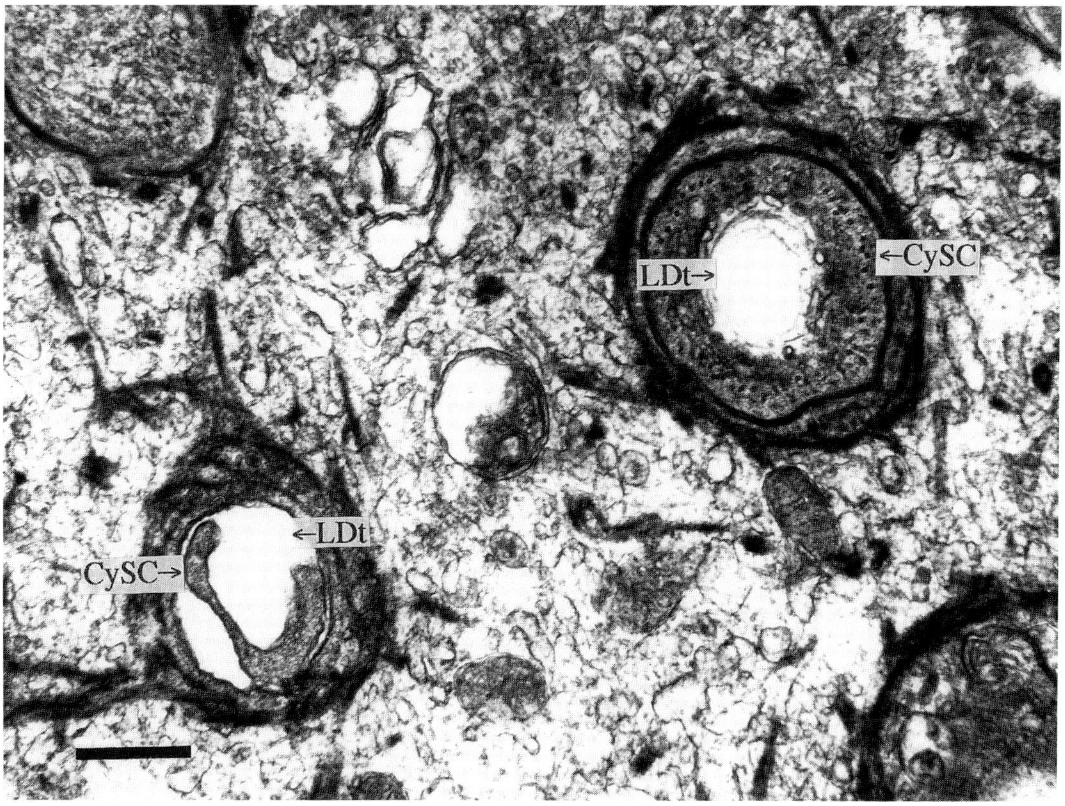

FIGURE 25 Cross-section of secretory cell processes surrounded by the central lobe of the central cell, forming ductules. See Fig. 23 for abbreviations. Transmission electron micrograph of the esophageal tegumental gland. Scale bar: 0.5 μm. (From Factor *et al.*, 1995)

(reviewed by Smith, 1978). This confusion extended to *Homarus americanus* and may be traced to an incorrect statement by Herrick (1909, p. 251): "The embryology of the animal shows that the inner wall of the intestine is primarily due to an ingrowth from the outside skin and in the early larvae an intestinal cuticle can be detected, but if the latter is present in the adult, it is reduced to a layer of extreme thinness." [Anderson's (1982) account of the origins of the midgut contradicts that of Herrick (see Section VII,A).] This may be the source for the mistaken statement in the widely used laboratory manual *Selected Invertebrate Types* (Lochhead, 1950, p. 438): "In larval lobsters the midgut is said to be short and the hindgut long, as in crayfishes."

Contrary to these statements, larval lobsters have a long midgut and a short hindgut (Fig. 28), just as in the adult (Fig. 1), and the division between the midgut and the hindgut occurs in the sixth abdominal segment in all stages (Factor, 1980, 1981b). There is an obvious distinction between the midgut (intestinal) and hindgut (rectal) epithelia. In all life history stages of *Homarus americanus*, the epithelia in all parts of the midgut lack a cuticle and have a brush border of microvilli, while the hindgut epithelium (like the foregut epithelium) is always lined with cuticle and lacks microvilli.

A thin, membranelike layer has been observed in the intestine of some stage I larvae (Factor, 1981b). Although it superficially resembles a delicate cuticle (e.g., it stains positively with Mallory's triple stain), and it may be the source of Herrick's confusion about an intestinal cuticle in larvae, it is not a cuticle and likely represents a peritrophic membrane. Its origin and constituents are not known.

VII. Development and Metamorphosis of the Digestive System

A. Embryonic Origins

The presumptive stomodeum is a patch of superficial cells along the ventral midline, which later invaginates to form the lining of the foregut in all crustacean embryos (Anderson, 1982). Because proctodeal development occurs late, the presumptive proctodeum is more difficult to trace; it is a small group of superficial

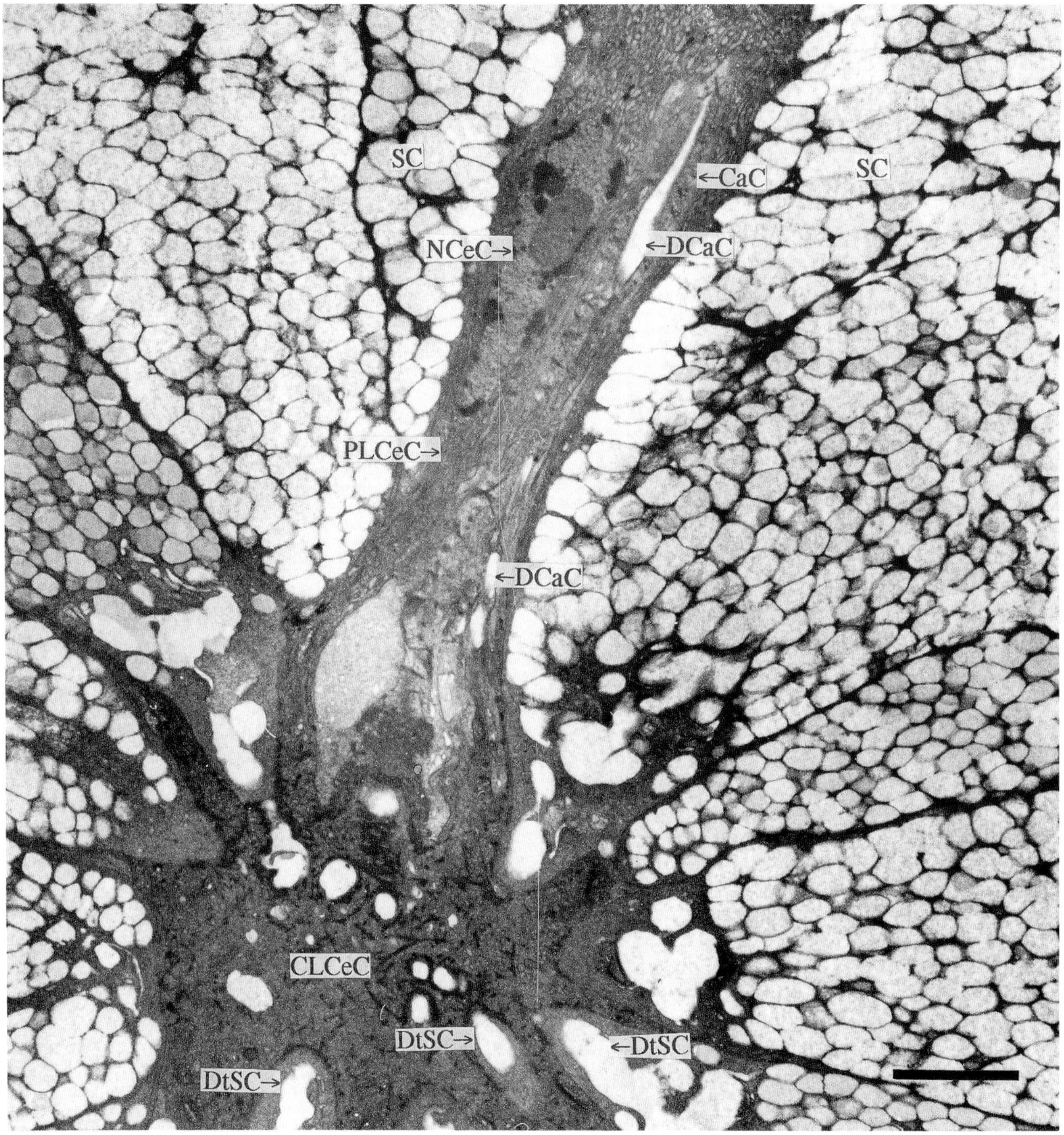

FIGURE 26 Longitudinal section of the canal cell, which forms the main duct exiting the tegumental gland, surrounded by the peripheral lobe of the central cell (with the nucleus). Note the extensions of apical cytoplasm of secretory cells, forming ductules that extend into the central cell. See Fig. 23 for abbreviations. Transmission electron micrograph of the rectal tegumental gland. Scale bar: 5.0 μm. (From Factor *et al.*, 1995.)

cells along the posteroventral midline of the presumptive ectoderm (Anderson, 1982).

Anderson (1982) provides the following description of the complex development of the decapod midgut. An anterior midgut rudiment proliferates, invades the yolk, and forms a vitellophage epithelium, while a posterior midgut rudiment proliferates posteroventrally as a plate of cells, contiguous with the vitellophage epithelium. This plate grows out into the caudal papilla and forms a long posterior midgut

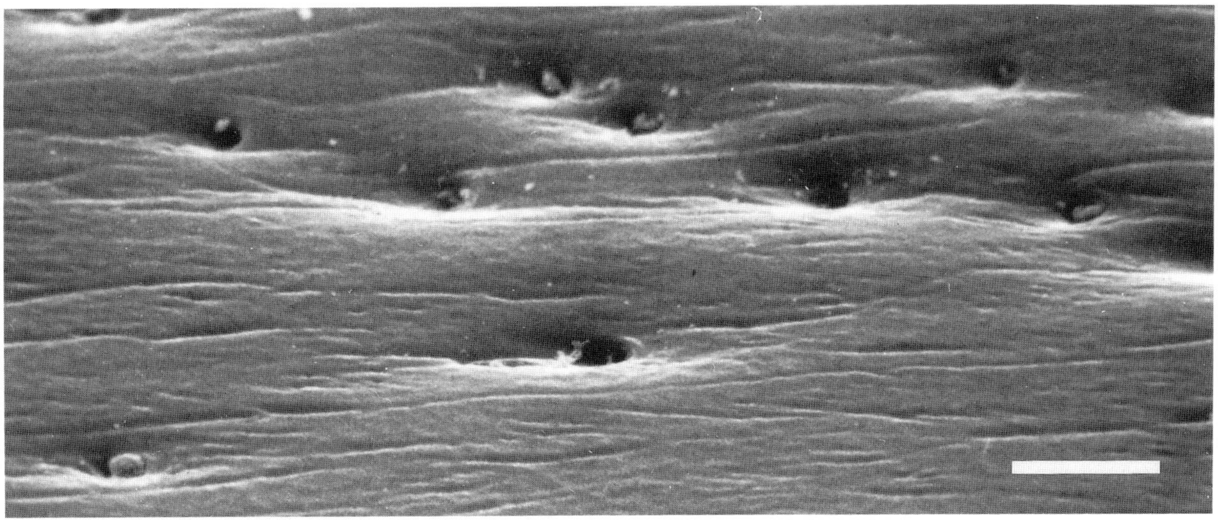

FIGURE 27 Scanning electron micrograph of pores in the luminal surface of the esophageal cuticle showing presumed termini of ducts from the tegumental glands. Scale bar: 5.0 μm. (From Factor *et al.*, 1995.)

tube, which connects with the short proctodeum. The vitellophage epithelium is transformed into definitive epithelium at two sites, anteriorly, on either side of the junction with the stomodeum, and posteriorly, on either side of the junction with the posterior midgut tube. At both sites, the forming epithelium bulges out as anterior and posterior digestive gland rudiments; eventually, the vitellophage epithelium is transformed almost entirely into digestive glands. The gut is completed by the joining together of the digestive glands, stomodeum, and posterior midgut.

B. Developmental Changes in the Foregut

Some features of the foregut are relatively well developed in stage I larvae and undergo only minor changes besides the obvious increase in size. [This account of the development of the digestive system is based primarily on Factor (1980, 1981b); Hinton and Corey (1979) have also described aspects of development and metamorphosis.] The esophagus in stage I has labral, lateral, and metastomal ridges (Fig. 30d and h) and is similar in structural organization to stage IV and the adult (Figs. 31a, 32c, and 33h). The cardiopyloric valve and the lower cardiac canals of the cardiac stomach (Figs. 29b and 30d and f) and the chitinous ridge and the gland filter of the pyloric stomach (Figs. 29c and 30c and g) are all present in stage I. Regardless of the type of food being consumed by the various stages, the degree of development of the gastric mill, or the mechanism of processing food, particles must still be sorted on the basis of size so that only appropriately small particles enter the digestive glands. This may account for the early development of a functional pyloric filter. Even though the filter apparatus appears to be functional from stage I on, the structure of the pyloric stomach increases somewhat in complexity during the early stages (Figs. 31e, 32d, and 33f).

The most dramatic changes in the foregut take place in the gastric mill of the cardiac stomach (Factor, 1981b). No substantial medial, lateral, or accessory lateral teeth are present in stage I; the only obvious elaborations of the wall of the cardiac stomach are setae, folds, and ridges (Figs. 29a, 30a, and 34a), which obviously participate in food processing and routing, but no heavily cuticularized teeth are present. Some of the ridges are in the positions of future teeth and may be their rudiments. Stage II larvae also lack the teeth of the gastric mill. Recognizable medial and lateral teeth first appear in the third stage. They are heavily cuticularized and are accompanied ventrally by ridges, which may be rudiments of the accessory lateral teeth. The dorsal ridge present in stages I and II is replaced by the medial tooth in later stages; it may represent a dorsal cardiopyloric valve and apparently assists the ventral cardiopyloric valve in retaining larger food particles in the cardiac stomach. By stage IV (Figs. 31b and c and 32b and d), the gastric mill has a full complement of well-developed medial, lateral, and accessory lateral teeth (cf. Figs. 34a and b). However, the stage IV mill is not identical to that of the adult; for example, the relatively sharp lateral teeth of stage IV, with pointed cusps arranged in two rows (Fig. 34c), give rise to teeth with a single row of rather blunt, molar-like processes in the adult (Fig. 34d).

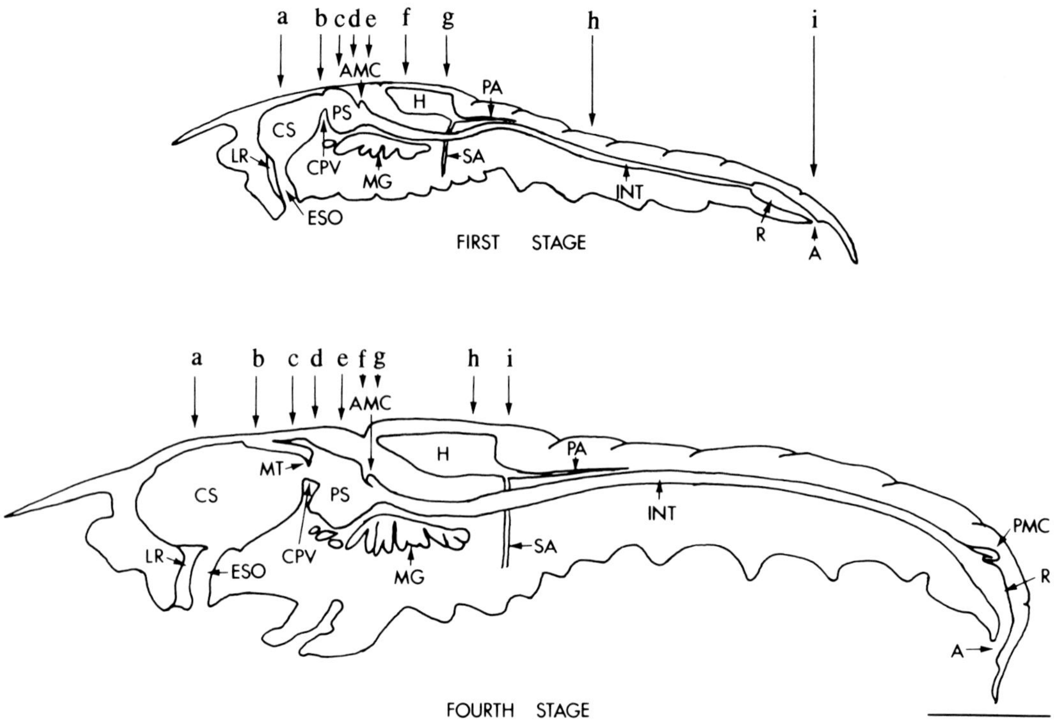

FIGURE 28 Digestive system of stage I and IV lobsters illustrating features that lie along the sagittal plane. Arrows indicate planes of the cross-sections illustrated in Figs. 29 and 31. Scale bar: 1.0 mm. (From Factor, 1981b, *J. Morphol.* with permission of Wiley–Liss. Copyright © 1981 Alan R. Liss, Inc.) **Abbreviations for Figs. 28–33:** A, Anus; AGM, anterior gastric muscle; AMC, anterior midgut caecum; AMG, anterior lobe of the midgut (digestive) gland; CEC, circumesophageal commissure; CPV, cardiopyloric valve; CS, cardiac stomach; DAMG, duct of the anterior lobe of the midgut (digestive) gland; DDMG, duct of the dorsal lobe of the midgut (digestive) gland; DMG, dorsal lobe of the midgut (digestive) gland; DPMG, duct of the posterior lobe of the midgut (digestive) gland; ESO, esophagus; G, gills; GF, gland filter; H, heart; INT, intestine; L, labrum; LAR, lateral ridge of the esophagus; LCC, lower cardiac canal; LE, lumen of the esophagus; LR, labral ridge of the esophagus; LT, lateral tooth; M, midgut; MAN, mandible; MG, midgut (digestive) gland; MGD, duct of the midgut (digestive) gland into the intestine; MR, metastomal ridge of the esophagus; MT, medial tooth; PA, posterior aorta; PMC, posterior midgut caecum; PMG, posterior lobe of the midgut (digestive) gland; PS, pyloric stomach; R, rectum; RC, ridge in the wall of the cardiac stomach; SA, sternal artery; VN, ventral nerve.

C. Developmental Changes in the Midgut

As in all other life history stages of *Homarus americanus*, the stage I midgut begins immediately posterior to the pyloric stomach, where the cuticular lining of the foregut ends, and continues to the sixth abdominal segment, where it joins the hindgut. The general form of the digestive gland in stage I is similar to that of latter stages; each of the three lobes (Fig. 35) comprises many small, branching tubules and has its own collecting duct (Figs. 30a, b, and g and 33f and g), which empties into the main duct (Figs. 29e, 30g, 31f, and 33f). The complexity of the digestive gland increases considerably from stage I to stage IV (Factor, 1981b). Stage I lobsters have fewer and shorter tubules than those present in stage IV, but apparently greater luminal volume relative to the size of the organ (cf. stage I and IV cross-sections in Figs. 29b and c, and 31d).

Yolk has been seen intercalated among the tubules of the digestive gland in embryos and first-stage larvae, but not in later stages (Biesiot, 1982). All cell types found in the adult digestive gland—embryonic (E cells), fibrillar (F cells), secretory (B cells), and resorptive (R cells)—are present in stages I–IV, but not all types are found in embryos (Biesiot, 1986; Sasaki *et al.*, 1986). The embryonic digestive gland comprises E, R, and F cells; B cells do not develop until a day or so before hatching (Biesiot and McDowell, 1995). Embryonic R cells, in contrast to those of adults, have only one or two lipid vacuoles, presumably representing resorbed yolk lipids. Larval R cells do not attain the adult morphology until stage VI (Sasaki *et al.*, 1986).

The anterior midgut caeca are present and well developed in stage I and are similar to those found in stage IV (Factor, 1981b). They bifurcate from a com-

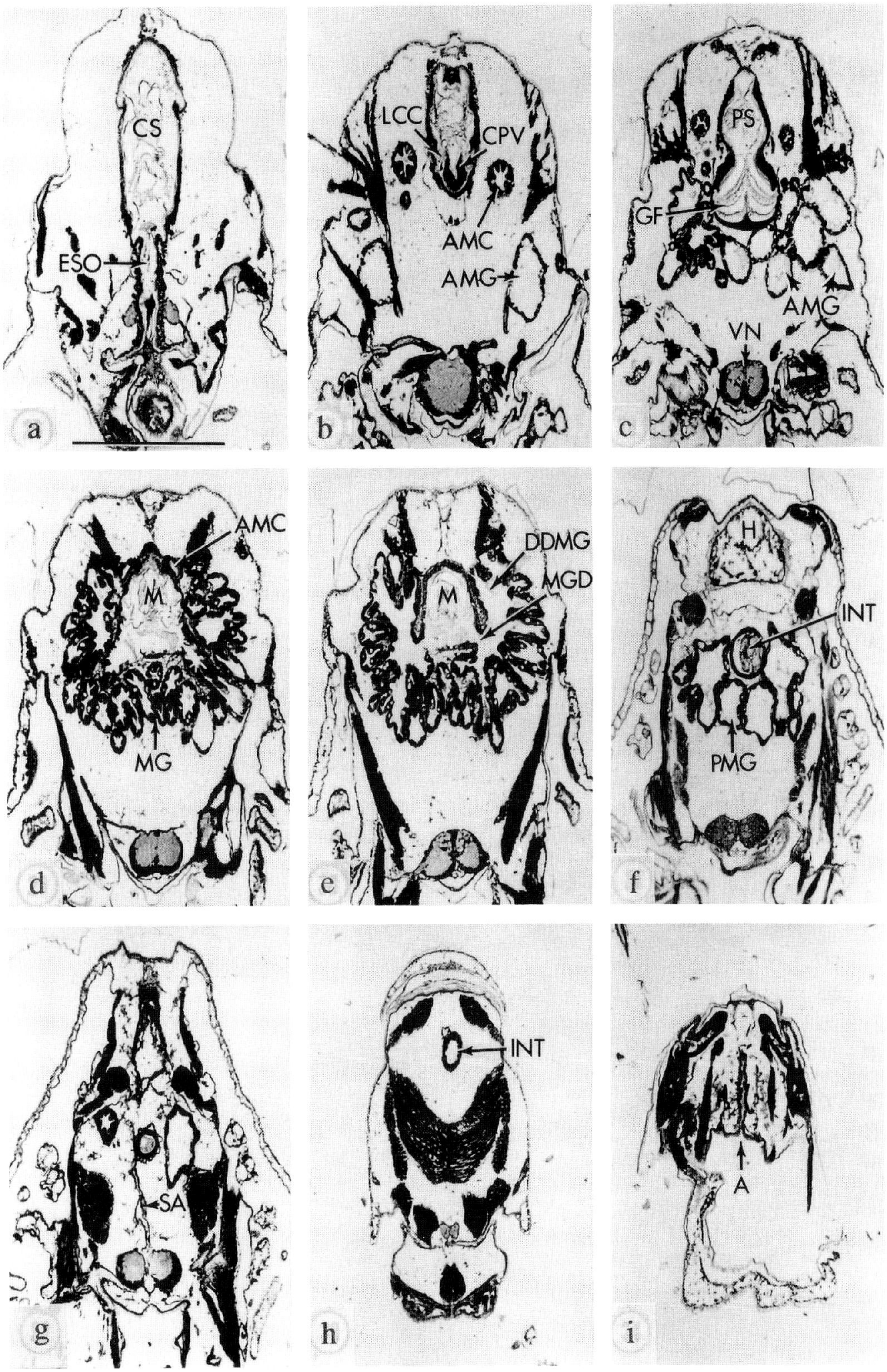

FIGURE 29 Selected serial cross-sections of a stage I larval lobster, progressing from anterior to posterior. See Fig. 28 for abbreviations and planes of section. Photomicrographs of paraffin sections stained with Mallory's triple stain. All are to the same scale; scale bar: 0.5 mm. (From Factor, 1981b, *J. Morphol.* with permission of Wiley–Liss. Copyright © 1981 Alan R. Liss, Inc.)

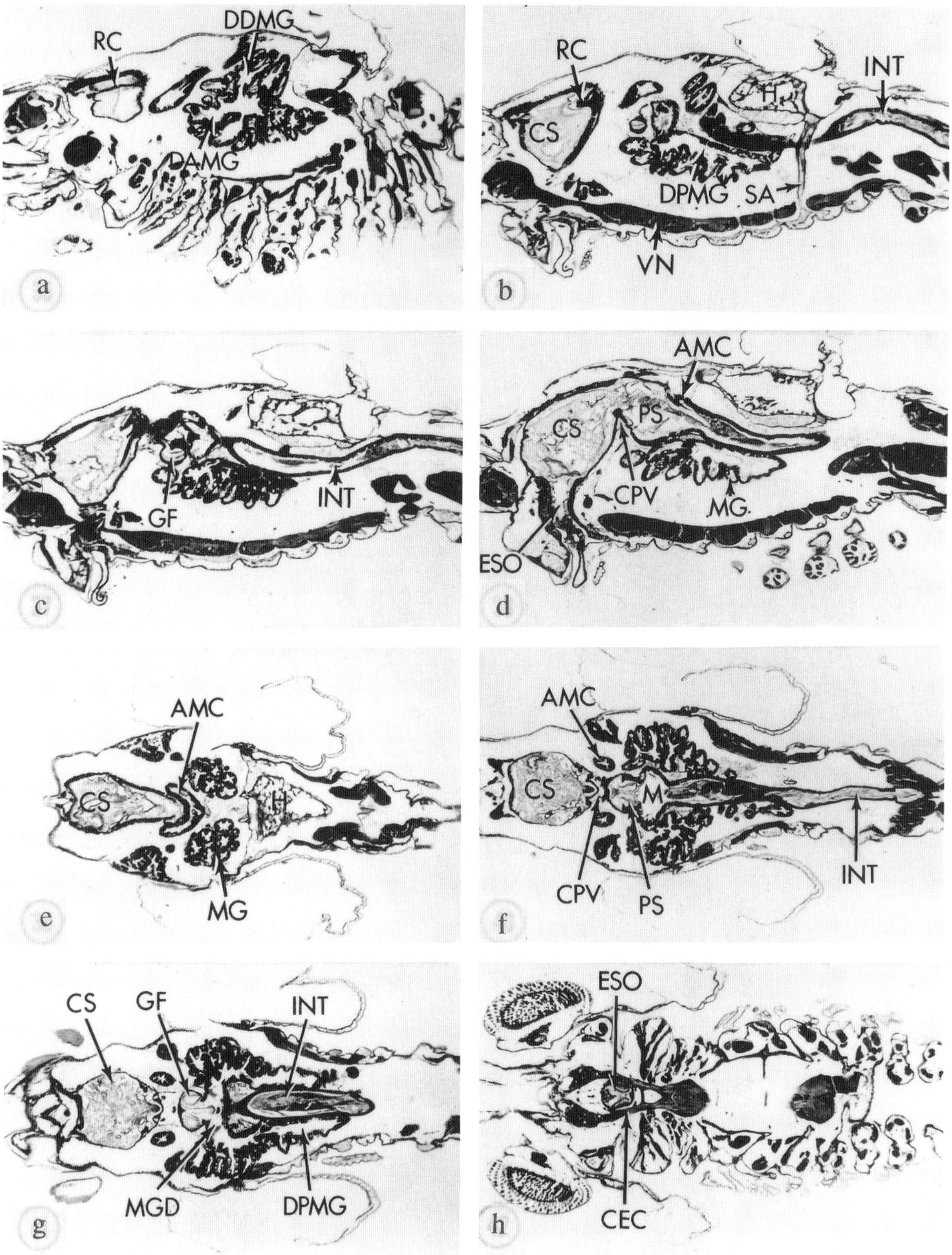

FIGURE 30 Selected serial sagittal and frontal sections of stage I larval lobsters. (a–d) Sagittal sections, progressing from the right side to the midline. (e–h) Frontal sections, progressing from dorsal to ventral. See Fig. 28 for abbreviations. Photomicrographs of paraffin sections stained with Mallory's triple stain. All are to the same scale; scale bar: 0.5 mm. (From Factor, 1981b, *J. Morphol.* with permission of Wiley–Liss. Copyright © 1981 Alan R. Liss, Inc.)

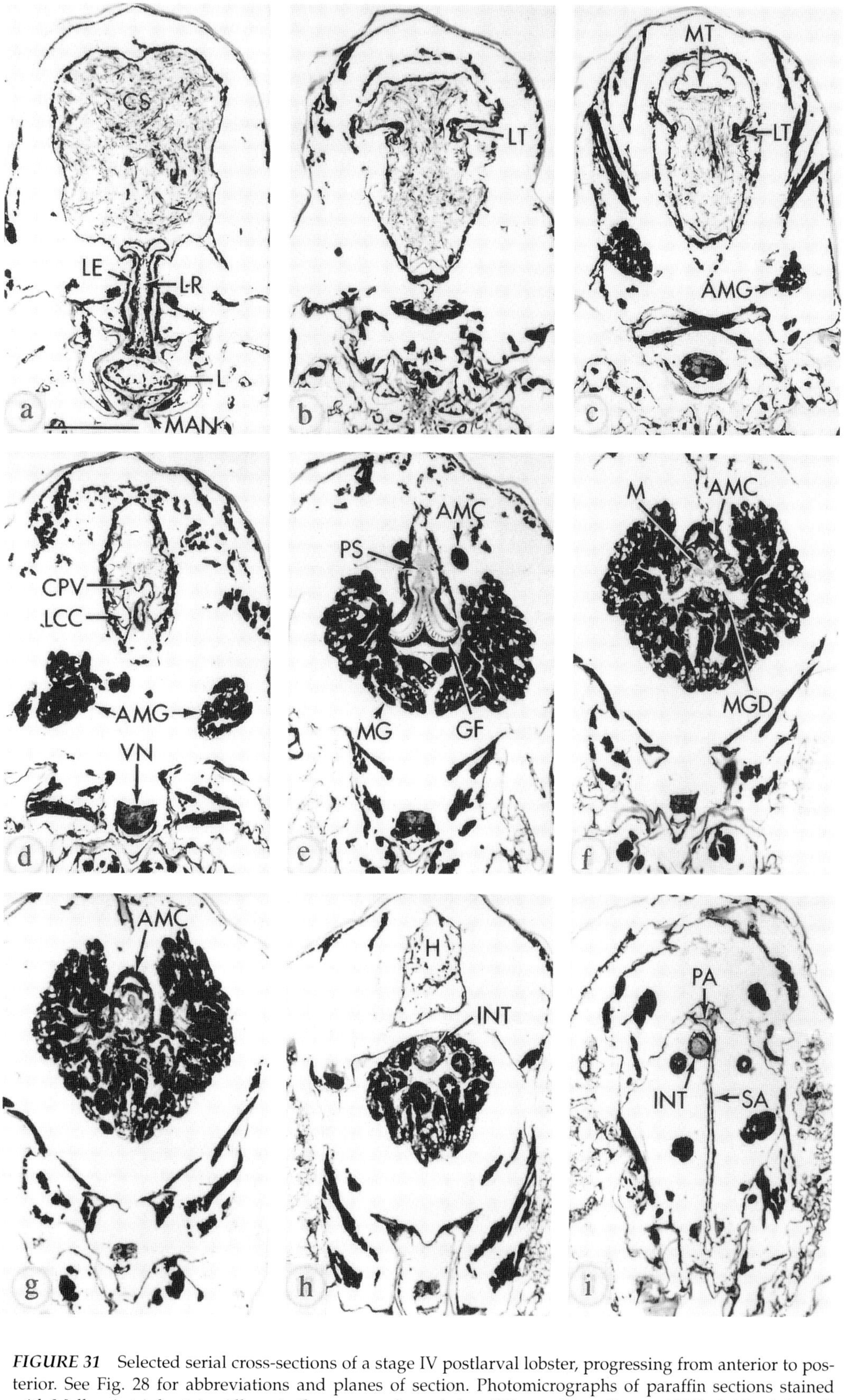

FIGURE 31 Selected serial cross-sections of a stage IV postlarval lobster, progressing from anterior to posterior. See Fig. 28 for abbreviations and planes of section. Photomicrographs of paraffin sections stained with Mallory's triple stain. All are to the same scale; scale bar: 0.5 mm. (From Factor, 1981b, *J. Morphol.* with permission of Wiley–Liss. Copyright © 1981 Alan R. Liss, Inc.)

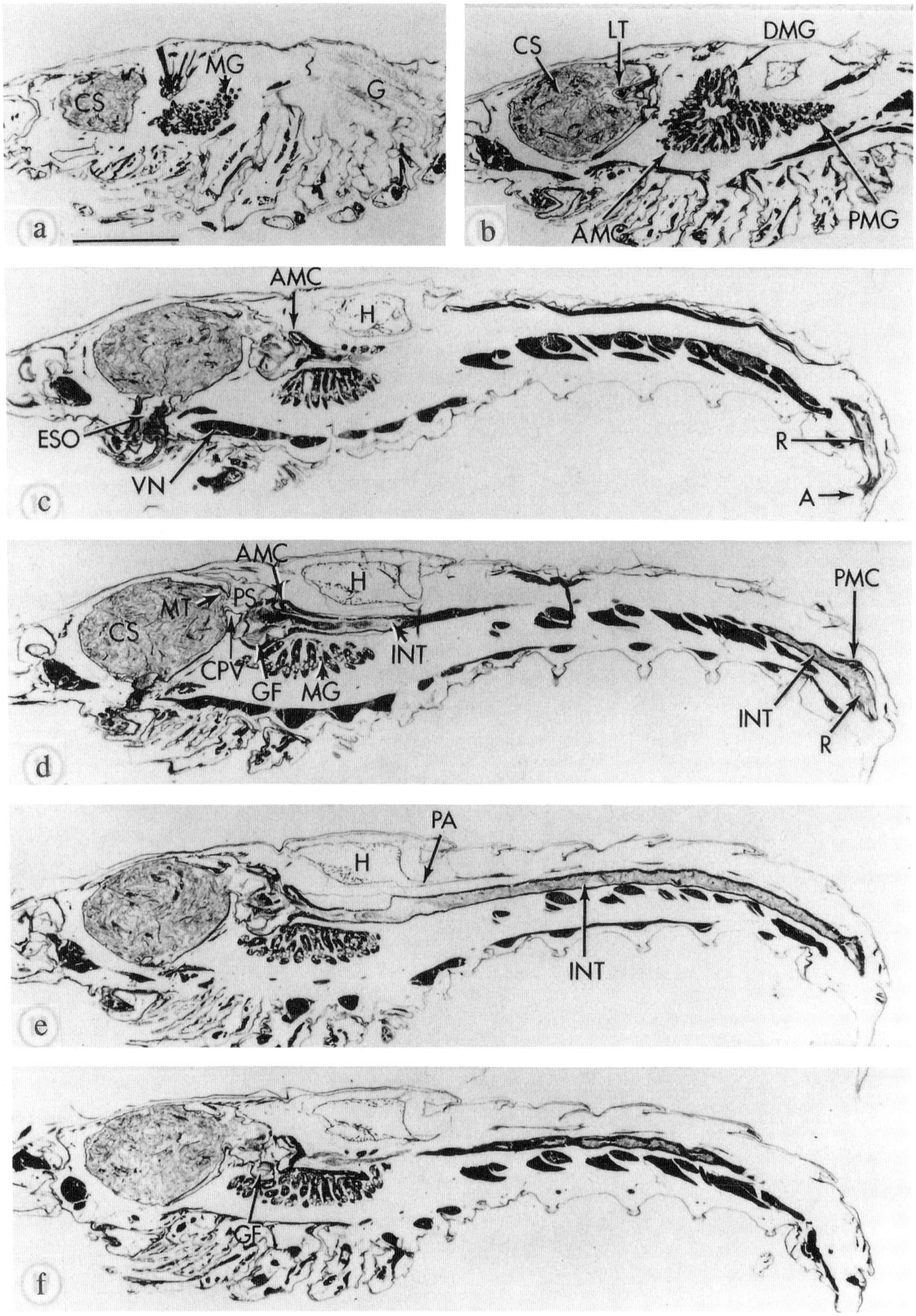

FIGURE 32 Selected serial sagittal sections of a stage IV postlarval lobster, progressing from the right side to the midline. See Fig. 28 for abbreviations. Photomicrographs of paraffin sections stained with Mallory's triple stain. All are to the same scale; scale bar: 1.0 mm. (From Factor, 1981b, *J. Morphol.* with permission of Wiley–Liss. Copyright © 1981 Alan R. Liss, Inc.)

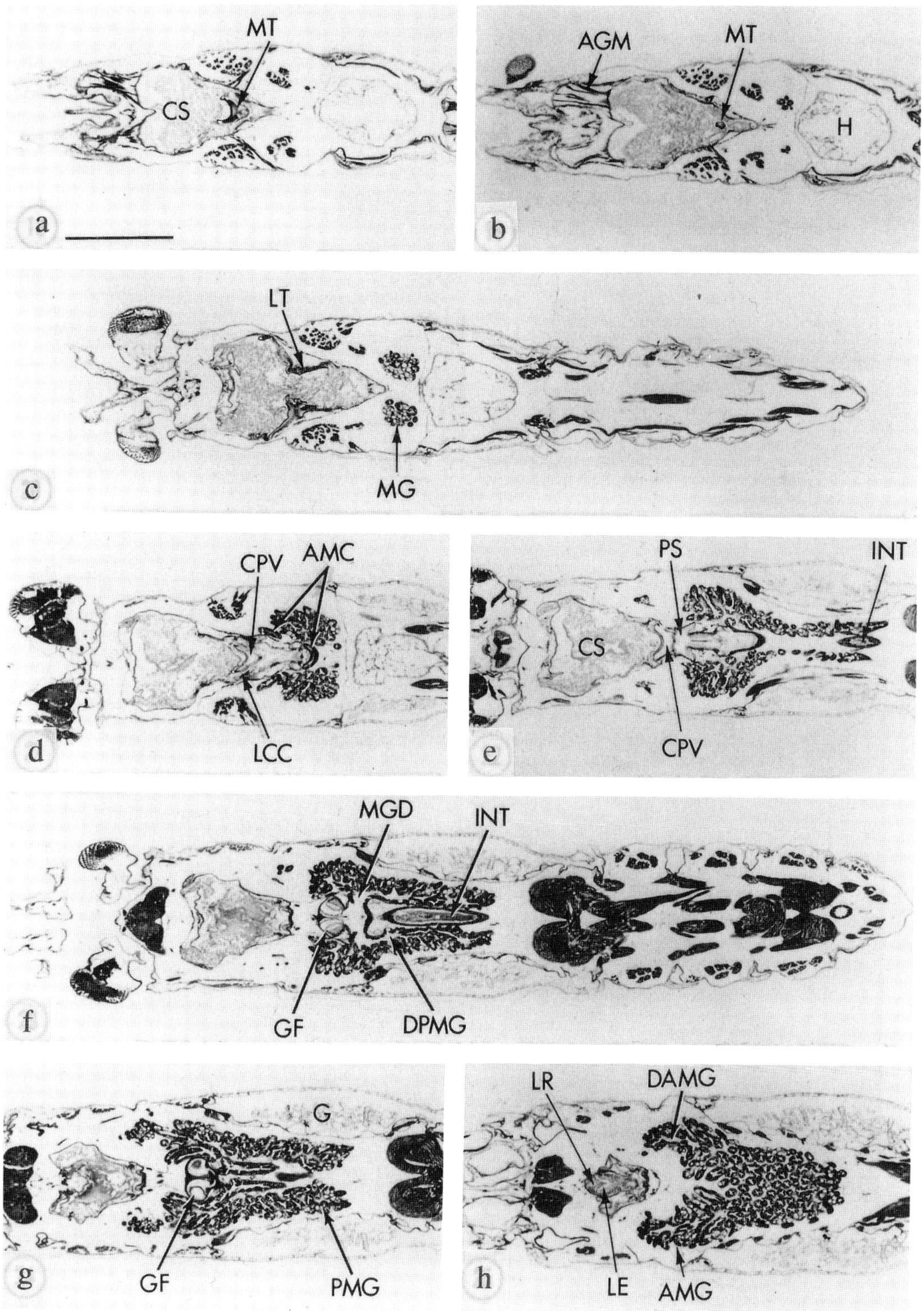

FIGURE 33 Selected serial frontal sections of a stage IV postlarval lobster, progressing from dorsal to ventral. See Fig. 28 for abbreviations. Photomicrographs of paraffin sections stained with Mallory's triple stain. All are to the same scale; scale bar: 1.0 mm. (From Factor, 1981b, *J. Morphol.* with permission of Wiley–Liss. Copyright © 1981 Alan R. Liss, Inc.)

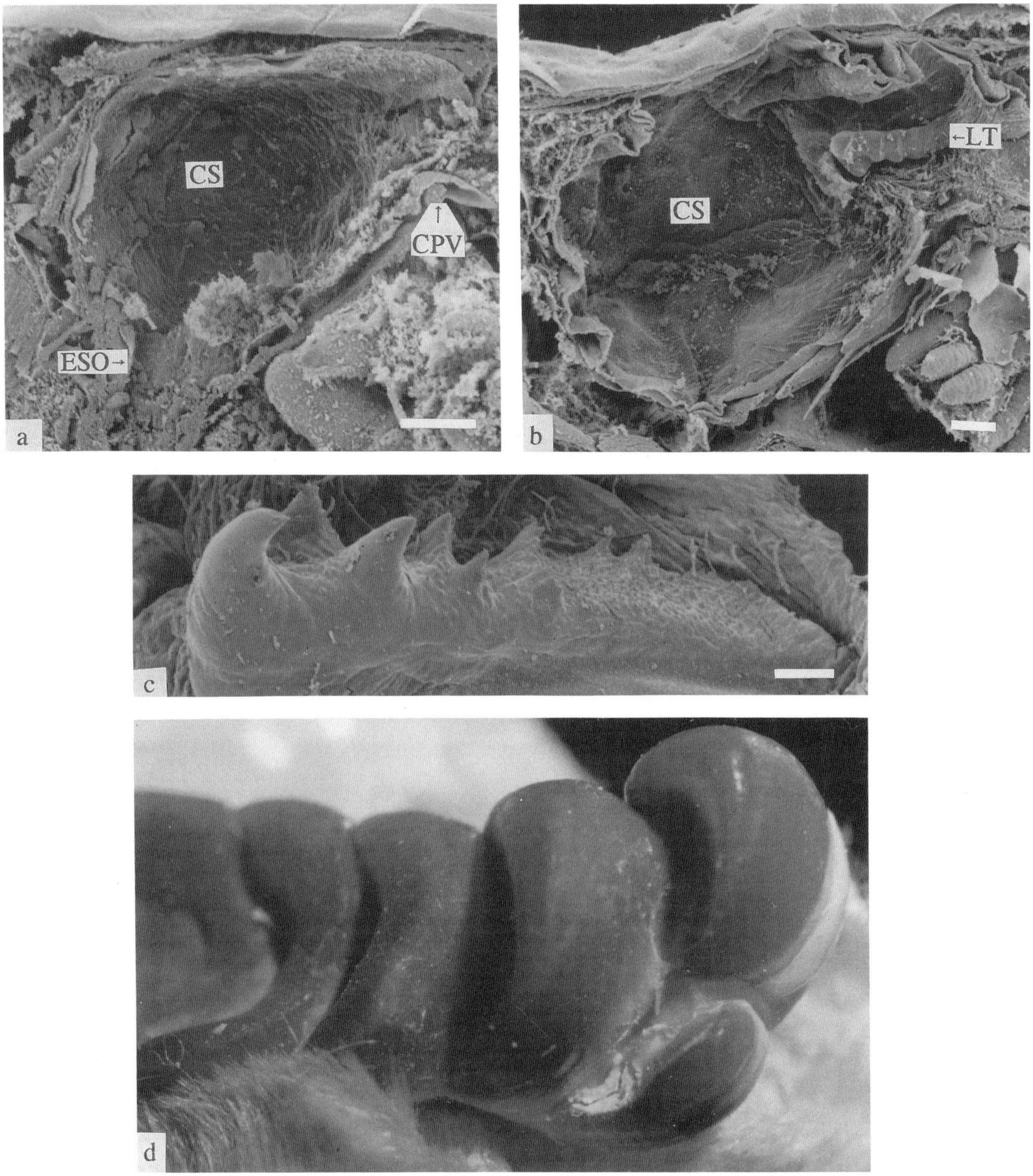

FIGURE 34 Cardiac stomach (CS) of stage I larva (a) and stage IV postlarva (b). Note the absence of teeth in stage I (a) and the well-developed lateral teeth (LT) in stage IV (b). Lateral tooth of stage IV (c, anterior is to the left) and the anterior end of a lateral tooth of an adult (d, anterior is to the right). Scanning electron micrographs (a–c) and photomacrograph (d) of dissected specimens. CPV, Cardiopyloric valve; ESO, esophagus. Scale bars: (a–c) 0.1 mm. [(a–c) from Factor, 1981b, *J. Morphol.* with permission of Wiley–Liss. Copyright © 1981 Alan R. Liss, Inc.]

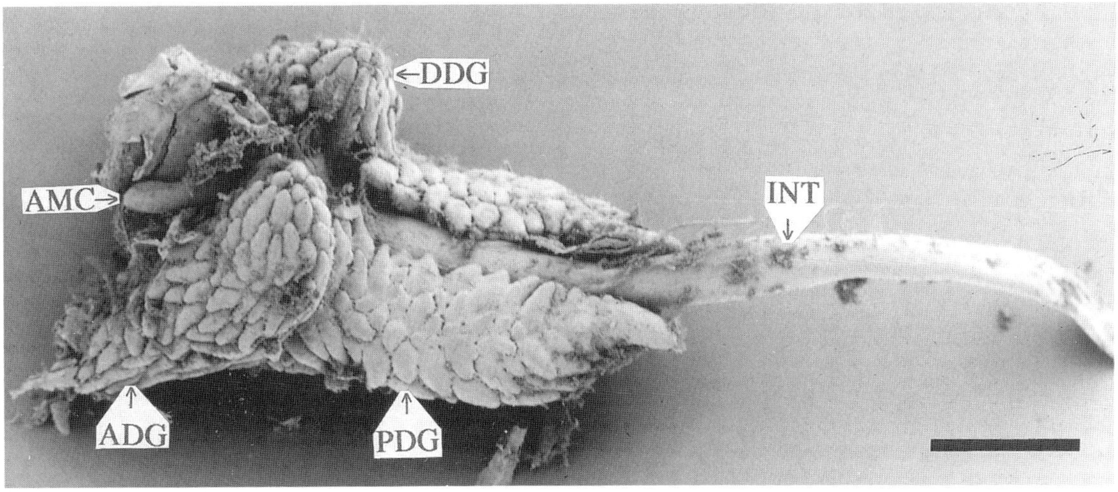

FIGURE 35 Scanning electron micrograph of the dissected digestive gland of a stage IV postlarva. ADG, Anterior lobe of the digestive gland; AMC, anterior midgut caecum; DDG, dorsal lobe of the digestive gland; INT, intestine; PDG, posterior lobe of the the digestive gland. Scale bar: 0.25 mm. (From Factor, 1981b, *J. Morphol.* with permission of Wiley–Liss. Copyright © 1981 Alan R. Liss, Inc.)

mon point along the dorsal midline of the anterior-most part of the intestine and send one caecum forward along either side of the foregut (Figs. 29b–d, 30d–g, 31e–g, 32c–e, and 33d).

The stage I midgut has no posterior midgut caecum. A simple outpocketing of the dorsal wall of the intestine appears to be a rudimentary caecum in stage II, but there is no lateral bend in the tube. By stage III, the caecum has the same form as in stage IV; the tube arises from the left side of the intestine, bends across to the right side, and lies above the rectum (Fig. 32d) (Factor, 1981b).

From stage I on, the epithelia of the intestine, the digestive gland, and the anterior and posterior midgut caeca all have a brush border of microvilli and lack a cuticular lining (contrary to Hinton and Corey, 1979). A peritrophic membrane may be present in the intestine (see Section VI).

D. Developmental Changes in the Hindgut

The rectum of all three larval stages is generally similar to that of the stage IV postlarva (Fig. 32c). They all have the same series of longitudinal ridges found in the adult and the same basic arrangement of tissues (Factor, 1981b).

E. Developmental Correlations

Aspects of development of the digestive system can be correlated with development of the mouthparts, with changes in habitat and feeding, and with the larger array of changes that accompanies metamorphosis.

Development of heavily cuticularized medial and lateral teeth in the gastric mill (the internal food-processing apparatus) can be correlated with development of the feeding appendages (the external food-processing apparatus), particularly the mouthparts. The mouthparts become heavily setose and gain more teeth and spines along their edges. Larvae have fine, pointed teeth on the medial edge of the incisor process and a pad of setae in the location of the future molar process. The delicate teeth develop into more solid, heavily cuticularized structures in the postlarva (stage IV), with thick, blunt teeth on the incisor process, and are eventually transformed into the solid, toothless cutting edge of the adult. The pad of setae is replaced by a solid molar process in stage IV, which is more like that of the adult (Factor, 1978; Lavalli and Factor, Chapter 14).

Developmental changes in the digestive system and feeding appendages also coincide with changes in behavior that result in the important transition from planktonic to benthic life, which typically occurs during the fourth stage (see Ennis, Chapter 3, and Lawton and Lavalli, Chapter 4). These developmental changes suggest that the lobster undergoes a dramatic change in diet. However, benthic postlarvae and shelter-restricted juveniles are also capable of feeding on plankton, as are the larvae, as well as on small benthic organisms (see Lavalli and Factor, Chapter 14). This appears to be a versatile intermediate feeding strategy. The transition in feeding habits and food as lobsters move from a planktonic to a benthic life appears to be more gradual than abrupt.

Nevertheless, the changes in the digestive system (as well as in the mouthparts) help prepare lobsters for their change of diet as they assume a benthic existence.

Finally, developmental changes in the digestive system coincide with the changes in external anatomy that occur at the metamorphic molt to stage IV (the postlarva). These anatomical changes include the loss of exopodites on the pereiopods, the development of functional pleopods, and the shift of the function of swimming from the thoracic to the abdominal appendages (Charmantier *et al.*, 1991; Factor, Chapter 1). Developmental changes that result in a well-formed gastric mill can also be considered to be metamorphic.

VIII. Directions for Further Research

Much remains to be learned about the digestive system of *Homarus americanus*, from both structural and functional viewpoints. It seems remarkable, for example, that we know virtually nothing about the role of two organs of the digestive system (the anterior and posterior midgut caeca). Our understanding of the other digestive organs is woefully incomplete. Although not meant to be an exhaustive list, several areas that would benefit from further study follow.

1. The detailed understanding of the stomatogastric neuromuscular system gained from studies of the spiny lobster should be extended to *Homarus americanus*, with the goal of understanding both neurological organization and details of the operation of the gastric mill and the pyloric filter.
2. Additional histological, ultrastructural, and physiological studies are needed for a more complete understanding of the cells and tissues of most of the digestive organs.
3. The function of the esophageal and rectal tegumental glands should be addressed, including the nature of their secretions and their possible roles in the physiology or mechanics of digestion and the formation of the integument.
4. The functional significance of the unusually complex basement membranes found in the organs of the midgut remains unknown.
5. Physiological studies are needed to determine the possible roles of the intestine and the anterior and posterior midgut caeca in food, water, and ion absorption and osmotic balance.
6. A better understanding is needed of food processing in the early larval stages, which appear to have neither masticating mandibles nor a functional gastric mill.

IX. Summary

The complex digestive system of *Homarus americanus* is divided into three regions (foregut, midgut, and hindgut), based on embryonic origins.

The foregut, which is lined with a thin layer of cuticle, comprises the mouth, esophagus (a relatively short, muscular tube), cardiac stomach, and pyloric stomach. Enzymes from the digestive gland are mixed with food in the cardiac stomach, a spacious, saclike chamber that contains the plates and teeth of the gastric mill. The pyloric stomach is a complex filtering apparatus. The function of this stomatogastric system of nerves, muscles, and ossicles is the mastication of ingested food into fine particles in the cardiac gastric mill and the sorting of these particles in the cardiac sac and the pyloric filter.

The midgut comprises the intestine, a gently curved tube, and several associated organs: the digestive gland (=hepatopancreas, or midgut gland), the anterior midgut caeca, and the posterior midgut caecum. The midgut organs lack a cuticular lining.

The wall of the intestine consists of a simple columnar epithelium, which is gently folded into longitudinal folds, has a brush border of microvilli, and is underlain by an unusually complex basement membrane. The connective tissue is a matrix of collagen fibers in which the other subepithelial elements are embedded, including circular and longitudinal muscles, hemal sinuses, and a variety of cell types.

The digestive gland, which is primarily responsible for the synthesis and secretion of digestive enzymes, the final digestion of food, and the absorption of nutrients, is organized as a series of blindly ending tubules of rather simple construction. Each tubule includes a digestive epithelium, a basement membrane, contractile cells, and tunica propria. The epithelium has embryonic (E cells), fibrillar (F cells), secretory (B cells), and resorptive cells (R cells). The nonepithelial elements are part of the extensive network of connective tissue. Processes of contractile cells, which form a network surrounding each tubule, are responsible for peristaltic contractions. Interspersed among the digestive tubules are numerous terminal hepatic arterioles. The arterioles include an endothelium, which forms the vessel; an acellular endothelial intima lining the lumen; circulating hemocytes within the lumen; fixed phagocytes on the periphery; and a perforated membrane covering the periphery. The fixed phagocytes are responsible for removing foreign particles from circulation and their exposed surfaces are bathed in hemolymph circulating through the digestive gland.

The hindgut, like the foregut, is lined with cuticle

and constitutes a muscular rectum that begins in the sixth abdominal segment and continues to the anus. The thick, muscular wall of the rectum has six substantial longitudinal ridges. It consists of a simple columnar epithelium that is underlain by a typical basement membrane. The connective tissue contains a fibrous matrix in which the other subepithelial elements are embedded, including hemal sinuses, circular and longitudinal muscles, and a variety of cell types.

Both the esophagus and the rectum have virtually identical tegumental glands embedded in the loose connective tissue beneath the epithelia. Each tegumental gland is roughly spherical; comprises secretory, central, and canal cells; and is surrounded by a thin connective tissue layer. A system of ducts collects secretory product and transports it out of the gland. The function of esophageal and rectal tegumental glands is uncertain.

The most dramatic developmental changes in the foregut take place in the gastric mill of the cardiac stomach. Stage I and II larvae lack the teeth characteristic of the masticating stomach of the adult. By stage IV, the gastric mill has a full complement of well-developed medial, lateral, and accessory lateral teeth. The complexity of the digestive gland increases considerably from stage I to stage IV. All cell types found in the adult digestive gland are present in stages I–IV. The stage I midgut has anterior midgut caeca, but not the posterior midgut caecum, which first appears in stage II. The rectum of the larvae is generally similar to that of the stage IV postlarva and the adult.

Development of the gastric mill coincides with development of the mouthparts, with the changes in behavior that result in the transition from planktonic to benthic life, and with the changes in external anatomy that occur at the metamorphic molt to stage IV (the postlarva).

Acknowledgments

Drs. John M. Anderson, Patricia Biesiot, Douglas Conklin, Phyllis T. Johnson, and Kari L. Lavalli provided valuable comments and suggestions.

References

Anderson, D. T. (1982). Embryology. *In* "The Biology of Crustacea" (D. E. Bliss, ed.), Vol. 2, "Embryology, Morphology, and Genetics" (L. G. Abele, ed.), pp. 1–41. Academic Press, New York.

Arsenault, A. L., Clattenburg, R. E., and Aiken, D. E. (1979). The morphology and secretory-transport mechanism of the tegumental glands of the lobster (*Homarus americanus*) as related to the molt cycle. *J. Submicrosc. Cytol.* 11, 193–207.

Barker, P. L., and Gibson, R. (1977). Observations on the feeding mechanism, structure of the gut, and digestive physiology of the European lobster *Homarus gammarus* (L.) (Decapoda: Nephropidae). *J. Exp. Mar. Biol. Ecol.* 26, 297–324.

Bayer, R. C., Gallagher, M. L., Leavitt, D. F., and Rittenburg, J. H. (1979). A radiographic study of the lobster (*Homarus americanus*) alimentary canal. *Proc. Annu. Meet.—World Maric. Soc.* 10, 561–564.

Biesiot, P. (1982). Development of the hepatopancreas in early stages of the American lobster *Homarus americanus*. *Am. Zool.* 22, 871.

Biesiot, P. M. (1986). "Changes in midgut gland morphology and enzyme activities associated with development in early stages of the American lobster *Homarus americanus*," Ph.D. dissertation, Joint Program in Oceanography, Woods Hole Oceanogr. Inst. Mass. Inst. Tech., WHOI-86-20.

Biesiot, P. M., and McDowell, J. E. (1995). Midgut gland development during early life history stages of the American lobster *Homarus americanus*. *J. Crust. Biol.* (in press).

Bryant, E. A. (1992). "Development and Three-dimensional Ultrastructure of Midgut Basement Membranes in the Lobster (*Homarus americanus*)," baccalaureate thesis. State University of New York at Purchase.

Charmantier, G., Charmantier-Daures, M., and Aiken, D. E. (1991). Metamorphosis in the lobster *Homarus* (Decapoda): A review. *J. Crust. Biol.* 11, 481–495.

Claiborne, B. J., and Ayers, J. (1987). Functional anatomy and behavior. *In* "The Crustacean Stomatogastric System" (A. I. Selverston and M. Moulins, eds.), pp. 9–29. Springer-Verlag, Berlin.

Cochran, D. M. (1935). The skeletal musculature of the blue crab *Callinectes sapidus* Rathbun. *Smithson. Misc. Collect.* 92, 1–76.

Conklin, D. E. (1980). Nutrition. *In* "The Biology and Management of Lobsters" (J. S. Cobb and B. F. Phillips, eds.), Vol. I, pp. 277–300. Academic Press, New York.

Cuénot, L. (1903). L'organe phagocytaire des Crustacés Décapodes. *C. R. Hebd. Seances Acad. Sci.* 137, 619–620.

Cuénot, L. (1905). L'organe phagocytaire des Crustacés Décapodes. *Arch. Zool. Exp. Gen., Ser.* 4 3, 1–16 + plate i.

Factor, J. R. (1978). Morphology of the mouthparts of larval lobsters, *Homarus americanus* (Decapoda: Nephropidae), with special emphasis on their setae. *Biol. Bull. (Woods Hole, Mass.)* 154, 383–408.

Factor, J. R. (1980). "Studies on the Digestive System of the Lobster, *Homarus americanus*: Larval Development and Aspects of Adult Histology and Ultrastructure," Diss. Abstr. 41B, 95-B (Order No. 80156-70). Ph.D. dissertation. Cornell University, Ithaca, New York.

Factor, J. R. (1981a). Unusually complex basement membranes in the midgut of two decapod crustaceans, the stone crab (*Menippe mercenaria*) and the lobster (*Homarus americanus*). *Anat. Rec.* 200, 253–258.

Factor, J. R. (1981b). Development and metamorphosis of the digestive system of larval lobsters, *Homarus americanus* (Decapoda: Nephropidae). *J. Morphol.* 169, 225–242.

Factor, J. R., and Beekman, J. (1990). The digestive system of the lobster, *Homarus americanus*: III. Removal of foreign particles from the blood by fixed phagocytes of the digestive gland. *J. Morphol.* 206, 293–302.

Factor, J. R., and Naar, M. (1985). The digestive system of the lobster, *Homarus americanus*: I. Connective tissue of the digestive gland. *J. Morphol.* 184, 311–321.

Factor, J. R., and Naar, M. (1990). The digestive system of the lobster, *Homarus americanus*: II. Terminal hepatic arterioles of the digestive gland. *J. Morphol.* 206, 283–293.

Factor, J. R., Manzo, P., and Hall, H. (1995). The digestive system of the lobster, *Homarus americanus*: IV. Tegumental glands of the esophagus and rectum. (in preparation.)

Felgenhauer, B. E. (1992). Internal anatomy of the Decapoda: An overview. *In* "Microscopic Anatomy of Invertebrates" (F.W. Harrison, ed.), Vol. 10, "Decapod Crustacea" (F. W. Harrison and A. G. Humes, eds.), pp. 45–75. Wiley–Liss, New York.

Gibson, R., and Barker, P. L. (1979). The decapod hepatopancreas. *Oceanogr. Mar. Biol.* 17, 285–346.

Harris-Warrick, R. M., Marder, E., Selverston, A. I., and Moulins, M., eds. (1992). "Dynamic Biological Networks. The Stomatogastric Nervous System." MIT Press, Cambridge, Massachusetts.

Herrick, F. H. (1909). Natural history of the American lobster. *Bull. U.S. Bur. Fish.* 29, 149–408 + 20 plates.

Hinton, D. J., and Corey, S. (1979). The mouthparts and digestive tract in the larval stages of *Homarus americanus*. *Can. J. Zool.* 57, 1413–1423.

Huxley, T. H. (1880). "The Crayfish. An Introduction to the Study of Zoology." Appleton, New York.

Icely, J. D., and Nott, J. A. (1992). Digestion and absorption: Digestive system and associated organs. *In* "Microscopic Anatomy of Invertebrates" (F. W. Harrison, ed.), Vol. 10, "Decapod Crustacea" (F. W. Harrison and A. G. Humes, eds.), pp. 147–201. Wiley–Liss, New York.

Johnson, B. R., and Hooper, S. L. (1992). Overview of the stomatogastric nervous system. *In* "Dynamic Biological Networks. The Stomatogastric Nervous System" (R. M. Harris-Warrick, E. Marder, A. I. Selverston, and M. Moulins, eds.), pp. 1–30. MIT Press, Cambridge, Massachusetts.

Johnson, B., and Talbot, P. (1987). Ultrastructural analysis of the pleopod tegumental glands in male and female lobsters, *Homarus americanus*. *J. Crust. Biol.* 7, 288–301.

Johnson, P. T. (1980). "Histology of the Blue Crab, *Callinectes sapidus*. A Model for the Decapoda." Praeger, New York.

Johnson, P. T. (1987). A review of fixed phagocytic and pinocytotic cells of decapod crustaceans, with remarks on hemocytes. *Dev. Comp. Immunol.* 11, 679–704.

Johnson, P. T., Stewart, J. E., and Arie, B. (1981). Histopathology of *Aerococcus viridans* var *homari* infection (gaffkemia) in the lobster *Homarus americanus*, and a comparison with histological reactions to a gram-negative species, *Pseudomonas perolens*. *J. Invertebr. Pathol.* 38, 127–148.

Katz, P. S., and Tazaki, K. (1992). Comparative and evolutionary aspects of the crustacean stomatogastric system. *In* "Dynamic Biological Networks. The Stomatogastric Nervous System." (R. M. Harris-Warrick, E. Marder, A. I. Selverston, and M. Moulins, eds.), pp. 221–261. MIT Press, Cambridge, Massachusetts.

Leavitt, D. F., and Bayer, R. C. (1982). A description of the muscle net surrounding the digestive epithelium in the midgut gland of the lobster *Homarus americanus*. *J. Crust. Biol.* 2, 40–44.

Lochhead, J. H. (1950). Crayfishes (and *Homarus*). *In* "Selected Invertebrate Types" (F. A. Brown, Jr., ed.), pp. 422–447. Wiley, New York.

Maynard, D. M., and Dando, M. R. (1974). The structure of the stomatogastric neuromuscular system in *Callinectes sapidus*, *Homarus americanus* and *Panulirus argus* (Decapoda Crustacea). *Philos. Trans. R. Soc. London, Ser. B* 268, 161–220.

Milne Edwards, H. (1834). "Histoire Naturelle des Crustacés. Comprenant l'Anatomie, la Physiologie et la Classification de ces Animaux," Vol. I. Librairie Encyclopédique de Roret, Paris.

Milne Edwards, H. (1840). "Histoire Naturelle des Crustacés. Comprenant l'Anatomie, la Physiologie et la Classification de ces Animaux. Atlas." Librairie Encyclopédique de Roret, Paris.

Mocquard, M. F. (1883). Recherches Anatomiques sur l'Estomac des Crustacés Podophthalmaires. *Ann. Sci. Nat. Zool.* 16, 1–309.

Mykles, D. L. (1979). Ultrastructure of alimentary epithelia of lobsters, *Homarus americanus* and *H. gammarus*, and crab, *Cancer magister*. *Zoolomorphologie* 92, 201–215.

Patwardhan, S. S. (1935a). On the structure and mechanism of the gastric mill in Decapoda. I. The structure of the gastric mill in *Paratelphusa querini* (M. Edw.). *Proc.—Indian Acad. Sci., Sect. B* 1B, 183–195.

Patwardhan, S. S. (1935b). On the structure and mechanism of the gastric mill in Decapoda. II. A comparative account of the gastric mill in Brachyura. *Proc.—Indian Acad. Sci., Sect. B* 1B, 359–375.

Patwardhan, S. S. (1935c). On the structure and mechanism of the gastric mill in Decapoda. III. Structure of the gastric mill in Anomura. *Proc.—Indian Acad. Sci., Sect. B* 1B, 405–413.

Patwardhan, S. S. (1935d). On the structure and mechanism of the gastric mill in Decapoda. IV. The structure of the gastric mill in reptantous Macrura. *Proc.—Indian Acad. Sci., Sect. B* 1B, 414–422.

Patwardhan, S. S. (1935e). On the structure and mechanism of the gastric mill in Decapoda. V. The structure of the gastric mill in natantous Macrura—Caridae. *Proc.—Indian Acad. Sci., Sect. B* 1B, 693–704.

Patwardhan, S. S. (1935f). On the structure and mechanism of the gastric mill in Decapoda. VI. The structure of the gastric mill in natantous Macrura—Penaeidea and Stenopidae; conclusion. *Proc.—Indian Acad. Sci., Sect. B* 2B, 155–174.

Sasaki, G. C., Capuzzo, J. M., and Biesiot, P. (1986). Nutritional and bioenergetic considerations in the development of the American lobster *Homarus americanus*. *Can J. Fish. Aquat. Sci.* 43, 2311–2319.

Selverston, A. I., and Moulins, M., eds. (1987). "The Crustacean Stomatogastric System." Springer-Verlag, Berlin.

Smith, R. I. (1978). The midgut caeca and the limits of the hindgut of Brachyura: A clarification. *Crustaceana* 35, 195–205.

Talbot, P., and Demers, D. (1993). Tegumental glands of Crustacea. *In* "The Crustacean Integument, Morphology and Biochemistry" (M. N. Horst and J. A. Freeman, eds.), pp. 151–191. CRC Press, Boca Raton, Florida.

Turrigiano, G. C., and Heinzel, H.-G. (1992). Behavioral correlates of stomatogastric network function. *In* "Dynamic Biological Networks. The Stomatogastric Nervous System" (R. M. Harris-Warrick, E. Marder, A. I. Selverston, and M. Moulins, eds.), pp. 197–220. MIT Press, Cambridge, Massachusetts.

van Weel, P. B. (1974). Hepatopancreas? *Comp. Biochem. Physiol.* 47S, 1–9.

Wales, W. (1982). Control of mouthparts and gut. *In* "The Biology of Crustacea" (D. E. Bliss, ed.), Vol. 4, "Neural Integration and Behavior" (D. C. Sandeman and H. L. Atwood, eds.), pp. 165–191. Academic Press, New York.

Wiens, T. J. (1982). Small systems of neurons: Control of rhythmic and reflex activities. *In* "The Biology of Crustacea" (D. E. Bliss, ed.), Vol. 4, "Neural Integration and Behavior" (D. C. Sandeman and H. L. Atwood, eds.), pp. 193–240. Academic Press, New York.

Williams, L. W. (1907). The stomach of the lobster and the food of larval lobsters. *Annu. Rep.—R.I. Comm. Inland Fish.* 37, 153–180.

Yonge, C. M. (1924). Studies on the comparative physiology of digestion. II. The mechanism of feeding, digestion, and assimilation in *Nephrops norvegicus*. *J. Exp. Biol.* 1, 343–389.

Yonge, C. M. (1932). On the nature and permeability of chitin. I. The chitin lining the foregut of decapod Crustacea and the function of the tegumental glands. *Proc. R. Soc. London, Ser. B* 111, 298–329.

Yonge, C. M. (1945). Crayfish and lobsters. *Sch. Sci. Rev.* 26, 333–342.

CHAPTER

16

Digestive Physiology and Nutrition

DOUGLAS E. CONKLIN
Department of Animal Science
University of California at Davis
Davis, California 95616

I. Introduction

The nutritional needs of the American lobster *Homarus americanus* are of interest to a variety of individuals. Most immediately obvious is the need of both fishers and those charged with regulating the fishery to understand the relationship between food availability and catch. Surprisingly, given the long history of the fishery, dietary intake studies of lobsters in the wild were, until recently, rare. Although early fisheries biologists occasionally noted the types of animals consumed by lobsters, it was not until the 1970s that specific studies aimed at defining the impact of food resources on lobster populations were initiated. These recent studies often have attempted to define feeding of lobsters within an ecological framework of predator–prey relationships. Most studies, however, must still rely on the enumeration of distinguishable remains of prey items in the gut. Such analyses of gut contents leave open questions relating to the total nutritional intake and specific nutrient content of the natural diet. Thus, while information from field-oriented dietary studies has increased dramatically over the last two decades, many implications are still in dispute and more needs to be done.

Distribution of live lobsters to the marketplace brought this hardy species to the attention of a wide range of individuals and stimulated research into both laboratory and commercial culture. Lobsters are important to crustacean specialists and as biological models in such diverse fields as neurobiology and degenerative muscle diseases. Typically, laboratory diets presently consist of the occasional provision of chopped fish or clams. As these studies become increasingly more sophisticated and rigorous, however, the need to conscientiously control dietary input becomes increasingly important.

While the expense of culture is typically a secondary consideration for the laboratory researcher, costs are paramount for entrepreneurs aspiring to market lobsters from farms rather than the sea. As food may account for more than half the projected cost in aquaculture ventures, potential lobster farmers seek to maximize growth while reducing nutrient input to a minimum. Interest in lobster aquaculture, particularly strong during the 1970s and early 1980s, provided much of the emphasis on specific nutrient requirements. (See Aiken and Waddy, Chapter 8, on aquaculture.) Unfortunately, research efforts dwindled as interest in the commercial culture of crustaceans shifted from lobsters to marine shrimp, and information on the specific nutrient requirements of lobsters remains fragmentary.

In spite of their disparate goals and methods, each group brings to the subject a variety of valuable insights. Studies on both natural and laboratory diets are considered in this chapter along with selected aspects of feeding behavior and digestive physiology. Comparisons with the closely related European lobster, *Homarus gammarus* (=*H. vulgaris*), are made where relevant. In addition, relevant information derived from other crustacean species, particularly marine shrimp, will be used to frame the information on *H. americanus* within the context of an emerging crustacean nutritional model.

II. Digestive Physiology

Morphological and functional details of the digestive system of *Homarus americanus*, as well as larval development and metamorphosis, are presented by Factor (Chapter 15). In addition, several extensive reviews of the digestive system of crustaceans in general are available (Gibson and Barker, 1979; Dall and Moriarty, 1983; Icely and Nott, 1992). The following section, therefore, highlights those specific aspects of digestive physiology that provide insight into natural diets and are of value in developing efficient formulated diets for culture.

A. Overview of Digestion

Postmetamorphic lobsters locate and stalk prey at some distance through chemoreceptive setae located on the antennules (see Atema and Voigt, Chapter 13, on sensory biology). Nitrogenous compounds such as amino acids are generally the most effective attractants (McLeese, 1970; Ache, 1972; Carter and Steele, 1982a).

In postmetamorphic animals, food is torn into pieces small enough to pass through the esophagus by combined action of the mandibles and the third maxillipeds, but external mastication is minimal (see Lavalli and Factor, Chapter 14). Passage of food is lubricated by mucus secretions from tegumental glands located in the walls of the esophagus. There is no evidence of amylase production by the esophageal tegumental glands; thus, enzymatic breakdown of food does not begin until it enters the cardiac stomach.

In postmetamorphic animals, reduction of food particle size in the cardiac chamber is accomplished by action of the gastric mill, which grinds ingested food into fine particles. The complex structure and movement of the stomatogastric system (see Factor, Chapter 15) foster circulation patterns that efficiently combine digestive enzymes from the digestive gland with the food mass while it is being repeatedly passed to the gastric mill for trituration. Circulation of the fluidized mass also enhances the selective separation of fine particles from the food mass. Channels on the floor of the cardiac stomach, which are covered by setal screens, facilitate movement of a slurry containing finely ground food particles in the pyloric stomach, which provides further screening of particles before they move into the digestive gland. A radiographic study using barium sulfate shows the size of particles passing into the digestive gland of the adult to be less than 1 μm (Bayer *et al.*, 1979).

Although the larval mandible and maxillipeds do not have the elaboration of teeth found in older animals, this does not imply that larval lobsters consume their prey whole. The adult brine shrimp used in laboratory larval culture are comparatively quite large but easily torn into pieces for ingestion by all the larval stages.

Lack of a gastric mill in larval lobsters is puzzling in that, in nature, the larvae prey on a variety of planktonic organisms. While most of these are soft-bodied, some grinding would seem necessary to reduce food effectively. In a review of the mouthparts of crustaceans, Factor (1989) noted that brachyuran and anomuran larvae typically have massive molar processes for grinding of food when no gastric mill is present. Looking for an analogous structure in homarid larvae, Lavalli and Factor (1992) suggested that the "pad of setae" on the inner surface of the mandibles might be used by the larvae to masticate food. If this is true, much of the grinding of food by lobster larvae would have to be accomplished externally by this series of pads and ridges. This may limit the type of food larvae can successfully use and should be considered in the development of artificial larval feeds.

The final phase of digestion and nutrient uptake takes place in the digestive gland. Both by volume and complexity, the digestive gland (i.e., midgut gland, or hepatopancreas) dominates the midgut region of the crustacean digestive tract (see Factor, Chapter 15, for the structural organization of the complex organ). The digestive gland is the production site of digestive enzymes that initiate digestion upon delivery to the stomach, as well as the location for the completion of the digestive process. In addition, most, if not all, of the absorption of nutrients takes place in the tubules of the digestive gland. Studies on the uptake kinetics of glucose and amino acids suggest that transport rates for the digestive gland are several orders of magnitude greater than those of the intestine (Ahearn and Clay, 1987a,b, 1988; Ahearn *et al.*, 1983, 1985).

The four principal cell types found in the crustacean digestive gland (Fig. 1) are typically referred to by initials, following the classification proposed by Hirsch and Jacobs (1930; Jacobs, 1928): E cells (*embryonalenzellen*, or embryonic cells); R cells (*resorptionzellen*, or resorptive cells); F cells (*fibrillenzellen*, or fibrillar cells); and B cells (*blasenzellen*, or blisterlike cells). The relationship and exact function of these four cell types in crustaceans are matters of some controversy and may vary depending on species. However, based on studies of *Homarus gammarus* (Barker and Gibson, 1977), the most likely sequence of differentiation is E cells generating both R and F cells, with F cells maturing into B cells (Gibson, 1983). An additional cell type, the small M cell (midget cell) found in penaeid shrimp (Al-Mohanna *et al.*, 1985), has not been noted in lobsters.

According to the functional pattern proposed by Gibson and Barker (1979), each cell type has a distinct role in the digestive gland. The E cells, at the distal end of the tubule and the only cell type showing cell division, are undifferentiated progenitors of F and R cells. The F cells contain abundant rough endoplasmic reticulum, which produces the fibrillar appearance at the light microscopy level. Digestive enzymes produced in the rough endoplasmic reticulum are packaged in the Golgi complex of the F cells and stored in cytoplasmic vacuoles. As the vacuoles expand and coalesce, the appearance of the cells distorts into the characteristic blisterlike form of B cells. A thin layer of cytoplasm encompasses a single large vacuole that compresses the other cellular structures to the basal region of the cell. In response to feeding, the vacuoles are released in periodic surges. Normally, secretion is thought to be by merocrine or apocrine discharge; however, the stimulation of food intake following a period of starvation may induce holocrine secretion (Gibson and Barker, 1979). B cells that are not lost to the tubule lumen through holocrine secretion undergo a restitution cycle, during which the cells appear as F cells before developing again into B cells. R cells, the most abundant cell type in the digestive gland, are the primary site of uptake and storage of nutrients, although F and B cells also take up some nutrients through pinocytosis. R cells have a dense distinct brush border of microvilli that promotes contact digestion and transport of digested nutrients into the cells. Abundant lipid inclusions represent the primary storage material of the R cells, although small particles of glycogen also occur in the cytoplasm (Barker and Gibson, 1977).

Undigested residues from the digestive gland pass back to the intestine to be combined with the coarser material that was diverted away from the openings to the digestive gland. Compaction of this residual material, a process initiated in the pyloric stomach, and its passage as feces through the intestine to the hindgut for eventual elimination complete the digestion process. As the residual material moves through,

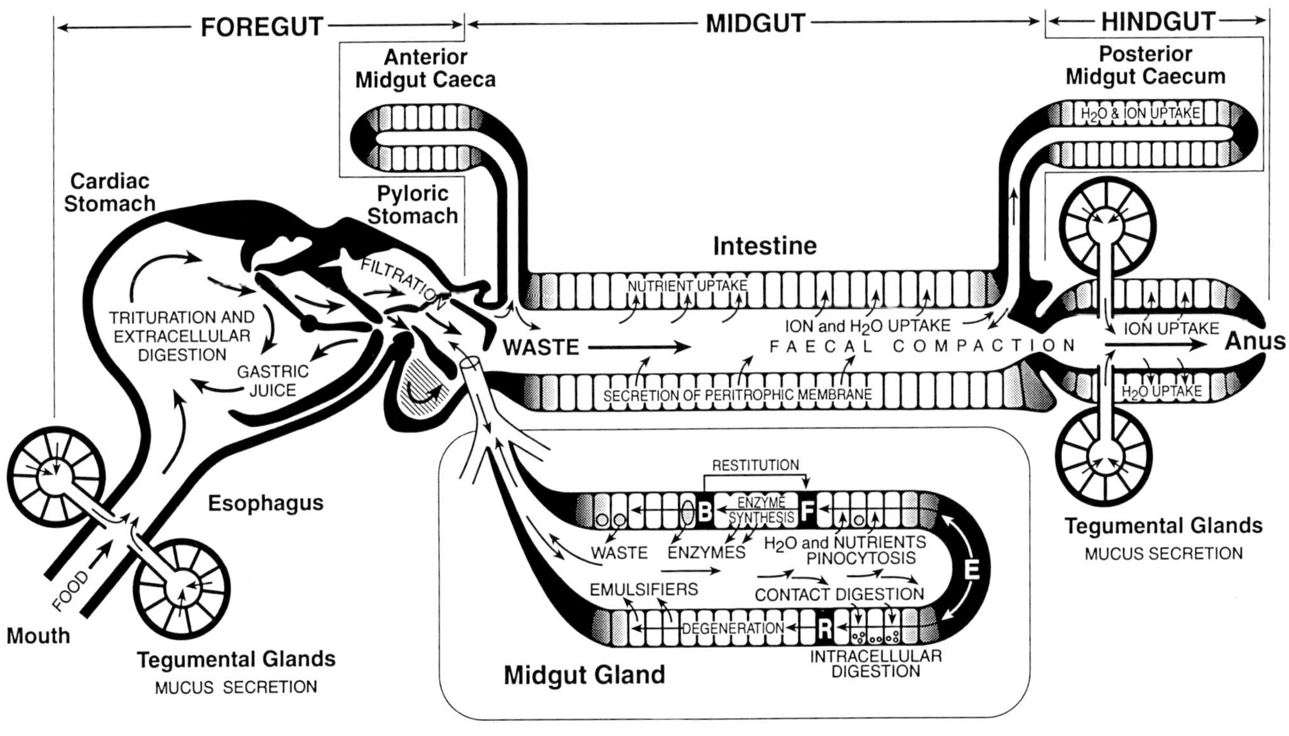

FIGURE 1 Diagram of presumptive digestive tract function (see text for details) in *Homarus americanus*. (Adapted from Gibson, 1983, with permission.)

water and ions are removed and transported across the epithelial borders (Mykles, 1979, 1980). Epithelial secretions enclose the concentrated residual material in a peritrophic membrane as it passes through the intestine (Forster, 1953; Dall, 1967; Mykles, 1979, 1980). The epithelial cells of the posterior midgut caecum are specialized for water uptake and the caecum is thought to have a minor role in digestion (Mykles, 1979). Water uptake by the caecum is particularly important in molting in that the majority of water absorbed during ecdysis in the lobster occurs in the intestine and the caeca (Mykles, 1980).

The cells lining the short hindgut appear to be specialized for cation transport (Mykles, 1979, 1980), but their specific functional role with regard to nutrition has yet to be defined. Infolding of the basal membrane in the hindgut epithelial cells as well as folding of the entire epithelium allow for stretching of the hindgut lining under fecal loading.

B. *Digestive Enzymes*

The literature on digestive enzymes is difficult to interpret for several reasons, not the least of which is a general paucity of studies (Dall and Moriarty, 1983). Comparison among available studies is hampered by the use of seemingly innocuous assay differences, such as the nature of the substrate, which may give strikingly different results. For example, Brockerhoff *et al.* (1970) originally concluded that amylase activity in the adult lobster is insignificant, but later, using a soluble form of starch as a substrate, found appreciable activity (Wojtowicz and Brockerhoff, 1972). Terminology is an additional problem. Researchers rely on enzyme activity parameters derived from mammalian systems as a guide in the isolation and classification of crustacean enzymes. However, the digestive system of crustaceans appears to differ significantly, for example, the unique protease identified in the crayfish *Astacus fluviatilis* (Zwilling *et al.*, 1981; Titani *et al.*, 1987). Thus, reports that crustaceans lack this or that enzyme should be "treated with caution" (Dall and Moriarty, 1983).

The general pattern of digestive enzyme activity in *Homarus americanus* is defined in some detail for both larvae (Biesiot and Capuzzo, 1990a and b) and adults (Brockerhoff *et al.*, 1970; Wojtowicz and Brockerhoff, 1972; Hoyle, 1973). Although there is considerable agreement, some differences exist, particularly with respect to pH optima. Differing isolation techniques, ranging from using fresh stomach fluid to using pH-adjusted, dialyzed, freeze-dried preparations, may account for some of these anomalies.

The digestive fluid is acidic in both larvae (pH 5.5, Biesiot and Capuzzo, 1990a, b) and adults (pH 4.6–4.8, Hoyle, 1973; pH ~5.0, Brockerhoff *et al.*, 1970), although the mechanism for establishing and maintaining the acidic state is unknown. The acidic pH of the digestive fluid is of interest because many enzymes examined have an activity optimum near pH 8.0. While it has been suggested that the pH optimum of the various enzymes could be altered by various ionic components in the digestive fluid or that the pH at the actual site of digestion within the food mass may be less acidic (Brockerhoff *et al.*, 1970; Hoyle, 1973), there is no evidence to support these suggestions. Many of the enzymatic fractions examined have two activity peaks, with the acidic peak near the pH of the digestive fluid. Other activity peaks, such as those for amylase, are quite broad. Although the optimum pH for amylase activity in digestive extracts from larval lobsters is 6.5, two thirds of the broad activity curve is at the pH of the digestive extract, 5.5 (Biesiot and Capuzzo, 1990b).

In adults, enzymatic activity is stimulated by feeding; however, enzymatic adaptation to differing levels of dietary components appears to be minor in *Homarus americanus* (Hoyle, 1973). *Homarus gammarus* adults undergo cyclic secretory bursts of enzymatic activity following the initial stimulation of feeding (Barker and Gibson, 1977). There appears to be an ontogenetic increase in enzyme activities among embryos and larvae of *H. americanus* (Biesiot and Capuzzo, 1990a). In *H. americanus* larvae, overall enzymatic activity tends to be constant, not cyclical, perhaps reflecting the continuous input of small amounts of food rather than the large intermittent meals of adults (Biesiot and Capuzzo, 1990b).

Protease activity in the stomach juice of adults has two pH optima (5.5 and 7.5) when assayed with a collagen (Azocoll) substrate (Hoyle, 1973). These are similar to the two activity peaks (pH 5.3 and 6.5) noted in crude digestive tract extracts of stage I larvae tested with Azocoll (Biesiot and Capuzzo, 1990b). A wider range of peak activity (pH 4.0 and 8.0 optima) occurs when Azocoll is used as a substrate for freeze-dried preparations of adult gastric fluid (Brockerhoff *et al.*, 1970). Despite the use of differing preparations, the acidic pH reported for the gastric fluid is appropriate for at least one of the pH optima associated with proteolytic activity.

The specific identity of the enzymes responsible for protease activity is not known. At least seven proteolytic enzymes have been noted in *Homarus americanus* adults (Brockerhoff *et al.*, 1970). Interestingly, the pH optimum of most of these proteolytic enzymes is basic and thus outside the normal acidic pH range of the digestive fluid (Brockerhoff *et al.*,

1970). One of these enzymes is classified as "trypsin" based on its molecular weight (25,000) and pH optimum of 8.0. Trypsin, an endopeptidase that cleaves peptide bonds in the inner portion of the protein molecule where the carboxyl group is contributed by the basic amino acids arginine and lysine, is found in a variety of crustaceans (DeVillez and Buschlen, 1967).

While functionally indistinguishable, crustacean "trypsins" differ from mammalian trypsin in several ways, most importantly in their ability to attack undenatured protein and their resistance to autodigestion (Zwilling *et al.*, 1969; Eisen *et al.*, 1973). The amino acid sequence of crayfish trypsin shows a 43.6% sequence identity with bovine trypsin (Titani *et al.*, 1983). Interestingly, resistance to autodigestion by the crustacean trypsin was a function of a single amino acid difference at the point of autocatalytic cleavage. While the purported trypsin in lobsters has not been sufficiently characterized as yet, it is likely to prove to be similar to that of the crayfish.

The digestive fluid of adult *Homarus americanus* shows "chymotryptic" activity, but this cannot be separated from the trypsin fraction and it may be that both types of enzymatic activity result from a single protein (Brockerhoff *et al.*, 1970). The trypsin enzyme isolated from the crab *Uca pugilator* has both tryptic and chymotryptic activity (Eisen *et al.*, 1973). Trypsin has yet to be identified in larval lobsters, although it has been reported in the zoeal stages of the shrimp *Penaeus japonicus* (Galgani and Benyamin, 1985) and the spider crab *Hyas araneus* (Hirche and Anger, 1987).

"Carboxypeptidase A" activity (pH optimum of 7.5) has been reported in *Homarus americanus*, but not carboxypeptidase B activity (Brockerhoff *et al.*, 1970). Because *H. americanus* appears to lack both a dipeptidase and a leucine aminopeptidase, digestion of small peptides may take place intracellularly (Brockerhoff *et al.*, 1970; Gibson and Barker, 1979).

Lipids are an important energy source for crustaceans (Capuzzo, 1983) and lobsters require several specific fatty acids (Conklin *et al.*, 1983). A lipase found in the gastric fluid of adult lobsters acts on olein to produce fatty acids, diglycerides, and β-monoglycerides (Brockerhoff *et al.*, 1967). Intriguingly, the β-monoglyceride is preferentially retained following lipolytic action and absorption (Brockerhoff and Hoyle, 1967). This preferential treatment of the β-position fatty acid leads to an enrichment of polyunsaturated acids in the β-position of tissue lipids. The lipase activity of the freeze-dried digestive fluid from adults is reported to be maximal at a pH of 7.0 (Brockerhoff *et al.*, 1970). In larvae, the optimum lipase activity for the hydrolysis of a olein emulsion occurs at a pH of 5.5, a value more closely allied with the gastric fluid pH of lobsters (Biesiot and Capuzzo, 1990b). Although lipid metabolism is important to larval lobsters (Sasaki *et al.*, 1986), lipase activity in stage I larvae is surprisingly much lower than in adults. Lipid digestion in lobsters is aided by the emulsifying and solubilizing action of fatty aryl–taurine complexes (Holwerda and Vonk, 1973), in combination with dietary lecithin (Lester *et al.*, 1975). The emulsifiers may be liberated when degenerative R cells rupture (Gibson and Barker, 1979).

Amylase activity in lobsters often shows a broad pH optimum rather than a sharp peak: ~pH 6.0–7.0 in larvae (Biesiot and Capuzzo, 1990b) and pH 5.0–6.0 in adults (Hoyle, 1973). A pH optimum of 5.2 has been found in an isolated amylase fraction from adults, with sharply increased activity at 0.05–0.1 *M* NaCl concentrations (Wojtowicz and Brockerhoff, 1972), an effect not seen with larval amylase.

Several other potential enzymes in the digestive fluid of *Homarus americanus* have been described based on substrate activity (Brockerhoff *et al.*, 1970): a ribonuclease, a phosphatase, and a chitobiase (with strong activity); and an α- and β-glucosidase, a β-galactosidase, and a chitinase (with weak activities). Three exochitinases and two chitobiases have also been isolated from gastric juice (Lynn, 1990). Chitinoclastic activity in the gut of the shrimp *Penaeus setiferus* is due to a combination of enzymes produced by the shrimp digestive gland, a chitinase and a chitobiase enzyme, and bacterial production of these enzymes (Hood and Meyers, 1977). Nothing is known about the microbiological component of the lobster's digestive tract and its possible contribution of digestive enzymes.

While there is much yet to be learned, in general, postmetamorphic lobsters appear to be well equipped to consume and digest a wide variety of organisms. The combined action of the mouthparts and the gastric mill quickly and efficiently reduces food to a slurry of fine particles that are moved to the digestive gland for further digestion and absorption. To date, studies on digestive enzymes have tended to be descriptive, but it appears that lobsters have a full complement of digestive enzymes to effectively utilize prey the organisms consumed in the wild.

III. Nutritional Parameters of Natural Diets

With regard to natural diets, lobsters can be categorized into three groups: (1) planktonic stages, (2) shelter-restricted and emergent juveniles, and (3) foraging juveniles and adults. (See Factor, Chapter 1, and Lawton and Lavalli, Chapter 4, for overviews of life history.) These categories reflect distinctly different feeding strategies, resulting in differing sets of prey items, but not necessarily obligatory changes in nutritional requirements. For example, in the wild, the three larval stages feed on planktonic prey species, while juveniles and adults eat benthic organisms. However, a diet of live adult brine shrimp (*Artemia* sp.) that supports rapid growth and high survival of cultured larvae also serves postmetamorphic animals through the first 3 months of growth. This is consistent with the role planktonic organisms continue to play in the diet during the early benthic period (see Lawton and Lavalli, Chapter 4, and Lavalli and Factor, Chapter 14). Thus, while the specific nutrient requirements are probably similar, these dietary groupings reflect distinct combinations of morphological and behavioral characteristics that lead to the consumption of different prey items in nature.

A. Planktonic Stages

The first three larval stages presumably feed on a similar array of available planktonic food organisms. Lavalli and Factor (1992) compiled a list of food organisms consumed by larval lobsters that includes copepods, amphipods, other decapod larvae, echinoderm larvae, fish larvae, worms, mollusks, and diatoms. There is some, mostly anecdotal, evidence of selection of specific or larger prey items (Williamson, 1905; Williams, 1907; Harding *et al.*, 1983; Cobb *et al.*, 1983), but the extent to which selection occurs in nature, particularly for the three larval stages, has yet to be characterized. (See Ennis, Chapter 3, and Lawton and Lavalli, Chapter 4, on larval and postlarval ecology, and Lavalli and Factor, Chapter 14, on feeding appendages.)

The postlarva (stage IV) is an interesting transitional stage. The molt between stages III and IV is metamorphic (Neil *et al.*, 1976; Charmantier *et al.*, 1991; Factor, Chapter 1), resulting in a postlarva with the general appearance of the adult. Although the postlarva has the morphological characteristics of the benthic stages that follow, it may remain in the water column, feeding on planktonic organisms, for much of the instar period. Its prey is similar to that of the earlier larval stages, but there is a preference for the larger planktonic organisms, such as adult copepods and the larger species of copepods, despite generally higher densities of smaller copepods in the water column (Juinio and Cobb, 1992).

Several authors (Phillips and Sastry, 1980; Harding *et al.*, 1983; Harding and Trites, 1988; Castell and Kean, 1986; Ennis, 1986) have speculated that diet may be a critical limiting factor for the planktonic larval stages of *Homarus americanus*. Because "larval survival appears to be greatly affected by changes in the time to complete development" (Ennis, 1986), factors slowing developmental rates would subject the larvae to increasing periods of predation. This general idea is supported by analysis of temperature and larval survival, suggesting that development slows at lower temperature regimes with a corresponding decrease in larval survival (Caddy, 1979). In that diet presumably also could limit growth rates, an impact on larval survival by both food quantity and quality has been proposed.

Rising water temperatures in the spring may couple the hatch with water currents fostering inshore movement to allow lobster larvae to intercept temporal prey assemblages of small copepods (Harding *et al.*, 1983). This would allow "rapid growth in warm waters past a vulnerable size in its life cycle." While intuitively attractive, evidence supporting the hypothesis that food influences larval development rates is contradictory. Often cited are the results of studies carried out by Templeman (1936), who clearly showed that food restriction dramatically lengthens larval development times. On the other hand, in experiments done by Carlberg and Van Olst (1976) and Eagles *et al.* (1986), the effect of feeding level is limited to survival, with only negligible effects on larval development times. These strikingly contrasting results are seemingly irreconcilable.

Another way to look at the issue is the examination of field data for possible links between plankton productivity and the fishery. There was a striking fit between freshwater runoff and *Homarus americanus* landings 6 years later in Quebec (Sutcliffe, 1972), suggesting that river runoff is likely to enhance biological productivity in near-shore areas (see Bugden *et al.*, 1982). However, continuing work specifically examining larval abundance produced more mixed results. A firm correlation between larval abundance and runoff existed in one area of the Gulf of St. Lawrence, but not in others (Sutcliffe, 1973). Since 1984, there has been no correlation between freshwater runoff and subsequent lobster landings in Quebec, leaving the issue unresolved (reviewed by Drinkwater *et al.*, 1991).

Available data on larval feeding behavior, while admittedly scanty, suggest that food resources for larval lobsters may be more than ample. The significant numbers of newly ingested prey items that were larger than would be expected based on availability suggest preferential consumption (Juinio and Cobb, 1992). In that feeding by lobster larvae appears to be by tactile stimulation only (*Homarus americanus*, Juinio and Cobb, 1992), such preferential ingestion implies rejection of smaller prey items after they are encountered, an unlikely event under conditions of food limitation. An abundance of prey also is suggested by the fact that only a very few larvae have empty guts (Juinio and Cobb, 1992).

Besides the quantitative aspects of diet, there is also the possibility of variation in quality of diet affecting larval survival (Castell and Kean, 1986). The effect of food quality can be shown in the laboratory. Feeding "poorer-quality" frozen *Artemia* to cultured lobster larvae results in slower development, decreased total weight gain, and molting problems (Eagles *et al.*, 1986). Batches of commercially frozen *Artemia* with a high content of fragmented and colorless adults are considered of lower quality.

Because of the importance of long-chain polyunsaturated fatty acids for marine fish and crustaceans (Castell and Kean, 1986), the possibility of an essential fatty acid deficiency has been a common focus of research. The dietary importance of polyunsaturated fatty acids in crab larvae has been shown in experiments using various prey species (Sulkin, 1978; Sulkin and Van Heukelem, 1980; Levine and Sulkin, 1984; Staton and Sulkin, 1991). The marine rotifer *Brachionus plicatilis* is deficient in polyunsaturated fatty acids compared to brine shrimp nauplii and larval development of several species of crabs is correspondingly arrested when only the rotifer is provided as food. While such laboratory experiments are useful in understanding some facets of larval nutrition, extrapolation to larval food quality in nature should be done with caution. Little is known as yet about the biosynthetic abilities of various species of larvae in comparison to the juvenile and adult stages. Single-prey dietary studies also provide limited insight into larval diets in nature, where the larvae feed on a variety of organisms. Lobster larvae are particularly well developed morphologically and raptorial feeding on zooplankton can commence immediately upon hatching. The large size range of acceptable prey is impressive and is reflected in the wide array of items noted in the gut of wild-caught lobster larvae. Development of a specific nutrient deficiency under such conditions seems improbable.

The ability to accumulate substantial energy reserves during the larval stages (Sasaki *et al.*, 1986) may facilitate a delay of settlement when necessary to find appropriate substrate (see Ennis, Chapter 3, and Lawton and Lavalli, Chapter 4). Such energy reserves also argue against the idea of food limitation.

B. Shelter-Restricted and Emergent Juveniles

Where the substrate is appropriate, newly settled lobsters construct burrows for shelter (Lawton and Lavalli, Chapter 4). During this shelter-restricted phase of the lobster's life history (Lawton and Lavalli, in Chapter 4, provide definitions of life history phases), feeding is accomplished by generating a current through the burrow using the pleopods and capturing organisms carried in by the current (Barshaw and Bryant-Rich, 1988; Barshaw, 1989). Prey found within the burrow or just within reach of the burrow entrance serves as an additional source of food (Barshaw and Bryant-Rich, 1988). Suspension feeding on the mesoplankton diet available to shelter-restricted lobsters supports growth rates comparable to those found in other laboratory studies at ambient seawater temperatures (Hughes and Matthiessen, 1962) and roughly that predicted from field data (Hudon, 1987). This would indicate that suspension feeding on available suprabenthic plankton and additional food found within the burrows could account for the growth of lobsters in the field over the first 2 years, even though they are shelter bound.

C. Foraging Juveniles and Adults

Transition to active foraging is gradual. Even vagile juveniles (>25-mm carapace length) continue to spend much of their time in burrows (Lawton and Lavalli, Chapter 4). Hudon (1987) reported that lobsters between 20- and 30-mm carapace length were only occasionally seen in the open by divers and quickly retreated to shelter. The behavior of older juvenile lobsters (>40-mm carapace length) is markedly different from that of the younger animals. Rather than swimming away or retreating with a tail flip, larger juveniles tend to stand their ground with a defensive posture of raised claws (Hudon, 1987).

Identifying the prey of foraging lobsters and estimating the contribution of the various species to the total diet is a daunting challenge because of the shredding of prey during ingestion and the extensive mastication of food in the stomach. Identification of dietary components is difficult and heavily dependent on fragments of indigestible skeletal structures. Consequently, the value of food items lacking in persistent skeletal elements or those rapidly digested tends to be underestimated. Visual estimates are typically used to determine gut fullness. The volume contribution of various prey items within the gut has to be calculated using various subjective estimates. While admitting that interpretation is fraught with potential problems (see Hyslop, 1980), considerable information on the natural diet of foraging lobsters is available and a number of conclusions can be drawn from the aggregate. The most salient is the diverse nature of the diet of larger juveniles and adults. Carter and Steele (1982b) list the identifiable remains of 33 taxa of organisms; Elner and Campbell (1987), 65 taxa; and Hudon and Lamarche (1989), 53 taxa. This diverse diet consists chiefly of animal prey, generally over 90%; however, plant material is a small (<5%), but consistent, component. The multiplicity of acceptable prey species helps to ensure that foraging lobsters are seldom without food and animals found with empty stomachs are rare (9%, Scarratt, 1980; 7.8%, Carter and Steele, 1982b; 2%, Hudon and Lamarche, 1989). As expected, stomach fullness varies seasonally, with the average being around 25% during the cold early months of the year and rising to 50–80% during the summer, when the lobsters are more active (Ennis, 1973; Scarratt, 1980; Elner and Campbell, 1987). Feeding is not completely related to temperature, as Ennis (1973) reported that feeding activity remained high in the fall despite decreasing temperature. Carter and Steele (1982b) reported similar results, noting no significant difference in stomach fullness over the period from June through November.

While foraging lobsters are clearly opportunistic feeders, they are not indiscriminate. Although dietary components change both year to year and seasonally and are influenced by habitat, there is an overall consistency in principal prey organisms, mussels, crabs, polychaetes, and sea urchins. Although lobster remains are found in the stomach, cannibalism is a minor consideration (Weiss, 1970; Carter and Steele, 1982b; Elner and Campbell, 1987; Hudon and Lamarche, 1989). The exact level of cannibalism (<10%) is obscured by consumption of the cast exuviae after molting. Habitat differences are reflected in dietary composition primarily through changes in the consumption of less important prey items (Elner and Campbell, 1987; Hudon and Lamarche, 1989). The general consistency of diet, in spite of differences in prey abundance among habitats, is seen as a result of the selection of available prey with higher caloric value (Scarratt, 1980; Hudon and Lamarche, 1989). The resulting diet is high in protein (Leavitt *et al.*, 1979), supporting tissue synthesis during the intermolt. A shift to a calcium-rich diet in the fall (Weiss, 1970; Ennis, 1973; Leavitt *et al.*, 1979; Scarratt, 1980; Carter and Steele, 1982b) has been interpreted to reflect an increased requirement for calcium for shell hardening following the molt in late summer. Thus, while a select few species are always taken, the total spectrum varies year to year and seasonally with availability of prey and the molt cycle of the lobster.

Selection of prey by foraging lobsters also relates to size of the lobster; the various size classes within a population may concentrate on different components of available prey within a particular habitat (Scarratt, 1980; Carter and Steele, 1982b; Elner and Campbell, 1987). The stronger chelae of larger lobsters permit them to feed on larger and more heavily armored species (Elner and Campbell, 1981); however, prey size is not strictly correlated with lobster size, as consumption of smaller armored prey appears to be energetically advantageous (Juanes, 1992). Thus, larger size results in a wider choice and a more varied diet.

IV. Ration Formulation and Feeding

The commercial culture of lobsters, first envisioned by Lord in 1867, remains an enticing but elusive prospect, hindered, in large part, by the lack of a suitable diet. The intensive culture approach envisioned for commercial culture dictates a complete ration providing all of needed nutrients. While lobsters can be easily cultured in a laboratory setting using live or fresh foodstuffs, rations appropriate for commercial culture situations still await development. Reviews on the laboratory culture of both larvae (Chang and Conklin, 1993) and postmetamorphic animals (Conklin and Chang, 1993) are available and Aiken and Waddy (Chapter 8) provide an extensive discussion of commercial aquaculture. The primary focus of this section is the specific nutritional requirements needed for the formulation of complete rations.

A. *Laboratory Feeding Studies*

The most widely used larval feed for *Homarus americanus* is adult brine shrimp (*Artemia* sp.), either live or frozen (Carlberg and Van Olst, 1976; Schuur *et al.*, 1976; Logan and Epifianio, 1978; Capuzzo and Lancaster, 1979b; Eagles *et al.*, 1986; Chang and Conklin, 1993); however, nauplii have also been used successfully (Castell, 1977; MacKenzie, 1987; Snyder and Chang, 1986). A high density of food is necessary to prevent cannibalism by the larvae. A brine shrimp-to-larvae ratio of 4:1 is recommended (Schuur *et al.*, 1976). Actual consumption has been observed to be 2.5 adult brine shrimp per day for stage I larvae, which increases steadily to 8.8 per day for stage IV lobsters (postlarvae) (Logan and Epifanio, 1978), with a reported overall average larval consumption of approximately eight adult brine shrimp per larva per day (Carlberg and Van Olst, 1976).

Live brine shrimp are preferable from a nutritional standpoint (Eagles *et al.*, 1986) and are easier to use. As live brine shrimp adults can avoid entrapment on the overflow screen of the *kreisel* larval rearing chamber (see Aiken and Waddy, Chapter 8), they are usually fed only once a day in excess. As heavy concentrations of frozen brine shrimp tend to clog the overflow screen, an automatic feeding device (Serfling *et al.*, 1974) that periodically dispenses additional amounts is needed for frozen brine shrimp.

Initial studies with formulated artificial diets for lobster larvae are disappointing (Kurmaly *et al.*, 1990). Survival on all of the microparticulate and microencapsulated diets tested is discouragingly poor. Further work is necessary before the potential of artificial formulations for larval culture can be seriously evaluated. For the foreseeable future, live brine shrimp will continue to be the best diet for the culture of larval *Homarus americanus*.

As with larval culture, a diet of live adult brine shrimp supports high survival and excellent growth in postlarvae (stage IV) and early juveniles. Growth rates of juveniles fed live adult brine shrimp will vary with temperature and other culture parameters, but are consistent if these parameters are maintained at their known optima (Van Olst *et al.*, 1980). At a temperature of 20°C, a rate of growth equal to a 0.1-mm/day increase in carapace length is used as a "brine shrimp standard" for *Homarus americanus* (Conklin *et al.*, 1977). Growth rate on a diet of frozen adult brine shrimp is only about 60% of that of the live standard. Other organisms, such as fouling organisms (D'Agostino, 1980), amphipods (D'Agostino, 1980; Good *et al.*, 1982), barnacle larvae (Daniel *et al.*, 1985), or mesoplankton (Lavalli, 1991), are effective to a degree. Such alternative diets typically provide lower growth rates than live brine shrimp, and as they are not commercially available, they are unlikely to be used widely.

Larger lobsters require larger food items and the live brine shrimp diet becomes less effective after about 4–6 months of growth. The larger juveniles and adults are typically fed small shrimp or chopped mollusks, such as mussels or squid. While a diet of chopped fresh food is effective for the limited amount of food needed for most laboratory situations, such a diet would not be practical for large-scale commercial culture. To meet the needs of the commercial facilities sometimes envisioned (production of 80,000 1-lb lobsters per month), an estimated 24 metric tons of fresh foodstuffs would be required daily (Conklin *et al.*, 1983). Pelletized rations that could be dispensed easily appear to be essential for commercial culture. Of course, effective formulation of such rations requires detailed information on the nutritional requirements of lobsters. Delineating these requirements has been the focus of lobster nutritionists since the mid-1970s.

B. Nutritional Requirements

Our understanding of the nutritional needs of *Homarus americanus* has been greatly expanded over the last decade, particularly with respect to selected aspects of lipid nutrition. However, in other areas, research is still lacking and many gaps identified in earlier reviews (Conklin, 1980; Conklin *et al.*, 1983) remain. The study of nutrition in *H. americanus* has occurred in the context of advances in other commercially important, culturable crustaceans. Reviews are available for penaeid shrimp (Akiyama *et al.*, 1992; Chen, 1993), freshwater prawns (D'Abramo and Sheen, 1994), crayfish (D'Abramo and Robinson, 1989), and spiny lobsters (Kanazawa, 1994).

Three dominant characteristics of lobster nutritional research are: a narrow focus with respect to size class, the use of live adult brine shrimp (*Artemia* sp.) as a control diet, and the use of purified ingredients as a foundation for test diets. Lobster nutritional studies have been almost wholly carried out with early juveniles younger than 120 days of age. It is assumed that the specific nutrient requirements for lobsters do not change significantly throughout the growth phase and that nutritional needs identified for these early juveniles are applicable to older animals. It is important to remember, however, that this assumption has yet to be adequately tested.

Two purified test diets have been developed for lobster nutritional studies (Table 1). Although both diets consist of purified ingredients, they differ significantly in composition. The diet called HFX CRD 84, developed by Castell and co-workers (described in Boghen *et al.*, 1982) uses an isolated crab concentrate as a protein source, in contrast to the casein used as the primary protein in the BML 81S diet, developed by Conklin *et al.* (1983). A second prominent difference is the need for lecithin in the BML 81S diet. For reasons that are not yet well understood (Baum *et al.*, 1990), phospholipids are not required in the HFX CRD 84 diet. Other differences in the two diets are thought to be minor and both support good growth of juvenile lobsters, although neither is as good as a diet of live brine shrimp.

The composition of brine shrimp is known (reviewed by Simpson *et al.*, 1983; Watanabe *et al.*,

TABLE 1 **Standard Reference Diets for Crustacean Research**

Ingredients	Reference diet composition (%)	
	HFX CRD 84	BML 81S
Crab protein concentrate	40.0	—
Casein (vitamin free)	—	31.0
Egg white (spray-dried)	—	4.0
Wheat gluten	5.0	5.0
Corn starch	15.0	24.0
Dextrin	5.0	—
Celufil (nonnutritive bulk)	17.8	12.1
Cod liver oil	6.0	4.0
Corn oil	3.0	2.0
Soy lecithin	—	10.0
Cholesterol	1.0	0.5
Mineral mix (modified Bernhart–Tomarelli)	4.0	3.0
Vitamin premix CRD	2.0	—
Vitamin premix BML-2	—	4.0
Vitamin E (DL-α-tocopherol)	0.2	0.2
Vitamin A acetate (50,000 IU/g)	—	0.1
Cholecalciferol (D_3) (4 million IU/g)	—	0.1
Choline chloride (70%)	1.0	—
Totals	100.0	100.0

1983; Léger *et al.*, 1986) and, in one sense, nutrient levels meeting the specific dietary requirements of larval and early juvenile lobsters can be identified. However, mimicking this standard has proven to be difficult. Early juvenile lobsters reared on the best artificial formulation grow at only 60–70% of the rate achieved on a diet of brine shrimp. The reason for this deficit in growth response is not known, but it is likely due to less than optimum ratios of the various nutrients rather than unknown deficiencies.

1. Proteins

Brine shrimp, as well as preferred natural prey, are high in protein (Table 2). As protein is the most expensive component of formulated feeds, defining optimal levels of dietary protein is a major goal of aquaculture nutrition. Reported optimum protein levels for lobsters fed artificial formulations vary widely (60%, Castell and Budson, 1974; 53%, Gallagher *et al.*, 1976a; 37%, Lucien-Brun *et al.*, 1985; 35%, Boghen and Castell, 1981; 30%, Conklin *et al.*, 1975; D'Abramo *et al.*, 1981b). The variable results may be influenced to an extent by differing experimental conditions, but the most significant factors probably relate to protein quality and the level of nonprotein energy in the diet.

Using casein, a milk protein, as a protein source in experimental diets is convenient, as it is commercially available in a purified vitamin-free form, but it is not the most efficient protein for lobster growth. The addition of shrimp meal (D'Abramo *et al.*, 1981b; Cruz-Suárez *et al.*, 1993) or the use of proteins extracted from crustaceans such as shrimp (Boghen *et al.*, 1982) or crabs (Castell *et al.*, 1989a,b) generally produces more effective results (rapid growth at lower protein levels). The reason for the superior response with crustacean proteins is not understood. In general, crustaceans seem to have little problem digesting protein, although Norman-Boudreau and Conklin (1985) reported problems with a trypsin inhibitor in egg white. The total protein digestion in juvenile lobsters for two formulated diets containing several different proteins is over 90% (Bordner *et al.*, 1983).

Protein quality relates to the amino acid composition of the protein. Ideally, it should match the essential amino acid needs of the animal. Information as to essential amino acid requirements in crustaceans comes primarily from radiolabeling experiments delineating a lack of synthesis of specific amino acids. Amino acids not synthesized by *Homarus americanus* (Gallagher and Brown, 1975b) mirror those found with other crustaceans. All show the same underlying pattern and the 10 amino acids essential for crustaceans—arginine, histidine, isoleucine, leucine, lysine, methionine, phenylalanine, threonine, tryptophan, and valine—are the same 10 typically required by vertebrates (Wilson, 1989).

Besides the classic 10 essential amino acids, additional amino acids may be required for some crustaceans. Both asparagine and taurine have been suggested (Dadd, 1983). Asparagine synthesis from aspartic acid, as occurs in vertebrates, is lacking in some crustacean species (*Astacus astacus*, Zandee, 1966b; *A. leptodactylus*, Van Marrewijk and Zandee, 1975; *Macrobrachium rosenbergii*, Smith *et al.*, 1987). The biosynthesis of asparagine has yet to be examined in *Homarus americanus*. The free amino acid taurine is found in high concentrations in crustaceans; it is thought to serve as an osmoregulatory agent (Allen and Garrett, 1971) and may also be involved in reproduction (Pochon-Masson *et al.*, 1984). While taurine is essential for some animals, crustaceans appear to be capable of its synthesis and there is evidence of taurine synthesis in *H. americanus* (Gilles and Schoffeniels, 1969; Finney, 1978).

Unfortunately, radiolabeling provides little infor-

TABLE 2 **Proximate Composition of Brine Shrimp and Lobster Larvae Expressed as a Percentage of Dry Weight**

	Brine shrimp, adults		Lobster larvae[c]		
Constituent	Live[a]	Frozen[b]	Stage I	Stage IV	Stage V
Moisture (% of wet weight)	90.0	90.0	~90.0[d]	—	—
Crude protein (CP)	58.0	50.2	68.8	62.0	62.6
Amino acid protein	43.0	—	—	—	—
Crude fat (CF)	5.1	2.4	5.0	4.2	3.2
Cholesterol	0.46	—	—	—	—
Carbohydrates (CHO)	16.3[e]	17.2	7.8[e]	9.6[e]	11.7[e]
Crude fiber	3.5	2.9	—	—	—
Chitin	—	—	7.8	9.8	12.3
Ash	20.6	29.2	18.4	24.2	22.5
Totals	100.0	99.0	100.0	100.0	100.0

[a]From Gallagher and Brown (1975a).
[b]From Good *et al.* (1982).
[c]From Capuzzo and Lancaster (1979b).
[d]From Watanabe *et al.* (1978a, 1983).
[e]Calculated as CHO = 100.0 - (CP + CF + Ash).

mation on the amount of synthesis in relation to metabolic needs. Information is lacking on the quantitative amino acid requirements supporting rapid growth. The use of isolated amino acids to ascertain quantitative requirements for lobsters (Mason and Castell, 1980) has not been a fruitful approach. Rapid leaching of crystalline amino acids from test diets makes quantifying actual consumption difficult. Not only are leaching losses for the entire mixture of added free amino acids high, but the various individual amino acids are lost in differing proportions (Lim, 1993). In that nitrogenous compounds also can serve as attractants (McLeese, 1970; Carter and Steele, 1982a), a growth response may reflect increased consumption rather than a superior amino acid composition. Beyond the problems associated with leaching, the use of free amino acids appears to disrupt normal patterns of uptake. For *Penaeus japonicus*, tissue incorporation of added free arginine is very low in contrast to arginine coming from dietary protein (Deshimaru, 1983).

In formulated diets, arginine levels approximating those in brine shrimp are particularly hard to reach without the use of protein from crustaceans (Table 3) or other marine sources. A mixture of proteins having an amino acid composition similar to that of *Penaeus japonicus* or the control diet of short-necked clam *Venerupis philippinarum* produces superior growth rates in *P. japonicus* in comparison to alternative amino acid compositions (Deshimaru and Shigeno, 1972; Deshimaru, 1983). Although similar studies have yet to be carried out with lobsters, the high levels of arginine in the isolated crab protein used in the HFX CRD 84 diet may account for its ability to support good growth at relatively low protein levels.

Inefficient use of protein also may result from the inappropriate levels of nonprotein energy components in the diet. As with most other animals, consumption in crustaceans is controlled by energy levels of the diet (Sedgwick, 1979; Davis and Arnold, 1993). Excessive dietary energy can limit food consumption and, consequently, diminish protein intake and growth. If nonprotein energy is too low, part of the dietary protein will be oxidized to make up the deficit, again reducing potential growth. Ideally, protein intake is at a maximum and all the protein goes into tissue synthesis.

2. Protein and Energy Ratios

Lack of information on appropriate dietary energy levels for *Homarus americanus* makes defining optimum protein levels difficult. The potential for protein sparing by adding carbohydrates to the diets of juvenile lobsters has been shown by Capuzzo and Lancaster (1979a), but these results have not been incorporated into more recent studies with lobsters. The apparent protein requirement of *Penaeus monodon* can be reduced from 40 to 30% by increasing dietary starch levels from 20 to 30% (Shiau and Peng, 1992).

TABLE 3 **Amino Acid Composition (Percentage of Total Weight of Amino Acids) of Brine Shrimp in Comparison to the Two Standard Reference Diets Used in the Culture of Lobsters**[a]

Amino acid	Brine shrimp Adults Live[b]	Brine shrimp Adults Meal[c]	Brine shrimp Nauplii[d]	HFX CRD 84[e]	BML 81S[f]
Arginine	6.5	6.8	7.3	8.1	3.9
Histidine	1.8	2.2	1.9	3.0	2.9
Isoleucine	5.3	5.1	3.8	4.4	5.2
Leucine	8.0	8.6	8.9	7.7	8.9
Lysine	7.6	7.4	8.9	7.6	7.0
Methionine[g]	2.7	2.3	1.3	2.7	2.9
Phenylalanine	4.7	5.3	4.7	5.1	4.7
Threonine	4.6	4.9	2.5	5.0	3.7
Tryptophan[g]	1.0	—	1.5	—	1.4
Valine	5.4	5.3	4.7	4.8	6.5
Alanine	6.9	5.2	6.0	5.2	3.2
Aspartic acid	9.2	10.1	11.0	10.8	6.8
Cystine[g]	2.2	1.3	0.6	2.3	0.5
Glutamic acid	14.2	14.6	12.9	14.7	22.3
Glycine	5.3	4.9	5.0	4.9	2.2
Proline	5.2	4.7	6.9	4.2	9.9
Serine	4.8	5.2	6.7	4.6	4.8
Tyrosine	4.5	4.6	5.4	4.2	4.9

[a]Amino acids identified as essential are shown in boldface type.

[b]From Gallagher and Brown (1975a).

[c]From Deshimaru and Shigueno (1972).

[d]From Watanabe *et al.* (1978a).

[e]From Castell *et al.* (1989b).

[f]BML 81S amino acid composition was calculated as the total of 0.775 of casein values plus 0.1 of egg white values plus 0.125 of gluten values provided in Castell *et al.* (1989a).

[g]Partially or totally destroyed by acid hydrolysis. Nitrogen from these amino acids is recorded as NH_3 with conventional amino acid analysis. As the amount of NH_3 is not listed above, the sum of the amino acids may not equal 100%.

Similarly, dietary protein levels for *P. japonicus* can be reduced from 61 to 42% without reducing growth by increasing levels of dextrin and pollack liver oil (Koshio *et al.*, 1993). It is likely that the protein requirement for lobsters is around the 30% level, given an appropriate protein source and sufficient nonprotein energy in the diet. However, this remains supposition because of the many questions that need to be answered regarding protein quality and energy sources.

3. Carbohydrates

The nonprotein energy sources in diets come from either digestible carbohydrates or lipids. It is assumed, but has not been rigorously tested, that lobsters and other decapod crustaceans are incapable of cellulose digestion. Starch, the typical carbohydrate source in formulated diets for crustaceans, is poorly digested by lobsters, approximately 60% compared to more than 90% for proteins and lipids (Bordner *et al.*, 1983). The digestibility and utilization of other digestible carbohydrates are not known for lobsters. Survival rates of the prawn (*Penaeus monodon*) fed starch and dextrin were higher than those fed glucose, with the best protein utilization arising from diets containing starch, intermediate for diets with dextrin, and poor for diets with glucose (Shiau and Peng, 1992).

4. Lipids

Unsaturated fatty acids are thought to be key elements in the diet of crustaceans. As with other animals, crustaceans lack the ability to synthesize the unsaturated fatty acids, linoleic (18:2n-6) and linolenic (18:3n-3) acids (Zandee, 1966c; Kanazawa *et al.*, 1979a–d). In addition, bioconversion by marine decapods of 18:3n-3 to other fatty acids, especially the highly unsaturated fatty acids, eicosapentaenoic (20:5n-3) and docosahexaenoic (22:6n-3) acids, is seemingly quite poor (Teshima *et al.*, 1992a). Brine shrimp, particularly adults, contain high levels of these polyunsaturated fatty acids (Table 4), although levels can vary depending on their diet. The addition of 5.0% cod liver oil (high in the n-3 polyunsaturated fatty acids) to a formulated diet improved the growth of *Homarus americanus* adults when compared to a diet containing only corn oil, in which linoleic acid predominates (Castell and Covey, 1976). The value of polyunsaturated fatty acids in the diet of juvenile lobsters has also been noted (D'Abramo *et al.*, 1980).

Uptake and utilization of these fatty acids from the diet are favored if they are provided as phospholipid. The addition of soybean lecithin to purified diets dramatically improves the survival rate of *Homarus americanus* juveniles; lack of the lecithin fraction results in heavy mortalities at ecdysis when the lobsters are unable to free themselves from the cast molt (Conklin *et al.*, 1980). The key component of this fraction is phosphatidylcholine (D'Abramo *et al.*, 1981a) and the most effective phosphatidylcholine molecules are those containing polyunsaturated fatty acids.

Phospholipids play an important role in the transport of cholesterol via the hemolymph (D'Abramo *et al.*, 1982, 1985a). A phospholipid deficiency in the diet of juvenile lobsters inhibits mobilization of cholesterol from the digestive gland to the hemolymph. While the role of phosphatidylcholine in the transport of cholesterol in *Homarus* is clearly evident, the overall effect of phospholipids in the diet of the lobster is still muddled. Other components of the diet can complicate the response to lecithin. For example, juvenile lobsters do not require dietary soy lecithin when a purified protein derived from the rock crab (*Cancer irroratus*) is substituted for the casein (Kean *et al.*, 1985a). However, the addition of lecithin increases serum cholesterol whichever the source of protein, milk or crab (Baum *et al.*, 1990).

Lobsters, like all arthropods (Teshima, 1983), are incapable of *de novo* sterol synthesis (Zandee, 1966a) and require a dietary source for normal growth and survival. Diets for *Homarus americanus* should contain 0.5% dietary cholesterol (Castell *et al.*, 1975), although the requirement can be satisfied by as little as 0.12% cholesterol in diets containing lecithin (D'Abramo *et al.*, 1984). In contrast to crayfish (D'Abramo *et al.*, 1985b) and, to some extent, penaeid shrimp (Teshima *et al.*, 1989), lobsters are unable to use phytosterols as a replacement for cholesterol (D'Abramo *et al.*, 1984).

Crustaceans also require a dietary source of carotenoids for normal pigmentation. A variety of carotenoids can be transformed by the lobster into astaxanthin, the primary pigment that gives the characteristic red color upon boiling. Besides pigmentation, it has been suggested that carotenoids play other vital roles in the nutrition of other crustaceans (Gillchrist and Lee, 1972). However, these suggestions are generally unsupported (reviewed by Nelis *et al.*, 1989). Only a slight increase in growth occurs in juvenile lobsters fed a carotenoid-supplemented diet compared to carotenoid-free diets (D'Abramo *et al.*, 1983) and survival was unaffected.

Reduced production of viable larvae from lobster females with pale-colored eggs containing presumably lower than normal levels of astaxanthin has been noted (Castell and Kean, 1986). Lobster larvae have lower than normal levels of carotenoids when fed on frozen brine shrimp batches with low levels of carotenoids (Eagles *et al.*, 1986); the ensuing postlarvae also are smaller and suffer from cheliped loss. While these situations are of interest, a specific metabolic role of carotenoids in lobster reproduction or larval survival has yet to be established.

5. Vitamins

Defining the vitamin requirements for lobsters has been surprisingly difficult (Conklin *et al.*, 1977; Castell *et al.*, 1975) and little specific information is available. Concern over rapid leaching of these water-soluble compounds (Goldblatt *et al.*, 1979; Slinger *et al.*, 1979) encouraged the high levels of supplementation in the two test diets (Table 1), particularly BML 81S. Both use similar vitamin premixes derived directly from formulations used in vertebrate feeds, which are of unknown efficacy for lobsters. Included in the premixes are all of the factors typically considered vitamins, plus *p*-aminobenzoic acid. The B-complex compounds—thiamin, riboflavin, niacin, pantothenic acid, vitamin B_6 (pyridoxine), biotin, folic acid, and vitamin B_{12} (cobalamin)—are all water soluble, as are choline, *myo*-inositol (inositol), vitamin C (ascorbic acid), and *p*-aminobenzoic acid. Vitamins A, D, E, and K are fat soluble. The vitamin requirements of *Homarus americanus* have yet to be delineated, with the possible exception of vitamin C. However, the vitamin requirements of shrimp,

TABLE 4 **Fatty Acid Composition (Percentage of Total Lipid) of Brine Shrimp Showing Variation between Adults and Larvae, with Source of Eggs (Year), and Previous Nutritional History**

Adults[a]		Nauplii				
			1975[c]	1977[d]	Hatch[e]	Fed[f]
Fatty acid[b]	%	Fatty acid[b]	%	%	%	%
14:0	1.4	14:0	1.1	0.9		
14:1	2.3					
15:0	0.7					
15:1	0.8					
16:0	13.5	16:0	11.2	9.5	15.5	14.0
16:1	13.8	16:1n-7	4.3[h]	12.0[h]	8.4	6.5
17:0	1.3	17:0	1.5[i]	0.9[i]	—	3.0
17:1	0.9				1.4	0.5
18:0	5.9	18:0	4.4	6.8	4.2	12.3
18:1	35.6	18:1n-9	25.1[h]	36.1[h]	37.8	22.6
18:2	6.2	18:2n-6	6.1	3.4	7.7	4.2
18:3		18:3n-3	28.4	10.3	19.7	23.4
20:0	2.0	20:0	4.5[j]	1.2[j]	—	2.7
20:1	0.1				—	0.6
		20:4n-6	1.0[k]	0.4[k]		
20:4	2.2[g]	20:4n-3	0.3	0.4	—	5.8
20:5	12.0	20:5n-3	3.1	9.5		

[a]From Gallagher and Brown (1975a).

[b]Fatty acids are identified by use of the shorthand designation. The first number is the number of carbon atoms in the molecule. The number following the colon is the number of double bonds and the number n-*x* (replacing the earlier ω designation) indicates the position of the first double bond from the methyl end of the molecule. The discrimination and detection of individual fatty acids depend on the column and solvent system used in fatty acid analysis.

[c]From Watanabe *et al.* (1978b); *Artemia salina* (San Francisco Brand, 1975) nauplii hatched at the Laboratory of Fish Nutrition, Tokyo University of Fisheries.

[d]From Watanabe *et al.* (1978b); nauplii (*Artemia salina* from San Francisco, 1977) just after hatching.

[e]From Claus *et al.* (1979); San Francisco Bay strain of *Artemia salina*, instar I.

[f]From Claus *et al.* (1979); San Francisco Bay strain of *Artemia salina*, 48-hour-old larvae alga fed (*Scenedesmus* sp.).

[g]From Gallagher and Brown (1975a); includes 20:3 and 20:4.

[h]From Watanabe *et al.* (1978b); small amounts of the other monoenes included.

[i]From Watanabe *et al.* (1978b); includes 16:2 and 17:0.

[j]From Watanabe *et al.* (1978b); includes 18:4n-3 and 20:0.

which are better known, are likely to bear some relevance to lobsters. These should serve as a rough guide until specific studies are carried out with *H. americanus*. The vitamin requirements for shrimp are summarized in Table 5 and compared with the levels used in the two lobster reference diets and vitamin analysis of brine shrimp adults.

For shrimp, specific requirements for vitamins in the B complex have been shown or suggested for thiamin, riboflavin, niacin, pyridoxine (B_6), and, surprisingly, vitamin B_{12}. A dietary level for thiamin of approximately 120 mg/kg of diet is thought to satisfy the requirement of shrimp (Deshimaru and Kuroki, 1979), and 50 mg/kg has been recommended for commercial shrimp diets (Akiyama *et al.*, 1992). Levels of vitamin supplementation in commercial diets can be lower than in refined laboratory diets because of the additional vitamins contained in other component feedstuffs. While the thiamin needs of *Penaeus monodon* are met with 20 mg/kg of diet

(Chen *et al.*, 1991), *P. japonicus*, considered to be the most carnivorous of the cultured shrimp, requires 60 mg/kg for maximum growth and 120 mg/kg for tissue saturation (Deshimaru and Kuroki, 1979). The riboflavin requirement for *P. monodon* is just over 20 mg/kg of diet (Chen and Hwang, 1992); juveniles fed a riboflavin-deficient diet exhibit several signs of deficiency, including a decrease in tissue concentration. Riboflavin supplementation of 40 mg/kg is recommended for commercial shrimp feeds (Akiyama *et al.*, 1992). Niacin supplementation at just under 10 mg/kg produces maximum weight gain in *P. monodon* (Shiau and Suen, 1994). As levels of niacin supplementation above 70 mg/kg significantly reduce weight gain, the recommended levels of 200 mg/kg for commercial shrimp feeds (Akiyama *et al.*, 1992) should be viewed with caution. Appropriate dietary levels of pantothenic acid have yet to be established, but 75 mg/kg have been recommended for commercial shrimp feeds (Akiyama *et al.*, 1992). The pyridoxine (vitamin B_6) requirement for animals is commonly higher than that for thiamin (McDowell, 1989), and the recommended level of 50 mg/kg of feed (Akiyama *et al.*, 1992) should be viewed as a minimum; survival and growth of *P. japonicus* are best at 60 mg/kg (Deshimaru and Kuroki, 1979). Recommended biotin and folic acid supplementation levels in commercial shrimp feeds are 1 mg and 10 mg/kg of diet, respectively (Akiyama *et al.*, 1992); confirming studies with cultured crustaceans have yet to be carried out. *p*-Aminobenzoic acid is a component of the folacin molecule and is used by bacteria in its synthesis. *p*-Aminobenzoic acid is unlikely to have a direct role in crustacean nutrition. The demonstration of a vitamin B_{12} requirement of 0.2 mg/kg of diet for *P. monodon* is surprising in that this vitamin is generally required in such minute amounts (Shiau and Lung, 1993).

Besides the four identified B vitamins, crustaceans probably require both choline and inositol, which are components of phospholipids. Compared to other vitamins, they are found in substantial amounts in animals. Phospholipids are key elements of biological membranes and in crustaceans are the primary lipid component of the lipoprotein transport system. The addition of choline chloride at 600 mg/kg of diet has been reported both to improve growth and survival of *Penaeus japonicus* juveniles (Kanazawa *et al.*, 1976) and, paradoxically, to have no effect (Deshimaru and Kuroki, 1979). For maximum growth and survival, at least 400 mg/kg of diet of inositol is required (Kanazawa *et al.*, 1976; Deshimaru and Kuroki, 1979). The recommended level for commercial diets is 400 mg/kg choline and 300 mg/kg inositol (Akiyama *et al.*, 1992).

All penaeids tested to date require a dietary source of ascorbic acid (vitamin C) (reviewed by Conklin, 1990). Suggested dietary levels vary widely depending

TABLE 5 A Comparison of Vitamin Levels in Brine Shrimp with Levels in the Standard Lobster Reference Diets, as Well as the Recommended and Required Feed Levels for Penaeid Shrimp[a]

Vitamin	Brine shrimp	HFX CRD 84	BML 81S	Shrimp feed	Required shrimp
Water soluble					
Thiamin	127	64	200	50	120
Riboflavin	17	144	320	40	20
Niacin	130	520	1040	200	10
Vitamin B_6	8	48	120	50	60
Pantothenic acid	68	286	600	75	10
Biotin	1	1.6	40	1	?
Folic acid	7	19.4	200	10	?
Vitamin B_{12}	3	54	40	0.1	0.2
Choline	6100	7000	3600	400	600
Inositol	1200	2540	2200	300	400
p-Aminobenzoic acid	—	404	1200	—	—
Ascorbic acid	220	1220	4840	100	120
Fat soluble					
Vitamin A (IU)	6650	51,000	50,000	10,000	4800
Vitamin E	—	2000	2000	300	100
Vitamin D (IU)	—	340	4 million	5000	8000
Vitamin K	—	16	—	5	40

[a]Values are expressed as milligrams per kilogram of dry weight unless otherwise noted. See text for further details.

on the source of ascorbic acid. Ascorbic acid itself is easily oxidized to an inactive form and thus the apparent feed requirement can appear much higher than the biological requirement. Use of chemically stable forms of ascorbate, such as L-ascorbyl-2-sulfate, L-ascorbyl-2-phosphate, and L-ascorbyl-2-polyphosphate, greatly reduces the feed requirement. The requirement is 100 mg/kg of diet for *Penaeus japonicus* when using Mg-L-ascorbyl-2-phosphate (Shigueno and Itoh, 1988) and 120 mg/kg for *P. vannamei* when using L-ascorbyl-2-polyphosphate (He and Lawrence, 1993a). Interestingly, *Homarus americanus* may not require ascorbic acid at all. There is evidence of ascorbate synthesis in the lobster (Desjardins *et al.*, 1985) and dietary trials show no effect on growth and survival of juveniles fed diets lacking ascorbic acid (Kean *et al.*, 1985b).

There have been hints that both *Homarus americanus* (Stewart and Castell, 1979) and larval shrimp (Kanazawa, 1983, 1985) might require some of the fat-soluble vitamins. Identifying the specific dietary need for these fat-soluble factors has been difficult, however, because fish oil, used as a source of the required long-chain polyunsaturated fatty acids, contains significant levels of these vitamins. The use of the fatty acid fraction extracted from menhaden oil in experiments with *Penaeus vannamei* circumvents this obstacle (He *et al.*, 1992). Addition of the four fat-soluble vitamins at the following levels to a control diet significantly improves growth: vitamin A acetate, 240 mg/kg of diet (4800 IU); vitamin D_3, 20 mg/kg (8000 IU); vitamin E acetate, 1600 mg/kg (400 IU); and menadione (vitamin K), 40 mg/kg. Poorer growth resulted when vitamin A, vitamin D, and vitamin E are removed from the mixture, but no effect is seen with the removal of menadione. Lack of vitamin E also significantly reduces survival. The requirement for vitamin E is suggested to be 100 mg/kg of diet (25 IU) (He and Lawrence, 1993b). The primary role of vitamin E in the diet is antioxidant protection; a reduction in the growth of shrimp fed a vitamin E-deficient diet can be prevented by the addition of the synthetic antioxidant butylated hydroxytoluene (He and Lawrence, 1993b). *Penaeus monodon* diets supplemented with 30–40 mg/kg of menadione (vitamin K) promote higher rates of growth and better conversion of protein than unsupplemented controls (Shiau and Liu, 1994). The levels of fat-soluble vitamins (A, D, and E) suggested by recent laboratory work (He *et al.*, 1992; He and Lawrence, 1993b) are noticeably lower than those recommended for commercial shrimp diets (Akiyama *et al.*, 1992): vitamin A, 10,000 IU/kg of diet; vitamin D, 5000 IU/kg of diet; and vitamin E, 300 mg/kg of diet. Conversely, the recommended feed level for vitamin K, 5 mg/kg of diet, is much lower than that suggested by the work with *P. monodon* (Shiau and Liu, 1994).

6. Minerals

Because crustaceans are capable of absorbing ions from water and minerals are particularly abundant in seawater, defining the mineral requirements of marine crustaceans has been particularly difficult. The exoskeleton of *Homarus americanus* contains significant amounts of calcium carbonate, and calcium needs have received the most attention. Although marine crustaceans can take calcium from the surrounding seawater (Hayes *et al.*, 1962; Deshimaru *et al.*, 1978), the addition of calcium to the diet of juvenile lobsters appears to improve mineralization of the exoskeleton; however, there is no impact on survival or growth (Conklin *et al.*, 1975). A calcium-to-phosphorus ratio of 1:2 has been suggested as optimum for juvenile lobsters; however, growth and survival on all of the formulated diets used in the experiment were poor (Gallagher *et al.*, 1976b). This early work should be reexamined now that better diets are available. Because greater calcium deposition in *Penaeus monodon* was noted with increased vitamin K in the diet (Shiau and Liu, 1994), studies should incorporate this vitamin as one of the factors. Such studies should also include vitamin D, as increasing the calcium level in a diet containing lecithin, but without added vitamins A and D, improves survival (Baum *et al.*, 1991). Dietary requirements of *H. americanus* for other minerals have not been determined, although suggested levels for a variety of minerals have been proposed for penaeid shrimp (Table 6).

V. Directions for Further Research

Research needs for crustacean nutrition through the end of the century have been extensively detailed by D'Abramo and Lovell (1991). Their suggestions covering nutrient requirements, digestion, broodstock and larval diets, and feed formulation are all applicable to studies with *Homarus americanus*; however, several are particularly important.

Primarily because crystalline amino acid supplementation has proved ineffective for crustaceans, quantitative amino acid requirements of crustaceans, including the lobster, remain unknown. Development of techniques to effectively deliver individual amino acid supplementation would be extremely useful in optimizing the protein component of diets. Techniques for linking amino acids to proteins (Teshima *et al.*, 1992b), encapsulating them in a non–water-soluble

TABLE 6 **Mineral Content of Brine Shrimp in Comparison with Recommended Feed Levels for Penaeid Shrimp**

Mineral	Brine shrimp (% of dry wt)[a]	Shrimp feed (%/kg of diet or ppm)[b]
Calcium	0.100	< 2.3
Phosphorus	0.930	0.8
Magnesium	0.222	0.2
Sodium	5.11	0.6
Potassium	0.832	0.9
Cobalt	—	10 ppm
Copper	0.001	35 ppm
Iron	0.275	300 ppm
Manganese	0.013	20 ppm
Selenium	—	1 ppm
Zinc	0.008	110 ppm

[a]From Gallagher and Brown (1975a).
[b]From Akiyama *et al.* (1992).

form (Chen *et al.*, 1992) or some other mechanism to provide specific amino acid enrichment of diets, are important to develop fully. Beyond allowing for the accurate determination of quantitative essential amino acid requirements, such techniques will ultimately be required on a commercial basis to provide needed flexibility in formulating cost-effective rations for aquaculture of lobsters and other crustaceans.

Digestibility studies are a second key element to improving existing diets. Although chromic oxide as an inert internal marker has been used extensively in animal nutrition studies, it is segregated from other components of the diet by the digestive process in lobsters (Bordner *et al.*, 1983; Leavitt, 1985). Because chromic oxide is not an effective marker for *Homarus americanus*, other techniques for measuring digestibility should be investigated. Both the gravimetric method (total collection) and the ash ratio method have been evaluated with the lobster (Leavitt, 1985); however, both have significant problems. Use of indigestible fiber (cellulose or lignin) appears, from studies with fish (Anderson *et al.*, 1991), to have some potential and its applicability to studies with lobsters should be investigated. Accurate determination of feedstuff digestibility is important in defining optimum protein–energy ratios and in formulating commercial feeds, and is needed for the application of a bioenergetics approach to nutrition (reviewed by Capuzzo, 1983) both in culture and in the field.

Unfortunately, it will be some time before formulated diets have been improved to the level where they will become useful to lobster biologists probing nutritional questions in the field. Even then, some situations, such as examining the feeding of shelter-restricted juveniles in the laboratory, may require the use of live feed. For this type of study with larvae and early juveniles, a diet of live brine shrimp should be used, as several studies indicate that frozen brine shrimp are deficient (Barshaw and Bryant-Rich, 1988; Eagles *et al.*, 1986). While brine shrimp is not available to lobsters in the wild, its nutrient composition is similar to that of natural zooplankton prey (Watanabe *et al.*, 1983). There is also the possibility of manipulating, to a fair degree, various aspects of this live diet. Both larvae and early juvenile lobsters will feed on all sizes of brine shrimp from freshly hatched nauplii to mature adults; however, the energetic cost of feeding on the various sizes should be quite different. The nutrient quality of brine shrimp nauplii can be diminished by starving the nauplii or enhanced in various ways, depending on what they are fed (see Léger *et al.*, 1986). Similar modification of nutrient quality also should be possible when using adult brine shrimp. While this type of approach lacks the flexibility of nutrient variation possible with formulated diets, it should prove to be a useful technique for exploring a number of questions relating to the nutrition of lobsters in the wild.

VI. Summary

The fishery for *Homarus americanus* is the most valuable of all the lobster fisheries and represents an important component of the economy along the northern Atlantic coast of the United States and Canada. Questions about the relationship between food availability and catch are of vital interest to the fishery. The high market value has also stimulated a continuing interest in the commercial culture of lobsters. The potential of lobster aquaculture has encouraged further interest into the nutritional needs of lobsters and the problems of feed development.

Food items are torn by the mouthparts and passed through the esophagus to the cardiac stomach. Knowledge of the digestive enzymes of lobsters is still quite limited, but an array of enzymes, including proteinases, lipases, and various carbohydrases, is found in both larval and adult lobsters. Food masticated into fine particles by the gastric mill is mixed with digestive enzymes that initiate the digestive process and reduced to a slurry. Suitably small particles are passed to the digestive gland for completion of the digestive process. The digestive gland contains a variety of cell types specialized for the production of digestive enzymes and the absorption and storage of nutrients. Undigested material and wastes from the digestive gland are compacted in the intestine and passed out through the hindgut and the anus.

The type of prey consumed by lobsters varies dramatically with life history phase, from planktonic organisms during the larval stages to a broadening array of both mobile and sessile benthic organisms for juvenile and adult animals. The wide array of prey items greatly complicates the critical assessment of the impact of food resources on the growth and survival of lobsters in the wild.

There is little evidence to support the idea that prey densities have a significant impact on the survival of the larvae. Larvae appear to have little trouble encountering adequate food. It is also unlikely that larval development would be hindered by specific nutrient deficiencies. The short premetamorphic developmental period, the existence of substantial energy reserves, and the ability to feed successfully on a variety of planktonic organisms all argue against diet as a critical limiting factor for larval survival, and therefore recruitment to the fishery.

Recently settled postlarvae have a preference for burrows and other protected habitats and only feed within these shelters. The importance of suspension feeding in burrows has only been recently realized; information relating growth and survival of postlarvae and early juveniles to food levels in the field is lacking. The potential contribution of algae to the nutrition of shelter-restricted juveniles is intriguing. Diatoms themselves are not sufficient; however, their contribution to the total diet could be important. For larger lobsters, plant material is generally not an important component of the diet, but ingestion does not appear to be inadvertent and thus it may have a specific role in the diet.

As lobsters grow and begin to forage outside their shelters, the range of available food continues to be extended, making it easier to maintain a protein-rich diet supportive of growth. Foraging lobsters are not indiscriminate feeders and are able to maintain a rough consistency of diet, although prey availability varies both seasonally and year to year. This adaptability again lessens the chances that diet becomes limiting in either quality or quantity.

Although significant progress has been made, information on the specific nutrient requirements of lobsters remains fragmentary. Most of the research has focused on juveniles and consequently the culture of both larvae and adults is still dependent on live or fresh food items. Lobster larvae typically are fed live adult brine shrimp (*Artemia* sp.) and the adults are fed a combination of chopped fish and molluscs, supplemented with small shrimp or crabs. Purified test diets have been developed for juvenile culture, but growth rates are still limited in comparison to what can be achieved with a diet of live adult brine shrimp.

Despite the obvious importance of protein in the diet, knowledge of optimum dietary levels is still lacking. The inability to determine quantitative amino acid requirements with crystalline amino acids accounts for much of the problem. It is suspected that the amino acid composition of casein, a common protein source in experimental diets, does not correspond well with the amino acid requirements of the lobster. Consequently, estimates of dietary protein requirements of more than 35% are probably inaccurate. A lack of information as to the appropriate protein–energy ratio is an additional constraint in finding optimum dietary protein levels for the lobster.

Crustaceans, including *Homarus americanus*, lack the ability to biosynthesize linoleic and linolenic acids, as well as cholesterol, and appear to have limited ability to convert these fatty acids to other important fatty acids, especially eicosapentaenoic and docosahexaenoic acids. Fish oils are commonly used to provide sufficient amounts of these essential fatty acids in formulated lobster diets. In addition, phospholipids are necessary for the effective utilization of dietary cholesterol by juveniles. Carotenoids are needed for normal pigmentation, but do not appear to have a role in growth or general well-being.

For the most part, vitamin and mineral requirements have not been delineated for the lobster. The lobster seems unique among cultured crustaceans in that it can biosynthesize ascorbic acid and does not need a dietary source of vitamin C. Until vitamin and mineral requirements of the lobster are specifically identified, premixes based on the requirements found for shrimp and other animals must be used in the formulation of rations.

References

Ache, B. W. (1972). Amino acid receptors in the antennules of *Homarus americanus*. *Comp. Biochem. Physiol. A* 42A, 807–811.

Ahearn, G. A., and Clay, L. P. (1987a). Membrane-potential–sensitive, Na^+-independent lysine transport by lobster hepatopancreatic brush border membrane vesicles. *J. Exp. Biol.* 127, 373–387.

Ahearn, G. A., and Clay, L. P. (1987b). Na^+-Cl^-–glutamate cotransport by lobster hepatopancreatic brush border membrane vesicles. *J. Exp. Biol.* 130, 175–191.

Ahearn, G. A., and Clay, L. P. (1988). Electroneutral, Na^+-$2Cl^-$–leucine cotransport by lobster hepatopancreatic brush border membrane vesicles. *J. Exp. Biol.* 136, 363–381.

Ahearn, G. A., Monckton, E. A., Henry, A. E., and Botfield, M. C. (1983). Alanine transport by lobster hepatopancreatic cell suspensions. *Am. J. Physiol.* 244, R150–R162.

Ahearn, G. A., Grover, M. L., and Dunn, R. E. (1985). Glucose transport by lobster hepatopancreatic brush-border membrane vesicles. *Am. J. Physiol.* 248, R133–R141.

Akiyama, D. M., Dominy, W. G., and Lawrence, A. L. (1992). Penaeid shrimp nutrition. *In* "Marine Shrimp Culture" (A. W. Fans and L. J. Lester, eds.), pp. 535–568. Elsevier, Amsterdam.

Allen, J. A., and Garrett, M. R. (1971). Taurine in marine invertebrates. *Adv. Mar. Biol.* 9, 205–253.

Al-Mohanna, S. Y., Nott, J. A., and Lane, D. J. W. (1985). M-"midget" cells in the hepatopancreas of the shrimp *Penaeus semisulcatus* de Haan 1844 (Decapoda: Natantia). *Crustaceana* 48, 260–268.

Anderson, J., Capper, B. S., and Bromage, N. R. (1991). Measurement and prediction of digestible energy values in feedstuffs for the herbivorous fish tilapia (*Oreochromis iloticus* Linn.). *Br. J. Nutr.* 66, 37–48.

Barker, P. L., and Gibson, R. (1977). Observations on the feeding mechanism, structure of the gut, and digestive physiology of the European lobster *Homarus gammarus* (L.) (Decapoda: Nephropidae). *J. Exp. Mar. Biol. Ecol.* 26, 297–324.

Barshaw, D. E. (1989). Growth and survival of early juvenile American lobsters, *Homarus americanus*, on a diet of plankton. *Fish. Bull.* 87, 366–370.

Barshaw, D. E., and Bryant-Rich, D. R. (1988). A long-term study on the behavior and survival of early juvenile American lobsters, *Homarus americanus*, in three naturalistic substrates: Eelgrass, mud, and rocks. *Fish. Bull.* 86, 789–796.

Baum, N. A., Conklin, D. E., and Chang, E. S. (1990). Effect of dietary lecithin in combination with casein or crab protein on cholesterol uptake and transport in the lobster *Homarus americanus*. *J. World Aquacult. Soc.* 21, 277–287.

Baum, N. A., Conklin, D. E., Castell, J. D., and Boston, L. D. (1991). Nutritionally induced molt death syndrome in aquatic crustaceans: III. The effect of varying levels of calcium in the reference diet, BML 81S for juvenile (*Homarus americanus*). *Crust. Nutr. Newsl.* 7, 115–118.

Bayer, R. C., Gallagher, M. L., Leavitt, D. F., and Rittenburg, J. H. (1979). A radiographic study of the lobster (*Homarus americanus*) alimentary canal. *Proc. Annu. Meet.—World Maric. Soc.* 10, 561–564.

Biesiot, P. M., and Capuzzo, J. M. (1990a). Changes in digestive enzyme activities during early development of the American lobster *Homarus americanus* Milne Edwards. *J. Exp. Mar. Biol. Ecol.* 136, 107–122.

Biesiot, P. M., and Capuzzo, J. M. (1990b). Digestive protease, lipase and amylase activities in stage I larvae of the American lobster, *Homarus americanus*. *Comp. Biochem. Physiol. A* 95A, 47–54.

Boghen, A. D., and Castell, J. D. (1981). Nutritional value of different dietary proteins to juvenile lobsters, *Homarus americanus*. *Aquaculture* 22, 343–351.

Boghen, A. D., Castell, J. D., and Conklin, D. E. (1982). In search of a reference protein to replace "vitamin-free casein" in lobster nutrition studies. *Can. J. Zool.* 60, 2033–2038.

Bordner, C. E. D'Abramo, L. R., and Conklin, D. E. (1983). Assimilation of nutrients by cultured hybrid lobsters (*Homarus* sp.) fed experimental diets. *J. World Maric. Soc.* 14, 11–24.

Brockerhoff, H., and Hoyle, R. J. (1967). Conversion of a dietary triglyceride into depot fat in fish and lobster. *Can. J. Biochem.* 45, 1365–1370.

Brockerhoff, H., Stewart, J. E., and Tacreiter, W. (1967). Digestion of triglycerides by lobster. *Can. J. Biochem.* 45, 421–422.

Brockerhoff, H., Hoyle, R. J., and Hwang, P. C. (1970). Digestive enzymes of the American lobster (*Homarus americanus*). *J. Fish. Res. Board Can.* 27, 1357–1370.

Bugden, G. L., Hargrave, B. T., Sinclair, M. M., Tang, C. L., Terriasult, J.-C., and Yeats, P. A. (1982). Freshwater Runoff Effects in the Marine Environment: The Gulf of St. Lawrence Example. *Can. Tech. Rep. Fish. Aquat. Sci.* 1078, 1–89.

Caddy, J. F. (1979). The influence of variations in the seasonal temperature regime on survival of larval stages of the American lobster (*Homarus americanus*) in the southern Gulf of St. Lawrence. *Rapp. P.-V. Run. Cons. Int. Explor. Mer* 175, 204–216.

Capuzzo, J. M. (1983). Crustacean bioenergetics: The role of environmental variables and dietary levels of macronutrients on energetic efficiencies. *In* "Proceedings of the Second International Conference on Aquaculture Nutrition: Biochemical and Physiological Approaches to Shellfish Nutrition" (G. Pruder, C. J. Langdon, and D. E. Conklin, eds.), pp. 71–86. Louisiana State University, Baton Rouge.

Capuzzo, J. M., and Lancaster, B. A. (1979a). The effects of dietary carbohydrate levels on protein utilization in the American lobster (*Homarus americanus*). *Proc. Annu. Meet.—World Maric. Soc.* 10, 689–700.

Capuzzo, J. M., and Lancaster, B. A. (1979b). Some physiological and biochemical considerations of larval development in the American lobster, *Homarus americanus* Milne Edwards. *J. Exp. Mar. Biol. Ecol.* 40, 53–62.

Carlberg, J. M., and Van Olst, J. C. (1976). Brine shrimp (*Artemia salina*) consumption by the larval stages of the American lobster (*Homarus americanus*) in relation to food chemistry and water temperature. *Proc. Annu. Meet.—World Maric. Soc.* 7, 379–389.

Carter, J. A., and Steele, D. H. (1982a). Attraction to and selection of prey by immature lobsters (*Homarus americanus*). *Can. J. Zool.* 60, 326–336.

Carter, J. A., and Steele, D. H. (1982b). Stomach contents of immature lobsters (*Homarus americanus*) from Placentia Bay, Newfoundland. *Can. J. Zool.* 60, 337–347.

Castell, J. D. (1977). Production of juvenile lobsters (*Homarus americanus*) for nutrition research. *Actes Colloq.—Cent. Natl. Exploit. Oceans* 4, 277–281.

Castell, J. D., and Budson, S. D. (1974). Lobster nutrition: The effect on *Homarus americanus* of dietary protein levels. *J. Fish. Res. Board Can.* 31, 1363–1370.

Castell, J. D., and Covey, J. F. (1976). Dietary lipid requirements of adult lobsters, *Homarus americanus*. *J. Nutr.* 106, 1159–1165.

Castell, J. D., and Kean, J. C. (1986). Evaluation of the role of nutrition in lobster recruitment. *Can. J. Fish. Aquat. Sci.* 43, 2320–2327.

Castell, J. D., Mason, E. G., and Covey, J. F. (1975). Cholesterol requirements of juvenile American lobster (*Homarus americanus*). *J. Fish. Res. Board Can.* 32, 1431–1435.

Castell, J. D., Kean, J. C., D'Abramo, L. R., and Conklin, D. E. (1989a). A standard reference diet for crustacean nutrition research. I. Evaluation of two formulations. *J. World Aquacult. Soc.* 20, 93–99.

Castell, J. D., Kean, J. C., McCann, D. G. C., Boghen, A. D., Conklin, D. E., and D'Abramo, L. R. (1989b). A standard reference diet for crustacean nutrition research. II. Selection of a purification procedure for production of the rock crab *Cancer irroratus* protein ingredient. *J. World Aquacult. Soc.* 20, 100–106.

Chang, E. S., and Conklin, D. E. (1993). Larval culture of the American lobster (*Homarus americanus*). *In* "Handbook of Mariculture" (J. P. McVey, ed.), Vol. 1, 2nd ed., pp. 489–495. CRC Press, Boca Raton, Florida.

Charmantier, G., Charmantier-Daures, M., and Aiken, D. E. (1991). Metamorphosis in the lobster *Homarus* (Decapoda): A review. *J. Crust. Biol.* 11, 481–495.

Chen, H.-Y. (1993). Recent advances in nutrition of *Penaeus monodon*. *J. World Aquacult. Soc.* 24, 231–240.

Chen, H.-Y., and Hwang, G. (1992). Estimation of the dietary riboflavin required to maximize tissue riboflavin concentration in juvenile shrimp (*Penaeus monodon*). *J. Nutr.* 122, 2474–2478.

Chen, H.-Y., Wu, F.-C., and Tang, S.-Y. (1991). Thiamin requirement of juvenile shrimp (*Penaeus monodon*). *J. Nutr.* 121, 1984–1989.

Chen, H.-Y., Leu, Y.-T., and Roelants, I. (1992). Quantification of arginine requirements of juvenile marine shrimp, *Penaeus monodon*, using microencapsulated arginine. *Mar. Biol.* 114, 229–233.

Claus, C., Benijts, F., and Vandeputte, G. (1979). The biochemical composition of the larvae of two strains of *Artemia salina* (L.) reared on two different algal foods. *J. Exp. Mar. Biol. Ecol.* 36, 171–183.

Cobb, J. S., Gulbrasen, T., Phillips, B. F., Wand, D., and Syslo, M. (1983). Behavior and distribution of larval and early juvenile *Homarus americanus*. *Can. J. Fish. Aquat. Biol.* 40, 2184–2188.

Conklin, D. E. (1980). Nutrition. *In* "The Biology and Management of Lobsters" (J. S. Cobb and B. F. Phillips, eds.), Vol. 1, pp. 277–300. Academic Press, New York.

Conklin, D. E. (1990). Vitamin requirements of juvenile penaeid shrimp. Advances in Tropical Aquaculture, Workshop in Tahiti, 1989. IFREMER. *Actes Colloq.* 9, 287–308.

Conklin, D. E., and Chang, E. S. (1993). Culture of juvenile lobster (*Homarus americanus*). *In* "Handbook of Mariculture" (J. P. McVey, ed.), Vol. 1, 2nd ed., pp. 497–510. CRC Press, Boca Raton, Florida.

Conklin, D. E., Devers, K., and Shleser, R. A. (1975). Initial development of artificial diets for the lobster, *Homarus americanus*. *Proc. Annu. Meet.—World Maric. Soc.* 6, 237–243.

Conklin, D. E., Devers, K., and Bordner, C. (1977). Development of artificial diets for the lobster, *Homarus americanus*. *Proc. Annu. Meet.—World Maric. Soc.* 8, 841–852.

Conklin, D. E., D'Abramo, L. R., Bordner, C. E., and Baum, N. A. (1980). A successful purified diet for the culture of juvenile lobster: The effect of lecithin. *Aquaculture* 21, 243–249.

Conklin, D. E., D'Abramo, L. R., and Norman-Boudreau, K. (1983). Lobster nutrition. *In* "Handbook of Mariculture" (J. P. McVey, ed.), Vol. 1, 2nd ed., pp. 413–423. CRC Press, Boca Raton, Florida.

Cruz-Suárez, L. E., Ricque-Marie, D., Martinez-Vega, J. A., and Wesche-Ebeling, P. (1993). Evaluation of two shrimp by-product meals as protein sources in diets for *Penaeus vannamei*. *Aquaculture* 115, 53–62.

D'Abramo, L. R., and Lovell, R. T. (1991). Aquaculture research needs for the year 2000: Fish and crustacean nutrition. *World Aquacult.* 22, 57–62.

D'Abramo, L. R., and Robinson, E. H. (1989). Nutrition of crayfish. *Rev. Aquat. Sci.* 1, 711–728.

D'Abramo, L. R., and Sheen, S.-S. (1994). Nutritional requirements, feed formulation, and feeding practices for intensive culture of the freshwater prawn *Macrobrachium rosenbergii*. *Rev. Fish. Sci.* 2, 1–21.

D'Abramo, L. R., Bordner, C. E., Daggett, G. R., Conklin, D. E., and Baum, N. A. (1980). Relationship among dietary lipids, tissue lipids, and growth in juvenile lobsters. *Proc. Annu. Meet.—World Maric. Soc.* 11, 335–345.

D'Abramo, L. R., Bordner, C. E., Conklin, D. E., and Baum, N. A., (1981a). Essentiality of dietary phosphatidylcholine for the survival of juvenile lobsters. *J. Nutr.* 111, 425–431.

D'Abramo, L. R., Conklin, D. E., Bordner, C. E., Baum, N. A., and Norman-Boudreau, K. A. (1981b). Successful artificial diets for the culture of juvenile lobsters. *J. World Maric. Soc.* 12, 325–332.

D'Abramo, L. R., Bordner, C. E., and Conklin, D. E. (1982). Relationship between dietary phosphatidylcholine and serum cholesterol in the lobster *Homarus* sp. *Mar. Biol.* 67, 231–235.

D'Abramo, L. R., Baum, N. A., Bordner, C. E., and Conklin, D. E. (1983). Carotenoids as a source of pigmentation in juvenile lobsters fed a purified diet. *Can. J. Fish. Aquat. Sci.* 40, 699–704.

D'Abramo, L. R., Bordner, C. E., Conklin, D. E., and Baum, N. A. (1984). Sterol requirement of juvenile lobsters, *Homarus* sp. *Aquaculture* 42, 13–25.

D'Abramo, L. R., Baum, N. A, Bordner, C. E., Conklin, D. E., and Chang, E. S. (1985a). Diet-dependent cholesterol transport in the American lobster. *J. Exp. Mar. Biol. Ecol.* 87, 83–96.

D'Abramo, L. R., Wright, J. S., Wright, K. H., Bordner, C. E., and Conklin, D. E. (1985b). Sterol requirement of cultured juvenile crayfish, *Pacifastacus leniusculus*. *Aquaculture* 49, 245–255.

Dadd, R. H. (1983). Insect nutrition: Relevance to marine invertebrates. *In* "Proceedings of the Second International Conference on Aquaculture Nutrition: Biochemical and Physiological Approaches to Shellfish Nutrition" (G. Pruder, C. J. Langdon, and D. E. Conklin, eds.), pp. 33–55. Louisiana State University, Baton Rouge.

D'Agostino, A. (1980). Growth and color of juvenile lobsters (*Homarus americanus*) kept on diets of natural and artificial foodstuffs. *In* "Proceedings of the 1980 Lobster Nutrition Workshop," Maine Sea Grant Tech. Rep. 58, pp. 41–45. University of Maine, Orono.

Dall, W. (1967). The functional anatomy of the digestive tract of shrimp *Metapenaeus bennettae* Racek and Dall (Crustacea: Decapoda: Penaeidae). *Aust. J. Zool.* 15, 699–714.

Dall, W., and Moriarty, D. J. W. (1983). Functional aspects of nutrition and digestion. *In* "The Biology of Crustacea" (D. E. Bliss, ed.), Vol. 5, "Internal Anatomy and Physiological Regulation" (L. H. Mantel, ed), pp. 215–261. Academic Press, New York.

Daniel, P. C., Bayer, R. C., and Chapman, S. (1985). Barnacle larvae (*Balanus* spp.) as a potential diet for juvenile lobsters (*Homarus americanus*). *Aquaculture* 46, 67–70.

Davis, D. A., and Arnold, C. R. (1993). Evaluation of five carbohydrate sources for *Penaeus vannamei*. *Aquaculture* 114, 285–292.

Deshimaru, O. (1983). Protein and amino acid nutrition of the prawn *Penaeus japonicus*. *In* "Proceedings of the Second International Conference on Aquaculture Nutrition: Biochemical and Physiological Approaches to Shellfish Nutrition" (G. Pruder, C. J. Langdon, and D. E. Conklin, eds.), pp. 106–122. Louisiana State University, Baton Rouge.

Deshimaru, O., and Kuroki, K., (1979). Requirement of prawn for dietary thiamine, pyridoxine, and choline chloride. *Bull. Jpn. Soc. Sci. Fish.* 45, 363–367.

Deshimaru, O., and Shigueno, K. (1972). Introduction to the artificial diet for prawn, *Penaeus japonicus*. *Aquaculture* 1, 115–133.

Deshimaru, O., Kuroki, K., Sakamoto, S., and Yone, Y. (1978). Absorption of labelled calcium-^{45}Ca by prawn from sea water. *Bull. Jpn. Soc. Sci. Fish.* 44, 975–977.

Desjardins, L. M., Castell, J. D., and Kean, J. C. (1985). Synthesis of dehydroascorbic acid by subadult lobsters (*Homarus americanus*). *Can. J. Fish. Aquat. Biol.* 42, 370–373.

DeVillez, E., and Buschlen, K. (1967). Survey of tryptic digestive enzymes in various species of Crustacea. *Comp. Biochem. Physiol.* 21, 541–546.

Drinkwater, K. F., Harding, G. C., Vass, W. P., and Gauthier, D. (1991). The relationship of Quebec lobster landings to freshwater runoff and wind storms. *Can. Spec. Publ. of Fish. Aquat. Sci.* 113, 179–187.

Eagles, M. D., Aiken, D. E., and Waddy, S. L. (1986). Influence of light and food on larval American lobsters, *Homarus americanus*. *Can. J. Fish. Aquat. Sci.* 43, 2303–2310.

Eisen, A. Z., Henderson, K. O., Jeffrey, J. J., and Bradshaw, R. A. (1973). A collagenolytic protease from the hepatopancreas of the fiddler crab, *Uca pugilator*. Purification and properties. *Biochemistry* 12, 1814–1822.

Elner, R. W., and Campbell, A. (1981). Force, function and mechanical advantage in the chelae of the American lobster *Homarus americanus* (Decapoda: Crustacea). *J. Zool.* 193, 269–286.

Elner, R. W., and Campbell, A. (1987). Natural diets of lobster *Homarus americanus* from barren ground and macroalgal habitats off southwestern Nova Scotia, Canada. *Mar. Ecol. Prog. Ser.*

37, 131–140.

Ennis, G. P. (1973). Food, feeding, and condition of lobster, *Homarus americanus*, throughout the seasonal cycle in Bonavista Bay, Newfoundland. *J. Fish. Res. Board Can.* 30, 1905–1909.

Ennis, G. P. (1986). Stock definition, recruitment variability, and larval recruitment processes in the American lobster, *Homarus americanus*: A review. *Can. J. Fish. Aquat. Sci.* 43, 2072–2084.

Factor, J. R. (1989). Development of the feeding apparatus in decapod crustaceans. *In* "Crustacean Issues" (F. R. Schram, ed.), Vol. 4, "Functional Morphology of Feeding and Grooming in Decapod Crustacea" B. E. Felgenhauer, L. Watling, and A. B. Thistle, eds.), pp. 185–203. Balkema, Rotterdam, The Netherlands.

Finney, C. M. (1978). Isotopic labeling of taurine; implications for its synthesis in selected tissues of *Homarus americanus*. *Comp. Biochem. Physiol. B* 61B, 409–413.

Forster, G. R. (1953). Peritrophic membrane in Caridea (Crustacea Decapoda). *J. Mar. Biol. Assoc. U.K.* 32, 344–345.

Galgani, F., and Benyamin, Y. (1985). Radioimmunoassay of shrimp trypsin: Application to the larval development of *Penaeus japonicus* Bate, 1888. *J. Exp. Mar. Biol. Ecol.* 87, 145–151.

Gallagher, M. L., and Brown, W. D. (1975a). Composition of San Francisco bay brine shrimp (*Artemia salina*). *Agric. Food Chem.* 23, 630–632.

Gallagher, M. L., and Brown, W. D. (1975b). Amino acid requirements of the lobster (*Homarus americanus*). *Fed. Proc., Fed. Am. Soc. Exp. Biol.* 34, 880.

Gallagher, M. L., Conklin, D. E., and Brown, W. D. (1976a). Development of artificial diets for the lobster, *Homarus americanus*. *Proc. Annu. Meet.—World Maric. Soc.* 7, 363–378.

Gallagher, M. L., Brown, W. D., Conklin, D. E., and Sifri, M. (1976b). Effects of varying calcium phosphorus ratios in diets fed to juvenile lobsters (*Homarus americanus*). *Comp. Biochem. Physiol. A* 60A, 467–471.

Gibson, R. (1983). Feeding and digestion in decapod crustaceans. In "Proceedings of the Second International Conference on Aquaculture Nutrition: Biochemical and Physiological Approaches to Shellfish Nutrition" (G. Pruder, C. J. Langdon, and D. E. Conklin, eds.), pp. 59–70. Louisiana State University, Baton Rouge.

Gibson, R., and Barker, P. L. (1979). The decapod hepatopancreas. *Oceanogr. Mar. Biol.* 17, 285–346.

Gillchrist, B. M., and Lee, W. L. (1972). Carotenoid pigments and their possible role in reproduction in the sand crab, *Emerita analoga* (Stimpson, 1857). *Comp. Biochem. Physiol. B* 42B, 263 291.

Gilles, R., and Schoffeniels, E. (1969). Metabolism of arabinose in the ventral nerve cord of the lobster *Homarus vulgaris* (M. Edw.). *Comp. Biochem. Physiol.* 28, 1145–1152.

Goldblatt, M. J., Brown, W. D., and Conklin, D. E. (1979). Nutrient leaching from pelleted rations. *In* "Finfish Nutrition and Fishfeed Technology" (J. E. Halver and K. Tiews, eds.), Vol. 2, pp. 117–129. Heenemann Verlangs-Gesellschaft, Berlin.

Good, L. K., Bayer, R. C., Gallagher, M. L., and Rittenberg, J. H. (1982). Amphipods as a potential diet for juveniles of the American lobster (*Homarus americanus* Milne Edwards). *J. Shellfish Res.* 2, 183–187.

Harding, G. C., and Trites, R. W. (1988). Dispersal of *Homarus americanus* larvae in the gulf of Maine from Browns Bank. *Can. J. Fish. Aquat. Sci.* 45, 416–425.

Harding, G. C., Drinkwater, K. F., and Vass, W. P. (1983). Factors influencing the size of American lobster (*Homarus americanus*) stocks along the Atlantic coast of Nova Scotia, Gulf of St. Lawrence, and the Gulf of Maine: A new synthesis. *Can. J. Fish. Aquat. Sci.* 40, 168–184.

Hayes, D. K., Singer, L., and Armstrong, W. D. (1962). Calcium homeostatic mechanisms and uptake of radioisotopes in the lobster. *Am. J. Physiol.* 202, 383–386.

He, H., and Lawrence, A. (1993a). Vitamin C requirements of the shrimp *Penaeus vannamei*. *Aquaculture* 114, 305–316.

He, H., and Lawrence, A. (1993b). Vitamin E requirement of *Penaeus vannamei*. *Aquaculture* 118, 245–255.

He, H., Lawrence, A., and Liu, R. (1992). Evaluation of dietary essentiality of fat-soluble vitamins, A, D, E and K, for penaeid shrimp (*Penaeus vannamei*). *Aquaculture* 103, 177–185.

Hirche, H. J., and Anger, K. (1987). Digestive enzyme activities during larval development of *Hyas araneus* (Decapoda, Majidae). *Comp. Biochem. Physiol. B* 87B, 297–302.

Hirsch, G. C., and Jacobs, W. (1930). Der Arbeitsrhythmus der Mitteldarmdrüse von *Astacus leptodactylus* II. Teil: Waschstum als primärer Faktor des Rhythmus eines polyphasichen organigen Sekretionssystems. *Z. Vergl. Physiol.* 12, 524–557.

Holwerda, D. A., and Vonk, H. F. (1973). Emulsifiers in the intestinal juice of Crustacea. Isolation and nature of surface-active substances from *Astacus leptodactylus* Esch. and *Homarus vulgaris* L. *Comp. Biochem. Physiol. B* 45B, 51–58.

Hood, M. A., and Meyers, S. P. (1977). Microbiological and chitinoclastic activities associated with *Penaeus setiferus*. *J. Oceanogr. Soc. Jpn.* 33, 235–241.

Hoyle, R. J. (1973). Digestive enzyme secretion after dietary variation in the American lobster (*Homarus americanus*). *J. Fish. Res. Board Can.* 30, 1647–1653.

Hudon, C. (1987). Ecology and growth of postlarval and juvenile lobster, *Homarus americanus*, off Îles de la Madeleine (Quebec). *Can. J. Fish. Aquat. Sci.* 44, 1855–1869.

Hudon, C., and Lamarche, G. (1989). Niche segregation between American lobster *Homarus americanus* and rock crab *Cancer irroratus*. *Mar. Ecol. Prog. Ser.* 52, 155–168.

Hughes, J. T., and Matthiessen, G. C. (1962). Observations on the biology of the American lobster *Homarus americanus*. *Limnol. Oceanogr.* 7, 414–421.

Hyslop, E. J. (1980). Stomach content analysis—a review of methods and their application. *J. Fish Biol.* 17, 411–429.

Icely, J. D., and Nott, J. A. (1992). Digestion and absorption: Digestive system and associated organs. *In* "Microscopic Anatomy of Invertebrates" (F. W. Harrison, ed.), Vol. 10, "Decapod Crustacea" (F. W. Harrison and A. G. Humes, eds.), pp. 147–201. Wiley–Liss, New York.

Jacobs, W. (1928). Untersuchungen über die cytologie der sekretbildungin der mitteldarmdrüse von *Astacus leptodactylus*. *Z. Zellforsch. Mikrosk. Anat.* 8, 1–62.

Juanes, F. (1992). Why do decapod crustaceans prefer small-sized molluscan prey? *Mar. Ecol. Prog. Ser.* 87, 239–249.

Juinio, M. A. R., and Cobb, J. S. (1992). Natural diet and feeding habits of the postlarval lobster *Homarus americanus*. *Mar. Ecol. Prog. Ser.* 85, 83–91.

Kanazawa, A. (1983). Penaeid nutrition. *In* "Proceedings of the Second International Conference on Aquaculture Nutrition: Biochemical and Physiological Approaches to Shellfish Nutrition" (G. Pruder, C. J. Langdon, and D. E. Conklin, eds.), pp. 87–105. Louisiana State University, Baton Rouge.

Kanazawa, A. (1985). Nutrition of penaeid prawns and shrimps. *In* "Proceedings of the First International Conference on the Culture of Penaeid Prawns/Shrimps" (Y. Taki, J. H. Primavera, and J. A. Llobrera, eds.), pp. 123–130. SEAFDEC Aquaculture Department, Iloilo City, the Philippines.

Kanazawa, A. (1994). Nutrition and food. *In* "Spiny Lobster Management" (B. F. Phillips, J. S. Cobb, and J. Kittaka, eds.), pp. 483–494. Fishing News Books, Blackwell, London.

Kanazawa, A., Teshima, S., and Tanaka, N. (1976). Nutritional requirements of prawn—V: Requirements for choline and inositol. *Mem. Fac. Fish., Kagoshima Univ.* 25, 47–51.

Kanazawa, A., Teshima, S., Ono, I., and Chalayondeja, K. (1979a). Biosynthesis of fatty acids from acetate in prawns, *Penaeus monodon* and *Penaeus merguiensis*. *Mem. Fac. Fish., Kagoshima Univ.* 28, 21–26.

Kanazawa, A., Teshima, S., and Tokiwa, S. (1979b). Biosynthesis of fatty acids from palmitic acid in the prawn, *Penaeus japonicus*. *Mem. Fac. Fish., Kagoshima Univ.* 28, 17–20.

Kanazawa, A., Teshima, S., and Endo, M. (1979c). Requirements of prawn, *Penaeus japonicus* for essential fatty acids. *Mem. Fac. Fish., Kagoshima Univ.* 28, 27–33.

Kanazawa, A., Teshima, S., and Ono, K. (1979d). Relationship between essential fatty acid requirements of aquatic animals and the capacity for bioconversion of linolenic acid to highly unsaturated fatty acids. *Comp. Biochem. Physiol. B* 63B, 295–298.

Kean, J. C., Castell, J. D., Boghen, A. G., D'Abramo, L. R., and Conklin, D. E. (1985a). A re-evaluation of the lecithin and cholesterol requirements of juvenile lobster (*Homarus americanus*) using crab protein-based diets. *Aquaculture* 47, 143–149.

Kean, J. C., Castell, J. D., and Trider, D. J. (1985b). Juvenile lobster (*Homarus americanus*) do not require dietary ascorbic acid. *Can. J. Fish. Aquat. Sci.* 42, 368–370.

Koshio, S., Teshima, S., Kanazawa, A., and Watase, T. (1993). The effect of dietary protein content on growth, digestion efficiency and nitrogen excretion of juvenile kuruma prawns, *Penaeus japonicus*. *Aquaculture* 113, 101–114.

Kurmaly, K., Jones, D. A., and Yule, A. B. (1990). Acceptability and digestion of diets fed to larval stages of *Homarus gammarus* and the role of dietary conditioning behaviour. *Mar. Biol.* 106, 181–190.

Lavalli, K. L. (1991). Survival and growth of early-juvenile American lobsters *Homarus americanus* through their first season while fed diets of mesoplankton, microplankton, and frozen brine shrimp. *Fish. Bull.* 89, 61–68.

Lavalli, K. L., and Factor, J. R. (1992). Functional morphology of the mouthparts of juvenile lobsters, *Homarus americanus* (Decapoda: Nephropidae), and comparison with the larval stages. *J. Crust. Biol.* 12, 467–510.

Leavitt, D. F. (1985). An evaluation of gravimetric and inert marker techniques to measure digestibility in the American lobster. *Aquaculture* 47, 131–142.

Leavitt, D. F., Bayer, R. C., Gallagher, M. L., and Rittenburg, J. H. (1979). Dietary intake and nutritional characteristics in wild American lobsters, *Homarus americanus*. *J. Fish. Res. Board Can.* 36, 965–969.

Léger, P., Bengtson, D. A., Simpson, K. L., and Sorgeloos, P. (1986). The use and nutritional value of *Artemia* as a food source. *Oceanogr. Mar. Biol.* 24, 521–623.

Lester, R., Carey, M. C., Cooperstein, L. A., and Dowd, S. R. (1975). Crustacean intestinal detergent promotes sterol solubilization. *Science* 189, 1098–1100.

Levine, D. M., and Sulkin, S. D. (1984). Nutritional significance of long-chain polyunsaturated fatty acids to the zoeal development of the brachyuran crab, *Eurypanopeus depressus* (Smith). *J. Exp. Mar. Biol. Ecol.* 81, 211–223.

Lim, C. (1993). Effect of dietary pH on amino acid utilization by shrimp (*Penaeus vannamei*). *Aquaculture* 114, 293–303.

Logan, D. T., and Epifanio, C. E. (1978). A laboratory energy balance for the larvae and juveniles of the American lobster *Homarus americanus*. *Mar. Biol.* 47, 381–389.

Lord, W. B. (1867). "Crab, Shrimp, and Lobster Lore Gathered amongst the Rocks at the Sea-shore, by the Riverside, and in the Forest." Routledge, London.

Lucien-Brun, H., Van Wormhoudt, A., Lachaux, A., and Ceccaldi, H. J. (1985). Effets de regimes composes sur la croissance de homards juveniles, *Homarus gammarus* L.: Estimation biochimique de la composition optimale du regime alimentaire en proteines. *Aquaculture* 46, 97–109.

Lynn, K. R. (1990). Chitinases and chitobiases from the American lobster (*Homarus americanus*). *Comp. Biochem. Physiol. B* 96B, 761–766.

MacKenzie, B. R. (1987). Larval lobster (*Homarus americanus* Milne Edwards) development with Great Salt Lake, Utah and reference I strains of *Artemia* nauplii. *J. World Aquacult. Soc.* 18, 6–10.

Mason, E. G., and Castell, J. D. (1980). The effects of supplementing purified proteins with limiting essential amino acids on growth and survival of juvenile lobsters (*Homarus americanus*). *Proc. Annu. Meet.—World Maric. Soc.* 11, 346–354.

McDowell, L. R. (1989). "Vitamins in Animal Nutrition." Academic Press, San Diego.

McLeese, D. W. (1970). Detection of dissolved substances by the American lobster (*Homarus americanus*) and olfactory attraction between lobsters. *J. Fish. Res. Board Can.* 27, 1371–1378.

Mykles, D. L. (1979). "The Mechanism of Fluid Absorption at Ecdysis in Lobster, *Homarus americanus* and *H. gammarus*, and Dungeness Crab, *Cancer magister*," Ph.D. thesis. University of California, Berkeley.

Mykles, D. L. (1980). The mechanism of fluid absorption at ecdysis in the American lobster, *Homarus americanus*. *J. Exp. Biol.* 84, 89–101.

Neil, D. M., Macmillan, D. L., Robertson, R. M., and Laverack, M. S. (1976). The structure and function of thoracic exopodites in the larvae of the lobster *Homarus gammarus* (L.). *Philos. Trans. R. Soc. London, Ser. B* 274, 53–68.

Nelis, H. J., Lavens, P., Moens, L., Sorgeloos, P., and De Leenheer, A. P. (1989). Carotenoids in relation to *Artemia* development. *In* "Biochemistry and Cell Biology of *Artemia*" (T. H. MacRae, J. C. Bagshaw, and A. H. Warner, eds.), pp. 159–190. CRC Press, Boca Raton, Florida.

Norman-Boudreau, K., and Conklin, D. E. (1985). Antinutritional properties of egg white in diets for lobsters. *Bull. Jpn. Soc. Sci. Fish.* 51, 953–957.

Phillips, B. F., and Sastry, A. N. (1980). Larval ecology. *In* "The Biology and Management of Lobsters" (J. S. Cobb and B. F. Phillips, eds.), Vol. 1, pp. 2–82. Academic Press, New York.

Pochon-Masson, J., Payen, G. G., Portemer, C., and Chatagner, F. (1984). Variations of taurine concentration in the ovaries and hepatopancreas of the female crab *Carcinus maenas* L. (Decapoda Brachyura) during the different phases of its genital activity. *Int. J. Invertebr. Reprod. Dev.* 7, 127–133.

Sasaki, G. C., Capuzzo, J. M., and Biesiot, P. (1986). Nutritional and bioenergetic considerations in the development of the American lobster *Homarus americanus*. *Can. J. Fish. Aquat. Sci.* 43, 2311–2319.

Scarratt, D. J. (1980). The food of lobsters. *Can. Tech. Rep. Fish. Aquat. Sci.* 954, 66–91.

Schuur, A., Fisher, W. S., Van Olst, J. C., Carlberg, J. M., Hughes, J. T., Shleser, R. A., and Ford, R. F. (1976). "Hatchery Methods for the Production of Juvenile Lobsters *Homarus americanus*," Sea Grant Publ. 48. Institute of Marine Resources, University of California, La Jolla.

Sedgwick, R. W. (1979). Influence of dietary protein and energy on growth, food consumption and food conversion efficiency in *Penaeus merguiensis* de Man. *Aquaculture* 16, 7–30.

Serfling, S. A., Van Olst, J. C., and Ford, R. F. (1974). An automatic feeding device and the use of live and frozen *Artemia* for culturing larval stages of the American lobster, *Homarus americanus*. *Aquaculture* 3, 311–314.

Shiau, S.-Y., and Liu, J.-S. (1994). Quantifying the vitamin K requirement of juvenile marine shrimp (*Penaeus monodon*) with menadione. *J. Nutr.* 124, 277–282.

Shiau, S.-Y., and Lung, C.-Q. (1993). Estimation of the vitamin B_{12} requirement of the grass shrimp, grass prawn, *Penaeus monodon*. *Aquaculture* 117, 157–163.

Shiau, S.-Y., and Peng, C.-Y. (1992). Utilization of different carbohydrates at different dietary protein levels in grass prawn, *Penaeus monodon*, reared in seawater. *Aquaculture* 101, 241–250.

Shiau, S.-Y., and Suen, G.-S. (1994). The dietary requirement of juvenile grass shrimp (*Penaeus monodon*) for niacin. *Aquaculture* 125, 139–145.

Shigueno, K., and Itoh, S. (1988). Use of Mg-L-ascorbyl-2-phosphate as a vitamin C source in shrimp diets. *J. World Aquacult. Soc.* 19, 168–174.

Simpson, K. L., Klein-MacPhee, G., and Beck, A. D. (1983). Zooplankton as a food source. *In* "Proceedings of the Second International Conference on Aquaculture Nutrition: Biochemical and Physiological Approaches to Shellfish Nutrition" (G. Pruder, C. J. Langdon, and D. E. Conklin, eds.), pp. 180–201. Louisiana State University, Baton Rouge.

Slinger, S. J., Razzaque, A., and Cho, C. Y. (1979). Nutrient leaching from pelleted rations. *In* "Finfish Nutrition and Fishfeed Technology" (J. E. Halver and K. Tiews, eds.), Vol. 2, pp. 425–434. Heenemann Verlangs-Gesellschaft, Berlin.

Smith, B. R., Miller, G. C., Mead, R. W., and Taylor, R. E. L. (1987). Biosynthesis of asparagine and taurine in the freshwater prawn, *Macrobrachium rosenbergii* (De Man). *Comp. Biochem. Physiol. B* 87B, 827–831.

Snyder, J. J., and Chang, E. S. (1986). Effects of eyestalk ablation on larval molting rates and morphological development of the American lobster, *Homarus americanus*. *Biol. Bull. (Woods Hole, Mass.)* 170, 232–243.

Staton, J. L., and Sulkin, S. D. (1991). Nutritional requirements and starvation resistance in larvae of the brachyuran crabs *Sesarma cinereum* (Bosc) and *S. reticulatum* (Say). *J. Exp. Mar. Biol. Ecol.* 152, 271–284.

Stewart, J. E., and Castell, J. D. (1979). Various aspects of culturing the American lobster, *Homarus americanus*. *In* "Advances in Aquaculture" (T. V. R. Pillay and W. A. Dill, eds.), pp. 314–319. Fishing News Books, Farnham, Surrey, England.

Sulkin, S. D. (1978). Nutritional requirements during larval development of the portunid crab, *Callinectes sapidus* Rathbun. *J. Exp. Mar. Biol. Ecol.* 34, 29–41.

Sulkin, S. D., and Van Heukelem, W. F. (1980). Ecological and evolutionary significance of nutritional flexibility in planktonic larvae of the deep sea red crab *Geryon quinquedens* and the stone crab *Menippe mercenaria*. *Mar. Ecol. Prog. Ser.* 2, 91–95.

Sutcliffe, W. H., Jr. (1972). Some relations of land drainage, nutrients, particulate material and fish catch in two eastern Canadian Bays. *J. Fish. Res. Board Can.* 29, 357–362.

Sutcliffe, W. H., Jr. (1973). Correlations between seasonal river discharge and local landings of American lobster (*Homarus americanus*) and Atlantic halibut (*Hippoglossus hippoglossus*) in the Gulf of St. Lawrence. *J. Fish. Res. Board Can.* 30, 856–859.

Templeman, W. (1936). The influence of temperature, salinity, light and food conditions on the survival and growth of the larvae of the lobster. *J. Biol. Board Can.* 2, 485–497.

Teshima, S. (1983). Sterol metabolism. *In* "Proceedings of the Second International Conference on Aquaculture Nutrition: Biochemical and Physiological Approaches to Shellfish Nutrition" (G. Pruder, C. J. Langdon, and D. E. Conklin, eds.), pp. 205–216. Louisiana State University, Baton Rouge.

Teshima, S., Kanazawa, A., Koshio, S., and Kondo, N. (1989). Nutritive value of sitosterol for prawn, *Penaeus japonicus*. *Nippon Suisan Gakkaishi* 55, 153–157.

Teshima, S., Kanazawa, A., and Koshio, S. (1992a). Ability for bioconversion of n-3 fatty acids in fish and crustaceans. *Oceanis* 18, 67–75.

Teshima, S., Kanazawa, A., and Koshio, S. (1992b). Supplemental effects of methionine-enriched plastein in *Penaeus japonicus* diets. *Aquaculture* 101, 85–93.

Titani, K., Sasagawa, T., Woodbury, R. G., Ericsson, L. H., Dörsam, H., Kraemer, M., Neurath, H., and Zwilling, R. (1983). Amino acid sequence of crayfish (*Astacus fluviatilis*) trypsin I_f. *Biochemistry* 22, 1459–1465.

Titani, K., Torff, H.-J., Hormel, S., Kumar, S., Walsh, K. A., Rödl, J., Neurath, H., and Zwilling, R. (1987). Amino acid sequence of a unique protease from the crayfish (*Astacus fluviatilis*). *Biochemistry* 26, 222–226.

Van Marrewijk, W. J. A., and Zandee, D. I. (1975). Amino acid metabolism of *Astacus leptodactylus* (Esch.)—454 454II. Biosynthesis of the non-essential amino acids. *Comp. Biochem. Physiol. B* 50B, 449–455.

Van Olst, J. C., Carlberg, J. M., and Hughes, J. T. (1980). Aquaculture. *In* "The Biology and Management of Lobsters" (J. S. Cobb and B. F. Phillips, eds.), Vol. 2, pp. 333–384. Academic Press, New York.

Watanabe, T., Arakawa, T., Kitajima, C., Fukusho, K., and Fujita, S. (1978a). Proximate and mineral composition of living feeds used in seed production of fish. *Bull. Jpn. Soc. Sci. Fish.* 44, 973–984.

Watanabe, T., Oowa, F., Kitajima, C., and Fujita, S. (1978b). Nutritional quality of brine shrimp, *Artemia salina*, as a living feed from the viewpoint of essential fatty acids for fish. *Bull. Jpn. Soc. Sci. Fish.* 44, 1115–1121.

Watanabe, T., Kitajima, C., and Fujita, S. (1983). Nutritional values of live organisms used in Japan for mass propagation of fish: A review. *Aquaculture* 34, 115–143.

Weiss, H. M. (1970). "The Diet and Feeding Behaviour of the Lobster, *Homarus americanus*, in Long Island Sound," Ph.D. thesis. University of Connecticut, Storrs.

Williams, L. W. (1907). The stomach of the lobster and the food of larval lobsters. *Annu. Rep.—R.I. Comm. Inland. Fish.* 37, 153–180.

Williamson, H. C. (1905). A contribution to the life-history of the lobster (*Homarus vulgaris*). *Annu. Rep. Fish. Board Scot.* (1904) 23, 65–107.

Wilson, R. P. (1989). Amino acids and proteins. *In* "Fish Nutrition" (J. E. Halver, ed.), 2nd ed., pp. 111–151. Academic Press, San Diego.

Wojtowicz, M. B., and Brockerhoff, H. (1972). Isolation and some properties of the digestive amylase of the american lobster (*Homarus americanus*). *Comp. Biochem. Physiol. B* 42B, 295–302.

Zandee, D. I. (1966a). Absence of cholesterol synthesis as contrasted with the presence of fatty acid synthesis in some arthropods. *Comp. Biochem. Physiol.* 20, 811–822.

Zandee, D. I. (1966b). Metabolism in the crayfish *Astacus astacus* (L.). I. Biosynthesis of amino acids. *Arch. Int. Physiol. Biochim.* 74, 35–44.

Zandee, D. I. (1966c). Metabolism in the crayfish *Astacus astacus* (L.). IV. The fatty acid composition and the biosynthesis of the fatty acids. *Arch. Int. Physiol. Biochim.* 74, 614–626.

Zwilling, R., Pfleiderer, G., Sonneborn, H., Kraft, V., and Stucky, I. (1969). Evolution of endopeptidases: Common and different traits of bovine and crayfish trypsin. *Comp. Biochem. Physiol.* 28, 1275–1287.

Zwilling, R., Dorsam, H., Torff, H., and Rodl, J. (1981). Low molecular mass protease: Evidence for a new family of proteolytic enzymes. *FEBS Lett.* 127, 75–78.

CHAPTER

17

Circulation, the Blood, and Disease

GARY G. MARTIN *and* JO ELLEN HOSE
Department of Biology
Occidental College
Los Angeles, California 90041

I. Introduction

The circulatory system distributes nutrients, respiratory gases, metabolic wastes, and hormones throughout the body. In *Homarus americanus*, the circulatory system is open; that is, the blood (hemolymph) and its circulating cells (hemocytes) directly perfuse all tissues. Although there is an elaborate system of vessels carrying hemolymph from the heart to the tissues, the vessels eventually terminate. There are no beds of capillaries for exchange. Generally, the hemolymph then flows into large open spaces (sinuses), bathing the tissues directly, with no intervening capillary walls. The open circulation also rapidly disseminates pathogens. Much of the basic anatomy and physiology of the system was reviewed by Maynard (1960).

Open circulatory systems have traditionally been considered sluggish, with low pressure and little regulation of flow patterns. However, it has recently been recognized that open circulatory systems in some higher Malacostraca are "complex, highly efficient, and tightly regulated systems capable of a degree of tissue perfusion that rivals that of vertebrate closed systems" (McMahon and Burnett, 1990). It is timely to review the open circulatory system in *Homarus americanus* because of these advances in our understanding of circulatory physiology. Furthermore, because hemolymph is easy to sample and is an integration of the physiological, nutritional, and immunological status, hemolymph parameters could be used to assess the health of individuals and populations. The goal of this chapter is to summarize the current knowledge of (1) the morphology and function of the structures comprising the open circulatory system in *H. americanus*, including the heart, vessels, and sinuses; (2) the cellular and acellular components of the hemolymph and their functional roles; and (3) hemolymph-borne diseases and immunological defense mechanisms available to the lobster.

II. Circulation

A. General Pattern of Hemolymph Flow

In *Homarus americanus*, hemolymph flows from the heart to the body, then to the gills, and back to the heart. The beating heart may be seen by flexing the abdomen downward and looking beneath the dorsal carapace of the cephalothorax. Contraction of the heart pumps hemolymph through seven arteries: five directed anteriorly, one ventrally, and one posteriorly (Fig. 1). The anteriorly directed arteries include the ophthalmic (=cephalic, median, or anterior aorta), the paired antennal, and the paired hepatic arteries. The ophthalmic artery runs along the midline just beneath the shell to the brain and the eyestalks. This vessel becomes expanded into the cor frontale above the stomach; its contractions assist in propelling the blood anteriorly. The paired antennal arteries originate just lateral to the ophthalmic artery and pass forward with branches to the digestive gland, stomach, gonads, thoracic muscles, and branchial muscles. The paired hepatic arteries leave the ventral surface of the heart and also supply the digestive gland. The posterior end of the heart tapers into the bulbus arteriosus, which divides into the dorsal abdominal artery (=superior abdominal artery) and sternal artery. The former carries hemolymph posteriorly, serving the abdominal musculature and the abdominal appendages. The sternal artery descends to the ventral surface of the lobster, where it branches into the ventral abdominal artery (=inferior abdominal artery), which extends posteriorly, and the ventral thoracic artery, which extends anteriorly. The ventral abdominal artery provides hemolymph to some of the ventral abdominal musculature, but surprisingly none of the pleopods. The ventral thoracic artery lies beneath the nerve cords and supplies the legs and the mouthparts of the cephalothorax.

The major arteries branch to varying degrees to form vessels with increasingly smaller diameters. The smallest vessels, often with lumen diameters of less than 8 μm, have been called capillaries (Maynard, 1960). We propose that they be called arterioles, because they appear to be important in the distribution of hemolymph rather than the exchange of metabolites between hemolymph and tissues. The branching pattern of these smaller vessels in *Homarus americanus* has not been described in the detail known for *Astacus fluviatilis* (Baumann, 1921; Maynard, 1960); however, the general pattern is assumed to be similar. In most cases, the arterioles terminate (as terminal arterioles) and the hemolymph flows into irregularly shaped sinuses, directly perfusing the organs and tissues. Sinuses have received little attention, perhaps because of their irregular shape. In a few tissues, such as the cerebral ganglion, the optic lobe, and some of the musculature, the arterioles do not give rise to sinuses, but instead branch to form capillaries that appear to be the sites of gas and nutrient exchange (Sandeman, 1967; Steinacker, 1975; Govind and Guchardi, 1986).

Deoxygenated hemolymph from the sinuses and the capillaries eventually merges in the sternal sinus. Hemolymph then enters the afferent branchial sinuses of the gills. Blood flows through the gills, which are subdivided into filaments. Each filament contains a vascular loop with an outer afferent and inner efferent vessel. These basic vessels branch to form a complex vasculature before an efferent vessel from each gill delivers oxygenated hemolymph into the branchiocardiac sinus and then into the pericardial cavity.

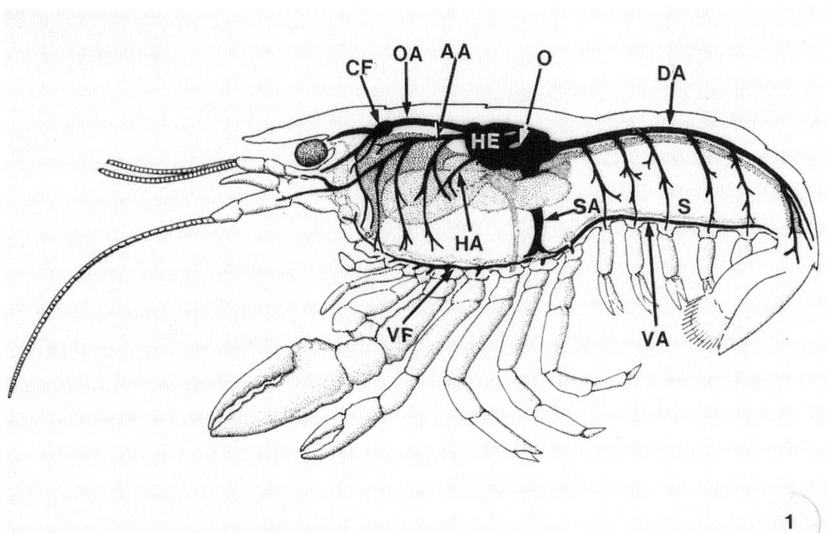

FIGURE 1 Schematic drawing of *Homarus americanus* highlighting the circulatory system. AA, Antennal artery; CF, cor frontale; DA, dorsal abdominal artery (=superior abdominal artery); HA, hepatic artery; HE, heart; O, ostium; OA, ophthalmic artery; S, segmental artery; SA, sternal artery; VA, ventral abdominal artery (=inferior abdominal artery); VF, ventral thoracic artery (=inferior thoracic artery). (Redrawn after McLaughlin, 1980. From "Comparative Morphology of Recent Crustacea" by Patsy A. McLaughlin. Copyright © 1980 by W.H. Freeman and Company. Reprinted by permission.)

(McMahon, in Chapter 18, provides a discussion of physiology.) Hemolymph enters the heart through the six ostia (paired dorsal, lateral, and ventral) (Herrick, 1909; Maynard, 1960; McMahon and Burnett, 1990); valves prevent backflow of hemolymph during contraction. Following contraction, the heart is refilled with hemolymph from the pericardial cavity and the rate of refilling varies with the pressure gradient across the ostial valves.

B. Morphology of the Circulatory System

1. Heart

The heart is a single, muscular chamber making up 0.1–0.15% of total body weight. A thin pericardium, composed of spongy and fibrous connective tissue (Figs. 2 and 3), surrounds the bulk of the heart (myometrium), which is composed of striated muscle arranged into a complex series of trabecular strands. Cardiac muscle fibers (Fig. 4) are composed of multinucleate, cross-striated cells connected by intercalated disks. Thick and thin filaments (Fig. 5), as well as two separate membrane systems, have been found in cardiac muscle (Smith and Anderson, 1972). The first system is composed of a series of invaginations of the plasma membrane that occur at the Z level and are analogous to the transverse tubules of other skeletal muscles. From these arise the second set of membranes, longitudinally oriented tubules with a small diameter, which form a collar surrounding the fibril at the H level. It is thought that this network of smaller tubules is analogous to the transverse tubular system of other fibrils (Smith and Anderson, 1972; see Govind, Chapter 12, on muscles and their innervation).

Satellite cells, reserve inclusion (RI) cells, and phagocytes are also present in the heart. Satellite cells may function in regeneration of cardiac cells (Martynova *et al.*, 1986). Although they have not been studied in *Homarus americanus*, satellite cells are present in 21 other species of decapod crustaceans, wedged between the intima and the plasma membrane of cardiac muscle fibers (Midsukami, 1981). They are less than 2 μm wide through the nucleus and extend up to 30 μm in length oriented parallel to the muscle fibers. RI cells (cyanocytes) both produce and store hemocyanin (see Section IV,B) and have been observed within the myocardium of *H. americanus*. Phagocytic cells have been observed in the heart of *H. americanus* (Cornick and Stewart, 1968) and other decapods (Foster *et al.*, 1981; Johnson, 1980, 1987). Although RI cells reportedly possess some phagocytic capabilities (Johnson, 1980), it is unclear whether the cardiac phagocytes of *H. americanus* are hemocytes or RI cells.

The heart is suspended within a pericardial cavity by three pairs of alary ligaments. Although the alary ligaments themselves contain little or no muscle, their attachment to the body wall contains elastic material, muscle, and neural innervation (McMahon and Burnett, 1990).

2. Vessels

Homarus americanus has an extensive set of vessels conducting hemolymph away from the heart. Although the distribution of these vessels has been described, a satisfactory functional classification of the vessels is still lacking (Maynard, 1960; Burnett, 1975). Therefore, at present only three categories of vessels are commonly identified: arteries, large vessels with relatively thick walls (Figs. 6, 7, and 9); arterioles, small vessels with thinner walls that open into sinuses or capillaries (Fig. 11); and capillaries, small vessels with thin walls that are morphologically similar to arterioles, but appear to be involved in gas and nutrient exchange between hemolymph and tissue because they do not give rise to sinuses (Sandeman, 1967; Steinacker, 1975; Govind and Guchardi, 1986). Elucidation of the functional differences between small distributional arteries, arterioles, terminal arterioles, capillaries, and sinuses requires further study. However, the layers of all vessel walls include the following, progressing outward from the lumen: (1) intima, an acellular layer; (2) endothelium, a cellular layer; (3) a loose connective tissue layer; and (4) external lamina, a thin, outer, acellular layer. Burnett (1975) included the last two layers as his *fibrous outer layer* or *external lamina*. In large arteries, the endothelial layer is thick and clusters of cells are separated by layers of elastinlike material (Fig. 6). As vessels decrease in diameter, the connective tissue layer thins and the endothelial layer is reduced to a single, continuous layer of cells (Fig. 11); finally, the endothelium is lost. Maynard (1960) referred to capillaries (our *arterioles*) as small-caliber vessels with a single endothelial layer and without a visible intima. The intima is always visible using transmission electron microscopy (TEM), although highly reduced in thickness in the smallest vessels.

The acellular intima (Figs. 6–8 and 11) has been called a basal lamina or basement membrane, although it differs from the more thoroughly studied basement membranes of vertebrates in the following ways: (1) it is not separated from the plasma by an endothelial layer; (2) it is relatively thick, 12 μm compared to 50–80 nm for vertebrates (Bloom and Fawcett, 1975); and (3) it is not divided into zones, such as lamina densa and rara, as described for vertebrates. In histological sections, it is eosinophilic,

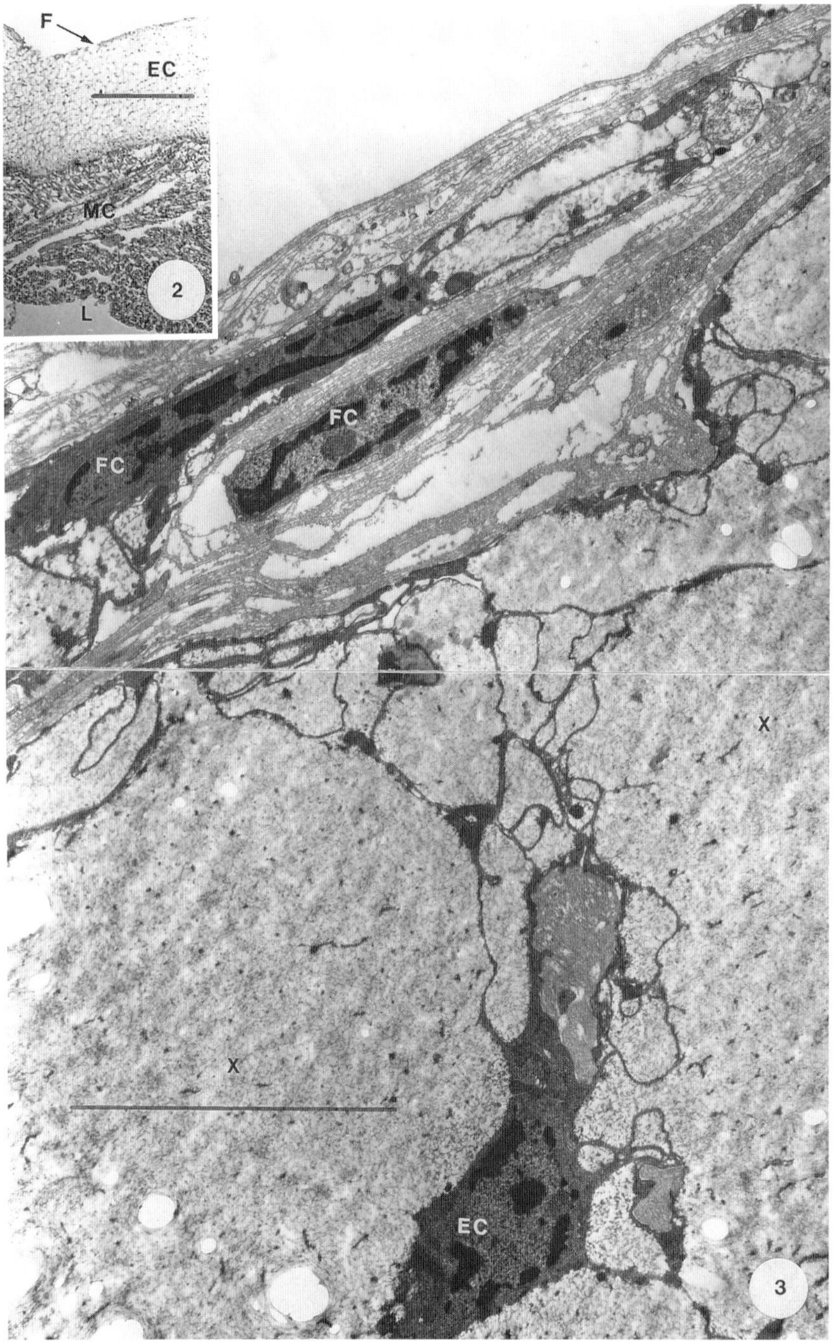

FIGURE 2 Cross-section through the wall of the heart. Note the outer fibrous layer (F), netlike epicardium (EC), myocardium (MC), and lumen (L). Photomicrograph of paraffin section stained with hematoxylin and eosin. Scale bar: 0.5 μm.

FIGURE 3 Transmission electron micrograph of the outer heart wall. Note the cells of the outer fibrous layer (FC) and the epicardial cell (EC) surrounded by polygonal extracellular areas (X). Scale bar: 10.0 μm.

periodic acid–Schiff positive, and Alcian blue negative (at pH 2.5) and stains with a variety of "elastin" stains. Reviewing the phylogenetic distribution of elastin, based on amino acid composition and the presence of characteristic desmosine, Elder (1973) and Sage and Gray (1979) concluded that invertebrate "elastin" is different from its vertebrate counterpart. Shadwick *et al.* (1990) demonstrated that invertebrate elastic tissue lacks the amorphous appearance of vertebrate elastin and is instead composed of fine fibrils. In *Homarus americanus*, the intima is composed of fine, 25-nm-wide fibrils (Fig. 8) and a tannic acid–glutaraldehyde fixative reveals a banding pattern that is unlike that of collagen; both features are consistent with some descriptions of invertebrate elastic tissue. There is no information on the presence of type IV collagen, laminin, or fibronectin in this layer, molecules characteristic of the basement membranes of vertebrates and insects (Ashurst, 1982; Fessler and Fessler, 1989; Kefalides *et al.*, 1979; Martin and Timpl, 1987).

In arteries, the filamentous material of the intima extends into the wall, forming a meshwork that separates clusters of endothelial cells (Fig. 6). The cells have an irregular shape and do not form distinct layers (Fig. 7). The cytoplasm contains rough endoplasmic reticulum, mitochondria, and numerous microtubules. Muscle is thought to be absent from the wall of major arteries except adjacent to valves. However, muscle cells have been observed interspersed between the endothelial cells and mixed orientation in the dorsal abdominal artery (Fig. 7) (G. G. Martin and J. E. Hose, unpublished observations). In the outer part of the vessel wall, the elastin staining decreases and the extracellular material is presumably collagen. Two types of cells are present (Fig. 9). One type is large, characterized by numerous vacuoles (perhaps the result of inadequate fixation), and embedded in the outermost layer of the elastinlike material. The second type of cell resembles a fibroblast; it has an irregular shape with extensive cell processes and the most abundant organelles include rough endoplasmic reticulum and mitochondria. The extracellular matrix around these cells appears more filamentous and has less amorphous material between the filaments. This material is oriented primarily parallel to the surface of the vessel and

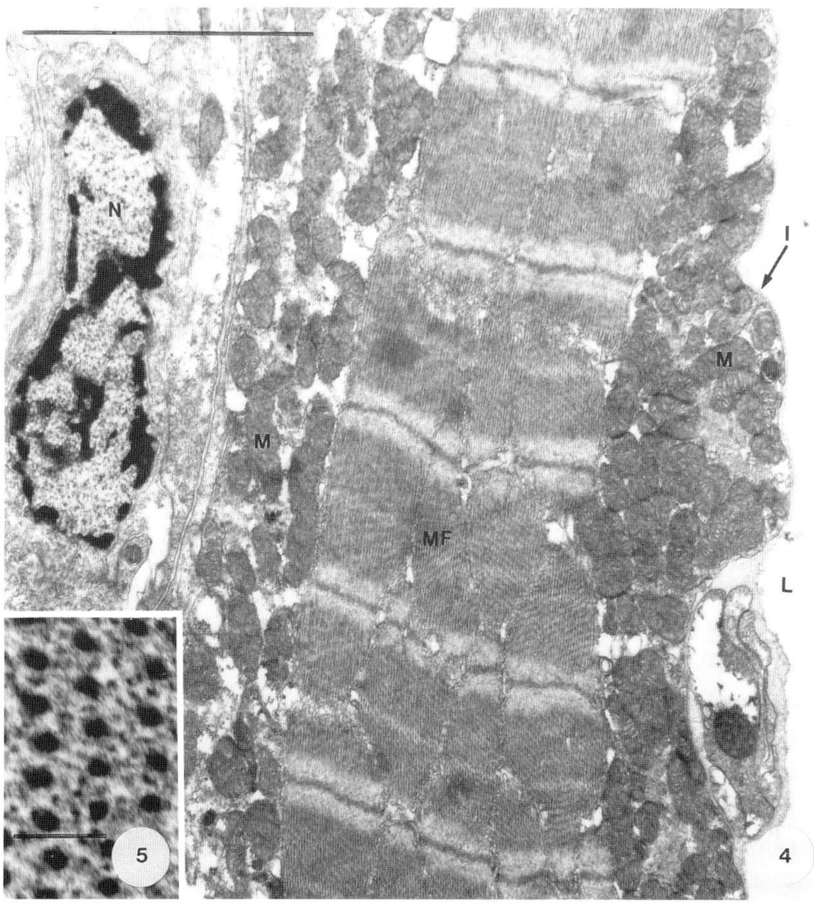

FIGURE 4 Transmission electron micrograph of a myocardial cell. Note the myofibers (MF), mitochondria (M), nucleus (N), intima (I), and heart lumen (L). Scale bar: 5.0 μm.

FIGURE 5 Transmission electron micrograph of a cross-section through myofibers showing both thick and thin filaments. Scale bar: 0.1 μm.

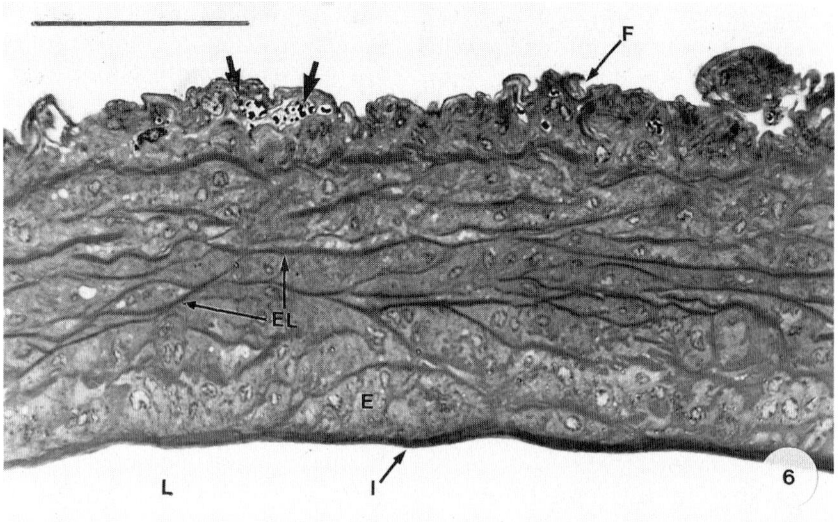

FIGURE 6 Cross-section through the wall of the dorsal abdominal artery. Note the vessel lumen (L), intima (I), and clusters of endothelial cells (E) separated by strands of elastinlike material (EL). Arrows point to *Aerococcus viridans* in small vessels in the outer fibrous layer (F). Photomicrograph of a plastic section stained with methylene blue. Scale bar: 0.1 mm.

becomes the outermost layer of the vessel (Fig. 10). The outer layer of the vessel wall has been called a collagenous adventitia; the presence of collagenous material is consistent with measurements of nonlinear elasticity exhibited by artificially inflated vessels (Shadwick *et al.*, 1990). Ultrastructurally, this outermost layer appears similar to the lining of all hemolymph channels, but it is not clear whether it is the same as the intima lining the vessel lumen and this should be resolved. Foreign materials in the hemolymph, such as bacteria, may be recognized as nonself because they are not covered by intima (Salt, 1970).

C. Circulatory Physiology

The heart of a 450-g lobster beats at a rate of 50–136 beats per minute and has an estimated hemolymph volume turnover time of 3–8 minutes (Burger and Smythe, 1953). Blood pressure in the heart ranges from 9 to 22 mm Hg at systole and 0 to 5 mm Hg at diastole. There is no correlation between body size and heart rate or blood pressure. Pressure in the ventral thoracic sinus is approximately 2–6 mm Hg. Blood flow has been described as intermittent in the ventral thoracic sinus, moving at a rate of 0.3–0.5 cm/sec and increasing to 5 cm/sec with abdominal flexion.

The heart begins to pulse rhythmically in 4- to 5-week-old embryos, long before nervous innervation is established (Herrick, 1909). However, in the adult, a neurogenic pacemaker system located in the cardiac ganglion in the dorsal wall of the heart regulates the initiation of contraction, control of heart rate, and contractility of heart muscle (Maynard, 1960; Hartline, 1979). In addition, it is likely that excitatory and inhibitory nerves from the central nervous system (Bullock and Horridge, 1965; Wilkens *et al.*, 1974; Young and Coyer, 1979) and release of cardioactive substances from the pericardial organ (Cooke and Sullivan, 1982) also regulate heart activity. *Homarus americanus* responds to hypoxia by increasing its cardiac output (McMahon and Wilkens, 1975). (See McMahon, Chapter 18, for a comprehensive account of the physiology of *H. americanus*.)

Hemolymph flow to the major arteries could be regulated by opening or closing valves in the heart; these are sensitive to octopamine and proctolin, but not serotonin (Kuramoto and Ebara, 1984). Muscles and valves in the dorsal abdominal artery may pre-

FIGURE 7 Transmission electron micrograph of a cross-section through the inner wall of the dorsal abdominal artery (=superior abdominal artery) in a region containing muscle (MU). Note the endothelial cells (E), elastinlike material (EL), and thick intima (I) adjacent to the vessel lumen (L). Scale bar: 10.0 μm.

FIGURE 8 Transmission electron micrograph showing fibers in the intima. Scale bar: 1.0 μm.

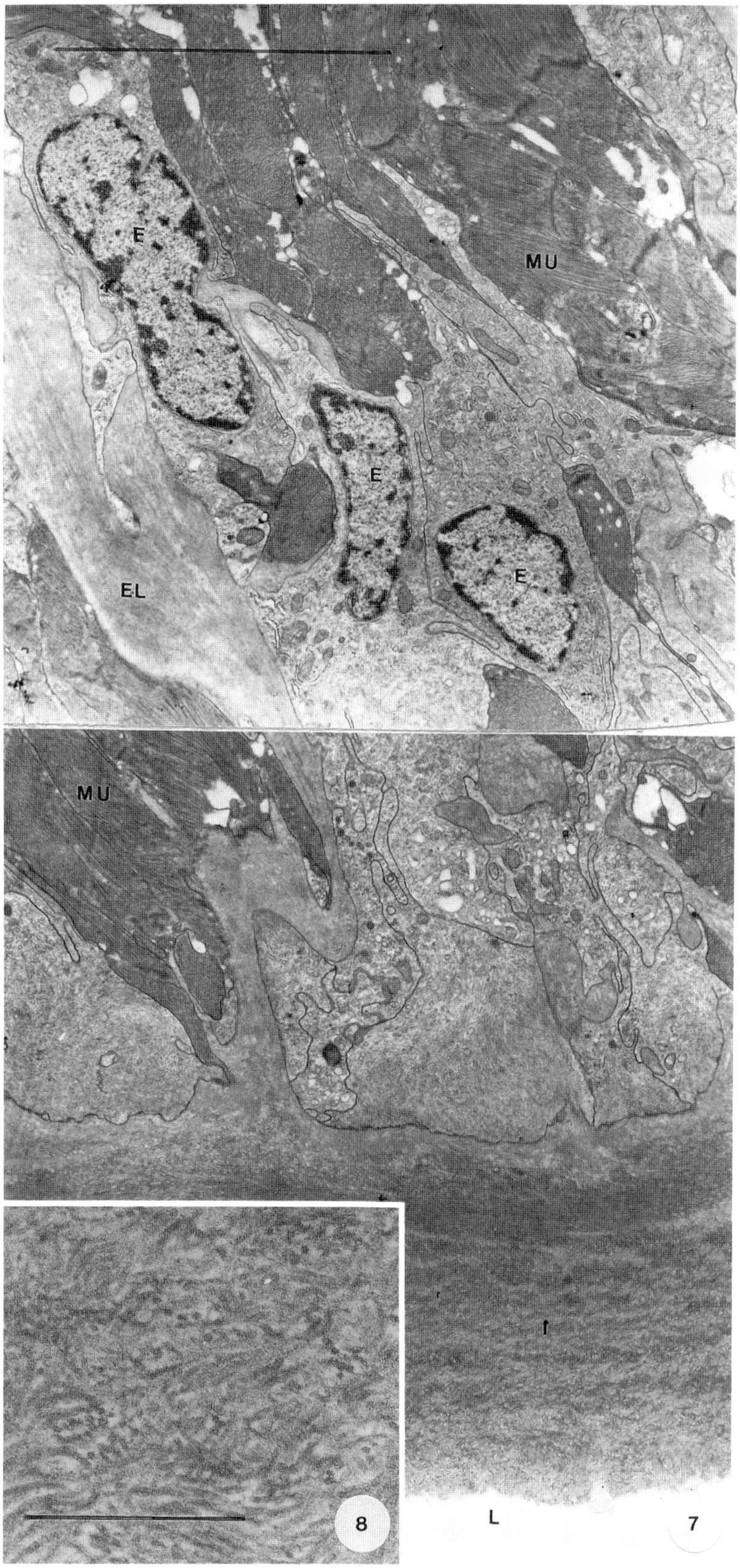
E
MU
E
E
EL
MU
L
8
7

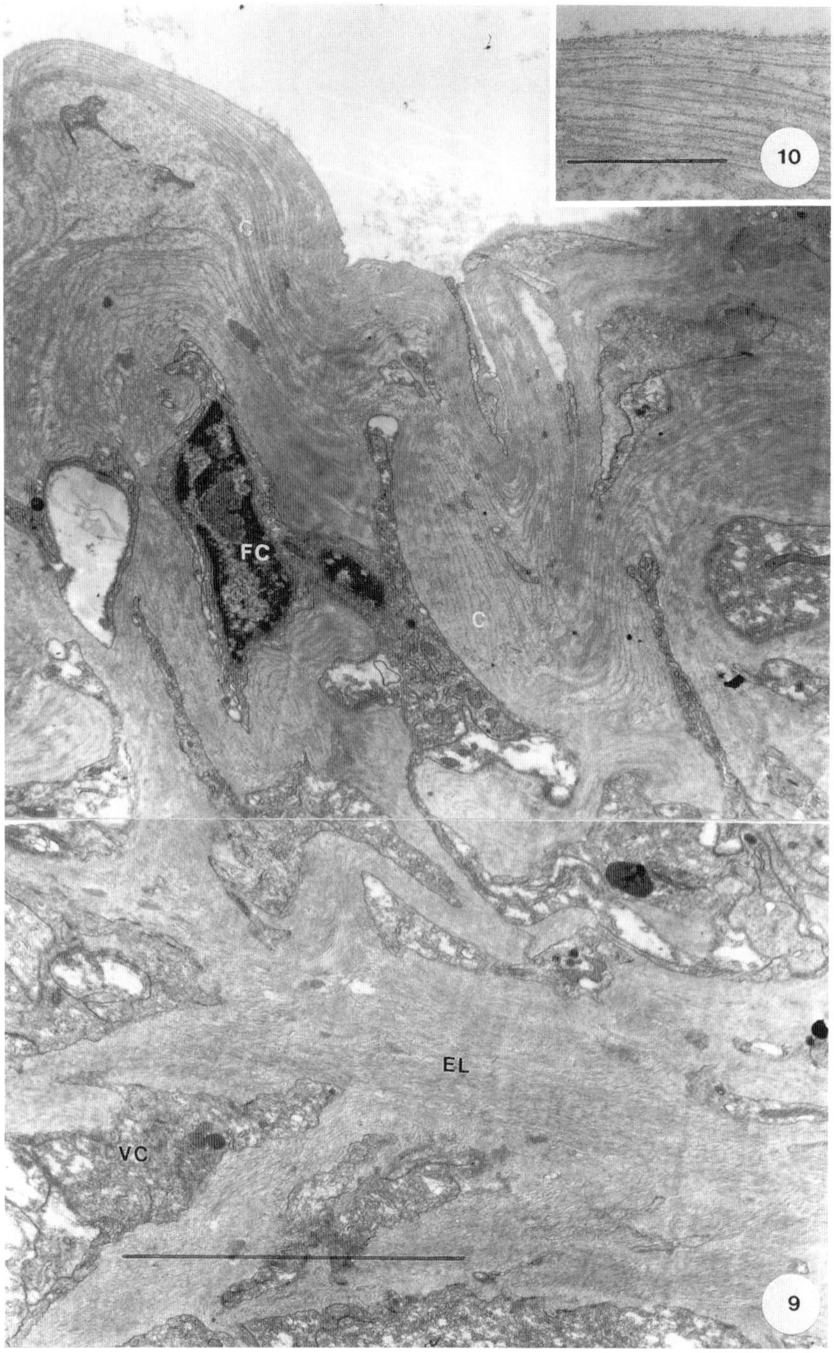

FIGURE 9 Transmission electron micrograph of a cross-section through the outer wall of the dorsal abdominal artery showing the outer fibrous layer. Note the fibroblastlike cells (FC) surrounded by possible collagenous fibers (C) and vacuolated cells (VC) embedded in the elastinlike fibers (EL). Scale bar: 10.0 μm.

FIGURE 10 Transmission electron micrograph showing layered fibers covering the outer wall of the dorsal abdominal artery. Scale bar: 1.0 μm.

vent backflow and facilitate blood propulsion and distribution (Maynard, 1960). However, most of the major arteries of decapods like the lobster lack muscle and therefore seem unable to modify hemolymph flow.

Major arteries, such as the dorsal abdominal artery of the lobster, are elastic (Shadwick *et al.*, 1990). The vessels distend during systole, storing energy from the heart contraction in the wall of the stretched vessel. Elastic recoil of the wall assists in blood propulsion and is an effective pulse-smoothing system. The walls are easily distended at low pressures, suggesting an extensible component (analogous to elastin) in the wall. Since they become stiffer at high pressure, a rigid component such as collagen may also be present. A large artery in *Homarus americanus* is therefore similar in design to a vertebrate artery.

III. Hemolymph

Hemolymph, the fluid circulating throughout the body, has a pH of 7.6 (Stewart *et al.*, 1966). It transports gases, nutrients, cellular waste, hormones, and molecules associated with internal defense, as well as the circulating hemocytes.

A. Volume

The volume of hemolymph in the lobster varies with the molt cycle. (See Waddy *et al.*, Chapter 10, for a discussion of the molt cycle and molt stages.) During molt stage D, hemolymph volume is approximately 30% of the animal's wet weight. It rises to 55% immediately after molting (stages A_1–A_2) and then decreases to initial levels by intermolt (stage C) (Mykles, 1980). Hemolymph volume as well as tissue and cell water content are also affected by eyestalk ablation and may vary with the size of the lobster (Jackson *et al.*, 1987).

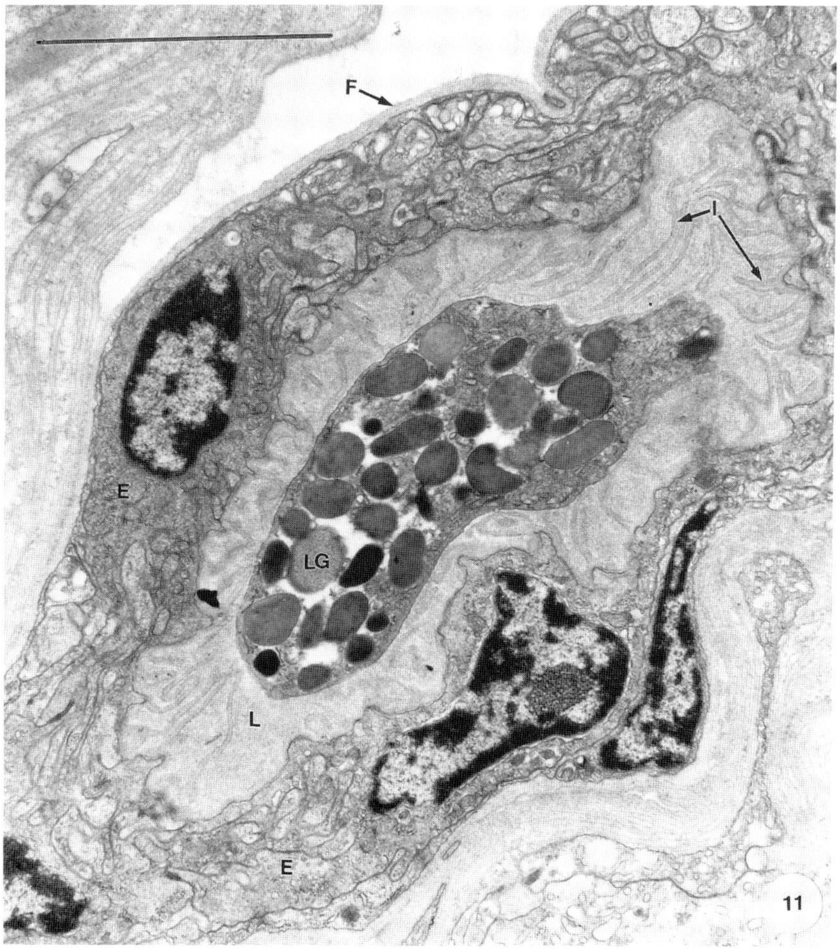

FIGURE 11 Transmission electron micrograph of a cross-section through an arteriole showing folded intima (I), a single endothelial layer (E), and the outer fibrous layer (F). A large-granule hemocyte (LG) fills the vessel lumen (L). Scale bar: 5.0 μm.

B. *Major Hemolymph Proteins*

1. Hemocyanin

Hemocyanin is the oxygen-carrying protein in the hemolymph. (See McMahon, Chapter 18, on physiology.) Freshly drawn blood is colorless and develops a blue tinge upon exposure to air. Hemocyanin is a large, copper-containing molecule composed of a minimum of six subunits, each approximately 75 kDa (Mangum, 1993). It is produced primarily in the digestive gland by RI cells (Senkbeil and Wriston, 1981). [RI cells and the cyanocytes described by Ghiretti-Magaldi *et al.* (1977) are identical.] Production of hemocyanin is stimulated by chronic hypoxia and may indeed be affected by water quality (Senkbeil and Wriston, 1981; Engel and Brouwer, 1991). Only a small fraction of the hemocyanin present may be essential for respiratory function. Hemocyanin does not appear to function in osmoregulation, since hemocyanin production is not enhanced when lobsters are placed into water of abnormally low salinity (50% seawater) (Senkbeil and Wriston, 1981).

2. Coagulogen

The crustacean plasma clotting protein analogous to vertebrate fibrinogen is coagulogen. It has a molecular mass of approximately 400 kDa and is composed of two identical subunits. The site of synthesis of plasma coagulogen is unknown, but it does not appear to be produced by hemocytes. Plasma coagulogen may account for up to 20% of the total blood protein in lobsters (Durliat, 1985). Cross-linking of coagulogen to form a clot is initiated by transglutaminase, an enzyme released from hyaline hemocytes (Doolittle and Fuller, 1972; Lorand, 1972; Lorand and Conrad, 1984; Martin *et al.*, 1991). The intensity of gelation is directly correlated with the number of circulating hemocytes, particularly the hyaline hemocytes. Various anticoagulants have been used to prevent, or at least delay, the rapid clotting of hemolymph. Sodium citrate (10%) and EDTA (5%) do not inhibit cellular lysis, but prevent plasma clotting (Durliat and Vranckx, 1981). Amines, including histamine and serotonin, delay the clotting time in *Homarus americanus* (Lorand *et al.*, 1964). *N*-Ethylmalemide (0.02 *M* in 3% NaCl) prevents cellular aggregation and hemolysis and maintains hemocyte morphology (Bang, 1970; Durliat and Vranckx, 1981; Martin *et al.*, 1991). Coagulation is dependent on calcium ions that may be regulated by cAMP. A highly specific octopamine-stimulated increase in cAMP in hemolymph raises the rate of coagulation and changes the nature of the clotting reaction (Battelle and Kravitz, 1978).

A second form of coagulogen (cellular) has been described within the hemocytes of some crustaceans, yet it is not abundant, has a lower molecular mass (about 70 kDa) than plasma coagulogen, and its function is not clear (Madaras *et al.*, 1981; Durliat, 1985; Durliat and Vranckx, 1989). Its presence in *Homarus americanus* has not been determined.

3. Defensive Proteins

Like other invertebrates, humoral antibodies are not present in the hemolymph of *Homarus americanus*. Instead, a variety of molecules, such as agglutinins, lysins, and opsonins, effect the recognition and elimination of foreign materials.

Agglutinins are molecules capable of agglutinating foreign cells (tests include bacteria and vertebrate erythrocytes) and possibly enhancing their removal by phagocytosis. A natural agglutinin in lobster hemolymph has been identified with a pH stability between 6.0 and 9.0, a heat-inactivation temperature of 56–65°C, and a requirement for calcium (Cornick and Stewart, 1973). Two similar agglutinins have also been found; one is specific for *N*-acetylneuraminic acid and the second, for *N*-acetylgalactosamine (Campbell *et al.*, 1982). The agglutinins may be produced by hemocytes and agglutinin levels may be correlated with changes in the number of circulating hemocytes (Cornick and Stewart, 1973, 1978).

Bacteriocidins and inducible bacteriocidins have been demonstrated in the hemolymph of *Homarus americanus* (Acton *et al.*, 1969; Stewart and Zwicker, 1972). They require the interaction of plasma and cellular components. The plasma portion is assumed to be the bacteriocidin in an inactive form that requires activation by some material released from hemocytes. Elevated temperature increases the time of response. The normally low level of bactericidal activity in hemolymph of untreated lobsters appears to be an effective growth inhibitor against a wide variety of bacteria; this inhibition may be enhanced by vaccination (Stewart and Zwicker, 1972). Heat stability tests suggest that several bacteriocidins may be present in *H. americanus* (Stewart and Zwicker, 1972).

There is some evidence for lysins in *Homarus americanus*. Lobster hemolymph was tested for hemolysins after these molecules were identified in crabs and sea stars (Hall *et al.*, 1972). This ability to lyse red blood cells may show similarities to the vertebrate complement system. Investigations of lobster hemolysins have utilized an activator, such as cobra venom factor. Caution has been raised in interpreting the results of such experiments because the cobra venom factor contains low levels of phospholipase activity, which actually may be the cause of cell lysis (Hall *et al.*, 1972).

Prophenoloxidase is an important enzyme in the recognition and elimination of foreign materials from the hemolymph of crustaceans (reviewed by Söderhäll *et al.*, 1988). It has long been observed that cellular defense reactions such as encapsulation result in melanization. Phenoloxidase is present in the hemolymph as well as in hemocytes, and it is a key enzyme in the production of the pigment melanin. The inactive proenzyme is activated by β-1,3-glucans of fungi, endotoxin of gram-negative bacteria, and peptidoglycans of gram-positive bacteria. Phenoloxidase may play two additional roles in the defense against foreign materials: (1) intermediates in melanin production are cytotoxic and may cause the death of invading cells; and (2) phenoloxidase is sticky and appears to stimulate phagocytosis, encapsulation, and hemocyte locomotion, all important in cellular defense.

The molecule α_2-macroglobulin may also be involved in the humoral response in *Homarus americanus*. This high-molecular-weight protease inhibitor was initially found in the plasma of vertebrates, where it has been shown to inhibit proteases (primarily endopeptidases). Lobster macroglobulin has a molecular mass of about 342 kDa and is composed of two units connected by disulfide bonds (Spychert *et al.*, 1987). The amino acid sequence shows considerable variation from that of vertebrate macroglobulin. Although it is able to inhibit a wide variety of proteases, its specific role is unknown.

IV. Hemocytes

The first studies of *Homarus* hemocytes (blood cells) were carried out on the European lobster, *H. gammarus*, nearly 100 years ago (Cuénot, 1903). Hemocytes have been popular material for study ever since (Fischer-Piette, 1931; Toney, 1958; Hearing and Vernick, 1967). However, only recently have the hemocytes been reassessed within the context of a comprehensive scheme of classification of decapod hemocytes (Hose *et al.*, 1990; Martin and Hose, 1992). The two major categories of blood cells of *H. americanus*—hyaline hemocytes and granulocytes—can be distinguished by a set of morphological features and have distinct functions, as discussed in detail in the sections that follow.

Mean total hemocyte counts (THCs) for *Homarus americanus* range from 19 to 25×10^3 cells per milliliter (Yeager and Tauber, 1935; Stewart *et al.*, 1983); these counts are within the range of measurements for other decapods. Using a mean hemolymph volume of 300 ml, the total number of hemocytes in a 1.0-kg animal varies from 5.7 to 7.5×10^6 cells. Molting, development, reproductive status, nutritional condition, and disease have been shown to influence decapod hemocyte abundance, although the first three factors generally have received little attention in *H. americanus*. Prolonged starvation of lobsters results in a reduction in THC to 40% of normal, which is assumed to be reversible (Stewart *et al.*, 1967). The lethal bacterial disease gaffkemia causes profound and irreversible hemocytopenia (see Section VI,A).

A. Hemocyte Classification

Traditionally, hyaline hemocytes and granulocytes were distinguished within arthropods by the absence or presence of cytoplasmic granules (see Ravindranath, 1980). The term *hyaline hemocyte* was reserved for small, agranular hemocytes, which were infrequently found in *Homarus americanus* (Cornick and Stewart, 1978). Examination of cells responsible for initiating coagulation in the lobster yielded a group that did not fit the historical classification; despite their relatively small size, they contained numerous cytoplasmic granules. The identification of hyaline hemocytes merely by the absence of granules clearly was misleading, so the following suite of morphological criteria was developed to distinguish between hyaline cells and granulocytes (Table 1).

1. Hyaline Hemocytes

Hyaline hemocytes can be recognized by their generally ovoid shape; size range from 8 to 16 μm in the longest dimension (Figs. 12 and 13), high nucleocytoplasmic (N:C) ratio of 35–45%, and relatively small number of granules (eight to 20 per section, similar to that found in some small-granule hemocytes). In differential counts of fixed smears and unclotted hemocyte preparations, hyaline hemocytes compose approximately 22% of the total in intermolt lobsters (range, 12–34%). This percentage is much lower than the 66% hyalocytes seen by Cornick and Stewart (1978); because of the overlap in cell size and the presence of granules in some hyaline cells, their estimate undoubtedly contains a large proportion of granulocytes. Their prohyalocyte category probably represents immature, agranular hyaline cells, which are infrequently observed in healthy, intermolt animals [in some decapods, these may become more

TABLE 1 **Characteristics of Hematopoietic Cells and Hemocytes of *Homarus americanus***

	Location[a]	Size (μm)[a]	N:C ratio (%)	No. of granules per cell[b]	Granule type[c]	% of Total[d]
Hyaline cells						
Hyaline stem cell	HPT	13 × 9	40	<5	ed, h, or s	7.4
Hyaline hemocyte	HPT	11 × 6	40	10–15	ed, h, or s	2.1
	C	14 × 11	40	14	ed, h, or s	22.4
Granulocytes						
Granulocyte stem cell	HPT	15 × 10	30	<2	el, f	20.9
Small-granule stem cell	HPT	12 × 9	30	3–20	el, f; ed, h	52.8
Small-granule hemocyte	HPT	10 × 7	40	>20	ed, h	10.0
	C	21 × 14	27	27	ed, h	60.2
Large-granule hemocyte	HPT	14 × 10	20	>20	ed, h	6.3
	C	23 × 12	20	73	ed, h	16.4

[a]HPT, Hematopoietic tissue, sectioned cells, measurements from transmission electron microscopy; C, circulation, hemolymph smear, measurements from phase contrast microscopy.

[b]Number of granules per sectioned cell for HPT; number per whole cell for C.

[c]ed, Electron dense; h, homogeneous; s, striated; el, electron lucent; f, fibrillar.

[d]Percentage of the total number of cells by location (HPT or C).

numerous prior to molting (Hose *et al.*, 1992) and during some disease states.]

When stained with Giemsa, the dark blue nucleus is surrounded by a thin rim of pink cytoplasm. Using methylene blue, the cytoplasm stains darker than that of the granulocytes. This tinctorial quality probably corresponds to the enhanced electron density of hyaline cells resulting from the presence of numerous, small (50-nm-diameter) cytoplasmic deposits (Figs. 13–15). These cytoplasmic deposits stain slightly with Sudan black B, but the intensity is much less than in penaeid shrimp (Martin and Hose, 1992). Although the identity of these deposits has not yet been determined, they are presumably a clotting factor that initiates hemolymph gelation (Omori *et al.*, 1989), possibly a "cellular coagulogen" or another coagulation factor such as transglutaminase (Martin *et al.*, 1991).

The cytoplasm of hyaline hemocytes can contain no granules at all (prohyalocyte) or up to 20 small granules (0.2–0.4 μm in length), ovoid to cylindrical in shape. Ultrastructurally, there are two main types of granules: (1) contents homogeneous and electron dense; and (2) contents forming a reticular or striated (concentric) pattern (Fig. 14). The reticular/striated granules are more abundant than the homogeneous granules; they seem to be analogous to the striated granules characteristic of shrimp hyaline hemocytes (Omori *et al.*, 1989). Other organelles include mitochondria, rough endoplasmic reticulum, Golgi bodies, free ribosomes, and a circumferential band of 50–60 microtubules (Cohen *et al.*, 1983).

2. Granulocytes

Lobster granulocytes are larger than hyaline hemocytes (>20 μm in the longest dimension) and usually contain more cytoplasmic granules. The N:C ratio is low (<30%) and does not overlap with the ratio of hyaline hemocytes. Although they represent a continuum of differentiation, the granulocytes can be subdivided into small-granule hemocytes (sometimes called semigranulocytes) (Figs. 16 and 17) and large-granule (or refractile) hemocytes (Figs. 18 and 19). As in all decapods, the small-granule hemocytes are more numerous, constituting 60% of the differential count. Large-granule hemocytes represent about 16% of the total; they probably correspond to the circulating chromophobic granulocytes of Cornick and Stewart (1978).

Granulocytes vary in shape from elliptical to teardrop shaped. The nucleus is central in elliptical cells, but is eccentrically placed in other cells, primarily large-granule hemocytes. The N:C ratio progressively decreases and the amount of condensed, marginal chromatin increases with maturation into large-granule hemocytes. The cytoplasm is pink in Giemsa-stained smears and appears more electron lucent than in hyaline hemocytes using TEM (Figs. 6 and 17).

The number of granules increases as small-granule hemocytes mature into large-granule cells concomitant with an increase in granule diameter. Small-granule hemocytes are identified by the presence of only small (0.8–1.0 μm long), ovoid to cylindrical granules (Fig. 6). At the EM level (Fig. 17), most of the granules

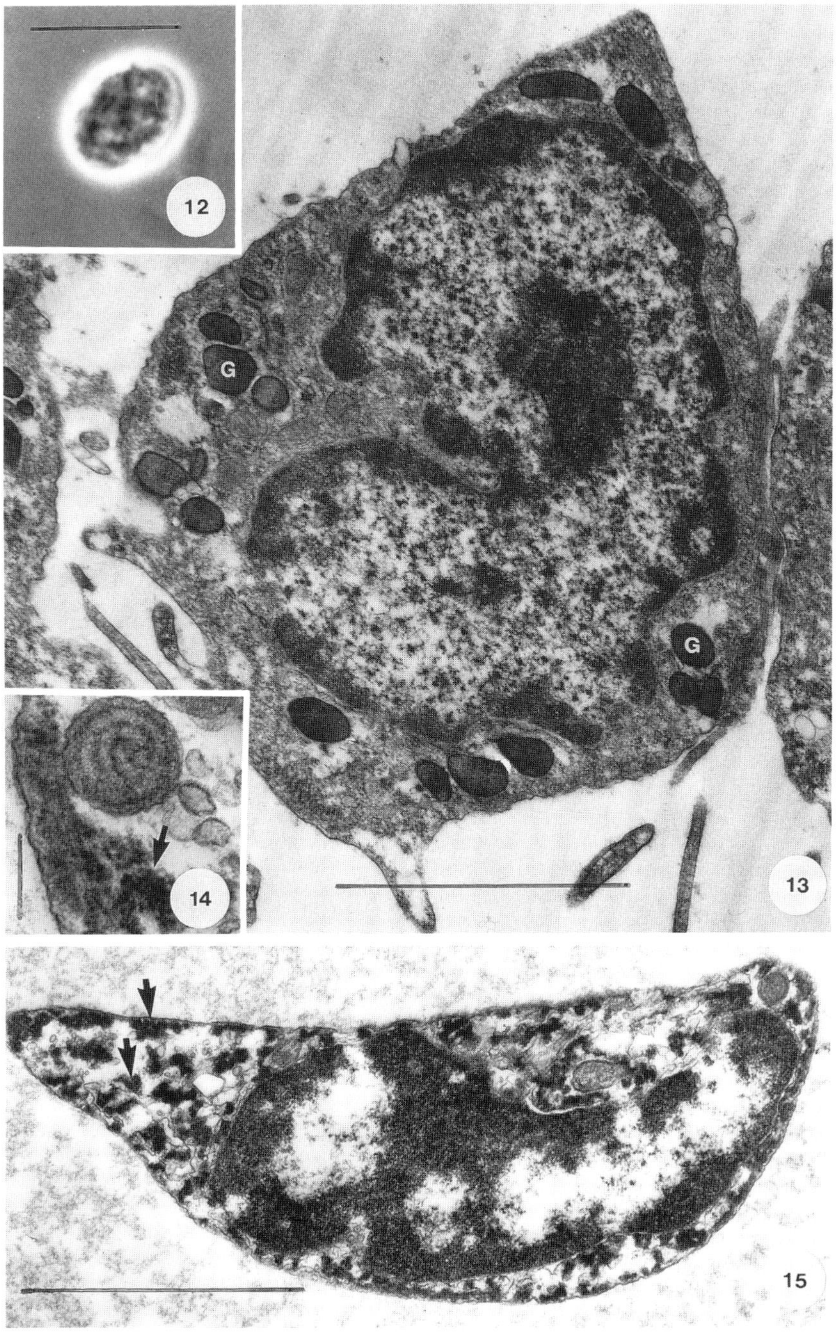

FIGURE 12 Hyaline hemocyte. Phase contrast photomicrograph of a fixed cell. Scale bar: 15.0 μm.

FIGURE 13 Transmission electron micrograph of a hyaline hemocyte showing electron-dense cytoplasm and granules (G). Scale bar: 3.0 μm.

FIGURE 14 Transmission electron micrograph of cytoplasm from a hyaline hemocyte containing deposits (arrow) and granule with a substructure of concentric rings. Scale bar: 0.5 μm.

FIGURE 15 Transmission electron micrograph of a hyaline cell in the early stage of lysis. Note the clumping of the cytoplasmic deposits (arrows) and leached cytoplasm. Scale bar: 3.0 μm.

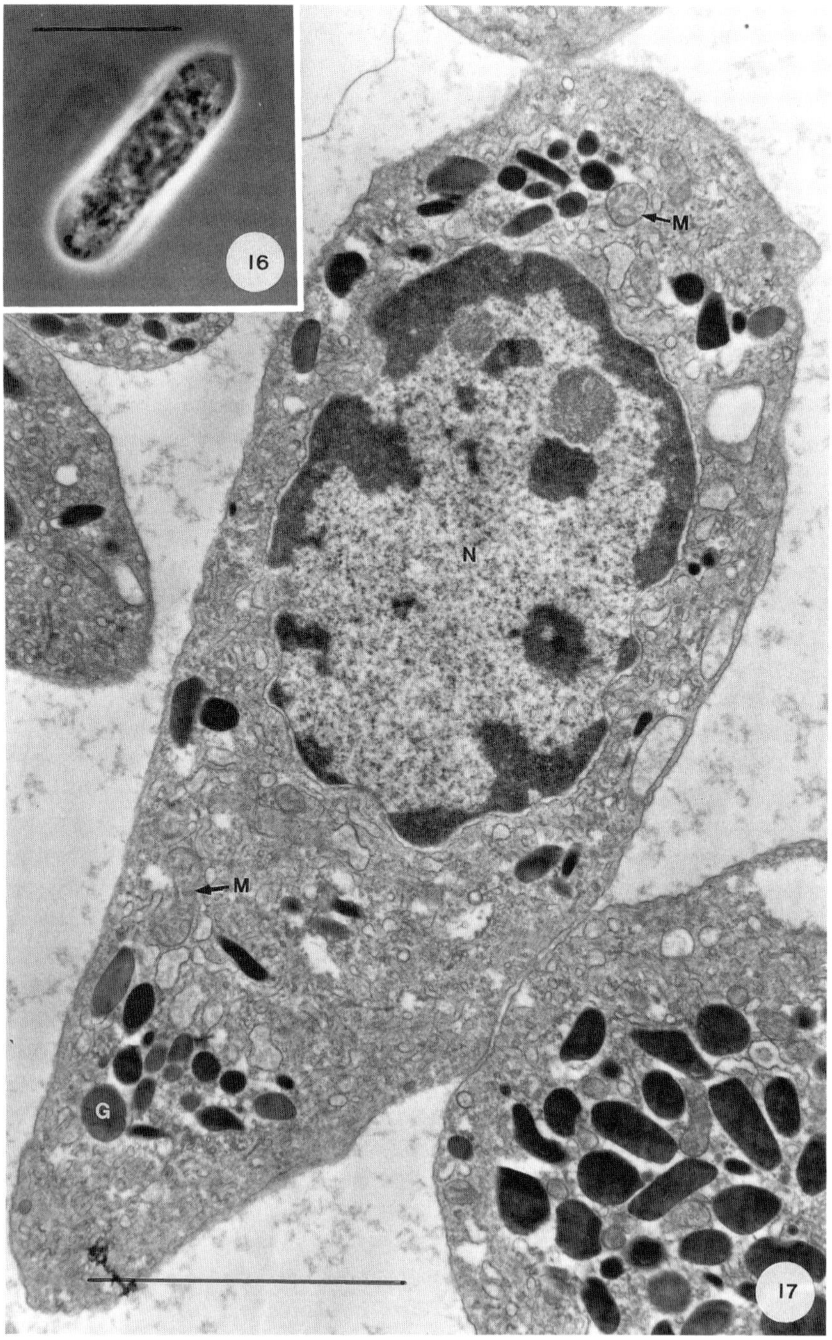

FIGURE 16 Small-granule hemocyte. Phase-contrast photomicrograph of a fixed cell. Scale bar: 15.0 μm.

FIGURE 17 Transmission electron micrograph of a small-granule hemocyte showing the nucleus (N), cytoplasmic granules (G), and mitochondria (M). Scale bar: 5.0 μm.

are homogeneous and electron dense and are frequently surrounded by a clear halo (possibly an artifact). One or two Golgi complexes are present along with numerous vesicles, rough endoplasmic reticulum, mitochondria, free ribosomes, and a marginal band of 50–60 microtubules (Cohen *et al.*, 1983).

Large-granule hemocytes are slightly larger than small-granule hemocytes and can be distinguished from them by the presence of many (usually >20) large (1.2- to 1.4-μm-diameter), spheroid granules. On phase contrast microscopy, they are so numerous that they obscure the nucleus (Fig. 18). Because the large granules are frequently unstained by Giemsa, previous researchers have termed these "chromophobic granulocytes" (Cornick and Stewart, 1978). Using TEM, most of these granules have a homogeneous, electron-dense substructure, although rarely a granule with a punctate substructure is present (Fig. 19). Small (0.8- to 1.0-μm-diameter), spheroidal, electron-dense granules are also numerous.

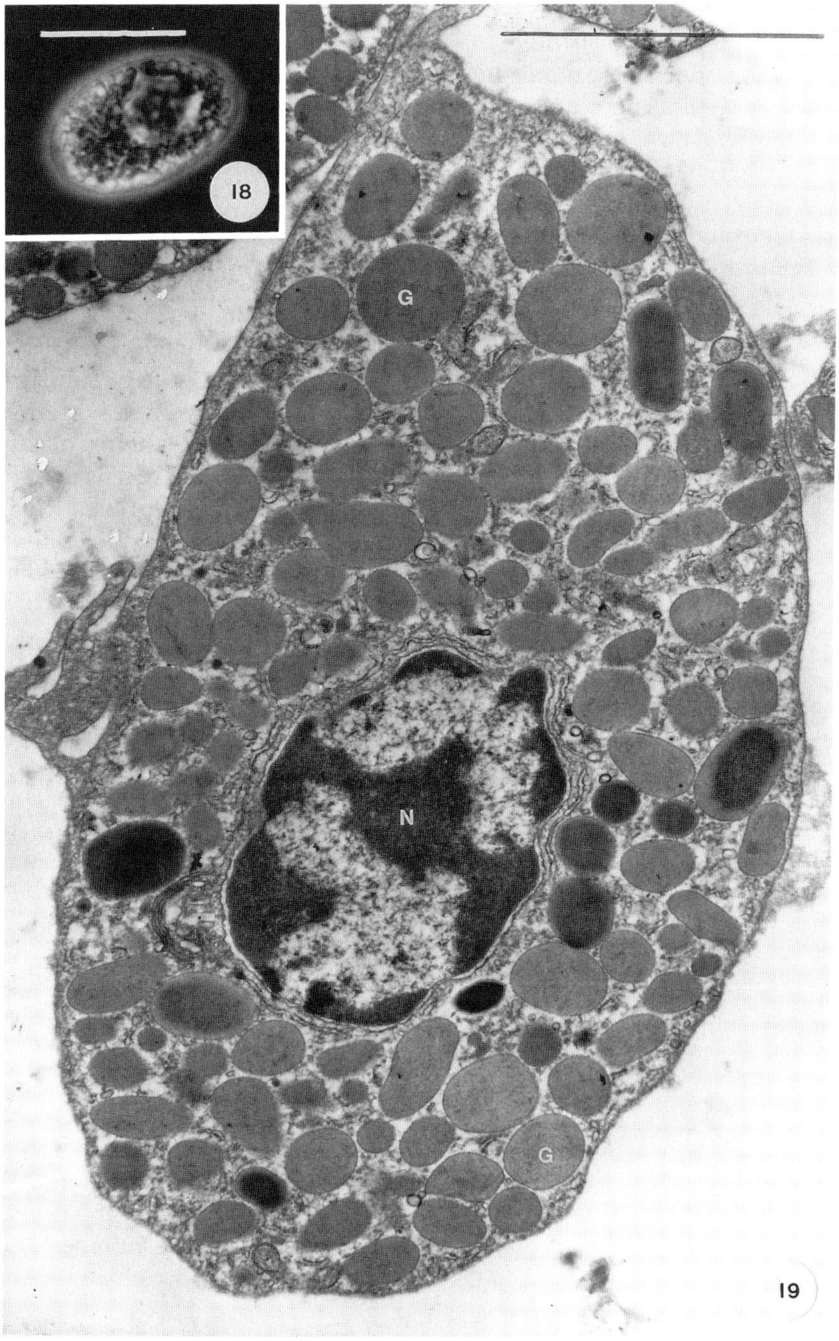

FIGURE 18 Large-granule hemocyte. Note the refractile granules around the circumference of the cell. Phase-contrast photomicrograph of a fixed cell. Scale bar: 15.0 μm.

FIGURE 19 Transmission electron micrograph of a large-granule hemocyte showing the nucleus (N) and cytoplasm filled with granules (G). Scale bar: 5.0 μm.

B. Cytochemical Correlations

Cytochemistry can facilitate identification of the hemocytes. Granulocytes, which are active in phagocytosis and encapsulation, contain numerous reaction sites for lysosomal enzymes such as acid phosphatase and β-glucuronidase. These reaction sites are located primarily in tiny vesicles and the trans-cisternae of Golgi, and less frequently in small granules. About 25% of the small-granule hemocytes of *Homarus americanus* and 50% of the large-granule cells are positive for these reaction sites; since typical percentages are 70–90% for decapods, lobsters have lower figures than any other decapod species studied (Hose *et al.*, 1990). Small-granule hemocytes have more reactive sites than do large-granule cells. Hyaline hemocytes rarely contain lysosomal enzymes (0–3%).

Encapsulation of large foreign bodies involves prophenoloxidase (proPO), an enzyme present in both plasma and granulocytes. ProPO is located within the granules of most large-granule hemocytes (>70% of cells), is less abundant in small-granule hemocytes (5–14%), and is absent from hyaline hemocytes.

Hyaline hemocytes, by virtue of their involvement in hemolymph coagulation, probably contain transglutaminase or other clotting factors that cross-link plasma coagulogen. When hyaline cells lyse, the tiny cytoplasmic deposits give rise to long, filamentous strands, which seem to be the focus for clot initiation. The chemical identity of these deposits remains unclear, but likely is a lipoprotein clotting factor since they stain with Sudan black B.

C. Hemocyte Function

1. Clotting

Hemocyte agglutination and hemolymph coagulation both contribute to hemolymph clotting in *Homarus americanus* (Tait type B coagulation, with an intermediate percentage of lysing hyaline hemocytes; Tait, 1911) (Fig. 20).

Upon an initiating event, lobster hyaline hemocytes undergo a series of rapid morphological changes, culminating in cytolysis and release of cytoplasmic factors that trigger coagulation of the hemolymph. These are the "explosive corpuscles" described by Hardy (1892). The earliest alterations observed are aggregation of the tiny cytoplasmic deposits (Fig. 15) and bleb formation. Granules become concentrated around the nuclear envelope. The plasma membranes surrounding the blebs then rupture, releasing the filamentous deposit strands (Figs. 20–22) and disrupted organelles until all the cytoplasm is gone. Within a few minutes, only an electron-lucent nucleus remains (Fig. 21), highlighting the difficulty early researchers encountered when attempting to define the hemocyte type that initiates clotting. The filamentous strands expand to form typical clot material, present as a sphere around each hyaline hemocyte; the expansion pushes outward on the surrounding granulocytes, forcing them into a compacted mass around the periphery (Fig. 20). The strands may be the sites where a transglutaminase causes cross-linking of plasma coagulogen to form a gel (Fuller and Doolittle, 1971; Ghidalia *et al.*, 1981; Lorand and Conrad, 1984).

Granulocytes, which are morphologically unchanged during hyaline cell lysis, agglutinate in the gelled hemolymph to form a seal at the site of injury.

2. Exoskeleton Hardening

Little is known about the role of hemocytes in exoskeleton hardening in *Homarus americanus*. It is assumed, however, to be similar to the pattern seen in other decapods, as follows. Both types of hemocytes participate in the tanning of the exoskeleton during early postecdysis. There is an increased abundance of circulating hemocytes just after molting and a subsequent movement into the epithelium of the exoskeleton (Kollman, 1908; Vacca and Fingerman, 1983; Tsing *et al.*, 1989; Hose *et al.*, 1992). Large numbers of hemocytes accumulate near the epidermis during ecdysis and early postecdysis and appear to release cytoplasmic constituents that harden the exoskeleton (Vacca and Fingerman, 1983). Granulocytes reportedly predominate prior to ecdysis and exocytose granules containing basic amino acids, contributing to proteins necessary for the formation of the new exoskeleton. Some granulocytes may also contribute cytoplasmic, nonphenolic reducing substances, which are cyclically reactive for basic amino acids (lysine, arginine, and histidine). Hyaline hemocytes are most abundant during postecdysis. They release diphenolic tanning substances (catecholamines such as norepinephrine, DOPA, dopamine, *N*-acetyldopamine, and *N*-acetylnorepinephrine) into the hemolymph at ecdysis, which bind to two large hemolymph proteins (1.5 kDa and >4 kD). Subsequently, these molecules are transported to the cuticle, attach to exoskeleton glucosides, and cross-link the protein matrix. Following sclerotinization, the number of hemocytes declines, both in the hemolymph and in the epidermis.

3. Clearance of Foreign Material

Granulocytes are responsible for the recognition of nonself material, phagocytosis of small particles (in conjunction with the fixed phagocytes), and the encapsulation of particles too large for phagocytosis

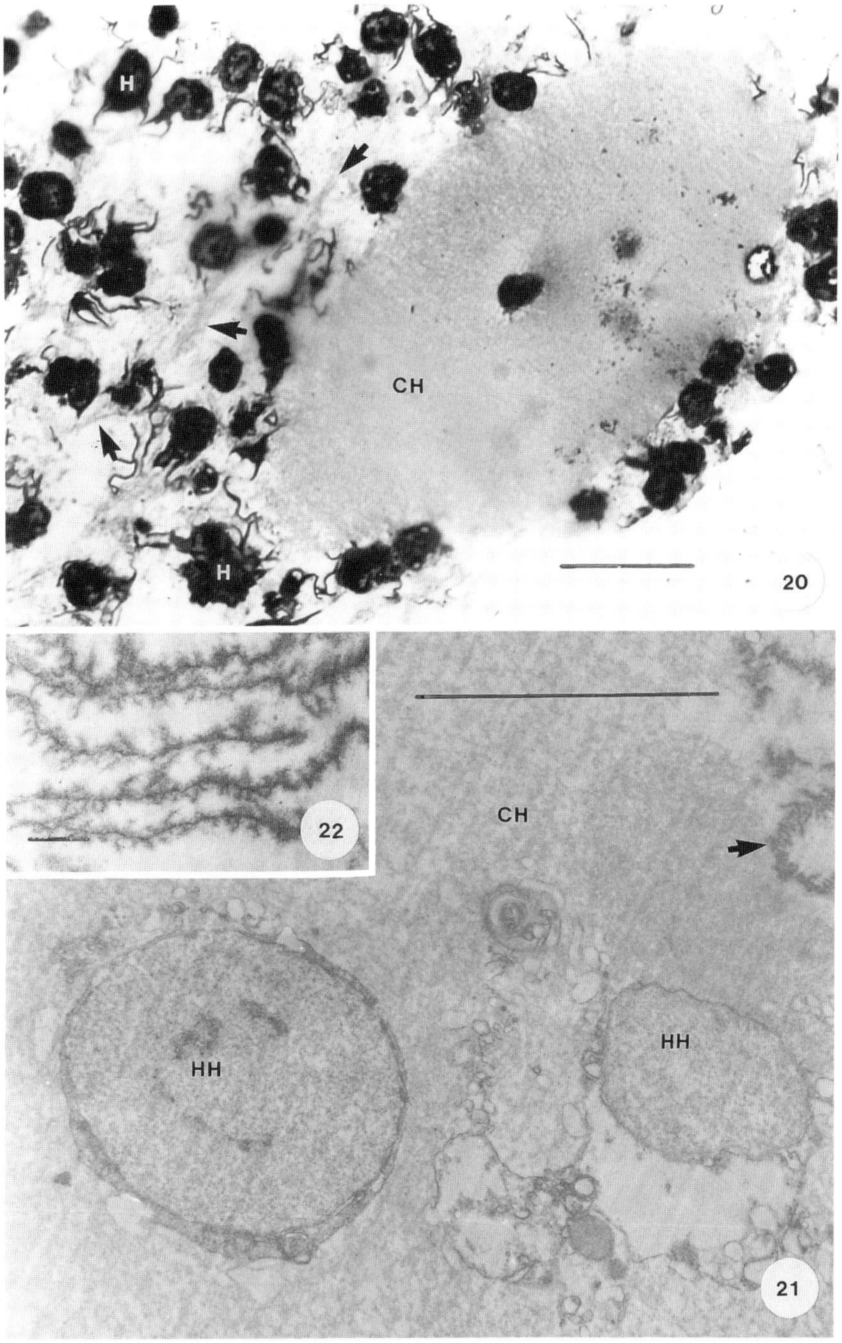

FIGURE 20 Clot showing coagulated hemolymph (CH) surrounded by hemocytes (H) and filamentous strands (arrows). Photomicrograph of live cells. Scale bar: 50.0 μm.

FIGURE 21 Transmission electron micrograph of two lysed hyaline hemocytes (HH) embedded in coagulated hemolymph (CH). Note the filamentous strand (arrow) before it expands to form the coagulated hemolymph. Scale bar: 10.0 μm.

FIGURE 22 Transmission electron micrograph of filamentous strands shown at higher magnification. Scale bar: 10.0 μm.

(Hose *et al.*, 1990). The recognition of foreign material has been studied in other decapods (reviewed by Söderhäll *et al.*, 1988), but the mechanisms in *Homarus americanus* are expected to be similar. Most of the phenoloxidase (PO) exists as a hemolymph protein, but large-granule hemocytes contain the inactivated form (proPO) (Hose *et al.*, 1990). Upon stimulation of crayfish *in vivo* by β-1,3-glucans, endotoxin, or peptidoglycans, or *in vitro* using a calcium ionophore (A23187), degranulation occurs and proPO is released into the hemolymph (Söderhäll *et al.*, 1988). The activated proPO then covers the foreign cells. An as yet unidentified component of the PO cascade acts as an opsonin to enhance phagocytosis.

Phagocytosis of bacteria by hemocytes in *Homarus americanus* has been widely studied and is accomplished both *in vitro* and *in vivo* primarily by small-granule hemocytes and, to a lesser extent, the large-granule hemocytes (Hose *et al.*, 1990). *In vitro* experiments utilizing hemocytes adherent to surfaces (Goldenberg *et al.*, 1984, 1986) show that phagocytic activation is preceded by flattening and spreading (the formation of filopodia and pseudopodia), granular darkening and dispersion, and vacuolation (Fig. 23). During spreading, the marginal bands of microtubules become twisted and subsequently disappear (Cohen *et al.*, 1983). A glycoprotein "adhesion factor" is present in both types of granulocytes which allows adherence to the substrate and spreading (Söderhäll et al., 1988). Hyaline hemocytes, in contrast, never display spreading behavior *in vitro*, instead rapidly lysing in the presence of bacteria or sheep red blood cells (SRBCs) and releasing clotting factors (Goldenberg *et al.*, 1986; Hose *et al.*, 1990). This reaction aids the immobilization of foreign material.

Initial attachment of bacteria and SRBCs to granu-

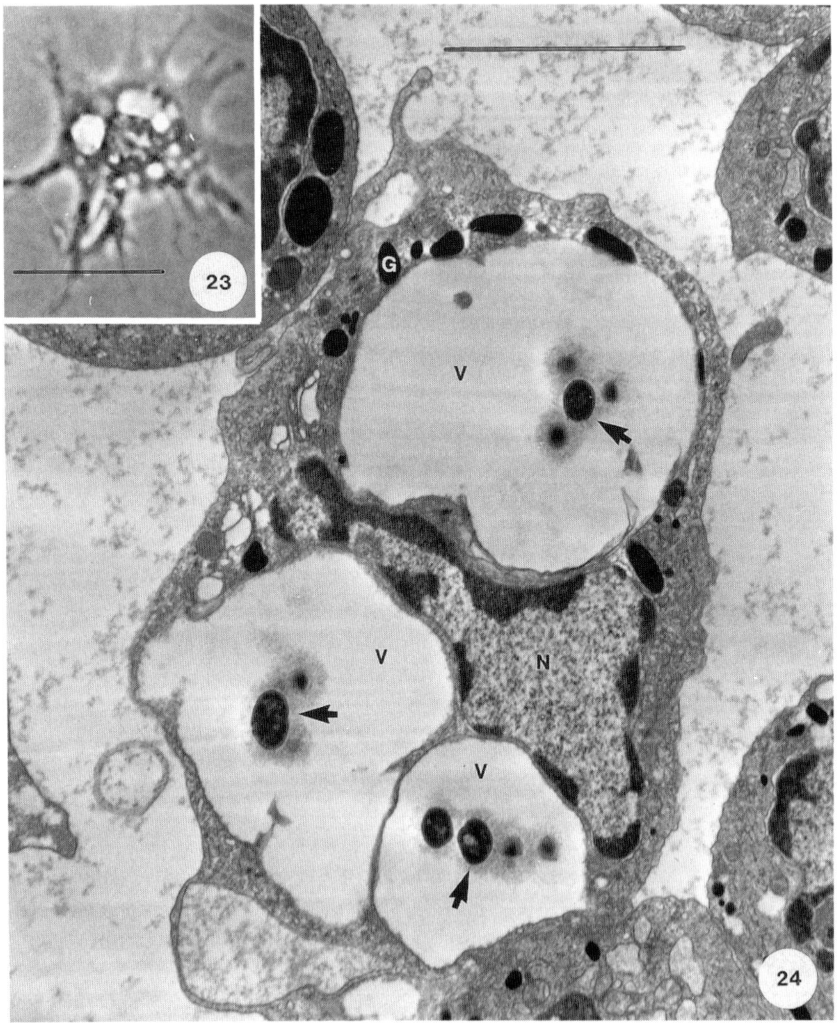

FIGURE 23 Small-granule hemocyte spread on a cover slip. Photomicrograph of a live cell. Scale bar: 15.0 μm.

FIGURE 24 Transmission electron micrograph of a small-granule hemocyte showing three large phagocytic vacuoles (V) containing *Aerococcus viridans* (arrows). G, Granule; N, nucleus. Scale bar: 5.0 μm.

locytes occurs on the plasma membrane of granulocytes, possibly by lectin binding sites (Goldenberg *et al.*, 1984; Söderhäll *et al.*, 1988). The plasma membrane extends around the particle and encloses it within a vacuole (Hose and Martin, 1989). Adjacent phagosomes may fuse, producing large areas containing up to 30 bacteria (Fig. 24). Vesicles or cytoplasmic granules containing lysosomal enzymes then fuse with the phagocytic vacuole and the bacteria are degraded. Hemocytes containing many secondary lysosomes ultimately lyse. The granulocytes also demonstrate cytotoxic responses against certain foreign cells (SRBCs, tumor cells, and non-tumor cells) through mechanisms independent of the PO system (Söderhäll *et al.*, 1988). Although the cell type is unidentified, hemocytes are thought to produce hemolymph agglutinins that facilitate recognition and attachment (Cornick and Stewart, 1973).

In vivo, the introduction of foreign particles is uniformly accompanied by a rapid and profound decrease in the number of both foreign particles and circulating hemocytes. Hemocytes appear to settle out on the lining of the open circulatory system, where they have been observed phagocytosing bacteria adherent to the intima of the hemal spaces of various organs. Foreign particles are also removed from circulation in the digestive gland by fixed phagocytes (FPs) lining the vessels (Figs. 25 and 26) (Johnson, 1987; Factor and Beekman, 1990; Factor and Naar, 1990). Factor (Chapter 15) considers the structure of the FPs of the digestive gland. The physiological importance of the FPs exceeds that of the hemocytes for small particles such as viruses. Johnson (1987) theorized that the FPs of the digestive gland arise from certain undifferentiated hemocytes. Finally, foreign material is removed from circulation in the gills. For larger particles such as bacteria, this may result from physical effects rather than phagocytosis. The podocytes of the gills are not phagocytic cells, but can remove foreign proteins (and perhaps small viruses) by pinocytosis (Johnson, 1987).

Foreign material too large to be phagocytized is sequestered within loose aggregations of hemocytes (nodule formation) or by successive layers of hemocytes (encapsulation). These processes appear to be similarly initiated by the proPO system and granulocyte agglutination (Söderhäll, 1982; Johansson and Söderhäll, 1989a,b). Most of the hemocytes adhering *in vitro* to fungal hyphae are large-granule hemocytes, about 25% are small-granule cells (Hose *et al.*, 1990). As granulocytes continue to adhere, they frequently degranulate, forming cellular capsules in which it is difficult to identify the responsible cell types. Three layers are typically present in decapod capsules: a central zone, usually melanized, of foreign material and lysed hemocytes; a thick layer of flattened hemocytes bound together by junctions; and an outer layer of loosely attached hemocytes retaining their discoid shape (Krol *et al.*, 1989). Initiation of the proPO cascade culminates in melanin production, thus producing the brown–black specks apparent upon dissection of gaffkemic lobsters (Cornick and Stewart, 1968). Some of the melanin intermediates in crayfish are toxic or at least inhibitory to the growth of microorganisms (Söderhäll and Ajaxson, 1982). Released granules also contain a degranulating factor and a cell-adhesion factor promoting encapsulation in crayfish (Smith and Söderhäll, 1983; Kobayashi *et al.*, 1990).

4. Other Functions

Hemocytes are involved in many other physiological processes in a variety of decapods, although the responsible cell types have not yet been identified and the processes have not yet been studied in *Homarus americanus*. The role of hemocytes during wound healing and autotomy has been described (Bittner, 1973; Lumb *et al.*, 1991); exoskeleton regeneration is certainly necessary and participation by both the hyaline hemocytes and granulocytes appears essential. The involvement of melanin and the proPO system in wound healing in a crab implicates granulocytes; indeed, "melanin" granules attach to the first-formed repair layer of the cuticle (Halcrow and Smith, 1986). Hemocytes have also been implicated in the repair of skeletal muscle damage in crayfish (Uhrik *et al.*, 1989).

V. Hematopoietic Tissue

A. General Morphology

The hematopoietic tissue (HPT) of *Homarus americanus* is a thin (40- to 800-μm) layer of tissue that is loosely bound to the dorsal surface of the foregut and composed of numerous, ovoid lobules ranging in size from 30 × 20 μm to 180 × 80 μm (Figs. 27–29). Each lobule contains densely packed stem cells and maturing hemocytes in a loose, fibrillar material; each is surrounded by a connective tissue capsule (Fig. 29). The lobules are most abundant over the dorsal surface of the foregut and become gradually replaced by connective tissue and striated muscle fibers toward the anterior and lateral margins of the foregut.

The capsule is composed of an outer, fibrillar layer (0.2 μm thick) and an inner layer of collagen fibers containing an occasional elongate, fibroblastlike cell.

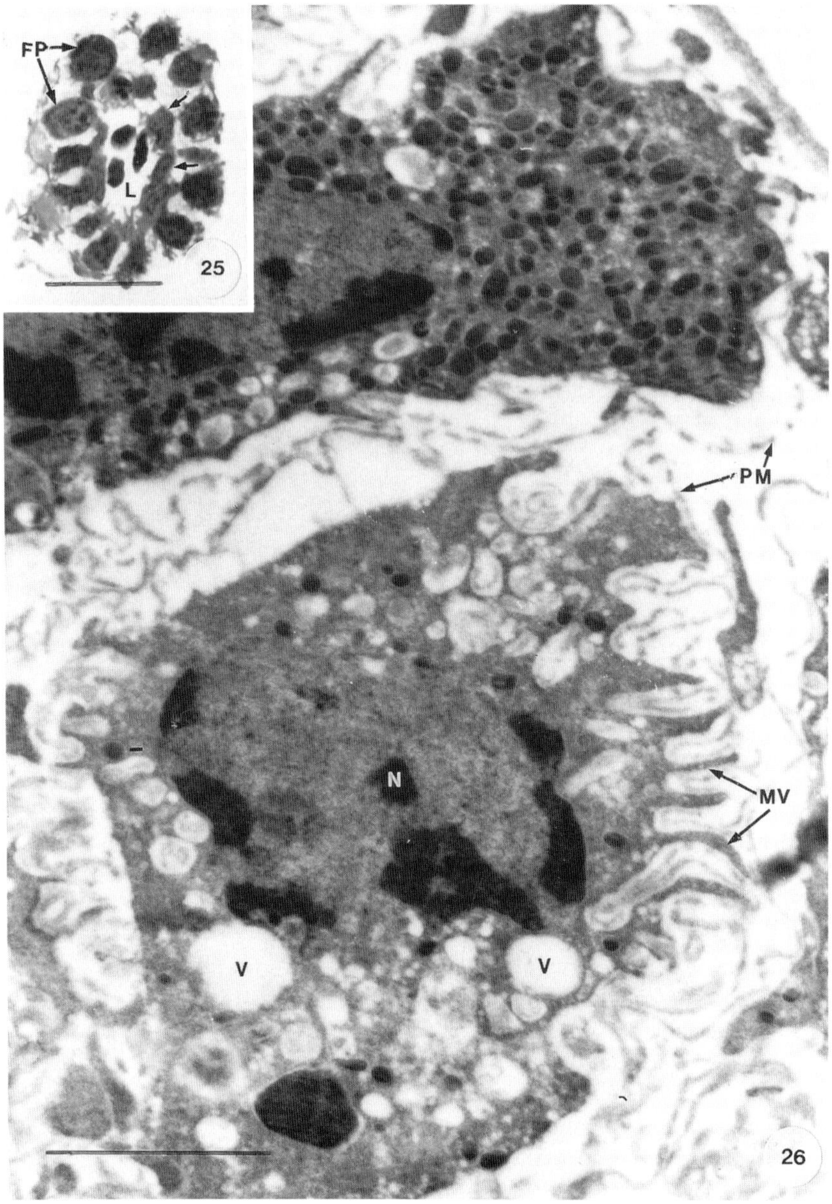

FIGURE 25 Cross-section through a vessel in the digestive gland. Note the three hemocytes in the vessel lumen (L), the endothelial cells (arrows), and the outer layer of fixed phagocytes (FP). Photomicrograph of a plastic section stained with methylene blue. Scale bar: 20.0 μm.

FIGURE 26 Transmission electron micrograph of two fixed phagocytes from a vessel in the digestive gland. The upper cell is filled with granules, whereas the lower one displays cell processes and (presumably) endocytic vacuoles (V). Note the perforated membrane that covers these cells (PM). MV, Microvilli; N, nucleus. Scale bar: 2.5 μm.

In some lobules, the encapsulating connective tissue layer has been seen to rupture, with hemocytes being released into adjacent hemal spaces (Fig. 29) or into the dorsal hemocoel. Diapedesis of hemocytes through the encapsulating connective tissue has never been observed.

B. Stem Cells and Maturing Hemocytes

The proportions of differentiating granulocytes and hyaline hemocytes reflect the percentages present in circulation. Granulocytes form the bulk of the intermolt hematopoietic tissue (mean, 90%), with

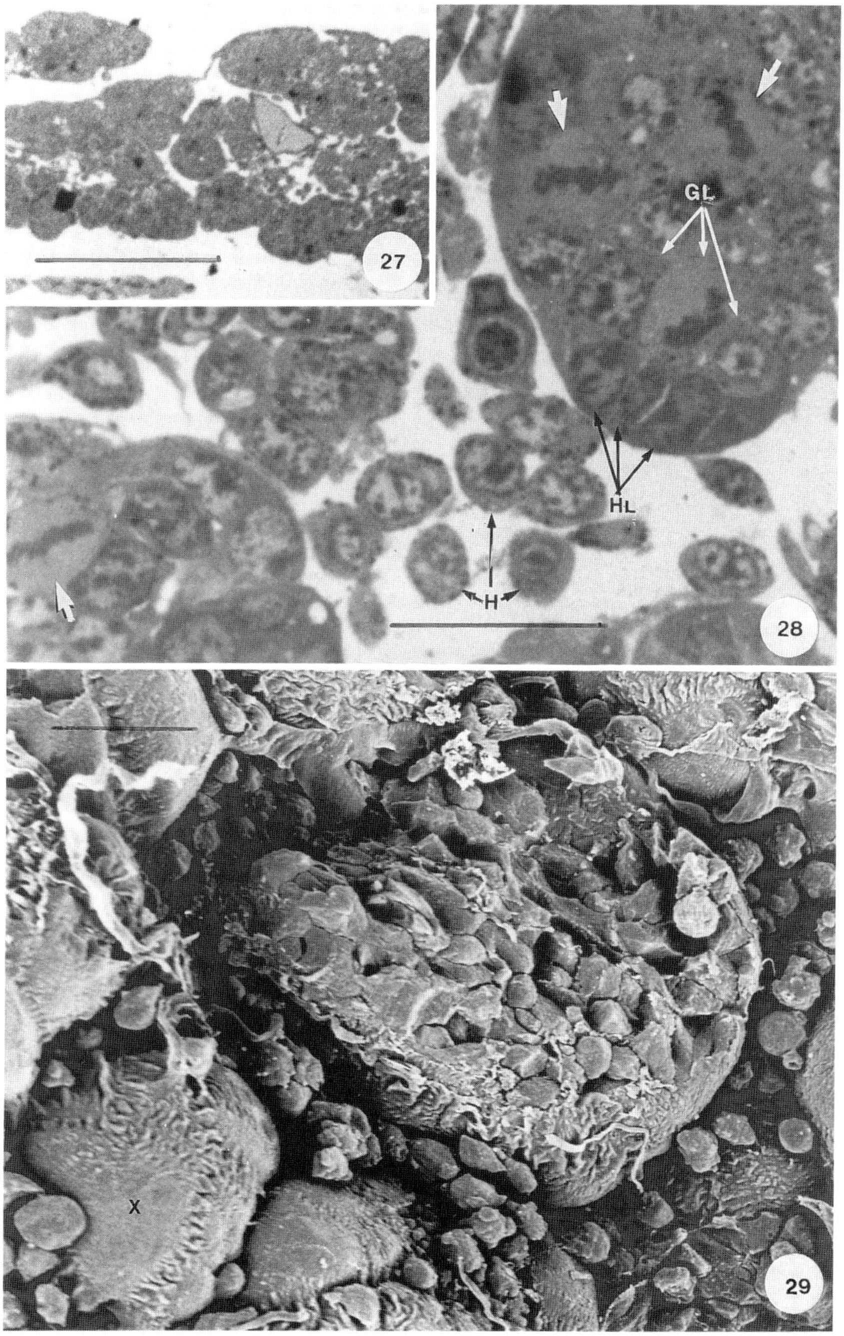

FIGURE 27 Cross-section through the hematopoietic tissue showing the aggregation of lobules. Photomicrograph of a plastic section stained with methylene blue. Scale bar: 250 μm.

FIGURE 28 Hemocytes (H) in hemolymph between two nodules of the hematopoietic tissue. Note the mitotic cells (short arrows), and stem and maturing cells of the hyaline (HL) and granulocyte (GL) lineages. Photomicrograph of a plastic section stained with methylene blue. Scale bar: 50.0 μm.

FIGURE 29 Scanning electron micrograph of hematopoietic tissue. Note that the small nodule in the lower left (X) is covered by a fibrous layer. The large nodule in the center lacks an investing fibrous coat and appears to be releasing hemocytes. Scale bar: 2.0 μm.

hyaline cells averaging 10% (range, 1–35%). Rarely, cells resembling small RI cells (Johnson, 1980) are observed. Although some lobules are composed solely of granulocytes, both cell types are usually present as small, interspersed clusters. Frequently, a gradient of maturation can be observed, with stem cells at one end of the lobule, immature hemocytes intermediate in position, and detaching hemocytes grouped at the opposite end.

Cells of the hyaline line can be differentiated from granulocytes by their darkly staining cytoplasm. The youngest hyaline stem cells are elliptical, averaging 13 × 9 μm (Fig. 30). The oval nucleus is composed primarily of heterochromatin and is surrounded by a thin band of dark blue cytoplasm (electron dense at the TEM level) containing a few small (0.6-μm), ovoid, electron-dense granules. The tiny, round, cytoplasmic deposits and striated granules characteristic of shrimp hyaline stem cells are not present. Hyaline hemocytes (Fig. 31 and Table 1) are slightly smaller than hyaline stem cells (11 × 6 μm) and range as a continuum from stem cells to cells indistinguishable from circulating hyaline hemocytes. During maturation, the cell size decreases slightly and the N:C ratio generally increases. The number of cytoplasmic granules increases from the zero to three per section through the center of each stem cell to the 14 or 15 present in mature cells within the hemal spaces and in circulating hyaline hemocytes.

Maturing granulocytes can be grouped into a developmental series of four distinct cell types, although intermediates are observed. The earliest is the granulocyte stem cell, followed by the small-granule stem cell, the small-granule hemocyte, and finally the large-granule hemocyte. Cells within the HPT, but mostly detached from adjacent cells, are termed hemocytes and may be released into circulation. Increasing numbers of cytoplasmic granules and a decreasing N:C ratio mark the progressive maturation of the small-granule stem cell into the large-granule hemocyte. Small-granule hemocytes can mature into large-granule hemocytes in the HPT, within the dorsal hemocoel, and in circulation.

Granulocyte stem cells (Fig. 32) are recognizable by their large size (15 × 10 μm) and elliptical nucleus with one to three prominent (2.5-μm) nucleoli. The cytoplasm contains numerous mitochondria, abundant rough endoplasmic reticulum, and up to two small, ovoid, electron-lucent cytoplasmic granules. Small-granule stem cells (Fig. 33) are smaller (12 × 9 μm; Table 1), with many (three to 20 per cross-section) small (< 1.0-μm), ovoid–cylindrical cytoplasmic granules. In addition to granules with a fibrillar substructure, the cylindrical granules are composed of a homogeneous, electron-dense material. The nucleus contains a thick band of marginal chromatin and usually a single nucleolus (1.8 μm). Small-granule stem cells are usually the most frequent category observed, with a mean of 53% (range, 32–80%).

Small-granule hemocytes (Fig. 34) are detached from the adjacent cells and resemble their circulating counterparts in size (10 × 8 μm), nuclear characteristics, and granule number. Large-granule hemocytes (Fig. 35) are infrequently present, but are distinguished by their larger (1.3- to 2.0-μm), cylindrical, electron-dense cytoplasmic granules; numerous small granules; lower N:C ratio; larger size (12 × 8 μm); and more eccentrically placed nucleus.

A pluripotential cell with features common to the hyaline cell and the granulocyte has not been identified in decapod HPT. The least differentiated hyaline and granulocyte stem cells both lack granules, but the cytoplasm of the former is always more electron dense and has distinctive chromatin patterns. The inability to identify a common precursor lends support to the idea that there are two distinct lines of hemocyte maturation in the decapods (Hose *et al.*, 1992; Martin *et al.*, 1993; Clare and Lumb, 1995).

Decapod RI cells (Fig. 36) store and probably also manufacture hemocyanin, which is contained within the cytoplasm and is identified as a crystalline lattice with a band periodicity of 11 nm (Ghiretti-Magaldi *et al.*, 1977). The precursors of RI cells (also called cyanoblasts) are thought by several investigators to arise from hemocytes because of their location in the adjacent connective tissue near the hematopoietic tissue (Ghiretti-Magaldi *et al.*, 1977; G. G. Martin and J. E. Hose, unpublished observations). In *Homarus americanus*, RI cells are present throughout the spongy connective tissue, in the myocardium, and lining all hemal spaces. They are never seen free in the hemolymph. Cells resembling immature RI cells (Fig. 37) are uncommon, but were observed in hemal spaces adjacent to the HPT in one apparently healthy intermolt lobster. Although the putative cyanoblasts have the darkly staining cytoplasm of hyaline cells, they are enormous (25-μm diameter) compared to hemocytes and most hema-topoietic cells. These cells have a high N:C ratio with a single large, central nucleolus. However, the typical cytoplasmic inclusions of RI cells are not visible using light microscopy or TEM. Because of the lack of supportive experimental evidence that RI cells arise in the HPT or are derived from hemocytes, the genesis of RI cells is not yet decided. Further complicating the discussion of the RI cells of *H. americanus* are the large cells with

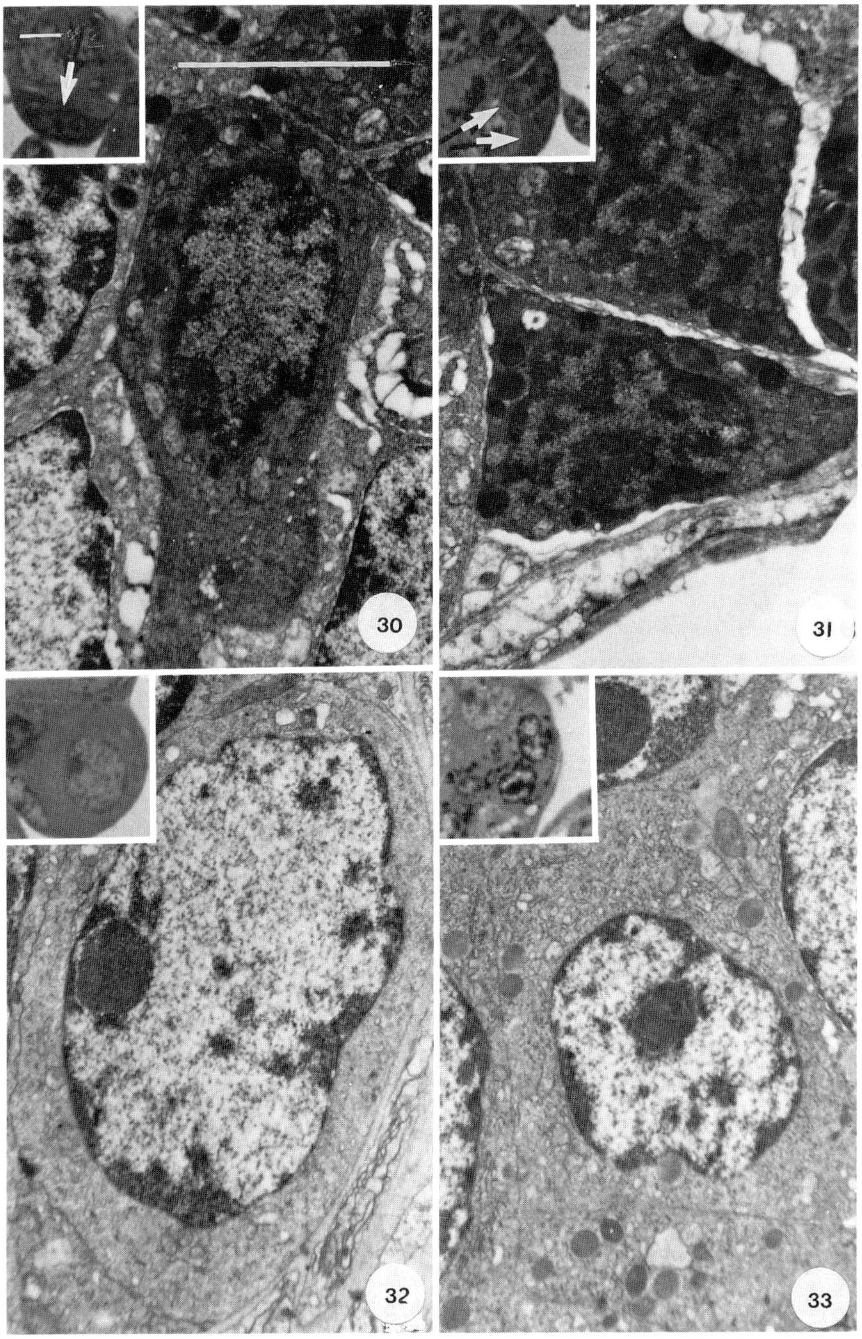

FIGURES 30–35 Tranmission electron micrographs of stem and maturing hemocytes from the hematopoietic tissue. Insets are photomicrographs of plastic sections of the same cell types, stained with methylene blue. All scale bars: 5.0 μm. *FIGURE 30* Hyaline stem cell. *FIGURE 31* Hyaline hemocyte. *FIGURE 32* Granulocyte stem cell. *FIGURE 33* Small-granule stem cell. *FIGURE 34* Small-granule hemocyte. *FIGURE 35* Large-granule hemocyte. The darker cytoplasmic staining of the hyaline cells compared to that of the granulocytes is a key differentiating characteristic. Stem cells contain fewer granules than do the hemocytes, which can also be identified by the detachment of their plasma membranes from the surrounding cells.

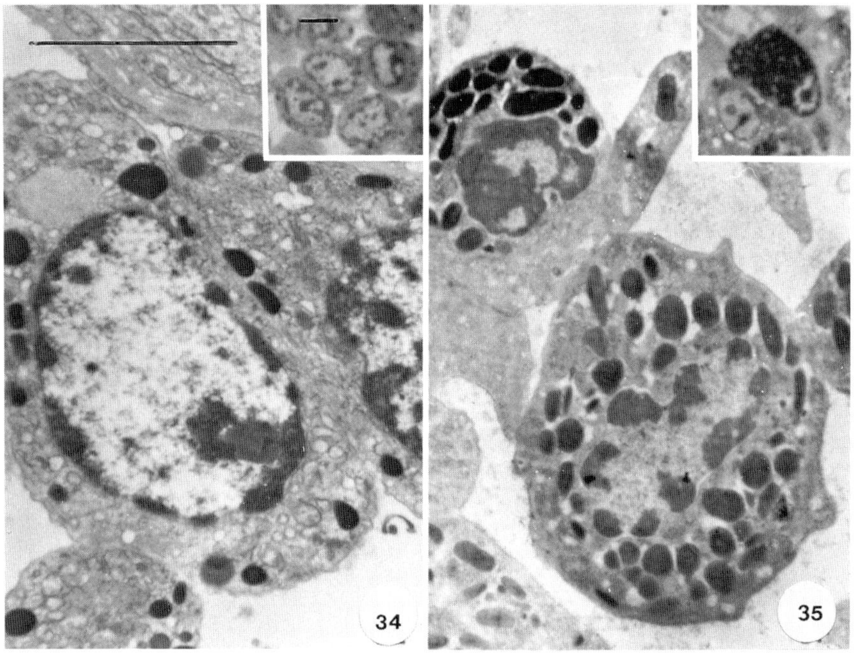

FIGURES 30–35 Continued

neutrophilic or basophilic granules observed by Johnson (1980) within the hepatopancreas and similar cells with basophilic granules in the midgut and the anterior midgut caeca. The identity of these cells remains unresolved, although Johnson suggests that they may be related to the circulating basophilic or chromophobic granulocytes described by Cornick and Stewart (1978).

Notably lacking from the literature on *Homarus americanus*, indeed for all of the Decapoda, are investigations on hematopoiesis using modern cell culture and molecular probe techniques. Probably the most fruitful investigations will involve situations in which hematopoiesis is stimulated, such as during late stages of gaffkemia (Johnson *et al.*, 1981), or as the animal prepares to molt, progressing from molt stage C (Hose *et al.*, 1992).

C. Mitotic Index

The average mitotic rate of stem cells in intermolt lobsters is 5.1%, ranging from 0.7 to 15.8% (Martin *et al.*, 1993). Hyaline stem cells, granulocyte and small-granule stem cells, and, rarely, small-granule hemocytes are capable of division. The preponderance of the dividing cells are cells of the granulocyte line (90%), mostly small-granule stem cells (80%). Hyaline cells are generally uncommon in intermolt hematopoietic tissue and they exhibit low mitotic rates.

In crustaceans, the molt cycle controls hemocyte production and release (Marrec, 1944; Johnson, 1980; Tsing *et al.*, 1989; Hose *et al.*, 1992). Granulopoiesis is maximal during the intermolt phase, while hyaline cell production is greatest during ecdysis. This synchrony reflects the integral role played by hyaline hemocytes during molting (Vacca and Fingerman, 1983). Regulation of hematopoiesis by the extracellular matrix of the HPT is important in vertebrates (Zuckerman and Rhodes, 1985; Zipori, 1988), but has not been studied in decapods.

VI. Disease

Adult lobsters are relatively resistant to disease, with the exceptions of gaffkemia, a bacterial blood disease, and shell disease, erosion of the exoskeleton by bacteria and fungi. Unlike the case with other decapods, there have been no reported viral epizootics in *Homarus americanus*. Pathogenic protozoans, fungi, and metazoan parasites are also rare in adults (reviewed by Fisher *et al.*, 1978), although several species of eukaryotic parasites have been reported (Boghen, 1978; Brattey and Campbell, 1985).

Disease resistance is related to the effective cellular and humoral clearance mechanisms of the lobster. In all decapods, the exoskeleton is the first barrier to infection; it contains PO and an inhibitor against proteolytic enzymes (Söderhäll *et al.*, 1988). Upon entry of pathogens, a clot rapidly forms in healthy animals which prevents loss of hemolymph, seals the wound, and immobilizes microorganisms (see Section IV,C,1).

Humoral factors with antimicrobial action include agglutinins, bactericidins, and a protease inhibitor such as α_2-macroglobulin (see Section III,B,3). The existence of several additional antimicrobial factors can be theorized based on their presence in insect hemolymph (Söderhäll *et al.*, 1988). Cellular clearance mechanisms of foreign material include hemocyte responses (phagocytosis, nodule formation, and encapsulation), pinocytosis by podocytes in the gills and the antennal glands, and, most importantly, removal by the fixed phagocytes (see Section IV,C,3) (Figs. 25 and 26), which line the vessels of the digestive gland. The fixed phagocytes of decapods effectively sequester and phagocytose many pathogens. A notable exception is the bacterium causing gaffkemia, which is resistant to degradation and probably multiplies within the fixed phagocytes of the lobster (Johnson, 1987).

A. Gaffkemia

Heavy mortalities in lobsters result from infection with virulent strains of the gram-positive bacterium *Aerococcus viridans* var. *homari* (formerly *Gaffkya homari*). Because the bacterium does not contain

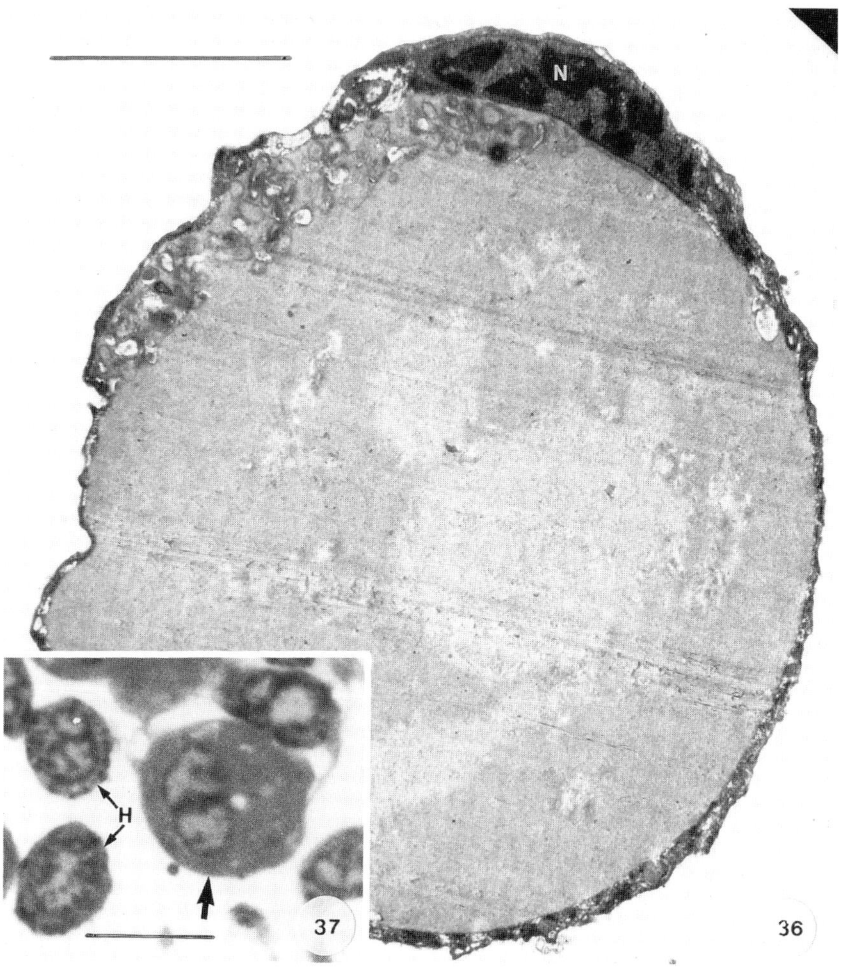

FIGURE 36 Transmission electron micrograph of reserve inclusion (RI) cells from the heart. Note the nucleus (N) and the thin rim of cytoplasm surrounding the large central space. Scale bar: 5.0 μm.

FIGURE 37 Unidentified cell type (thick arrow) observed in hemal spaces adjacent to hematopoietic tissue. Note the large size of the cell and nucleus as compared to adjacent hemocytes (H). Photomicrograph of a plastic section stained with methylene blue. Scale bar: 18.0 μm.

exoenzymes, it can only enter through breaks in the integument (Stewart *et al.*, 1969). A surprisingly low number of bacteria (five) is sufficient to cause death in 90% of infected lobsters within 17 days. This bacterium is "unique in its ability to overcome the otherwise highly effective intrinsic defense mechanisms of the lobster" (Cornick and Stewart, 1968). *In vitro*, avirulent strains are effectively phagocytized; the phagocytosis rate for virulent strains is reduced by half, with fewer bacteria ingested per hemocyte (Schapiro *et al.*, 1977). Virulent strains continue to divide after phagocytosis, within the hemocytes, and are not killed within the phagosomes (Fig. 24); presumably, this is because of the acidic polysaccharide capsule, which is absent from avirulent strains (Cornick and Stewart, 1968; Paterson and Stewart, 1974; Johnson *et al.*, 1981). Two other factors enhance the pathogenicity of *A. viridans*: resistance to agglutinins and stimulation (rather than inhibition) of growth by hemolymph serum (Cornick and Stewart, 1968).

The clotting ability of the hemolymph is also lost in infected animals and wounded lobsters may exsanguinate. Despite the premature release of differentiating hemocytes from the HPT (Johnson *et al.*, 1981), the THC declines by almost 90%. Granulocytes are preferentially lost until late in the course of infection; after 5 days, only hyaline hemocytes remain (Stewart *et al.*, 1983). It has been hypothesized that small-granule hemocytes (eosinophils) contribute to the enhanced phagocytic capability following administration of vaccines. Utilization of the lobster's glycogen and ATP reserves by the bacterium lead to hepatopancreatic dysfunction and death (Stewart and Arie, 1973).

Recently, some progress has been made in the development of an effective vaccine against gaffkemia (Keith *et al.*, 1992; Paterson and Keith, 1992). Injection of an inactivated *Aerococcus viridans* bacterium confers protection against disease beginning 6 and lasting until 90 days after administration. The percentage of phagocytic hemocytes and the number of ingested bacteria increase concomitant with the induction of immunity (Paterson and Stewart, 1979). The response lacks specificity, however, since phagocytosis of SRBCs is also enhanced in vaccinated animals. Because of the week-long latent period and extended duration, immunity could result from production of new phagocytes. Injection of the vaccine also delays molting, a beneficial effect for the commercial fishery (Keith *et al.*, 1992). Although immersion delivery confers less disease protection than does injection, both methods result in enhanced survival of lobsters in the field.

B. Shell Disease

Erosion of the exoskeleton (shell disease) is common throughout the crustaceans, grossly appearing as pitting, melanization, and necrotic lesions of the claws, tail, and carapace (Rosen, 1970; Sindermann, 1991). Because higher prevalences of shell disease are found in captive lobsters than in wild populations, environmental stressors, such as low oxygen concentrations, extreme temperatures, and poor water quality, are thought to be contributory factors (Getchell, 1991). The involvement of environmental pollutants has been hypothesized (Sindermann, 1991). In natural populations, shell disease is more prevalent and most severe in large animals, which molt infrequently, and in ovigerous females. The cuticular lesions are caused by lipolytic and chitinoclastic microbes, primarily bacteria of the genera *Pseudomonas*, *Vibrio*, and *Aeromonas*, which are normal shell epibionts (Estrella, 1991). Sindermann (1991) has theorized that "shell disease is an external indication of metabolic disturbance or trauma which results in failure of an important defense mechanism—chitin deposition—to keep pace with activities of chitin degrading microorganisms."

Although wounding frequently precedes shell lesions, a single initiating factor in shell disease still has not been defined, since large lesions may result without disruption of the epicuticle (Sindermann, 1991). Chitinolytic microorganisms may be introduced by wounding or mechanical abrasion and degrade the chitinous layers of the shell; chitin destruction can occur without disruption of the epicuticle. Lipolytic bacteria may also initiate shell disease by breaking down the lipoidal epicuticle. Affected areas of the shell lack calcium and phosphorus (Bayer *et al.*, 1989) and hemocytic involvement is minimal (Noga, 1991). Melanization is typically found at the site of infection; in extreme cases, portions of the exoskeleton are absent (Sindermann, 1991). Large lesions can cause death during ecdysis due to adhesions between the exoskeleton and underlying tissues.

C. Other Microbial Diseases

Unlike the situation with other decapods, systemic bacterial infections caused by gram-negative bacteria are rare in *Homarus americanus*. The sole report, despite years of commercial holding and rearing, is of a *Vibrio*-like bacterium infecting juvenile lobsters subjected to nutritional stress (Bowser *et al.*, 1980). Recently, a ciliate (*Mugardia*, =*Anophrys*) has been implicated in sporadic mortalities of adults in holding pounds; the ciliate was recovered from the hemolymph and the hepatopancreas and caused damage to the intestine (Sherburne and Bean, 1991). Only one serious epizootic of the shrimp pathogen *Fusarium solani* has been reported in juvenile and adult lobsters (Lightner and Fontaine, 1975). The fungus causes a chronic, necrotizing condition in the gills and/or exoskeleton and may expand into adjacent tissues. Death results from tissue destruction, fungal toxins, or secondary bacterial infections. Encapsulated hyphae, some melanized, are prominent histologically.

D. Diseases of Eggs and Larvae

Microbial fouling with primarily filamentous epibionts occurs on the surfaces of the eggs and larvae of *Homarus americanus*, and heavy infestations on the gills can cause mortalities in young juveniles as well (Nilson *et al.*, 1976; Fisher *et al.*, 1978). The filamentous forms involved are typically cyanophytes (*Oscillatoria* and *Anabaena*), protozoans (*Vorticella*), and a bacterium (*Leucothrix mucor*). Nonfilamentous bacteria and diatoms may also be involved.

Two fungi, *Lagenidium callinectes* and *Haliphthoros milfordensis*, infect crustacean eggs and larvae, including those of *Homarus americanus* (Fisher *et al.*, 1976, 1978). The fungi spread over the surface of the egg mass, then slowly penetrate inside. Infected eggs do not hatch. Larval infections spread more rapidly and may be accelerated by the presence of epibionts. The fungi then grow throughout the tissues. While no host defense is apparent in *L. callinectes*-infected larvae, *H. milfordensis* mycelia are heavily melanized (Fisher *et al.*, 1978). Death occurs from tissue destruction or, in *H. milfordensis*-infected larvae, during molting from adhesions formed between inflammatory cells and adjacent tissues.

VII. Directions for Further Research

Surprisingly little original work has been performed on *Homarus americanus*, given its experimental advantages over smaller, less easily maintained decapods. Its size makes it amenable to physiological studies requiring repetitive bleedings and biochemical studies in which large volumes of hemolymph are necessary. *Homarus americanus* is particularly attractive for studying hemocytes and their maturation. The lobster holds promise as a model for continuing research in the following areas of crustacean circulation, hematology, and mechanisms against disease.

1. Physiology of open circulatory systems.
2. Characterization of the materials lining all surfaces exposed to the hemolymph; this topic may be central to recognition of nonself material by hemocytes and other phagocytes.
3. Existence of inducible defense proteins in decapods similar to the defensins and cecropins of insects.
4. Determination of the health status of individuals and populations using hemolymph measurements.
5. Development of culturing techniques for hemocytes and HPT, as initiated by Brody and Chang (1989). *In vitro* systems would facilitate (a) verification of the hemocyte maturation scheme presented here; (b) elucidation of factors influencing hematopoiesis; (c) comparison of roles of the crustacean and vertebrate extracellular matrix in hematopoiesis; and (d) study of the effects of environmental pollutants on the production and activity of hemocytes.
6. Mechanisms of immune modulation by vaccines against *Aerococcus viridans*.

The use of molecular and immunochemical techniques to answer these questions should yield information applicable to many areas of invertebrate biology, as well as to the biology of *Homarus americanus*.

VIII. Summary

The blood of *Homarus americanus* performs two essential functions: transport of gas, nutrients, and metabolites, and defense against foreign materials. Blood leaves the dorsal heart through several large-diameter vessels that branch, become progressively narrower, and distribute the blood throughout the body. In most cases, the vessels terminate and the blood flows into irregularly shaped sinuses, where gas and nutrient exchange takes place. From the sinuses, the blood flows through the gills and back to the heart. The blood contains two types of circulating cells: hyaline cells, which lyse readily to initiate coagulation of the hemolymph; and granulocytes, which phagocytose or encapsulate foreign materials. Blood cells are produced in a thin layer of tissue that lies on top of the foregut. The tissue is composed of a series of ovoid lobules with an average mitotic rate of 5%. The blood also contains hemocyanin (respiratory pigment that is not packaged inside circulating cells), coagulogen (clotting protein analogous to fibrinogen), and a variety of defensive proteins (including agglutinins, bacteriocidins and bacteriolysins, and proPO). Proteins that agglutinate foreign cells presumably make them easier to phagocytose, while bacteriocidins and lysins prevent the rapid growth of bacteria. ProPO is a major enzyme in the recognition and elimination of foreign materials. It is activated by cell wall molecules of fungi as well as gram-positive and -negative bacteria. The activity of hemocytes, defensive proteins, and FPs in the gills and digestive gland provides an effective system to fight infection and disease. The major diseases of the lobster include shell disease and gaffkemia. Shell disease is caused by chitin-degrading microorganisms that result in the erosion of the exoskeleton. Gaffkemia is a lethal disease caused by the bacterium *Aerococcus viridans*. Because of its importance in aquaculture and its large size, *H. americanus* is an excellent system for further physiological, cellular, and molecular studies on circulation, immune function, and diseases in Crustacea.

Acknowledgments

We wish to thank our students for their help with our contribution to the work on *Homarus americanus*. Special thanks to Maria Pereyra for drawing Fig. 1 and for support from National Science Foundation grant DCB-8502150 (to G.G.M. and J.E.H.).

References

Acton, R. T., Weinheimer, P. F., and Evan, E. E. (1969). A bactericidal system in the lobster *Homarus americanus*. *J. Invertebr. Pathol.* 13, 463–464.

Ashurst, D. (1982). The structure and development of insect connective tissue. *In* "Insect Ultrastructure" (R. King and H. Akai, eds.), Vol. 2, pp. 313–350. Plenum, New York.

Bang, F. (1970). Cellular aspects of blood clotting in the seastar and the hermit crab. *J. Reticuloendothel. Soc.* 7, 161–172.

Battelle, B. A., and Kravitz, E. A. (1978). Targets of octopamine action in the lobster: Cyclic nucleotide changes and physiological effects in hemolymph, heart and exoskeletal muscle. *J. Pharmacol. Exp. Ther.* 205, 438–448.

Baumann, H. (1921). Das Gefäßsystem von *Astacus fluviatilis* (*Potamobus astacus* L.). *Z. Wiss. Zool.* 118, 246–312.

Bayer, R. C., Prince, D. L., Waltz, C. D., Corey, A. R., and Getchell, R. G. (1989). Scanning electron microscopy and X-ray analysis of shell disease lesions in the American lobster. *J. Shellfish Res.* 8, 481–482.

Bittner, G. D. (1973). Degeneration and regeneration in crustacean neuromuscular systems. *Am. Zool.* 13, 379–408.

Bloom, W., and Fawcett, D. W. (1975). "A Textbook of Histology." Saunders, Philadelphia.

Boghen, A. D. (1978). A parasitological survey of the American lobster *Homarus americanus* from the Northumberland Strait, southern Gulf of St. Lawrence. *Can. J. Zool.* 56, 2460–2462.

Bowser, P. R., Rosemark, R., and Reiner, C. R. (1980). A preliminary report of vibriosis in cultured American lobsters, *Homarus americanus*. *J. Invertebr. Pathol.* 37, 80–85.

Brattey, J., and Campbell, A. (1985). A survey of parasites of the American lobster, *Homarus americanus* (Crustacea: Decapoda), from the Canadian Maritimes. *Can. J. Zool.* 64, 1998–2003.

Brody, M. D., and Chang, E. S. (1989). Development and utilization of crustacean long-term primary cell cultures: Ecdysteroid effects *in vitro*. *Invertebr. Reprod. Dev.* 16, 141–147.

Bullock, T. H., and Horridge, G. A. (1965). "Structure and Function in the Nervous Systems of Invertebrates." Freeman, San Francisco.

Burger, J. W., and Smythe, C. (1953). The general form of circulation in the lobster, *Homarus*. *J. Cell. Comp. Physiol.* 42, 369–383.

Burnett, B. R. (1975). "Structure of the Eumalacostracan Arterial Wall," M.A. thesis. University of California, San Diego.

Campbell, P. A., Hartman, A. L., and Abel, C. A. (1982). Stimulation of B cells but not T cells by sialic acid-specific lectin. *Immunology* 45, 155–162.

Clare, A. S., and Lumb, G. (1995). Haemocyte development in the blue crab, *Callinectes sapidus* Rathbun. *Mar. Biol.* (in press).

Cohen, W. D., Nemhauser, I., and Cohen, M. F. (1983). Marginal bands of lobster blood cells: Disappearance associated with changes in cell morphology. *Biol. Bull. (Woods Hole, Mass.)* 164, 50–61.

Cooke, I. M., and Sullivan, R. E. (1982). Hormones and neurosecretion. *In* "The Biology of Crustacea" (D. E. Bliss, ed.), Vol. 3, "Neurobiology: Structure and Function" (H. L. Atwood and D. C. Sandeman, eds.), pp. 205–290. Academic Press, New York.

Cornick, J. W., and Stewart, J. E. (1968). Interaction of the pathogen *Gaffkya homari* with natural defense mechanisms of *Homarus americanus*. *J. Fish. Res. Board Can.* 25, 695–709.

Cornick, J. W., and Stewart, J. E. (1973). Partial characterization of a natural agglutinin in the hemolymph of the lobster, *Homarus americanus*. *J. Invertebr. Pathol.* 21, 255–262.

Cornick, J. W., and Stewart, J. E. (1978). Lobster (*Homarus americanus*) hemocytes: Classification, differential counts and associated agglutinin activity. *J. Invertebr. Pathol.* 31, 194–203.

Cuénot, L. (1903). Études physiologiques sur les Crustacés Décapodes. *Arch. Biol. Lieges* 13, 245–303.

Doolittle, R. F., and Fuller, G. M. (1972). Sodium dodecyl sulfate-polyacrylamide gel electrophoresis studies on lobster fibrinogen and fibrin. *Biochim. Biophys. Acta* 263, 805–809.

Durliat, M. (1985). Clotting processes in Crustacea Decapoda. *Biol. Bull. (Woods Hole, Mass.)* 60, 473–498.

Durliat, M., and Vranckx, R. (1981). Action of various anticoagulants on hemoymphs of lobsters and spiny lobsters. *Biol. Bull. (Woods Hole, Mass.)* 160, 55–68.

Durliat, M., and Vranckx, R. (1989). Relationships between plasma and hemocyte proteins in Decapoda. *Comp. Biochem. Physiol. B* 92B, 595–603.

Elder, H. Y. (1973). Distribution and function of elastic fibers in the invertebrates. *Biol. Bull. (Woods Hole, Mass.)* 144, 43–63.

Engel, D. W., and Brouwer, M. (1991). Hemocyanin concentrations in the hemolymph of blue crabs, *Callinectes sapidus*, from Eastern N.C. and lobsters, *Homarus americanus*, from Massachusetts. *Can. Tech. Rep. Fish. Aquat. Sci.* 1774, 592–593.

Estrella, B. T. (1991). Shell disease in American lobster (*Homarus americanus*, H. Milne Edwards, 1837) from Massachusetts coastal waters with considerations for standardizing sampling. *J. Shellfish Res.* 10, 483–488.

Factor, J. R., and Beekman, J. (1990). The digestive system of the lobster, *Homarus americanus*: III. Removal of foreign particles from the blood by fixed phagocytes of the digestive gland. *J. Morphol.* 206, 293–302.

Factor, J. R., and Naar, M. (1990). The digestive system of the lobster, *Homarus americanus*: II. Terminal arterioles of the digestive gland. *J. Morphol.* 206, 283–291.

Fessler, J., and Fessler, L. (1989). *Drosophila* extracellular matrix. *Annu. Rev. Cell Biol.* 5, 309–339.

Fischer-Piette, E. (1931). Culture de tissue de crustacés. La glande lymphatique du homard. *Arch. Zool. Exp. Gen.* 74, 33–52.

Fisher, W. S., Nilson, E. H., Follett, L. F., and Shleser, R. A. (1976). Hatching and rearing lobster larvae (*Homarus americanus*) in a disease situation. *Aquaculture* 7, 75–80.

Fisher, W. S., Nilson, E. H., Steenbergen, J. F., and Lightner, D. V. (1978). Microbial diseases of cultured lobsters: A review. *Aquaculture* 14, 115–140.

Foster, C. A., Farley, C. A., and Johnson, P. T. (1981). Virus-like particles in cardiac cells of the brown shrimp, *Penaeus aztecus* Ives. *J. Submicrosc. Cytol.* 13, 723–726.

Fuller, G. M., and Doolittle, R. F. (1971). Studies of invertebrate fibrinogen. II. Transformation of lobster fibrinogen into fibrin. *Biochemistry* 10, 1311–1315.

Getchell, R. G. (1991). Lobster shell disease survey. *J. Shellfish Res.* 10, 489–490.

Ghidalia, W., Vendrely, R., Montmory, C., Coirault, Y., and Brouard, M. O. (1981). Coagulation in decapod Crustacea. *J. Comp. Physiol.* 142, 473–478.

Ghiretti-Magaldi, A., Milanese, C., and Tognon, G. (1977). Hemopoiesis in Crustacea Decapoda: Origin and evolution of hemocytes and cyanocytes of *Carcinus maenas*. *Cell Differ.* 6, 167–186.

Goldenberg, P. Z., Huebner, E., and Greenberg, A. H. (1984). Activation of lobster hemocytes for phagocytosis. *J. Invertebr. Pathol.* 43, 77–88.

Goldenberg, P. Z., Greenberg, A. H., and Gerrard, J. M. (1986). Activation of lobster hemocytes: Cytoarchitectural aspects. *J. Invertebr. Pathol.* 47, 143–154.

Govind, C. K., and Guchardi, J. (1986). Vascularization of a lobster (*Homarus americanus*) muscle. *J. Morphol.* 188, 327–334.

Halcrow, K., and Smith, J. C. (1986). Wound closure in the crab *Carcinus maenas* (L.). *Can. J. Zool.* 64, 2770–2778.

Hall, J. L., Rowlands, D. T., Jr., and Nilsson, U. R. (1972). Complement-like hemolytic activity in lobster hemolymph. *J. Immunol.* 109, 816–823.

Hartline, D. K. (1979). Integrative physiology of the lobster cardiac ganglion. *Am. Zool.* 19, 53–65.

Hearing, V. J., and Vernick, S. H. (1967). Fine structure of the blood cells of the lobster, *Homarus americanus*. *Chesapeake Sci.* 8, 170–186.

Herrick, F. H. (1909). Natural history of the American lobster. *Bull. U.S. Bur. Fish.* 29, 149–408 + 20 plates.

Hose, J. E., and Martin, G. G. (1989). Defense functions of granulocytes in the ridgeback prawn, *Sicyonia ingentis* Burkenroad 1938. *J. Invertebr. Pathol.* 53, 335–346.

Hose, J. E., Martin, G. G., and Gerard, A. S. (1990). A decapod hemocyte classification scheme integrating morphology, cytochemistry and function. *Biol. Bull. (Woods Hole, Mass.)* 178, 33–45.

Hose, J. E., Martin, G. G., Tiu, S., and McKrell, N. (1992). Patterns of hemocyte production and release throughout the molt cycle in the penaeid shrimp *Sicyonia ingentis*. *Biol. Bull. (Woods Hole, Mass.)* 183, 185–199.

Jackson, S. A., Bruce, M. J., Chang, E. S., and Clegg, J. S. (1987). Effects of eyestalk ablation upon water relations in the American lobster, *Homarus americanus*. *J. Exp. Zool.* 244, 389–393.

Johansson, M. W., and Söderhäll, K. (1989a). Cellular immunity in crustaceans and the proPO system. *Parasitol. Today* 5, 171–176.

Johansson, M. W., and Söderhäll, K. (1989b). A cell adhesion factor from crayfish haemocytes has degranulating activity towards crayfish granular cells. *Insect Biochem.* 19, 183–190

Johnson, P. T. (1980). "Histology of the Blue Crab, Callinectes sapidis: A Model for the Decapoda." Praeger, New York.

Johnson, P. T . (1987). A review of fixed phagocytic and pinocytotic cells of decapod crustaceans, with remarks on hemocytes. *Dev. Comp. Immunol.* 11, 679–704.

Johnson, P. T., Stewart, J. E., and Arie, B. (1981). Histopathology of *Aerococcus viridans* var. *homari* infection (Gaffkemia) in the lobster, *Homarus americanus*, and a comparison with histological reactions to a gram-negative species, *Pseudomonas perolens*. *J. Invertebr. Pathol.* 38, 127–148.

Kefalides, N. A., Alper, R., and Clark, C. C. (1979). Biochemistry and metabolism of basement membranes. *Int. Rev. Cytol.* 61, 167–228.

Keith, I. R., Paterson, W. D., Airdrie, D., and Boston, L. D. (1992). Defense mechanisms of the American lobster, *Homarus americanus*: Vaccination provided protection against gaffkemia infections in laboratory and field trials. *Fish Shellfish Immunol.* 2, 109–119.

Kobaysahi, M., Johansson, M. W., and Söderhäll, K. (1990). The 76kD cell-adhesion factor from crayfish haemocytes promotes encapsulation *in vitro*. *Cell Tissue Res.* 260, 13–18.

Kollman, M. (1908). Recherches sur les leucocytes de le tissue lymphoide des Invertébrés. *Ann. Sci. Nat. Zool. Ser. 9* 8, 1–240.

Krol, R. A., Hawkins, W. E., Vogelbein, W. K., and Overstreet, R. M. (1989). Histopathology and ultrastructure of the hemocytic response to bacterial infection in cultured *Penaeus vannamei*. *J. Aquat. Anim. Health* 1, 37–42.

Kuramoto, T., and Ebara, A. (1984). Neurohormonal modulation of the cardiac outflow through the cardioarterial valve in the lobster. *J. Exp. Biol.* 111, 123–130.

Lightner, D. V., and Fontaine, C. T. (1975). A mycosis of the American lobster, *Homarus americanus*, caused by *Fusarium* sp. *J. Invertebr. Pathol.* 25, 94–99.

Lorand, L. (1972). Fibrinoligase: The fibrin-stabilizing factor system of blood plasma. *Ann. N.Y. Acad. Sci.* 202, 1–348.

Lorand, L., and Conrad, S. M. (1984). Transglutaminases. *Mol. Cell. Biochem.* 58, 9–35.

Lorand, L., Jacobsen, A., and Schuel, A. (1964) Inhibitors of fibrin polymerization. *Biol. Bull. (Woods Hole, Mass.)* 129, 413.

Lumb, G., Clare, A. S., and Costlow, J. D. (1991). Cheliped regeneration in the megalops of the mud crab, *Rithropanopeus harrisii* (Gould). *Invert. Reprod. Dev.* 20, 87–96.

Madaras, F., Chew, M. Y., and Parkin, J. D. (1981). Purification and characterization of the sand crab (*Ovalipes bipustulatus*) coagulogen (fibrinogen). *Thromb. Haemostasis* 45, 77–81.

Mangum, C. (1993). Hemocyanin subunit composition and oxygen binding in two species of the lobster genus *Homarus* and their hybrids. *Biol. Bull. (Woods Hole, Mass.)* 184, 105–113.

Marrec, M. (1944). L'organe lymphocytogène des Crustacés Decapodés. Son activite cyclique. *Bull. Inst. Oceanogr.* 867, 1–4.

Martin, G., and Timpl, R. (1987). Laminin and other basement membrane components. *Annu. Rev. Cell Biol.* 3, 57–86.

Martin, G. G., and Hose, J. E. (1992). Vascular elements and blood (hemolymph). *In* "Microscopic Anatomy of Invertebrates" (F. W. Harrison, ed.), Vol. 10, "Decapod Crustacea" (F. W. Harrison and A. G. Humes, eds.), pp. 117–146. Wiley–Liss, New York.

Martin, G. G., Hose, J. E., Omori, S., Chong, C., Hoodbhoy, T., and McKrell, N. (1991). Localization and roles of coagulogen and transglutaminase in hemolymph coagulation in decapod crustaceans. *Comp. Biochem. Physiol. B* 100B, 517–522.

Martin, G. G., Hose, J. E., Choi, M., Provost, R., Omori, G., McKrell, N., and Lam, G. (1993). Organization of hematopoietic tissue in the intermolt lobster, *Homarus americanus*. *J. Morphol.* 216, 65–78.

Martynova, M. G., Økland, S., Rumyantsev, P. P., and Tjønneland, A. (1986). The ultrastructure of the myocardium and the myocardial satellite cells in *Hyas araneus* L., *Carcinus maenas* L. and *Cancer pagurus* L. (Crustacea, Decapoda). *Zool. Anz.* 216, 357–366.

Maynard, D. M. (1960). Circulation and heart function. *In* "The Physiology of Crustacea" (T. H. Waterman, ed.), Vol. 1, pp. 161–226. Academic Press, New York.

McLaughlin, P. A. (1980). "Comparative Morphology of Recent Crustacea." Freeman, San Francisco.

McMahon, B. R., and Burnett, L. E. (1990). The crustacean open circulatory system: A reexamination. *Physiol. Zool.* 63, 35–71.

McMahon, B. R., and Wilkens, J. L. (1975). Respiratory and circulatory responses to hypoxia in the lobster *Homarus americanus*. *J. Exp. Biol.* 62, 637–655.

Midsukami, M. (1981). The structure and distribution of satellite cells of cardiac muscle in decapod crustaceans. *Cell Tissue Res.* 219, 69–83.

Mykles, D. L. (1980). The mechanism of fluid absorption at ecdysis in the American lobster, *Homarus americanus*. *J. Exp. Biol.* 84, 89–101.

Nilson, E. H., Fisher, W. S., and Shleser, R. A. (1976). A new mycosis of larval lobster (*Homarus americanus*). *J. Invertebr. Pathol.* 27, 177–183.

Noga, E. J. (1991). Shell disease in marine crustaceans: Concluding remarks. *J. Shellfish Res.* 10, 505–506.

Omori, S., Martin, G. G., and Hose, J. E. (1989). Morphology of hemocyte lysis and clotting in the ridgeback prawn, *Sicyonia ingentis*. *Cell Tissue Res.* 255, 117–123.

Paterson, W. D., and Keith, I. R. (1992). Disease and defense mechanisms of the American lobster, *Homarus americanus*. *In* "Diseases in Asian Aquaculture" (M. Shariff, R. P. Subasinghe, and J. R. Arthur, eds.), pp. 81–87. Asian Fish Society, Manila, Philippines.

Paterson, W. D., and Stewart, J. E. (1974). *In vitro* phagocytosis by hemocytes of the American lobster (*Homarus americanus*). *J. Fish. Res. Board Can.* 31, 1051–1056.

Paterson, W. D., and Stewart, J. E. (1979). Rate and duration of phagocytic increase in lobsters induced by *Pseudomonas perolens* endotoxin. *Dev. Comp. Immunol.* 3, 353–357.

Paterson, W. D., Stewart, J. E., and Zwicker, B. M. (1976). Phagocytosis as a cellular immune response mechanism in the American lobster, *Homarus americanus*. *J. Invertebr. Pathol.* 27, 95–104.

Ravindranath, M. H. (1980). Haemocytes in haemolymph coagulation of arthropods. *Biol. Rev.* 55, 139–170.

Rosen, B. (1970). Shell disease of aquatic crustaceans. *Am. Fish. Soc., Spec. Publ.* 5, 409–415.

Sage, H., and Gray, W. R. (1979). Studies on the evolution of elastin. 1. Phylogenetic distribution. *Comp. Biochem. Physiol. B* 64B, 313–327.

Salt, G. (1970). The Cellular Defense Reactions of Insects. *Cambridge Monogr. Exp. Biol.* 16.

Sandeman, D. C. (1967). The vascular circulation in the brain, optic lobes and thoracic ganglion of the crab *Carcinus*. *Proc. R. Soc. London Ser. B* 168, 82–90.

Schapiro, H. C., Steenbergen, J. F., and Fitzgerald, Z. A. (1977). Hemocytes and phagocytosis in the American lobster *Homarus americanus*. *In* "Comparative Pathobiology" (L. A. Bulla, Jr., and T. C. Cheng, eds.), pp. 127–133. Plenum, New York.

Senkbeil, E. G., and Wriston, J. C., Jr. (1981). Hemocyanin synthesis in the American lobster, *Homarus americanus*. *Comp. Biochem. Physiol. B* 68B, 163–171.

Shadwick, R. E., Pollock, C. M., and Stricker, S. A. (1990). Structure and biomechanical properties of crustacean blood vessels. *Physiol. Zool.* 63, 90–101.

Sherburne, S. W., and Bean, L. L. (1991) Mortalities of impounded and feral Maine lobsters, *Homarus americanus* H. Milne-Edwards, 1837, caused by the protozoan ciliate *Mugardia* (formerly *Anophrys* = *Paranophrys*), with initial prevalence data from ten locations along the Maine coast and one offshore area. *J. Shellfish Res.* 10, 315–326.

Sindermann, C. J. (1991). Shell disease in marine crustaceans—A conceptual approach. *J. Shellfish Res.* 10, 491–494.

Smith, D. S., and Anderson, M. E. (1972). The disposition of membrane systems in cardiac muscle of a lobster *Homarus americanus*. *Tissue Cell* 4, 629–645.

Smith, V. J., and Söderhäll, K. (1983). Induction of degranulation and lysis of haemocytes in the freshwater crayfish, *Astacus* by components of the prophenoloxidase activating system *in vitro*. *Cell Tissue Res.* 233, 295–303.

Söderhäll, K. (1982). Prophenoloxidase activating system and melanization—A recognition mechanism of arthropods? A review. *Dev. Comp. Immunol.* 6, 601–611.

Söderhäll, K., and Ajaxon, R. (1982). Effects of quinones and melanin on mycelial growth of *Aphanomyces* spp. and extracellular proteases of *Aphanomyces astaci*, a parasite on crayfish. *J. Invertebr. Pathol.* 39, 105–109.

Söderhäll, K., Johansson, M. W., and Smith, V. J. (1988). Internal defense mechanisms. *In* "Freshwater Crayfish: Biology, Management and Exploitation" (D. M. Holdich and R. S. Lowery, eds.), pp. 213–235. Croom Helm, London.

Spychert, S. E., Arya, S., Isenman, D. D., and Painter, R. H. (1987). A functional, thioester-containing α_2-macroglobulin homologue isolated from the hemolymph of the American lobster (*Homarus americanus*). *J. Biol. Chem.* 262, 14606–14611.

Steinacker, A. (1975). Perfusion of the central nervous system of decapod crustaceans. *Comp. Biochem. Physiol. A* 52A, 103–104.

Stewart, J. E., and Arie, B. (1973). Depletion of glycogen and adenosine triphosphate as major factors in the death of lobsters (*Homarus americanus*) infected with *Gaffkya homari*. *Can. J. Microbiol.* 19, 1103–1110.

Stewart, J. E., and Zwicker, B. M. (1972). Natural and induced bactericidal activities in the hemolymph of the lobster, *Homarus americanus*: Products of hemocyte–plasma interaction. *Can. J. Microbiol.* 18, 1499–1509.

Stewart, J. E., Dingle, J. R., and Odense, P. H. (1966). Constituents of the hemolymph of the lobster, *Homarus americanus* Milne Edwards. *Can. J. Biochem.* 44, 1447–1459.

Stewart, J. E., Cornick, J. W., and Dingle, J. R. (1967). An electronic method for counting lobster (*Homarus americanus*) hemocytes and the influence of diet on hemocyte numbers and hemolymph proteins. *Can. J. Zool.* 45, 291–304.

Stewart, J. E., Arie, B., Zwicker, B. M., and Dingle, J. R. (1969). Gaffkemia, a bacterial disease of the lobster, *Homarus americanus*: Effects of the pathogen, *Gaffkemia homari*, on the physiology of the host. *Can. J. Microbiol.* 15, 925–932.

Stewart, J. E., Arie, B., and Marks, L. (1983). Hemocyte patterns during gaffkemia infections and induction of resistance in *Homarus americanus*. *Rapp. P.-V. Reun., Cons. Int. Explor. Mer* 182, 126–129.

Tait, J. (1911). Types of crustacean blood coagulation. *J. Mar. Biol. Assoc. U.K.* 9, 191–198.

Toney, M. E., Jr. (1958). Morphology of the blood cells of some crustacea. *Growth* 22, 35–50.

Tsing, A., Arcier, J.-M., and Brehelin, M. (1989). Hemocytes of penaeid and palaemonid shrimps: Morphology, cytochemistry, and hemograms. *J. Invertebr. Pathol.* 53, 64–77.

Uhrik, B., Rýdlová, K., and Zacharová, D. (1989). The role of haemocytes during degeneration and regeneration of crayfish muscle fibres. *Cell Tissue Res.* 255, 443–449.

Vacca, L. L., and Fingerman, M. (1983). The roles of hemocytes in tanning during the molting cycle: A histochemical study of the fiddler crab, *Uca pugilator*. *Biol. Bull. (Woods Hole, Mass.)* 165, 758–777.

Wilkens, J. L., Wilkens, L. A., and McMahon, B. R. (1974). Central control of cardiac and ventilatory pacemakers in the crab *Cancer magister*. *J. Comp. Physiol.* 90, 89–104.

Yeager, J. F., and Tauber, O. E. (1935). On the hemolymph cell counts of some marine invertebrates. *Biol. Bull. (Woods Hole, Mass.)* 69, 66–70.

Young, R. E., and Coyer, P. E. (1979). Phase coordination in the ventilatory and cardiac rhythms of the lobster *Homarus americanus*. *J. Exp. Biol.* 82, 53–74.

Zipori, D. (1988). Hemopoietic microenvironments. *In* "Hematopoiesis" (N. Testa and R. Gale, eds.), pp. 27–62. Dekker, New York.

Zuckerman, K., and Rhodes, R. (1985). The hematopoietic extracellular matrix. *In* "Hematopoietic Stem Cell Physiology" (J. Palek and P. Quesenbery, eds.), pp. 257–266. Liss, New York.

CHAPTER

18

The Physiology of Gas Exchange, Circulation, Ion Regulation, and Nitrogenous Excretion: An Integrative Approach

BRIAN R. McMAHON
Department of Biological Sciences
University of Calgary
Calgary, Alberta, Canada T2N 1N4

I. Introduction

Despite the huge economic importance of the American lobster, *Homarus americanus*, study of its basic physiology has fallen behind that of many other crustaceans. A few systems, such as the functioning of the cardiac ganglion, have been extensively studied. Others of similar apparent importance, including many of the physiological areas included in this chapter, are poorly understood. In order to provide a better overall understanding of the physiology of *H. americanus*, this chapter cements our fragmentary knowledge of the lobster together with a more general matrix of the physiology of other lobsters and of closely related decapod crustaceans. Although gas exchange, circulation, water and salt regulation, acid–base balance, and nitrogenous excretion are listed separately, all are complexly interrelated and cannot be dealt with in isolation. This chapter is designed to reveal these interrelationships and their integration within the overall responses of *H. americanus* to environmental perturbations and changes in activity level.

II. Respiration: Gas Exchange and Transport

A. Structure and Function of the Gills

Filamentous gills of the trichobranchiate pattern (Fig. 1A) arise from the epipodites of the thoracic appendages. In *Homarus americanus*, 20 paired gills arise from the seven appendages (Fig. 1B). They are located in a well-developed branchial cavity protected by the branchiostegites on either side. Histology of the lobster gill is not well known but appears similar to that of crayfish (Dunel-Erb *et al.*, 1982), except that the filaments are shorter. Lobster gill filaments are tubular in form, unlike the lamellate (phyllobranchiate) pattern of crab gills. Gill surface area is not known for *H. americanus*, but is probably similar per unit mass to that of marine brachyuran decapods (Gray, 1957). The filaments are extremely numerous

BIOLOGY OF THE LOBSTER *Homarus americanus*
Jan Robert Factor, Editor

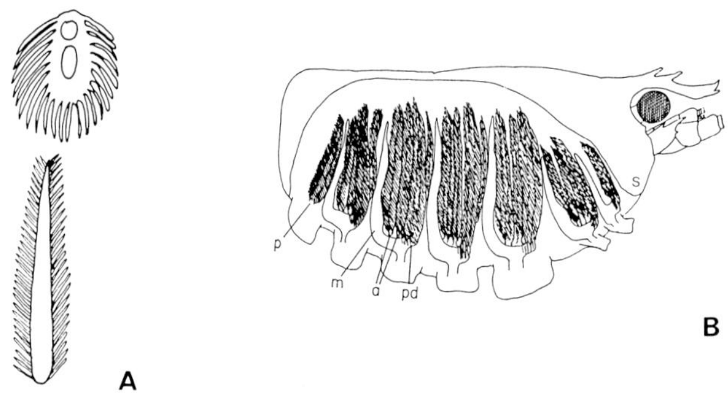

FIGURE 1 *Homarus americanus* gills. (A) Filament and filament cross section to show trichobranchiate gill structure. (B) Location and type of gills within the right branchial cavity. a, Arthrobranch; m, mastigobranch; p, pleurobranch; pd, podobranch; S, position of scaphognathite. (From McLauglin, 1983.)

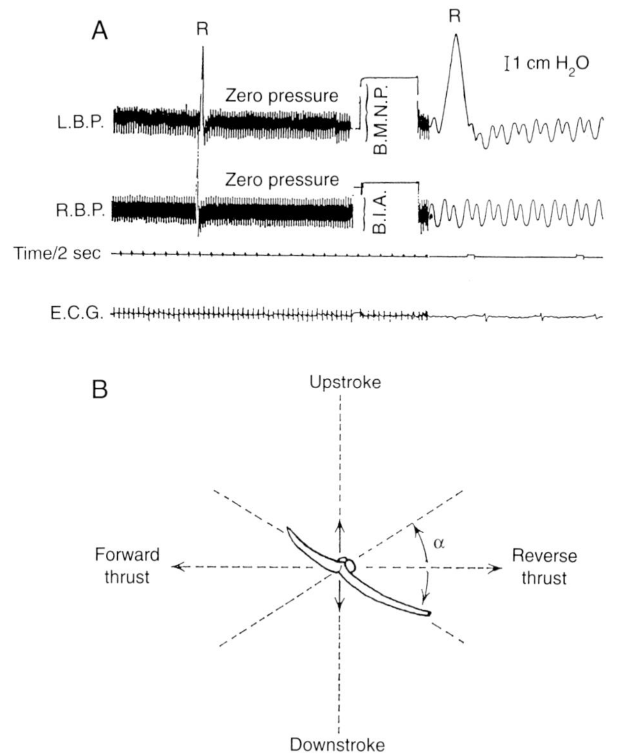

FIGURE 2 (A) Hydrostatic pressure recordings from the branchial chambers of *Homarus americanus* to show pressure oscillations developed during forward and reversed scaphognathite pumping. The faster record clearly shows negative pressure resulting from both up and down strokes of pressures from both left (L.B.P.) and right (R.B.P.) branchial chambers. B.M.N.P., Branchial mean negative pressure; B.I.A., branchial irrigation amplitude (branchial pulse pressure). (From McMahon and Wilkens, 1975.) (B) Changes in attack angle of the scaphognathite blade which result in forward and reversed pumping. (From Wilkens and McMahon, 1972.)

and clearly incorporate a large diffusion area. The pattern of hemolymph[1] perfusion of the branchiae is presented by Martin and Hose (Chapter 17).

B. Gill Ventilation

Ventilation (irrigation) of the branchial cavities occurs by the action of the scaphognathites, which are lateral, blade-like structures of the second maxillae thought to represent combined exopodite and epipodite (see Lavalli and Factor, Chapter 14). These oscillate within the confines of narrow prebranchial chambers (Patterson, 1968) to cause water movements through the branchial chambers.

The movements of the scaphognathites of *Homarus americanus* and their neural control have been studied in detail (Wilkens and McMahon, 1972; McMahon and Wilkens, 1983). Contraction of intrinsic elevator and depressor muscles causes rhythmic oscillations of the scaphognathite blade. Both upward and downward deflections cause water to be pumped out of the prebranchial chambers (Fig. 2A) and drawn through the branchial cavity (Fig. 1). Water can be pumped out of the prebranchial cavity in either direction by changing the attack angle of the pump from positive to negative (Fig. 2B). The normal mode (positive attack angle, forward beating) generates water flow out of the scaphognathite channel anteriorly and the resultant subambient pressure causes water to enter the branchial cavity ventrally, between the limb bases. This water is channeled to flow in between the gill filaments and thus forms the ventilatory current. Reversed pumping (negative attack angle) occurs frequently in many brachyuran decapods (McMahon and Wilkens, 1983), but in *H. americanus* is limited to occasional single reversed beats (Fig. 2A). These occur in conjunction with contraction of the epimeral muscles, which depress the branchiostegites. The result is a rapid pulse of water directed posteriorly through the branchial cavity and out ventrally between the limb bases (Wilkens and McMahon, 1972). This backflow probably serves to remove particles which otherwise could clog the input filters, which are setae on the limb bases and edges of the carapace, and/or the gill sieve, although other functions have also been proposed (McMahon and Wilkens, 1983).

The actions of the scaphognathites generate subambient pressure throughout the branchial cavities [Wilkens and McMahon, 1972, *Homarus americanus*; Butler *et al.*, 1978, *H. gammarus*[2] (=*H. vulgaris*)]. The gill arrays associated with each limb (Fig. 1B) are loosely compartmentalized by the epipodites, which are flattened plates interleaved with and separating

[1] The term hemolymph is used here to describe the circulating fluid of the open circulatory system of the lobster, which has the functions of both blood and lymph.

the sets of gills. The level of subambient pressure varies slightly between compartments (Wilkens and McMahon, 1972), but flow through each compartment seems similar. In the majority of active aquatic animals, the relationship between water flow across and hemolymph flow through a gas-exchange surface is countercurrent. That is, water and hemolymph flow are opposed at the exchange surface, allowing greater efficiency of gas transfer than a cocurrent model in which the two media flow in the same direction (reviewed by McMahon and Wilkens, 1983, for various crustaceans). Other systems which incorporate elements of both cross- and cocurrent, such as the multicapillary system described for the gills of the dogfish (Piiper and Scheid, 1975) and the crayfish (Burggren *et al.*, 1974), also increase efficiency. Details of water flow-through at the filament level have not been published for *H. americanus*, but flow patterns similar to the multicapillary pattern of crayfish (Burggren *et al.*, 1974) appear likely on structural grounds.

Levels of water flow through the branchial cavities of *Homarus americanus* vary considerably with factors which affect oxygen demand or supply. In quiescent, fettered animals, average flow rates can reach 1163 ml/kg/min, a level which confirms that the scaphognathite is a highly efficient pump capable of generating ventilatory flows at a level comparable to those of other aquatic organisms (McMahon and Wilkens, 1983). Lower levels are reported from unfettered animals (Butler *et al.*, 1978). The lowest levels (<200 ml/kg/min) are associated with intermittent scaphognathite pumping patterns, where the scaphonathites of one or both sides may cease pumping often and for extended periods (Fig. 3; see Section II,G of this chapter). Although such intermittent ventilation is commonly observed in quiescent, aquatic animals, there is some evidence from other decapods that the scaphognathite pump becomes much less effective at low pumping rates (reviewed by McMahon and Wilkens, 1983).

C. Oxygen Uptake

The percentage efficiency of removal of oxygen from the ventilatory water stream ($E_w\%$) can be assessed by the following equation:

$$E_W\% = \frac{Pi_{O_2} - Pe_{O_2}}{100}, \quad (1)$$

where Pi_{O_2} and Pe_{O_2} are the oxygen partial pressures (in mm Hg)[3] of inhalant and exhalant water, respectively. Values obtained for *Homarus americanus* (Table 1) are within the range for other aquatic animals with similar activity levels. Oxygen uptake levels are tightly correlated with oxygen demand and thus vary markedly with many factors, including physiological state, behavioral state, activity level, and changes in such environmental factors as temperature and oxygen level (see Section IV,A). Animals are often labeled as oxygen conformers or oxygen regulators, depending on their ability to steadily maintain oxygen consumption levels while external oxygen levels fall (Mangum and van Winkle, 1973). In fact, the ability of an animal to "regulate" oxygen consumption varies with many factors (Herried, 1980), such that almost all animals are capable of a degree of regulation depending on the oxygen level and the precise circumstances; thus, the concept really has little practical value. *H. americanus* has been termed an oxygen conformer (Amberson *et al.*, 1924; Thomas, 1954), but actually regulates well (see Section IV,A).

D. Hemolymph Gas and Acid–Base Levels

The levels of ventilation described in Section II,C and Table 1 are sufficient to well oxygenate the hemolymph of quiescent lobsters (Table 2). In restrained lobsters showing signs of disturbance, mean levels of oxygen pressure are 69 mm Hg for postbranchial (i.e., arterialized) hemolymph (Pa_{O_2}) and 18 mm Hg for prebranchial (i.e., venous) hemolymph (Pv_{O_2}) in well-aerated water (McMahon and Wilkens, 1975). These levels (Table 2) are a little higher than those reported for *Homarus gammarus* using animals that were less disturbed by the experimental protocol (Butler *et al.*, 1978). Circulating oxygen levels fall rapidly in undisturbed animals which exhibit unilateral pumping and pausing (McMahon and Wilkens, 1983) and it is likely that mean oxygen levels of undisturbed quiescent lobsters under natural conditions would be substantially lower (Forgue *et al.*, 1992).

At these enhanced oxygen levels, the circulating oxygen-carrier molecule hemocyanin (Hc) is fully saturated in postbranchial hemolymph, and prebranchial hemolymph carries a substantial oxygen reserve (Table 2). As is typical of most aquatic decapod crustaceans, the amount of oxygen carried (oxygen capacity, C_{O_2}) in lobster hemolymph is (by vertebrate standards) relatively low (Table 2) and oxygen dissolved in the hemolymph (i.e., not bound to Hc) makes up a substantial proportion of the oxygen

---

[2]*Homarus gammarus* is the valid species name for the European lobster (see Williams, Chapter 2) and is used throughout this chapter. The invalid synonym *H. vulgaris* is commonly found in the literature, including many of the papers cited in this chapter.

[3]1 mm Hg = 1 torr = 1.33×10^{-2} *Pa*.

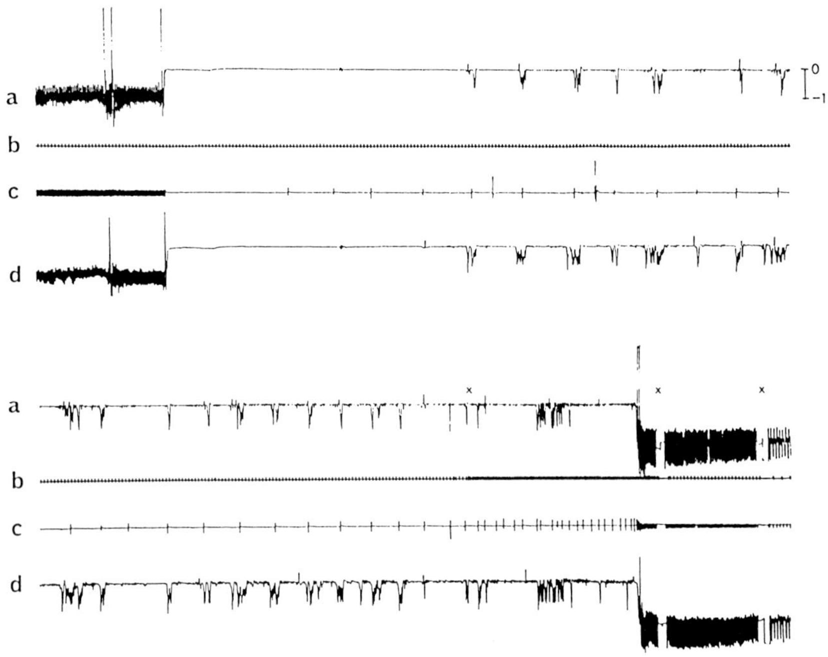

FIGURE 3 Extended period of cessation of heart and ventilatory pumping (13 min) in *Homarus americanus,* quiescent but restrained in the laboratory. a,d, Right and left branchial chamber pressure; b, time = 10 seconds; c, heart rate (electrocardiogram); x, periods when chart was stationary. Bar in upper trace, 1 cm of water pressure.

TABLE 1 **Respiratory and Circulatory Variables in Normoxic and Hypoxic *Homarus americanus***

Ambient oxygen tension range (mm Hg)	BRR[a] (bt/min)	BMNP[b] (cm H_2O)	$\dot{V}_G$[c] (ml/kg/mm)	Ut[d] (%)	ΔP_G[e] (mm Hg)	T_{O_2}[f] $\dot{V}_{O_2}/\text{min}/\dot{V}_G$	E_w[g] (%)	$\dot{V}_{O_2}$[h] (ml/kg/min)	E.C.G.[i] (bt/min)
100–150	101	1.5	487	23	67	0.007	27	0.50	96
70–90	116	1.8	595	29	39	0.013	38	0.88	82
30–50	137	2.3	666	41	15	0.032	59	0.53	47
10–20	88	1.1	394	43	5	0.030	80	0.15	30
30–50	141	(2.2)	(563)	(44)	—	—	—	0.98	(36)
70–90	126	2.3	641	37	—	—	—	1.10	(70)
100–150	103	1.7	579	32	—	—	—	1.63	90

Note. The data above are averaged from experiments (nine on six animals) characterized by ~2-hour periods of adjustment to the experimental chamber and exposure to several maintained levels of oxygen depletion. Data in parentheses are where $n = 5$ or less data points. The last three lines are from animals recovering from hypoxia.

[a]BRR, scaphognathite rate.

[b]BMNP, maintained negative pressure in the right branchial cavity.

[c]$\dot{V}_G$, volume of water pumped through both branchial cavities.

[d]Ut, utilization of oxygen expressed as the percentage of oxygen content removed from the branchial water flow.

[e]ΔP_G, average tension difference between inspired water and prebranchial blood in mm Hg.

[f]T_{O_2}, transfer factor (diffusing capacity of the gills for O_2) expressed as ml O_2/min/kg/mm Hg = $\dot{V}_{O_2}/\Delta P_G$

[g]E_w, effectiveness (%) of oxygen transfer from water to blood = $[(Pi_{O_2} - Pe_{O_2})/(Pi_{O_2} - Pv_{O_2})] \times 100$.

[h]$\dot{V}_{O_2}$, oxygen uptake from the branchial water expressed as ml O_2/kg body wt/min.

[i]E.C.G., electrocardiogram.

TABLE 2 **Oxygen Transport in Normoxic and Hypoxic *Homarus americanus***

Ambient tension range (P_{O_2})	Pa_{O_2} [a] (mm Hg)	Pv_{O_2} [b] (mm Hg)	P_{O_2} difference $a-v$ (mm Hg)	Sa_{O_2} [c] (%)	Sv_{O_2} [c] (%)	Ca_{O_2} [d] (vol %)	Cv_{O_2} [d] (vol %)	Total O_2 delivered to tissues (vol %)	O_2 delivered by Hc[e] (vol %)	O_2 delivered from solution (vol %)	$\dot{Q}$ [f] (ml/min)	$\dot{V}_{O_2}$ [g] (ml/kg/min)	E.C.G. (bt/min)
100–150	69	18	51	100	93	0.86	0.80	0.248	0.06	0.188	60.5	0.50	92
70–90	42	17	25	100	91	0.86	0.78	0.173	0.08	0.093	88.2	0.50	84
30–50	22	12	10	96	75	0.82	0.64	0.217	0.18	0.037	59.9	0.47	50
10–20	8	7	1	32	24	0.275	0.206	0.073	0.069	0.004	54.8	0.13	29

[a] Pa_{O_2}, postbranchial blood oxygen tension.

[b] Pv_{O_2}, prebranchial blood oxygen tension.

[c] Sa_{O_2}, Sv_{O_2}, percentage oxygen saturation of post- and prebranchial blood calculated from the dissociation curve at pH 7.50 and 15°C.

[d] Ca_{O_2}, Cv_{O_2}, oxygen content of post- and prebranchial blood based on hemocyanin oxygen capacity of 0.86 vol % = 100% saturation.

[e] Hc, hemocyanin.

[f] $\dot{Q}$, cardiac output in ml/min calculated using the Fick principle.

[g] $\dot{V}_{O_2}$, oxygen uptake from the branchial water expressed as ml O_2/kg body wt/min.

available for delivery to tissues. In fact, in restrained *Homarus americanus*, only 24% of O_2 delivered to tissues under normal oxygen (normoxic) conditions is supplied from the hemocyanin (Fig. 4) (McMahon and Wilkens, 1975). Under stressful conditions (e.g., hypoxic exposure), however, this proportion increases markedly (as discussed in Section IV,A). Oxygen and carbon dioxide pressure profiles from environment to tissues have been recorded using a mass spectrometer (McMahon and Burggren, 1990). The results displayed in Fig. 5 are for masked and rigidly restrained lobsters with a window cut from the branchial chamber; they are thus severely disturbed, but nonetheless provide the first such data for any crustacean or indeed any invertebrate animal. The values presented for tissue oxygen pressure (Pt_{O_2}) show the oxygen pressure in an artificial hemolymph space created by the catheter on insertion into the tissues and may not accurately reflect the levels found in tissue lacunae (blood sinuses). Intracellular oxygen levels are not known. This oxygen profile (Fig. 5A) reveals that the total partial pressure gradient for oxygen from external water to tissue space is 102 torr, which corresponds with a total diffusion pressure required for CO_2 of approximately 10 torr (Fig. 5B). This difference reflects the much lower solubility, and hence diffusion rate, of oxygen in biological fluids and suggests that O_2 uptake rather than CO_2 elimination is the limiting factor for gas exchange in *H. americanus*, as in other aquatic animals.

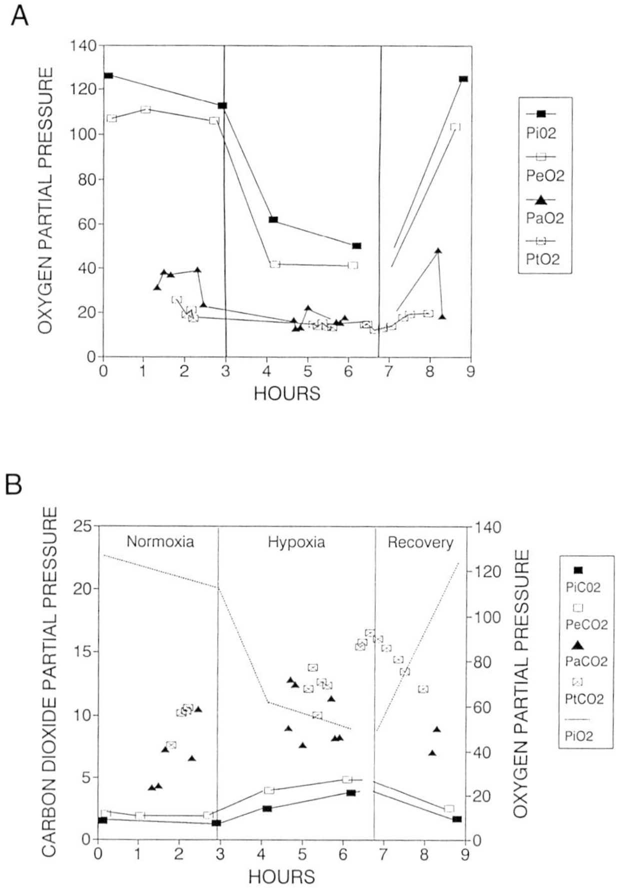

FIGURE 5 Effects of hypoxia. Oxygen (A) and CO_2 (B) pressure gradients recorded during normoxia and hypoxic exposure (hours 4–7) from restrained lobsters by a mass spectrometer catheter inserted into abdominal muscle. Pi_{O_2} Pe_{O_2} Pa_{O_2}, and Pt_{O_2}, Inhalant, exhalant, arterialized, and tissue oxygen partial pressure, respectively. Pi_{CO_2}, etc., Partial pressures of carbon dioxide. (From McMahon and Burrgren, 1990.)

E. Functioning of Hemocyanin

The functioning of crustacean Hc has received a great deal of recent attention (reviewed by Mangum, 1983a; McMahon, 1985; Morris, 1990; and Truchot, 1992) and is reasonably well known. Less specific information is available for *Homarus* Hc O_2 binding. *H. americanus* Hc is usually in the form of dodecamers composed of two hexameric units (Mangum, 1983b). Several structurally different subunits occur and could allow construction of physiologically different functional molecules. Hemocyanin O_2-binding properties have been described for *H. americanus* (Mangum, 1983b, 1993). The molecule is highly cooperative (n = 2.8–4.8) and has typically high (for a crustacean) oxygen affinity [pressure for 50% O_2 saturation (P_{50}) = 9 mm Hg at pH 7.85]. As is common with crustacean oxygen-carrier molecules, P_{50} is quite labile, being decreased by increases in lactate (Mangum, 1983b) and other ions (Na^+, Cl^-, and Ca^{2+}) and quite strongly influenced by changes in ambient H^+ (Bohr shift) (Butler *et al.*, 1978; Mangum, 1983a,b).

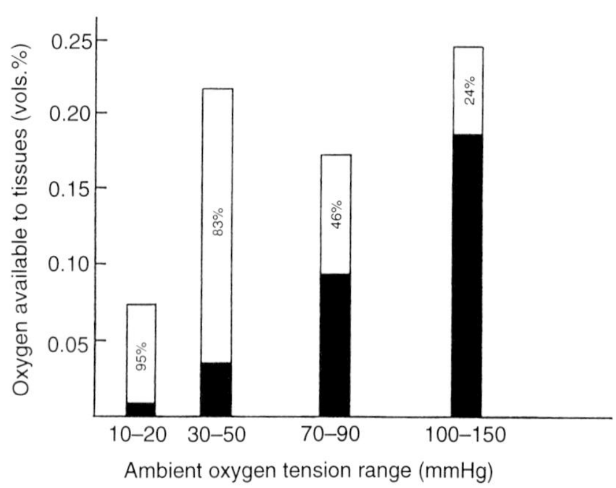

FIGURE 4 Oxygen delivery from dissolved and hemocyanin-bound O_2 fractions in normoxic and hypoxic lobsters. Black bars, percent of O_2 delivered from O_2 in solution. White bars, amount (in %) delivered from Hc-bound O_2. (From McMahon and Wilkens, 1975.)

Similar properties for Hc from both hypoxic and normoxic *H. gammarus* were presented by Butler *et al.* (1978) and McMahon *et al.* (1978). A detailed comparative study of Hc of *H. americanus*, *H. gammarus*, and their natural hybrids revealed significant but small differences in subunit composition, O_2 affinity, cooperativity (Fig. 6), and sensitivity to ions and lactate between species, with the hybrids occupying a position intermediate between the two but more similar to *H. gammarus* (Mangum, 1993).

F. Carbon Dioxide Elimination and Acid–Base Balance

In addition to the transport of O_2, the circulatory system in *Homarus americanus* is also responsible for the reverse transport of CO_2 which is eliminated across the gills. The majority of CO_2 carried by the hemolymph is carried as bicarbonate ion (HCO_3^-). A smaller, but unknown, fraction may be bound to protein (carbamino-bound CO_2), while the remainder travels as free CO_2 (i.e., in solution). Partial pressures of CO_2 within the environment and hemolymph of *H. americanus* are presented in Fig. 5B. The amount of CO_2 carried in the hemolymph at any one time depends on the rate of production, but also on many other factors, such as pH, temperature, ionic composition, and ambient CO_2 level. In quiescent, nonrestrained lobsters, circulating CO_2 levels in arterial and venous hemolymph (Fig. 15) are essentially similar to those of other marine crustaceans under similar conditions. Of the total CO_2 present in lobster hemolymph (C_{CO_2}, i.e., bound plus dissolved), more than 90% is present as HCO_3^-. Total CO_2 in prebranchial hemolymph (Cv_{CO_2} = 5.95 ±0.06 m*M*) exceeds that in postbranchial (arterialized) hemolymph (Ca_{CO_2} = 5.84 ±0.2 m*M*) by 0.11 m*M*, which represents the loss of CO_2 across the gill surface. Circulating CO_2 partial pressures (Pa_{CO_2} = 1.97 ±0.06 mm Hg; Pv_{CO_2} = 2.52 ±0.17 mm Hg) are much lower than those for oxygen, reflecting the much greater solubility and hence diffusibility of this gas in body fluids. While it is generally assumed that the majority of bicarbonate is reconverted to CO_2 by the action of carbonic anhydrase on the branchial epithelium prior to excretion (Burnett and McMahon, 1987), both free CO_2 and HCO_3^- can be eliminated at the outer gill surface. The ratio of these substances eliminated varies complexly with the demands of branchial ion and acid–base regulatory processes (see Section V of this chapter). It is generally assumed that most excretion of CO_2 occurs at the gill surface, but possible contributions of the antennal glands and body surface cannot be discounted.

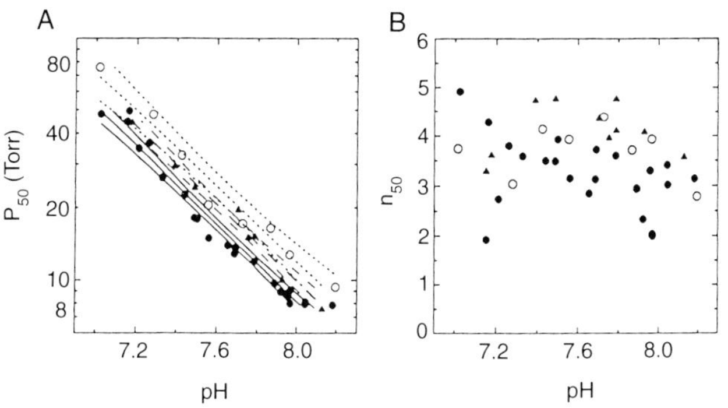

FIGURE 6 (A) Affinity (P_{50}), and (B) cooperativity (n_{50}) of oxygen binding by *Homarus americanus* hemocyanin (closed circles, solid lines) in comparison with that of *H. gammarus* (open circles, dotted lines) and their hybrids (triangles, dashed lines). Curves are fitted regression lines ±95% confidence limits. (From Mangum, 1983b.)

Protons (H^+) are thought to be a continual by-product of metabolism. Excess protons are neutralized by an extremely complex series of interrelated processes conveniently, but loosely, referred to as acid–base balance. In *Homarus americanus*, as in other crustaceans, this balance involves many factors; these include production and elimination of CO_2 and H^+ and regulation of respiratory and hemolymph flows, ions and nitrogenous end-products, and most other bodily processes in minor or less well-understood ways. Acid–base balance occurs in both the intra- and extracellular compartments. Most extracellular acid–base balance occurs at the gill surface, where it is a function of balance between the rates of the elimination of CO_2 and NH_4^+ and the rates of loss and uptake of many ions. All of these processes are separate and may vary independently, but they are also interrelated in such a way that the requirements of acid–base balance can be met in a variety of ways (see Section V of this chapter). In *Homarus* and other poikilotherms, intra- or extracellular pH is not regulated, but can vary with temperature and other factors. What is regulated is the H^+:OH^- ratio or the degree of dissociation of alpha-imidazole, which, in turn, affects the ionization and activity states of the metabolic proteins. In effect, however, we can think of the regulation of pH, or better of H^+ or OH^-, as long as other major conditions are held constant. At 15°C, the pH of circulating hemolymph is 7.800 (16 n*M* H^+) for arterialized and 7.704 (19.8 n*M* H^+) for venous hemolymph in *H. gammarus* (McMahon *et al.*, 1978).

G. Pauses

The periods of intermittent ventilation (pauses) mentioned above are typical of decapod crustaceans.

Periods of virtually complete cessation of ventilation and acardia (Fig. 3) have been noted in *Homarus americanus* (McMahon and Wilkens, 1972). These can be as long as 20 or more minutes in undisturbed, but fettered, lobsters. Such pauses may serve to conserve resources or may simply be a function of insufficient stimulation of the respiratory and cardiac drivers due to high O_2 availability under quiescent experimental conditions. Ventilation thus can be seen as an activity which is used only when necessary, i.e., when O_2 stores are depleted. In support, these pause periods are seen less frequently during exercise (Guirguis and Wilkens, 1995) or hypoxia when oxygen need is greatest (McMahon and Wilkens, 1975).

III. Circulation

The anatomy of the open circulatory system of crustaceans is reviewed by McLauglin (1983) and is described for *Homarus americanus* by Martin and Hose (Chapter 17). McLauglin's (1983) proposed unified terminology is followed here. Briefly, hemolymph flow throughout the body is powered by a single, condensed, muscular ventricle which is suspended within a reservoir chamber, the pericardial sinus. The ventricle pumps hemolymph through seven separate arterial systems which supply specific body regions (see Martin and Hose, Chapter 17, Section II,A and Fig. 1). The anterior aorta (=ophthalmic artery, cephalic artery) delivers hemolymph to the cephalic area; paired anterior lateral (=antennal) arteries supply the anterior viscera, antennal glands, etc., and paired lateral (=hepatic) arteries supply the digestive gland. A large single sternal artery descends ventrally and splits to form ventral thoracic and ventral abdominal arteries supplying the limbs, some anterior appendages, and the ventral nervous system. Finally, a large posterior aorta (=dorsal abdominal artery) supplies the abdominal musculature and swimmerets via paired, segmental, abdominal arteries. The fine details of the vascular distribution system have not been determined for *H. americanus*, but a complex system of distribution vessels (arterioles) of similar size to vertebrate capillaries are found in many tissues (see illustrations by Renata Sandeman, as published in McMahon and Burnett, 1990; see Martin and Hose, Chapter 17). Hemolymph is collected from these vessels by a tightly organized system of sinusoids and sinuses which deliver hemolymph back to the pericardial sinus via the gills and the branchio-pericardial "veins." An alternate return route for hemolymph passing through the branchiostegal walls is known for some brachyuran decapods and likely also occurs in *H. americanus*. Hemolymph reenters the ventricle via three pairs of muscular ostial valves (Fig. 7).

A. Heart

1. Structure

The heart is composed of multinucleate, striated, cardiac muscle fibers which make end-to-end contact at intercalated discs to form a loosely arranged network with several irregular layers of cross-connecting muscle strands (Maynard, 1960; Martin and Hose, Chapter 17). Muscle cell innervation occurs from the motor cells of the cardiac ganglion and is both multiterminal and polyneuronal. Excitatory junction potentials that show both summation and facilitation have been reported for *Homarus americanus* (Anderson and Cooke, 1971). The ostial valves have separately innervated musculature (Alexandrowicz, 1932), but active closure has not been reported. The heart is suspended within the pericardial sinus by the arteries and by the elastic alary ligaments which are important in diastolic filling (see Section III,A,6).

2. Excitation

In *Homarus americanus*, excitation of the heart is neurogenic with the pacemaker information being

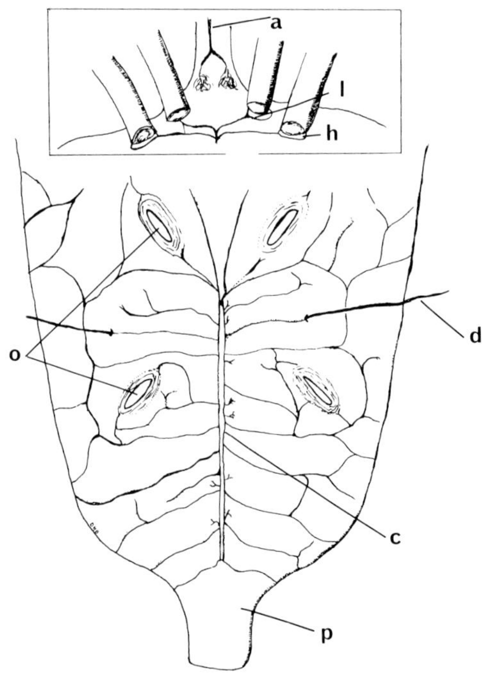

FIGURE 7 Diagram of the lobster heart showing arterial outputs (a,h,l,p), ostial valves (o), nerve dorsalis (d), and the location and innervation (N. dorsalis) of the cardiac ganglion and its cells (c) and the location and innervation of the anterior cardiac valves. (Modified from Alexandrowicz, 1932.)

initiated within the cardiac ganglion located medially on the inner dorsal wall of the heart (Fig. 7) (Alexandrowicz, 1932). Input from the central nervous system can modify heart rate, but is not needed to initiate beating. The cardiac ganglion contains nine neurons of two types. These have been termed small (pacemaker) and large (follower) cells. The four pacemaker cells have traditionally been associated with initial excitation and the five follower cells with innervation of the myocardium; however, both are autorhythmic and it is unlikely that their functions are so clearly differentiated (Berlind, 1985). Notwithstanding the above, the output of the cardiac ganglion is a neural burst transmitted to the myocardium via the axonal tree of the follower cells. As noted above, innervation is both polyneuronal and multiterminal, ensuring efficient, rapid contraction of the entire muscle mass.

3. Autoregulation

Under appropriate conditions, hearts isolated from the central nervous system (CNS) still beat but at a much slower rate. It has long been reported that perfusion of the isolated *Homarus* heart increases the beat rate (Maynard, 1960). It was thus thought that this heart exhibits a stretch response similar to the Frank–Starling response, well known for mammalian hearts. Inflation of a naturally beating, semi-isolated heart of *H. americanus* with well-oxygenated Ringers increases heart rate (Wilkens, 1993); inflation with oxygen-depleted Ringers solution, however, does not, indicating that the apparent "Starling effect" is associated more with reoxygenation of the myocardium rather than with stretch of the muscle fibers. Further work is needed to resolve this interesting question.

4. Neural Control of Heart Performance

In *Homarus americanus*, the cardiac ganglion is innervated by two pairs of cardioaccelerator (CA) and a single pair of inhibitor (CI) nerves which arise from the subesophageal ganglion (Fig. 7) (Alexandrowicz, 1932). By analogy with other decapods (reviewed by Wilkens, 1989, and McMahon *et al.*, 1995), it seems likely that these cardioregulatory nerves are at least partially driven by command interneurons arising from the circumesophageal connectives from the cephalic ganglia. Tonic activity is normally present and increased firing has been associated with a variety of stimuli, including increased ventilatory rate and stimulation of sensory receptors, and with activity. The inhibitory transmitter in *Homarus* is GABA (gamma-amino-butyric acid); the excitatory transmitter is not known, but a variety of substances, including acetylcholine, are known to excite the cardiac ganglion cells. Direct effects of neural stimulation of *H. americanus* cardioregulatory nerves has not been reported, but in the semi-isolated heart of the crayfish *Procambarus clarkii* stimulation of the CI nerves causes bradycardia and even short periods of cardiac arrest (Wilkens and Walker, 1992). The effects of CI stimulation are usually fast adapting, which is interesting since long periods (20 min) of cardiac arrest are seen in *H. americanus* under quiescent laboratory conditions (McMahon and Wilkens, 1972, 1975). Stimulation of the CA nerves increases both heart rate and contractility (Wilkens and Walker, 1992), but, in contrast with the effects of CI stimulation, the effects of CA stimulation are quite long-lived. In part, this may be due to stimulation of neurohormonal release from the pericardial organs, which also receive innervation from the CA nerves (McMahon *et al.*, 1995). The CA nerves may also play a role in autoadjustment of heart function in activity in *H. americanus*. Rapid increase in heart rate accompanying activity is dependent on intact CA innervation, although slow and incomplete rate increase still occurs after denervation (Fig. 14) (Guirguis and Wilkens, 1995). Heart rate also increases in compensation for decrease in stroke volume following cutting some of the alary ligaments, and this is similarly dependent on innervation from the CNS.

5. Neurohormonal Control of the Heart

Many neurohormones, including a number known to have effects on the heart, are present in the pericardial organs, a major neurohemal release site situated in the pericardial sinus adjacent to the heart. Substances known to affect either rate or amplitude of beating in semi-isolated *Homarus americanus* hearts include acetylcholine, dopamine, 5-hydroxytryptamine, octopamine, and the peptides proctolin, CCAP, and several FMRFamide-like peptides (Kuramoto *et al.*, 1995). Studies on the effect of some of these substances on the heart of *H. americanus* and other decapods *in vivo*, however, indicate that much caution must be used in the interpretation of the results of experiments on isolated hearts. Although the majority of these substances are cardio-excitatory on hearts maintained *in vitro*, *none* have cardio-excitatory effects when injected *in vivo* (McGaw *et al.*, 1994). Furthermore, addition of these substances only increases the *in vitro* rate to levels normally seen *in vivo*. It thus seems likely that the heart rate *in situ* is the product of a complex blend of neurohormonal and neural input. Variations in the blend are used to adjust cardiac performance to meet the demands of variation in activity level or environment (as discussed in Section IV).

6. Mechanical Performance of the Heart

Contraction of the ventricle drives hemolymph out through the cardio-arterial valves at the bases of each arterial system (Fig. 7). Pressure waveforms recorded from the arterial system of *Homarus americanus* (Fig. 8) (Jorgenson and Farrell, personal communication) are essentially similar to those of other decapods (reviewed by McMahon and Burnett, 1990). In early systole, ventricular pressure rises rapidly since the ostial valves close immediately and the cardio-arterial valves remain closed until pressure in the ventricle exceeds that in the artery (Fig. 8). Pressures developed by the ventricle are 27/6 cm H_20 (2.65/0.6 kP*a*), systolic over diastolic. Pressure in the posterior aorta is approximately 26/19 cm H_20 (2.56/1.85 kP*a*). Note that ventricular pressure exceeds arterial pressure during much of systole as hemolymph is forced into the artery from the heart. Following ventricular ejection and the closure of the arterial valves, intraventricular pressure falls rapidly to diastolic levels as the myocardium relaxes. These pressure measurements, made using sensitive pressure transducers, lie within the range measured more crudely by Burger and Smythe (1953). The rather square shape of the pressure waveform is thought to be associated with rapid runoff through the relatively low-pressure circuits found in crustacean distribution systems (G. B. Bourne, personal communication).

Some of the energy of systolic contraction is transferred to elastic tissues in the arterial trunks and the alary ligaments as these are stretched by the decrease in ventricular volume. In diastole, this elastic energy is essential to return the ventricle to its presystolic volume. This volume change can only occur as hemolymph enters the heart through the open ostial valves. Pressure within the ventricle thus falls below that in the surrounding pericardial sinus during early diastole, but this pressure differential is small (approximately 0.1 kP*a*), reflecting the low resistance of the three pairs of ostial valves. Pressures within the pericardial sinus are maintained above zero, suggesting a degree of overall body inflation resulting from cardiac activity. This "turgor" pressure is presumably important in maintenance of the hydrostatic components of the skeleton.

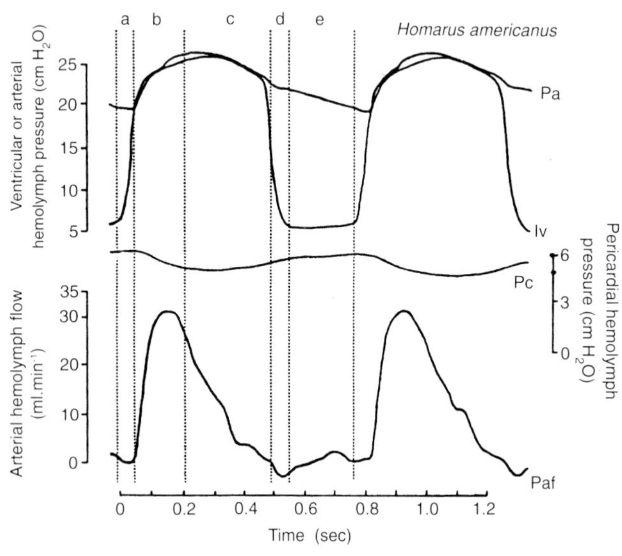

FIGURE 8 Pressure profiles recorded via catheters from the ventricle and cardiovascular system of *Homarus americanus*. (From Jorgenson and Farrell, personal communication.)

7. Cardiac Output and Its Control

Until recently, cardiac output has been measured only indirectly by use of the Fick principle (McMahon and Wilkens, 1983). Such estimates have been reported for both *Homarus americanus* (22–63 ml/min, Burger and Smythe, 1953; 61 ml/min, McMahon and Wilkens, 1975) and *H. gammarus* (43 ml/kg/min, McMahon *et al.*, 1978). At least during sustained high heart rates, typical of disturbed animals, pressure in several posterior arteries is maintained substantially above pericardial sinus pressure throughout diastole (see sternal artery traces, Figs. 10 and 11). This is a function of the elasticity of the vessel walls (see Martin and Hose, Chapter 17) and, particularly in the case of these arteries, of the bulbous arteriosus, which can be observed to pulsate markedly with each heart beat in *H. americanus* (B. R. McMahon, unpublished observations). A similar "windkessel" effect has been described for the bulbous arteriosus of the crayfish *Procambarus clarkii* (Reiber, 1994). Arterial pressures recorded for *H. americanus* are similar to those reported for the spiny lobster *Panulirus interruptus* (Belman, 1975). New methods of flow measurement, using miniature pulsed Doppler transducers implanted adjacent to each of the arteries leaving the heart, have been used in *H. americanus* to allow more direct measurement of cardiac output (Reiber *et al.*, 1992; McMahon and Reiber, 1991). Where these flows are recorded simultaneously, summation of each individual flow yields cardiac output. Comparison of these values (Table 3) with the earlier Fick estimates is surprisingly good (Reiber *et al.*, 1995), but, unlike the earlier estimates, these can be made without substantial disturbance and allow us to monitor rapid changes occurring in response to natural stimuli. Based on these direct measurements, stroke volume of quiescent *H. americanus* is 0.7 ml and cardiac output is 54 ml/min (or 94 ml/kg/min) at 15°C.

Cardiac function is not constant in *Homarus americanus*, however, and heart rate and especially stroke

TABLE 3 **Arterial Flows and Cardiac Output in *Homarus americanus***

	Flow rate	Percentage of cardiac output
Anterior aorta	7.76 ±0.76 ml/min	12.8
Sternal artery	38.90 ±4.1 ml/min	64.0
Ventral abdominal artery	0.28 ±0.04 ml/min	0.5
Ventral thoracic artery	3.98 ±0.10 ml/min	6.5
Lateral artery (right)	0.57 ±0.04 ml/min	1.0
Posterior aorta	12.61 ±1.0 ml/min	20.7
Cardiac output	60.8 ±4.4 ml/mm	
Heart frequency	82.5 ±2.9 beats/min	
Stroke volume	0.69 ±0.05 ml/beat	

volume can vary considerably with activity and other factors, leading to marked changes in cardiac output (as discussed in Section IV).

B. Vascular System

1. Structure

The fine structure of the arterial distribution system vessels is well reviewed by Martin and Hose (Chapter 17). Generally, crustacean arteries are highly elastic structures which distend markedly at low pressure, but become quite stiff at higher pressure (Shadwick *et al.*, 1990). Videographic analysis of dissected sections of the arterial system of *Homarus americanus* shows that marked distension occurs with each heart beat in all arterial systems tested [Table 4; only the hepatic (lateral) arteries were not tested]. It is important to reemphasize here that the distribution system is arranged so that each arterial system perfuses a particular body region.

2. Arterial Flow

Previous thinking concerning hemolymph flow in crustaceans suggested that the hearts of these animals are capable of generating only low (sluggish) flow. The use of microflow transducers has also allowed the first assessment of arterial flow patterns in *Homarus americanus*. In fact, such measurements of both cardiac output and arterial flow indicate that these are similar to or larger than those of poikilothermic vertebrates at equivalent temperature. Flow varies predictably among arterial systems (Fig. 9) and with a variety of internal and external conditions. In quiescent but tethered lobsters, the majority of the flow passes ventrally through the sternal artery and via the ventral thoracic and ventral abdominal arteries to supply the limbs. The posterior aorta passes 21% and the anterior aorta passes 13% of the total cardiac

TABLE 4 **Arterial Diameters in *Homarus americanus***

Vessel	Systolic diameter (D_s)	Diastolic diameter (D_d)	ΔD (D_s–D_d)	Mean diameter (D_{XA})
Anterior aorta	1.16 mm (±0.11)	0.94 mm (±0.11)	0.24 mm	1.1 mm
Sternal artery	1.92 mm (±0.09)	1.67 mm (±0.09)	0.25 mm	1.83 mm
Ventral abdominal artery	0.39 mm (±0.02)	0.27 mm (±0.03)	0.12 mm	0.35 mm
Ventral thoracic artery	1.44 mm (±0.21)	1.23 mm (±0.16)	0.25 mm	1.39 mm
Lateral artery (right)	0.93 mm (±0.03)	0.76 mm (±0.01)	0.17 mm	0.87 mm
Posterior aorta	1.98 mm (±0.08)	1.61 mm (±0.14)	0.38 mm	1.86 mm
Hepatic artery	n.d.	n.d.	n.d.	0.45 mm

Note. Mean arterial diameters were calculated from observations of six lobsters and are expressed as mean ± 1 SE. The hepatic artery was not readily accessible to obtain a pulsatile diameter; the value given indicates a visual determination of mean diameter (n.d., not determined). (From Reiber, 1994.)

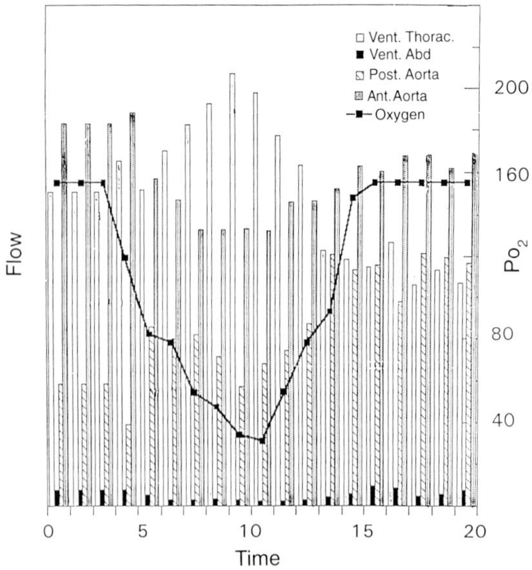

FIGURE 9 Histogram of arterial flow distribution in *Homarus americanus* under normoxic and hypoxic conditions and during recovery. P_{O_2}, Oxygen partial pressure in mm Hg. Flow in arbitrary units. Time in hours. (From McMahon, 1992.)

output. Only 2% of flow passes through the paired anterolateral (antennal) arteries. Flow through the paired hepatic arteries has not been measured, but these arteries are small (Table 4). By analogy with flows measured in the crab *Cancer magister* (Airriess and McMahon, 1994), hepatic arterial flow is assumed to be rather small, at least in unfed animals.

Within each arterial system, flow appears highly variable both with time and during various behavior patterns. Variability in flow in a particular arterial system often seems to occur without concurrent variation in heart rate, suggesting that *Homarus americanus* is able to adjust flow through each arterial system independently. Studies confirm this and, further, indicate that special circulatory "reflexes" are able to shunt hemolymph flow into particular arterial systems. Due to the highly regional nature of the arterial distribution, this effectively allows adjustment of hemolymph flow to certain body regions. These circulatory reflexes can be well illustrated by the response of *H. americanus* to hypoxia (see Section IV,A).

3. Regulation of Regional Flow

The mechanism by which regional flow is controlled in lobsters is not well understood. There is usually no muscle in the walls of the distribution vessels (see Martin and Hose, Chapter 17) and thus control by change in peripheral resistance, which is typical of vertebrates, cannot occur here. Part of the answer probably lies in the cardio-arterial valves which are located at the origin of each arterial tree. These are similar to vertebrate cardio-arterial valves in that they can prevent reflux of hemolymph during diastole, but differ in that they contain muscle (Alexandrowicz, 1932). These muscles respond to both neural and neurohormonal stimulation to alter the degree of tension on the flap wave, which has been shown to affect flow in the sternal artery and posterior aorta of *Homarus americanus* (Kuramoto *et al.*, 1995).

There is evidence for both neural and neurohormonal control of arterial flow patterns in *Homarus americanus*. Suggestive evidence for neural control is shown in Fig. 10. Here, a sudden startling stimulus causes an initial short bradycardia, followed by marked alteration in flow patterns in several arteries. The rapidity of the changes in both heart rate and arterial flow strongly suggest neural effects on both the cardiac ganglion and cardio-arterial valves. Direct effects of stimulation of the nerve of the posterior valve of *H. americanus* have been demonstrated (Kuramoto *et al.*, 1995).

Neurohormonal control of regional hemolymph flow has also been demonstrated in *Homarus americanus*. The peptides proctolin (Starrat and Brown, 1975) and the FMRFamide-like peptide F1 (Kobierski *et al.*, 1987) are both found in the pericardial organs of *H. americanus*. Substances released from these organs would enter the hemolymph immediately prior to its entry into the heart, a prime situation for influencing cardiac function. Experimental *in vivo* infusion of either proctolin or F1 into the pericardial sinus of *H.*

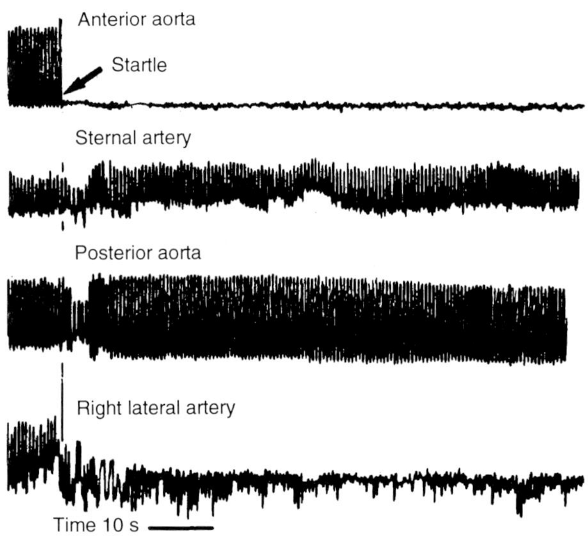

FIGURE 10 Effects of a sudden startling stimulus on patterns of arterial flow distribution in *Homarus americanus*. Stimulation occurred at arrow. Flow resumed in anterior aorta and right lateral artery only after 1 minute. (Modified slightly from McMahon, 1992, to increase trace clarity.)

americanus produces dramatic dose-related effects on flow in several arterial systems (McMahon and Reiber, 1991; McMahon, 1992). Proctolin (Fig. 11) causes dramatic decrease in flow through the anterior aorta and anterolateral arteries and some decrease in the posterior aorta, but markedly increased flow through the sternal artery (Figs. 9 and 12). Conversely, F1 has virtually opposite effects, decreasing sternal arterial flow and increasing flow into the other two systems. Since the sternal artery largely supplies structures associated with locomotion and ventilation and the anterolateral artery supplies visceral structures, the effects of these peptide hormones appear functionally analogous to those of the autonomic nervous system in terms of the control of regional hemolymph flow in vertebrates. It is important to note, however, that although both peptides have been described as cardioaccelerators in isolated *H. americanus* hearts (Kobierski *et al.*, 1987), they have no excitatory effect on heart rate *in vivo* (Fig. 11). Wilkens and McMahon (1992) introduced a semi-isolated preparation in which the heart retained its normal mechanical connections; thus, it maintained normal mechanical function, but was isolated from normal neural and neurohormonal stimulation. Aminergic substances enhance both rate and force of beating in these semi-isolated *H. americanus* hearts (Kuramoto *et al.*, 1995), but their effects have not been tested *in vivo*. Serotonin, octopamine, and dopamine, however, are all known to affect both heart and arterial hemolymph flow patterns *in vivo* in the brachyuran decapod *Cancer magister* (Airriess and McMahon, 1992) and would probably be active in *H. americanus* also. At least in the posterior aorta, more precise regional control is provided by muscular valves at the origin of segmental lateral arteries (J. L. Wilkens, personal communication).

IV. Integrated Respiratory and Circulatory Responses and Physiological Compensation

A. Hypoxia

Variations in the external environment inevitably have major effects on either oxygen supply or demand or both. In crustaceans, complex integrated responses involving several systems allow physiological compensation by which oxygen supply is matched to oxygen demand. For *Homarus americanus*, these mechanisms can be illustrated by the response to hypoxia, which has been reasonably well studied. American lobsters are occasionally subjected to hypoxic conditions within their natural habitat, for example, when bottom water is trapped beneath surface layers and stagnates. Initial exposure to hypoxic conditions is associated with increase in ventilatory (scaphognathite) pumping, which, in turn, increases the volume of water flowing across the gills (Table 1) (McMahon and Wilkens, 1975). This could serve to restore oxygen delivery, provided that the animal can increase the efficiency of oxygen extraction from the water. This is achieved in *H. americanus* by a variety of factors (McMahon and Wilkens, 1975; Butler *et al.*, 1978; McMahon *et al.*, 1978). First, oxygen levels in both arterial and venous hemolymph decrease (Table 2, Fig. 5). This increases the diffusion gradient for oxygen across the gills and also increases the transfer factor for oxygen diffusion by increasing the participation of Hc O_2 binding in gas exchange (Table 2, Fig. 4). Second, oxygen affinity increases which also increases the diffusivity of oxygen across the gills. Two factors are involved in this increase in affinity. The first involves an increase in pH, which occurs as additional CO_2 is lost across the gills during initial hyperventilation (reasoning from Butler *et al.*, 1978).

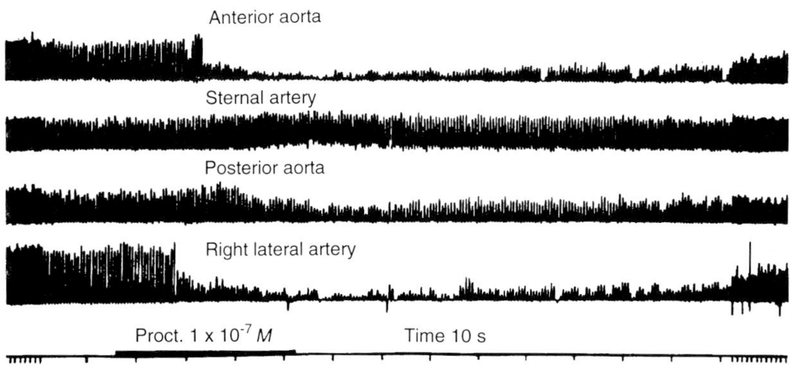

FIGURE 11 Effects of infusion of proctolin into the pericardial sinus on arterial flow patterns in *Homarus americanus*. Period of proctolin infusion indicated by the bar on time trace.

The second occurs as a result of increased production of lactate from an increase in anaerobic metabolism and its release into the hemolymph. Both factors cause an increase in oxygen binding affinity of Hc, the first via the Bohr effect and the second via a specific action of lactate on oxygen binding by lobster hemocyanin (B. R. Mangum, 1983b; Truchot, 1992).

The rate of beating of the swimmerets (abdominal pleopods) also increases dramatically in hypoxic exposure (B. R. McMahon, unpublished observations). The amount of oxygen uptake occurring over the pleopod surfaces is likely to be minor due to the relatively small area and thick wall of these structures when compared to the gills. More probably, the movements of the pleopods either function to cause local mixing, which refreshes the partially O_2-depleted water near the lobster, or may be part of a locomotor response (Cattaert and Clarac, 1983) designed to move the animal to a new, potentially less-depleted area.

Concomitant changes are also observed in the cardiovascular system. Unlike the situation for ventilation rate, heart rates decrease progressively in hypoxia (Table 2) (Reiber *et al.*, 1992). Cardiac stroke volume, however, is maintained or increased, limiting the decrease in cardiac output. Coupled with this in severe hypoxia, there is shunting of hemolymph away from the more anterior arteries and into the sternal artery and particularly into the ventral thoracic artery (Fig. 9) (McMahon, 1992). This allows increased perfusion of the limbs and other appendages, including the scaphognathites (McMahon, 1992). More importantly, increased flow through this ventral route could increase return flow through the gills at the expense of flow returning through the branchiostegites, which preferentially receive hemolymph from the anterior lateral arteries (Von Raben, 1934). Similar, but more marked, changes accompany hypoxic exposure in the crayfish (Reiber *et al.*, 1992) and crab (Airriess and McMahon, 1994). Taken together, these modifications increase both water and hemolymph flow through the gills and also further increase the effectiveness of hemolymph oxygenation.

Surprisingly, in the face of the above, Amberson *et al.* (1924) and Thomas (1954) reported that *Homarus americanus* and *H. gammarus* are both "oxyconformers," i.e., they have little or no regulatory ability and allow oxygen consumption to fall progressively with external oxygen level. Despite these early opinions, however, recent studies confirm that both species are capable of maintaining oxygen consumption in quite severely hypoxic water (Fig. 12) (McMahon and Wilkens, 1975; Butler *et al.*, 1978). The disagreement probably stems from differences in experimental

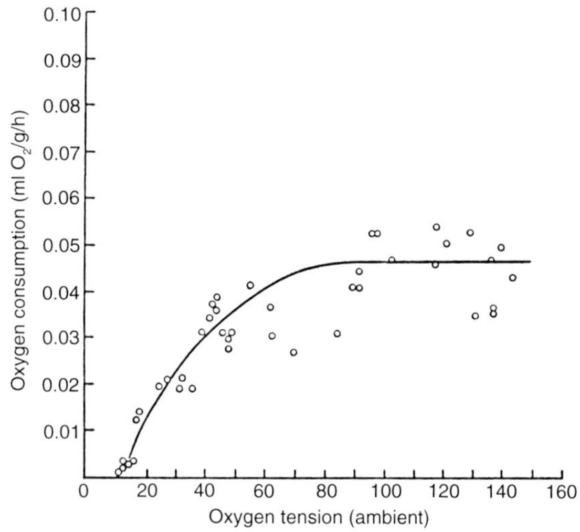

FIGURE 12 Oxyregulation: Oxygen consumption as a factor of ambient oxygen depletion in *Homarus americanus*. (From McMahon and Wilkens, 1975.)

regime. In the earlier studies, ambient oxygen levels were allowed to decline progressively and unreasonably rapidly. The more recent studies utilized step changes, where oxygen was depleted rapidly but then maintained at a constant level. The former regime does not allow sufficient time for compensatory responses to develop.

B. Activity

Homarus americanus is a relatively active species with both juveniles and adults capable of migrating approximately 2 km/day (Cooper and Uzman, 1980). Several studies have examined the physiological responses during increased activity. An examination of energetics of walking in *H. americanus* showed that this lobster could maintain oxygen consumption throughout 40 minutes of sustained walking at moderate speed (Fig. 13) (Houlihan *et al.*, 1985). There was little evidence of oxygen debt following sustained walking, suggesting that this activity is largely aerobic. Oxygen consumption is linearly related to walking speed over the range 1–4 m/min. Factorial scope for sustained increase in oxygen consumption during walking in *H. americanus* (3–4) is similar to that measured for activity in other crustaceans (McMahon and Wilkens, 1983) and is not significantly changed with increase in mass over the weight range used. The demonstrated capabilities of *H. americanus* for sustained aerobic activity are correlated with a preponderance of slow, mitochondria-rich muscle fibers in the limb depressor muscles (Houlihan *et al.*, 1985).

The increase in cardiac output that generally accompanies activity in crustaceans (McMahon and

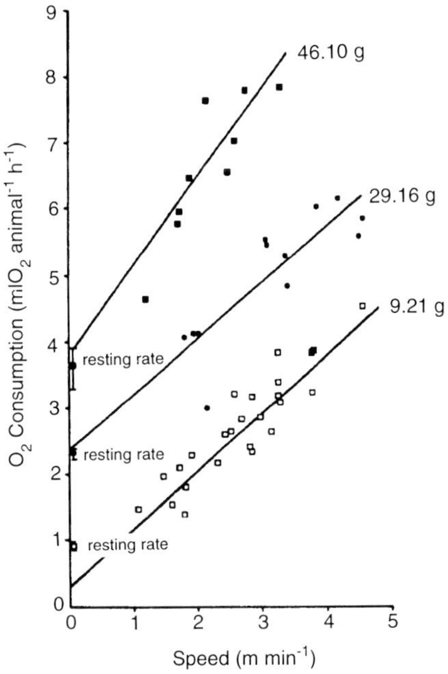

FIGURE 13 Rates of oxygen consumption as a function of walking speed in three size classes of *Homarus americanus* at 22°C. Mean values for inactive lobsters are included against the ordinate. Lines are linear regressions plotted in each case without the data for inactive animals. (From Houlihan *et al.*, 1985.)

Wilkens, 1983) can result from increase in either heart rate or stroke volume. In intact *Homarus americanus*, heart rate increases approximately 60% at the onset of treadmill walking (Fig. 14) (Guirguis and Wilkens, 1995). Cutting the cardioregulatory nerves abolishes this rapid increase, but heart rate continues to increase slowly during activity. These data support the conclusion that the initial rapid increase is caused by input from the CNS, but that a slower component, possibly associated with neurohormonal stimulation, may be important in maintaining elevated heart rates during sustained activity (Guirguis and Wilkens, 1995). Data from several brachyuran crustaceans (McMahon and Burnett, 1990; De Wachter and McMahon, unpublished observations) indicate that patterns of hemolymph distribution also change during activity, with increased flow diverted into the sternal artery as noted for hypoxia (see Section IV,A). The importance of this step is clear since increase in flow in this artery increases O_2 supply to the limbs, as well as to the scaphognathites and the ventral nervous system, which coordinates locomotor responses. The swimmerets often also beat during walking in *H. americanus* (McMahon and Wilkens, unpublished observations) and in *H. gammarus* (Cattaert and Clarac, 1983). Fast swimmeret movement seen at the onset of locomotor periods may be associated with startle responses, but a second slow rhythm is often seen in extended walking periods. The extent to which this response contributes to the locomotor process is not known.

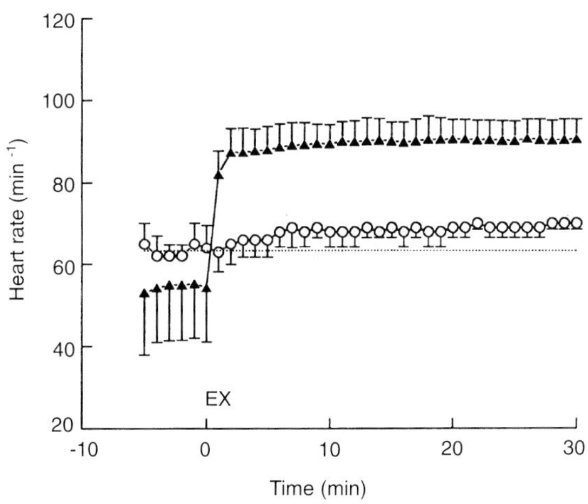

FIGURE 14 Effects of severing cardioaccelerator innervation on heart rate response to treadmill walking in *Homarus americanus*. EX, Start of treadmill walking; (▲), control; (O), denervated. Bars, ±SEM (standard error around the mean). (From Guirguis and Wilkens, personal communication.)

C. Effects of Disturbance on Respiratory and Circulatory Performance

Disturbance of any kind has strong effects on ventilatory performance and thus on circulating levels of oxygen, CO_2, and pH. Enhanced ventilatory performance is associated with preexperimental procedures involving masking, catheter implantation, and air exposure in *Homarus gammarus* (McMahon *et al.*, 1978). The increased ventilation volume is associated with higher O_2 and CO_2 partial pressures, but lower contents, and with decreased pH (increased [H^+]). Levels do not fully stabilize until 48 hours after the handling stress (Fig. 15). While possibly not part of natural responses of lobsters, disturbance is a usual consequence of laboratory experiments. In light of this, it is important to minimize all handling procedures and to allow sufficient time for the animals to recover from preexperimental procedures before making physiological measurements. Failure to do this risks complicating or confusing responses to the desired test condition with additional effects associated with operator or other disturbance. Effects of air exposure per se have not been investigated for *H. americanus*, but can be expected to be similar to those reported for *H. gammarus* (Taylor and Whitley, 1993).

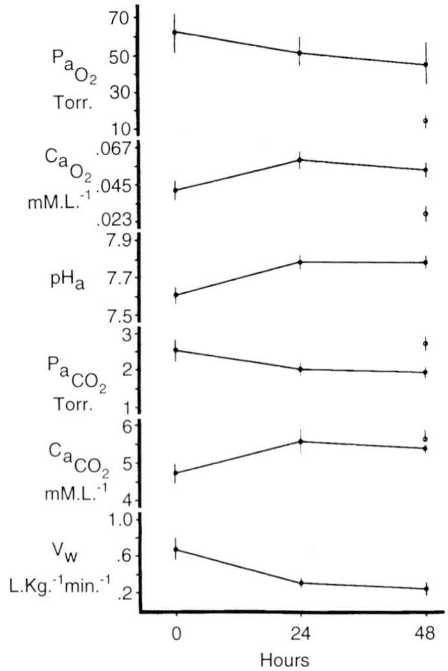

FIGURE 15 Effects of handling and disturbance on ventilation and hemolymph acid–base parameters in *Homarus americanus*. V_w, ventilation volume; Pa_{O_2}, O_2 pressure; Pa_{CO_2}, CO_2 pressure; Ca_{O_2}, O_2 content; Ca_{CO_2}, CO_2 content; closed circles, pH*a*, pH of arterialized (postbranchial) hemolymph; open circles, pH*v*, pH of venous (prebranchial) hemolymph at the 48-hour mark. All ±SEM, temperature at 15°C. (From McMahon *et al.*, 1978.)

V. Osmotic and Ionic Regulation and Nitrogen Excretion

It is not possible to treat ionic and osmotic regulation and the excretion of nitrogenous end-products in *Homarus americanus* in isolation since they are all very closely interrelated. In the case of water and salt regulation, the interrelationship is obvious—a change in water balance will affect salt balance and vice versa. In the case of nitrogenous excretion, the interrelationship is less obvious, but equally important. In an aquatic animal such as *H. americanus*, the vast majority of excess nitrogen arises from deamination of amino acids. The NH_3 so formed is highly toxic and must be eliminated or converted into less toxic excretory products, such as urea and uric acid. Depending on pH, a major fraction of NH_3 is converted into NH_4^+ by reaction with protons, but both are usually present in body fluids at normal pH. Both are osmotically active and thus must be taken into account when considering water balance. Additionally, NH_4^+ is a charged particle and acts as an important counterion in many ion-regulatory processes. Amino acids themselves are also important osmotic effectors in water balance. For these reasons, these topics are considered together.

Three organs, the gut, the antennal glands, and the gills are implicated in excretory and ion-regulatory function in *Homarus americanus*. The renal system differs radically from the other two in that its input fluid is entirely internal (i.e., hemolymph), while the gills and, to a lesser extent, the gut can interact directly with both the internal and external media. The structure of the intestine of *H. americanus* is considered by Mykles (1979), from the perspective of the transporting epithelium, and by Factor (Chapter 15). The structure of the antennal gland is considered in Section V,A. The structure of the gills is considered in Section II,A.

Homarus americanus lives in coastal waters and has been taken from waters of below 25 ppt salinity (Burger, 1957; Cole, 1940). The limits of salinity tolerance are a complex function involving several variables (Fig. 16) (McLeese, 1956). Despite a reasonable level of tolerance, the ion-regulatory capability has been described as limited and *H. americanus* is largely an osmoconformer in dilute media (Robertson, 1960). Ion concentrations in animals in normal-strength seawater, and the extent of osmotic regulation at salinities below 25 ppt (Fig. 17), have been determined by McLeese (1956) and Dall (1970). Hemolymph ion levels are not massively different from seawater on the one hand and urine on the other, confirming that little regulation occurs. Hemolymph, however, is always slightly hyperosmotic, even in undiluted seawater, and this presumably allows the maintenance of turgor and also allows a flow of urine through the renal organs (antennal glands). Below 26 ppt seawa-

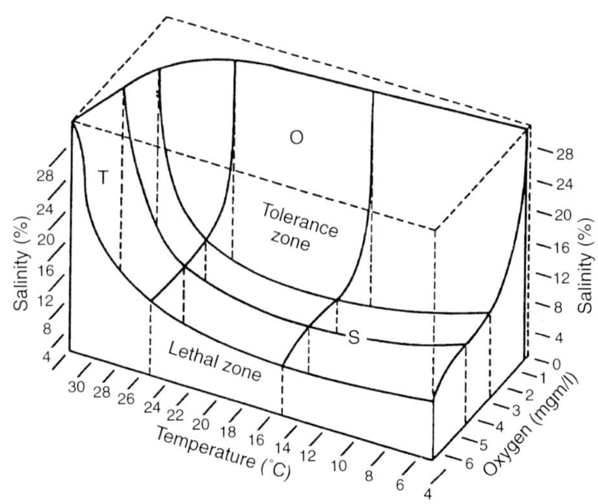

FIGURE 16 Tolerance of variation in environmental salinity (S), oxygen concentration (O), and temperature (T) in *Homarus americanus* in the laboratory. (From McLeese, 1956.)

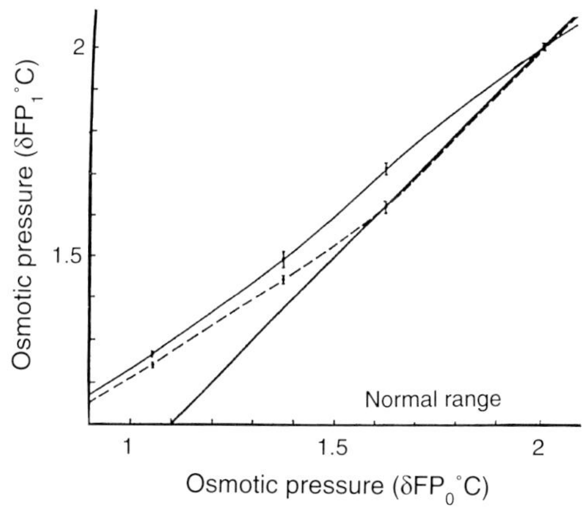

FIGURE 17 Limited osmoregulatory capability in *Homarus americanus* exposed to a range of decreased salinities. Δ*FPi* and Δ*FPo*, depression of freezing point (osmolarity) of body fluid or urine and of the medium respectively. Heavy line, isosmotic level; light line, blood; dashed line, urine. Variation at each point expressed as ±SEM. (Modified from McLeese, 1956.)

ter, clear regulation occurs which is associated with the removal of osmolytes (presumably Na^+ and Cl^-) from urine.

A. Antennal Glands

The excretory organs of adult decapod crustaceans are a pair of antennal glands located in the anterior cephalothorax near the base of the antennae. Each gland consists of a blindly ending coelomosac, which is well perfused with hemolymph, and a tightly coiled tubular section, which is divided into several anatomically distinct regions (Fig. 18). The tubule terminates in a large bladder, in which urine is held before disposal through the excretory pore.

The antennal gland differs fundamentally from the ion-regulatory sites in either the gut or gills in that exchange occurs between the hemolymph and its filtrate, thus allowing secondary regulation of filtered substances. In *Homarus americanus*, ultrafiltration occurs across the coelomosac (end sac) membrane using the hydrostatic pressure of the antennary artery, producing potential urine at a rate of 2 ml/kg/hr (Burger, 1957). In crayfish, filtration occurs through spaces between podocyte cells (Fig. 18C) and through a molecular filter in the epithelial basement membranes; a similar process is thought to occur in *H. americanus*. The structure of this ultrafiltration system is remarkably like that of the vertebrate nephron (Reigel, 1972) and is functionally similar in that it retains large molecules in the hemolymph while allowing smaller molecules to pass through. Filtered fluid then passes into a glandular region, the labyrinth, which appears to function in the secretion of wastes. Following this, the potential urine enters a tubular region which functions in the active regulation of urinary ions and active uptake of other filtered substances (e.g., glucose) as required. Reabsorption of ions from the urine has been reported for *H. americanus* during acclimation to dilute environments (McLeese, 1956; Dall, 1970). The tubular region, however, is relatively shorter in *H. gammarus* than in freshwater decapods such as the crayfish (Fig. 18A), in which the much greater requirement for urinary salt conservation is associated with a much extended tubule (Peters, 1935). The epithelium of this tubular region (Fig. 18E) shows the extensive membrane infolding and high number of mitochondria which are characteristic of ion-transporting epithelia. Measurements of *H. gammarus* indicate some regulation of the divalent ions Ca^{2+} and Mg^{2+}, but little or no regulation of Na^+, K^+, or Cl^- (Robertson, 1949). It is likely that the antennal gland also plays a minor role in acid–base balance and in elimination of the nitrogenous end-products, NH_3 and NH_4^+ (Wheatly, 1985).

B. Gut

A great deal of exchange with the environment occurs across the wall of the lobster alimentary tract (Mykles, 1979, 1981). The epithelia of *Homarus* midgut and midgut caeca have structural characteristics of ion-transporting epithelia (Mykles, 1979) and can transport large volumes of fluid (Mykles, 1981). Water flux of 32.4 μl/cm/hr occurs in isolated perfused *H. americanus* midgut in lobster Ringer's solution; water transport is strongly linked to Na^+ but not Cl^- transport and is cyanide- but not ouabain-sensitive (Mykles, 1981). Movement of water across the lobster midgut thus depends on active transport, probably of Na^+, which likely serves to create a standing osmotic potential in the epithelium, causing a transepithelial water flux. Ions entering the hemolymph from the gut thus can clearly originate from ingested seawater, but a potentially large, but variable, contribution must also come from food sources.

C. Gills

The majority of movement of water, ions, other acid–base equivalents, and NH_3 and NH_4^+ in crustaceans doubtless occurs across the gill surface. At least in the case of ions, the gills are selectively permeable and have selective transport processes of many kinds. Ions are often exchanged, for example K^+ is exchanged for Na^+ at most cell membranes and

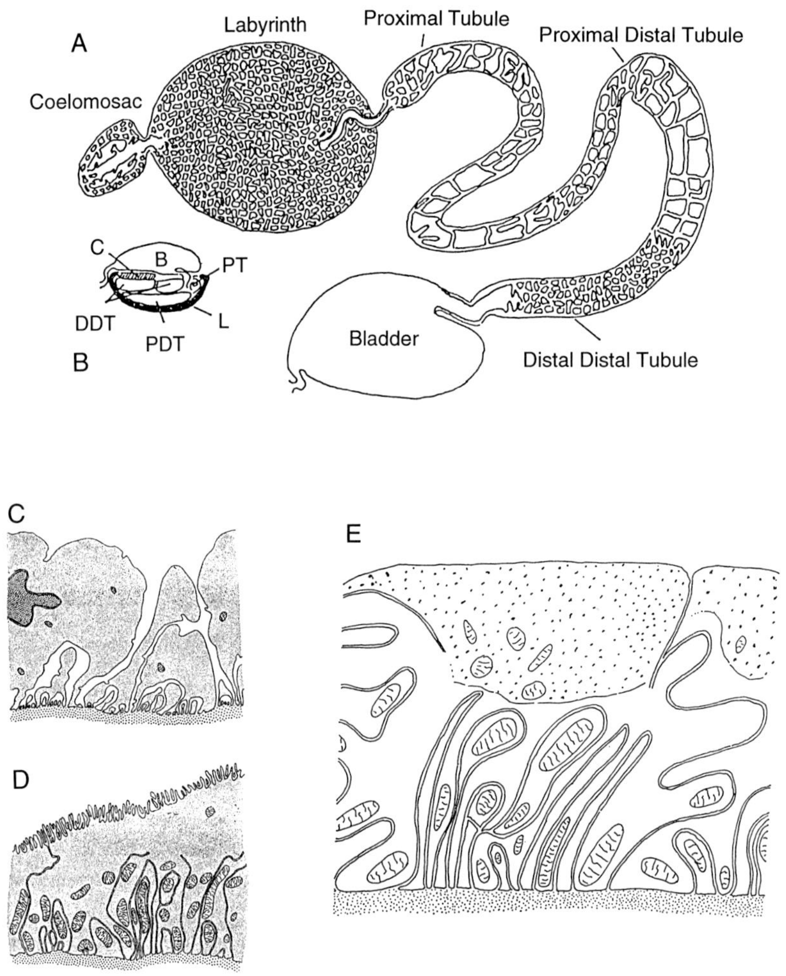

FIGURE 18 Structure of the antennal gland of the crayfish. (A) Dissected tubule to show regions. (B) Spatial relationships *in situ*. Histology of epithelial cells: (C) from the coelomosac showing podocytes; (D) from the labyrinth and proximal tubule; (E) from the distal part of the distal tubule. B, bladder; C, coelomosac; DDT, distal region of the distal tubule; L, labyrinth; PDT, proximal part of distal tubule; PT, proximal tubule. (Modified from Reigel, 1972.)

often the exchange is for an ion that is normally excreted, e.g., Na^+/NH_4^+ exchange. Little specific data are available for *Homarus americanus*, indeed ion exchange in primarily marine forms is generally very poorly understood. It is important to note the extremely integrated nature of these exchange mechanisms. Movement of Na^+, for instance, is associated with movements of K^+, H^+, NH_4^+, Cl^-, and perhaps organic substances such as glucose and amino acids (Kirschner, 1992). Movement of Cl^- can be associated with Na^+, K^+, and HCO_3^-. All are intimately associated with the movement of water and, therefore, volume regulation at both the cellular and whole-animal levels. Ion regulation is thus intimately associated with both water balance and nitrogenous excretion, with acid–base balance via both H^+ and HCO_3^-, and with CO_2 elimination via branchial carbonic anhydrase-mediated transport of HCO_3^-. It is likely that the balance between the various mechanisms varies complexly to suit the immediate needs of all the regulatory systems.

VI. Integrated Responses and Physiological Compensation

It is apparent that the respiratory and circulatory systems in *Homarus americanus* are strongly coregulated. For example, if ventilation stops, as in the "pauses" discussed in Section II,G, heart beat also ceases (see Fig. 3). This is logical, since once ventilation has ceased, the water bathing the gill filaments rapidly becomes deoxygenated and there is a danger of the animal losing oxygen from the hemolymph to the environment. Brief periods of cardiac arrest also accompany ventilatory reversals and could again protect the animal from poor gas exchange as spent (exhalant) water refluxes over the gill filaments. In the other direction, both heart and particularly ventilatory flow rates are increased very rapidly at the onset of an activity period (McMahon and Wilkens, 1983; Guirguis and Wilkens, 1995) as part of the compensatory responses designed to increase oxygen supply to tissues (Houlihan *et al.*, 1985). The strongly integrated nature of physiological responses to environmental stress can be well illustrated by reference to the overall responses of *H. americanus* to hypoxic exposure. The initial increase in water and hemolymph flow across the gills not only brings additional O_2, but also causes increased CO_2 efflux. The loss of CO_2 causes hemolymph alkalosis which, in turn, (a) increases Hc oxygen affinity to increase the efficiency of oxygen uptake, and (b) stimulates the retention of HCO_3^- to allow acid–base regulation. If hypoxia is severe enough for increased anaerobic metabolism, increased lactate release into the hemolymph also contributes to increase the affinity of Hc O_2 binding and further improves O_2 uptake from the depleted water. Increased water flow across the gills is also thought to increase passive ion loss, which could further complicate the picture presented above (Hassall and McMahon, unpublished observations). Physiological compensation to even a simple stepchange in the environment thus involves a large number of interrelated changes in many physiological systems, all of which are integrated to allow effective compensation.

Neural and neurohormonal control mechanisms thus must affect many systems simultaneously. Hormones that elevate cardiovascular function also elevate ventilatory water and gill hemolymph flow, but also probably decrease hemolymph flow to visceral organs (McGaw *et al.*, 1994; McMahon, 1992) and probably also affect many other bodily systems. Thus it is probably less useful to study any physiological system in isolation and future work should attempt to ascertain the roles of each individual system within the overall compensatory response.

VII. Directions for Further Research

This account of the respiratory, circulatory, and ion-regulatory physiology of *Homarus americanus* is clearly incomplete and there is still much to learn. It is hoped that this chapter will stimulate further physiological research on this scientifically interesting and economically important species. The following list is not exhaustive, but outlines some areas in which further study would be profitable.

Only the most basic aspects of oxygen uptake and delivery to tissues are understood. Both the structure and function of the gills of *Homarus americanus* need investigation using both light and electron microscopy. Hemolymph flow through the gill has also not been studied. Almost nothing is known of the mechanisms involved in CO_2 transport or acid–base balance.

The structure and function of the cardiac ganglion of *Homarus americanus* are relatively well known, but very little is known of the manner in which cardiac performance is controlled either by the CNS or via the neurohormonal system. Control of heart performance by oxygen level is poorly known. The structure, innervation, and pharmacological responses of the cardio-arterial valves is described for *Homarus* heart preparations *in vitro* and a role in control of regional hemolymph flow has been proposed (McMahon, 1992); yet, the mechanisms that coordinate these valves to adjust regional blood flow to suit varying metabolic demands are poorly understood. Muscular valves are also present in the peripheral circulation (Alexandrowicz, 1932). While these certainly function in local control of hemolymph perfusion (McMahon *et al.*, 1995), knowledge of the precise mechanisms involved is urgently required. Details of the microcirculation are unknown for any *Homarus* tissue.

Few studies have focused on ion regulation. Fine structure of the gills and renal organs is unknown. Studies on ion movements across the gills of *Homarus americanus* are unknown at any level for animals under "normal" conditions or during compensation for any environmental change. The latter is also true for the vast majority of marine invertebrates. Details of acid–base regulation at any level are not understood.

These are examples of information that is required to understand the functioning of individual systems. Physiological systems do not work in isolation, but are complexly integrated into a series of overall responses. Details of this integration or its control are almost completely unknown.

In addition, two key areas almost completely lack physiological understanding: the molting process, in which all of the above systems are crucially involved; and larval development, in which all physiological systems can be expected to change radically with both increase in size and at metamorphosis.

VIII. Summary

This chapter focuses on three interrelated physiological processes. *Gas exchange* encompasses the structure and function of gills, the generation of the ventilatory stream, oxygen uptake and transport, CO_2 transport and elimination, acid–base balance, and respiratory control. *Circulation* encompasses heart structure, heart excitation and dynamics, control of heart performance, vascular function in an open circulatory system, and control of both heart performance and regional blood flow. *Osmotic and ionic regulation and nitrogenous excretion* involves discussion of the structure and function of the three major regulatory systems—the gills, the antennal glands, and the alimentary tract.

The functioning of these systems is discussed for quiescent animals under natural conditions, in activity, and in response to environmental and other stressors. These systems do not, however, act independently; rather, they are complexly integrated into an overall compensatory response that involves the coordinated control of many (possibly all) physiological systems. This level of control is not well known in *Homarus americanus*, but some aspects of these integrated responses are outlined. Finally, areas in which understanding of these aspects of lobster physiology are lacking, and which require additional study, are identified.

References

Airriess, C. N., and McMahon, B. R. (1992). Aminergic modulation of circulatory performance in the crab *Cancer magister*. *In* "Comparative Physiology" (R. K. H. Kinney, E. Kinne Saffran, and K. W. Beyenbach, eds.), Vol. 11, "Phylogenetic Models in Functional Coupling of the CNS and Cardiovascular System" (R. B. Hill, K. Kuwasawa, B. R. McMahon, and T. Kuramoto, eds.), pp. 123–121. Karger, Basel.

Airriess, C. N., and McMahon, B. R. (1994). Cardiovascular adaptations enhance tolerance of environmental hypoxia in the crab *Cancer magister*. *J. Exp. Biol.* 190, 23–41.

Alexandrowicz, J. S. (1932). The innervation of the heart of the Crustacea. I. Decapoda. *Q. J. Microsc. Sci.* 75, 181–249.

Amberson, W. R., Mayerson, H. S., and Scott, W. J. (1924). The influence of oxygen tension on metabolic rate in invertebrates. *J. Gen. Physiol.* 7, 171–176.

Anderson, M., and Cooke, I. M. (1971). Neural activation of the heart of the lobster *Homarus americanus*. *J. Exp. Biol.* 55, 449–468.

Belman, B. (1975). Some aspects of the circulatory physiology of the spiny lobster *Panulirus interruptus*. *Mar. Biol.* 29, 295–305.

Berlind, A. (1985). Endogenous burst-organizing potentials in two classes of neurons in the lobster cardiac ganglion respond differently to alterations in divalent ion concentration. *J. Comp. Physiol. (A)* 157A, 845–856.

Burnett, L. E., and McMahon, B. R. (1987). Gas exchange, hemolymph acid–base status and the role of branchial water stored during air exposure in three littoral crab species. *Physiol. Zool.* 60, 27–36.

Burger, J. W. (1957). The general form of excretion in the lobster *Homarus*. *Biol. Bull. (Woods Hole, Mass.)* 113, 207–223.

Burger, J. W., and Smythe, C. McC. (1953). The general form of circulation in the lobster *Homarus*. *J. Cell. Comp. Physiol.* 42, 369–383.

Burggren, W. W., McMahon, B. R., and Costerton, J. W. (1974). Branchial blood and water-flow patterns and the structure of the gill of the crayfish *Procambarus clarkii*. *Can. J. Zool.* 52, 1511–1518.

Butler, P. J., Taylor, E. W., and McMahon, B. R. (1978). Respiratory and circulatory changes in the lobster (*Homarus vulgaris*) during long term exposure to moderate hypoxia. *J. Exp. Biol.* 73, 131–146.

Cattaert, D., and Clarac, F. (1983). Influence of walking on swimmeret beating in the lobster *Homarus gammarus*. *J. Neurobiol.* 14, 421–439.

Cole, W. H. (1940). The composition of fluids and sera of some marine animals and of the seawater in which they live. *J. Gen. Physiol.* 23, 575–584.

Cooper, R. A., and Uzman, J. R. (1980). Ecology of juvenile and adult *Homarus*. *In* "The Biology and Management of Lobsters" (J. S. Cobb and B. R. Phillips, eds.), Vol. 2, pp. 97–142. Academic Press, New York.

Dall, W. (1970). Osmoregulation in the lobster *Homarus americanus*. *J. Fish. Res. Board Can.* 27, 1123–1130.

Dunel-Erb, S., Massabuau, J.-C., and Laurent, P. (1982). Organisation fonctionelle de la branchie d'Ecrevisse. *C. R. Séances Soc. Biol. Ses Fil.* 1763, 248–258.

Forgue, J., Truchot, J.-P., and Massabuau, J.-C. (1992). Low arterial PO_2 in resting crustaceans is independent of oxygen affinity. *J. Exp. Biol.* 170, 257–264.

Gray, I. E. (1957). A comparative study of the gill area of crabs. *Biol. Bull. (Woods Hole, Mass.)* 112, 43–42.

Guirguis, M. S., and Wilkens, J. L. (1995). The role of the cardioregulatory nerves in mediating heart rate responses to locomotion and neurohormones in *H. americanus*. *Biol. Bull. (Woods Hole, Mass.)* 188, 179–185.

Herried, C. F. (1980). Hypoxia in invertebrates. *Comp. Biochem. Physiol. (A)* 67A, 311–320.

Houlihan, D. F., Govind, C. K., and El Haj, A. (1985). Energetics of swimming in *Callinectes sapidus* and walking in *Homarus americanus*. *Comp. Biochem. Physiol. (A)* 82A, 267–279.

Kirchner, L. B. (1992). Water and ions. *In* "Comparative Animal Physiology: Environmental and Metabolic Physiology," 4th Ed. (C. Ladd Prosser, ed.), Chap. 2. Wiley-Liss, New York.

Kobierski, L. B., Beltz, B., Trimmer, B. A., and Kravitz, E. D. (1987). FMRFamide-like peptides of *Homarus americanus*: Distribution, immunocytochemical mapping and ultrastructural localization in terminal varicosities. *J. Comp. Neurol.* 266, 1–15.

Kuramoto, T., Wilkens, J. L., and McMahon, B. R. (1995). Neural control of cardiac outflow through the sternal valve in the lobster *H. americanus*. *Physiol. Zool.* 68, 443–452.

Mangum, C. P. (1983a). Oxygen transport in the blood. *In* "The Biology of Crustacea" (D. E. Bliss, ed.), Vol. 5, "Internal Anatomy and Physiological Regulation" (L. H. Mantel, ed.), pp. 373–429. Academic Press, New York.

Mangum, C. P. (1983b). On the distribution of lactate sensitivity among the hemocyanins. *Mar. Biol. Lett.* 4, 139–149.

Mangum, C. P. (1993). Hemocyanin subunit composition and oxygen binding in two species of the lobster genus *Homarus* and their hybrids. *Biol. Bull. (Woods Hole, Mass.)* 184, 105–113.

Mangum, C. P., and van Winkle, W. (1973). Responses of aquatic invertebrates to declining oxygen conditions. *Am. Zool.* 13, 513–518.

Maynard, D. M. (1960). Circulation and heart function. *In* "The Physiology of Crustacea" (T. H. Waterman, ed.), Vol. 1, pp. 161–226. Academic Press, New York.

McGaw, I. J., Airriess, C. N., and McMahon, B. R. (1994). Peptidergic modulation of cardiovascular dynamics in the Dungeness crab *Cancer magister*. *J. Comp. Physiol. (B)* 164B, 103–111.

McLauglin, P. (1983). Internal anatomy. *In* "The Biology of Crustacea" (D. E. Bliss, ed.), Vol. 5, "Internal Anatomy and Physiological Regulation" (L. H. Mantel, ed.), pp. 1–52. Academic Press, New York.

McLeese, D. W. (1956). Effects of temperature, salinity and oxygen on the survival of the American Lobster. *J. Fish. Res. Board Can.* 13, 247–272.

McMahon, B. R. (1985). Functions and functioning of crustacean hemocyanin. *In* "Respiratory Pigments in Animals: Relation Structure and Function" (J. N. Lamy, J. P. Truchot, and R. Gilles, eds.), pp. 35–58. Springer, New York.

McMahon, B. R. (1992). Factors controlling the distribution of cardiac output in decapod crustaceans. *In* "Comparative Physiology" (R. K. H. Kinney, E. Kinne Saffran, and K. W. Beyenbach, eds.), Vol. 11, "Phylogenetic Models in Functional Coupling of the CNS and Cardiovascular System" (R. B. Hill, K. Kuwasawa, B. R. McMahon, and T. Kuramoto, eds.), pp. 51–61. Karger, Basel.

McMahon, B. R., and Burggren, W. W. (1990). O_2 and O_2 profiles in normoxic and hypoxic lobsters. *Physiologist* 33, 70A.

McMahon, B. R., and Burnett, L. E. (1990). The crustacean open circulatory system: A reexamination. *Physiol. Zool.* 63, 35–71.

McMahon, B. R., and Rieber, C. L. (1991). Effects of proctolin in controlling the distribution of cardiac output in the lobster. *FASEB J.* 5, A1057.

McMahon, B. R., and Wilkens, J. L. (1972). Simultaneous apnoea and bradycardia in the lobster *Homarus americanus*. *Can. J. Zool.* 50, 165–170.

McMahon, B. R., and Wilkens, J. L. (1975). Respiratory and circulatory responses to hypoxia in the lobster *Homarus americanus*. *J. Exp. Biol.* 62, 637–655.

McMahon, B. R., and Wilkens, J. L. (1983). Ventilation, perfusion and oxygen uptake. *In* "The Biology of Crustacea" (D. E. Bliss, ed.), Vol. 5, "Internal Anatomy and Physiological Regulation" (L. H. Mantel, ed.), pp. 289–372. Academic Press, New York.

McMahon, B. R., Butler, P. J., and Taylor, E. W. (1978). Acid–base changes during recovery from disturbance and during long-term exposure to hypoxia *Homarus vulgaris*. *J. Exp. Zool.* 205, 361–370.

McMahon, B. R., Smith, P. S., and Wilkens, J. L. (1995). Physiology of invertebrate circulatory systems. *In* "Handbook of Comparative Physiology" (W. H. Danztler, ed.). Am. Physiol. Soc., Washington, DC. (in press.)

Morris, S. (1990). Organic ions as modulators of respiratory pigment function during stress. *Physiol. Zool.* 63, 253–287.

Mykles, D. L. (1979). Ultrastructure of alimentary epithelia of lobsters *Homarus americanus* and *H. gammarus* and the crab *Cancer magister*. *Zoolomorphologie* 92, 201–215.

Mykles, D. L. (1981). Ionic requirements of transepithelial potential difference and net water flux in the perfused midgut of the American lobster *Homarus americanus*. *Comp. Biochem. Physiol. (A)* 69A, 317–320.

Paterson, N. F. (1968). The anatomy of the Cape rock lobster, *Jasus lalandii*. *Ann. S. Afr. Mus.* 51, 1–232

Peters, H. (1935). Ueber den Einfluss des Salzgehaltes in Aussenmedium auf den Bau und die Funcktion der Excretionsorgane dekapoder Crustaceen (nach Untershungen au *Potamobius fluviatilis* und *Homarus vulgaris*). *Z. Morphol. Oekol. Tiere* 30, 355–381.

Piiper, J., and Scheid, P. (1975). Gas transport efficiency of gilles, lungs and skin: Theory and experimental data. *Respir. Physiol.* 23, 209–221.

Reiber, C. L. (1994). Hemodynamics of the crayfish *Procambarus clarkii*. *Physiol. Zool.* 67, 449–467.

Reiber, C. L., McMahon, B. R., and Burggren, W. W. (1992). Redistribution of cardiac output in response to hypoxia: A comparison of the freshwater crayfish *Procambarus clarkii* and the lobster *Homarus americanus*. *In* "Comparative Physiology" (R. K. H. Kinney, E. Kinne Saffran, and K. W. Beyenbach, eds.), Vol. 11, "Phylogenetic Models in Functional Coupling of the CNS and Cardiovascular System" (R. B. Hill, K. Kuwasawa, B. R. McMahon, and T. Kuramoto, eds.), pp. 22–28. Karger, Basel.

Reiber, C. L., McMahon, B. R., and Burggren, W. W. (1995). Arterial blood flow and cardiac output in decapod Crustacea (*Procambarus clarkii* and *Homarus americanus*). *J. Comp. Physiol.* (in press.)

Reigel, J. A. (1972). "Comparative Physiology of Renal Excretion." Oliver & Boyd, Edinburgh.

Robertson, J. D. (1949). Ionic regulation in some marine invertebrates. *J. Exp. Biol.* 26, 182–200.

Robertson, J. D. (1960). Osmotic and ionic regulation. *In* "The Physiology of Crustacea" (T. H. Waterman, ed.), Vol. 1, pp. 317–339. Academic Press, New York.

Shadwick, R. E., Pollock, C. A., and Strycker, S. A. (1990). Structure and biomechanical properties of crustacean blood vessels. *Physiol. Zool.* 63, 90–101.

Starrat, A. N., and Brown, B. E. (1975). Structure of the pentapeptide proctolin, a proposed neurotransmitter in insects. *Life Sci.* 17, 1253–1256.

Taylor, E. W., and Whitely, N. M. (1993). Oxygen transport and acid–base balance in the haemolymph of the lobster *Homarus gammarus* during aerial exposure and resubmersion. *J. Exp. Biol.* 144, 417–436.

Thomas, H. J. (1954). The oxygen uptake of the lobster *Homarus vulgaris* Edw. *J. Exp. Biol.* 31, 228–251.

Truchot, J.-P. (1992). Respiratory function of arthropod haemocyanins. *In* "Advances in Comparative and Environmental Physiology," Vol. 13, pp. 377–408. Springer-Verlag, Berlin.

Von Raben, K. (1934). Veranderungen im Kiemendeckel und im den Kiemeneineger Brachyuren (Decapoden) in Verlauf der Anpassung an die Feuchtluftatmung. *Z. Wiss. Zool.* 145, 425–461.

Wheatly, M. G. (1985). The role of the antennal gland in ion and acid–base regulation during hyposaline exposure of the Dungeness crab *Cancer magister* (Dana). *J. Comp. Physiol.* 155, 445–454.

Wilkens, J. L. (1989). Cardiac and circulatory control in decapod Crustacea with comparisons to molluscs. *Experientia* 43, 990–993.

Wilkens, J. L. (1993). Re-evaluation of the stretch sensitivity hypothesis of crustacean hearts: Hypoxia not lack of stretch causes reduction in heart rate of isolated hearts. *J. Exp. Biol.* 176, 223–232.

Wilkens, J. L., and McMahon, B. R. (1972). Aspects of branchial irrigation in the lobster *Homarus americanus*. *J. Exp. Biol.* 56, 469–479.

Wilkens, J. L., and McMahon, B. R. (1992). Intrinsic properties and extrinsic neurohormonal control of crab hemodynamics. *Experientia* 48, 827–834.

Wilkens, J. L., and Walker, R. L. (1992). Nervous control of crayfish cardiac hemodynamics. *In* "Comparative Physiology" (R. K. H. Kinney, E. Kinne Saffran, and K. W. Beyenbach, eds.), Vol. 11, "Phylogenetic Models in Functional Coupling of the CNS and Cardiovascular System" (R. B. Hill, K. Kuwasawa, B. R. McMahon, and T. Kuramoto, eds.), pp. 115–122. Karger, Basel.

Index

I

J

T

ISBN 0-12-247570-4
90018
9 780122 475702